# Methods in Enzymology

Volume 248
PROTEOLYTIC ENZYMES
ASPARTIC AND METALLO PEPTIDASES

# METHODS IN ENZYMOLOGY

EDITORS-IN-CHIEF

## John N. Abelson     Melvin I. Simon

DIVISION OF BIOLOGY
CALIFORNIA INSTITUTE OF TECHNOLOGY
PASADENA, CALIFORNIA

FOUNDING EDITORS

## Sidney P. Colowick and Nathan O. Kaplan

*Methods in Enzymology*

*Volume 248*

# Proteolytic Enzymes: Aspartic and Metallo Peptidases

EDITED BY

## *Alan J. Barrett*

DEPARTMENT OF BIOCHEMISTRY
STRANGEWAYS RESEARCH LABORATORY
CAMBRIDGE, ENGLAND

## ACADEMIC PRESS

San Diego  New York  Boston  London  Sydney  Tokyo  Toronto

Copyright © 1995 by ACADEMIC PRESS, INC.

Academic Press, Inc.
A Division of Harcourt Brace & Company
525 B Street, Suite 1900, San Diego, California 92101-4495

*United Kingdom Edition published by*
Academic Press Limited
24-28 Oval Road, London NW1 7DX

International Standard Serial Number: 0076-6879

International Standard Book Number: 0-12-182149-8

PRINTED IN THE UNITED STATES OF AMERICA
95  96  97  98  99  00  MM  9  8  7  6  5  4  3  2  1

# Table of Contents

## Section I. Methods

## Section II. Aspartic Peptidases

# Section III. Metallopeptidases

# Contributors to Volume 248

Article numbers are in parentheses following the names of contributors.
Affiliations listed are current.

ANGELA ANASTASI (43), *Department of Pathology, The Medical School, St. Luke's Hospital, Msida, Malta*

DAVID S. AULD (14), *Center for Biochemical and Biophysical Sciences and Medicine, and Department of Pathology, Harvard Medical School and Brigham and Women's Hospital, Boston, Massachusetts 02115*

HÉLÈN BARELLI (36), *Institut de Pharmacologie Moléculaire et Cellulaire, CNRS, Université de Nice-Sophia Antipolis, F-06560 Valbonne, France*

ALAN J. BARRETT (7, 13, 32, 43), *Department of Biochemistry, Strangeways Research Laboratory, Cambridge CB1 4RN, United Kingdom*

ANDREW B. BECKER (44), *Department of Molecular Pharmacology, Stanford University School of Medicine, Stanford, California 94305*

JUDD BERMAN (3), *Department of Medicinal Chemistry, Glaxo Inc. Research Institute, Research Triangle Park, North Carolina 27709*

D. MARK BICKETT (3), *Department of Biochemistry, Glaxo Inc. Research Institute, Research Triangle Park, North Carolina 27709*

JOSEPH G. BIETH (5), *INSERM Unité 392, Laboratoire d'Enzymologie, Université Louis Pasteur de Strasbourg, F-67400 Illkirch, France*

JÓN B. BJARNASON (21, 22), *The Science Institute, University of Iceland, IS-107 Reykjavík, Iceland*

JUDITH S. BOND (20), *Department of Biochemistry and Molecular Biology, College of Medicine, Pennsylvania State University, Hershey, Pennsylvania 17033*

JACQUES BOUVIER (37), *Animal Health Department, Ciba-Geigy Ltd., CH-1566 St. Aubin, Switzerland*

MOLLY A. BROWN (32), *Department of Biochemistry, Strangeways Research Laboratory, Cambridge CB1 4RN, United Kingdom*

MICHAEL BRUNNER (46), *Institut für Physiologische Chemie, Universität München, D-80336 München 2, Germany*

DAVID J. BUTTLE (4), *Department of Biochemistry, Strangeways Research Laboratory, Cambridge CB1 4RN, United Kingdom*

PAUL CANNON (25), *Institute of Biochemistry and Cell Biology, Syntex Discovery Research, Palo Alto, California 94303*

NIAMH X. CAWLEY (9), *Laboratory of Developmental Neurobiology, National Institute of Child Health and Human Development, NIH and Department of Biochemistry, Uniformed Services, University of the Health Sciences, Bethesda, Maryland 20892*

FRÉDÉRIC CHECLER (36), *Institut de Pharmacologie Moleculaire et Cellulaire, CNRS, Université de Nice-Sophia Antipolis, F-06560 Valbonne, France*

VALÉRIE CHESNEAU (45), *Laboratoire de Biochimie des Signaux Régulateurs, Cellulair et Moléculaires, Unité de Recherches Associée au Centre National de la Scientifique, Université Pierre et Marie Curie, 75006 Paris, France*

PAUL COHEN (45), *Laboratoire de Biochimie des Signaux Régulateurs, Cellulair et Moléculaires, Unité de Recherches Associée au Centre National de la Scientifique, Université Pierre et Marie Curie, 75006 Paris, France*

CHRISTOPHER A. CONLIN (34), *Department of Biological Sciences, Mankato State University, Mankato, Minnesota 56002*

PIERRE CORVOL (18), *Institut National de la Santé et de la Recherche Medicale, Collége de France, 75005 Paris, France*

THOMAS CRABBE (28), *Celltech Therapeutics Ltd., Slough SL1 4EN, United Kingdom*

PHILIPPE CRINE (17), *Département de Biochimie, Faculté de Médecine, Université de Montréal, Montréal, Canada H3C 3J7*

PAMELA M. DANDO (32), *Department of Biochemistry, Strangeways Research Laboratory, Cambridge CB1 4RN, United Kingdom*

PASCALE DAUCH (36), *Institut de Pharmacologie Moleculaire et Cellulaire, CNRS, Université de Nice-Sophia Antipolis, F-06560 Valbonne, France*

PETER A. DEDDISH (41), *Departments of Pharmacology and Anesthesiology, University of Illinois College of Medicine, Chicago, Illinois 60612*

MARIANNA DIOSZEGI (25), *Institute of Biochemistry and Cell Biology, Syntex Discovery Research, Palo Alto, California 94303*

VINCENT DIVE (36), *Départment d'Ingénierie et d'Etude des Protéines, C.E.N. de Saclay, Laboratoire de Structure des Protéines en Solution, 91191 Gif/Yvette, France*

ROBERT ETGES (37), *Department of Biochemistry, University of Puerto Rico, San Juan, Puerto Rico 00936*

STEPHAN FISCHER (51), *Boehringer Mannheim GmbH, D-82372 Penzberg, Germany*

THIERRY FOULON (45), *Laboratoire de Biochimie des Signaux Régulateurs, Cellulair et Moléculaires, Unité de Recherches Associée au Centre National de la Scientifique, Université Pierre et Marie Curie, 75006 Paris, France*

MARIE-CLAUDE FOURNIÉ-ZALUSKI (17), *Département de Pharmacochimie Moléculaire et Structurale, Institut National de la Sante et de la Recherche Medicale, Université René Descartes, 75270 Paris Cedex 06, France*

JAY W. FOX (21, 22), *Department of Microbiology, University of Virginia, Health Sciences Center, Charlottesville, Virginia 22908*

URSULA GEUSS (51), *Boehringer Mannheim GmbH, D-68305 Mannheim, Germany*

PAUL GLYNN (23), *Medical Research Council Toxicology Unit, University of Leicester, Leicester LE1 9HN, United Kingdom*

MICHAEL GREEN (3), *Department of Medicinal Chemistry, Glaxo Inc. Research Institute, Research Triangle Park, North Carolina 27709*

MARIE-LUISE HAGMANN (51), *Boehringer Mannheim GmbH, D-82372 Penzberg, Germany*

CHRISTOPHER J. HANDLEY (4), *Department of Biochemistry and Molecular Biology, Monash University, Clayton Victoria 3168, Australia*

LOUIS B. HERSH (16), *Department of Biochemistry, College of Medicine, University of Kentucky, Lexington, Kentucky 40536*

LINDA HOWARD (23), *Department of Cell Biology and Lombardi Cancer Center, Georgetown University School of Medicine, Washington 20007*

GRAZIA ISAYA (33), *Department of Genetics, Yale University School of Medicine, New Haven, Connecticut 06510*

KARL E. KADLER (49, 50), *Department of Biological Sciences, Research Division of Biochemistry, University of Manchester, Manchester M13 9PT, United Kingdom*

TAKASHI KAGEYAMA (8), *Department of Cellular and Molecular Biology, Primate Research Institute, Kyoto University, Inuyama, Aichi 484, Japan*

FRANTISEK KALOUSEK (33), *Department of Genetics, Yale University School of Medicine, New Haven, Connecticut 06510*

CHIH-MIN KAM (1), *School of Chemistry and Biochemistry, Georgia Institute of Technology, Atlanta, Georgia 30332*

EFRAT KESSLER (48), *Maurice and Gabriela Goldschleger Eye Research Institute, Sackler Faculty of Medicine, Tel-Aviv University, Tel-Hashomer 52621, Israel*

C. GRAHAM KNIGHT (2, 6, 32), *Department of Cell Adhesion and Signalling, Strangeways Research Laboratory, Cambridge CB1 4RN, United Kingdom*

GEORG-BURKHARD KRESSE (51), *Boehringer Mannheim GmbH, D-82372 Penzberg, Germany*

CHINGWEN LI (16), *Department of Biochemistry, College of Medicine, University of Kentucky, Lexington, Kentucky 40536*

SAMANTHA J. LIGHTFOOT (49), *Department of Biological Sciences, Research Division of Biochemistry, University of Manchester, Manchester M13 9PT, United Kingdom*

XINLI LIN (11), *Protein Studies Program, Oklahoma Medical Research Foundation, Oklahoma City, Oklahoma 73104*

REGGIE Y.C. LO (47), *Department of Microbiology, University of Guelph, Guelph, Ontario, Canada N1G 2W1*

Y. PENG LOH (9), *Section on Cellular Neurobiology, Laboratory of Developmental Neurobiology, National Institute of Child Health and Human Development, and Department of Biochemistry, Uniformed Services, University of the Health Sciences, Bethesda, Maryland 20892*

HIROSHI MAEDA (24), *Department of Microbiology, Kumamoto University Medical School, Kumamoto 860, Japan*

GERARD M. MCGEEHAN (3), *Department of Biochemistry, Glaxo Inc. Research Laboratories, Research Triangle Park, North Carolina 27709*

NORMAN MCKIE (32), *Department of Biochemistry, Strangeways Research Laboratory, Cambridge CB1 4RN, United Kingdom*

ALAN MELLORS (47), *Department of Chemistry and Biochemistry, Guelph-Waterloo Centre for Graduate Work in Chemistry, University of Guelph, Guelph, Ontario, Canada N1G 2W1*

CHARLES G. MILLER (34), *Department of Microbiology, University of Illinois at Urbana-Champaign, Urbana, Illinois 61801*

VÉRONIQUE MONNET (35), *INRA Centre De Recherches De Jouy-en-Josas, Station De Recherches Laitières, Domaine de Vilvert, 78352 Jouy-en-Josas, Cedex, France*

CESARE MONTECUCCO (39), *Centro CNR Biomembrane and Dipartimento di Scienze Biomediche, Università di Padova, 75-35100 Padova, Italy*

KAZUYUKI MORIHARA (15), *Institute of Applied Life Sciences, Graduate School, University of East Asia, Yamaguchi 751, Japan*

GILLIAN MURPHY (28, 30), *Department of Cell and Molecular Biology, Strangeways Research Laboratory, Cambridge CB1 4RN, United Kingdom*

HIDEAKI NAGASE (27), *Department of Biochemistry and Molecular Biology, University of Kansas Medical Center, Kansas City, Kansas 66160*

WALTER NEUPERT (46), *Institut für Physiologische Chemie, Universität München, D-80336 München 2, Germany*

FLORENCE NOBLE (17), *Département de Pharmacochimie Moléculaire et Structurale, Institut National de la Sante et de la Recherche Medicale, Centre , National de la Recherche Scientifique, Université René Descartes, 75270 Paris Cedex 06, France*

ADRIAN R. PIEROTTI (45), *Department of Biological Sciences, Glasgow Caledonian University, Glasgow GL4 OBA, Scotland*

ANDREW G. PLAUT (38), *Department of Medicine, Gastroenterology Division, Tufts University School of Medicine and New England Medical Center, Boston, Massachusetts 02111*

JAMES C. POWERS (1), *School of Chemistry and Biochemistry, Georgia Institute of Technology, Atlanta, Georgia 30332*

ANNIK PRAT (45), *Laboratoire de Biochimie des Signaux Régulateurs, Cellulair et Moléculaires, Unité de Recherches Associée au Centre National de la Scientifique, Université Pierre et Marie Curie, 75006 Paris, France*

NEIL D. RAWLINGS (7, 13, 32), *Department of Biochemistry, Strangeways Research Laboratory, Cambridge CB1 4RN, United Kingdom*

BERNARD P. ROQUES (17), *Département de Pharmacochimie Moléculaire et Structurale, Institut National de la Santé et de la Recherche Medicale, Centre National de la Recherche Scientifique, Université René Descartes, 75270 Paris Cedex 06, France*

RICHARD A. ROTH (44), *Department of Molecular Pharmacology, Stanford University School of Medicine, Stanford, California 94305*

KRISHNAN SANKARAN (12), *Department of Microbiology and Immunology, Uniformed Services, University of the Health Sciences, Bethesda, Maryland 20814*

GIAMPIETRO SCHIAVO (39), *Centro CNR Biomembrane and Dipartimento di Scienze Biomediche, Università di Padova, 75-35100 Padova, Italy*

PASCAL SCHNEIDER (37), *Department of Biochemistry, University of Dundee, Dundee DD1 4HN, Scotland*

ATSUSHI SERIZAWA (32), *Sapporo Research Laboratory, Snow Brand Milk Products Co., Ltd., Sapporo 065, Japan*

RANDAL A. SKIDGEL (40, 41), *Department of Pharmacology and Anesthesiology, University of Illinois College of Medicine, Chicago, Illinois 60612*

FLORENT SOUBRIER (18), *Institut National de la Santé et de la Recherche Medicale, Collége de France, 75005 Paris, France*

VALENTIN M. STEPANOV (42), *Protein Chemistry Laboratory, Institute of Microbial Genetics, Moscow 113545, Russia*

WALTER STÖCKER (19), *Zoologisches Institut der Universität Heidelberg, Physiologie, Im Neuenheimer Feld, D-69120 Heidelberg, Germany*

KENJI TAKAHASHI (10), *Department of Biophysics and Biochemistry, Faculty of Science, The University of Tokyo, Tokyo 113, Japan*

FULONG TAN (41), *Departments of Pharmacology and Anesthesiology, University of Illinois College of Medicine, Chicago, Illinois 60612*

JORDAN TANG (11), *Oklahoma Medical Research Foundation, Protein Studies Program, Oklahoma City, Oklahoma 73104*

HARALD TSCHESCHE (26), *Biochemistry Department, University Bielefeld, D-33615, Bielefeld Germany*

HAROLD E. VAN WART (25), *Institute of Biochemistry and Cell Biology, Syntex Discovery Research, Palo Alto, California 94303*

BRUNO VINCENT (36), *Institut de Pharmacologie Moléculaire et Cellulaire, CNRS, Université de Nice-Sophia Antipolis, F-06560 Valbonne, France*

JEAN PIERRE VINCENT (36), *Institut de Pharmacologie Moléculaire et Cellulaire, CNRS, Université de Nice-Sophia Antipolis, F-06560 Valbonne, France*

ROD B. WATSON (49), *Department of Biological Sciences, Research Division of Biochemistry, University of Manchester, Manchester M13 9PT, United Kingdom*

FRANCES WILLENBROCK (30), *Department of Biochemistry, Queen Mary and Westfield College, University of London, London E1 4NS, United Kingdom*

TRACY A. WILLIAMS (18), *Institut National de la Santé et de la Recherche Medicale, Collége de France, 75005 Paris, France*

JEFFREY S. WISEMAN (3), *Department of Biochemistry, Glaxo Inc. Research Institute, Research Triangle Park, North Carolina 27709*

J. FREDERICK WOESSNER, JR. (29, 31), *Department of Biochemistry and Molecular Biology, University of Miami, School of Medicine, Miami, Florida 33101*

RUSSELL L. WOLZ (20), *Department of Biochemistry and Molecular Biology, College of Medicine, Pennsylvania State University, Hershey, Pennsylvania 17033*

ANDREW WRIGHT (38), *Department of Molecular Biology and Microbiology, Tufts University School of Medicine, Boston, Massachusetts 02111*

HENRY C. WU (12), *Department of Microbiology and Immunology, Uniformed Services, University of the Health Sciences, Bethesda, Maryland 20814*

ROBERT ZWILLING (19), *Zoologisches Institut der Universität Heidelberg, Physiologie Im Neuenheimer Feld, D-69120 Heidelberg, Germany*

# Preface

Through the earlier, general volumes on proteolytic enzymes (Volumes 19, 45, and 80), *Methods in Enzymology* made available over 200 authoritative articles on these enzymes and their inhibitors. Since the appearance of the latest of these volumes, however, there have been many profound advances in this field of study. The biomedical importance of proteolytic enzymes, suspected for so long, has been established beyond reasonable doubt for a number of groups, including the matrix metalloproteinases, the viral polyprotein-processing enzymes, and the prohormone-processing peptidases. The more recent, specialized Volumes 222, 223, and 241 have dealt with some of these areas, but others have remained to be covered.

The resurgence of excitement about proteolytic enzymes has inevitably resulted in an information explosion, but some of the new understanding has also helped us develop novel approaches to the management of the mass of data. As a result, we can now "see the forest for the trees" a little more clearly. Like other proteins, the proteolytic enzymes have benefited from the recent advances in molecular biology, and amino acid sequences are now available for many hundreds of them. These can be used to group the enzymes in families of evolutionarily related members. Also, there has been a major overhaul of the recommended nomenclature for peptidases by the International Union of Biochemistry and Molecular Biology. In Volume 244 on peptidases of serine and cysteine type and in this volume on aspartic, metallo, and other peptidases, the chapters on specific methods, enzymes, and inhibitors are organized within the rational framework of the new systems for classification and nomenclature.

The peptidases of the aspartic and metallo types dealth with in this volume depend for their activity on the nucleophilic activity of an ionized water molecule, unlike the enzymes, described in Volume 244, in which the nucleophilic character of a serine or cysteine residue is at the heart of the catalytic mechanism. A wide variety of specificities of peptide bond hydrolysis is represented in each set of peptidases, together with an equally wide range of biological functions.

ALAN J. BARRETT

# METHODS IN ENZYMOLOGY

VOLUME XXXVIII. Hormone Action (Part C: Cyclic Nucleotides)
*Edited by* JOEL G. HARDMAN AND BERT W. O'MALLEY

VOLUME XXXIX. Hormone Action (Part D: Isolated Cells, Tissues, and Organ Systems)
*Edited by* JOEL G. HARDMAN AND BERT W. O'MALLEY

VOLUME XL. Hormone Action (Part E: Nuclear Structure and Function)
*Edited by* BERT W. O'MALLEY AND JOEL G. HARDMAN

VOLUME XLI. Carbohydrate Metabolism (Part B)
*Edited by* W. A. WOOD

VOLUME XLII. Carbohydrate Metabolism (Part C)
*Edited by* W. A. WOOD

VOLUME XLIII. Antibiotics
*Edited by* JOHN H. HASH

VOLUME XLIV. Immobilized Enzymes
*Edited by* KLAUS MOSBACH

VOLUME XLV. Proteolytic Enzymes (Part B)
*Edited by* LASZLO LORAND

VOLUME XLVI. Affinity Labeling
*Edited by* WILLIAM B. JAKOBY AND MEIR WILCHEK

VOLUME XLVII. Enzyme Structure (Part E)
*Edited by* C. H. W. HIRS AND SERGE N. TIMASHEFF

VOLUME XLVIII. Enzyme Structure (Part F)
*Edited by* C. H. W. HIRS AND SERGE N. TIMASHEFF

VOLUME XLIX. Enzyme Structure (Part G)
*Edited by* C. H. W. HIRS AND SERGE N. TIMASHEFF

VOLUME L. Complex Carbohydrates (Part C)
*Edited by* VICTOR GINSBURG

VOLUME LI. Purine and Pyrimidine Nucleotide Metabolism
*Edited by* PATRICIA A. HOFFEE AND MARY ELLEN JONES

VOLUME LII. Biomembranes (Part C: Biological Oxidations)
*Edited by* SIDNEY FLEISCHER AND LESTER PACKER

VOLUME LIII. Biomembranes (Part D: Biological Oxidations)
*Edited by* SIDNEY FLEISCHER AND LESTER PACKER

VOLUME LIV. Biomembranes (Part E: Biological Oxidations)
*Edited by* SIDNEY FLEISCHER AND LESTER PACKER

VOLUME LV. Biomembranes (Part F: Bioenergetics)
*Edited by* SIDNEY FLEISCHER AND LESTER PACKER

VOLUME LVI. Biomembranes (Part G: Bioenergetics)
*Edited by* SIDNEY FLEISCHER AND LESTER PACKER

VOLUME 233. Oxygen Radicals in Biological Systems (Part C)
*Edited by* LESTER PACKER

VOLUME 234. Oxygen Radicals in Biological Systems (Part D)
*Edited by* LESTER PACKER

VOLUME 235. Bacterial Pathogenesis (Part A: Identification and Regulation of Virulence Factors)
*Edited by* VIRGINIA L. CLARK AND PATRIK M. BAVOIL

VOLUME 236. Bacterial Pathogenesis (Part B: Integration of Pathogenic Bacteria with Host Cells)
*Edited by* VIRGINIA L. CLARK AND PATRIK M. BAVOIL

VOLUME 237. Heterotrimeric G Proteins
*Edited by* RAVI IYENGAR

VOLUME 238. Heterotrimeric G-Protein Effectors
*Edited by* RAVI IYENGAR

VOLUME 239. Nuclear Magnetic Resonance (Part C)
*Edited by* THOMAS L. JAMES AND NORMAN J. OPPENHEIMER

VOLUME 240. Numerical Computer Methods (Part B)
*Edited by* MICHAEL L. JOHNSON AND LUDWIG BRAND

VOLUME 241. Retroviral Proteases
*Edited by* LAWRENCE C. KUO AND JULES A. SHAFER

VOLUME 242. Neoglycoconjugates (Part A: Synthesis)
*Edited by* Y. C. LEE AND REIKO T. LEE

VOLUME 243. Inorganic Microbial Sulfur Metabolism
*Edited by* HARRY D. PECK, JR., AND JEAN LEGALL

VOLUME 244. Proteolytic Enzymes: Serine and Cysteine Peptidases
*Edited by* ALAN J. BARRETT

VOLUME 245. Extracellular Matrix Components
*Edited by* E. RUOSLAHTI AND E. ENGVALL

VOLUME 246. Biochemical Spectroscopy
*Edited by* KENNETH SAUER

VOLUME 247. Neoglycoconjugates (Part B: Biomedical Applications)
*Edited by* Y. C. LEE AND REIKO T. LEE

VOLUME 248. Proteolytic Enzymes: Aspartic and Metallo Peptidases
*Edited by* ALAN J. BARRETT

VOLUME 249. Enzyme Kinetics and Mechanism (Part D: Developments in Enzyme Dynamics)
*Edited by* DANIEL L. PURICH

VOLUME 250. Lipid Modifications of Proteins
*Edited by* PATRICK J. CASEY AND JANICE E. BUSS

VOLUME 251. Biothiols (Part A: Monothiols and Dithiols, Protein Thiols, and Thiyl Radicals)
*Edited by* LESTER PACKER

VOLUME 252. Biothiols (Part B: Glutathione and Thioredoxin; Thiols in Signal Transduction and Gene Regulation) (in preparation)
*Edited by* LESTER PACKER

VOLUME 253. Adhesion of Microbial Pathogens (in preparation)
*Edited by* RON J. DOYLE AND ITZHAK OFEK

VOLUME 254. Oncogene Techniques (in preparation)
*Edited by* PETER K. VOGT AND INDER M. VERMA

VOLUME 255. Small GTPases and Their Regulators (Part A: Ras Family) (in preparation)
*Edited by* W. E. BALCH, CHANNING J. DER, AND ALAN HALL

VOLUME 256. Small GTPases and Their Regulators (Part B: Rho Family) (in preparation)
*Edited by* W. E. BALCH, CHANNING J. DER, AND ALAN HALL

VOLUME 257. Small GTPases and Their Regulators (Part C: Proteins Involved in Transport) (in preparation)
*Edited by* W. E. BALCH, CHANNING J. DER, AND ALAN HALL

VOLUME 258. Redox-Active Amino Acids in Biology (in preparation)
*Edited by* JUDITH P. KLINMAN

VOLUME 259. Energetics of Biological Macromolecules (in preparation)
*Edited by* MICHAEL L. JOHNSON AND GARY K. ACKERS

VOLUME 260. Mitochondrial Biogenesis and Genetics, (Part A) (in preparation)
*Edited by* GIUSEPPE M. ATTARDI AND ANNE CHOMYN

VOLUME 261. Nuclear Magnetic Resonance and Nucleic Acids (in preparation)
*Edited by* THOMAS L. JAMES

VOLUME 262. DNA Replication (in preparation)
*Edited by* JUDITH L. CAMPBELL

# Section I

# Methods

# [1] Peptide Thioester Substrates for Serine Peptidases and Metalloendopeptidases

*By* JAMES C. POWERS and CHIH-MIN KAM

## Introduction

Synthetic peptide substrates are widely used in biochemical and physiological studies of proteolytic enzymes. Synthetic peptide substrates can be used to detect enzyme during isolation, to assay enzyme activity, to determine enzyme concentrations, to investigate enzyme specificity, and to determine inhibitor potency. The three most commonly used synthetic substrates are peptide 4-nitroanilides, peptide thioesters, and peptide derivatives of 7-amino-4-methylcoumarin. Amino acid and peptide thioesters are sensitive substrates for serine peptidases and metalloendopeptidases because the substrates have high $k_{cat}/K_m$ values for enzymatic hydrolysis rates and low background hydrolysis rates, and the thiol-leaving group can be easily detected at low concentrations. Cleavage of the thioester bond can be monitored continuously by reaction with a thiol reagent such as 4,4'-dithiodipyridine or 5,5'-dithiobis (2-nitrobenzoic acid) contained in the assay mixture to produce a chromogenic compound.[1,2] Alternately, the thiol detection reagent can be used after the reaction has been concluded. Synthetic peptide thioester substrates have been used for substrate mapping of elastases, chymotrypsin-like enzymes, coagulation enzymes, and complement proteins, and they are also useful for detecting various new serine peptidase activities in cell extracts such as lymphocyte and natural killer cell granules. Several synthetic peptide thioesters have been used to monitor enzyme activities of various metalloendopeptidases such as collagenases, stromelysin, gelatinase, and thermolysin.

---

[1] D. R. Grassetti and J. F. Murray, Jr., *Arch. Biochem. Biophys.* **119,** 41 (1967).
[2] D. A. Farmer and J. H. Hageman, *J. Biol. Chem.* **250,** 7366 (1975).

Synthetic Methods

*Materials.* CDI,[3] DCC, HOBt, and benzyl mercaptan can be obtained from Aldrich Chemical Company, Inc. (Milwaukee, WI). All Boc amino acids can be obtained from Chemical Dynamics Corp. (South Plainfield, NJ), Bachem Bioscience, Inc. (Philadelphia, PA), and numerous other sources.

*Boc-Ala-Ala-AA-SBzl.* Boc-AA-SBzl derivatives are prepared by coupling Boc-AA-OH and benzyl mercaptan using the DCC/HOBt method. Boc-AA-SBzl can then be deblocked with HCl in dioxane or ethyl acetate to give HCl · AA-SBzl which is further coupled with Boc-Ala-Ala-OH to give the final product Boc-Ala-Ala-AA-SBzl.[4,5] When the amino acid AA is glutamic acid or aspartic acid, the side-chain carboxyl group is protected with a *tert*-butyl group, which along with the Boc group can be removed by trifluoroacetic acid or HCl in ethyl acetate.[5]

*Boc-Ala-Ala-Nva-SBzl.* To prepare Boc-Ala-Ala-Nva-SBzl,[4] Boc-Nva-OH (2.17 g, 10 mmol) is dissolved in dry THF (10 ml), CDI (1.62 g, 10 mmol) is added, and the reaction is stirred at 0° for 45 min. Benzyl mercaptan (1.16 ml, 10 mmol) is added, and the reaction mixture is allowed to warm to 25° overnight. The solvent is removed under reduced pressure, ethyl acetate (20 ml) is added, and the organic solution is washed with 10% (w/v) citric acid, 4% (w/v) NaHCO$_3$, and saturated aqueous NaCl. The ethyl acetate solution is dried over MgSO$_4$, then filtered, and the solvent is removed. Boc-Nva-SBzl is solidified with hexane or petroleum ether (80% yield), produces one spot on thin-layer chromatography (TLC) [CHCl$_3$: methanol (9:1, v/v)], and is used for subsequent reaction without further purification.

Boc-Nva-SBzl (0.97 g, 3 mmol) is treated with 10 equivalents of 2.2 *N* HCl in dioxane and allowed to stir at 25° for 45 min. The solvent is then removed by evaporation, and ether is added to solidify the product. The resulting Nva-SBzl · HCl is dried *in vacuo* and used for subsequent steps

---

[3] AA, Amino acid residue; Abz, 2-aminobenzoyl; AMC, 7-amino-4-methylcoumarin; Boc, *tert*-butyloxycarbonyl; Bu-*i*, isobutyl; Bzl, benzyl; CDI, 1,1'-carbonyldiimidazole; DCC, *N,N*'-dicyclohexylcarbodiimide; DMF, dimethylformamide; EDC, 1-ethyl-3-[3-(dimethylamino)-propyl]carbodiimide hydrochloride; HEPES, 4-(2-hydroxyethyl)-1-piperazineethanesulfonic acid; HOBt, *N*-hydroxybenzotriazole; MES, 2-(*N*-morpholino)ethanesulfonic acid; Mu, *N*-morpholinocarbonyl; Nba, 4-nitrobenzylamine; NNap-OCH$_3$, 1-methoxy-3-naphthylamine; SBzl(Cl), SCH$_2$C$_6$H$_4$-4-Cl; Suc, succinyl; THF, tetrahydrofuran; Tricine, *N*-[tris(hydroxymethyl)methyl]glycine; Z, benzyloxycarbonyl.

[4] J. W. Harper, R. R. Cook, J. Roberts, B. J. Mclaughlin, and J. C. Powers, *Biochemistry* **23,** 2995 (1984).

[5] S. Odake, C.-M. Kam, L. Narasimhan, M. Poe, J. T. Blake, O. Krahenbuhl, J. Tschopp, and J. C. Powers, *Biochemistry* **30,** 2217 (1991).

without further purification. Nva-SBzl · HCl (0.78 g, 3 mmol), Boc-Ala-Ala-OH (0.78 g, 3.0 mmol), HOBt (0.61 g, 4.5 mmol), and triethylamine (0.42 ml, 3.0 mmol) are dissolved in 10 ml of DMF, and the solution is cooled to $- 10°$ in an ice–water–salt bath. DCC (0.68 g, 3.3 mmol) is added, and the reaction mixture is stirred at $- 10°$ for 2 hr and at 25° overnight. Dicyclohexylurea is removed by filtration, DMF is removed by evaporation, and the residue is dissolved in ethyl acetate and washed as described above. The final product is recrystallized from $CHCl_3$–petroleum ether with cooling, mp 140°–141°; TLC $R_f$ of 0.61 [$CHCl_3$ : methanol (9 : 1, v/v)]. Analysis calculated for $C_{23}H_{35}N_3O_5S$: C, 59.38; H, 7.58; N, 9.02. Found: C, 59.58; H, 7.62; N, 9.14.

*Boc-Ala-Ala-Asp-SBzl.* To prepare Boc-Ala-Ala-Asp-SBzl,[5] to a THF solution (60 ml) of Boc-Asp(O-*tert*-Bu)-OH (11.6 g, 40 mmol) is added HOBt hydrate (3.1 g, 20 mmol), benzyl mercaptan (5.2 ml, 44 mmol), and DCC (9.9 g, 48 mmol) in 15 ml of THF at $-5°$. After stirring at 0° for 12 hr and at room temperature for 24 hr, the reaction mixture is filtered and the solvent removed by evaporation. Ethyl acetate is added to the residue, and the solution is washed successively with 1 N HCl, 10% (w/v) $Na_2CO_3$, and a saturated NaCl solution. The organic layer is dried over $MgSO_4$, and the solvent is removed by evaporation. The crude product is purified by silica gel chromatography using ethyl acetate : *n*-hexane (1 : 10, v/v) as an eluant to give Boc-Asp(O-*tert*-Bu)-SBzl (13.5 g, 85% yield) as a pale yellow oil. An ethyl acetate solution (70 ml) saturated with HCl is added to Boc-Asp(O-*tert*-Bu)-SBzl (4.0 g, 8.8 mmol) at 0°, and the solution is stirred at 25° for 2.5 hr. The solvent is removed, and ethyl acetate is added to the residue to give a white precipitate of H-Asp-SBzl · HCl which is filtered and dried *in vacuo* (2.1 g, 87% yield).

To a THF solution (20 ml) of Boc-Ala-Ala-OH (0.52 g, 2.0 mmol) is added successively N-methylmorpholine (0.22 ml, 2.0 mmol) and isobutyl chloroformate (0.26 ml, 2.0 mmol) at $-15°$. After stirring for 2 min and addition of a cold THF solution (3 ml) of triethylamine (0.56 ml, 4 mmol), the mixture is added to a DMF solution (1 ml) of H-Asp-SBzl · HCl (0.55 g, 2 mmol) at $-15°$. After stirring for 1 hr, the reaction mixture is quenched by the addition of 1 N HCl (4 ml) and is then concentrated *in vacuo*. An ethyl acetate solution of the residue is washed with water and dried over $MgSO_4$. The solvent is removed by evaporation, and the crude product is purified by silica gel chromatography using $CHCl_3$ : methanol (50 : 1, v/v) as an eluant and solidified with *n*-hexane to give the final product (0.67 g, 70% yield) as a white powder; mp 72°–79°; TLC $R_f$ of 0.52 ($CHCl_3$ : methanol : $CH_3COOH$, 80 : 10 : 5, v/v/v). Analysis calculated for $C_{22}H_{31}O_7N_3S$ · 0.25$H_2O$ · 0.25$C_6H_{14}$ : C, 55.61; H, 6.95; N, 8.28. Found: C, 55.71; H, 7.00; N, 7.94.

*Lysine- or Arginine-Containing Thioesters.* Z-Lys-SBzl is prepared by coupling Z-Lys-OH with benzyl mercaptan using the DCC method.[6] Simple N-blocked arginine thioesters are synthesized by coupling an N-blocked arginine derivative with a thiol using either the DCC–HOBt or DCC methods.[7] Di- and tripeptide thiol esters are then prepared by deblocking the Boc-Arg-SR with HCl in dioxane followed by coupling with the appropriate peptide acid using the pentachlorophenyl active ester method. The arginine side chain is protonated with HCl during the synthesis of arginine thioesters; therefore, it is not necessary to protect the guanidino group with other reagents.

*Z-Arg-SBzl · HCl.* To prepare Z-Arg-SBzl · HCl,[8] Z-Arg-OH · HCl (3.45 g, 10 mmol) is dissolved in DMF, HOBt (1.35 g, 1.0 mmol) is added, and the solution is cooled to 0°. Benzyl mercaptan (1.24 g, 10 mmol) and DCC (2.16 g, 10.5 mmol) are then added. The reaction mixture is stirred overnight at 0°, followed by removal of the dicyclohexylurea by filtration and the solvent by evaporation. The crude product is purified by flash column chromatography on silica gel (32–64 $\mu$m) using 15% methanol in CHCl$_3$ as an eluant. The product is obtained as a white foam after trituration with petroleum ether and drying *in vacuo* (20% yield); TLC $R_f$ of 0.64 [CHCl$_3$ : methanol : CH$_3$COOH (10 : 3 : 1, v/v/v)]. Analysis calculated for C$_{21}$H$_{27}$N$_4$O$_3$Cl · 0.5H$_2$O: C, 54.83; H, 6.14; N, 12.18. Found: C, 54.76, H, 6.10; N, 12.40.

*Thioester Substrates for Metalloendopeptidases.* Thioester substrates for metalloendopeptidases typically contain a thioester bond in the interior of a peptide sequence and require additional synthetic steps. The synthesis usually involves coupling of an N-terminal peptide fragment with the thiol-containing fragment using standard peptide condensation methods to form the thioester bond. For example, CH$_3$CO-Pro-Leu-Gly-$^S$Leu-Leu-Gly-OC$_2$H$_5$ is prepared from CH$_3$CO-Pro-Leu-Gly-OH and HSCH[CH$_2$CH(CH$_3$)$_2$]-CO-Leu-Gly-OC$_2$H$_5$ using the DCC–HOBt method. The required HSCH[CH$_2$CH(CH$_3$)$_2$]-CO-Leu-Gly-OC$_2$H$_5$ is synthesized by coupling L-$\alpha$-mercaptoisocaproic acid and Leu-Gly-OC$_2$H$_5$ using EDC–HOBt.[9]

*Boc-Abz-Gly-Pro-Leu-SCH$_2$CO-Pro-Nba.* Boc-Abz-Gly-Pro-OH and Leu-SCH$_2$CO-Pro-Nba · HCl are synthesized by standard peptide coupling methods. Boc-Abz-Gly-Pro-OH (0.78 g, 2 mmol) and Leu-SCH$_2$CO-Pro-Nba · HCl (0.94 g, 2 mmol) are then coupled using the mixed anhydride

[6] G. D. Green and E. Shaw, *Anal. Biochem.* **93**, 223 (1979).
[7] B. J. McRae, K. Kurachi, R. L. Heimark, K. Fujikawa, E. W. Davie, and J. C. Powers, *Biochemistry* **20**, 7196 (1981).
[8] R. R. Cook, B. J. McRae, and J. C. Powers, *Arch. Biochem. Biophys.* **234**, 82 (1984).
[9] H. Weingarten, R. Martin, and J. Feder, *Biochemistry* **24**, 6730 (1985).

method with isobutyl chloroformate and $N$-methylmorpholine in THF–DMF (1:1, v/v) to give Boc-Abz-Gly-Pro-Leu-SCH$_2$CO-Pro-Nba,[10] which is purified by flash column chromatography on silica gel (32–63 $\mu$m) using methanol–ether (1:20, v/v) as an eluant (0.61 g, 37% yield); TLC [CHCl$_3$:methanol (5:1, v/v)] $R_f$ of 0.79; mp 120°–130° (decomposition). Analysis calculated for C$_{39}$H$_{51}$N$_7$O$_{10}$S · H$_2$O: C, 56.56; H, 6.46; N, 11.84; S, 3.87. Found: C, 56.68; H, 6.47; N, 11.28; S, 3.84.

## Assay Methods

*Reagents.* The thiol reagents 4,4'-dithiodipyridine (Aldrithiol-4) and 5,5'-dithiobis(2-nitrobenzoic acid) (Ellman's reagent), and Tricine can be obtained from Aldrich and other sources. *HEPES* is purchased from Research Organics, Inc. (Cleveland, OH). Several amino acid and peptide thioester substrates are commercially available, and some are listed in Table I. Chymotrypsin, trypsin, porcine pancreatic elastase (PPE), and *Clostridium histolyticum* collagenase can be obtained from Sigma Chemical Co. (St. Louis, MO). Human leukocyte elastase (HLE) and human cathepsin G can be obtained from Athens Research and Technology, Inc. (Athens, GA). Proteinase 3 (human neutrophil) was provided by Dr. Koert Dolman of the University of Amsterdam. Rat mast cell peptidase I and II (RMCP I and RMCP II) were obtained from Drs. Richard Woodbury and Hans Neurath of the University of Washington. Bovine thrombin, factor IXa, factor Xa, factor XIa, and factor XIIa were provided by Drs. Kotoku Kurachi, Kazuo Fujikawa, and Earl Davie of University of Washington (Seattle, WA). Human factor VIIa was provided by Dr. George Vlasuk of Merck, Sharp & Dohme Research Laboratories (West Point, PA). Human factor D, C2a, and Bb were obtained from Dr. John Volanakis of University of Alabama (Birmingham, AL). Human C1̄s and C1̄r were provided by Dr. T.-Y. Lin of Merck (Rohway, NJ).

*Assay for Serine Peptidases.* The rates of enzymatic hydrolysis of synthetic peptide thioester substrates catalyzed by serine peptidases are measured in various pH 7.5–8.0 buffers containing 8–10% dimethyl sulfoxide (DMSO) and at 25° in the presence of the thiol reagent 4,4'-dithiodipyridine[1] or 5,5'-dithiobis(2-nitrobenzoic acid).[2] The buffers used are 0.1 M HEPES, 10 m$M$ CaCl$_2$, pH 7.5, for bovine trypsin and the coagulation enzymes (thrombin, factor IXa, factor Xa, factor XIa, and factor XIIa); 0.1 $M$ HEPES, 0.5 $M$ NaCl, pH 7.5, for chymotrypsin, cathepsin G, RMCP I, RMCP II, HLE, PPE, proteinase 3, factor D, C2a, and Bb; 0.1 $M$ Tris-HCl, 20 m$M$ CaCl$_2$, 0.1 $M$ NaCl for C1̄r and C1̄s. Stock solutions of substrates (5–

[10] C. F. Vencill, D. Rasnick, K. V. Crumbley, N. Nishino, and J. C. Powers, *Biochemistry* **24,** 3149 (1985).

TABLE I
COMMERCIALLY AVAILABLE THIOESTER SUBSTRATES[a]

| Lys or Arg derivatives | Phe derivatives | Val or Nva derivatives | Others |
|---|---|---|---|
| Z-Lys-SBzl[a,b,c,d] | Phe-SBzl | Boc-Ala-Ala-Val-SBzl | Z-Tyr-SBzl |
| Z-Arg-SBzl | $CH_3$O-Suc-Phe-SBzl | Boc-Ala-Pro-Nva-4-Cl-SBzl[b] | Z-Trp-SBzl |
| Z-Arg-Arg-SBzl | Mu-Phe-SBzl | $CH_3$O-Suc-Ala-Ala-Pro-Val-SBzl | Z-Gly-Pro-SBzl |
| Z-Gly-Arg-SBzl | Boc-Ala-Ala-Phe-SBzl | Mu-Ala-Ala-Pro-Val-SBzl | Boc-Ala-Ala-Asp-SBzl |
| Z-Phe-Arg-SBzl | Suc-Ala-Pro-Phe-SBzl | | Boc-Ala-Ala-Met-SBzl |
| Z-Trp-Arg-SBzl | $CH_3$O-Suc-Gly-Leu-Phe-SBzl | | $CH_3$O-Suc-Ala-Ala-Pro-Ala-SBzl |
| D-Val-Leu-Lys-SBzl | Mu-Gly-Leu-Phe-SBzl | | Mu-Ala-Ala-Pro-Ala-SBzl |
| | Suc-Phe-Leu-Phe-SBzl[b,c] | | |
| | Suc-Val-Pro-Phe-SBzl[b] | | |
| | Suc-Ala-Ala-Pro-Phe-SBzl[c] | | |
| | $CH_3$O-Suc-Ala-Ala-Pro-Phe-SBzl | | |
| | Mu-Ala-Ala-Pro-Phe-SBzl | | |

[a] The substrates which are available from Enzyme System Products (6497 Sierra Lane, Dublin, CA 94568) have no footnote, whereas those available from other sources are indicated by a footnote.
[b] Bachem Bioscience Inc., 3700 Horizon Drive, Renaissance at Gulph Mills, King of Prussia, PA 19406.
[c] Sigma Chemical Co., P.O. Box 14508, St. Louis, MO 63178.
[d] Novabiochem., P.O. Box 12087, La Jolla, CA 92039-2087.

10 mM) are prepared in DMSO and stored at $-20°$. Most substrate solutions are stable in DMSO for several months at $-20°$.

The Arg- or Lys-containing substrates are less stable, and some hydrolysis occurs after long storage times. An accurate concentration of the substrates can be determined by measuring the absorbance increase at 324 nm after adding 25 $\mu$l of a trypsin solution ($>10^{-5}$ M) and 10 $\mu$l of the substrate stock solution ($<5$ mM) to 2 ml of buffer containing 4,4'-dithiodipyridine (0.34 mM). After trypsin has completely hydrolyzed the substrate, the original concentration of the thioester substrate is calculated using an $\varepsilon_{324}$ of 19,800 $M^{-1}$ cm$^{-1}$. Boc-Ala-Ala-Asp-SBzl also partially hydrolyzes on storage in DMSO after a few days and has fast hydrolysis rates in pH 7.5–8.0 buffer. Thus, solutions of the substrate must be freshly prepared.

The initial rates of thioester hydrolysis are measured at 324 nm ($\varepsilon_{324}$ of 19,800 $M^{-1}$ cm$^{-1}$ for 4-thiopyridone) using a UV–visible spectrometer when 10–50 $\mu$l of an enzyme stock solution (0.01–2 $\mu$M) is added to a cuvette containing 2.0 ml of buffer, 150 $\mu$l of 4,4'-dithiodipyridine (5 mM), and 25 $\mu$l of substrate (0.004–10 mM). For Lys- or Arg-containing thioester substrates or Boc-Ala-Ala-Asp-SBzl, the same volumes of substrate and 4,4'-dithiodipyridine are added to the reference cell in order to compensate for the background hydrolysis rates of the substrates. The enzymatic hydrolysis rates can also be measured at 405 nm ($\varepsilon_{405}$ of 13,260 $M^{-1}$ cm$^{-1}$ for 2-nitro-5-thiobenzoate anion) using a microplate reader when 10–25 $\mu$l of an enzyme stock solution (0.01–2 $\mu$M) is added to a well containing 0.2 ml of buffer, 10 $\mu$l of 5,5'-dithiobis(2-nitrobenzoic acid) (5 mM), and 10 $\mu$l of substrate ($<5$ mM). For the blank reaction, the same volumes of substrate and thiol reagent, but no enzyme, are added to buffer in the well. Initial hydrolysis rates are measured in duplicate or triplicate for each substrate concentration and are averaged in each case. The kinetic constants $k_{cat}$, $K_m$, and $k_{cat}/K_m$ can be obtained from Lineweaver–Burk plots. Correlation coefficients are usually greater than 0.99.

*Assay for Metalloendopeptidases.* The enzymatic hydrolysis rates of thioesters by metallopeptidases are measured similarly to those for serine peptidases, and a typical assay for *Clostridium histolyticum* collagenase follows.[10] The hydrolysis rate is measured in 50 mM Tricine, pH 7.5, containing 10 mM CaCl$_2$, 0.5% (v/v) DMSO, and 2.4% (v/v) methanol at 25°. To a cuvette containing 2.0 ml of buffer is added 10 $\mu$l of 4,4'-dithiodipyridine (56 mM) in DMSO, 50 $\mu$l of the substrate Z-Gly-Pro-Leu-SCH$_2$CO-Pro-Nba in methanol (1–4 mM), and 10 $\mu$l of collagenase in buffer. The increase in absorbance at 324 nm ($\varepsilon_{324}$ of 19,800 $M^{-1}$ cm$^{-1}$) is monitored with a UV–visible spectrometer. Background hydrolysis of the substrate is negligible in most cases, but it can be corrected when significant by including the substrate and thiol reagent in the reference cell. The kinetic constants are determined by either Hanes or Lineweaver–Burk plots. Correlation coefficients are usually greater than 0.99.

## Thioester Substrates for Serine Peptidases

A variety of amino acid and peptide thioesters have been described as synthetic substrates for various serine peptidases. The first thioester substrate described is the leucine aminopeptidase substrate Leu-SC$_2$H$_5$,[11] whereas the first thioester substrate used with serine peptidases is Bz-Tyr-SBzl which is effectively hydrolyzed by $\alpha$-chymotrypsin and subtilisin BPN'.[2] The tetrapeptide CH$_3$O-Suc-Ala-Ala-Pro-Val-SBzl is a very sensitive substrate for HLE and PPE,[12] and the lysine derivative Z-Lys-SBzl is the first thioester substrate described for trypsin-like serine peptidases.[6]

Peptide thioesters which contain a large hydrophobic group at the P1 site[13] are good substrates for chymotrypsin-like enzymes,[4,14,15] and kinetic constants for representative thioesters are shown in Table II. Thioesters containing Phe or Met at the P1 site usually have $k_{cat}/K_m$ values of $10^5$–$10^7$ $M^{-1}$ sec$^{-1}$ with chymotrypsin, cathepsin G, RMCP I, and RMCP II. Suc-Val-Pro-Phe-SBzl, Suc-Phe-Leu-Phe-SBzl, and Suc-Ala-Ala-Pro-Phe-SBzl are particularly good substrates for chymases and have very low detection limits.[14,15]

Elastase prefers longer peptide sequences than chymotrypsin and smaller aliphatic residues at the P1 site.[4] Tripeptide thioesters have frequently been used as synthetic substrates of HLE, PPE, and proteinase 3.[4,12,16] The kinetic constants for representative thioesters are shown in Table III. Boc-Ala-Ala-AA-SBzl containing Val, Nva, Leu, Ile, or Nle in the P1 site are good substrates of PPE and HLE with $k_{cat}/K_m$ values of $10^5$–$10^6$ $M^{-1}$ sec$^{-1}$. Proteinase 3 hydrolyzes the substrates at slower rates than PPE and HLE by approximately 2- to 65-fold. CH$_3$O-Suc-Ala-Ala-Pro-Val-SBzl, Boc-Ala-Ala-Val-SBzl, and Boc-Ala-Pro-Nva-SBzl (Cl) are particularly useful substrates for HLE.[4,12] The reactivity of the substrate improves when the thiobenzyl group is substituted with Cl, NO$_2$, and CH$_3$O

---

[11] R. M. Metrione, *Biochim. Biophys. Acta* **268**, 518 (1972).

[12] M. J. Castillo, K. Nakajima, M. Zimmerman, and J. C. Powers, *Anal. Biochem.* **99**, 53 (1979).

[13] The nomenclature for individual amino acid residues (P2, P1, P1', etc.) of a substrate and for the subsites (S2, S1, S1', etc.) of the enzyme is that of I. Schechter and A. Berger, *Biochem. Biophys. Res. Commun.* **27**, 157 (1967).

[14] J. W. Harper, G. Ramirez, and J. C. Powers, *Anal. Biochem.* **118**, 382 (1981).

[15] J. C. Powers, T. Tanaka, J. W. Harper, Y. Minematsu, L. Barker, D. Lincoln. K. V. Crumley, J. E. Fraki, N. M. Schechter, G. G. Lazarus, K. Nakajima, K. Nakashino, H. Neurath, and R. G. Woodbury, *Biochemistry* **24**, 2048 (1985).

[16] C.-M. Kam, J. E. Kerrigan, K. M. Dolman, R. Goldschmeding, A. Von dem Borne, and J. C. Powers, *FEBS Lett.* **297**, 119 (1992).

TABLE II

Kinetic Constants for Hydrolysis of Thioester Substrates by Chymotrypsin-like Enzymes

| Substrate | Parameter | Chymotrypsin | Human leukocyte cathepsin G | RMCP I | RMCP II |
|-----------|-----------|--------------|------------------------------|--------|---------|
| Boc-Ala-Ala-Phe-SBzl[a] | $k_{cat}$ | 135 | 44 | 320 | 100 |
| | $K_m$ | 9.1 | 43 | 6.9 | 4.8 |
| | $k_{cat}/K_m$ | 15 | 1.0 | 46 | 2.1 |
| Boc-Ala-Ala-Nva-SBzl[a] | $k_{cat}$ | 12 | 6.9 | 71 | 9.6 |
| | $K_m$ | 7.6 | 33 | 6.1 | 13 |
| | $k_{cat}/K_m$ | 1.6 | 0.21 | 12 | 0.74 |
| Boc-Ala-Ala-Met-SBzl[a] | $k_{cat}$ | 22 | 7.0 | 190 | 21 |
| | $K_m$ | 6.0 | 34 | 6.2 | 25 |
| | $k_{cat}/K_m$ | 3.7 | 0.20 | 31 | 0.84 |
| Suc-Val-Pro-Phe-SBzl[b] | $k_{cat}$ | 39 | 22 | 130 | 13 |
| | $K_m$ | 21 | 19 | 25 | 26 |
| | $k_{cat}/K_m$ | 1.9 | 1.2 | 5.2 | 0.5 |
| Suc-Phe-Leu-Phe-SBzl[c] | $k_{cat}$ | 11 | 24 | 1.9 | 31 |
| | $K_m$ | 6.0 | 32 | 1.5 | 21 |
| | $k_{cat}/K_m$ | 1.9 | 0.78 | 1.3 | 1.5 |
| Suc-Ala-Ala-Pro-Phe-SBzl[c] | $k_{cat}$ | 31 | 14 | — | 390 |
| | $K_m$ | 2.0 | 48 | — | 83 |
| | $k_{cat}/K_m$ | 15 | 0.30 | 4.7 | 4.7 |

[a] Kinetic constants were measured in 0.1 $M$ HEPES, 0.5 $M$ NaCl buffer, pH 7.5, 10% DMSO, and at 25°. The units for $k_{cat}$, $K_m$, and $k_{cat}/K_m$ are sec$^{-1}$, $\mu M$, and $10^6$ $M^{-1}$ sec$^{-1}$. Data were obtained from J. W. Harper, R. R. Cook, J. Roberts, B. J. Mclaughlin, and J. C. Powers, *Biochemistry* **23**, 2995 (1984). Boc-Ala-Ala-Ile-SBzl and Boc-Ala-Ala-Val-SBzl are not hydrolyzed by the four enzymes. Boc-Ala-Ala-Ala-SBzl is hydrolyzed only by chymotrypsin ($k_{cat}/K_m$ of 3500 $M^{-1}$ sec$^{-1}$) but not by the other three enzymes.

[b] For RMCP I and RMCP II, 50 m$M$ phosphate, 10% DMSO, pH 8.0, was used. Data were obtained from J. C. Powers, T. Tanaka, J. W. Harper, Y. Minematsu, L. Barker, D. Lincoln, K. V. Crumley, J. E. Fraki, N. M. Schechter, G. G. Lazarus, K. Nakajima, K. Nakashino, H. Neurath, and R. G. Woodbury, *Biochemistry* **24**, 2048 (1985).

[c] For RMCP I and RMCP II, 50 m$M$ phosphate, 10% DMSO, pH 8.0, was used. Data were obtained from J. W. Harper, G. Ramirez, and J. C. Powers, *Anal. Biochem.* **118**, 382 (1981).

groups.[4] The 4-chloro derivative Boc-Ala-Pro-Nva-SBzl (Cl) has a higher $k_{cat}/K_m$ value than any other investigated tripeptide by at least 14-fold.

Arginine- or lysine-containing thioesters have been used as substrates for various trypsin-like enzymes,[5–8,17–25] and representative kinetic data obtained with bovine trypsin,[6,7] several coagulation enzymes,[6–8,17,18] and

[17] R. P. Link and F. J. Castellino, *Biochemistry* **22**, 999 (1983).

[18] R. Lottenberg, U. Christensen, C. M. Jackson, and P. L. Coleman, this series, Vol. 80, p. 341.

[19] B. J. McRae, T.-Y. Lin, and J. C. Powers, *J. Biol. Chem.* **256**, 12362 (1981).

[20] C.-M. Kam, B. J. McRae, J. W. Harper, M. M. Niemann, J. E. Volanakis, and J. C. Powers, *J. Biol. Chem.* **262**, 3444 (1987).

TABLE III
KINETIC CONSTANTS FOR HYDROLYSIS OF THIOESTER SUBSTRATES BY PORCINE PANCREATIC
ELASTASE, HUMAN LEUKOCYTE ELASTASE, AND PROTEINASE $3^a$

| Substrate | Parameter | $PPE^b$ | $HLE^b$ | Proteinase $3^c$ |
|---|---|---|---|---|
| Boc-Ala-Ala-Ala-SBzl | $k_{cat}$ | 73 | 63 | 10.1 |
| | $K_m$ | 100 | 87 | 106 |
| | $k_{cat}/K_m$ | 0.74 | 0.73 | 0.095 |
| Boc-Ala-Ala-Val-SBzl | $k_{cat}$ | 63 | 9.0 | 10.4 |
| | $K_m$ | 95 | 3.0 | 28 |
| | $k_{cat}/K_m$ | 0.66 | 3.0 | 0.38 |
| Boc-Ala-Pro-Nva-SBzl (4-Cl) | $k_{cat}$ | 230 | 10 | — |
| | $K_m$ | 100 | 0.08 | — |
| | $k_{cat}/K_m$ | 2.2 | 130 | — |
| Boc-Ala-Ala-Nva-SBzl | $k_{cat}$ | 570 | 20 | 63.3 |
| | $K_m$ | 84 | 2.2 | 63 |
| | $k_{cat}/K_m$ | 6.8 | 9.2 | 1.0 |
| Boc-Ala-Ala-Met-SBzl | $k_{cat}$ | 71 | 82 | 12.2 |
| | $K_m$ | 42 | 96 | 61 |
| | $k_{cat}/K_m$ | 1.7 | 0.86 | 0.2 |
| Boc-Ala-Ala-Ile-SBzl | $k_{cat}$ | 7.5 | 5.4 | 1.9 |
| | $K_m$ | 7.6 | 1.4 | 31 |
| | $k_{cat}/K_m$ | 0.98 | 3.9 | 0.06 |
| $CH_3O$-Suc-Ala-Ala-Pro-Val-SBzl$^d$ | $k_{cat}$ | 23 | 18 | — |
| | $K_m$ | 31 | 15 | — |
| | $k_{cat}/K_m$ | 0.74 | 1.2 | — |

$^a$ Kinetic constants were measured in 0.1 $M$ HEPES, 0.5 $M$ NaCl buffer, pH 7.5, 9–10%
DMSO, and at 25°. The units for $k_{cat}$, $K_m$, and $k_{cat}/K_m$ are sec$^{-1}$, $\mu M$, and $10^6$ $M^{-1}$ sec$^{-1}$.
Boc-Ala-Ala-Phe-SBzl is not hydrolyzed by HLE and PPE.
$^b$ Data were obtained from J. W. Harper, R. R. Cook, J. Roberts, B. J. Mclaughlin, and
J. C. Powers, *Biochemistry* **23**, 2995 (1984).
$^c$ Data were obtained from C.-M. Kam, J. E. Kerrigan, K. M. Dolman, R. Goldschmeding,
A. Von dem Borne, and J. C. Powers, *FEBS Lett.* **297**, 119 (1992).
$^d$ Data were obtained from M. J. Castillo, K. Nakajima, M. Zimmerman, and J. C. Powers,
*Anal. Biochem.* **99**, 53 (1979).

[21] K. Cho, T. Tanaka, R. R. Cook, W. Kiesiel, K. Fujikawa, K. Kurachi, and J. C. Powers,
*Biochemistry* **23**, 644 (1984).
[22] T. Tanaka, B. J. McRae, K. Cho, R. Cook, J. E. Fraki, D. E. Johnson, and J. C. Powers,
*J. Biol. Chem.* **258**, 13552 (1983).
[23] M. Orlowski, M. Lesser, J. Ayala, A. Lasdun, C.-M. Kam, and J. C. Powers, *Arch. Biochem.
Biophys.* **269**, 125 (1989).
[24] J. C. Powers, B. J. McRae, T. Tanaka, K. Cho, and R. R. Cook, *Biochem. J.* **220**, 569 (1984).
[25] C.-M. Kam, G. P. Vlasuk, D. E. Smith, K. E. Arcuri, and J. C. Powers, *Thromb. Haemostasis*
**64**, 133 (1990).

complement peptidases[19,20] are shown in Tables IV and V. The Lys- or Arg-containing thioesters are hydrolyzed very rapidly by trypsin with $k_{cat}/K_m$ values of $10^6$ $M^{-1}$ $sec^{-1}$, and Z-Arg-SBzl and Z-Lys-SBzl are particularly useful substrates for most trypsin-like enzymes including enzymes such as the lymphocyte granzyme A.[5–7]

Thrombin is the most active coagulation serine peptidase toward thioester substrates, with factor IXa and VIIa being much less active. Thrombin hydrolyzes most of the Arg-containing thioesters with $k_{cat}/K_m$ values of $10^5$–$10^6$ $M^{-1}$ $sec^{-1}$.[7,8] Factor IXa is much more specific and hydrolyzes only a handful of thioester substrates, with $k_{cat}/K_m$ values of $10^3$–$10^4$ $M^{-1}$ $sec^{-1}$. The most useful substrate for factor IXa is Z-Trp-Arg-SBzl.[7,17] Factor VIIa also effectively hydrolyzes Z-Arg-SBzl ($k_{cat}/K_m$ of 1600 $M^{-1}$ $sec^{-1}$).[25]

The complement enzymes factor D, Bb, and C2 are less active and more specific thioesterases than $C\overline{1s}$ ($k_{cat}/K_m$ of $10^5$–$10^6$ $M^{-1}$ $sec^{-1}$).[19,20] Factor D, Bb, and C2a hydrolyze only a few peptide thioesters with $k_{cat}/K_m$ values of $10^2$–$10^4$ $M^{-1}$ $sec^{-1}$. Interestingly, factor I is one of a handful of serine peptidases for which we have not been able to find an effective thioester substrate. The enzymatic hydrolysis of various peptide thioester substrates by human lung tryptase,[22] human skin tryptase,[22] murine and human granzyme A,[5] urokinase,[6] crotalase,[6] plasma and pancreatic kallikrein,[6,7,24] and

TABLE IV

KINETIC CONSTANTS FOR HYDROLYSIS OF THIOESTER SUBSTRATES BY BOVINE TRYPSIN AND BLOOD COAGULATION ENZYMES[a]

| Substrate | Parameter | Trypsin | Thrombin | Factor IXa | Factor Xa | Factor XIa | Factor XIIa |
|---|---|---|---|---|---|---|---|
| Z-Lys-SBzl[b] | $k_{cat}$ | 75 | 35 | — | — | — | — |
| | $K_m$ | 50 | 40 | — | — | — | — |
| | $k_{cat}/K_m$ | 1.5 | 0.88 | — | — | — | — |
| Z-Arg-SBzl[c] | $k_{cat}$ | 94 | 7.0 | — | 85 | — | — |
| | $K_m$ | 5.3 | 1.9 | — | 190 | — | — |
| | $k_{cat}/K_m$ | 18 | 3.7 | — | 0.45 | — | — |
| Z-Gly-Arg-SBu-$i$[d] | $k_{cat}$ | 95 | 17 | NH | 100 | 110 | 7.1 |
| | $K_m$ | 21 | 30 | — | 750 | 960 | 320 |
| | $k_{cat}/K_m$ | 4.5 | 0.56 | — | 0.13 | 0.12 | 0.02 |
| Boc-Phe-Phe-Arg-SBzl[d] | $k_{cat}$ | 32 | 13 | — | 64 | 40 | — |
| | $K_m$ | 8.2 | 11 | — | 1600 | 760 | — |
| | $k_{cat}/K_m$ | 3.9 | 1.2 | 0.03 | 0.04 | 0.05 | — |

[a] Kinetic constants were measured in 0.1 $M$ HEPES, 10 m$M$ CaCl$_2$ buffer, pH 7.5, 9% DMSO, and at 25° for all substrates except for Z-Lys-SBzl for which 0.1 $M$ Tris, 0.1 $M$ NaCl, buffer, pH 8.0, was used. The units for $k_{cat}$, $K_m$, and $k_{cat}/K_m$ are $sec^{-1}$, $\mu M$, and $10^6$ $M^{-1}$ $sec^{-1}$. NH, No hydrolysis.
[b] G. D. Green and E. Shaw, *Anal. Biochem.* **93**, 223 (1979).
[c] R. R. Cook, B. J. McRae, and J. C. Powers, *Arch. Biochem. Biophys.* **234**, 82 (1984).
[d] B. J. McRae, K. Kurachi, R. L. Heimark, K. Fujikawa, E. W. Davie, and J. C. Powers, *Biochemistry* **20**, 7196 (1981).

TABLE V
KINETIC CONSTANTS FOR HYDROLYSIS OF THIOESTER SUBSTRATES BY HUMAN COMPLEMENT ENZYMES[a]

| Substrate | Parameter | Factor D[b] | C2a[b] | Bb[b] | $\overline{\text{C1s}}$[c] | $\overline{\text{C1r}}$[c] |
|---|---|---|---|---|---|---|
| Z-Arg-SBzl | $k_{cat}$ | 0.64 | — | — | — | — |
| | $K_m$ | 3700 | — | — | — | — |
| | $k_{cat}/K_m$ | 170 | — | — | — | — |
| Z-Lys-Arg-Bu-$i$ | $k_{cat}$ | 1.0 | NH | SH | 39 | — |
| | $K_m$ | 2.5 | | | 240 | — |
| | $k_{cat}/K_m$ | 2500 | | | 160,000 | — |
| Z-Leu-Ala-Arg-Bzl | $k_{cat}$ | NH | 1.66 | 0.57 | 9.7 | 1.3 |
| | $K_m$ | | 240 | 110 | 16 | 160 |
| | $k_{cat}/K_m$ | | 7000 | 5400 | 600,000 | 8200 |
| Z-Gly-Leu-Ala-Arg-Bzl | $k_{cat}$ | NH | 2.34 | 0.59 | 4.1 | 3.4 |
| | $K_m$ | | 300 | 60 | 70 | 350 |
| | $k_{cat}/K_m$ | | 7900 | 9200 | 60,000 | 9500 |

[a] Kinetic constants were measured in 0.1 $M$ HEPES, 0.5 $M$ NaCl buffer, pH 7.5, 10% DMSO, and at 25° for factor D, C2a, and Bb, and 0.1 $M$ HEPES, 10 m$M$ CaCl$_2$, buffer, pH 7.5, 9% DMSO, and 30° for $\overline{\text{C1r}}$ and $\overline{\text{C1s}}$. The units for $k_{cat}$, $K_m$, and $k_{cat}/K_m$ are sec$^{-1}$, $\mu M$, and $M^{-1}$ sec$^{-1}$. NH, No hydrolysis; SH, slow hydrolysis.
[b] C.-M. Kam, B. J. McRae, J. W. Harper, M. M. Niemann, J. E. Volanakis, and J. C. Powers, *J. Biol. Chem.* **262**, 3444 (1987).
[c] B. J. McRae, T.-Y. Lin, and J. C. Powers, *J. Biol. Chem.* **256**, 12362 (1981).

sheep lymph capillary injury-related protease[23] are reported in the literature and not shown here.

One thioester substrate has been reported as an active site titrant of trypsin-like enzymes and is useful for determining enzyme active-site concentrations. Benzyl *p*-guanidinothiobenzoate, like *p*-nitrophenyl *p'*-guanidinobenzoate (NPGB), has high acylation rates with serine peptidases and forms stable acyl–enzyme derivatives which deacylate slowly. The compound is an active-site titrant for trypsin, thrombin, factor Xa, factor XIIa, activated protein C, and lung tryptase, and is more specific than NPGB.[26]

*Specificity.* Owing to high intrinsic reactivities, peptide thioester substrates are not as specific as amide substrates such as 4-nitroanilides, but they are often quite selective toward various serine peptidases. For example, the tripeptides Boc-Ala-Ala-AA-SBzl (where AA is Ala, Val, or Ile) are good substrates for PPE, HLE, and proteinase 3, but they are either slowly hydrolyzed or not hydrolyzed by chymotrypsin, cathepsin G, RMCP I, and RMCP II. Similarly, Boc-Ala-Ala-Phe-SBzl is a good substrate for chymotrypsin but is not hydrolyzed by the two elastases. Z-Arg-SBzl is a good substrate for most trypsin-like enzymes but is not hydrolyzed by

[26] R. R. Cook and J. C. Powers, *Biochem. J.* **215**, 287 (1983).

TABLE VI
KINETIC CONSTANTS FOR HYDROLYSIS OF PEPTIDE THIOESTERS BY HUMAN
AND BACTERIAL METALLOPROTEINASES

| Enzyme | Substrate | $k_{cat}$ (sec$^{-1}$) | $K_m$ ($\mu M$) | $k_{cat}/K_m$ ($M^{-1}$ sec$^{-1}$) |
|---|---|---|---|---|
| Fibroblast collagenase[a] | CH$_3$CO-Pro-Leu-Gly-$^S$Leu-Leu-Gly-OC$_2$H$_5$ | 100 | 3900 | 26,000 |
| Recombinant stromelysin[b] | CH$_3$CO-Pro-Leu-Gly-$^S$Leu-Leu-Gly-OC$_2$H$_5$ | 2.1 | 270 | 7800 |
| *Clostridium* collagenase[c] | Z-Gly-Pro-Leu-$^S$Gly-Pro-NH$_2$ | — | — | 4200 |
| *Clostridium* collagenase[c] | Boc-Abz-Gly-Pro-Leu-$^S$Gly-Pro-Nba | — | — | 63,000 |
| *Clostridium* collagenase[c] | Abz-Gly-Pro-Leu-$^S$Gly-Pro-Nba | 8.1 | 260 | 32,000 |
| Fibroblast collagenase[d] | CH$_3$CO-Pro-Leu-Ala-$^S$Nva-Trp-NH$_2$ | 170 | 1400 | 123,000 |
| Gelatinase[d] | CH$_3$CO-Pro-Leu-Ala-$^S$Nva-Trp-NH$_2$ | 7.5 | 45 | 167,000 |
| Stromelysin[d] | CH$_3$CO-Pro-Leu-Ala-$^S$Nva-Trp-NH$_2$ | 15 | 590 | 25,000 |
| Thermolysin[d] | CH$_3$CO-Pro-Leu-Ala-$^S$Nva-Trp-NH$_2$ | 1.7 | 360 | 4600 |

[a] Kinetic constants were measured in 50 m$M$ HEPES, 10 m$M$ CaCl$_2$ buffer, pH 6.5–7.0, and at 25°. Data were obtained from H. Weingarten and J. Feder, *Anal. Biochem.* **147**, 437 (1985); H. Weingarten, R. Martin, and J. Feder, *Biochemistry* **24**, 6730 (1985).

[b] Kinetic constants were measured in 50 m$M$ MES, 10 m$M$ CaCl$_2$ buffer, pH 6.0, and at 22°. Data were obtained from Q.-Z. Ye, L. L. Johnson, D. J. Hupe, and V. Baragi, *Biochemistry* **31**, 11231 (1992).

[c] Kinetic constants were measured in 50 m$M$ Tricine, 10 m$M$ CaCl$_2$, 0.5% DMSO, 2.4% methanol buffer, pH 7.5, and at 25°. Data were obtained from C. F. Vencill, D. Rasnick, K. V. Crumbley, N. Nishino, and J. C. Powers, *Biochemistry* **24**, 3149 (1985).

[d] Kinetic constants were measured in 0.1 $M$ MES, 10 m$M$ CaCl$_2$, 3.3% DMSO buffer, pH 6.0, and at 25°. Data were obtained from R. L. Stein and M. Izquierdo-Martin, *Arch. Biochem. Biophys.* **308**, 274 (1994).

elastase or chymotrypsin-like enzymes. Boc-Ala-Ala-Asp-SBzl is effectively hydrolyzed by human and murine lymphocyte granzyme B with $k_{cat}/K_m$ values of $10^4$–$10^5$ $M^{-1}$ sec$^{-1}$ but is not effectively hydrolyzed by several other serine peptidases tested.[5,27]

## Thioester Substrates for Metallopeptidases

The compound Leu-SC$_2$H$_5$ is the first reported thioester substrate for a metallopeptidase, leucine aminopeptidase.[11] Other peptide thioesters that are hydrolyzed by human and bacterial metallopeptidases are shown in Table VI.[9,10,28–30] CH$_3$CO-Pro-Leu-Gly-$^S$Leu-Leu-Gly-OC$_2$H$_5$ is the first reported thioester substrate for fibroblast collagenase[9,28] and has been used

[27] M. Poe, J. T. Blake, D. A. Boulton, M. Gammon, N. H. Sigal, J. K. Wu, and H. J. Zweerink, *J. Biol. Chem.* **266**, 98 (1991).

[28] H. Weingarten and J. Feder, *Anal. Biochem.* **147**, 437 (1985).

[29] Q.-Z. Ye, L. L. Johnson, D. J. Hupe, and V. Baragi, *Biochemistry* **31**, 11231 (1992).

[30] R. L. Stein and M. Izquierdo-Martin, *Arch. Biochem. Biophys.* **308**, 274 (1994).

to detect the activity of a recombinant stromelysin catalytic domain.[29] Peptides that contain the Gly-Pro-Leu-$^S$Gly-Pro, Pro-Leu-Gly-$^S$Leu-Leu-Gly, or Pro-Leu-Ala-$^S$Nva-Trp sequences are hydrolyzed by metalloendopeptidases such as collagenase and stromelysin. $CH_3CO$-Pro-Leu-Ala-$^S$Nva-Trp-$NH_2$ has been shown to be hydrolyzed not only by fibroblast collagenase but also by other metalloendopeptidases such as stromelysin, gelatinase, and thermolysin.[30] Several peptide thioesters such as Boc-Abz-Gly-Pro-Leu-$^S$Gly-Pro-Nba have been used to monitor the activity of *Clostridium histolyticum* collagenase.[10] *Clostridium* collagenase has a distinct preference for two prolines at the P2 and P2′ subsites, and substrates lacking those structural features such as Z-Gly-Pro-Leu-$SCH_3$ and $CH_3CO$-Leu-$SCH_2CO$-Pro-Nba are not hydrolyzed or are poorly cleaved.[10]

### Thioester Substrates for Other Enzymes

In addition to peptidases, thioester substrates have been developed for other enzymes in particular esterases. For example, acetylcholinesterase activity can be measured with the substrate acetylthiocholine in the presence of Ellman's reagent.[31] Several thioester-containing phospholipids are hydrolyzed by phospholipase $A_1$, phospholipase $A_2$, and lysophospholipases in continuous spectrometric assays in the presence of a thiol-reactive reagent.[32–34]

### Advantages and Limitations

One of the advantages of peptide thioester substrates for peptidases is the high sensitivity. This is the result of low background hydrolysis rates in neutral pH buffers and extremely high $k_{cat}/K_m$ values for enzymatic hydrolysis rates. In the case of elastase and chymotrypsin, sensitivity comparisons have been made between thioester substrates and substrates with the identical peptide sequence and different detector groups.[12,14] With the elastases PPE and HLE, $CH_3O$-Suc-Ala-Ala-Pro-Val-SBzl has a higher $k_{cat}/K_m$ value than either the corresponding 4-nitroanilide, ethyl ester, AMC, or NNap-$OCH_3$ substrates. The thioester substrate is capable of detecting picomolar concentrations of active-site titrated HLE or PPE using either Ellman's reagent or 4,4′-dithiodipyridine.[12] The substrate detected lower

[31] G. L. Ellman, K. D. Courtney, V. Andres, Jr., and R. M. Featherstone, *Biochem. Pharmacol.* **7,** 88 (1961).

[32] G. L. Kucera, C. Miller, P. J. Sisson, R. W. Wilcox, Z. Wiemer, and M. Waite, *J. Biol. Chem.* **263,** 12964 (1988).

[33] L. Yu and E. A. Dennis, *Methods Enzymol.* **197,** 65 (1991).

[34] L. J. Reynolds, L. L. Hughes, L. Yu, and E. A. Dennis, *Anal. Biochem.* **217,** 25 (1994).

concentrations of elastase than the fluorogenic substrates even though fluorescence detection is orders of magnitude more sensitive than chromogenic detection using a spectrometer. Similar substrate sensitivities are observed with chymotrypsin-like enzymes when using Suc-Phe-Leu-Phe-SBzl and MeO-Suc-Ala-Ala-Pro-Phe-SBzl.[14]

Owing to the high sensitivity, thioester substrates are also useful tools for substrate mapping of newly isolated peptidases. Using small amounts of enzyme, it is possible to learn quickly significant information on the substrate preference for a newly isolated enzyme or for an enzyme with limited availability. For example, factor IXa and factor VIIa have the lowest enzymatic activity among the coagulation serine peptidases toward synthetic substrates, and subsite mapping studies with peptide thioesters lead to the discovery of the substrates Z-Trp-Arg-SBzl and Z-Arg-SBzl. Until relatively recently, those were the only effective thioester substrates for the two enzymes[7,17,25] and probably are as effective as the described fluorescent substrates.[35] In the case of complement enzymes, similar subsite mapping studies lead to the development of substrates for factor D, C2a, Bb, C1r, and C1s.[18] Thioester substrates are still the only effective synthetic substrates for enzymes such as factor D. Interestingly, no synthetic thioester substrate has yet been reported for factor I.

Thioester substrates can be used to detect new serine peptidase activities in crude cell preparations such as the granules isolated from cytolytic lymphocytes. The granules have been found to contain several trypsin-like enzymes (granzyme A), several chymases, and peptidases specific for Asp, Met, and Ser in P1.[36,37] The subsite mapping studies with thioesters were important in the eventual characterization of a Met-specific peptidase from rat natural killer cells which hydrolyzes Boc-Ala-Ala-Met-SBzl effectively.[38] Similarly, Boc-Ala-Ala-Asp-SBzl has been discovered to be the only effective substrate for lymphocyte granzyme B[5,27] and lymphocyte fragmentin 2.[39] Interestingly, no effective thioester substrate has been discovered for granzymes C, D, and E from murine cytotoxic T lymphocytes, but this simply may indicate that the enzymes are not yet available in an active form.[5]

[35] S. Butenas, T. Orfeo, J. H. Lawson, and K. G. Mann, *Biochemistry* **31,** 5399 (1992).
[36] D. Hudig, N. J. Gregg, C.-M. Kam, and J. C. Powers, *Biochem. Biophys. Res. Commun.* **149,** 882 (1987).
[37] D. Hudig, N. J. Allison, T. M. Pickett, U. Winkler, C.-M. Kam, and J. C. Powers, *J. Immunol.* **147,** 1360 (1991).
[38] M. J. Smyth, T. Wiltrout, J. A. Trapani, K. S. Ottaway, R. Sowder, L. E. Henderson, C.-M. Kam, J. C. Powers, H. A. Young, and T. J. Sayers, *J. Biol. Chem.* **267,** 24418 (1992).
[39] L. Shi, C.-M. Kam, J. C. Powers, R. Aebersold, and A. H. Greenberg, *J. Exp. Med.* **176,** 1521 (1992).

The use of thioester substrates has been limited by the commercial availability of only a small number of compounds (Table I). Most derivatives are easily synthesized, but some derivatives such as Arg and Asp thioesters are more difficult. The Arg and Asp derivatives also have higher background hydrolysis rates. Thioester substrates are less specific than other peptide substrates such as peptide 4-nitroanilides and 7-amino-4-methylcoumarin derivatives, but this is compensated by the high reactivities. High concentrations of serum albumin, glutathione, or other free thiol compounds would interfere with single point assays by producing high background absorbance but would have little effect on rate assays.

## Summary

Peptide thioesters are sensitive substrates of various serine peptidases and metalloendopeptidases. Thioester substrates generally have high enzymatic hydrolysis rates and low background hydrolysis rates, and the hydrolysis rates can be easily monitored in the presence of thiol reagents such as 4,4'-dithiodipyridine or 5,5'-dithiobis (2-nitrobenzoic acid). Peptide thioester substrates have been invaluable for the study of enzyme specificity and enzyme inhibitors, especially in cases where no other practical synthetic substrates are available. Tripeptide substrates of the type Boc-Ala-Ala-AA-SBzl, where AA is nearly all of the 20 common amino acids, have now been synthesized and should be useful for the subsite mapping of new serine peptidases and the study of crude cell preparations containing serine peptidases.

## Acknowledgment

This work has been supported by grants from the National Institutes of Health (GM42212 and HL34035).

# [2] Fluorimetric Assays of Proteolytic Enzymes

## By C. Graham Knight

## Introduction

Fluorogenic substrates of proteolytic enzymes may be divided into three groups. The first group employs aromatic amines, such as 7-amino-4-methylcoumarin, which change fluorescence characteristics on acylation by an amino acid. Cleavage of the amide bond by a peptidase is accompanied by an increase in fluorescence, when monitored at the appropriate wavelengths. In most substrates of this type, the fluorescent group is conjugated to a short peptide which confers selectivity and specificity. As the fluorescent leaving group occupies the S1' specificity pocket, the substrates are most useful for peptidases whose specificities are dominated by the S subsites. Many proteinases, however, require both S and S' sites to be occupied by amino acids, and the enzymes will not cleave efficiently substrates of the peptidyl 4-methyl-7-coumarylamide (NH-Mec) type. In the second group of substrates this limitation is overcome by using an appropriate peptide sequence to separate the fluorophore from another group that quenches the fluorescence via intramolecular contacts. Such quenching is often inefficient when the length of the intervening peptide is longer than 3 or 4 residues. The third group of substrates is related to the second, but employs the principle of resonance (or radiationless) energy transfer, by which the fluorescent donor can be efficiently quenched by a suitable acceptor even when the groups are separated by 10 or more residues. Each of these classes of substrates will be discussed in more detail below.

## Peptidyl Derivatives of Fluorescent Amines

### 2-Naphthylamides

Peptidyl 2-naphthylamides and 4-methoxy-2-naphthylamides have been used most commonly as chromogenic substrates, the liberated 2-naphthylamine being converted to an intensely colored azo dye.[1] The fluorescence of naphthylamine can be detected with far higher sensitivity, but concerns about the carcinogenic nature of 2-naphthylamine have led most laboratories equipped with fluorimeters to replace naphthylamides with the analogous substrates derived from 7-amino-4-methylcoumarin. The latter are not only safer, but also more fluorogenic. The peptidyl naphthylamides remain valuable histochemical reagents, but routine use for fluorimetric assays cannot be recommended on safety grounds.

### 4-Methyl-7-coumarylamides

The favorable properties of 7-amino-4-methylcoumarin as a fluorescent leaving group were recognized first by Zimmerman et al.,[2] who synthesized glutaryl-Phe-NH-Mec as a substrate for chymotrypsin. Both the substrate and the product are intensely fluorescent, but with different excitation and

[1] A. J. Barrett and H. Kirschke, this series, Vol. 80, p. 535.
[2] M. Zimmerman, E. Yurewicz, and G. Patel, Anal. Biochem. **70**, 258 (1976).

emission wavelengths. Thus, glutaryl-Phe-NH-Mec has a $\lambda_{ex}$ of 325 nm and $\lambda_{em}$ of 395 nm, whereas 7-amino-4-methylcoumarin has a $\lambda_{ex}$ of 345 nm and $\lambda_{em}$ of 445 nm. When assays are made at a $\lambda_{ex}$ of 370 nm and $\lambda_{em}$ of 460 nm, the liberation of 7-amino-4-methylcoumarin can be measured against a low background fluorescence of unhydrolyzed substrate.

## Assay Methods: General Principles

Peptidyl 4-methyl-7-coumarylamide substrates for the more familiar serine and cysteine proteinases are widely available. Stock solutions (1–5 m$M$) of the substrate and 7-amino-4-methylcoumarin are prepared in dry dimethyl sulfoxide (DMSO) and stored at 4° in dark bottles (exposure to bright light should be avoided). The peptide content of the substrates can be less than 100%, and it is recommended that for kinetic studies the concentration of the substrate stock should be checked by total enzymatic hydrolysis, comparing the final fluorescence to that of a standard solution of 7-amino-4-methylcoumarin. Assay procedures will differ among enzymes, but the general principles may be illustrated, based on specific procedures described for chymotrypsin,[2] elastase,[3] and cathepsin B.[1]

Fluorescence is very temperature-dependent, so the fluorimeter should have a thermostatted cell holder, preferably of the electronic type. For continuous rate assays it is convenient to have an electronic stirrer. The buffer is filtered through a clean glass sinter and equilibrated to the assay temperature in a separate water bath. The fluorimeter is first calibrated with 7-amino-4-methylcoumarin, using a $\lambda_{ex}$ of 370 nm and $\lambda_{em}$ of 460 nm, so that the fluorescence increase during the assay may be converted to a molar concentration.

The following procedure is used in this laboratory with Perkin-Elmer (Norwalk, CT) fluorimeters controlled by the FLUSYS software.[4] Pre-warmed buffer (2.45 ml) is placed in the cuvette and 2 or 3 min allowed for thermal equilibration. Substrate solution (25 $\mu$l of 0.5 m$M$) is added to give a final concentration of 5 $\mu M$, and the instrument reading is set to zero. 7-Amino-4-methylcoumarin (25 $\mu$l of 0.05 m$M$) is added and the reading set to an arbitrary full scale deflection of 1000 units corresponding to 10% hydrolysis. The free coumarin is about 200 times more fluorescent than the substrate, so more sensitive assays can be made with the full scale deflection scaled to 2% hydrolysis. The cuvette is then washed, filled with fresh substrate solution, and allowed to equilibrate, and the assay is started by the addition of enzyme. That order of addition will detect residual enzyme adsorbed to the cuvette, and we routinely include 0.05% (w/v) Brij

---

[3] M. J. Castillo, K. Nakajima, M. Zimmerman, and J. C. Powers, *Anal. Biochem.* **99**, 53 (1979).
[4] N. D. Rawlings and A. J. Barrett, *Comput. Appl. Biosci.* **6**, 118 (1990).

35 in our buffers to decrease the adsorption. On occasion, it is necessary to wash the cuvette thoroughly with aqueous 0.05% (w/v) Brij 35 between runs. Although for kinetic studies the use of a continuously recording spectrofluorimeter is strongly recommended, the release of 7-amino-4-methylcoumarin during a fixed time interval can be measured after stopping the reaction with a suitable reagent.[1]

### Synthesis of 7-Amido-4-methylcoumarins of Side-Chain-Protected Amino Acids

For an enzyme with an unusual specificity, it may be necessary to synthesize a novel peptidyl-NH-Mec substrate, as we did for the asparagine-specific endopeptidase of legume seeds.[5] Although almost all the amino acid 7-amido-4-methylcoumarins are commercially available (Bachem Feinchemikalien AG, CH-4416 Bubendorf, Switzerland), those with unprotected side chains are unsuitable for synthesis owing to the possibility of side reactions. We wished to prepare a peptidyl-Glu-NH-Mec, but synthesis of the protected Boc-Glu(OBn)-NH-Mec (where Boc and OBn stand for *tert*-butyloxycarbonyl and benzyl ester, respectively) by standard methods[6] gave very poor yields. Such reactions are known to be difficult,[7] but we found the yield to be greatly increased by the addition of the acylation catalyst 1-hydroxy-7-azabenzotriazole[8] (Millipore Corporation, Bedford, MA 01730-9135). The experimental protocol that follows is likely to be of general applicability.

*Synthesis of Boc-Glu(OBn)-NHMec.* Boc-Glu(OBn)-OH (1.77 g, 5.25 mmol), 7-amino-4-methylcoumarin (0.88 g, 5.0 mmol), and 1-hydroxy-7-azabenzotriazole (0.68 g, 5.0 mmol) are dissolved in dry dimethylformamide (10 ml) at 0°. Dicyclohexylcarbodiimide (1.03 g, 5.0 mmol) in dimethylformamide (3 ml) is added, and the mixture is stirred at 0° for 2.5 hr and then at room temperature for 48 hr. Insoluble dicyclohexylurea is filtered and the filtrate rotary evaporated to leave an oil. The oil is dissolved in ethyl acetate (100 ml) and extracted with 30% (v/v) saturated $K_2CO_3$ (two times, 100 ml each), 10% (w/v) $KHSO_4$ (two times, 100 ml each), water (100 ml), and saturated NaCl (100 ml). After drying over $MgSO_4$, the ethyl acetate is evaporated to leave an off-white solid. The solid is triturated with 90%

[5] A. A. Kembhavi, D. J. Buttle, C. G. Knight, and A. J. Barrett, *Arch. Biochem. Biophys.* **303**, 208 (1993).
[6] T. Sasaki, T. Kikuchi, N. Yumoto, N. Yoshimura, and T. Murachi, *J. Biol. Chem.* **259**, 12489 (1984).
[7] S. Khammungkhune, *Synthesis* 614, (1980).
[8] L. A. Carpino, *J. Am. Chem. Soc.* **115**, 4397 (1993).

aqueous methanol and recrystallized twice from hot ethanol/water (yield 2.15 g, 87%). Boc-Glu(OBn)-NH-Mec ($M_r$ 494.55) has mp 151°–152° (found: C, 65.65; H, 6.14; N, 5.66; $C_{27}H_{30}N_2O_7$ requires C, 65.57; H, 6.11; N, 5.66%).

### 6-Amino-1-naphthalenesulfonamides

A disadvantage of the peptidyl 7-amino-4-methylcoumarylamides is that the structure occupying the S1′ subsite is not amenable to systematic variation and so the influence of P1′ on enzyme specificity cannot be explored. The problem has been addressed with the design of peptidyl 6-amino-1-naphthalesulfonamides, in which the sulfonamide group can be substituted with different alkyl chains to confer greater selectivity.[9,10] The substrates, designed for serine proteinases involved in coagulation and fibrinolysis, have less than 0.1% of the fluorescence ($\lambda_{ex}$ 352 nm, $\lambda_{em}$ 470 nm) of the parent sulfonamide. The fluorescence characteristics suggest that the 6-amino-1-naphthalenesulfonamides may have wider applications as substrates for routine screening for enzymes and inhibitors with fluorescent plate readers.[11] At present, they are not available commercially.

### Rhodamine 110

The xanthene dye rhodamine 110 is the most fluorescent amine leaving group so far described, having very high values of both molar absorptivity and quantum yield.[12] A disadvantage of rhodamine 110 is that the syntheses of peptidyl derivatives are quite difficult, as the xanthene ring carries two amine groups and both must be acylated to eliminate the fluorescence. Relative to rhodamine 110, the monosubstituted Z-Arg-NH-rhodamine (where Z is benzyloxycarbonyl) is about one-tenth as fluorescent, whereas (Z-Arg-NH)$_2$-rhodamine is 3000-fold less fluorescent again.[12] Thus for the most sensitive assays, stringent purification is necessary to eliminate all traces of the free amine. Although kinetic studies with peptidyl-rhodamine 110 have been described,[13,14] the substrates are probably most suited to the characterization of proteinase activities in living cells. Rhodamine 110 has the advantages that it is retained within the cell on release and its fluorescence is excited by visible light at 495 nm. Some rhodamine 110

[9] S. Butenas, T. Orfeo, J. H. Lawson, and K. G. Mann, *Biochemistry* **31,** 5399 (1992).
[10] S. Butenas, N. Ribarik, and K. G. Mann, *Biochemistry* **32,** 6531 (1993).
[11] D. M. Bickett, M. D. Green, J. Berman, M. Dezube, A. S. Howe, P. J. Brown, J. T. Roth, and G. M. McGeehan, *Anal. Biochem.* **212,** 58 (1993).
[12] S. P. Leytus, L. L. Melhado, and W. F. Mangel, *Biochem. J.* **209,** 299 (1983).
[13] S. P. Leytus, W. L. Patterson, and W. F. Mangel, *Biochem. J.* **215,** 253 (1983).
[14] S. P. Leytus, D. L. Toledo, and W. F. Mangel, *Biochim. Biophys. Acta* **788,** 74 (1984).

substrates are commercially available (Molecular Probes Inc., Eugene, OR 97402-0414).

## Quenched Fluorescent Peptide Substrates

Assays with substrates of the peptidyl-NH-Mec type give misleading results if the enzyme cleaves an amide bond in the peptide portion of the molecule. This may occur, for example, if the S1' subsite is of such restricted accessibility, or so favors a specific electrostatic interaction, that aromatic amine leaving groups are strongly excluded. Examples are the cleavages (at the position marked +) of Suc-Leu+Tyr-NH-Mec by the ClpP protein-ase (see Volume 244 in this series, [24]) and of Z-Arg+Arg-NH-Mec by omptin (see Volume 244, [28]). The alternative cleavages were detected only when the reaction mixtures were examined by high-performance liquid chromatography (HPLC). Many other peptidases cleave only bonds be-tween amino acid residues or within an extended peptide sequence, and HPLC provides an invaluable method of monitoring such activities, al-though it is less suitable for routine assays.

Fluorimetric approaches to monitoring such cleavages were developed initially for carboxypeptidase A[15] and trypsin,[16] using substrates in which a short peptide sequence separated a fluorescent donor from another group that acted as a quencher of fluorescence. This quenching can only be re-lieved when an intervening bond is cleaved, thus allowing the donor and acceptor to diffuse apart and so fluorescence to appear. Quenching can occur during normal random intramolecular contacts, but a more versatile approach employs resonance energy transfer, by which the excited state of the fluorophore can be deactivated via dipole–dipole interactions over separations of up to 50 Å (5 nm).

## Contact-Quenched Substrates

The first peptidase substrates in which a fluorescent $o$-aminobenzoyl (Abz) group was quenched by contact with an adjacent nitrophenyl chromo-phore were designed for leucine aminopeptidase. The close proximity is essential for quenching, and cleavage of the substrates Lys(Abz)+Ala-OBnNO$_2$ and Lys (Abz)+Ala-Ala-OBnNO$_2$ (where OBnNO$_2$ denotes a 4-nitrobenzyl ester) is accompanied by 27- and 18-fold increases in fluores-cence, respectively.[17] Quenching of the Abz group is equally effective with

[15] S. A. Latt, D. S. Auld, and B. L. Vallee, *Anal. Biochem.* **107**, 595 (1972).
[16] A. Carmel, M. Zur, A. Yaron, and E. Katchalski, *FEBS Lett.* **30**, 11 (1973).
[17] A. Carmel, E. Kessler, and A. Yaron, *Eur. J. Biochem.* **73**, 617 (1977).

4-nitrophenylalanine, and Abz-Gly+Phe($NO_2$)-Pro [where Phe($NO_2$) is 4-nitrophenylalanyl], designed as a substrate for dipeptidyl carboxypeptidase, is cleaved with a 71-fold increase in fluorescence.[18] It was recognized by Yaron et al.[19] that the quenching mechanism in the short peptide substrates is collisional, and the large fluorescence increases reflect the minimal separation of fluorophore and quencher.

The cleavage of longer peptides is accompanied by smaller increases in fluorescence, although these depend on the structure of the peptide. For example, the cleavage of Abz-Gly-Pro-Leu+Gly-Pro-NH-BnNO$_2$ (where NH-BnNO$_2$ is 4-nitrobenzylamide) by clostridial collagenase[20] and of Abz-Ala-His-Gln-Val-Tyr+Phe($NO_2$)-Val-Arg-Lys-Ala by human immunodeficiency virus (HIV) proteinase[21] results in about a 10-fold increase in fluorescence. By contrast, only a 2- to 3-fold increase is observed when bovine factor XIa cleaves Abz-Glu-Phe-Ser-Arg+Val-Val-Gly-NH-BnNO$_2$[22] and hepatitis A virus 3C proteinase cleaves Dns-Leu-Arg-Thr-Gln+Ser-Phe($NO_2$)-Ser-NH$_2$ (where Dns represents 5-dimethylaminonaphthalene-1-sulfonyl).[23]

o-Aminobenzoyl derivatives of peptidyl 4-nitroanilides have been proposed as substrates having both chromogenic and fluorogenic properties.[24] Although assays for both metalloproteinases[25] and subtilisin[26] have been described, more effective quenching of the o-aminobenzoyl group can be achieved by resonance energy transfer, as described below.

### Quenching by Resonance Energy Transfer

As contact-quenched peptides become longer and the increases in fluorescence on cleavage become smaller, the assays become less sensitive because the release of product is measured against a significant background fluorescence. For that reason the synthesis of new substrates based on collisional quenching cannot be recommended. When the fluorophore is

[18] A. Carmel and A. Yaron, Eur. J. Biochem. **87,** 265 (1978).

[19] A. Yaron, A. Carmel, and E. Katchalski-Katzir, Anal. Biochem. **95,** 228 (1979).

[20] C. F. Vencill, D. Rasnick, K. V. Crumley, N. Nishino, and J. C. Powers, Biochemistry **24,** 3149 (1985).

[21] Y.-S. E. Cheng, F. H. Yin, S. Foundling, D. Blomstrom, and C. A. Kettner, Proc. Natl. Acad. Sci. U.S.A. **87,** 9660 (1990).

[22] M. J. Castillo, K. Kurachi, N. Nishino, I. Ohkubo, and J. C. Powers, Biochemistry **22,** 1021 (1983).

[23] D. A. Jewell, W. Swietnicki, B. M. Dunn, and B. A. Malcolm, Biochemistry **31,** 7862 (1992).

[24] E. K. Bratovanova and D. D. Petkov, Anal. Biochem. **162,** 213 (1987).

[25] I. Y. Filippova, E. N. Lysogorskaya, E. S. Oksenoit, E. P. Troshchenkova, and V. M. Stepanov, Bioorg. Khim. **14,** 467 (1988).

[26] N. A. Stambolieva, I. P. Ivanov, and V. M. Yomtova, Arch. Biochem. Biophys. **294,** 703 (1992).

Abz, only a minor structural change, Phe(NO$_2$) to Tyr(NO$_2$) [Tyr(NO$_2$ is 3-nitrotyrosyl], provides the analogous substrates quenched by resonance energy transfer, which can be 80–90% efficient over separations of 12 to 14 residues.[27]

The size of the fluorescence increase on cleavage of the quenched fluorescent substrate depends on a number of physical and experimental factors. Of prime importance is the efficiency of resonance energy transfer between the fluorescent donor and the quenching acceptor. Förster[28] derived the relationship

$$(F_0 - F)/F_0 = R_0^6/(R^6 + R_0^6) \tag{1}$$

where $F$ and $F_0$ are the fluorescence intensities of the donor in the presence and absence of the acceptor, respectively, and $R$ is the distance between them. $R_0$, the distance at which the energy transfer efficiency is 50%, is given (in Å) by

$$R_0 = 9.79 \times 10^3 (\kappa^2 Q J n^{-4})^{1/6} \tag{2}$$

where $\kappa^2$ is an orientation factor having an average value close to 0.67 for freely mobile donors and acceptors, $Q$ is the quantum yield of the unquenched fluorescent donor, $n$ is the refractive index of the intervening medium, and $J$ is the overlap integral, which expresses in quantitative terms the degree of spectral overlap,

$$J = \int_0^\infty \varepsilon_\lambda F_\lambda \lambda^4 d\lambda \Big/ \int_0^\infty F_\lambda d\lambda \tag{3}$$

where $\varepsilon_\lambda$ is the molar absorptivity of the acceptor and $F_\lambda$ is the donor fluorescence at wavelength $\lambda$. Equation (2) predicts that quenching will be efficient over wide separations when the fluorophore has a high quantum yield and a fluorescence emission spectrum that is exactly overlapped by a strongly absorbing acceptor, although ideally the quencher should have little absorbance at the excitation wavelength. In practice, these conditions are seldom met exactly, and many useful substrates have been described that employ relatively weakly fluorescent donors and quenchers with moderate degrees of spectral overlap.

*Quenched Fluorescence Assays: General Principles*

Assays with quenched fluorescent peptides are made under conditions very similar to those for peptidyl-NH-Mec substrates. They require clean solutions, stirred at a constant temperature, and continuous monitoring of

[27] M. Meldal and K. Breddam, *Anal. Biochem.* **195**, 141 (1991).
[28] T. Förster, *Ann. Phys.* (*Leipzig*) **2**, 55 (1948).

the released fluorescence is preferred. Careful calibration of the assays is essential, as light absorption by the quenching group can lead to substantial errors at substrate concentrations greater than 10 to 20 $\mu M$. The following method of calibration is used in this laboratory.[29,30] An appropriate volume of buffer, prewarmed to the assay temperature, is placed in the cuvette and a small portion of substrate stock added. The solution is stirred over 2 to 3 min, until a constant fluorescence signal is obtained (it may be necessary to decrease the slit widths on the excitation side to avoid photochemical decomposition). The output is set to zero, and an amount of the fluorescence standard is added so that the full scale deflection corresponds to between 2 and 10% hydrolysis of the substrate.

The standard should correspond as closely as possible to the structure of the fluorophore in the substrate. Acetyl-Trp-NH$_2$, for example, is an appropriate standard for peptides containing an internal tryptophan or -Trp-NH$_2$. When the fluorescent group is conjugated at the N terminus, it is recommended that the standard should contain at least the first and preferably the second amino acid, as in Mcc-Pro-Leu-OH (where Mcc is 7-methoxycoumarin-3-carboxylyl).[31] Simple amides may give an incorrect calibration, as in the case of Abz-NH$_2$, which we find to be far more fluorescent than Abz-Pro-OH. A similar effect occurs with the proline derivative of the $N$-methylanthraniloyl group.[11] The fluorescence of acetyl-Lys(Mca)-OH (where Mca is 7-methoxycoumarin-4-acetyl), however, is comparable to that of Mca-NH$_2$.[32]

In studies made at a single substrate concentration, one calibration is sufficient for a number of experiments, although it is advisable to repeat the calibration periodically to compensate for any instrumental drift. In experiments in which the substrate concentration is varied, as in the measurement of $k_{cat}$ and $K_m$, it is essential to calibrate at each concentration. If this is not done, increasing absorptive quenching at higher substrate concentrations will result in spuriously low rate values, with correspondingly low kinetic constants.

The extent of trivial quenching should be measured routinely before first using a quenched fluorescent substrate, so that concentrations do not exceed those at which the correction is about 2-fold. One method is to measure the self-quenching of the residual fluorescence of the substrate. The fluorimeter is zeroed with buffer in the cuvette, and the fluorescence of substrate solutions of increasing concentration is measured. The values

[29] C. G. Knight, *Biochem. J.* **274,** 45 (1991).
[30] C. G. Knight, F. Willenbrock, and G. Murphy, *FEBS Lett.* **296,** 263 (1992).
[31] U. Tisljar, C. G. Knight, and A. J. Barrett, *Anal. Biochem.* **186,** 112 (1990).
[32] A. Anastasi, C. G. Knight, and A. J. Barrett, *Biochem. J.* **290,** 601 (1993).

are plotted versus substrate concentration and the initial linear portion of the curve extrapolated. The difference between the extrapolated and experimental points at any concentration shows the extent of quenching. In another method, the fluorimeter is calibrated first with 1 $\mu M$ fluorescence standard. The cuvette is then refilled with substrate solution and the fluorescence set to zero, before adding 1 $\mu M$ standard. The fluorescence is recorded and the procedure repeated at higher substrate concentrations. The extent of quenching is shown by a plot of the apparent fluorescence of the standard versus substrate concentration.

With the thimet oligopeptidase substrate Dnp-Pro-Leu-Gly-Pro-DL-Amp-D-Lys-OH [where Dnp is 2,4-dinitrophenyl and DL-Amp is 3-(7-methoxy-4-coumaryl)-DL-2-aminopropionyl], both methods showed that the extent of quenching was 50% at 10 $\mu M$ substrate.[29] In this case, the value of $K_m$ (1.23 $\mu M$) could be accurately determined by measurements of rates in the concentration range 0.2 to 8 $\mu M$. When $K_m$ is high, it may be possible to determine the individual kinetic constants with stopped assays, but it is preferable to measure only $k_{cat}/K_m$ at substrate concentrations well below $K_m$.[30]

### Dabcyl–Edans Substrates

A continuous fluorimetric assay for the human immunodeficiency virus 1 proteinase (retropepsin) was developed by Matayoshi et al.[33] based on a natural processing site for retropepsin in the Pr$^{gag}$ precursor polyprotein. The peptide Ser-Gln-Asn-Tyr + Pro-Ile-Val-Gln was labeled at the N terminus with 4-(4-dimethylaminophenylazo)benzoic acid (Dabcyl) and at the C terminus with 5-[(2-aminoethyl)amino]naphthalene-1-sulfonic acid (Edans). The Dabsyl absorption band overlaps almost exactly the fluorescence emission spectrum of Edans, so that a 40-fold fluorescence enhancement occurs on cleavage. A disadvantage of the Dabcyl–Edans substrate is that it must be made using time-consuming solution-phase, rather than solid-phase, peptide synthesis.[34] A side-chain-labeled amino acid, Glu(Edans), has been described,[35] that overcomes this objection.

The fluorescence yield from the Edans group, as defined by the product of the molar absorptivity $\varepsilon_{max}$ and quantum yield $\Phi_F$,[36] is small, Edans having an $\varepsilon_{max}$ of 5400 $M^{-1}$ cm$^{-1}$ at 336 nm and a $\Phi_F$ of 0.13.[33] Although

[33] E. Matayoshi, G. T. Wang, G. A. Krafft, and J. Erickson, *Science* **247**, 954 (1990).

[34] G. T. Wang, E. Matayoshi, H. J. Huffaker, and G. A. Krafft, *Tetrahedron Lett.* **31**, 6493 (1990).

[35] L. L. Maggiora, C. W. Smith, and Z. Y. Zhang, *J. Med. Chem.* **35**, 3727 (1992).

[36] I. D. Campbell and R. A. Dwek, "Biological Spectroscopy," Benjamin-Cummings, Menlo Park, California, 1984.

substrates made with the Dabcyl–Edans donor–acceptor pair are likely therefore to be rather insensitive, they have been found useful in studies of the cysteine proteinases.[37,38]

### Dansyl–Tryptophan Substrates

Energy transfer from tryptophan to a dansyl (5-dimethylaminonaphthalene-1-sulfonyl) group was used in the first quenched fluorescent peptidase substrates, designed for carboxypeptidase A.[15] Energy transfer is 50% efficient at a separation of 21 Å.[39] An advantage of the Dns–Trp system is that it also allows the formation and turnover of enzyme–substrate complexes to be followed by energy transfer from enzyme tryptophans to dansyl peptides lacking a tryptophan.[40] Such versatility has been exploited in a number of elegant studies of proteinase–substrate interactions employing stopped-flow techniques.[41] Although useful substrates have been described with 5 amino acid residues between Dns and Trp,[42,43] the extent of quenching decreased to 60% in a peptide with 7 intervening residues,[44] which suggests that other fluorophore–quencher pairs may be more suitable for longer peptides.

### 2,4-Dinitrophenyl–Tryptophan Substrates

A more effective quencher of tryptophan fluorescence is the $N$-2,4-dinitrophenyl (Dnp) group, with transfer efficiencies greater than 50% when the donor and acceptor are separated by less than about 30 Å.[45] Thus the cleavage of short peptide substrates, such as Dnp-Pro-Leu+Gly-Pro-Trp-D-Lys-OH by thimet oligopeptidase[46] and Dnp-Pro-Leu-Gly+Leu-Trp-Ala-D-Arg-NH$_2$ by vertebrate collagenase and gelatinase[47] is accompanied by large fluorescence increases. Although Dnp–Trp substrates have been

[37] C. García-Echeverría and D. H. Rich, *FEBS Lett.* **297**, 100 (1992).
[38] C. García-Echeverría and D. H. Rich, *Biochem. Biophys. Res. Commun.* **187**, 615 (1992).
[39] I. Z. Steinberg, *Annu. Rev. Biochem.* **40**, 83 (1971).
[40] S. A. Latt, D. S. Auld, and B. L. Vallee, *Biochemistry* **11**, 3015 (1972).
[41] D. S. Auld, *in* "Enzyme Mechanisms" (M. I. Page and A. Williams, eds.), p. 240. Royal Society of Chemistry, London, 1987.
[42] M. Ng and D. S. Auld, *Anal. Biochem.* **183**, 50 (1989).
[43] W. Stöcker, M. Ng, and D. S. Auld, *Biochemistry* **29**, 10418 (1990).
[44] K. F. Geoghegan, R. W. Spencer, D. E. Danley, L. G. Contillo, Jr., and G. C. Andrews, *FEBS Lett.* **262**, 119 (1990).
[45] N. M. Green, *Biochem. J.* **90**, 564 (1964).
[46] A. J. Barrett, C. G. Knight, M. A. Brown, and U. Tisljar, *Biochem. J.* **260**, 259 (1989).
[47] M. S. Stack and R. D. Gray, *J. Biol. Chem.* **264**, 4277 (1989).

valuable tools in studies of the matrix metalloproteinases,[48,49] there are disadvantages to the use of tryptophan as a fluorophore. The fluorescence yield is small, tryptophan having an $\varepsilon_{max}$ of 5600 $M^{-1}$ cm$^{-1}$ at 280 nm and a $\Phi_F$ of 0.2. Tryptophan is also abundant in proteins, so that assays made with crude enzyme preparations, or with protein inhibitors, will be less sensitive owing to a high fluorescence background.

*2,4-Dinitrophenyl–7-Methoxycoumarin Substrates*

Derivatives of 7-methoxycoumarin are also quenched by the 2,4-dinitrophenyl group, and compared to tryptophan they have the advantages of higher molar absorptivities and quantum yields. For example, 7-methoxycoumarin-4-acetic acid in water has an $\varepsilon$ of 14,500 $M^{-1}$ cm$^{-1}$ at 325 nm and $\Phi_F$ of 0.49,[50] and it is thus about 10- and 6-fold more fluorescent than Edans and tryptophan, respectively. In the first substrate of this type, we used 7-methoxycoumarin-3-carboxylic acid (Mcc-OH) to prepare the peptide Mcc-Pro-Leu+Gly-Pro-D-Lys(Dnp)-OH as a substrate for thimet oligopeptidase.[31] The fluorescence of the product Mcc-Pro-Leu is excited at 345 nm, and the fluorescence emission at 405 nm is overlapped by a prominent shoulder in the Lys(Dnp) absorption spectrum at 410 nm. The cleavage is accompanied by a 27-fold increase in fluorescence, and as expected the assay was unaffected by the addition of bovine serum albumin in concentrations up to 4 mg/ml.

The more favorable fluorescence characteristics of the coumarins were also clearly shown in a substrate designed for the matrix metalloproteinases.[30] We used 7-methoxycoumarin-4-acetic acid (Mca-OH) and 3-(2,4-dinitrophenyl-L-2,3-diaminopropionic acid (Dpa-OH), a shorter side-chain homolog of Lys(Dnp), to prepare Mca-Pro-Leu-Gly+Leu-Dpa-Ala-Arg-NH$_2$. Not only was the cleavage accompanied by a 190-fold increase in fluorescence ($\lambda_{ex}$ 328 nm, $\lambda_{em}$ 390 nm), but assays made with the substrate were 50 to 100 times more sensitive than those made with the closely related Dnp-Pro-Leu-Gly+Leu-Trp-Ala-D-Arg-NH$_2$.[47] Although part of the difference reflects the favorable substitution of Dpa for Trp in P2′, the assays were more sensitive than could be accounted for by differences in $k_{cat}/K_m$, reflecting the higher intrinsic fluorescence of the Mca group. We have designed   Mca-Nle-Val-Lys-Lys-Tyr+Leu-Asn+Ser-Lys(Dnp)-Leu-Asp-D-Lys-OH as a substrate for the bacterial endopeptidase pitrilysin.[32] Al-

[48] S. Netzel-Arnett, S. K. Mallya, H. Nagase, H. Birkedal-Hansen, and H. E. Van Wart, *Anal. Biochem.* **195**, 86 (1991).

[49] L. Niedzwiecki, J. Teahan, R. K. Harrison, and R. L. Stein, *Biochemistry* **31**, 12618 (1992).

[50] R. P. Haugland, "Handbook of Fluorescent Probes and Research Chemicals," p. 24. Molecular Probes, Eugene, Oregon, 1985.

though the Mca and Dnp groups are separated by 9 amino acid residues, the cleavage is accompanied by a 20-fold increase in fluorescence.

The coumarin-derived peptides described above were labeled with the fluorescent group by acylation of the amino terminus. In some circumstances it may be necessary to leave the terminus as a free amine or to use it for the attachment of a solubilizing group. A coumarin label can be incorporated in such peptides via the unnatural amino acid DL-2-amino-3-(7-methoxy-4-coumaryl)propionic acid, an analog of tryptophan in which the indole ring is replaced by 7-methoxycoumarin.[29] The fluorescent properties are similar to those of the Mca group.

*2,4-Dinitrophenyl–o-Aminobenzoyl Substrates*

The 2,4-dinitrophenyl group is also an effective quencher of *o*-aminobenzoyl (Abz) fluorescence, as was demonstrated by Juliano *et al.*[51] who made Abz-Gly-Gly-Phe-Leu+Arg-Arg-Val-NHC$_2$H$_4$NH-Dnp as a substrate for endooligopeptidase A. Cleavage of the substrate was accompanied by a 16-fold increase in fluorescence ($\lambda_{ex}$ 320 nm, $\lambda_{em}$ 420 nm). When similarly quenched substrates were developed for the kallikreins,[52,53] it was found that the degree of quenching was highly dependent on the peptide sequence. Thus Abz-Phe-Arg+Arg-Val-NHC$_2$H$_4$NH-Dnp was cleaved with a 7-fold increase in fluorescence, whereas hydrolysis of the related Abz-Phe-Arg+Ser-Val-NHC$_2$H$_4$NH-Dnp was accompanied by a 28-fold increase.[52] This effect may reflect a more extended peptide backbone arising from electrostatic repulsion in the Arg-Arg substrate and consequently greater spatial separation of the donor and acceptor groups. However, such electrostatic effects may be balanced by greater conformational freedom in longer peptides, as shown by the tricationic renin substrate Abz-Tyr-Ile-His-Pro-Phe-His-Leu-Val+Ile-His-NHC$_2$H$_4$NH-Dnp which is also cleaved with a 7-fold increase in fluorescence.[54]

A disadvantage of substrates containing the -NHC$_2$H$_4$NH-Dnp group is that they cannot be prepared by the usual methods of solid-phase peptide synthesis, but that objection does not apply to the equally effective quenchers of Abz fluorescence, Lys(Dnp)[32] and Dpa.[30] A complementary synthetic strategy is to introduce the fluorescent group as Lys(Abz) and to label the amino terminus with Dnp.[11]

[51] L. Juliano, J. R. Chagas, I. Y. Hirata, E. Carmona, M. Sucupira, E. S. Oliveira, E. B. Oliveira, and A. C. M. Camargo, *Biochem. Biophys. Res. Commun.* **173,** 647 (1990).

[52] J. R. Chagas, L. Juliano, and E. S. Prado, *Anal. Biochem.* **192,** 419 (1991).

[53] J. R. Chagas, I. Y. Hirata, M. A. Juliano, W. Xiong, C. Wang, J. Chao, L. Juliano, and E. S. Prado, *Biochemistry* **31,** 4969 (1992).

[54] M. C. F. Oliveira, I. Y. Hirata, J. R. Chagas, P. Boschcov, R. A. S. Gomes, A. F. S. Figueiredo, and L. Juliano, *Anal. Biochem.* **203,** 39 (1992).

*3-Nitrotyrosine–o-Aminobenzoyl Substrates*

3-Nitrotyrosine is another quencher of *o*-aminobenzoyl fluorescence, although less efficient than Dnp, and is also compatible with solid-phase peptide synthesis.[27] In a study of substrates for subtilisin Carlsberg, peptides with up to 15 residues between the Abz and Tyr(NO$_2$) groups were effectively quenched.[27] Quenching was attributed to long-range energy transfer; however, spectral overlap between Abz and Tyr(NO$_2$) is only modest, and the high degree of quenching more probably reflects the flexibility of the peptides described that allows shorter end-to-end distances.[55] Peptides containing Tyr(NO$_2$) and Abz have been valuable tools in characterizing the substrate specificity of the subtilisins.[56,57]

Choosing Peptide and Quencher–Fluorophore Pair

How does one design a quenched fluorescent substrate? In many cases, there is an existing assay employing a chromogenic peptide or in which the cleavage of a specific peptide is followed by HPLC and the peptides form the basis for the new substrate. When this is not the case, HPLC can be used to screen the wide variety of commercially available bioactive peptides for susceptibility to cleavage. Ideally, the cleavage should occur quite rapidly at a single site, which is identified by amino acid analysis of the products. The peptide sequence is then examined for residues which can be replaced by the fluorescent and quenching groups. Commonly, the N terminus is labeled, and the other member of the donor–acceptor pair replaces a hydrophobic residue on the C-terminal side of the scissile bond. If some flexibility in placement is possible, the position should be chosen to minimize the structural change. For example, a tryptophan may be replaced by a coumarin[29] or a Dnp-amino acid.[30] These guidelines should not be applied too rigidly, however, as it is often observed that quenched fluorescent peptides bind more tightly than the parent sequences.

With such a variety of quenchers and fluorophores available, many different substrates can be derived from the same peptide sequence. Although all will be effective substrates to some extent, some will be more suited to particular investigations. For example, if one wishes to measure individual $k_{cat}$ and $K_m$ values, it is necessary to cover as wide a concentration range as possible. To minimize trivial quenching, weakly absorbing quencher–fluorophore pairs, such as Dns–Trp and Tyr(NO$_2$)–Abz, are

---

[55] T. E. Creighton, *in* "Proteins: Structures and Molecular Properties," p. 171. Freeman, New York, 1993.
[56] H. Grøn, M. Meldal, and K. Breddam, *Biochemistry* **31,** 6011 (1992).
[57] L. M. Bech, S. B. Sørensen, and K. Breddam, *Biochemistry* **32,** 2845 (1993).

more suitable in continuous rate[27] and stopped-flow assays.[43] By contrast, to compare $k_{cat}/K_m$ values, it is necessary to measure rates at substrate concentrations well below $K_m$. Under those conditions, more efficient fluorophores, such as the coumarins, and more absorbent quenching groups, such as Dnp, will give appropriately sensitive assays. If stopped assays are used, the substrates may also be used to determine the individual kinetic constants.[30] Comparisons of different quencher–fluorophore pairs have seldom been made, but we found Mca-Gly-Gly-Phe-Leu-Arg+Arg-Ala-Lys(Dnp)-OH to be a 3-fold more sensitive substrate for insulysin than the Abz analog (A. Anastasi, C. G. Knight, and A. J. Barrett, unpublished work, 1993).

An important feature of any substrate is the ability to prepare it by solid-phase synthesis, and the sequence should be chosen with that in mind. Multiple column methods are ideal for preparing structural analogs for specificity mapping,[56] although defined substrate mixtures have also been used.[58]

Pitfalls and Precautions

Many of the cautionary points have been covered above but bear repeating. Before using a quenched fluorescent substrate, it is necessary to characterize the extent of trivial quenching over the concentration range to be employed in the assays. Failure to do that may lead to large corrections being applied at the higher substrate concentrations, with a corresponding loss of accuracy. It follows that if the concentration of substrate is varied, as during the determination of $K_m$ values, it is necessary to correct for trivial quenching at each substrate concentration to avoid a spurious apparent fall in rate at the higher concentrations. Accurate temperature control is essential, and stirring is helpful in maintaining the temperature and decreasing photobleaching. The sensitivity of some fluorimeters may be too high, so that fluorophore concentrations in the micromolar range are off scale. This is most conveniently controlled by decreasing the slit width in the excitation light path.

As cleavage at any peptide bond in a quenched fluorescent substrate leads to increasing fluorescence, the products of the hydrolysis reaction should be characterized. This is most conveniently done by separating the individual products by HPLC and determining the amino acid composition. We have found that synthetic standards do not give reliable indications of the site of cleavage. Ideally, the substrate should exhibit a single cleavage,

[58] J. Berman, M. Green, E. Sugg, R. Anderegg, D. S. Millington, D. L. Norwood, J. McGeehan, and J. Wiseman, *J. Biol. Chem.* **267**, 1434 (1992).

but substrates with two cleavages can be useful in the earlier stages of an investigation. In such a case, it may be possible to suppress one cleavage by introducing an $N$-methyl or a $\beta$-branched amino acid at the scissile bond.

## Notes Added in Proof

### Synthetic Peptide Libraries

Perhaps the most exciting new development has been the application of synthetic peptide libraries[59,60] to the study of peptidase specificity. Meldal and co-workers[61] used a "portion mixing" approach to prepare a family of substrates for subtilisin Carlsberg with the general structure H-Tyr(NO$_2$)-X$^5$-X$^4$-P-X$^3$+X$^2$-X$^1$-Lys(Abz)-OH. All the peptides had proline in P2 to direct the cleavage to the X$^3$+X$^2$ bond.[56] The starting H-Lys(Abz)-resin was distributed between 20 reaction vessels, each with a single activated amino acid. After coupling, the resins were recombined and redistributed among the wells. This procedure was repeated five times, and finally Tyr(NO$_2$) was coupled. Each resin bead had a unique sequence and on treatment with subtilisin the most susceptible became fluorescent as the Tyr(NO$_2$)-containing product was released. These beads, which contained about 5 pmol of substrate, were subjected to gas-phase sequence analysis. The development of a highly permeable polyethylene glycol-polyacrylamide resin[62] contributed greatly to the success of the method. Although Meldal *et al.*[61] observed some discrepancies between the bead and the classical solution kinetic approaches to defining specificity, the library method is clearly of great promise.

### New Fluorogenic Substrates

Macromolecular fluorogenic substrates, in which fluorescein-labeled peptides are conjugated to gluconoylated polylysine, have been developed by Mayer[63,64] After proteolysis the polymer is precipitated and the fluorescence released into the supernatant quantified. The method is highly sensitive and a substrate mimicking a cleavage site in the HIV-1 *gag–pol* precursor allowed the detection of a few fmol of retropepsin.[63] The method is

---

[59] M. A. Gallop, R. W. Barrett, W. J. Dower, S. P. A. Fodor and E. M. Gordon, *J. Med. Chem.* **1233**, 1251 (1994).

[60] P. J. Schatz, *Curr. Opin. Biotechnol.* **5**, 487 (1994).

[61] M. Meldal, I. Svendsen, K. Breddam and F.-I. Auzanneau, *Proc. Natl. Acad. Sci. USA* **91**, 3314 (1994).

[62] F.-I. Auzanneau, M. Meldal and K. Bock, *J. Peptide Sci.* **1**, 31 (1995).

[63] F. Anjuère, M. Monsigny, Y. Lelièvre and R. Mayer, *Biochem. J.* **291**, 869 (1993).

[64] F. Anjuère, M. Monsigny and R. Mayer, *Anal. Biochem.* **198**, 342 (1991).

compatible with fluorescence plate readers and so suitable for rapid screening of multiple samples.

New substrates for papain based on conserved sequences of natural inhibitors of the cystatin family have been described by Gauthier and co-workers.[65,66] They found that quenched fluorescent substrates spanning the consensus sequences were far more efficient than those with 7-amino-4-methylcoumarin in P1'.

By contrast, a quenched fluorescent substrate for the interleukin-1$\beta$ converting enzyme was no more sensitive than the corresponding peptidyl-NH-Mec.[67]

The solid-phase synthesis of substrates containing Abz and $NH(CH_2)_2NHDnp$ has been developed by Nishino[68] using the nucleophile-sensitive Kaiser oxime resin.[69]

New substrates have also been described for neprilysin,[70] renin,[71] thimet oligopeptidase,[72] and stromelysin 1,[73] the latter substrate allowing for the first time some discrimination between members of the matrix metalloproteinase family.

An interesting new principle is the application of bioluminescence to proteinase assays with the $\alpha$-chymotrypsin substrate 6-(N-acetyl-L-Phe)aminoluciferin.[74]

[65] F. Gauthier, T. Moreau, G. Lalmanach, M. Brillard-Bourdet, M. Ferrer-Di Martino and L. Juliano, *Arch. Biochem. Biophys.* **306,** 304 (1993).

[66] C. Serveau, L. Juliano, P. Bernard, T. Moreau, R. Mayer and F. Gauthier, *Biochimie* **76,** 153 (1994).

[67] M. W. Pennington and N. A. Thornberry, *Peptide Res.* **7,** 72 (1994).

[68] N. Nishino, Y. Makinose and T. Fujimoto, *Chem. Lett.* **1992,** 77 (1992).

[69] M. A. Findeis and E. T. Kaiser, *J. Org. Chem.* **54,** 3478 (1989).

[70] N. Goudreau, C. Guis, J.-M. Soleilhac and B. P. Roques, *Anal. Biochem.* **219,** 87 (1994).

[71] G. T. Wang, C. C. Chung, T. F. Holzman and G. A. Krafft, *Anal. Biochem.* **210,** 351 (1993)

[72] C. G. Knight, P. M. Dando and A. J. Barrett, *Biochem. J.* in press, (1995).

[73] H. Nagase, C. G. Fields and G. B. Fields, *J. Biol. Chem.* **269,** 20952 (1994).

[74] T. Monsees, W. Miska and R. Geiger, *Anal. Biochem.* **221,** 329 (1994).

# [3] Defined Substrate Mixtures for Mapping of Proteinase Specificities

By Gerard M. McGeehan, D. Mark Bickett,
Jeffrey S. Wiseman, Michael Green, and Judd Berman

## Introduction

The optimization of substrate design is an initial step toward the development of robust enzyme assays. In addition, the information gained from the optimization procedure can serve as a basis for inhibitor design. We previously reported a method which allows rapid optimization of substrates by screening pools of specific peptide mixtures.[1] Optimal substrates can be directly identified in each pool by using a combined high-performance liquid chromatography/mass spectrometry (HPLC/MS) technique. Herein we describe the optimization of substrates derived from the fluorogenic substrate Dnp-Pro-Leu-Gly-Leu-Trp-Ala-D-Arg-NH$_2$ (where Dnp represents the 2,4-dinitrophenyl group).[2] The substrate is cleaved specifically at the Gly-Leu linkage by the Zn$^{2+}$ metalloproteinases interstitial collagenase (EC 3.4.24.7) and gelatinase B (EC 3.4.24.35), yielding Dnp-Pro-Leu-Gly-OH and H-Leu-Trp-Ala-D-Arg-NH$_2$. Using the template, four separate substrate mixtures were synthesized at each subsite, P$_2$ Leu through P$_2'$ Trp. The mixtures contained either naturally occurring L-amino acids, D-amino acids, or either of two distinct sets of miscellaneous amino acids. Combined, 88 unique substitutions were surveyed at each position, and, over the four subsites, 352 potential substrates were evaluated. Using the approach we have identified and characterized a new peptide substrate which has a 35-fold improvement in $k_{cat}/K_m$ with interstitial collagenase.

Other approaches to peptide mixture analyses have been reported, and these are briefly described. These include (1) N-terminal sequence analysis of peptide mixtures,[3–6] (2) a novel, phage-based strategy for examining

[1] J. Berman, M. Green, E. E. Sugg, R. Anderegg, D. S. Millington, D. L. Norwood, G. M. McGeehan, and J. S. Wiseman, *J. Biol. Chem.* **267**, 1434 (1992).
[2] M. S. Stack and R. D. Gray, *J. Biol. Chem.* **264**, 4277 (1989).
[3] M. S. Rangeheard, G. Langrand, C. Triantaphylides, and J. Baratti, *Biochim. Biophys. Acta* **1004**, 20 (1989).
[4] A. J. Birkett, D. F. Soler, R. L. Wolz, J. S. Bond, J. Wiseman, J. Berman, and R. B. Harris, *Anal. Biochem.* **196**, 137 (1991).
[5] J. R. Petithory, F. R. Masiarz, J. F. Kirsch, D. V. Santi, and B. A. Malcolm, *Proc. Natl. Acad. Sci. U.S.A.* **88**, 11510 (1991).
[6] V. Schellenberger, R. A. Siegel, and W. J. Rutter, *Biochemistry* **32**, 4344 (1993).

METHODS IN ENZYMOLOGY, VOL. 248

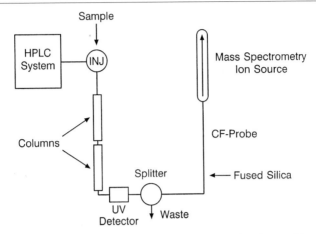

FIG. 1. Experimental configuration for mixture analysis using the combined HPLC/MS technique. Tandem HPLC columns are used to optimize separation of the components within the mixtures. Details of the separation and MS analysis are provided in the text.

proteinase substrate specificity,[7] and (3) the use of electrospray ionization mass spectrometry in mixture analysis.[8] Those methods complement our approach by identifying novel peptide sequences that can be optimized further using chemical strategies described here.

## Substrate Mapping by High-Performance Liquid Chromatography/ Mass Spectrometry

### Procedures

The method of substrate mapping is a straightforward process. The steps involved depend on having the capacity to (i) synthesize mixtures with substitutions at defined positions, (ii) establish a reliable enzyme assay to follow turnover of standard substrates in a mixture, (iii) achieve satisfactory chromatographic resolution of mixture components to analyze turnover, and (iv) obtain strong enough ionization signals to make mass assignments. The setup for mixture analysis used in the studies is shown in Fig. 1. The procedures outlined below describe each experimental step in more detail.

*Peptide Synthesis.* The protected peptides with mixtures at defined positions are prepared starting with 0.5 mmol *p*-methylbenzhydrylamine resin

[7] D. J. Matthews and J. A. Wells, *Science* **260,** 1113 (1993).
[8] D. B. Kassel, M. D. Green, R. Wiethe, R. Swanstrom, and J. Berman, manuscript in preparation.

(1.1 mmol/g) (Peptides International, Louisville, KY). The residues with fixed assignments are coupled using double couplings (preformed hydroxybenzotriazole esters in N-methylpyrrolidone) followed by capping with acetic anhydride. The mixed residues are coupled using extended single couplings (preformed hydroxybenzotriazole esters in N-methylpyrrolidone, 1 hr) followed by exhaustive capping with acetic anhydride. The mixture used for coupling the mixed residues consists of an equimolar amount of each of the amino acids (0.25 mmol total). After the final coupling and $N^{\alpha}$-Boc deprotection, the resins are neutralized and treated with 2,4-dinitrofluorobenzene (0.55 mmol) and diisopropylethylamine (0.55 mmol) in N-methylpyrrolidone (10 ml) for 3 hr. This places a dinitrophenyl group at the amino terminus of full-length peptides which is a useful UV tag for detection during chromatography. The resins are then treated with thiophenol [His(Dnp) removal] and are cleaved using 2 ml anisole in 10 ml HF at $-10°$ for 40 min. The HF is removed *in vacuo*, and the residues are diluted with 50 ml trifluoroacetic acid (TFA) and filtered. The residues are then additionally rinsed two times with 50 ml TFA each time. The TFA solutions are concentrated under reduced pressure, diluted with 150 ml of a 60% $CH_3CN$ solution (v/v) in water, and lyophilized to yellow solids.

*Mixture Digests.* Peptide mixtures are dissolved by warming in dimethyl sulfoxide (DMSO). Insoluble material is pelleted by centrifugation. Concentrations of Dnp-containing peptides are determined by measuring the absorbance at 375 nm (Dnp has $\varepsilon_{375}$ of 16,000 $M^{-1}$ cm$^{-1}$). Substrate mixtures are diluted to around 250 $\mu M$ final Dnp concentration (10–15 $\mu M$ per substrate) in assay buffer (200 m$M$ NaCl, 50 m$M$ Tris, 5 m$M$ CaCl$_2$, 0.05% Brij, pH 7.6). Standard substrate, Dnp-Pro-Leu-Gly-Leu-Trp-Ala-D-Arg-NH$_2$, is added to the D and miscellaneous amino acid mixtures to a final concentration of 15 $\mu M$. The hydrolyses are run at 37° and are initiated by adding recombinant human fibroblast collagenase or collagenase/EDTA (control) to each mixture. The reactions are followed by monitoring the increases in trytophan fluorescence ($\lambda_{ex}$ 280 nm, $\lambda_{em}$ 346 nm) and are quenched with EDTA after 30–50% hydrolysis.

*Substrate Analysis.* Relative $k_{cat}/K_m$ values of individual substrates are determined in competition assays at 37°. Gelatinase and collagenase are adjusted to 1.1 and 15 n$M$, respectively, in a 2-ml reaction volume. Reactions are initiated with substrate (10 $\mu M$ final concentration). Control digests contain excess EDTA or SC 43937, a potent inhibitor of both collagenase and gelatinase.[9] Fluorescence increase for the parent substrate Dnp-Pro-Leu-Gly-Leu-Trp-Ala-D-Arg-NH$_2$ is monitored ($\lambda_{ex}$ 280 nm, $\lambda_{em}$ 346 nm) using a Perkin-Elmer (Norwalk, CT) LS-5B luminescence spectrometer.

[9] J. P. Dickens, D. K. Donald, G. Keen, and W. P. McKay, U.S. Patent 4,599,361, July 8, 1986.

Hydrolyses are followed to 30–50% conversion of the parent substrate, then quenched with EDTA. HPLC analysis of the Dnp substrate and product peaks is monitored at 375 nm. Under competitive conditions, turnover of each substrate is first order with respect to other substrates, and relative $k_{cat}/K_m$ values can be determined directly using Eq. (1).[10]

$$[\ln(S/S_0)]/[\ln(S'/S_0')] = (k_{cat}/K_m)/(k'_{cat}/K'_m) \qquad (1)$$

*HPLC/CF-FAB-MS.* Chromatography is performed using a Waters (Milford, MA) 600 solvent delivery system and two Waters Delta Pak $C_{18}$ columns connected in series (150 × 3.9 mm i.d.). Solvent A consists of 0.1% trifluoroacetic acid with 1% glycerol, and solvent B consists of 0.1% trifluoroacetic acid, 1% glycerol, and 60% $CH_3CN$. Gradients are developed over 90 min using a concave slope at a flow rate of 1 ml/min. The effluent is split, and 10 $\mu$l/min is directed into the probe of a VG 70S double-focusing mass spectrometer. The magnetic field is scanned repetitively once every 5 sec from $m/z$ 500 to $m/z$ 1500. The ion source temperature is maintained at 50°, and the accelerating voltage is 5 kV.

Alternatively, 1% thioglycerol (v/v) is used in place of glycerol in both solvent A and B. The chromatography effluent is split, and 10 $\mu$l/min is directed into the frit-FAB (fast atom bombardment) probe of a JEOL SX-102 mass spectrometer. The magnetic field is scanned once every 6 sec from $m/z$ 100 to $m/z$ 1000. The ion source temperature is maintained at 55°, and the accelerating voltage is 10 kV.

*Estimates versus Measured Rates*

Mixtures (D, L, M1, M2) are prepared at each of the $P_2$–$P_2'$ positions (Table I). The results obtained with collagenase and gelatinase at $P_1'$ and $P_2'$ for the *L*, M1 , and M2 mixtures are shown in Fig. 2 to illustrate the typical data obtained. Analysis of the data shows that the specificity at $P_2'$ is very broad for both enzymes, and most substrates preferred by collagenase show similar enhancements with gelatinase. However, the specificity becomes more stringent around the scissile bond. The best $P_1'$ substrates incorporate straight-chain aliphatic residues at that position, there is notable selectivity for $\gamma$-S substituted cysteines, and gelatinase tolerates some branching (2-thienylalanine). There is also strict selectivity at $P_1$, with L-$\alpha$-aminobutyric acid (Abu) and Met being the only functional substitutes for Gly. The selectivity broadens again at $P_2$, with large hydrophobic groups like cyclohexylalanine (Cha) being optimal for collagenase (data not shown).

[10] R. H. Abeles, W. R. Frisell, and C. G. MacKenzie, *J. Biol. Chem.* **235**, 853 (1960); correction, *J. Biol. Chem.* **235**, (6) (1960).

TABLE I

PEPTIDE SUBSTRATE MIXTURES SYNTHESIZED BASED ON FLUOROGENIC SUBSTRATE
Dnp-Pro-Leu-Gly~Leu-Trp-Ala-D-Arg-NH$_2$

| Site | Peptide substrate[a] | Mixture |
|------|---------------------|---------|
| P$_2'$ | Dnp-Pro-Leu-Gly~Leu-<u>Xaa</u>-Ala-D-Arg-NH$_2$ | D, L, M1, M2 |
| P$_1'$ | Dnp-Pro-Leu-Gly~<u>Xaa</u>-Trp-Ala-D-Arg-NH$_2$ | D, L, M1, M2 |
| P$_1$ | Dnp-Pro-Leu-<u>Xaa</u>~*Cys(Me)*-*His*-Ala-D-Arg-NH$_2$ | D, L, M1, M2 |
| P$_2$ | Dnp-Pro-<u>Xaa</u>-Gly~*Cys(Me)*-*His*-Ala-D-Arg-NH$_2$ | D, L, M1, M2 |

[a] Positions incorporating the synthetic mixtures are designated <u>Xaa</u>. Positions changed from the original substrate are designated in italics. The D and L mixtures contain the natural amino acid side chains. The M1 and M2 mixtures contain 24 and 26 unnatural amino acids, respectively, all with the L configuration at the α carbon.

An accurate determination of the true $k_{cat}/K_m$ ratio must be made with individual peptide substrates to validate the mapping results. Thus, substrates are synthesized that incorporate the optimal substitutions found at P$_2$–P$_2'$ for collagenase. The respective $k_{cat}/K_m$ enhancements measured for the individual peptides (1–7) are shown in Table II. Overall, there is a 35-fold enhancement in $k_{cat}/K_m$ for the fully substituted substrate 7 over the published peptide substrate 1, representing a greater than 1.5 order of magnitude increase in turnover. Rates for peptides 5–7 are also faster than those for the physiological substrate collagen.[11] Turnover changes of that magnitude can have a significant impact on assay development by drastically cutting the time an assay takes or by allowing a precious enzyme to be conserved.

Figure 3 compares the relative $k_{cat}/K_m$ values determined for the individual substitutions at P$_2$ through P$_2'$ (Table II) with the expected enhancements calculated from the mixture analysis. There is good agreement at most of the positions. Histidine at P$_2'$ (peptide 2) gave a measured enhancement of 2.9 which correlates well with the value of 3.1 estimated from the mixtures. Substitution of $S$-methylcysteine at P$_1'$ (peptide 3) gave a 4.8-fold increase, which was greater than the estimate from the mixture analysis (3.2). Combination of the two adjacent substitutions gave a peptide (4) with $k_{cat}/K_m$ enhancement of 9.6 over the original substrate. The rate enhancement was more than the additive sum (7.7) but less than the multiplicative product (13.9) of the values at P$_1'$ and P$_2'$. Incorporation of L-α-aminobutyric acid at P$_1$ resulted in a 2-fold increase in peptide turnover with substrate 5, as expected. Incorporation of cyclohexylalanine alone at

[11] H. G. Welgus, J. J. Jeffrey, and A. Z. Eisen, *J. Biol. Chem.* **256,** 9511 (1981).

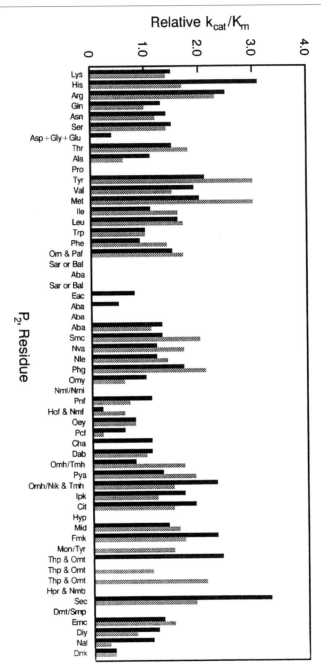

Fig. 2

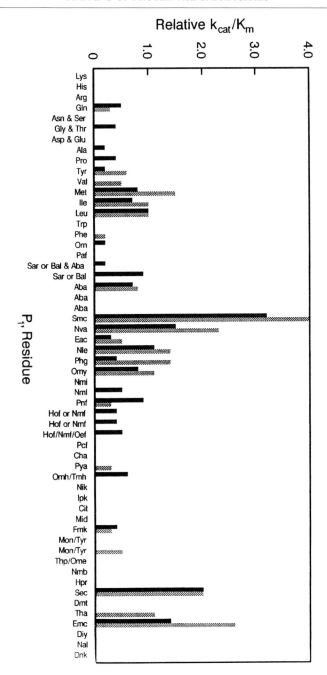

FIG. 2 (continued)

TABLE II
OPTIMIZED SUBSTRATES IDENTIFIED FROM ANALYSIS OF $P_2$–$P_2'$
MIXTURES FOR COLLAGENASE

| Substrate | Relative $k_{cat}/K_m{}^a$ |
|---|---|
| Type I collagen | 10.0 |
| 1 Dnp-Pro-Leu-Gly~Leu-Trp-Ala-D-Arg-NH$_2$ | 1.0 |
| 2 Dnp-Pro-Leu-Gly~Leu-His-Ala-D-Arg-NH$_2$ | 2.9 |
| 3 Dnp-Pro-Leu-Gly~Cys(Me)-Trp-Ala-D-Arg-NH$_2$ | 4.8 |
| 4 Dnp-Pro-Leu-Gly~Cys(Me)-His-Ala-D-Arg-NH$_2$ | 9.6 |
| 5 Dnp-Pro-Leu-Abu~Cys(Me)-His-Ala-D-Arg-NH$_2$ | 19.0 |
| 6 Dnp-Pro-Cha-Gly~Cys(Me)-His-Ala-D-Arg-NH$_2$ | 30.0 |
| 7 Dnp-Pro-Cha-Abu~Cys(Me)-His-Ala-D-Arg-NH$_2$ | 35.0 |

$^a$ Relative $k_{cat}/K_m$ values were determined under competitive conditions for each substrate.

$P_2$ (peptide 6) gave a 3.1-fold enhancement, again in close agreement with the value of 3.4 determined from the mixture analyses. The results underscore the fact that the competitive assay method can give reasonably accurate estimates of turnover of components in complex mixtures.

FIG. 2. Peptide substrate mixtures were hydrolyzed with collagenase (solid bars) or gelatinase (stippled bars) until 30–50% parent substrate was consumed. Reactions were quenched with EDTA (25 m$M$ final concentration). Analysis of mixture components was performed using the HPLC/MS procedure.[1] Under the competitive conditions of the assay, turnover of each substrate is first order relative to other substrates, allowing direct ranking of $k_{cat}/K_m$ values. Data presented represent turnover of substrates with amino acid substitutions of the L configuration. No turnover was observed for D-amino acid mixtures. Components that were hydrolyzed at less than 20% of control rates were treated as no turnover. The three-letter designations for the amino acids are used as follows: Aba, aminobutyric acid; Ala, alanine, Arg, arginine; Asn, asparagine; Asp, aspartate; Bal, $\beta$-alanine, Cha, cyclohexylalanine; Cit, citrulline; Dab, 2,4-diamino butyric acid; Diy, 3,5-diiodotyrosine; Dmt, 5,5,-dimethylthiazolidine-4-carboxylic acid; Dnk, $\varepsilon$-$N$-2,4-dinitrophenyllysine; Eac, 8-aminooctanoic acid; Emc, $S$-mercaptoethylcysteine; Fmk, $\varepsilon$-$N$-formyllysine; Gln, glutamine; Glu, glutamate; Gly, glycine; His, histidine; Hof, homophenylalanine; Hpr, pipecolic acid; Hyp, 4-hydroxyproline; Ile, isoleucine; Ipk, $\varepsilon$-$N$-isopropyllysine; Leu, leucine; Lys, lysine; Met, methionine; Mid, methionine sulfoxide; Mon, methionine sulfone; Nal, naphthylalanine; Nik, $\varepsilon$-$N$-nicotinoyllysine; Nle, norleucine; Nmb, $\alpha$-$N$-methylbutyric acid; Nmf, $\alpha$-$N$-methylphenylalanine; Nmi, $\alpha$-$N$-methylisoleucine; Nml, $\alpha$-$N$-methylleucine; Nva, norvaline; Oey, $O$-ethyltyrosine; Ome, $O$-methylthreonine; Omh, 1-methylhistidine; Omy, $O$-methyltyrosine; Orn, ornithine, Paf, $p$-aminophenylalanine; Pcf, $p$-chlorophenylalanine; Phe, phenylalanine; Phg, phenylglycine; Pnf, $p$-nitrophenylalanine; Pro, proline; Pya, pyridylalanine; Sar, sarcosine; Sec, $S$-ethylcysteine; Ser, serine; Smc, $S$-methylcysteine; Tha, thienylalanine; Thp, thioproline; Thr, threonine; Tmh, 3-methylhistidine; Trp, tryptophan; Tyr, tyrosine; Val, valine.

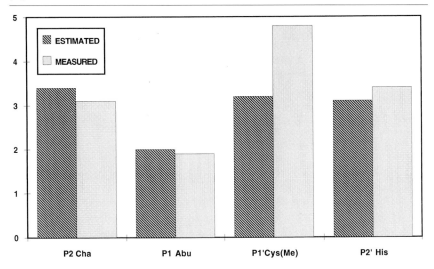

FIG. 3. Comparison of the relative $k_{cat}/K_m$ values predicted from analysis of the proteolysis mixture (hatched bars) and values determined from direct measurements with the purified peptide substrates (stippled bars). Assays were conducted under competitive conditions, and all the data were calculated based on the log (%S) remaining at the time of the quench. Abu, $\alpha$-Aminobutyric acid; Cha, cyclohexylalanine; Cys(Me), S-methylcysteine; His, histidine.

## Protocol Advantages

The essential feature of substrate design is optimization of turnover of target peptides (increased $k_{cat}/K_m$). The HPLC/MS technique described above has the capacity to determine simultaneously the relative $k_{cat}/K_m$ and the structures of each component in a mixture. The utility of the technique described here is then 2-fold. First, the procedure gives a reasonable, semiquantitative estimate of turnover within a limited dynamic range. The advantage derives from the competitive conditions of the assay under which turnover of each substrate can be treated as first order with respect to other substrates simultaneously. That obviates the need to synthesize and evaluate large numbers of peptides, individually. The dynamic range of the assay presented is about 20, representing turnover from 0.2 to 4.0 of the control peptide. The range can be extended by performing repetitive analysis of a mixture over an extended period to generate progress curves for all components, especially the slowly hydrolyzed components.[6] Second, the chemical approach has the capacity to incorporate many diverse unnatural amino acids in peptide mixtures which cannot be obtained through biological approaches. The functional enzyme assay then will determine what the optimal substitutions are at the various subsites. This is an attractive approach for generating novel, selective substrates for target enzymes. Indeed, of the positions surveyed in this study, the optimal substitutions at three of four sites were unnatural amino acids. The functional data also can represent a good starting point for the design of peptide-based enzyme inhibitors.[12]

Alternative Approaches

*Amino-Terminal Sequencing*

Other approaches to peptide mixture analyses have been published which complement the HPLC/MS method. Amino-terminal sequence analysis has been performed on peptide substrate mixtures to map the specificities of several enzymes.[3–6] A short experimental protocol is described for the approach.[5] Briefly, an amino-terminally acylated substrate mixture is prepared using chemistry similar to that described above. Mixtures are resuspended at 400–600 $\mu M$ in a buffer appropriate for the target proteinase. Digests are initiated by addition of enzyme, and the extent of proteolysis is quantitated by fluorescamine labeling.[13] Aliquots are removed at timed intervals and quenched. Samples containing 5–20 nmol of amine are submitted to automated Edman sequencing using standard phenylthiohydantoin (PTH) chemistry to characterize the cleavage products. The yields of the PTH–amino acids are used to quantitate the peptide products. Sequencing is continued to confirm the specificity of the cleavage site.

Again, analysis of mixtures is conducted under competitive conditions, and there is generally good agreement between the $k_{cat}/K_m$ estimates and $k_{cat}/K_m$ values determined for individually synthesized substrates using the procedure.[5] As with the chemical strategy described above, the approach allows the incorporation of many amino acids into the substrate mixtures. A distinct advantage of the method is the chromatographic resolution achieved with the PTH–amino acid conjugates, which determines the accuracy of the $k_{cat}/K_m$ estimates. The major drawback of the approach is that PTH standards of each of the mixture components must be prepared and analyzed prior to initiating any digests. If mixtures incorporate large numbers of unnatural amino acids, this characterization can be rate limiting.

*Phage Libraries*

A novel, phage display-based strategy for examining proteinase substrate specificity has been described.[7] The phage library approach has a

---

[12] G. M. McGeehan, C. J. Aquino, D. M. Bickett, P. J. Brown, M. Green, E. E. Sugg, J. S. Wiseman, and J. Berman, submitted for publication.
[13] T. Hammerle, C. U. T. Helen, and E. Wimmer, *J. Biol. Chem.* **266,** 5412 (1991).

distinct advantage over the other methods in the capacity to survey a large number of subsites using a combinatorial approach. The experimental details are beyond the scope of this chapter, but the general strategy is shown in Fig. 4. In the first step, a phagemid vector is generated which encodes a ligand/antigen attached by a linker to a phage coat protein. The linker sequence encodes a peptide sequence that contains fixed positions (Aaa) and multiply degenerate positions (Xaa). After transfection into *Escherichia coli*, the phage are harvested and captured on a plate coated with the appropriate antibody or receptor. The target proteinase is applied, and phage with susceptible linkers are liberated into the medium. The soluble phage are harvested, individual phage are plaque purified, and the susceptible cleavage sites are determined by nucleotide sequencing. In the second step, each susceptible cleavage sequence (e.g., Baa--Caa in Fig. 4B) is incorporated into an individual phagemid linking a reporter enzyme to the phage. Each construct is transfected into *E. coli* and harvested. The phage are then captured on a plate coated with an antibody to M13. The target proteinase is applied and liberates the marker enzyme from phage

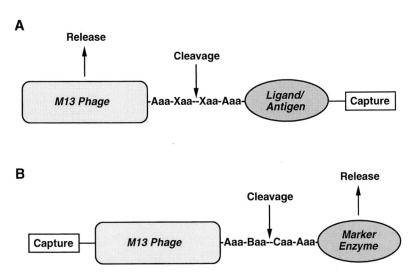

FIG. 4. Two-stage approach to define proteinase specificities using filamentous phage M13. (A). Phage pools are captured with an antibody or receptor on a solid-phase support. Treatment with the target proteinase generates soluble phage particles during the cleavage of susceptible bonds (Xaa--Xaa). (B) The defined cleavage sequence (Baa--Caa) is incorporated into a construct which links the phage to a marker enzyme. The phage is captured on a solid-phase support with an M13-specific antibody. Proteinase treatment releases the marker enzyme, giving an indication of the rate of peptide cleavage.

with susceptible linkers. The rate of release of marker enzyme gives an indication of turnover ($k_{cat}/K_m$) for the peptide cleavage.

The major limitations of the approach are that the analysis is restricted to the natural L-amino acids and there is no direct means of measuring relative turnover. However, the phage-based method nicely complements the chemical approaches by identifying novel peptide sequences that can be used as templates for optimization. The chemical diversity can be amplified by preparing and analyzing complex mixtures based on the templates.

## Electrospray Ionization Mass Spectrometry

A major limitation to all the methods described above is the need to separate mixture components prior to analysis. Advances in instrumentation may obviate the need to perform the separations in all cases. Methods are under development[8] to use electrospray ionization mass spectrometry to characterize peptide mixtures. The major advantages of the method are the ability to ionize mixtures for analysis without extensive chromatography and the ability to quantitate product as well as substrate turnover. As long as the size and physical properties of the mixture components are related, most peptides will ionize with similar efficiency. As with the other chemically based approaches, the method uses competitive conditions for enzyme hydrolysis, allowing simultaneous determination of the relative $k_{cat}/K_m$ and the structures of components in the mixtures. The major limitations of the procedure are that (1) isobaric compounds cannot be incorporated in a mixture and (2) the ionization of a mixture may not be efficient enough to generate a strong signal.

## Conclusion

The methods described above are not mutually exclusive and, in many ways, are complementary. The chemical approaches are similar in the nature of peptide mixture synthesis and in the competitive nature of the mixture analysis. Great chemical diversity can be achieved using the array of unnatural amino acids currently available. The development of new biological approaches is also exciting, as these will help identify novel templates for further peptide substrate optimization.

# [4] Assay of Proteoglycan Degradation

*By* CHRISTOPHER J. HANDLEY and DAVID J. BUTTLE

## Introduction

Proteoglycans comprise a large number of distinct molecules, all of which contain a protein core to which are covalently attached sulfated glycosaminoglycans (SGAG).[1] Proteoglycans are structurally important components of extracellular matrices. The highest concentration is found in cartilage, where proteoglycan molecules, such as aggrecan, are arranged into enormous macromolecular aggregates via binding to hyaluronic acid, a noncovalent association that is stabilized by link protein. The polyanionic nature of the aggregates, conferred by the presence of large numbers of sulfate and carboxyl groups, is responsible for the binding of water, and it is the high concentration of bound water that accounts for the property of compressive stiffness so important to the correct functioning of articular joints and intervertebral discs.[2] In basement membranes, such as those of the kidney, polyanionic proteoglycans act as a selective barrier to the passage of charged solutes and macromolecules, and on the external surfaces of cells they are thought to be important in many physiological processes that require cell–cell and cell–matrix interactions.[1]

The turnover of proteoglycans is a normal physiological process of removal and replacement. Removal appears to involve the fragmentation of the protein core by the action of proteolytic enzymes or oxygen free radicals, with hydrolysis of the glycosaminoglycan side chains occurring later. Any increase in the rate of removal or decrease in biosynthesis of proteoglycans may have pathological consequences, such as in the development of arthritis. Thus, proteinases involved in the degradation of proteoglycans could be contributing to the pathology of disease, and methods for determining rates of proteoglycan breakdown are important indicators of the process.

Owing to the extended structure of the proteoglycan molecule (at least of the monomer when in solution), the core protein is quite accessible to the action of endopeptidases. This, coupled with the availability of convenient

---

[1] D. Heinegård and Y. Sommarin, this series, Vol. 144, p. 305.

[2] A. Maroudas, J. Mizrahi, E. P. Katz, E. J. Wachtel, and M. Soudry, *in* "Articular Cartilage Biochemistry" (K. Kuettner, R. Schleyerbach, and V. C. Hascall, eds.), p. 311. Raven, New York, 1986.

methods for the estimation of SGAG content, makes the proteoglycan monomer a useful substrate for the assay of proteolytic activity.

## Methods

### Assay of Sulfated Glycosaminoglycans with Dimethylmethylene Blue

*Principle.* 1,9-Dimethylmethylene blue (DMB) was initially synthesized and used as a metachromatic histological stain.[3] The method was adapted for the spectrophotometric determination of SGAG in solution, where the metachromatic shift that occurs on binding of the basic dye to acidic polymers was quantitated. The method was limited, however, because of interference by proteins.[4] It gained wider applicability after it was found that the interference by proteins could be eliminated by treatment of the sample with papain, and that cartilage proteoglycan could be quantitated directly, without papain digestion. However, other polyanions, such as DNA and hyaluronic acid, interfered positively with the assay.[5] The interference could be removed by lowering the pH of the assay buffer and including NaCl in the buffer formulation.[6] The latter method is described below, for measurement in a spectrophotometer cuvette. The method has also been adapted for use in microwell plates, but in this case it is important to ensure that the samples are all read within a short time period (60 sec) after addition of the color reagent.[7]

*Color Reagent.* 1,9-Dimethylmethylene blue (obtainable from Serva Feinbiochemica, Heidelberg, Germany, or Aldrich Chemical Co., Milwaukee, WI) (16 mg) is dissolved in 1000 ml of water containing 3.04 g glycine, 2.37 g NaCl, and 95 ml of 0.1 $M$ HCl. The solution should be pH 3.0 and have an $A_{525}$ of 0.31. It is stable for at least 3 months when stored in a brown bottle at room temperature. Storage at 4° leads to a reversible loss of color.

*Procedure.* One hundred microliters of sample, containing up to 5 $\mu$g of SGAG, is placed in a polystyrene tube, and 2.5 ml of the color reagent is added. The mixture is poured into a disposable spectrophotometer cuvette, and $A_{525}$ is read immediately or after a fixed time of less than 60 sec. The assay is calibrated by the use of a reagent blank (without SGAG) and standards containing up to 5 $\mu$g of an appropriate SGAG.

[3] K. B. Taylor and G. M. Jeffree, *Histochem. J.* **1,** 199 (1969).
[4] R. Humbel and S. Etringer, *Rev. Roum. Biochim.* **11,** 21 (1974).
[5] R. W. Farndale, C. A. Sayers, and A. J. Barrett, *Connect. Tissue Res.* **9,** 24 (1982).
[6] R. W. Farndale, D. J. Buttle, and A. J. Barrett, *Biochim. Biophys. Acta* **883,** 173 (1986).
[7] M. Poe, R. L. Stein, and J. K. Wu, *Arch. Biochem. Biophys.* **298,** 757 (1992).

*Comments.* The SGAG–DMB complex is unstable. Aggregates form and precipitate with time, and it is therefore advisable to read the absorption at the earliest convenient fixed time period. Vigorous mixing, as achieved by vortexing, shaking, or aspirating through sampling devices, accelerates the formation of precipitates. The complexes adhere to the walls of tubes and cuvettes so that the use of disposable plastic cuvettes is recommended. Glass cuvettes should be rinsed with methanol followed by water before reuse.

The difference spectrum for SGAG–DMB complexes[6] shows a positive peak at 525 nm. A larger negative peak is seen at 590 nm, and the sensitivity of the assay could be increased by the use of that wavelength. However, interference from other polyanions is more pronounced than at 525 nm.

Different SGAG–DMB complexes give differing degrees of metachromasia.[5] Interference at 525 nm by other polyanions is, however, minimal (Fig. 1). An indication of the amounts of particular SGAG present can be obtained by digestion with specific polysaccharide lyases. For instance, the proportion of SGAG that is chondroitin sulfate, keratan sulfate, or heparan sulfate can be estimated from the reduction in metachromasia following

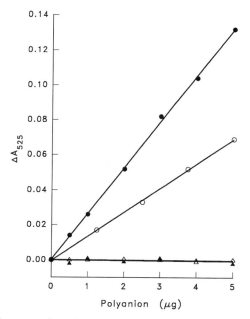

Fig. 1. Calibration curves for polyanions with DMB. The polyanions tested are as follows: ●, whale condroitin 4-sulfate; ○, keratan sulfate; ▲, calf thymus DNA; △, human umbilical hyaluronic acid. Reproduced with permission from Farndale *et al.*[6]

complete digestion with chondroitin AC lyase, keratanase, or heparitinase, respectively.[6,8,9]

*Assay of Degrading Activity by Release of Proteoglycan from Insoluble Matrix and Assay for Sulfated Glycosaminoglycans*

*Principle.* Proteoglycan breakdown can be conveniently assayed by measuring the release of proteoglycan from an insoluble matrix. Soluble proteoglycan fragments are easily separated from insoluble material by centrifugation or filtration, and levels of SGAG can be determined radiochemically or by the DMB assay. The insoluble matrix may simply be a piece of killed (frozen and thawed) and washed cartilage.[10,11] A more uniform and structurally less complex matrix can be made by entrapping purified proteoglycan monomers in a cross-linked polyacrylamide gel through which proteoglycan breakdown products can diffuse. The method was introduced by Dingle *et al.*,[12] who synthesized 10% (w/v) acrylamide beads containing $^{35}$S-labeled proteoglycan. Nagase and Woessner[13] modified and simplified the method of synthesis of the proteoglycan-containing polyacrylamide gel, as described below.

*Proteoglycan Monomer.* Extraction of proteoglycan monomer from slices of bovine nasal septum cartilage is achieved by treatment in 4 $M$ guanidine hydrochloride (50 m$M$ sodium acetate/acetic acid, pH 5.8, 23°, 36 hr) and dissociative density gradient centrifugation.[14] The product is dialyzed against water and freeze-dried.

*Proteoglycan Particles.* Solution A contains 1 $M$ Tris-HCl, pH 8.5, containing 0.2% (w/v) *N,N,N',N'*-tetramethylethylenediamine. Solution B contains 3 g acrylamide and 61 mg *N,N'*-methylenebisacrylamide made to 10 ml in water. Solution C contains 0.112 g ammonium persulfate in 20 ml water. To prepare proteoglycan particles, dissolve 480 mg proteoglycan monomer in 28 ml of solution A. Add and mix 8 ml of solution B and 12 ml of solution C. Allow 30 min at room temperature for polymerization to take place. Mince the gel with a pair of scissors in 300 ml water, homogenize, and then allow the gel to settle. Decant and retain the fines, and repeat the homogenization and decanting step with the gel fragments until no more rapidly settling gel remains. Combine the fines in 5 liters of water

[8] F. A. G. Reubsaet, J. P. M. Langeveld, and J. H. Veerkamp, *Biochim. Biophys. Acta* **838**, 144 (1985).

[9] J. R. Jenner, D. J. Buttle, and A. K. Dixon, *Ann. Rheum. Dis.* **45**, 441 (1986).

[10] D. J. Buttle, J. Tudor, and A. J. Barrett, *Br. J. Radiol.* **57**, 475 (1984).

[11] L. J. Ignarro, A. L. Oronsky, and R. J. Perper, *Clin. Immunol. Immunopathol.* **2**, 36 (1973).

[12] J. T. Dingle, A. J. Barrett, A. M. J. Blow, and P. E. N. Martin, *Biochem. J.* **167**, 775 (1977).

[13] H. Nagase and J. F. Woessner, *Anal. Biochem.* **107**, 385 (1980).

[14] V. C. Hascall and S. W. Sajdera, *J. Biol. Chem.* **245**, 4920 (1970).

and allow them to settle at 4° for 2 hr. Twice more, replace the water and allow the fines to settle again. Pour off 90% of the water and add with mixing an equal volume of acetone. Repeat the process until the particles turn opaque and settle into a thin layer. The particles are then resuspended in acetone and the suspension filtered over Whatman (Clifton, NJ) No. 1 paper. The remaining acetone is evaporated. This should leave the particles dry and free flowing. If they are at all sticky they should be rehydrated and the acetone extraction repeated. Freeze-dry the particles and store dry at room temperature. Papain digestion and the DMB assay can be used to estimate the recovery of proteoglycan, which should be greater than 50%.

*Papain.* The papain reagent is a suspension of 2× crystallized papain from Sigma Chemical Co. (St. Louis, MO).

*Procedure.* The particles are distributed into a series of 1.5-ml minicentrifuge tubes. This can be achieved by weighing (about 4 mg/tube), but a great deal of time can be saved by the construction and use of a small dispensing cup. The cup can be made from the tip of a glass Pasteur pipette. Heat the pipette tip over a Bunsen burner, then bend to an angle slightly more than 90° about 8 mm from the tip, and seal the base of the cup with a twist. Fill the cup (tap the contents down a few times and refill), then pour the contents into the minicentrifuge tube. An appropriate buffer and the endopeptidase under test are added, and the tubes are incubated with agitation (on a shaking platform or an end-over-end mixer) for a fixed amount of time. At the end of the incubation the particles are separated from the liquid phase by centrifugation (10,000 $g$, 2 min, 4°). The amount of SGAG in the supernatant is determined by the DMB assay (above).

The amount of residual SGAG in the particles is determined by the DMB assay of the product of papain digestion. Add 1 ml of 0.1 $M$ NaH$_2$/Na$_2$HPO$_4$, pH 6.8, containing 1 m$M$ EDTA, 4 m$M$ cysteine, and 0.4 mg papain. Incubation with agitation for 60 min at 40° leads to total solubilization of the remaining proteoglycan, which can be assayed with DMB following sedimentation of the particles by centrifugation. The amount of SGAG released during the first digestion is expressed as a percentage of total SGAG in the particles.

*Comments.* The method is appropriate for assaying the activity of endopeptidases (or polysaccharide lyases) that are small enough to enter the particles. This is not usually a consideration as, apart from a few exceptions such as the multicatalytic endopeptidase complex (EC 3.4.99.46), all endopeptidases will be able to enter the 5% (w/v) polyacrylamide gel. Endopeptidases vary in the ability to cleave the proteoglycan monomer into fragments small enough to escape from the gel, but enzymes that can do this can be assayed with sensitivity. For instance, nanogram quantities of trypsin (EC 3.4.21.4), thermolysin (EC 3.4.24.27), papain (EC 3.4.22.2), pepsin A (EC

3.4.23.1),[13] cathepsin B (EC 3.4.22.1),[15] matrilysin (EC 3.4.24.23), gelatinases A (EC 3.4.24.24) and B (EC 3.4.24.35), stromelysin 1 (EC 3.4.24.17), and stromelysin 2 (EC 3.4.24.22)[16] have all been detected by the method. Dose–response or time curves for the release of proteoglycan from an insoluble matrix are not linear, but they can be approximated to a straight line when enzyme concentration or time are plotted logarithmically. The phenomenon was first reported by Dingle et al.[12] and confirmed[13] for the use of proteoglycan entrapped in polyacrylamide beads. The same relationship is found for release from dead cartilage in vitro[10] and from live cartilage explants following stimulation of resorption with interleukin 1.[17] Although the reasons for the semilogarithmic relationship are not fully understood, it is one of the factors that makes the assay sensitive at low enzyme concentrations and short time periods.

The DMB assay is a convenient and robust method for the investigation of proteoglycan release from solid tissues in model systems of physiology and pathology. For instance, the method has been used in the implication of matrix metalloendopeptidases and the cysteine endopeptidase cathepsin B in cartilage proteoglycan breakdown.[18–21] For endopeptidases that break down proteoglycan in a highly specific manner, different methods for the analysis and characterization of breakdown products are required.

*Isolation and Characterization of Core Protein Fragments Originating from Catabolism of Proteoglycans Present in Culture Medium and Biological Fluids*

*Principle.* The increased availability of micro-sequencing technology and especially the ability to obtain N-terminal amino acid sequence data from polypeptides trapped on polyvinylidene difluoride membranes from electroblots of polyacrylamide gels have allowed detailed information to be obtained about the molecular mechanisms involved in the extracellular

---

[15] D. J. Buttle, M. Abrahamson, D. Burnett, J. S. Mort, A. J. Barrett, P. M. Dando, and S. Hill, *Biochem. J.* **276,** 325 (1991).

[16] G. Murphy, M. I. Cockett, R. V. Ward, and A. J. P. Docherty, *Biochem. J.* **277,** 277 (1991).

[17] J. Saklatvala, L. M. C. Pilsworth, S. J. Sarsfield, J. Gavrilovic, and J. K. Heath, *Biochem. J.* **224,** 461 (1984).

[18] J. S. Nixon, K. M. K. Bottomley, M. J. Broadhurst, P. A. Brown, W. H. Johnson, G. Lawton, J. Marley, A. D. Sedgwick, and S. E. Wilkinson, *Int. J. Tissue React.* **13,** 237 (1991).

[19] H. J. Andrews, T. A. Plumpton, G. P. Harper, and T. E. Cawston, *Agents Actions* **37,** 147 (1992).

[20] C. B. Caputo, L. A. Sygowski, D. J. Wolanin, S. P. Patton, R. G. Caccese, A. Shaw, R. A. Roberts, and G. DiPasquale, *J. Pharmacol. Exp. Ther.* **240,** 460 (1987).

[21] D. J. Buttle, C. J. Handley, M. Z. Ilic, J. Saklatvala, M. Murata, and A. J. Barrett, *Arthritis Rheum.* **36,** 1709 (1993).

catabolism of aggrecan, the major proteoglycan of cartilage.[22] Prior to the use of this technology, studies of the mechanisms of aggrecan catabolism had been restricted to measurement of the kinetics of degradation and release of proteoglycan from cartilage and the determination of changes in the chemical and physical properties of the products. Nevertheless, those studies established that the critical molecular event involved in the extracellular catabolism of aggrecan is the proteolytic cleavage of the core protein.[23,24]

Amino-terminal amino acid sequence data can be obtained from aggrecan core protein fragments appearing in the medium of cartilage explant cultures and from biological fluids such as serum and synovial fluid. After separation of the polypeptide fragments by electrophoresis on polyacrylamide gels and electroelution to polyvinylidene difluoride membranes, the resulting polypeptide bands can be subjected to N-terminal amino acid sequence analysis and probed with monoclonal antibodies that interact with specific epitopes within the aggrecan macromolecule. The N-terminal sequence data can be compared with the published amino acid sequence of the core protein of aggrecan.[25] These data, combined with the reactivity of the core protein fragments with specific antibodies, can determine the sites of proteolytic cleavage within the core protein of aggrecan and thus indicate the type of proteinase that is involved.[26]

When undertaking such an analysis, it is important to consider the nature of the system and the amount of core protein fragments present. Levels of core protein fragments originating from the catabolism of aggrecan in cartilage explant cultures can be very low, and it is therefore necessary to isolate the polypeptides free from other proteins, if analysis of the N-terminal amino acid sequence is to be undertaken. Where the analysis of aggrecan core protein fragments is to be carried out using specific antibodies, sufficient amounts of the core protein polypeptides need to be used, and this may require some purification of products to allow for the application of sufficient material to polyacrylamide gels. The presence of exogenous proteins can be tolerated.

In the following sections the isolation and analysis of aggrecan fragments present in synovial fluid or serum, or lost to the medium of explant cultures of articular cartilage, will be described.

[22] P. Matsudaira, *J. Biol. Chem.* **262**, 10038 (1987).
[23] C. J. Handley and D. A. Lowther, *Biochim. Biophys. Acta* **582**, 234 (1979).
[24] M. A. Campbell, C. J. Handley, V. C. Hascall, R. A. Campbell, and D. A. Lowther, *Arch. Biochem. Biophys.* **234**, 275 (1984).
[25] K. Doege, M. Sasaki, T. Kimura, and Y. Yamada, *J. Biol. Chem.* **266**, 894 (1987).
[26] M. Z. Ilic, C. J. Handley, H. C. Robinson, and M. T. Mok, *Arch. Biochem. Biophys.* **294**, 115 (1992).

*Procedure: Isolation of Proteolytic Products of Aggrecan Catabolism Present in Medium of Cartilage Explant Cultures and in Other Biological Fluids.* Aggrecan fragments can be isolated from the medium of explant cultures of cartilage and other biological fluids such as synovial fluid and serum by ion-exchange chromatography followed by density gradient centrifugation. In all cases, consideration should be given to the addition of proteinase inhibitors to the fluid of interest, to control proteolytic activity during storage and in the initial purification steps.[27] No pretreatment of the biological fluid is necessary, except in the case of synovial fluid where the viscosity needs to be reduced by the specific degradation of hyaluronan. In this case, synovial fluid (3 ml) is mixed with 1 ml of 0.2 $M$ sodium acetate buffer, pH 6.0, containing proteinase inhibitors[28] and 2 units of *Streptomyces* sp. hyaluronidase (EC 4.2.2.1). The digest is incubated at 37° for 10 hr.

Aggrecan fragments are best isolated from culture medium or biological fluids by ion-exchange chromatography using buffers containing 6 $M$ urea. This approach allows for the concentration as well as the purification of core protein fragments from the fluids. In the case of aggrecan fragments present in the medium from explant cultures of articular cartilage, up to 400 ml of the pooled medium can be applied to a column of Q-Sepharose (Pharmacia, Uppsala, Sweden) (1.0 × 20 cm) equilibrated with 50 m$M$ sodium acetate buffer, pH 6, containing 6 $M$ urea and 0.15 $M$ sodium chloride. The column is then washed at a flow rate of 30 ml/hr with at least 100 ml of the same urea buffer prior to elution with a linear sodium chloride gradient of 0.15–1.5 $M$ (total volume of 400 ml) in urea buffer. If the fragments are radiolabeled, aliquots of fractions from the column are assayed for radioactivity; if fragments are unlabeled, fractions are assayed chemically for hexuronate or SGAG. Figure 2 shows a typical elution profile from Q-Sepharose for aggrecan core protein fragments from the medium of explant cultures of articular cartilage, where the macromolecules elute at a sodium chloride concentration of approximately 0.75 $M$.

Instead of using a linear gradient, the Q-Sepharose column can be eluted with a stepwise gradient. After applying the sample of culture medium or biological fluid to the column, the column is washed with sodium acetate buffer, pH 6, containing 6 $M$ urea and 0.15 $M$ sodium chloride, then with 5–10 bed volumes of the same buffer but containing 0.5 $M$ sodium chloride. The aggrecan fragments are then eluted from the ion-exchange medium with 4 bed volumes of 4 $M$ guanidinium chloride in 0.1 $M$ sodium acetate, pH 6. Experience has shown that greater than 95% recovery of aggrecan fragments is obtained from Q-Sepharose ion-exchange columns.

[27] T. R. Oegema, V. C. Hascall, and R. Einsenstein, *J. Biol. Chem.* **259,** 1720 (1979).
[28] Y. Oike, K. Kimata, T. Shinomura, and S. Suzuki, *Biochem. J.* **191,** 203 (1980).

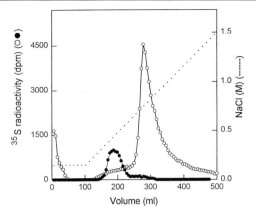

FIG. 2. Elution profiles of aggrecan core protein fragments from Q-Sepharose before (○) and after (●) digestion with chondroitin ABC lyase and keratanase. Reproduced with permission from Ilic et al.[26]

Analysis by sodium dodecyl sulfate (SDS)–polyacrylamide gel electrophoresis of fractions from Q-Sepharose containing aggrecan fragments showed that link protein was also present. If N-terminal amino acid sequencing of fragments is to be carried out, it is necessary to remove link protein from the preparation, which is achieved by subjecting fractions containing the aggrecan fragments to density gradient centrifugation. Fractions from Q-Sepharose that contain aggrecan fragments are pooled and concentrated by precipitation with 3 volumes of ethanol at 4° for 12 hr. The precipitate is then resuspended in 4 $M$ guanidinium chloride and 50 m$M$ sodium acetate, pH 6. Alternatively, the pooled fractions from Q-Sepharose are diluted with water so that the ionic strength is below 0.5 $M$ and then reapplied to Q-Sepharose eluted with sodium acetate buffer, pH 6, containing 0.5 $M$ sodium chloride. The column is washed with the same buffer and the aggrecan fragments eluted with 4 bed volumes of 4 $M$ guanidinium chloride in 0.1 $M$ sodium acetate, pH 6. The resulting solution of aggrecan fragments in 4 $M$ guanidinium chloride is adjusted to a density of 1.5 g/ml with solid cesium chloride and subjected to centrifugation at 100,000 g for 40 hr at 10°. The resulting density gradients are fractionated into three equal fractions. Aggrecan fragments are recovered in the bottom fraction of each gradient, and 2-ml aliquots of the fraction are applied to Sephadex G-50 (1.4 × 17 cm) equilibrated in 50 m$M$ ammonium hydrogen carbonate and the excluded volume fractions from the column collected and lyophilized.

To obtain a separation of aggrecan core protein fragments by electrophoresis on SDS–polyacrylamide gels, the aggrecan core proteins need to be deglycanated using chondroitin ABC lyase and keratanase. Lyophilized samples of aggrecan core protein fragments are resuspended in 2 ml of 0.1 $M$ Tris and 0.1 $M$ sodium acetate, pH 7, containing proteinase inhibitors,[28] 0.12 units of chondroitin ABC lyase (EC 4.2.2.4), and 0.24 units of keratanase (EC 3.2.1.103). The choice of the grade of chondroitinase ABC is important. Proteinase-free grade (obtainable from Seikagaku Kogyo Co., Tokyo, Japan, or ICN Biomedicals Inc., Costa Mesa, CA) should be used because the preparation is highly pure and does not contain added carrier protein which complicates subsequent separation and analysis of the aggrecan fragments.

After incubation for 10 hr at 37° the digest is subjected to ion-exchange chromatography in order to remove the endoglycosidases, which is necessary if N-terminal amino acid sequence analysis of aggrecan core proteins is to be carried out. The digest is applied to a column of Sephadex G-50 equilibrated with 6 $M$ urea and 50 m$M$ sodium acetate, pH 6, containing 0.15 $M$ sodium chloride, and the excluded volume fractions are pooled, then applied to a column of Q-Sepharose, and eluted with a linear sodium chloride gradient as described above. Figure 2 shows the elution profile of chondroitin ABC lyase- and keratanase-digested aggrecan core protein fragments from Q-Sepharose, where the polypeptides elute at a sodium chloride concentration of approximately 0.47 $M$. The core protein fragments are recovered from the fractions by ethanol precipitation or by being exchanged into ammonium hydrogen carbonate solution by size-exclusion chromatography followed by lyophilization.

*Separation and Characterization of Aggrecan Core Protein Fragments by Electrophoresis on Sodium Dodecyl Sulfate–Polyacrylamide Gels.* The separation of deglycosylated core protein fragments originating from aggrecan is best carried out by electrophoresis on gradient polyacrylamide–SDS gels. Electrophoresis gives a better separation of the core protein polypeptides than is achieved with ion-exchange, size-exclusion, and reversed-phase chromatography and allows for the convenient characterization of the core protein products by either N-terminal amino acid sequencing or determining reactivity to antibodies. Peptides derived from the culture medium are subjected to electrophoresis on 4–10% gradient polyacrylamide–SDS gels (1.5 mm thick) at a loading concentration of approximately 2 mg per well. After electrophoresis the gels are stained with Coomassie blue or subjected to electroelution (250 mA; 18 hr; 4°) using polyvinylidene difluoride membranes in a buffer consisting of 10 m$M$ CAPS [3-(cyclohexylamino)propanesulfonic acid], pH 11, in 10% (v/v) aqueous methanol solution.[22] The polyvinylidene difluoride membranes are washed in water, then stained with Coomassie blue, and bands corresponding to peptides are cut out and subjected to amino acid sequencing in a gas-phase peptide sequenator. As an alternative to staining the polyvinylidene difluoride membranes to visualize the core protein fragments, the membranes can be viewed while drying. Regions of the membranes containing peptides dry slower, thus enabling the core protein bands to be marked.

Figure 3 (lane d) shows a Coomassie blue-stained electroblot of aggrecan fragments isolated from the medium of articular cartilage explant cultures. It is evident that six peptides are present. Peptide 2 has the same N-terminal sequence as the bovine aggrecan core protein, VEVXEP. Peptides 3–5 have the common sequence ARGSVILXA that corresponds to a sequence starting at residue 374 (residue 1 is the N-terminal valine residue of the core protein of the secreted aggrecan macromolecule) in the human aggrecan core protein. Peptide 6 has the sequence AGEGPXGILE and peptide 7 the sequence LGQRPPVT that correspond to sequences starting at residues 1820 and 1920, respectively, in the human aggrecan sequence.[25] Amino acid yields from the first cycle of N-terminal amino acid sequence analysis of each core protein band originating from 2 mg of purified aggrecan core protein fragments ranged from 2 to 18 pmol.

After electrophoresis and electroelution onto polyvinylidene difluoride membranes, the membranes can be probed with antibodies which react with known epitopes present in the aggrecan macromolecule. This is performed on 4–10% gradient polyacrylamide–SDS gels (0.75 mm thick) requiring approximately 1 $\mu$g of core protein fragments to be loaded in each

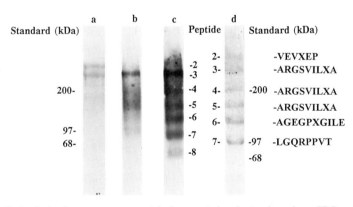

FIG. 3. Analysis of aggrecan core protein fragments by electrophoresis on SDS–polyacrylamide gradient gels. After electrophoresis the blots were reacted with the following: lane a, antibody 12/21/1-C-6 which reacts with a common peptide present in both G1 and G2 domains of the aggrecan core protein; lane b, antibody 1/20/5-D-4 which reacts with keratan sulfate disaccharides; lane c, antibody 5/6/3-B-3 which reacts with terminal chondroitin 6-sulfate disaccharides, Lane d shows a gel stained with Coomassie blue and the corresponding N-terminal amino acid sequence of each core peptide band. Reproduced with permission from Ilic et al.[26]

well. After electroelution to polyvinylidene difluoride membranes (200 mA; 1.5 hr; 4°), the membranes are soaked in 5% (w/v) skim milk in phosphate-buffered saline to block protein binding sites. The membranes can then be probed with monoclonal antibodies that react with either protein or carbohydrate epitopes present in aggrecan macromolecules. The polyvinylidene difluoride membranes are then incubated with a second antibody conjugated with either horseradish peroxidase or alkaline phosphatase.

Examples of monoclonal antibodies that have been used in work elucidating the mechanism of catabolism of aggrecan by mature articular cartilage are shown in Fig. 3. Lane a (Fig. 3) shows that peptides 2 and 3 react with the monoclonal antibody 12/21/1-C-6 that recognizes a common epitope within the two N-terminal globular domains (G1 and G2) of aggrecan,[29] lane b (Fig. 3) shows that peptides 2–7 are positive for the antibody 1/20/5-D-4 which reacts with regions of keratan sulfate containing disulfated disaccharides.[29] Lane c (Fig. 3) demonstrates that all the peptides react with the antibody 5/6/3-B-3 which recognizes terminal 6-sulfated chondroitin sulfate disaccharides.[29] The monoclonal antibodies 1/20/5-D-4 and 5/6/3-B-3 are commercially available from Seikagaku Kogyo Co. or ICN Biomedicals Inc. This use of antibodies allows for the detection of nanogram levels of core protein peptides present and is more sensitive than direct staining methods.

*Comments.* The above experimental approach has the potential to be used to investigate the catabolism of other species of proteoglycans by tissue or cell cultures of connective tissues. However, some modification of the methods may be required, especially in the use of density gradients to isolate the fragments where proteoglycans of low buoyant density are present and in the selection of antibodies. The separation of core protein fragments by electrophoresis on SDS–polyacrylamide gels and subsequent analysis by N-terminal sequence analysis and reactivity with antibodies have been used to characterize the mechanism of degradation of a proteoglycan by a specific protease. In this work, the nature of products arising from the action of human leukocyte elastase on an isolated aggrecan complex preparation was investigated under conditions of partial and total digestion.[30] The sites of proteolytic cleavage within the aggrecan core protein by leukocyte elastase were determined, as well as the way in which the proteinase modified the structure of link protein that was also present in the substrate.

---

[29] B. Caterson, J. E. Christner, J. R. Baker, and J. R. Couchman, *Fed. Proc.* **44,** 386 (1985).
[30] M. T. Mok, M. Z. Ilic, C. J. Handley, and H. C. Robinson, *Arch. Biochem. Biophys.* **292,** 442 (1992).

# [5] Theoretical and Practical Aspects of Proteinase Inhibition Kinetics

*By* Joseph G. Bieth

## Introduction

Many physiological and pathological processes are controlled by proteinases whose activity is itself regulated by protein proteinase inhibitors. The endogenous inhibitors usually exhibit a good proteinase class specificity, for example, serine proteinase inhibitors do not inhibit cysteine proteinases and vice versa. However, when tested with a rough *in vitro* assay, a given proteinase inhibitor will usually inhibit many proteinases belonging to the same class. The only way to define the specificity of such inhibitors is to assess the rate and equilibrium constants for the interaction with proteinases. This may show that an inhibitor can react with two proteinases with widely different $K_i$ values although it yields similar "percent of inhibition" values with them when a rough all-or-none assay is used. In such work as the characterization of a new protein proteinase inhibitor, the search for a physiological function of an inhibitor, the study of the effect of site-directed mutagenesis on inhibitory potency, the investigation of proteinase specificity using a set of inhibitors, and the rational design of synthetic proteinase inhibitors, the kinetic parameters describing the proteinase–inhibitor interaction should be determined in a proper way. This chapter selects essential kinetic equations underlying this determination, provides practical advice, and attempts to correlate kinetic parameters with the *in vivo* function of inhibitors.

## Reversible and Irreversible Inhibition

### Simplest Reaction Schemes

A proteinase inhibitor (I) may react with the target proteinase (E) to form either a reversible or an irreversible enzyme–inhibitor complex (EI). If the binding is reversible, the simplest reaction is

$$E + I \underset{k_{diss}}{\overset{k_{ass}}{\rightleftharpoons}} EI$$

Scheme I. Reversible inhibition.

where $k_{ass}$ and $k_{diss}$ are the rate constants of complex formation and dissociation, respectively. The equilibrium concentrations of E, I, and EI are related to each other by $K_i$, the equilibrium dissociation constant of the enzyme–inhibitor complex, also called the inhibition constant:

$$K_i = \frac{[E][I]}{[EI]} \tag{1}$$

When the equilibrium between E, I, and EI is reached, the rate of EI formation, namely, $k_{ass}[E][I]$, is equal to the rate of EI dissociation, $k_{diss}[EI]$:

$$k_{ass}[E][I] = k_{diss}[EI] \tag{2}$$

From Eqs. (1) and (2) it follows that

$$K_i = \frac{k_{diss}}{k_{ass}} \tag{3}$$

Equation (3) shows that $K_i$ may be determined using either measurements at equilibrium or rate determinations.

To define rigorously irreversible inhibition (Scheme II), one sets $k_{diss} = 0$ in Scheme I:

$$E + I \xrightarrow{k_{ass}} EI$$

SCHEME II. Irreversible inhibition.

*Multistep Inhibition*

Schemes I and II are sometimes valid only if the inhibition is studied with low enzyme or inhibitor concentrations (say, $\leq 0.1 \ \mu M$). With higher concentrations, an intermediate species EI* may be detected so that more complex reaction schemes must be used. Most proteinase inhibitors probably interact with the target proteinases via such multistep mechanisms. The relationship between the microscopic rate constants depicted in Schemes III and IV and the macroscopic constants shown in Schemes I and II will be given later.

$$E + I \underset{k_{-1}}{\overset{k_1}{\rightleftharpoons}} EI^* \underset{k_{-2}}{\overset{k_2}{\rightleftharpoons}} EI$$

SCHEME III. Reversible inhibition.

$$E + I \underset{k_{-1}}{\overset{k_1}{\rightleftharpoons}} EI^* \xrightarrow{k_2} EI$$

SCHEME IV. Irreversible inhibition.

## Reversal of Inhibition

If free and bound reactants are in thermodynamic equilibrium, there is a continuous exchange between the free and the bound molecules. The exchange is possible only if the free enzyme and the free inhibitor which are released from the complex are chemically identical with the enzyme and the inhibitor that have been mixed together at the start of the reaction. Therefore, to demonstrate that the enzyme–inhibitor binding is truly reversible, one should examine the structure of the free enzyme and inhibitor following dissociation of the complex. Such an analysis has shown that some protein proteinase inhibitors interact with the cognate enzymes according to Scheme V,[1]

$$E + I \rightleftharpoons L \rightleftharpoons C \rightleftharpoons L^* \rightleftharpoons E + I^*$$

SCHEME V

where I is the virgin inhibitor that was initially reacted with the enzyme, I* is a modified but still reactive inhibitor with a peptide bond cleaved at the active center, L and L* are loose enzyme–inhibitor complexes, and C is the stable complex. Although all species depicted in Scheme V are in equilibrium with one another, the whole sequence of reactions does not correspond to a true thermodynamic equilibrium because part of the free inhibitor released from the stable complex C is chemically different from the inhibitor initially reacted with the enzyme.

Irreversible inhibitors that belong to the serpin superfamily of proteins[2] react with proteinases to form complexes that slowly decompose into active enzyme and inactive inhibitor as shown in Scheme VI,

$$E + I \xrightarrow{k_{ass}} EI \xrightarrow{k_{deg}} E + I^*$$

SCHEME VI

where I* is now an inactive inhibitor species and $k_{deg}$ is the first-order rate constant for the breakdown of EI. If EI forms rapidly on mixing E and I and if $k_{deg}$ is high enough (say, $k_{deg} \geq 10^{-3}$ sec$^{-1}$, $t_{1/2\,deg} = 0.693/k_{deg} \leq 11$ min), active enzyme will be present after a reasonable time of incubation.

[1] M. Laskowski, Jr., and I. Kato, *Annu. Rev. Biochem.* **49**, 593 (1980).
[2] J. Travis and G. S. Salvesen, *Annu. Rev. Biochem.* **52**, 655 (1983).

Hence, reversible inhibition might be claimed although all steps of the reaction pathway are irreversible.

These two examples show that the presence of a residual enzymatic activity in a mixture of enzyme and inhibitor does not unambiguously prove that the inhibition is reversible from a thermodynamic point of view. Experimental evidence for the reversal of inhibition must be obtained. Failure to reverse inhibition is, however, not definitive evidence for irreversible inhibition. For instance, if $k_{diss}$ is low (e.g., $10^{-5}$ $sec^{-1}$), the half-time of reversal of inhibition ($t_{1/2(diss)} = 0.693/k_{diss}$) is about 20 hr. As a consequence, no reversal of inhibition can be observed if the back reaction is followed only for a few hours. The inhibition may be erroneously classified as irreversible. Reversible inhibitors with low $k_{diss}$ values may also have low $K_i$ values (i.e., high affinities), as can be inferred from Eq. (3). Such inhibitors are called reversible tight-binding inhibitors. Most reversible proteinase inhibitors belong to that class. The next section analyzes the binding behavior of such enzyme–inhibitor systems.

## Reversible Tight-Binding Inhibition

### Classical versus Tight-Binding Reversible Inhibitors

With classical reversible inhibitors, inhibition takes place only if a large molar excess of inhibitor is reacted with the enzyme, say, $10^{-5}$ $M$ inhibitor plus $10^{-8}$ $M$ enzyme to yield $7 \times 10^{-9}$ $M$ enzyme–inhibitor complex. In this example, the concentration of bound inhibitor [EI] is negligible with respect to the concentration of total inhibitor $[I]_o$. Hence, the concentration of free inhibitor [I] may be assumed to be equal to $[I]_o$; for a classical inhibitor,

$$[I]_o = [I]$$
$$\text{Total} = \text{free} \tag{4}$$

With tight-binding reversible inhibitors significant inhibition takes place if equimolar concentrations of enzyme and inhibitor are reacted, say, $10^{-8}$ $M$ inhibitor plus $10^{-8}$ $M$ enzyme to yield $7 \times 10^{-9}$ $M$ enzyme–inhibitor complex. Now the concentration of bound inhibitor [EI] is no longer negligible with respect to that of total inhibitor $[I]_o$. Hence, the concentration of free inhibitor [I] is no longer equal to $[I]_o$; for a tight-binding inhibitor,

$$[I]_o = [I] + [EI]$$
$$\text{Total} = \text{free} + \text{bound} \tag{5}$$

Most equations for steady-state velocities of inhibitor interactions found in textbooks assume that $[I]_o = [I]$ and can therefore not be used for tight-binding inhibitors such as protein proteinase inhibitors.

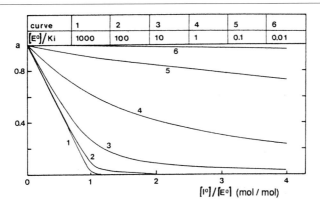

| curve | 1 | 2 | 3 | 4 | 5 | 6 |
|-------|------|-----|----|---|-----|------|
| $[E^o]/K_i$ | 1000 | 100 | 10 | 1 | 0.1 | 0.01 |

FIG. 1. Theoretical plots of fractional free enzyme ($a$) as a function of molar ratios of inhibitor to enzyme for variable $[E]_o/K_i$ ratios. [From J. G. Bieth, *in* "Bayer-Symposium V: Proteinase Inhibitors" (H. Fritz, H. Tchesche, L. J. Greene, and E. Truscheit, eds.), p. 463. Springer-Verlag, Berlin and New York, 1974.]

*Importance of $[E]_o/K_i$*

Tight-binding inhibition does not only take place if $K_i$ is "low," that is, if the inhibitor is a tight binder, but also if $[E]_o$ is "high." To demonstrate this, Eq. (5) can be rewritten as follows[3,4]:

$$[I]_o = K_i \frac{1-a}{a} + [E]_o(1-a)$$

(6)

$$\text{Total} = \quad \text{free} + \text{bound}$$

where $a$ is the fraction of total enzyme that is not bound to the inhibitor:

$$a = 1 - \frac{[EI]}{[E]_o}$$

(7)

Equation (6) clearly shows that both $K_i$ and $[E]_o$ are important factors for tight-binding inhibition. As a matter of fact, the binding behavior of an enzyme–inhibitor system depends on $[E]_o/K_i$. This is illustrated in Fig. 1, which shows the influence of increasing inhibitor concentrations on constant enzyme concentrations for a series of values of $[E]_o/K_i$. For each curve, $a$ has been computed using Eq. (8) which is derived from Eq. (6):

[3] L. H. Easson and E. Stedman, *Proc. R. Soc. London B* **121**, 141 (1936).
[4] J. G. Bieth, *in* "Bayer-Symposium V: Proteinase Inhibitors" (H. Fritz, H. Tchesche, L. J. Greene, and E. Truscheit, eds.), p. 463. Springer-Verlag, Berlin and New York, 1974.

$$a = 1 - \frac{([E]_o + [I]_o + K_i) - \{([E]_o + [I]_o + K_i)^2 - 4[E]_o[I]_o\}^{1/2}}{2[E]_o} \qquad (8)$$

It can be seen (Fig. 1) that if $[E]_o/K_i \geq 100$ the binding curves are almost linear and intercept the abscissa at $[I]_o/[E]_o = 1$. In that case, almost all the inhibitor is bound to the enzyme, that is, Eq. (6) may be written as $[I]_o \approx [E]_o(1 - a)$ or

$$a \approx 1 - \frac{[I]_o}{[E]_o} \qquad (9)$$

Under these conditions, the binding is so tight that the inhibitor titrates the enzyme. On the other hand, if $[E]_o/K_i \leq 0.01$, there is virtually no EI complex present at $[I]_o = [E]_o$. Hence, the inhibition will be "classical," and Eq. (6) will be written as $[I]_o \approx K_i (1 - a)/a$. If $0.01 < [E]_o/K_i < 100$, then $[I]_o = [I] + [EI]$ whatever the absolute values of $[E]_o$ and $K_i$.

*Influence of Substrate*

Let $v_o$ and $v_i$ be the steady-state rates of substrate hydrolysis in the absence and presence of inhibitor, respectively. Because the enzymatic rate is proportional to the concentration of noninhibited enzyme, the fractional rate, $v_i/v_o$, will be equal to $a$, the fractional free enzyme defined in Eq. (7):

$$a = \frac{v_i}{v_o} \qquad (10)$$

If the substrate and the inhibitor bind at the same site of the enzyme (competitive inhibition) and if enzyme, substrate, and inhibitor are in true equilibrium with the respective complexes, $K_{i(app)}$ will replace $K_i$ in Eqs. (6) and (8). $K_{i(app)}$ is the substrate-dependent equilibrium dissociation constant:

$$K_{i(app)} = K_i(1 + [S]_o/K_m) \qquad (11)$$

Now the enzyme–inhibitor binding behavior depends on $[E]_o/K_{i(app)}$. It is therefore possible to shift from tight-binding inhibition to classic inhibition just by measuring $v_i$ with a large excess of substrate over $K_m$.

For many reversible protein proteinase inhibitors $v_i/v_o$ does not vary with the substrate concentration. This may be interpreted to mean that the inhibition is purely noncompetitive. However, the knowledge that such inhibitors bind proteinases in a substrate-like fashion[1] makes noncompetitive inhibition unlikely. It is more reasonable to assume that $v_i/v_o$ does not increase with the substrate concentration because substrate-induced partial dissociation of EI is too slow to be detectable during the assay time. The behavior of such systems is described by Eqs. (6) and (8).

Determination of $K_i$

*Preliminary Experiments*

Add increasing amounts of inhibitor to constant amounts of proteinase. Let the buffered mixtures incubate for a fixed period of time (say, 5 min) at a constant temperature. Add a small volume of a stock solution of a synthetic substrate ($\leq 50$ $\mu$l/ml reaction mixture) and measure the initial rates of substrate hydrolysis in the presence of inhibitor ($v_i$) and in its absence ($v_o$). The resulting inhibition curve may either be linear or concave.

*Linear Inhibition Curve.* A linear inhibition curve that intercepts the $x$ axis at $[I]_o/[E]_o \approx 1$ (e.g., curves 1 and 2 of Fig. 1) indicates that the 5-min incubation time is sufficient to ensure maximal binding of inhibitor to enzyme and suggests either irreversible or reversible tight-binding inhibition with $[E]_o/K_i \geq 100$ or $[E]_o/K_{i(app)} \geq 100$. In any case, the experiment yields the equivalence between proteinase and inhibitor, a parameter that is particularly important if the active-site titer of a proteinase and/or an inhibitor might not be known precisely.

Even if reversal of inhibition can be demonstrated (see section below), that is, even if inhibition is reversible, $K_i$ cannot be determined from the linear inhibition curve [$K_i$ does not appear in Eq. (9)]. Because the shape of the inhibition curve depends on $[E]_o/K_i$ or $[E]_o/K_{i(app)}$, the lower the enzyme concentration and the higher the substrate concentration, the more the inhibition curve will tend to be concave (i.e., to be usable for the calculation of $K_i$). The inhibition experiments must therefore be run again using a 10-fold or a 100-fold lower enzyme concentration and a much more sensitive substrate. It is important to emphasize that the rate at which the equilibrium between E, I, and EI is reached depends on $[E]_o$ and $[I]_o$. The 5-min preincubation time used for the first inhibition experiment may therefore not be sufficient to allow the more diluted partners to be in equilibrium. The time dependence of inhibition must therefore be tested again.

Figure 2 shows the results of an experiment where the inhibition of porcine pancreatic elastase by eglin c, an 8-kDa reversible proteinase inhibitor, was tested using two widely different enzyme concentrations. With 5 $\mu M$ elastase, the inhibition curve was found to be linear up to about 80% inhibition (Fig. 2, curve 1). In contrast, with 30 n$M$ enzyme (and a much more sensitive substrate), a concave inhibition curve was obtained (Fig. 2, curve 2). The experiment clearly illustrates that it is possible to shift from a linear inhibition curve to a concave curve by just lowering $[E]_o$. Note that the E + I incubation time had to be increased from 1 min ($[E]_c = 5$ $\mu M$) to 20 min ($[E]_o = 30$ n$M$).

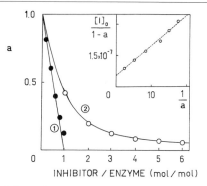

Fig. 2. Experimental illustration of the influence of $[E]_o$ on the shape of the inhibition curve. The effect of increasing concentrations of recombinant eglin c on the activity of constant concentrations of porcine pancreatic elastase was studied using two widely different concentrations of enzyme. The inhibition curve was linear with 5 $\mu M$ elastase (curve 1) and concave with 30 n$M$ elastase (curve 2). Other conditions were as follows: for curve 1, E + I preincubation time, 1 min; substrate, 6 m$M$ Suc-Ala$_2$-$p$-nitroanilide; for curve 2, E + I preincubation time, 20 min; substrate, 0.5 m$M$ Suc-Ala$_3$-$p$-nitroanilide. Curve 2 is theoretical and has been calculated using Eq. (8) and 14.5 nM, the best estimate of $K_{i(app)}$ obtained by nonlinear regression analysis. The substrate-independent $K_i$ [Eq. (11)] was found to be 9.6 ± 0.3 nM. The $[E]_o/K_{i(app)}$ ratio for curve 1 is therefore approximately 520, which explains why the curve is linear.

If the decrease in $[E]_o$ is not possible because there is no hypersensitive substrate available or if the decrease fails to yield a concave inhibition curve, $K_i$ must be determined through separate measurements of $k_{ass}$ and $k_{diss}$ [see Eq. (3)] as shown in the next section.

*Concave Inhibition Curve.* The inhibition curve may be concave if the 5-min incubation time is not sufficient to achieve complete association of the two partners or if the association is complete but the inhibition is reversible with $K_i \geq 0.1[E]_o$ (see Fig. 1). To decide between the two alternatives, the enzyme–inhibitor mixtures should be incubated until the residual activity shows no further change with time. If the resulting inhibition curve is linear, the above rules should be applied. If, on the other hand, the inhibition curve is still concave one has to check whether the substrate displaces the E + I $\rightleftharpoons$ EI equilibrium toward the left. This is best done by recording the time dependence of substrate hydrolysis after addition of a small aliquot of substrate to an equilibrium mixture of E and I. An exponential release of product with time is then unambiguous proof of substrate-induced dissociation of EI. The new equilibrium will be obtained once the release of product is linear with time.

If, immediately after addition of substrate, there is a linear release of product with time, either there is no substrate-induced dissociation of EI or the disssociation is so fast that the equilibrium between E, I, S, and the

respective complexes is attained within the time required to mix the reagents. To choose between the two alternatives, the fractional activity $a$
[Eq. (10)] should be measured using two widely different substrate concentrations (say, $0.5K_m$ and $3K_m$) and constant enzyme and inhibitor concentrations. If $a$ increases with $[S]_o$, the inhibition is substrate-dependent and E,
I, and S are in fast equilibrium. In that case there is good evidence that E,
I, and EI are in true equilibrium. On the other hand, if $a$ does not vary
with $[S]_o$, either the inhibition is not reversible from a thermodynamic
viewpoint (see *"Reversal of Inhibition,"* p. 61) or $k_{diss}$ is so low that substrate-induced dissociation of EI does not occur during the assay of enzy-
matic activity. For example, if $k_{diss} = 10^{-4}$ sec$^{-1}$ ($t_{1/2(diss)} \approx 2$ hr), substrate-
induced dissociation cannot be detected if the assay lasts 15 min. In that
case it is mandatory to measure $k_{diss}$ separately (see later section) to make
sure that the system is truly reversible, that is, that $K_i$ has a true physical
meaning.

*Calculation of $K_i$*

Let us emphasize again that calculation of $K_i$ is only possible if the inhibi-
tion is truly reversible and if the inhibition curve is reasonably concave. The
Easson–Stedman plot[3] is a graphical method that allows a rough estimate of
$K_i$. The plot is based on Eq. (12), obtained by rearranging Eq. (6):

$$\frac{[I]_o}{1 - a} = \frac{K_i}{a} + [E]_o \tag{12}$$

If substrate-induced dissociation of EI occurs, that is, if the inhibition is
competitive, $K_{i(app)}$ [Eq. (11)] replaces $K_i$ in Eq. (12). Thus, a plot of $[I]_o/$
$(1 - a)$ as a function of $1/a$ yields a straight line whose slope is $K_i$ (or
$K_{i(app)}$) and which intercepts the $y$ axis at $[E]_o$ (see inset to Fig. 2).

The Easson–Stedman plot has been widely used in the past. It must,
however, be handled with great care to avoid extraction of erroneous
parameters. The user of the plot should be aware that Eq. (12) is a linearized
form of an equation that describes the nonlinear effect of $[I]_o$ on $a$ [Eq.
(8)], in the same way as the familiar Lineweaver–Burk plot is based on
a linearized form of the Michaelis–Menten equation that describes the
nonlinear dependence of $v$ versus $[S]_o$. Owing to the linearization, $a$ appears
on both abscissa and ordinate so that any error in the parameter will be
reflected twice, thus leading to scattered data points. The scatter may be
all the more serious because $a$ is the ratio of two independent measurements,
$v_o$ and $v_i$. The natural tendency of the experimenter is to overlay the data
points with a ruler to draw the "best line" through them or to calculate
the best line by linear regression analysis. This assumes that the errors

are equally distributed throughout the data, which is not case because linearization of a nonlinear equation will lead to a distortion of the error distribution. The Easson–Stedman plot should, therefore just be used to get a rough estimate of $K_i$ or $K_{i(app)}$. The refined value of the parameter as well as the error should be calculated by fitting the data to Eq. (8) by nonlinear regression. This is now easy to do with the ENZFITTER software (Biosoft, Cambridge CB21LR, UK). The user should also print out both the experimental points and the theoretical curve calculated using Eq. (8) and $K_i$. If, and only if, the data points are evenly distributed along the theoretical curve, Eq. (8) adequately describes the data and the estimated $K_i$ has a physical meaning. The ENZFITTER program helpfully substitutes this visual analysis by plotting the residuals [e.g., $a$(observed) $-$ $a$(calculated)] as a function of $[I]_o$ to check the appropriateness of Eq. (8). Diagram 1A shows the equation editor screen for Eq. (8) in ENZFITTER.

Figure 2, curve 2, shows the inhibition of 30 n$M$ elastase by increasing concentrations of eglin c. The enzyme activity was measured with $[S]_o = 0.5K_m$. Immediately after mixing the reagents, the release of product was linear with time, suggesting either that the substrate did not dissociate EI or that the dissociation occurred during mixing. To decide between the two possibilities we have done the same experiment using $[S]_o = 3K_m$. The new inhibition curve was much less concave, suggesting competitive

```
═══════════════════════════════════Equation Name═══════════════════════════════
 Tight-binding inhibition Eq.(8)

═══════════Y Axis Name═══════════      ═════════X Axis Name═══════════  ═══VAR═══
 a                                      [Inhibitor]                      Io

═══════════Variable Name═══════  ═VAR═  ═════════Variable Name═══════  ═══VAR═══
 1  Inhibition constant    Ki      5
 2                                  6
 3                                  7
 4                                  8

═════════Prompted Constant═════  ═VAR═  ═══════Prompted Constant═════  ═══VAR═══
 1  Enzyme concentration   Eo      3
 2                                  4

═══════════════════════════════════Equation Definition═════════════════════════
 Temp1  SQRT(SQR(Eo+Io+Ki) - (4*Eo*Io))
 Temp2

 Y =    1 - ((Eo+Io+Ki) - Temp1)/(2*Eo)
```

DIAGRAM 1A

inhibition (data not shown). The Easson–Stedman plot shown in the inset to Fig. 2 gave a rough estimate of $K_{i(app)}$ that was used by the ENZFITTER program for the first iteration of the nonlinear regression analysis. The refined value of $K_{i(app)}$ was $(1.45 \pm 0.02) \times 10^{-8}\ M$. The substrate-independent value of $K_i$, calculated using Eq. (11), was $(0.96 \pm 0.03) \times 10^{-8}\ M$. Note that the absolute error in $K_i$ is larger than that in $K_{i(app)}$ because the error in $K_m$ was taken into account. Identical calculations were done for $[S]_o = 3K_m$ and gave $K_i = (1.2 \pm 0.2) \times 10^{-8}\ M$. The two $K_i$ values are identical within the limits of experimental error, confirming that the inhibition is indeed competitive. Curve 2 of Fig. 2 is a theoretical line generated using Eq. (8) and $K_{i(app)} = 1.45 \times 10^{-8}\ M$. The curve fits the experimental points well, confirming that the inhibition data are adequately described by Eq. (8).

## Measurement of Rate Constants of Inhibition

From a kinetic point of view, reversible inhibitors that follow a one-step mechanism are characterized by the macroscopic rate constants $k_{ass}$ and $k_{diss}$ (see Scheme I). Measurement of those constants is often the only way to get $K_i$ [see Eq. (3)] and provides useful information for delineating the physiological function of proteinase inhibitors (see last section). In addition, measurement of the rate of inhibition using variable inhibitor concentrations may help detect intermediates (see Schemes III and IV).

In previous articles we have recommended measuring the association rate constant $k_{ass}$ in a discontinuous way that consists of mixing equimolar concentrations of E and I and waiting for a given time before adding substrate to measure the enzyme activity remaining at that time. The experiment was then repeated using variable incubation times. If the E + I reaction is irreversible, the data are analyzed in a straightforward manner using the simplified form of the second-order rate equation. If the inhibition is reversible, the method assumes that during the course of the experiment the reverse reaction does not take place to a significant extent. This must of course be ascertained. Because mixing of the reagents takes some time, the half-life of the reaction should not be less than 2 min. Therefore, if $k_{ass}$ or $[E]_o$ is too high, $k_{ass}$ will be difficult to measure. On the other hand, the substrate should virtually stop the E + I association. This requires $[S]_o \geq 5\ K_m$ which is not always easy to get. Last, if a reversible inhibitor has a high $k_{diss}$, the E + I association will reach an equilibrium position which renders the data difficult to analyze. The discontinuous method may therefore yield $k_{ass}$ values with poor physical significance. We now prefer to use a continuous method outlined below.

*Progress Curve Method: Theory*

In the progress curve method the enzyme E is added to a mixture of inhibitor I and substrate S and the release of product P is continuously recorded.[5-7] In the case of reversible competitive inhibition, a one-step enzyme–inhibitor binding is described by Scheme VII. If the $E + I \rightleftharpoons EI$

$$E + S \rightleftharpoons ES \longrightarrow E + P$$
$$+$$
$$I$$
$$k_{ass} \big\Vert k_{diss}$$
$$EI$$

SCHEME VII

equilibrium is reached at a significantly slower rate than the $E + S \rightleftharpoons ES$ equilibrium, there will be an exponential release of P with time until the E, I, S system reaches steady state, where the release of P becomes linear with time (see Fig. 3).

If $[I]_o \geq 10\ [E]_o$ and if $[S]_o$ is not significantly depleted during the reaction (say, $\leq 5\%$ substrate hydrolysis) the biphasic release of product with time (progress curve) is described by the following relationship (see also Diagram 1B):

$$[P] = v_s t + \frac{v_z - v_s}{k}(1 - e^{-kt}) \tag{13}$$

where $v_z$ is the initial velocity at $t = 0$, $v_s$ is the final steady-state velocity, and $k$ is the pseudo-first-order rate constant for the approach to steady state. The constant $k$ is related to $k_{ass}$ and $k_{diss}$ through Eq. (14):

$$k = \frac{k_{ass}[I]_o}{1 + [S]_o/K_m} + k_{diss} \tag{14}$$

On the other hand, the initial and the steady-state rates $v_z$ and $v_s$ are given by

$$v_z = \frac{V_m}{1 + K_m/[S]_o} \tag{15}$$

$$v_s = \frac{V_m}{(1 + K_m/[S]_o)(1 + [I]_o/K_i)} \tag{16}$$

[5] S. Cha, *Biochem. Pharmacol.* **24**, 2177 (1975).
[6] J. F. Morrisson and C. T. Walsh, *Adv. Enzymol. Relat. Areas Mol. Biol.* **61**, 201 (1988).
[7] W.-X. Tian and C.-L. Tsou, *Biochemistry* **21**, 1028 (1982).

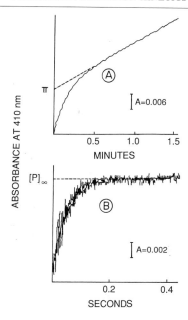

FIG. 3. Experimental illustration of the progress curve method used to measure the kinetics of proteinase inhibition. The enzyme–inhibitor system was the same as in Fig. 2. Elastase was added to a mixture of inhibitor and substrate, and the release of product (absorbance at 410 nm) was recorded as a function of time. The substrate was 2 m$M$ Suc-Ala$_3$-$p$-nitroanilide (=2$K_m$). (A) Progress curve recorded with 50 n$M$ elastase and 500 n$M$ eglin c. A manually driven stopped-flow accessory mounted on an ordinary spectrophotometer was used to mix the reagents. The biphasic curve characterizes reversible inhibition where an exponential pre-steady-state release of product is followed by a steady-state (linear) release. The curve was analyzed by nonlinear regression using the ENZFITTER software (see Diagram 1B) and a variant of Eq. (13) where $\pi$ (see $y$ axis) stands for $(v_z - v_s)/k$. Rough estimates of $k_{ass}$ ($9.0 \times 10^5\ M^{-1}\ sec^{-1}$) and $k_{diss}$ ($1.4 \times 10^{-2}\ sec^{-1}$) were calculated using Eqs. (14) and (24). (B) Progress curve recorded with 2 $\mu M$ elastase and 60 $\mu M$ eglin c. A stopped-flow apparatus was used to monitor the fast reaction. The exponential progress curve normally diagnoses irreversible inhibition. However, if a reversible inhibitor is tested at a concentration much higher than $K_i$, as is the case here ($[I]_o/K_i = 6000$), it will behave like an irreversible inhibitor for which only $k_{ass}$ can be calculated. The exponential was analyzed using Eq. (21), and $k_{ass}$ was found to be $10^6\ M^{-1}\ sec^{-1}$. The rough estimate is in good agreement with that determined from the curve shown in (A).

where $V_m$ and $K_m$ are the maximal velocity and the Michaelis constant, respectively, and $K_i = k_{diss}/k_{ass}$.

Equation (16) is the equation for "classical" competitive inhibition which assumes that $[I]_o = [I]$, that is, that the enzyme does not deplete the inhibitor. This assumption is also valid for tight-binding inhibition provided that $[I]_o$ is at least 10-fold higher than $[E]_o$, which is the case here.

```
╔══════════════════════════Equation Name══════════════════╗
║ Approach to the steady-state Eq.(13)                     ║
╚══════════════════════════════════════════════════════════╝
╔═════════Y Axis Name═════════╗ ╔════════════X Axis Name════════╤══VAR══╗
║ Product absorbance          ║ ║ Time                          │ t     ║
╚═════════════════════════════╝ ╚═══════════════════════════════╧═══════╝
╔═══════════Variable Name══════════╤══VAR══╗ ╔═══════Variable Name══════╤══VAR══╗
║ 1│π = (Vz-Vs)/k                  │ PI    ║5║                          │       ║
║ 2│Rate constant                  │ k     ║6║                          │       ║
║ 3│Steady-state velocity          │ Vs    ║7║                          │       ║
║ 4│Offset at t=0                  │ b     ║8║                          │       ║
╚══════════════════════════════════╧═══════╝ ╚══════════════════════════╧═══════╝
╔═══════════Prompted Constant══════╤══VAR══╗ ╔═════Prompted Constant════╤══VAR══╗
║ 1│                               │       ║3║                          │       ║
║ 2│                               │       ║4║                          │       ║
╚══════════════════════════════════╧═══════╝ ╚══════════════════════════╧═══════╝
╔══════════════════════════Equation Definition════════════════════╗
║ Temp1│                                                           ║
║ Temp2│                                                           ║
║      │                                                           ║
║ Y =  │ ((Vs*t) + PI*(1-EXP(-k*t))) + b                           ║
╚══════════════════════════════════════════════════════════════════╝
```

DIAGRAM 1B

If a competitive inhibitor reacts with an enzyme via a two-step mechanism, Scheme VIII applies.

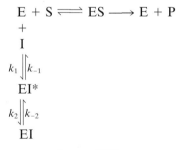

SCHEME VIII

If the $E + I \rightleftharpoons EI^*$ equilibrium is reached rapidly and at about the same rate as the $E + S \rightleftharpoons ES$ equilibrium but at a significantly higher rate than the $EI^* \rightleftharpoons EI$ equilibrium, there will again be an exponential release of P with time. If the condition $[I]_o \geq 10[E]_o$ is fulfilled and if there is less than 5% substrate hydrolysis during the reaction, the progress curve will be described by the same equation as before [Eq. (13)]. The rate constant $k$ will, however, have a different significance:

$$k = \frac{k_2[I]_o}{[I]_o + K_i^*(1 + [S]_o/K_m)} + k_{-2} \tag{17}$$

where $K_i^* = k_{-1}/k_1$ (see Diagram 1C). The initial rate $v_z$ will also have a different meaning:

$$v_z = \frac{V_m}{(1 + K_m/[S]_o)(1 + [I]_o/K_i^*)} \tag{18}$$

Thus, $v_z$ is no longer the rate of the uninhibited reaction as it was in the case of the one-step mechanism but the rate of a "classical" competitive inhibition as was $v_s$ in the case of the one-step inhibition mechanism [Eq. (16)]. This is so because the E + I $\rightleftharpoons$ EI* and the E + S $\rightleftharpoons$ ES equilibria are reached rapidly and with similar rates so that, at the beginning of the reaction, the inhibitor behaves like a classical competitive inhibitor. On the other hand, the steady-state velocity $v_s$ will be given by

$$v_s = \frac{V_m}{(1 + K_m/[S]_o)(1 + [I]_o/K_i)} \tag{19}$$

an equation that is apparently identical to Eq. (16) of the one-step mechanism but in which $K_i$ has a different meaning:

```
╔════════════════════Equation Name════════════════════════╗
║ Ki* for reversible inhibition Eq.(17)                    ║
╚══════════════════════════════════════════════════════════╝

╔═══════Y Axis Name══════╗╔═══════X Axis Name═══════╦═VAR═╗
║ Rate constant k        ║║ [Inhibitor]             ║ Io  ║
╚════════════════════════╝╚═════════════════════════╩═════╝

╔═══════Variable Name════╦═VAR═╦═══════Variable Name═══════╦═VAR═╗
║1│Rate constant k2      │ k2  │5│                         │     ║
║2│Equilm. constant Ki*  │ Ki  │6│                         │     ║
║3│Rate constant k-2     │ k   │7│                         │     ║
║4│                      │     │8│                         │     ║
╚═══════════════════════╩═════╩═══════════════════════════╩═════╝

╔═══════Prompted Constant═══╦═VAR═╦═══════Prompted Constant═══╦═VAR═╗
║1│[Substrate]              │ So  │3│                         │     ║
║2│Michaelis constant       │ Km  │4│                         │     ║
╚═════════════════════════╩═════╩═══════════════════════════╩═════╝

╔════════════════════Equation Definition════════════════════╗
║ Temp1                                                      ║
║ Temp2                                                      ║
║                                                            ║
║ Y =  ((k2*Io) / (Io + (Ki*(1+So/Km)))) + k                ║
╚═══════════════════════════════════════════════════════════╝
```

DIAGRAM 1C

$$K_i = K_i^* \frac{k_{-2}}{k_2 + k_{-2}} \tag{20}$$

If the inhibition is irreversible and if substrate and inhibitor compete for the same site of the enzyme, the progress curve method may again be used, a one-step enzyme–inhibitor binding being described by Scheme IX.

$$E + S \rightleftharpoons ES \longrightarrow E + P$$
$$+$$
$$I$$
$$\downarrow k_{ass}$$
$$EI$$

SCHEME IX

If ES forms much faster than EI, there will again be an exponential release of P with time. Once all E has been transformed into EI, [P] will be constant with time (i.e., $v_s = 0$). Hence Eq. (13) becomes

$$[P] = \frac{v_z}{k}(1 - e^{-kt}) \tag{21}$$

The entry of Eq. (21) in ENZFITTER is shown in (Diagram 1D).

| ═══════════Equation Name═══════════ |
| --- |
| Rate of irreversible inhibition Eq.(21) |

| ═══Y Axis Name═══ | ═══X Axis Name═══ | ═VAR═ |
| --- | --- | --- |
| Product absorbance | Time | t |

| | ═══Variable Name═══ | ═VAR═ | | ═══Variable Name═══ | ═VAR═ |
| --- | --- | --- | --- | --- | --- |
| 1 | Initial velocity | Vz | 5 | | |
| 2 | Rate constant | k | 6 | | |
| 3 | Offset at t=0 | b | 7 | | |
| 4 | | | 8 | | |

| | ═Prompted Constant═ | ═VAR═ | | ═Prompted Constant═ | ═VAR═ |
| --- | --- | --- | --- | --- | --- |
| 1 | | | 3 | | |
| 2 | | | 4 | | |

| | ═══════════Equation Definition═══════════ |
| --- | --- |
| Temp1 | |
| Temp2 | |
| Y = | ((Vz/k)*(1 - EXP(-k*t))) + b |

DIAGRAM 1D

that is, the equation of a simple exponential. Because $k_{diss} = 0$, Eq. (14) becomes

$$k = \frac{k_{ass}[I]_o}{1 + [S]_o/K_m} \qquad (22)$$

On the other hand, $v_z$ is still the velocity in the absence of inhibitor [Eq. (15)].

If the enzyme and the irreversible inhibitor combine via a two-step mechanism, Scheme X applies.

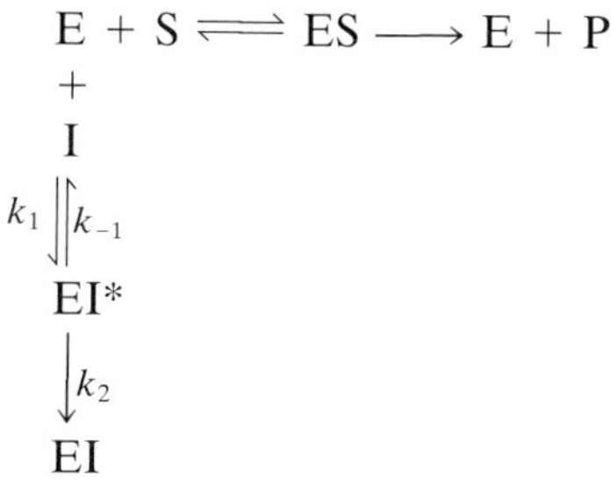

$$E + S \rightleftharpoons ES \longrightarrow E + P$$

$$+$$

$$I$$

$$k_1 \left\Vert k_{-1} \right.$$

$$EI^*$$

$$\downarrow k_2$$

$$EI$$

SCHEME X

For reasons given above, the progress curve will be a simple exponential [Eq. (21)]. The rate constant $k$ will, however, have a different meaning:

$$k = \frac{k_2[I]_o}{[I]_o + K_i^*(1 + [S]_o/K_m)} \qquad (23)$$

[The entry of Eq. (23) in ENZFITTER is shown in Diagram 1E.] On the other hand, $v_z$ will have the same meaning as in the case of two-step reversible inhibition [Eq. (18)].

*Progress Curve Method: Practical Aspects*

Add an aliquot of enzyme solution to a buffered and thermostatted mixture of substrate and inhibitor contained in a spectroscopic cuvette. Mix the solution as rapidly as possible, and record immediately the absorbance or fluorescence versus time. Use a modern spectrophotometer or spectrofluorimeter on-line with a microcomputer which is loaded with the ENZFITTER software. In the preliminary experiment it is recommended to use $[I]_o/[E]_o = 10$ (e.g., $[E]_o = 10\,nM$ and $[I]_o = 100\,nM$) and $[S]_o = K_m$ and to follow the reaction for 30 min. The resulting progress curve may have one of the following shapes.

```
╔══════════════════════Equation Name══════════════════════╗
║Ki* for irreversible inhibition Eq.(23)                   ║
╠═══════════════════════╦══════════════════════════════════╣
║══════Y Axis Name══════║════════X Axis Name═══════╤══VAR══ ║
║Rate constant k        ║[Inhibitor]               │Io     ║
╠═══════════════════════╩══════════════════════════╧════════╣
║══════Variable Name══════╤══VAR══╤══════Variable Name══════╤══VAR══║
║1│Rate constant k2       │k2     │5                        │      ║
║2│Equilm. constant Ki*   │Ki     │6                        │      ║
║3│                       │       │7                        │      ║
║4│                       │       │8                        │      ║
╠═══════Prompted Constant═══════╤══VAR══╤═══Prompted Constant═══════╤══VAR══╣
║1│[Substrate]                  │So     │3                          │      ║
║2│Michaelis constant           │Km     │4                          │      ║
╠══════════════════════Equation Definition══════════════════╣
║Temp1                                                       ║
║Temp2                                                       ║
║                                                            ║
║Y =   (k2*Io) / (Io + (Ki*(1+So/Km)))                       ║
╚════════════════════════════════════════════════════════════╝
```

DIAGRAM 1E

*Case 1: Flat Progress Curve.* A flat progress curve means that the inhibition is so fast that virtually no substrate is turned over, that is, the pseudo-first-order rate constant $k$ [Eqs. (13) or (21)] is too high. The constant may be lowered by decreasing $[I]_o$ and/or increasing $[S]_o$ [see Eqs. (14), (17), (22), or 23)]. The decrease in $[I]_o$ must be accompanied by a decrease in $[E]_o$ so that the $[I]_o/[E]_o$ ratio remains equal to 10. A better signal-producing substrate should therefore be used (e.g., fluorescent or thiobenzyl ester substrates). Increasing $[S]_o$ not only decreases $k$ but also increases the spectroscopic signal by increasing $v_z$ [Eqs. (15) or (18)] and $v_s$ [Eqs. (16) or (19)]. If the concentration changes yield a significant optical signal but a progress curve whose pre-steady-state phase is barely visible, the observation time should be shortened from 30 min to a few minutes or less. In that case, the mixing time becomes dramatically important. If an acceptable progress curve may be recorded in a few minutes, an ordinary spectrophotometer or spectrofluorimeter may still be used but mixing of the reagents must be done with a manually driven stopped-flow device whose dead-time is about 50 msec. If the presteady-state is still not clearly visible under these conditions, a stopped-flow apparatus with a dead-time of about 1 msec and a short observation time must then be used.

*Case 2: Simple Exponential Progress Curve.* The progress curve may be a simple exponential as shown, for example, in Fig. 3B. Such a curve normally characterizes irreversible inhibition [Schemes IX or X and Eq.

(21)]. However, it may also be observed with very high concentrations of a reversible inhibitor as is the case with the example shown in Fig. 3B. It may be demonstrated that if $[I]_o \gg K_{i(app)}$, Eqs. (14) and (17) which characterize reversible inhibition shift to Eqs. (22) and (23) which are valid for irreversible inhibition. This is the pseudo-irreversible behavior of reversible inhibitors. Figure 3 illustrates the shift from reversible inhibition behavior to pseudo-irreversible inhibition behavior on increasing $[I]_o$. To decide between true irreversible inhibition and pseudo-irreversible inhibition one must therefore run progress curves using lower $[I]_o$ and higher $[S]_o$. One should also run separate dissociation experiments (see below).

Whether the inhibition is truly irreversible or not, it is possible to determine $k_{ass}$ or $k_2$ and $K_i^*$ by running progress curves with variable inhibitor concentrations. To derive $k$ from a simple exponential progress curve it is not necessary to supply the ENZFITTER program (see Diagram 1D, p. 74) with initial values of $k$ and $v_z$. The program calculates $k$ and $[P]_\infty$, the amplitude of the exponential (Fig. 3B), which is equal to $v_z/k$. If $k$ varies linearly with $[I]_o$, $k_{ass}$ will be calculated by linear regression analysis. If the variation is hyperbolic, nonlinear regression analysis based on Eq. (23) (see Diagram 1E, p. 76) will yield $k_2$ and $K_i^*$. The largest experimental value of $k$ is taken as the initial estimate of $k_2$, whereas $[I]_o$ corresponding to the middle of the curve is taken as the initial estimate of $K_i^* (1 + [S]_o/K_m)$.

*Case 3: Biphasic Progress Curve.* A biphasic progress curve is shown in Fig. 3A. Such a curve characterizes reversible inhibition (Scheme VII or VIII) and is described by Eq. (13). To derive $v_z$, $v_s$, and $k$ by nonlinear regression analysis, it is convenient to set $\pi = (v_z - v_s)/k$ in Eq. (13) (see Diagram 1B, p. 72). It may be demonstrated that $\pi$ is the intercept of the linear tail of the progress curve with the ordinates (see Fig. 3A). The initial estimate of $k$ is then calculated using the time corresponding to $\pi/2$ which approximates $t_{1/2}$ of the pre-steady state ($k = 0.693/t_{1/2}$). The initial estimate of $v_s$ is calculated by the built-in software of the spectrometer. Thus, feeding the ENZFITTER program with $\pi$, $v_s$, and $k$ allows the refined values of $v_z$, $v_s$, and $k$ to be calculated. The following relationship allows a rough estimate of $k_{diss}$ (or $k_{-2}$):

$$k_{diss} = kv_s/v_z \tag{24}$$

from which an estimate of $k_{ass}$ (one-step mechanism, scheme VII) or $k_{ass(app)}$ (two-step mechanism) (Scheme VIII) may be calculated. The latter constant is given by

$$k_{ass(app)} = \frac{k_2}{[I]_o + K_i^*} \tag{25}$$

Precise values of $k_{ass}$ and $k_{diss}$ (one-step mechanism) or $K_i^*$, $k_2$, and $k_{-2}$ (two-step mechanism) are obtained by measuring $k$ as a function of $[I]_o$. As outlined in case 2 the relationship is either linear or hyperbolic. In the latter case the data are fitted to Eq. (17) by nonlinear regression (see Diagram 1C, p. 73). The initial estimates of $k_2$ and $K_i^*$ are derived as shown above for Eq. (23), whereas the initial estimate of $k_{-2}$ is taken as the average value of $k_{-2}$ calculated using Eq. (24).

*Case 4: Linear Progress Curve.* A linear progress curve means that the pre-steady-state phase has occurred during the time required to mix the reagents. To see kinetically the pre-steady state one must therefore decrease the mixing time by using a fast-kinetic accessory. To illustrate this we used the porcine pancreatic elastase/eglin c pair. The system yields linear progress curves if the reagents are mixed manually and if the release of product is followed for about 10–15 min. The $K_i$ value inferred from the progress curves was found to be $0.96 \times 10^{-8}$ $M$ (see section on determination of $K_i$). With a manually operated stopped-flow device and an observation time in the minute range, a biphasic progress curve was observed (Fig. 3A). This experiment yielded $k_{ass} = 9 \times 10^5$ $M^{-1}$ sec$^{-1}$ and $k_{diss} = 1.4 \times 10^{-2}$ sec$^{-1}$, that is, a calculated $K_i$ of $1.5 \times 10^{-8}$ $M$, which is in fair agreement with the experimentally determined $K_i$.

*Other Practical Aspects.* If chromogenic or fluorogenic substrates are used, the absorbances or fluorescence intensities need not be transformed into molar concentrations of product in order to exploit the progress curves. In addition, one should always watch for depletion of substrate by calculating $[P]_{end}$, the concentration of product at the end of the progress curve. The condition $[P]_{end} \leq 0.05[S]_o$ should always be fulfilled. This is easier to do with $[S]_o > K_m$ than with $[S]_o < K_m$.

The value of $K_m$ should be as accurate as possible because it enters into the calculation of $k_{ass}$ [Eqs. (14) and (22)] and $K_i^*$ [Eqs. (17) and (23)]. Never take a $K_m$ value from the literature but redetermine it under the buffer and temperature conditions used in experiments. The error in $K_m$ should be included in the error in $k_{ass}$ or $K_i^*$.

Most proteinase–inhibitor systems probably involve a reaction intermediate EI*. The intermediate, however, is not seen kinetically if $[I]_o < K_i^*$ $(1 + [S]_o/K_m)$, because that inequality transforms Eq. (17), characteristic of a two-step mechanism, into Eq. (26):

$$k = \frac{k_2[I]_o}{K_i^*(1 + [S]_o/K_m)} + k_{-2} \qquad (26)$$

which is identical to Eq. (14), characteristic of a one-step mechanism, if one sets $k_2/K_i^* = k_{ass}$ and $k_{-2} = k_{diss}$. Thus, a kinetic behavior characteristic of a one-step mechanism does not necessarily mean that the system does not involve a reaction intermediate. The search for EI* usually requires high concentrations of inhibitor and a fast kinetic apparatus.

*Direct Measurement of $k_{diss}$ or $k_{-2}$*

The progress curve method is not able to distinguish an irreversible inhibitor from a reversible tight-binding inhibitor that behaves in a pseudo-irreversible way under the concentration conditions imposed by the methodology. It is therefore safe to check the reversible nature of the binding by shifting the E + I ⇌ EI equilibrium toward the left. This may be achieved by trapping free E or I by tight-binding ligands and measuring the appearance of free E or free I as a function of time.

$\alpha_2$-Macroglobulin is a particularly useful dissociating ligand because (i) it forms irreversible complexes with almost any proteinase, (ii) $\alpha_2$-macroglobulin–proteinase complexes have substantial activity on synthetic substrates, and (iii) complex formation stabilizes the proteinase so that reactions may be followed for days.[8] The inhibitor should, however, be larger than 20 kDa; otherwise, it will slowly reassociate with the $\alpha_2$-macroglobulin-bound proteinase.[9] $\alpha_1$-Proteinase inhibitor may also be used because it forms irreversible complexes with a number of serine proteinases. If the dissociation process lasts for days it may be followed electrophoretically.[10] Other dissociating agents include chromogenic or fluorogenic active-site titrants of proteinases, very high concentrations of substrate, or another proteinase.[11]

The EI complex should be prepared by mixing equimolar amounts of highly concentrated E and I (say, $10^{-5}$ M) to ensure full association of the two partners. The mixture should then be diluted 100- to 1000-fold into a buffered solution of dissociating agent. In the ideal case the E + I → EI reassociation will be negligible so that the decay of EI will be a first-order process:

$$[EI] = [EI]_o \, e^{-\lambda t} \tag{27}$$

where $\lambda = k_{diss}$ or $k_{-2}$. If reassociation is substantial, the system will reach a new equilibrium after a time. In that case, only the initial part of the dissociation reaction can be taken into account and the value of $\lambda$ will be tentative. However, such an experiment is still very useful because it ascertains that the binding is reversible.

[8] M. Aubry and J. G. Bieth, *Biochim. Biophys. Acta* **438**, 221 (1976).
[9] B. Faller, S. Dirrig, M. Rabaud, and J. G. Bieth, *Biochem. J.* **270**, 639 (1990).
[10] B. Faller and J. G. Bieth, *Biochem. J.* **280**, 27 (1991).
[11] J. G. Bieth, *Bull. Eur. Physiopathol. Respir.* 16(Suppl.), 183 (1980).

## Percent Inhibition and $IC_{50}$

To define the class to which a new proteinase belongs one usually uses a set of class-specific reversible or irreversible inhibitors. For each of the compounds one should study the time dependence of the inhibition process in order to find out the optimal inhibition time. This is rarely done. The incubation time is either not indicated or is the same for the whole set of inhibitors. This may lead to ambiguous results. Also, "percent inhibition," a common way to express results of such studies, has a poor significance. Everybody is hoping to construct a table where the percentages of inhibition are either equal or close to 0% or 100%! Percentages that are outside those all-or-none limits usually generate useless speculations instead of prompting further investigations. One should be aware that a 10% inhibition may be shifted toward a 95% inhibition just by increasing $[I]_o$ and/or the time of incubation of enzyme and inhibitor.

The $IC_{50}$, or inhibitor concentration for which 50% enzyme inhibition is observed, is commonly used in inhibitor screening. It should be stressed that the parameter has little physical meaning. First of all, it should never be used for irreversible inhibitors which always yield 100% inhibition if the incubation time is sufficient. For reversible inhibitors, $IC_{50}$ is given by one of the following:

Classical competitive inhibition: $\quad IC_{50} = K_i(1 + [S]_o/K_m)$ (28)

Classical noncompetitive inhibition: $\quad IC_{50} = K_i$ (29)

Tight-binding competitive inhibition: $\quad IC_{50} = K_i(1 + [S]_o/K_m) + 0.5[E]_o$ (30)

Tight-binding noncompetitive inhi- $\quad IC_{50} = K_i + 0.5[E]_o$ (31)
bition:

Tight-binding inhibition with $\quad IC_{50} = 0.5[E]_o$ (32)
$[E]_o/K_i$ or $[E]_o/K_{i(app)} \geq 100$:

Equations (28)–(32) illustrate the very poor significance of $IC_{50}$ values, which not only depend on the enzyme–inhibitor affinity but may also depend on $[S]_o$ and/or $[E]_o$. Thus, the higher the E–I affinity the poorer the significance of $IC_{50}$, with an extreme situation occurring where $IC_{50}$ is independent of the affinity!

### In Vivo Inhibition

Endogenous proteinase inhibitors have at least two physiological functions. First, they may have a proteolysis-preventing function that consists of preventing or at least minimizing the degradation of endogenous substrates by proteinases liberated at sites where they do not normally play a physiological role (e.g., release of neutrophil elastase at sites of inflamma-tion). Mucus proteinase inhibitor and $\alpha_1$-proteinase inhibitor, the physiolog-ical elastase inhibitors, are thought to act in such a manner. Second, inhibi-tors may have a proteolysis-regulating function where an inhibitor allows a proteinase to transform a substantial part of a biological substrate during the inhibition process. Here, partial proteolysis is a physiological event, whereas in the first case it is a deleterious action. The thrombin/antithrom-bin III/fibrinogen system illustrates this function.

Both functions require that proteolysis stops after a time, which implies that the inhibition is irreversible and permanent (Schemes IX or X). How-ever, reversible inhibitors (Schemes VII or VIII) which exhibit a pseudo-irreversible inhibition behavior may also meet this requirement. That be-havior has been discussed and illustrated in a preceding section (see Fig. 3). Pseudo-irreversible inhibition takes place *in vivo* if $[I]_{vivo} \geq 10[E]_{vivo}$ and $[I]_{vivo} \geq 10^3 K_i$, where $[I]_{vivo}$ and $[E]_{vivo}$ are the *in vivo* inhibitor and proteinase concentrations, respectively.[12]

*In Vivo Inhibition Frequency*

To predict the *in vivo* function of irreversible (or pseudo-irreversible) proteinase inhibitors, one must consider the relationship between $[P]_\infty$, the concentration of product at the end of the inhibition process, and the various parameters describing the enzyme–substrate–inhibitor interaction. If $[I]_{vivo} \geq 10[E]_{vivo}$ and if $[P]_\infty \leq 0.1[S]_{vivo}$, $[P]_\infty/[S]_{vivo}$ is given by[7,12]

$$\frac{[P]_\infty}{[S]_{vivo}} = \frac{k_{cat}[E]_{vivo}}{K_m} \times \frac{1}{k_{ass}[I]_{vivo}} \tag{33}$$

for one-step irreversible inhibition (Scheme IX) or

$$\frac{[P]_\infty}{[S]_{vivo}} = \frac{k_{cat}[E]_{vivo}}{K_m} \times \frac{K_i^*}{k_2[I]_{vivo}} \tag{34}$$

for two-step irreversible inhibition (Scheme X). The only difference be-tween the two relationships is that $k_{ass}$ in Eq. (33) becomes $k_2/K_i^*$ in Eq. (34). The two constants have the same meaning as already mentioned above.

Let us now arbitrarily assume that a proteolysis-preventing function requires $[P]_\infty/[S]_{vivo} \leq 0.02$; that is, less than 2% of the biological substrate should be cleaved during the irreversible inhibition process. Because $k_{cat}/$

[12] J. G. Bieth, *Biochem. Med.* **32,** 387 (1984).

$K_m$ as well as $[E]_{vivo}$ are usually unknown, we shall arbitrarily set their highest limit to $10^5 \, M^{-1} \, \sec^{-1}$ and $10^{-6} \, M$, respectively.[12] This leads to

$$k_{ass}[I]_{vivo} \left( \text{or} \frac{k_2[I]_{vivo}}{K_i^*} \right) \geq 5 \, \sec^{-1} \tag{35}$$

The above pseudo-first-order rate constant may be called the "*in vivo* inhibition frequency*.*" If the frequency is higher than $5 \, \sec^{-1}$, one may conclude that the inhibitor plays a proteolysis-preventing function even if one does not know the values of $[E]_{vivo}$ and $k_{cat}/K_m$. On the other hand, if the frequency is lower than $5 \, \sec^{-1}$, one should measure $[E]_{vivo}$ and $k_{cat}/K_m$ to ascertain whether the inhibitor plays a proteolysis-preventing function or a proteolysis-regulating function, that is, whether $[P]_\infty/[S]_{vivo}$ is lower or higher than 0.02.

*Special Features of in Vivo Two-Step Inhibition*

If irreversible inhibition follows a two-step mechanism (Scheme X) and if $[I]_o \geq 10[E]_o$, then $k$, the pseudo-first-order rate constant of release of product [Eq. (23)], is also equal to the pseudo-first-order constant of formation of EI in the presence of S:

$$k = \frac{k_2[I]_o}{[I]_o + K_{i(app)}^*} \tag{36}$$

where $K_{i(app)}^* = K_i^*(1 + [S]_o/K_m)$. If $[I]_o$, $k_2/K_i^*$, and $[S]_o/K_m$ are constant $K_i^*$ and $k_2$ increase, $k$ will increase and reach a limiting value $k_{lim 1}$ once $K_{i(app)}^* \gg [I]_o$:

$$k_{lim 1} = \frac{k_2[I]_o}{K_{i(app)}^*} \tag{37}$$

On the other hand, if $K_i^*$ and $k_2$ decrease, $k$ will decrease and reach a limiting value $k_{lim 2}$ once $K_{i(app)}^* \ll [I]_o$:

$$k_{lim 2} = k_2 \tag{38}$$

The analysis shows that inhibitors with identical *in vivo* inhibition frequencies $(k_2[I]_{vivo}/K_i^*)$, that is, inhibitors which yield identical $[P]_\infty/[S]_{vivo}$ values may form EI with widely different rates if the inhibition follows a two-step reaction (see also Fig. 4). In other words, the *in vivo* efficacy of a two-step inhibitor is not related to its rate of EI formation.

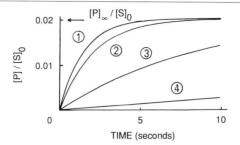

Fig. 4. Theoretical progress curves for proteinase-catalyzed breakdown of a substrate in the presence of an irreversible inhibitor whose reaction with the enzyme involves an intermediate EI* (Scheme X). All curves were calculated using $[E]_o = 0.1 \ \mu M$, $[I]_o = 1 \ \mu M$, $[S]_o = K_m = 0.1$ m$M$, $k_{cat} = 20$ sec$^{-1}$, and $k_2/K_i^* = 10^6 \ M^{-1}$ sec$^{-1}$. Under those conditions all curves have an asymptotic value $[P]_\infty = 2 \ \mu M$, that is, $[P]_\infty/[S]_o = 0.02$, suggesting that the inhibitor plays a proteolysis-preventing function (see text). The curves differ only in the absolute values of $k_2$ and $K_i^*$ and, hence, in the initial concentration of EI*, that is, the concentration of EI* at the start of the reaction when [EI] is negligible: $[EI^*]_{initial} = [E]_o/(1 + K_{i(app)}^*/[I]_o)$. For curve 1, $k_2 = 10$ sec$^{-1}$, $K_i^* = 10 \ \mu M$, $[EI^*]_{initial} = 4.7$ n$M$; for curve 2, $k_2 = 1$ sec$^{-1}$, $K_i^* = 1 \ \mu M$, $[EI^*]_{initial} = 33$ n$M$; for curve 3, $k_2 = 0.1$ sec$^{-1}$, $K_i^* = 0.1 \ \mu M$, $[EI^*]_{initial} = 83$ n$M$; and for curve 4, $k_2 = 0.01$ sec$^{-1}$, $K_i^* = 0.01 \ \mu M$, $[EI^*]_{initial} = 98$ n$M$. It can be seen that the lower $K_i^*$ (i.e., the higher $[EI^*]_{initial}$), the slower the inhibition process ($t_{1/2} = 1.4$ and 693 sec for curves 1 and 4, respectively). Thus, two-step inhibitors with identical *in vivo* inhibition frequencies (see text) may form the final EI complex with widely different rates; that is, the rate of EI formation *in vivo* is not a determinant of inhibitor potency for such inhibitors. On the other hand, the lower $K_i^*$, the lower [P], that is, the more efficient the proteolysis-preventing function of the inhibitor during the initial stage of the inhibition process.

This is not the case with one-step inhibitors for which the rate of *in vivo* EI formation depends only on $k_{ass}[I]_{vivo}$, the *in vivo* inhibition frequency. In that case, inhibitors with identical frequencies also have identical rates of EI formation. The delay time of inhibition,[11,12] the very popular concept of *in vivo* efficacy of one-step inhibitors, therefore does not apply to two-step inhibitors.

Last, it is noteworthy that at the start of the inhibition reaction, a two-step inhibitor whose $[I]_{vivo}$ is higher than $K_{i(app)}^*$, will be more efficient in preventing proteolysis than a one-step inhibitor with an identical inhibition frequency. For example, if $[I]_{vivo} = 100 \ K_{i(app)}^*$, 99% of the *in vivo* liberated proteinase will be taken up as an inactive EI* complex. Therefore, during the initial stage of the inhibition process, the irreversible two-step inhibitor will behave like a one-step tight-binding reversible inhibitor with a pseudo-irreversible behavior:

$$\text{E} \underset{k_{-1}}{\overset{k_1[\text{I}]_{vivo}}{\rightleftharpoons}} \text{EI*}, \qquad k_1[\text{I}]_{vivo} \gg k_{-1}$$

On the other hand, the presence of EI* requires $k_{-1} \gg k_2$ from which it follows that $k_1 \gg k_2/K_i^*$. The formation of EI* will therefore be very fast. The special efficiency of two-step inhibitors at the start of the inhibition process is illustrated in Fig. 4, which shows that during the first seconds of the inhibition reaction, [P] greatly decreases as $[I]_{vivo}/K_i^*$ increases.

### Discussion

Most proteinase inhibitors probably form EI* complexes with their target enzymes. Whether the intermediates are detected *in vitro* depends on the inhibitor concentration range used in the progress curve experiments: if $[I]_o$ is not of the order of magnitude of $K_{i(app)}^*$, EI* is not seen kinetically. *In vitro* experiments with two-step inhibitors therefore yield either $k_2$ (and $k_{-2}$) and $K_i^*$ separately or the $k_2/K_i^*$ ratio. In any case, the *in vivo* inhibition frequency $k_2[I]_{vivo}/K_i^*$ may be inferred from the *in vitro* measurements.

We have suggested that reversible inhibitors may have the physiological function of preventing or regulating proteolysis if they have a pseudo-irreversible behavior *in vivo*, that is, if $[I]_{vivo}/K_i \geq 10^3$. Reversible inhibitors might, however, also act *in vivo* as "proteinase reservoirs"[13] or as intracellular proteinase activity amplifiers[14] as suggested earlier. Furthermore, the above *in vivo* inhibition concept may also be used to help predict the *in vivo* potency of antiproteinase drugs.[13]

The present analysis should be considered as a rough attempt to correlate *in vitro* determined kinetic parameters with *in vivo* function of endogenous or exogenous inhibitors. In the living organism, proteinase inhibition is obviously more complex than what may be anticipated from the oversimplified equations given in this section. Our *in vivo* inhibition theory can only suggest a potential *in vivo* function.

### Acknowledgments

The data in Figs. 2 and 3 have been taken from a postuniversity course on tight-binding inhibition held each year in our laboratory and have been gathered by Christophe Adam and Martine Cadène. The author thanks Christian Boudier, Martine Cadène, and Philippe Mellet for helpful comments regarding the manuscript.

[13] J. G. Bieth, *in* "Pulmonary Emphysema and Proteolysis" (C. Mittman and J. C. Taylor, eds.), Vol. 2, p. 93. Academic Press, New York, 1987.
[14] J. G. Bieth, *in* "Cysteine Proteinases and Their Inhibitors" (V. Turk, ed.), p. 217. de Gruyter, Berlin, 1986.

# [6] Active-Site Titration of Peptidases

## *By* C. GRAHAM KNIGHT

### Introduction

Measurement of the active-site molarity of a peptidase is a valuable step toward defining the molecular properties of the enzyme. In the absence of such a method, catalytic activity (micromoles of substrate cleaved per second) can only be defined in terms of protein concentration. Protein concentration may be measured directly or derived from an established assay for which a specific activity was calculated on the basis that the purified protein was fully active. This assumption is likely to be erroneous, for even with rapid and mild methods of protein isolation, a substantial proportion of the enzyme protein in a purified preparation may be inactive, perhaps as a result of modifications *in vivo* prior to isolation. Although affinity purification may overcome this objection, enzymes vary widely in lability, and a fully active preparation cannot be assumed. If only small quantities are available, the concentration may have been determined by comparison to a standard protein such as albumin which can cause further errors. Although measurements of absorbance can be related to the residual weight of exhaustively dried solutions, tightly bound water or nonvolatile impurities will lead to error.

In studies with recombinant enzymes, it is essential to measure active-site concentrations, so that the effect of amino acid substitutions can be correctly ascribed to effects on catalysis rather than effects on the stability of the protein. Active-site titration, by placing the enzyme activity on a molar basis, also allows the absolute comparison of kinetic constants obtained on different occasions and in different laboratories. In contrast, comparisons of activity are likely to be influenced by differences in the assay conditions or the presence of inhibitory impurities.

The criteria which a titrant should fulfill remain those defined for the first reagents of this type.[1] The titrant should be a stable, well-characterized molecule that combines rapidly only with the active site with 1 : 1 stoichiometry to yield either an inactive or a slowly turned-over complex. Three classes of reagents are potential titrants: substrates that form a stable intermediate acyl–enzyme with the release of an easily quantified first product,

---

[1] M. L. Bender, M. L. Begué-Cantón, R. L. Blakely, L. J. Brubacher, J. Feder, C. R. Gunter, F. J. Kézdy, J. V. Killheffer, Jr., T. H. Marshall, C. G. Miller, R. W. Roeske, and J. K. Stoops, *J. Am. Chem. Soc.* **88,** 5890 (1966).

irreversible inhibitors that react with a single active-site residue, and tight-binding reversible inhibitors. Each of the classes will be considered in turn and representative examples presented.

## Burst Titrations with Specific Substrates

*Kinetic Analysis*

The first titrants were developed for serine and cysteine proteinases following the discovery of the acyl–enzyme mechanism.[2] The kinetics of substrate turnover by such enzymes is described by Eq. (1), where $K_s$ is

$$E + S \underset{}{\overset{K_s}{\rightleftharpoons}} E \cdot S \overset{k_2}{\longrightarrow} \underset{+ P_1}{ES'} \overset{k_3}{\longrightarrow} E + P_2 \qquad (1)$$

the dissociation constant of the initial noncovalent $E \cdot S$ complex, $k_2$ is the first-order acylation rate constant, and $k_3$ the deacylation constant. If $k_3 \ll k_2$, there will be an initial rapid release, or "burst," of the first product $P_1$ on mixing enzyme and substrate. This accompanies the buildup of the steady-state concentration of the acyl–enzyme, $ES'$, and is followed by a slower rate of turnover. A classic example of biphasic behavior is the burst of $p$-nitrophenol released during the reaction of trypsin with a large molar excess of $p$-nitrophenyl guanidinobenzoate.[3] When $[S]_o \gg [E]_o$, the final steady-state portion of the curve can be described by

$$[P_1] = \pi + vt \qquad (2)$$

where $\pi$ is the initial burst of 4-nitrophenol and $v$ is the final steady-state velocity. The magnitude of $\pi$ is given by

$$\pi = [E]_o \{k_2/(k_2 + k_3)\}^2/(1 + K_m/[S]_o)^2 \qquad (3)$$

Inspection of Eq. (3) reveals that $\pi = [E]_o$ only when $k_2 \gg k_3$ and $[S]_o \gg K_m$, where $\gg$ means "at least 100 times greater than." When these conditions are met, most commonly by using chemically reactive phenolic esters as substrates, the size of the burst is independent of titrant concentration. If the condition that $[S]_o \gg K_m$ cannot be met, the burst should be measured over a range of titrant concentrations at a constant value of $[E]_o$. The data are plotted as $1/\pi^{1/2}$ versus $1/[S]_o$, and the intercept on the ordinate gives $(k_2 + k_3)/(k_2 [E]_o^{1/2})$. It must be emphasized that there is no comparable procedure to derive $[E]_o$ if $k_3 \approx k_2$ and for this reason burst titrations

---

[2] B. S. Hartley and B. A. Kilby, *Biochem. J.* **56**, 288 (1954).
[3] T. Chase, Jr., and E. Shaw, this series, Vol. 19, p. 20.

TABLE I

BURST TITRANTS OF SERINE PROTEINASES

| Titrant | Proteinase | Ref. |
|---|---|---|
| Acetyl-Ala-azaPhe-ONp | Cathepsin G, chymotrypsin, subtilisin | a |
| Acetyl-Ala-Ala-azaNle-ONp | Leukocyte elastase, pancreatic elastase | b |
| Benzyl p-guanidinothiobenzoate | Thrombin, trypsin, tryptase | c |
| trans-Cinnamoylimidazole | Chymotrypsin | d |
| | Subtilisin | e |
| Fluorescein mono-p-guanidinobenzoate | Factor IXa, factor Xa, kallikrein | f |
| | Plasmin, trypsin, thrombin, urokinase | g |
| 2-Hydroxy-5-nitro-α-toluenesulfonic acid sultone | Chymotrypsin | h |
| 4-Methylumbelliferyl p-guanidino-benzoate | Factor Xa, plasmin, thrombin, trypsin | i |
| | Factor XIIa | j |
| | Tryptase | k |
| 4-Methylumbelliferyl p-trimethyl-ammonium cinnamate | Chymotrypsin | i |
| p-Nitrophenyl guanidinobenzoate | Plasmin, thrombin, trypsin | l |

[a] B. F. Gupton, D. L. Carroll, P. M. Tuhy, C.-M. Kam, and J. C. Powers, *J. Biol. Chem.* **259**, 4279 (1984).

[b] J. C. Powers, R. Boone, D. L. Carroll, B. F. Gupton, C.-M. Kam, N. Nishino, M. Sakamoto, and P. Tuhy, *J. Biol. Chem.* **259**, 4288 (1984).

[c] R. R. Cook and J. C. Powers, *Biochem. J.* **215**, 287 (1983).

[d] G. R. Schonbaum, B. Zerner, and M. L. Bender, *J. Biol. Chem.* **236**, 2930 (1961).

[e] M. L. Bender, M. L. Begué-Cantón, R. L. Blakely, L. J. Brubacher, J. Feder, C. R. Gunter, F. J. Kézdy, J. V. Killheffer, Jr., T. H. Marshall, C. G. Miller, R. W. Roeske, and J. K. Stoops, *J. Am. Chem. Soc.* **88**, 5890 (1966).

[f] P. E. Bock, P. A. Craig, S. T. Olson, and P. Singh, *Arch. Biochem. Biophys.* **273**, 375 (1989).

[g] L. L. Melhado, S. W. Peltz, S. P. Leytus, and W. F. Mangel, *J. Am. Chem. Soc.* **104**, 7301 (1982).

[h] F. J. Kézdy and E. T. Kaiser, this series, Vol. 19, p. 3.

[i] G. W. Jameson, D. V. Roberts, R. W. Adams, W. S. A. Kyle, and D. T. Elmore, *Biochem. J.* **131**, 107 (1973).

[j] M. M. Bernado, D. E. Day, S. T. Olson, and J. D. Shore, *J. Biol. Chem.* **268**, 12468 (1993).

[k] N. M. Schechter, G. Y. Eng, and D. R. McCaslin, *Biochemistry* **32**, 2617 (1993).

[l] T. Chase, Jr., and E. Shaw, this series, Vol. 19, p. 20.

cannot be recommended for mutant enzymes whose catalytic mechanism may be perturbed.[4]

*Chemical Considerations*

Some representative burst titrants for the serine proteinases are given in Table I. The majority are based on specific substrates, but all have small

[4] M. Rheinnecker, G. Baker, J. Eder, and A. R. Fersht, *Biochemistry* **32**, 1199 (1993).

values of $k_3$ so that the size of the burst can be measured accurately. Although trypsin and chymotrypsin have been titrated successfully with Z-Lys-ONp and Z-Tyr-ONp (where Z and ONp stand for benzyloxycarbonyl and $p$-nitrophenyl ester, respectively),[1] those substrates are turned over rapidly after the initial burst, and better reagents are now available. Similarly, although cysteine proteinases of the papain family can be titrated with nitrophenyl esters, spontaneous hydrolysis is a problem, and the use of the irreversible inhibitor E-64 [$trans$-epoxysuccinyl-L-leucylamido(4-guanidino)butane] is preferred (see below). By contrast, product release is slow after the initial stoichiometric reactions between trypsin and $p$-nitrophenyl guanidinobenzoate,[3] and chymotrypsin and $p$-nitrophenyl trimethylacetate.[5]

Slow turnover is also a feature of the reaction of chymotrypsin with $p$-nitrophenyl 3-acetyl-2-benzyl carbazate, in which the $\alpha$-methine (CH) group of $C_2H_3O$-Phe-ONp has been replaced by a nitrogen atom.[6] Such aza-esters form stable carbamates, R-NH-CO-OR', on reaction with the active-site serine residue. That principle has been used by Powers and co-workers[7,8] in the design of selective azapeptidyl titrants, in which the P1 residue is an azaamino acid, for enzymes with more extended active sites, such as human leukocyte elastase and cathepsin G. Improved methods for azapeptide synthesis[9,10] suggest that the approach could be exploited more fully.

*Sensitivity of Titration*

Because the molar concentration of the detectable product is equal to that of the enzyme, the sensitivity of titrations depends on the sensitivity of detection of the product. For compounds releasing $p$-nitrophenol ($\varepsilon$ of 12,600 $M^{-1}$ cm$^{-1}$ at pH 7.5[11]), the minimal enzyme concentration is 1–10 $\mu M$. Far greater sensitivity is available in principle with fluorogenic titrants, such as those derived from 4-methylumbelliferone (7-hydroxy-4-methylcoumarin),[12] although the esters are rather labile and prone to photolytic

[5] M. L. Bender, F. J. Kézdy, and F. C. Wedler, *J. Chem. Ed.* **44,** 84 (1967).

[6] D. T. Elmore and J. J. Smyth, *Biochem. J.* **107,** 103 (1968).

[7] J. C. Powers, R. Boone, D. L. Carroll, B. F. Gupton, C.-M. Kam, N. Nishino, M. Sakamoto, and P. Tuhy, *J. Biol. Chem.* **259,** 4288 (1984).

[8] B. F. Gupton, D. L. Carroll, P. M. Tuhy, C.-M. Kam, and J. C. Powers, *J. Biol. Chem.* **259,** 4279 (1984).

[9] M. Quibell, W. G. Turnell, and T. Johnson, *J. Chem. Soc. Perkin Trans. 1,* 2843 (1993).

[10] J. Gante, *Synthesis,* 405 (1989).

[11] R. R. Cook and J. C. Powers, *Biochem. J.* **215,** 287 (1983).

[12] G. W. Jameson, D. V. Roberts, R. W. Adams, W. S. A. Kyle, and D. T. Elmore, *Biochem. J.* **131,** 107 (1973).

decomposition. The requirement that $[S]_o \gg K_m$ also restricts the range of suitable substrates, as fluorimetric assays are commonly made at concentrations below 20 $\mu M$ to avoid absorptive quenching effects. The acyl–enzyme mechanism [Eq. (1)] predicts

$$K_m = K_s k_3/(k_2 + k_3) \tag{4}$$

so $K_m$ will be very low only if acylation is rapid and turnover is negligible. If the condition that $[S]_o \gg K_m$ cannot be met, $[E]_o$ can be obtained from measurements of the burst at different substrate concentrations, as described above, provided that $k_2 \gg k_3$.

Titrants that fulfill these conditions are 4-methylumbelliferyl $p$-guanidinobenzoate for trypsin and 4-methylumbelliferyl $p$-trimethylammonium cinnamate for chymotrypsin,[12] which allow the measurement of enzyme concentrations in the range 0.02 to 0.2 $\mu M$. A further 20-fold increase in the sensitivity of titration can be achieved with fluorescein mono-$p$-guanidinobenzoate,[13] the most sensitive titrant currently available for proteinases with trypsin-like specificity. The advantages of fluorescein compared to 4-methylumbelliferone are a higher molar absorptivity and quantum yield, and increased photostability.

It could be argued that many of the reagents discussed above are redundant and of merely academic interest, as they were developed for pancreatic enzymes available in abundance. Although the titrants are also applicable to some of the rarer enzymes of the clotting cascade, for example, there is a problem with the titration of serine proteinases of different specificities and available in small amounts. A possible way forward is illustrated by the titration of the secreted, soluble kexin with a highly specific fluorogenic substrate.[14] Measurements of 7-amino-4-methylcoumarin (NH-Mec) release from Acetyl-Pro-Met-Tyr-Lys-Arg-NH-Mec ($k_{cat}/K_m$ of $1.1 \times 10^7$) were made in a quench-flow apparatus which allowed the initial burst to be measured on a millisecond time scale with less than 20 $\mu g$ of enzyme. Although it was not rigorously established that $k_2 \gg k_3$, the low value of $K_m$ (2.2 $\mu M$) suggests that this is a reasonable assumption, and indeed the enzyme concentration agreed well with that determined by amino acid analysis. The method should be applicable to other serine proteinases if specific fluorogenic substrates with $k_2 \gg k_3$ are available.

[13] L. L. Melhado, S. W. Peltz, S. P. Leytus, and W. F. Mangel, *J. Am. Chem. Soc.* **104,** 7301 (1982).
[14] C. Brenner and R. S. Fuller, *Proc. Natl. Acad. Sci. U.S.A.* **89,** 922 (1992).

*Titration of Cysteine Proteinases with Disulfides*

Compared to the case for serine proteinases, burst titration of the cysteine enzymes is not easy. The active-site SH group is very prone to oxidation, and low molecular weight thiols, such as 2-mercaptoethanol, are required to maintain the reduced state. In designing a titrant, the exogenous thiols must be taken into account, as they accelerate the hydrolysis of specific esters. Although burst titrants have been described,[1,15] another approach is to measure selectively the concentration of active-site thiols in the presence of the low $M_r$ thiols. Fortunately, a unique property of the enzymes of the papain family is the enhanced reactivity of the active-site SH group at low pH, owing to the formation of a stable thiolate–imidazolium ion pair with the adjacent histidine.[16] Reactivity is particularly marked toward 2,2′-dipyridyl disulfide and allowed Brocklehurst and Little[17] to titrate papain at pH 3.8 in the presence of a 100-fold excess of 2-mercaptoethanol. However, the method requires large amounts of enzyme.

*Experimental Protocols for Burst Titrants*

Protocols for the titration of trypsin[3] and chymotrypsin[18] have been described in an earlier volume in this series, and procedures for other *p*-nitrophenyl ester titrants are similar. Although the high reactivity of *p*-nitrophenyl esters leads to rapid enzyme acylation, it also favors spontaneous hydrolysis and catalytic cleavage. These effects are minimized at acidic pH values, but the molar absorptivity of *p*-nitrophenol is greatly decreased ($pK_a$ 7.04). The burst of *p*-nitrophenol from the azapeptide titrants[7,8] under acidic conditions was measured near an isobestic point at 345 nm ($\varepsilon$ of 6250 $M^{-1}$ cm$^{-1}$) and the spectrophotometer calibrated with each substrate and buffer used. For enzymes with trypsin-like specificity, Cook and Powers[11] have proposed titration with benzyl 4-guanidinothiobenzoate, detecting the released benzyl mercaptan with 4,4′-dithiopyridine. The molar absorptivity of the liberated thiopyridone is high ($\varepsilon$ of 19,800 $M^{-1}$ cm$^{-1}$ at 324 nm) and is hardly affected by pH.

In contrast to the case for absorbance, the fluorescence of a sample cannot be placed on an absolute scale, but can only be expressed relative to an arbitrary standard, which introduces an additional source of error. Thus, fluorogenic titrants require more controlled conditions and careful

[15] P. Ascenzi, P. Aducci, G. Amiconi, and A. Ballio, *Gazz. Chim. Ital.* **117,** 421 (1987).
[16] K. Brocklehurst, F. Willenbrock, and E. Salih, *in* "Hydrolytic Enzymes" (A. Neuberger and K. Brocklehurst, eds.), p. 39. Elsevier, Amsterdam, 1987.
[17] K. Brocklehurst and G. Little, *Biochem. J.* **133,** 67 (1973).
[18] F. J. Kézdy and E. T. Kaiser, this series, Vol. 19, p. 3.

calibration to ensure accuracy. Fluorescence is very sensitive to environment, and care must be taken to maintain a constant temperature and buffer composition. In the case of 4-methylumbelliferone, it is important to measure the fluorescence intermittently, or to decrease the slit widths if recording continuously, to avoid photobleaching. For Perkin–Elmer (Norwalk, CT) fluorimeters running the FLUSYS software,[19] which controls the instrument via an IBM-compatible computer and collects the data continuously in a disk file, it is easy to calibrate the response with 4-methylumbelliferone as a standard.

Useful fluorophores have $\varepsilon$ values exceeding 10,000 $M^{-1}$ cm$^{-1}$, so there will be self-quenching owing to absorption of the exciting light except at concentrations less than 10 $\mu M$. Moreover, the fluorescence of the released product is measured in the presence of the unhydrolyzed titrant, which may have a significant absorbance, so the calibration should be made under similar conditions. Alternatively, a standard curve can be made using an independent fluorescence standard, such as quinine[12] or a fluorescent plastic block,[13] or the method can be calibrated *in situ*[20] using, for example, trypsin previously titrated with *p*-nitrophenyl guanidinobenzoate.

## Titration with Irreversible Inhibitors

Active-site-directed irreversible inhibitors are attractive candidates in principle as titrants for proteinases, as they would allow the determination of enzymes at the concentrations used in assays. For use as a titrant, the inhibitor should react exclusively with the active enzyme with a 1 : 1 stoichiometry, and it should be chemically stable and unreactive toward other functional groups on the protein or present in the assay mixture. In practice, those conditions are fulfilled quite rarely, and very few of the reagents are suitable titrants. This probably reflects the fact that the active sites of aspartic, serine, and metalloproteinases contain groups that are only weakly nucleophilic in an aqueous environment. For example, a wide variety of peptidylchloromethanes have been described as selective inhibitors of serine proteinases. The reagents have been invaluable in kinetic and structural studies, but they are unsuitable as titrants. When the soluble, secreted kexin of yeast was titrated with Phe-Ala-Lys-Arg-CH$_2$Cl[21] the apparent enzyme concentration was twice that determined by initial burst titration.[14] The discrepancy was reasonably ascribed to the hydration and instability of the inhibitor.

[19] N. D. Rawlings and A. J. Barrett, *Comput. Appl. Biosci.* **6**, 118 (1990).
[20] N. M. Schechter, G. Y. Eng, and D. R. McCaslin, *Biochemistry* **32**, 2617 (1993).
[21] H. Angliker, P. Wikstrom, E. Shaw, C. Brenner, and R. S. Fuller, *Biochem. J.* **293**, 75 (1993).

*Serine Proteinase Titration with* $^3H$*-Labeled Diisopropyl Fluorophosphate*

Although chloromethanes are unsatisfactory titrants of serine protein-ases, the active-site serine is reactive toward organophosphorus inhibitors, and diisopropyl fluorophosphate (DFP) allowed the first active-site titration of a proteolytic enzyme.[22] The method may be illustrated by the titration of $\alpha$-lytic proteinase.[23] The enzyme is treated with a 5- to 10-fold molar excess of [$^3$H]DFP at room temperature until no activity remains (the minimum time may be determined with unlabeled DFP). [$^3$H]DFP is avail-able with a specific radioactivity of about 185 GBq/mmol (5 Ci/mmol), suitable for the titration of 5–10 $\mu$g of enzyme. The mixture is applied to a column of Sephadex G-25, and the total radioactivity in the protein-containing fractions is determined. From this the concentration of protein-bound DFP can be calculated knowing the specific radioactivity. If each molecule of DFP reacts exclusively with an active-site serine residue, that corresponds to the active-site molarity. Side reactions of DFP are known, and the method is probably most suitable for purified enzymes. When chymase[24] was titrated with lima bean trypsin inhibitor standardized with chymotrypsin, the specific activity was 40% lower than that determined with [$^3$H]DFP and a less pure chymase preparation.

*Cysteine Proteinase Titration with Epoxypeptides*

The enhanced reactivity of the active-site cysteine in papain and related cysteine proteinases has been exploited in the design of inhibitors, such as peptidyldiazomethanes,[25] that are essentially unreactive toward low $M_r$ thiols. Despite promising early indications, the diazomethanes are now known to react with some serine proteinases.[25] However, E-64, L-*trans*-epoxysuccinyl-L-leucylamido(4-guanidino)butane, originally isolated from cultures of *Aspergillus japonicus*,[26] is not only unreactive toward low $M_r$ thiols, but also does not react with aspartic, serine, and metalloproteinases.[27] The epoxides, therefore, are very useful titrants of cysteine proteinases of the papain family.

[22] A. K. Balls and E. F. Jansen, *Adv. Enzymol.* **13,** 321 (1952).
[23] D. M. Epstein and R. H. Abeles, *Biochemistry* **31,** 11216 (1992).
[24] N. M. Schechter, J. L. Sprows, O. L. Schoenberger, G. S. Lazarus, B. S. Cooperman, and H. Rubin, *J. Biol. Chem.* **264,** 21308 (1989).
[25] E. N. Shaw, this series, Vol. 244, p. 649.
[26] K. Hanada, M. Tamai, M. Yamagishi, S. Ohmura, J. Sawada, and I. Tanaka, *Agric. Biol. Chem.* **42,** 523 (1978).
[27] A. J. Barrett, A. A. Kembhavi, M. A. Brown, H. Kirschke, C. G. Knight, M. Tamai, and K. Hanada, *Biochem. J.* **201,** 189 (1982).

*Kinetics of Irreversible Inhibition*

In considering titration with irreversible inhibitors, it is helpful to understand the kinetic basis of the application. The kinetics can be analyzed according to

$$E + I \underset{}{\overset{K_i}{\rightleftharpoons}} E \cdot I \xrightarrow{k_2} EI \tag{5}$$

where $K_i$ is the dissociation constant of the noncovalent $E \cdot I$ complex and $k_2$ is the first-order rate constant for inactivation. When $[I]_o \gg [E]_o$, the disappearance of activity will follow Eq. (6)[28]:

$$2.3 \log (a_t/a_o) = -k_{obs} \, t \tag{6}$$

where $a_o$ is the activity at zero time, $a_t$ is the activity at time $t$, and $k_{obs}$ is the pseudo-first-order rate constant for inactivation. According to the scheme in Eq. (5),

$$k_{obs} = k_2/(1 + K_i/[I]_o) \tag{7}$$

If $K_i \gg [I]_o$, $k_{obs}/[I]_o$ is the apparent second-order rate constant $k_2'$ for the reaction

$$E + I \xrightarrow{k_2'} EI \tag{8}$$

The parameter $k_2'$ is very useful, as it allows the calculation of the extent of reaction at any time and $[E]_o/[I]_o$ ratio during a titration. When equivalent concentrations of enzyme and inhibitor are mixed, the decrease in activity with time is described by Eq. (9):

$$t = (1/k_2')[(a_o - a_t)/a_o a_t] \tag{9}$$

For very rapidly reacting inhibitors, it is convenient to determine $k_2'$ at equivalence, using Eq. (9).[27] At other relative concentrations, the loss of activity with time is given by

$$\frac{a_t}{a_o} = 1 - \frac{[I]_o \exp\{k_2't([E]_o - [I]_o)\} - [I]_o}{[E]_o \exp\{k_2't([E]_o - [I]_o)\} - [I]_o} \tag{10}$$

Figure 1 shows that the time taken to reach 99% reaction, calculated using Eq. (10), increases markedly in the region of the equivalence point. Because the concentrations of both enzyme and inhibitor are greatly depleted here, the probability of a bimolecular encounter becomes very low indeed, and the reaction slows right down. The practical consequence is that titration curves will show a deviation from linearity around the equiva-

[28] R. Kitz and I. B. Wilson, *J. Biol. Chem.* **237**, 3245 (1962).

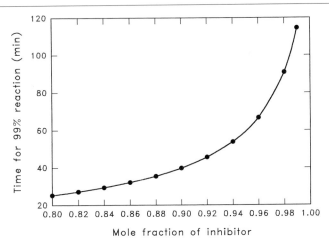

FIG. 1. Slowing of the reaction as the concentrations of enzyme and inhibitor approach equivalence during a titration. The data points were calculated using Eq. (10), assuming 99% reaction between the inhibitor and 20 n$M$ proteinase, with $k_2' = 500{,}000\ M^{-1}\ sec^{-1}$.

lence point if insufficient time has been allowed for reaction. We had thought previously[27] that the deviations could be ignored, provided the initial part of the titration curve was linear. However, when titration curves differing only in the incubation time are calculated using Eq. (10), as shown in Fig. 2, it is apparent that this conclusion is incorrect. When the reaction

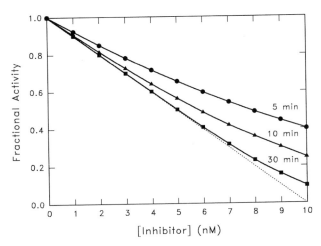

FIG. 2. Simulated titration curves, calculated using Eq. (10), showing the effect of incubation times between 5 and 30 min on the reaction between an inhibitor and 20 n$M$ proteinase, with $k_2' = 500{,}000\ M^{-1}\ sec^{-1}$. The stoichiometric titration line (---) is also shown.

time is too short, extrapolation of the initial part of the titration curve will give an incorrectly high value of $[E]_o$. It is strongly recommended that if the deviation at the equivalence point exceeds 10–20%, the titration be repeated with double the incubation time, or with increased concentrations of enzyme and inhibitor.

When the value of $k_2'$ is known, Eq. (10) allows appropriate enzyme and inhibitor concentrations to be estimated. For example, if 30 min is considered a convenient reaction time, a 10 n$M$ enzyme solution will undergo 98% reaction with 0.7 equivalents of titrant in that time if $k_2' = 500,000\ M^{-1}\ sec^{-1}$. If $k_2' = 5000\ M^{-1}\ sec^{-1}$, 1 $\mu M$ enzyme will be required, as 10 n$M$ enzyme will need 2 days to react to the same extent. In practice, the reaction is accelerated by allowing the enzyme and inhibitor to react in a very small volume, from which a sample is taken to measure residual activity (see below).

*Experimental Protocols*

The suitability of an active-site thiol-reactive inhibitor as a titrant can be assessed by measuring $k_{obs}$, the pseudo-first-order rate constant for inactivation. The procedures are straightforward and may be illustrated with E-64.[27] Dissolve the titrant in dry dimethyl sulfoxide to give a 5 m$M$ solution (stable for several months at 4°). Dilute the solution further in assay buffer to give a conveniently measurable rate of inactivation of the enzyme. Equilibrate the titrant and enzyme (10–100 n$M$) stock solutions separately at an appropriate temperature. At zero time, mix the solutions and remove samples (10–25 $\mu$l) at suitable time intervals (30–60 sec). Dilute immediately into an assay mixture (at least 1.0 ml) and measure the remaining activity. When $[I]_o \gg [E]_o$, the kinetics of inactivation will follow Eq. (6) and $k_{obs}$ can be calculated. The kinetics of inactivation may also be followed continuously if an appropriate substrate is available.[29]

The basic titration procedure may also be illustrated with E-64.[27] Dilute the stock solution of titrant to give solutions 1, 2, 3, ... , 10 $\mu M$ in assay buffer. Add 25 $\mu$l of each solution to 75 $\mu$l of enzyme solution and allow to stand for 30 min at 30°. The enzyme concentration should be such that the residual activity can be measured by adding a 10- to 25-$\mu$l portion of each mixture to 1–2 ml of a peptidyl-7-amino-4-methylcoumarin substrate. Plot the activity versus inhibitor concentration and extrapolate the initial linear portion to cut the baseline at the enzyme concentration. Ideally, the equivalence point should be 70–80% of the maximum inhibitor concentration, so that the initial slope is well-defined. If the equivalence point falls

---

[29] C. Crawford, R. W. Mason, P. Wilkstrom, and E. Shaw, *Biochem. J.* **253,** 751 (1988).

outside that range, the concentration of one of the reactants should be adjusted. For example, if no activity remains after the first 10–20% of the titration, dilute the inhibitor 10-fold.

## Titration with Tight-Binding Proteinase Inhibitors

The kinetic properties of tight-binding inhibitors have been extensively reviewed[30,31] and are not discussed in detail here. Briefly, tight-binding inhibitors bind to the target enzymes with dissociation constants such that under assay conditions inhibition occurs at comparable concentrations of enzyme and inhibitor. This complicates the kinetic analysis, as the concentration of free inhibitor cannot be equated with the total concentration because a significant fraction is bound to the enzyme. Moreover, in many cases, the rates of formation of the final equilibrium complex, and of its dissociation, are slow. However, in the context of active-site titration, we are concerned only with the final equilibrium state, and those difficulties can be ignored provided sufficient time has been allowed for equilibration. If in doubt, the approach to equilibrium should be followed by assaying the mixture at intervals until the residual activity reaches a constant value.

Assay methods differ widely in sensitivity, so how can one judge whether an inhibitor will bind tightly enough to give a valid titration? The parameter determining whether the binding will be stoichiometric is the ratio $[E]_o/K_i$.[32] Provided the data are analyzed by curve-fitting (see below), valid titration curves will be obtained when $[E]_o/K_i > 2$ (Fig. 3).

### Experimental Protocol

The experimental approach is straightforward and may be illustrated by the titration of subtilisin BPN' by chymotrypsin inhibitor 2.[4] Varying volumes of a stock solution of inhibitor (in this case 1.84 $\mu M$ in assay buffer) in the range 0 to 50 $\mu$l are added to a 100-$\mu$l portion of enzyme solution (approximately 250 n$M$) and the volume made up to 150 $\mu$l with assay buffer. After incubation for 30 min at room temperature, a 50-$\mu$l portion of each mixture is added to 950 $\mu$l of substrate solution, and the initial velocities are measured. Velocities are plotted against inhibitor concentration, and the initial linear part of the plot is extrapolated to cut the baseline at the equivalence point corresponding to the concentration of active enzyme. For subtilisin BPN' and chymotrypsin inhibitor 2[4], the value

[30] J. F. Morrison, *Trends Biochem. Sci.* **7**, 102 (1982).
[31] J. F. Morrison and C. T. Walsh, *Adv. Enzymol.* **61**, 201 (1987).
[32] A. Goldstein, *J. Gen. Physiol.* **27**, 529 (1944).

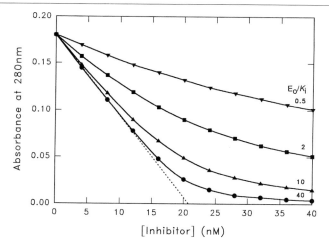

FIG. 3. Simulated tight-binding titrations of 20 n$M$ cathepsin D by pepstatin analogs, calculated using Eq. (11), at different values of $[E]_o/K_i$. When the titration is analyzed by curve-fitting, an $[E]_o/K_i$ ratio of 2 is probably sufficient to give accurate values of $[E]_o$ and $K_i$. Much higher values of the ratio are necessary for a manual extrapolation to be accurate.

obtained was the same as that determined with *N-trans*-cinnamoylimidazole,[33] showing that the inhibitor was 100% active.

An assumption implicit in the protocol is that dissociation of the preformed enzyme–inhibitor complex can be neglected on dilution into the assay, so that the observed activity reflects the concentration of free active enzyme in the incubation mixture. If dissociation does occur, the progress curve will show an upward curvature, and the titration procedure should be reversed. That is to say, the enzyme and inhibitor are preincubated in the larger volume and the rate measured after the addition of a small volume of substrate stock. However, the more dilute system will take longer to reach equilibrium. With very slow, tight-binding inhibitors, dissociation on dilution can be neglected, and it is necessary to preincubate at enzyme concentrations up to 100 times greater than that used in the assay in order to reach equilibrium in a reasonable time. Thus, in the subtilisin BPN′ titration by chymotrypsin inhibitor 2, where $K_i = 2.9$ p$M$,[34] the enzyme concentration before dilution was approximately 250 n$M$. In another example, Murphy and Willenbrock (this volume, [30]) preincubated gelatinase B and the tissue inhibitor of metalloproteinases (TIMP), $K_i < 1$ p$M$, for 6 hr at an enzyme concentration of 100 n$M$, before diluting 100-fold into the assay.

[33] G. R. Schonbaum, B. Zerner, and M. L. Bender, *J. Biol. Chem.* **236**, 2930 (1961).
[34] C. Longstaff, A. F. Campbell, and A. R. Fersht, *Biochemistry* **29**, 7339 (1990).

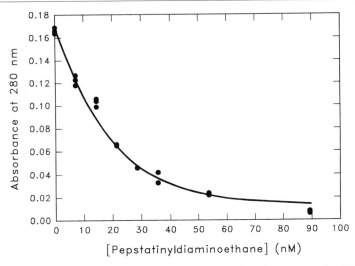

[Pepstatinyldiaminoethane] (nM)

FIG. 4. The inhibition of cathepsin D by pepstatinyldiaminoethane was analyzed by fitting to Eq. (11), giving $[E]_0 = 22.5 \pm 0.3$ n$M$ and $K_i = 5.1 \pm 0.4$ n$M$.

## Analyzing Titration Curves

The data to be analyzed are a set of activities and inhibitor concentrations. Although the data may be plotted by hand, that is not a satisfactory procedure, even if the titration has an extended initial linear portion, as is shown, for example, by cathepsin D and pepstatin.[35] If the data are scattered or fall on a curve almost from the start, it is necessary to fit the data to an equation for equilibrium binding[36]:

$$v = SA(E_o - 0.5\,\{(E_o + I + K_i) - [(E_o + I + K_i)^2 - 4E_oI]^{1/2}\}) \quad (11)$$

where $SA$ is specific activity (rate per unit enzyme concentration); $E_o$, enzyme concentration; $I$, inhibitor concentration; and $K_i$, the dissociation constant. Another form of Eq. (11)[37] is

$$v = [v_o/2E_o]\,\{E_o - I - K_i + [(I + K_i - E_o)^2 + 4K_iE_o]^{1/2}\} \quad (12)$$

Figure 4 shows data from a titration of cathepsin D by pepstatinyldiaminoethane fitted to Eq. (11) by the ENZFITTER program (Elsevier-

[35] C. G. Knight and A. J. Barrett, *Biochem. J.* **155,** 117 (1976).
[36] W. G. Gutheil and W. W. Bachovchin, *Biochemistry* **32,** 8723 (1993).
[37] J. W. Williams and J. F. Morrison, this series, Vol. 63, p. 437.

| Equation Name | | |
|---|---|---|
| Tight-binding titration Eq.(11) | | |

| Y Axis Name | X Axis Name | VAR |
|---|---|---|
| Rate | [Inhibitor] | I |

| | Variable Name | VAR | | Variable Name | VAR |
|---|---|---|---|---|---|
| 1 | Inhibition constant | Ki | 5 | | |
| 2 | Enzyme concentration | Eo | 6 | | |
| 3 | | | 7 | | |
| 4 | | | 8 | | |

| | Prompted Constant | VAR | | Prompted Constant | VAR |
|---|---|---|---|---|---|
| 1 | Specific activity | SA | 3 | | |
| 2 | | | 4 | | |

| | Equation Definition |
|---|---|
| Temp1 | |
| Temp2 | |
| $Y =$ | SA*(Eo-0.5*((Eo+I+Ki)-SQRT(SQR(Eo+I+Ki)-4*I*Eo))) |

FIG. 5. Equation editor screen when Eq. (11) is correctly entered into the program ENZFIT-TER. If the specific activity $SA$ is unknown, it should be entered as variable 3 and the Prompted Constant box left empty.

Biosoft, Cambridge CB2 1LA, UK). Entry of equations into the program is straightforward, and the equation editor screen for Eq. (11) is shown in Fig. 5. When fitting the data, it is necessary to supply initial estimates for $SA$, $E_o$, and $K_i$ (although with a well-established assay $SA$ can be treated

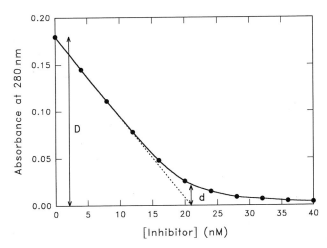

FIG. 6. Graphical estimation of $[E]_o$ and $K_i$ from a titration of cathepsin D by pepstatin (see text for details). Note that the approach of the titration curve to the baseline is asymptotic.

## TABLE II
### TIGHT-BINDING INHIBITOR TITRANTS

| Titrant | Proteinase | Ref. |
|---|---|---|
| Chymotrypsin inhibitor 2 | Subtilisin BPN' | a |
| Lima bean trypsin inhibitor | Chymase | b |
| $\alpha_1$-Proteinase inhibitor ($\alpha$-PI) | Cathepsin G, chymotrypsin, neutrophil elastase | c |
| Tissue inhibitor of metallo- proteinases (TIMP) | Collagenase, gelatinase, matrilysin, stromelysin | d |
| Pepstatin | Cathepsin D | e |
|  | Pepsin | f |
| Phosphoramidon | Thermolysin | f |
| Roche Ro-8959 | Human immunodeficiency virus proteinase | g |
| Ala-Ala-Phe$\psi$[CH(OH)CH$_2$]Gly-Val-Val-OCH$_3$ | Human and simian immunodeficiency virus proteinases | h |

[a] M. Rheinnecker, G. Baker, J. Eder, and A. R. Fersht, *Biochemistry* **32**, 1199 (1993).

[b] N. M. Schechter, J. L. Sprows, O. L. Schoenberger, G. S. Lazarus, B. S. Cooperman, and H. Rubin, *J. Biol. Chem.* **264**, 21308 (1989).

[c] G. Salvesen and H. Nagase, *in* "Proteolytic Enzymes: A Practical Approach" (R. J. Beynon and J. S. Bond, eds.), p. 83. IRL Pess, Oxford, 1989.

[d] G. Murphy and F. Willenbrock, this volume, [30].

[e] C. G. Knight and A. J. Barrett, *Biochem. J.* **155**, 117 (1976).

[f] T. S. Angeles, G. A. Roberts, S. A. Carr, and T. D. Meek, *Biochemistry* **31**, 11778 (1992).

[g] J. T. Griffiths, L. H. Phylip, J. Konvalinka, P. Strop, A. Gustchina, A. Wlodawer, R. J. Davenport, R. Briggs, B. M. Dunn, and J. Kay, *Biochemistry* **31**, 5193 (1992).

[h] S. K. Grant, I. C. Deckman, M. D. Minnich, J. Culp, S. Franklin, G. B. Dreyer, T. A. Tomaszek, Jr., C. Debouck, and T. D. Meek, *Biochemistry* **30**, 8424 (1991).

as a constant). It is convenient to estimate the parameters graphically.[38] Plot the rate versus [I] data and extrapolate the initial part to obtain $[E]_o$, as shown schematically in Fig. 6. The specific activity, $SA$, is given by $D/[E]_o$. With Eq. (13) calculate the value of the parameter $y$:

$$y = d/D \tag{13}$$

which approximates the fraction of active enzyme remaining unbound at the equivalence point. An estimate of the dissociation constant $K_i$ is given by

$$K_i = y^2[E]_o/(1 - y) \tag{14}$$

Unless the titrant has an exceptionally low value of $K_i$, the titration curve will not approximate the intersection of two straight lines. The natural

[38] N. M. Green and E. Work, *Biochem. J.* **54**, 347 (1953).

deviation from linearity in the region of the equivalence point reflects the stoichiometric nature of the interaction. As the inhibitor concentration increases past equivalence, any remaining free enzyme becomes bound and activity disappears.

*Examples of Tight-Binding Titrants*

Some tight-binding titrants are listed in Table II. The most convenient titrants are those of low $M_r$, such as pepstatin, which can be rigorously purified and characterized. The protein proteinase inhibitors are also valuable, but the protein and inhibitory concentrations cannot be equated[39] (this is especially true for mutant inhibitors[40]). Before the reagents can be used for titration, it is necessary to standardize them with a proteinase whose active-site concentration has been determined with a chemical titrant. For example, before titration of elastase, chymotrypsin, and cathepsin G,[41] $\alpha_1$-proteinase inhibitor ($\alpha_1$-PI) is standardized with trypsin previously titrated with *p*-nitrophenyl guanidinobenzoate.[3] $\alpha_1$-PI freshly prepared from human plasma is about 80% active.[42] In the case of the matrix metalloproteinases and TIMP, there is at present no independent method of standardization, but the development of low $M_r$ tight-binding inhibitors of the enzymes[43,44] seems likely to provide a chemical titrant shortly.

It has also been noted that active-site titrants are useful reagents for determining whether the activity of a proteinase of low specific activity is due to a contaminant. For example, when the sweet protein thaumatin, an apparent cysteine proteinase, was titrated with E-64, the equivalence point occurred at 1/100 of the molar quantity of thaumatin,[45] indicating the presence of a trace amount of the enzyme thaumatopain.[46]

[39] Q.-L. Ying and S. R. Simon, *Biochemistry* **32**, 1866 (1993).
[40] M. O'Shea, F. Willenbrock, R. A. Williamson, M. I. Cockett, R. B. Freedman, J. J. Reynolds, A. J. P. Docherty, and G. Murphy, *Biochemistry* **31**, 10146 (1992).
[41] G. Salvesen and H. Nagase, *in* "Proteolytic Enzymes: A Practical Approach" (R. J. Beynon and J. S. Bond, eds.), p. 83. IRL Press, Oxford, 1989.
[42] D. A. Lomas, D. L. Evans, S. R. Stone, W.-S. W. Chang, and R. W. Carrell, *Biochemistry* **32**, 500 (1993).
[43] D. Grobelny, L. Poncz, and R. E. Galardy, *Biochemistry* **31**, 7152 (1992).
[44] D. J. Buttle, C. J. Handley, M. Z. Ilic, J. Saklatvala, M. Murato, and A. J. Barrett, *Arthritis Rheum.* **36**, 1709 (1993).
[45] M. Cusack, R. J. Beynon, and P. B. Rodgers, *Biochem. Soc. Trans.* **15**, 880 (1987).
[46] M. Cusack, A. G. Stephen, R. Powls, and R. J. Beynon, *Biochem. J.* **274**, 231 (1991).

# Section II

# Aspartic Peptidases

# [7] Families of Aspartic Peptidases, and Those of Unknown Catalytic Mechanism

*By* NEIL D. RAWLINGS and ALAN J. BARRETT

## Introduction

Endopeptidases directly dependent on aspartic acid residues for catalytic activity represent by far the simplest sub-subclass of peptidases in that only three families are known, and crystallographic data are available to show that at least two of these are distantly related. The two families known to be related are those of pepsin (A1) and retropepsin (A2). The enzymes from pararetroviruses such as the cauliflower mosaic virus that form family A3 show some signs of relationship to retropepsins, and provisionally we include them also in clan AA. All of the known members of clan AA are endopeptidases.

Crystallographic studies have shown that the enzymes of the pepsin family are bilobed molecules with the active-site cleft located between the lobes, and each lobe contributing one of the pair of aspartic acid residues that is responsible for the catalytic activity.[1,2] The lobes are homologous to one another, having arisen by gene duplication. A retropepsin molecule, on the other hand, represents just one lobe and carries only one catalytic aspartic residue, so that activity requires the formation of a noncovalent homodimer.[2,3]

In clan AA, the catalytic Asp residues occur within the motif Asp-Xaa-Gly, in which Xaa can be Ser or Thr (Fig. 1). The presence of the motif in a protein has often been taken as evidence that it is an aspartic peptidase, but this is quite unsound. The Asp residue in the catalytic triad of the serine peptidases of the subtilisin family (S8) occurs in such a motif (this series, Volume 244, [2]), and the motif is also present in many proteins that are not peptidases.

Members of clan AA are all more or less strongly inhibited by pepstatin by a mechanism that is known to be similar at least for families A1 and

---

[1] A. R. Sielecki, M. Fujinaga, R. J. Read, and M. N. G. James, *J. Mol. Biol.* **219,** 671 (1991).
[2] T. L. Blundell, J. B. Cooper, A. Šali, and Z. Zhu, *in* "Structure and Function of the Aspartic Proteinases" (B. M. Dunn, ed.), p. 443. Plenum, New York, 1991.
[3] M. Miller, M. Jaskólski, J. K. M. Rao, J. Leis, and A. Wlodawer, *Nature (London)* **337,** 576 (1989).

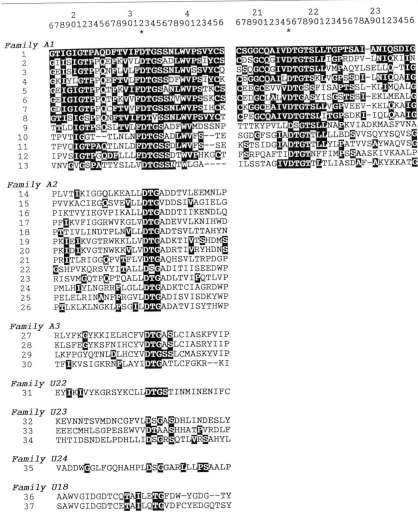

FIG. 1. Conservation of sequences around the catalytic residues in aspartic endopeptidases. Sequences are numbered according to that of pig pepsin A [P. Sepulveda, J. Marciniszyn, D. Liu, and J. Tang, *J. Biol. Chem.* **250**, 5082 (1975)]. One cDNA sequence for the gene of a variant of pig pepsin A (PEPA_PIG) has an extra residue (Ile-229A) not seen in pepsins A from other species. Asterisks indicate the catalytic residues, and the presumed catalytic residues in families U28, U32, U33, and U18. Residues identical to those in pig pepsin A are shown as white letters on black. Key to sequences: 1, pig pepsin A; 2, rabbit pepsin F; 3, rat gastricsin; 4, cattle chymosin; 5, mouse submandibular renin; 6, human renin; 7, mouse cathepsin D; 8, rabbit cathepsin E; 9, yeast barrierpepsin; 10, *Pencillium janthinellum* penicillopepsin; 11, endothiapepsin; 12, *Rhizomucor miehei* mucorpepsin; 13, polyporopepsin; 14, human immunodeficiency virus 1 retropepsin; 15, human immunodeficiency virus 2 retropepsin; 16, African green monkey immunodeficiency virus retropepsin; 17, cattle immunodeficiency virus retropepsin; 18, equine infectious anemia virus retropepsin; 19, visna lentivirus retropepsin; 20,

A2.[4,5] Leader peptidase II (family U11) and thermopsin (family U16) also are sensitive to this inhibitor, however. Covalently reacting inhibitors of pepsin, namely, diazoacetylnorleucine methyl ester (DANLME), 1,2-epoxy-3-(p-nitrophenoxy)propane (EPNP), and p-bromophenacyl bromide, have also been used as diagnostic reagents for aspartic endopeptidases.

An acidic pH optimum is certainly no indication of relationship to pepsin. Unrelated peptidases that are maximally active at acidic pH include the lysosomal cysteine endopeptidases (family C1), serine carboxypeptidases (S10), thermopsin (U16), and the enzymes of the scytalidopepsin (U18) and pseudomonapepsin (U25) families (see below).

In this chapter we take the opportunity to give brief mention of a number of other families of peptidases for which there is as yet no clear indication of the type of catalytic mechanism.

Pepsin Family (A1)

All members of the pepsin family have been found in eukaryotes. The family includes animal enzymes from the digestive tract, such as pepsin and chymosin, lysosomal enzymes such as cathepsin D, and enzymes involved in posttranslational processing such as renin (this series, Volume 80, [35]) and yeast aspartic protease 3 (this volume, [9]). There are also examples from protozoa (e.g., *Eimeria, Plasmodium*), fungi, and a plant. Members of the family are listed in Table I.

Family A1 contains many enzymes that enter the secretory pathway, and it is probable that all the proteins are synthesized with signal peptides and propeptides. Unusually, barrierpepsin from yeast has a long C-terminal extension, which can be removed without affecting activity.[6]

Many three-dimensional structures are now known for members of the pepsin family (Table I), and they show bilobed molecules in which the two

---

[4] P. M. D. Fitzgerald, B. M. McKeever, J. F. VanMiddlesworth, J. P. Springer, J. C. Heimbach, C. T. Leu, W. K. Herber, R. A. F. Dixon, and P. L. Darke, *J. Biol. Chem.* **265**, 14209 (1990).

[5] J. B. Cooper, S. I. Foundling, T. L. Blundell, J. Boger, R. A. Jupp, and J. Kay, *Biochemistry* **28**, 8596 (1989).

[6] V. L. Mackay, S. K. Welch, M. Y. Insley, T. R. Manney, J. Holly, G. C. Saari, and M. L. Parker, *Proc. Natl. Acad. Sci. U.S.A.* **85**, 55 (1988).

---

ovine lentivirus retropepsin; 21, feline leukemia virus retropepsin; 22, Rous avian sarcoma virus retropepsin; 23, human T-cell leukemia virus type II retropepsin; 24, mouse mammary tumor virus retropepsin; 25, sheep pulmonary adenomatosis virus retropepsin; 26, squirrel monkey retrovirus retropepsin; 27, cauliflower mosaic virus endopeptidase; 28, figwort mosaic virus endopeptidase; 29, carnation etched ring virus endopeptidase; 30, soybean chlorotic mottle virus endopeptidase; 31, *Drosophila melanogaster* transposon 297 retropepsin-like sequence; 32, *D. melanogaster* transposon *copia* endopeptidase; 33, tobacco transposon Tnt1 endopeptidase; 34, yeast Ty1-17 transposon endopeptidase; 35, maize transposon *bs1* retropepsin-like sequence; 36, scytalidopepsin; 37, aspergillopepsin II.

## TABLE I
MEMBERS OF FAMILIES OF ASPARTIC PEPTIDASES[a]

| Family | EC | Database code | Structure |
|---|---|---|---|
| **Family A1: Pepsin** | | | |
| Aspartic endopeptidase P111 | - | PIR JT0398 | |
| Aspartic proteinase (barley) | - | (X56136) | |
| Aspartic proteinase (cattle) | - | (L06151) | |
| Aspartic proteinase (*Eimeria*) | - | (Z24676) | |
| Aspartic proteinase 3 (yeast) | - | YAP3_YEAST | |
| Aspergillopepsin I | 3.4.23.18 | PEPA_ASPAW, (D13894) | |
| Barrierpepsin (*Saccharomyces*) | - | BAR1_YEAST | |
| Candidapepsin | 3.4.23.24 | CAR1_CANAL, CARP_CANTR, CARL_CANPA, (L22358), (X62289) | |
| Cathepsin D | 3.4.23.5 | CATD_*, (M88822), (M95187), (S49650) | b |
| Cathepsin E | 3.4.23.34 | CATE_*, (L08418) | |
| Chymosin | 3.4.23.4 | CHYM_* | b |
| Endothiapepsin | 3.4.23.22 | CARP_CRYPA | b |
| Gastricsin | 3.4.23.3 | PEPC_* | |
| Mucorpepsin | 3.4.23.23 | CARP_RHIMI, CARP_RHIPU CAR2_CANAL, CAR2_CANAL, CARR_CANPA | b |
| Penicillopepsin | 3.4.23.20 | PENP_PENJA | b |
| Pepsin A | 3.4.23.1 | PEPA_*, PEP1_MACFU, PEP2_MACFU, PEP4_MACFU | b |
| Pepsin F | 3.4.23.1 | PEPF_RABIT | |
| Pepsin II | 3.4.23.1 | PEP1_RABIT, PEP2_RABIT, PEP4_RABIT | |
| Pepsin III | 3.4.23.1 | PEP3_RABIT | |
| Embryonic pepsin (chicken) | - | PEPE_CHICK | |
| Polyporopepsin | 3.4.23.29 | CARP_POLTU | |
| Renin, renal | 3.4.23.15 | RENI_* | b |
| Renin, submandibular | 3.4.23.15 | RENS_MOUSE | |
| Rhizopuspepsin | 3.4.23.21 | CARP_RHICH, CARP_RHINI | b |
| Rhizopuspepsin II | 3.4.23.21 | CRP2_RHINI, (D13939) | |
| Rhizopuspepsin III | 3.4.23.21 | CRP3_RHINI | |
| Rhizopuspepsin IV | 3.4.23.21 | (X56992) | |
| Saccharopepsin | 3.4.23.25 | CARP_SACFI, CARP_YEAST, SAX1_SCHPO, SAX2_SCHPO | |
| **Family A2: Retropepsin** | | | |
| Retropepsin | 3.4.23.16 | GAG_AVIMA, GAG_RSVP, POL_BAEVM, POL_BIV06, POL_EIAV, POL_FIVPE, POL_FLV, POL_GALV, POL_HIV1A, POL_HIV2D, POL_MLVAV, POL_OMVVS, POL_SIVMK, POL_VILV, VPRT_BLV, VPRT_JSRV, VPRT_MPMV, VPRT_MMTVB, VPRT_HTL1A, VPRT_SMRVH, VPRT_SRV1, VPRT_HUMAN | b |
| **Family A3: Cauliflower mosaic virus peptidase** | | | |
| Pararetropepsin endopeptidase | | POL_CAMVD, POL_CERV, | |

TABLE I (*continued*)

| Family | EC | Database code | Structure |
|---|---|---|---|
| Family U22: *Drosophila* transposon 297 endopeptidase | | POL_FMVD, POL_SOCMV | |
| Endopeptidase (*Drosophila* transposon 297) | - | POL2_DROME | |
| Endopeptidase (*Drosophila* transposon 17.6) | - | POL3_DROME, | |
| Endopeptidase (*Drosophila* transposon 412) | - | POL4_DROME | |
| Retrovirus-related endopeptidase (*Schizosaccharomyces*) | - | (L10324) | |
| Family U23: *Drosophila* transposon *copia* endopeptidase | | | |
| Endopeptidase (*Drosophila* transposon *copia*) | - | COPI_DROME | |
| Endopeptidase (tobacco transposon Tnt1) | - | POLX_TOBAC | |
| Endopeptidase (yeast transposon Ty1-17) | - | YCB9_YEAST | |
| Family U24: Maize transposon *bs1* endopeptidase | | | |
| Endopeptidase (maize transposon *bs1*) | - | POLB_MAIZE | |
| Family U16: Thermopsin | | | |
| Thermopsin | 3.4.99.43 | THPS_SULAC | |
| Family U18: Scytalidopepsin | | | |
| Scytalidopepsin B | 3.4.23.32 | PRTB_SCYLI | |
| Aspergillopepsin II | 3.4.23.19 | PRTA_ASPNG | |
| Family U25: Pseudomonapepsin | | | |
| Pseudomonapepsin | 3.4.23.37 | (D37970) | |
| Family U11: Leader peptidase II | | | |
| Leader peptidase II | 3.4.99.35 | LPSA_* | |
| Family U4: Sporulation sigma$^E$ factor processing peptidase | | | |
| Sporulation sigma factor$^E$ processing peptidase (*Bacillus subtilis*) | - | SP2G_* | |

[a] EC is the enzyme nomenclature number (Nomenclature Committee of the International Union of Biochemistry and Molecular Biology, "Enzyme Nomenclature 1992," Academic Press, Orlando, Florida, 1992, and Supplement);—indicates that no EC number has been assigned. Literature references to the individual proteins are generally to be found in the database entries for which the codes are given.
[b] Structure is included in the Brookhaven database.

lobes have similar structures, strongly indicating that they are homologous, despite very little amino acid sequence similarity. It seems clear that the ancestral enzyme evolved by gene duplication followed by gene fusion. There is a short hexapeptide connecting the two lobes, and each lobe contributes one of the pair of catalytic Asp residues.[7] There are indications that dimers of the N-terminal lobe of pepsinogen can express some catalytic activity.[8] The catalytic mechanism of pepsin and pepsin homologs has been reviewed by Polgár.[9]

In almost all members of the pepsin family, the catalytic Asp residues

[7] J. Tang, M. N. G. James, I. N. Hsu, J. A. Jenkins, and T. L. Blundell, *Nature* (*London*) **271**, 618 (1978).
[8] X. Lin, Y. Lin, G. Koelsch, A. Gustchina, A. Wlodawer, and J. Tang, *J. Biol. Chem.* **267**, 17257 (1992).
[9] L. Polgár, "Mechanisms of Protease Action." CRC Press, Boca Raton, Florida, 1989.

are contained in an Asp-Thr-Gly-Xaa motif in both N- and C-terminal lobes of the enzyme (Fig. 1). With few exceptions, Xaa is Ser or Thr, the side chains of which can hydrogen bond to the Asp. One exception to the rule occurs in the C-terminal lobe of human renin, in which Xaa is Ala, as it also is in almost all the retropepsins (Fig. 1). Human renin and the retropepsins also have higher pH optima than typical members of the pepsin family. Ido and co-workers have suggested that the reason is the effect of the hydrogen bond on the acidity of the Asp residue, and they described a site-directed mutagenesis experiment in support of that idea.[10]

There is marked conservation of cysteine residues within the pepsin family. In pepsin, there are three disulfide loops, one following the first catalytic Asp (Asp-32, mature pig pepsin A numbering), one preceding the second (Asp-215), and one toward the C terminus. The first loop is present in all members of the family except several fungal enzymes; it is generally small, containing five residues in most members of the family, but is larger in candidapepsin and barrierpepsin. The second loop is commonly of five or six residues, and is found only in the animal enzymes. The third and largest disulfide loop is conserved in all members of the family except polyporopepsin, which does not contain any cysteine residues. The fact that the two domains have different disulfide bonding patterns indicates that the loops were introduced after the internal duplication of the molecule. Cathepsin D has an additional loop, from Cys-26 to Cys-91B, which contains the first catalytic Asp and the first of the loops that occur generally throughout the family. Cathepsin E is unusual among the members of the pepsin family in that the proenzyme is a disulfide-bonded homodimer, and the active enzyme can be either dimeric or monomeric (see [8] in this volume).

The mechanism by which the 44-residue propeptide of pepsinogen inactivates the proenzyme has been elucidated from the crystal structure of pepsinogen.[1] In the precursor, the first 11 residues of mature pepsin are displaced by residues of the propeptide, in a six-stranded $\beta$ sheet. The propeptide contains two helices that block the active-site cleft, and the conserved Asp-11 hydrogen bonds a conserved arginine in the propeptide. This stabilizes the propeptide conformation and is probably responsible for the acid triggering of the conversion of pepsinogen to pepsin. Two tyrosine residues block the S1 and S1' binding sites.

The aspartic proteinase from barley is unusual in that a 104-residue insert is present in the C-terminal lobe. The insert is homologous (RDF score 13.22 this series, Volume 244, [1]) to the precursor of human saposins (SAP_HUMAN), which contains four saposin peptides. The individual saposin molecules are proteolytically excised and act as activators for lysosomal enzymes including $\beta$-galactosylceramidase and $\beta$-galactosidase. The barley

---

[10] E. Ido, H. Han, F. J. Kezdy, and J. Tang, *J. Biol. Chem.* **266**, 24359 (1991).

insert does not correspond to a single saposin domain, but rather overlaps the C-terminal third of one domain and the N-terminal two-thirds of the next. Not only are cysteine residues (and presumably disulfides) conserved, but also a potential N-linked glycosylation site. Thus the barley aspartic proteinase is the only mosaic protein so far recognized in the pepsin family.

The barley enzyme is most closely related to cathepsin D, and both are processed to two-chain forms. In the barley enzyme two such forms exist (32 plus 16 kDa, and 29 plus 11 kDa), both being cleaved within the 104-residue insert.[11] The processing of mammalian cathepsin D to the two-chain molecule occurs at quite a different location (this series, Volume 80, [35]).

The pepsins as well as cathepsin D and the majority of other members of the family show specificity for the cleavage of bonds in peptides of at least six residues with hydrophobic amino acids in both the P1 and P1' positions.[12] The crystal structures have given some insight into the basis of the specificity of catalysis. The binding cleft takes the form of a groove running across the junction of the two lobes, with an extended loop projecting across the cleft to form a flexible, eleven-residue "flap" that encloses substrates and inhibitors in the active site. The specificity subsites are formed by several hydrophobic residues surrounding the catalytic Asp residues and by three residues on the flap, Tyr-75, Gly-76, and Asp-77. Some fungal aspartic endopeptidases are able to activate trypsinogen by cleavage of a Lys+Ile bond, showing an affinity for the anionic lysine side chain in S1 (this series, Volume 45, [35]; Ref. 13). In one enzyme, penicillopepsin, the anionic binding site was shown to be Asp-77.[14] That residue is conserved in all family members able to activate trypsinogen, but it is replaced by Thr or Ser in those unable to do so (pepsin A, mucorpepsin, chymosin, renin). Yeast aspartic proteinase 3 and the mammalian proopiomelanocortin-processing enzyme require basic amino acids in both the S1 and S1' subsites (this volume, [9]). Yeast aspartic proteinase 3 retains Asp-77, and the specificity for a basic residue in S1' might be mediated by an Asp residue (448 in the yeast preproaspartic protease sequence) that corresponds to uncharged residues in penicillopepsin and pepsin.

The gene structures of several pepsin homologs are known. Each gene contains nine exons, and the pattern and phase of exon/intron junctions are strikingly conserved. Comparison of the positions and phases of the intron junctions in the parts of the gene encoding the two homologous lobes reveals little similarity, however, which implies that the introns were inserted

[11] P. Sarkkinen, N. Kalkkinen, C. Tilgmann, J. Siuro, J. Kervinen, and L. Mikola, *Planta* **186,** 317 (1992).

[12] B. Keil, "Specificity of Proteolysis." Springer-Verlag, Berlin, 1992.

[13] T. Hofmann, R. S. Hodges, and M. N. G. James, *Biochemistry* **23,** 635 (1984).

[14] M. N. G. James, A. R. Sielecki, and T. Hofmann, *in* "Aspartic Proteinases and Their Inhibitors" (V. Kostka, ed.), p. 165. de Gruyter, Berlin, 1985.

into the ancestral gene after the initial duplication that formed the two lobes. The conservation of introns through the many gene duplications that have given rise to the individual enzymes in the pepsin family contrasts with the situation in the chymotrypsin family (S1), in which there has been more variation (this series, Volume 244, [2]), and the papain family (C1), in which insertion of introns appears to have continued after the separation of the individual enzymes to the extent that the ancestral pattern is obscured (this series, Volume 244, [33]). What evolutionary pressure may have led to the conservation of structure in the genes of the pepsin family remains unclear.

### Retropepsin Family (A2)

The hypothesis put forward by Tang *et al.*[7] for the evolution of the pepsin family implied the existence at some time of a small molecule analogous to a single lobe of pepsin, which would have been active as a homodimer. Unexpectedly, confirmation of the feasibility of the proposal came from the crystal structure of retropepsin, the polyprotein processing enzyme from the human immunodeficiency virus.[2,3] Structures have now been elucidated also for retropepsins from Rous sarcoma virus and avian myeloblastosis virus, and they show that all of the retropepsins are similar in structure to one lobe of pepsin. The viral enzyme is probably not truly primitive, but rather derived from the N-terminal domain of a homolog of pepsin, since it contains the flap that is an important part of the structure of the substrate binding site.[8]

Retroviruses are enveloped, single-stranded RNA viruses (this series, Volume 244, [2]). Retroviral RNA is transcribed into double-stranded DNA (by a virus-encoded reverse transcriptase), which is then incorporated into the host genome. The retroviral RNA includes at least three genes, in the order *gag, pol, env*. The *gag* and *env* genes encode the structural proteins of the virus. The *pol* gene encodes the three enzymes necessary for transcribing the viral RNA to DNA and incorporating it into the host genome, and also encodes retropepsin. The *pol* gene is translated as part of a *gag–pol* fusion, and the retropepsin is usually the N-terminal protein in the *pol* polyprotein. Unusually, it is the *gag* gene that encodes the retropepsin (protein p15) in avian retroviruses such as Rous sarcoma virus.[15] In oncoviruses (e.g., simian Mason–Pfizer virus), the endopeptidase gene is a separate open reading frame between the *gag* and *pol* genes.[16]

Retropepsin is required for processing of all three viral polyproteins, although initial stages of *env* polyprotein cleavage are performed by cellular enzymes. Processing occurs at a very late stage in virion assembly, usually after budding of virus particles from the cell membrane; inactive virions

---

[15] H.-G. Kräusslich and E. Wimmer, *Annu. Rev. Biochem.* **57,** 701 (1988).

[16] P. Sonigo, C. Barker, E. Hunter, and S. Wain-Hobson, *Cell (Cambridge, Mass.)* **45,** 375 (1986).

containing only *gag* polyprotein can be formed. Processing seems to be essential for RNA dimerization within the virion, and hence for infectivity. There is therefore intense interest in the development of inhibitors of retropepsins as antiretroviral agents.[17]

A subset of the retropepsins from oncoviruses and avian retroviruses are larger proteins with N-terminal extensions, and in some representatives the N-terminal domain is homologous to dUTPases (e.g., DUTP_ECOLI). Examples of such oncoviruses are mouse mammary tumour virus (VPRT_MMTVB), simian Mason–Pfizer virus (VPRT_MPMV), sheep pulmonary adenomatosis virus (VPRT_JSRV), squirrel monkey retrovirus (VPRT_SMRVH), and a human retrovirus (VPRT_HUMAN). Homologs of *Escherichia coli* dUTPase are also found in visna and other ovine lentiviruses as well as in feline immunodeficiency virus, but these are located elsewhere in the *pol* polyprotein.

## Cauliflower Mosaic Virus Peptidase Family (A3)

Cauliflower mosaic virus belongs to a group of plant viruses known as pararetroviruses. Although the viral genome is double-stranded DNA, it contains an open reading frame (ORF V) which is analogous to the *pol* gene of retroviruses. ORF V encodes a polyprotein which includes a reverse transcriptase, which is homologous to that of retroviruses and, on the basis of an Asp-Thr-Gly triplet near the N terminus, was suggested to include an aspartic proteinase as well.[18] The existence of an endopeptidase was confirmed by mutational studies, which implicated Asp-45 in catalysis (Fig. 1). There was also weak inhibition by pepstatin. The peptidase is larger than retropepsin, but because it contains only one Asp-Thr-Gly sequence, it is assumed to be active only as a dimer.[19] Other pararetroviruses contain sequences homologous to the cauliflower mosaic virus peptidase (Table I).

The pararetrovirus endopeptidase releases itself from the ORF V polyprotein and is responsible for some of the cleavages required for processing of the ORF IV polyprotein, which encodes coat proteins.[19] Although no crystallographic data are available, we suggest that pararetrovirus peptidases are distant homologs of pepsin and retropepsin, and form a separate family in clan AA.

## Putative Aspartic Peptidases

We now come to a group of families of peptidases (U22, U33, U34, U23, U24, U25, U11, and U4) that are strictly categorized as of unknown catalytic type but for which there are some indications of aspartic type.

---

[17] C. U. T. Hellen and E. Wimmer, *Experientia* **48,** 201 (1992).
[18] J. Fuetterer and T. Hohn, *Trends Biochem. Sci.* **12,** 92 (1987).
[19] M. Torruella, K. Gordon, and T. Hohn, *EMBO J.* **8,** 2819 (1989).

*Putative Transposon Endopeptidases: Families of Drosophila Transposon
297 Endopeptidase (U22), copia Endopeptidase (U23), and Maize
Transposon bs1 Endopeptidase (U24)*

Transposons are transposable genetic elements that are capable of moving (by means of RNA transcription and reverse transcription) from one place to another in the genome, and they contain one or two open reading frames that code for polyproteins.[20] The polyproteins contain all the enzymes needed to accomplish the transpositions, including a reverse transcriptase and, it is believed, a peptidase that processes the polyproteins. Transposons resemble retroviruses in many respects, with the open reading frames being analogous to the *gag* and *pol* genes. No sequence that is clearly that of a retropepsin is present, but the proteins do contain Asp-Thr-Gly or Asp-Ser-Gly tripeptides (or Asp-Thr-Ala, in the tobacco Tnt1 sequence). Little enzymological work has been done on the putative peptidases, and there are no inhibition data.

The putative endopeptidases of family U22 are encoded by open reading frames that correspond to the *pol* gene of retroviruses,[21] whereas that of the *copia* transposon (U23) is encoded by the open reading frame corresponding to the Rous sarcoma virus *gag* gene.[22] The *gypsy* transposon from *Drosophila* is homologous to *Drosophila* transposons in family U22; it has been suggested to contain a peptidase,[23] but has no Asp-Thr-Gly motif.

For family U23, Adams *et al.* have shown that deletion of part of the 5' end of yeast *TyB* (including the Asp-Ser-Gly tripeptide) prevents processing,[20] and consistent results were obtained by another group subsequently.[23] For the product of the *copia* transposon in *Drosophila*, autocatalytic processing has been shown to be necessary for the release of the protein VPL from the polyprotein precursor, and mutation of the putative catalytic Asp prevents processing.[22] The autocatalytic processing of the *copia* transposon of *Drosophila*, thought to be mediated by the transposon endopeptidase, was not significantly inhibited by pepstatin, however.[22]

Family U24 comprises only the putative endopeptidase of maize transposon *bs1*.[24]

*Thermopsin family (U16)*

Among the few peptidases that are known from archaebacteria, there is a subtilisin (family S8) and a multicatalytic endopeptidase complex (S25),

---

[20] S. E. Adams, J. Mellor, K. Gull, R. B. Sim, M. F. Tuite, S. M. Kingsman, and A. J. Kingsman, *Cell (Cambridge, Mass.)* **49**, 111 (1987).
[21] S. Inouye, S. Yuki, and K. Saigo, *Eur. J. Biochem.* **154**, 417 (1986).
[22] K. Yoshioka, H. Honma, M. Zushi, S. Kondo, S. Togashi, T. Miyake, and T. Shiba, *EMBO J.* **9**, 535 (1990).
[23] D. J. Garfinkel, A. M. Hedge, S. D. Youngren, and T. D. Copeland, *J. Virol.* **65**, 4573 (1991).
[24] M. A. Johns, M. S. Babcock, S. M. Fuerstenberg, S. I. Fuerstenberg, M. Freeling, and R. B. Simpson, *Plant Mol. Biol.* **12**, 633 (1989).

but thermopsin from the thermophilic *Sulfolobus acidocaldarius* shows no relationship to any other protein (this volume, [11]). The enzyme has a pH optimum of 2, is maximally active at 70°, and is covalently attached to the cell membrane. Thermopsin is apparently synthesized as a precursor with a 41-residue prepropeptide. Sensitivity to inhibition by pepstatin suggests a possible distant relationship to the pepsin clan; however, the typical Asp-Xaa-Gly motif is not present, and there is no evidence of an internal duplication.

## Scytalidopepsin B Family (U18)

Pepstatin-insensitive peptidases active at very low pH values are known from a variety of fungi (*Aspergillus, Scytalidium*) and bacteria (*Xanthomonas, Pseudomonas, Bacillus*). Some of the enzymes can be inhibited by EPNP (this volume, [10]; Ref. 25); EPNP has been shown to modify a catalytic aspartic residue in pepsin.[26] The bacterial enzymes are inhibited by the carboxyl-specific carbodiimides (this volume, [10]) and also by the peptide aldehyde tyrostatin.[27] Unlike enzymes from the pepsin family, these endopeptidases are thermostable.

Scytalidopepsin B from *Scytalidium* and aspergillopepsin II (*Aspergillus* proteinase A; this volume, [10]) have been sequenced and are homologous. The catalytic residue of scytalidopepsin B was reported[28] to be Glu-53, which reacted with EPNP and occurs within a segment similar to that around Asp-32 of pepsin (Fig. 1). However, the corresponding residue is Gln in aspergillopepsin II.[29] The amino acid sequence originally reported for scytalidopepsin B also had Gln at that position.[30]

Aspergillopepsin II is a secreted enzyme, synthesized as a precursor. Activation involves not only removal of the 59-residue prepropeptide, but also excision of an internal 11-residue peptide to produce a two-chain molecule. The gene for the proteinase does not include introns.[29]

## Leader Peptidase II Family (U11)

Bacterial cell walls contain large quantities of murein lipoprotein. This is a small protein (7.2 kDa) that has the N-terminal cysteine substituted on

[25] K. Oda, T. Nakazima, T. Terashita, K.-I. Suzuki, and S. Murao, *Agric. Biol. Chem.* **51,** 3073 (1987).

[26] J. Tang, *J. Biol. Chem.* **246,** 4510 (1971).

[27] K. Oda, Y. Fukuda, S. Murao, K. Uchida, and M. Kainosho, *Agric. Biol. Chem.* **53,** 405 (1989).

[28] D. Tsuru, S. Shimada, S. Maruta, T. Yoshimoto, K. Oda, S. Murao, T. Miyata, and S. Iwanaga, *J. Biochem.* (*Tokyo*) **99,** 1537 (1986).

[29] H. Inoue, T. Kimura, O. Makabe, and K. Takahashi, *J. Biol. Chem.* **266,** 19484 (1991).

[30] T. Maita, S. Nagata, G. Matsuda, S. Maruta, K. Oda, S. Murao, and D. Tsuru, *J. Biochem.* (*Tokyo*) **95,** 465 (1984).

sulfur with the $CH_2(OOCR_1)CH(OOCR_2)CH_2$- group, and the C-terminal lysine is bound to the membrane peptidoglycan (murein) through the ε-amino group. Secretion of the lipoprotein from the cytoplasm is mediated by a leader peptide, which is cleaved by a specialized peptidase of the inner membrane known as leader peptidase II (this volume, [12]).

Leader peptidase II is strongly inhibited by the antibiotic globomycin, but it is also inhibited by pepstatin, which suggests that the enzyme may be an aspartic endopeptidase. No catalytic residues have been identified, but there are four aspartic residues conserved throughout the four known sequences (Asp-23, Asp-114, Asp-123, and Asp-141 in *E. coli* leader peptidase II). No serine, cysteine, or histidine residue is conserved in the alignment, so it is likely that the aspartic residues are responsible for catalysis.

*Pseudomonapepsin Family (U25)*

Pseudomonapepsin, an acid endopeptidase from a *Pseudomonas* species, is not inhibited by pepstatin, DANLME, or EPNP, standard inhibitors for the pepsin family (A1), but is inhibited by tyrostatin (*N*-isovaleryl-tyrosyl-leucyl-tyrosinal). It is a single-chain molecule with 372 amino acids and one disulfide bond, showing no relationship to known aspartic endopeptidases.[30a] Pseudomonapepsin is a secreted enzyme, synthesized as a precursor with a signal peptide and a large propeptide, which is autocatalytically activated by cleavage at a Leu–Ala bond.

We report here that pseudomonapepsin is related to a gene from the slime mold *Physarum polycephalum*, the product of which has not been characterized. The putative *Physarum* protein[30b] overlaps the entire length of the mature *Pseudomonas* endopeptidase, and possesses a signal peptide, but apparently no propeptide. Eight aspartate residues are conserved between the sequences of pseudomonapepsin and the *Physarum* gene. One of these occurs in the sequence Asp-Ser-Gly in *Physarum* (resembling a catalytic residue in family A1), but the corresponding sequence is Asp-Glu-Gly in pseudomonapepsin.

*Sporulation Sigma$^E$ Factor Processing Peptidase Family (U4)*

Bacilli produce spores under the direction of a protein known as σ (sigma)$^E$, which switches on the genes necessary for sporulation. The σ factor is produced as a precursor, and the peptidase believed to be responsible for processing it is the product of the *spoIIGA* gene. Because of the presence of an Asp-Ser-Gly motif in the processing peptidase, the enzyme

[30a] K. Oda, T. Takahashi, Y. Tokuda, Y. Shibano, and S. Takahashi, *J. Biol. Chem.* **269**, 26518 (1994).
[30b] M. Benard, D. Pallotta, and G. Pierron, *Exp. Cell Res.* **201**, 506 (1992).

TABLE II

MEMBERS OF FAMILIES OF UNKNOWN CATALYTIC TYPE[a]

| Family | EC | Database code |
|---|---|---|
| **Family U7: Endopeptidase IV** | | |
| Endopeptidase IV (*Escherichia coli*) | - | SPPA_ECOLI, LICA_HAEIN |
| Minor capsid protein precursor C | - | VCAC_LAMBD |
| (bacteriophage λ) | | |
| *sohB* gene product (*E. coli*) | - | (M73320) |
| **Family U2: Aminopeptidase iap** | | |
| Alkaline phosphatase isozyme conversion | - | IAP_ECOLI |
| protein (*Escherichia coli*) | | |
| **Family U5: Tail-specific protease** | | |
| Tail-specific protease (*Escherichia coli*) | - | (M75634) |
| OrfX (*Agmenellum*) | - | (X63049) |
| **Family U6: Murein endopeptidase** | | |
| Penicillin-insensitive murein endopeptidase | - | MEPA_ECOLI |
| (*Escherichia coli*) | | |
| **Family U8: Bacteriophage murein endopeptidase** | | |
| Murein endopeptidase (bacteriophage) | - | ENPP_* |
| **Family U9: Prohead endopeptidase** | | |
| Prohead endopeptidase (bacteriophage T4) | - | PCPP_BPT4 |
| **Family U3: Spore endopeptidase** | | |
| Spore endopeptidase (*Bacillus*) | - | (M55262), {9124} |
| **Family U20: γ-D-Glutamyl-L-diamino acid** | | |
| **endopeptidase II** | | |
| γ-D-glutamyl-L-diamino acid endopeptidase | - | (X64809) |
| II (*Bacillus sphaericus*) | | |
| **Family U26: *Enterococcus* D-Ala-D-Ala** | | |
| **carboxypeptidase** | | |
| D-Ala-D-Ala carboxypeptidase (*Enterococcus*) | - | (M90647) |
| **Family U29: Encephalomyelitis virus** | | |
| **proteinase 2A** | | |
| Proteinase 2A (Theiler's murine | | |
| encephalomyelitis virus) | - | POLG_TMEVD |
| Proteinase 2A (Encephalomyocarditis virus) | - | POLG_EMCV |
| **Family U27: *Lactococcus* ATP-dependent** | | |
| **proteinase** | | |
| ATP-dependent proteinase (*Lactococcus*) | - | (X67821) |
| **Family U28: Aspartyl dipeptidase** | | |
| Aspartyl dipeptidase (*Salmonella*) | - | [b] |

[a] See Table I for explanation.
[b] C. A. Conlin, K. Håkensson, A. Liljas, and C. G. Miller, *J. Bacteriol.* **176,** 166 (1994).

has been assumed to be an aspartic proteinase. The peptidase is located in the inner membrane and possesses five membrane-spanning domains and a large cytoplasmic domain that contains the putative catalytic Asp.[31]

## Families of Peptidases of Unknown Catalytic Mechanism

There are a number of incompletely characterized peptidases that cannot yet be assigned to any catalytic type and show no homology to peptidases of known type. These are listed in Table II.

[31] P. Stragier, C. Bonamy, and C. Karmazyn-Campelli, *Cell (Cambridge, Mass.)* **52,** 697 (1988).

*Family U7. Escherichia coli* endopeptidase IV is probably the best known peptidase that is still of unknown catalytic type. It is found in the periplasmic membrane and is involved in the initial stages of degradation of leader peptides after cleavage from the preproteins. Endopeptidase IV has been shown to cleave the leader peptides released from prelipoproteins to give fragments that are further digested by oligopeptidase A (family M3; see [18] in this volume).[32] Endopeptidase IV has a requirement for at least three residues either side of the scissile bond, and it is presumably an oligopeptidase.[32] Another *E. coli* enzyme, a product of the *sohB* gene, is homologous.[33] The C protein from bacteriophage λ shares a C-terminal homologous domain with endopeptidase IV, and it may be an endopeptidase involved in the processing of the bacteriophage prohead B protein and degradation of the scaffold protein around which the prohead is formed.[34]

*Family U2.* The product of the *iap* gene from *E. coli* may be the arginyl aminopeptidase that converts alkaline phosphatase from one isozyme to another by removal of successive N-terminal arginines, or it may be an activator of the aminopeptidase. The *iap* protein has a leader peptide and is known to be proteolytically processed; however, it is not secreted or found in the periplasm and is assumed to be a membrane protein.[35]

*Family U5.* An *E. coli* peptidase that removes C-terminal hydrophobic peptides from proteins has been named tail-specific protease and may be identical to protease Re,[36] which is a serine peptidase. A large N-terminal fragment of a homolog has been sequenced from the cyanobacterium *Synechococcus* (or *Agmenellum*) *quadruplicatum.* There are five serine residues conserved between the two sequences, both of which are homologous to the repeated domains from mammalian interphotoreceptor retinoid-binding protein (IRBP_BOVIN, IRBP_HUMAN).[36]

*Family U6.* Family U6 consists solely of *E. coli* murein endopeptidase,[37] which is responsible for cleaving D-alanyl-γ-*meso*-2,6-diaminopimelyl peptide bonds connecting the peptidoglycan units of murein in the bacterial cell wall.[37] The 31-kDa enzyme has a leader peptide and is located in the periplasm. Murein endopeptidase is insensitive to penicillin, unlike penicillin-binding protein 4 from *E. coli,* which is a penicillin-sensitive

---

[32] P. Novak and I. K. Dev, *J. Bacteriol.* **170,** 5067 (1988).

[33] L. Baird, B. Lipinska, S. Raina, and C. Georgopoulos, *J. Bacteriol.* **173,** 5763 (1991).

[34] J. Kochan and H. Murialdo, *Virology* **131,** 100 (1983).

[35] Y. Ishino, H. Shinagawa, K. Makino, M. Amemura, and A. Nakata, *J. Bacteriol.* **169,** 5429 (1987).

[36] K. R. Silber, K. C. Keiler, and R. T. Sauer, *Proc. Natl. Acad. Sci. U.S.A.* **89,** 295 (1992).

[37] W. Keck, A. M. van Leeuwen, M. Huber, and E. W. Goodell, *Mol. Microbiol.* **4,** 209 (1990).

enzyme with similar activity (see family S13, Volume 244 in this series, [20]).

*Family U8.* Bacteriophages are viral parasites of bacteria. Phage particles need to escape from the host cell in order to infect other cells, and bacteriophage-encoded lytic enzymes, including a lysozyme, make escape possible by causing rupture of the bacterial cell wall. In bacteriophage P22, gene 15 was found to be essential for lysis under some conditions.[38] The gene product is believed to be an endopeptidase with specificity similar to that of *E. coli* murein endopeptidase, cleaving the bacterial cell wall cross-links. An *E. coli* homolog has been cloned on the complementary strand to the plasmid *iss* gene.[39] The only conserved residue likely to be catalytic is Asp-85.

*Family U9.* Bacteriophages also require a peptidase for processing of the proteins that are assembled into the phage prohead and for degradation of the scaffold protein around which the prohead is built.[34] In bacteriophage λ, protein C may perform both functions (see family U7 above). In bacteriophage T4, the product of gene 21 is the processing endopeptidase, usually cleaving Glu + Ala or Glu + Gly bonds, and has been located in the center of the prohead.[40] The gp21 proteinase is itself synthesized as a precursor, and activation occurs by self-cleavage. The enzyme is not inhibited by serine or cysteine proteinase inhibitors,[40] and it shows no sequence relationship to other peptidases (including protein C from bacteriophage λ).

*Family U3.* The product of the *gpr* gene in *Bacillus* is an endopeptidase involved in spore germination, initiating the degradation of small, acid-soluble proteins. The endopeptidase is synthesized as a precursor, which is processed by the GPR endopeptidase itself.[41] The sequences from two species of *Bacillus* show no relationship to other peptidases.

*Family U20.* Another *Bacillus* enzyme associated with sporulation is γ-D-glutamyl-L-diamino acid endopeptidase II, which cleaves the cross-linked peptide side chains of peptidoglycans. The enzyme is thiol-dependent, being inhibited by *N*-ethylmaleimide, iodoacetamide, and *p*-hydroxymercuribenzoate, and clearly may be a cysteine peptidase.[42] The peptidase is synthesized with a leader peptide and shows homology with a C-terminal domain of p54 protein from *Enterococcus* (P54_ENTFC).

*Family U26.* The *vanY* gene from *Enterococcus* is believed to encode a penicillin-insensitive D-Ala-D-Ala carboxypeptidase involved in the removal of terminal D-Ala residues from cell wall peptidoglycan precursors.

[38] S. Casjens, K. Eppler, R. Parr, and A. R. Poteete, *Virology* **171**, 588 (1989).
[39] S. Chuba Leon Banerjee Palchaudhuri, *Mol. Gen. Genet.* **216**, 287 (1989).
[40] B. Keller and T. A. Bickle, *Gene* **49**, 245 (1986).
[41] J.-L. Sanchez-Salas and P. Setlow, *J. Bacteriol.* **175**, 2568 (1993).
[42] T. Bourgogne, M.-J. Vacheron, M. Guinand, and G. Michel, *Int. J. Biochem.* **24**, 471 (1992).

It is not known for certain whether the enzyme is a serine or metallo-type carboxypeptidase; there is no sequence similarity to other D-Ala-D-Ala carboxypeptidases (families S11, S12, S13, and M15), but the Ser-Xaa-Xaa-Lys motif that contains two of the active-site residues of serine D-Ala-D-Ala carboxypeptidases is present.[43]

*Family U29.* Cardioviruses such as encephalomyelitis virus are picorna-viruses and like others in the group produce a polyprotein that is processed by two endopeptidases which are initially contained in the polyprotein as proteins 3C and 2A. Cardiovirus protein 3C is homologous to picornain 3C and is included in family C3 (this series, Volume 244, [41]). Protein 2A, however, shows no relationship to picornain 2A and is responsible for a single cleavage in the polyprotein, that of a Gly + Pro bond between proteins 2A and 2B.[44]

*Family U27.* A transposon-like sequence with two open reading frames occurs in a *Lactococcus lactis* plasmid. One of the open-reading frames encodes a homolog of clpA that has been named clpL, whereas the other encodes a 310-residue protein with no known homologs. Because increased ATP-dependent proteolysis occurs in *E. coli* cells that are expressing both transposon genes, it has been assumed[45] that the 310-residue protein is a peptidase analogous to clpP (family S14; see Volume 244 in this series, [24]).

*Family U28.* The sequence of the 25-kDa *Salmonella* aspartyl dipepti-dase, which seems to be essential for the cleavage of Asp-Pro by the organism, shows no similarity to any known peptidase.[46] The enzyme, active at neutral pH, is unaffected by EDTA[47] and so is evidently not a metallopep-tidase, unlike the majority of known dipeptidases.

[43] M. Arthur, C. Molinas, and P. Courvalin, *Gene* **120,** 111 (1992).
[44] A. C. Palmenberg, G. D. Parks, D. J. Hall, R. H. Ingraham, T. W. Seng, and P. V. Pallai, *Virology* **190,** 754 (1992).
[45] D. C. Huang, X. F. Huang, G. Novel, and M. Novel, *Mol. Microbiol.* **7,** 957 (1993).
[46] C. A. Conlin, K. Håkensson, A. Liljas, and C. G. Miller, *J. Bacteriol.* **176,** 166 (1994).
[47] T. H. Carter and C. G. Miller, *J. Bacteriol.* **159,** 453 (1984).

# [8] Procathepsin E and Cathepsin E

*By* TAKASHI KAGEYAMA

## Historical Background

Cathepsin E (EC 3.4.23.34) is a nonsecretory intracellular proteinase, and its precursor is known as procathepsin E. The enzyme is a member of the pepsin family of aspartic proteinases.[1] In 1962 Lapresle and Webb[2] gave the name cathepsin E to a new proteolytic enzyme that they had purified from rabbit bone marrow and that was different from the tissue cathepsins A, B, C, and D known at that time. Subsequent acceptance of the designation cathepsin E, however, required a considerable length of time because cathepsins E from different sources seemed to have different properties[3-9] and, hence, to be different enzymes, and also because the proteinase was postulated to be a form of cathepsin D (EC 3.4.23.5), a common aspartic proteinase found in lysosomes.[10] From 1989 to 1993 the primary structures of human,[11] guinea pig,[12] and rabbit[13] cathepsins E were determined consecutively by molecular cloning of the respective cDNAs, and it was proved that the enzymes belong to an independent group within the family of aspartic proteinases. The enzyme was shown to be synthesized as a prepro form and to be processed to an active form, a phenomenon common to other aspartic proteinases.

In the past, a variety of names was given to (pro)cathepsin E obtained from different sources such as gastric mucosa,[3-6] thymus,[7] spleen,[8] and blood cells.[9] The names included slow moving proteinase (SMP),[3] non-pepsin proteinase,[4] cathepsin D-like acid proteinase,[5,6] cathepsin D-type protease,[7] cathepsin E-like acid proteinase,[8] and erythrocyte membrane acid proteinase (EMAP).[9] However, the identification of all of the enzymes as (pro)cathepsin E has now been confirmed.[14-16]

[1] A. J. Barrett, in "Proteinases in Mammalian Cells and Tissues" (A. J. Barrett, ed.), p. 209. Elsevier/North-Holland, Amsterdam, 1977.
[2] C. Lapresle and T. Webb, Biochem. J. 84, 455 (1962).
[3] I. M. Samloff, R. T. Taggart, T. Shiraishi, T. Branch, W. A. Reid, R. Heath, R. W. Lewis, M. J. Valler, and J. Kay, Gastroenterology 93, 77 (1987).
[4] N. B. Roberts and W. H. Taylor, Biochem. J. 169, 617 (1978).
[5] T. Kageyama and K. Takahashi, J. Biochem. (Tokyo) 87, 725 (1980).
[6] N. Muto, K. Murayama-Arai, and S. Tani, Biochim. Biophys. Acta 745, 61 (1983).
[7] N. Yago and W. E. Bowers, J. Biol. Chem. 250, 4749 (1975).
[8] K. Yamamoto, N. Katsuda, and K. Kato, Eur. J. Biochem. 92, 499 (1978).
[9] K. Yamamoto, M. Takeda, H. Yamamoto, M. Tatsumi, and Y. Kato, J. Biochem. (Tokyo) 97, 821 (1985).
[10] V. Turk, I. Kregar, F. Gubensek, and D. Lebez, Enzymologia 36, 182 (1969).
[11] T. Azuma, G. Pals, T. K. Mohandas, J. M. Couvreur, and R. T. Taggart, J. Biol. Chem. 264, 16748 (1989).
[12] T. Kageyama, M. Ichinose, S. Tsukada, K. Miki, K. Kurokawa, O. Koiwai, M. Tanji, E. Yakabe, S. B. P. Athauda, and K. Takahashi, J. Biol. Chem. 267, 16450 (1992).
[13] T. Kageyama, Eur. J. Biochem. 216, 717 (1993).
[14] R. A. Jupp, A. D. Richards, J. Kay, B. M. Dunn, J. B. Wyckoff, I. M. Samloff, and K. Yamamoto, Biochem. J. 254, 895 (1988).

Assay Methods

*Assay with Hemoglobin*

*Principle.* Cathepsin E catalyzes the hydrolysis of proteins at extremely low pH values, such as pH 2.0. The hemoglobin-digestion method developed by Anson and Mirsky[17] for assaying pepsin (EC 3.4.23.1) is suitable for the assay of cathepsin E. The assay is performed by measuring increases in levels of trichloroacetic acid-soluble peptides as a function of time. The trichloroacetic acid-soluble peptides are quantitated spectrophotometrically or fluorometrically. The method allows determination of the potential activity of procathepsin E as well as the activity of cathepsin E, as procathepsin E is rapidly converted to cathepsin E at acidic pH. The method is appropriate for estimations of the activity of cathepsin E in crude homogenates of tissue since cathepsin D and other intracellular proteinases have negligible activity at pH 2.0. In gastric homogenates, however, the activity is the sum of the activities of pepsin and cathepsin E.

*Reagents*

HCl, 1 $N$

Trichloroacetic acid, 5% (w/v)

Dialyzed hemoglobin, 8% (w/v): Sixteen grams of hemoglobin substrate powder (Worthington Diagnostic Systems Inc., Freehold, NJ) is dissolved in about 100 ml of distilled water, and the solution is dialyzed against three lots of 5 liters of distilled water, for 12–24 hr each, at 4°; the solution is diluted with distilled water to bring the concentration to 8% and is stored at −20°

Sodium borate buffer, pH 8.5, 0.5 $M$

Fluorescamine solution: The solution is prepared by dissolving 10 mg of fluorescamine (Hoffmann-La Roche Inc., Nutley, NJ) in 100 ml of acetone; the solution is kept at room temperature in the dark

Standard solution of leucine, 100 $\mu M$: The standard solution of leucine is prepared by dissolving 1.31 mg of L-leucine in 100 ml of 3.3% trichloroacetic acid

*Procedure.* The 8% hemoglobin solution is diluted about 3-fold with distilled water, then acidified to pH 2.0 with 1 $N$ HCl, and the final concen-

[15] S. Yonezawa, T. Tanaka, N. Muto, and S. Tani, *Biochem. Biophys. Res. Commun.* **144,** 1251 (1987).
[16] N. I. Tarasova, P. B. Szecsi, and B. Foltmann, *Biochim. Biophys. Acta* **880,** 96 (1986).
[17] M. L. Anson and A. E. Mirsky, *J. Gen. Physiol.* **16,** 59 (1932).

tration of hemoglobin is brought to 2% with distilled water. This substrate solution can be kept at 4° for a few weeks. The amount of trichloroacetic acid-soluble material increases slightly during such storage. One milliliter of the substrate solution is incubated with an appropriate amount of enzyme solution (up to 50 $\mu$l) at 37°. The volume of enzyme solution added is so small, compared to the volume of the hemoglobin solution, that no correction for a change in reaction volume is necessary. No increase in the volume of enzyme solution is recommended because the buffering power of the substrate solution is weak. When the enzyme is dissolved in a buffer of a high concentration at neutral or alkaline pH, the pH of the reaction mixture must be checked carefully. The reaction is stopped by the addition of 2 ml of 5% trichloroacetic acid. After standing for about 30 min, the mixture is filtered through filter paper (No. 2 filter paper, Toyo Roshi Co., Tokyo, Japan; or other equivalent filter paper) and the absorbance of the filtrate is measured at 280 nm. The "blank" is prepared by adding the enzyme solution after the hemoglobin has been precipitated with trichloroacetic acid.

When a crude homogenate of tissue is used as the source of enzyme, an appreciable increase in absorbance is occasionally not obtained after an incubation for even a few hours. In this case, the trichloroacetic acid-soluble peptides are quantitated fluorometrically[18] as follows. One hundred microliters of the filtrate is mixed with 1.0 ml of 0.5 $M$ sodium borate buffer, pH 8.5, and then 0.2 ml of fluorescamine solution is added with vigorous shaking. Measurement of fluorescence is carried out with excitation at 390 nm and emission at 475 nm. The leucine solution is used as the standard for a primary amine.

*Units of Activity.* There is no international agreement about units obtained by the hemoglobin-digestion method. Tyrosine-equivalent amounts of trichloroacetic acid-soluble peptides have been used by some investigators, including Anson and Mirsky.[17] The present author uses an arbitrary system whereby the amount of enzyme that causes an increase in the absorbance of the hemoglobin filtrate of 1.0 at 280 nm in 1 min is taken as 1 unit. The fluorometric analysis of the hemoglobin filtrate gives a rather accurate measure of the amount of released peptides since fluorescamine reacts with the terminal amino group of released peptides. One unit of enzyme corresponds to the amount that causes release of about 10 $\mu$mol of leucine equivalents of trichloroacetic acid-soluble peptides per minute.

---

[18] S. de Bernardo, M. Weigele, V. Toome, K. Manhart, W. Leimgruber, P. Böhlen, S. Stein, and S. Udenfriend, *Arch. Biochem. Biophys.* **163,** 390 (1974).

## Assay with Substance P

*Principle.* Cathepsin E rapidly hydrolzyes various biologically active peptides at weakly acidic pH.[13] Substance P is the best peptide substrate. The hydrolysis is highly specific for the cleavage of the $Phe^8$-$Phe^9$ bond of substance P. The released peptides are quantitated by high-performance liquid chromatography (HPLC) on a reversed-phase column.

### Reagents

Substance P, 0.5 m$M$: Substance P (0.54 mg) is dissolved in 0.8 ml of 0.1% (v/v) acetic acid, and the solution is stored at $-20°$
Sodium acetate buffer, 0.2 $M$, pH 5.0
Sodium formate buffer, 0.2 $M$, pH 4.0
Tris-HCl buffer, 0.2 $M$, pH 8.0
Trichloroacetic acid, 5% (w/v)
E-64 (*trans*-epoxysuccinyl-L-leucylamido(4-guanidino)butane, an inhibitor of cysteine proteinases), 10 m$M$: E-64 (3.57 mg) is dissolved in 1 ml of dimethyl sulfoxide, and the solution is stored at $-20°$

*Procedure.* The reaction mixture is composed of 17 $\mu$l of 0.2 $M$ sodium acetate buffer, pH 5, 2 $\mu$l of the 0.5 m$M$ solution of substance P, and 1 $\mu$l of enzyme solution. Incubation is carried out at 37° for an appropriate time. The hydrolysis of substance P proceeds linearly until about 50% of the initial amount of substance P is degraded. Thirty minutes is sufficient for assays of 0.25 ng of cathepsin E. The reaction is stopped by the addition of 60 $\mu$l of 0.2 $M$ Tris-HCl buffer, pH 8.0, and the entire mixture is subjected to HPLC. A column (0.48 cm i.d. $\times$ 25 cm) of ODS-120T (Tosoh Corporation, Tokyo, Japan) is equilibrated with 0.1% (v/v) trifluoroacetic acid. Substance P and the products of hydrolysis are eluted by a linear gradient of acetonitrile from 0 to 70% (v/v) over the course of 28 min in the presence of 0.1% trifluoroacetic acid at a flow rate of 0.8 ml/min. Peptides are monitored at 215 nm. Three peaks, corresponding to substance P(1–7), substance P(8–11), and intact substance P, are eluted in that order within 20–26 min.

When a crude tissue homogenate is used as the source of enzyme, the following changes are made. The reaction is carried out in 0.2 $M$ sodium formate buffer, pH 4.0, to determine the potential activity of procathepsin E as well as the activity of cathepsin E. The buffer contains 1 $\mu$$M$ E-64 to inhibit cysteine proteinases. The reaction is stopped by the addition of 20 $\mu$l of 5% trichloroacetic acid. Precipitated proteins are removed by centrifugation, and the supernatant is subjected to HPLC.

*Units of Activity.* The increase in the amount of substance P(1–8) is determined from the area of the peak on the chart after HPLC. One unit

TABLE I
PURIFICATION OF PROCATHEPSIN E FROM GASTRIC MUCOSA OF GUINEA PIGS[a]

| Step | Protein (mg) | Activity[b] (units) | Specific activity (units/mg protein) | Yield (%) |
|---|---|---|---|---|
| 1. Supernatant of crude homogenate | 176 | 161 | 0.91 | 100 |
| 2. DEAE-Sephacel | | | | |
|    Procathepsin E | 6.5 | 31 | 4.8 | 19 |
|    Progastricsin[c] | 11 | 85 | 7.7 | 53 |
| 3. Purification of procathepsin E | | | | |
|    Sephadex G-100 | 2.0 | 20 | 10 | 12 |
|    FPLC | 0.67 | 15 | 22 | 9.3 |
| 4. Purification of progastricsin[c] | | | | |
|    Sephadex G-100 | 4.5 | 49 | 11 | 30 |
|    FPLC 1 | 1.4 | 27 | 19 | 17 |
|    FPLC 2 | 0.92 | 16 | 17 | 9.9 |

[a] T. Kageyama, M. Ichinose, S. Tsukada, K. Miki, K. Kurokawa, O. Koiwai, M. Tanji, E. Yakabe, S. B. P. Athauda, and K. Takahashi, *J. Biol. Chem.* **267,** 16450 (1992).
[b] Determined by the hemoglobin-digestion method.
[c] Purification of progastricsin (pepsinogen C) is also shown because activities of both procathepsin E and progastricsin are determined by the hemoglobin-digestion method.

is defined as the amount of enzyme that degrades 1 $\mu$mol of substance P per minute under the experimental conditions.

## Purification of Procathepsin E

Cathepsin E is present as procathepsin E in certain types of animal tissue, and it is present at a high level in mammalian gastric mucosa.[3-6,12,13,19] Purification of procathepsin E from gastric mucosa of guinea pig[12] is described here, using a modified version of procedures applied to the purification of other mammalian gastric procathepsins E (Table I). All procedures except for FPLC (fast protein liquid chromatography) are performed at 0°–4°. Chromatography and gel filtration are carried out in 10 m$M$ sodium phosphate buffer, pH 7.0 (referred to as the buffer). Because procathepsin E is easily converted to active cathepsin E under weakly acidic conditions,[12,20] a neutral or weakly alkaline pH is maintained throughout the purification.

*Step 1: Preparation of Mucosal Extract.* Gastric mucosa (total weight 2.4 g) is stripped from the frozen stomachs (total weight 4.5 g) of two adult

[19] O. Matsuzaki and K. Takahashi, *Biomed. Res.* **9,** 515 (1988).
[20] S. B. P. Athauda, T. Takahashi, T. Kageyama, and K. Takahashi, *Biochem. Biophys. Res. Commun.* **175,** 152 (1991).

guinea pigs and homogenized at 0°C with a Waring blender in 40 ml of the buffer. The homogenate is centrifuged at 15,000 $g$ for 60 min, and the supernatant is used for subsequent purification.

*Step 2: Chromatography on DEAE-Sephacel.* The supernatant is applied to a column (1.2 cm i.d. × 25 cm) of DEAE-Sephacel (Pharmacia LKB Biotech., Uppsala, Sweden) (Fig. 1). Proteins are eluted with a 500-ml linear gradient of 0–0.5 $M$ NaCl. Proteolytic activity against hemoglobin is detected as two peaks. The first and second peaks are those of procathepsin E and progastricsin, respectively. Fractions containing procathepsin E are combined and concentrated to about 10 ml with a membrane filter.

*Step 3: Gel Filtration.* The concentrated solution is subjected to gel filtration on a column (2 cm i.d. × 160 cm) of Sephadex G-100. Fractions with proteolytic activity are combined.

*Step 4: Fast Protein Liquid Chromatography.* The concentrated protein solution is subjected to FPLC on a Mono Q column (HR 5/5, Pharmacia LKB Biotech.) for final purification. Proteins are eluted with a linear gradient of NaCl from 0 to 0.5 $M$ over the course of 30 min at a flow rate of 1.0 ml/min. Procathepsin E is eluted as a single peak in the range of 0.1 to 0.2 $M$ NaCl.

*Purification of Other Mammalian Gastric Procathepsins E.* Procathepsins E from other mammals, such as humans[5] and rabbits,[13] can be purified

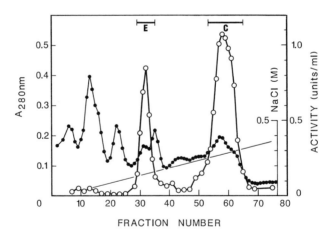

FIG. 1. Chromatography of the supernatant of a crude extract of guinea pig gastric mucosa on a column of DEAE-Sephacel. The fraction size was 10 ml. ○, Proteolytic activity determined by the hemoglobin-digestion method; ●, absorbance at 280 nm. E, Procathepsin E; C, progastricsin. [Reproduced from T. Kageyama, M. Ichinose, S. Tsukada, K. Miki, K. Kurokawa, O. Koiwai, M. Tanji, E. Yakabe, S. B. P. Athauda, and K. Takahashi, *J. Biol. Chem.* **267**, 16450 (1992).]

essentially by the same procedure as described above. However, because the level of the proenzyme in the gastric mucosa of those mammals is lower than that of guinea pig, the introduction of an additional step between Steps 3 and 4 is appropriate. After partial purification of procathepsin E by chromatography on DEAE-Sephacel and gel filtration on Sephadex G-100, the enzyme preparation is applied to a column (roughly 1 g wet weight of resin per 20 mg protein) of concanavalin A (Con A)-Sepharose that has been equilibrated with the buffer supplemented with 0.5 $M$ NaCl. Many of the contaminating proteins, including pepsinogens, pass unadsorbed through the column. Procathepsin E is adsorbed and eluted with 10 column volumes of the buffer supplemented with 0.5 $M$ NaCl and 0.4 $M$ α-D-methylglucoside. The eluate is then dialyzed against the buffer. Final purification is achieved by FPLC on a Mono Q column as described above in Step 4.

## Activation of Procathepsin E and Isolation of Cathepsin E and Released Propeptides

Activation of procathepsin E occurs autocatalytically under acidic conditions.[12,20] In the case of guinea pig procathepsin E, the rate of activation is maximal at pH 4.0, and the rate is still significant at lower pH values. Procathepsin E (0.4 mg) is dissolved in 4 ml of 10 m$M$ sodium phosphate buffer, pH 7.0, and mixed with 1 ml of 0.1 $M$ HCl to lower the pH to 2.0. After incubation at 14° for 30 min, 5 ml of 0.2 $M$ sodium acetate buffer, pH 5.5, is added. The mixture is applied to a column (1 cm i.d. × 2 cm) of SP-Sephadex that has been equilibrated with 0.1 $M$ sodium acetate buffer, pH 5.5. The flow-through fraction contains cathepsin E. Activation peptides released from procathepsin E, which are very basic in character, are adsorbed to the column and eluted with 0.1 $M$ sodium acetate buffer, pH 5.5, supplemented with 1 $M$ NaCl. Because, after the release of the prosegment by cleavage at the Leu[32]-Asn[33] bond, some peptide bonds in the prosegment are secondarily cleaved, several peptides are included in the high-salt eluate. The eluate is loaded onto a reversed-phase column (0.46 cm i.d. × 25 cm) of ODS-120T that has been equilibrated with 0.1% trifluoroacetic acid for HPLC. Each peptide is eluted separately with a linear gradient of acetonitrile from 0 to 50% over the course of 60 min in the presence of 0.1% trifluoroacetic acid at a flow rate of 0.8 ml/min.

## Enzymatic Properties

*Purity and Molecular Mass.* The procedures described above yield preparations of procathepsin E and cathepsin E that appear to be homogeneous when examined by polyacrylamide gel electrophoresis (PAGE) on nondenaturing gels by the methods of Ornstein[21] and Davis[22] and by sodium dodecyl sulfate (SDS)–PAGE by the method of Laemmli.[23] The molecular mass of guinea pig procathepsin E has been estimated to be 82 and 43 kDa by SDS–PAGE under nonreducing and reducing conditions, respectively. Thus, procathepsin E is a dimer. The molecular mass of guinea pig cathepsin E has been estimated to be 76 and 40 kDa by SDS–PAGE under nonreducing and reducing conditions, respectively.[12] The dimeric form is maintained in cathepsin E. Human[19] and rabbit[13] (pro)cathepsins E give similar results. The molecular masses cited above are slightly larger than those expected from the primary structures of the respective (pro)enzyme. The discrepancy is due to the attachment of a carbohydrate side chain to the cathepsin E moiety which accounts for 3–4% of the total molecular mass.[5,24]

*Stability.* Procathepsin E can be stored in neutral or weakly alkaline solutions that contain 50% glycerol at −20° for several months without significant loss of activity. The proenzyme can also be stored at 4° in a saturated solution of ammonium sulfate, if the solution is adjusted to neutral or weakly alkaline pH. Cathepsin E is stable under similar conditions provided that the pH of the storage solution is kept at around pH 5.5.

*Interconversion of Dimeric (Pro)cathepsin E and Monomeric Form.* Dimeric (pro)cathepsin E is converted to monomers by incubation with 10 m$M$ reducing reagent, such as 2-mercaptoethanol, L-cysteine, or reduced glutathione, at 37° for 20 min, in an appropriate buffer, namely, 25 m$M$ sodium phosphate buffer, pH 7.0, for procathepsin E and 25 m$M$ sodium acetate buffer, pH 5.5, for cathepsin E.[12,25,26] The dimer is regenerated after removal of the reducing reagent from the solution by precipitation with ammonium sulfate or dialysis. Inactivation of the enzyme is not observed during the course of interconversion. The hydrolytic activity of the dimer and that of the monomer are nearly identical, although a slightly higher activity with respect to digestion of hemoglobin at pH 5 is found with monomeric cathepsin E than with dimeric cathepsin E.[12] It must be noted

[21] L. Ornstein, *Ann. N.Y. Acad. Sci.* **121**, 321 (1964).
[22] J. Davis, *Ann. N.Y. Acad. Sci.* **121**, 404 (1964).
[23] U. K. Laemmli, *Nature (London)* **227**, 680 (1970).
[24] S. B. P. Athauda, O. Matsuzaki, T. Kageyama, and K. Takahashi, *Biochem. Biophys. Res. Commun.* **168**, 878 (1990).
[25] K. Yamamoto, H. Yamamoto, M. Takeda, and Y. Kato, *Biol. Chem. Hoppe-Seyler* **369**(Suppl.), 315 (1988).
[26] S. Yonezawa, K. Fujii, Y. Maejima, K. Tamoto, Y. Mori, and N. Muto, *Arch. Biochem. Biophys.* **267**, 176 (1988).

TABLE II
HYDROLYSIS OF PROTEIN SUBSTRATES BY CATHEPSIN E

|  | Relative activity[a] (%) | | | | |
|---|---|---|---|---|---|
| | Rat | | | Human | Guinea pig |
| Substrate | Stomach[b] (pH 2.5) | Spleen[c] (pH 3.1) | Erythrocyte[d] (pH 3.8) | erythrocyte[d] (pH 3.8) | stomach[e] (pH 5.0) |
| Hemoglobin | 100 | 100 | 100 | 100 | 100 |
| Bovine serum albumin | 4.3 | 17 | 16 | 19 | 43 |
| Human serum albumin | | 3 | 20 | 28 | |
| Ovalbumin | | | | | 12 |
| Bovine γ-globulin | 10 | | | | 6 |
| Casein | 6.3 | | | | 110 |

[a] Activity of cathepsin E with hemoglobin as substrate is taken as 100% in each case.
[b] N. Muto, K. Murayama-Arai, and S. Tani, *Biochim. Biophys. Acta* **745**, 61 (1983).
[c] K. Yamamoto, N. Katsuda, and K. Kato, *Eur. J. Biochem.* **92**, 499 (1978).
[d] M. Takeda, E. Ueno, Y. Kato, and K. Yamamoto, *J. Biochem. (Tokyo)* **100**, 1269 (1986).
[e] T. Kageyama, *Eur. J. Biochem.* **216**, 717 (1993).

that the monomeric cathepsin E is much less stable at weakly alkaline pH, for example, pH 8.5, than the dimeric cathepsin E.[12,13]

*Hydrolysis of Protein and Peptide Substrates, and Specificity.* Cathepsin E hydrolyzes various protein and peptide substrates. Protein substrates, such as hemoglobin, serum albumin, and casein, are hydrolyzed efficiently (Table II). Hemoglobin is the best protein substrate, and its hydrolysis proceeds most rapidly at around pH 2.5 and at a significantly lower rate at higher pH values such as pH 4.0 and pH 5.0.[3–6,12,13] The low pH optimum for the hydrolysis of protein substrates by cathepsin E resembles that of pepsins but differs from that of cathepsin D, which has maximal proteolytic activity at pH 3.5 and is almost inactive at pH 2.0.

Peptide substrates, such as biologically active peptides and chromogenic synthetic substrates, are also hydrolyzed efficiently by cathepsin E (Table III). The best peptide substrates are substance P and related tachykinins. Chromogenic substrates, such as Pro-Pro-Thr-Ile-Phe-Phe(NO$_2$)-Arg-Leu and Lys-Pro-Ile-Glu-Phe-Phe(NO$_2$)-Arg-Leu, are less suitable substrates than substance P and other tachykinins.[14,27] The rates of hydrolysis of many peptide substrates are highest at around pH 5 and are quite low at around pH 2.0. The rates of hydrolysis of substance P and other

[27] J. Kay and B. M. Dunn, *Scand. J. Clin. Lab. Invest.* **52**(Suppl. 210), 23 (1992).

TABLE III
RELATIVE HYDROLYTIC ACTIVITIES OF CATHEPSIN E AND CATHEPSIN D
AGAINST BIOLOGICALLY ACTIVE PEPTIDES[a]

| Peptide | Activity[b] (nmol/min/μg protein) | | |
|---|---|---|---|
| | Rabbit cathepsin E | Guinea pig cathepsin E | Bovine cathepsin D |
| Substance P | 45.0 | 12.5 | 0.097 |
| Neurokinin A | 22.5 | 11.8 | 0.019 |
| Eledoisin | 11.5 | 5.3 | 0.16 |
| Kassinin | 3.2 | 3.3 | 0.010 |
| Cholecystokinin-8 | 0.009 | 0.009 | 0.003 |
| Neuromedin C | UC[c] | UC | 0.001 |
| Bombesin | UC | UC | 0.007 |
| Neurotensin | UC | UC | 0.018 |
| Acidic FGF (102–111)[d] | 2.7 | 1.0 | 0.012 |
| Basic FGF (106–120) | 1.5 | 1.0 | 0.18 |
| β-Endorphin | 1.8 | 0.47 | 18.6 |
| Dynorphin A | 0.065 | 0.046 | 0.005 |
| Porcine renin substrate | 15.3 | 13.6 | 4.2 |
| Human renin substrate | 0.63 | 0.55 | 0.066 |
| Big endothelin-1 | 0.11 | 0.14 | 0.057 |

[a] T. Kageyama, *Eur. J. Biochem.* **216**, 717 (1993).
[b] Hydrolysis of each peptide was determined by a method similar to the standard assay of the hydrolysis of substance P. The rate of hydrolysis refers to the rate at the major site of cleavage of each peptide substrate.
[c] UC, Uncleaved.
[d] FGF, Fibroblast growth factor.

tachykinins by cathepsin E at pH 5 are very high, being several hundred-fold higher than rates of hydrolysis by cathepsin D (Table III). $K_m$ and $k_{cat}$ values for peptide substrates have been measured and are shown in Table IV.

It is notable that the pH optima for hydrolysis of various protein and peptide substrates differ significantly. The differences may be due to several factors, such as changes in the conformation and extent of ionization of the various substrates at different pH values. In the case of protein substrates, it is reasonable to assume that destruction of the tertiary structure of each protein at an extremely low pH may be partly responsible for the maximum rate of hydrolysis under such acidic conditions.

The specificity of cleavage sites has been rather well defined on the basis of studies of the hydrolysis of various peptides,[13] as well as the B

TABLE IV
KINETIC CONSTANTS FOR HYDROLYSIS OF PEPTIDES BY CATHEPSIN E

| Substrate[a] | Assay pH | $K_m$ ($\mu M$) | $k_{cat}$ ($sec^{-1}$) | $k_{cat}/K_m$ ($\mu M^{-1} sec^{-1}$) | Ref. |
|---|---|---|---|---|---|
| Substance P | 5 | 3.9 | 47 | 12.1 | b |
| Arg-modified substance P | 5 | 2.3 | 64 | 27.8 | b |
| [Tyr$^8$]Substance P | 5 | 7.1 | 40 | 5.63 | b |
| [His$^{10}$]Substance P | 5 | 11 | 23 | 2.09 | b |
| Ac-substanceP | 5 | 9.3 | 26 | 2.80 | b |
| Substance P (2–11) | 5 | 42 | 22 | 0.53 | b |
| Ac-substance P (2–11) | 5 | 83 | 19 | 0.23 | b |
| Substance P (3–11) | 5 | 660 | 70 | 0.11 | b |
| Ac-substance P (3–11) | 5 | 400 | 58 | 0.15 | b |
| Substance P (4–11) | 5 | 24 | 0.74 | 0.03 | b |
| Substance P (1–9) | 5 | >500 | —$^f$ | — | b |
| Neurokinin A | 5 | 25 | 42 | 1.68 | b |
| Eledoisin | 5 | 5.9 | 10 | 1.70 | b |
| Acidic FGF (102–111) | 5 | 143 | 18 | 0.13 | b |
| Porcine renin substrate | 5 | 3.4 | 13 | 3.82 | b |
| Big endothelin-1 | 4.3 | 9 | 0.13 | 0.014 | c |
| Lys-Pro-Ala-Glu-Phe-Phe(NO$_2$)-Arg-Leu | 3.1 | 160 | 135 | 0.84 | d |
| Pro-Thr-Glu-Phe-Phe(NO$_2$)-Arg-Leu | 3.1 | 145 | 27 | 0.19 | d |
| Lys-Pro-Ile-Glu-Phe-Phe(NO$_2$)-Arg-Leu | 3.5 | 70 | 170 | 2.4 | e |
| Pro-Pro-Thr-Ile-Phe-Phe(NO$_2$)-Arg-Leu | 3.5 | 30 | 75 | 2.5 | e |

$^a$ Ac-, Acetyl; FGF, fibroblast growth factor.
$^b$ T. Kageyama, *Eur. J. Biochem.* **216**, 717 (1993).
$^c$ P. S. Robinson, W. E. Lees, J. Kay, and N. D. Cook, *Biochem. J.* **284**, 407 (1992).
$^d$ I. M. Samloff, R. T. Taggart, T. Shiraishi, T. Branch. W. A. Reid, R. Heath, R. W. Lewis, M. J. Valler, and J. Kay, *Gastroenterology* **93**, 77 (1987).
$^e$ R. A. Jupp, A. D. Richards, J. Kay, B. M. Dunn, J. B. Wyckoff, I. M. Samloff, and K. Yamamoto, *Biochem. J.* **254**, 895 (1988).
$^f$ Not measured.

chain of insulin (Fig. 2).[13,19,28,29] The P1 and P1' positions are occupied almost exclusively by hydrophobic and aromatic amino acids. This specificity is commonly observed in other aspartic proteinases, such as pepsin and cathepsin D, although the rates of hydrolysis of various proteins and peptides differ significantly between cathepsin E and other aspartic proteinases. Such differences may be caused by variations in the interactions among

[28] S. Yonezawa, T. Tanaka, and T. Miyauchi, *Arch. Biochem. Biophys.* **256**, 499 (1987).
[29] S. B. P. Athauda, T. Takahashi, H. Inoue, M. Ichinose, and K. Takahashi, *FEBS Lett.* **292**, 53 (1991).

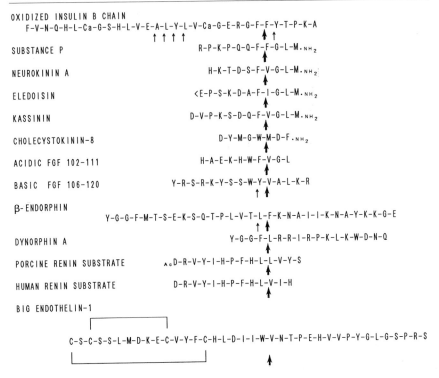

Fig. 2. Sites of cleavage of various biologically active peptides by cathepsin E. The arrows indicate sites of cleavages that produce major peptides ( ↑ ) and intermediate or minor peptides ( ↑ ). NH$_2$ in small letters indicates an amide group at the carboxyl terminus of the peptide. Ac, Acetyl; Ca, cysteic acid; E, pyroglutamic acid.

positions other than the P1 and P1′ positions, as well as differences among the active sites of the respective aspartic proteinases. In the case of cathepsin E, peptides with the Lys-Pro combination at the P5 and P4 positions have been shown to be optimal substrates.[27]

*Activators and Inhibitors.* ATP and analogs are known to stimulate the activity of cathepsin E at weakly acidic and neutral pH.[30] Pepstatin,[5,12,13] tripeptide analogs,[31] and the protein inhibitor from *Ascaris lumbricoides*[3,14] inhibit cathepsin E very effectively. Diazoacetyl-DL-norleucine methyl es-

---

[30] D. J. Thomas, A. D. Richards, R. A. Jupp, E. Ueno, K. Yamamoto, I. M. Samloff, B. M. Dunn, and J. Kay, *FEBS Lett.* **243,** 145 (1989).

[31] J. E. Bird, T. L. Waldron, D. K. Little, M. M. Asaad, C. R. Dorso, G. DiDonato, and J. A. Norman, *Biochem. Biophys. Res. Commun.* **182,** 224 (1992).

ter[5] and 1,2-epoxy-3-($p$-nitrophenoxy)propane[5] inactivate cathepsin E irreversibly via modification of the aspartic acid residues at the active site.

## Primary Structure

Complete primary structures are known for human, guinea pig, and rabbit preprocathepsins E (Fig. 3). They consist of three regions, namely, the prepeptide (signal peptide), the propeptide (activation segment), and the active enzyme. The extent of the similarities among the structures is more than 80% at both the nucleotide and the amino acid level. The amino

```
            -10        -1 1      10        20        30        40        50        60
            :           : :       :         :         :         :         :         :
Rabbit E   MKTLPLLLLLLLDLGQA  GTLDRVPLRRQPSLRKKLRAQGQLSEPWKAHKVDM/QYTETCTMEQSANEPLINYLDME
Guinea E   MKTFLLLLLVLLELGQA PG/LHRVPLSRRESLRKKLRAQGQLTELWKSQNL/MDQ----CSTIQSANEPLINYLDME
Human  E   MKTLLLLLLVLLELGEA  GSLHRVPLRRHPSLKKKLRARSQLSEPWKSHNLDM/QF7ESCSMDQSAKEPLINYLDME

            70 CHO    80        90       100       110       120       130       140
            :  |       :         :         :         :         :         :         :
RE         YFGTISIGSPPQNFTVIFT VSSNLWVPSVYCTSPACQMHPQFRPSQSNTYSEVGTPFSIAYGTGSLTGIIGADQVSVQG
GE         YFGTISIGSPPQNFTVIFT GSSNLWVPSVYCTSPACQTHPVFHPSLSSTYREVGNSFSIQYGTGSLTGIIGADQVSVEG
HE         YFGTISIGSPPQNFTVIFT GSSNLWVPSVYCTSPACKTHSRFQPSQSSTYSQPGQSFSIQYGTGSLSGIIGADQVSVEG

            150       160       170       180       190       200       210       220
            :         :         :         :         :         :         :         :
RE         LTVVGQQFGESVKEPGQTFV NAEFDGILGLGYPSLAAGGVTPVFDNMMAQNLVSLPMFSVYMSSNPEGGSGSELTFGGYD
GE         LTVVGQQFGESVQEPGKTFV HAEFDGILGLGYPSLAAGGVTPVFDNMMAQNLVALPMFSVYMSSNP-GGSGSELTFGGYD
HE         LTVVGQQFGESVTEPGQTFV DAEFDGILGLGYPSLAVGGVTPVFDNMMAQNLVDLPMFSVYMSSNPEGGAGSELIFGGYD

            230       240       250       260       270       280       290       300
            :         :         :         :         :         :         :         :
RE         SSHFSGSLNWVPVTKQGYWQ IALDEIQVGGSPMFCPEGCQAIVTGTGSLITGPSDKIIQLQAAIGATPMDGEYAVECENL
GE         PSHFSGSLNWVPVTKQAYWQ IALDGIQVGDSVMFCSEGCQAIVTGTGSLITGPPGKIKQLQEALGATYVDEGYSVQCANL
HE         HSHFSGSLNWVPVTKQAYWQ IALDNIQVGGTVMFCSEGCQAIVTGTGSLITGPSDKIKQLQNAIGAAPVDGEYAVECANL

            310       320       330       340       350       360       370
            :         :         :         :         :         :         :
RE         NIMPDVTFVINGVPYTLSAT AYTLPDFVDGMQFCGSGFQGLDIQPPAGPLWILGDVFIRQFYSVFDRGSNRVGLAPAVP
GE         NMMLDVTFIINGVPYTLNPT AYTLLDFVDGMQVCSTGFEGLEIQPPAGPLWILGDVFIRQFYAVFDRGNNRVGLAPAVP
HE         NVMPDVTFTINGVPYTLSPT AYTLLDFVDGMQFCSSGFQGLDIHPPAGPLWILGDVFIRQFYSVFDRGNNRVGLAPAVP
```

FIG. 3. Comparison of the amino acid sequences of procathepsin E from rabbit, guinea pig, and human. Common residues are shown by shading. Residues in dark boxes are Asp residues at the active site. Bars represent spaces that have been inserted to maximize alignment. The italic letters at positions 1 and 40 approximately indicate the N-terminal amino acids of procathepsin E and cathepsin E, respectively. Sequence for rabbit from T. Kageyama, *Eur. J. Biochem.* **216,** 717 (1993); for guinea pig from T. Kageyama, M. Ichinose, S. Tsukada, K. Miki, K. Kurokawa, O. Koiwai, M. Tanji, E. Yakabe, S. B. P. Athauda, and K. Takahashi, *J. Biol. Chem.* **267,** 16450 (1992); and for humans from T. Azuma, G. Pals, T. K. Mohandas, J. M. Couvreur, and R. T. Taggart, *J. Biol. Chem.* **264,** 16748 (1989). CHO indicates the site of N-glycosylation in human procathepsin E [S. B. P. Athauda, O. Matsuzaki, T. Kageyama, and K. Takahashi, *Biochem. Biophys. Res. Commun.* **168,** 878 (1990)].

acid compositions also resemble one another (Table V). In the guinea pig proenzyme, the absence of five residues is noted, and four of those are positioned in the NH$_2$-terminal region of the cathepsin E moiety. The extent of the similarity between the primary structure of cathepsin E and other aspartic proteinases is less than 60%. The evolutionary relationships among various aspartic proteinases, based on nucleotide sequences, indicate that procathepsin E is closer to the pepsinogens than are procathepsin D and prorenin.[12]

Structural features common to species of procathepsin E but absent from other aspartic proteinases are the occurrence of a Cys residue in the

TABLE V
AMINO ACID COMPOSITION OF PROCATHEPSIN E

| Amino acid | Number of residues per molecule | | |
|---|---|---|---|
| | Guinea pig[a] | Rabbit[b] | Human[c] |
| Asp | 16 | 18 | 20 |
| Asn | 17 | 14 | 15 |
| Thr | 24 | 25 | 22 |
| Ser | 33 | 34 | 40 |
| Glu | 15 | 16 | 13 |
| Gln | 23 | 26 | 24 |
| Pro | 24 | 28 | 25 |
| Gly | 40 | 41 | 39 |
| Ala | 21 | 21 | 20 |
| Cys | 7 | 7 | 7 |
| Val | 31 | 31 | 30 |
| Met | 10 | 12 | 10 |
| Ile | 19 | 20 | 20 |
| Leu | 33 | 27 | 29 |
| Tyr | 14 | 14 | 13 |
| Phe | 19 | 21 | 22 |
| Lys | 7 | 7 | 9 |
| His | 5 | 3 | 7 |
| Arg | 9 | 9 | 9 |
| Trp | 5 | 5 | 5 |
| Total | 372 | 379 | 379 |
| $M_r$ | 40,080 | 40,831 | 40,928 |

[a] T. Kageyama, M. Ichinose, S. Tsukada, K. Miki, K. Kurokawa, O. Koiwai, M. Tanji, E. Yakabe, S. B. P. Athauda, and K. Takahashi, *J. Biol. Chem.* **267,** 16450 (1992).
[b] T. Kageyama, *Eur. J. Biochem.* **216,** 717 (1993).
[c] T. Azuma, G. Pals, T. K. Mohandas, J. M. Couvreur, R. T. Taggart, *J. Biol. Chem.* **264,** 16748 (1989).

$NH_2$-terminal region of the cathepsin E moiety and the absence of Lys-37 (numbering based on human pepsinogen A) in the pro-segment. The Cys residue is repsonsible for the generation of the dimer. The absence of Lys-37 may be responsible for the rapid activation of procathepsin E under weakly acidic conditions, for example, at pH 4, because in other aspartic proteinases, such as pepsinogens, the Lys residue has been shown to be essential for stabilization of the prosegment and a very low pH, for example, pH 2.0, is needed for induction of the conformational change responsible for activation.[32]

(Pro)cathepsin E is a glycoprotein.[5] A mannose-rich carbohydrate chain is attached to Asn-73 in human procathepsin E.[24] The carbohydrate chain has been estimated to be composed of 2 glucosamine and 8 mannose residues per monomeric form in human procathepsin E.[5] The site of the carbohydrate chain is thought to be the same in guinea pig and rabbit procathepsins E as the consensus sequence for a glycosylation site, Asn-X-Thr, is common to all three procathepsins E (Fig. 3).

The gene for human procathepsin E consists of nine exons, resembling other mammalian genes for aspartic proteinases.[33] It is located on chromosome 1.[34]

## Distribution

(Pro)cathepsin E is distributed in various types of tissues: contractile tissues, such as the stomach,[3–6,12,13] colon,[13] and urinary bladder[35,36]; and immune-associated tissues.[7,8,37,38] Appreciable amounts are found in human[9] and rat[39] erythrocytes and rat neutrophils,[28] but no activity has been detected in human neutrophils.[40] The occurrence in blood cells has been reported to be species-specific.[41] Wide distribution has been recognized in

[32] M. N. G. James and A. R. Sielecki, Nature (London) 319, 33 (1986).
[33] T. Azuma, W. Liu, D. J. Vander Laan, A. M. Bowcock, and R. T. Taggart, J. Biol Chem. 267, 1609 (1992).
[34] J. M. Couvreur, T. Azuma, D. A. Miller, M. Rocchi, T. K. Mohandas, F. A. Boudi, and R. T. Taggart, Cytogenet. Cell Genet. 53, 137 (1990).
[35] N. Muto, M. Yamamoto, S. Tani, and S. Yonezawa, J. Biochem. (Tokyo) 103, 629 (1988).
[36] H. Sakai, T. Saku, Y. Kato, and K. Yamamoto, Biochim. Biophys. Acta 991, 367 (1989).
[37] G. Finzi, M. Cornaggia, C. Capella, R. Fiocca, F. Bosi, E. Solcia, and I. M. Samloff, Histochemistry 99, 201 (1993).
[38] E. Solcia, M. Paulli, E. Silini, R. Fiocca, G. Finzi, S. Kindl, E. Boveri, F. Bosi, M. Cornaggia, C. Capella, and I. M. Samloff, Eur. J. Histochem. 37, 19 (1993).
[39] M. Takeda, E. Ueno, Y. Kato, and K. Yamamoto, J. Biochem. (Tokyo) 100, 1269 (1986).
[40] E. Ichimaru, H. Sakai, T. Saku, K. Kunimatsu, Y. Kato, I. Kato, and K. Yamamoto, J. Biochem. (Tokyo) 108, 1009 (1990).
[41] S. Yonezawa and K. Nakamura, Biochim. Biophys. Acta 1073, 155 (1991).

fetal tissues, with especially high levels in the liver.[13] Procathepsin E is expressed frequently in malignant tissues, such as gastric carcinoma[42,43] and ovarian mucinous tumors.[44] The expression of mRNA for cathepsin E has been shown to be closely correlated with the level of (pro)cathepsin E in various tissues.[11-13] The ratio of cathepsin E to procathepsin E in tissues has not been determined exactly. Procathepsin E is thought to be the predominant form in gastric mucosa, as judged from the results of SDS–PAGE immunoblot analysis.[45] The intracellular localization of (pro)cathepsin E has not been clarified. It has been postulated that the (pro)enzyme is localized in the cystosol and in light particulate fractions[26,46] or that it is distributed strategically in certain types of cells.[47]

[42] T. Saku, H. Sakai, N. Tsuda, H. Okabe, Y. Kato, and K. Yamamoto, *Gut* **31,** 1250 (1990).
[43] C. Furihata, M. Tatematsu, K. Miki, T. Katsuyama, K. Sudo, N. Miyagi, T. Kubota, S. Jin, K. Kodama, N. Ito, Y. Konishi, K. Suzuki, and T. Matsushima, *Cancer Res.* **44,** 727 (1984).
[44] P. Tenti, A. Aguzzi, C. Riva, L. Usellini, R. Zappatore, J. Bara, I. M. Samloff, and E. Solcia, *Cancer* **69,** 2131 (1992).
[45] S. Yonezawa, Y. Maejima, N. Hagiwara, T. Aratani, R. Shoji, T. Kageyama, S. Tsukada, K. Miki, and M. Ichinose, *Dev. Growth Differ.* **35,** 349 (1993).
[46] K. Bennett, T. Levine, J. S. Ellis, R. J. Peanasky, I. M. Samloff, J. Kay, and B. M. Chain, *Eur. J. Immunol.* **22,** 1519 (1992).
[47] T. Saku, H. Sakai, Y. Shibata, Y. Kato, and K. Yamamoto, *J. Biochem.* (*Tokyo*) **110,** 956 (1991).

# [9] Processing Enzymes of Pepsin Family: Yeast Aspartic Protease 3 and Pro-opiomelanocortin Converting Enzyme*

By Y. Peng Loh and Niamh X. Cawley

## Introduction

Neuropeptides and peptide hormones are synthesized as large inactive precursors which are cleaved by endopeptidases at paired or single basic residues to yield biologically active peptides.[1] The processing generally occurs in the trans-Golgi network or within secretory vesicles.[1] A number

* The opinions or assertions contained herein are the private ones of the authors and are not to be construed as official or reflecting the views of the U.S. Department of Defense or the Uniformed Serivces University of the Health Sciences.
[1] Y. P. Loh, M. C. Beinfeld, and N. P. Birch, *in* "Mechanisms of Intracellular Trafficking and Processing of Proproteins" (Y. P. Loh, ed.), p. 179. CRC Press, Boca Raton, Florida, 1993.

of processing enzymes which can cleave prohormones at paired/single basic residues have been described (for reviews, see Refs. 1–3). Many of the processing enzymes are subtilisin-like serine proteinases[2,3] or aspartic proteinases,[4–7] although ones belonging to the metallo- and thiol proteinase classes have also been reported.[1] In this chapter, two processing enzymes of the aspartic proteinase family, namely, yeast aspartic protease 3 and pro-opiomelanocortin converting enzyme (EC 3.4.23.17), are discussed. The two enzymes are immunologically related, and they have been the most extensively studied of the aspartic processing enzymes, especially with respect to specificity for various prohormones.

Yeast aspartic protease 3 (YAP3p), encoded by the *YAP3* gene, has been cloned and the amino acid sequence deduced from the nucleotide sequence.[7] YAP3p is induced in yeast mutants deficient in the *KEX-2* gene, which encodes the subtilisin-like serine proteinase, Kexin, that is normally involved in cleaving pro-α-mating factor, when this pheromone is overexpressed. YAP3p may be an alternative enzyme for processing pro-α-mating factor in yeast mutants lacking the *KEX-2* gene. YAP3p is a membrane-associated enzyme rendering it difficult to extract and purify. A recombinant *YAP3* gene has been engineered and expressed to produce a C-terminally, truncated form of YAP3p which is soluble. This enzyme has been purified and shown to process several mammalian prohormones specifically at paired or single basic residues.[8,9]

Proopiomelanocortin-converting enzyme (PCE) was the first prohormone processing enzyme identified as an aspartic proteinase. It is found in pituitary intermediate and anterior lobes where proopiomelanocortin (POMC) is synthesized.[1,4] PCE has been purified to apparent homogeneity from bovine intermediate lobe secretory vesicles, the intracellular site of POMC processing, and shown to cleave POMC at paired basic residues.[4] Treatment of mouse pituitary neurointermediate lobe with pepstatin A, an aspartic proteinase inhibitor, partially blocked the cleavage of endogenous POMC, supporting a physiological role of PCE in POMC processing *in vivo*.[10] PCE has also been purified from neural lobe secretory vesicles and

[2] D. F. Steiner, S. P. Smeekens, S. Ohagi, and S. J. Chan, *J. Biol. Chem.* **267**, 23435 (1992).
[3] N. G. Seidah, R. Day, M. Marcinkiewicz, and M. Chretien, *Ann. N.Y. Acad. Sci.* **680**, 135 (1993).
[4] Y. P. Loh, D. C. Parish, and R. Tuteja, *J. Biol. Chem.* **260**, 7194 (1985).
[5] T. J. Krieger and V. Y. H. Hook, *Biochemistry* **31**, 4223 (1992).
[6] R. Mackin, B. Noe, and J. Spiess, *Endocrinology* **129**, 1951 (1991).
[7] M. Egel-Mitani, H. P. Flygenring, and M. T. Hansen, *Yeast* **6**, 127 (1990).
[8] A. V. Azaryan, M. Wong, T. C. Friedman, N. X. Cawley, F. E. Estivariz, H.-C. Chen, and Y. P. Loh, *J. Biol. Chem.* **268**, 11968 (1993).
[9] N. X. Cawley, B. D. Noe, and Y. P. Loh, *FEBS Lett.* **332**, 273 (1993).
[10] Y. P. Loh, *FEBS Lett.* **238**, 142 (1988).

shown to process provasopressin.[11] It is likely that PCE is present in other endocrine and neuroendocrine tissues. PCE has been implicated as the mammalian homolog of YAP3p because it shares similar specificity with YAP3p and shows cross-reactivity with an antibody against YAP3p on Western blots.[12]

In the following sections, the methods for assay and purification, the physical properties, and the specificities of recombinant YAP3p and PCE are described. In addition, the role of glycosylation and the conformation of the substrate in influencing the specificity of the processing enzymes is discussed.

## Assay Methods for Yeast Aspartic Protease 3p and Proopiomelanocortin-Converting Enzyme

A rapid and sensitive assay has been developed for the purification of YAP3p and PCE on the basis of the principle that the enzymes cleave human $\beta$-lipotropin ($\beta_h$-LPH) at paired-basic residue sites to generate $\beta$-endorphin and $\beta$-melanocyte-stimulating hormone ($\beta$-MSH) and the finding that the products are soluble in 10% trichloroacetic acid (TCA) whereas the substrate is not.[8] This allows YAP3p and PCE enzymatic activity to be assayed as TCA-soluble counts per minute (cpm) generated from custom iodinated human $\beta$-lipotropin ([$^{125}$I]$\beta_h$-LPH). $\beta$-LPH is obtainable from the National Pituitary Agency (Baltimore, MD). Samples to be assayed are incubated in a final volume of 160 $\mu$l of 0.1 $M$ sodium citrate, pH 4.0, with [$^{125}$I]$\beta_h$-LPH ($\sim$10,000–20,000 cpm) for 30 min at 37°. The reaction is stopped by the sequential addition of 20 $\mu$l of ice-cold bovine serum albumin (BSA) (0.1% w/v final) and 20 $\mu$l of TCA (10% v/v final) and allowed to precipitate on ice for 30 min. After centrifugation at 10,000 $g$ for 10 min, two aliquots of 75 $\mu$l of the supernatant are counted in a gamma counter to determine TCA-soluble products generated.

For a more sensitive and quantitative assay for YAP3p, adrenocorticotropin hormone [ACTH(1–39)] can be used as a substrate. YAP3p has been shown to cleave ACTH at the tetrabasic residue site, Lys$^{15}$-Lys$^{16}$-Arg$^{17}$-Arg$^{18}$, to release ACTH(1–15) and the N-terminally extended corticotropin-like intermediate lobe peptide [CLIP(16–39)]. The products can be detected and quantitated by high-performance liquid chromatography (HPLC) analysis [C$_{18}$ column, with buffer A being 0.1% (v/v) trifluoroacetic acid (TFA), buffer B being 80% (v/v) acetonitrile/0.1% (v/v) TFA, and a

[11] D. C. Parish, R. Tuteja, M. Alstein, H. Gainer, and Y. P. Loh, *J. Biol. Chem.* **261**, 14392 (1986).
[12] Y. P. Loh, N. X. Cawley, T. C. Friedman, and L.-P. Pu, *in* "Aspartic Proteinases" (K. Takahashi, ed.), in press. Plenum, New York, 1995.

linear gradient from 20% B to 27% B in 30 min followed by 27% B to 37% B in 30 min being used to separate the products]. Ten micrograms of ACTH(1–39) is incubated with the enzyme in a volume of 100 $\mu$l of 0.1 $M$ sodium citrate, pH 4.0, for 30 min. The reaction is stopped with 10 $\mu$l of 1% TFA, and 50 $\mu$l is analyzed by HPLC. The products are detected and quantitated by absorbance at 214 nm, allowing specific activity to be calculated. PCE does not cleave ACTH(1–39), however, and therefore the assay cannot be used for PCE.

Enzyme Source

*Yeast Aspartic Protease 3p.* A plasmid, pYAP3LC, consisting of the 3′ truncated *YAP3* gene with a stop codon following base 1596[7], inserted into the pEMBLyex4 vector under the galactose promoter, was transformed into yeast cells, strain BJ3501.[8] The cells are grown in yeast nitrogen base (YNB) glucose medium to an $OD_{600-nm}$ of 0.8–1.0 at 30° with shaking, harvested by centrifugation at 2000 $g$ for 5 min, and resuspended in YNB galactose medium to an equivalent optical density. Both growth media lack amino acids but are supplemented with 0.02% (w/v) histidine. The cells are induced overnight, 12–20 hr, and then harvested as before. The pellet of yeast is resuspended in lysis buffer, 2 ml/g wet yeast (50 m$M$ HEPES, 1 m$M$ CaCl$_2$, pH 7.4), and lysed by at least six freeze–thaw cycles. The suspension of yeast cells is then centrifuged at 10,000 $g$ for 10 min to remove the cell debris. The supernatant obtained is used as the enzyme source for the purification of YAP3p. An alternate, richer source of YAP3p enzymatic activity is the growth medium of the induced cells.

*Proopiomelanocortin-Converting Enzyme.* A preparation of highly purified intermediate lobe secretory vesicles from 60 bovine pituitaries obtained by following the procedure in Loh *et al.*[13] is used as enzyme source. The vesicles are pelleted by ultracentrifugation at 100,000 $g$ for 30 min. The pellet is resuspended in 1 ml lysis buffer (10 m$M$ sodium succinate, pH 5.0) and lysed by six freeze–thaw cycles. The membranes are then pelleted by ultracentrifugation at 100,000 $g$ for 30 min. The supernatant obtained is used as the enzyme source for the purification of PCE.

Purification Procedure for Yeast Aspartic Protease 3p

A two-step procedure has been used to purify YAP3p to apparent homogeneity (Fig. 1).[8] The enzyme is first purified through a concanavalin A (Con A)-Sepharose affinity column and then subjected to chromatogra-

---

[13] Y. P. Loh, W. W. H. Tam, and J. T. Russell, *J. Biol. Chem.* **259**, 8238 (1984).

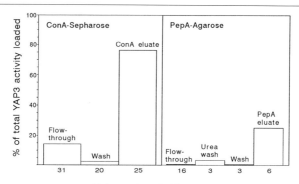

Volume of pooled fractions(ml)

Fig. 1. Typical enzymatic activity profiles of YAP3 from a yeast cell extract purified through Con A-Sepharose and pepstatin A-agarose columns. The bars show the enzymatic activity in each pooled fraction expressed as a percentage of total activity loaded on the column. The volume of each pooled fraction is indicated on the abscissa. Activity was assayed using [$^{125}$I]$\beta$-LPH as substrate. Note that 75–80% of the total loaded enzymatic activity is bound and recovered from the Con A column. Although the pepstatin A column typically gives only a 25–30% recovery of total loaded enzymatic activity, the eluted pepstatin A-bound YAP3 is highly pure.

phy on a pepstatin A (PepA)-agarose affinity column. Both steps are carried out at 4°. The following procedures contain some modifications which are an improvement to that published previously.[8]

*Concanavalin A Chromatography.* The soluble extract of the yeast cells or growth medium is diluted with 4 volumes of Con A equilibration buffer (10 m*M* Tris-HCl, 0.7 m*M* MgCl$_2$, 1 *M* NaCl, 0.1% (v/v) Triton X-100, pH 7.4). The enzyme solution is batch processed by addition to 1–5 ml of preequilibrated Con A-Sepharose beads (Pharmacia LKB Biotechnology Inc., Piscataway, NJ) followed by gentle mixing for 2 hr. The Con A-Sepharose beads are then used to pour a column, and the flow-through is collected. The column is washed with equilibration buffer until there is no detectable protein in the washes. For maximum recovery of enzymatic activity, the column is allowed to sit in 2–3 volumes of elution buffer (Con A equilibration buffer with 0.5 *M* methyl-$\alpha$-D-mannopyranoside) overnight. The eluent is then collected and assayed.

*Pepstatin A Chromatography.* The Con A eluate is brought to pH 4.0 by dialysis against 50 m*M* sodium citrate, pH 4.0, and then batch processed with 1–5 ml PepA-agarose beads (Pierce Chemical Co., Rockford, IL) equilibrated in 50 m*M* sodium citrate, pH 4.0. After 2 hr of gentle mixing, a column is poured and the flow-through is collected. Nonspecifically bound proteins are removed by washing with 3 volumes of equilibration buffer containing 6 $M$ urea. The urea is removed by washing the column with 3 volumes of equilibration buffer alone. The specifically bound protein is eluted with 50 m$M$ Tris-HCl, 2 $M$ LiBr, pH 8.6, and the pool of activity is desalted immediately on a PD10 (Sephadex G-25 M, Pharmacia LKB) column with 50 m$M$ ammonium bicarbonate/0.02% (v/v) Tween 20. The resulting desalted protein solution containing the purified YAP3p can be lyophilized and reconstituted without much loss of YAP3p activity.

## Purification Procedure for Pro-opiomelanocortin Converting Enzyme

The PCE activity can be purified to apparent homogeneity using the same procedure as for YAP3p above with a few modifications.[4,14] The Con A eluate is brought to pH 4.0 with 1 $M$ sodium citrate, pH 4.0, and immediately incubated with the PepA-agarose beads. Also, 1 $M$ NaCl and 0.1% Triton X-100 are present in the PepA equilibration and wash buffers.[14]

## Comments

The pepstatin A-agarose purchased from Pierce Chemical Co. has sometimes shown problems with the binding of YAP3p and PCE. This varied with lot numbers. An alternate procedure has recently been developed for YAP3p in which the pepstatin A column is replaced by a FPLC step using a MonoQ column (Pharmacia, Piscataway, NJ). The MonoQ column is run in 20 m$M$ Na phosphate buffer pH 7.0 with a linear salt gradient from 0–250 m$M$ NaCl in 20 min. YAP3p activity eluted between 100-200 m$M$ salt concentration.

## Properties of Yeast Aspartic Protease 3

*Physical Properties.* The YAP3p structure deduced from the nucleotide sequence of the *YAP3* gene consists of a 569-amino acid protein with a 21-residue signal peptide, a putative 45-amino acid proregion, a Ser/Thr-rich region, and 10 possible N-linked glycosylation sites (Fig. 2).[7] By alignment with other aspartic proteases, the two active-site aspartic acid residues in YAP3p are Asp-101 and Asp-371. The sequence around Asp-101 is typical for classic aspartic proteinases, Asp-Thr-Gly-(DTG), but in the case of Asp-371 the sequence is Asp-Ser-Gly (DSG). The *YAP3* sequence exhibits the highest degree of homology to the yeast *BAR1*-encoded "barrier" aspartic proteinase,[7] barrierpepsin.

---

[14] N. P. Birch, H. P. J. Bennett, F. E. Estivariz, and Y. P. Loh, *Eur. J. Biochem.* **201,** 85 (1991).

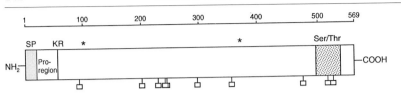

FIG. 2. Schematic representation of the structure of the YAP3 protein. SP, Signal peptide (amino acids 1–21); KR, potential zymogen activation site (Lys[66]-Arg[67]); Ser/Thr, serine/threonine-rich domain (amino acids 503–545); □, potential N-linked glycosylation sites; *, active-site aspartic residues (Asp[101], Asp[371]).

Recombinant YAP3p expressed in yeast is found in two forms, having $M_r$ values around 70,000 and 180,000 which are both enzymatically active. Both forms are present in yeast cell extracts although the relative amounts vary considerably depending on the length of the galactose induction time. However, the 180-kDa form is the predominant form found secreted into the growth medium. A minor 90-kD form has also sometimes been observed in the growth medium. The 70-, 90-, and 180-kDa forms are glycosylated, and upon endoglycosidase H treatment, they are reduced to a 65-kDa form which is close to the expected size of the protein backbone, although some O-linked sugars may be present (N. X. Cawley and Y. P. Loh, unpublished data). Presumably, the 180-kDa form is the mature form which has been extensively glycosylated.

*Stability.* In purified form YAP3p is an extremely stable enzyme. It can be stored in a neutral pH buffer at −20° for more than 10 months or for longer periods in a lyophilized form.

*pH Optimum.* Using [125I]$\beta_h$-LPH as substrate, YAP3p has a pH optimum between pH 4 and pH 5 for both forms of the enzyme.

*Inhibitor Profile.* The YAP3p enzyme is inhibited by pepstatin A, an aspartic proteinase inhibitor, but not by serine or thiol proteinase inhibitors (see Table I). The $K_i$ for pepstatin A determined using the 180-kDa form of YAP3p and Boc-Arg-Val-Arg-Arg-MCA as substrate is $4.8 \times 10^{-7}\ M$.

*Substrate Specificity.* The YAP3p enzyme is highly specific and selectively recognizes only those paired or monobasic sites within intact prohormones or prohormone fragments that are cleaved *in vivo*. YAP3p cleaves between or on the carboxyl side of paired basic residues, and on the carboxyl side of single basic residues. No difference in the substrate specificity of the 70- and 180-kDa forms of YAP3p were found with several substrates tested.

Yeast aspartic protease 3p can process POMC at paired basic residues to yield ACTH(1–39) and β-LPH; bovine N-POMC(1–77) to yield $\gamma_3$-MSH and Lys-$\gamma_3$-MSH; human ACTH(1–39) to yield ACTH(1–15) and Lys-Arg-CLIP; and human β-LPH to yield β-endorphin 1–31, β-endorphin 1–29,

TABLE I
EFFECT OF PROTEOLYTIC INHIBITORS ON ENZYME ACTIVITIES

| | | Inhibition[a] (%) | |
| Inhibitor | Concentration (mM) | YAP3p[b] | PCE[c] |
| --- | --- | --- | --- |
| p-Chloromercuribenzoate | 1.0 | 0 | 2 |
| Iodoacetamide | 1.0 | 0 | — |
| Dithiodipyridine | 0.1 | — | 2 |
| Phenylmethylsulfonyl fluoride | 1.0 | 0 | 2 |
| Diisopropyl fluorophosphate | 1.0 | — | 0 |
| $N^\alpha$-p-Tosyl-L-lysine chloromethyl ketone | 0.2 | 0 | 7 |
| Leupeptin | 1.0 | 5 | 58 |
| Pepstatin A | 0.1 | 85 | 100 |
| o-Phenanthroline | 0.1 | 0 | — |
| Captopril | 0.1 | 0 | — |

[a] Values are means from two experiments. Inhibition (%) = (cpm in products of control incubate − cpm in products of inhibitor-treated incubate)/(cpm in products of control incubate) × 100.
[b] YAP3p was assayed using human [$^{125}$I]β-LPH as substrate.
[c] PCE was assayed using [$^3$H]POMC as substrate.

β-endorphin 1–28, γ-LPH, and β-MSH (Fig. 3).[8] However, it does not cleave the Arg-Arg pair of γ$_3$-MSH to yield γ$_2$-MSH, which is consistent with the pattern of processing *in vivo* in the mammalian system.

The YAP3p enzyme can also cleave anglerfish prosomatostatin I (apSSI) at a pair of basic residues to yield anglerfish somatostatin-14 and can cleave

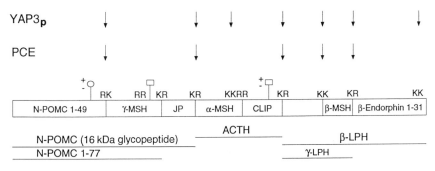

FIG. 3. Diagrammatic representation of bovine pro-opiomelanocortin. The arrows indicate the paired basic sites cleaved by PCE and YAP3p. □, N-linked glycosylation sites; ○, O-linked glycosylation site. MSH, melanocyte-stimulating hormone; JP, joining peptide; CLIP, corticotropin-like intermediate lobe peptide; LPH, lipotropin; END, endorphin; ACTH, adrenocorticotropin.

anglerfish prosomatostatin II (aPSSII) at a monobasic Arg residue, within a motif containing an Arg at the $-6$ position, to yield somatostatin$-28$.[9,15] Computer analysis of many prohormones has shown that single basic residue cleavage generally occurs within a motif which has another basic residue at the $-7$, $-6$, $-5$, $-4$, or $-3$ position.[1] YAP3p does not cleave the paired basic residue site of aPSSII to generate somatostatin-14, indicating that the enzyme is highly specific and recognizes not only the monobasic and paired basic residue sites but also the tertiary conformation in the two substrates, so as to cleave only the paired basic residue site in aPSSI and the monobasic residue site in aPSSII, as *in vivo*.

Porcine cholecystokinin-33 (CCK33) is cleaved by YAP3p at a specific monobasic Arg site which has an upstream Arg at the $-5$ position, to yield CCK 8.[15a] In addition, YAP3p can cleave the synthetic substrate Boc-Arg-Val-Arg-Arg-MCA between the Arg-Arg pair to yield Arg-MCA, but other peptides such as Boc-Gly-Lys-Arg-MCA and Boc-Gly-Arg-Arg-MCA are very poor substrates, suggesting that the upstream basic residue at the $-3$ position is important for cleavage specificity of the substrate.[8] The $K_m$ for Boc-Arg-Val-Arg-Arg-MCA was approximately 10-fold greater than the $K_m$ for ACTH(1–39), indicating that YAP3p prefers larger substrates. Such differences in substrate affinity are likely to be due to better conformation matching between the enzyme and a larger substrate than short synthetic substrates which lack inherent three-dimensional structure.

*Subcellular Localization.* Immunocytochemistry studies using an antibody against YAP3p show that the enzyme is localized in the periplasm of induced yeast cells.[12] That finding together with the presence of YAP3p in the growth medium indicates that YAP3p is localized in the secretory pathway of yeast cells and is not a vacuolar enzyme.

## Properties of Proopiomelanocortin-Converting Enzyme

*Physical Properties.* The enzyme PCE has been purified to apparent homogeneity from bovine intermediate lobe[4] and neural lobe[11] secretory vesicles and characterized as a glycoprotein of $M_r$ around 70,000 with an isoelectric point between 3.5 and 4.[4] PCE shows cross-reactivity with an antibody against YAP3p on Western blots suggesting that the two enzymes are immunologically related and likely share some structural similarity.[12]

*Stability.* Purified PCE can be stored at $-20°$ for 2–3 days.

[15] Y. Bourbonnais, J. Ash, M. Daigle, and D. Y. Thomas, *EMBO J.* **12,** 285 (1993).
[15a] N. X. Cawley, H-C. Chen, M. C. Beinfeld, and Y. P. Loh (1995), submitted.

*pH Optimum.* The enzyme has a pH optimum of pH 4.0–5.5, assayed using mouse POMC as substrate.

*Inhibitor Profile and Effects of Calcium Ions.* The activity of PCE is inhibited by two aspartic proteinase inhibitors, pepstatin A and diazocetyl-norleucine methyl ester, but not by thiol or serine proteinase inhibitors (Table I). The enzyme is stimulated by $Ca^{2+}$, although the extent of stimulation varies with different substrates.[14]

*Substrate Specificity.* Proopiomelanocortin converting enzyme is highly specific for paired basic residues of prohormones. The enzyme cleaves mouse POMC to form 21 to 23-kDa ACTH, ACTH(1–39), β-lipotropin, and β-endorphin 1–31[4]; bovine N-POMC(1–77) to yield (Lys) $\gamma_3$-MSH and N-POMC(1–49)[16]; and human β-lipotropin to yield β-MSH and β-endorphin 1–31 (see Fig. 3).[17] Analysis of the products generated from intact POMC and N-POMC(1–77) by PCE shows that PCE cleaves between and on the carboxyl side of paired basic residues. PCE, however, unlike YAP3p, does not cleave at the tetrabasic residues of ACTH(1–39) or the Lys-Lys pair of β-endorphin 1–31 to yield β-endorphin 1–28 or β-endorphin 1–29, or the Arg-Arg pair of $\gamma_3$-MSH, indicating selectivity in the paired basic residues cleaved. O-linked glycosylation at Thr-45 of N-POMC(1–77) greatly inhibits the ability of PCE to cleave the $Arg^{49}$-$Lys^{50}$ pair to form (Lys) $\gamma_3$-MSH and N-POMC(1–49).[16] This is perhaps due to stearic hindrance caused by formation of a salt bridge between the terminal sialic acid and Arg-49, since N-POMC(1–77) molecules that have a truncated sugar chain lacking sialic acids are cleaved by PCE. The cleavage specificity is consistent with the observation that N-POMC(1–49) found in the intermediate pituitary is never O-glycosylated at Thr-45.[18] This demonstrates that posttranslational modifications such as glycosylation of a substrate can play a critical role in determining the specificity of cleavage by PCE.

In addition, PCE can cleave provasopressin at a Lys-Arg pair to yield vasopressin-Lys-Arg[11] and proinsulin at Lys-Arg and Arg-Arg to yield the A and B chains of insulin, respectively.[4] The ability of PCE to cleave at monobasic residues has not been examined yet.

*Distribution and Subcellular Localization.* The PCE activity is found in the secretory vesicles of rat and bovine anterior, intermediate, and neural lobes of the pituitary. There is a soluble and membrane component of PCE

[16] N. P. Birch, F. E. Estivariz, H. P. J. Bennett, and Y. P. Loh, *FEBS Lett.* **290,** 191 (1991).
[17] Y. P. Loh, *J. Biol. Chem.* **261,** 11949 (1986).
[18] M. A. Seger and H. P. J. Bennett, *J. Steroid Biochem.* **25,** 703 (1986).

within the secretory vesicles, but there appears to be no difference in the specificity between the two forms.[19] PCE is secreted in a coordinately regulated manner with POMC-derived peptide hormones from intermediate lobe cells, further confirming the subcellular localization in secretory vesicles.[20] Presence of the enzyme in other endocrine/neuroendocrine tissues is likely but has not been studied thus far.

[19] F. E. Estivariz, N. P. Birch, and Y. P. Loh, *J. Biol. Chem.* **264**, 17796 (1989).
[20] M. G. Castro, N. P. Birch, and Y. P. Loh, *J. Neurochem.* **52**, 1619 (1989).

# [10] Proteinase A from *Aspergillus niger*

*By* KENJI TAKAHASHI

## Introduction

The mold *Aspergillus niger* var. *macrosporus* produces two kinds of extracellular acid endopeptidases namely, proteinases A and B (or Proctases A and B), and they are usually present as the minor and the major endopeptidase components, respectively, in the commercial crude enzyme powder (Proctase, Meiji Seika Kaisha Ltd., Tokyo).[1] Proteinase A is a nonpepsin-type acid proteinase, resistant to pepstatin and other specific inhibitors of pepsin family aspartic proteinases, and hence it is also called a pepstatin-insensitive acid proteinase and now has the recommended name aspergillopepsin II (EC 3.4.23.19). Proteinase B is a typical pepsin-type aspartic proteinase.[2] Proteinase A has a unique two-chain structure with no similarity in primary structure to ordinary aspartic proteinases.[3]

## Assay Method

*Principle.* The proteolytic activity of proteinase A is determined with bovine hemoglobin as the substrate essentially according to Anson[4] by measuring the absorbance at 280 nm of trichloroacetic acid-soluble hydroly-

[1] Y. Koaze, H. Goi, K. Ezawa, Y. Yamada, and T. Hara, *Agric. Biol. Chem.* **28**, 216 (1964).
[2] W.-J. Chang, S. Horiuchi, K. Takahashi, M. Yamasaki, and Y. Yamada, *J. Biochem. (Tokyo)* **80**, 975 (1976).
[3] K. Takahashi, H. Inoue, K. Sakai, T. Kohama, S. Kitahara, K. Takishima, M. Tanji, S. B. P. Athauda, T. Takahashi, H. Akanuma, G. Mamiya, and M. Yamasaki, *J. Biol. Chem.* **266**, 19480 (1991).
[4] M. L. Anson, *J. Gen. Physiol.* **20**, 565 (1937).

sis products. Casein can be also used as the substrate for the assay in a similar manner.

*Reagents*

Hemoglobin solution, 2.5% (w/v): Bovine hemoglobin (7.5 g) is dissolved in 100 ml of distilled water under magnetic stirring at 4° for several hours. The hemoglobin solution is then dialyzed against several changes of distilled water (4 liters each) at 4° for 2 days, and then filtered through filter paper. To the filtrate is added 30 ml of 1 *M* phosphoric acid buffer, pH 2.0, and the total volume is adjusted to 300 ml with distilled water. The 2.5% hemoglobin solution thus obtained is stored at 4°

Trichloroacetic acid (TCA), 5% (w/v)

*Procedure.* Hemoglobin solution (400 μl) is preincubated at 37° for 5 min, then mixed with 200 μl of an enzyme solution preincubated in the same way, and the mixture is kept at 37° for 30 min. To this is added 800 μl of 5% TCA, and the mixture is centrifuged at 12,000 rpm for 10 min. Blank sample is treated in the same manner using 200 μl of distilled water instead of enzyme solution. The absorbance at 280 nm of the supernatant is measured against the enzyme blank.

Purification Procedures

The enzyme is an acidic protein and therefore can be purified mainly by chromatography on various anion exchangers. The previous procedures included Sephadex G-75 gel filtration followed by ammonium sulfate fractionation and sulfoethyl-cellulose chromatography,[1] DEAE-Sephadex A-25 chromatography followed by Sephadex G-100 gel filtration,[5] and DEAE-Sephadex A-25 chromatography followed by ultrafiltration and DEAE-Sephadex A-50 chromatography.[6] The enzyme has also been purified by two steps of chromatography on DEAE-Toyopearl. Proteinase A can be separated from proteinase B by the first chromatography on DEAE-Toyopearl. At this stage, proteinase A is already fairly well purified, and it can be completely purified by the second chromatography on DEAE-Toyopearl. We use the following procedure with DEAE-Toyopearl for large-scale preparation of proteinase A, but it can be scaled down when necessary. All procedures (or at least Step 2) are performed at 4°.

*Step 1: First DEAE-Toyopearl Chromatography.* The crude enzyme powder (Proctase, 100 g) is dissolved in 900 ml of 50 m*M* HCl adjusted to

[5] K. Iio, Doctoral dissertation, University of Tokyo (1978).
[6] S. Kitahara, Doctoral dissertation, University of Tokyo (1982).

pH 3.0 with sodium acetate and applied to a column (11 × 50 cm) of DEAE-Toyopearl 650S equilibrated with the same solvent (Fig. 1a). The column is eluted with the same solvent at a flow rate of 1 liter/min. The proteinase A fractions are pooled, concentrated by ultrafiltration with concomitant 1 : 10 dilution with distilled water, and lyophilized. About 1.5 g of partially purified enzyme (crude Proctase A) can be obtained essentially free from proteinase B.

*Step 2: Second DEAE-Toyopearl Chromatography.* The crude proctase A (500 mg) is dissolved in 5 ml of 50 m$M$ sodium acetate adjusted to pH 4.5 with HCl (solvent A) and applied to a column (2.2 × 30 cm) of DEAE-Toyopearl 650S equilibrated with the same solvent. The column is eluted at a flow rate of 1 ml/min first with a linear gradient (200 ml) from solvent A to 50 m$M$ HCl adjusted to pH 3.0 with sodium acetate (solvent B), then with solvent B (Fig. 1b). The proteinase A fractions are pooled, dialyzed against several changes of distilled water adjusted to pH 4.5 with HCl at 4° for 2 days, and lyophilized. The yield of pure proteinase A is normally 350 mg.

Properties

*Purity and Physical Properties.* The molecular weight of the enzyme was estimated to be 19,200 by both sedimentation equilibrium and gel-

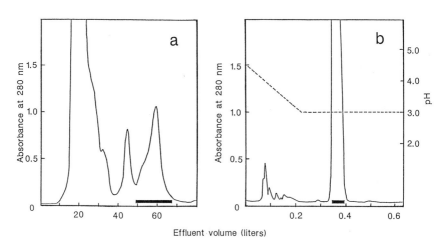

Fɪɢ. 1. Chromatographic purification of proteinase A on DEAE-Toyopearl. (a) First chromatography; (b) second chromatography. Solid line, absorbance at 280 mm; dashed line, pH. Fractions marked by bars were pooled.

TABLE I

AMINO ACID COMPOSITION OF PROTEINASE A

| | Residues per molecule | | |
|---|---|---|---|
| Amino acid | Proteinase A | L Chain | H Chain |
| Lys | 3 | 1 | 2 |
| His | 1 | 0 | 1 |
| Arg | 0 | 0 | 0 |
| Asp | 18 | 1 | 17 |
| Asn | 5 | 1 | 4 |
| Thr | 25 | 3 | 22 |
| Ser | 30 | 8 | 22 |
| Glu | 19 | 3 | 16 |
| Gln | 5 | 0 | 5 |
| Pro | 3 | 1 | 2 |
| Gly | 23 | 7 | 16 |
| Ala | 16 | 3 | 13 |
| Cys | 4 | 0 | 4 |
| Val | 21 | 4 | 17 |
| Met | 1 | 0 | 1 |
| Ile | 8 | 1 | 7 |
| Leu | 6 | 1 | 5 |
| Tyr | 11 | 3 | 8 |
| Phe | 8 | 1 | 7 |
| Trp | 5 | 1 | 4 |
| Total | 212 | 39 | 173 |
| Molecular weight | 22,265 | 3918 | 18,347 |

filtration methods,[7] whereas it was calculated from the amino acid sequence to be 22,265.[3] On sodium dodecyl sulfate (SDS)–polyacrylamide gel electrophoresis, in the absence of 2-mercaptoethanol, the enzyme gave two peptide bands, one stained densely and the other stained faintly with Coomassie Brilliant Blue, at positions corresponding to molecular weights of approximately 23,000 and 5000, respectively. The same results were obtained in the presence of 2-mercaptoethanol except that the former band appeared at a position corresponding to a molecular weight of approximately 33,000. The results indicate that the enzyme is composed of two polypeptide chains, a high molecular mass or heavy chain (H chain) and a low molecular mass or light chain (L chain), associated noncovalently with one another. As the molecular weights of the two polypeptide chains calculated from the amino acid sequences[3] are 18,347 and 3918, respectively (Table I), the migration

[7] S. Horiuchi, M. Honjo, M. Yamasaki, and Y. Yamada, *Sci. Pap. Coll. Gen. Educ. Univ. Tokyo* **19,** 127 (1969).

behavior of the H chain in the presence of 2-mercaptoethanol is apparently abnormal. The reason for this is not clear at present. The isoelectric point of the enzyme was estimated to be pH 3.3,[5] and the $E^{1\%}_{280\,nm,1\,cm}$ value of the enzyme solution was 13.3.[7]

The enzyme has a high solubility in aqueous solution and can be dissolved to a concentration of up to 10 m$M$ at pH 4.5. It can be crystallized into two different forms from an ammonium sulfate solution by the hanging-drop vapor diffusion method. One of the crystal forms belongs to the orthorhombic space group $P2_12_12_1$ with unit cell dimensions of $a$ = 54.7 Å, $b$ = 70.4 Å, and $c$ = 38.0 Å.[8] The solvent content was estimated to be 21%, which is among the lowest values for protein crystals reported.[9] The crystals diffracted X-rays to a resolution of at least 1.5 Å. The other crystal form also belongs to the orthorhombic space group $P2_12_12_1$ with unit cell dimensions of $a$ = 69.8 Å, $b$ = 87.6 Å, and $c$ = 60.8 Å.[10] The solvent content was estimated to be 41%, and the crystals diffracted to a resolution of at least 1.6 Å.

*Amino Acid Composition and Primary Structure.* Proteinase A is synthesized in the cell as a single-chain precursor form, prepro-proteinase A, of 282 amino acid residues (Fig. 2),[11] and secreted into the culture medium as a two-chain mature form. The two chains, L chain and H chain, were dissociated at pH 7 and above and separated by gel filtration, and the amino acid sequences were determined at the protein level (Fig. 2).[3] The amino acid compositions of the mature protein and constituent chains are given in Table I. The mature enzyme (212 residues) is composed of a 39-residue L chain noncovalently bound with a 173-residue H chain. Comparison of the primary structures of the precursor and the mature form indicates that three peptide bonds, Asn-Glu (residues 59–60), Tyr-Gly (residues 98–99), and Arg-Gln (residues 109–110), in the prepro form must be cleaved to produce the mature form, in addition to the signal peptide cleavage. The NH$_2$-terminal residue of the H chain is a pyroglutamic acid, which is thought to be formed from a glutamine residue after cleavage at the NH$_2$ side of the residue. The H chain contains two disulfide bonds. The mature protein is rich in acidic amino acids, containing 18 free aspartic and 19 free glutamic acid residues, and very low in basic amino acids, containing only 1 histidine

---

[8] M. Tanokura, H. Matsuzaki, S. Iwata, A. Nakagawa, T. Hamaya, T. Takizawa, and K. Takahashi, *J. Mol. Biol.* **223,** 373 (1992).

[9] B. W. Matthews, *J. Mol. Biol.* **33,** 491 (1968).

[10] M. Tanokura, H. Sasaki, T. Muramatsu, S. Iwata, T. Hamaya, T. Takizawa, and K. Takahashi, *J. Biochem. (Tokyo)* **114,** 457 (1993).

[11] H. Inoue, T. Kimura, O. Makabe, and K. Takahashi, *J. Biol. Chem.* **266,** 19484 (1991).

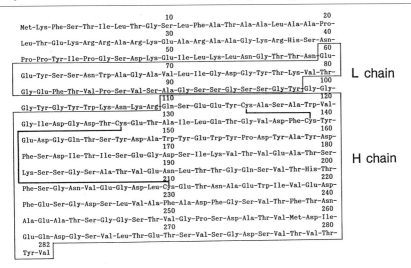

FIG. 2. Amino acid sequences of prepro-proteinase A and mature proteinase A. Prepro-proteinase A is a single chain protein of 282 residues. Mature proteinase A, shown as enclosed in the boxes, is composed of a 39-residue L chain (residues 60 to 98 in the preproenzyme) and a 173-residue H chain (residues 110 to 282 in the preproenzyme) associated noncovalently. The NH$_2$-terminal Gln (residue 110 in the preproenzyme) is converted to a pyroglutamic acid residue in the mature enzyme.

and 3 lysine residues per molecule. The high content of hydroxyamino acids (30 Ser and 25 Thr residues per molecule) is also notable.

The amino acid sequence of proteinase A has no similarity with those of the pepsin-type aspartic proteinases so far sequenced, and there is no Asp-Ser/Thr-Gly sequence that represents the consensus active-site sequences of the major families of aspartic proteinases. Sequence similarity is found only with another fungal enzyme, the acid proteinase B from *Scytalidium lignicola* (EC 3.4.23.32, scytalidopepsin B).[12] The sequence identity between the two proteinases is approximately 50%, indicating that the enzymes have evolved from the same ancestral protein; however, proteinase A is different from the *Scytalidium* enzyme in some important points.[3] First, proteinase A is a two-chain enzyme, whereas the *Scytalidium* enzyme is reported to be a one-chain enzyme. Second, proteinase A has two disulfide bonds, whereas the *Scytalidium* enzyme is reported to have three, and only one of the disulfide bonds is common to both enzymes.

[12] T. Maita, S. Nagata, G. Matsuda, S. Maruta, K. Oda, and D. Tsuru, *J. Biochem.* (*Tokyo*) **95**, 465 (1984).

Third, proteinase A has no similarity in the residues corresponding to the putative active-site residues of the *Scytalidium* enzyme.[13-15] Glutamic acid at position 53 and aspartic acid at position 98 in the *Scytalidium* enzyme are deduced to be active-site residues from chemical modification studies,[13-15] whereas the corresponding positions in the H chain of the present enzyme are occupied by a glutamine residue (at position 133 in Fig. 2) and a lysine residue (at position 181), respectively.[3]

*Secondary and Higher Structures.* From the circular dichroism (CD) spectrum of the enzyme, the contents of $\alpha$ helix, $\beta$ sheet, and $\beta$ turn were estimated to be approximately 2, 64, and 16%, respectively, at pH 5.5. The high content of $\beta$-sheet structure was also suggested from the two-dimensional nuclear magnetic resonance (NMR) spectrum of the enzyme. On incubation in $D_2O$ at pH 4.6 for 24 hr, about 100 amide protons remained unexchanged with deuterium, suggesting that the enzyme has a tightly packed core structure.[16]

*Stability.* The enzyme is fairly stable under weakly acidic conditions (especially between pH 3.0 and 5.5 and at low temperature), but autolysis takes place on prolonged standing in solution. The enzyme solution at pH 3.0–5.5 can be stored frozen for weeks and can be lyophilized without loss of activity. Above pH 6, the enzyme is very unstable, and rapid and irreversible inactivation takes place between pH 6 and 7 with concomitant dissociation of the two chains.[16] Similar inactivation with chain dissociation occurs in 3 *M* guanidine- hydrochloride at pH 5.5. So far, we have not been able to regenerate any of the activity or original conformation by mixing the two isolated chains. The enzyme activity is maximal at around 65° as measured using hemoglobin as the substrate at pH 2.5 and drops above that temperature; however, some activity is still observed at 80°.

*pH Profile.* The enzyme is optimally active at pH 1.1 toward hemoglobin as the substrate. About 35% of the maximum activity remains at pH 3.0, and a slight activity is still present at pH 4 to 6. A similar pH dependence was also observed with casein as the substrate, with an optimum of pH 1.7–1.8. The pH–activity profiles indicate that at least two carboxyl groups are involved in catalysis.

[13] D. Tsuru, S. Shimada, S. Maruta, T. Yoshimoto, K. Oda, S. Murao, T. Miyata, and S. Iwanaga, *J. Biochem.* (*Tokyo*) **99,** 1537 (1986).

[14] D. Tsuru, R. Kobayashi, N. Nakayama, and T. Yoshimoto, *Agric. Biol. Chem.* **53,** 1305 (1989).

[15] D. Tsuru, A. Naotuka, R. Kobayashi, T. Yoshimoto, K. Oda, and S. Murao, *Agric. Biol. Chem.* **53,** 2751 (1989).

[16] K. Takahashi, M. Tanokura, H. Inoue, M. Kojima, Y. Muto, M. Yamasaki, O. Makabe, T. Kimura, T. Takizawa, T. Hamaya, E. Suzuki, and H. Miyano, *in* "Structure and Function of the Aspartic Proteinases" (B. Dunn, ed.), p. 203. Plenum, New York and London, 1991.

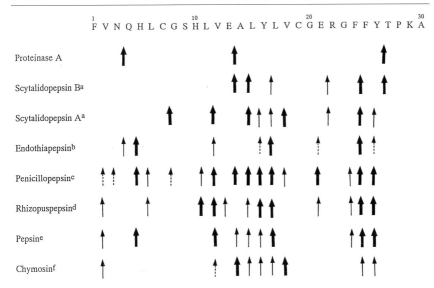

FIG. 3. Comparison of the cleavage specificity of proteinase A toward the B chain of oxidized insulin with those of some related acid proteinases. [a]K. Oda and S. Murao, *Agric. Biol. Chem.* **40**, 1221 (1976); [b]D. C. Williams, J. R. Whitaker, and P. V. Caldwell, *Arch. Biochem. Biophys.* **149**, 52 (1972); [c]G. Mains, M. Takahashi, J. Šodek, and T. Hofmann, *Can. J. Biochem.* **49**, 1134 (1971). (*S*-sulfo B chain was used as substrate); [d]D. Tsuru, A. Hattori, H. Tsuji, T. Yamamoto, and J. Fukumoto, *Agric. Biol. Chem.* **33**, 1419 (1969); [e]F. Sanger and H. Tuppy, *Biochem. J.* **49**, 481 (1951); [f]B. Foltmann, *C. R. Trav. Lab. Carlsberg* **34**, 319 (1964). The arrows indicate relative rates of cleavage for each peptide.

*Substrate Specificity.* The specificity of action of the enzyme toward several peptides, including oxidized insulin B chain and oxidized ribonuclease A, has been investigated, and the results are summarized in Figs. 3 and 4 and Table II. The enzyme hydrolyzes fairly specifically the three peptide bonds in oxidized insulin B chain (Fig. 3), namely, Asn-Gln (residues 3–4), Glu-Ala (residues 13–14), and Tyr-Thr (residues 26–27).[17] Among them, the Tyr-Thr bond is cleaved most rapidly, followed by cleavages at the Glu-Ala bond and the Asn-Gln bond in that order. On prolonged digestion, however, some additional cleavages may occur rather slowly. Essentially the same specificity was observed at pH 1.5 and 5.5. The cleavage specificity is quite different from that of porcine pepsin and related acid proteinases as shown in Fig. 3. Cleavages at the Asn-Gln and Tyr-Thr bonds are especially notable because those peptide bonds have not previously been found to be cleaved by pepsin-type aspartic proteinases except for the

[17] K. Iio and M. Yamasaki, *Biochim. Biophys. Acta* **429**, 912 (1976).

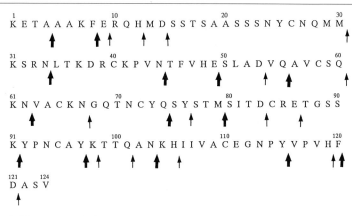

Fig. 4. Cleavage specificity of proteinase A toward oxidized ribonuclease A, measured by analysis of a 2-hr digestion mixture at pH 1.8 and 37° (K. Takahashi, in preparation). The arrows indicate relative rates of cleavage.

TABLE II

Specificity of Action of Proteinase A[a]

| Peptide substrate | Sequence and cleavage site | Half-life |
|---|---|---|
| 1. Tyr[8]-Substance P | R P K P Q Q F Y \| G L M (NH2) | <0.5 min |
| 2. Substance P | R P K P Q Q F F \| G L M (NH2) | 0.5 min |
| 3. pCl-Phe[7,8]-Substance P | R P K P Q QF* F* \| G L M (NH2) | 0.5 min |
| 4. Physalemin | ⟨ E A ↓ D P N K E Y \| G L M (NH2) | <2 min |
| 5. Substance P6–11 | Q F F \| G L M (NH2) | 3 min |
| 6. Oxidized insulin A chain | G I V E Q C C A S V C S L Y \| Q L E N Y C N | 3 min |
| 7. Substance P7–11 | F F \| G L M (NH2) | 20 min |
| 8. Neurotensin | ⟨ E L Y E N \| K P R R P Y I L | 40 min |
| 9. Eledoisin | ⟨ E P S K D A F \| I G L M (NH2) | 45 min |
| 10. Neurotensin 1–8[b] | ⟨ E L Y E ↓ N \| K P R | 2 hr |
| 11. Bradykinin | R P P G F \| S P F R | 3 hr |
| 12. Angiotensin II | D R V Y \| I H P F | 5 hr |
| 13. Val[5]-Angiotensin II | D R V Y \| V H P F | 16 hr |
| 14. Des-Asp[1]-Angiotensin II | R V Y \| I H P E | 50 hr |
| 15. Neurotensin 1–6 | ⟨ E L Y E ↓ N K | >50 hr |

[a] Data from E. Ido, T. Saito, and M.Yamasaki, *Agric. Biol. Chem.* **51**, 2855 (1987), and E. Ido, M. Yamasaki, and K. Takahashi, in preparation. Each peptide (0.9 m$M$) was digested at pH 1.5 and 30° with 0.8 $\mu M$ enzyme (peptides 1, 2, 5, 7, 12, 14) or 0.4 $\mu M$ enzyme (peptides 3, 4, 6, 8–11, 13, 15). F*, p-Chlorophenylalanine; ⟨E, pyroglutamic acid residue. Vertical arrows indicate sites of cleavage.
[b] The E–N and N–K bonds were cleaved in a mutually exclusive manner.

cleavage of the Asn-Gln bond by endothiapepsin.[18,19] It is also notable that the Tyr-Thr and Ala-Leu bonds, but not the Asn-Gln bond, were cleaved by the *Scytalidium* enzyme B, which also cleaved several other peptide bonds as well.[20] A synthetic peptide, $CH_3CO$-Arg-Gly-Phe-Phe-Tyr-Thr-Pro-Arg-Ala-OH, essentially corresponding to the COOH-terminal region of oxidized insulin B chain, was also cleaved specifically at the Tyr-Thr bond at pH 1.5 and 30° with $K_m = 0.42$ m$M$ and $k_{cat} = 0.14$ sec$^{-1}$.[6]

Table II summarizes the results obtained with proteinase A and some other smaller peptides. Most peptides were cleaved at only one site, and the rate of hydrolysis differed markedly among them. The results indicate that a stretch of at least four residues (positions P1 to P4) at the $NH_2$-terminal side of the cleavage site is recognized by the enzyme and that the residue at P1 position is most important. On the other hand, several peptidyl 4-methylcoumaryl-7-amide (MCA) derivatives have been tested, but in none has the X-MCA bond been cleaved so far, indicating that the residues at the COOH-terminal side of the cleavage site (P1′, P2′, etc.) are also important for hydrolysis.

Figure 4 summarizes the results of cleavage of oxidized ribonuclease A by the enzyme.[21] Various peptide bonds were cleaved to various extents. Among these, cleavages at Asn-X bonds are most notable, followed by those of Tyr-X, Phe-X, and Gln-X bonds. Other cleavages include Glu-X, Asp-X, His-X, Ala-X, Met($O_2$)-X, and Lys-X bonds, which were cleaved partially and/or rather slowly. No other peptide bonds, however, were cleaved to a detectable extent. Thus, the enzyme has a rather broad but still fairly restricted and distinct specificity.

*Inhibitors.* Water-soluble carbodiimides are potent inhibitors of the enzyme.[16] They are thought to react with the catalytic carboxyl groups at the active site. The enzyme is resistant to the usual aspartic proteinase inhibitors, such as diazoacetyl-DL-norleucine methyl ester in the presence of cupric ions, 1,2-epoxy-3-($p$-nitrophenoxy)propane, and pepstatin.[2] Furthermore, the enzyme seems to be resistant to all the ordinary proteinase inhibitors so far known.

[18] K. Borivoj, "Specificity of Proteolysis." Springer-Verlag, Berlin, Heidelberg, and New York, 1992.

[19] D. C. Williams, J. R. Whitaker, and P. V. Caldwell, *Arch. Biochem. Biophys.* **149,** 52 (1972).

[20] K. Oda and S. Murao, *Agric. Biol. Chem.* **40,** 1221 (1976).

[21] K. Takahashi, in preparation.

# [11] Thermopsin

## By XINLI LIN and JORDAN TANG

Introduction

Thermopsin is an extracellular acid proteinase from *Sulfolobus acido-caldarius,* a thermophilic archaebacterium which is best cultured in pH 2 at 70°, and, not surprisingly, thermopsin shows maximal activity under these conditions. In the native state, the enzyme is covalently attached to the bacterial cell wall. The biological function of thermopsin is assumed to be the digestion of protein substrates in the growth medium. The purification, characterization, and gene sequencing of thermopsin have been reported.[1,2] The heterologous expression of the enzyme has also been described.[3] In this chapter, we summarize the available information on the enzyme.

Assay for Thermopsin Activity

*Radiolabeled Bovine Hemoglobin as Substrate.* Routine assay of thermopsin activity is carried out using [14]C-methylated bovine hemoglobin as the substrate. Preparation of the substrate has been described elsewhere.[4] The principle of the assay is to digest radiolabeled hemoglobin, then determine the extent of digestion by the radioactivity from the peptides soluble in trichloroacetic acid. The amount of enzyme is proportional to the extent of the digestion.

The assay mixture contains 0.51% (w/v) substrate and thermopsin in 0.1 ml of 0.1 $M$ sodium formate, pH 3.2. After incubating the mixture in an Eppendorf tube at 80° for a period ranging from 5 to 30 min, depending on the level of activity used, an aliquot of 0.1 ml of 10% trichloroacetic acid is added to end the proteolysis. The tube is centrifuged for 2 min at full speed in an Eppendorf centrifuge. The radioactivity of an aliquot of the clear supernatant is determined in a scintillation counter. For each radioactive hemoglobin preparation, a standard activity curve should be prepared using different amount of enzyme in order to find out the linear range of the assay. Both the amount of enzyme and the incubation time can be adjusted to keep the assays in the linear range.

[1] X. Lin and J. Tang, *J. Biol. Chem.* **265,** 1490 (1990).
[2] M. Fusek, X. Lin, and J. Tang, *J. Biol. Chem.* **265,** 1496 (1990).
[3] X. Lin, M. Liu, and J. Tang, *Enzyme Microb. Technol.* **14,** 696 (1992).
[4] X. Lin, R. N. S. Wong, and J. Tang, *J. Biol. Chem.* **264,** 4482 (1989).

TABLE I
PURIFICATION OF THERMOPSIN FROM 400 LITERS OF *Sulfolobus Acidocaldarius*
GROWTH MEDIUM

| Step | Total protein[a] (mg) | Total enzyme[b] (mg) | Specific activity (mg enzyme/mg protein) | Yield (%) | Purification (-fold) |
|---|---|---|---|---|---|
| Cells | — | 41 | — | — | — |
| Medium | 7140 | 2.7 | $3.8 \times 10^{-4}$ | 100 | 1 |
| DEAE-Sepharose | 510 | 2.2 | $4.3 \times 10^{-3}$ | 81 | 11 |
| Phenyl-Sepharose | 70 | 1.4 | $2.0 \times 10^{-2}$ | 52 | 53 |
| Sephadex G-100 | 26 | 1.1 | $4.2 \times 10^{-2}$ | 41 | 110 |
| FPLC | 0.47 | 0.35 | 0.74 | 13 | 1947 |
| HPLC | 0.35 | 0.35 | 1 | 13 | 2632 |

[a] Measured by absorbance at 280 nm assuming 1 $A_{280}$ unit equals 1.2 mg protein/ml.
[b] Measured by proteolytic activity of thermopsin with purified thermopsin as standard.

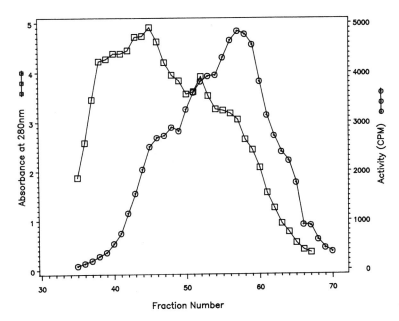

FIG. 1. DEAE-Sepharose CL-6B column chromatography of thermopsin. [14]C-Labeled hemoglobin was used for the enzyme assay. Fractions 51–63 were combined for further purification.

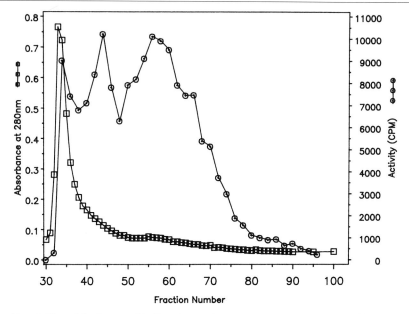

FIG. 2. Phenyl-Sepharose CL-4B column chromatography of thermopsin. [14]C Labeled hemoglobin was used for the enzyme assay. Fractions 50–74 were combined for further purification.

*Synthetic Substrates.* For kinetic studies, it is convenient to use synthetic peptide substrates. An appropriate substrate is Lys-Pro-Ala-Glu-Phe-Phe(NO$_2$)-Ala-Leu. Hydrolysis of the Phe-Phe(NO$_2$) bond by thermopsin is monitored spectrophotometrically at 300 nm. In a typical assay,[2] hydrolysis of the substrate (typically $1.9 \times 10^{-4}$ $M$ in 0.1 $M$ citric acid–HCl, pH 2.1) is followed in a Varian (Sunnyvale, CA) DMS 300 recording spectrophotometer equipped with a circulating water bath for temperature control of the reaction chamber (typically at 75°).

## Purification of Thermopsin

The majority of thermopsin activity is associated with the archaebacterial cell wall, with less than 10% of the enzyme activity being found in the culture medium. However, release of thermopsin from the cell wall has proved difficult, and the enzyme is therefore purified from the culture medium.

*Large-Scale Culture of Sulfolobus Acidocaldarius.* The thermophilic archaebacterium *Sulfolobus acidocaldarius* can be obtained from the Amer-

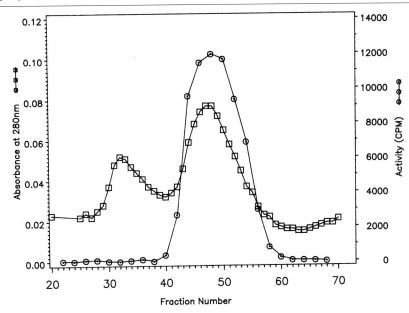

FIG. 3. Sephadex G-100 column chromatography of thermopsin. [14]C-Labeled hemoglobin was used in the enzyme assay. Fractions 42–56 were combined for further purification.

ican Type Culture Collection (ATCC, Rockville, MD). The following procedure is important for expanding the culture. On receiving the bacterium sample, an equal volume of ATCC medium 1256 is added and the bacteria incubated at 70° without shaking. After several days, the solution becomes turbid, indicating growth of the organisms. Further expansion of a 1:5 dilution and several 1:10 dilutions are done in sequence using the same medium. When a culture volume of 100 ml is reached, a 250-ml flask is used and gentle shaking is applied during the culture. For a 1-liter culture, a 3-liter flask is used.

A 4-liter culture of the archaebacterium is inoculated into 36 liters of autoclaved ATCC medium 1256, pH 2, in a 40-liter stainless steel container with temperature regulated at 70° (±2°). During the culture, gentle stirring is maintained by use of a motor-driven propeller, and a stream of air is bubbled through the medium. Growth is monitored spectrophotometrically at 540 nm. The cells are usually grown to sufficient density in 2 days. Because of the very low content of thermopsin in the growth medium, a very large amount of culture medium is needed for enzyme purification. By using two culture containers, about 250 liters of cell culture can be obtained each week.

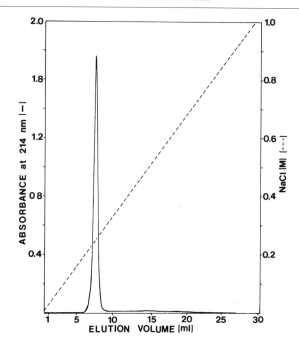

Fig. 4. Second FPLC Mono Q anion exchange of thermopsin.

To keep a productive bacterial stock, it is necessary to continue the small culture by 1 : 10 dilution every 3 days regardless of whether the large (40-liter) culture is being maintained. A 4-liter aliquot can be saved for inoculation of the next 40-liter culture. However, results appear inferior to those achieved by following the procedure above. Also, that alternative leads to progressive deterioration of the large culture.

*Purification of Thermopsin from Culture Medium.* A summary of a typical purification is shown in Table I. Four hundred liters of cell culture is filtered to separate medium from the cells in a Millipore (Bedford, MA) Pelican cassette system using a 0.45-$\mu$m filter. The clear filtrate, which usually contains 5–10% of the total proteolytic activity, is again filtered using the same apparatus fitted with a filter of 10,000 molecular weight cutoff. Toward the end of the filtering process, the solvent is also changed to 20 m$M$ Tris-HCl, pH 8.0, by several additions of the buffer. The final volume is about 1.5 liters.

The concentrated medium is centrifuged at 16,000 $g$ for 30 min at 4°, and the clear supernatant is applied to a 4.5 × 32 cm DEAE-Sepharose CL-6B column equilibrated with 20 m$M$ Tris-HCl, pH 8.0. The column is eluted with a linear gradient of 0 to 1 $M$ NaCl in 2 liters of the same buffer

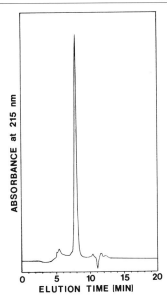

Fig. 5. Final HPLC gel filtration of thermopsin.

(Fig. 1). The fractions containing the thermopsin activity peak (eluted at about 0.4 $M$ NaCl) are pooled. A buffer of 1 $M$ sodium formate, pH 3.2, is added to the pooled enzyme solution to achieve a final sodium formate concentration of 0.25 $M$ and pH 3.2. The acidified crude enzyme solution is then incubated at 80° for 1 hr to destroy contaminating proteins by proteolysis.

The enzyme solution is applied to a 2.5 × 47 cm phenyl-Sepharose CL-4B column preequilibrated with 0.25 $M$ sodium formate, pH 3.2. The column is washed first with 4 liters of 0.1 $M$ sodium formate, pH 3.2, then eluted with 0.1 $M$ Tris-HCl, pH 8.0, to recover thermopsin (Fig. 2). The eluent is then concentrated by ultrafiltration to about 10 ml in an Amicon (Danvers, MA) apparatus fitted with a membrane to retain molecular weights above 10,000. The buffer is changed to 0.1 $M$ sodium formate, pH 3.2, by several additions of that buffer to the ultrafiltration apparatus.

The resulting acidic enzyme solution is again heated at 80° for 1 hr, cooled to room temperature, and applied to a 2.5 × 90 cm Sephadex G-100 column preequilibrated and eluted (flow rate 30 ml/hr) with a solution containing 20 m$M$ Tris-HCl, pH 8.0, 50 m$M$ NaCl, and 1% 2-propanol (Fig. 3). The active fractions from the gel-filtration chromatography are pooled and subjected to anion-exchange chromatography using a Mono Q column in a Pharmacia (Piscataway, NJ) FPLC (fast protein liquid chromatography)

```
CTCAATCTTCCGTGGGTGCTGTAACGAGACCTGTGATTAAGTGAATGCTTTATAGTTTATTGACGCTGTGATTGTTCGG   79
        .           .           .           .           .           .           .
                                                                  -41
                                                                  Met Asn Phe
TATCAAGCGATTTTTTCTTTATTTTTTAAAAGAAAGCTTATATACATGAAAATTATTTAAAAACTG ATG AAT TTT   154
        .           .           .           .           .           .           .
        -30                                                 -20
Lys Ser Ile Cys Leu Ile Ile Leu Leu Ser Ala Leu Ile Ile Pro Tyr Ile Pro Gln Asn
AAA TCC ATT TGT TTA ATA ATT TTG TTA AGT GCC TTA ATA ATA CCA TAT ATT CCA CAA AAT   214

            -10                                         1
Ile Tyr Phe Phe Pro His Arg Asn Thr Thr Gly Ala Thr Ile Ser Ser Gly Leu Tyr Val
ATC TAT TTT TTC CCT CAT CGT AAC ACA ACA GGA GCC ACA ATA AGC TCA GGT CTC TAT GTG   274

            10                                          20
Asn Pro Tyr Leu Tyr Tyr Thr Ser Pro Pro Ala Pro Ala Gly Ile Ala Ser Phe Gly Leu
AAT CCC TAT CTC TAC TAT ACA TCT CCT CCA GCT CCT GCA GGC ATT GCA TCA TTT GGA CTT   334

            30                                          40
Tyr Asn Tyr Ser Gly Asn Val Thr Pro Tyr Val Ile Thr Thr Asn Glu Met Leu Gly Tyr
TAT AAC TAC TCA GGT AAT GTG ACG CCA TAT GTA ATA ACT ACA AAC GAA ATG TTA GGA TAT   394

            50                                          60
Val Asn Ile Thr Ser Leu Leu Ala Tyr Asn Arg Glu Ala Leu Arg Tyr Gly Val Asp Pro
GTT AAT ATA ACC TCA TTA TTA GCG TAC AAT AGG GAA GCA CTA AGG TAT GGA GTA GAT CCC   454

            70                                          80
Tyr Ser Ala Thr Leu Gln Phe Asn Ile Val Leu Ser Val Asn Thr Ser Asn Gly Val Tyr
TAC AGT GCG ACT TTA CAG TTC AAC ATA GTA CTC TCT GTA AAC ACG AGC AAT GGA GTA TAT   514

            90                                          100
Ala Tyr Trp Leu Gln Asp Val Gly Gln Phe Gln Thr Asn Lys Asn Ser Leu Thr Phe Ile
GCG TAC TGG CTA CAG GAT GTT GGA CAG TTT CAA ACA AAC AAG AAT TCC TTA ACG TTT ATT   574

            110                                         120
Asp Asn Val Trp Asn Leu Thr Gly Ser Leu Ser Thr Leu Ser Ser Ser Ala Ile Thr Gly
GAT AAT GTA TGG AAT TTG ACA GGA AGT TTA TCA ACT TTA AGT TCA AGT GCA ATA ACA GGT   634

            130                                         140
Asn Gly Gln Val Ala Ser Ala Gly Gly Gly Gln Thr Phe Tyr Tyr Asp Val Gly Pro Ser
AAT GGG CAA GTT GCA TCC GCA GGT GGT GGA CAA ACC TTC TAT TAT GAT GTG GGA CCG TCT   694

            150                                         160
Tyr Thr Tyr Ser Phe Pro Leu Ser Tyr Ile Tyr Ile Ile Asn Met Ser Tyr Thr Ser Asn
TAC ACC TAT TCC TTC CCA CTC TCA TAT ATA TAC ATA ATA AAC ATG AGC TAC ACC AGT AAT   754

            170                                         180
Ala Val Tyr Val Trp Ile Gly Tyr Glu Ile Ile Gln Ile Gly Gln Thr Glu Tyr Gly Thr
GCA GTG TAT GTA TGG ATA GGT TAT GAG ATT ATA CAA ATA GGA CAA ACT GAA TAT GGT ACC   814

            190                                         200
Val Asn Tyr Tyr Asp Lys Ile Thr Ile Tyr Gln Pro Asn Ile Ile Ser Ala Ser Leu Met
GTA AAC TAC TAC GAT AAG ATA ACA ATA TAC CAG CCA AAC ATT ATC TCT GCA TCA CTT ATG   874

            210                                         220
Ile Asn Gly Asn Asn Tyr Thr Pro Asn Gly Leu Tyr Tyr Asp Ala Glu Leu Val Trp Gly
ATA AAC GGA AAC AAT TAC ACT CCA AAC GGT CTA TAT TAC GAT GCC GAA TTA GTT TGG GGT   934

            230                                         240
Gly Gly Gly Asn Gly Ala Pro Thr Ser Phe Asn Ser Leu Asn Cys Thr Leu Gly Leu Tyr
GGC GGA GGA AAC GGT GCA CCA ACA TCA TTT AAC TCA TTA AAT TGT ACA TTA GGG CTA TAC   994

            250                                         260
Tyr Ile Ser Asn Gly Ser Ile Thr Pro Val Pro Ser Leu Tyr Thr Phe Gly Ala Asp Thr
TAT ATC AGT AAT GGT AGC ATA ACT CCT GTT CCT TCA CTG TAC ACT TTT GGC GCC GAC ACA   1054

            270                                         280
Ala Glu Ala Ala Tyr Asn Val Tyr Thr Thr Met Asn Asn Gly Val Pro Ile Ala Tyr Asn
GCC GAG GCT GCA TAT AAC GTC TAC ACA ACA ATG AAC AAC GGA GTG CCT ATA GCA TAT AAT   1114

            290                                         299
Gly Ile Glu Asn Leu Thr Ile Leu Thr Asn Asn Phe Ser Val Ile Leu Ile ***
GGC ATT GAG AAT CTT ACC ATA TTG ACA AAT AAC TTC AGT GTT ATA CTA ATT TAA TTGTTGT   1175

CCAAGCCAAAAATAAGGAAGAGAAAAAAGGGAAATGTGCATTCATTTTTTACTTTTTCCCACTGTTCTTTACCTATTTC   1254

ATTGTAGCTGGCTCATTACTCTCTATAAATCAATATGATTTAGGAAGATTCAATATATGATTTGCAATGTATGATAAGA   1333

CTAAATTCTGAGAGACCGGTGCTACCTTGTAAAGCCTTGTCTCCCTCAACTTCCTCTCTATTCCAGTATCAACAGCATA   1412

TCCATACCCTCCGTAGGTATCCATGGCAA  1441
```

Fig. 6. Nucleotide and amino acid sequences of thermopsin. The thermopsin gene is coded between nucleotides 146 and 1168. Nucleotide numbers are shown on the right at the end of each line. The underlined triplets in the nucleotide sequence indicate the region recognized

apparatus. The column is equilibrated with 20 m$M$ 1-methylpiperazine, 1% (v/v) 2-propanol, pH 4.5. A linear gradient from 0 to 1 $M$ NaCl in 30 min with a flow rate of 1 ml/min is employed for thermopsin elution, which usually occurs at 0.25 $M$ NaCl (Fig. 4). The collected active thermopsin fractions are adjusted to pH 3.2 by the addition of 1 $M$ sodium formate, pH 3.1. The solution is heated at 80° for 1 hr. The heated enzyme is once more subjected to FPLC Mono Q purification. The active fractions are pooled and subjected to a final step of purification using high-performance liquid chromatography (HPLC) gel filtration in a 7.5 × 300 mm column (TSK G3000SW) equilibrated and eluted with 0.1 $M$ ammonium bicarbonate, pH 8.1 (Fig. 5).

The purified enzyme is homogeneous by the criteria of chromatographic behavior, sodium dodecyl sulfate–polyacrylamide gel electrophoresis (SDS–PAGE), and N-terminal sequence determination.[1]

Properties of Thermopsin

*Amino Acid Sequence.* The amino acid sequence of the enzyme was deduced from the DNA sequence of the thermopsin gene.[1] As shown in Fig. 6, the enzyme consists of 299 residues. A 41-residue stretch in front of the thermopsin sequence apparently contains the leader and possibly a "pro" sequence. The thermopsin sequence shows no similarity with other sequences in GenBank/EMBL and NBRF Databanks, including those for aspartic proteinases. There are 11 potential N-glycosylation sites. Two of them (Asn-24 and Asn-28) revealed no positive identification in Edman degradation studies, suggesting that they are glycosylated.

*Enzyme Properties.* The optimal activity is observed near pH 2 (Fig. 7). Under high sensitivity assay conditions, activity can be measured up to pH 12. Thermopsin loses activity during lyophilization, but it is stable in solution at 4° over several months. The temperature for maximal activity is about 90° (Fig. 8) although low activity can be detected at 13°. Thermopsin is competitively inhibited by pepstatin with a $K_i$ of 2 × $10^{-7}$ $M$ at 76°. The enzyme is also inactivated by the aspartic proteinase inhibitors diazoacetylnorleucine methyl ester (DAN) and 1,2-epoxy-3-(*p*-nitrophenoxy)propane (EPNP).[2] The facts suggest that thermopsin may have a catalytic site

---

by the synthetic oligonucleotide probe. The amino acid residue numbers are placed directly above the residues. The NH$_2$-terminal position of the mature enzyme is residue 1. The NH$_2$-terminal amino acid residues confirmed by Edman degradation are underlined. The amino acid residues that precede the thermopsin NH$_2$-terminal position are numbered in negative numbers in the reversed direction. Potential transcription termination signals and promoters are indicated by solid underlining. The potential ribosome binding site is boxed.

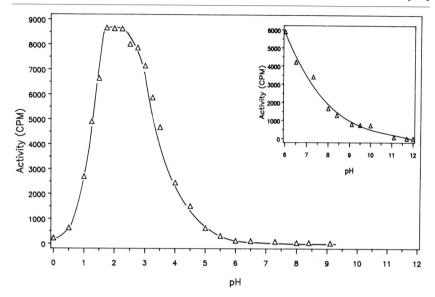

Fig. 7. Dependence of the proteolytic activity of thermopsin on pH. *Inset:* Residual activity of thermopsin between pH 6 and pH 12 using a higher sensitivity assay.

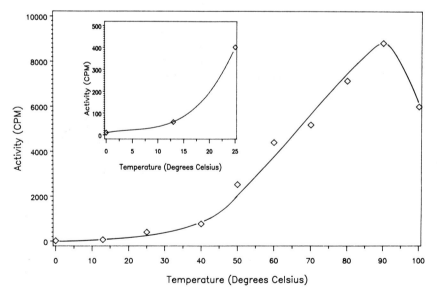

Fig. 8. Dependence of the proteolytic activity of thermopsin on temperature. *Inset:* Residual activity of thermopsin between 0° and 25° using a higher sensitivity assay as described in the text.

related to that of the aspartic proteinases. Inhibition by mercury acetate but not by iodoacetamide suggests that the inhibition of former is not mediated by the reaction to a sulfhydryl group on thermopsin.

The preference for residues at the cleavage sites appears to be for those with large hydrophobic side chains. The following bonds in insulin B chain are cleaved by thermopsin: Leu-Val, Leu-Tyr, Phe-Phe, Phe-Tyr, and Tyr-Thr (2). For hydrolysis of the synthetic peptide substrate described above, the $K_m$ is $5.3 \times 10^{-5}$ $M$ and the $k_{cat}$ is 14.3 sec$^{-1}$. Kinetic results also suggest that urea, acetamide, and phenylalaninamide are competitive inhibitors by virtue of structural similarity to a peptidic structure. Kinetic data collected at different temperatures revealed that the enzyme is most efficient (highest $k_{cat}/K_m$ values) between 50° and 65°. Above that range, the $K_m$ increases sharply.

## Synthesis and Purification of Recombinant Thermopsin

The heterologous expression of recombinant thermopsin in *Escherichia coli* or in yeast would permit the study of structure–function relationships by site-directed mutagenesis. Although we have succeeded in producing recombinant thermopsin, the yield is too low to be of practical use. The following description, however, provides a basis for further improvement of recombinant enzyme synthesis.

*Thermopsin Gene.* The thermopsin gene, about 1.1 kbp, has been cloned from *S. acidocaldarius*,[1] and the DNA sequence is shown in Fig. 6.

*Expression of Thermopsin as Fusion Protein.* Various constructs have been made in order to express thermopsin either in the cytosol or in the periplasmic space of *E. coli,* including the use of a strong T7 promoter.[5] However, none produced detectable thermopsin, probably owing to rapid degradation. We then used the strategy to express thermopsin as a fusion protein in *E. coli* as inclusion bodies, thus preventing it from being degraded. Recombinant thermopsin is refolded from urea and purified for further studies. Pepsinogen was chosen as the fusion partner because it can be expressed as inclusion bodies at a high level.

The fusion gene consists of the 5′ 1110-bp *Eco*RI–*Bam*HI fragment of the pepsinogen gene[4] connected in phase to the coding sequence of the mature thermopsin gene (Fig. 9). Construction of the expression vector for the pepsinogen–thermopsin (PP-TH) fusion protein is as follows. The pepsinogen gene in an expression vector, pGBT-T19-PP,[4] is cut with *Eco*RI, filled-in with Klenow, and blunt-end ligated into the *Bam*HI site of the Klenow filled vector pET-3b. This generates a recombinant pepsinogen

[5] F. M. Studier, A. H. Rosenberg, J. J. Dunn, and J. W. Dubendorff, this series, Vol. 185, p. 60.

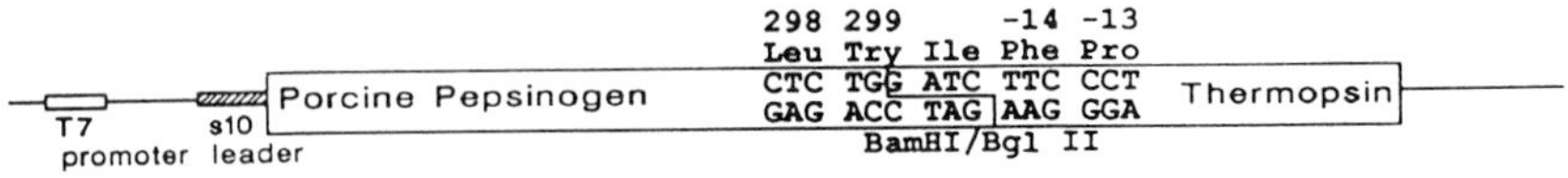

```
                                    298 299        -14 -13
                                    Leu Try Ile Phe Pro
─────────┬─────────────────────┐    CTC TGG ATC TTC CCT   ┌──────────┬───────
─────────┤ ▨▨▨▨ Porcine Pepsinogen │    GAG ACC TAG│AAG GGA   │ Thermopsin │
T7       s10└─────────────────────┘          BamHI/Bgl II  └──────────┴───────
promoter leader
```

FIG. 9. Gene construct for the expression of thermopsin as a fusion protein to pepsinogen.

that has 16 extra amino acids on the N terminus. When transfected into *E. coli* strain BL21(DE3),[5] the vector, pET-3b-PP, directs the synthesis of about 200 mg of pepsinogen per liter of cell culture in the form of inclusion bodies. The thermopsin gene is modified by oligonucleotide-directed mutagenesis for the addition of a starting codon before residue-14,[1] which is immediately preceded by a *Bgl*II site. Ligating of the *Bgl*II fragment of the thermopsin gene to the unique *Bam*HI site of the pepsinogen gene in pET-3b-PP results in pepsinogen–thermopsin fusion in which the correct reading frame is preserved (Fig. 9). The resulting hybrid gene, pET-3b-PP-TH, has the N-terminal 299 amino acids of pepsinogen followed by an Ile residue created in the fusion and the modified thermopsin gene. The fusion gene for thermopsin shown in Fig. 9 is transformed into an expression host, *E. coli* BL21(DE3),[5] for expression.

*Refolding of Thermopsin Fusion Protein.* A colony of *E. coli* BL21(DE3) harboring pET-PP-TH is picked and grown in 100 ml of ZB[5]/ampicillin (50 mg/liter) medium overnight. Ten milliliters of the overnight culture is inoculated into 1 liter of LB/ampicillin (50 × mg/liter) and grown at 37° to an optical density of 0.6 at 600 nm. Isopropylthiogalactoside (IPTG) is then added to a final concentration of 0.5 m$M$, and the culture is continued for another 3 hr.

The cells are harvested by centrifugation at 3500 $g$ for 30 min and resuspended in 20 ml of TN/lysozyme buffer (50 m$M$ Tris-HCl, 0.15 $M$ NaCl, 100 mg lysozyme, pH 7.4). After freeze and thaw, 0.6 ml of 1 $M$ MgCl$_2$ and 0.2 mg of DNase are added, and the cells are ruptured by sonication. The homogenized solution is diluted to 250 ml with TN buffer

TABLE II
YIELD OF RECOMBINANT THERMOPSIN AT DIFFERENT PURIFICATION STEPS

| Preparation | Total activity (cpm) | Activity relative to RD-1 (%) |
|-------------|----------------------|-------------------------------|
| RD-1        | 156,200              | 100                           |
| RD-2-C      | 660,362              | 420                           |
| RD-2-F      | 106,510              | 93                            |
| RD-3-C      | 624,373              | 400                           |
| RD-3-CH     | 546,738              | 350                           |

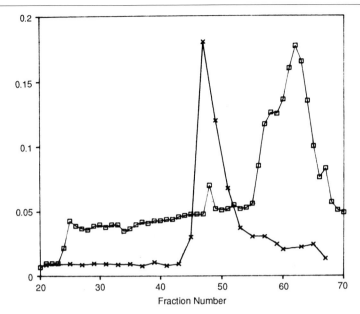

Fig. 10. Sephacryl S-300 column chromatography of recombinant thermopsin. The pepsino-gen–thermopsin fusion protein was expressed as inclusion bodies, refolded, partially purified, and incubated at pH 3.5, 70°, to digest the pepsinogen moiety, then loaded onto the column. □, Absorbance at 280 n$M$. x, Activity in cpm/300,000.

containing 1% Triton X-100 and stirred at 20° for 1 hr. The solution is centrifuged at 16,000 $g$ for 20 min. The resulting pellet is resuspended in TN/1% Triton buffer and washed once more. The final pellet is dissolved in 8 $M$ urea, 20 m$M$ Tris-HCl, 1 m$M$ EDTA, 1 m$M$ glycine, 100 m$M$ 2-mercaptoethanol, pH 8.0, to a volume of 20 ml. After stirring overnight at 4°, the residual insoluble material is removed by centrifugation at 289,000 $g$ for 1 hr at 4°. The clear supernatant was recovered.

The supernatant is mixed with 9 volumes of 8 $M$ urea/50 m$M$ CAPS [3-(cyclohexylamino)propanesulfonic acid], pH 11.5, and diluted, with efficient stirring, into 100 volumes of 10 m$M$ Tris, pH 9.0. The solution, RD-1, is kept at room temperature for 2 hr and then at 4° overnight. Solution RD-1 is concentrated in an Amicon ultrafiltration cell with a molecular mass cutoff of 10,000. The retained solution, about 5 ml, is referred to as RD-2-C, and the flow-through (filtrate) is referred to as RD-2-F. RD-2-C is mixed with 0.5 ml of 1 $M$ sodium formate, pH 3.2, and some insoluble material is removed by centrifugation. The resulting supernatant is referred to as RD-3-C. Glycerol is then added to RD-3-C to a final concentration of 10% (v/v). The solution is heated at 70° for 35 min and is termed RE-

3-CH. Thermopsin activity at each step is assayed at high temperature using the highly sensitive radioactive assay described above. RE-3-CH is then applied to a $3 \times 90$-cm column of Sephacryl S-300, which had been equilibrated and eluted with 20 m$M$ 1-methylpiperazine, 5% (v/v) glycerol, 0.1% (v/v) 2-propanol, 0.1 m$M$ EDTA, 0.1 $M$ NaCl, pH 4.5. The flow rate is 37 ml/hr, and 8.6-ml fractions are collected.

The vector pET-3b-PP-TH directs the synthesis of a 70-kDa fusion protein, which could be detected either by Coomassie blue staining of the purified inclusion bodies or by Western blot analysis using antipepsin antiserum as probe. Contributions to the fusion protein are about 35 kDa from pepsinogen and 35 kDa from thermopsin. The expression level of the fusion protein is about 4 mg per liter of culture, which is about 50 times lower than that of pepsinogen expression itself. The activities generated from the above-described refolding steps are shown in Table II. Final chromatography of the material on a column of Sephacryl S-300 (Fig. 10) produces an active thermopsin peak. The elution position of the peak corresponds to about 35 kDa, indicating that during the incubation at 70°, pH 3.2, pepsinogen polypeptide in the fusion protein was indeed digested and eliminated from the recombinant thermopsin molecule.

*Properties of Recombinant Thermopsin.* Because the refolded recombinant thermopsin has not been purified to homogeneity, only limited properties have been determined. The maximal activity of recombinant thermopsin was found in the range of 70° to 80°. The temperature–activity profile is similar to that of unpurified native thermopsin (Fig. 8).

Potential Industrial Use of Thermopsin

As the majority of industrial enzymes are thermostable ones, thermopsin appears to have potential for industrial use. An advantage of thermopsin is its unique stability under conditions of low pH and high temperature where most other proteins are denatured and thus will be digested by the enzyme. The fact that thermopsin is active in sodium dodecyl sulfate up to 1.1 $M$ may also enhance its usefulness.

## [12] Bacterial Prolipoprotein Signal Peptidase

*By* KRISHNAN SANKARAN and HENRY C. WU

### Introduction

Since the discovery and structural elucidation of the major outer membrane protein of *Escherichia coli* (also known as Braun's or murein lipoprotein),[1,2] more than 130 lipoproteins have been identified in bacteria.[3] The common feature shared by this structurally and functionally diverse group of membrane proteins in bacteria is the presence of an extensively lipid-modified N-terminal amino acid, namely, cysteine substituted on sulfur with the $CH_2(OOCR)$ $CH(OOCR')CH_2$- (diacylglyceryl) group. The lipoproteins are initially synthesized as prolipoproteins[4] with signal sequences typically consisting of a positively charged N-terminal region followed by a stretch of uncharged and hydrophobic amino acids, and ending in a sequence called the lipobox, which in the majority of known lipoproteins is -Leu-(Ser or Ala)-(Gly or Ala)-Cys-.[3,5] That sequence seems to determine the recognition site for the prolipoprotein modification and processing enzymes. As might be expected, the N-terminal cysteine of the mature proteins is invariant among all the lipoproteins.

According to the postulated pathway for lipoprotein maturation based on both *in vivo* and *in vitro* studies using *E. coli* murein prolipoprotein as the prototype model, diacylglyceryl modification of the cysteine residue in the lipobox of lipoprotein precursors precedes the processing of the lipid-modified prolipoproteins by a specific endopeptidase called prolipoprotein signal peptidase (or signal peptidase II, EC3.4.99.35).[6] From the ensuing biochemical evidence[7–9] it became clear that the lipid modification is actually a prerequisite for the cleavage of the signal peptide by signal peptidase II. Among the enzymes involved in the pathway, signal peptidase II has

[1] V. Braun and K. Rehn, *Eur. J. Biochem.* **10**, 426 (1969).
[2] K. Hantke and V. Braun, *Eur. J. Biochem.* **34**, 284 (1973).
[3] V. Braun and H. C. Wu, *in* "Bacterial Cell Wall" (New Comprehensive Biochemistry, Volume 27) (J.-M. Ghuysen and R. Hackenbeck, eds.), p. 319. Elsevier, Amsterdam, 1993.
[4] S. Inouye, S. Wang, J. Sekizawa, S. Halegoua, and M. Inouye, *Proc. Natl. Acad. Sci. U.S.A.* **74**, 1004 (1977).
[5] S. Hayashi and H. C. Wu, *J. Bioenerg. Biomembr.* **22**, 451 (1990).
[6] M. Tokunaga, H. Tokunaga, and H. C. Wu, *Proc. Natl. Acad. Sci. U.S.A.* **79**, 2255 (1982).
[7] M. Tokunaga, J. M. Loranger, and H. C. Wu, *J. Biol. Chem.* **259**, 3825 (1984).
[8] I. K. Dev and P. H. Ray, *J. Biol. Chem.* **259**, 11114 (1984).
[9] H. Yamada, H. Yamagata, and S. Mizushima, *FEBS Lett.* **166**, 179 (1984).

been studied in greatest detail. Its amino acid sequence has been deduced from the nucleotide sequence of the cloned gene[10,11]; it has been purified to apparent homogeneity[8] and some of its properties studied[7,8,12]; and its inner membrane localization[7] and membrane topology[13] have been established. Sequence comparison among signal peptidase II from *Escherichia coli* and other bacteria has revealed a similar membrane topology and the existence of highly conserved regions of the enzyme.[14–16]

## Assay for Signal Peptidase II

*Strain.* *Escherichia coli* B is the strain used for preparing diacylglyceryl-modified prolipoprotein, the substrate for signal peptidase II.

*Media.* Growth media used are Luria broth or M9 minimal medium.

### Reagents

Globomycin, a cyclic peptide antibiotic obtained from Dr. M. Arai of Sankyo Co. (Tokyo, Japan), is prepared as a stock solution (10 mg/ml) in dimethyl sulfoxide (DMSO) and stored in aliquots at −20°.

[35S]Methionine (Tran[35]S-label from ICN Biomedicals, Costa Mesa, CA), specific activity about 1000 Ci/mmol

Trichloroacetic acid (TCA), 50% (w/v)

Triton buffer[8]: 50 m$M$ Tris-HCl, pH 8.0, containing 0.1 m$M$ EDTA, 0.15 $M$ NaCl, 1 mg/ml bovine serum albumin (BSA) and 2% (v/v) Triton X-100

50 m$M$ Sodium phosphate buffer, pH 7.0, containing 1% (w/v) sodium dodecyl sulfate (SDS)

Immunoaffinity column: $\gamma$-globulin fraction is obtained from antilipoprotein antiserum by sodium sulfate precipitation or any one of the standard procedures, then coupled to CNBr-activated Sepharose CL-4B (Pharmacia, Piscataway, NJ) according to the procedure given by the manufacturer

0.2 $M$ Glycine hydrochloride buffer, pH 2.2, containing 0.1% (v/v) Triton X-100

[10] M. A. Innis, M. Tokunaga, M. E. Williams, J. M. Loranger, S. Y. Chang, S. Chang, and H. C. Wu, *Proc. Natl. Acad. Sci. U.S.A.* **81,** 3708 (1984).

[11] F. Yu, H. Yamada, K. Daishima, and S. Mizushima, *FEBS Lett.* **173,** 264 (1984).

[12] I. K. Dev, R. J. Harvey, and P. H. Ray, *J. Biol. Chem.* **260,** 5891 (1985).

[13] F. J. Muñoa, K. W. Miller, R. Beers, M. Graham, and H. C. Wu, *J. Biol. Chem.* **266,** 17667 (1991).

[14] L. Isaki, M. Kawakami, R. Beers, R. Hom, and H. C. Wu, *J. Bacteriol.* **172,** 469 (1990).

[15] L. Isaki, R. Beers, and H. C. Wu, *J. Bacteriol.* **172,** 6512 (1990).

[16] X. Zhao and H. C. Wu, *FEBS Lett.* **299,** 80 (1992).

Assay buffer: 50 m$M$ Tris-HCl (pH 7.4), containing 0.25% (w/v) Nikkol (octaethylene glycol mono-$n$-dodecyl ether, Nikko Chemical Co., Tokyo, Japan) or 0.25% (v/v) Triton X-100, and 1 m$M$ dithiothreitol (DTT)
All chemicals are of analytical grade.

*Principle of Assay.* Like a typical signal peptidase assay, the method is based on the separation of the substrate (the preproteins) and the product (the processed form) by sodium dodecyl sulfate–polyacrylamide gel electrophoresis (SDS–PAGE) by virtue of the molecular weight difference (2000–3000). In the case of prolipoprotein signal peptidase, the commonly used substrate is [35S]methionine-labeled diacylglyceryl-modified murein prolipoprotein, prepared from globomycin-treated *E. coli* B cells. The labeled substrate is purified from the crude cell envelope fraction either by immunoprecipitation[7] or by an immunoaffinity method[8] using antilipoprotein antibody. In the semiquantitative gel assay, the substrate (diacylglyceryl prolipoprotein) and the product (the apolipoprotein) are separated in a high-percentage polyacrylamide gel. In a much simpler assay developed by Dev and Ray,[8] the released [35S]methionine-labeled signal peptide is recovered in the acetone supernatants of the reaction mixtures and counted.

*Preparation of Substrate.* Globomycin blocks lipoprotein maturation at the cleavage step by specifically inhibiting signal peptidase II.[17,18] This results in the accumulation of diacylglyceryl-modified prolipoproteins. *Escherichia coli* B, a strain that is hypersensitive to globomycin, is grown in M9 medium supplemented with glucose (0.2%) and thiamin (1 $\mu$g/ml) to an $A_{600\,nm}$ of 0.6, when globomycin (10 $\mu$g/ml, final) is added. After 10 min, [35S]methionine (10 $\mu$Ci/ml of culture) is added and the incubation continued for 20 min. Labeling is stopped with either TCA (5%, final) or prechilled acetone (50%, v/v, final). The pellet is washed once with 50% acetone, resuspended in one-tenth the volume of the culture with 50 m$M$ sodium phosphate buffer (pH 7.0) containing 1% SDS, and heated at 100° for 5 min. The lysate is then diluted 1:10 with Triton buffer and clarified by centrifugation.

The clear supernatant is passed through an antilipoprotein antibody-Sepharose column (1 ml packed bed volume for each 5 ml of culture) equilibrated with Triton buffer. The column is washed with 10 bed volumes each of Triton buffer and 0.1% Triton X-100 in water to allow a sharp change in pH during elution. The lipoprotein is eluted with 0.2 $M$ glycine hydrochloride buffer (pH 2.2) containing 0.1% Triton X-100. The peak fractions are combined, neutralized with Tris base (200 $\mu$l of 1 $M$ solution for every 1 ml of fraction), and stored at $-20°$ in aliquots, or they are precipitated with

[17] M. Inukai, M. Takeuchi, K. Shimizu, and M. Arai, *J. Antibiot.* **31,** 1203 (1978).
[18] M. Hussain, S. Ichihara, and S. Mizushima, *J. Biol. Chem.* **255,** 3707 (1980).

acetone, resuspended in a suitable buffer, and stored. From a 25-ml culture about 10 $\mu$Ci of [$^{35}$S]methionine-labeled diacylglyceryl-modified prolipoprotein is obtained. The affinity gel can be reused several times if it is neutralized by passing through Triton buffer (5 bed volumes) immediately after the elution.

*Assay Procedure.* Purified [$^{35}$S]methionine-labeled modified prolipoprotein (20,000 cpm) is incubated with the enzyme preparation in 50 $\mu$l of assay buffer. After incubation at 37°, 200 $\mu$l of prechilled acetone is added to stop the reaction and to precipitate the processed and unprocessed forms. The supernatant is counted for the measurement of the released signal peptide, and the pellet is analyzed by polyacrylamide gel electrophoresis using either the Ito gel system[19] [which is quicker (4 hr)] or the Tricine gel system[20] [which is more time consuming (16 hr) but capable of separating all the intermediates in lipoprotein maturation]. The gels are dried and subjected to fluorography or Phosphor Imager (Molecular Dynamics, Sunnyvale, CA) analysis. For quantitation, the radioactive band from the dried gel is cut, crushed, eluted with 1% SDS (42°, overnight), and counted. Alternatively, the autoradiogram can be scanned or quantitated during Phosphor Imager analysis. Correction is made for the fact that there are three methionines in prolipoprotein and two in the apolipoprotein or mature lipoprotein.

Owing to the stoichiometric recovery of the signal peptide in the acetone supernatant, the assay[8] is simple and gives a quantitative measurement of the enzyme activity. Unit definition for signal peptidase II activity has been arbitrary. In the gel assay, 1 unit has been defined as the amount of enzyme required to process 50% of the substrate in 30 min at 37°. If the specific activity of the labeled substrate is known, the unit can be converted to the international unit.

## Purification of Signal Peptidase II

*Solubilization of Enzyme Activity.* Signal peptidase II, a minor integral membrane protein in the cytoplasmic membrane, can be solubilized from the crude cell envelope fraction obtained by French press with 1% Triton X-100, Nikkol, or octylglucoside.[7] Although octylglucoside solubilizes the enzyme, it does not support activity, whereas sarkosyl, a detergent commonly used for solubilizing inner membrane proteins, inactivates the enzyme even at 0.1% (w/v).[7]

*Purification of Enzyme.* The E. coli signal peptidase II has been purified to apparent homogeniety by Dev and Ray.[8] In the method, 200 g of frozen *E. coli* B cell pellet is suspended and thawed at 30° in 200 ml of 0.2 *M* Tris-

[19] K. Ito, T. Date, and W. Wickner, *J. Biol. Chem.* **255**, 2123 (1980).
[20] H. Schägger and C. von Jagow, *Anal. Biochem.* **166**, 368 (1987).

HCl (pH 8.0) containing 25% (w/v) sucrose and 1 m$M$ DTT. Spheroplasts are formed using the EDTA–lysozyme method.[21] Subsequent to osmotic lysis of the spheroplasts and removal of the unbroken cells, the cell envelope is collected by centrifugation and washed thoroughly. The washed pellet is extracted with 2% Triton X-100 in 100 m$M$ Tris–acetate, pH 4.0, containing 1 m$M$ DTT, 5 m$M$ MgCl$_2$, and 10% glycerol; extraction under acidic pH has been found to yield higher specific activity than that obtained at neutral or alkaline pH. The supernatant is heat treated at 65° for 7.5 min, and the resulting pellet is discarded. The supernatant is dialyzed to pH 5.5 and conductivity equivalent to 20 m$M$ in 10 m$M$ triethanolamine hydrochloride, pH 7.5, 5 m$M$ MgCl$_2$, 1 m$M$ DTT, 10% glycerol, and 1% Triton X-100. The dialyzed supernatant is then passed through a DE-52 (Whatman, Clifton, NJ) column (2.5 × 5 cm) at room temperature and the unbound fractions are collected.

The highly active fractions are dialyzed against chromatofocusing buffer (10 m$M$ imidazole hydrochloride, pH 8.1, 1 m$M$ DTT, 10% glycerol, and 1% Triton X-100) and fractionated on the chromatofocusing column PBE-94 (Pharmacia, 1.5 × 3.5 cm) using Polybuffer 74 (pH 4.5, Pharmacia) as eluent. The enzyme activity elutes as a single peak, and one fraction appears to be at least 95% pure as judged by polyacrylamide gel electrophoresis and Coomassie blue staining. The activity has been found to be unstable without glycerol and DTT. The purification procedure results in a 35,000-fold increase in the specific activity with an overall yield of 40%. With the availability of hyperexpressing clones,[10] the procedure could yield even larger amounts of enzyme.

Yu *et al.*[11] placed the gene for signal peptidase II (*lsp*) under control of the *trp* promoter; after inducing both the runaway replication of the plasmid at 37° and transcription of the cloned *lsp* gene with β-indoleacrylic acid, the enzyme is purified from *E. coli* cells using DEAE-cellulose chromatography in the presence of sodium EDTA. Subsequent purification using high-performance liquid chromatography (HPLC) gel filtration in the presence of SDS gives a preparation the amino acid composition of which matched the predicted values based on the nucleotide sequence of the cloned gene.

Cloning, Sequencing, and Hyperexpression of Gene Encoding Signal Peptidase II

The *E. coli lsp* gene has been identified by two independent strategies. Using either complementation of a temperature-sensitive mutant defective

---

[21] B. Witholt, M. Boekhout, M. Borck, J. Kingma, H. van Heerikhuizen, and L. de Leij, *Anal. Biochem.* **74,** 160 (1976).

in signal peptidase II activity[22,23] or selection for clones exhibiting increased globomycin resistance,[24] the same plasmid pLC3-13 from the Carbon–Clarke collection was identified to contain the structural gene for signal peptidase II (*lsp*). The clone and its subclones were shown to be overexpressing signal peptidase II activity, up to 75 times the wild-type activity.[10] The gene was sequenced[10,11] and the deduced amino acid sequence revealed *E. coli* signal peptidase II to comprise 164 amino acid residues. A combination of genetic mapping,[25] sequencing of neighboring genes,[26,27] and Northern blot analysis[28] showed that the *lsp* gene is one of the five genes in the *x-ileS-lsp-orf149-lytB* operon.[29]

## Properties of Signal Peptidase II

*Molecular Properties.* The molecular weight of the enzyme, 18,144, calculated from the deduced amino acid sequence,[10,11] is in excellent agreement with the reported molecular mass of about 18 kDa for the purified enzyme.[8] In addition, there is a fair agreement between the deduced and the calculated amino acid compositions.[11] The optimum pH for the purified preparation is 6.0,[8] whereas the crude enzyme exhibits a broad pH optimum between 6.0 and 9.0 with the maximum at 7.9.[7,18] The activity drops sharply below pH 6.0. The optimum temperature for activity under the assay conditions is 37°. The enzyme is thermally stable especially in the membrane; in the absence of detergent it is active even at 80°,[18] and in the presence of detergent (2% Triton X-100), it can withstand brief exposure up to 60°.[8]

Signal peptidase II can be solubilized from the membrane by a variety of detergents, the most effective ones being Triton X-100, Nikkol, and octylglucoside. The enzyme requires detergent for optimal activity *in vitro*; Triton X-100 and Nikkol support activity well, whereas octylglucoside (1%, w/v) inactivates the enzyme completely.[7] Although the enzyme is stable especially in the crude state, it is unstable during purification unless 10% glycerol, 1% Triton X-100, and 1 m*M* dithiothreitol are included in the

[22] H. Yamagata, C. Ippolite, M. Inukai, and M. Inouye, *J. Bacteriol.* **152**, 1163 (1982).
[23] H. Yamagata, K. Daishima, and S. Mizushima, *FEBS Lett.* **158**, 301 (1983).
[24] M. Tokunaga, J. M. Loranger, and H. C. Wu, *J. Biol. Chem.* **258**, 12102 (1983).
[25] M. Regue, J. Remenick, M. Tokunaga, G. A. Mackie, and H. C. Wu, *J. Bacteriol.* **158**, 632 (1984).
[26] M. Tokunaga, J. M. Loranger, S. Y. Chang, M. Regue, S. Chang, and H. C. Wu, *J. Biol. Chem.* **260**, 5610 (1985).
[27] Y. Kamio, C. K. Lin, M. Regue, and H. C. Wu, *J. Biol. Chem.* **260**, 5616 (1985).
[28] K. W. Miller, J. Bouvier, P. Stragier, and H. C. Wu, *J. Biol. Chem.* **262**, 7391 (1987).
[29] C. E. Gustafson, S. Kaul, and E. E. Ishiguro, *J. Bacteriol.* **175**, 1203 (1993).

buffers. Under these conditions the purified enzyme is stable for at least 1 month at 4° or −20°.[8] The molecular mass of the native enzyme has not been determined; it is not known whether it is active as a monomer, dimer, or oligomer.

*Kinetic Properties and Substrate Specificity.* Signal peptidase II is specific for lipid-modified prolipoproteins in bacteria.[9,30,31] The $K_m$ of the purified enzyme for diacylglyceryl-modified murein prolipoprotein has been shown to be $6 \pm 1 \, \mu M$.[8] The $K_m$ values for other modified prolipoproteins may vary considerably. It has been shown that unmodified prolipoproteins or other precursor proteins such as promaltose binding protein[8] or M13 pro-coat protein[30] (which are substrates for signal peptidase I) are not processed by the enzyme.

Signal peptidase II cleaves the glycyl diacylglyceryl–cysteine bond in murein prolipoprotein and in the majority of other lipoprotein precursors. Other prolipoproteins contain Ala- or Ser-diacylglyceryl–cysteine cleavage sites. Oligonucleotide-directed site-specific mutations of the cleavage site have clearly demonstrated the indispensability of the cysteine residue for modification and processing. Glycine at the −1 position can be replaced with alanine or serine; however, mutant prolipoproteins with leucine or isoleucine substitutions at the −1 position are not modified and conse-quently not processed,[32] whereas a glycine to threonine substitution allows modification at a slower rate but does not allow processing of the lipid-modified mutant prolipoprotein.[33] In the case of a mutant prepenicillinase containing a deletion of the lipobox (-Leu-Ala-Gly-Cys-Ala-), the Cys in an upstream sequence, -Leu-Phe-Ser-Cys-Val-, is recognized by the modifi-cation enzymes, albeit poorly, but not by signal peptidase II, probably because of the bulky phenylalanine residue in the sequence.[34]

The observations indicate that signal peptidase II recognizes the struc-ture surrounding the modification and cleavage sites, and this requirement may be more stringent for signal peptidase II than for modification enzymes. The lipobox sequence -Leu-Ala-Gly-Cys-, seen in the majority of lipopro-teins, is recognized by the modification enzymes and signal peptidase II even when placed internally in a protein, as in the cases of a fusion protein between the erythromycin resistance gene product and murein lipoprotein[35]

[30] M. Tokunaga, J. M. Loranger, P. B. Wolfe, and H. C. Wu, *J. Biol. Chem.* **257**, 9922 (1982).
[31] M. Tokunaga, J. M. Loranger, and H. C. Wu, *J. Cell. Biochem.* **24**, 113 (1984).
[32] J. Gennity, J. Goldstein, and M. Inouye, *J. Bioenerg. Biomembr.* **22**, 233 (1990).
[33] S. Pollitt, S. Inouye, and M. Inouye, *J. Biol. Chem.* **261**, 1835 (1986).
[34] S. Hayashi, S. Y. Chang, S. Chang, and H. C. Wu, *J. Biol. Chem.* **259**, 10448 (1984).
[35] S. Hayashi, S. Y. Chang, S. Chang, C. Z. Giam, and H. C. Wu, *J. Biol. Chem.* **260**, 5753 (1985).

and a *lacZ–lpp* fusion product in which the lipoprotein is fused C-terminally to β-galactosidase.[36]

It is not clear whether signal peptidase II can act on glyceryl-modified prolipoprotein, lacking the acyl groups. In *Klebsiella pneumoniae* strain K21, the secreted form of pullulanase is reported to be unacylated but processed by signal peptidase II.[37]

Globomycin is a potent, reversible, and noncompetitive inhibitor of signal peptidase II with a $K_i$ of 36 n$M$.[12] Among the many proteinase inhibitors tried, tosylarginylmethyl ester, $HgCl_2$, and pepstatin inhibited the enzyme significantly at concentrations below 1 m$M$; phenylmethylsulfonyl fluoride, phenylethyl alcohol, $N$-ethylmaleimide, and chymostatin, at 2–4 m$M$ concentrations, inhibited signal peptidase II activity up to 50%.[8] Unlike the microsomal signal peptidase from eukaryotic cells,[38,39] signal peptidase II does not require phospholipids,[7] nor does it require metal ions or other cofactors.[8]

## Membrane Topology of Signal Peptidase II

The hydropathy profile of the predominantly hydrophobic *E. coli* enzyme[10] exhibits four major hydrophobic domains each containing a stretch of at least 10 uncharged amino acids. These are flanked by charged residues in such a way that when the hydrophobic domains span the membrane the positive charges fall on one side, probably cytoplasmic, allowing interactions with the negatively charged cytoplasmic side, in accordance with the "positive inside" rule of von Heijine.[40] The enzyme is modeled hypothetically as an integral membrane protein with four membrane-spanning segments connected by two periplasmic loops and one positively charged cytoplasmic loop (Fig. 1); the polar N-terminal and the positively charged C-terminal segments lie on the cytoplasmic side.[10] Support for the model was obtained using PhoA and LacZ as reporter enzymes in a series of *lsp–phoA* and *lsp–lacZ* gene fusions.[13] Strains containing *phoA* fusions to the predicted periplasmic loops of signal peptidase II exhibited higher alkaline phosphatase activity than those fused to the predicted cytoplasmic domains. As expected, corresponding fusions with *lacZ* showed the opposite results: fusions to the cytoplasmic loops showed higher activity than those to the periplasmic loops of signal peptidase II.

[36] G. P. Vlasuk, J. Ghrayeb, and M. Inouye, *in* "The Enzymes of Biological Membranes" (A. N. Martonosi, ed.), Vol. 2, p. 309. Plenum, New York and London, 1985.
[37] M. G. Kornacker, A. Boyd, A. P. Pugsley, and G. S. Plastow, *Mol. Microbiol.* **3,** 497 (1989).
[38] R. C. Jackson and W. R. White, *J. Biol. Chem.* **256,** 2545 (1981).
[39] M. D. Lively and K. A. Walsh, *J. Cell. Biochem.* **7B,** 331 (1983).
[40] G. von Heijine, *EMBO J.* **5,** 3021 (1986).

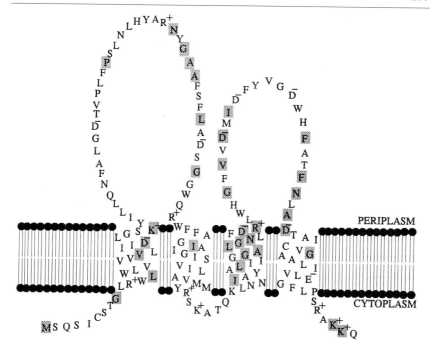

Fig. 1. Membrane topology of *E. coli* signal peptidase II based on the hydropathy profile and evidence from the orientation of LacZ and PhoA fusions. Residues that are identical in the four organisms studied (Fig. 2) have been highlighted.

## Structure–Function Relationship of Signal Peptidase II

The genes encoding signal peptidase II from *Enterobacter aerogenes*,[14] *Pseudomonas fluorescens*,[15] and *Staphylococcus aureus*[16] have been cloned and sequenced. In spite of the species diversity, the signal peptidases II possess similar hydropathy profiles indicating the same membrane topology. In addition, comparison of deduced amino acid sequences of the four enzymes revealed regions of high sequence similarity, suggestive of conserved domains involved in enzyme function (Fig. 2).

The C-terminal sequence containing a cluster of positively charged residues seems to be important for activity, as any alteration resulting in abolition of the positive charge affected the enzyme activity.[10] A mutant signal peptidase II from *E. coli* in which the last 5 amino acids containing three lysyl residues were replaced with 12 amino acids derived from the vector was shown to be 10% as active as the wild-type enzyme. Similarly, when the sequence was terminated 5 amino acids short of the full length, activity was reduced to 30% of the wild-type enzyme activity.

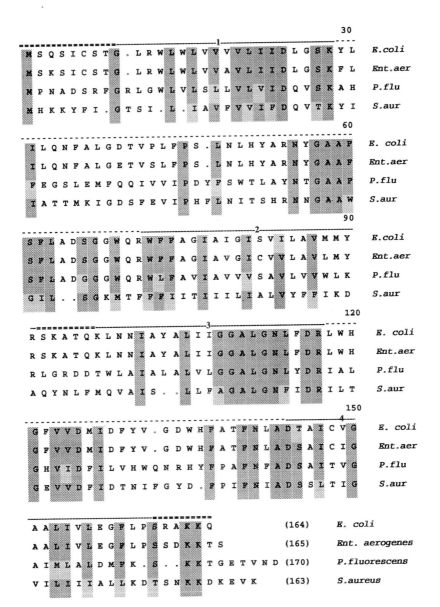

FIG. 2. Multiple sequence alignment among signal peptidase II sequences from *Escherichia coli, Enterobacter aerogenes, Pseudomonas fluorescens,* and *Staphylococcus aureus.* Regions of high similarity in all four sequences have been highlighted, allowing for only one conserved amino acid change, with variability [C. Sanders and R. Schneider, *Proteins* **9**, 56 (1991)] less than 20%. Identical residues are darkly shaded, and conserved substitutions are lightly shaded. The sequences on the cytoplasmic side (= = =), periplasmic side (– – –), and inside the membrane (---, domains 1–4) have been indicated for the *E. coli* sequence. Dots indicate gaps introduced to maximize alignment.

The globomycin-resistant and temperature-sensitive allele of the *E. coli lsp* gene[22] has been shown to produce protein defective in signal peptidase II activity both *in vitro* and *in vivo*. DNA sequencing of the polymerase chain reaction (PCR) product of the mutant allele has revealed a replacement of Asp-23 with Gly.[41] The Asp-23 residue, which is in the first membrane-spanning domain, is conserved among the four signal peptidases II so far sequenced.

The deduced amino acid sequence of signal peptidase II does not bear significant similarity to other proteins in databanks, including other proteinases. There is no conserved serine or cysteine residue, essential for serine and thiol proteinases, respectively. Because signal peptidase II does not require any metal ion for activity and is active in the presence of EDTA, it is not a metalloproteinase. It is, however, inhibited by pepstatin,[8] and there is an essential aspartic acid (position 23 in *E. coli* sequence, Fig. 1)[41] which is conserved in all four organisms studied. The sequence around that residue (-Ile-Asp-Leu-Gly-Ser- in *E. coli*) is similar to the conserved sequence -Phe-Asp-Thr-Gly-Ser- seen in the N-terminal domain of endopeptidases in the pepsin family (this volume, [7]). The sequence (-Leu-Ala-Asp-Thr-Ala-Ile, in *E. coli*) around another conserved aspartic acid, Asp-141, also resembles the sequence -(Ile, Leu, or Val)-Val-Asp-Thr-Gly-Thr seen at the C-terminal domain in members of the pepsin family.[42] Hence, it is tempting to suggest that signal peptidase II could be a novel aspartic proteinase that is active around neutral pH, possibly because of the changes in the residues around the active-site aspartic acid in a manner suggested by Ido *et al.* for viral proteinases.[43]

## Concluding Remarks

Bacterial prolipoprotein signal peptidase is a unique signal peptidase capable of cleaving the signal peptides of prolipoproteins after diacylglyceryl modification of a cysteine residue which becomes the N-terminal amino acid in the mature forms of all the known lipoproteins. The sequence present just in front of the cleavage site (lipobox) serves as the determinant not only for the lipid modification but also for subsequent processing by signal peptidase II. The enzyme, localized to the inner membrane, has been purified from a wild-type *E. coli* strain, and an overproducing clone and some of the biochemical properties of the enzyme have been studied.

[41] R. M. Dai and H. C. Wu, unpublished data (1991).
[42] B. Foltmann and V. B. Peterson, *in* "Advances in Experimental Medicine and Biology" (J. Tang, ed.), vol. 95, p. 3. Plenum, New York and London, 1977.
[43] E. Ido, H.-P. Han, F. J. Kezdy, and J. Tang, *J. Biol. Chem.* **266,** 24359 (1991).

Among notable features of the enzyme are its small size, thermostability, stringent structural requirements in the substrate, and noncompetitive inhibition by the potent peptide antibiotic globomycin. The deduced amino acid sequence shows that the protein is very hydrophobic and lacks signal sequence. Comparison of signal peptidase II sequences from *E. coli, Enterobacter aerogenes, Pseudomonas fluorescens,* and *Staphylococcus aureus* supports the proposed membrane topology of the enzyme and reveals conserved regions that may be important for enzyme function. Limited biochemical studies preclude conclusive statements on the mechanism of action, but available data suggest the importance of both the aspartic acid at position 23 in the *E. coli* sequence and the positively charged C-terminal region. A serious effort in studying the structure–function relationship of this essential enzyme will lead to a better understanding and establishment of the mechanism of action of this unique endopeptidase.

Acknowledgment

Research was supported by Grant GM 28811 from the National Institutes of Health.

# Section III

# Metallopeptidases

## [13] Evolutionary Families of Metallopeptidases

*By* NEIL D. RAWLINGS and ALAN J. BARRETT

### Introduction

Metallopeptidases form the most diverse of the catalytic types of peptidases, about 30 families being recognized by our criteria (see [1], Volume 244, this series, for definitions of the terms family and clan). About half of the families comprise enzymes containing the His-Glu-Xaa-Xaa-His (or HEXXH) motif that has been shown by X-ray crystallography to form part of the site for binding of the metal (normally zinc) atom in some families, and almost certainly does so in the others.

The HEXXH pentapeptide is very common in protein sequences, and only about one-fifth of the occurrences of the motif in the SwissProt database are in proteins that are known to be metallopeptidases. The motif is small enough to have arisen independently on several occasions, so additional evidence would be needed before one could suggest that all HEXXH-containing enzymes are related. Tertiary structures show only slight similarities between such HEXXH-containing endopeptidases as thermolysin and astacin.[1]

A consensus sequence containing the HEXXH pentapeptide can be defined somewhat more precisely as abXHEbbHbc, in which b is an uncharged residue, c is hydrophobic, and X can be almost any amino acid. Proline is never found in this region; all of the available tertiary structures for metallopeptidases containing HEXXH show the motif in a helix, which would be broken by proline. As can be seen from Figs. 1, 2, and 4 below, residue a in the abXHEbbHbc consensus is most commonly Val or Thr, and it is known to form part of the S1' subsite at least in thermolysin (family M4) and neprilysin (M13).[2]

We shall here divide the families of metallopeptidases into five groups, the first three of which contain the HEXXH motif. In the first set, a glutamic acid residue completes the metal-binding site, and because the families in this "HEXXH + E" set show signs of relationship to one another they form clan MA. In the second set, a third histidine residue is a ligand; these "HEXXH + H" families also are interrelated, and thus form clan MB.

[1] W. Bode, F. X. Gomis-Rüth R. Huber, R. Zwilling, and W. Stöcker, *Nature (London)* **358,** 164 (1992).
[2] J. Vijayaraghavan, Y.-A. Kim, D. Jackson, M. Orlowski, and L. B. Hersh, *Biochemistry* **29,** 8052 (1990).

The third set of HEXXH families is that in which additional metal ligands are as yet unidentified. A fourth group of metallopeptidase families is a heterogeneous one in which it is known that the essential metal atoms are bound at motifs other than HEXXH, and the fifth and final group of families contains those in which the metal ligands are quite unknown.

As we review the families of metallopeptidases, we shall note that many are represented in both prokaryotes and eukaryotes. Moreover, the prokaryotic and eukaryotic enzymes often have much in common, in terms of both structure and properties. We have suggested that this may be because the metallopeptidases of higher organisms were introduced into the eukaryotic cell relatively recently, at the time of the endosymbioses that led to the formation of mitochondria and other organelles, rather than having been derived from the ultimate common ancestor of prokaryotic and eukaryotic cells.[3,4] This point is discussed further below.

### HEXXH + E Metallopeptidase Clan (MA)

Clan MA contains the most thoroughly characterized of the metalloendopeptidases, thermolysin, and other enzymes from both prokaryotic and eukaryotic organisms (Table I; Fig. 1). The three-dimensional structure of thermolysin shows that, in the HEXXH motif, the His residues are zinc ligands and the Glu has a catalytic function. The Glu residue promotes the nucleophilic attack of a buried water molecule on the carbonyl carbon atom of the substrate peptide bond, forming a tetrahedral intermediate.[5] The additional zinc-binding site in thermolysin is another Glu residue, 20 residues C-terminal to the HEXXH motif.[6] Aspartate commonly occurs four residues C-terminal to the zinc-binding glutamate, and it can be seen in Fig. 1 that this residue (Asp-170) is conserved in families other than M1 of clan MA.

Despite exceedingly slight similarity in sequences, the peptidases in the families of thermolysin (M4) and neprilysin (M13) have several characteristics in common. Thus, the sequences show similar clustering of hydrophobic residues,[7] the main determinant of substrate specificity is a hydrophobic residue in P1′, and there is strong inhibition by phosphoramidon.

It may well be that the biological function of the ancestor of clan MA was the digestion of extracellular food proteins, in an early bacterium. Most

---

[3] A. J. Barrett and N. D. Rawlings, in "Innovations on Proteases and Their Inhibitors" (F. X. Avilés, ed.), Gruyter, Berlin, 1993.

[4] A. J. Barrett and N. D. Rawlings, in "Proteolysis and Protein Turnover" (J. S. Bond and A. J. Barrett, eds.), p. 73. Portland Press, London, 1993.

[5] L. H. Weaver, W. R. Kester, and B. W. Matthews, J. Mol. Biol. **114**, 119 (1977).

[6] B. W. Matthews, L. H. Weaver, and W. R. Kester, J. Biol. Chem. **249**, 8030 (1974).

[7] T. Benchetrit, V. Bissery, J. P. Mornon, A. Devault, P. Crine, and B. P. Roques, Biochemistry **27**, 592 (1988).

TABLE I
METALLOPEPTIDASES WITH HEXXH + E ZINC-BINDING MOTIF: CLAN MA[a]

| Family | EC | Database code |
|---|---|---|
| **Family M4: Thermolysin** | | |
| Bacillolysin[b] | 3.4.24.28 | THER_BACCE, THER_BACCL, NPRE_BACSU, NPRE_BACPO, (K02497), (M62845), (M64815), (X61286), (X61380), PIR B36706 |
| Coccolysin | 3.4.24.30 | (M37185) |
| Extracellular elastase (*Staphylococcus*) | - | (X69957) |
| Extracellular endopeptidase (*Erwinia*) | - | (M36651) |
| Extracellular endopeptidase (*Serratia*) | - | (M59854) |
| Metalloendopeptidase (*Legionella*) | - | PROA_LEGPN |
| Metalloendopeptidase (*Listeria*) | - | PRO1_LISMO, PRO2_LISMO |
| Pseudolysin[b] | 3.4.24.26 | ELAS_PSEAE |
| Thermolysin[b] | 3.4.24.27 | THER_BACST, THER_BACTH |
| Vibriolysin (*Vibrio*) | - | NPRV_VIBPR, HAPT_VIBCH, (I01814), (L02528) |
| **Family M5: Mycolysin** | | |
| Mycolysin | 3.4.24.31 | MYCO_STRCI |
| **Family M13: Neprilysin** | | |
| Endothelin-converting enzyme | - | (D29683) |
| Kell blood group protein | - | KELL_HUMAN |
| Neprilysin | 3.4.24.11 | NEP_*, (M81591) |
| Peptidase O (*Lactococcus*) | - | (L18760) |
| **Family M1: Membrane alanyl aminopeptidase** | | |
| Alanyl/arginyl aminopeptidase (*Saccharomyces*) | - | (L12542) |
| Aminopeptidase yscII (*Saccharomyces*) | - | APE2_YEAST |
| Glutamyl aminopeptidase | - | AMPE_MOUSE, (L14721) |
| Leukotriene A₄ hydrolase | 3.3.2.6 | LKHA_HUMAN, (M63848), (D16669), (M88793) |
| Lysyl aminopeptidase (*Lactococcus*) | 3.4.11.15 | (X61230), (Z21701) |
| Membrane alanyl aminopeptidase | 3.4.11.2 | AMPN_*, (M75750), (M91449) |
| **Family M2: Peptidyl-dipeptidase A** | | |
| Peptidyl-dipeptidase A | 3.4.15.1 | ACE_*, ACET_* |

[a] EC is the enzyme nomenclature number (Nomenclature Committee of the International Union of Biochemistry and Molecular Biology, "Enzyme Nomenclature 1992," Academic Press, Orlando, Florida, 1992, and Supplement); — indicates that no EC number has been assigned. Sequence codes are from the SwissProt database, except those in parentheses, which are from the EMBL database, and those prefixed "PIR," which are from the National Biomedical Research Foundation Protein Information Resource database. Literature references to the individual proteins are generally to be found in the database entries.
[b] Structure is included in the Brookhaven database.

of the enzymes of modern bacteria are secreted to act extracellularly, but peptidase O (family M13) is an integral protein of the plasma membrane of a *Lactococcus*, and thus acts to modify the composition of the pericellular environment (this volume, [35]). The eukaryotic integral membrane proteins of families M13, M1, and M2 similarly function outside the cell. Characteristically for such proteins, they are heavily glycosylated and presumably contain disulfide bonds. Perhaps as a result of glycosylation, as

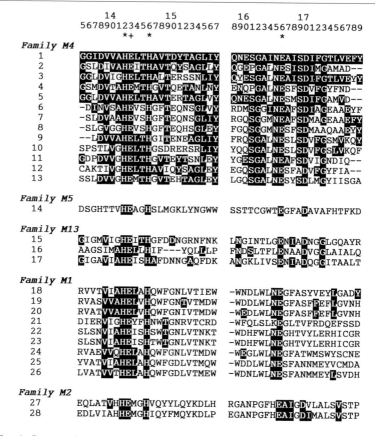

FIG. 1. Conservation of sequences around the zinc ligands and catalytic residues of clan MA. Residues are numbered according to thermolysin. Asterisks (*) indicate the zinc ligands, and + indicates the catalytic glutamic acid. Residues identical to those in the thermolysin sequence are shown in white on black. Key to sequences (gene names are given in parentheses where confusion with other metalloendopeptidases from the same species is possible): 1, thermolysin (*Bacillus* sp.); 2, *Bacillus subtilis* bacillolysin; 3, *Bacillus megaterium* bacillolysin; 4, *Bacillus amyloliquefaciens* bacillolysin; 5, *Bacillus brevis* metalloendopeptidase, 6, *Vibrio cholerae* hemagglutinin; 7, pseudolysin; 8, *Legionella pneumophila* metalloendopeptidase; 9, *Erwinia carotovora* metalloendopeptidase (*prt1*); 10, *Serratia marcescens* metalloendopeptidase (*smp*); 11, *Bacillus polymyxa* bacillolysin; 12, *Listeria monocytogenes* metalloendopeptidase; 13, coccolysin; 14, *Streptomyces cacaoi* mycolysin; 15, human neprilysin; 16, human Kell blood group protein; 17, *Lactococcus lactis* peptidase O; 18, human membrane alanyl aminopeptidase; 19, mouse glutamyl aminopeptidase; 20, human glutamyl aminopeptidase; 21, *Escherichia coli* aminopeptidase N; 22, human leukotriene-A₄ hydrolase; 23, guinea pig leukotriene-A₄ hydrolase; 24, yeast aminopeptidase yscII; 25, *Lactococcus lactis* lysyl aminopeptidase; 26, *Lactococcus delbrueckii* lysyl aminopeptidase; 27, mouse dipeptidyl-peptidase A repeat 1; 28, mouse dipeptidyl-peptidase A repeat 2.

well as their location, several members of the clan have been independently recognized as cell-surface antigens.

## Thermolysin Family (M4)

The thermolysin family contains only secreted eubacterial endopeptidases, from both gram-positive and gram-negative bacteria. The mature enzymes are all of moderate size, in the region of 35 kDa.

The tertiary structure of thermolysin shows the zinc ligands to be His-142, His-146, and Glu-166, and the catalytic residue Glu-143 (Fig. 1). The structure also shows four calcium-binding sites that contribute to the unusual thermal stability of the enzyme.[6] Other members of the family include enzymes from pathogens (e.g., *Legionella, Listeria, Pseudomonas, Vibrio*). Pseudolysin (*Pseudomonas* elastase) causes tissue damage by degrading collagens, elastin, and fibronectin. The structure of pseudolysin shows a close relationship to thermolysin (see [15] in this volume). Thermolysin and its precursor lack cysteine residues, but this is not the case for all members of the family. For example, pseudolysin has four cysteines forming two disulfide bridges.

The substrate specificity of most members of family M4 is primarily for an aromatic residue in P1'.[8] Analysis of the tertiary structures of three members of the family has led to the conclusion that the enzymes undergo hinge-bending motion during catalysis.[9] The broader specificity of pseudolysin, acting on such large native proteins as elastin and collagen, may be attributable to the wider active-site cleft ([15], this volume).

## Mycolysin Family (M5)

Mycolysin, the single member of the M5 family, is secreted by *Streptomyces griseus* (and is thus a component of pronase) and also is produced by several other *Streptomyces* spp. The enzyme has a specificity similar to that of thermolysin, and the residues directly involved in activity (Fig. 1) also are as in thermolysin.[10] The endopeptidase is synthesized as a precursor with a large propeptide (171 residues) that is autolytically processed by cleavage between Ala residues.[10] Mycolysin is subject to anomalous inhibition by serine proteinase inhibitors that are produced by *Streptomyces* spp.[11]

[8] K. Morihara and H. Tsuzuki, *Arch. Biochem. Biophys.* **146**, 291 (1971).
[9] D. R. Holland, D. E. Tronrud, H. W. Pley, K. M. Flaherty, W. Stark, J. N. Jansonius, D. B. McKay, and B. W. Matthews, *Biochemistry* **31**, 11310 (1992).
[10] P.-C. Chang and Y.-H. Wu Lee, *J. Biol. Chem.* **267**, 3952 (1992).
[11] K. Kajiwara, A. Fujita, H. Tsuyuki, T. Kumazaki, and S. Ishii, *J. Biochem (Tokyo)* **110**, 350 (1991).

*Neprilysin Family (M13)*

Neprilysin (EC 3.4.24.11), previously known as enkephalinase, common acute lymphocytic leukemia antigen (CALLA), and neutral endopeptidase 24.11, is a mammalian enzyme that has the capacity to degrade enkephalins and many other biologically active peptides (see [16] in this volume). That property of the enzyme has stimulated much pharmacologically oriented work on inhibitors ([17], this volume). The neprilysin family contains eukaryote and prokaryote oligopeptidases, and one of the mammalian blood group antigens, Kell. In neprilysin, zinc ligands are known (Fig. 1) and are analogous to those in thermolysin.[12] Neprilysin and homologs are much larger proteins than thermolysin, however (about 90 kDa as compared to 35 kDa).

Neprilysin and the other members of the family share with thermolysin the characteristic of potent inhibition by phosphoramidon, and in animals this inhibitor seems to be rather selective for this family. The known endopeptidases of the family appear to be oligopeptidases, acting on oligopeptides and polypeptides but not on proteins.

Neprilysin is most abundant as an integral protein of the kidney and intestinal microvillar membranes, but it is also present in smaller amounts elsewhere in the body. Neprilysin has a short, N-terminal cytoplasmic domain (27 residues), followed by a transmembrane domain (23 residues), and the rest of the molecule is extracellular. The cytoplasmic domain includes a conformationally restrained octapeptide which has been proposed to be a "stop transfer sequence" preventing proteolysis and secretion.[13] The membrane-spanning domain is homologous to that of human sucrase-isomalatase (RDF score 8.15 for the first 100 residues of both sequences, see Vol. 244 [1]). Neprilysin does not require proteolytic activation.

Short forms of neprilysin result from alternative splicing of the mRNA. The gene consists of 24 exons, with exon 3 encoding the initiation codon, transmembrane domain, and cytoplasmic domain and exons 4–23 encoding the extracellular portion of the enzyme. Exon 19 encodes the HEXXH motif.[14] A shortened form of the enzyme lacking exons 5–18 is known that retains the membrane-spanning domain and the active-site residues.[15] Another alternatively spliced form of neprilysin lacks exon 16 and seems to be essentially inactive.[16]

[12] H. Le Moual, B. P. Roques, P. Crine, and G. Boileau, *FEBS Lett.* **324**, 196 (1993).

[13] B. Malfroy, P. R. Schofield, W.-J. Kuang, P. H. Seeburg, A. J. Mason, and W. J. Henzel, *Biochem. Biophys. Res. Commun.* **144**, 59 (1987).

[14] L. D'Adamio, M. A. Shipp, E. L. Masteller, and E. L. Reinherz, *Proc. Natl. Acad. Sci. U.S.A.* **86**, 7103 (1989).

[15] C. Llorens-Cortes, B. Giros, and J.-C. Schwartz, *J. Neurochem.* **55**, 2146 (1990).

[16] H. Iijima, N. P. Gerard, C. Squassoni, J. Ewig, D. Face, J. M. Drazen, Y.-A. Kim, B. Shriver, L. B. Hersh, and C. Gerard, *Am. J. Physiol.* **262**, L725 (1992).

Kell blood group antigens are clinically important because incompatibility can cause severe hemolytic reactions in blood transfusions, and erythroblastosis in newborn infants.[17] The Kell blood group system comprises at least 24 antigenic determinants, all located on a 93-kDa glycoprotein that spans the erythrocyte membrane and is attached to the underlying cytoskeleton. Although the Kell protein is homologous to neprilysin, and possesses all of the potential catalytic residues, no activity has so far been described for it.

Endothelin-converting enzyme is a membrane-bound, phosphoramidon-sensitive peptidase, with N-terminal cytoplasmic and transmembrane domains in positions equivalent to those in neprilysin and the Kell blood group protein.[17a] It generates endothelin from the 38-residue precursor, big endothelin-1, by cleavage of a -Trp+Val- bond.

The eubacterial enzyme is peptidase O from *Lactococcus lactis*, which has a very similar specificity to neprilysin (see [35]).

## Alanyl Aminopeptidase Family (M1)

The alanyl aminopeptidase family contains only aminopeptidases, though from a wide variety of organisms including eubacteria (*Lactococcus, Escherichia, Caulobacter*), yeast, a nematode, and mammals (Table I). The aminopeptidases differ widely in specificity, hydrolyzing acidic, basic, or neutral N-terminal residues. Also a member of the family is leukotriene-$A_4$ hydrolase, an enzyme well known for its activity in the hydrolysis of the epoxide leukotriene-$A_4$ to form an inflammatory mediator. The leukotriene hydrolase was discovered to have aminopeptidase activity,[18] and site-directed mutagenesis led to the identification of the zinc ligands of the family (Fig. 1). The two activities of leukotriene-$A_4$ hydrolase are attributable to nonidentical but overlapping active sites. The zinc atom is essential for both activities, so that substitution of the ligand Glu-166 (Fig. 1 numbering) eliminates both. The Glu-143 residue is important only for the aminopeptidase activity, however.[19]

Membrane alanyl aminopeptidase and glutamyl aminopeptidase have N-terminal membrane-spanning domains that are not proteolytically removed. Both are typically homodimers and heavily glycosylated (reviewed

[17] S. Lee, E. D. Zambas, W. L. Marsh, and C. M. Redman, *Proc. Natl. Acad. Sci. U.S.A.* **88**, 6353 (1991).

[17a] D. Xu, N. Emoto, A. Giaid, C. Slaughter, S. Kaw, D. deWit, and M. Yanagisawa, *Cell (Cambridge, Mass.)* **78**, 473 (1994).

[18] J. Z. Haeggström, A. Wetterholm, R. Shapiro, B. L. Vallee, and B. Samuelsson, *Biochem. Biophys. Res. Commun.* **172**, 965 (1990).

[19] A. Wetterholm, J. F. Medina, O. Rådmark, R. Shapiro, J. Z. Haeggström, B. L. Vallee, and B. Samuelsson, *Proc. Natl. Acad. Sci. U.S.A.* **89**, 9141 (1992).

in Ref. 20, pp. 59, 72). The enzymes are quantitatively important components (of the order of 5% of total protein) in intestinal brush border membranes.[21] Membrane alanyl aminopeptidase is the myeloid leukemia marker, CD13,[22] and glutamyl aminopeptidase is a differentiation-related kidney antigen[23] and an immature B cell marker.[24] The cytoplasmic members of the family such as *Escherichia coli* aminopeptidase N[25] and leukotriene-$A_4$ hydrolase[26] lack membrane-spanning domains; they are correspondingly shorter at the N terminus and synthesized without prepropeptides.

*Peptidyl-Dipeptidase A Family (M2)*

Peptidyl-dipeptidase A (also known as angiotensin-converting enzyme) is a mammalian peptidase that cleaves off C-terminal dipeptides, notably to convert angiotensin I to angiotensin II ([18], this volume). It is unable to cleave bonds involving proline (Ref. 20, p. 227).

The enzyme exists in two molecular forms, both derived from the same gene but resulting from alternative modes of transcription. The best known form of the enzyme, from lung endothelium, contains two homologous domains that have arisen by gene duplication. The structure of the gene reflects the duplication, with number and size of exons conserved between domains, as well as the phase of exon–intron junctions.[27] This is indicative of a very recent duplication–fusion event, contrasting with the ancient duplication in the pepsin family (A1, see [7], this volume) in which no similarities are seen in terms of intron number, position, or junction phasing between the repeats. The date of the duplication in peptidyl-dipeptidase A can be estimated at 415 million years ago, about the time of the divergence of fish and amphibia. The testis-specific form of the enzyme contains only the more C-terminal of the repeats. There is evidence that the promoter of the gene was also duplicated, so that a second promoter exists within intron 12. It is this internal promoter that is responsible for the testicular

---

[20] J. K. McDonald and A. J. Barrett, "Mammalian Proteases: A Glossary and Bibliography. Volume 2: Exopeptidases." Academic Press, London, 1986.

[21] A. J. Kenny, *in* "Proteinases in Mammalian Cells and Tissues" (A. J. Barrett, ed.), p. 393. North-Holland, Amsterdam, 1977.

[22] J. Olsen, H. Sjöström, and O. Norén, *FEBS Lett.* **251,** 275 (1989).

[23] D. M. Nanus, D. Engelstein, G. A. Gastl, L. Gluck, M. J. Vidal, M. Morrison, C. L. Finstad, N. H. Bander, and A. P. Albino, *Proc. Natl. Acad. Sci. U.S.A.* **90,** 7069 (1993).

[24] Q. Wu, J. M. Lahti, G. M. Air, P. D. Burrows, and M. D. Cooper, *Proc. Natl. Acad. Sci. U.S.A.* **87,** 993 (1990).

[25] M. T. McCaman and J. D. Gabe, *Mol. Gen. Genet.* **204,** 148 (1986).

[26] C. D. Funk, O. Rådmark, J. Y. Fu, T. Matsumoto, H. Jörnvall, T. Shimizu, and B. Samuelsson, *Proc. Natl. Acad. Sci. U.S.A.* **84,** 6677 (1987).

[27] C. Hubert, A. M. Houot, P. Corvol, and F. Soubrier, *J. Biol. Chem.* **266,** 15377 (1991).

transcripts, and elements responsive to cAMP and steroid hormones have been identified that can account for the expression selectively in testis.[27] Both forms of the enzyme are heavily glycosylated, integral membrane proteins anchored by a C-terminal, transmembrane domain. In the testicular form, the N-terminal 72 residues are unique, not being found in the lung enzyme.

Both domains of the endothelial enzyme are catalytically active, but with different kinetic constants, and site-directed mutagenesis has identified some zinc ligands and catalytic residues in both domains (Fig. 1).[28] At present, additional zinc ligands for the two domains have not been positively identified, but, as shown in Fig. 1, Glu residues can plausibly be aligned with Glu-166 of thermolysin.

There is a bacterial enzyme with a specificity very similar to that of peptidyl-dipeptidase A, but it has a completely unrelated amino acid sequence (see family M3).

## HEXXH + H Metallopeptidase Clan (MB)

The elucidation of the tertiary structures of astacin[1] and adamalysin[29] (family M12) showed the essential zinc atom bound by three His residues, two from the HEXXH motif and a third only six residues C-terminal to this. It was readily apparent that the third His residue was conserved in the interstitial collagenase family (M10) and in autolysin from *Chlamydomonas* (family M11) (Fig. 2; Table II). Moreover, the tertiary structures subsequently elucidated for serralysin[30] and the catalytic domains of interstitial collagenase[31] and neutrophil collagenase[32] clearly confirmed the relationship of those proteins of family M10 to proteins of family M12. Indeed, the amino acid sequences of human interstitial collagenase and fibrolase reach a level of similarity (RDF score 5.94) almost sufficient to justify the merging of families M10 and M12.

A Met residue (Met-143 in Fig. 2) that is completely conserved throughout clan MB, and seems to serve a vital function in the structure of the active sites of the enzymes, has led to the term "metzincins" for the peptidases of the clan.[33] The Met-143 residue is not present in thermolysin, and what slight topological similarity exists between astacin and thermolysin is limited to four strands of $\beta$-pleated sheet and two long helices.[33] Thus there is little

[28] L. Wei, F. Alhenc-Gelas, P. Corvol, and E. Clauser, *J. Biol. Chem.* **266**, 9002 (1991).
[29] F.-X. Gomis-Rüth, L. F. Kress, and W. Bode, *EMBO J.* **12**, 4151 (1993).
[30] U. Baumann, S. Wu, K. M. Flaherty, and D. B. McKay, *EMBO J.* **12**, 3357 (1993).
[31] B. Lovejoy, A. Cleasby, A. M. Hassell, K. Longley, M. A. Luther, D. Weigl, G. McGeehan, A. B. McElroy, D. Drewry, M. H. Lambert, and S. R. Jordan, *Science* **263**, 375 (1994).
[32] P. Reinemer, F. Grams, R. Huber, T. Kleine, S. Schnierer, M. Piper, H. Tschesche, and W. Bode, *FEBS Lett.* **338**, 227 (1994).
[33] W. Bode, F.-X. Gomis-Rüth, and W. Stöcker, *FEBS Lett.* **331**, 134 (1993).

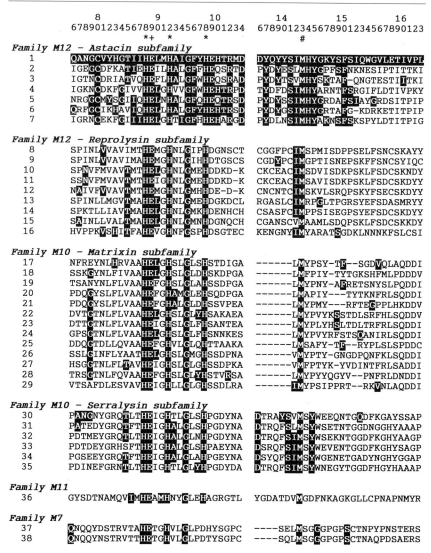

FIG. 2. Conservation of sequences around the zinc ligands and catalytic residues of clan MB. Residues are numbered according to astacin. Asterisks (*) indicate the zinc ligands, + indicates the catalytic Glu residue, and # indicates the Met turn. Residues identical to those in the astacin sequence are shown in white on black. Key to sequences: 1, astacin; 2, mouse meprin α subunit; 3, rat meprin ß subunit; 4, human bone morphogenetic protein 1; 5, *Oryzias latipes* hatching enzyme (HCE1); 6, *O. latipes* hatching enzyme (LCE); 7, *Drosophila melanogaster* tolloid protein; 8, trimerelysin I (HR1B); 9, jararhagin; 10, trimerelysin II (HR2); 11, trimerelysin II (HR2A); 12, green habu snake endopeptidase; 13, ruberlysin; 14, bushmaster snake hemorrhagic factor LHFII; 15, southern copperhead snake fibrolase; 16, cattle myelin-associated metalloendopeptidase; 17, human interstitial collagenase; 18, rat interstitial colla-

similarity between structures of proteins in clans MA and MB, although it could not be excluded that they are very distantly related.

The Gly residue (Gly-95) totally conserved in the extended active-site motif, HEXXHXXGXXH (Fig. 2), contributes to an important hairpin loop that permits the third His to bind the zinc atom. The *Streptomyces* neutral proteinase (family M7) is tentatively included in clan MB on the grounds that it retains Gly-95 and its hydrophobic surroundings. If indeed it is a relative, a very important change has taken place, in that the third zinc ligand (His) has been replaced by Asp (Fig. 2).

All of the enzymes of clan MB are endopeptidases; there are no known exopeptidases. Most of the enzymes are soluble, secreted proteins, but the meprins and some reprolysins (in family M12) are integral membrane proteins. Families M10, M11, and M12 contain mosaic proteins. In family M10, gene structures are available to suggest that the additional domains result from shuffling of phase 1-bounded exons, and the same may well be true of the other two families. This would be similar to the situation in the chymotrypsin family (this series, Volume 244, [2]), except that here the additional domains are located C-terminally rather than N-terminally to the peptidase domain.

Very few metalloendopeptidases act on arylamide substrates of the kinds hydrolyzed by most serine- and cysteine-type endopeptidases, that is to say, peptide nitroanilides and peptide aminomethylcoumarins (although metalloexopeptidases commonly cleave such substrates). Several endopeptidases in clan MB are active on arylamide substrates, however; these include astacin, meprin, and serralysin. The enzymes of clan MB are scarcely inhibited by phosphoramidon, unlike many in clan MA.

A novel type of propeptide mechanism arose early in the evolution of clan MB. In many of the eukaryotic members, a cysteine residue occurs in a conserved motif in the propeptide and interacts with the essential zinc atom to prevent it from binding the catalytic water molecule, maintaining the proenzyme in an inactive state (Fig. 3).[34] This "cysteine switch" mechanism is present in some but not all members of families M10, M11,

---

[34] E. B. Springman, E. L. Angleton, H. Birkedal-Hansen, and H. E. Van Wart, *Proc. Natl. Acad. Sci. U.S.A.* **87**, 364 (1990).

---

genase; 19, human neutrophil collagenase; 20, human gelatinase A; 21, human gelatinase B; 22, mouse stromelysin 1; 23, human stromelysin 1; 24, rat stromelysin 2; 25, human stromelysin 3; 26, human matrilysin; 27, human macrophage elastase; 28, *Paracentrotus lividus* envelysin; 29, soybean leaf metalloendopeptidase; 30, *Pseudomonas aeruginosa* serralysin; 31, *Serratia marcescens* serralysin; 32, *Erwinia chrysanthemi* serralysin D; 33, *E. chrysanthemi* serralysin B; 34, *E. chrysanthemi* serralysin C; 35, *E. chrysanthemi* serralysin G; 36, *Chlamydomonas reinhardtii* autolysin; 37, *Streptomyces lividans* metalloendopeptidase; 38, *Streptomyces coelicolor* metalloendopeptidase.

TABLE II
METALLOPEPTIDASES WITH HEXXH + H ZINC-BINDING MOTIF: CLAN MB[a]

| Family | EC | Database code |
|---|---|---|
| **Family M12: Astacin** | | |
| **Astacin subfamily** | | |
| Astacin[b] | 3.4.24.21 | ASTA_ASTFL |
| Blastula protease-10 (*Paracentrotus*) | - | (X56224) |
| Bone morphogenetic protein 1 | - | BMP1_HUMAN, (L12249) |
| Choriolytic enzyme (*Oryzias*) | - | HCE1_ORYLA, HCE2_ORYLA, LCE_ORYLA |
| Meprin $\alpha$-subunit | 3.4.24.18 | MEPA_MOUSE, (S43408) |
| Meprin $\beta$-subunit | 3.4.24.18 | (M88601) |
| Metalloendopeptidase (*Caenorhabditis*) | - | (M75746) |
| PABA-peptide hydrolase | 3.4.24.18 | (M82962) |
| SpAN protein (*Strongylocentrotus*) | - | (M84144) |
| *tolloid* gene product (*Drosophila*) | - | TLD_DROME |
| UVS.2 protein (*Xenopus*) | - | [g] |
| **Reprolysin subfamily** | | |
| Adamalysin[c] | 3.4.24.46 | ADAM_CROAD |
| Atrolysin A | 3.4.24.1 | [h] |
| Atrolysin C | 3.4.24.42 | HRTD_CROAT |
| Atrolysin D | 3.4.24.42 | (U01237) |
| Atrolysin E | 3.4.24.44 | HRTE_CROAT |
| Endopeptidase (green habu snake) | - | DISA_TRIGA, (X77089) |
| $\alpha$-Fibrinogenase | - | DISR_AGKRH |
| Fibrolase (southern copperhead snake) | - | FIBR_AGKCO |
| Haemorrhagic factor LHFII (bushmaster snake) | - | HRL2_LACMU |
| HR1B-endopeptidase (habu snake) | - | HR1B_TRIFL |
| HR2a-endopeptidase (habu snake) | - | HR2A_TRIFL |
| Jararhagin | - | DISJ_BOTJA |
| Myelin-associated metalloproteinase (cattle) | - | (Z21961) |
| ORF9 Metalloendopeptidase (human) | - | (D14665) |
| PH-30 $\alpha$ subunit | - | (Z11719) |
| Ruberlysin | 3.4.24.48 | HRT2_CRORU |
| Russellysin | 3.4.24.58 | [i] |
| Trimerelysin II | 3.4.24.53 | HR2_TRIFL |
| **Family M10: Interstitial collagenase** | | |
| **Serralysin subfamily** | | |
| Serralysin[d] | 3.4.24.40 | APRA_PSEAE, PRTB_ERWCH, PRTC_ERWCH, PRTX_ERWCH, PRZN_*, (X70011), (X71365) |
| **Matrixin subfamily** | | |
| Envelysin | 3.4.24.12 | HE_PARLI |
| Gelatinase A | 3.4.24.24 | COG2_* |
| Gelatinase B | 3.4.24.35 | COG9_HUMAN, (Z27231), (X78324), (D26514) |
| Interstitial collagenase[e] | 3.4.24.7 | COG1_*, COGZ_RAT |
| Macrophage elastase | - | COGM_MOUSE, (L23808) |
| Matrilysin | 3.4.24.23 | COG7_HUMAN, (L24374), (U04444) |
| Metalloproteinase (soybean) | - | MEP1_SOYBN |
| Neutrophil collagenase[f] | 3.4.24.34 | COG8_HUMAN |
| Stromelysin 1 | 3.4.24.17 | COG3_* |

TABLE II  (*continued*)

| Family | EC | Database code |
|---|---|---|
| Stromelysin 2 | 3.4.24.22 | COGX_*, (X76537) |
| Stromelysin 3 | - | COGY_*, (Z27093) |
| Membrane-bound metalloendopeptidase | - | (D26512) |
| **Family M11: Autolysin** | | |
| Autolysin | 3.4.24.38 | GLE_CHLRE |
| **Family M7: *Streptomyces* small neutral protease** | | |
| Small neutral protease (*Streptomyces*) | - | (M81703), (M86606), (Z11929) |

[a] See Table I for explanation. References to three-dimensional structures are footnoted *b–f*. References to sequences are footnoted *g–i*.
[b] F.-X. Gomis-Rüth, W. Stöcker, R. Huber, R. Zwilling, and W. Bode, *J. Mol. Biol.* **229**, 945 (1993).
[c] F.-X. Gomis-Rüth, L. F. Kress, and W. Bode, *EMBO J.* **12**, 4151 (1993).
[d] U. Baumann, S. Wu, K. M. Flaherty, and D. B. McKay, *EMBO J.* **12**, 3357 (1993).
[e] B. Lovejoy, A. Cleasby, A. M. Hassell, K. Longley, M. A. Luther, D. Weigl, G. McGeehan, A. B. McElroy, D. Drewry, M. H. Lambert, and S. R. Jordan, *Science* **263**, 375 (1994).
[f] N. S. Andreeva, A. S. Zdanov, A. E. Gustchina and A. A. Fedorov, *J. Biol. Chem.* **259**, 11353 (1984).
[g] S. M. Sato and T. D. Sargent, *Dev. Biol.* **138**, 135 (1990).
[h] L. A. Hite, L.-G. Jia, J. B. Bjarnason, and J. W. Fox, *Arch. Biochem. Biophys.* **308**, 182 (1994).
[i] H. Takeya, S. Nishida, T. Miyata, S. Kawada, Y. Saisaka, T. Morita, and S. Iwanaga, *J. Biol. Chem.* **267**, 14109 (1992).

and M12, and must therefore be very primitive, but it is not present in family M7.

*Astacin Family (M12)*

All known members of the astacin family (Table II) are from animals. A phylogenetic tree constructed to display the relationships between the sequences diagrammatically shows that the astacin family contains two subfamilies (Fig. 4). The divergence between the subfamilies is deep, but they are linked by the relationships of the peptidase domain of bone morphogenetic protein 1 to both ruberlysin (RDF score of 7.26) and atrolysin C (6.81), and that of *Oryzias latipes* hatching enzyme heavy component to ruberlysin (7.04). Both subfamilies contain mosaic proteins.

Astacin, the type enzyme of the astacin subfamily, is a digestive enzyme from the crayfish hepatopancreas ([19], this volume). The enzymes from the medaka fish (*Oryzias latipes*) are involved in hatching of the embryo.[35]

[35] S. Yasumasu, K. Yamada, K. Akasaka, K. Mitsunaga, I. Iuchi, H. Shimada, and K. Yamagami, *Dev. Biol.* **153**, 250 (1992),

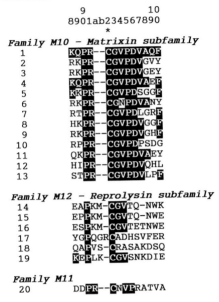

FIG. 3. Conservation of sequences around the activation cysteine in clan MB. Residues are numbered according to human interstitial preprocollagenase. The asterisk (*) indicates the activation cysteine. Residues identical to these in the human interstitial preprocollagenase sequence are shown in white on black. Key to sequences: 1, human interstitial procollagenase; 2, rat interstitial procollagenase; 3, mouse interstitial procollagenase; 4, pig interstitial procollagenase; 5, human neutrophil procollagenase; 6, human progelatinase A; 7, human progelatinase B; 8, rat prostromelysin 1; 9, human prostomelysin 1; 10, human prostromelysin 3; 11, human promatrilysin; 12, mouse macrophage proelastase; 13, proenvelysin; 14, projararhagin; 15, green habu snake proendopeptidase; 16, Malayan pit viper α-profibrinogenase; 17, cattle myelin-associated prometalloendopeptidase; 18, guinea pig pro-PH-30 α subunit; 19, human ORF9 putative prometalloendopeptidase; 20, proautolysin.

Meprin is a membrane component of rodents and humans that occurs in several forms constructed from two (α and β) subunits which are homologous and both contain catalytic sites ([20], this volume). The human bone morphogenetic protein 1, the *tolloid* protein from *Drosophila*, and proteins from sea urchins and *Xenopus* are involved in morphogenesis ([19], this volume).

The crystal structure of astacin shows that Tyr-145 (Fig. 2) acts as a fourth zinc ligand, and this is confirmed by the absorbance spectrum of the copper form of the enzyme. This Tyr residue occurs within the motif Ser-Xaa-Met-His-**Tyr** (where Xaa is an aliphatic hydrophobic residue), which is present in all members of the astacin subfamily. However, there is no

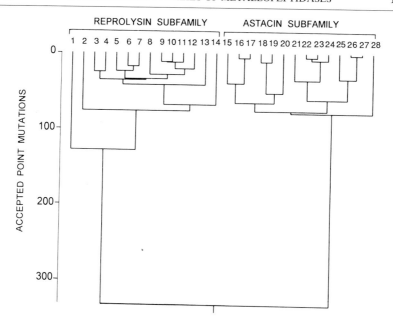

FIG. 4. Phylogenetic tree for the astacin family (M12). The alignment of peptidase domain sequences used for the phylogenetic tree was based on that produced by PILEUP [Genetics Computer Group, *in* "Program Manual for the GCG Package" (Anonymous, ed.). University of Wisconsin, Madison, 1991] using the default parameters. The tree was constructed with the help of the KITSCH program of the PHYLIP package [J. Felsenstein, *Evolution* **39**, 783 (1985)], which assumes a constant evolutionary rate. Because of the low sequence similarity between mammalian reprolysins and the astacin subfamily, the divergence between the two subfamilies may be more distant than depicted. Key: 1, cattle myelin-associated metalloendopeptidase; 2, guinea pig sperm protein PH-30 $\alpha$ subunit; 3, jararhagin; 4, western diamondback rattlesnake atrolysin E; 5, green habu snake endopeptidase; 6, habu snake trimerelysin II; 7, habu snake HR2a endopeptidase; 8, southern copperhead fibrolase; 9, red rattlesnake ruberlysin; 10, eastern diamondback rattlesnake adamalysin; 11, western diamondback rattlesnake atrolysin C; 12, bushmaster snake hemorrhagic factor LHFII; 13, Malayan pit viper $\alpha$-fibrinogenase; 14, human putative metalloendopeptidase; 15, *Drosophila melanogaster tolloid* protein; 16, *Xenopus laevis* bone morphogenetic protein 1; 17, human bone morphogenetic protein 1; 18, *Strongylocentrotus purpuratus* SpAN protein; 19, *Paracentrotus lividus* blastula endopeptidase 10; 20, *X. laevis* UVS.2 protein; 21, rat meprin $\beta$ subunit; 22, rat meprin $\alpha$ subunit; 23, mouse meprin $\alpha$ subunit; 24, human PABA-peptide hydrolase; 25, *Oryzias latipes* low choriolytic enzyme; 26, *Oryzias latipes* high choriolytic enzyme 1; 27, *Oryzias latipes* high choriolytic enzyme 2; 28, crayfish astacin.

such fourth ligand in the structure of adamalysin,[29] nor in any of the known sequences for the reprolysin subfamily.

The activation mechanism of astacin reveals remarkable parallelism to that of chymotrypsin (this series, Volume 244, [2]). The N-terminal Ala formed by cleavage of the propeptide is buried in a water-filled cavity, with the -NH$_3^+$ group forming a salt bridge to the carboxylate of Glu-99 (Fig. 2 numbering), which is the neighboring residue to the third zinc ligand, His-98. Presumably, there is a large conformational change when the proenzyme is processed.

As shown in Fig. 5, the *Oryzias* choriolytic enzymes are very simple, containing just a peptidase domain of about 200 residues in addition to pre- and propeptides. Astacin apparently is similar, but bone morphogenetic protein 1 has a C-terminal extension that includes complement component

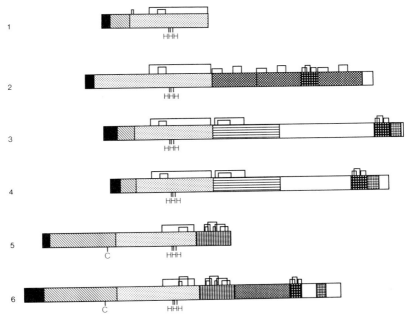

FIG. 5. Polypeptide chain structures of some members of the astacin family (M12). Lengths of chains are proportional to the number of amino acid residues. Brackets indicate known disulfide bonds, and H indicates a zinc ligand. Shaded segments from the top, and left to right correspond to the following: signal peptides (black), propeptides (diagonal lines), peptidase domains (stippled), complement component C1r-like repeats (crosshatched), epidermal growth factor-like domains (bold checks), transmembrane regions (light checks), MAM domains (horizontal lines), disintegrin domains (bricks), and cysteine-rich region (chicken wire). Unshaded areas are of unknown function. Key to structures: 1, medaka fish high choriolytic enzyme; 2, human bone morphogenetic protein 1; 3, mouse meprin α chain; 4, mouse meprin β chain; 5, atrolysin E; 6, guinea pig PH-30 α chain.

C1r-like segments.[36] Such segments are also present in *tolloid* protein from *Drosophila*, UVS.2 protein from *Xenopus*, and the blastula proteinases from the sea urchins *Paracentrotus* and *Strongylocentrotus* ([20], this volume). Apart from the UVS.2 protein, the proteins also contain epidermal growth factor (EGF)-like domains. The EGF-like domains, in turn, are also found in both meprin chains, as well as complement component C1r and other members of the chymotrypsin family. The meprin chains contain the "MAM" structure in the C-terminal extensions that is believed to act as a cell-surface adhesion domain. MAM domains ocur in other membrane-bound proteins, such as the receptor protein tyrosine phosphatase $\mu$ (PTPU _*), which is involved in cytosolic dephosphorylation, and the A5 protein (A5_XENLA) from the visual center of *Xenopus* brain, which also contains C1r-like repeats.[37] All the additional domains in the astacin family are appended C-terminally to the peptidase domain, in contrast to the situation in the chymotrypsin family where similar structures are N-terminal. Because no gene structures are available for members of the astacin family, it is not known how the mosaic organization has arisen, although the involvement of domains also found in the chymotrypsin family would suggest phase 1 exon shuffling.

Astacin contains two disulfide bridges, which are apparently conserved throughout the subfamily. Conservation of cysteine residues in the peptidase domains of other members indicates that they may contain a third disulfide loop.

The second subgroup of the astacin family is termed the reprolysin subfamily following a suggestion of Bjarnason and Fox (see [21] and [22] in this volume), and has also been described as the adamalysins.[33] Most of the known members of the subfamily (Table II) are snake venom metalloendopeptidases, but there are also some mammalian proteins among which are a myelin-associated endopeptidase from cattle ([23], this volume), a human homolog (EMBL:D14665), and protein PH-30, which is involved in sperm–egg fusion in the guinea pig.[38] Protein PH-30 is composed of two homologous subunits, $\alpha$ and $\beta$, but only the $\alpha$ subunit retains the active site.

The pattern of disulfide loops in the peptidase domains of the reprolysin subfamily is quite different to that of the astacin subfamily, emphasizing the deep divergence between the two branches. There are only two disulfide loops in adamalysin,[29] ruberlysin, and atrolysin C, whereas other members of the subfamily have two additional conserved Cys residues in the peptidase domain.

---

[36] N. D. Rawlings and A. J. Barrett, *Biochem. J.* **266,** 622 (1990).
[37] G. Beckmann and P. Bork, *Trends Biochem. Sci.* **18,** 40 (1993).
[38] T. G. Wolfsberg, J. F. Bazan, C. P. Blobel, D. G. Myles, P. Primakoff, and J. M. White, *Proc. Natl. Acad. Sci. U.S.A.* **90,** 10783 (1993).

The precursors of the endopeptidases in the reprolysin subfamily are mosaic proteins. They are synthesized with N-terminal signal peptides and propeptides, but they also have various additional domains attached C-terminally to the peptidase domain (as in the astacin subfamily) (Fig. 5). The propeptides contain a sequence motif similar to that which provides the "cysteine switch" controlling the activity of the matrixin subfamily of family M10 (Fig. 3), and it seems probable that this is functional ([22], this volume). The size of the mature enzymes varies widely, from 25 kDa (ruberlysin) to 60 kDa (trimerelysin I), depending on the extent to which C-terminal, nonpeptidase domains are cleaved off (see [21] and [22]).

The disintegrin domain from the green habu snake enzyme is known as trigramin, and it binds to platelet fibrinogen receptors, preventing platelet aggregation.[39] However, the disintegrin-like domains that are released during the activation of many other snake venom enzymes differ from the disintegrin peptides themselves in the lack of the Arg-Asp-Gly (RGD) motif that has been thought to be of fundamental importance in the activity of the peptides. The mammalian enzymes and jararhagin have a cysteine-rich domain following the disintegrin domain, and PH-30 α subunit also has an EGF-like domain and a potential transmembrane domain.[38]

The known mammalian members of the reprolysin subfamily mostly lack residues that would be essential for peptidase activity. These include mouse cyritestin (EMBL: X64227), a human breast cancer-associated protein, BRCA1 (EMBL: D17390), guinea pig sperm protein PH-30 β subunit (EMBL: Z11720), and AEP I (EMBL: X66139, EMBL: X66140), which is an androgen-regulated epididymal protein cloned from rat and monkey.[38] The proteins contain C-terminal disintegrin-like, EGF-like, and transmembrane domains. The group clearly does contain active peptidases as well, however, as is shown by the characterization of the cattle myelin-associated proteinase by Glynn and co-workers (see [23] in this volume). Other proteins that have the potential to show proteolytic activity, but have not yet been tested, include the α subunit of PH-30 and a human homolog. It may well be that the mammalian members of the reprolysin subfamily will prove to include a group of membrane-associated, cell-surface metalloproteinases with the potential for many important biological functions.

It is apparent that an ancestor of both modern reptiles and mammals must have possessed a reprolysin that contained the metallopeptidase and disintegrin-like domains. Whatever its function, that protein has been the evolutionary precursor not only of a diverse group of metallopeptidases, but also of the disintegrins and related proteins.

---

[39] M. S. Dennis, W. J. Henzel, R. M. Pitti, M. T. Lipari, M. A. Napier, T. A. Deisher, S. Bunting, and R. A. Lazarus, *Proc. Natl. Acad. Sci. U.S.A.* **87,** 2471 (1989).

*Interstitial Collagenase Family (M10)*

The interstitial collagenase family is one of the largest families of metallopeptidases and includes a variety of enzymes from organisms as diverse as bacteria (*Serratia, Erwinia, Pseudomonas*) ([24], this volume), plants (soybean), and animals (sea urchin, mammals). The known enzymes are all secreted endopeptidases, and those in mammals are involved in proteolytic degradation of collagens and other proteins of the extracellular matrix (see [25]–[29]).

The members of the interstitial collagenase family are listed in Table II, and a phylogenetic tree for the sequences (Fig. 6) shows a deep division between two subfamilies, one containing the eubacterial enzymes, conveniently termed the serralysin subfamily, and the other containing eukaryote enzymes such as the matrix metalloproteinases, which can be termed the matrixin subfamily.[40] The divergence between the two subfamilies is very ancient having occurred about 2500 million years ago.

Members of the serralysin subfamily have all been found in gram-negative bacteria, and at least five different genes are known from *Erwinia*. Serralysins are secreted as proproteins, requiring proteolytic activation. All serralysins lack cysteine residues, so proteolytic activation is not via a cysteine-switch mechanism, as is the case in the matrixin subfamily. The only sulfur-containing amino acid in *Pseudomonas* serralysin is Met-143, which occurs in the Met turn thought to be important for active-site structure (Fig. 2).

Serralysins contain six "glycine-rich" repeats (Gly-Gly-Xaa-Gly-Asn-Asp) in the C-terminal domain that are similar to sequences in the nonproteolytic hemolysins from *Actinobacillus, Bordetella, Escherichia*, and *Pasteurella* (HLYA_*, CYAA_BORDE, LKTA_PASHA). The tertiary structure of *Pseudomonas* serralysin shows that the repeats form a tight spiral in which calcium atoms are bound (Ref. 30; [24]). All known eubacterial proteins containing the repeating motif are secreted proteins, although they lack signal peptides. Instead, a special secretion pathway is employed involving the so-called ABC (ATP-binding cassette) family of transmembrane transport proteins. Mutational studies with *Erwinia* serralysin G have shown that a C-terminal 29-residue segment serves as a translocation signal and that the C-terminal Asp-Xaa-Xaa-Xaa motif (where Xaa is any hydrophobic amino acid), which is conserved throughout the subfamily, is essential for secretion.[41]

The catalytic domains in the matrixin subfamily each contain two zinc atoms and one or two calcium atoms. One of the zinc atoms is catalytic, but the other atoms contribute to the structure of the protein.[31] The noncata-

---

[40] J. F. Woessner, Jr., *FASEB J.* **5**, 2145 (1991).
[41] J.-M. Ghigo and C. Wandersman, *J. Biol. Chem.* **269**, 8979 (1994).

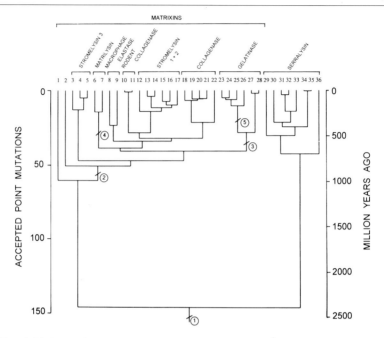

FIG. 6. Phylogenetic tree for the interstitial collagenase family (M10). The alignment of peptidase domain sequences and construction of the phylogenetic tree were as for Fig. 4. The tree has been calibrated for evolutionary distance in millions of years ago by assuming that the divergence between soybean leaf metalloendopeptidase and the other matrixins represents 1000 million years ago. Key: 1, soybean leaf metalloendopeptidase; 2, *Paracentrotus lividus* envelysin; 3, *Xenopus* stromelysin 3; 4, mouse stromelysin 3; 5, human stromelysin 3; 6, cat matrilysin; 7, human matrilysin; 8, mouse macrophage elastase; 9, human macrophage elastase; 10, mouse interstitial collagenase; 11, rat interstitial collagenase; 12, rat stromelysin 2; 13, mouse stromelysin 1; 14, rat stromelysin 1; 15, rabbit stromelysin 1; 16, human stromelysin 2; 17, human stromelysin 2; 18; cattle interstitial collagenase; 19, pig interstitial collagenase; 20, rabbit interstitial collagenase; 21, human interstitial collagenase; 22, human neutrophil collagenase; 23, cattle gelatinase B; 24, human gelatinase B; 25, rabbit gelatinase B; 26, mouse gelatinase B; 27, human gelatinase A; 28, mouse gelatinase A; 29, *Erwinia chrysanthemi* serralysin G; 30, *E. chrysanthemi* serralysin B; 31, *E. chrysanthemi* serralysin D; 32, *E. chrysanthemi* serralysin A; 33, *E. chrysanthemi* serralysin C; 34, *Serratia marcescens* serralysin; 35, *Serratia* sp. serralysin; 36, *Pseudomonas aeruginosa* serralysin. Notable events that occurred during the evolution of the family are marked by numbers in circles (see text).

lytic zinc atom is bound by one Asp and three His residues that are conserved throughout the subfamily. The peptidase domains of the matrixins lack disulfide loops, and, indeed, mature matrilysin, like serralysins, does not contain any cysteine residues at all and the soybean leaf metalloendopeptidase contains just one.[42] It seems that all members of the matrixin

---

[42] G. McGeehan, W. Burkhart, R. Anderegg, J. D. Becherer, J. W. Gillikin, and J. S. Graham, *Plant Physiol.* **99**, 1179 (1992).

subfamily (including that from soybean) are inhibited by the tissue inhibitors of metalloproteinases (TIMPs) ([30], this volume), which have not been found to affect any other metallopeptidases.

Matrilysin and the soybean leaf metalloendopeptidase are simple molecules, each containing only a pre- and propeptide in addition to the catalytic domain. The other members of the matrixin subfamily are mosaic proteins, as can be seen in Fig. 7. The collagenases, stromelysins, and gelatinases each have a C-terminal segment that consists of a short, proline-rich "hinge" sequence and a longer domain that is homologous to vitronectin, limunectin, and hemopexin (VTNC_*, EMBL: M64996, HEMO_*). An extended hinge region accounts for the larger size of the gelatinase B molecule. Vitronectin and limunectin are proteins that have affinity for connective tissue matrix

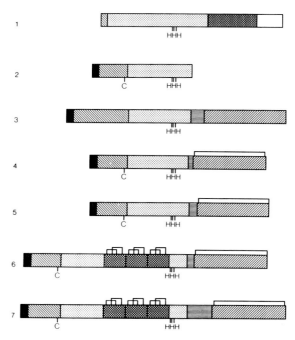

FIG. 7. Polypeptide chain structures of some members of the interstitial collagenase family (M10). Lengths of chains are proportional to the number of amino acid residues. Brackets indicate known disulfide bonds, C indicates the switch cysteine, and H indicates a zinc ligand. Shaded segments correspond to the following: signal peptides (black), propeptides (diagonal lines top left to bottom right), peptidase domains (stippled), calcium-binding domain (chicken wire), hinge regions (horizontal lines), vitronectin-like domains (diagonal lines bottom left to top right), and fibronectin type II-like domains (cross-hatched). Unshaded areas are of unknown function. Key to structures: 1, *Serratia* serralysin; 2, human matrilysin; 3, envelysin; 4, human interstitial collagenase; 5, human stromelysin 1; 6, human gelatinase A; 7, human gelatinase B.

components, and they are composed of divergent repeats of a unit of about 50 residues, which corresponds approximately to a single exon. Gelatinases A and B also each contain an insert in the peptidase domain that shows similarity to type II segments in fibronectin.[43] The recently determined sequence of a membrane-associated matrixin shows a C-terminal, 24 residue, membrane-spanning domain attached to the end of the vitronectin-like domain.[43a]

The structures of the genes known for several of the matrixins are remarkably similar.[44] Not only are comparable exons of similar size, but the exon–intron junctions are closely conserved, suggesting that the insertion of introns predated the formation of the separate enzymes. The fibronectin-like repeats in the gelatinase genes, which are composed of three exons of identical nucleotide number, and the vitronectin-like domains in collagenases, stromelysins, and gelatinases are preceded by phase 1 exon–intron junctions. This suggests that the mosaic character of these proteins arose by phase 1 exon shuffling, in much the same way as did that of serine peptidases of the chymotrypsin family (this series, Volume 244, [2]).

The relationships between the members of the matrixin subfamily and some of the events in their evolution are represented diagrammatically in the phylogenetic tree (Fig. 6). As was mentioned above, the acquisition of the cysteine switch (event ①) occurred very early in the evolution of clan MB, before the divergence of families M10, M11, and M12. It seems probable that the radiation of the matrixins coincided with the appearance of the extracellular matrix proteins that are their substrates, in the very earliest animals. The vitronectin-like segment was introduced into an ancestral matrixin resembling the soybean enzyme between 1000 million and 850 million years ago (event ②), and it was lost from matrilysin as a secondary event between 650 million and 250 million years ago ((④). The gelatinases form a single branch, and it can be reasoned that the insertion of a fibronectin-like segment (event ③) occurred between divergence of the gelatinases from the other matrixins (about 650 million years ago, during early animal evolution) and the gene duplication that gave rise to gelatinases A and B (about 450 million years ago, during the early evolution of vertebrates). The hinge region was extended in the evolution of gelatinase B (event ⑤ in Fig. 6). Thus, invertebrates would be expected to possess only one

[43] T. E. Petersen, H. C. Thøgersen, K. Skorstengaard, K. Vibe-Pedersen, P. Sahl, L. Sottrup-Jensen, and S. Magnusson, *Proc. Natl. Acad. Sci. U.S.A.* **80**, 137 (1983).

[43a] H. Sato, T. Takino, Y. Okada, J. Cao, A. Shinagawa, E. Yamamoto, and M. Seiki, *Nature* **370**, 61 (1994).

[44] M. Gaire, Z. Magbanua, S. McDonnell, L. McNeil, D. H. Lovett, and L. M. Matrisian, *J. Biol. Chem.* **269**, 2032 (1994).

gelatinase. Human stromelysin 3 is a divergent member of the family, with an origin by gene duplication about 800 million years ago.

## Autolysin Family (M11)

The only known member of family M11 is autolysin, from the acellular, biflagellated alga *Chlamydomonas*. Autolysin contains the three His residues and the Met that are conserved throughout the whole of clan MB (Fig. 2).

Autolysin is synthesized as a 70-kDa proenzyme with an N-terminal signal peptide and a 155-residue propeptide that appears to contain a cysteine switch (Fig. 3). The signal peptide targets the proenzyme to the periplasm of the gamete cells, where it is stored. Flagellar agglutination between gametes of the opposing mating types triggers the activation of autolysin by cleavage at a Lys+Glu bond.[45]

The 50-kDa mature enzyme specifically degrades the proline- and hydroxyproline-rich proteins of the algal cell wall, contributing to the degradation of the cell wall that releases the gametes, which then fuse. There are proline-rich regions in both the propeptide and mature enzyme that are characteristic of the hydroxyproline-rich glycoprotein network of *Chlamydomonas* and *Volvox* cell walls, and they may allow autolysin to associate with similar regions in cell wall proteins. These may therefore be functionally analogous to domains in the matrixins that cause them to interact with extracellular matrix components. The soybean leaf metalloendopeptidase (family M10) also can act on cell wall proteins.[42]

## Streptomyces Extracellular Neutral Proteinase Family (M7)

Again, only one member is known for the M7 family, from *Streptomyces* spp. The deduced sequence of the proenzyme shows a signal peptide, a propeptide, and an HEXXH motif.[46] The mature enzyme, which is capable of hydrolyzing milk proteins, is one of the smallest peptidases known, being about 16 kDa. The enzyme differs from thermolysin in several respects: it is stabilized by calcium, it is not inhibited by phosphoramidon (although 1,10-phenanthroline is inhibitory), and it contains a putative disulfide bridge. A Leu-Gly-Leu sequence near the HEXXH motif suggests a structural similarity to interstitial collagenase (Fig. 2). If the catalytic site is indeed similar to that of other members of clan MB, an Asp residue apparently replaces His-98 (Fig. 2 numbering) as the third ligand. Met-143 is the

---

[45] T. Kinoshita, H. Fukuzawa, T. Shimada, T. Saito, and Y. Matsuda, *Proc. Natl. Acad. Sci. U.S.A.* **89,** 4693 (1992).

[46] J. S Lampel, J. S. Aphale, K. A. Lampel, and W. R. Strohl, *J. Bacteriol.* **174,** 2797 (1992).

TABLE III
OTHER METALLOPEPTIDASES WITH HEXXH MOTIF[a]

| Family | EC | Database code |
|---|---|---|
| **Family M3: Thimet oligopeptidase** | | |
| Metalloendopeptidase (*Schizophyllum*) | - | (M97181) |
| Mitochondrial intermediate peptidase | 3.4.24.59 | PMIP_RAT |
| Oligopeptidase A (*Salmonella*) | - | OPDA_* |
| Oligopeptidase M | - | SABP_PIG, (D13310) |
| Peptidase F (*Lactococcus*) | - | (Z32522) |
| Peptidyl-dipeptidase, bacterial | - | DCP_* |
| Saccharolysin | 3.4.24.37 | YCF7_YEAST |
| Thimet oligopeptidase | 3.4.24.15 | MEPD_RAT, (D21871), [b] |
| **Family M6: Immune inhibitor A** | | |
| Immune inhibitor A (*Bacillus thuringiensis*) | - | INA_BACTL |
| **Family M8: Leishmanolysin** | | |
| Leishmanolysin | 3.4.24.36 | GP63_*, (M94364), (M94365), (X64394) |
| **Family M9: *Vibrio* collagenase** | | |
| Collagenase (*Vibrio*) | 3.4.24.3 | (X62635) |
| **Family M31: Clostridial collagenase** | | |
| Collagenase (*Clostridium*) | 3.4.24.3 | (D13791) |
| **Family M26: IgA-specific metalloendopeptidase** | | |
| IgA-specific metalloendopeptidase (*Streptococcus*) | 3.4.24.13 | [c] |
| **Family M27: Tetanus toxin** | | |
| Tetanus toxin | - | TETX_CLOTE |
| Botulinum neurotoxin | - | BXA_CLOBO, BXB_CLOBO, BXC1_CLOBO, BXD_CLOBO, BXE_CLOBO, BXF_CLOBO |
| **Family M30: *Staphylococcus* neutral protease** | | |
| Neutral protease (*Staphylococcus*) | - | (X73315) |
| **Family M32: Carboxypeptidase Taq** | | |
| Carboxypeptidase Taq (*Thermus*) | - | (D17669) |

[a] See Table I for explanation.
[b] H. F. Dovey, P. A. Seubert, S. Sinha, L. Conlogue, S. P. Little, and E. M. Johnstone, Patent WO 92/07068 (1992).
[c] J. V. Gilbert, A. G. Plaut, and A. Wright, *Infect. Immun.* **59**, 7 (1991).

only Met residue C-terminal to the HEXXH motif and may be part of a Met-turn, characteristic of clan MB.

## Other HEXXH Metallopeptidase Families

A number of families of metallopeptidases contain the HEXXH motif that is almost certainly part of a zinc-binding site, but additional ligands have not yet been identified with certainty (Table III).

### Thimet Oligopeptidase Family (M3)

Thimet oligopeptidase is a cytosolic, thiol-activated enzyme of higher animals that is capable only of cleaving peptides of 6–17 residues. As reported in [32] in this volume, a second, mitochondrial oligopeptidase,

oligopeptidase M, with different specificity has been discovered. Other homologs that share the oligopeptidase specificity are known from yeast (saccharolysin)[47] and eubacteria (see [34] and [35]). As described in [34], the family also contains a eubacterial peptidyl-dipeptidase with specificity and inhibitor characteristics similar to those of the mammalian angiotensin-converting enzyme.

The mammalian mitochondrial intermediate peptidase involved in the processing of nuclear-encoded mitochondrial proteins is also a member of family M3 ([33]). Proteins transported into the mitochondrion possess an N-terminal targeting signal that is removed by the mitochondrial processing peptidase ([46]), but some also contain an additional octapeptide that is removed by the mitochondrial intermediate peptidase.

The sequences of members of the thimet oligopeptidase family contain the HEXXH motif, but neither catalytic residues nor zinc ligands have been experimentally confirmed. There are two conserved Glu residues C-terminal to the HEXXH motif, Glu-172 and Glu-179 (Fig. 8 numbering). The conserved His-149 seems too close to the HEXXH motif to be a ligand. The thiol dependence of thimet oligopeptidase might be explained by the presence of a cysteine residue (Cys-152) close to the HEXXH motif, but the presence of the residue is not correlated with thiol dependence in other members of the family.

The mix of distinctive endopeptidase, oligopeptidase, and exopeptidase activities in family M3 makes it the most diverse of the metallopeptidase families, and the elucidation of the structural basis of such large differences in a homologous set of proteins will be awaited with the greatest interest. Like family M1, family M3 shows a surprising similarity between prokaryote and eukaryote sequences, suggesting a rather recent introduction of the family into eukaryotic cells.

*Immune Inhibitor A Family (M6)*

The single known member of family M6 is an endopeptidase from *Bacillus thuringiensis*, a bacterium that is a pathogen of lepidoptera and is used in their control. The immune inhibitor A, one of several active factors secreted by the bacterium, is named for its activity in the degradation of host antibacterial proteins attacin and cecropin.[48] The amino acid sequence includes the HEXXH motif (Fig. 8), but otherwise there is little similarity to other enzymes.

---

[47] M. Büchler, U. Tisljar, and D. H. Wolf, *Eur. J. Biochem.* **219**, 627 (1994).
[48] A. Lövgren, M. Zhang, A. Engström, G. Dalhammar, and R. Landén, *Mol. Microbiol* **4**, 2137 (1990).

```
              14        15        16        17        18
      345678901234567890123456789012345678901234567890123456789012345678
              *+    *                            *
Family M4
   1  LSGGIDVVAHELTHAVTDYTAGLIYQNESGAINEAISDIFGTLVEFYANKNPDWEI

Family M3
   2  QHDEVETYFHEFGHVMHQLCSQAEFAMFSG-THVERDFVEAPSQMLENWVWEKEPL
   3  RHDEVRTYFHEFGHVMHQICAQTDFARFSG-TNVETDFVEVPSQMLENWVWDIDSL
   4  TPGMMENLFHEMGHAMHSMLGRTRYQHVTG-TRCPTDFAEVPSILMEYFSNDYRVV
   5  KHNEIVTFFHEIGHGIHDLVGQNKESRFNGPGSVPWDFVEAPSQMLEFWTWNKNEL
   6  THDEVITLFHEFGHGLHHMLTRIETAGVSGISGVPWDAVELPSQFMENWCWEPEAL
   7  IWDDVITLFHEFGHTLHGLFARQRYATLSG-TNTPRDFVEFPSQINEHWATHPQVF
   8  DDLFTLVHETGHSMHSAFTRENQPYVYG--NYPIFLAEIASTTNEN

Family M6
   9  EDGAVGVNAHEYGHDLGLPDEYDTDYTGHGEPIQAWSVMSGGTWAGKIAGTTPTSF

Family M8
  10  DQLVTRVVTHEMAHALGFSGPFFEDARIVANVPNVRGKNEDVPVINSSTAVAKARE
  11  DQLVTRVVTHEMAH-VGFSGTFFTEILLVTQMMNIRGKDPNVSVINSSTAVAKARE
  12  DHLMVHAVTHEIAHSLGFSNAFFTNTGIGQFVTGVRGNPDTVPVINSPTVVAKARE

Family M9
  13  ADHFVWNLEHEYVHYLDGRFDLYGGFSHPTEKIVWWSEGIAEYVAQENDNQAALET

Family M31
  14  IYTLEELFRHEFTHYLQGRYVVPGMWGQEFYQEGVLTWYEEGTAEFFAGSTRTDG

Family M26
  15  DKDGAITYTHEMTHDSDNEIYLGGYGRRSGLGPEFFAKGLLQAPDHPDDATITVNS

Family M27
  16  FQDPALLLMHELIHVLHGLYGMQV--SSHEIIPSKQEIYMQHTYPISAEELFTFGG
  17  ATDPAVTLAHELIHAGHRLYGIAI-NPNRVFKVNTNAYYEMSGLEVSFEELRTFGG
  18  CMDPILILMHELNHAMHNLYGIAIPNDQTISSVTSNIFYSQYNVKLEYAEIYAFGG
  19  CMDPVIALMHELTHSLHQLYGINIPSDKRIRPQVSEGFFSQDGPNVQFEELYTFGG

Family M30
  20  EKKVYSTLAHEYQHMVNANQKLLKEQKEDGMDVWLDEAFAMASEHMYLQKPLDHRI
```

FIG. 8. Conservation of sequences around the zinc ligands and catalytic residues of other metallopeptidases with the HEXXH motif. Residues are numbered according to thermolysin (included for comparison). Asterisks (*) indicate the zinc ligands, and + indicates the catalytic Glu in thermolysin. Residues identical to those in the thermolysin sequence are shown in white on black. Key to sequences: 1, thermolysin; 2, rat thimet oligopeptidase; 3, rabbit "microsomal endopeptidase"; 4, rat mitochondrial intermediate peptidase; 5, yeast saccharolysin; 6, *Escherichia coli* oligopeptidase A; 7, *E. coli* peptidyl-dipeptidase; 8, *Lactococcus lactis* peptidase F; 9, *Bacillus thuringiensis* immune inhibitor A; 10, *Leishmania major* leishmanolysin; 11, *Leishmania mexicana* leishmanolysin; 12, *Crithidia fasciculata* leishmanolysin; 13, *Vibrio alginolyticus* collagenase; 14, *Clostridium perfringens* collagenase; 15, *Streptococcus sanguis* IgA-specific metalloendopeptidase; 16, tetanus toxin; 17, botulinum neurotoxin type A; 18, botulinum neurotoxin type C1; 19, botulinum neurotoxin type D; 20, *Staphylococcus hyicus* neutral metalloendopeptidase.

## Leishmanolysin Family (M8)

Leishmanolysin is an enzyme of *Leishmania* and some related parasitic protozoa in the trypanosomatid group (this volume, [37]). In the promastigote form of the parasite, the endopeptidase is the major cell-surface protein and is attached to the cell membrane by a glycosylphosphatidylinositol anchor. The amastigote resides in the lysosomes of host macrophages and produces forms of the enzyme that differ from the great majority of metallopeptidases in having acidic pH optima that may well be an adaptation to that environment. The 63-kDa protein has a predominantly $\beta$ structure and contains an HEXXH motif. The nearest of several potential zinc ligands is Glu-188 (Fig. 8).

## Vibrio Collagenase Family (M9)

Few endopeptidases have the ability to cleave the helical region of native collagen, but nevertheless the ability to cleave collagen has evolved independently on several occasions. Among serine peptidases, there are brachyurin and crustacean trypsins in family S1, and among the metallopeptidases there are the interstitial collagenases (M10) and microbial enzymes in at least two families. Unlike the interstitial collagenases, the microbial enzymes cleave collagen helices at multiple sites. The collagenases from *Vibrio* (family M9) and *Clostridium* (M31) show no similarity in sequence to vertebrate interstitial collagenase except for the HEXXH motif.

*Vibrio* collagenase cleaves the helical region of collagen at Xaa+Gly-Pro-Xaa bonds. It is secreted enzyme and is synthesized as a precursor with a signal peptide and apparently a propeptide. Activation occurs by cleavage of a Ser+Thr bond and possibly other bonds, because three different sized forms of the active enzyme are known.[49]

## Clostridium Collagenase Family (M31)

The collagenase from *Clostridium perfringens* is a large (120-kDa, 1104-residue), mosaic protein which shows no resemblance to the peptidase domains of *Vibrio* or mammalian collagenases.[50] However, both *Clostridium* and *Vibrio* collagenases share a domain C-terminal to the peptidase domain, which is also found in *Achromobacter* lysyl endopeptidase (family S5). At the C terminus of clostridial collagenase there are two repeats of a 90-residue domain that is homologous to incomplete open reading frames

[49] H. Takeuchi, Y. Shibano, K. Morihara, J. Fukushima, S. Inami, B. Keil, A.-M. Gilles, S. Kawamoto, and K. Okuda, *Biochem. J.* **281,** 703 (1992).
[50] O. Matsushita, K. Yoshihara, S. Katayama, J. Minami, and A. Okabe, *J. Bacteriol.* **176,** 149 (1994).

sequenced in conjunction with the genes for phospholipase C from *Bacillus thuringiensis* (EMBL: X12952) and *Bacillus cereus* (EMBL: M30809).

## *Immunoglobulin A-Specific Metalloendopeptidase Family (M26)*

The immunoglobulin A (IgA)-specific endopeptidase from *Streptococcus sanguis* cleaves the Pro+Thr bond in the hinge region of the heavy chain of immunoglobulin A (this volume, [38]). Unrelated serine peptidases with similar specificity are known (see this series, Volume 244, [11]). The *Streptococcus* enzyme of 1669 amino acids contains an HEXXH motif, and site-specific mutagenesis of His-142 or Glu-143 (Fig. 8 numbering) abolishes activity. It has also been postulated that Glu-166 is the third zinc ligand. There are 10 tandem repeats of an 18- to 20-residue peptide, VXPXQVXXXPEYXGXXXGAX, of unknown function near the N terminus (residues 139–336).

## *Tetanus Toxin Family (M27)*

Tetanus and botulism are caused by species of gram-positive, anaerobic bacteria in the genus *Clostridium*. In both conditions, neurotoxins cause motor paralysis by blocking acetylcholine release at the neuromuscular junction, and it has been discovered that the neurotoxins are metalloproteinases (this volume, [39]). The enzymes are inhibited by EDTA, 1,10-phenanthroline, and captopril.[51]

The botulinus toxin is synthesized as a 150-kDa precursor, which is proteolytically processed into a two-chain molecule (a 50-kDa, N-terminal light chain and a 100-kDa, C-terminal heavy chain), with the chains linked by a disulfide bridge. The protein accumulates in the bacterial cell until it lyses. The heavy chain has a high and specific affinity for neurons, and after binding it induces internalization of the protein into endosome-like vesicles. The protein then passes into the cytoplasm, and the free, catalytically active light chain is able to carry out a specific cleavage of synaptobrevin, a component of the neuroexocytosis apparatus. Botulinum toxin cleaves each of the two isoforms of synaptobrevin at $Gln^{60}$+Lys, whereas the tetanus toxin cleaves at the $Gln^{76}$+Phe bond found only in synaptobrevin 2.[51] The catalytic domains of the toxins contain a zinc-binding motif that conforms to the Jongeneel[52] consensus, and the botulinum toxin has been shown to contain one atom of zinc per molecule. A potential third ligand could be the conserved Glu-182 (Fig. 8 numbering), which is approximately the same distance from the HEXXH motif as the third ligand in thermolysin. The conserved His-149 (also seen in family M3) seems too close to be a ligand.

[51] G. Schiavo, C. C. Shone, O. Rossetto, F. C. G. Alexander, and C. Montecucco, *J. Biol. Chem.* **268**, 11516 (1993).

[52] C. V. Jongeneel, J. Bouvier, and A. Bairoch, *FEBS Lett.* **242**, 211 (1989).

*Staphylococcus hyicus Extracellular Metalloendopeptidase Family (M30)*

Family M30 is another one-member family, which contains a secreted, 38.5-kDa zinc endopeptidase. The enzyme is synthesized with a signal peptide and a 75-residue propeptide. It is inhibited by 1,10-phenanthroline but, unlike thermolysin (family M4), not inhibited by phosporamidon.[53] The sequence includes an HEXXH motif, and Glu-269 may be a third ligand.[53]

*Carboxypeptidase Taq Family (M32)*

Carboxypeptidase Taq is an enzyme from the thermophilic eubacterium *Thermus aquaticus* that is synthesized with a 28-residue prepropeptide.[53a] The carboxypeptidase contains one atom of zinc per molecule. The amino acid sequence shows no relationship to any other known protein, but does contain an HEXXH potential zinc-binding site, and carboxypeptidase Taq may therefore be the first carboxypeptidase to be discovered that binds zinc at such a motif.

Families of Metallopeptidases with Metal Ligands Other than HEXXH

Five families of metallopeptidases are known from crystallographic structures or site-directed mutagenesis to bind the essential metal atoms without the involvement of the HEXXH consensus sequence that is important in so many of the families. In the endopeptidases of the pitrilysin family, it seems that an HXXEH sequence binds zinc. Most of the ligands in the two families of carboxypeptidases (M14, M15) are His residues, whereas the ligands of the pairs of metal atoms in the aminopeptidase families (M17, M24) are predominantly carboxylic acids (Fig. 9).

*Pitrilysin Family (M16)*

The pitrilysin family contains two very different subfamilies of enzymes (Table IV). In the pitrilysin subfamily are pitrilysin, which is a periplasmic oligopeptidase from *E. coli*, and the animal enzymes insulysin and NRD-convertase. The second subfamily is that of mitochondrial processing peptidase (MPP), which also contains subunits of ubiquinol–cytochrome–$c$ reductase.

Mature pitrilysin (this volume, [43] and [44]) is synthesized as a large protein of about 950 residues including a 23-residue signal peptide.[54] Site-

[53] S. Ayora and F. Götz, *Mol. Gen. Genet.* **242,** 421 (1994).

[53a] S.-H. Lee, H. Taguchi, E. Yoshimura, E. Minagawa, S. Kaminogawa, T. Ohta, and H. Matsuzawa, *Biosci. Biotechnol. Biochem.* **58,** 1490 (1994).

[54] P. W Finch, R. E. Wilson, K. Brown, I. D. Hickson, and P. T. Emmerson, *Nucleic Acids Res.* **14,** 7695 (1986).

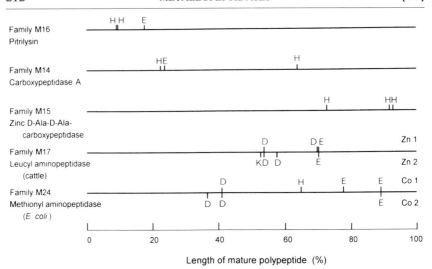

FIG. 9. Metal ligands in metallopeptidases lacking the HEXXH consensus sequence. The location of each ligand residue (as a percentage of the distance from the N to C terminus in the mature protein) is indicated, using the one-letter code. The aminopeptidases (families M17 and M24) bind two metal ions per moelcule.

directed mutagenesis has identified His-88, His-92, and Glu-169 as zinc ligands, and Glu-91 as the catalytic residue.[55,56] These residues are conserved in insulysin and other members of the subfamily (Fig. 10). The HXXEH motif is apparently a functional inversion of the HEXXH zinc-binding motif found in so many other families. It seems that only the N-terminal 50 kDa or so of the large molecules of pitrilysin and insulysin are directly involved in catalysis, as all members of the MPP subfamily are smaller.

Pitrilysin-like proteins are found in other bacteria, including the product of the *pqqF* gene from *Klebsiella pneumoniae*.[57] Although no enzymology has been done on that protein, it retains the zinc ligands and the catalytic residue, and is probably a peptidase. The sequence is more divergent from pitrilysin than would be expected for bacterial species that are so closely related (both are enterobacters) and may represent a different enzyme.

Insulysin, unlike pitrilysin, is a thiol-dependent enzyme (this volume, [44]), and it occurs in *Drosophila* as well as mammals.[58] At the C terminus,

[55] A. B. Becker and R. A. Roth, *Proc. Natl. Acad. Sci. U.S.A.* **89,** 3835 (1992).

[56] A. B. Becker and R. A. Roth, *Biochem. J.* **292,** 137 (1993).

[57] J. J. M. Meulenberg, E. Sellink, W. A. M. Loenen, N. H. Riegman, M. van Kleff, and P. W. Postma, *FEMS Microbiol. Lett.* **71,** 337 (1990).

[58] W.-L. Kuo, B. D. Gehm, and M. R. Rosner, *Mol. Endocrinol.* **4,** 1580 (1990).

TABLE IV
OTHER METALLOPEPTIDASES WITH KNOWN METAL LIGANDS[a]

| Family | EC | Database code |
|---|---|---|
| **Family M16: Pitrilysin** | | |
| **Pitrilysin subfamily** | | |
| Insulysin | 3.4.24.56 | IDE_*, (X67269), (Z17458) |
| Metalloendopeptidase (*Bacillus*) | - | YPP_BACSU |
| Pitrilysin | 3.4.24.55 | PTR_ECOLI |
| *pqqF* gene product (*Klebsiella*) | - | PQQF_KLEPN |
| Sporozoite developmental protein (*Eimeria*) | - | (M98842) |
| **Mitochondrial processing peptidase subfamily** | | |
| Mitochondrial processing peptidase α subunit | 3.4.99.41 | MPP1_* |
| Mitochondrial processing peptidase β subunit | 3.4.99.41 | MPP2_*, (L12965) |
| **Family M14: Carboxypeptidase A** | | |
| **Carboxypeptidase A subfamily** | | |
| Carboxypeptidase A[b] | 3.4.17.1 | CBPA_*, CBPC_*, CBP1_*, CBP2_RAT, (M27717) |
| Carboxypeptidase B | 3.4.17.2 | CBPB_*, (M75106) |
| Carboxypeptidase (*Simulium*) | - | (L08481) |
| Carboxypeptidase (*Streptomyces*) | - | CBPS_STRRA, (U00619) |
| Carboxypeptidase T (*Thermoactinomyces*) | 3.4.17.18 | CBPT_THEVU |
| **Carboxypeptidase H subfamily** | | |
| Carboxypeptidase H | 3.4.17.10 | CBPH_*, [c] |
| Carboxypeptidase M | 3.4.17.12 | CBPM_HUMAN |
| Carboxypeptidase (*Drosophila*) | - | (U03883) |
| Lysine carboxypeptidase | 3.4.17.3 | CBPN_HUMAN |
| **Family M15: Zinc D-Ala-D-Ala carboxypeptidase** | | |
| Muramoyl-pentapeptide carboxypeptidase | 3.4.17.8 | CBPM_STRGR |
| Zinc D-Ala-D-Ala carboxypeptidase | 3.4.17.14 | (X55794) |
| **Family M17: Leucyl aminopeptidase** | | |
| Aminopeptidase A (*Escherichia coli*) | - | AMPA_ECOLI |
| Leucyl aminopeptidase[b] | 3.4.11.1 | AMPL_ARATH, AMPL_BOVIN, AMPL_RICPR, AMPL_SOLTU, YOJ6_CAEEL |
| **Family M24: Methionyl aminopeptidase** | | |
| Methionyl aminopeptidase[b] | 3.4.11.18 | AMPM_* |
| X-Pro aminopeptidase (*Escherichia coli*) | 3.4.11.9 | AMPP_ECOLI, (M91546) |
| X-Pro dipeptidase | 3.4.13.9 | PEPQ_ECOLI, PEPD_HUMAN, (M88837) |

[a] See Table I for explanation.
[b] Structure is included in the Brookhaven database.
[c] W. W. Roth, R. B. Mackin, J. Spiess, R. H. Goodman, and B. D. Noe, *Mol. Cell. Endocrinol.* **78,** 171 (1991).

insulysin bears the tripeptide -(Ala or Ser)-Lys-Leu, a peroxisomal targeting signal.[59]
Described in [45], this volume, is a homologous enzyme termed N-arginine dibasic convertase (NRD-convertase) that cleaves on the amino

[59] F. Authier, R. A. Rachubinski, B. I. Posner, and J. J. M. Bergeron, *J. Biol. Chem.* **269,** 3010 (1994).

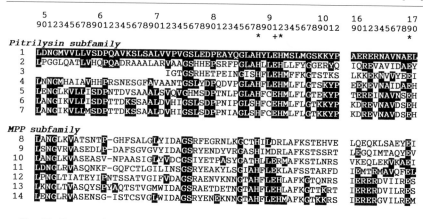

FIG. 10. Conservation of sequences around the zinc ligands and catalytic residues of the pitrilysin family (M16). Residues are numbered according to pitrilysin. Asterisks (*) indicate the zinc ligands, and + indicates the catalytic Glu residue. Residues identical to those in the pitrilysin sequence are shown in white on black. Key to sequences: 1, pitrilysin; 2, *pqqF* gene product from *Klebsiella pneumoniae*; 3, orfP gene product from *Bacillus subtilis*; 4, *Eimeria bovis* sporozoite protein; 5, *Drosophila melanogaster* insulysin; 6, rat insulysin; 7, human insulysin; 8, yeast mitochondrial processing peptidase (MPP) α subunit; 9, *Neurospora crassa* MPP α subunit; 10, potato MPP α subunit; 11, rat MPP α subunit; 12, yeast MPP β subunit; 13, *Neurospora crassa* MPP β subunit; 14, rat MPP β subunit.

side of Arg residues in the pairs of basic amino acids that so commonly mark the points of posttranslational processing. The enzyme differs from other members of the family in having an Asp/Glu-rich segment inserted very close to the catalytic site.

In the second subfamily, MPP is responsible for removing the mitochondrial target sequence attached to the N terminus of mitochondrial proteins that are synthesized in the cytoplasm ([46] in this volume). This is not the only processing enzyme in the mitochondrion; there is a mitochondrial leader peptidase, which removes leader peptides from proteins synthesized in the mitochondrion (family S26), and also the mitochondrial intermediate peptidase (family M3; [33]), which completes the processing of some imported mitochondrial proteins after the action of MPP. Although some cleavage sites for MPP are known, the rules governing the specificity remain unclear.[60]

Mitochondrial processing peptidase is a heterodimer, the subunits being homologous but not identical. The subunit now known as β-MPP[61] has the zinc ligands and the catalytic Glu residue of pitrilysin conserved, whereas α-MPP has His-92 (Fig. 10 numbering) replaced by Arg (or Lys in the rat

[60] B. S. Glick, E. M. Beasley, and G. Schatz, *Trends Biochem. Sci.* **17**, 453 (1992).
[61] F. Kalousek, W. Neupert, T. Omura, G. Schatz, and U. K. Schmitz, *Trends Biochem. Sci.* **18**, 249 (1993).

enzyme) and Glu-91 replaced by Asp. Thus $\beta$-MPP appears to be the catalytic subunit, but neither subunit is active alone.[62] $\alpha$-MPP and $\beta$-MPP are each about one-half the size of pitrilysin, so that the heterodimer is about the same size as pitrilysin. However, this appears to be coincidental, as there is no evidence that pitrilysin has evolved by gene duplication. Both $\alpha$-MPP and $\beta$-MPP are nuclear-encoded proteins synthesized in the cytoplasm and have N-terminal mitochondrial targeting signals.

The $\alpha$-MPP subunit has a long glycine-rich conserved motif, (Gly)$_4$-Ser-Phe-Ser-Ala-(Gly)$_2$-Pro-Gly-Lys-Gly-Met-Tyr-Ser-Arg-Leu-Tyr, in the C-terminal segment that is not found in other members of the family or any other protein. *Neurospora* $\beta$-MPP is also subunit I, and the potato $\alpha$-MPP is also subunit III, of the ubiquinol–cytochrome-*c* reductase complex. In *Saccharomyces* and mammals the reductase complex subunits and MPP subunits are different proteins, though subunits I (UCR1_YEAST) and II UCR2_*) of the cytochrome-*c* reductase are distant homologs of MPP.[63] Cytochrome-*c* reductase is a multisubunit enzyme (9 subunits in fungi, 10 in plants, and 11 in mammals); however, the actual mechanism involves only three subunits, and the functions of the other subunits are unknown.[64] The product of an open reading frame from *Plasmodium falciparum*, cloned in association with a multidrug resistance gene, is also a homolog of MPP but lacks the zinc ligands and catalytic Glu residue.[65]

A dated phylogenetic tree (not shown) gives a time of 1400 million years ago for the divergence of pitrilysin and insulysin. This is too recent to represent the eukaryote/eubacterium divergence, which suggests that insulysin probably reached eukaryotes by transfer of a gene from an endosymbiont. The divergence of the pitrilysin and MPP subfamilies is calculated as 3500 million years ago, however.

*Carboxypeptidase A Family (M14)*

The carboxypeptidase A family contains two subfamilies of carboxypeptidases (Table IV; Fig. 11). Members of the carboxypeptidase H (or "regulatory") subfamily have sequences that are longer at the C terminus than those of the carboxypeptidase A ("digestive") subfamily.[66] Carboxypeptidase M (this volume, [41]) is membrane bound via a glycosylphosphatidylinositol anchor, but most other carboxypeptidases such as lysyl carboxypeptidase

[62] V. Géli, *Proc. Natl. Acad. Sci. U.S.A.* **90,** 6247 (1993).

[63] U. Schulte, M. Arretz, H. Schneider, M. Tropschug, E. Wachter, W. Neupert, and H. Weiss, *Nature (London)* **339,** 147 (1989).

[64] H. Weiss, K. Leonard, and W. Neupert, *Trends Biochem. Sci.* **15,** 178 (1990).

[65] T. Triglia, S. J. Foote, D. J. Kemp, and A. F. Cowman, *Mol. Cell. Biol.* **11,** 5244 (1991).

[66] A. L. Osterman, N. V. Grishin, S. V. Smulevitch, M. V. Matz, O. P. Zagnitko, L. P. Revina, and V. M. Stepanov, *J. Protein Chem.* **11,** 561 (1992).

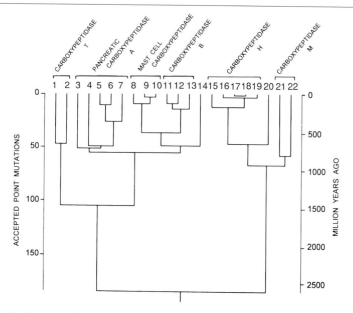

FIG. 11. Phylogenetic tree for the carboxypeptidase A family (M14). The alignment of peptidase sequences and construction of the phylogenetic tree were as for Fig. 4. Key: 1, *Thermoactinomyces* carboxypeptidase T; 2, *Streptomyces* carboxypeptidase; 3, crayfish pancreatic carboxypeptidase B; 4, *Simulium* pancreatic carboxypeptidase A; 5, human pancreatic carboxypeptidase A; 6, cattle pancreatic carboxypeptidase A; 7, rat pancreatic carboxypeptidase A; 8, human mast cell carboxypeptidase; 9, rat mast cell carboxypeptidase; 10, mouse mast cell carboxypeptidase; 11, cattle pancreatic carboxypeptidase B; 12, human pancreatic carboxypeptidase B; 13, rat carboxypeptidase B; 14, human arginine carboxypeptidase; 15, anglerfish carboxypeptidase H; 16, human carboxypeptidase H; 17, rat carboxypeptidase H; 18, mouse carboxypeptidase H; 19, cattle carboxypeptidase H; 20, human lysine carboxypeptidase; 21, *Drosophila* carboxypeptidase; 22, human carboxypeptidase M.

([40]) and carboxypeptidase T ([42]) are soluble. The carboxypeptidase A subfamily includes enzymes from animals and actinomycetes, and the calculated time of divergence between cattle carboxypeptidase A and *Thermoactinomyces* carboxypeptidase is about 1300 millions years ago, suggesting an endosymbiont origin for the family in eukaryotes.

As shown in Fig. 12, zinc ligands in human carboxypeptidase A are His-69, Glu-72, and His-196.[67] The catalytic residue is Glu-270. Those residues are conserved in all members of the family. The S1' subsite binds the C terminus of the substrate and is the primary recognition site in carboxypeptidases. In rat mast cell carboxypeptidase A2, which releases bulky C-terminal residues, Thr-268 is replaced by Ala and Leu-203 by Met, enlarging the

[67] D. C. Rees, M. Lewis, and W. N. Lipscomb, *J. Mol. Biol.* **168**, 367 (1983).

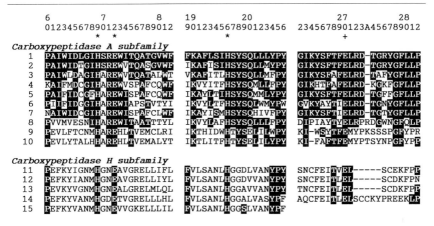

FIG. 12. Conservation of sequences around the zinc ligands and catalytic residues of the carboxypeptidase A family (M14). Residues are numbered according to cattle carboxypeptidase A. Asterisks (*) indicate the zinc ligands, and + indicates the catalytic Glu residue. Residues identical to those in the cattle carboxypeptidase A sequence are shown in white on black. Key to sequences: 1, cattle carboxypeptidase A; 2, human carboxypeptidase A1; 3, rat carboxypeptidase A2; 4, human carboxypeptidase A; 5, rat carboxypeptidase B; 6, crayfish carboxypeptidase B; 7, human plasma carboxypeptidase B; 8, *Simulium vittatum* carboxypeptidase; 9, *Streptomyces capreolus* carboxypeptidase; 10, *Thermoactinomyces vulgaris* carboxypeptidase T; 11, human carboxypeptidase H; 12, anglerfish carboxypeptidase H; 13, human lysine carboxypeptidase; 14, human carboxypeptidase M; 15, *Drosophila melanogaster* carboxypeptidase fragment.

specificity pocket.[68] The conservation of other residues involved in catalysis and the specificity of the enzymes have been discussed fully by Osterman *et al.*[66]

The enzymes of the carboxypeptidase A family, including those from bacteria, are synthesized as inactive precursors with propeptides. Three-dimensional structures of procarboxypeptidases A and B show that in each the propeptide, which has a globular domain followed by an extended α-helical segment, shields the catalytic site without making contact with it, whereas the substrate-binding site is blocked by specific contacts.[69]

### Zinc D-Ala-D-Ala Carboxypeptidase Family (M15)

The zinc D-Ala-D-Ala carboxypeptidase differs from most serine-type D-Ala-D-Ala carboxypeptidases (families S11–S13; this series, Volume 244, [2]) in that the metalloenzyme does not perform the transpeptidation reac-

[68] Z. Faming, B. Kobe, C.-B. Stewart, W. J. Rutter, and E. J. Goldsmith, *J. Biol. Chem.* **266**, 24606 (1991).

[69] A. Guasch, M. Coll. F. X. Avilés, and R. Huber, *J. Mol. Biol.* **224**, 141 (1992).

tion and has no structural relationship to $\beta$-lactamases.[70] The enzyme is synthesized with an N-terminal extension of 42 residues which includes a signal peptide; whether a propeptide exists is uncertain.[71]

The crystallographic structure of the enzyme from *Streptomyces* has been determined and the zinc ligands identified as His-153, His-194, and His-196.[72] The tertiary structure shows no relation to those of other zinc peptidases, and the arrangement of ligands is unique in that the short–long spacing[73] is reversed (Fig. 9). The molecule comprises two domains, of which the more N-terminal contains 80 residues and is homologous with the C-terminal third of *Bacillus* N-acetylmuramoyl-L-alanine amidase (ALYS_BACSU). Thus, the zinc D-Ala-D-Ala carboxypeptidase is a mosaic protein.

*Leucyl Aminopeptidase Family (M17)*

The enzymes of family M17 contain "cocatalytic" zinc atoms as defined in Ref. 74. In other words, two metal atoms are located very close together, bound mainly by carboxylic amino acids, with at least one ligand residue shared by the two zinc atoms.

The family contains aminopeptidases from bacteria, animals, and plants (Table IV). A crystallographic structure has been determined for the cattle lens enzyme, showing it to be a homohexamer.[75] Two zinc atoms are bound per subunit by residues clustered in the relatively short segment at residues 250–334. For one atom, the ligands are Asp-255, Asp-332, and Glu-334, and for the other, Lys-250, Asp-255, Asp-273, and Glu-334 (Figs. 9 and 13). Also present in the active site are Lys-262 and Arg-336. All those residues are conserved in all known members of the family. The roles of the two metal atoms in catalysis have been discussed by Kim and Lipscomb.[76] The monomeric unit of cattle leucyl aminopeptidase has a two-domain structure with the catalytic site restricted to the C-terminal domain, and the N-terminal 137 residues can be removed without affecting activity.[77] In the native hexamer, the six active sites are located in the interior.[76]

Bacterial members of family M17 include homologs from *E. coli* and

[70] B. J. Sutton, P. J. Artymiuk, A. E. Cordero-Borboa, C. Little, D. C. Phillips, and S. G. Waley, *Biochem. J.* **248**, 181 (1987).
[71] C. Duez, B. Lakaye, S. Houba, J. Dusart, and J.-M. Ghuysen, *FEMS Microbiol. Lett.* **71**, 215 (1990).
[72] O. Dideberg, P. Charlier, G. Dive, B. Joris, J. M. Frère, and J. M. Ghuysen, *Nature* (*London*) **299**, 469 (1982).
[73] B. L. Vallee and D. S. Auld, *Biochemistry* **29**, 5647 (1990).
[74] B. L. Vallee and D. S. Auld, *Proc. Natl. Acad. Sci. U.S.A.* **90**, 2715 (1993).
[75] H. Kim and W. N. Lipscomb, *Biochemistry* **32**, 8465 (1993).
[76] H. Kim and W. N. Lipscomb, *Proc. Natl. Acad. Sci. U.S.A.* **90**, 5006 (1993).
[77] S. K. Burley, P. R. David, A. Taylor, and W. N. Lipscomb, *Proc. Natl. Acad. Sci. U.S.A.* **87**, 6878 (1990).

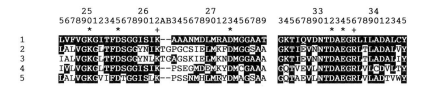

FIG. 13. Conservation of sequences around the metal ligands and catalytic residues of the cattle leucyl aminopeptidase family (M17). Residues are numbered according to cattle leucyl aminopeptidase. Asterisks (*) indicate the metal ligands, and + indicates the putative catalytic Lys and Arg residues. Residues identical to those in the cattle leucyl aminopeptidase sequence are shown in white on black. Key to sequences: 1, cattle leucyl aminopeptidase; 2, *Arabidopsis thaliana* leucyl aminopeptidase; 3, potato leucyl aminopeptidase; 4, *E. coli* aminopeptidase A; 5, *Rickettsia prowazekii* leucyl aminopeptidase.

*Rickettsia prowazekii.* Plant homologs are known from potato and *Arabidopsis.* The percentages of identical amino acids between the bacterial, plant, and mammalian enzymes in family M17 are exceptionally high, considering the evolutionary diversity of those groups of organisms. Finally, analysis of the structures of cattle leucine aminopeptidase and carboxypeptidase A has led to the suggestion that family M17 is distantly related to family M14.[78]

## Methionyl Aminopeptidase Family (M24)

Family M24 represents a second set of aminopeptidases that contain "cocatalytic" metal atoms,[74] but those are not necessarily zinc atoms. The known members of the family are methionyl aminopeptidase from bacteria and fungi, X-Pro dipeptidase from bacteria and animals, and *E. coli* X-Pro aminopeptidase (aminopeptidase P). The X-Pro aminopeptidase can also act on dipeptides and tripeptides, and it has a specificity intermediate between those of the other two enzymes.

Methionyl aminopeptidase is a universally important enzyme because it removes the N-terminal Met residues from newly synthesized proteins, acting cotranslationally, presumably in association with ribosomes.[79] The three-dimensional structure of *E. coli* methionyl aminopeptidase is known.[80] The enzyme binds two cobalt atoms, the ligands being Asp-97, Asp-108, His-177, Glu-204, and Glu-235 (Figs. 9 and 14). However, it should be noted that Glu-235 is absent from yeast methionyl aminopeptidase. The X-ray structure of the molecule reveals an internal homology no longer apparent in the amino acid sequence, suggesting that the two halves of the molecule are the result of a gene duplication. This is a situation analogous

[78] P. J. Artymiuk, H. M. Grindley, J. E. Park, D. W. Rice, and P. Willett, *FEBS Lett.* **303,** 48 (1992).

[79] Y.-H. Chang, U. Teichert, and J. A. Smith, *J. Biol. Chem.* **267,** 8007 (1992).

[80] S. L. Roderick and B. W. Matthews, *Biochemistry* **32,** 3907 (1993).

FIG. 14. Conservation of sequences around the cobalt ligands and catalytic residues of the methionyl aminopeptidase family (M24). Residues are numbered according to *Escherichia coli* methionyl aminopeptidase. Asterisks (*) indicate the cobalt ligands. Residues identical to those in the *E. coli* methionyl aminopeptidase sequence are shown in white on black. Key to sequences: 1, *E. coli* methionyl aminopeptidase; 2, *Bacillus subtilis* methionyl aminopeptidase; 3, yeast methionyl aminopeptidase; 4, *E. coli* X-Pro aminopeptidase; 5, *Streptomyces lividans* X-Pro aminopeptidase; 6, human X-Pro dipeptidase; 7, *E. coli* X-Pro dipeptidase.

to that of pepsin (see [7] in this volume), and we would postulate that the ancestral enzyme was a homodimer.

Methionyl aminopeptidases from *E. coli, Salmonella typhimurium,* and *Bacillus subtilis* do not have propeptides, whereas long N-terminal extensions are found in other members of the family, with *E. coli* X-Pro dipeptidase having a long C-terminal extension as well. Yeast methionyl aminopeptidase differs from the prokaryotic enzymes in consisting of two functional domains: a unique N-terminal domain containing two motifs resembling zinc fingers that may allow the protein to interact with ribosomes and a catalytic C-terminal domain resembling prokaryotic methionyl aminopeptidases.[79] Disruption of the gene for yeast methionyl aminopeptidase is not lethal, which suggests that there is an alternative processing enzyme.[79]

X-Pro dipeptidase and X-Pro aminopeptidase are cytosolic manganese-dependent metallopeptidases that are also thiol-dependent. X-Pro aminopeptidase is a homotetramer.

A number of other proteins that are not peptidases have been shown to be homologous to methionyl aminopeptidase.[81] Two that retain the metal-binding residues are a hypothetical protein from the Asn-tRNA 5′ region of the archaebacterium *Methanothermus fervidus* (YSYN_METFE; RDF score 9.29) and a rat initiation factor 67-kDa associated protein (EMBL: L10652; RDF score 8.25). Perhaps surprisingly, two other homologs lack metal-binding ligands and express enzymatic activity that is not metal-dependent. These are *Flavobacterium* creatinase (CREA_FLASU; RDF score 11.43), which has also been shown to have a tertiary fold similar to that of methionyl aminopeptidase, and *Agrobacterium* agropine synthesis cyclase (AGS_AGRRA; RDF score 6.68).

[81] J. F. Bazan, L. H. Weaver, S. L. Roderick, R. Huber, and B. W. Matthews, *Proc. Natl. Acad. Sci. U.S.A.* **91,** 2473 (1994).

Metallopeptidase Families with Unknown Metal Ligands

Inevitably, there are a number of families of peptidases for which there is evidence of metal-dependence but no clear information as to which of the amino acids serve as ligands for the presumed metal atoms. This kind of information is usually derived from X-ray crystallography, site-directed mutagenesis, or analysis of the presence of familiar motifs in the sequence. Without such data, the conclusion that the peptidase is truly a metalloenzyme must be examined critically, as inhibition by chelating agents and even analytical evidence of tightly associated metal atoms have been reported for serine and cysteine peptidases. Because the metallopeptidases with unknown metal ligands have generally been studied less intensively than those in the other groups of families above, there is a possibility that some may prove not to be metallopeptidases at all.

Useful generalizations about the amino acids that most commonly form the metal ligands in metalloenzymes, which may be helpful in the identification of metal ligands in these families, have been made in Ref. 82.

*Yeast Aminopeptidase I Family (M18)*

Family M18 contains only one enzyme, a leucyl aminopeptidase from the yeast *Saccharomyces cerevisiae*. This is the product of the *LAP4* gene, and it has been named aminopeptidase I, aminopeptidase III, and aminopeptidase yscI.[83] The enzyme, located in the yeast vacuole, is a glycoprotein that is inactivated by chelating agents and can then be reactivated by zinc and chloride ions. It is synthesized as a precursor with a 45-residue prepropeptide and is activated by cleavage of a Leu+Glu bond to give a mature enzyme of 424 residues.

*Membrane Dipeptidase Family (M19)*

Membrane dipeptidase is abundant in association with renal brush border membranes and has also been termed renal dipeptidase. It is a zinc enzyme with broad specificity for dipeptides. The enzyme is known from a number of mammalian species, and it has been cloned and expressed in several laboratories (Table V). It is synthesized with a signal peptide and a C-terminal hydrophobic domain, both of which are removed in posttranslational processing. Membrane dipeptidase is then anchored to the membrane by covalently bound glycosylphosphatidylinositol.[84]

His-219 in pig renal membrane dipeptidase has been shown by directed

---

[82] B. L. Vallee and D. S. Auld, *Biochemistry* **32**, 6493 (1993).
[83] Y. H. Chang and J. A. Smith, *J. Biol. Chem.* **264**, 6979 (1989).
[84] I. A. Brewis, M. A. J. Ferguson, A. J. Turner, and N. M. Hooper, *Biochem. Soc. Trans.* **21**, 46S (1993).

TABLE V
METALLOPEPTIDASES WITH UNKNOWN METAL LIGANDS[a]

| Family | EC | Database code |
|---|---|---|
| **Family M18: Yeast aminopeptidase I** | | |
| Aminopeptidase I (*Saccharomyces*) | - | AMPL_YEAST, LAP4_YEAST |
| **Family M19: Membrane dipeptidase** | | |
| Gene R product (*Acinetobacter*) | - | YPQR_ACICA |
| Membrane dipeptidase | 3.4.13.19 | MDP1_*, (D13139), (L27113) |
| Open reading frame X product (*Klebsiella*) | - | YPQQ_KLEPN |
| Renal brush border membrane protein (rabbit) | - | MDP4_RABIT |
| **Family M20: Glutamate carboxypeptidase** | | |
| Glutamate carboxypeptidase (*Pseudomonas*) | 3.4.17.11 | CBPG_PSES6 |
| Gly-X carboxypeptidase (*Saccharomyces*) | 3.4.17.4 | CBPS_YEAST |
| Peptidase T (*Salmonella*) | - | PEPT_SALTY |
| Peptidase V (*Lactococcus*) | - | (L27596) |
| **Family M22: O-Sialoglycoprotein endopeptidase** | | |
| O-Sialoglycoprotein endopeptidase (*Pasteurella*) | 3.4.24.57 | (M62364) |
| OrfX (*Escherichia coli*) | - | YRUX_ECOLI |
| OrfX (*Haloarcula*) | - | (X70117) |
| OrfX (*Salmonella*) | - | (M14427) |
| Orf YKR038c (*Saccharomyces*) | - | (Z28263) |
| **Family M23: β-lytic endopeptidase** | | |
| β-Lytic endopeptidase | 3.4.24.32 | PRLB_* |
| LasA protein (*Pseudomonas*) | - | LASA_PSEAE |
| **Family M25: X-His-dipeptidase** | | |
| X-His dipeptidase | 3.4.13.3 | PEPD_ECOLI |
| **Family M28: *Vibrio* leucyl aminopeptidase** | | |
| Leucyl aminopeptidase (*Vibrio*) | 3.4.11.10 | AMPX_VIBPR |
| **Family M29: Thermophilic aminopeptidase** | | |
| Thermophilic aminopeptidase (*Thermus*) | 3.4.11.12 | AMPT_THEAQ, (D13386) |
| Aminopeptidase II (*Bacillus*) | - | (D13385) |
| **Family M33: Aminopeptidase Y** | | |
| Aminopeptidase Y (*Saccharomyces*) | - | (L31635) |
| ipa-8r protein (*Bacillus*) | - | (X73124) |

[a] See Table I for explanation.

mutagenesis to be essential for activity. A pentapeptide somewhat similar to the HXXEH motif of the pitrilysin family (M16), Asp-His-Leu-Asp-His, which is present only in the mammalian dipeptidases, has been suggested as a zinc-binding motif, but has now been shown not to be essential for activity.[84a]

The two eubacterial members of family M19 are known only as fragmentary nucleotide sequences. Both sequences, from *Klebsiella* and *Acinetobacter*, are adjacent to genes encoding enzymes of the pyrroloquinoline quinone biosynthetic pathway.[85,86]

Proteins in a group of bacterial hemolysin activators including that of *Serratia marcescens* (HLYB_SERMA) are distantly related to the peptidases of family M19. There is no evidence that the proteins are peptidases, however.

[84a] S. Keynan, N. M. Hooper, and A. J. Turner, *FEBS Lett.* **349**, 50 (1994).

[85] J. J. M. Meulenberg, E. Sellink, N. H. Riegman, and P. W. Postma, *Mol. Gen. Genet.* **232**, 284 (1992).

[86] A. M. Cleton-Jansen, H. P. Horsman, R. G. Huinen, and P. van de Putte, *J. Bacteriol.* **171**, 447 (1989).

*Glutamate Carboxypeptidase Family (M20)*

The glutamate carboxypeptidase family contains two peptidases which we had previously placed in separate families.[87] Glutamate carboxypeptidase from *Pseudomonas* sp. was known to be related to two amidases, succinyldiaminopimelate desuccinylase (DAPE_ECOLI) and acetylornithine deacetylase (ARGE_ECOLI),[88] but the elucidation of the sequence of aminoacylase[89] has provided a link between the bacterial enzymes and Gly-X carboxypeptidase from yeast (Table V).

Succinyldiaminopimelate desuccinylase and acetylornithine deacetylase are both cobalt-dependent metalloenzymes and are involved in the biosynthesis of lysine and arginine, respectively.[88] Acetylornithine deacetylase is also thiol-dependent. Glutamate carboxypeptidase is a homodimer in which each subunit binds two zinc atoms.[90] Aminoacylase also is a zinc enzyme.[89] Peptidase V from *Lactococcus* is a dipeptidase that hydrolyses $\beta$-Ala+His (carnosine).[90a] Metal ligands for the enzymes are not known, but potential metal-binding residues conserved in our alignment of the sequences are Asp-141, Asp-174, Glu-175, Glu-176, His-229, and Glu-390 (glutamate carboxypeptidase numbering).

Two enzymes with activities related to that of aminoacylase are also distant members of family M20. These are hydantoin utilization protein C from *Pseudomonas* (HYUC_PSESN), which converts *N*-carbamyl-L-amino acids to L-amino acids, and a thermostable aminoacylase from *Bacillus stearothermophilus* (EMBL:X74289).

*O-Sialoglycoprotein Endopeptidase Family (M22)*

Family M22 contains an enzyme secreted by the causative agent of bovine shipping fever, *Pasteurella haemolytica*, which only cleaves proteins that are heavily *O*-sialoglycosylated. Substrates include glycophorin A from erythrocytes and the leukocyte surface antigens CD34, CD43, CD44, and CD45 (this volume, [47]). Treatment of glycophorin A with neuraminidase destroys its suceptibility to hydrolysis.[91] The enzyme has an $M_r$ of 35,000. The sequence His[111]-Met-Glu-Gly-His has been suggested to be a zinc-binding site.[92] The enzyme is inhibited by EDTA and prolonged treatment

[87] N. D. Rawlings and A. J. Barrett, *Biochem. J.* **290**, 205 (1993).
[88] A. Boyen, D. Charlier, J. Charlier, V. Sakanyan, I. Mett, and N. Glansdorff, *Gene* **116**, 1 (1992).
[89] M. Jakob, Y. E. Miller, and K.-H. Röhm, *Biol. Chem. Hoppe-Seyler* **373**, 1227 (1992).
[90] L. F. Lloyd, C. A. Collyer, and R. F. Sherwood, *J. Mol. Biol.* **220**, 17 (1991).
[90a] K. F. Vongerichten, J. R. Klein, H. Matern, and R. Plapp, *J. Gen. Microbiol.* **140**, 2591 (1994).
[91] D. R. Sutherland, K. M. Abdullah, P. Cyopick, and A. Mellors, *J. Immunol.* **148**, 1458 (1992).
[92] K. M. Abdullah, R. Y. C. Lo, and A. Mellors, *J. Bacteriol.* **173**, 5597 (1991).

with 1,10-phenanthroline, but not by phosphoramidon or inhibitors of serine and cysteine peptidases. Cleavage of glycophorin A occurs at $Arg^{31}+Asp$ primarily but also at $Glu^{60}+Arg$, $Ala^{65}+His$, and $Tyr^{34}+Ala$ after prolonged treatment.[93] Homologous sequences are known from other eubacteria (*Escherichia coli* and *Salmonella typhimurium*[92]), the archaebacterium *Haloarcula marismortui*,[93a] and yeast,[93b] but the enzymes have not been isolated. We report here that a homolog also exists in the cyanobacterium *Synechocystis*. The partial sequence of the homolog occurs on the complementary strand to the gene for photosystem I subunit III,[94] although the reading frames do not overlap, and is interrupted, possibly as the result of a sequencing error.

### β-Lytic Endopeptidase Family (M23)

Family M23 contains bacterial enzymes. These are the β-lytic endopeptidases of *Lysobacter* and *Achromobacter*, the *Pseudomonas* LasA protein, and a fibrinogenolytic enzyme from *Aeromonas* (Table V; see also [48] in this volume). All the enzymes have specificity for Gly bonds, especially in -Gly-Gly+Xaa- sequences. They lyse the cell walls of gram-positive bacteria in which the peptidoglycan cross-links contain Gly (or multiple Gly) residues. The action of the LasA enzyme on elastin is also attributable to the cleavage of glycyl bonds, and the *Aeromonas* enzyme expresses its fibrinogenolytic activity by cleavage of a -Gly-Gly+Ala- bond located near the cross-link site in fibrin.[95]

The enzymes in family M23 are very resistant to inhibition by EDTA (10 mM), but they are inhibited by 1,10-phenanthroline and thiol compounds. None has been reported to interact with $\alpha_2$-macroglobulin.

The enzymes contain zinc, but zinc ligands have not been identified. However, there is a conserved His-Xaa-His motif, resembling that which has been shown to contain two zinc ligands in the zinc D-Ala-D-Ala carboxypeptidase family (M15; see above), the members of which also hydrolyze peptidoglycan components. All the other conserved His residues are N-terminal to this motif, just as is the third zinc ligand in the D-Ala-D-Ala carboxypeptidase.

*Lysobacter* β-lytic endopeptidase is synthesized with a 171-residue propeptide at the N terminus. The propeptide contains several clusters of

[93] K. M. Abdullah, E. A. Udoh, P. E. Shewen, and A. Mellors, *Infect. Immun.* **60**, 56 (1992).
[93a] E. Arndt and C. Steffens, *FEBS Lett.* **314**, 211 (1992).
[93b] B. Dujon, *et al.*, *Nature* **369**, 371 (1994).
[94] P. R. Chitnis, D. Purvis, and N. Nelson, *J. Biol. Chem.* **266**, 20146 (1991).
[95] A. G. Loewy, U. V. Santer, M. Wieczorek, J. K. Blodgett, S. W. Jones, and J. C. Cheronis, *J. Biol. Chem.* **268**, 9071 (1993).

Arg residues, as do the propeptides of α-lytic endopeptidase (family S2), subtilisin (family S8), and lysyl endopeptidase (family S5). It has been postulated that the basic residues assist in the correct folding of the enzyme by providing hydrophilic and hydrophobic regions, and they may help transport the mature enzyme through the outer membrane of a gram-negative bacterium.[96]

### X-His Dipeptidase Family (M25)

The only known member of family M25 is a cytoplasmic exopeptidase from *E. coli* that cleaves the substrate β-Ala-His. The enzyme is synthesized without pre- or propeptides.[97]

### Vibrio Leucyl Aminopeptidase Family (M28)

The leucyl aminopeptidase from *Vibrio* is synthesized as a 54-kDa proenzyme but is secreted as a 30-kDa protein. Although showing a preference for hydrolysis of leucyl bonds, the enzyme has a broad specificity. A 99-residue C-terminal domain as well as an 85-residue N-terminal propeptide are believed to be removed during maturation of the *Vibrio* aminopeptidase. The C-terminal domain is homologous to similarly located domains in otherwise unrelated peptidases of *Vibrio*, *Xanthomonas*, and *Helicobacter*, in families S8 (this series, Volume 244, [2]) and M4.[98] That domain has been experimentally excised from a *Vibrio* subtilisin homolog without affecting activity.[99] The zinc ligands in *Vibrio* leucyl aminopeptidase have not been identified, but they have been postulated to be His-211, Glu-214, and His-336, forming a pattern similar to that of carboxypeptidase A (family M14).[98]

### Aminopeptidase T Family (M29)

Broad specificity, cobalt-dependent aminopeptidases stable to high temperatures have been isolated from thermophilic bacteria (*Bacillus stearothermophilus*,[100] *Thermus aquaticus*,[101] and *Thermus thermophilus*). All the enzymes are oligomeric, composed of subunits of about 47 kDa, and the three known sequences (Table V) are very similar. The enzymes from *Thermus*[101] and aminopeptidases II and III of *B. stereothermophilus* are

---

[96] S. L. Li, S. Norioka, and F. Sakiyama, *J. Bacteriol.* **172**, 6506 (1990).

[97] B. Henrich, U. Monnerjahn, and R. Plapp, *J. Bacteriol.* **172**, 4641 (1990).

[98] G. Van Heeke, S. Denslow, J. R. Watkins, K. J. Wilson, and F. W. Wagner, *Biochim. Biophys. Acta* **1131**, 337 (1992).

[99] S. M. Deane, F. T. Robb, S. M. Robb, and D. R. Woods, *Gene* **76**, 281 (1989).

[100] G. Roncari, E. Stoll, and H. Zuber, this series, Vol. 45, p. 522.

[101] H. Motoshima, N. Azuma, S. Kaminogawa, M. Ono, E. Minagawa, H. Matsuzawa, T. Ohta, and K. Yamauchi, *Agric. Biol. Chem.* **54**, 2385 (1990).

homodimers. *Bacillus stereothermophilus* also produces a distinct amino-peptidase I that is a larger molecule, reported to contain 12 subunits of two nonidentical but homologous types; it is probably unrelated to family M29.

### Aminopeptidase Y Family (M33)

Yeast aminopeptidase Y is a glycosylated, vacuolar enzyme able to release a variety of amino acids, but preferentially Arg and Lys. The amino-peptidase contains zinc, but no HEXXH motif. It is synthesized with a 56-residue prepropeptide, and processing was shown with purified cerevi-sin.[101a,101b] We report here that the amino acid sequence is homologous to the product of the *ipa-8r* gene from *Bacillus subtilis*[101c] and contains domains similar to some in three families of otherwise unrelated peptidases. The mature aminopeptidase Y polypeptide can be thought of as comprising four segments. The second of these resembles a sequence inserted between the active site His and Ser residues of several bacterial members of the subtilisin family of serine peptidases (S8), including the cell-wall associated endopeptidases of *Lactococcus* and *Lactobacillus,* the C5a peptidase from *Streptococcus,* and an intracellular endopeptidase from *Bacillus.* The third segment of aminopeptidase Y is similar to one found in a metal-dependent leucyl aminopeptidase from *Vibrio* (family M29) and an *E. coli* arginyl aminopeptidase of unknown catalytic mechanism (family U2). The catalytic residues of aminopeptidase Y seem likely to be located in the first or fourth segments of the enzyme, which are not detectably related to other proteins.

### Discussion

The 30 families of metallopeptidases that we have recognized in this chapter comprise 12 that contain only exopeptidases, 17 that contain only endopeptidases, and a single family, M3, that clearly contains both exopepti-dases and endopeptidases. Among the serine and cysteine peptidases there is similarly a tendency for related enzymes to be either exopeptidases or endopeptidases, but again, a few families are exceptional in containing some of each (this series, Volume 244, [2] and [33]). It seems that the evolution of any exopeptidase into an endopeptidase, or vice versa, is a rare event.

One can also generalize by saying that peptidases of all catalytic types

[101a] T. Yasuhara, T. Nakai, and A. Ohashi, *J. Biol. Chem.* **269,** 13644 (1994).
[101b] M. Nishizawa, T. Yasuhara, T. Nakai, Y. Fujiki, and A. Ohashi, *J. Biol. Chem.* **269,** 13651 (1994).
[101c] P. Glaser, *et al., Molecular Microbiology* **10,** 371 (1993).

commonly are synthesized with signal peptides directing secretion and propeptides that hold the enzyme in an inactive state until it is in the environment in which the potentially damaging activity can safely be expressed. Among the metallopeptidases, families M4–M12, M14, M18, M23, M28, and M30 are known to contain peptidases that are synthesized with signal peptides and propeptides, although some of the families are mixed. Thus, the serralysins in family M10 lack signal peptides (being secreted by a different mechanism). Also, bone morphogenetic protein 1 in family M12 and lysine carboxypeptidase in family M14 are believed to differ from other members of their families in lacking propeptides. The lengths of propeptides range from about 12 residues in immune inhibitor A (family M6) to 171 residues in $\beta$-lytic endopeptidase (M23). There are several families which are believed to have signal peptides but not propeptides: M2, M16, M19, M20, and M31. Mitochondrial processing peptidase (family M16) has an N-terminal, mitochondrial targeting signal, rather than a secretory signal. The signal peptide of *Clostridium* collagenase (M31) is very long at 86 residues, and may also include a propeptide. The peptide contains the sequence Pro-Leu-Gly-Pro that conforms to the substrate specificity of the enzyme, and it may be a site of autolytic processing.

There are several families whose members do not contain signal or propeptides: M1, M3, M13, M17, M22, and M24–M27. Peptidases that are integral membrane proteins with an N-terminal transmembrane segment do not require cleavable signal peptides, it seems. Examples of membrane peptidases lacking such sequences are membrane alanyl aminopeptidase

TABLE VI
DATES OF EUBACTERIAL/EUKARYOTIC DIVERGENCE IN METALLOPEPTIDASE FAMILIES[a]

| Family | Calibration divergence | | Dated divergence | | |
|--------|------------|------|------------|------|------|
|        | Comparison | Mya  | Comparison | Mya  | Rate |
| M1  | Human/yeast        | 1000 | Human/*Lactococcus*  | 1162 | 0.84 |
| M3  | Human/yeast        | 1000 | Human/*Escherichia*  | 1445 | 0.65 |
| M10 | Human/soybean      | 1000 | Human/*Serratia*     | 2500 | 0.60 |
| M14 | Bovine/*Simulium*  | 700  | Bovine/*Streptomyces*| 1344 | 0.66 |
| M16 | Human/*Drosophila* | 700  | Human/*Escherichia*  | 1424 | 0.66 |
| M17 | Human/potato       | 1000 | Potato/*Escherichia* | 937  | 0.81 |

[a] Phylogenetic trees with contemporary tips were prepared for the six families of metallopeptidases by use of the KITSCH program [J. Felsenstein, *Evolution* **39,** 783 (1985)]. The trees were calibrated by use of the geological dates of divergence (Mya, millions of years ago) of mammals from fungi, higher plants, and insects as shown. The dates of divergence of prokaryotic and eukaryotic members of the families were then read from the tree. Of course, the analysis depends on the assumption that the rate of evolution within any one family has been constant. Other studies have given support to that general assumption, and, in the case of metallopeptidases, the rather uniform calculated rates of evolution of the families (given in mutations/residue/$10^9$ years) are reassuring.

and neprilysin. In contrast, peptidases with C-terminal transmembrane domains such as peptidyl-dipeptidase A (M2) and meprin (M12) do possess N-terminal signal peptides. In the tetanus and botulinum toxins (M27) activation requires an internal cleavage in the molecule, C-terminal to the peptidase domain. The O-sialoglycoprotein endopeptidase (M22) is a secreted protein despite lacking a signal peptide, and secretion must occur by a special mechanism (perhaps via the ABC pathway whereby serralysins are secreted).

## Origins of Metallopeptidases in Eukaryotes

Many families of metallopeptidases contain both prokaryotic and eukaryotic members, and several have yielded enough sequences to allow the construction of dated phylogenetic trees. From those it is possible to deduce the dates of divergence of the prokaryotic and eukaryotic members of the families (Table VI). With the single exception of family M10, the dates of divergence appear to be far too recent to represent the divergence of the kingdoms, about 3000 million years ago, and suggest horizontal transfer of genes. Presumably, the genes for a number of bacterial metallopeptidases were transferred into the eukaryotic cell by the endosymbiotic organisms that gave rise to organelles in primitive eukaryotes. The exception is family M10, that of interstitial collagenase, in which the members of the serralysin subfamily are sufficiently different from those of the matrixin subfamily to suggest that the ancestor of the family (and probably of the whole of clan MB) was present in the organism that gave rise to both modern bacteria and eukaryotes. Family M22 also was present in the ultimate ancestor, as is shown by the fact that members occur in eubacteria, archaebacteria, and eukaryotes.

# [14] Removal and Replacement of Metal Ions in Metallopeptidases

*By* DAVID S. AULD

## Removal of Metal

### Role of Chelators

Metal chelators contain anionic or neutral oxygen, nitrogen, and sulfur atoms spatially arranged to give bi-, tri-, or tetradentate ligation to a metal atom. Chelators (C) can inhibit metalloenzyme (EM) catalysis either by removal of the metal (M) from the enzyme or by directly binding to it. The former mechanism may be written

$$EM \xrightleftharpoons{K_D} E + M \tag{1}$$

followed by

$$M + C \xrightleftharpoons{K_1} MC \xrightleftharpoons{K_2} MC_2 \xrightleftharpoons{K_3} MC_3 \tag{2}$$

where $K_1 = [MC]/([M][C])$, $K_2 = [MC_2]/([MC][C])$, $K_3 = [MC_3]/([MC_2][C])$, and $K_D = [E][M]/[EM]$, and the apparent stability constants for di- and trichelator·metal complexes are $\beta_2 = K_1K_2 = [MC_2]/([M][C]^2)$ and $\beta_3 = K_1K_2K_3$. The latter mechanism is written

$$EM + C \xrightleftharpoons{K_{EMC}} EMC \tag{3}$$

$$EMC \xrightarrow{k_T} E + MC \tag{4}$$

where $K_{EMC} = [EM][C]/[EMC]$ and $k_T$ is the rate constant for the breakdown of the ternary complex into apoenzyme (E) and MC.

The complexes of metals and proteins have been classified as metalloproteins or metal–protein complexes by Vallee on the basis of the apparent stability constants governing the associations.[1] The definition of a metalloprotein was given as follows: "The metal atoms of metalloproteins are bound so firmly that they are not removed from the protein by the isolation procedure." Volume V of *The Proteins* is an excellent source of information on the characteristics of metalloproteins.[2] The series of books for stability constants of metal–ion complexes by Martell and collaborators are also excellent sources of values for $K_n$ and $\beta_n$ for a wide range of chelators.[3–5] The tabulation of such constants is beyond the scope of this chapter, but some comments on the frequently used chelators and their characteristics will be made in the hopes it will help investigators of metallopeptidase catalysis. Because the overwhelming majority of the metallopeptidases are zinc enzymes, this chapter will generally focus on the chemistry of zinc and related transition metals.

[1] B. L. Vallee, *Adv. Protein Chem.* **10**, 401 (1955).

[2] B. L. Vallee and W. E. Wacker, *in* "The Proteins" Vol. 5, p. 1. Academic Press, New York, 1970.

[3] R. M. Smith and A. E. Martell, *in* "Critical Stability Constants," Vol. 6, 2nd Suppl. Plenum, New York, 1989.

[4] L. G. Sillén and A. E. Martell, *in* "Stability Constants of Metal–Ion Complexes," Special Publ. No. 25. The Chemical Society, London, 1971.

[5] L. G. Sillén and A. E. Martell, *in* "Stability Constants of Metal–Ion Complexes," Special Publ. No. 17. The Chemical Society, London, 1964.

TABLE I
VALUES OF log $K_1$ AND log $\beta_n$ FOR CHELATORS BINDING TO TRANSITION METALS[a]

| Chelator | Ni$^{2+}$ | Cu$^{2+}$ | Fe$^{2+}$ | Zn$^{2+}$ | Cd$^{2+}$ | Co$^{2+}$ | Mn$^{2+}$ |
|---|---|---|---|---|---|---|---|
| OP | 8.6(24.3) | 9.1(20.9) | 5.0(21.3) | 5.9(17.0) | 6.4(15.8) | 7.3(19.9) | 3.9(10. |
| HQSA | 9.0(22.9) | 12(21.5) | 8.4(21.8) | 7.5(14.3) | 7.7(14.2) | 8.8(15.9) | 5.7(10. |
| EDTA | 18.6 | 18.7 | 14.3 | 16.4 | 16.4 | 16.2 | 13.8 |
| EGTA | 13.6 | 17.7 | 11.9 | 12.5 | 16.1 | 12.5 | 12.1 |
| DPA | 8.0(14.1) | 8.9(17.1) | 5.7(10.4) | 6.3(10.8) | 5.7(10.0) | 6.7(12.7) | 5.0(8.5 |

[a] Conditions for determination of constants were generally 25° or 20°, 0.1 $M$ ionic strength. T values of log $\beta_n$ are given in parentheses. The value of $n$ is 3 for all OP complexes and for t Ni$^{2+}$·HSQA complex. In all other cases it is 2. Equilibrium parameters are taken from Refs. 3-

Table I lists the values of the stability constants, $K_1$ and $\beta_n$, for some commonly used chelators binding to transition metals. The metals chosen can lead to active metallopeptidases (see below). For the planar aromatic bidentate chelators 1,10-phenanthroline (OP) and 8-hydroxyquinoline-5-sulfonic acid (HQSA), the stability constants fall into roughly three groups. The binding strength decreases in the order (Ni$^{2+}$, Cu$^{2+}$, and Fe$^{2+}$) > (Zn$^{2+}$, Cd$^{2+}$, and Co$^{2+}$) > Mn$^{2+}$. There are some shifts in the order for the polydentate carboxylic acid chelators ethylenediamine-$N,N,N',N'$)-tetra-acetic acid (EDTA), ethylene glycol bis($\beta$-aminoethyl ether)-$N,N,N',N'$-tetraacetic acid (EGTA), and pyridine-2,6-dicarboxylic acid (DPA). The most striking change is that the stability constant for Fe$^{2+}$ falls to a value close to that of Mn$^{2+}$.

The apparent stability constant (generally $\geq 10^{11}$) for a 1 m$M$ concentration of any of the chelators should be sufficient to remove the metal from most metallopeptidases at equilibrium conditions. However, the characteristics of the metal binding site and the chelator structure likely dictate which chelator will be most effective. Thus EDTA and OP have essentially the same apparent stability constant, 2.5 $\times$ 10$^{16}$ and 1.0 $\times$ 10$^{17}$, respectively (Table I), at a 1 m$M$ chelator concentration, but they have markedly different inhibition characteristics toward metalloenzymes (see below).

*Kinetic Parameters for Chelator Inhibition*

Analysis of the inhibition of enzyme activity can be made using the function[6]:

$$\log(V_c/V_i - 1) = -\log K_i + \bar{n}[\mathrm{I}] \tag{5}$$

[6] T. Coombs, J.-P. Felber, and B. L. Vallee, *Biochemistry* **1**, 899 (1962).

TABLE II
VALUES OF $pK_i$ AND $\bar{n}$ FOR METALLOPEPTIDASE INHIBITION BY
1,10-PHENANTHROLINE[a]

| Enzyme | $pK_i$ | $\bar{n}$ | Ref. |
|---|---|---|---|
| Carboxypeptidase A | 4.2 | 2.3 | 6 |
| *Streptomyces griseus* carboxypeptidase | 3.6 | 1.6 | 8 |
| Angiotensin-converting enzyme | 5.2 | 1.9 | 9 |
| Thermolysin | 4.4 | 1.8 | 10 |
| *Bacillus cereus* neutral protease | 4.5 | 1.9 | 11 |
| *Clostridium histolyticum* γ collagenase | 4.0 | 1.8 | 12 |
| *Clostridium histolyticum* ζ collagenase | 4.4 | 1.8 | 12 |
| Astacin | 4.3 | 2.1 | 13 |
| Yeast aminopeptidase | 4.9 | 2.2 | 14 |

[a] See Eq. (5) for definition of parameters.

where $V_c$ and $V_i$ are the velocities in the absence and presence of inhibitor
(I). When $\bar{n}$ is 1 it is likely that a stable ternary enzyme · metal · chelator
complex is formed. A value of $\bar{n}$ much greater than 1 implies that the
chelator is removing the metal from the enzyme and/or there is more than
one chelator binding site on the enzyme.[7]
   The formation constant for the enzyme · metal complex, $1/K_D$, will usually be greater than the value of $K_1$ for the formation of the chelator · metal
complex but not greater than $\beta_2$ or $\beta_3$ (Table I). Therefore the concentration
of the chelating agent must be in the range where the di- and trichelator ·
metal complexes are favored in order for the chelator to effectively compete with the apoenzyme for the metal. Table II[8-14] lists results for the
inhibition of a number of metallopeptidases by OP. The values of $\bar{n}$ are
frequently greater than 1, indicating more than one molecule of OP is
involved in the inhibition. The $pK_i$ values are in the range of 3.6 to 5.2.
This means inhibition is occurring over the concentration range where the
$Zn(OP)_2$ and $Zn(OP)_3$ complexes predominate.[7] Under these conditions
OP can readily remove the metal from the enzyme. If the mechanism of

[7] D. S. Auld, this series, Vol. 158, p. 110.
[8] K. Breddam, T. J. Bazzone, B. Holmquist, and B. L. Vallee, *Biochemistry* **18**, 1563 (1979).
[9] P. Bünning and J. F. Riordan, *J. Inorg. Biochem.* **24**, 183 (1985).
[10] B. Holmquist and B. L. Vallee, *J. Biol. Chem.* **249**, 4601 (1974).
[11] B. Holmquist, *Biochemistry* **16**, 4591 (1977).
[12] E. L. Angleton and H. E. Van Wart, *Biochemistry* **27**, 7413 (1988).
[13] W. Stöcker, R. L. Wolz, R. Zwilling, D. A. Strydom, and D. S. Auld, *Biochemistry* **27**, 5026 (1988).
[14] G. Metz and K.-H. Röhm, *Biochim. Biophys. Acta* **429**, 933 (1976).

TABLE III
EQUILIBRIUM CONSTANTS FOR $Ca^{2+}$, $Mg^{2+}$, AND $Zn^{2+}$ BINDING
TO CHELATORS[a]

| Chelator | $\log K_1$ ($\log \beta_n$) | | | Ref. |
| | $Ca^{2+}$ | $Mg^{2+}$ | $Zn^{2+}$ | |
| --- | --- | --- | --- | --- |
| DPA | 4.36(7.4) | 2.34(3.0) | 6.32(11.8) | 15, 16 |
| HQSA | 3.52 | 4.06(7.6) | 7.54(14.3) | 17, 18 |
| EDTA | 10.7 | 8.9 | 16.4 | 19 |
| EGTA | 11.0 | 5.2 | 12.9 | 20 |
| OP | 1.05 | 1.51 | 6.43(17.0) | 21, 22 |

[a] Formation constants are given at 25°, 0.1 $M$ ionic strength. $\beta$ values are for $\beta_2$, except for OP, where $n = 3$.

zinc removal is the same for all the enzymes, the differences in $pK_i$ may reflect the strength of binding zinc.

## Calcium- and Magnesium-Dependent Metalloenzyme Systems

Peptidases that require an alkaline earth metal, e.g., calcium, for activation present a number of problems in interpreting the results of chelator inhibition studies. If the chelator concentration chosen is high enough to reduce the effective free calcium concentration, loss of enzymatic activity can be due to decreased calcium concentration. If the chelator binds as tightly to calcium as to a group IIB or transition element metal such as zinc, then a high concentration of calcium can decrease the concentration of free chelator and thus prevent inhibition that is due to chelator binding to the enzyme zinc site or removal of zinc from the protein.

The choice of chelator for such studies is therefore critical for the evaluation of results. Table III[15–22] gives the first stability constant and the highest $\beta$ value for binding of the alkaline earth metals calcium and magnesium and the group IIB metal zinc to a number of commonly used chelators. Only OP displays a high binding constant for zinc and a weak

[15] R. Aruga, Bull. Soc. Chim. Fr. I-79 (1983).
[16] B. Szpoganicz and A. E. Martell, Inorg. Chem. 23, 4442 (1984).
[17] R. Näsänen and E. Uisitalo, Acta Chem. Scand. 8, 112 and 835 (1954).
[18] C. F. Richard, R. L. Gustafson, and A. E. Martell, J. Am. Chem. Soc. 81, 1033 (1959).
[19] B. D. Sarma and P. Ray, J. Indian Chem. Soc. 33, 841 (1956).
[20] G. Anderegg, Helv. Chim. Acta 47, 1801 (1964).
[21] S. Capone, A. De Robertis, C. De Stefano, and R. Scarella, Talanta 32, 675 (1985).
[22] I. M. Kolthoff, D. Leussing, and T. S. Lee, J. Am. Chem. Soc. 73, 390 (1951).

one for calcium and magnesium. It is therefore the best choice when there is no prior information available.

The chelator DPA may be of use for zinc enzymes that are activated by magnesium since there is about a $10^9$ difference in the affinity of zinc and magnesium for DPA. For a calcium-activated zinc peptidase, HQSA should be useful since there is approximately a $10^{10}$ difference in the affinity of $Ca^{2+}$ and $Zn^{2+}$ toward HQSA. EDTA cannot distinguish between nonmetallo- and metalloenzymes that are activated by calcium or magnesium since it binds tightly to $Ca^{2+}$, $Mg^{2+}$, and $Zn^{2+}$. The differences in the stability constants of HQSA and EDTA for zinc and calcium likely explain the results of chelator inhibition studies with the *Clostridium histolyticum* collagenases.[23] All six collagenases are totally inhibited by HQSA and EDTA. However, zinc restores activity only to the HQSA-inhibited enzymes. Because EDTA has a strong affinity for calcium (Table III), it likely removes $Ca^{2+}$ ions that are critical to the tertiary structure of the protein, resulting in an enzyme more susceptible to autoproteolysis and to difficulties in refolding. EGTA, although providing some difference in affinity for $Ca^{2+}$ and $Mg^{2+}$, does not differentiate between $Ca^{2+}$ and $Zn^{2+}$.

### Mechanisms of Chelator Action

Inhibition can be due to formation of a stable ternary chelator · metal · enzyme complex or can arise from metal removal by an $S_N1$ or $S_N2$ type of mechanism.[7] In an $S_N1$ type mechanism, the metal spontaneously dissociates from the metalloenzyme in a rate-determining step, and the free metal rapidly reacts with the chelator in solution [Eqs. (1) and (2)]. The rate of inactivation is therefore independent of chelator concentration. This type of mechanism has frequently been observed for EDTA as exemplified by carbonate dehydratase (CA),[24] angiotensin-converting enzyme (ACE),[25] and carboxypeptidase A (CPD-A).[26] In these cases the first-order rate constant for zinc release and inhibition is independent of EDTA concentration over a wide range ($5 \times 10^{-4}$ to $5 \times 10^{-2}$ $M$ for CA, $1 \times 10^{-7}$ to $5 \times 10^{-4}$ $M$ for ACE, and $5 \times 10^{-4}$ to $5 \times 10^{-3}$ $M$ for CPD-A).

In contrast, an $S_N2$ type reaction includes the formation of a ternary enzyme · metal · chelator (EMC) complex [Eqs. (3) and (4)]. In most cases this is followed by dissociation of the chelator-bound metal, but a stable

[23] M. D. Bond and H. E. Van Wart, *Biochemistry* **23**, 3085 (1984).
[24] Y. Kidani and J. Hirose, *J. Biochem.* (*Tokyo*) **81**, 1383 (1977).
[25] S. G. Kleemann, W. M. Keung, and J. F. Riordan, *J. Inorg. Biochem.* **26**, 93 (1986).
[26] J. E. Billo, K. K. Brito, and R. G. Wilkins, *Bioorg. Chem.* **8**, 461 (1978).

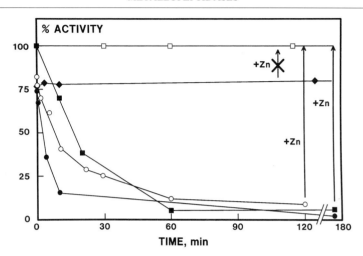

Fig. 1. Time-dependent inhibition of astacin by 1 m$M$ OP ($\bigcirc$), 2 m$M$ OP ($\bullet$), and 5 m$M$ DPA ($\blacksquare$) at 25° in 20 m$M$ HEPES, pH 7.8.[13] The time course in the absence of inhibitors ($\square$) and in the presence of the nonchelating isomer of OP, MP ($\blacklozenge$), is also shown. Activity is completely restored by addition of zinc in the case of the chelators OP and DPA but not for MP.

EMC complex has been observed for horse liver alcohol dehydrogenase.[27] The observed rate constant for inactivation, $k$, should have a hyperbolic dependence on the chelator concentration for the simplest form of this type of mechanism [Eq. (5)].

$$k = \frac{k_{\mathrm{T}} + [\mathrm{C}]}{K_{\mathrm{EMC}} + [\mathrm{C}]} \qquad (6)$$

Saturation kinetics have been observed in the removal of metals from CA by DPA[28] and cobalt CPD-A by OP.[29] However, if the concentration of chelator is greater than 10-fold below $K_{\mathrm{EMC}}$, then a linear dependence on metal removal and/or inhibition is expected. This behavior has been found using OP, HQSA, and DPA with a number of metalloenzymes. In contrast, if the chelator should bind in more than one mode simultaneously to the protein, then a higher order dependence on the chelator is expected.[30] Another criterion that is consistent with some form of an $S_N2$ mechanism is the stoichiometry of the reaction as given by $\bar{n}$. Values of $\bar{n}$ in the range

[27] D. E. Drum and B. L. Vallee, *Biochemistry* **9**, 4078 (1970).
[28] Y. Kidani, J. Hirose, and H. Koike, *J. Biochem.* (*Tokyo*) **79**, 43 (1976).
[29] R. J. Rogers and J. E. Billo, *J. Inorg. Biochem.* **12**, 335 (1980).
[30] R. L. Wolz and R. Zwilling, *J. Inorg. Biochem.* **35**, 157 (1989).

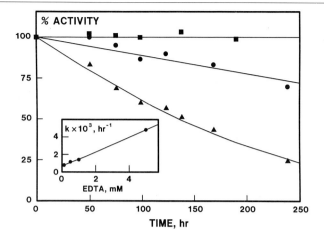

FIG. 2. Time-dependent inhibition of astacin by 1 m$M$ (●) and 5 m$M$ (▲) EDTA. Activity in the absence of EDTA is also shown (■).[13] *Inset:* Dependence of the first-order rate of inactivation, $k$, on EDTA concentration.

1.5 to 3.0 indicate that two or three chelator molecules are involved in the inhibition.

*Time-Dependent Nature of Chelator Action*

Chelators can inhibit the enzyme either instantaneously or in a time-dependent fashion. The nature of the inhibition is likely related to whether the chelator can catalyze the removal of the metal and how tightly the protein binds the metal.

Inhibition of astacin catalysis by chelators serves to demonstrate this and a number of other features. Astacin is inhibited by the metal-binding agents OP, EDTA, DPA, HQSA, and 2,2'-bipyridyl (BP).[13] In all cases the inhibition is time dependent and reversed by dilution. Figure 1 shows the time course of inactivation of astacin in solutions containing millimolar concentrations of OP and DPA. Inhibition occurs within a few minutes for those chelators. In contrast to OP, the nonchelating isomer 1,7-phenanthroline (MP) displays only a slight instantaneous inhibition that is not reversed by the addition of zinc. Further analysis reveals that OP inhibits in both an instantaneous and a time-dependent manner. The $K_i$ and $\bar{n}$ values for the instantaneous inhibition by OP ($3.09 \times 10^{-3}$, 1.15) are essentially the same as those obtained for MP ($2.95 \times 10^{-3}$, 1.33), whereas after preincubation for 1 hr with the protein only the OP values change ($5 \times 10^{-4}$, 2.12).[13] These results are consistent with a model of synergistic binding of OP to

TABLE IV
HALF-LIVES FOR ZINC DISSOCIATION FROM
ZINC ENZYMES[a]

| Enzyme | $t_{1/2}$ | Ref. |
|--------|-----------|------|
| Angiotensin-converting enzyme | 36 sec | 25 |
| Carboxypeptidase A | 29 min | 26 |
| Astacin | 39 days | 13 |
| Carbonate hydratase | 3 years | 31 |

[a] Measured at 25°, pH 7–7.5.

the enzyme catalyzing the release of zinc in an $S_N2$ manner.[30] In marked contrast, inactivation by EDTA is extremely slow, occurring over a period of days (Fig. 2). The rate constant for EDTA inhibition depends linearly on the concentration of EDTA up to 5 m$M$, suggestive of a weakly formed ternary complex. The intercept yields the rate constant for the spontaneous dissociation of zinc from the enzyme. The value of $k$ ($2.1 \times 10^{-7}$ sec$^{-1}$) indicates that the half-life for noncatalyzed zinc release is 39 days. That parameter can vary from seconds to years for zinc enzymes (Table IV).

## Metal Replacement

### Methods

There are three fundamentally different approaches to replacing the native metal in a metalloenzyme[32]: (1) preparation of the metal-free protein (apoprotein) by removal of the native metal by a chelator and/or competition with hydrogen ions at low pH, followed by insertion of the new metal ion, (2) direct displacement of the native metal by a second metal, and (3) biosynthesis of the protein under native metal-deficient and new metal-enriched conditions. The latter method has been used successfully to replace the two zinc atoms of the catalytic domain of stromelysin by cobalt.[33] However for many metallopeptidases method 1 is still the preferred one. This approach usually uses the solution state of the enzyme, although excellent results can be obtained with the crystalline enzyme if that is available.[32,34]

[31] A. Y. Romans, M. E. Graichen, C. H. Lochmuller, and R. W. Henkens, *Bioorg. Chem.* **9,** 217 (1978).
[32] D. S. Auld, this series, Vol. 158, p. 71.
[33] S. P. Salowe, A. I. Marcy, G. C. Cuca, C. K. Smith, I. E. Kopka, W. K. Hagmann, and J. D. Hermes, *Biochemistry* **31,** 4537 (1991).
[34] D. S. Auld and B. Holmquist, *Biochemistry* **13,** 4355 (1974).

TABLE V
PROCEDURE FOR METAL REPLACEMENT BY EQUILIBRIUM DIALYSIS OR ULTRAFILTRATION

| Step | Metal-free buffer[a] | Chelator | New metal | Number of exchanges | Time (hr) |
|------|------------------|----------|-----------|---------------------|-----------|
| 1 | + | + | − | 2–3 | 6–9 |
| 2 | + | − | − | 2–3 | 6–9 |
| 3 | + | − | + | 2 | 6 |
| 4 | + | − | − | 2 | 3–6 |

[a] Metal-free buffer should be prepared either by heterogeneous phase extraction with diphenylthiocarbazone (dithizone) in $CCL_4$ or $CHCl_3$ or by Chelex 100 treatment.[35]

The apoprotein can be prepared by removing the native metal with a chelator during equilibrium dialysis, ultrafiltration, or gel-permeation chromatography. Of the three methods ultrafiltration gives the best environment for preparation of the new metallopeptidase because the device used is small and readily protected from contaminating metal ions. The general procedure for ultrafiltration and equilibrium dialysis is outlined in Table V. An assumption is that the chelator can remove metal from the enzyme in less than 1 hr. Inhibition studies can yield information on which inhibitor to use and at what concentration, so that rapid removal of the metal is achieved (see above). The containers are cleansed of small amounts of metal ions in the first step of the procedure. Special care must be taken in steps 2–4 (Table V) not to reintroduce the native metal ion or other adventitious metal ions. Zinc is particularly prevalent in many buffers as well as in metals that are not spectroscopically pure. Other precautions and technical details have been documented elsewhere in this series.[32]

In designing the conditions to remove metal from the enzyme, the pH and temperature chosen are particularly important variables to consider. The ligands to catalytic, cocatalytic, and structural zinc sites are predominately His > Glu, Asp > Cys.[36,37] All these amino acids have side chains that ionize in the range pH 4 to 8. The basic forms of the side chains are the likely ligands to the metal. Lowering the pH should therefore favor metal removal owing to competition between $H^+$ ions and the metal for the ligands. The value of pH chosen for metal removal should be on the acid side of neutrality as long as the apoenzyme is stable.

Temperature is also an important factor. Because the activation energies of metal dissociation are reasonably high, higher temperatures favor metal

[35] B. Holmquist, this series, Vol. 158, p. 6.
[36] B. L. Vallee and D. S. Auld, Acc. Chem. Res. 26, 543 (1993).
[37] B. L. Vallee and D. S. Auld, Biochemistry 32, 6493 (1993).

removal. Thus the time-dependent inhibition of carboxypeptidase A increases from 10 min at 25° to 1 hr at 0°.[38] Once again, the choice of temperature must be evaluated in the light of the temperature stability of the native and apoenzyme. If the apoenzyme is very unstable at any reasonable temperature and pH, then gel permeation or direct displacement of the metal may be the method of choice for metal exchange.[32]

*Inhibition of Enzymatic Activity by Excess Metal Ion*

Several zinc peptidases are known to be inhibited by excess zinc.[39,40] The mechanism of inhibition has been determined only in the case of carboxypeptidase A.[39,41] The enzyme is inhibited competitively by zinc, lead, and cadmium with apparent inhibition constants, $K_I$, of 24 $\mu M$, 48 $\mu M$, and $1.1 \times 10^{-2}$ $M$, respectively, in 0.5 $M$ NaCl at pH 7.5 and 25°. Negligible inhibition is found for manganese, copper, cobalt, nickel, and mercury.

The state(s) of hydroxylation of zinc and the possible site(s) of interaction with the enzyme were investigated by determining the strength of zinc inhibition over the range pH 4.6–10.5 (Fig. 3). The pH dependence of $pK_I$ follows a pattern which indicates that the enzyme is selectively inhibited by zinc monohydroxide, ZnOH$^+$ ($K_I$ of 0.71 $\mu M$).[41] The formation of the inhibitory ZnOH$^+$ complex from fully hydrated Zn$^{2+}$ is characterized by an ionization constant of 9.05, and the consecutive conversion of ZnOH$^+$ to Zn(OH)$_2$, Zn(OH)$_3^-$, and Zn(OH)$_4^{2-}$ complexes takes place with ionization constants of 9.75, 10.1, and 10.5, respectively. Ionization of a ligand, LH, in the inhibitory site of the enzyme ($pK_{LH}$ 5.8) is obligatory for binding of the ZnOH$^+$ complex. The enzymatic activity ($k_{cat}/K_m$) is influenced by three ionizable groups: $pK_{EH_2}$ = 5.78, $pK_{EH}$ = 8.60, and $pK_E$ = 10.2. Because the values of $pK_{LH}$ and $pK_{EH_2}$ are virtually identical, it is likely that the inhibitory ZnOH$^+$ complex interacts with the group responsible for $pK_{EH_2}$.

The ionizable ligand LH is assigned to Glu-270 because chemical modification of that residue decreases the affinity of the native enzyme for zinc and lead by approximately 100-fold.[39] In addition, previous studies have suggested that $pK_{EH_2}$ reflects the ionization of Glu-270 and its interaction with a water molecule coordinated to the catalytic zinc ion.[40] A bridging interaction between the Glu-270-coordinated metal hydroxide and the cata-

[38] J.-P. Felber, T. Coombs, and B. L. Vallee, *Biochemistry* **1**, 231 (1962).
[39] K. S. Larsen and D. S. Auld, *Biochemistry* **30**, 2613 (1991).
[40] D. S. Auld, *in* "Enzyme Mechanisms" (M. I. Page and A. Williams, eds.), p. 256. Royal Society of Chemistry, Herts, U.K., 1987.
[41] K. S. Larsen and D. S. Auld, *Biochemistry* **28**, 9620 (1989).

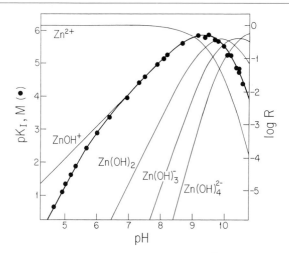

Fig. 3. Dependence on pH of the $pK_I$ values for zinc inhibition (●) of carboxypeptidase A and the relative distribution ($R$) of mononuclear zinc complexes (thin lines).[4] The limiting inhibition constant (0.71 $\mu M$) and the ionization constants affecting activity ($pK_{LH}$ = 5.80, $pK_1$ = 9.05, $pK_2$ = 9.75, $pK_3$ = 10.10, and $pK_4$ = 10.30) were used to calculate the theoretical curve.

lytic metal ion is implicated since lead and zinc induce a perturbation in the electronic absorption spectrum of cobalt carboxypeptidase. This mode of inhibition may occur for other metallopeptidases as a number of them have a similar arrangement of a zinc binding site and a nearby general base catalyst. In fact, inhibition by metal monohydroxide multidentate complexes could be important to the regulatory and/or toxicology process of zinc enzymes and possibly zinc proteins in general.[39]

*Replacement of Zinc with Other Metals*

A number of metallopeptidases have had the catalytic zinc ions replaced with other transition metals to yield catalytically active enzymes (Table VI[42–50]). Evaluation of which metals can replace zinc should be done once

[42] D. S. Auld and B. L. Vallee, *Biochemistry* **9,** 602 (1970).
[43] R. C. Davis, J. R. Riordan, D. S. Auld, and B. L. Vallee, *Biochemistry* **7,** 1090 (1968).
[44] N. Zisapel and M. Sokolovsky, *Biochem. Biophys. Res. Commun.* **53,** 722 (1973).
[45] J. M. Prescott, F. W. Wagner, B. Holmquist, and B. L. Vallee, *Biochemistry* **24,** 5350 (1985).
[46] F. H. Carpenter and J. M. Vahl, *J. Biol. Chem.* **248,** 294 (1973).
[47] M. P. Allen, A. H. Yamanda, and F. H. Carpenter, *Biochemistry* **22,** 3778 (1983).
[48] H. E. Van Wart and S. H. Lin, *Biochemistry* **20,** 5682 (1981).
[49] A. Schäffer and D. S. Auld, *Biochemistry* **25,** 2476 (1986).
[50] H. E. Van Wart and B. L. Vallee, *Biochemistry* **17,** 3385 (1978).

TABLE VI
REPLACEMENT OF ZINC BY OTHER METALS IN METALLOPEPTIDASES

| Enzyme | Substrate[a] | High activity[b] | Low activity, <3% | Ref. |
|---|---|---|---|---|
| | | **Catalytic sites** | | |
| Carboxypeptidase A | Bz-GGF | $Zn^{2+}(100)$, $Co^{2+}(500)$, $Mn^{2+}(20)$, $Ni^{2+}(100)$, $Cd^{2+}(5)$, $Cu^{2+}(4)$ | $Hg^{2+}$, $Fe^{3+}$, $Co^{3+}$ | 42, 49, 50 |
| | Bz-G-OF | $Zn^{2+}(100)$, $Co^{2+}(130)$, $Mn^{2+}(200)$, $Ni^{2+}(100)$, $Cd^{2+}(220)$, $Hg^{2+}(25)$ | $Co^{3+}$ | 43, 50 |
| Carboxypeptidase B | Bz-GR | $Zn^{2+}(100)$, $Co^{2+}(110)$, $Mn^{2+}(230)$ | $Cd^{2+}$ | 44 |
| | Bz-G-OR | $Zn^{2+}(100)$, $Co^{2+}(500)$, $Mn^{2+}(20)$, $Cd^{2+}(500)$ | | 44 |
| | Cbz-AAF | $Zn^{2+}(100)$, $Co^{2+}(65)$, $Mn^{2+}(85)$, $Cd^{2+}(75)$ | | 44 |
| *Streptomyces griseus* carboxypeptidase | Bz-GR | $Zn^{2+}(100)$, $Co^{2+}(1100)$, $Mn^{2+}(60)$, $Ni^{2+}(80)$ | $Cu^{2+}$, $Ca^{2+}$, $Co^{3+}$, $Cd^{2+}$ | 8 |
| | Bz-G-OR | $Zn^{2+}(100)$, $Co^{2+}(110)$, $Mn^{2+}(120)$, $Ni^{2+}(10)$, $Cd^{2+}(80)$ | $Cu^{2+}$, $Ca^{2+}$, $Co^{3+}$ | 8 |
| Angiotensin-converting enzyme | Fa-FGG | $Zn^{2+}(100)$, $Co^{2+}(55)$, $Mn^{2+}(25)$ | $Fe^{2+}$, $Ni^{2+}$, $Cu^{2+}$, $Cd^{2+}$, $Hg^{2+}$, $Mg^{2+}$, $Ca^{2+}$ | 9 |
| Astacin | S-AAA-pNA | $Zn^{2+}(100)$, $Co^{2+}(50)$, $Cu^{2+}(50)$ | $Mn^{2+}$, $Ni^{2+}$, $Fe^{2+}$, $Cd^{2+}$, $Ca^{2+}$, $Mg^{2+}$ | 13 |
| Thermolysin | Fa-GL-NH$_2$ | $Zn^{2+}(100)$, $Co^{2+}(200)$, $Mn^{2+}(10)$, $Fe^{2+}(60)$ | $Ni^{2+}$, $Cu^{2+}$, $Hg^{2+}$, $Cd^{2+}$, $Mg^{2+}$, $Pb^{2+}$, $Fe^{3+}$ | 10 |
| *Bacillus cereus* neutral proteinase | Fa-GL-NH$_2$ | $Zn^{2+}(100)$, $Co^{2+}(200)$, $Mn^{2+}(20)$ | | 11 |
| γ Collagenase | Fa-GLPA | $Zn^{2+}(100)$, $Co^{2+}(500)$, $Ni^{2+}(50)$, $Cu^{2+}(270)$ | $Hg^{2+}$, $Cd^{2+}$ | 12 |
| ζ Collagenase | Fa-GLPA | $Zn^{2+}(100)$, $Co^{2+}(120)$, $Ni^{2+}(110)$ | $Hg^{2+}$, $Cd^{2+}$, $Cu^{2+}$ | 12 |
| | | **Cocatalytic sites** | | |
| *Aeromonas* aminopeptidase | A-pNA | $diZn^{2+}(100)$, $diCo^{2+}(620)$, $diCu^{2+}(1230)$, $diNi^{2+}(1200)$, $diCd^{2+}(20)$ | $Mn^{2+}$, $Mg^{2+}$, $Ca^{2+}$ | 45 |
| Bovine aminopeptidase | L-NH$_2$ | $diZn^{2+}(100)$, $diCo^{2+}(200)$, $Mg^{2+}Zn^{2+}(1000)$, $Mn^{2+}Zn^{2+}(2500)$ | $Mn^{2+}$, $Mg^{2+}$, $Ca^{2+}$ | 46, 47 |
| Porcine aminopeptidase | L-pNA | $diZn^{2+}(100)$, $Mg^{2+}Zn^{2+}(1000)$, $Mn^{2+}Zn^{2+}(1200)$, $Cu^{2+}Zn^{2+}(50)$, $Ni^{2+}Zn^{2+}(75)$ | $Hg^{2+}$, $Cd^{2+}$ | 48 |

[a] The one-letter code is used for amino acids, together with standard abbreviations for substituents. Fa, Furylacryloyl; OF, phenyllactate; OR, argininic acid; pNA, *p*-nitroanilide; Bz, benzoyl; Cbz, carbobenzoxy; S, succinyl.

[b] Relative activities (%) are given in parentheses.

both the native zinc ion and the chelator are removed. If metals are added to the enzyme in the presence of a chelator, there are many explanations for the results obtained. For example, the inhibition caused by the inhibitor may have been the result of its hydrophobic or ionic properties and not because of its metal-binding characteristics. Addition of excess metal may reverse the inhibition by preventing the binding of the chelator to the enzyme. In addition, some of the metal · chelator complexes may be inhibitory, leading to the wrong conclusion about such metals. Even if the inhibition is due to a zinc · chelator action, addition of excess new metal ions may only allow the enzyme to regain its zinc binding capacity, and the resulting activity will not be directly due to the action of the new metal ion. In compiling the list given in Table VI, cases were chosen only where an apoenzyme free of chelator was prepared before adding new metal ions. The addition of new metal ions should be done at different metal concentrations to reduce the possibility of binding the metal to an inhibitory site.

Cobalt is the one metal that substitutes for zinc in the carboxypeptidases, endopeptidases, and aminopeptidases. Manganese frequently yields active carboxypeptidases and endopeptidases but likely plays a supportive role in the cocatalytic aminopeptidases.[37] Nickel, although producing active carboxypeptidases[8,42,43] and *Aeromonas* aminopeptidase,[45] does not generally yield an active endopeptidase[10,11,13] or dipeptidyl-peptidase.[9] Although copper is generally thought to be unable to substitute for zinc, it yields active metallopeptidases in several instances.[12,13,42,45]

The carboxypeptidases serve to remind us that the results obtained on one class of substrates need not pertain to another (Table VI). Thus, the cadmium-substituted enzymes, although frequently having low peptidase activity, have elevated esterase activity.[8,42,43] This observation also applies to the catalysis of the hydrophobic CPD-A substrate by carboxypeptidase B, whose specificity is generally considered to be for basic C-terminal amino acids.[44]

In evaluating the results of activity values less than 5% special precautions must be taken. The assays must allow the use of enzyme concentrations well above adventitious zinc concentrations ($\sim 10$ n$M$ under the best metal-free conditions). Both zinc and the substituted metal content of the protein and the buffer should be determined under the conditions employed to ensure that the activity is due to the new metal and not contaminating zinc. Stopped-flow based assays are very useful for such studies because a higher protein concentration can be used with rapidly turned-over substrates.[50] In this manner we were able to demonstrate that copper carboxypeptidase has 1.5 to 4% peptidase activity toward dansylated tripeptides and from 8 to 17% activity toward the ester analogs. The results for a number of the listed low activity metals have not been obtained under such conditions, so it is not clear whether these metals might yield an active enzyme with, for example, 1% activity. That activity may seem small with respect to the native zinc enzyme activity, but it would still be orders of magnitude above a nonenzymatic reaction.

# [15] Pseudolysin and Other Pathogen Endopeptidases of Thermolysin Family

*By* Kazuyuki Morihara

Introduction

*Pseudomonas aeruginosa* is an opportunistic pathogen which can cause fatal infection in vulnerable hosts.[1] Several products of the organism that are related to virulence have been identified and characterized.[2] Among them, elastase has long been thought to be of major significance.[3] The elastase is now called pseudolysin (EC 3.4.24.26) by recommendation of the IUBMB. Pseudolysin is a Zn-metalloendopeptidase belonging to the thermolysin family (or family MH, see [13] in this volume).[3] Similar pathogenic endopeptidases are also produced by *Vibrio cholerae*,[4] *Legionella pneumophila*,[5] and other bacteria.[6] This chapter mainly concerns pseudolysin, with some mention of related enzymes.

Assay Methods

*Principle*

Three assay methods are described, namely, casein digestion, elastin digestion, and hydrolysis of synthetic peptides. The first method is based on that of Kunitz,[7] which involves enzymatic digestion of casein under defined conditions. The digestion mixture is treated with trichloroacetic acid

---

[1] R. D. Feigin and W. T. Shearer, *J. Pediatr.* **87**, 677 (1975).

[2] O. Pavlovskis and B. Wretlind, *in* "Medical Microbiology" (C. S. F. Easman and J. Jeljaszewicz, eds.), Vol. 1, p. 97. Academic Press, London, 1982.

[3] K. Morihara and J. Y. Homma, *in* "Bacterial Enzymes and Virulence" (I. A. Holder, ed.), p. 41. CRC Press, Boca Raton, Florida, 1985.

[4] B. A. Booth, M. Boesman-Finkelstein, and R. A. Finkelstein, *Infect. Immun.* **42**, 639 (1983).

[5] L. A. Dreyfus and B. A. Iglewski, *Infect. Immun.* **51**, 736 (1986).

[6] P. L. Mäkinen, D. B. Clewell, F. An, and K. K. Mäkinen, *J. Biol. Chem.* **264**, 3325 (1989).

[7] M. Kunitz, *J. Gen. Physiol.* **30**, 291 (1947).

(TCA) to remove undigested casein. The extent of digestion is measured by the amount of unprecipitated products, which contain tyrosine, according to the method of Lowry et al.[8] The method is not specific for pseudolysin, but it is usually used for routine assay. The second method is based on the colorimetric procedure of Sachar et al.,[9] in which orcein–elastin is used as substrate. Enzymatic digestion of elastin results in dissolution of insoluble elastin. The extent of digestion is then measured by the amount of colored soluble products. The third method is based on that of Feder,[10] which gives a precise measure of peptidase activity using a chromophoric substrate, furylacryloyl-Gly-Leu-NH$_2$. The hydrolysis is assayed by following the decrease in absorbance at 345 nm that accompanies hydrolysis of the Gly-Leu bond in the substrate.

*Casein-Digestion Method*

*Reagents*

Substrate, 2% casein: the substrate solution is made by dissolving 2 g of casein (Hammrsten, Merck, Darmstadt, Germany) in 70 ml of 0.1 $N$ NaOH; the solution is then adjusted to pH 7.4 by addition of 1 $M$ KH$_2$PO$_4$, and the total volume is adjusted to 100 ml with distilled water

Tris-HCl buffer, 10 m$M$, pH 7.4, with 10 m$M$ CaCl$_2$

Trichloroacetic acid (TCA), 10% (w/v)

Na$_2$CO$_3$, 0.4 $M$

Folin–Ciocalteu phenol reagent (Sigma, St. Louis, MO), diluted with an equal volume of distilled water

*Procedure.* The casein solution and pseudolysin solution suitably diluted with 10 m$M$ Tris-HCl buffer are incubated in a water bath at 40° for at least 10 min prior to the assay. The reaction is carried out in 1 ml of 2% casein solution with 1 ml of pseudolysin solution (enzyme content, 0.4–4 μg) at 40° for 10 min. The reaction is terminated by the addition of 2 ml of 10% TCA, and, after shaking, the reaction mixture is kept in the water bath at 40° for about 30 min. The precipitate is filtered through Whatman (Clifton, NJ) No. 1 filter paper. One milliliter of the filtrate is mixed with 5 ml of 10% Na$_2$CO$_3$, followed by the addition of 1 ml of diluted phenol reagent. After 15 min, the absorbance at 670 nm is read for determination of the amount of liberated tyrosine. The blank is prepared by first mixing

[8] O. H. Lowry, N. J. Rosebrough, A. L. Farrand, and R. J. Randall, *J. Biol. Chem.* **193**, 265 (1951).
[9] L. A. Sachar, K. K. Winter, N. Sicher, and S. Frankel, *Proc. Soc. Exp. Biol. Med.* **90**, 323 (1955).
[10] J. Feder, *Biochem. Biophys. Res. Commun.* **32**, 326 (1968).

the casein solution with TCA and then adding the enzyme solution to the casein–TCA mixture. The specific activity is expressed as milligrams of tyrosine released per minute (units) per milligram of enzyme. Purified psuedolysin exhibits approximately 20 units/mg.

In variants of the above procedure, the TCA-soluble products can be assayed directly at 280 nm,[11] or a chromophoric substrate such as azocasein (Serva, Heidelberg, Germany) is used,[12] and the extent of digestion is determined by reading the absorbance of the TCA-soluble products at 370 nm.

### Elastin-Digestion Method

#### Reagents

Orcein–elastin: Orcein (50 mg) is dissolved in 5 ml of 70% (v/v) ethanol containing 1 ml of concentrated HCl, and 1 g of elastin (Calbiochem, La Jolla, CA, C grade) is added. The suspension is stirred at room temperature for several hours. The insoluble product is collected by centrifugation (4000 rpm, 10 min) and washed with a large quantity of 70% ethanol until the supernatant becomes colorless. The precipitate is suspended in distilled water and lyophilized

Tris-HCl buffer, 30 m$M$, pH 8.0, with 2 m$M$ CaCl$_2$

Sodium phosphate buffer, 0.7 $M$, pH 6.0

*Procedure.* The enzyme solution is dialyzed against 30 m$M$ Tris-HCl buffer containing 2 m$M$ CaCl$_2$ for 2 days (two buffer changes a day) in the cold to remove inhibitory ions prior to the assay, and the enzyme content is suitably adjusted by dilution with the same buffer or by concentration using an Amicon (Danvers, MA) miniconcentrator. The reaction mixture containing 3 ml of enzyme solution (25–100 $\mu$g of purified pseudolysin) and 20 mg of orcein–elastin is shaken rapidly at 40° for 3 hr. The reaction is terminated by the addition of 2 ml of 0.7 $M$ phosphate buffer, pH 6.0, and the mixture is then filtered. The absorbance of the filtrate is read at 590 nm. An increase in absorbance of 0.42 results from the complete hydrolysis of 20 mg orcein–elastin. Activity is expressed as milligrams of elastin dissolved per hour (units) per milligram of enzyme. Purified pseudolysin exhibits approximately 35 units/mg.

Elastin–Congo Red (ICN Biochemicals, Costa Mesa, CA) can also be used as a chromophoric substrate in place of orcein–elastin. When native elastin is used as substrate, the protein content of the soluble product is assayed according to the method of Lowry et al.[8]

---

[11] H. Matsubara, this series, Vol. 19, p. 642.

[12] E. Kessler, M. Safrin, N. Landshman, A. Chechick, and S. Blumberg, *Infect. Immun.* **38,** 716 (1982).

*Hydrolysis of Synthetic Substrate*

*Reagents*

Furylacryloyl (FA)-Gly-Leu-NH$_2$ (Sigma), 10 m$M$, in 50 m$M$ Tris-HCl buffer, pH 7.5: 30.68 mg of FA-Gly-Leu-NH$_2$ is dissolved in 0.1 ml of spectrophotometrically pure dimethylformamide, to which approximately 7 ml of distilled water and 1.0 ml of 0.5 $M$ Tris-HCl buffer (pH 7.5) are added. After final adjustment to pH 7.5, the volume is brought to 10 ml. The stock solution is stored at $-20°$. Tris-HCl buffer, 50 m$M$, pH 7.5, with 10 m$M$ CaCl$_2$.

*Procedure.* A solution of 1 m$M$ FA-Gly-Leu-NH$_2$ is prepared by a 1/10 dilution of the 10 m$M$ stock solution with 50 m$M$ Tris-HCl buffer (pH 7.5). The reaction is initiated by adding 20 $\mu$l of enzyme solution containing pseudolysin (20–100 $\mu$g) to 1 ml of 1 m$M$ FA-Gly-Leu-NH$_2$ (0.870 absorbance units) in a 1.5-ml cuvette of 1-cm light path, maintained at 25°. The decrease in absorbance at 345 nm is plotted on a chart recorder. Under these conditions, the recorder trace is linear for approximately 3 min, during which time less than 10% of the substrate is hydrolyzed; the initial velocity is determined from that portion of the trace. Complete hydrolysis of a 1 m$M$ solution results in an absorbance change of 0.345 at 345 nm. One unit of activity is defined as the amount of enzyme that gives an initial hydrolytic activity of 1 $\mu$mol of FA-Gly-Leu-NH$_2$ per minute. Purified pseudolysin has a specific activity of approximately 3 units/mg.

The peptidase activity can also be assayed by the same general procedure using any of several other FA-tripeptides such as FA-Gly-Leu-Ala or FA-Ala-Leu-Ala.[13] The activity on the FA-tripeptides can be more than 50 times that on FA-Gly-Leu-NH$_2$, as it is with simple peptides (see Table I, below).

## Production and Isolation of Pseudolysin

*Organism*

Most cultures of *Pseudomonas aeruginosa* isolated from human infections exhibit marked elastinolytic activity,[14] but then show a greater or lesser reduction in elastinolytic activity by repeated transfer on nutrient agar over a 6-month period.[15] Therefore, it has been necessary to select a

---

[13] J. M. Saulnier, F. M. Curtil, M.-C. Duclos, and J. M. Wallach, *Biochim. Biophys. Acta* **995,** 285 (1989).

[14] W. Scharmann, *Zentralbl. Bakteriol. Hyg. Abt. 1: Orig. A* **220,** 435 (1972).

[15] S. Goto, M. Ogawa, T. Takita, Y. Kaneko, and S. Kuwahara, *in* "Abstract of Proc. 9th Meet. Japan *Pseudomonas aeruginosa* Society," p. 22. Tokyo, Japan, 1975.

particularly suitable strain for elastase production, which continues to produce the enzyme even through many subcultures. The strain IFO 3455 is suitable for that purpose, and it has been used for industrial production of pseudolysin for over 10 years at the Nagase Biochemical Co. (Fukuchiyamashi, Kyoto, Japan).

*Culture*

A nutrient medium (1% w/v meat extract, 1% peptone, 1% yeast extract, 0.5% glucose, and 0.5% NaCl, pH 7.0) is usually used for the production.[16] A 500-ml flask containing 100 ml of the medium is shaken (130 rpm, 7 cm amplitude) for 1 or 2 days at 30°. A 5-liter jar fermenter containing 3 liters of medium and a small amount of antifoamant can also be used. Two shaking flasks (200 ml broth) of fresh culture are used as the inoculum, and the culture is grown with aeration at 2 liters/min and stirring at 400 rpm at 30° for 7–8 hr.

*Purification*

*Procedure 1.* The culture broth is centrifuged at 5000 rpm for 20 min to remove bacterial cells, and the clear supernatant is used as the starting material.[16] The supernatant is brought to 60% saturation with solid ammonium sulfate and allowed to stand overnight at 4°. The precipitate is then collected by centrifugation (5000 rpm, 20 min) and dissolved in distilled water. After a second centrifugation, a clear supernatant is obtained. Acetone is then added to the solution in the cold (below 4°) until the concentration reaches 65% (v/v). The precipitate formed by acetone treatment is collected by centrifugation (4°), dissolved in 20 m$M$ phosphate buffer at pH 8, and then dialyzed against the same buffer for 2 days at 4°. The retentate is applied to a column of DEAE-cellulose (Serva) which was previously washed with 1 $M$ NaOH and 1 $M$ Na$_2$HPO$_4$, and then equilibrated with 20 m$M$ phosphate at pH 8.0. Column chromatography is performed at 4°. After washing the column with 20 m$M$ phosphate buffer (pH 8.0), purified pseudolysin can be obtained by stepwise elution with the same buffer containing 50 m$M$ NaCl. The total yield of activity is about 25%, and the specific activity increases about 300-fold with this procedure.

Pseudolysin-containing fractions are pooled, concentrated by ultrafiltration, and stored at −70°. Pseudolysin remains stable and active (retaining >90% activity) for more than 1 year when stored under these conditions.

*Procedure 2.* Pseudolysin can be purified without separation of the bacterial cells,[17] which may be useful for large-scale production. The cul-

[16] K. Morihara, H. Tsuzuki, T. Oka, H. Inoue, and M. Ebata, *J. Biol. Chem.* **240,** 3295 (1965).
[17] K. Morihara, Jpn. Pat. Appl., Sho-40-8114 (1965).

ture broth is adjusted to pH 5.0 with acetic acid and then mixed with a 1/30 volume of wet resin (Amberlite CG 50), which has previously been equilibrated with 0.5 $M$ acetate buffer (pH 5.0) containing 1 m$M$ CaCl$_2$ and 0.2 m$M$ ZnCl$_2$. The suspension is gently stirred for 1 hr at 4°, and the resin is collected by centrifugation and washed with distilled water several times. Crude pseudolysin can be eluted from the resin by the addition of 2 $M$ sodium acetate (pH 10) containing 1 m$M$ CaCl$_2$ and 0.2 m$M$ ZnCl$_2$. The eluted enzyme solution is concentrated by ultrafiltration. The activity yield is 80–90%.

Further purification can be achieved by affinity chromatography using Sepharose-ε-aminocaproyl-D-phenylalanine methyl ester as the ligand.[18] The column is equilibrated with 20 m$M$ glycine containing 5 m$M$ CaCl$_2$. The crude enzyme preparation for affinity chromatography (such as the product of fractionation or adsorption on the ion-exchange resin, as above) is dialyzed against the buffer used for equilibration of the column before being applied to the column. After the column has been further washed with the buffer, purified pseudolysin is eluted with 0.1 $M$ Tris-HCl buffer (pH 8.0).

The enzyme is readily crystallized by the addition of ammonium sulfate to give a concentrated solution of the enzyme prepared by either Procedure 1 or 2.

Properties of Pseudolysin

*Enzymatic Properties*

The enzymatic properties of pseudolysin have been studied.[16] The optimum pH for hydrolysis of either casein or elastin is pH 7 to 8, and the enzyme is stable at pH 6 to 10 (16 hr at 4°) or at 70° (10 min at pH 7.0) in the presence of Ca$^{2+}$. The enzymatic activity is inhibited by metal chelators (1 m$M$) such as EDTA and 1,10-phenanthroline but is not affected by other inhibitors of endopeptidases such as diisopropyl fluorophosphate, tosyl-L-phenylalanine chloromethyl ketone, and $p$-chloromercuribenzoate. The enzyme contains 1 g atom Zn$^{2+}$ per molecule, which is essential for activity.[18] The metal can be removed, reversibly, by treatment with metal chelators in the presence of Ca$^{2+}$.

Pseudolysin extensively digests various protein substrates such as casein, hemoglobin, ovalbumin, and fibrin, when they have previously been denatured. The specificity for cleavage of peptide bonds by pseudolysin has been

---

[18] K. Morihara and H. Tsuzuki, *Agric. Biol. Chem.* **39**, 1123 (1975).

TABLE I
RATE OF HYDROLYSIS OF SYNTHETIC SUBSTRATES
BY PSEUDOLYSIN[a]

| Substrate[b] | Rate ($\mu$mol/min/mg enzyme) |
|---|---|
| Z-Ala-Gly+Leu-Ala | 5280 |
| Z-Ala+Leu-Ala | 1130 |
| Z-Gly+Leu-Ala | 410 |
| Z-Gly+Leu-Gly | 30 |
| Z-Gly+Leu-NH$_2$ | 8 |

[a] The reaction mixture contained 2.5 m$M$ peptide,
0.1 $M$ Tris-HCl (pH 7.0), 2.5 m$M$ CaCl$_2$, and a suitable
amount of enzyme, which was kept at 40°.
[b] + shows the bond split.

studied using oxidized insulin B chain and various synthetic peptides.[18–20]
The results indicate that pseudolysin cleaves on the amino side of hydrophobic or aromatic amino acid residues (i.e., with such residues in the P1′
position). This is typical of endopeptidases belonging to the thermolysin
family. There are some quantitative differences between pseudolysin and
thermolysin, however. For example, with synthetic substrates of the general
structure Z-Phe-Xaa-Ala, in which Z is benzyloxycarbonyl and the Phe-
Xaa bond was cleaved, the order of preference for Xaa was as follows: Phe
(1130) > Leu (380), Tyr (300) > Val (240) > Ile (130) for pseudolysin.
For thermolysin the order was Leu (490) > Phe (310), Ile (310) > Val
(190) ≫ Tyr (10) (the value in parentheses being micromoles peptide
hydrolyzed/min/mg enzyme at pH 7.0 and 40°).

The activity of pseudolysin against compounds based on Z-Gly-Leu-
NH$_2$ was markedly accelerated when the amino acid residues at either P2′
or P1 or both were replaced with Ala. Also, elongation of the peptide bond
to the P2 position of peptide substrates results in a marked increase in
peptidase activity (Table I). A fluorogenic substrate, 2-aminobenzoyl-Ala-
Gly+Leu-Ala-4-nitrobenzylamide ($K_m$ = 0.11 m$M$, $k_{cat}$ = 100 sec$^{-1}$, at pH
7.2, 25°; +, cleavage site), has been synthesized for assay of pseudolysin,[21]
in which the rate of hydrolysis can be conveniently assayed by the increase
in fluorescence caused by separation of the fluorogenic and the quenching
group (see [2] in this volume).

Phosphoramidon (N-$\alpha$-L-rhamnopyranosyloxy(hydroxyphosphinyl)-L-
leucyl-L-tryptophan), an inhibitor of thermolysin, is also an efficient com-

[19] K. Morihara and H. Tsuzuki, *Arch. Biochem. Biophys.* **114,** 158 (1966).
[20] K. Morihara and H. Tsuzuki, *Arch. Biochem. Biophys.* **146,** 297 (1971).
[21] N. Nishino and J. C. Powers, *J. Biol. Chem.* **255,** 3482 (1980).

petitive inhibitor of pseudolysin with a $K_i$ value of 40 n$M$.[22] Pseudolysin is also inhibited competitively[21] by synthetic peptides containing hydroxamic acid [HONHCOCH(CH$_2$C$_6$H$_5$)-CO-Ala-Gly-NH$_2$, $K_i$ = 44 n$M$; HONHCOCH[CH$_2$CH(CH$_3$)$_2$]CO-Ala-Gly-NH$_2$, $K_i$ = 57 n$M$], and thiol functional groups [HSCH$_2$CH(CH$_2$C$_6$H$_5$)-CO-Ala-Gly-NH$_2$, $K_i$ = 64 n$M$]. These inhibitors contain a ligating group which interacts with the active-site Zn$^{2+}$ of pseudolysin. These inhibitors (which are also good competitive inhibitors of thermolysin) can be used as ligands for affinity chromatography of pseudolysin.

Thermolysin is irreversibly inhibited by ClCH$_2$CO-HOLeu-OCH$_3$, but pseudolysin is not[21] (HOLeu is $N$-hydroxyleucine). The tripeptide analog ClCH$_2$CO-HOLeu-Ala-Gly-NH$_2$ can inhibit pseudolysin irreversibly) $k_3/K_i$ = 0.092 $M^{-1}$ sec$^{-1}$),[3] as well as thermolysin ($k_3/K_i$ = 40 $M^{-1}$ sec$^{-1}$).[3] This may indicate that the binding pocket of pseudolysin is bigger than that of thermolysin.

*Structure*

The $M_r$ of pseudolysin is 33,000, and the isoelectric point is 5.9.[16,23] The sequence of 301 amino acid residues has been deduced from the nucleotide sequence of the pseudolysin gene.[23,24] Considerable homology can be seen between the amino acid sequence of pseudolysin and that of thermolysin, as shown below.

The three-dimensional structure of pseudolysin has been solved at 1.6 Å resolution.[25] The overall tertiary structures of pseudolysin and thermolysin are very similar, both having an upper and a lower domain separated by a cleft (Fig. 1). The secondary structure of pseudolysin is summarized in Fig. 2, where a comparison is made with that of thermolysin.[26] The sequence alignment of pseudolysin and thermolysin implied by the structural superposition is also shown in Fig. 2. The pattern of similarity between the proteins is reflected in conserved amino acid residues. In the amino-terminal domain, residues 1–135 (1–137 in thermolysin), 19% of the residues are identical. In the region encompassing the two-active site helices, residues 136–180 (138–182 in thermolysin), the identity in sequence is 48%. In the remainder of the two proteins, residues 181–301 (183–316 in thermolysin), 29% of the residues are identical. In total, 28% of the residues are identical.

[22] K. Morihara and H. Tsuzuki, *Jpn. J. Exp. Med.* **48**, 81 (1978).
[23] R. A. Bever and B. H. Iglewski, *J. Bacteriol.* **170**, 4309 (1988).
[24] J. Fukushima, S. Yamamoto, K. Morihara, Y. Atsumi, H. Takeuchi, S. Kawamoto, and K. Okuda, *J. Bacteriol.* **171**, 1698 (1989).
[25] M. M. Thayer, K. M. Flaherty, and D. B. McKay, *J. Biol. Chem.* **266**, 2864 (1991).
[26] B. W. Matthews, J. N. Jansonius, P. M. Colman, B. P. Schoenborn, and D. Duporque, *Nature (London) New Biol.* **238**, 37 (1972).

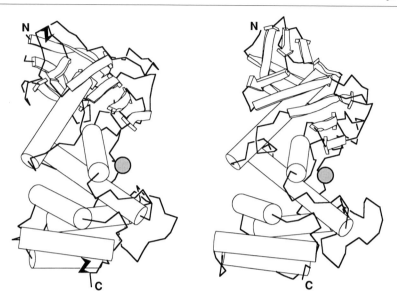

FIG. 1. Schematic drawings of pseudolysin (left) and thermolysin (right), oriented to show the active-site clefts. Helices are drawn as cylinders, strands as ribbons; the active-site zinc is shown as a sphere. (From Thayer et al.,[25] with permission.)

The area of greatest similarity (residues 136–180 in pseudolysin) spans the active-site cleft of pseudolysin which includes the three $Zn^{2+}$ ligands of His-140, His-144, and Glu-164; the active site Glu-141; and the three amino acids Tyr-155, His-223, and Asp-221, which are thought to comprise the substrate-binding region of pseudolysin. All these amino acids are conserved in thermolysin. Site-directed mutagenesis of pseudolysin at Glu-141[27] and His-223[27,28] indicated that both amino acid residues are key ones for catalysis.

The major difference between the two enzymes in tertiary structure is that the substrate-binding cleft is more open in pseudolysin than in thermolysin. This may account for the difference in specificity against amino acid residues at the P1′ position, and in the effects of inhibitors as mentioned above. Other differences include the presence of four cysteine residues in pseudolysin at positions 30, 58, 270, and 297, whereas thermolysin contains no cysteine. The crystal structure shows that in pseudolysin the cysteines

[27] S. Kawamoto, Y. Shibano, J. Fukushima, N. Ishii, K. Morihara, and K. Okuda, Infect. Immun. 61, 1400 (1993).
[28] K. McIver, E. Kessler, and D. E. Ohman, J. Bacteriol. 173, 7781 (1991).

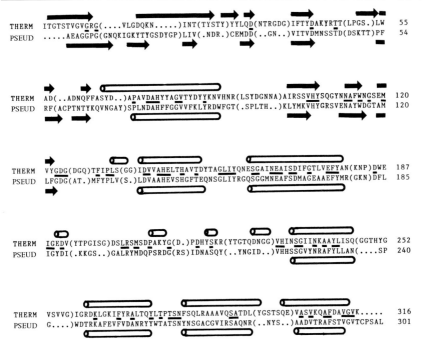

FIG. 2. Sequence alignment of thermolysin (THERM) and pseudolysin (PSEUD) implied by superposition of the structures. β strands are denoted by arrows, helices by cylinders; identical residues are highlighted with bars between the sequences. For segments of sequence included in parentheses, the structures do not superimpose; hence, sequences in these regions are not aligned. (From Thayer et al.,[25] with permission.)

form disulfide bonds with their nearest neighbors. In addition, thermolysin has four calcium binding sites, whereas pseudolysin has only one.

*Pathogenic Activity*

It is now clear that pseudolysin has the potential to contribute in many ways to pathogenesis by *P. aeruginosa*.[3,29–31] The evidence comes from studies with purified pseudolysin, as well as comparative studies using protease-deficient strains. Nonetheless, pseudolysin is relatively nontoxic, with an $LD_{50}$ (50% lethal dose) for mice in the range of 60–400 μg depending on the route of inoculation. Of primary interest, however, is the fact that

[29] L. W. Heck, P. A. Alarcon, R. M. Kulahavy, K. Morihara, M. W. Russell, and J. F. Mestecky, *J. Immunol.* **144**, 2253 (1990).
[30] L. W. Heck, K. Morihara, W. B. McRae, and E. J. Miller, *Infect. Immun.* **51**, 115 (1986).
[31] D. R. Galloway, *Mol. Microbiol.* **5**, 2315 (1991).

purified pseudolysin is capable of degrading several molecules of biological significance to the host. These include various complement components, laminin, fibrin, human collagens, human $\alpha_1$-proteinase inhibitor, and immunoglobulins of the IgG, IgA, and secretory IgA classes. Furthermore, it is now clear that pseudolysin contributes to the activation of Hageman factor and inactivates human $\gamma$-interferon. At the tissue level, it has been shown that pseudolysin is capable of damaging intact pulmonary and corneal tissue and can also destroy epithelial cell junctions. Thus, it seems clear that pseudolysin could easily destroy the host defense mechanism in a number of ways. This may be explained by an "aggressin activity" of pseudolysin, seen when the virulence of protease-deficient strains is increased by the addition of a minute amount of pseudolysin to the experimental animal model of *Pseudomonas* infection.[32,33]

Although there seems to be little question that the proteolytic activity of pseudolysin contributes to the pathogenesis associated with *Pseudomonas* infections, the biochemical basis of elastin degradation remains unresolved. Elastin degradation is of particular significance in pathogenesis since several tissues are composed of elastin and require elastic properties for carrying out their function in the body. For example, lung tissue is composed of approximately 28% elastin and relies on that protein component for expansion and contraction. Furthermore, vascular tissue also contains elastin, which is an important factor in resilience. Consequently, the ability of pseudolysin to destroy elastin is likely to be particularly damaging. Evidence for the destruction of vascular tissue is cited in several studies that suggest that elastase may be responsible for internal organ (including lung) hemorrhages.

Other Pathogen Endopeptidases of Thermolysin Family

The sequences of the structural genes encoding the Zn-metalloendopeptidases from *Vibrio cholerae*,[34] *Legionella pneumophila*,[35] and *Enterococcus* (formerly *Streptococcus*) *faecalis*[36] have been determined, from which the amino acid sequences of the three enzymes were deduced. The sequences show extensive amino acid identity with pseudolysin, especially in the enzymes from *V. cholerae* (identity, 62%) and *L. pneumophila* (identity, 52%). The structural identity between each of the three enzymes and pseudolysin

[32] K. Kawaharajo and J. Y. Homma, *Jpn. J. Exp. Med.* **45,** 515 (1975).
[33] I. A. Holder and C. G. Haidaris, *Can. J. Microbiol.* **25,** 593 (1979).
[34] C. C. Häse and R. A. Finkelstein, *J. Bacteriol.* **173,** 3311 (1991).
[35] W. J. Black, F. D. Quinn, and L. S. Tompkins, *J. Bacteriol.* **172,** 2608 (1990).
[36] Y. A. Su, M. C. Sulavik, P. He, K. K. Mäkinen, P.-L. Mäkinen, S. Fiedler, R. Wirth, and D. B. Clewell, *Infect. Immun.* **59,** 415 (1991).

is most pronounced in the regions forming the enzymatic active site of pseudolysin. The residues constituting the active site of pseudolysin have been mentioned above, and all the residues are identical in each of the three enzymes from *Vibrio, Legionella,* and *Enterococcus,* except Tyr-215 in the *E. faecalis* enzyme, coccolysin (EC 3.4.24.30). It is therefore considered that the mechanism of catalysis by these enzymes is the same as that of thermolysin or pseudolysin (see [13] in this volume). It is suggested that each of the three metalloendopeptidases from *V. cholerae,*[37] *L. pneumophila,*[38] and *E. faecalis*[39] play a role in the pathogenesis which occurs by infection with the respective bacteria.

[37] B. A. Booth, T. J. Dyer, and R. A. Finkelstein, *in* "Advances in Research on Cholera and Related Diarrheas" (R. B. Sack and Y. Zinnaka, eds.), p. 19. KTK Scientific Publ., Tokyo, 1990.
[38] L. A. Dreyfus and B. H. Iglewski, *Infect. Immun.* **51**, 736 (1986).
[39] P.-L. Mäkinen, D. B. Clewell, F. An, and K. K. Mäkinen, *J. Biol. Chem.* **264**, 3325 (1989).

# [16] Neprilysin: Assay Methods, Purification, and Characterization

*By* CHINGWEN LI and LOUIS B. HERSH

## Introduction

Neprilysin (neutral endopeptidase 24.11, EC 3.4.24.11) is a plasma membrane-bound zinc-containing enzyme which degrades and inactivates a number of bioactive peptides. Neprilysin is one of the many zinc metallopeptidases (see [13] in this volume). The enzyme was first isolated from porcine kidney brush border as an endopeptidase hydrolyzing the B chain of insulin.[1,2] It is an ectoenzyme composed of a 23-amino acid N-terminal cytoplasmic domain, a 28-amino acid membrane-spanning domain, and an approximately 700-amino acid extracellular domain which contains the active site. The enzyme was rediscovered as an enkephalin degrading peptidase in rat brain and given the trivial name "enkephalinase".[3,4] Neprilysin was subsequently shown to be the protein described as CD10 or "CALLA"

[1] M. A. Kerr and A. J. Kenny, *Biochem. J.* **137**, 477 (1974).
[2] M. A. Kerr and A. J. Kenny, *Biochem. J.* **137**, 489 (1974).
[3] B. Malfroy, J. P. Swerts, A. Guyon, B. Roques, and J. C. Schwartz, *Nature (London)* **276**, 523 (1978).
[4] S. Sullivan, H. Akil, and J. D. Barchas, *Commun. Psychopharmacol.* **2**, 525 (1978).

(common acute lymphoblastic leukemia antigen) which is present on a group of pre-B cell leukemias.

The substrate specificity of neprilysin is directed toward cleavage of peptide substrates on the amino side of aromatic residues. There are a number of physiological peptides which serve as substrates for the enzyme *in vitro*. These include the enkephalins, substance P, atrial natriuretic factor (ANF), gastrin releasing peptide (GRP), and endothelin, for which evidence exists that these are substrates *in vivo*. Other substrates which are less well documented as being physiologically relevant include melanocyte-stimulating hormone, neurotensin, oxytocin, angiotensins, and calcitonin. The broad substrate specificity of neprilysin would at first seem inconsistent with its role in regulating the physiological action of specific peptides such as enkephalins in the brain, substance P in the lung, and endothelin in the endometrium. However, these and other peptides, rather than the enzyme, have a more discrete distribution, and thus the function of neprilysin in different tissues is determined by substrate availability rather than by enzyme specificity.

## Cellular Localization and Tissue Distribution

Neprilysin exhibits a wide tissue distribution, but activity of the enzyme is quite variable among different tissues. Kidney was found to have the highest neprilysin concentration, and the enzyme is located primarily on the brush border membranes of the kidney proximal tubule.[5] The presence of neprilysin in purified glomeruli from dog kidney suggests a role as a regulator of plasma peptides in that part of the nephron.[6] A relatively high concentration of neprilysin is found in membranes of the brush border epithelial cells of intestine, lymph nodes, and placenta, whereas significant levels of enzyme are localized to muscle cells in the stomach, small intestine, and colon.[7] Neprilysin is found in lower concentrations in adrenal glands, testis, prostate, fibroblasts, neutrophils, and lung.[7,8] In the upper respiratory tract the enzyme is thought to occur in epithelial cells, serous cells of submucosal glands, and vessel walls.[9] The finding of neprilysin in human

[5] A. J. Kenny, *Trends Biochem. Sci.* **11**, 40 (1986).
[6] C. Landry, P. Santagata, W. Bawab, M. C. Fournie-Zaluski, B. P. Roques, P. Vinay, and P. Crine, *Biochem. J.* **291**, 773 (1993).
[7] A. J. Kenny, S. L. Stephenson, and A. J. Turner, *in* "Mammalian Ectoenzymes" (A. J. Kenny and A. J. Turner, eds.), p. 169. Elsevier, Amsterdam, 1987.
[8] N. Sales, I. Dutriez, B. Maziere, M. Ottaviani, and B. P. Roques, *Regul. Pept.* **33**, 209 (1991).
[9] J. N. Baraniuk, K. Ohkubo, O. J. Kwon, J. Mak, J. Rohde, M. A. Kaliner, S. R. Durham, and P. J. Barnes, *J. Appl. Physiol.* **74**, 272 (1993).

endometrium, where it is confined to the stromal cells, suggests a possible role in regulating the ovulatory cycle.[10,11]

In the mammalian brain neprilysin was found in relatively high concentrations in the choroid plexus, substantia nigra, caudate putamen, globus pallidus, olfactory tubercle, nucleus accumbens, and the substantia gelatinosa of the spinal cord.[12] The enzyme was also detected in the meninges of rat and human spinal cord.[13]

In the immune system, CALLA/neprilysin has long been used as a immunological marker in leukemia and lymphoma diagnosis. It was found that CALLA/neprilysin is expressed in the non-T acute lymphoblastic leukemia, disappears at the stage of B cell chronic lymphocytic leukemia, and reappears in a significant proportion of B cell lymphomas.[14] The CALLA/neprilysin antigen was initially thought to be a leukemia-specific antigen, but it was subsequently demonstrated that the enzyme is also expressed in normal cells of bone marrow, spleen, and fetal liver, suggesting that it plays a role in metabolizing some regulatory peptides in normal lymphoid differentiation.[15]

## Assay Methods

There are a number of assay methods used to follow neprilysin activity. These include a radiometric assay measuring the hydrolysis of tritiated [*tyrosyl*-³H]D-Ala²-Leu-enkephalin, fluorometric assays following the hydrolysis of synthetic peptides containing a fluorophor, and assays following the cleavage of internally quenched substrates. Each assay has particular advantages and limitations.

### Radiolabeled Enkephalin as Substrate

#### Reagents

[*tyrosyl*-³H]D-Ala²-Leu-enkephalin (available from Amersham, Arlington Heights, IL, Cat. No. TRK.602, at a specific radioactivity of

[10] M. L. Casey, J. W. Smith, K. Nagai, L. B. Hersh, and P. C. MacDonald, *J. Biol. Chem.* **266,** 23041 (1991).

[11] J. R. Head, P. C. MacDonald, and M. L. Casey, *J. Clin. Endocrinol. Metab.* **76,** 769 (1993).

[12] G. Waksman, R. Bouboutou, J. Devin, R. Besselievre, M. C. Fournie-Zaluski, and B. P. Roques, *Biochem. Biophys. Res. Commun.* **131,** 262 (1985).

[13] J. M. Zajac, Y. Charnay, J. M. Soleilhac, N. Sales, and B. P. Roques, *FEBS Lett.* **216,** 118 (1987).

[14] K. C. Anderson, M. P. Bates, B. L. Slaughenhoupt, G. Pinkus, S. F. Schlossman, and L. M. Nadler, *Blood* **63,** 1424 (1984).

[15] T. W. LeBien and R. T. McCormack, *Blood* **73,** 625 (1989).

25–55 Ci/mmol); the specific radioactivity can be adjusted depending on the required sensitivity
200 m$M$ Tris-HCl buffer, pH 7.4
0.5 $M$ HCl
Porapak Q beads (Waters, Milford, MA)
Ethanol
10 m$M$ Sodium phosphate buffer, pH 7.5
Scintillation fluid

*Procedure.* The radiometric assay involves measuring the release of tritiated Tyr-D-Ala-Gly ([*tyrosyl*-³H]D-Ala-Gly) from the substrate [*tyrosyl*-³H]D-Ala²-Leu-enkephalin. The reaction mixture containing 12.5 μl of Tris buffer, sufficient [*tyrosyl*-³H]D-Ala²-Leu-enkephalin to give a 20 n$M$ solution, and enzyme in a 50-μl volume is incubated at 37° for an appropriate time (15–60 min). The reaction is terminated by adding 5 μl of 0.5 $M$ HCl. The [*tyrosyl*-³H]Tyr-D-Ala-Gly formed is separated from [*tyrosyl*-³H]D-Ala²-Leu-enkephalin by chromatography on Porapak Q beads.[16] The substrate [*tyrosyl*-³H]D-Ala²-Leu-enkephalin is retained on the column, whereas the product Tyr-D-Ala-Gly is not. D-Ala²-Leu-enkephalin can be eluted from the column with 50% (v/v) ethanol. For convenience Pasteur pipettes can be prepared containing 80 mg of dry Porapak Q beads and equilibrated with ethanol for 1 hr or longer. Prior to use the columns are washed with a minimum of 4 ml of ethanol, 50% ethanol, and twice with 10 m$M$ sodium phosphate buffer. (The Porapak Q beads can be cleaned with absolute ethanol and reused.) The quenched reaction mixture is applied to the column and the [*tyrosyl*-³H]Tyr-D-Ala-Gly eluted with 4 ml of the sodium phosphate buffer. A 1-ml aliquot is mixed with an appropriate scintillation fluid and counted in a scintillation counter. Controls in which enzyme is omitted should be used to correct for any [*tyrosyl*-³H]D-Ala²-Leu-enkephalin which does not bind to the beads. It should be cautioned that [³H]Tyr and [*tyrosyl*-³H]Tyr-D-Ala, which are products of aminopeptidase and dipeptidyl aminopeptidase reactions, respectively, are also not retained on the column, so the use of a specific neprilysin inhibitor is recommended to verify neprilysin activity.

An alternative is to use thin-layer chromatography to separate reactants and products. An aliquot of the reaction mixture is mixed with peptide standards (Tyr, Tyr-Ala, Tyr-Ala-Gly, and Ala²-Leu-enkephalin), spotted on a silica gel thin-layer plate, and run in chloroform–methanol–acetic acid–water (45:30:6:9, v/v) as solvent. After development of the chromatogram, the plate is dried and sprayed with 0.1% fluorescamine in acetone, if necessary followed by spraying immediately thereafter with 0.5% pyridine

---

[16] Z. Vogel and M. Alstein, *FEBS Lett.* **80**, 332 (1977).

in acetone. The standards are located with a long wavelength hand-held UV lamp. The area of the silica gel plate containing Tyr-D-Ala-Gly and D-Ala$^2$-Leu-enkephalin is cut from the plate, placed into a scintillation vial, and shaken with 0.5 ml of 0.5 $M$ NaCl for 1 hr. Scintillation fluid (10 ml) is added and the radioactivity determined.

The radiometric assay is simple to use, but in some tissues and cell lines the levels of interfering aminopeptidase and dipeptidyl aminopeptidase activities are so great they make assaying the enzyme difficult. The use of the D-Ala analog of enkephalin serves to reduce the interfering activities, but it is recommended that assays be conducted in the presence or absence of a specific neprilysin inhibitor, such as phosphoramidon, and the difference between the two taken as neprilysin activity.

*Synthetic Chromogenic Substrates*

Use of synthetic peptides of the type glutaryl-Ala-Ala-Phe-4-methoxy-2-naphthylamide to measure enzyme activity is probably the most widely applied assay for neprilysin.[17] With this substrate neprilysin releases glutaryl-Ala-Ala and Phe-4-methoxy-2-naphthylamide as products [Eq. (1)]. The 4-methoxy-2-naphthylamide can be measured fluorometrically after liberation by the action of an aminopeptidase [Eq. (2)]. The free 4-methoxy-2-naphthylamide can also be measured spectrophotometrically after conversion to a diazonium salt.[17]

$$\text{Glut-Ala-Ala-Phe-4-MeO-2-NA} \xrightarrow{\text{neprilysin}} \text{glut-Ala-Ala} + \text{Phe-4-MeO-2-NA} \quad (1)$$

$$\text{Phe-4-MeO-2-NA} \xrightarrow{\text{aminopeptidase}} \text{Phe} + \text{4-MeO-2-NA} \quad (2)$$

*Reagents*

200 m$M$ MES buffer, pH 6.5

40 m$M$ Glutaryl-Ala-Ala-Phe-4-methoxy-2-naphthylamide (Sigma, St. Louis, MO, Cat. No. G 3769), prepared in dimethylformamide

Buffer/substrate mixture: add 50 $\mu$l of 40 mM glutaryl-Ala-Ala-Phe-4-methoxy-2-naphthylamide to 10 ml of 200 m$M$ MES buffer; the stock glutaryl-Ala-Ala-Phe-4-methoxy-2-naphthylamide should be added slowly with shaking to the buffer

Phosphoramidon: 0.5 m$M$ stock solution prepared in 10 m$M$ MES buffer, pH 6.5

Microsomal leucine aminopeptidase (Sigma, Cat. No. L 5006), supplied as 2 mg/ml suspension in ammonium sulfate

[17] M. Orlowski and S. Wilk, *Biochemistry* **20**, 4942 (1981).

Aminopeptidase/phosphoramidon mix: mix 200 $\mu$l of phosphoramidon stock solution and 50 $\mu$l of leucine aminopeptidase in 1 ml of 0.1 $M$ MES buffer, pH 6.5

*Procedure.* The assay, although somewhat less sensitive than the radiometric assay, is simple and is not subject to interference by dipeptidyl aminopeptidase or aminopeptidase activities, although other activities, particularly chymotrypsin-like enzymes, will react with the substrate. Thus it is recommended that assays be conducted in the presence and absence of a specific neprilysin inhibitor such as phosphoramidon.

Reaction mixtures contain 0.5 ml of buffer/substrate mixture, enzyme, and water in a final volume of 1.0 ml. Following incubation at 37° for the desired time, 100 $\mu$l of the phosphoramidon/aminopeptidase mix is added and the reaction permitted to proceed for an additional 20 min. The fluorescence is then measured at an excitation wavelength of 340 nm and an emission wavelength of 425 nm. A standard curve is constructed with free 4-methoxy-2-naphthylamide. It is crucial that the phosphoramidon be premixed with the leucine aminopeptidase because the enzyme is contaminated with small amounts of neprilysin, which is effectively inhibited by the phosphoramidon. An incubation in which 10 $\mu M$ phosphoramidon is added at zero time serves as a control for the presence of other peptidases which may act on the substrate.

*Quenched Fluorescent Substrates*

The use of the internally quenched fluorogenic substrate dansyl-D-Ala-Gly-Phe(NO$_2$)-Gly, developed by Florentin *et al.*,[18] offers the advantage that the reaction can be monitored in a continuous fashion, but it suffers from decreased sensitivity. With that substrate cleavage occurs at the Gly-Phe(NO$_2$) bond, relieving quenching of the fluorescent dansyl group by Phe(NO$_2$) and resulting in an increase in fluorescence [Eq. (3)].

$$\text{Dansyl-D-Ala-Gly-Phe(NO}_2)\text{-Gly} \xrightarrow{\text{neprilysin}}$$
$$\text{dansyl-D-Ala-Gly} + \text{Phe(NO}_2)\text{-Gly} \quad (3)$$

*Reagents*

200 m$M$ MES buffer, pH 6.5

10 m$M$ Dansyl-D-Ala-Gly-Phe(NO$_2$)-Gly (Sigma, Cat. No. D 2155), prepared in methanol

Buffer/substrate mixture: add 100 $\mu$l of 10 m$M$ dansyl-D-Ala-Gly-Phe(NO$_2$)-Gly to 10 ml of 200 m$M$ MES buffer

*Procedure.* Reaction mixtures contain 0.5 ml of the buffer/substrate mixture, enzyme, and water in a total volume of 1.0 ml. The reaction is

---

[18] D. Florentin, A. Sassi, and B. P. Roques, *Anal. Biochem.* **141,** 62 (1984).

initiated by the addition of enzyme, and the increase in fluorescence (excitation wavelength 342 nm, emission wavelength 562 nm) is monitored continuously using a strip-chart recorder. As dansyl-Ala-Gly is not commercially available, the assay can be standardized by adding an excess of enzyme to known amounts of dansyl-D-Ala-Gly-Phe($NO_2$)-Gly and permitting the reaction to proceed to completion.

## Purification Methods

Neprilysin can be purified either by detergent solubilization followed by classic purification procedures conducted in the presence of detergent or as a papain-solubilized enzyme which is purified and stable in the absence of detergent.

### Purification of Papain-Solubilized Neprilysin

A modification of the procedure of Almenoff and Orlowski (1983) is described for purification of the enzyme from rat kidney.[19] Unless noted all purification steps are performed at 4°.

#### Reagents

Buffer A: 50 m$M$ Tris-HCl buffer, pH 7.8 containing 0.2 $M$ NaCl
Papain: Twice-crystallized suspension from Sigma (Cat. No. P 4762) at 25 mg/ml
0.25 $M$ Dithiothreitol (DTT), prepared by dissolving 77.1 mg DTT in 2 ml buffer A

*Homogenization.* Rat kidneys (100 g) are homogenized in 7 volumes of buffer A using either a Waring Blendor (New Hartford, CT) or a Tissuemizer (Tekmar Co., Cincinnati, OH). The homogenate is centrifuged at 10,000 $g$ for 15 min, and the supernatant is carefully decanted and centrifuged again at 40,000 $g$ for 1 hr. The resulting pellet is resuspended in around 200 ml of buffer A with the aid of a glass homogenizer and recentrifuged at 40,000 $g$ for 1 hr. The procedure is repeated, and the final pellet is resuspended in 200 ml of buffer A containing 2 g of deoxycholic acid. The suspension is slowly stirred overnight in the cold. The next day the excess deoxycholic acid is precipitated by the dropwise addition of around 9 ml of a freshly prepared 20% (w/v) streptomycin sulfate solution. The suspension is stirred for 2 min after all the streptomycin sulfate has been added and then centrifuged at 40,000 $g$ for 25 min. The supernatant is concentrated to approximately 50 ml in 1 or 2 Amicon (Danvers, MA) concentrators, and 2 ml DTT is added followed by 0.4 ml papain to solubilize the enzyme. The suspension is brought to 37° and slowly stirred at that temperature for

---

[19] J. Almenoff and M. Orlowski, *Biochemistry* **22**, 590 (1983).

1.5 hr. The suspension is then centrifuged at 40,000 g for 20 min and the precipitate discarded.

*Molecular Sieve Chromatography.* The solubilized enzyme is chromatographed on a 500-ml column (4.5 × 70 cm) of Sephacryl S-200 equilibrated and run with buffer A. The active fractions are pooled and concentrated to 10–20 ml using an Amicon concentrator.

*Hydrophobic Chromatography.* Saturated ammonium sulfate (neutralized to around pH 7) is added to the concentrated enzyme to give a 40% solution. The sample is loaded onto a 10-ml Toyo Pearl Phenyl 650S column (TosoHaas, Montgomeryville, PA). The column is washed first with 5 volumes of buffer A containing 40% saturated ammonium sulfate, then with the same volume of buffer containing 25% saturated ammonium sulfate. All ammonium sulfate solutions were prepared from a saturated (100%) stock solution. The enzyme is then eluted with a 100-ml gradient of decreasing ammonium sulfate from 25 to 7.5% in buffer A. Active fractions are pooled, dialyzed against 50 mM Tris-HCl buffer, pH 7.8, and then concentrated on an Amicon concentrator.

In most cases the enzyme obtained after hydrophobic chromatography appears homogeneous as judged by sodium dodecyl sulfate–polyacrylamide gel electrophoresis (SDS–PAGE). However, an additional ion-exchange chromatography step can be employed if needed. The concentrated fraction from the hydrophobic column is chromatographed on a Pharmacia (Piscataway, NJ) Mono Q column in 50 mM Tris-HCl buffer, pH 7.8, using a gradient of NaCl from 0 to 0.50 M to elute the enzyme.

The above procedure yields 0.5 to 1 mg of enzyme. The overall recovery is low (~10%) mainly as a result of the relatively inefficient solubilization of the enzyme with papain. However, this is in part compensated by the high tissue content of the enzyme and the relative ease with which it can be purified. The purified enzyme retains activity when frozen or can be kept at 4° for several months without loss of activity.

*Purification of Detergent-Solubilized Neprilysin*

*Reagents*

Buffer B: 25 mM Tris-HCl buffer, pH 7.4, containing 0.25 M sucrose

Buffer C: 25 mM Tris-HCl buffer, pH 7.4, containing 0.1% (v/v) Triton X-100

Buffer D: 25 mM Tris-HCl buffer, pH 7.4, containing 0.1% Triton X-100, 1 mM $MgCl_2$, 1 mM $CaCl_2$, 1 mM $MnCl_2$, and 0.5 M NaCl

Chromatofocusing resin: Polybuffer exchanger 94 (Pharmacia)

Histidine buffer: 25 mM Histidine hydrochloride, pH 6.3, containing 0.1% Triton X-100

*Homogenization.* Rat kidneys (50 g) are homogenized in 5 volumes of buffer B using a Polytron (Brinkman Instruments Inc., Westbury, NY) homogenizer. The homogenate is centrifuged at 8000 *g* for 15 min, and the supernatant is filtered through glass wool and centrifuged again at 40,000 *g* for 30 min. The resulting pellet is resuspended in around 200 ml of buffer B and recentrifuged at 40,000 *g* for 30 min. This step is repeated twice, and the final pellet is resuspended in 80 ml of buffer B. Triton X-100 is added to a final concentration of 1% (v/v), and the suspension is stirred for 1 hr at room temperature. [The detergent octaethylene glycol dodecyl ether ($C_{12}E_8$), at a final concentration of 0.1%, can be substituted for Triton. That detergent is as efficient in solubilizing the enzyme as Triton and gives a solution which is optically clear at 280 nm.] The solubilized membranes are centrifuged at 100,000 *g* for 75 min.

*Ion-Exchange Chromatography.* The solubilized membranes are applied to a 50-ml (2 × 14.5 cm) column of DEAE cellulose (DE-52, Whatman, Clifton, NJ) equilibrated with buffer C. The column is washed with 500 ml of buffer C, and then the enzyme is eluted in a broad peak by washing the column with buffer C containing 25 m*M* NaCl. (An additional peak of activity, corresponding to ~50% of that eluted with 25 m*M* NaCl, can be eluted with higher salt. However, this fraction is of considerably lower specific activity and thus more difficult to purify.)

*Lectin Affinity Chromatography.* The pooled fractions from the DEAE column are adjusted to 0.5 *M* NaCl and applied to a 100-ml column of concanavalin A (Con A)-Sepharose preequilibrated with buffer D. After washing with 5 column volumes of starting buffer, the enzyme is eluted with a 1-liter linear gradient of 0 to 1.0 *M* 1-*O*-methyl-D-glucose in buffer D. The enzyme, which elutes in a broad peak, is concentrated to approximately 5 ml with an Amicon stirred cell using a YM30 membrane.

*Molecular Sieve Chromatography.* The concentrated enzyme is chromatographed on a 500-ml column (4.5 × 70 cm) of Sephacryl S-200 equilibrated and run with buffer C containing 0.15 *M* NaCl. The active fractions are pooled and concentrated to around 20 ml as above and dialyzed against histidine buffer.

*Chromatofocusing.* The dialyzed enzyme is applied to a 10-ml chromatofocusing column equilibrated with the histidine buffer. The enzyme is eluted with 150 ml of Polybuffer 74 diluted 1/12 containing 0.1% Triton X-100 and adjusted to pH 4.0 with HCl. The eluted enzyme is concentrated to approximately 2 ml, and Polybuffer is removed by passage of the enzyme over a 40-ml (1.5 × 25 cm) Sephadex G-75 column equilibrated with buffer C.

As with the purified papain-solubilized enzyme, the purified detergent-solubilized enzyme retains activity when frozen or can be kept at 4° for several months without loss in activity.

## Identification of Functional Residues in Neprilysin

The use of chemical modification as well as site-directed mutagenesis has led to the elucidation of several critical active-site residues in neprilysin. The similarities in catalytic properties of neprilysin and the bacterial endopeptidase thermolysin, whose structure has been determined by X-ray crystallography, facilitated such studies. Although the similarities between the enzymes is in general not extended to sequence homology, an exception is a conserved $\alpha$-helical segment, amino acids 580 to 587. The helical region contains the consensus sequence VxxHExxH. In thermolysin the two histidines serve as zinc ligands, the glutamate serves to facilitate nucleophilic attack of a zinc-bound water molecule, whereas the valine forms a part of the binding site which determines the specificity of the enzyme. Evidence for a similar function of these residues in neprilysin has been obtained by site-directed mutagenesis. Changing histidine-583 and histidine-587 to phenylalanine or glutamate-584 to valine or aspartate led to a loss in enzyme activity.[20,21] The latter mutation occurred without a loss of inhibitor binding, suggesting that the amino acid substitution does not disrupt the structure of the active site. Changing valine-580 to a leucine resulted in a change in the substrate specificity of the enzyme, confirming the importance of that residue in substrate binding.[22] A third active-site histidine (his-711) was identified by its reaction with diethyl pyrocarbonate and shown to be an essential residue by site-directed mutagenesis.[23] The role of histidine-711 is believed to be analogous to that of histidine-231 in thermolysin which serves to stabilize the transition state. Using thermolysin as a model, other active-site residues which have been identified by mutagenesis include Glu-646, which is proposed to serve as the third zinc ligand, and Arg-747, which is thought to interact with the carbonyl amide group of the P1' residue.[24,25]

An active-site residue unique to neprilysin is arginine-102, initially identified by its reaction with radiolabeled phenylglyoxal and studied by site-

[20] A. Devault, V. Sales, C. Nault, A. Beaumont, B. Roques, P. Crine, and G. Boileau, *FEBS Lett.* **231**, 54 (1988).

[21] A. Devault, C. Nault, M. Zollinger, M. C. Fournie-Zaluski, B. Roques, P. Crine, and G. Boileau, *J. Biol. Chem.* **263**, 4033 (1988).

[22] J. Vijayaragahavan, Y.-A. Kim, D. Jackson, M. Orlowski, and L. B. Hersh, *Biochemistry* **29**, 8052 (1990).

[23] R. C. Bateman, Jr., and L. B. Hersh, *Biochemistry* **26**, 4237 (1987).

[24] H. LeMoual, A. Devault, B. Roques, P. Crine, and G. Boileau, *J. Biol. Chem.* **266**, 15670 (1991).

[25] A. Beaumont, H. LeMoual, G. J. Boileau, P. Crine, and B. Roques, *J. Biol. Chem.* **266**, 214 (1991).

directed mutagenesis.[26] The residue is believed to form an ionic interaction with the free carboxylate of substrates, although such an interaction is not essential because peptide amides can serve as substrates for the enzyme. Changing the residue by mutagenesis effects only the reaction of substrates containing a free carboxylate.

Thus, the identification and study of active-site residues in neprilysin has supported the proposal that the enzyme is mechanistically and structurally related to the thermolysin family of bacterial neutral metalloendopeptidases.

[26] R. C. Bateman, Jr., D. Jackson, C. Slaughter, S. Unnithan, Y. G. Chai, C. Moomaw, and L. B. Hersh, *J. Biol. Chem.* **264,** 6151 (1989).

## [17] Inhibitors of Neprilysin: Design, Pharmacological and Clinical Applications

*By* Bernard P. Roques, Florence Noble, Philippe Crine, and Marie-Claude Fournié-Zaluski

### Introduction

There has been considerable interest in ectoenzymes (membrane-bound peptidases) since the discovery that inhibition of angiotensin-converting enzyme (ACE), the enzyme implicated in the formation of angiotensin II from angiotensin I, led to antihypertensive effects,[1] and that the inhibition of another membrane-bound Zn-metallopeptidase, involved in the inactivation of the opioid enkephalins in the brain, induced analgesic responses.[2] The latter enzyme was shown to be identical to the neutral endopeptidase EC 24.11 (NEP, also known as neprilysin, EC 3.4.24.11),[3] a well-characterized enzyme demonstrated to be present in brush border cells of the intestine and the kidney, where it inactivates the atrial natriuretic peptide (ANP), and at the surface of B cells as a marker of common acute lymphoblastic leukemia antigen (CALLA) (reviewed in Roques *et al.*[4]) (see also [16] in this volume).

[1] M. A. Ondetti, B. Rubin, and D. W. Cushman, *Science* **196,** 441 (1977).
[2] B. P. Roques, M. C. Fournié-Zaluski, E. Soroca, J. M. Lecomte, B. Malfroy, C. Llorens, and J. C. Schwartz, *Nature (London)* **288,** 286 (1980).
[3] M. A. Kerr and A. J. Kenny, *Biochem J.* **137,** 489 (1974).
[4] B. P. Roques, F. Noble, V. Daugé, M. C. Fournié-Zaluski, and A. Beaumont, *Pharmacol. Rev.* **45,** 87 (1993).

In contrast to classic neurotransmitters, activation or interruption of the responses induced by regulatory peptides is ensured by ectoenzymes which cleave the peptide into inactive fragments. Therefore, the pharmacological effects resulting from inhibition of the enzymatic processes will appear only in tissues where the peptide substrate is tonically or phasically released. This was expected to avoid, or at least to minimize, the side effects resulting from excessive and ubiquitous stimulation of receptors by exogenously administered agonists or antagonists.[5] The results summarized here show that this hypothesis has been verified in particular in the case of mixed inhibitors of NEP/APN (aminopeptidase N) or NEP/ACE, leading to new analgesics and antihypertensive agents that are now undergoing clinical trials. Moreover, thiorphan, Tiorfan, the first synthetic inhibitor of NEP,[2] is on the market as a novel antidiarrheal agent.

## Structure, Substrate Specificity, and Mechanism of Action of Neprilysin, a Pharmacologically Relevant Multisubstrate Metabolizing Enzyme

The primary structure of rabbit NEP was elucidated by cloning and sequencing the cDNA to the mRNA coding for the kidney enzyme.[6] The 94-kDa N-glycosylated ectoenzyme contains 749 amino acid residues. The protein has an $NH_2$-terminal cytoplasmic domain of 23 residues, anchoring the protein in the plasma membrane, and a large extracellular domain that contains the active site.

Although the sequence of NEP shows only a weak similarity with those of other Zn-metallopeptidases, some of the most important amino acids in the active site of thermolysin (TLN), a Zn-metallopeptidase that has been crystallized alone and with various inhibitors, appear to have been conserved.[7] Thus, the ectoenzymes contain a consensus sequence VxxHExxH, which has been found in other Zn-endopeptidases also (see this volume, [13]).

The histidines (His-583, His-587 in NEP) are two of three Zn-coordinating ligands, and the glutamate (Glu-584) plays a role in catalysis by polarizing a water molecule (Fig. 1). The third Zn ligand was identified by site-directed mutagenesis as Glu-646, and replacement of that residue by Val led to a complete loss of enzyme activity and [³H]HACBO-Gly [HONH-

---

[5] B. P. Roques, *Kidney Int.* **34,** S27 (1988).

[6] A. Devault, C. Lazure, C. Nault, H. Le Moal, N. G. Seidah, M. Chretien, P. Kahn, J. Powell, J. Mallet, A. Beaumont, B. P. Roques, P. Crine, and C. Boileau, *EMBO J.* **6,** 1317 (1987).

[7] T. Benchetrit, V. Bissery, J. P. Mornon, A. Devault, P. Crine, and B. P. Roques, *Biochemistry* **27,** 592 (1988).

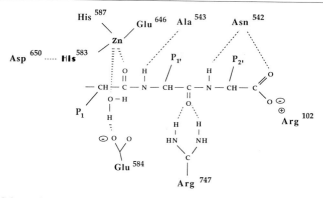

FIG. 1. Schematic representation of binding of a substrate to the active site of NEP.

$CH_2$-$CH(CH_2\Phi)$-$CONH$-$CH_2$-$COOH$] (selective NEP inhibitor) binding affinity.[8]

This third ligand was shown to belong to a short consensus sequence ExxxD[9] allowing to assign its position in the sequence of ACE and endothelin converting enzyme (ECE),[9a] another physiologically important member of the TLN-derived class of zinc metalloenzymes.

All the critical amino acids present in the active site of TLN were also found in NEP (reviewed in Roques *et al.*[4]). Unlike TLN, however, NEP possesses an Arg-102 residue located at the edge of the active site which is probably responsible for the dipeptidyl carboxypeptidase activity of NEP as observed in the release of Phe-Leu from Leu-enkephalin (Tyr-Gly-Gly-Phe-Leu). The replacement of Arg-102 by Glu has allowed the specificity of the enzyme to be changed, by charge–polarity reversal.[9b] This result has been confirmed by the replacement of Arg-102 with several other amino acids.[10] The Glu[102]-NEP can be used to separate unusual amino acid enantiomers and to position inhibitors in the NEP active site.[11]

All the Zn-metallopeptidases have similarities in the mechanisms of action. Schematically, the hydrolysis occurs through the formation of a pentacoordinated complex of the metal which includes the three Zn-coordi-

[8] H. Le Moual, A. Devault, B. P. Roques, P. Crine, and G. Boileau, *J. Biol. Chem.* **266,** 15670 (1991).

[9] B. P. Roques, *Biochem. Soc. Trans.* **21,** 678 (1993).

[9a] K. Shimida, M. Takahashi, and K. Tanjawa, *J. Biol. Chem.* **269,** 18275 (1994).

[9b] A. Beaumont, B. Barbe, H. Le Moual, G. Boileau, P. Crine, M. C. Fournié-Zaluski, and B. P. Roques, *J. Biol. Chem.* **267,** 2138 (1992).

[10] Y. A. Kim, B. Shriver, T. Quay, and L. B. Hersh, *J. Biol. Chem.* **267,** 12330 (1992).

[11] I. Gomez-Monterrey, S. Turcaud, E. Lucas, L. Bruetschy, B. P. Roques, and M. C. Fournié-Zaluski, *J. Med. Chem.* **36,** 87 (1993).

nating amino acids of the peptidase, the oxygen of the scissile bond, and the water molecule that is initially bound to the Zn atom.[12]

The specificity of the Zn-metallopeptidases is essentially ensured by van der Waals and ionic interactions between the S2, S1, S1', and S2' subsites and the lateral chains of the corresponding P2, P1, P1', and P2' moieties of the substrate. NEP has a broad selectivity and can cleave various short linear or cyclic peptides as well as insulin B chain[13] or interleukin (IL)-$\alpha$1.[14] The active site of NEP, especially the catalytic site, has been shown to be large,[15] accounting for the ability of the enzyme to cut the Cys-Phe bond of ANP (SLRRSSCFGGRMDRIGAQSGLGCNSFRY).

The enzyme also behaves as an efficient endopeptidase, cleaving various linear peptides *in vitro*, such as substance P and neurokinins,[16–18] gastrin and cholecystokinin-8 (CCK8),[17,19–21] and neurotensin.[22]

Aside from acting on the enkephalins and Met-enkephalin-Arg-Phe, NEP shows little activity toward other opioid peptides (dynorphin, $\beta$-endorphin). This is an important observation because it indicates that the "opioid" pharmacological effects induced by NEP inhibitors, and probably by mixed APN/NEP inhibitors, are mainly due to the protection of enkephalins.

Development of Selective Inhibitors of Neprilysin

Several classes of potent and selective inhibitors of NEP have been developed, characterized by the nature of the Zn-coordinating moiety (Table I) (reviewed in Roques *et al.*[4]). The compounds have been rationally designed by taking into account the structure and the mechanisms of action of NEP deduced from those of TLN.

The thiol inhibitor thiorphan [HS-CH$_2$-CH(CH$_2$Φ)-CONH-CH$_2$-COOH] was the first described synthetic potent NEP inhibitor ($K_1$ 4 n$M$),

[12] B. W. Matthews, *Acc. Chem. Res.* **21**, 333 (1988).
[13] M. A. Kerr and A. J. Kenny, *Biochem. J.* **137**, 477 (1974).
[14] M. E. Pierart, T. Najidovski, T. E. Appelboom, and M. M. Deschot-Lanckman, *J. Immunol.* **140**, 3808 (1988).
[15] M. C. Fournié-Zaluski, P. Chaillet, E. Soroca-Lucas, J. Costentin, and B. P. Roques, *J. Med. Chem.* **26**, 60 (1983).
[16] R. Matsas, I. S. Fulcher, A. J. Kenny, and A. J. Turner, *Proc. Natl. Acad. Sci. U.S.A.* **80**, 3111 (1983).
[17] R. Matsas, A. J. Kenny, and A. J. Turner, *Biochem. J.* **223**, 433 (1984).
[18] S. L. Stephenson and A. J. Kenny, *Biochem. J.* **241**, 237 (1987).
[19] C. Durieux, B. Charpentier, E. Fellion, G. Gacel, D. Pélaprat, and B. P. Roques, *Peptides* **6**, 495 (1985).
[20] C. Durieux, B. Charpentier, D. Pélaprat, and B. P. Roques, *Neuropeptides* **7**, 1 (1986).
[21] K. A. Zuzel, C. Rose, and J. C. Schwartz, *Neuroscience (Oxford)* **15**, 149 (1985).
[22] F. Checler, J. P. Vincent, and P. Kitabgi, *J. Neurochem.* **41**, 375 (1983).

which is only about 50-fold more potent in inhibiting NEP than ACE. To try to increase NEP selectivity, various structural modifications of the P1' and/or P2' moieties of thiorphan were made. However, this generally led to highly potent mixed inhibitors of NEP and ACE, such as ES34 [HS-CH$_2$-CH(CH$_2\Phi$)-CONH-CH(CH$_2$CH(CH$_3$)$_2$)-COOH] (NEP, $K_1$ 4.5 n$M$; ACE, $K_1$ 55 n$M$),[23] subsequently designated SQ 28,133 (Table I).[24] Retroinversion of the amide bond of thiorphan led to complete differentiation of NEP and ACE inhibition, and the resulting retrothiorphan [HS-CH$_2$-CH(CH$_2\Phi$)-NHCO-CH$_2$-COOH] is almost as potent as thiorphan, having a $K_1$ of 6 n$M$, but displays a drastic loss of potency for ACE ($K_1$ >10,000 n$M$).[25] Thiorphan and retrothiorphan cocrystallized with TLN are identically localized in the active site of the enzyme.[26] This result was in good agreement with computer simulations[27] and emphasizes the great structural analogy in the active sites of both enzymes. In agreement with the flexibility of the side chain of the arginine residue (Arg-102) located at the surface of NEP and at the edge of the S2' subsite,[9] a clear increased selectivity, without affinity loss, was also obtained by replacing the glycine moiety of thiorphan with longer aminoalkylcarboxylic acids, such as aminoheptanoic acid in SQ 29,072 [HS-CH$_2$-CH(CH$_2\Phi$)-CONH-(CH$_2$)$_6$-COOH][28] or by a heterocyclic hydrazide as in RU 44,004 [(R,S)HS-CH$_2$-CH(CH$_2\Phi$)-CONH-NC$_4$H$_8$O].

Introducing a carboxyl group at the N terminus of pseudodipeptides, capable of interacting with the S1' and S2' subsites, led to modest NEP inhibitors with $K_1$ values in the micromolar range.[29] Three carboxyl NEP inhibitors have been extensively studied, SCH 32,615, SCH 39,370, and UK 69,578 (Table I). In the latter compound, a cyclopentyl group and a $p$-aminocyclohexane carboxyl moiety were introduced in the P1' and P2' positions, respectively, to improve the selectivity for NEP.[23] Carboxyl inhibitors of NEP are significantly less potent than the corresponding thiol inhibitors, a result not found in the case of ACE.

The optimization of the interactions of hydroxamate inhibitors belong-

[23] M. C. Fournié-Zaluski, E. Lucas, G. Waksman, and B. P. Roques, *Eur. J. Biochem.* **139**, 267 (1984).

[24] A. A. Seymour, J. N. Swerdel, and B. Abboa-Offei, *J. Cardiovasc. Pharmacol.* **17**, 456 (1991).

[25] B. P. Roques, E. Lucas-Soroca, P. Chaillet, J. Costentin, and M. C. Fournié-Zaluski, *Proc. Natl. Acad. Sci. U.S.A.* **80**, 3178 (1983).

[26] S. L. Roderick, M. C. Fournié-Zaluski, B. P. Roques, and B. W. Matthews, *Biochemistry* **28**, 1493 (1989).

[27] T. Benchetrit, M. C. Fournié-Zaluski, and B. P. Roques, *Biochem. Biophys. Res. Commun.* **147**, 1034 (1987).

[28] A. A. Seymour, S. A. Fennell, and J. N. Swerdel, *Hypertension* (*Dallas*) **14**, 87 (1989).

[29] M. C. Fournié-Zaluski, E. Soroca-Lucas, G. Waksman, J. C. Schwartz, and B. P. Roques, *Life Sci.* **31**, 2947 (1982).

TABLE I

Selective Neprilysin Inhibitors and Mixed Neprilysin/Angiotensin-Converting Enzyme Inhibitors

| Inhibitor structure[a] | Inhibitor | $K_I$ (nM) | |
|---|---|---|---|
| | | NEP | ACE |
| — ENDOPEPTIDASE-24.11 ACTIVE SITE — | | | |
| | | $K_I$ (nM) | |
| | | NEP | ACE |
| $^-S$ - $CH_2$ - CH - C - N - $CH_2$ - COO$^-$ | Thiorphan | S 4 <br> R 4 | S 140 <br> R 860 |
| $^-S$ - $CH_2$ - CH - N - C - $CH_2$ - COO$^-$ | Rétrothiorphan | S 210 <br> R 2.3 | > 10,000 |
| $^-S$ - $CH_2$ - CH - C - N - $(CH_2)_6$ - COO$^-$ | SQ 29 072 | 26 | > 10,000 |
| $COO^-$ ... $CH_2$ - CH - NH - CH - C - N - $CH_2$ - CH - COO$^-$ | SCH 39,370 | 11 | > 10,000 |
| $COO^-$ ... $CH_2$ - CH - NH - CH - C - N - $CH_2$ - $CH_2$ - COO$^-$ | SCH 32,615 | 20 | > 10,000 |

| 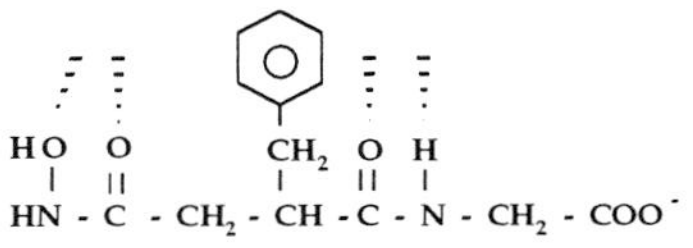 | 1.7 | > 10,000 |
| HACBO-Gly | | |
| 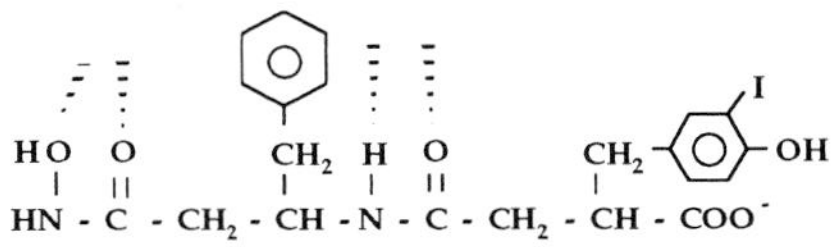 | 0.03 | > 10,000 |
| RB 104 | | |
| | 200 | > 10.000 |
| RU 44004 | | |
| | 28 | 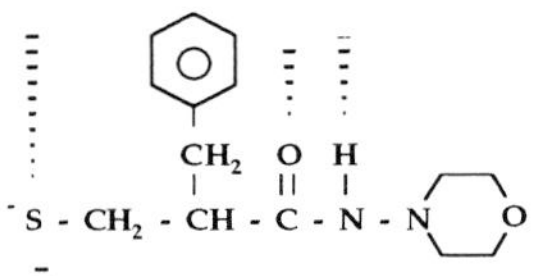 > 10.000 |
| 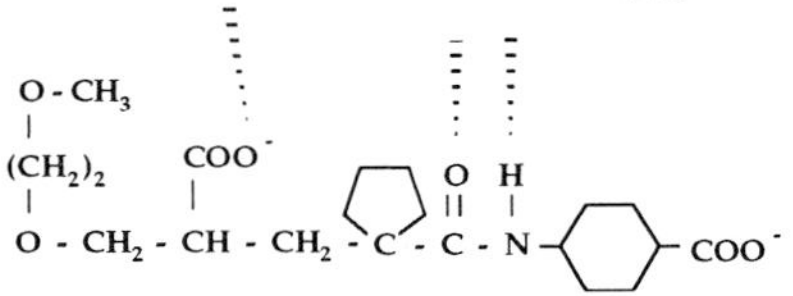 UK 69578 | | |

HO — O == O H
HN - C - CH₂ - CH - CH - C - N - CH₂ - COO⁻
CH₂
Φ

HO — O == H O
HN — C - CH₂ - CH - N - C - CH₂ - CH - COO⁻
CH₂
Φ
OH
I

O
N
S - CH₂ - CH - C - N
CH₂
Φ

O - CH₃
(CH₂)₂ COO⁻
O - CH₂ - CH - CH₂ - C - N
H O
C - N — COO⁻

$^a$ The symbol Φ represents a phenyl group.

ing to the HACBO-Gly series[30-32] has shown that the absolute configuration of the P1' residue, as well as the size and the hydrophobicity of the P2' residue, does not greatly influence enzyme recognition. The replacement of Gly in retro-HACBO-Gly by a highly hydrophobic aromatic moiety led to the inhibitor RB 104. [125]I-Labeled RB 104 [HONH-CH$_2$-

$$CH(CH_2\Phi)\text{-}NH\text{-}CO\text{-}CH_2\text{-}CH(CH_2\underset{\text{I}}{\overset{\text{I}}{\langle\bigcirc\rangle}}\text{-}OH)\text{-}COOH]$$ is the most po-

tent NEP inhibitor described so far ($K_I$ 0.03 n$M$), a property that has been used to visualize NEP directly in crude membrane fractions after gel electrophoresis.[33]

Another interesting series of inhibitors are the phosphorus-containing dipeptides,[34,34a] among which is the natural competitive inhibitor of NEP, phosphoramidon.

All the inhibitors generally have a limited ability to penetrate the gastrointestinal and blood–brain barriers. Protection of the thiol and/or carboxyl groups by lipophilic moieties led to an improvement in bioavailability. The resulting prodrugs, such as acetorphan, are rapidly transformed by esterases in blood and the brain.

## Development of Mixed Inhibitors of Neprilysin and Aminopeptidase N, and of Neprilysin and Angiotensin-Converting Enzyme

*In vivo,* the enkephalins are rapidly metabolized both by NEP and by aminopeptidase N (APN). Interestingly, both enzymes are Zn-metallopeptidases (see [13] in this volume), leading to the development of the concept of mixed inhibitors.[35] This was achieved using hydroxamate-containing inhibitors. The loss of binding affinity, arising from a relative inability of the lateral chains of the inhibitors to fit adequately the respective subsites

[30] M. C. Fournié-Zaluski, A. Coulaud, R. Bouboutou, P. Chaillet, J. Devin, G. Waksman, J. Costentin, and B. P. Roques, *J. Med. Chem.* **28,** 1158 (1985).

[31] J. Xie, J. M. Soleilhac, N. Renwart, J. Peyroux, B. P. Roques, and M. C. Fournié-Zaluski, *Int. J. Pept. Protein Res.* **34,** 246 (1989).

[32] J. Xie, J. M. Soleilhac, C. Schmidt, J. Peyroux, and M. C. Fournié-Zaluski, *J. Med. Chem.* **32,** 1497 (1989).

[33] M. C. Fournié-Zaluski, J. M. Soleilhac, S. Turcaud, R. Lai-Kuen, P. Crine, A. Beaumont, and B. P. Roques, *Proc. Natl. Acad. Sci. U.S.A.* **89,** 6388 (1992).

[34] R. L. Elliot, N. Marks, M. J. Berg, and P. S. Portoghese, *J. Med. Chem.* **28,** 1208 (1985).

[34a] S. de Lombaert, L. Blanchard, J. Tan, Y. Sakane, C. Berry, and R. D. Ghai, *Bioorg. Med. Chem. Lett.,* in press (1995).

[35] M. C. Fournié-Zaluski, P. Chaillet, R. Bouboutou, A. Coulaud, P. Chérot, G. Waksman, J. Costentin, and B. P. Roques, *Eur. J. Pharmacol.* **102,** 525 (1984).

of the different enzymes, could be counterbalanced by the strength of coordination to the Zn atom. Accordingly, bidentate inhibitors such as kelatorphan and RB 38A (Table II) are highly potent NEP/APN mixed inhibitors.[35] The mixed inhibitors were shown to inhibit completely enkephalin degradation *in vitro*[36] and *in vivo*.[37] Using this new approach a large number of bidentate inhibitors with nanomolar affinities both for NEP and for APN have been synthesized. However, their high water solubility prevents them from crossing the gastrointestinal and blood–brain barriers. Therefore, new lipophilic systemically active prodrugs such as RB 101 were developed by linking highly potent and hydrophobic thiol-containing APN and NEP inhibitors via a disulfide bond. One of the main advantages of the mixed inhibitors is the relative stability of the disulfide bond in plasma, as well as its rapid breakdown in the brain.[38]

The diuretic and vasorelaxant peptide ANP is a physiological substrate for NEP at the renal[39,40] and at the vascular levels.[41] In addition to ANP clearance occurring through the C-receptors, enzymatic degradation by NEP has been proposed to be responsible for the short half-life of ANP in the circulation. Selective NEP inhibitors, including the *S* isomer of thiorphan, lead to an increase in natriuresis and diuresis by the protection of ANP from degradation, but without any significant antihypertensive effect.[42,43] The concept of mixed inhibitors has therefore been extended to NEP/ACE blockers, to obtain a new class of antihypertensives.[44] Relatively few mixed NEP/ACE inhibitors have so far been synthesized, but the number will probably grow in the near future.

Based on the analogies between the active sites of both enzymes, mixed inhibitors such as ES34 (also designated SQ 28,133), PC57, and mixanpril I have been designed (Table II).[23,24,45] Conformationally constrained thiol

[36] G. Waksman, R. Bouboutou, J. Devin, S. Bourgoin, F. Cesselin, M. Hamon, M. C. Fournié-Zaluski, and B. P. Roques, *Eur. J. Pharmacol.* **117**, 233 (1985).
[37] S. Bourgoin, D. Le Bars, F. Artaud, A. M. Clot, R. Bouboutou, M. C. Fournié-Zaluski, B. P. Roques, M. Hamon, and F. Cesselin, *J. Pharmacol. Exp. Ther.* **238**, 360 (1986).
[38] M. C. Fournié-Zaluski, P. Coric, S. Turcaud, E. Lucas, F. Noble, R. Maldonado, and B. P. Roques, *J. Med. Chem.* **35**, 2473 (1992).
[39] S. L. Stephenson and A. J. Kenny, *Biochem. J.* **243**, 183 (1987).
[40] C. Landry, P. Santagata, W. Bawab, M. C. Fournié-Zaluski, B. P. Roques, P. Vinay, and P. Crine, *Biochem. J.* **291**, 773 (1993).
[41] J. M. Soleilhac, E. Lucas, A. Beaumont, S. Turcaud, J. B. Michel, D. Ficheux, M. C. Fournié-Zaluski, and B. P. Roques, *Mol. Pharmacol.* **41**, 609 (1992).
[42] D. B. Northridge, C. T. Alabaster, J. M. C. Connell, S. G. Dilly, A. F. Lever, A. G. Jardine, P. L. Barclay, H. J. Dargie, I. N. Findlay, and G. M. R. Samuels, *Lancet* **2**, 591 (1989).
[43] J. C. Kahn, M. Patey, J. L. Dubois-Rande, P. Merlet, A. Castaigne, C. Lim-Alexandre, J. M. Lecomte, D. Duboc, C. Gros, and J. C. Schwartz, *Lancet* **335**, 118 (1990).
[44] B. P. Roques and A. Beaumont, *Trends Pharmacol. Sci.* **11**, 245 (1990).
[45] M. C. Fournié-Zaluski, P. A. Coric, S. Turcaud, N. Rousselet, W. Gonzalez, B. Barbe, I. Pham. N. Jullian, J. P. Michel, and B. P. Roques, *J. Med. Chem.* **37**, 1070 (1994).

TABLE II
## Selected Mixed Inhibitors of Neprilysin/Aminopeptidase N

|  | Inhibitor | $K_1 (nM)$ | |
| | | NEP | APN |

| Inhibitor structure | Inhibitor | NEP | APN |
|---|---|---|---|
| ———— Mixed inhibitors NEP/APN ———— | | | |

S₁ H⁵⁸⁷ E⁶⁴⁶ S′₁ S′₂ (schematic of active site with H⁵⁸³, Zn²⁺, (−), (+) R¹⁰²)

K₁ (nM): NEP, APN

| Structure | Inhibitor | NEP | APN |
|---|---|---|---|
| ⁻O O CH₂ CH₃ <br> \| \|\| \| \| <br> HN - C - CH₂ - CH - CONH - CH - COO⁻ | Kélatorphan | 1.8 | 380 |
| ⁻O O CH₂ CH₂ <br> \| \|\| \| \| <br> HN - C - CH₂ - CH - CONH - CH - COO⁻ | RB 38A | 0.9 | 120 |
| SCH₃ <br> \| <br> (CH₂)₂ <br> \| <br> ⁺H₃N - CH - CH₂ - S⁻ | PC 18 | >10.000 | 8 |
| CH₂ CH₂ <br> \| \| <br> ⁻S - CH₂ - CH - CONH - CH - COO⁻ | ST43 | 1.5 | >10.000 |
| SCH₃ CH₂ CH₂ <br> \| \| \| <br> (CH₂)₂ <br> \| <br> ⁺H₃N - CH - CH₂ - S - S - CH₂ - CH - CONH - CH - COO⁻ | RB 101 (prodrug) | >10.000 | >10.000 |

|  |  | NEP | ACE |
|---|---|---|---|
| CH₃ <br> \| <br> CH - CH₃ <br> \| <br> CH₂ O H CH₂ <br> \| \|\| \| \| <br> S⁻ - CH₂ - CH - C - N - CH - COO⁻ | ES 34 (SQ 28,133) | 4.5 | 55 |
| CH₃ — CH O H CH₃ <br> \| \|\| \| \| <br> S⁻ - CH₂ - CH - C - N - CH - COO⁻ | Mixanpril | 1.7 | 4.2 |
| CH₃ <br> \| <br> CH₃ CH₂ <br> CH₃ S⁻ O H CH \| O H CH₂—⟨O⟩—OH <br> \| \| \|\| \| \| \| \|\| \| \| <br> CH₃ - CH - CH - C - N - CH - C - N - CH - COO⁻ | PC 57 | 1.4 | 0.2 |

inhibitors synthesized in pharmaceutical industries were also reported to have potent inhibitory potencies against NEP and ACE.[46,46a,46b]

## Analgesic Responses Induced by Enkephalin-Degrading Enzyme Inhibitors

### Antinociceptive Effects of Selective or Mixed Inhibitors

Inhibitors of NEP or APN are able to potentiate the analgesic effects of exogenous enkephalins and, more interestingly, possess a weak intrinsic opioidergic action after intracerebroventricular (i.c.v.) injection. This has been established for thiorphan or bestatin alone, or in association, retrothiorphan, and phosphoryl- or carboxyl-containing inhibitors. Different analgesic tests have been employed (hot plate test, tail flick test, writhing test, paw pressure test, all electric stimulation test, tail withdrawal test, formalin test), and different routes of administration have also been used. All the responses observed were antagonized by prior administration of naloxone (reviewed in Roques et al.[4]).

Owing to the complementary roles of NEP and APN in enkephalin inactivation, selective inhibition of only one of the peptidases gives weak antinociceptive effects. As expected, mixed inhibitors are much more effective. Complete inhibition of enkephalin metabolism by intracerebroventricular kelatorphan or RB 38A induced naloxone-antagonized antinociceptive responses in all the various assays commonly used to select analgesics,[35,47] which were greater than the responses obtained after coadministration of NEP and APN inhibitors. However, even at very high concentrations (150 $\mu$g i.c.v.), at which they have been shown to inhibit completely enkephalin metabolism,[36,37] mixed inhibitors were unable to produce the maximum analgesic effect induced by morphine, except in the hot plate and writhing tests.[47] This suggests that the local increase in enkephalin concentration produced by the mixed inhibitors remains too low to saturate opioid recep-

---

[46] G. A. Flynn, D. G. Beight, S. Mehdi, J. R. Koehl, E. L. Giroux, J. F. French, P. W. Hake, and R. C. Dage, J. Med. Chem. **36**, 2420 (1993).

[46a] J. L. Stanton, D. M. Sperbeck, A. J. Trapani, D. Cote, Y. Sakane, C. J. Berry, and R. D. Ghai, J. Med. Chem. **36**, 3829 (1993).

[46b] J. Das, J. A. Robl, J. A. Reid, C. Q. Sun, R. N. Misra, B. R. Brown, D. E. Ryono, M. M. Asaad, J. E. Bird, N. C. Trippodo, E. W. Petrillo, and D. S. Karanewsky, Bioorg. Med. Chem. Lett. **4**, 2193 (1994).

[47] C. Schmidt, J. Peyroux, F. Noble, M. C. Fournié-Zaluski, and B. P. Roques, Eur. J. Pharmacol. **192**, 253 (1991).

tors, a conclusion that is in agreement with the results of *in vivo* binding experiments.[48,49]

Unfortunately, owing to high water solubility, the bidentate inhibitors are unable to cross the blood–brain barrier, preventing the effects resulting from administration of mixed inhibitors to be investigated by a clinically relevant route. Therefore, as previously discussed, new lipophilic thiol-containing APN and NEP inhibitors linked by mercapto groups have been synthesized.[38] RB 101, the first systemically active prodrug, generates, through a biologically dependent cleavage of the disulfide bond, the potent APN [(*S*)2-amino-1-mercapto-4-methylthiobutane, $K_I$ 11 n$M$] and NEP {*N*-[(*R,S*)-2-mercaptomethyl-1-oxo-3-phenylpropyl]-L-phenylalanine, $K_I$ 2 n$M$} inhibitors. RB 101 easily crosses the blood–brain barrier, as shown by the complete inhibition of cerebral NEP following intravenous (i.v.) injection in mice. The prodrug induces strong, dose-dependent antinociceptive responses in the hot plate ($ED_{50}$ 9 mg/kg i.v.) and writhing tests in mice ($ED_{50}$ 3.25 mg/kg i.v.), and the tail flick and tail electric stimulation tests in rats. In all of the tests used, the pain-alleviating effect of RB 101 was suppressed by naloxone.[50]

At the spinal level, electrophoretic administration of kelatorphan in the substantia gelatinosa of spinal cats led to naloxone-reversible inhibition of nociceptive responses and marked potentiation of coadministered Met-enkephalin.[51] When an acute noxious stimulus was induced by transcutaneous electrical stimulation or when a more prolonged chemical noxious stimulus was elicited by a subcutaneous (s.c.) injection of 5% (v/v) formalin, the direct spinal application of bestatin, thiorphan, or kelatorphan abolished C-fiber-evoked activity in a dose-dependent manner. Kelatorphan was considerably more effective than APN or NEP selective inhibitors, and it was significantly more efficient than a combination thereof.[52,53]

Several studies have used the expression of immediate early genes as markers for neuronal activity in an attempt to differentiate the pain modulatory effects of exogenously administered opioids from tonically released endogenous opioid peptides. When administered intravenously before heat stimulation, both morphine and, to a lesser extent, kelatorphan and RB 101

[48] E. Meucci, P. Delay-Goyet, B. P. Roques, and J. M. Zajac, *Eur. J. Pharmacol.* **171,** 167 (1989).
[49] M. Ruiz-Gayo, A. Baamonde, S. Turcaud, M. C. Fournié-Zaluski, and B. P. Roques, *Brain Res.* **571,** 306 (1992).
[50] F. Noble, J. M. Soleilhac, E. Soroca-Lucas, S. Turcaud, M. C. Fournié-Zaluski, and B. P. Roques, *J. Pharmacol. Exp. Ther.* **261,** 181 (1992).
[51] C. R. Morton, Z. Q. Zhao, and A. W. Duggan, *Eur. J. Pharmacol.* **140,** 195 (1987).
[52] A. H. Dickenson, A. F. Sullivan, M. C. Fournié-Zaluski, and B. P. Roques, *Brain Res.* **408,** 185 (1987).
[53] A. F. Sullivan, A. H. Dickenson, and B. P. Roques, *Br. J. Pharmacol.* **98,** 1039 (1989).

reduced the induction of immediate early genes, such as c-*fos*, in the superficial dorsal horn and the deep dorsal horn of rats.[54,55]

Promising preliminary clinical results after intrathecal administration of either kelatorphan (J. Meynadier, M. C. Fournié-Zaluski, and B. P. Roques, unpublished results, 1988) in morphine-tolerant patients or the association of bestatin and thiorphan in normal patients[56] have been obtained.

Enkephalin-degrading enzyme inhibitors are particularly efficient in arthritic rats, which serve as a model of chronic pain. Kelatorphan, at doses as low as 2.5 mg/kg i.v., at which acetorphan was ineffective, produced potent naloxone-reversible antinociceptive responses in normal rats that were comparable to those induced by 1 mg/kg i.v. morphine; however, at higher doses (2, 10, 15 mg/kg i.v.) the effects of kelatorphan were no more pronounced than those of acetorphan.[57] Unlike acetorphan, kelatorphan was found to be much more effective in arthritic than in normal rats in raising the vocalization threshold, even at 5 mg/kg i.v. (244% in arthritic versus 144% in normal rats). Given the very weak passage of kelatorphan into the brain, its strong antinociceptive effects in inflammatory pain raise the question of a possible action at the level of peripheral nociceptors, where all opioid targets including NEP seem to be present. On a model of unilateral inflammatory pain (intraplantar injection of Freund's complete adjuvant), a greater elevation of paw pressure threshold in inflamed compared to noninflamed paws was observed in rats that received an intraplantar injection of thiorphan plus bestatin.[58] As expected, slightly higher effects were observed in the same model following systemic administration of RB 38A or RB 101.[59]

*Interactions between Cholecystokinin and Enkephalin Systems in Control of Pain*

Anatomical studies have shown that the distribution of both cholecystokinin (CCK) and CCK receptors parallels that of endogenous opioid receptors in the pain-processing regions of both the brain and spinal cord.[60,61]

[54] T. R. Tölle, J. Schadrack, J. M. Castro-Lopes, G. Evan, B. P. Roques, and W. Zieglgansberger, *Pain* **56**, 103 (1994).

[55] C. Abaddie, P. Honoré, M. C. Fournié-Zaluski, B. P. Roques, and J. M. Besson, *Eur. J. Pharmacol.* **258**, 215 (1994).

[56] J. Meynadier, S. Dalmas, J. M. Lecomte, C. Gros, and J. C. Schwartz, *Pain Clin.* **2**, 201 (1988).

[57] V. Kayser, M. C. Fournié-Zaluski, G. Guilbaud, and B. P. Roques, *Brain Res.* **497**, 94 (1989).

[58] C. G. Parsons and A. Herz, *J. Pharmacol. Exp. Ther.* **255**, 795 (1990).

[59] R. Maldonado, O. Valverde, S. Turcaud, M. C. Fournié-Zaluski, and B. P. Roques, *Pain* **58**, 77 (1994).

[60] C. Gall, J. Lauterborn, D. Burks, and K. Seroogy, *Brain Res.* **403**, 403 (1987).

[61] M. Pohl, J. J. Benoliel, S. Bourgoin, M. C. Lombard, A. Mauborgne, H. Taquet, A. Carayon, J. M. Besson, F. Cesselin, and M. Hamon, *J. Neurochem.* **55**, 1122 (1990).

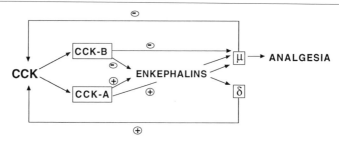

FIG. 2. Hypothetical model of interactions between CCK, via CCK$_A$ and CCK$_B$ receptors, and the opioid system via $\delta$-opioid and $\mu$-opioid receptors.

This overlapping distribution has focused investigations on the role of CCK in nociception. The existence of regulatory mechanisms between CCK and enkephalin systems in the control of pain have been proposed. Schematically, activation of CCK$_B$ receptors could negatively modulate the opioid system, either directly (via binding of opioid agonists or C-fiber-evoked activity) or indirectly (via release of endogenous enkephalins), whereas stimulation of CCK$_A$ receptors would enhance opioid release[62,63] (Fig. 2). This is in agreement with results obtained following the blockade of CCK$_B$ binding sites by selective antagonists, which significantly increased the antinociceptive responses induced by RB 101.[64] Thus the antinociception observed after the association of PD 134,308 with RB 101 was about 800% higher than that observed with RB 101 given alone in the tail flick test in rats (Fig. 3A).[65]

## Tolerance, Dependence, and Side Effects of Selective and Mixed Inhibitors of Neprilysin and Aminopeptidase N

In addition to serious drawbacks such as respiratory depression, the development of tolerance analgesia and physical and psychic dependence observed during chronic administration limits the use of morphine and surrogates. In contrast, kelatorphan, ionophoretically applied into the nucleus ambigus of cats, produces a low, partially naloxone-reversible reduction of respiratory frequency.[66] Furthermore, no signs of withdrawal were observed after administration of naloxone in animals chronically treated

[62] M. Derrien, F. Noble, R. Maldonado, and B. P. Roques, *Neurosci. Lett.* **160**, 193 (1993).
[63] F. Noble, M. Derrien, and B. P. Roques, *Br. J. Pharmacol.* **109**, 1064 (1993).
[64] R. Maldonado, M. Derrien, F. Noble, and B. P. Roques, *Neuroreport* **4**, 947 (1993).
[65] O. Valverde, R. Maldonado, M. C. Fournié-Zaluski, and B. P. Roques, *J. Pharmacol. Exp. Ther.* **270**, 77 (1994).
[66] M. P. Morin-Surun, E. Boudinot, M. C. Fournié-Zaluski, J. Champagnat, B. P. Roques, and M. Denavit-Saubié, *Neurochem. Int.* **20**, 103 (1992).

with RB 101 (Fig. 3B).[67] Moreover, chronic administration of the mixed inhibitor prodrug did not induce tolerance or cross-tolerance with morphine,[68] and unlike morphine it did not induce psychic dependence (Fig. 3C).[69]

Addiction, following chronic treatment with opiates, is probably due to multiple cellular events. The moderate degree or lack of tolerance or physical dependence observed after chronic treatment with the mixed inhibitors (RB 38A or RB 101) could be explained by a weaker, but more specific, stimulation of the opioid-binding sites by the tonically released endogenous opioids, thus minimizing receptor desensitization or down-regulation that usually occurs after the general stimulation of opioid receptors by exogenously administered agonists.[70,71] Moreover, it has been demonstrated, in the locus coeruleus, which is the most critical structure implicated in the development of dependence,[72,73] that there is little or no tonic endogenous opioid action.[74] Another possible reason could be a weaker adenylate cyclase expression[75] and protein phosphorylation,[70] or differences in the efficiency of triggering the release of endogenous antiopiate peptides. On the other hand, because of higher intrinsic efficacy,[76,77] enkephalins need to occupy less opioid receptors than morphine to give the same pharmacological responses, which could also explain the lack of or moderate side effects observed after chronic treatment with mixed inhibitors compared to opiates.

The biochemical mechanisms involved in the rewarding effects of morphine are unknown, but several studies have shown that dopaminergic neurons, particularly those that project from the ventral tegmental area (VTA) to the nucleus accumbens, may play a major role in the euphorogenic properties of opiates.[78,79] The failure of RB 101 to induce dependence syndrome measured by a conditioned place preference assay probably results from a lower recruitment of opioid receptors and from the relatively

[67] F. Noble, P. Coric, M. C. Fournié-Zaluski, and B. P. Roques, *Eur. J. Pharmacol.* **223**, 94 (1992).
[68] F. Noble, S. Turcaud, M. C. Fournié-Zaluski, and B. P. Roques, *Eur. J. Pharmacol.* **223**, 83 (1992).
[69] F. Noble, M. C. Fournié-Zaluski, and B. P. Roques, *Eur. J. Pharmacol.* **230**, 139 (1993).
[70] E. J. Nestler and J. F. Tallman, *Mol. Pharmacol.* **33**, 127 (1988).
[71] B. J. Morris and A. Herz, *Neuroscience* (*Oxford*) **29**, 433 (1989).
[72] R. S. Duman, J. F. Tallman, and E. J. Nestler, *J. Pharmacol. Exp. Ther.* **246**, 1033 (1988).
[73] R. Maldonado, L. Stinus, L. H. Gold, and G. F. Koob, *J. Pharmacol. Exp. Ther.* **261**, 669 (1992).
[74] J. T. Williams, M. J. Christie, R. A. North, and B. P. Roques, *J. Pharmacol. Exp. Ther.* **243**, 397 (1987).
[75] I. Matsuoka, R. Maldonado, N. Defer, F. Noel, J. Hanoune, and B. P. Roques, *Eur. J. Pharmacol.* **268**, 215 (1994).
[76] F. Porreca, D. Lopresti, and S. J. Ward, *Eur. J. Pharmacol.* **179**, 129 (1990).
[77] F. Noble and B. P. Roques, *Neurosci. Lett.*, in press.
[78] A. G. Phillips and F. G. Le Piane, *Behav. Brain Res.* **5**, 225 (1982).
[79] M. A. Bozarth, *Behav. Brain Res.* **22**, 107 (1986).

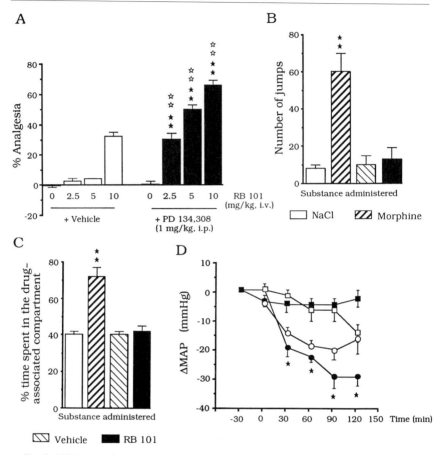

FIG. 3. (A) Antinociceptive effects of RB 101 in the tail flick test in rats and potentiation by the selective $CCK_B$ antagonist PD 134,308 (1 mg/kg i.p.) . ★★, $p < 0.01$ compared to control group: ☆☆, $p < 0.01$ compared to the same dose of RB 101 without PD 134,308. (B) Comparison of the withdrawal symptoms induced by naloxone after chronic treatment with morphine (6 mg/kg i.p.) or RB 101 (160 mg/kg i.p.) injected twice daily for 5 days. ★★, $p < 0.01$ compared with other groups. (C) Comparison of the psychic dependence induced by chronic morphine (6 mg/kg i.p.) or RB 101 (160 mg/kg i.p.) in the place preference test. ★★, $p < 0.01$ compared with other groups. (D) Hypotensive effects of selective inhibitors of NEP (□, retrothiorphan) and ACE (○, enalapril) and of the mixed NEP/ACE inhibitor (●, mixanpril) (25 mg/kg i.v. for each compound) in the spontaneous hypertensive rat. ΔMAP, Variation of mean arterial pressure; ★, $p < 0.05$ compared with enalapril; ■, control.

poor capability of endogenous enkephalins to modify intracellular events such as protein phosphorylation with subsequent modulation of the dopaminergic transmission in the nucleus accumbens. This hypothesis is supported by the minimal changes in dopamine release in the nucleus accumbens after local administration of kelatorphan into the VTA[80] and by the apparent absence of any effects on the levels of dopamine and metabolites in the nucleus accumbens following intravenous administration of acetorphan.[81] Accordingly, when injected in the VTA of rats, kelatorphan, unlike morphine, had either no effect or slightly decreased the rate of intracranial stimulation.[82]

The enkephalin-degrading enzyme inhibitors minimize the severity of naloxone-precipitated withdrawal syndrome in rats (reviewed in Roques *et al.*[4]). The greater efficiency of the mixed inhibitors compared to selective ones[83] is probably due to the resulting greater increase in enkephalins in certain brain regions, especially those enriched in $\mu$-opioid receptors, such as periaqueductal gray matter, which also contains high levels of NEP[84] and could be an important site of action for the development of physical morphine dependence.[85] Mixed inhibitors such as RB 101 could therefore represent more efficient compounds than methadone in the treatment of opioid addiction.[85a]

## Behavioral Effects of Neprilysin and Neprilysin/Aminopeptidase N Inhibitors

Dopamine receptor antagonists have been found to facilitate the behavioral effects induced by acute administration of mixed inhibitors into the nucleus accumbens, which was maximal in the third week after the beginning of treatment.[86] The delay corresponds to the appearance of the antipsychotic effect of the neuroleptics, suggesting that alterations in the opioidergic system, very likely through interrelations with the dopaminergic pathway, could be taking place in a neuronal system critically involved in the control of mood.[87] Various pharmacological and biochemical studies have shown that enkephalins are involved in the control of behavior such as arousal, locomotion, self-administration, self-stimulation, learning, and memory functions, through modulation of the motor (nigrostriatal) and limbic cortical (mesocorticolimbic) dopaminergic systems. A link between

[80] V. Daugé, P. W. Kalivas, T. Duffy, and B. P. Roques, *Brain Res.* **599**, 209 (1992).
[81] N. Dourmap, A. Michael-Titus, and J. Costentin, *Eur. J. Neurosci.* **2**, 783 (1990).
[82] C. Heidbreder, M. Gewiss, S. Lallemand, B. P. Roques, and P. De Witte, *Neuropharmacology* **31**, 293 (1992).
[83] R. Maldonado, V. Daugé, J. Callebert, J. M. Villette, M. C. Fournié-Zaluski, J. Féger, and B. P. Roques, *Eur. J. Pharmacol.* **165**, 199 (1989).
[84] G. Waksman, E. Hamel, M. C. Fournié-Zaluski, and B. P. Roques, *Proc. Natl. Acad. Sci. U.S.A.* **83**, 1523 (1986).
[85] R. S. Laschka, H. Teschemacher, P. Mehrain, and A. Herz, *Psychopharmacology* (*Berlin*) **46**, 141 (1976).
[85a] B. P. Roques and F. Noble, *NIDA Res. Monograph*, in press (1994).
[86] R. Maldonado, V. Daugé, J. Féger, and B. P. Roques, *Neuropharmacology* **29**, 215 (1990).
[87] B. P. Roques, V. Daugé, G. Gacel, and M. C. Fournié-Zaluski, *Biol. Psychiatry* **7**, 287 (1985).

opioid and dopaminergic systems has been demonstrated by the clear anti-depressant-like effects that are observed after intravenous administration of RB 101. These effects, which were shown to be related to $\delta$-receptor and $D_1$-receptor activation, produced an increase in dopamine turnover in the striatum.[88] Furthermore, it has been demonstrated that the antidepressant-like effects of RB 101 are potentiated by the administration of $CCK_B$ antagonists, which have their own antidepressant-like effect after systemic administration in the suppression of motility and Porsolt tests.[89,90,90a]

The following results also support a relationship between the protection of endogenous enkephalins and behavioral modifications: (1) naloxone-reversible potentiation of the stereotypes induced by phencyclidine after intracerebroventricular administration of bestatin and thiorphan[91]; (2) potentiation by thiorphan or bestatin of the antidepressant actions of a subeffective dose of imipramine, measured using the forced swimming test[92]; (3) slight reduction in immobility time induced in the same test by either thiorphan or bestatine[93]; (4) suppression of motility in a conditioned emotional response to an environment associated previously with foot shock following intracerebroventricular administration of either thiorphan and bestatin[94]; and (5) reduced behavioral responses triggered by learned helplessness after intracerebroventricular treatment with RB 38A.[95]

### Gastrointestinal Effects of Enkephalin-Degrading Enzyme Inhibitors and Clinical Use as Antidiarrheal Agents

The well-known antidiarrheal effect of opioids has been therapeutically exploited for many years. The possible tonic participation of endogenous enkephalins at different levels of the gastrointestinal tract has been investigated essentially through inhibition of NEP by thiorphan (reviewed by Checler[96]). The antidiarrheal effects of thiorphan and acetorphan have

[88] A. Baamonde, V. Daugé, M. Ruiz-Gayo, I. G. Fulga, S. Turcaud, M. C. Fournié-Zaluski, and B. P. Roques, *Eur. J. Pharmacol.* **216,** 157 (1992).

[89] M. Derrien, C. Durieux, and B. P. Roques, *Br. J. Pharmacol.* **111,** 956 (1994).

[90] F. Hernando, J. A. Fuentes, B. P. Roques, and M. Ruiz-Gayo, *Eur. J. Pharmacol.* **261,** 275 (1994).

[90a] C. Smadja, R. Maldonado, S. Turcaud, M. C. Fournié-Zaluski, and B. P. Roques, *Psychopharmacology,* submitted.

[91] M. Hiramatsu, T. Nabeshima, H. Fukaya, H. Furukawa, and T. Kameyama, *Eur. J. Pharmacol.* **120,** 69 (1986).

[92] M. D. C. De Felipe, I. Jimenez, A. Castro, and J. A. Fuentes, *Eur. J. Pharmacol.* **159,** 175 (1989).

[93] L. Ben Natan, P. Chaillet, J. M. Lecomte, H. Marçais, G. Uchida, and J. Costentin, *Eur. J. Pharmacol.* **97,** 301 (1984).

[94] T. Nabeshima, A. Katoh, and T. Kameyama, *J. Pharmacol. Exp. Ther.* **244,** 303 (1988).

[95] P. Tejedor-Real, J. A. Mico, R. Maldonado, B. P. Roques, and J. Gibert-Rahola, *Biol. Psychiatry* **34,** 100 (1993).

[96] F. Checler, *in* "Neuropeptide Function in the Gastrointestinal Tract" (E. E. Daniel, ed.), p. 274. CRC Press, Boston, Massachusetts, 1991.

been compared to those of loperamide in a model of castor oil-induced diarrhea in rats.[97] The inhibitors delayed the onset of diarrhea with no reduction in gastrointestinal transit, in contrast to loperamide. The naloxone-antagonized antidiarrheal effect of thiorphan and its prodrug seem to result from an antisecretory effect, probably due to the stimulation of peripheral δ-opioid receptors. Thus, the main advantages of the use of inhibitors is, as in the case of analgesia, a more physiological and selective antisecretory effect, thereby reducing the risk of tolerance and rebound effects leading to a severe reduction in gastrointestinal propulsion induced by stimulation of $\mu$-receptors by loperamide.[98] For this reason, acetorphan is now on the market as an antidiarrheal agent under the registered trademark Tiorfan.[99]

## Inhibition of Neprilysin-Induced Inactivation of Atrial Natriuretic Peptide: Clinical Perspectives of Selective Neprilysin Inhibitors and Mixed Neprilysin/Angiotensin-Converting Enzyme Inhibitors

In addition to ANP clearance occurring through the C-receptors, enzymatic degradation, probably both in the kidney and in the vasculature, was proposed to be responsible for the short half-life of ANP in the circulation. Cleavage of ANP, hypothesized to be due to NEP in the kidney,[100] was shown to modify renal secretion.[101] In addition, ANP was shown to be cleaved in vascular tissue by thiorphan-sensitive metallopeptidase,[102] which has since been identified as NEP by gel electrophoresis using the highly potent inhibitor [125]I-labeled RB 104.[41] Numerous experiments in animals and in humans have shown that, especially under conditions of heart loading, inhibition of NEP potentiates the effects of endogenous ANP, thereby inducing a significant increase in natriuresis, diuresis, and cyclic GMP excretion, without changing potassium elimination.[43,103–105] However, the change in blood pres-

[97] H. Marçais-Collado, G. Uchida, J. Costentin, J. C. Schwartz, and J. M. Lecomte, *Eur. J. Pharmacol.* **144**, 125 (1987).

[98] P. Baumer, E. D. Danquechin-Dorval, J. Bertrand, J. M. Vetel, J. C. Schwartz, and J. M. Lecomte, *Gut* **33**, 753 (1992).

[99] J. F. Bergman, S. Chaussade, D. Couturier, P. Baumer, J. C. Schwartz, and J. M. Lecomte, *Aliment. Pharmacol. Ther.* **6**, 305 (1992).

[100] N. Ura, O. A. Carretero, and E. G. Erdös, *Kidney Int.* **32**, 507 (1987).

[101] A. A. Seymour, J. N. Swerdel, N. G. Delaney, M. Rom, D. W. Cushman, and J. M. Deforrest, *Fed. Proc.* **46**, 1296 (1987).

[102] P. P. Tamburini, J. A. Koehn, J. P. Gilligan, D. Charles, R. Palmesino, R. Sharif, C. McMartin, M. D. Erion, and M. J. S. Miller, *J. Pharmacol. Exp. Ther.* **251**, 956 (1989).

[103] A. M. Richards, E. Espiner, C. Frampton, H. Ikram, T. Handle, M. Sopwitch, and N. Cussans, *Hypertension (Dallas)* **16**, 269 (1990).

[104] A. M. Richards, G. Wittert, E. A. Espiner, T. G. Yandle, C. Frampton, and H. Ikram, *J. Clin. Endocrinol. Metab.* **72**, 1317 (1991).

[105] I. Pham, A. I. K. El Amrani, M. C. Fournié-Zaluski, P. Corvol, B. P. Roques, and J. B. Michel, *J. Cardiol. Res.* **20**, 847 (1992).

sure is slight in the spontaneously hypertensive rat and in patients with congestive heart failure in spite of a large increase in circulating ANP.[106,107]

On the basis of these data, we proposed to extend the concept of mixed inhibitors that we initially developed to block both NEP and APN also to dual inhibition of NEP and ACE.[44] In the spontaneously hypertensive rat, the maximal depressor responses induced either by the combination of selective NEP inhibitor and captopril or by mixed inhibitors of NEP/ACE, such as SQ 28,133 or mixanpril I, were greater than the responses to any of the inhibitors given alone (Fig. 3D).[24,108,109] This potentiating effect was also observed in conscious dogs with pacing-induced heart failure.[110] Moreover, the dual inhibition of NEP and ACE will eliminate the administration of diuretics that are required to potentiate the action of ACE inhibitors but which trigger hypokalemia and increases in plasma renin and in aldosterone. Clinical trials with NEP inhibitors, and probably in the near future with mixed NEP/ACE inhibitors, should confirm the advantages and the limits of these approaches in the treatment of essential hypertension, chronic cardiac failure, angor,[110a] and sodium retention states.

It is interesting to observe that in all the studies performed with NEP inhibitors, no major side effects have been observed. This suggests that the cough which is sometimes associated with chronic treatment with ACE inhibitors is not potentiated by NEP inhibitors.

## Conclusions and Perspectives

Neprilysin became an interesting pharmacological "target" after the discovery that it acts to regulate the physiological functions of the enkephalins in the central nervous system, where NEP was found to be mainly localized, by autoradiography using the tritiated inhibitor [$^3$H]HACBO-Gly, in areas involved in pain control (substantia gelatinosa of the spinal cord, periaqueductal gray matter) but also in behavioral control (nucleus accumbens, substantia nigra).[84] Interest was further heightened when NEP was shown to degrade ANP in the periphery.

[106] D. B. Northridge, A. G. Jardine, I. N. Findlay, M. Archibald, S. Dilly, and H. J. Dargie, *Am. J. Hypertens.* **3,** 682 (1990).

[107] E. P. Kromer, D. Elsner, H. W. Kahles, and G. A. Riegger, *Am. J. Hypertens.* **4,** 460 (1991).

[108] K. B. Margulies, M. A. Perrella, L. J. McKinley, and J. C. Burnett, *J. Clin. Invest.* **88,** 1636 (1991).

[109] I. Pham, W. Gonzales, A. I. K. El Amrani, M. C. Fournié-Zaluski, M. Philippe, I. Laboulandine, B. P. Roques, and J. B. Michel, *J. Pharmacol. Exp. Ther.* **109,** 1064 (1993).

[110] A. A. Seymour, M. M. Asaad, V. M. Lanoce, K. M. Langenbacher, S. A. Fennell, and W. L. Rogers, *J. Pharmacol. Exp. Ther.* **266,** 872 (1993).

[110a] G. Piedimonte, J. A. Nadel, C. S. Long, and J. I. E. Hoffman, *Circ. Res.* **75,** 770 (1994).

The main advantage of modifying the concentration of endogenous peptides by the use of NEP inhibitors is that pharmacological responses are induced only at receptors tonically or phasically stimulated by the natural effectors. In contrast to exogenous agonists or antagonists, chronic administration of mixed inhibitors did not induce changes in the synthesis of the clearing peptidases, in the synthesis of target peptide precursors, or in the secretion of active peptides.[5,111] The goal of discovering analgesics endowed with a potency similar to that of morphine, but devoid of major side effects, may now have been reached with the mixed NEP/APN inhibitors, although the compounds have yet to be evaluated in clinical trials. In addition to its antidiarrheic property, acetorphan was shown to induce natriuresis and diuresis in humans. The marketing of acetorphan will give important information concerning the possible extension of the clinical indications for NEP inhibitors, for instance, as new antidepressive agents. Moreover, the facilitating effect of $CCK_B$ antagonists on the antinociceptive and antidepressive-like responses induced by the endogenous enkephalins protected from their catabolism by inhibitors is a promising perspective in the management of pain and in the treatment of depression. The coadministration of the selective $CCK_B$ antagonist and mixed inhibitors could also considerably improve the treatment of drug abusers. Likewise, it seems from initial studies that the use of NEP inhibitors, and especially NEP/ACE inhibitors, as new antihypertensive agents has a promising future.

[111] P. Delay-Goyet, J. M. Zajac, and B. P. Roques, *Neurosci. Lett.* **103**, 197 (1989).

# [18] Peptidyl Dipeptidase A: Angiotensin I-Converting Enzyme

By PIERRE CORVOL, TRACY A. WILLIAMS, and FLORENT SOUBRIER

## Introduction

Peptidyl dipeptidase A (EC 3.4.15.1) is a $Zn^{2+}$-metallopeptidase which acts on many substrates, primarily by releasing a C-terminal dipeptide. Its broad substrate specificity explains its successive denominations as angiotensin I-converting enzyme (ACE) since it cleaves the C-terminal dipeptide from angiotensin I (AI) to produce the potent vasopressor octapeptide angiotensin II,[1] and as kininase II because it inactivates bradykinin (BK)

[1] L. T. Skeggs, J. R. Kahn, and N. P. Shumway, *J. Exp. Med.* **103**, 295 (1956).

by the sequential removal of two C-terminal dipeptides.[2,3] In addition to acting on those two physiological substrates which are involved in blood pressure regulation and water and salt metabolism, ACE cleaves C-terminal dipeptides from various oligopeptides with a free C terminus. ACE is also able to act as an endopeptidase on certain substrates which are amidated at the C termini by cleaving a C-terminal dipeptide amide. Finally, ACE can cleave a C-terminal tripeptide amide from substance P[4] and luteinizing hormone-releasing hormone (LH-RH),[5] and, perhaps more surprisingly, the N-terminal tripeptide from this last substrate. Clearly ACE displays a wide substrate specificity, and none of the names given to the enzyme adequately describe its varied actions. The term ACE chosen in this chapter simply reflects its main recognized physiological function in the renin–angiotensin system.

The role of ACE in blood pressure control and water and salt metabolism has been defined mainly by the use of highly specific ACE inhibitors which have been developed since the early 1980s. These drugs clearly show efficacy in the treatment of hypertension and congestive heart failure. Although the physiological importance of ACE is undebated, it is only relatively recently that its structure and function have been studied in detail by the application of molecular biological techniques.

Several reviews have been published on the biochemistry of ACE,[6–9] on its main enzymatic characteristics, and on ACE inhibitors.[10,11] The purpose of the present review is to discuss new aspects of the structure and function of ACE as revealed from discoveries made in ACE molecular biology.

General Properties

Angiotensin I-converting enzyme is an ectoenzyme anchored to the plasma membrane with the bulk of its mass exposed at the extracellular surface of the cell. There are two ACE isoforms which are glycosylated: a

---

[2] H. Y. T. Yang and E. G. Erdös, *Nature* (*London*) **215**, 1402 (1967).
[3] H. Y. T. Yang, E. G. Erdös, and Y. Levin, *Biochim. Biophys. Acta* **214**, 374 (1970).
[4] R. A. Skidgel, S. Engelbrecht, A. R. Johnson, and E. G. Erdös, *Peptides* **5**, 769 (1984).
[5] R. A. Skidgel and E. G. Erdös, *Proc. Natl. Acad. Sci. U.S.A.* **82**, 1025 (1985).
[6] R. L. Soffer, *Annu. Rev. Biochem.* **45**, 73 (1976).
[7] M. R. W. Ehlers and J. F. Riordan, *Biochemistry* **28**, 5311 (1989).
[8] E. G. Erdös, *Hypertension* (*Dallas*) **16**, 363 (1990).
[9] N. M. Hooper, *Int. Biochem. J.* **23**, 641 (1991).
[10] J. W. Ryan, this series, Vol. 163, p. 194.
[11] G. Lawton, P. H. Paciorek, and J. F. Waterfall, *Adv. Drug Res.* **23**, 161 (1992).

somatic form of around 150–180 kDa is found in endothelial, epithelial, and neuronal cells, and a smaller isoform (90–110 kDa) is present in germinal cells. The structural relationship between the two isoenzymes has been elucidated by molecular cloning of the corresponding cDNAs and of the ACE gene, and will be discussed later.

The metalloenzyme nature of ACE was inferred when the enzyme was first identified owing to inhibition of activity by the metal-chelating agent EDTA.[1] Subsequent studies using atomic absorption spectroscopy determined that the associated metal was zinc.[12] Zinc is essential for the catalytic activity of ACE and functions exclusively in the catalytic step of peptide hydrolysis. The role of the zinc ion in ACE catalysis is probably analogous to that of the active-site zinc ion in thermolysin, where the zinc ion functions directly in the catalytic step of peptide hydrolysis by polarizing or ionizing the zinc-bound water molecule which thus becomes activated to initiate the nucleophilic attack on the substrate carbonyl scissile bond.[13] Therefore, ACE belongs to the thermolysin branch of the zinc metallopeptidases[14] (see [13] in this volume).

*Substrate Specificity*

The primary specificity of ACE is to cleave C-terminal dipeptides from oligopeptide substrates with a free C terminus in the absence of a penultimate proline residue. It is via this action that ACE hydrolyzes both AI and BK. BK is in fact the most favorable substrate for ACE, having a $k_{cat}/K_m$ ratio of 3900–5000 $sec^{-1}$ m$M^{-1}$ which is considerably higher than that for AI hydrolysis ($k_{cat}/K_m$ of 147–189 $sec^{-1}$ m$M^{-1}$) (reviewed in Ehlers and Riordan[7]). At neutral pH, ACE hydrolyzes a wide range of substrates in addition to AI and BK by virtue of its classic specificity described above.

Other substrates include neurotensin,[4] [Met$^5$]enkephalin, [Met$^5$]enkephalin-Arg$^6$-Gly$^7$-Leu$^8$, $\beta$-neoendorphin, and dynorphins.[15] ACE also hydrolyzes C-terminal amidated peptides such as cholecystokinin-8 and various gastrin analogs.[16] However, it should be noted that although all of the above substrate hydrolyses have been observed *in vitro*, no convincing evidence exists to support a physiological role for such hydrolyses, with the exception of angiotensin I and bradykinin.

---

[12] P. Bünning and J. F. Riordan, *Biochemistry* **22**, 110 (1983).
[13] B. L. Vallee and D. S. Auld, *Biochemistry* **29**, 5647 (1990).
[14] W. Jiang and J. S. Bond, *FEBS Lett.* **312**, 110 (1992).
[15] R. A. Skidgel and E. G. Erdös, *Clin. Exp. Hypertens. Part A* **9**, 243 (1987).
[16] P. Dubreuil, P. Fulcrand, M. Rodriguez, H. Fulcrand, J. Laur, and J. Martinez, *Biochem. J.* **262**, 125 (1989).

*Anion Activation*

An interesting feature of the catalytic activity of ACE is the activating effect of monovalent anions. The requirement of chloride for the hydrolysis of angiotensin I has long been recognized.[1] In fact, ACE appears to be unable to hydrolyze that substrate in the absence of chloride, and the maximal rate of hydrolysis is achieved at around 30 m$M$ chloride.[17] However, the presence of chloride is not essential for the hydrolysis of all substrates; for example, bradykinin is hydrolyzed in the absence of chloride, with a maximal rate of hydrolysis occurring at only 20 m$M$ chloride.[12,18]

The kinetics of chloride activation are rather complex, being both substrate and pH dependent, but in all conditions chloride appears to activate ACE by enhancing substrate binding.[19] An analysis of the chloride activation of ACE activity for the hydrolysis of synthetic ACE substrates led to the description of distinct classes of ACE substrates[19-21] Shapiro *et al.*[19] showed that the hydrolysis of some substrates (class I) was greatly stimulated by chloride, and the degree of chloride stimulation increased with increasing pH. They suggested the existence of an essential activator mechanism which was ordered and bireactant with anion binding being prerequisite for substrate binding. In contrast, the hydrolysis of class II and class III substrates appeared to follow a nonessential activator mechanism in which anion and substrate bound randomly to the enzyme, with the possibility of reaction products forming in the absence of a bound anion. However, class II and class III substrates could be distinguished by the degree of chloride activation, which was slightest for class II substrates. According to this classification, bradykinin is a class II substrate whose hydrolysis follows a nonessential activator mechanism, with its quite efficient hydrolysis in the absence of chloride being a reflection of the very low $K_m$ of ACE for bradykinin (0.85–1.00 $\mu M$).

*Assays for Angiotensin I-Converting Enzyme*

Several methods have been described for assaying ACE activity and have been reviewed in this series[10] and elsewhere.[22] They consist of (1) spectrophotometric assays using $N$-benzoylglycyl-L-histidyl-L-leucine (Hip-

[17] L. Wei, F. Alhenc-Gelas, P. Corvol, and E. Clauser, *J. Biol. Chem.* **266**, 9002 (1991).
[18] F. E. Dorer, J. R. Kahn, K. E. Lentz, M. Levine, and L. T. Skeggs, *Circ. Res.* **34**, 824 (1974).
[19] R. Shapiro, B. Holmquist, and J. F. Riordan, *Biochemistry* **22**, 3850 (1983).
[20] H. S. Cheung, F. L. Wang, M. A. Ondetti, E. F. Sabo, and D. W. Cushman, *J. Biol. Chem.* **255**, 401 (1980).
[21] R. Shapiro and J. F. Riordan, *Biochemistry* **22**, 5315 (1983).
[22] A. J. Turner, N. M. Hooper, and J. A. Kenny, *in* "Neuropeptides: A Methodology" (G. Fink and A. J. Harmar, eds.), p. 189. Wiley, New York, 1989.

His-Leu); (2) fluorimetric assays with Hip-His-Leu, angiotensin I, or Hip-Gly-Gly as substrates, in which the released dipeptide is usually detected by its fluorescence when conjugated to fluorescamine; (3) radiometric assays using [³H]Hip-Gly-Leu as substrate, in which the liberated [³H]Hip is selectively extracted and counted; and (4) direct radioimmunoassay.[23]

Several comments can be made. First, the nature of the substrate used must be taken into consideration for determining the optimal conditions of the assay. As already discussed, some synthetic substrates require a high chloride concentration for optimum hydrolysis, such as Hip-His-Leu (300 m$M$ NaCl), whereas a much lower chloride concentration is needed for natural peptides such as AI (30 m$M$) or BK (5 m$M$). Second, the sensitivity of the assay can be increased by high-performance liquid chromatography (HPLC), which separates the products of the reaction, allowing the measurement of the absorbance of hippuric acid. Third, the specificity of the assay can be confirmed by a specific ACE inhibitor. Finally, as discussed later, the two active sites of ACE exhibit somewhat different catalytic properties toward various substrates. Some substrates are more readily catalyzed by the C domain of ACE, such as Hip-His-Leu at a high chloride concentration, whereas some are more efficiently hydrolyzed by the N domain, such as the N-terminal endoproteolytic cleavage of LH-RH. However, no selective substrates for either domain have yet been reported.

*Purification*

The enzyme can be purified by affinity chromatography employing the ACE inhibitor lisinopril {$N$-[($S$)-1-carboxy-3-phenylpropyl]-Lys-Pro} as the ligand.[24,25] An improvement in the purification yield can be achieved using a 28-Å spacer arm between ligand and matrix[26] rather than the 14-Å spacer initially described. The same procedure can be used to purify ACE from various tissues, including testis, and the soluble form of ACE.

Structure of Angiotensin I-Converting Enzyme

It has long been known that two molecular forms of ACE exist in mammalian tissues: a somatic form found in endothelial cells, epithelial tissues, and in the brain, and a testicular form present exclusively in germinal

[23] F. Alhenc-Gelas, J. Richard, D. Courbon, J. M. Warnet, and P. Corvol, *J. Lab. Clin. Med.* **117**, 33 (1991).
[24] H. G. Bull, N. A. Thornberry, and E. H. Cordes, *J. Biol. Chem.* **260**, 2963 (1985).
[25] J. J. Lanzillo, J. Stevens, Y. Dasarathy, H. Yotsumoto, and B. L. Fanburg, *J. Biol. Chem.* **260**, 14938 (1985).
[26] N. M. Hooper, J. Keen, D. J. C. Pappin, and A. J. Turner, *Biochem. J.* **247**, 85 (1987).

cells (round spermatids and spermatozoa). Both forms exhibited similar enzymatic activities but differed by two criteria, namely, molecular size and immunological properties. The somatic and the testicular enzymes have apparent molecular masses of 150–180 kDa and 90–110 kDa, respectively. Both testicular and somatic ACE share common epitopes. However, the existence of different epitopes between the two isoforms was demonstrated by the differential preabsorbtion of an endothelial ACE antibody to the pulmonary and testicular proteins.[27,28] cDNA cloning of the somatic form of ACE expressed in endothelial cells revealed that the enzyme is composed of two homologous domains, suggesting a gene duplication event in the course of evolution. The hypothesis was supported by the elucidation of the structure of the germinal form of ACE expressed in the testes, which contains only one of the two homologous domains of somatic ACE and therefore corresponds to the putative ancestral form of the enzyme.

*Somatic Form*

Complete cDNAs for the somatic form of ACE were first isolated from human endothelial[29] and mouse kidney cDNA libraries.[30] The human endothelial ACE mRNA is 4.3 kb,[29] whereas the rabbit pulmonary ACE mRNA is 5.0 kb.[31] Two ACE mRNAs have been detected in mice of 4.9 and 4.15 kb in both the kidney and lung.[32] A high level of sequence similarity has been found between the different mammalian ACE cDNA sequences cloned to date. The human,[29] mouse,[30] rabbit,[31] bovine,[33] and rat[34] sequences are 80–90% identical in both nucleotide and amino acid sequences. Human endothelial ACE comprises 1306 amino acid residues, of which 14 are cysteine residues, and contains 17 potential N-linked glycosylation sites but no Ser/Thr-rich region indicative of O-glycosylation. Comparison of the amino terminus of the human ACE precursor, as deduced from the

[27] T. Berg, J. Sulner, C. Y. Lai, and R. L. Soffer, *J. Histochem. Cytochem.* **34,** 753 (1986).

[28] T. A. Williams, K. Barnes, A. J. Kenny, A. J. Turner, and N. M. Hooper, *Biochem. J.* **288,** 878 (1992).

[29] F. Soubrier, F. Alhenc-Gelas, C. Hubert, J. Allegrini, M. John, G. Tregear, and P. Corvol, *Proc. Natl. Acad. Sci. U.S.A.* **85,** 9386 (1988).

[30] K. E. Bernstein, B. M. Martin, A. S. Edwards, and E. A. Bernstein, *J. Biol. Chem.* **264,** 11945 (1989).

[31] T. J. Thekkumkara, I. W. Livingston, R. S. Kumar, and G. C. Sen, *Nucleic Acids Res.* **20,** 683 (1992).

[32] K. E. Bernstein, B. M. Martin, E. A. Bernstein, J. Linton, L. Striker, and G. Striker, *J. Biol. Chem.* **263,** 11021 (1988).

[33] S. Y. Shai, R. S. Fishel, B. M. Martin, B. C. Berk, and K. B. Bernstein, *Circ. Res.* **70,** 1274 (1992).

[34] G. Koike, J. E. Krieger, H. J. Jacob, M. Mukoyama, R. E. Pratt, and V. J. Dzau, *Biochem. Biophys. Res. Commun.* **198,** 380 (1994).

cDNA sequence, with that of the purified mature human enzyme indicated the presence of a hydrophobic signal peptide of 29 amino acids. The primary sequence of ACE reveals a second hydrophobic sequence of 17 amino acids located near the C terminus of the enzyme which is responsible for membrane anchorage, as discussed later.

Molecular cloning of the somatic form of ACE also demonstrated that the enzyme is composed of two homologous domains, referred to here as the N and C domains, indicative of the duplication of an ancestral gene. The overall sequence similarity between the two domains is 60% in both nucleotide and amino acid sequences but increases to 89% in a sequence of 40 amino acids in each domain containing essential residues of the active site. Therefore, each domain of ACE contains a putative catalytic site. There is no sequence similarity between the carboxyl- and amino-terminal extremities of the molecule, and the central region which separates the two homologous domains is also unique. In human ACE, the number and position of the cysteine residues are identical in the two domains, indicating conserved possibilities for establishing disulfide bridges. The positions of the disulfide bridges are still unknown as well as whether the seventh cysteine in each domain is free or involved in an interdomain disulfide bridge.

Each domain contains the zinc-binding motif His-Glu-Xaa-Xaa-His identified in many zinc metallopeptidases including thermolysin and neutral endopeptidase 24.11 (neprilysin).[13] In the consensus sequence, the two histidine residues provide two of the three zinc-coordinating ligands, and the glutamate residue is the base donor in the catalytic reaction. The third zinc ligand has been identified in both thermolysin[35] and neprilysin[36] as a glutamate. That residue is separated from the second histidine zinc-binding ligand on the C-terminal side by 19 and 58 amino acid residues in thermolysin and neprilysin, respectively. The putative glutamate third zinc-coordinating residues in ACE have been proposed as Glu-389 and Glu-987, in the N and C domains of ACE, respectively.[29] Studies using site-directed mutagenesis have confirmed this hypothesis[36a] (Fig. 1).

*Germinal Form*

The presence of two homologous domains in somatic ACE was highly suggestive of a gene duplication. Because germinal ACE is approximately one-half the molecular size of the somatic isoenzyme and the germinal

[35] W. R. Kester and B. W. Matthews, *Biochemistry* **16**, 2506 (1977).
[36] H. LeMoual, A. Devault, B. P. Roques, P. Crine, and G. Boileau, *J. Biol. Chem.* **266**, 15670 (1991).
[36a] T. A. Williams, P. Corvol, and F. Soubrier, *J. Biol. Chem.* **269**, 29430 (1994).

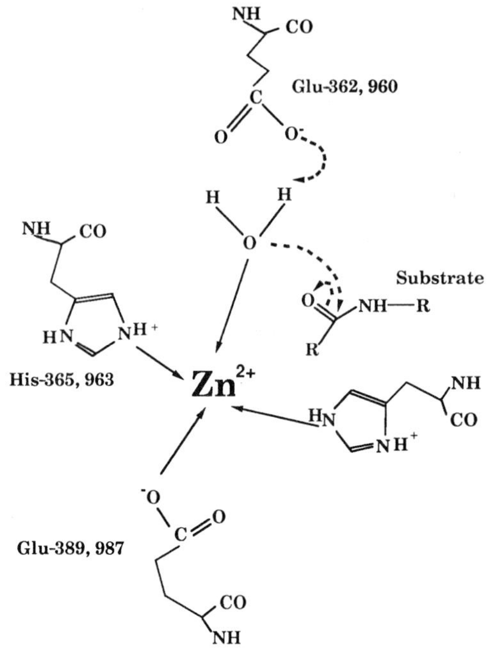

Fig. 1. Proposed mechanism for the action of ACE. Solid arrows represent zinc coordination bonds, and dashed arrows represent the first steps of the enzyme mechanism. Numbers indicate the amino acid position of the residue in the N and C domains, respectively (human ACE numbering). The mechanism is based on that proposed for thermolysin [P. M. Coleman, J. N. Jansonius, and B. W. Matthews, *J. Mol. Biol.* **70**, 701 (1972)].

ACE mRNA is 3 kb long, compared to 4.3 kb for the somatic enzyme, it was hypothesized that germinal ACE would correspond to a single domain, either the N or C domain. The primary structure of the germinal form of ACE found in testes has been deduced from the cDNA sequence from human,[37,38] rabbit,[39] and mouse.[40] Germinal ACE is comprised of 732 amino acids, compared to 1306 residues in somatic ACE, and corresponds to the C domain of somatic ACE; therefore, it contains a single active site. In fact, amino acid residues 68–732 of human germinal ACE are identical

[37] A.-L. Lattion, F. Soubrier, J. Allegrini, C. Hubert, P. Corvol, and F. Alhenc-Gelas, *FEBS Lett.* **252**, 99 (1989).
[38] M. R. W. Ehlers, E. A. Fox, D. J. Strydom, and J. F. Riordan, *Proc. Natl. Acad. Sci. U.S.A.* **86**, 7741 (1989).
[39] R. S. Kumar, J. Kusari, S. N. Roy, R. L. Soffer, and G. C. Sen, *J. Biol. Chem.* **264**, 16754 (1989).
[40] T. E. Howard, S.-Y. Shai, K. G. Langford, B. M. Martin, and K. E. Bernstein, *Mol. Cell. Biol.* **10**, 4294 (1990).

to residues 642–1306 of human somatic ACE. Therefore, germinal ACE contains the same hydrophobic sequence located near the carboxy terminus of somatic ACE proposed as the membrane-anchoring domain. The N-terminal 67-amino acid sequence of germinal ACE is specific to the isoenzyme and contains the signal peptide and a Ser/Thr-rich region for O-glycosylation. Germinal ACE is heavily O-glycosylated, with 90% of the O-linked carbohydrates added to a region covering 36 amino acids in the N-terminal germinal ACE-specific sequence. The function of the O-linked carbohydrates has not been determined; however, it has been demonstrated by site-directed mutagenesis that there is no relation between the O-linked glycosylation and enzyme activity or stability.[41] There are 7 cysteine residues in germinal ACE in all species from which the isoenzyme has been cloned, and the relative positions of cysteines have been conserved.

## Structure–Function Relationships of Angiotensin I-Converting Enzyme

The discovery by molecular cloning of the existence of two homologous ACE domains each containing the consensus sequence found in the active site of other $Zn^{2+}$ metallopeptidases was quite unexpected. Indeed, previous studies strongly suggested the presence of a single active site per ACE molecule: analysis of the zinc content of somatic ACE indicated the presence of a single zinc atom per molecule of ACE,[42] and studies with radiolabeled ACE inhibitors detected a single apparent high-affinity binding site.[43,44] As similar findings were observed for testicular ACE, which was found to correspond to the C domain of somatic ACE, this raised the question of whether the N domain was functional.[38]

### Functionality of N and C Domains

To establish whether both putative active sites of the somatic enzyme were functional, a series of ACE mutants, each containing only one intact domain, were constructed by deletion or point mutations of putative critical residues of the other domain.[17] Wild-type ACE and the different mutants were expressed in heterologous Chinese hamster ovary (CHO) stable cell lines. As negative controls, ACE constructs were designed in which either

[41] M. R. W. Ehlers, Y.-N. P. Chen, and J. F. Riordan, *Biochem. Biophys. Res. Commun.* **183**, 199 (1992).

[42] P. Bünning and J. F. Riordan, *J. Inorg. Biochem.* **24**, 183 (1985).

[43] S. M. Strittmatter and S. H. Snyder, Mol. Pharmacol. **29**, 142 (1986).

[44] F. Cumin, V. Vellaud, P. Corvol, and F. Alhenc-Gelas, *Biochem. Biophys. Res. Commun.* **163**, 718 (1989).

TABLE I
KINETIC PARAMETERS FOR HYDROLYSIS OF Hip-His-Leu AND ANGIOTENSIN I BY N AND C DOMAINS
OF ANGIOTENSIN I-CONVERTING ENZYME[a]

| | Hip-His-Leu | | | | Angiotensin I | | | |
|---|---|---|---|---|---|---|---|---|
| Enzyme | $K_m$ $(\mu M)$ | $k_{cat}$ $(sec^{-1})$ | $[Cl^-]_{opt}$ $(mM)$ | $K_a$ $(mM)$ | $K_m$ $(\mu M)$ | $k_{cat}$ $(sec^{-1})$ | $[Cl^-]_{opt}$ $(mM)$ | $K_a$ $(mM)$ |
| Wild type | 1540 | 408 | 800 | 180 | 16 | 40 | 30 | 2.9 |
| N domain | 2000 | 40 | 10 | 1.1 | 15 | 11 | 10 | 1.2 |
| C domain | 1590 | 364 | 800 | 217 | 18 | 34 | 30 | 6.4 |

[a] Compared to the wild-type enzyme. Modified from Wei et al.[17]

the zinc-binding histidines or the catalytic glutamic acid of the two domains was mutated.

The recombinant wild-type ACE was structurally, immunologically, and enzymatically identical to affinity-purified human kidney ACE.[45] Conversely, the two mutants, which possessed critical mutations in both the N and C domains were devoid of enzymatic activity. The mutants with only the N domain intact were able to hydrolyze the synthetic substrate Hip-His-Leu and angiotensin I. For those mutants, the Michaelis constant ($K_m$) for both substrates, Hip-His-Leu and angiotensin I, was similar to that of the wild-type enzyme, but the catalytic constant ($k_{cat}$) was 10 and 4 times lower for Hip-His-Leu and angiotensin I, respectively. The mutants containing only the C domain intact displayed similar $K_m$ values as for the N domain and the wild-type enzyme toward Hip-His-Leu and angiotensin I although, in this case, the $K_{cat}$ value for angiotensin I was 3 times higher than that for the N-domain active site and the $k_{cat}$ for Hip-His-Leu was 9 times higher (Table I). Both the N and the C domains have an absolute zinc requirement for activity and are sensitive to competitive ACE inhibitors. In all the experiments, both domains appeared to function independently, as the activity of the wild-type enzyme was equal to the sum of the activities of the N and C domains. There is no indication of cooperativity between the two domains. These studies established that ACE has two functional catalytic sites, both dependent on a zinc cofactor. In agreement with these findings, the zinc content of somatic ACE has been reinvestigated and shown to be two atoms of zinc per molecule.[28,46]

[45] L. Wei, F. Alhenc-Gelas, F. Soubrier, A. Michaud, P. Corvol, and E. Clauser, J. Biol. Chem. 266, 5540 (1991).
[46] M. R. W. Ehlers and J. F. Riordan, Biochemistry 30, 7118 (1991).

*Effect of Chloride on Catalytic Activity*

As discussed above, ACE is an unusual $Zn^{2+}$-metallopeptidase because the rate at which it catalyzes the hydrolysis of different substrates is specifically altered by monoanions such as chloride. It has been shown by several authors[18,20] that the rate of angiotensin I hydrolysis was more chloride-dependent compared to that of bradykinin. The chloride concentration has a similar effect on the hydrolysis of synthetic tripeptide substrates, which mimic the carboxydipeptide of a naturally occurring substrate, compared to their natural counterparts. For example, the hydrolysis of Hip-**His-Leu** (which mimics the carboxydipeptide of angiotensin I) was more dependent on chloride concentration than that of Hip-**Phe-Arg**[20] (which mimics the carboxydipeptide of bradykinin): the tripeptide Hip-Ala-Pro did not require chloride for optimal hydrolysis. These observations raise two questions concerning the sensitivity of the two active sites of somatic ACE to chloride activation and the amino acids involved in chloride binding.

From experiments conducted on wild-type somatic ACE and on mutants bearing either a functional N or C domain, it is clear that chloride has different effects on the catalytic activity of the two domains.[17,47] For Hip-His-Leu hydrolysis, the ACE functional N domain exhibits a low level of activity in the absence of chloride and an optimal activity at a concentration of 10 mmol/liter chloride. In contrast, the C domain is essentially inactive in the absence of chloride and requires much higher chloride concentrations for optimal activity ($\geq$800 mmol/liter).[17] Angiotensin I hydrolysis is also chloride dependent, the optimal chloride concentration being 10 and 30 m$M$ for the N and C domains, respectively (Table I). Bradykinin is hydrolyzed efficiently by both active sites, with similar $k_{cat}$ values in the presence of sodium chloride, but, again, the two active sites have different chloride activation profiles.[47]

Therefore, both domains act as peptidyl dipeptidases toward AI and BK but display different $k_{cat}$ values for the substrates and different chloride activation profiles. There is no indication of the amino acid residue which interacts with the chloride, although it has been proposed that a lysine residue near the active site may serve this functin. Chloride could induce a conformational change of the active site, resulting in an improved efficiency for angiotensin I hydrolysis or for class I substrates. The improvement of the rate of hydrolysis of the substrates, as well as the increase in the binding affinity of some ACE inhibitors, could be due to a more favorable interaction with the chloride-induced conformational change of the C domain rather than differential hydrolysis by the N- or C-domain active sites.

---

[47] E. Jaspard, L. Wei, and F. Alhenc-Gelas, *J. Biol. Chem.* **268**, 9496 (1993).

*Amino Acids Involved in Catalytic Activity*

A tyrosine and a lysine residue have been identified as essential active-site residues in the C domain of the wild-type rabbit somatic and testicular forms of ACE.[48,49] The residues were identified by chemical modification using 1-fluoro-2,4-dinitrobenzene (Dnp-F) which inactivated both forms of rabbit ACE. The loss of activity could be partially prevented by the presence of a competitive ACE inhibitor, suggesting that the inactivated amino acids were part of the catalytic site. A tyrosine residue was much more reactive (>90%) under these conditions than the lysine. Amino acid analysis and sequencing of tryptic digests of the modified enzymes, along with comparison of the peptide sequences with the known primary sequences deduced from the cDNA cloning of human somatic and testicular ACE, defined the amino acid positions of the lysine and tyrosine residues as Lys-118 and Tyr-200, respectively (human testicular ACE numbering).[49] The functional role of the tyrosine residue in human testicular ACE was subsequently investigated by mutating the residue to a phenylalanine.[50] The mutant enzyme displayed a decreased specific activity with $k_{cat}$ values 15- and 7-fold lower, compared to the wild-type enzyme, for the hydrolysis of the synthetic substrate furanacryloyl-Phe-Gly-Gly and AI, respectively. The change in $K_m$ for binding of the substrates was quite minor, and the $K_a$ for chloride activation was similar to that of the wild-type enzyme. In contrast, the mutant displayed a 100-fold decrease in affinity for the specific ACE inhibitor lisinopril, which is believed to act partly as a transition state analog. Thus, the results indicated that the tyrosine residue was not involved in substrate or chloride binding and was not crucial for catalysis; rather, it influences the rate of catalysis by stabilizing the transition state complex, in a manner analogous to Tyr-198 in carboxypeptidase A.[51]

However, the functional role of the tyrosine residue is unclear since a subsequent study, on rabbit testicular ACE, produced quite different results. In that study the lysine and tyrosine residues previously identified by chemical labeling with Dnp-F were mutated to produce two single mutants, and in addition a double mutant was constructed with the lysine and tyrosine residues mutated in the same mutant enzyme.[52] The authors reported that the Lys and Tyr single mutants exhibited the same enzymatic properties

[48] P. Bünning, S. G. Kleemann, and J. F. Riordan, *Biochemistry* **29**, 10488 (1990).
[49] Y.-N. P. Chen and J. F. Riordan, *Biochemistry* **29**, 10493 (1990).
[50] Y.-N. P. Chen, M. R. W. Ehlers, and J. F. Riordan, *Biochem. Biophys. Res. Commun.* **184**, 306 (1992).
[51] S. J. Gardell, D. Hilvert, J. Barnett, E. T. Kaiser, and W. J. Ruter, *J. Biol. Chem.* **262**, 576 (1987).
[52] I. Sen, S. Kasturi, M. A. Jabbar, and G. C. Sen, *J. Biol. Chem.* **268**, 25748 (1993).

as the wild-type enzyme. In contrast, the double mutant displayed distinctly different properties. In this instance, a 20-fold reduction in $k_{cat}$ and a 6-fold increase in $K_m$ was observed for the synthetic substrate Hip-His-Leu. Furthermore, the double mutant displayed a 100-fold increase in $K_i$ for lisinopril inhibition and decreased stimulation of enzyme activity by chloride. These observations led to the proposal that the tyrosine and lysine residues are not involved directly in the enzymatic functions of testicular ACE so that the single replacement of either residue can be tolerated without disturbing the usual functional properties of the enzyme. In contrast, the enzymatic properties of the double mutant imply an indirect role for both residues in catalysis, inhibitor binding, and chloride activation. However, because the double mutant displayed altered enzymatic properties for each measured parameter, the possibility exists that the observed activity of the double mutant was due to subtle changes in the three-dimensional structure of the protein.

*Interaction with Competitive Inhibitors*

Previous studies have been undertaken to examine the molecular mechanisms of ACE–inhibitor interactions.[44] The studies suggested the presence of a single class of affinity binding sites, in agreement with the previous observation of one zinc atom per ACE molecule. The finding that somatic ACE bears two functional sites, each of them being sensitive to ACE inhibitors, required a reexamination of the interaction of different ACE competitive inhibitors with the wild-type somatic ACE and mutants bearing only the N or the C domain.

Binding of the highly potent inhibitor [³H]trandolaprilat to wild-type ACE revealed the presence of two classes of high-affinity binding sites and a stoichiometry of two molecules of inhibitor bound per enzyme molecule. Using ACE mutants containing a single functional domain, it was found that both domains contain a single high-affinity binding site for trandolaprilat and that chloride enhances inhibitor potency essentially by stabilizing the enzyme–inhibitor complex and slowing the dissociation rate.[53] Interestingly, the effect of chloride on the action of inhibitors was more marked for the C domain compared to the N domain. Furthermore, the order of potency of the specific ACE inhibitors captopril, enalaprilat, and lisinopril was lisinopril {N-[(S)-1-carboxy-3-phenylpropyl]-L-lysyl-L-proline} > enalaprilat {N-[(S)-1-carboxy-3-phenylpropyl]-Ala-Pro} > captopril (D-2-methyl-3-mercaptopropanoyl-L-proline) for the C domain; however, that order was

---

[53] L. Wei, E. Clauser, F. Alhenc-Gelas, and P. Corvol, *J. Biol. Chem.* **267**, 13398 (1992).

TABLE II
INHIBITION OF N AND C DOMAINS OF ANGIOTENSIN I-CONVERTING ENZYME BY CAPTOPRIL, ENALAPRILAT, LISINOPRIL, AND TRANDOLAPRILAT[a]

| Enzyme | [Cl⁻]$_{opt}$ (m$M$) | $K_i$ ($\times 10^{10}$ $M$) | | | |
|---|---|---|---|---|---|
| | | Captopril | Enalaprilat | Lisinopril | Trandolaprilat |
| Wildtype | 300 | 13 | 6.5 | 3.9 | 0.45 |
| N domain | 300 | 8.9 | 26 | 44 | 3.1 |
| | 20 | 9.1 | 31 | 42 | 3.3 |
| C domain | 300 | 14 | 6.3 | 2.4 | 0.29 |
| | 20 | 111 | 78 | 27 | 2.2 |

[a] Using Hip-His-Leu as the substrate. From Wei et al.,[53] with permission.

reversed for the N domain (Table II). Such observations provide further evidence for the existence of structural differences between the two active sites of ACE.

The binding of various ACE inhibitors has been used to probe possible structural differences existing between the N- and C-domain active sites.[54,55] The binding of ACE inhibitors to purified rat lung and testis ACE was studied, and the binding parameters for the two inhibitor-binding sites in lung ACE were determined from the displacement of the ACE inhibitor [125]I-labeled RO 31-8472 {9-[1-carboxy-3-(4-hydroxyphenyl)propylamino]octahydro-10-oxo-6$H$-pyridazo[1,2-$a$][1,2]diazepine-1-carboxylic acid} with either unlabeled compound or another compound 351A, an analog of lisinopril. Similar $K_d$ vlaues for RO 31-8472 were determined for the two binding sites in lung ACE; however, with 351A, the $K_d$ for the N domain was more than 100-fold higher compared to that for the C domain. As the main structural difference between the two inhibitors is the length of the inhibitor side chain which interacts with the active-site zinc atom in each domain of ACE, it was proposed that the N domain of ACE contains a deeply recessed active site, and, therefore, the N-domain active site zinc atom is less accessible to the inhibitor 351A compared to that in the C domain.

The finding that each ACE domain interacts differently with competitive inhibitors is of substantial interest both for structure–function studies of the two active sites and for the possible design of an inhibitor more specific or even exclusively specific for one of the domains. Such a finding might

[54] R. B. Perich, B. Jackson, F. Rogerson, F. A. O. Mendelsohn, D. Paxton, and C. I. Johnston, Mol. Pharmacol. 42, 286 (1992).
[55] R. B. Perich, B. Jackson, and C. I. Johnston, Eur. J. Pharmacol. 266, 201 (1994).

be of therapeutic importance if the N or the C domain appears to act preferentially or even exclusively on a specific substrate(s).

### Substrate Specificity of N- and C-Domain Catalytic Sites

The finding that both the N and the C domains of ACE are enzymatically active toward Hip-His-Leu and AI, with higher catalytic constants for the C domain than for the N domain, and that the two domains differ in the affinities to bind several competitive ACE inhibitors, raises the question of the respective roles of the N and C domains in the hydrolysis of various substrates. Of particular interest is the catalytic rate of each domain on physiological substrates like BK, and on other substrates like substance P, where ACE acts as an endopeptidase by removing carboxyamidated di- and tripeptides, and LH-RH, where ACE cleaves an N-terminal tripeptide.

Using recombinant wild-type somatic ACE and mutants containing a single functional active site, it was found that both active sites converted BK to BK(1–7) and BK(1–5) with similar kinetics, although only the C-domain activity was stimulated by chloride.[47] Similarly, both domains hydrolyze substance P into a C-terminal di- and tripeptide amide. For the hydrolysis of that substrate, the C-domain active site was more readily activated by chloride and hydrolyzed substance P at a faster rate compared to the N domain.

Somatic ACE has been reported to perform the endopeptidase cleavage from the amino terminus of LH-RH ($Trp^3$-$Ser^4$) at a faster rate compared to testicular ACE, thus implying that the N-domain active site is primarily responsible for the cleavage.[46] In support of that observation, a subsequent study using the ACE mutants described above containing only a single intact domain demonstrated that the N-domain active site of ACE was able to cleave the $Trp^3$-$Ser^4$ bond of LH-RH 30 times faster compared to that of the C domain.[47] In addition, both the N and the C domains were able to cleave the carboxy-terminal tripeptide from the same substrate. The results suggest that the N domain may be preferentially involved in the N-terminal endoproteolytic cleavage of LH-RH, but it has not been demonstrated that ACE actually cleaves the peptide in the pituitary, particularly in view of its high $K_m$ and low catalytic rate.

Further evidence for a distinct specificity of the N domain comes from studies on the hydrolysis of acetyl-Ser-Asp-Lys-Pro, a peptide involved in medullary stem cell differentiation.[56] ACE inactivates the peptide by releasing the C-terminal dipeptide, and the effect seems to be mainly medi-

---

[56] K. J. Rieger, N. Saez-Servent, M. P. Papet, J. Wdzieczak-Bakala, J. L. Morgat, J. Thierry, W. Voelter, and M. Lenfant, *Biochem. J.* **296**, 373 (1993).

ated by the N domain, which hydrolyzes the peptide 40 times faster than does the C domain.[56a]

All the results indicate important functional and structural differences between the two subsites of the active catalytic sites despite the high degree of sequence homology. ACE cannot be considered as a true bifunctional enzyme, especially since an exclusive substrate has not been identified for either domain. Other enzymes contain two active sites and are truly bifunctional becuase they exhibit distinct functions, such as the brush border hydrolases sucrase–isomaltase[57] and lactase–phlorizin hydrolase.[58] For those enzymes, the duplication of an ancestral gene occurred in the course of evolution to form two active sites, but subsequent divergent evolution resulted in a different catalytic activity for each active site. Another example of a truly bifunctional zinc metalloenzyme is leukotriene-A4 hydrolase which is able to act both as a hydrolase in the biosynthesis of leukotrienes and as a peptidase.[59] Even if, at the present time, ACE cannot be considered a bifunctional enzyme, it appears that either the N or the C domain may exhibit a marked catalytic preference according to the substrate, its site of cleavage, and the chloride concentration. This may be of importance in physiological conditions where several potential substrates are present and may be preferentially cleaved by either domain. Similarly, the ACE activity of either domain might be influenced in the intracellular medium, where the chloride concentration is low. Another explanation could account for the specific properties of each active site, such as a lower selection pressure on the N domain which might have been responsible for mutational changes in key residues involved in the catalytic mechanism.

## Mechanism of Anchorage and Solubilization of Angiotensin I-Converting Enzyme

It has long been known from conventional studies on the purification of ACE that the enzyme was bound to the membrane of the cell and that a solubilization step, either by proteolytic cleavage with trypsin or by detergent, was required to purify the enzyme.[26] The molecular sizes of the soluble and the membrane-bound forms were similar, indicating that the

---

[56a] A. Rousseau, A. Michaud, M. T. Chauvet, M. Lenfant, and P. Corvol, *J. Biol. Chem.*, in press (1995).

[57] W. Hunziker, M. Spiess, G. Semenza, and H. F. Lodish, *Cell (Cambridge, Mass.)* **46**, 227 (1986).

[58] H. Wacker, P. Keller, R. Falchetto, G. Leofer, and G. Semenza, *J. Biol. Chem.* **267**, 18744 (1992).

[59] A. Wetterholm, J. F. Medina, O. Radmark, R. Shapiro, J. Z. Haeggstrom, B. L. Vallee, and B. Samuelson, *Proc. Natl. Acad. Sci. U.S.A.* **89**, 9141 (1992).

solubilization process did not greatly affect the size of the enzyme. Immuno-histochemical studies located the enzyme on the surface of various epithelial and endothelial cells. In addition to the membrane-bound form of the enzyme, ACE has also been found as a soluble form in plasma, semen, cerebrospinal fluid, and other body fluids. The relationship between the membrane-bound and the soluble forms of the enzyme has been elucidated to a large extent by molecular biology.

*Molecular Mechanism of Anchorage*

The mode of anchorage of ACE to the plasma membrane and the mechanism of its solubilization was first studied by Hooper *et al.*,[26] who demonstrated that ACE was inserted via the carboxyl terminus because the trypsin-solubilized (hydrophilic) and detergent-solubilized (amphipathic) enzymes had identical amino-terminal sequences.

The study[26] was further supported by determination of the primary structure of ACE, which revealed the presence of a highly hydrophobic sequence at the carboxyl-terminal extremity of the molecule (residues 1231–1247).[29] That this sequence constitutes the membrane anchor was subsequently demonstrated by site-directed mutagenesis studies on human endothelial ACE.[45] Transfection of CHO cells with the wild-type recombinant ACE cDNA resulted in 95% of newly synthesized enzyme bound to the cell membrane, whereas 5% was secreted into the cell supernatant. However, when a stop codon was introduced into the cDNA on the 5' site immediately before the sequence encoding the putative membrane anchor, ACE was recovered only in the cell supernatant. Among zinc ectopepti-dases, ACE is one of the enzymes to be inserted in plasma membranes via the carboxyl terminus (class I integral membrane protein), in contrast to neutral endopeptidase 24.11 (neprilysin) and aminopeptidase N which retain the membrane-inserted signal peptide (class II integral membrane proteins).

*Solubilization*

The plasma enzyme is probably derived from vascular endothelial cells and lacks the carboxyl terminus of ACE, as it is not recognized by a specific antibody directed against a peptide corresponding to that part of the enzyme. Therefore, the secreted circulating form of ACE is derived from the membrane-bound form by a posttranslational event, a mechanism which has been described for the solubilization of several ectoproteins. Posttranslational processing of the membrane-bound form of ACE into a soluble form has been observed during the culture of fibroblastic and epithelial cells which express recombinant somatic and germinal ACE.[45,60,61]

The relationship between the membrane-bound form of ACE and the circulating form has been elucidated in two different cell culture systems expressing either somatic or germinal ACE. The site of cleavage of human somatic ACE has been studied in CHO cells, which provide a convenient model system to study the solubilization process.[62] The carboxyl terminus of the secreted recombinant ACE, Ala-Gly-Gln-Arg, was established by carboxyl-terminal sequncing and corresponds to a cleavage site between Arg-1137 and Leu-1138. However the mutation of Arg-1137 to a glutamine residue did not prevent the secretion of recombinant ACE, indicating that the solubilizing enzyme can accommodate that change or can use an altenative cleavage site. In addition, the carboxyl-terminal sequencing of purified human plasma ACE showed the same cleavage site as above, and, therefore, it would appear that a similar mechanism for ACE secretion exists in CHO cells and in human vascular cells from which circulating ACE is derived. The site of cleavage of recombinant rabbit testicular ACE has been studied in a permanent epithelial cell line by Ramchandran et al.[61] Large amounts of ACE are secreted by the cells, and the cleavage site occurs at a monobasic site between Arg-663 and Ser-664. Interestingly, the secretion process could be enhanced by treatment of cells with phorbol esters. The enzyme involved in the secretion of ACE from pig kidney microvillar membranes has been partially characterized. The ACE-secreting enzyme is tightly associated with the membrane and is EDTA-sensitive, with the activity being substantially restored by the addition of $Mg^{2+}$ and to a lesser extent by $Zn^{2+}$ and $Mn^{2+}$.[63]

Several questions are still unanswered concerning the mechanism of ACE solubilization and its potential physiological role. The cleavage sequence found in human somatic ACE, Gln-Arg-Leu, is similar to the Gln-Lys-Leu cleavage site of the amyloid precursor protein (APP), which hydrolyzes APP within the ß-amyloid protein (AßP) and prevents deposition of AßP, the principal component of plaques in Alzheimer's disease. This might indicate the presence of a general posttranslational proteolytic mechanism in the case of ACE, as discussed by Ehlers et al.[60] for membrane-bound proteins. It is not known whether the vascular endothelial cell solubilizing enzyme is a rate-limiting factor for the production of plasma ACE; however,

[60] M. R. W. Ehlers, Y. N. P. Chen, and J. F. Riordan, *Proc. Natl. Acad. Sci. U.S.A.* **88,** 1009 (1991).
[61] R. Ramchandran, G. C. Sen, K. Misono, and I. Sen, *J. Biol. Chem.* **269,** 2125 (1994).
[62] V. Beldent, A. Michaud, L. Wei, M. T. Chauvet, and P. Corvol, *J. Biol. Chem.* **268,** 26428 (1993).
[63] S. Y. Oppong and N. M. Hooper, *Biochem. J.* **292,** 597 (1993).

in CHO cells, the secretion of soluble ACE is proportional to the level of cellular ACE, implying that the solubilizing enzyme is not rate-limiting, at least in that system. Finally, it cannot be excluded that, in some species other than human, soluble ACE results from the transcription of another gene or is produced from an alternative splicing eliminating the hydrophobic anchor domain.

### Structure of Gene for Angiotensin I-Converting Enzyme

The complete intron–exon structure of the human ACE gene was determined from restriction mapping of genomic clones and by sequencing the intron–exon boundaries.[64] The human ACE gene contains 26 exons: the somatic ACE mRNA is transcribed from exon 1 to 26, but exon 13 is removed by splicing from the primary RNA transcript; the germinal ACE mRNA is transcribed from exon 13 to 26. The structure of the human ACE gene provides further support for the hypothesis of duplication of an ancestral gene. Exons 4–11 and 17–24, which encode the two homologous domains of the enzyme, are highly similar in size and in sequence. In contrast, the intron sizes are not conserved.

The human somatic and germinal forms of ACE mRNA are transcribed from a single gene,[64] and thus the two species of ACE mRNA could be generated either by differential splicing of a common primary RNA transcript or by initiation of transcription from alternative start sites under the control of separate promoters. Several studies have demonstrated the presence of two functional promoters in the ACE gene, a somatic promoter and a testicular promoter,[64] by primer extension and RNase protection assays on mouse, rabbit, and human germinal mRNA.[40,65] The somatic ACE promoter is located on the 5′ side of the first exon of the gene. Fusion of fragments of the somatic ACE promoter of various sizes to a reporter gene and subsequent expression of the reporter gene demonstrated the presence of positive regulatory elements inside the 132-bp region upstream from the transcription start site and also indicated the presence of negative regulatory elements between positions $-132$ and $-343$ and between $-472$ and $-754$ from the transcription start site.[66]

The germinal ACE promoter is located on the 5′ side of the 5′ end of the germinal ACE mRNA with a transcription initiation site located inside the ACE gene. Thus, intron 12 which corresponds to the genomic sequence 5′ to the germinal specific exon 13, as deduced from the complete analysis

[64] C. Hubert, A.-M. Houot, P. Corvol, and F. Soubrier, *J. Biol. Chem.* **266,** 15377 (1991).
[65] R. S. Kumar, T. J. Thekumkara, and G. Sen, *J. Biol. Chem.* **266,** 3854 (1991).
[66] P. Testut, P. Corvol, and C. Hubert, *Biochem. J.* **293,** 843 (1993).

of the ACE gene in humans, was proposed as the putative germinal ACE promotor.[64] Unequivocal evidence for the promoter function of the sequence was obtained by using intron 12 to drive the transcription of a reporter gene in a germinal specific fashion.[67] A 689-bp mouse genomic fragment containing intron 12, exon 12, and part of intron 11 of the mouse ACE gene was fused to the ß-galactosidase coding sequence and microinjected into pronuclei of single-cell stage embryos. A histochemical analysis of the transgenic mice demonstrated that ß-galactosidase together with the ACE fragment was solely expressed in elongating spermatozoa. In another series of transgenic mice, a 91-bp fragment of intron 12 of the ACE gene was used as the promoter and was sufficient to confer a germinal cell-restricted pattern of transcription to the transgene.[68] Further mapping of the elements controlling transcription was achieved by DNase footprinting and gel mobility shift assays which demonstrated that a sequence between positions −42 and −62 specifically binds nuclear factors from testes extracts and contains a consensus cAMP responsive element (CRE).[68] The two alternative promoters of the ACE gene exhibit highly contrasting cell specificities as the somatic ACE promoter is active in several cell types in contrast to the germinal ACE promoter, which is only active in a stage-specific manner in male germinal cells.[69]

## Distribution in Tissues and Cells

The distribution of somatic and testicular ACE has been studied by using various techniques: enzymatic assays, immunohistochemistry, autoradiography using radiolabeled specific ACE inhibitors, and *in situ* hybridization. Elucidation of the human ACE gene structure allowed the production of antibodies directed against synthetic peptides specific for the two isoforms, as well as specific riboprobes for *in situ* hybridization studies.[70]

Angiotensin I-converting enzyme is found in the plasma membrane of vascular endothelial cells, with high levels being present at the vascular endothelial surface of the lung such that the active sites of ACE face the circulating substrates angiotensin I and bradykinin. In addition to the endothelial location of ACE, the enzyme is also expressed in the brush border of absorptive epithelia of the small intestine and the renal proximal

---

[67] K. G. Langford, S.-Y. Shai, T. E. Howard, M. J. Kovac, P. A. Overbeek, and K. E. Bernstein, *J. Biol. Chem.* **266,** 15559 (1991).

[68] T. Howard, R. Balogh, P. Overbeek, and K. E. Bernstein, *Mol. Cell. Biol.* **13,** 1 (1993).

[69] S. Nadaud, A. M. Houot, C. Hubert, P. Corvol, and F. Soubrier, *Biochem. Biophys. Res. Commun.* **189,** 134 (1992).

[70] M. Sibony, J.-M. Gasc, F. Soubrier, F. Alhenc-Gelas, and P. Corvol, *Hypertension* (*Dallas*) **21,** 827 (1993).

convoluted tubule. ACE is also found in plasma monocytes, macrophages, and T lymphocytes.[71] ACE has been found to enhance presentation of certain endogenously synthesized peptides to major histocompatibility complex class I-restricted cytotoxic T lymphocytes. In this case, ACE appears to function only in an intracellular secretory compartment of antigen-presenting cells.[72]

*In vitro* autoradiography, employing radiolabeled specific ACE inhibitors, and immunohistochemical studies have mapped the principal locations of ACE in brain.[73–75] The enzyme is found primarily in the choroid plexus, which may be the source of ACE in cerebrospinal fluid, ependyma, subfornical organ, basal ganglia (caudate putamen and globus pallidus), substantia nigra, and pituitary. The colocalization of ACE and the mammalian tachykinin peptide substance P to the striatonigral neuronal peptide led to suggestions of a physiological role for ACE in substance P hydrolysis in brain.[76,77] Neuronal ACE is smaller in size compared to the more widely distributed form of the enzyme, a difference which has been attributed to differences in the extent of N-glycosylation.[78] Neuronal ACE and the endothelial form of the enzyme display the same neuropeptide specificities and therefore do not represent distinct isoenzymes.[79]

The testes express both the somatic ACE in vascular endothelial cells and the germinal form.[27] The latter isoenzyme is specifically regulated as discussed above and is found in mature round spermatids but not in immature forms of germinal cells.[27,70] Its physiological role is still unknown.

Phylogenesis

The ancestral ACE gene has been duplicated during evolution, an event which seems to have occurred early in evolution. In all mammalian species where the gene has been cloned, that is, rabbits, mice, and humans, it

[71] O. Costerousse, J. Allegrini, M. Lopez, and F. Alhenc-Gelas, *Biochem. J.* **290**, 33 (1993).
[72] L. C. Eisenlohr, I. Bacik, J. R. Benninck, K. Bernstein, and J. W. Yewdell, *Cell (Cambridge, Mass.)* **71**, 963 (1992).
[73] R. Defendini, E. A. Zimmerman, J. A. Weare, F. Alhenc-Gelas, and E. G. Erdos, *Neuroendocrinology* **37**, 32 (1983).
[74] S. M. Strittmatter, M. M. S. Lo, J. A. Javitch, and S. H. Snyder, *Proc. Natl. Acad. Sci. U.S.A.* **81**, 1599 (1984).
[75] K. Barnes, R. Matsas, N. M. Hooper, A. J. Turner, and A. J. Kenny, *Neuroscience (Oxford)*, **27**, 799 (1988).
[76] S. M. Strittmatter, E. A. Thiele, M. S. Kapiloff, and S. H. Snyder, *J. Biol. Chem.* **260**, 9825 (1985).
[77] S. M. Strittmatter and S. H. Snyder, *Neuroscience (Oxford)* **21**, 407 (1987).
[78] N. M. Hooper and A. J. Turner, *Biochem. J.* **241**, 625 (1987).
[79] T. A. Williams, N. M. Hooper, and A. J. Turner, *J. Neurochem.* **57**, 193 (1991).

appears to be duplicated. A peptidyl dipeptidase, affinity-purified from the electric organ of *Torpedo marmorata*,[80] was recognized by a polyclonal antiserum raised against pig kidney ACE and was activated by chloride. The molecular mass of *Torpedo marmorata* ACE is 190 kDa, and therefore the enzyme also appears to be transcribed from a duplicated ACE gene.

An ACE-like enzyme has also been characterized in the housefly, Musca domestica.[81] The enzyme is able to hydrolyze the synthetic ACE substrate Hip-His-Leu and is inhibited by the specific ACE inhibitor captopril with an $IC_{50}$ of 0.4 $\mu M$. In addition, a cDNA encoding an enzyme with a high degree of sequence similarity to the C domain of ACE has also been cloned in *Drosophila melanogaster*.[82] The amino acid sequence homology reaches 65% in the region containing the zinc-binding motif. Interestingly, the ACE-like enzyme from *Drosophila melanogaster* does not appear to be duplicated as the full-length cDNA is 2.1 kb, an observation which is in agreement with the molecular mass of 86 kDa for the ACE-like enzyme from *Musca domestica*.

## Molecular Genetics of Angiotensin I-Converting Enzyme

The concentration of ACE in human blood serum is stable within a given subject but varies markedly from individual to individual. In healthy humans, no hormonal or environmental factors have been identified which affect serum ACE levels.[23] However, elevated serum ACE levels are observed in some pathological conditions, such as granulomatous diseases, hyperthyroidism, and diabetes.[83]

A family study of serum ACE levels, first performed in normal nuclear families, showed intrafamilial correlations of serum ACE levels. A segregation analysis of serum ACE levels revealed that the familial resemblance was due to a major gene effect which modulates the levels of serum ACE.[84] The cloning of the human ACE cDNA enabled the detection of a polymorphism in the ACE gene. The polymorphism consists of the presence or absence of a 287-bp DNA fragment.[85] The insertion polymorphism is located in intron 16 of the ACE gene and corresponds to an *Alu* repetitive sequence.[86] The polymorphism was subsequently analyzed in a group of

[80] A. J. Turner, J. Hryszko, N. M. Hooper, and M. J. Dowdall, *J. Neurochem.* **48,** 910 (1987).
[81] N. Lamango and R. E. Isaac, *Biochem. J.* **299,** 651 (1994).
[82] M. J. Cornell, D. Coates, and R. E. Isaac, *Biochem. Soc. Trans.* **21,** 243 (1993).
[83] J. Lieberman, *Am. J. Med.* **59,** 365 (1975).
[84] F. Cambien, F. Alhenc-Gelas, B. Herbeth, J. L. Andre, R. Rakotovao, M. F. Gonzales, J. Allegrini, and C. Bloch, *Am. J. Hum. Genet.* **43,** 774 (1988).
[85] B. Rigat, C. Hubert, F. Alhenc-Gelas, F. Cambien, P. Corvol, and F. Soubrier, *J. Clin. Invest.* **86,** 1343 (1990).
[86] B. Rigat, C. Hubert, P. Corvol, and F. Soubrier, *Nucleic Acids Res.* **20,** 1433 (1992).

80 healthy individuals along with the concentration of serum ACE. The frequency of the insertion allele (*I*) was 0.44, and the frequency of the deletion allele (*D*) was 0.56. In this study, the genetic background accounted for 47% of the total serum ACE variance. Using the *I*/*D* polymorphism as a DNA marker, a strong association was detected between the alleles of the ACE gene polymorphism and the level of serum ACE. Individuals homozygous for the *D* allele (*DD*) displayed serum ACE levels almost twice as high as those in individuals homozygous for the *I* allele (*II*).[8] Therefore, the ACE gene locus itself may be responsible for variations in the genetic control of serum ACE. The observations have been extended to the membrane-bound form of ACE in human circulating T lymphocytes. In these cells, ACE levels are also genetically determined, with comparatively high levels associated with the *DD* polymorphism.[71] This raises the possibility that a genetic determinant for the expression of ACE occurs in many tissues.

Because ACE is active in the metabolism of two major vasoactive peptides, several clinical and genetic studies have been performed to ascertain the role of ACE in cardiovascular diseases. In a case–control study, the frequency of the *DD* genotype was compared in matched subjects with and without a history of myocardial infarction. An increased frequency of the genotype was detected in individuals who had suffered a myocardial infarction,[87] suggesting that the ACE *DD* genotype is associated with an increased relative risk of myocardial infarction. The risk was more pronounced in patients with a history of myocardial infarction who where at low risk according to body mass index or plasma apolipoprotein B levels. Further epidemiological studies are underway to determine precisely the possible involvement of the ACE gene in coronary and heart diseases.

[87] F. Cambien, O. Poirier, L. Lecerf, A. Evans, J. P. Cambou, D. Arveiller, L. Gerald, J. M. Bard, L. Bara, S. Ricard, L. Tiret, P. Amouyel, F. Alhenc-Gelas, and F. Soubrier, *Nature* (*London*) **359**, 641 (1992).

# [19] Astacin

### By Walter Stöcker and Robert Zwilling

## Introduction

Astacin is a zinc endopeptidase from the digestive tract of the freshwater crayfish *Astacus astacus* L. (i.e., *Astacus fluviatilis* Fabr.) and other decapod

crustaceans.[1-3] The enzyme was recognized as distinct from classified peptidases because of its unusual cleavage specificity,[4,5] its resistance to a variety
of inhibitors,[6] and its amino acid sequence.[7,8] Astacin is the prototype for a
family of extracellular metalloproteinases which are widespread throughout
the animal kingdom from lower invertebrates to humans. In addition to
astacin, this protein family comprises membrane-bound zinc enzymes as
well as morphogenetically active molecules involved in hatching, em-
bryogenesis, and tissue remodeling (for overviews, see Refs. 9 and 10).
Most of these proteins have a multidomain structure including one astacin-
like catalytic domain that exhibits at least about 30% sequence identity
with the crayfish proteinase.[10] This chapter focuses on the purification and
the physical, chemical, and enzymatic properties of astacin in the light of
its three-dimensional structure.[11,12] In separate chapters, other branches of
the astacin-family are discussed (in this volume, per [20]–[23]).[13]

### Source and Site of Biosynthesis of Astacin

Astacin is synthesized in the F cells (fibrillar cells) of the crayfish hepatopancreas (midgut gland).[14] However, in starved animals, no accumulation
of intracellular proteases in F cells is detectable by immunohistochemistry
as would normally be seen in the pancreatic acinus cells of a vertebrate.
In *Astacus*, only after a stimulus (e.g., removal of gastric fluid or feeding)

[1] G. Pfleiderer, R. Zwilling, and H. H. Sonneborn, *Hoppe-Seyler's Z. Physiol. Chem.* **348,**
1319 (1967).
[2] R. Zwilling and H. Neurath, this series, Vol. 80, p. 633.
[3] W. Stöcker, R. L. Wolz, R. Zwilling, D. J. Strydom, and D. S. Auld, *Biochemistry* **27,**
5026 (1988).
[4] H. H. Sonneborn, R. Zwilling, and G. Pfleiderer, *Hoppe-Seyler's Z. Physiol. Chem.* **350,**
1097 (1969).
[5] E. Krauhs, H. Dörsam, M. Little, R. Zwilling, and H. Ponstingl, *Anal. Biochem.* **119,**
153 (1982).
[6] H.-J. Torff, H. Dörsam, and R. Zwilling, *Protides Biol. Fluids* **28,** 107 (1980).
[7] R. Zwilling, H. Dörsam, H.-J. Torff, and J. Rödl, *FEBS Lett.* **127,** 75 (1981).
[8] K. Titani, H.-J. Torff, S. Hormel, S. Kumar, K. A. Walsh, J. Rödl, H. Neurath, and
R. Zwilling, *Biochemistry* **26,** 222 (1987).
[9] E. Dumermuth, E. E. Sterchi, W. Jiang, R. L. Wolz, J. S. Bond, A. V. Flannery, and R. J.
Beynon, *J. Biol. Chem.* **266,** 21381 (1991).
[10] W. Stöcker, F.-X. Gomis-Rüth, W. Bode, and R. Zwilling, *Eur. J. Biochem.* **214,** 215 (1993).
[11] W. Bode, F.-X. Gomis-Rüth, R. Huber, R. Zwilling, and W. Stöcker, *Nature (London)* **358,**
164 (1992).
[12] F.-X. Gomis-Rüth, W. Stöcker, R. Huber, R. Zwilling, and W. Bode, *J. Mol. Biol.* **229,**
945 (1993).
[13] R. L. Wolz and J. S. Bond, this volume [20].
[14] G. Vogt, W. Stöcker, V. Storch, and R. Zwilling, *Histochemistry* **91,** 373 (1989).

can the transient appearance of massive intracellular immunofluorescence be observed.[14] Subsequently, astacin is released from F cells and stored extracellularly as active proteinase in the stomach (cardia) at a concentration of about 1 mg/ml.[14] Because of this mode of biosynthesis, an inactive zymogen (proenzyme) of astacin has not been observed. However, the nucleotide sequence of a complementary DNA contains a 5'-elongation corresponding to a possible pro part.[10,15] Hence, there are indications for a zymogen, which, however, may only show up transiently along the biosynthesis pathway.

## Assays

Generally, precautions are taken to avoid metal contamination.[16] Kinetic parameters are determined from Lineweaver–Burk plots or from nonlinear regression analysis of the Michaelis–Menten equation[17] in the programs ENZPACK3 or ENZFITTER (Elsevier/Biosoft, Cambridge, UK). The $K_i$ values are determined as briefly outlined in Table III.

### STANA Assay

Astacin activity can be assayed with heat-denatured casein or with the synthetic peptide succinylalanylalanylalanyl-4-nitroanilide (STANA[18]) as substrates.[2,19] Stock solutions of STANA (Bachem, Heidelberg, Germany) are prepared in 50 m$M$ HEPES–NaOH, pH 8.0. The peptide is readily soluble in that buffer up to 30 m$M$. Routine assays for astacin activity are conducted at 25° in plastic cuvettes containing 1 m$M$ STANA and 0.1–20 $\mu M$ enzyme in a total volume of 750 $\mu$l of the same buffer. The increase in absorbance at 405 nm, owing to the release of a nitroaniline product, is monitored for 10 min in a computer-interfaced spectrophotometer.[20] The concentration of product is based on the extinction coefficient $\varepsilon_{405\,nm}$ of 10,200 $M^{-1}$ cm$^{-1}$.[19] Substrate concentrations are determined by weight or

---

[15] E. Jacob, S. Reyda, W. Stöcker, R. Zwilling, unpublished (1995).
[16] B. Holmquist, this series, Vol. 158, p. 6.
[17] G. N. Wilkinson, *Biochem. J.* **80**, 324 (1961).
[18] Abbreviations: CBZ, carbobenzoxy; DNS, dansyl, 5-(dimethylamino)naphthalene-1-sulfonyl; FMOC, 9-fluorenylmethyloxycarbonyl; HEPES, 4-(2-hydroxyethyl)-1-piperazineethanesulfonic acid; MES, 2-(*N*-morpholino)ethanesulfonic acid; MP, 1,7-phenanthroline; OP, 1,10-phenanthroline; pNA, 4-nitroanilide; STANA, succinylalanylalanylalanyl-4-nitroanilide; Suc, succinyl; TFA, trifluoroacetic acid; TIMP, tissue inhibitor of metalloproteinases; Tris, trishydroxymethylaminomethane.
[19] J. Bieth, B. Spiess, and C. G. Wermuth, *Biochem. Med.* **11**, 350 (1974).
[20] W. Stöcker, B. Sauer, and R. Zwilling, *Biol. Chem. Hoppe-Seyler* **372**, 385 (1991).

spectrophotometrically at 405 nm after complete hydrolysis on addition of 100 µl concentrated NaOH.

### Assays with Fluorescent Oligopeptide Substrates

More sensitive assays use fluorescent, N-terminally dansylated substrates.[21,22] The turnover of these substrates is monitored either spectrofluorimetrically or by high-performance liquid chromatography (HPLC) or simply by thin-layer chromatography.

*Quenched Fluorescent Substrates.* Dansylated peptide substrates like DNS-PKR*APWV (dansyl-Pro-Lys-Arg*Ala-Pro-Trp-Val) are prepared by solid-phase peptide synthesis as described.[21] Cleavage of the central Arg*Ala bond causes an at least 10-fold increase of tryptophan fluorescence at 340 nm on excitation at 280 nm as compared to the intact substrate, where quenching by the dansyl group decreases the detectable tryptophan emission.[21,22] Substrate concentrations are based on the molar extinction coefficient of the dansyl group ($\varepsilon_{340\,nm}$ 4300 $M^{-1}$ cm$^{-1}$). Assays are run at 25° in 300-µl quartz cuvettes in a Perkin-Elmer (Norwalk, CT) LS 50 fluorescence photometer. One hundred microliters substrate at a final concentration of 0.1 m$M$ is mixed with 200 µl astacin solution (0.1–4 n$M$) in 50 m$M$ HEPES–NaOH buffer, pH 8.0. On excitation at 280 nm (slit width 2.5 mm) tryptophan fluorescence is detected at 350 nm (slit width 5.0 mm) through a 350-nm bandpass filter for 10 min. After a 5-min run time, 5 µl of astacin is added to a final concentration of 0.1 µ$M$, and recording is continued until the substrate is completely turned over within the next few minutes. The initial rate is taken from the first 5 min corresponding to a turnover of less than 10% of the substrate and calculated on the basis of the fluorescence yield after complete conversion of substrate to product.

Alternatively, stopped-flow assays are performed with 0.1 µ$M$ enzyme in 20 m$M$ HEPES–NaOH buffer, pH 7.8, at 25°, using a Durrum-Gibson stopped-flow instrument (Dionex Corporation, Sunnyvale, CA) equipped with a fluorescence detector and interfaced to a Digital Equipment Corporation (Maynard, MA) computer (Fig. 1a).[21–23]

*Spectrofluorimetric Assay with N-Dansylated Substrates.* N-Dansylated peptide substrates lacking a tryptophan residue can be analyzed under radiationless energy transfer conditions (RET) in stopped-flow assays, in 20 m$M$ HEPES–NaOH buffer, pH 8.0, at 25°. Substrate concentrations are between 2.0 × 10$^{-5}$ and 5 × 10$^{-4}$ $M$ and enzyme in the range 1.0 × 10$^{-7}$ to 5.0 × 10$^{-7}$ $M$. Under these conditions, formation and breakdown of

[21] M. Ng and D. S. Auld, *Anal. Biochem.* **183,** 50 (1989).
[22] W. Stöcker, M. Ng, and D. S. Auld, *Biochemistry* **29,** 10418 (1990).
[23] R. R. Lobb and D. S. Auld, *Biochemistry* **19,** 5297 (1980).

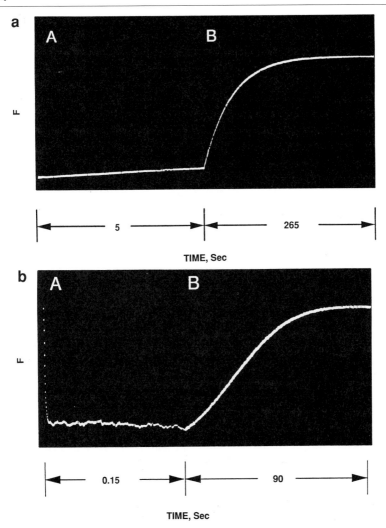

FIG. 1. Stopped-flow kinetics of astacin with dansylated oligopeptide substrates.[22] (a) Fluorescence change during hydrolysis of the quenched fluorescent substrate DNS-VKRAPWV ($3.25 \times 10^{-4}$ M) by astacin ($1.0 \times 10^{-7}$ M) at pH 7.8 and 25° (excitation of tryptophan at 285 nm and detection of Trp emission at 366 nm). The initial rate is calculated from the linear trace corresponding to the first 5 sec of the reaction (A), and complete turnover is achieved within the next 265 sec (B). (b) Observation of the formation and breakdown of the enzyme–substrate complex of DNS-GKRAPLV ($1.1 \times 10^{-4}$ M) and astacin ($1.0 \times 10^{-7}$ M) by excitation at 285 nm and emission of enzyme tryptophans at 366 nm. Rapid formation of the enzyme–substrate complex is shown (A), as well as steady-state conversion of the complex to enzyme plus products (B).

the enzyme–substrate complex can be observed on excitation of enzyme tryptophan residues at 285 nm either by monitoring the tryptophan emission at 366 nm or by detecting the dansyl emission at 550 nm arising from RET between enzyme tryptophans and the substrate dansyl group (Fig. 1b).[21–23]

*Chomatographic Assay with Dansylated Peptide Substrates.* For HPLC analysis, assays are conducted in 20 mM HEPES–NaOH buffer, pH 7.8, at room temperature.[22] A 95-μl aliquot of substrate ($5 \times 10^{-6}$ to $1 \times 10^{-3}$ $M$) is transferred to a 1.5-ml polypropylene tube, and the reaction is initiated by the addition of 5 μl of enzyme to give a final concentration of $5 \times 10^{-9}$ to $1.5 \times 10^{-7}$ $M$. For each substrate concentration, five samples are prepared, and the reaction is stopped after 1, 10, 30, 60, or 90 sec by addition of 5 μl glacial acetic acid. The samples are analyzed by HPLC on a reversed-phase column (e.g., Waters NovaPak $C_{18}$, $3.9 \times 150$ mm, Milford, MA). Samples (10–50 μl) are eluted with a linear acetonitrile gradient (20–60% (v/v)) in 0.06% (v/v) trifluoroacetic acid (TFA) at a flow rate of 1.5 ml/min. Peak areas are determined at 225 or 260 nm. The concentration of products at a given time, $P_t$, is determined according to the equation $P_t = [(P_a + P_b)/(P_a + P_b + S_a)]S_0$, where $P_a$, $P_b$, and $S_a$ are the integrated areas of product A and B and substrate peaks, respectively, and $S_0$ is the initial substrate concentration.[22]

## Preparation and Purification

A procedure for the purification of astacin has been described, yielding about 50–60 mg of lyophilized protein from 100 ml of crude digestive juice from the crayfish.[2] Astacin prepared in this way consists of immunologically identical multiple forms which are resolved further by anion-exchange chromatrography on a HR 5/5 Mono Q column (Pharmacia, Piscataway, NJ) (Fig. 2).[20] With this procedure, three peaks are obtained, corresponding to multiple forms termed *a*, *b*, and *c* and comprising approximately 75, 22, and 3% of the total protein, respectively. The kinetic properties of the *a* and *b* forms are almost identical (Table I).

## General Properties

*Stability and Solubility.* Astacin is remarkably stable against proteolytic digestion and autolysis. It is soluble at concentrations at least up to 20 mg/ml in dilute buffers. Below pH 4, however, the enzyme precipitates.[2]

*Absorbance Coefficient.* Astacin concentrations are based on the absorbance coefficient $\varepsilon_{280 \text{ nm}}$ of 42,800 $M^{-1}$ cm$^{-1}$.[3]

*pH Optimum.* The optimal pH region for the catalytic function of astacin is around neutrality, depending on the substrate used. With alanylalanyl-

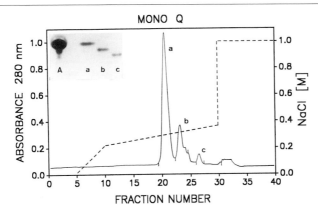

MONO Q

FRACTION NUMBER

FIG. 2. Separation of multiple forms of astacin by Mono Q anion-exchange chromatography.[20] The column was loaded with 4 mg astacin after DEAE- and Sephadex G-50 column chromatography, dissolved in 20 m$M$ Tris-HCl buffer, pH 8.0. Elution was achieved with a linear gradient of 0.2–0.35 $M$ NaCl within 20 min at a flow rate of 1 ml/min. *Inset*: Native electrophoresis [U. K. Laemmli, *Nature* (*London*) **227**, 680 (1970)] at pH 8.8 in the absence of reducing agents and sodium dodecyl sulfate (SDS). The acrylamide concentration was 12.5%, and the anode is at the bottom. A, Astacin starting material; *a, b, c*, peak fractions.

alanyl-4-nitroanilide the optimum is about pH 6.0, whereas with STANA it is about pH 8.0.[20]

*Isoelectric Point.* For the native (undenatured) protein an isoelectric point of pH 3.5 has been determined using Servalyte Precotes, pH 3–7 (Serva, Heidelberg, Germany) according to the manufacturer's recommendations. On the other hand, the isoelectric point of urea-denatured astacin, determined in the two-dimensional system of O'Farrell,[24] is around pH 4.8 (ampholines, pH 3–10, pH 3–5, and pH 5–7, from Serva), which is close to the calculated value of pH 4.54 obtained by the program HUSAR (Version 3.0, German Cancer Research Center at Heidelberg, Germany).

*Metal Content.* As determined by atomic absorption spectrometry, astacin contains 1 gram-atom zinc per mole. Other metals could not be detected.[3]

*Molecular Weight.* The relative molecular mass determined by sodium dodecyl sulfate (SDS)–polyacrylamide gel electrophoresis is 23,000 Da,[14] and a size of 22,614 Da was deduced from the amino acid sequence.[8] Earlier determinations leading to the designation "low molecular weight protease" arose from an extreme retention effect on gel-filtration columns.[2,25]

[24] P. H. O'Farrell, *J. Biol. Chem.* **250**, 4007 (1975).
[25] A. Feige, Diploma Thesis, Heidelberg University (1986).

TABLE I
CLEAVAGE OF NITROANILIDE SUBSTRATES BY ASTACIN

| Substrate | | | | | | | | $k_{cat}$ (sec$^{-1}$) | $K_m$ ($10^{-4}$ M) | $k_{cat}/K_m$ ($M^{-1}$ sec$^{-1}$) |
|---|---|---|---|---|---|---|---|---|---|---|
| P6 | P5 | P4 | P3 | P2 | P1 | * | P1' | | | |
| | | | A | A | A | * | pNA | 0.028 ± 0.0006 | 0.52 ± 0.04 | 53.8 |
| | | | | A | A | * | pNA | 0.172 ± 0.03 | 23.2 ± 0.7 | 7.4 |
| | | | | | A | * | pNA | | | 0.01[a] |
| | | | ↓ b | | ↓ c | | | | | |
| Suc | A | A | A | A | A | * | pNA | 118.0 ± 0.3 | 2.1 ± 0.18 | 56,190.0[b] |
| | | | | | | | | 0.026 ± 0.0006 | 0.56 ± 0.04 | 46.4[c] |
| | | Suc | A | A | **A** | * | pNA | 0.25 ± 0.01 | 9.7 ± 0.8 | 25.8[d] |
| | | | | | | | | 0.28 ± 0.02 | 12.5 ± 0.5 | 22.4[e] |
| | | | Suc | A | A | * | pNA | 0.012 ± 0.0003 | 7.1 ± 0.5 | 1.69 |
| | | Suc | A | A | V | * | pNA | 0.0003 ± 0.00008 | 11.4 ± 0.3 | 0.03[f] |
| | | Suc | A | A | F | * | pNA | 0.19 ± 0.009 | 4.2 ± 0.42 | 45.2[g] |
| | | Suc | P | A | A | * | pNA | 0.05 ± 0.0009 | 1.3 ± 0.1 | 38.5[g] |
| | | Suc | A | P | A | * | pNA | 0.008 ± 0.0003 | 4.1 ± 0.4 | 1.95[g] |
| | | Suc | A | A | P | * | pNA | 0.0008 ± 0.00003 | 9.7 ± 0.8 | 0.08[f] |

[a] Estimate, calculated from the rate of product appearance divided by the concentrations of enzyme (7 μM) and substrate (3.14 mM).
[b] Peptide bond cleavage monitored by HPLC using 2.5 × 10$^{-8}$ M astacin in 50 mM HEPES, pH 8.25, and Suc-Ala-Ala-Ala-Ala-Ala-pNA ranging from 0.16 to 4.95 mM.
[c] Amide bond cleavage by 1.1 μM astacin monitored spectrophotometrically at 405 nm.
[d] Astacin isozyme a.
[e] Astacin isozyme b.
[f] Enzyme concentration: 19 μM. Data from Ref. 20.
[g] Enzyme concentration: 1.1 μM. Data from Ref. 20.

*Primary Structure.* Astacin is a single-chain protein of 200 residues corresponding to a molecular mass of 22,614 Da, as deduced from amino acid sequence analysis.[8] The covalent structure contains two disulfide bridges from Cys-42 to Cys-198 and from Cys-64 to Cys-84.[8] There is no indication of attached carbohydrate, either from the sequence analysis or from X-ray structure analysis (see below). Translation of the nucleotide sequence of an astacin-encoding complementary DNA, derived from crayfish hepatopancreatic mRNA, revealed an N-terminal extension of 49 amino acid residues. The extension exhibits features of a prepro sequence typical for eukaryotic secreted proteins, including a signal peptide-like hydrophobic stretch and also potential processing sites for the maturation of the active enzyme.[10,15]

Cleavage Specificity

The primary specificity of astacin is dominated by short, uncharged amino acid side chains in position P1′ [26] of the substrate, and, in addition, by a proline residue in P2′.[4] An inspection of the 71 cleavage sites of astacin in the denatured chains of $\alpha$- and $\beta$-tubulin revealed a more extended substrate binding site beyond P1′ and P2′, preferring Ala in P1′, Pro in P2′ and P3, hydrophobic residues in P3′ and P4′, and Lys, Arg, Asn, and Tyr in P1 and P2.[5,22]

*Cleavage of Peptide Nitroanilides.* Because of the extended binding site, the specificity constants, $k_{cat}/K_m$, for the release of nitroaniline from substrates of the general structure Suc-Ala$_n$-pNA ($n$ = 2, 3, 5) and Ala$_n$-pNA ($n$ = 1, 2, 3) increase with the number of alanine residues.[20] Substitutions in peptides of the structure Suc-Ala-Ala-X-pNA (X = Ala, Phe, Pro, Val) cause a drop in activity following the order Phe > Ala > > Pro > Val (Table I). This is the same order seen for the frequency of these residues to occur in the P1 position of the tubulin cleavage pattern. The kinetic parameters obtained for peptides of the composition Suc-(Ala,Ala,Pro)-pNA also are in accord with this pattern, as a proline residue is well accepted in P3 but less tolerated in P2 and excluded from P1 (Table I).

*Cleavage of Dansylated Oligopeptide Substrates.* A series of fluorescent, N-dansylated oligopeptides developed based on the tubulin cleavage pattern are turned over up to $10^4$ times faster than the nitroanilides (Table II).[22] Lys and Arg in positions P1 and P2 yield high-turnover substrates, and in P3 the enzyme prefers Pro > Val > Leu > Ala > Gly. The substitution of Lys by Gly in P2 causes a 330-fold increase in $K_m$, demonstrating the importance of that interaction for substrate binding. The presence of a small side chain in P1′ is absolutely essential, as the activity drops by 30,000-fold if alanine is replaced by leucine (Table II). Also crucial is a proline in P2′, whose substitution by Ser decreases $k_{cat}$ 50-fold, whereas $K_m$ stays about the same (Table II). As indicated by the tubulin pattern, astacin accepts hydrophobic side chains including tryptophan in position P3′ (Table II), which has been convenient for the design of a sensitive quenched fluorescence assay (see above).[22] Investigation of substrates varying in length revealed that an optimal substrate for astacin comprises seven or more amino acid residues and minimally requires about five (Table II), which is in accord with the three-dimensional structure (see below).

---

[26] P1, P2, P3, etc., and P1′, P2′, P3′ designate substrate/inhibitor residues amino-terminal and carboxy-terminal to the scissile bond, respectively, and S1, S2, S3, etc., and S1′, S2′, S3′, the corresponding subsites of the proteinase [I. Schechter and A. Berger, *Biochem. Biophys. Res. Commun.* **27,** 157 (1967)].

TABLE II
KINETIC CONSTANTS FOR HYDROLYSIS OF FLUORESCENT PEPTIDE SUBSTRATES BY ASTACIN

| Substrate | | | | | | | | | | | $k_{cat}$ (sec$^{-1}$) | $K_m$ ($10^{-4}$ M) | $k_{cat}/K_m$ (M$^{-1}$ sec$^{-1}$) |
|---|---|---|---|---|---|---|---|---|---|---|---|---|---|
| P6 | P5 | P4 | P3 | P2 | P1 | * | P1' | P2' | P3' | P4' | | | |
| | | Dns | P | K | R | * | A | P | W | V$^a$ | 380 | 3.7 | 1.0 × 10$^6$ |
| | | Dns | V | K | R | * | A | P | W | V$^a$ | 190 | 2.5 | 7.6 × 10$^5$ |
| | | Dns | L | K | K | * | A | P | W | V$^a$ | 210 | 2.9 | 7.2 × 10$^5$ |
| | | Dns | L | K | R | * | A | P | W | V$^a$ | 210 | 3.3 | 6.4 × 10$^5$ |
| | | Dns | L | K | R | * | A | P | L | V$^b$ | 120 | 2.5 | 4.8 × 10$^5$ |
| | | Dns | A | A | R | * | A | P | L | V$^c$ | 200 | 4.8 | 4.2 × 10$^5$ |
| | | Dns | L | K | Y | * | A | P | W | V$^a$ | 67 | 2.8 | 2.4 × 10$^5$ |
| | | Dns | L | R | R | * | A | P | L | G$^b$ | 130 | 5.8 | 2.2 × 10$^5$ |
| | | Dns | L | K | N | * | A | P | L | V$^b$ | 180 | 10.0 | 1.8 × 10$^5$ |
| | | Dns | G | K | Y | * | A | P | W | V$^a$ | 40 | 2.7 | 1.5 × 10$^5$ |
| | | Dns | G | K | R | * | A | P | W | V$^a$ | 2.2 | 0.15 | 1.5 × 10$^5$ |
| | | Dns | G | K | R | * | A | P | L | V$^b$ | 2.3 | 0.17 | 1.3 × 10$^5$ |
| | | Dns | G | K | N | * | A | P | L | V$^c$ | 56 | 5.2 | 1.1 × 10$^5$ |
| | | Dns | G | K | N | * | A | P | L | V$^b$ | 39 | 4.0 | 9.6 × 10$^4$ |
| | | Dns | G | R | R | * | A | P | L | G$^c$ | 34 | 3.9 | 8.7 × 10$^4$ |
| | | Dns | G | P | R | * | A | P | L | V$^b$ | 20 | 5.3 | 3.8 × 10$^4$ |
| Dns | H | H | L | K | R | * | A | P | W | V$^a$ | 310 | 2.3 | 1.4 × 10$^6$ |
| | | Dns | L | K | R | * | A | P | L | V$^b$ | 120 | 2.5 | 4.8 × 10$^5$ |
| | | Dns | L | K | R | * | A | P | L$^b$ | | 43 | 8.7 | 4.9 × 10$^4$ |
| | | | Dns | K | R | * | A | P | L$^b$ | | 40 | 20.0 | 2.0 × 10$^4$ |
| | | | | Dns | R | * | A | P | L$^b$ | | 0.035 | 560.0 | 0.6 |
| | | Dns | L | R | R | * | A | S | L | G$^b$ | 2.7 | 6.7 | 4.0 × 10$^3$ |
| | | Dns | G | R | R | * | A | S | L | G$^b$ | 0.7 | 2.3 | 3.0 × 10$^3$ |
| | | Dns | G | G | R | * | A | P | W | V$^b$ | 20.0 | 50.0 | 3.9 × 10$^3$ |
| | | Dns | L | K | R | * | L | P | W | V$^b$ | | | 0.9$^c$ |

$^a$ Analyzed by tryptophan fluorescence changes between substrate and product.
$^b$ HPLC analysis.
$^c$ Analyzed by tryptophan fluorescence changes between enzyme–substrate complex and enzyme plus products. Data from Ref. 22.

*Cleavage of Type I Collagen.* Astacin does not readily cleave undenatured, globular proteins.[27] The tubulin chains, for example, are cleaved only in the denatured state.[5] However, astacin exhibits a specificity for short, uncharged side chains, which are positioned between proline residues. This is a sequence motif typical for the triple helix of type I collagen, a constituent of skin and tendon.[28] Astacin, in fact, cleaves

[27] W. Stöcker, *Habilitationsschrift,* Heidelberg University (1992).
[28] P. P. Fietzek and K. Kühn, *Int. Rev. Connect. Tissue Res.* **7,** 1 (1976).

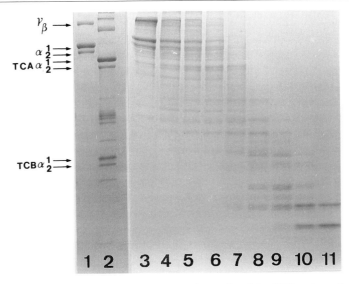

Fig. 3. Cleavage of type I collagen by astacin, analyzed by SDS–polyacrylamide gel electrophoresis [10–18% acrylamide, omitting reducing agents; U. K. Laemmli, *Nature* (*London*) **227**, 680 (1980)]. Lane 1 contained type I collagen (13 μg), and lane 2 contained collagen incubated with human interstitial collagenase for 18 hr at 20°. Arrows indicate collagen $\alpha_1$, $\alpha_2$, $\beta$, and $\gamma$ chains and the collagenase cleavage products TCA$\alpha_1$/$\alpha_2$ and TCB$\alpha_1$/$\alpha_2$. For lanes 3–11, collagen (13 μg) was incubated with astacin (2.0 × 10$^{-7}$ *M*) for 0.5 min (lane 3), 1 min (lane 4), 2 min (lane 5), 5 min (lane 6), 15 min (lane 7), 30 min (lane 8), 60 min (lane 9), 120 min (lane 10), and 18 hr (lane 11).

the triple helix of type I collagen.[27] Acid-solubilized calf skin collagen (1.5 mg/ml) is incubated with astacin (0.2 μ*M*) at 20° in phosphate buffer (5.0 m*M* Na$_2$HPO$_4$, 1.2 m*M* KH$_2$PO$_4$, 0.48 *M* NaCl, pH 7.4) or Tris buffer (20 m*M* Tris, 17 m*M* NaCl, 0.01% NaN$_3$, 5 m*M* CaCl$_2$, 50 m*M* ε-aminocaproic acid, pH 7.6) and subsequently resolved on SDS–polyacrylamide gradient gels (Fig. 3). Under the chosen conditions, astacin degrades the collagen substrate completely to small fragments within 15 min. This is in contrast to human fibroblast collagenase, which was used as a control (Fig. 3, lane 2).[27] The human enzyme cuts the $\alpha_1$ and $\alpha_2$ strands of type I collagen only at a single site, yielding fragments of one-quarter TCB$\alpha_1$/TCB$\alpha_2$ and three-quarters (TCA$\alpha_1$/TCA$\alpha_2$) of the native triple helix, respectively.[29] The collagen sample

[29] H. Birkedal-Hansen, this series, Vol. 144, p. 140.

and a preparation of human fibroblast collagenase were kindly provided by Dr. B. C. Adelmann-Grill (Martinsried, Germany).

### Protein Inhibitors

So far, the best characterized protein inhibitors of zinc endopeptidases are TIMP1 and TIMP2 (tissue inhibitors of metalloproteinases).[30] Of these, TIMP1 (a gift of Dr. T. E. Cawston, Cambridge, UK) has been tested with astacin and did not display inhibitory activity.

$\alpha_2$-*Macroglobulin.* Astacin is effectively inhibited by mammalian (bovine) $\alpha_2$-macroglobulin and also by crayfish $\alpha_2$-macroglobulin, a homolog inhibitor from the hemolymph of *Astacus astacus*.[31] It irreversibly forms a covalent 1 : 1 complex with astacin, which is inactive against the substrate STANA (Table III).[31,32]

*Inhibitors from Seeds of Ecballium elaterium.* Specific, natural inhibitors of astacin were detected in extracts from seeds of the mediterranean plant *Ecballium elaterium* (Table III).[33] Two peptides (termed EEAPI 1 and EEAPI 2, *Ecballium elaterium Astacus* protease inhibitor) could be purified and sequenced; they exhibit faint similarity with Bowman–Birk-type inhibitors of serine proteinases.[34] The $K_i$ values for astacin inhibition are in the micromolar range.

*PoI-d, Astacin Inhibitor from Potatoes.* Another astacin inhibitor (PoI-d, potato inhibitor d) has been purified from potato tubers (*Solanum tuberosum*).[35,36] Sequence analysis revealed 50 amino acids corresponding to a calculated molecular mass of 5533 Da. The sequence is very similar to polypeptide inhibitors of the inhibitor II family found in members of the Solanaceae.[37] PoI-d shows nonlinear, predominantly uncompetitive inhibition of astacin ($K_i$ of 0.1 $\mu M$; Table III), which indicates that it does not bind directly to the active site.

*Transition State Analog Inhibitors.* Replacement of the scissile peptide bond of a protease substrate by a phosphonamidate (-$PO_2$-NH-) or phosphi-

[30] J. F. Woessner, Jr., *FASEB J.* **5**, 2145 (1991).

[31] W. Stöcker, S. Breit, L. Sottrup-Jensen, and R. Zwilling, *Comp. Biochem. Physiol.* **98B**, 501 (1991).

[32] W. Stöcker, I. Yiallouros, J. R. Harris, and R. Zwilling, *Proc. Ger. Zool. Soc.* **85**, 171 (1992).

[33] A. Favel, H. Mattras, M. A. Coletti-Previero, R. Zwilling, E. A. Robinson, and B. Castro, *Int. J. Pept. Protein Res.* **33**, 202 (1989).

[34] D. Nalis, Thesis, University of Montpellier, France (1990).

[35] M. Herkert, W. Stöcker, and R. Zwilling, *Biol. Chem. Hoppe-Seyler* **371**, 760 (1990).

[36] M. Herkert, W. Stöcker, and R. Zwilling, submitted for publication (1995).

[37] H. M. Greenblatt, C. A. Ryan, and M. N. G. James, *J. Mol. Biol.* **205**, 201 (1989).

TABLE III
NATURAL AND SYNTHETIC INHIBITORS OF ASTACIN

| Inhibitor | $K_i$, $IC_{50}$ $(M)$ | Remarks |
|---|---|---|
| Protein inhibitors | | |
| $\alpha_2$-Macroglobulin | $nd^a$ | |
| EEAPI I,II | $nd^b$ | |
| PoI-d | $1.1 \times 10^{-7}$ | $NL^c$ |
| Synthetic inhibitors | | |
| 1,10-Phenanthroline, instantaneous | $3.1 \times 10^{-3}$ | $\bar{n} = 1.15^d$ |
| 1,10-Phenanthroline, 1 hr preincubation | $5.0 \times 10^{-4}$ | $\bar{n} = 2.12^d$ |
| 1,7-Phenanthroline, instantaneous* | $3.0 \times 10^{-3}$ | $\bar{n} = 1.33^d$ |
| Tyrosine hydroxamate | $1.7 \times 10^{-4}$ | $NC^e$ |
| CBZ-Phe$\psi$(PO$_2$H)-Ala-Pro-Phe-NH$_2$ | $2.7 \times 10^{-5}$ | $C^f$ |
| FMOC-Pro-Lys-Phe$\psi$(PO$_2$HCH$_2$)Ala-Pro-Leu-Val | $2.7 \times 10^{-8}$ | $C^g$ |

[a] Analyzed with STANA as the substrate, showing 1:1 stoichiometry (trap mechanism) and irreversible inhibition (covalent)[31,32]; nd, not determined.

[b] EEAPI, *Ecballium elaterium Astacus* protease inhibitor; preliminary $K_i$ value in the micromolar range.[34]

[c] PoI-d, Potato inhibitor, analyzed with DNS-VKRAPWV; $K_i$ calculated according to the equation for nonlinear inhibition (NL) [Refs. 35 and 36; A. Baici, *Eur. J. Biochem.* **119**, 9 (1981)].

[d] $\bar{n}$ gives the stoichiometry of the reaction (see text); the asterisk (*) signifies no change in activity on preincubation.[3]

[e] Analyzed with STANA as the substrate; $IC_{50}$ determined from Dixon plots (reciprocal velocity versus inhibitor concentration [R. L. Wolz, C. Zeggaf, W. Stöcker, and R. Zwilling, *Arch. Biochem. Biophys.* **281**, 275 (1990)]. NC, Noncompetitive inhibition.

[f] Analyzed with Dns-PKRAPWV as the substrate (0.5 $\mu M$–0.5 mM) after a 15-min preincubation of enzyme (4.2 nM) and inhibitor; $K_i$ determined from nonlinear regression analysis of the Michaelis–Menten equation in the absence and presence of 18.0 $\mu M$ inhibitor; apparent $K_i'$ values are corrected for competitive inhibition (c) by the equation $K_i = K_i'/(1 + [S]/K_m)$.

[g] Analyzed with Dns-PKRAPWV as the substrate (0.1 mM) after a 22-hr preincubation of enzyme (1.0 nM) and inhibitor (46 nM–4.6 $\mu M$); $K_i$ determined from nonlinear regression analysis of plots of the fractional activity, i.e., the quotient of the velocities in the presence and absence of inhibitor, respectively, $v_i/v_0$, versus the inhibitor concentration of the general equation for tight-binding inhibition. [J. Bieth, *Biochem. Med.* **32**, 387 (1984)].

nate moiety (-PO$_2$-CH$_2$-) has often resulted in effective protease inhibitors, in which the phosphorous group mimics the tetrahedral transition state during substrate turnover.[38–40a] Astacin is not inhibited by phosphorami-

[38] B. Holmquist, *Biochemistry* **16**, 4591 (1977).

[39] P. A. Bartlett and C. K. Marlowe, *Biochemistry* **22**, 4618 (1983).

[40] D. Grobelny, U. B. Goli, and R. E. Galardy, *Biochemistry* **28**, 4948 (1989).

[40a] A. Yiotakis, A. Lecoq, A. Nicolaou, J. Labadie, and V. Dive, *Biochem. J.* **303**, 323 (1994).

don, a natural phosphonamidate inhibitor of thermolysin-like enzymes.[41] In this property, the crayfish enzyme resembles meprin A and B[42] and the PABA-peptide hydrolase.[43] However, on the basis of good astacin substrates, pseudopeptides of this kind have been synthesized by Frank Grams (Martinsried) and by Dr. Vincent Dive (Gif-sur-Yvette, France), and some of them proved to be potent inhibitors of astacin, showing slow binding characteristics (Table III).

*Metal-Directed Inhibitors.* The zinc ion is indispensable for the catalytic activity of astacin as metal-binding agents like 1,10-phenanthroline (*o*-phenanthroline, OP), dipicolinic acid, 8-hydroxyquinoline-5-sulfonic acid, 2,2′-bipyridyl, and ethylenediaminetetraacetic acid (EDTA) inactivate the enzyme in a time- and concentration-dependent manner. Inhibition by the chelators is reversible on dilution or on addition of zinc(II) ions.[3,44] In addition, a series of amino acid hydroxamates and certain thiol-containing compounds, including 4-mercaptoaniline and cysteine, showed weak inhibition of astacin[45] owing to the metal-binding properties of the hydroxamate and thiol groups. Among the hydroxamates Tyr-NHOH is most effective, with an $IC_{50}$ of 0.17 m$M$ (Table III), followed by derivatives with other aromatic side chains, sulfur-containing residues, negatively charged residues, aliphatic residues, and positively charged side chains.[45] Except for Tyr-NHOH, this order deviates from the tubulin cleavage pattern, indicating that the small compounds may not bind in the same stereochemical orientation as extended substrates. Lower $K_i$ values in the micromolar range for the inhibition of astacin are reported for hydroxamates with longer backbones.[13,46a,60]

*Kinetics of Metal Removal.* The mechanism of metal removal by chelators like OP has been assessed by experiments in which astacin is assayed with STANA either instantaneously or after a 1-hr preincubation of enzyme and inhibitor.[3] The $IC_{50}$ values (i.e., inhibitor concentrations causing 50% inhibition) for OP inhibition of astacin are 3.1 m$M$ in the instantaneous assay and 0.5 m$M$ in the preincubated assay (Table III). Interestingly, the nonchelating isomer 1,7-phenanthroline (*m*-phenanthroline, MP) also shows inhibition ($IC_{50}$ 2.95 m$M$) (Table III). However, inhibition by MP is not time-dependent and cannot be reversed on addition to zinc. The time dependence of inhibition by OP indicates that in a first (fast) step,

[41] J. C. Powers and J. W. Harper, *in* "Proteinase Inhibitors" (A. J. Barrett and G. Salvesen, eds.), p. 220. Elsevier, Amsterdam, New York, and Oxford, 1986.

[42] R. J. Beynon, J. D. Shannon, and J. S. Bond, *Biochem. J.* **199**, 591 (1981).

[43] E. E. Sterchi, H. Y. Naim, M. J. Lentze, P. Hauri, and J. A. M. Fransen, *Arch. Biochem. Biphys.* **265**, 105 (1988).

[44] R. L. Wolz and R. Zwilling, *J. Inorg. Biochem.* **35**, 157 (1989).

[45] R. L. Wolz, C. Zeggaf, W. Stöcker, and R. Zwilling, *Arch. Biochem. Biophys.* **281**, 275 (1990).

[46a] R. L. Wolz, *Arch. Biochem. Biophys.* **310**, 144 (1994).

TABLE IV
ENZYMATIC ACTIVITY OF METAL-SUBSTITUTED
ASTACIN DERIVATIVES

| Enzyme Derivative | $\mu M$ | Substrate (m$M$) | Activity$^a$ (%) |
|---|---|---|---|
| Apoastacin$^b$ | 1.0 | 1.0 | 3 |
| Zn(II)-astacin | 1.0 | 4.0–18.0 | 100 |
| Co(II)-astacin | 1.0 | 0.2–8.0 | 140 |
| Cu(II)-astacin | 1.0 | 0.4–18.0 | 37 |
| Ni(II)-astacin | 1.3 | 1.0–24.0 | 3.2 |
| Hg(II)-astacin | 2.0 | 0.9–18.0 | 5.1 |

$^a$ Enzymatic activities were measured in the STANA assay. Percent values are based on the specificity constants ($k_{cat}/K_m$) with the zinc enzyme ($k_{cat}/K_m = 24.1 \pm 1.7\ M^{-1}\ sec^{-1}$ [20]) corresponding to 100%.
$^b$ The zinc-free apoastacin was assayed at 1 m$M$ substrate concentration and compared to the rate of the zinc enzyme under the same conditions. To avoid background metal contamination, kinetic experiments were run at enzyme concentrations above 1 $\mu M$. Data from Refs. 20 and 47.

the chelator binds to the active site, which is reflected by the MP inhibition.[3] There is evidence for cooperative binding of two OP molecules during metal removal.[44] In a second step, OP removes the metal to form higher bis- and tris-OP–zinc complexes in solution. This assumption is corroborated by the ligand/metal stoichiometry, $\bar{n}$,[46] which is about 1 for instantaneous OP inhibition and for MP but about 2 for preincubated OP (Table III). The inactivation by OP is on the order of a few minutes, whereas EDTA requires days.[3]

## Properties of Metal-Substituted Astacin Derivatives

*Preparation of Apoastacin.* Zinc-free apoastacin is produced by dialysis of 0.3-ml aliquots of enzyme (50–200 $\mu M$) dissolved in 50 m$M$ HEPES–NaOH buffer, pH 8.0, for 4 days at 4° versus four changes of 100 ml of the same buffer containing 10 m$M$ OP.[3] Subsequently, excess chelator is removed by dialysis against metal-free buffer until the OP concentration is below 0.001 m$M$ as judged from the extinction at 265 nm.[46] The dialysis procedure yields apoastacin containing less than 0.007 gram-atoms of zinc per mole of protein, as determined by atomic absorption spectroscopy, and exhibiting about 3% of the catalytic activity of zinc [Zn(II)]-astacin when assayed with 1 $\mu M$ enzyme, which probably reflects the background of zinc (Table IV).[3]

[46] D. S. Auld, this series, Vol. 158, p. 110.

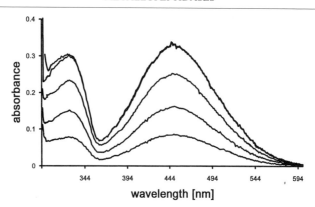

FIG. 4. Electronic absorption spectrum of Cu(II)-astacin.[47] Approximately 400 μl of 0.2 mM apoenzyme in 50 mM HEPES buffer, pH 8.0, was titrated with increasing amounts of copper(II) sulfate (0.037, 0.075, 0.11, 0.19, and 0.3 mM, from bottom to top). Saturation was reached at a molar ratio of 1:1.

*Metal-Substituted Astacin Derivatives.* Addition of stoichiometric amounts of zinc(II) ions to apoastacin fully restores the catalytic activity. Also copper(II) and cobalt(II) ions, but not mercury(II), nickel(II), or calcium(II) (among others), reactivate the apoenzyme.[3] To assess the role of the metal for catalysis, apoastacin (1.0 $\mu M$) is mixed with equimolar concentrations of $HgCl_2$, $CoCl_2$ or $CoSO_4$, $ZnSO_4$, $NiCl_2$, and $CuSO_4$, then analyzed in the STANA assay (Table IV). Based on the specificity constants ($k_{cat}/K_m$), Co(II)- and Cu(II)-astacin display 140 and 37% activity of the zinc enzyme, respectively, whereas Hg(II)-astacin and Ni(II)-astacin are almost inactive (Table IV).[47] The relative activities deviate from those of an earlier report[3] where Zn(II)-, Co(II)-, and Cu(II)-astacin were compared only at a single substrate concentration.

*Spectroscopy.* For spectroscopic analysis, apoastacin (50–200 $\mu M$) dissolved in 50 mM HEPES–NaOH buffer, pH 8.0, is titrated in a stepwise fashion with substoichiometric amounts of $CuSO_4$ or $CoSO_4$. Co(II)-astacin exhibits an absorption maximum at 514 nm ($\varepsilon$ 77 $M^{-1}$ cm$^{-1}$) with shoulders at 505 and 550 nm, similar to the spectra of Co(II)-thermolysin (EC 3.4.24.27) and Co(II)-carboxypeptidase A (EC 3.4.17.1).[3] By contrast, the spectrum of Cu(II)-astacin is very different from those of Cu(II)-thermolysin and Cu(II)-carboxypeptidase, as it displays an intense absorption band at 445 nm ($\varepsilon$ 1900 $M^{-1}$ cm$^{-1}$) and an additional band at 325 nm ($\varepsilon$ 1600 $M^{-1}$ cm$^{-1}$) (Fig. 4).[47] In these features, Cu(II)-astacin resembles Cu(II)-transferrin, Cu(II)-lactoferrin, and Cu(II)-conalbumin, which, unlike ther-

[47] F.-X. Gomis-Rüth, F. Grams, I. Yiallouros, H. Nar, U. Küsthardt, R. Zwilling, W. Bode, and W. Stöcker, *J. Biol. Chem.* **269**, 17111 (1994).

molysin[48] or carboxypeptidase,[49] contain tyrosine–metal ligands and whose respective spectral properties have been interpreted as phenol (O)–metal charge transfer bands (for further details, see Ref. 47). As discussed below, the metal coordination in astacin also involves a tyrosine ligand.[11,12]

The electron paramagnetic resonance (EPR) spectrum of copper(II)-astacin[47] can be simulated with parameters $g_\parallel = 2.29$, $g_\perp = 2.05$, and $A_\parallel = 139 \times 10^{-4}$ cm$^{-1}$, which are close to "normal," axial type II copper values. In the $g_\perp$ region, however, the spectrum reveals a fine structure indicating some rhombic distortion. This is consistent with the crystallographically determined copper coordination sphere (see below), exhibiting an almost ideal trigonal-bipyramidal geometry.

## Crystallization and X-Ray Structure of Astacin

*Crystallization.* A drop containing 3.5–5.0 μl native astacin (0.3–0.6 mM) in 1.0 mM MES–NaOH, pH 7.0) and 1.5 μl ammonium sulfate (1.0 M, in 0.1 M sodium phosphate, pH 7.0) is equilibrated against 1.0 M ammonium sulfate, pH 7.0, at 4° according to the hanging drop vapor diffusion method.[11,12,47] Trigonal crystals (space group $P3_121$) are harvested in 1.5 M ammonium sulfate buffer, pH 7.0. Details of the X-ray structure analysis via the multiple isomorphous replacement technique have been published, using one native, one apo, and six heavy atom derivative crystals, including cobalt(II), copper(II), mercury(II), and nickel(II).[12,47]

*Three-Dimensional Structure of Astacin.* The X-ray crystal structure analysis of astacin to 1.8 Å resolution revealed a kidney-shaped molecule, which is subdivided by a deep active-site cleft into two domains.[11,12] The N-terminal, upper part (Fig. 5) consists of a twisted five-stranded β-pleated sheet formed by four parallel strands (strands I, II, III, and V), one antiparallel strand (strand IV), and two long α helices (helices A and B) (Fig. 5). The lower, C-terminal domain is more irregularly organized except for a short 3$_{10}$ helix (helix C) and a long α helix (helix D), which is fixed to the N domain via the disulfide bond from Cys-42 to Cys-198. The other disulfide (Cys-64 to Cys-84) clamps β strand IV to the loop between strand V and helix B.

The catalytically active zinc ion is located at the bottom of the active-site cleft in the center of the molecule, ligated by three histidine residues, a tyrosine residue, and a water molecule in a trigonal-bipyramidal coordination sphere (Figs. 5 and 6). The zinc ligands His-92 and His-96, at distances of 2.0 and 2.2 Å, are part of the central helix B separated by a single helix turn (Figs. 5 and 6). The third zinc ligand, His-102 (2.0 Å), follows only six residues upstream of the second. The fourth and the fifth zinc ligands

[48] B. W. Matthews, *Acc. Chem. Res.* **21**, 333 (1988).
[49] D. N. Christiansen and W. N. Lipscomb, *Acc. Chem. Res.* **22**, 62 (1989).

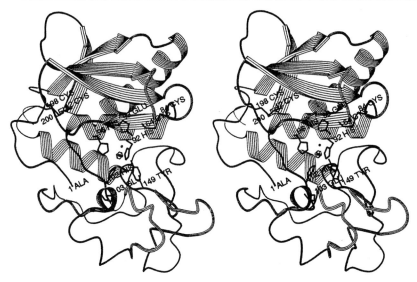

FIG. 5. Stereo pair of a ribbon plot of the astacin structure,[12] showing the secondary structure elements in a view perpendicularly to the active-site cleft. Only a few important side chains are labeled, including the zinc ligands, the disulfide-bonded cysteines, and Glu-103 which is salt bridged via a water molecule to the N-terminal alanine.

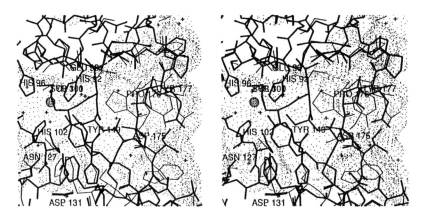

FIG. 6. Substrate binding sites of astacin and meprin $\alpha$.[10] Stereo plot of the putative substrate binding regions of a model of meprin (thick lines) superimposed on the astacin structure (thin lines) around the S1'/S2' subsite overlaid with the Connolly dot surface of the region. Also visible (left) are the zinc ion in its trigonal bipyramidal coordination sphere formed by His-92 (2.0 Å), His-96 (2.2 Å), His-102 (2.0 Å), Tyr-149 (2.5 Å), and a water molecule, Sol-300 (2.0 Å) (distances are in parentheses).

are supplied by the hydroxyl oxygen of Tyr-149 (2.5 Å) and a water molecule (Sol-300) (2.0 Å), also linked to Glu-93, respectively (Fig. 6). A model for the action of zinc peptidases implies that the metal-bound water is polarized between the zinc(II) ion, acting as a Lewis acid, and the glutamic acid residue, acting as the general base, for nucleophilic attack on the peptide bond.[48-50]

The three imidazole zinc ligands and the presumably catalytically important glutamate are combined in the consensus motif **HEXXHXXGXXH**, seen in all astacins. This conserved stretch of sequence is shared with the mammalian matrix metalloproteinases (matrixins), snake venom proteinases (adamalysins), and certain bacterial enzymes (serralysins).[10] These four superfamilies of zinc peptidases are related and form a family which has been called the metzincins,[51,51a,61] based on the motif **HEXXHXXGXXH**, on a conserved methionine-containing 1,4-$\beta$ turn (Met turn) beneath the active site, and, furthermore, on topologically equivalent five-stranded $\beta$ sheets and helices. Most remarkably, there is also some topological similarity between astacin and the archetypical zinc proteinase thermolysin, confined to four strands of the $\beta$-pleated sheet and to two long helices.[12,51,51a,61] One of the helices carries two of the zinc-binding histidines in a more general consensus motif **HEXXH** shared by the majority of metalloproteinases.[3,52-54]

As another important feature, astacin contains a water-filled cavity, in which the two N-terminal residues Ala-1 and Ala-2 are completely buried (Fig. 5). The ammonium group of Ala-1 is connected by a water-mediated salt bridge with the carboxylate of Glu-103, the direct neighbor of the zinc ligand His-102. This is a scenario reminiscent of the trypsin-like serine proteinases (EC 3.4.21.4),[55] which become activated after proteolytic removal of a propeptide from the inactive zymogen, through the interaction of the new N terminus with the carboxylate of Asp-194, neighboring the catalytic Ser-195. In astacin, the arrangement of the N terminus indicates an analogous activation mechanism, as N-terminally extended proforms could not exhibit such a conformation.[11] Evidence of proprotein processing in the astacins has been provided for meprin from mouse and rat kidney and for the PABA-peptide hydrolase from human intestine.[56-61]

[50] D. S. Auld, in "Enzyme Mechanisms" (M. I. Page and A. Williams, eds.), p. 240. Royal Society of Chemistry, Letchworth, U.K., 1987.

[51] W. Bode, F.-X. Gomis-Rüth, and W. Stöcker, *FEBS Lett.* **331**, 134 (1993).

[51a] W. Stöcker, F. Grams, U. Baumann, P. Reinemer, F.-X. Gomis-Rüth, D. B. McKay, and W. Bode, *Protein Sci.,* submitted (1995).

[52] C. V. Jongeneel, J. Bouvier, and A. Bairoch, *FEBS Lett.* **242**, 211 (1989).

[53] B. L. Vallee and D. S. Auld, *Acc. Chem. Res.* **26**, 543 (1993).

[54] N. D. Rawlings and A. J. Barrett, *Biochem. J.* **290**, 205 (1993).

[55] R. Huber and W. Bode, *Acc. Chem. Res.* **11**, 114 (1978).

[56] M. Z. Kounnas, R. L. Wolz, C. M. Gorbea, and J. S. Bond, *J. Biol. Chem.* **266**, 17350 (1991).

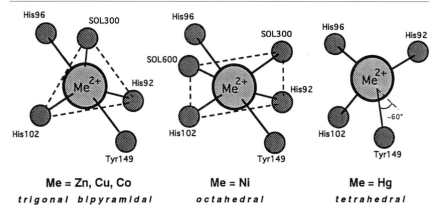

Me = Zn, Cu, Co            Me = Ni            Me = Hg
*trigonal bipyramidal*      *octahedral*       *tetrahedral*

FIG. 7. Active-site metal coordination in Zn(II)-, Co(II)-, Cu(II)-, Hg(II)-, and Ni(II)-astacin structures.[47] Schematic representation of the coordination geometries. Me, Metal; SOL, solvent ($H_2O$).

The cleavage specificities of astacin and meprin A are in part similar. Both enzymes prefer extended substrates with bulky residues in P1 and P2 and proline residues in P3 and P2'. However, they differ markedly in the preference for residues in P1', where astacin tolerates only small uncharged residues but meprin A also hydrolyzes substrates with Arg, Phe, or Lys residues in that position.[13,22,58]

The striking difference in cleavage specificities has been investigated by a computer model of the α chain of mouse meprin based on the astacin structure.[10] According to the model, the backbones of the two proteins match almost perfectly, if only five single-residue deletions, one double deletion, and three single-residue insertions are introduced. However, in the presumed S1' subsite, a substitution of Pro-176 by a Gly and the concomitant deletion of Tyr-177 lead to a more open configuration of the S1' subsite in meprin α (Fig. 6). The upper border of the S1'/S2' subsite is formed by Cys-64 and Thr-89 in both astacin and meprin α. However, at its bottom, Asp-175 of astacin is replaced by Ile or Leu in meprin and other astacins. Binding of aromatic side chains like Trp in S3' of astacin might be promoted by Tyr-177 and/or Trp-158, which are deleted and replaced

[57] J. Grünberg, E. Dumermuth, J. A. Eldering, and E. E. Sterchi, *FEBS Lett.* **335**, 376 (1993).
[58] R. L. Wolz, R. B. Harris, and J. S. Bond, *Biochemistry* **30**, 8488 (1991).
[59] C. M. Gorbea, P. Marchand, W. Jiang, N. G. Copeland, D. J. Gilbert, N. A. Jenkins, and J. S. Bond, *J. Biol. Chem.* **268**, 21035 (1993).
[60] D. Corbeil, P.-M. Milhiet, V. Simon, J. Ingram, A. J. Kenny, G. Boileau, and P. Crine, *FEBS Lett.* **335**, 361 (1993).
[61] G. D. Johnson and L. B. Hersh, *J. Biol. Chem.* **269**, 7682 (1994).

by Asn-158 in meprin, respectively. When the findings are taken together, there are significant differences in the S' subsites of astacin compared to those of the other members of the astacin family, whereas the S regions appear to be more similar.[10]

The X-ray crystal structures of native Zn(II)-, Cu(II)-, Co(II)-, Hg(II)-, and Ni(II)-astacin exhibit an identical overall framework.[47] However, the modes of metal coordination in the active sites differ considerably. The active compounds Cu(II)- and Co(II)-astacin contain trigonal bipyramidally coordinated metal as is seen in the zinc enzyme (Fig. 7).[47] By contrast, in the inactive mercury-astacin, the catalytically important water ligand is missing, leaving tetracoordinated metal,[12] and in the nickel derivative, an additional water ligand has entered the central plane, expanding it from trigonal to tetragonal and resulting in an octahedral ligand geometry (Fig. 7).[47] Thereby, the nucleophilicity of the general base should be reduced. Hence, both inactive derivatives presumably lack a sufficiently polarized water molecule for the attack of the scissile peptide bond to proceed.

### Acknowledgments

We thank Monika Dörhöfer and Wolfgang Forster for excellent technical assistance. This work was supported by the Deutsche Forschungsgemeinschaft (Sto 185/1-1,2,3 and Sto 185/3-1).

# [20] Meprins A and B

By RUSSELL L. WOLZ and JUDITH S. BOND

### Introduction

Meprins are oligomeric metalloendopeptidases with subunit molecular weights of 85,000–110,000.[1–3] They are disulfide-linked glycoproteins that are bound to the plasma membrane via the meprin $\beta$ subunit.[3–5] Secreted forms of the meprin $\alpha$ subunit have been reported.[6] Meprins are highly

[1] M. Z. Kounnas, R. L. Wolz, C. M. Gorbea, and J. S. Bond, *J. Biol. Chem.* **266**, 17350 (1991).

[2] W. Jiang, C. M. Gorbea, A. Flannery, R. J. Beynon, G. A. Grant, and J. S. Bond, *J. Biol. Chem.* **267**, 9185 (1992).

[3] C. M. Gorbea, P. Marchand, W. Jiang, N. G. Copeland, D. J. Gilbert, N. A. Jenkins, and J. S. Bond, *J. Biol. Chem.* **268**, 21035 (1993).

[4] S. S. Craig, J. F. Reckelhoff, and J. S. Bond, *Am. J. Physiol.* **253** (*Cell Physiol.* **22**), C535 (1987).

[5] P. Marchand, J. Tang, and J. S. Bond, *J. Biol. Chem.* **269**, 15388 (1994).

[6] A. V. Flannery, G. N. Dalzell, and R. J. Beynon, *Biochim. Biophys. Acta* **1041**, 64 (1990).

expressed in kidney and intestinal brush border membranes, and most of the information about the enzymes derives from studies of these tissues in mice, rats, and humans.[1,7-12] Meprin isoforms contain two types of subunits, referred to as $\alpha$ and $\beta$. Meprin A (EC 3.4.24.18) contains $\alpha$ subunits, and it can exist as homooligomers ($\alpha_2$) or heterooligomers ($\alpha\beta$). Meprin B (EC 3.4.24.63) contains only $\beta$ subunits and consists of homooligomers of that subunit.[13] Mouse meprin A has "high azocaseinase activity" and was first isolated from kidneys of BALB/c mice.[14] Meprin B, which has latent azocaseinase activity, was isolated from C3H/He mice, a strain in which kidney tissue expresses only the $\beta$ subunit.[15] Cloning and sequencing of the subunits from mice and rats have revealed that $\alpha$ and $\beta$ are approximately 45% identical, and that meprins are members of a family of metalloendopeptidases called the astacin family.[2,3,16] The prototype of this family is the crayfish digestive enzyme, astacin, which is discussed in the previous chapter in this volume ([19]; see also [13] in this volume).

In this chapter, we describe the purification, assay, structural characteristics, substrate specificity, and inhibitors of meprins A and B isolated from different mouse strains. Some discussion of the rat and human forms of the enzymes is also included.

## Purification of Meprin A

Meprin A is present in the kidney of random bred and at least 35 inbred, recombinant, and congenic strains of mice.[17] It has been purified from a random bred mouse strain (ICR), as well as from inbred mouse strains (BALB/c and C57BL/6).[1,14,15] It is most commonly purified from frozen male

[7] W. Jiang, P. M. Sadler, N. A. Jenkins, D. J. Gilbert, N. G. Copeland, and J. S. Bond, *J. Biol. Chem.* **268**, 10380 (1993).

[8] A. J. Kenny and J. Ingram, *Biochem. J.* **245**, 515 (1987).

[9] G. D. Johnson and L. B. Hersh, *J. Biol. Chem.* **267**, 13505 (1992); erratum *J. Biol. Chem.* **268**, 17647 (1993).

[10] D. Corbeil, F. Gaudoux, S. Wainwright, J. Ingram, A. J. Kenny, G. Boileau, and P. Crine, *FEBS Lett.* **309**, 203 (1992).

[11] E. E. Sterchi, H. Y. Naim, M. J. Lentze, H. P. Hauri, and J. A. M. Fransen, *Arch. Biochem. Biophys.* **265**, 105 (1988).

[12] E. Dumermuth, J. A. Eldering, J. Grünberg, W. Jiang, and E. E. Sterchi, *FEBS Lett.* **335**, 367 (1993).

[13] C. M. Gorbea, A. V. Flannery, and J. S. Bond, *Arch. Biochem. Biophys.* **290**, 549 (1991).

[14] R. J. Beynon, J. D. Shannon, and J. S. Bond, *Biochem. J.* **199**, 591 (1981).

[15] P. E. Butler and J. S. Bond, *J. Biol. Chem.* **263**, 13419 (1988).

[16] E. Dumermuth, E. E. Sterchi, W. Jiang, R. L. Wolz, J. S. Bond, A. V. Flannery, and R. J. Beynon, *J. Biol. Chem.* **266**, 21381 (1991).

[17] J. F. Reckelhoff, J. S. Bond, R. J. Beynon, S. Savarirayan, and C. S. David, *Immunogenetics* **22**, 617 (1985).

ICR mouse kidneys, following the method of Beynon et al.[14] as modified by
Kounnas et al.[1] (Table I). Kidneys (50 g) are homogenized in 200 ml of 20
mM Tris buffer, pH 7.5, using a Polytron (Brinkman Instruments, Westbury,
NY) homogenizer. The homogenate is centrifuged at 4° for 15 min at 600 g
and the supernatant fraction saved. The sediment is resuspended in 75 ml
of Tris-HCl buffer, pH 7.5, homogenized, and centrifuged at 4° for 15 min
at 600 g. The supernatant fraction is combined with that from the first
homogenization and subjected to ultracentrifugation at 4° for 60 min at
100,000 g.

The ultracentrifuge sediment containing the brush border membranes
is suspended in 200 ml of 50 mM potassium phosphate buffer, pH 6.2. To
release meprin from the membranes, 20 mg papain, activated by preincubation with 5 mM cysteine, is added to the suspended membranes and stirred
for 90 min at 37°. Papain is then inactivated with iodoacetamide (1 mM),
and the mixture is incubated for 20 min at 37°. The papain-treated membranes are sedimented by centrifugation at 100,000 g at 4° for 60 min. The
supernatant fluid, containing solubilized meprin, is subjected to ammonium
sulfate precipitation to 40% of saturation, then centrifuged at 4° for 20 min
at 8000 g. The 0 to 40% precipitate is discarded, and the ammonium sulfate

TABLE I
PURIFICATION OF MEPRIN A FROM ICR MICE AND MEPRIN B FROM C3H/He MICE[a]

| Step | Total proteinase activity (azocasein units) | Total protein (mg) | Specific activity (units/mg) | Purification (-fold) |
|---|---|---|---|---|
| Meprin A | | | | |
| Homogenate | 91,700 | 8800 | 10 | 1 |
| Membranes | 46,000 | 2340 | 19 | 1.9 |
| Papain/precipitate[b] | 15,100 | 163 | 93 | 9.3 |
| Mono Q | 5530 | 5.0 | 1100 | 110 |
| Superose | 1350 | 0.6 | 2250 | 225 |
| Meprin B | | | | |
| Homogenate | 4390 | 3900 | 1.1 | 1 |
| Membranes | 2510 | 1010 | 2.5 | 2.3 |
| Papain/precipitate | 910 | 64 | 14 | 12.7 |
| Mono Q | 867 | 9 | 97 | 88 |
| Superose | 148 | 1.1 | 134 | 120 |
| Mono Q | 46 | 0.2 | 210 | 190 |
| ACE affinity | 39 | 0.09 | 433 | 394 |

[a] Data from M. Z. Kounnas, R. L. Wolz, C. M. Gorbea, and J. S. Bond, J. Biol. Chem.
**266,** 17350 (1991).
[b] Ammonium sulfate precipitate after papain solubilization.

concentration in the supernatant fluid is raised to 80% saturation. The 40–80% precipitate, containing meprin, is sedimented by centrifugation at 4° for 20 min at 8000 $g$, resuspended in 6 ml of 20 m$M$ Tris-HCl buffer, pH 8.5, and desalted on Bio-Rad (Richmond, CA) 10DG columns with 20 m$M$ Tris-HCl buffer, pH 8.5.

The resulting 8 ml of desalted effluent is then applied to a Pharmacia (Piscataway, NJ) preparative Mono Q anion-exchange column (10 $\times$ 100 mm). Proteins are eluted at a flow rate of 4 ml/min with 20 m$M$ Tris-HCl buffer, pH 8.5, containing NaCl according to the following gradient program: no salt 0–12 min; 0–100 m$M$ from 12 to 16 min; 100–350 m$M$ from 16 to 49 min; and 350–1000 m$M$ from 49 to 53 min. The active fractions elute near 200 m$M$ NaCl and are combined and concentrated to 300 $\mu$l using Amicon (Danvers, MA) Centriprep and Centricon concentrators. The final step is gel-filtration chromatography on a Pharmacia Superose 12 column (10 $\times$ 300 mm) with 20 m$M$ Tris-HCl buffer, pH 8.5. Pure meprin A elutes in a volume of approximately 2 ml at a concentration near 1 mg/ml. The enzyme is stable in solution at 4° for at least 2 weeks or frozen at $-20°$ for more than 1 year with little loss of activity.

Trypsin has also been used to release meprin from the membrane.[14] The procedure is the same as detailed above except that the membranes are incubated for 18 hr with trypsin/toluene in Tris-HCl buffer, pH 7.5, instead of the 90-min incubation with papain. Soybean trypsin inhibitor is then used to inactivate trypsin.

Final yields of meprin A vary from 0.5 to 2 mg with specific activities in the range of 2500 to 10,000 azocasein units/mg. Purity is judged by electrophoresis and final specific azocaseinase activity. Final preparations also are checked routinely for the absence of contaminating activities of membrane alanyl aminopeptidase (EC 3.4.11.2) and angiotensin I-converting enzyme (EC 3.4.15.1; ACE, peptidyl-dipeptidase A).[1]

Purification of Meprin B

More than 20 inbred, recombinant, and congenic mouse strains have been identified in which kidney expresses meprin B but not meprin A.[17] Meprin B has also been identified in the small intestine of all strains of mice.[3] Meprin B has been purified from male C3H/He mouse kidneys.[1,15] The purification procedure is initially the same as that for meprin A, but two additional steps are required: anion-exchange chromatography with an extended gradient to remove alanyl aminopeptidase and affinity chromatography for the removal of ACE. After the steps detailed above for meprin A, the active fractions from gel filtration are combined and applied to a Mono Q column (5 $\times$ 50 mm). The column is then washed with 20 m$M$

Tris-HCl buffer, pH 8.5, for 5 min at a flow rate of 1 ml/min. Elution is achieved with a linear salt gradient from 0 to 200 m$M$ NaCl in 20 m$M$ Tris-HCl, pH 8.5, over a period of 35 min. Meprin B elutes at approximately 125 m$M$ salt. The active fractions are then subjected to affinity chromatography on a column of immobilized lisinopril {$N^{\alpha}$-[($S$)-1-carboxy-3-phenylpropyl]-L-lysyl-L-proline}.[18] Samples are applied to the lisinopril column in 20 m$M$ Tris-HCl, pH 7.5, containing 300 m$M$ NaCl. The ACE remains bound to the column, and the eluate contains pure meprin B. The lisinopril column can be regenerated by washing with 50 m$M$ sodium borate, pH 10, to remove the bound ACE.

Papain-purified preparations of meprin B have a final yield of 0.1 to 0.3 mg, and the specific activity against azocasein ranges from 200 to 600 units/mg. Papain-purified meprin B preparations can be activated 5- to 20-fold by limited trypsin treatment. For activation, meprin is incubated in 20 m$M$ Tris-HCl, pH 7.5, with tosyl phenylalanyl chloromethyl ketone (TPCK)-treated trypsin (Sigma, St. Louis, MO) at a meprin : trypsin ratio of 10 : 1 (w/w) for 2 hr at 37°; trypsin is then inhibited with a 2-fold excess of soybean trypsin inhibitor. This treatment results in meprin B activities comparable to those of meprin A. Meprin B can also be purified using toluene/trypsin solubilization; in this instance, final specific activities are similar to meprin A (2000 to 10,000 azocaseinase units/mg).[15]

## Assays

Meprins were originally discovered using azocasein as a substrate at pH 9.5.[14] Azocasein is an excellent substrate for mouse meprin A and trypsin-activated mouse meprin B. It is, however, a poorer substrate for rat and human meprins.[19] For enzyme from the latter two species, [125]I-iodinated insulin B chain,[20] which is cleaved at multiple sites, and N-benzoyltyrosyl-p-aminobenzoate,[11] which is cleaved at the arylamide bond, have been used as substrates. Experiments have also shown that arylamides such as succinyl-Ala-Ala-Ala-p-nitroanilide are substrates for meprin A.[21] Bradykinin is a substrate for meprin A from all species, but it is not hydrolyzed by meprin B. Assays for meprin A using azocasein and chromogenic and fluorogenic derivatives of bradykinin are described below.

[18] M. R. W. Ehlers, E. A. Fox, D. J. Strydom, and J. F. Riordan, *Proc. Natl. Acad. Sci. U.S.A.* **86,** 7741 (1989).
[19] K. Barnes, J. Ingram, and A. J. Kenny, *Biochem. J.* **264,** 335 (1989).
[20] A. J. Kenny, in "Proteinases in Mammalian Cells and Tissues" (A. J. Barrett, ed.), p. 393. North Holland Publishing Co., Amsterdam, 1977.
[21] R. L. Wolz, *Arch. Biochem. Biophys.* **310,** 144 (1994).

*Azocasein Assay*

Azocasein is casein that has been derivatized at multiple sites with sulfanilamide to produce a protein with an intense red color. After degradation of azocasein with a protease, product peptides are released which are soluble in dilute trichloroacetic acid (TCA), whereas undigested substrate is precipitated by TCA. Thus, the solubilized peptides can be separated from precipitated starting material by centrifugation and quantified colorimetrically. Although azocasein is a nonspecific substrate for meprin, it is very practical because large numbers of samples can be assayed simultaneously and the assay can be performed with samples such as crude homogenates or brush border membrane preparations which contain substances that would interfere with UV or fluorescence measurements. To determine the contribution of meprin in tissue samples, a spectrum of protease inhibitors must be used (see section on inhibitors below).

To follow meprin A or B activity during purification, a simplified azocasein hydrolysis assay is used.[17,22] Azocasein is available commercially (Sigma). Enzyme (0.5–5 $\mu$g of mouse kidney meprin A) is added to 20 m$M$ ethanolamine buffer, pH 9.5, containing 140 m$M$ NaCl, to give a total volume of 125 $\mu$l. The reaction is initiated by addition of 125 $\mu$l of azocasein solution (22 mg/ml in ethanolamine buffer, pH 9.5), and the mixture is incubated at 37° for 20 min. Addition of 1 ml of 5% (w/v) TCA then stops the reaction and precipitates undigested azocasein. The mixture is centrifuged at 25° for 4 min at 10,000 $g$. The absorbance of the digested azocasein fragments in the supernatant fluid is measured at 340 nm and compared to that of a blank reaction mixture containing no enzyme. The assay is most accurate when the final absorbance is between 0.1 and 0.5. Depending on the preparation, this corresponds to approximately 0.5–5 $\mu$g of mouse kidney meprin A per 20-min assay. A change in absorbance of 0.001 per minute (representing solubilization of 1.3 $\mu$g of azocasein per minute) is defined as 1 unit.

In instances where the absorbance change with time is nonlinear or where greater accuracy is required, essentially the same procedure is used except that a larger incubation mixture is prepared with 4- to 6-fold increases in the volumes of sample and azocasein.[17] Samples of 250 $\mu$l are removed from the incubation at various time points, and the reaction is stopped by addition of the sample to 1 ml of 5% (w/v) TCA. Absorbance is measured as before, and activity is calculated from the slope of a plot of absorbance against time.

---

[22] R. L. Wolz, unpublished work (1989).

*Assay of Meprin A with Nitrobradykinin*

Nitrobradykinin[23] [Arg-Pro-Pro-Gly-Phe(NO₂)-Ser-Pro-Phe-Arg; Table IV, peptide VIII] is a derivative which differs from the naturally occurring vasoactive peptide only by the addition of a nitro group to Phe-5. The basis for its use as a chromogenic substrate is a shift in the UV spectrum of nitrophenylalanine which can occur when the peptide bond on either side of the internal residue is cleaved.[24,25] The nitrophenylalanine residue has an absorbance maximum at 279 nm when at an internal position in a peptide. Hydrolysis adjacent to the residue produces a blue shift of the maximum to 273 nm when an N-terminal side cleavage creates a positively charged $\alpha$-amino group (the $pK_a$ of the $\alpha$-amino group is 6.65)[26] or a red shift to 282 nm when a C-terminal side cleavage creates a negatively charged $\alpha$-carboxyl group (the $pK_a$ of the $\alpha$-carboxyl group is 3.5).[24] The maximum absorbance change caused by such shifts is seen on the steep side of the peak at 310 nm where the absorbance decreases with the blue shift or increases with the red shift.

Meprin A cleaves nitrobradykinin on the C-terminal side of nitrophenylalanine, which at pH 8.7 causes a red shift and an increase in absorbance at 310 nm. Cleavages at other sites in the peptide at basic pH values are chromogenically silent, thereby conferring a specificity of colorimetric detection of hydrolysis only for enzymes which cleave the same bond as meprin A. In addition to the closely related mouse kidney,[23] rat kidney[27] and human intestinal[11] meprins A, the homologous crayfish digestive enzyme astacin also cleaves bradykinin and/or nitrobradykinin at the Phe-Ser bond.[21] The only other known enzyme, unrelated to meprins, which cleaves that bond in nitrobradykinin is thimet oligopeptidase (EC 3.4.24.15).[22]

Hydrolysis of nitrobradykinin and other chromogenic substrates of this type (Table IV, peptides VIII–XIII) can be monitored continuously by following the absorbance change at 310 nm. This is best done in a spectrophotometer (such as a Shimadzu UV160-A, Kyoto, Japan) in which the gain can be amplified 10-fold, allowing the measurement of absorbance to $10^{-4}$. Typically, a 1-ml reaction mixture contains 100 $\mu M$ nitrobradykinin in 50 m$M$ ethanolamine buffer, pH 8.7, and the reaction is initiated with 1 $\mu$g of meprin (12 n$M$). Product [Arg-Pro-Pro-Gly-Phe(NO₂)] concentra-

[23] R. L. Wolz and J. S. Bond, *Anal. Biochem.* **191**, 314 (1990).
[24] R. Salesse and J. Garnier, *J. Dairy Sci.* **59**, 1215 (1976).
[25] B. M. Dunn, B. Kammerman, and K. R. McCurry, *Anal. Biochem.* **138**, 68 (1984).
[26] T. Hofmann and R. S. Hodges, *Biochem. J.* **203**, 603 (1982).
[27] S. L. Stephanson and A. J. Kenny, *Biochem. J.* **255**, 45 (1988).

tion is calculated based on the measured $\Delta\varepsilon_{310}$ of 750 $M^{-1}$ cm$^{-1}$. Initial rates are calculated from the absorbance change during the first 10% or less of substrate hydrolysis and activity expressed as change in product concentration per unit time.

In a typical assay at 30° with 100 $\mu M$ substrate, a 10-min reaction time is sufficient to measure 1 n$M$ enzyme accurately (85 ng/ml). Sensitivity can be increased by using higher reaction temperatures, longer reaction times, and/or higher substrate concentrations. The initial rates of hydrolysis of nitrobradykinin are linearly dependent on the concentration of meprin A over a range of 85–2200 ng/ml (1–25 n$M$).[23]

### Assay of Meprin A with Fluorogenic Peptides

Hydrolysis of intramolecularly quenched fluorogenic substrates of the general structure 2ABz-Arg-Pro-Xxx-Xxx-Ser-Pro-Phe(NO$_2$)-Arg (Table IV, peptides I–VII) is followed fluorimetrically.[28,29] The 2-aminobenzoyl (2ABz) group attached to the N terminus of the peptide fluoresces when excited at 320 nm, but the fluorescence is quenched by the nitrophenylalanine at the C terminus of the peptide. Cleavage of the peptide between the groups relieves the quenching, and the increase in fluorescence at 420 nm can be monitored continuously. Typically, the increase in fluorescence after complete hydrolysis is approximately 2.5-fold relative to the uncleaved peptide.

For meprin A reactions, substrate is buffered to pH 8.7 with 50 m$M$ ethanolamine hydrochloride, and the reaction is initiated by the addition of 1.4–10.6 $\mu$g/ml of meprin A (16–125 n$M$ at the subunit molecular weight of 85,000). The maximum fluorescence change for each concentration of each substrate is determined by allowing the reaction to proceed to completion. Initial velocities are calculated from the fluorescence change during the first 10% or less of substrate hydrolysis and expressed as the change in concentration per unit time.

Molecular Properties

### Molecular Weight

The molecular weight of papain- or trypsin-solubilized meprin A isolated from BALB/c or ICR mice was estimated to be 270,000–320,000 based on gel filtration on Sepharose 6B or sodium dodecyl sulfate–polyacrylamide gel electrophoresis (SDS–PAGE) in the absence of 2-mercaptoethanol.[1,14]

---

[28] J. F. Soler and R. B. Harris, *Int. J. Pept. Protein Res.* **32,** 35 (1988).
[29] R. L. Wolz, R. B. Harris, and J. S. Bond, *Biochemistry* **30,** 8488 (1991).

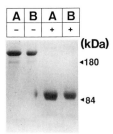

FIG. 1. Analysis by SDS–PAGE of meprin A and B in the absence and presence of 2-mercaptoethanol. Purified preparations (5 μg) of papain-solubilized meprin A (lanes 1 and 3) and meprin B (lanes 2 and 4) were boiled for 3 min with 1% (w/v) SDS in the absence (−) or presence (+) of 2-mercaptoethanol. Samples were subjected to SDS–PAGE on a 9% polyacrylamide gel at constant current of 60 mA. Gels were stained with Coomassie blue. Data from M. Z. Kounnas, R. L. Wolz, C. M. Gorbea, and J. S. Bond, *J. Biol. Chem.* **266,** 17350 (1991).

In the presence of reducing agents the subunit molecular weights appeared to be 85,000 to 90,000 (Fig. 1). Thus, it was proposed that meprin A and B are oligomers, probably tetramers. Studies on meprin A, using a variety of enzyme forms (protease-solubilized, detergent-solubilized, membrane-associated, or deglycosylated) and separation systems (native gel electrophoresis, gradient gel electrophoresis, or analytical centrifugation), indicate that the enzyme is composed of disulfide-linked dimers.[30] The dimers associate noncovalently to form tetramers and sometimes higher order oligomers in solution.

There is evidence that rat and human meprin A are also oligomers containing disulfide-linked dimers.[19] The subunits of the papain-purified rat enzyme are 82 and 72 kDa as determined by mobility on SDS–PAGE. The human meprin subunits have been estimated to be 100 kDa.[11]

*Carbohydrate*

Studies of trypsin- or papain-solubilized meprin A and B indicated that the subunits contain approximately 20% N-linked carbohydrate on the basis of mobility on SDS–PAGE before and after treatment with endoglycosidase F.[1,15] Chemical deglycosylation studies of kidney brush border membranes indicate that approximately 30% carbohydrate is associated with the subunits. On deglycosylation of membrane-bound α subunits, there is a decrease in molecular mass from 93 to 67 kDa; for β, there is a decrease from 115 to 79 kDa.[31] There are 10 potential N-glycosylation sites in the

[30] P. Marchand, J. Tang, and J. S. Bond, *J. Biol. Chem.* **269,** 15388 (1994).
[31] P. Marchand and J. S. Bond, unpublished work (1994).

α subunit, and at least 3 of those are glycosylated.[2] There are 9 potential N-glycosylation sites in the β subunit.[3]

Lectin blotting experiments indicate that both subunits contain high-mannose and/or complex biantennary oligosaccharides. The α subunit appears to contain tri- and tetraantennary N-linked oligosaccharides, whereas the β subunit seems to lack these more complex sugars, indicating that the posttranslational glycosylation of the subunits differs. No sialic acid has been found to be associated with either subunit. Chemical analyses of mouse meprin A indicate the presence of N-acetylglucosamine, mannose, galactose, fucose, and N-acetylgalactosamine.[30]

In addition to differences in oligosaccharides between the α and β subunits, there are sex-linked differences in N-linked glycosylation of meprin A from adult male and female mice.[32] These appear to be due to posttranslational modifications that are affected by estrogens.

*Isoelectric Point*

When trypsin-solubilized meprin A is subjected to chromatofocusing, the enzyme elutes between pH 4 and 5.[33] When the purified protein is subjected to isoelectric focusing, it separates into several evenly spaced bands in the range of pH 4 to 5. Because the subunits have no detectable sialic acid, the charge differences are likely due to modifications of amino acids or sugars, such as phosphorylation, sulfation, deamidation, or methylation.

*Metal Requirements*

One of the first properties of meprin to be recognized was its inhibition by metal-chelating agents such as EDTA and 1,10-phenanthroline.[14] No inhibition of meprin was observed with a variety of inhibitors of the other mechanistic classes of peptidases. Thus, meprin was characterized as a metalloproteinase. Subsequent analyses by atomic absorption spectroscopy revealed the presence of 1.1 mol zinc and 2.75 mol calcium per mole of meprin A subunit at the subunit molecular weight of 85,000.[33] Further confirmation that meprins A and B are metalloproteinases was found when the amino acid sequences were deduced from the respective cDNAs.[2,3,9,10,16] Within the proteinase domains, all of the residues which have been identified by X-ray crystallography as zinc-coordinating residues in astacin[34] are

[32] S. T. Stroupe, S. S. Craig, C. M. Gorbea, and J. S. Bond, *Am. J. Physiol.* **261**, E354 (1991).

[33] P. E. Butler, M. J. McKay, and J. S. Bond, *Biochem. J.* **241**, 229 (1987).

[34] W. Bode, F.-X. Gomis-Rüth, R. Huber, R. Zwilling, and W. Stöcker, *Nature (London)* **358**, 164 (1992).

strictly conserved in meprin as well as other members of the astacin family. Thus, inhibition by chelating agents, association of zinc with the protein, and sequence homology with a zinc binding site all implicate zinc as a requirement for meprin activity.

The requirement of metal for meprin activity has been demonstrated directly by showing that stoichiometric amounts of $Zn^{2+}$ or $Co^{2+}$ can reactivate the inactive apoenzyme.[22] In contrast to the case of astacin,[35] $Cu^{2+}$ cannot reactivate apomeprin.

*Primary Sequence and Domain Structure Deduced from cDNAs*

Mouse, rat, and human meprin subunits have been cloned and sequenced.[2,3,9–12] The primary translation products for mouse and rat α subunits are composed of 760 and 748 amino acids, respectively. There is an 86.8% identity between rat and mouse α subunits. Mouse and rat β subunits both consist of 704 amino acids and are 89% identical. The approximately 10% difference between rat and mouse for a given subunit is likely due to interspecies variation. Alignment of the mouse α and β subunits, however, reveals an overall identity in the amino acid sequence of only 42%. Thus, the apparent gene duplication which led to the two subunit types likely occurred before the speciation. Interestingly, there is very little identity between subunits in the N-terminal amino acids (approximately 40 residues) and in the approximately 50 C-terminal amino acids. In addition, there is a segment of 56 amino acids (residues 628 to 683 of the α subunit) that is missing in the β subunit.

The domain structures of the α and β subunits are similar (Fig. 2). Both contain a hydrophobic sequence near the N terminus which most likely corresponds to a signal peptide that is not present in the mature protein. A prosequence follows. For the mouse, the prosequence is not present in the mature membrane-bound α subunit, but it is present in the mature form of the β subunit. The evidence indicates that the difference in N-terminal processing of the subunits accounts for the high proteolytic activity of meprin A compared to meprin B in the kidney. When the prosequence is removed by trypsin treatment of β subunits, that subunit becomes active toward proteins such as azocasein. Increased proteolytic activity owing to removal of the prosequence is consistent with the information obtained from the crystal structure of astacin. In astacin, the N-terminal alanine is buried in a water-filled cavity, and its amino group forms a water-mediated salt bridge with Glu-103 adjacent to the active site (see below). Removal of the prosequence from meprin would be necessary to allow insertion of

---

[35] W. Stöcker, R. L. Wolz, R. Zwilling, D. J. Strydom, and D. S. Auld, *Biochemistry* **27**, 5026 (1988).

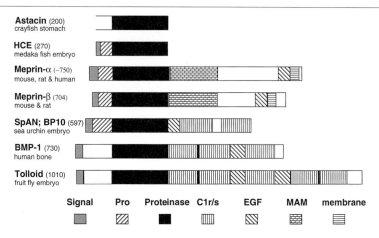

FIG. 2. Domain structure of astacin family members. The values in parentheses next to the names are numbers of amino acids for the complete primary translation products. For astacin, however, the number represents the amino acid content of the mature protein. References for the sequences are as follows: astacin, K. Titani, H.-J. Torff, S. Hormel, S. Kumar, K. A. Walsh, J. Rödl, H. Neurath, and R. Zwilling, *Biochemistry* **26**, 222 (1987); HCE (high choriolytic enzyme), S. Yasumasu, K. Yamada, K. Akasaka, K. Mitsunaga, I. Iuchi, H. Shimada, and K. Yamagami, *Dev. Biol.* **153**, 250 (1992); meprin α (mouse), W. Jiang, C. M. Gorbea, A. Flannery, R. J. Beynon, G. A. Grant, and J. S. Bond, *J. Biol. Chem.* **267**, 9185 (1992); meprin α (rat), D. Corbeil, F. Gaudoux, S. Wainwright, J. Ingram, A. J. Kenny, G. Boileau, and P. Crine, *FEBS Lett.* **309**, 203 (1992); meprin α (human), E. Dumermuth, J. A. Eldering, J. Grünberg, W. Jiang, and E. E. Sterchi, *FEBS Lett.* **335**, 367 (1993); meprin β (mouse), C. M. Gorbea, P. Marchand, W. Jiang, N. G. Copeland, D. J. Gilbert, N. A. Jenkins, and J. S. Bond, *J. Biol. Chem.* **268**, 21035 (1993); meprin β (rat), G. D. Johnson and L. B. Hersh, *J. Biol. Chem.* **267**, 13505 (1992); erratum *J. Biol. Chem.* **268**, 17647 (1993); BP10 (blastula protease-10), T. Lepage, C. Ghiglione, and C. Gache, *Development (Cambridge, UK)* **114**, 147 (1992); SpAN, S. D. Reynolds, L. M. Angerer, J. Palis, A. Nasir, and R. C. Angerer, *Development (Cambridge, UK)* **114**, 769 (1992); BMP-1 (bone morphogenetic protein-1), J. M. Wozney, V. Rosen, A. J. Celeste, L. M. Mitstock, M. J. Whitters, R. W. Kriz, R. M. Hewick, and E. A. Wang, *Science* **242**, 1528 (1988); *tolloid*, M. J. Shimell, E. L. Ferguson, S. R. Childs, and M. B. O'Connor, *Cell (Cambridge, Mass.)* **67**, 469 (1991). Abbreviations for the domains are: Signal, putative signal sequence; Pro, pro-peptide; C1r/s, complement proteins C1r and C1s; EGF, epidermal growth factor-like; MAM, receptor adhesive domain (G. Beckmann and P. Bork, *TIBS* **18**, 40 (1993); membrane, putative membrane-spanning region.

the mature N terminus (Asn in the α and β subunits of meprin) into the cavity and allow full activity. It should be noted, however, that this insertion is not an absolute requirement for function as meprin B, which still includes the prosequence, does have some proteolytic activities (Table II).

The proteinase or "astacin" domain of the α and β subunits contains about 200 amino acids. The sequences include, as all astacin family mem-

TABLE II
ACTION OF MEPRIN A AND B ON SUBSTRATES AND EFFECTS OF TRYPSIN TREATMENT OF ENZYMES[a]

| Enzyme | Trypsin treatment | Azocasein (units/mg) | Insulin B ($\mu M$/min/mg) | Nitrobradykinin ($\mu M$/min/mg) | YLVC(SO$_3^-$)GERG ($\mu M$/min/mg) |
|---|---|---|---|---|---|
| Meprin A | 0 | 4240 (100) | 0.73 (100) | 4.54 (100) | 0.095 (100) |
| Meprin A | + | 6590 (155) | 1.38 (189) | 4.16 (92) | 0.095 (100) |
| Ratio | 0/+ | 1.55 | 1.89 | 0.92 | 1.0 |
| Meprin B | 0 | 396 (9) | 0.09 (12) | <0.01 (0) | 0.467 (492) |
| Meprin B | + | 2230 (53) | 1.82 (249) | <0.01 (0) | 0.367 (386) |
| Ratio | 0/+ | 5.6 | 20.2 | — | 0.79 |

[a] Data from M. Z. Kounnas, R. L. Wolz, C. M. Gorbea, and J. S. Bond, *J. Biol. Chem.* **266,** 17350 (1991). Values in parentheses are percent activity relative to untreated meprin A. For the trypsin treatment, papain-purified meprin A or B (5–10 μg) was incubated with (+) or without (0) trypsin (0.5–1 μg) in 25 μl of 20 m$M$ Tris-HCl buffer, pH 7.5, for 1 hr at 37°. A 2-fold excess of soybean trypsin inhibitor was added. Assays with azocasein and nitrobradykinin were performed as described in the text. Hydrolysis rates for insulin B chain and YLVC(SO$_3^-$)GERG were determined by HPLC. C(SO$_3^-$) represents cysteic acid.

bers, a signature sequence of HEXXHXXGFXHEXXRXDRD.[16,36,37] The involvement of those residues in zinc binding and catalysis is discussed below.

Following the proteinase domain in meprin subunits is a sequence referred to as the MAM domain.[38] This sequence of about 170 amino acids has been proposed to act as an adhesion domain of cell-surface proteins. An epidermal growth factor-like domain is present near the C-terminal end of meprin subunit translation products. This is followed by a hydrophobic stretch of amino acids and a short hydrophilic tail of only 6 amino acids for $\alpha$ and 26 for $\beta$.

*Tertiary Structure of Proteinase Domain of Meprin*

The proteinase domains of the $\alpha$ and $\beta$ subunits of meprin have amino acid sequences which are approximately 30% identical to that of astacin. The active-site residues and disulfide bonds in the proteinase domains of meprin $\alpha$ and $\beta$ subunits are completely conserved in relation to astacin, and only eight single amino acid insertions and deletions are necessary for alignment of the remainder of the primary sequences.[16,39] Thus, on the

[36] W. Jiang and J. S. Bond, *FEBS Lett.* **312,** 110 (1992).
[37] W. Bode, F.-X. Gomis-Rüth, and W. Stöcker, *FEBS Lett.* **331,** 134 (1993).
[38] G. Beckman and P. Bork, *Trends Biochem. Sci.* **18,** 40 (1993).
[39] W. Stöcker, F.-X. Gomis-Rüth, W. Bode, and R. Zwilling, *Eur. J. Biochem.* **214,** 215 (1993).

basis of the X-ray crystallographic structure of astacin,[34,40] it was possible to construct computationally a three-dimensional model of the proteinase domain of the $\alpha$ subunit of meprin.[39]

The proteinase domain of meprin $\alpha$ is kidney-shaped with a deep active-site cleft separating the N- and C-terminal halves of the molecule (see [19] in this volume). The C-terminal "bottom" half has little defined secondary structure except for a three-turn $\alpha$ helix just before a disulfide bond ($Cys^{42}$-$Cys^{198}$, astacin numbering) which ends the domain. The N-terminal "top" half is organized with regard to secondary structure into a five-stranded $\beta$ sheet curving around the top and front surfaces of the molecule (as oriented with the cleft horizontally across the front) and two $\alpha$ helices. The second $\alpha$ helix forms the top of the active site and includes the zinc binding motif $His^{92}$-Glu-Ile-Leu-$His^{96}$ (astacin numbering). The catalytically essential zinc is deep in the center of the cleft and, in contrast to the expected tetracoordinated zinc as in thermolysin or carboxypeptidase, is pentacoordinated. The coordination geometry is a trigonal bipyramid involving the side chains of three histidine residues (His-92, His-96, and His-102), a tyrosine (Tyr-149), and a water molecule. Homologous histidine residues have also been proposed to bind zinc in the serralysin, adamalysin (snake venom), and matrixin families of metalloendopeptidases (see [13], this volume).[36,37] The water acts as a bridge between the zinc and the carboxylate side chain of a glutamic acid (Glu-93) and is thereby activated as a likely catalytic base in the reaction mechanism. There are some localized differences in the structures of astacin and meprin-$\alpha$ which are discussed below in relation to kinetic differences between the two enzymes.

### Kinetic Characteristics

#### Meprin A and B Endopeptidase Activity and Effects of Trypsin Treatment

The peptide bond specificity of mouse meprins A and B was compared using the oxidized insulin B chain as a substrate. Meprin A cleaves the insulin B chain at 10 sites (3 major and 7 minor),[33] whereas meprin B cleaves at only 4 sites (2 major and 2 minor).[1] The major sites of cleavage for meprin A are $Gly^{20}$-$Glu^{21}$, $Phe^{24}$-$Phe^{25}$, and $Phe^{25}$-$Tyr^{26}$. For meprin B, a unique site of cleavage is $Cys(SO_3^-)^{19}$-$Gly^{20}$. Three common sites of hydrolysis for meprin A and B are $His^5$-$Leu^6$, $Leu^6$-$Cys(SO_3^-)^7$, and $Ala^{14}$-$Leu^{15}$.

The activities of meprin A and B against various substrates and the effects of limited trypsin treatment on the activities of the two enzymes are different (Table II).[1] Papain-solubilized preparations of both meprin A and B hydrolyze both azocasein and insulin B chain. Prior to trypsin treatment, however, meprin A has a 5- to 10-fold higher specific activity against those substrates than does meprin B. Trypsin treatment activates meprin A 1.5- to 2-fold with azocasein and insulin B as substrates, whereas meprin B is activated 5- to 20-fold, giving it specific activities comparable to those of meprin A. The nonapeptide nitrobradykinin is hydrolyzed readily by meprin A but not detectably by meprin B. Trypsin activation has no effect on either enzyme with nitrobradykinin. The octapeptide Tyr-Leu-Val-Cys($SO_3^-$)-Gly-Glu-Arg-Gly was prepared as a substrate for meprin B based on a unique cleavage site within the insulin B chain. The synthetic peptide was hydrolyzed by both meprin A and B, but, in contrast with other substrates, the specific activity was 4- to 5-fold higher with meprin B. Trypsin treatment did not affect the activity of either enzyme on the octapeptide. Thus, both meprin A and B are activated against larger polypeptides, but not against the smaller ones tested.

*Meprin A Hydrolysis of Biologically Active Peptides*

A variety of biologically active peptides are cleaved by meprin A. In addition to the peptides listed in Table III, the rat enzyme has been reported to cleave [Met[5]]enkephalin-Arg[6]-Phe[7], substance P, neurotensin,[8] $\alpha$-atrial natriuretic peptide (human),[27] human transforming growth factor-$\alpha$,[41] endothelin I, and big endothelin I.[42]

The interaction of mouse meprin A with the peptides in Table III was characterized by two methods: (1) binding was assessed by inhibition of nitrobradykinin hydrolysis and (2) hydrolysis rates were determined directly by high-performance liquid chromatography (HPLC). A wide range of binding and catalytic constants is observed. $\alpha$-Melanocyte-stimulating hormone has the highest $k_{cat}/K_i$ ratio (equivalent to the specificity ratio) of $10^6$ $M^{-1}$ $sec^{-1}$, and neurotensin and luteinizing hormone releasing hormone (LHRH; gonadotropin) are also good substrates for meprin A. Bradykinin has the highest specificity ratio of the peptides with a single cleavage site and was used as the basis for the kinetic studies described below.

One common feature among most peptides cleaved by meprin A is the presence of proline. In particular, a proline in the P2' or P3' position seems to be important in directing the enzyme to the scissile bond. In bradykinin, for example, replacement of Pro-7 with any other residue results in multiple cleavages or prevents cleavage.[43]

[41] Y. Choudry and A. J. Kenny, *Biochem. J.* **280**, 57 (1991).
[42] T. Yamaguchi, H. Kido, and N. Katunuma, *Eur. J. Biochem.* **204**, 547 (1992).
[43] A. J. Birket, J. F. Soler, R. L. Wolz, J. S. Bond, J. Wiseman, J. Berman, and R. B. Harris, *Anal. Biochem.* **196**, 137 (1991).

TABLE III
MEPRIN A ACTION ON BIOLOGICALLY ACTIVE PEPTIDES[a]

| Peptide and sequence | $K_i$ ($\mu M$) | $k_{cat}$ (sec$^{-1}$) | $k_{cat}/K_i$ ($M^{-1}$ sec$^{-1}$) |
|---|---|---|---|
| α-Melanocyte-stimulating hormone, acetyl-Ser-Tyr-Ser⇓Met-Glu-His-Phe-Arg-Trp-Gly⇓Lys-Pro-Val | 36 | 34.5[b] | 9.6 × 10^5 |
| Neurotensin, pyroGlu-Leu-Tyr-Glu⇓Lys-Pro-Arg-Arg-Pro-Tyr-Ile-Leu | 68 | 7.6[b] | 1.1 × 10^5 |
| Luteinizing hormone-releasing hormone, pyroGlu-His-Trp⇓Ser⇓Tyr-Gly⇓Leu-Arg-Pro-Gly-NH₂ | 292 | 30.4[b] | 1.0 × 10^5 |
| Bradykinin, Arg-Pro-Pro-Gly-Phe⇓Ser-Pro-Phe-Arg | 425 | 21.8[b] | 5.1 × 10^4 |
| Angiotensin I, Asp-Arg-Val-Tyr⇓Ile-His-Pro-Phe-His⇓Leu | 552 | 0.64[c] | 1.2 × 10^3 |
| [desAsp]Angiotensin I, Arg-Val-Tyr⇓Ile-His-Pro-Phe-His⇓Leu | 246 | 0.14[c] | 5.6 × 10^2 |
| Angiotensin II, Asp-Arg-Val-Tyr⇓Ile-His-Pro-Phe | 740 | 0.15[c] | 2.0 × 10^2 |
| Angiotensin III, Arg-Val-Tyr⇓Ile-His-Pro-Phe | 477 | 0.10[c] | 2.0 × 10^2 |

[a] Data from R. L. Wolz, R. B. Harris, and J. S. Bond, *Biochemistry* **30**, 8488 (1991). $K_i$ values were determined from inhibition of nitrobradykinin hydrolysis. Initial velocities for peptide hydrolysis were determined by HPLC, then $k_{cat}$ was calculated by substituting $K_i$ for $K_m$ in the Michaelis–Menten equation. ⇓, Cleavage site.
[b] Rate constant for substrate disappearance.
[c] Rate constant for product appearance.

*Mouse Meprin A Substrate Specificity and Comparison with Astacin*

A series of bradykinin analogs was used to map kinetically the active site of meprin A (Table IV)[29] and to compare kinetically mouse meprin A with the homolog astacin.[21] Most of the peptides were cleaved by both enzymes at the site corresponding to the bond between Phe and Ser in native bradykinin, and hydrolyses followed simple Michaelis–Menten kinetics. Peptide I was the best substrate for both meprin A and astacin, and the other fluorogenic substrates had similar kinetics for both enzymes. Variations in the P1 to P3 positions affected both $K_m$ and $k_{cat}$ for both enzymes in similar ways, and, for any given peptide, the $k_{cat}/K_m$ values for the two enzymes were within a factor of 6. Thus, meprin A and astacin, two of the more distantly related enzymes in the astacin family,[16] share some common substrates and kinetic properties. This indicates that some of the peptides tested could also be substrates for the other putative proteases in the family.

Significant differences between meprin A and astacin were seen in peptides X–XII with P1′ residues Arg, Phe, and Lys, respectively. All three peptides were reasonably good substrates for meprin A, but cleavage by astacin was barely detectable. For those peptides, binding to astacin was assessed by competitive inhibition of nitrobradykinin hydrolysis, and it was found that the binding affinities were nevertheless in the same relatively narrow range (100–300 $\mu M$) as all the hydrolyzable peptides in this series.

Thus, for both meprin A and astacin, variations in the P1′ position affect mainly the rate of hydrolysis and not the equilibrium binding. This indicates a minor role for the S1′ subsite during the initial enzyme–substrate interaction. These initial binding affinities are not, however, a measure of the peptide–enzyme binding during the transition state. The effects of P1′ substitutions on the catalytic rate constant indicates that the S1′ subsite could play an important role in transition state binding or product release. The kinetic differences between meprin A and astacin indicate that this site is likely larger and less hydrophobic in meprin A than in astacin.

*Inhibitors of Meprin A and B*

The inhibitor profiles for meprin A and B are very similar.[1,15] Neither enzyme is affected by inhibitors of other mechanistic classes of proteinases including phenylmethylsulfonyl fluoride (PMSF), 3,4-dichloroisocoumarin, soybean trypsin inhibitor, pepstatin, and iodoacetic acid. Both meprin A and B can be totally inhibited by metal-chelating agents such as EDTA and 1,10-phenanthroline. Other general inhibitors of metalloproteinases which inhibit both meprin A and B include thiol-containing compounds, such as 2-mercaptoethanol and cysteine. Some inhibitors of other metalloproteinases which do not inhibit either meprin A or B include phosphorami-

TABLE IV
MEPRIN A AND ASTACIN HYDROLYSIS OF FLUOROGENIC AND CHROMOGENIC BRADYKININ ANALOGS[a]

| Peptide[b] | Site | | | | | | | | | Enzyme | $K_m$ ($\mu M$) | $k_{cat}$ (sec[-1]) | $k_{cat}/K_m$ ($M^{-1}$ sec[-1]) |
| | P5 | P4 | P3 | P2 | P1 | P1' | P2' | P3' | P4' | | | | |
|---|---|---|---|---|---|---|---|---|---|---|---|---|---|
| I | 2ABz | Arg | Pro | Ile | Phe ⇓←↑ | Ser | Pro | Npa | Arg | Mep | 296 | 132.7 | $4.48 \times 10^5$ |
| | | | | | | | | | | Ast | 29 | 78.4 | $2.74 \times 10^6$ |
| II | 2ABz | Arg | Hyp | Gly | Phe ⇓←↑ | Ser | Pro | Npa | Arg | Mep | 183 | 26.7 | $1.46 \times 10^5$ |
| | | | | | | | | | | Ast | 44 | 15.1 | $3.42 \times 10^5$ |
| III | 2ABz | Arg | Gly | Pro | Phe ⇓← | Ser | Pro | Npa | Arg | Mep | 220 | 1.1 | $4.81 \times 10^3$ |
| | | | | | | | | | | Ast | 327 | 7.1 | $2.18 \times 10^4$ |
| IV | 2ABz | Arg | Pro | Gly | Ala ⇓ 1↑ ←2 | Ser | Pro | Npa | Arg | Mep | 1380 | 98.5 | $6.92 \times 10^4$ |
| | | | | | | | | | | Ast | 154 | 44.4[d] | $2.88 \times 10^5$ |
| V | 2ABz | Arg | Pro | Gly | Glu ⇓←↑ | Ser | Pro | Npa | Arg | Mep | 1220 | 4.9 | $4.05 \times 10^3$ |
| | | | | | | | | | | Ast | 954 | 2.8 | $2.92 \times 10^3$ |
| VI | 2ABz | Arg | Pro | Gly3⇓ Lys5⇓ | Ser | Pro | Npa | Arg | | Mep | 402 | 2.4[d] | $6.07 \times 10^3$ |
| | | | | | | | | | | Ast | 281 | 3.6 | $1.29 \times 10^4$ |
| VII | 2ABz | Arg | Pro | Gly1⇓ Leu5⇓ | Ser | Pro | Npa | Arg | | Mep | 2460 | 12.0[d] | $4.87 \times 10^3$ |
| | | | | | | | | | | Ast | 867 | 14.3 | $1.65 \times 10^4$ |
| VIII | Arg | Pro | Pro | Gly | Npa ↑ | Ser | Pro | Phe | Arg | Mep | 290 | 40.9 | $1.41 \times 10^5$ |
| | | | | | | | | | | Ast | 85 | 78.7 | $9.22 \times 10^5$ |
| IX | Arg | Pro | Pro | Gly | Npa ↑ | Ala | Pro | Phe | Arg | Mep | 331 | 51.5 | $1.56 \times 10^5$ |
| | | | | | | | | | | Ast | 306 | 834 | $2.73 \times 10^6$ |
| X | Arg | Pro | Pro | Gly | Npa ↑ | Arg | Pro | Phe | Arg | Mep | 174 | 19.6 | $1.13 \times 10^5$ |
| | | | | | | | | | | Ast | 209[c] | 0.002 | $1.0 \times 10^1$ |
| XI | Arg | Pro | Pro | Gly | Npa ↑ | Phe | Pro | Phe | Arg | Mep | 226 | 7.6 | $3.34 \times 10^4$ |
| | | | | | | | | | | Ast | 194[c] | 0.004 | $2.0 \times 10^1$ |
| XII | Arg | Pro | Pro | Gly | Npa ↑ | Lys | Pro | Phe | Arg | Mep | 182 | 5.0 | $2.75 \times 10^4$ |
| | | | | | | | | | | Ast | 116[c] | 0.001 | 6.4 |
| XIII | Arg | Pro | Pro | Gly | Npa | Glu | Pro | Phe | Arg | Mep | 339[c] | <0.001 | <3 |
| | | | | | | | | | | Ast | 900[c] | 0.001 | 1.4 |

[a] Peptides I–VII are fluorogenic and VIII–XIII are chromogenic. Meprin A data are from R. L. Wolz, R. B. Harris, and J. S. Bond, *Biochemistry* 30, 8488 (1991), and astacin data are from R. L. Wolz, *Arch. Biochem. Biophys.* 310, 144 (1994).
[b] Abbreviations: 2ABz, 2-aminobenzoyl; Hyp, hydroxyproline; Npa, 4-nitrophenylalanine. ⇓, Meprin A cleavage; ↑, astacin cleavage. Numbers preceding the arrows indicate proportion of cleavage.
[c] ID$_{50}$ value.
[d]

don (an inhibitor of thermolysin and neprilysin) and captopril (an inhibitor of ACE).

Meprin A is inhibited by amino acid hydroxamates.[29] The potency of inhibition is dependent on the chemical nature of the side chains. Amino acid hydroxamates with aromatic side chains are the most effective followed, in order of decreasing affinity, by those with side chains containing sulfur, basic groups, aliphatic groups, and acidic groups. By comparison, affinities of amino acid hydroxamates for astacin follow a somewhat different order: aromatic, sulfur containing, acidic, aliphatic, and basic.[44]

Hydroxamate derivatives of peptides have also been characterized as to their affinity and mechanism of inhibition of meprin A and astacin (Table V). Peptide XIV was designed with a structure similar to the N-terminal cleavage product of bradykinin and has been characterized as a predominantly noncompetitive inhibitor of meprin A cleavage of nitrobradykinin.[29] Astacin is also inhibited by that peptide in a predominantly noncompetitive manner, but with an 8-fold weaker affinity compared to meprin A. Peptide XV incorporates the metal-complexing hydroxamate at the N terminus and includes amino acids patterned after the sequence of the C-terminal half of bradykinin. That peptide hydroxamate inhibits both enzymes by competitive mechanisms but has a 9-fold stronger affinity for astacin than for meprin A.

Preliminary reports[41] indicated that actinonin (peptide XVI), a naturally occurring peptide hydroxamate, is a very potent inhibitor of rat meprin A ($ID_{50}$ $3.5 \times 10^{-8}$ $M$). Actinonin is also a potent inhibitor of mouse meprin A hydrolysis of nitrobradykinin with $K_{is}$ and $K_{ii}$ values of 0.135 and 1.57 $\mu M$, respectively, indicating a predominantly competitive mechanism. Actinonin inhibition of astacin is essentially purely competitive but is 1000-fold weaker than its inhibition of meprin A, thus revealing another significant difference between the enzymes.

The 1000-fold difference in inhibition of meprin A and astacin by actinonin is likely due to differences in the size or shape of subsites in the respective extended active sites of the proteins. Actinonin has a bulky pentyl group adjacent to the metal-ligating hydroxamate. Hence if the pentyl group were situated in the transition state S1′ subsite discussed above for substrate peptides X–XII, the differing affinities could be explained.

In the three-dimensional model of meprin $\alpha$ subunit, the space on the right-hand side of the active-site zinc (as oriented with the N-terminal lobe on top and the active-site cleft in front) is bounded by a glycine residue (Gly-176). In the astacin crystal structure, the homologous space is partially filled by replacement of the glycine with a proline (Pro-176) and the insertion of a tyrosine residue (Tyr-177) which meprin lacks. By correlating the

[44] R. L. Wolz, C. Zeggaf, W. Stöcker, and R. Zwilling, *Arch. Biochem. Biophys.* **281,** 275 (1990).

TABLE V
Peptide Hydroxamate Inhibitors of Meprin A and Astacin[a]

| Inhibitor | Structure | Enzyme | $K_{is}$ $(\mu M)^b$ | $K_{ii}$ $(\mu M)^b$ | Mechanism |
|---|---|---|---|---|---|
| XIV | Acetyl-RPGY-NHOH | Meprin A | 2.35 | 8.42 | NC/C |
| | | Astacin | 18.9 | 122 | NC/C |

| | | | | | |
|---|---|---|---|---|---|
| XV | HONH-succinyl-PFR | Meprin A | 146 | 1750 | C/NC |
| | | Astacin | 16.1 | 850 | C |

| | | | | | |
|---|---|---|---|---|---|
| XVI | Actinonin | Meprin A | 0.135 | 1.57 | C/NC |
| | | Astacin | 130 | 9500 | C |

[a] Data from R. L. Wolz, *Arch. Biochem. Biophys.* **310**, 144 (1994).
[b] $K_{is}$ is the binding constant in the absence of substrate; $K_{ii}$ is the binding constant in the presence of substrate.
[c] Mechanisms: NC/C, noncompetitive predominant over competitive; C, competitive; C/NC, competitive predominant over noncompetitive.

above kinetic data with these structures, this space can be identified as the likely S1' subsite.

## Acknowledgment

This work was supported by National Institutes of Health Grant DK19691.

# [21] Snake Venom Metalloendopeptidases: Reprolysins

By JÓN B. BJARNASON and JAY W. FOX

## Introduction

Throughout the course of recorded history of human civilization mankind has shown a unique fascination for venomous snakes, both admiring and detesting the awesome powers residing within their serpentine embrace. The destructive effects of snake venoms on living organisms have been obvious to all, but their potential healing powers have long been suspected by some. Thus ancient emblems of some medical professions frequently have a snake in their design drawn from the caduceus, a winged staff with two serpents entwined around it, carried by Hermes of Greek mythology. In more recent times of modern scientific investigation, the study of snake venoms has yielded a vast body of important information on biological systems and insights into medical problems. Indeed, some venom components are thought to hold promise as agents in treatment of diseases and medical complications, or to hold the key to the design of pharmaceuticals. One of the most pronounced effects of snake envenomation is the occurrence of hemorrhage, localized or evasive throughout a substantial area of the involved extremity. Since the purification of five zinc-containing hemorrhagic metalloproteinases from the venom of *Crotalus atrox*,[1] snake venom-induced hemorrhage has increasingly been found to be associated with metal-dependent proteinases.[2-4]

In this chapter we review one of the most destructive agents of snake venoms, namely, the snake venom metalloproteinases (SVMPs). We choose to refer to these as the reprolysins because of their reptilian origin as well as their relationship with mammalian reproductive proteins (MRPs; ADAMs, disintegrin and metalloproteinase proteins or fertilins), which we

[1] J. B. Bjarnason and A. T. Tu, *Biochemistry* **17**, 3395 (1978).
[2] J. B. Bjarnason and J. W. Fox, *J. Toxicol. Toxin Rev.* **7**, 121 (1988, 1989).
[3] R. M. Kini and H. J. Evans, *Toxicon* **30**, 265 (1992).
[4] J. B. Bjarnason and J. W. Fox, *Pharmacol. Ther.* **62**, 325 (1994).

TABLE I
Reprolysins with Trivial Names and Enzyme Commission Numbers[a]

| Snake venom | Toxin | Trivial name | EC number |
|---|---|---|---|
| Crotalus atrox | Hemorrhagic toxin a | Atrolysin A | 3.4.24.1 |
| Agkistrodon piscivorus leucostoma | Leucostoma peptidase A | Leucolysin | 3.4.24.6 |
| Crotalus atrox | Hemorrhagic toxin b | Atrolysin B | 3.4.24.41 |
| Crotalus atrox | Hemorrhagic toxin c and d | Atrolysin C | 3.4.24.42 |
| Crotalus atrox | Atroxase | Atroxase | 3.4.24.43 |
| Crotalus atrox | Hemorrhagic toxin e | Atrolysin E | 3.4.24.44 |
| Crotalus atrox | Hemorrhagic toxin f | Atrolysin F | 3.4.24.45 |
| Crotalus adamanteus | Proteinase I and II | Adamalysin | 3.4.24.46 |
| Crotalus horridus horridus | Hemorrhagic protease IV | Horrilysin | 3.4.24.47 |
| Crotalus ruber ruber | Hemorrhagic toxin 2 | Ruberlysin | 3.4.24.48 |
| Bothrops jararaca | Bothropasin | Bothropasin | 3.4.24.49 |
| Bothrops jararaca | Jararaca protease | Bothrolysin | 3.4.24.50 |
| Ophiophagus hannah | Ophiophagus metalloendopeptidase | Ophiolysin | 3.4.24.51 |
| Trimeresurus flavoviridis | HR-1A (and 1B) | Trimerelysin I | 3.4.24.52 |
| Trimeresurus flavoviridis | Proteinase H2 | Trimerelysin II | 3.4.24.53 |
| Trimeresurus mucrosquamatus | Mucrotoxin A | Mucrolysin | 3.4.24.54 |
| Vipera russelli russelli | RVV-X | Russellysin | 3.4.24.58 |

[a] Recommended by the International Union of Biochemistry and Molecular Biology.

also consider to be members of the reprolysin subfamily. The term reprolysin is short and convenient in use, rolls well off the tongue, and is not species associated. It seems particularly relevant to review the reprolysins at this time. The sequence of 14 SVMPs are now known,[4,5] their domain compositions are becoming understood, and the X-ray crystal structures of two of them, adamalysin and atrolysin C, have been analyzed.[6,7] Seventeen SVMPs have been assigned trivial names and EC numbers by the nomenclature committee of the International Union of Biochemistry and Molecular Biology (Table I).

The term reprolysin strictly applies only to 14 of the snake venom metalloproteinases, whose total sequences are known. These are homologs which define the reprolysin subfamily of the M12 family of metalloproteinases (see [13] in this volume). However, in this chapter and in [22], we shall for convenience and euphony use the term reprolysin for all the metalloproteinases from snake venoms, and indeed assume that they all belong to this subfamily. Various direct and indirect evidence seems to justify this assumption, in particular, characteristics such as molecular

[5] L. A. Hite, L.-G. Jia, J. B. Bjarnason, and J. W. Fox, Arch. Biochem. Biophys. 308, 182 (1994).
[6] F.-X. Gomis-Rüth, L. F. Kress, E. Meyer, and W. Bode, EMBO J. 12, 4151 (1993).
[7] D. Zhang, I. Botos, F.-X. Gomis-Rüth, R. Doll, C. Blood, F. G. Njorge, J. W. Fox, W. Bode, and E. Meyer, Proc. Acad. Sci. USA 91, 8447 (1994).

masses, substrate specificities, chemical and biological activities, amino acid compositions and partial sequences, as well as the logic of living systems. Each time a new sequence for the enzymes has been presented, it has fallen into the category of reprolysins. If, on the other hand, an unrelated sequence would emerge from any of the remaining unsequenced snake venom metalloendopeptidases, those enzymes should not be considered reprolysins. In [22] of this volume we give a more exhaustive description of the chemical, physical, and functional characteristics of the reprolysins which have been most extensively studied, namely, the atrolysins.

## Overview of Snake Venom Metalloproteinases: Reprolysins

To date there are reports of at least 102 reprolysins (snake venom metalloproteinases, SVMPs) from 36 species of snakes, of which 13 come from the venom of the western diamondback rattlesnake (*Crotalus atrox*) alone (Table II).[8-96] This is not to say that over 100 distinct and different reprolysins exist. On the contrary, they are probably far fewer in number,

[8] J. B. Bjarnason and J. W. Fox, *Biochim. Biophys. Acta* **911,** 356 (1987).

[9] T. Kurecki, M. S. Laskowski, and L. F. Kress, *J. Biol. Chem.* **253,** 8340 (1978).

[10] D. C. Gowda, C. M. Jackson, P. Hensley, and E. A. Davidson, *J. Biol. Chem.* **269,** 10644 (1994).

[10a] I. LeHueron, C. Wicker, P. Guilloteau, R. Toullec, and A. Puigserver, *Eur. J. Biochem.* **193,** 767 (1990).

[11] C. Ouyang and T. Huang, *Biochim. Biophys. Acta* **439,** 146 (1976).

[12] C. Ouyang and T. Huang, *Toxicon* **15,** 161 (1977).

[13] N. Mori, T. Nikai, and H. Sugihara, *Toxicon* **22,** 451 (1984).

[14] T. Nikai, H. Ishisaki, A. T. Tu, and H. Sugihara, *Comp. Biochem. Physiol., C: Comp. Pharmacol.* **C72,** 103 (1982).

[15] X. Xu, C. Wang, J. Liu, and Z. Lu, *Toxicon* **19,** 633 (1981).

[16] K. Imai, T. Nikai, H. Sugihara, and C. L. Ownby, *Int. J. Biochem.* **21,** 667 (1989).

[17] G. Oshima, S. Iwanaga, and T. Suzuki, *J. Biochem* (*Tokyo*) **64,** 215 (1968).

[18] T. Omori, S. Iwanaga, and T. Suzuki, *Toxicon* **2,** 1 (1964).

[19] G. Oshima, T. Omori-Satoh, S. Iwanaga, and T. Suzuki, *J. Biochem.* (*Tokyo*) **72,** 1483 (1972).

[20] M. Satake, Y. Murata, and T. Suzuki, *J. Biochem.* (*Tokyo*) **53,** 438 (1965).

[21] S. Iwanaga, T. Omori, G. Oshima, and T. Suzuki, *J. Biochem.* (*Tokyo*) **57,** 392 (1965).

[22] B. Oshima, Y. Matsuo, S. Fwanaga, and T. Suzuki, *J. Biochem.* (*Tokyo*) **64,** 227 (1968).

[23] G. Oshima, S. Iwanga, and T. Suzuki, *Biochim. Biophys. Acta* **250,** 416 (1971).

[24] N. K. Ahmed, K. D. Tennant, F. S. Markland, and J. P. Lacz, *Haemostasis* **20,** 147 (1990).

[25] J. B. Moran and C. R. Geren, *Biochim. Biophys. Acta* **659,** 161 (1981).

[26] A. M. Spiekerman, K. K. Fredericks, F. W. Wagner, and J. M. Prescott, *Biochim. Biophys. Acta* **293,** 464 (1973).

[27] F. S. Markland, *Thromb. Haemostasis* **65,** 438 (1991).

[28] C. Ouyang, L. Hwang, and T. F. Huang, *Toxicon* **21,** 25 (1983).

[29] S. J. Van der Walt and F. J. Joubert, *Toxicon* **9,** 153 (1971).

[30] D. J. Strydom, F. J. Joubert, and N. L. Howard, *Toxicon* **24,** 247 (1986).

[31] C. H. Lawrence and B. J. Morris, *Biochim. Biophys. Acta* **657,** 13 (1981).

[32] D. Mebs and F. Panholzer, *Toxicon* **20,** 509 (1982).

[33] F. Aragon-Ortiz and F. Gubensek, *Toxicon* **25,** 759 (1987).

[34] M. T. Assakura, A. P. Reichl, and F. R. Mandelbaum, *Toxicon* **24,** 943 (1986).
[35] M. M. Tanizaki, R. B. Zingoli, H. Kawazaki, S. Imajoh, S. Yamazaki, and K. Suzuki, *Toxicon* **27,** 747 (1989).
[36] F. R. Mandelbaum, A. P. Reichl, and M. T. Assakura, *Toxicon* **30,** 955 (1982).
[37] M. J. I. Paine, H. P. Desond, R. D. G. Theakston, and J. M. Crampton, *J. Biol. Chem.* **247,** 22869 (1992).
[38] M. Maruyama, M. Sugiki, E. Yoshida, M. Mihara, and N. Nakajima, *Toxicon* **30,** 853 (1992).
[39] M. T. Assakura, A. P. Reichl, M. C. A. Asperti, and F. R. Mandelbaum, *Toxicon* **23,** 691 (1985).
[40] A. P. Reichl and F. R. Mandelbaum, *Toxicon* **31,** 187 (1993).
[41] S. M. T. Serrano, C. A. M. Sampaio, and F. R. Mandelbaum, *Toxicon* **31,** 483 (1993).
[42] F. R. Mandelbaum, M. T. Assakura, and A. P. Reichl, *Toxicon* **22,** 193 (1984).
[43] E. Bando, T. Nikai, and H. Sugihara, *Int. J. Biochem.* **23,** 1193 (1991).
[44] G. Ponnudurai, M. C. M. Chung, and N.-H. Tan, *Toxicon* **31,** 997 (1993).
[45] E. Daoud, A. T. Tu, and M. F. El-Asmar, *Thromb. Res.* **42,** 55 (1986).
[46] E. Daoud, A. T. Tu, and M. F. El-Asmar, *Thromb. Res.* **41,** 791 (1986).
[47] T. Kurecki and M. S. Laskowski, *in* "Toxicon Supplement Number 1, Toxins, Animal, Plant and Microbiol," p. 311. Pergamon, Elmsford, New York, 1978.
[48] T. Kurecki and L. F. Kress, *Toxicon* **23,** 657 (1985).
[49] J. B. Bjarnason, D. Hamilton, and J. W. Fox, *Biol. Chem. Hoppe-Seyler* **369,** 121 (1988).
[50] J. W. Fox, R. Campbell, L. Beggerly, and J. B. Bjarnason, *Eur. J. Biochem.* **156,** 65 (1986).
[51] J. D. Shannon, E. N. Baramova, J. B. Bjarnason, and J. W. Fox, *J. Biol. Chem.* **264,** 11575 (1989).
[52] J. B. Bjarnason and J. W. Fox, *Biochemistry* **22,** 3770 (1983).
[53] L. A. Hite, J. D. Shannon, J. B. Bjarnason, and J. W. Fox, *Biochemistry* **31,** 6203 (1992).
[54] T. Nikai, N. Mori, M. Kishida, H. Sugihara, and A. T. Tu, *Arch. Biochem. Biophys.* **231,** 309 (1984).
[55] T. Nikai, N. Mori, M. Kishida, M. Tsuboi, and H. Sugihara, *Am. J. Trop. Med. Hyg.* **34,** 1167 (1985).
[56] T. Willis and A. T. Tu, *Biochemistry* **27,** 4769 (1988).
[57] B. V. Pandya and A. A. Budzynski, *Biochemistry* **23,** 460 (1984).
[58] B. S. Hong, *Toxicon* **20,** 535 (1982).
[59] T. Nikai, R. Kito, N. Mori, H. Sugihara, and A. T. Tu, *Comp. Biochem. Physiol., B: Comp. Biochem.* **76B,** 679 (1983).
[60] M. Kruzel and L. F. Kress, *Anal. Biochem.* **151,** 471 (1985).
[61] O. Molina, R. K. Seriel, M. Martinez, M. L. Sierra, A. Varela-Ramirez, and E. D. Rael, *Int. J. Biochem.* **22,** 253 (1990).
[62] D. J. Civello, H. L. Duong, and C. R. Geren, *Biochemistry* **22,** 749 (1983).
[63] D. J. Civello, J. B. Moran, and C. R. Geren, *Biochemistry* **22,** 755 (1983).
[64] G. A. Ramirez, P. L. Fletcher, and L. D. Possani, *Toxicon* **228,** 285 (1990).
[65] N. Mori, T. Nikai, H. Sugihara, and A. T. Tu, *Arch. Biochem. Biophys.* **253,** 108 (1987).
[66] H. Takeya, A. Onikura, T. Nikai, H. Sugiharar, and S. Iwanaga, *J. Biochem. (Tokyo)* **108,** 711 (1990).
[67] M. Martinez, E. D. Rael, and N. L. Maddux, *Toxicon* **28,** 685 (1990).
[68] Q. Li, T. R. Colberg, and C. L. Ownby, *Toxicon* **31,** 711 (1993).
[69] E. F. Sanchez, C. R. Diniz, and M. Richardson, *FEBS Lett.* **282,** 178 (1991).
[70] E. F. Sanchez, A. Magalhaes, and C. R. Diniz, *Toxicon* **25,** 611 (1987).
[71] T. Takahashi and A. Ohsaka, *Biochim. Biophys. Acta* **207,** 65 (1970).

perhaps fewer than 10. Some of the reprolysins shown in Table II are now known to be isozymes from the same species of snake. This is exemplified by HT-c and HT-d (atrolysin C) from *C. atrox* venom. Those enzymes have a difference of only one amino acid (one charge) out of a total of 203.[5,8] Proteinase I and proteinase II (adamalysin) from the venom of *Crotalus adamanteus*[6,9] are probably also examples of such isozymes. There also appear to exist more distantly related isozymes from the same species of snake. Thus HT-b (atrolysin B) is most likely a cationic isozyme of the anionic atrolysin C (HT-c/d) isozymes. They share identical cleavage sites on the insulin B chain (bonds 5, 10, 14, 16, 23, see Table II), and there is a 78% sequence identity between atrolysin B and C[5]; for comparison, the anionic and cationic forms of bovine trypsinogen share 65% sequence identity.[10a] Finally, homologous enzymes probably exist in most of the different

[72] T. Nikai, M. Niikawa, Y. Komori, S. Sekoguchi, and H. Sugihara, *Int. J. Biochem.* **19**, 221 (1987).

[73] H. Takeya, M. Arekawa, T. Miyata, S. Iwanaga, and T. Omori-Satoh, *J. Biochem. (Tokyo)* **106**, 151 (1989).

[74] Y. Yamakawa, T. Omori-Satoh, and S. Sadahiro, *Biochim. Biophys. Acta* **925**, 124 (1987).

[75] K. Yonaha, M. Iha, Y. Tomihara, M. Nozaki, and M. Yamakawa, *Toxicon* **29**, 703 (1991).

[76] A. Ohsaka, H. Ikezawa, H. Kondo, and S. Kondo, *Jpn. J. Med. Sci. Biol.* **13**, 73 (1960).

[77] T. Omori-Satoh, A. Ohsaka, S. Kondo, and H. Kondo, *Toxicon* **5**, 17 (1967).

[78] T. Omori-Satoh and S. Sadahiro, *Biochim. Biophys. Acta* **580**, 392 (1979).

[79] H. Takeya, K. Oda, T. Miyata, T. Omori-Satoh, and S. Iwanaga, *J. Biol. Chem.* **265**, 16068 (1990).

[80] C. Ouyang and T. F. Huang, *Biochim. Biophys. Acta* **571**, 270 (1979).

[81] T. F. Huang, J. H. Chang, and C. Ouyang, *Toxicon* **22**, 45 (1984).

[82] T. Nikai, N. Mori, M. Kishida, K. Yuko, C. Takenaka, T. Muramkami, S. Shigezane, and H. Sugihara, *Biochim. Biophys. Acta* **838**, 122 (1985).

[83] H. Sugihara, M. Moriura, and T. Nikai, *Toxicon* **21**, 247 (1983).

[84] C. Ouyang and C. M. Teng, *Biochim. Biophys. Acta* **420**, 298 (1976).

[85] H. Sugihara, N. Mori, T. Nikai, M. Kishida, and M. Akagi, *Comp. Biochem. Physiol., B: Comp. Biochem.* **82B**, 29 (1985).

[86] J. W. Fox, B. Nikolov, and S. Shkenderov, manuscript in preparation (1994).

[87] Y. Komori and H. Sugihara, *Int. J. Biochem.* **20**, 1417 (1988).

[88] M. Samel and J. Siigur, *Comp. Biochem. Physiol., C: Comp Pharmacol. Toxicol.* **97C**, 209 (1990).

[89] E. Siigur and J. Siigur, *Biochim. Biophys. Acta* **1074**, 223 (1991).

[90] A. Gasmi, M. Karoni, Z. Benlosfar, H. Karoni, M. El Ayeb, and K. Dellagi, *Toxicon* **29**, 827 (1991).

[91] M. Ovadia, *Toxicon* **16**, 479 (1978).

[92] H. J. Evans, *Biochim. Biophys. Acta* **660**, 219 (1981).

[93] H. J. Evans and A. J. Barrett, *Hematology* **7**, 213 (1988).

[94] N. H. Tan and M. N. Saifuddin, *Toxicon* **28**, 385 (1990).

[95] Y. Yamakawa and T. Omori-Satoh, *Toxicon* **26**, 1145 (1988).

[96] M. Ovadia, *Toxin* **25**, 621 (1987).

TABLE II
CLASSIFICATION OF REPROLYSINS ISOLATED FROM SNAKE VENOM: SNAKE VENOM METALLOENDOPEPTIDASES

| Venom | Name or abbreviation | Molecular mass (kDa) | Specificity on oxidized insulin B chain[a] | Hemorrhagic activity | Proposed class assignment | Refs. |
|---|---|---|---|---|---|---|
| *Agkistrodon* | | | | | | |
| | | | Crotalid and viperid snake venoms | | | |
| *A. acutus* | FP | 24 | 10, 16 | Yes | I-B | Ouyang and Huang[11,12] |
| *A. acutus* | Ac-1 | 24.5 | | Yes | I-A | Mori et al.[13], Nikai et al.[14] |
| *A. acutus* | Ac-2 | 25 | 10, 14, 16, 23 | Yes | I-A | Mori et al.[13] |
| *A. acutus* | Ac-3 | 57 | | Yes | III | Mori et al.[13] |
| *A. acutus* | Ac-4 | 33 | | Yes | I | Mori et al.[13] |
| *A. acutus* | Ac-5 | 24 | | Yes | I-A | Mori et al.[13] |
| *A. acutus* | AaH-I | 22 | | Yes | I-A | Xu et al.[15] |
| *A. acutus* | AaH-II | 22 | | Yes | I-A | Xu et al.[15] |
| *A. acutus* | AaH-III | 22 | 3, 10, 14, 16, 22, 24 | Yes | I-B | Xu et al.[15] |
| *A. bilineatus* | Bilitoxin | 48 | 9, 14 | Yes | II | Imai et al.[16] |
| *A. halys blomhoffii* | Proteinase a | 50 | 5, 14, 16 | No | II | Oshima et al.[17] |
| *A. halys blomhoffii* | HR-1 | 85 | | Yes | IV | Omori et al.[18]; Oshima et al.[19] |
| *A. halys blomhoffii* | Proteinase c | 70 | 10, 14, 23 | No | III | Oshima et al.[17] |
| *A. halys blomhoffii* | HR-II (proteinase b) | 95 | 5, 9, 10, 14, 23 | Yes | IV | Omori et al.[18], Satake et al.[20]; Iwanaga et al.[21]; Oshima et al.[22]; Oshima et al.[23] |
| *A. contortrix contortrix* | Fibrolase | 23 | 14 | No | I | Ahmed et al.[24] |
| *A. contortrix mokasen* | Fibrinogenase | 23 | | n.d.[b] | I | Moran and Geren[25] |
| *A. piscivorus leucostoma* | *Leucostoma* peptidase A | 22.5 | 1, 5, 10, 14, 20, 23, 24 | n.d. | I | Spiekerman et al.[26] |
| *A. piscivorus conanti* | | 23.5 | | n.d. | I | Markland[27] |
| *A. rhodostoma* | α-Fibrinogenase | 25.4 | | n.d. | I | Ouyang et al.[28] |
| *Bitis* | | | | | | |
| *B. arietans* | Protease A | 25 | | No | I | Van der Walt and Joubert[29]; Strydom et al.[30] |
| *B. arietans* | Protease A | 24 | | n.d. | I | Lawrence and Morris[31] |
| *B. arietans* | Protease A | 80 | | n.d. | IV | Lawrence and Morris[31] |
| *B. arietans* | Protease A | 60 | | n.d. | III | Lawrence and Morris[31] |
| *B. arietans* | Protease A | 30 | | n.d. | I | Lawrence and Morris[31] |

| Species | Enzyme | $M_r$ | Substrate/activity | Inhibited | Class | Reference |
|---|---|---|---|---|---|---|
| B. arietans | HT-1 | n.d. | | Yes | | Mebs and Panholzer[32] |
| B. asper | Proteinase G | 18 | 4, 10, 14 | No | I | Aragon-Ortiz and Gubensek[33] |
| *Bothrops* | | | | | | |
| B. jararaca | HF-1 | n.d. | 5, 10, 14, 16, 24 | Yes | | Assakura et al.[34] |
| B. jararaca | Bothrolysin (*Jararaca* proteinase) | 23 | 10, 14, 16, 24 | No | I | Tanizaki et al.[35] |
| B. jararaca | Bothropasin | 48 | | Yes | II | Mandelbaum et al.[36] |
| B. jararaca | HF-2 | 50 | | Yes | III | Assakura et al.[34] |
| B. jararaca | HF-3 | 62 | | Yes | III | Assakura et al.[34] |
| B. jararaca | Jarahagin | 52 | 5, 9, 10, 14, 16, 20, 21, 23, 24 | Yes | III | Paine et al.[37] |
| B. jararaca | Jarafibrase I | 47 | | Yes | II | Maruyama et al.[38] |
| B. jararaca | Jarafibrase II | 21.4 | | Yes | I | Maruyama et al.[39]; Reichl and Mandelbaum[40] |
| B. moojeni | MP-A | 22.5 | | Yes | I-B | Serrano et al.[41] |
| B. moojeni | MP-B | 84 | | No | IV | Mandelbaum et al.[42] |
| B. neuwiedi | NHF-a | 46 | | Yes | II | Mandelbaum et al.[42] |
| B. neuwiedi | NHF-b | 58 | | Yes | III | Mandelbaum et al.[42] |
| *Calloselasma* | | | | | | |
| Calloselasma rotostoma | HP-I | 38 | | Yes | II | Bando et al.[43] |
| Calloselasma rhodostoma | Rhodostoxin | 34 | | Yes | I | Ponnudurai et al.[44] |
| *Cerastes* | | | | | | |
| Cerastes cerastes | Cerastase F-4 | 22.5 | | Yes | I-B | Daoud et al.[45,46] |
| *Crotalus* | | | | | | |
| C. adamanteus | Adamalysin (proteinase I) | 24.6 | 1, 5, 10, 14, 15, 16 | Yes | I-B | Kurecki and Laskowski[47]; Kurecki et al.[9] |
| C. adamanteus | Adamalysin (proteinase II) | 23.7 | 1, 5, 10, 14, 15, 16 | Yes | I-B | Kurecki and Laskowski[47]; Kurecki et al.[9] |
| C. adamanteus | Proteinase H | 85.7 | | Yes | III | Kurecki and Kress[48] |
| C. atrox | Atrolysin B (HT-b) | 24 | 5, 10, 14, 16, 23 | Yes | I-B | Bjarnason and Tu[1]; Bjarnason et al.[49]; Hite et al.[5] |
| C. atrox | Atrolysin C (HT-c) | 24 | 5, 10, 14, 16, 23 | Yes | I-B | Bjarnason and Tu[1]; Fox et al.[50]; Shannon et al.[51] |
| C. atrox | Atrolysin C (HT-d) | 24 | 5, 10, 14, 16, 23 | Yes | I-B | Bjarnason and Tu[1]; Shannon et al.[51] |
| C. atrox | Atrolysin E (HT-e) | 25.7 | 3, 9, 14 | Yes | I-A | Bjarnason and Tu[1]; Bjarnason and Fox[52]; Bjarnason and Fox[8]; Hite et al.[53] |
| C. atrox | Atrolysin F (HT-f) | 64 | 2, 4, 6, 10, 14, 16 | Yes | III | Nikai et al.[54] |
| C. atrox | HT-g | 60 | | Yes | III | Nikai et al.[55] |

*(continued)*

TABLE II (continued)

| Venom | Name or abbreviation | Molecular mass (kDa) | Specificity on oxidized insulin B chain[a] | Hemorrhagic activity | Proposed class assignment | Refs. |
|---|---|---|---|---|---|---|
| C. atrox | Atrolysin A (HT-a) | 68 | 3, 5, 10, 14, 16 | Yes | III | Bjarnason and Tu[1]; Bjarnason et al.[49]; Hite et al.[5] |
| C. atrox | Atroxase | 23.5 | 5, 9, 10, 14, 16 | No | I | Willis and Tu[56] |
| C. atrox | Protease I (HT-b) | 20 | | n.d. | I | Pandya and Budzynski[57] |
| C. atrox | Protease IV | 46 | | n.d. | II | Pandya and Budzynski[57] |
| C. atrox | Collagenase | 58 | | No | III | Hong[58] |
| C. atrox | Fibrinogenase | 31 | | No | I | Nikai et al.[59] |
| C. atrox | New α-protease | 26.7 | | No | I | Kruzel and Kress[60] |
| C. basiliscus basiliscus | B-1 | 27 | | Yes | I | Molina et al.[61] |
| C. basiliscus basiliscus | B-2 | 27.5 | | Yes | I | Molina et al.[61] |
| C. horridus horridus | Horrilysin (HP-IV) | 57 | 14 | Yes | III | Civello et al.[62,63] |
| C. m. nigrescens | Proteinase E | 21.4 | | n.d. | I | Ramirez et al.[64] |
| C. ruber ruber | Ruberlysin (HT-2) | 25 | 10, 14, 16, 23 | Yes | I | Mori et al.[65], Takeya et al.[66] |
| C. ruber ruber | HT-3 | 25 | 5, 10, 14, 16, 23 | Yes | I | Mori et al.[65], Takeya et al.[66] |
| C. ruber ruber | HT-1 | 60 | 10, 14, 16 | Yes | III | Mori et al.[65] |
| C. scutulatus scutulatus | P-13 | 27 | | Yes | I-A | Martinez et al.[67] |
| C. viridis viridis | HT-1 | 62 | | Yes | III | Li et al.[68] |
| C. viridis viridis | HT-2 | 68 | | Yes | III | Li et al.[68] |
| *Lachesis* | | | | | | |
| L. muta muta | LHF-II | 23.5 | | Yes | I | Sanchez et al.[69] |
| L. muta muta | LHF-I | 100 | | Yes | IV | Sanchez et al.[70] |
| *Trimeresurus* | | | | | | |
| T. flavoviridis | HR-2a | 23 | | Yes | I-A | Takahashi and Ohsaka et al.[71]; Nikai et al.[72] |
| T. flavoviridis | HR-2b | 24 | 1, 9, 10, 14 | Yes | I-A | Takahashi and Ohsaka et al.[71]; Nikai et al.[72] |
| T. flavoviridis | Trimerelysin II (proteinase H2) | 23 | 8, 10, 14, 16 | No | I | Takahashi and Ohsaka et al.[71]; Takeya et al.[73] |
| T. flavoviridis | Basic | 24 | 3, 10, 14 | No | I | Yamakawa et al.[74] |
| T. flavoviridis (Okinawa) | HR-2a | 24 | | Yes | I | Yonaha et al.[75] |
| T. flavoviridis (Okinawa) | HR-2b | 19 | 10, 14, 16, 23, 24 | Yes | I | Yonaha et al.[75] |

| Species | Name | | | | | Reference |
|---|---|---|---|---|---|---|
| T. flavoviridis (Okinawa) | HR-1 | 46 | | Yes | II | Yonaha et al.[75] |
| T. flavoviridis | Trimerelysin (HR-1A) | 60 | 10, 14 | Yes | III | Ohsaka et al.[76]; Omori-Satoh et al.[77]; Omori-Satoh and Sadahiro[78] |
| T. flavoviridis | HR-1B | 60 | | Yes | III | Omori-Satoh and Sadahiro[78]; Takeya et al.[79] |
| T. gramineus | HR-1 | 24 | | Yes | I | Ouyang and Huang[80] |
| T. gramineus | HR-2 | 82 | | Yes | III | Huang et al.[81] |
| T. mucrosquamatus | HR-a | 15 | 10, 16, 22 | Yes | I | Nikai et al.[82] |
| T. mucrosquamatus | HR-b | 27 | 14 | Yes | I | Nikai et al.[82] |
| T. mucrosquamatus | Mucrolysin | 94 | 9, 10, 14, 15, 16 | n.d. | IV | Sugihara et al.[83] |
| T. mucrosquamatus | α-Fibrinogenase | 22.4 | | No | I | Ouyang and Teng[84] |
| T. mucrosquamatus | P-1 | 23 | | No | I | Sugihara et al.[85] |
| T. mucrosquamatus | P-2 | 23.5 | | No | I | Sugihara et al.[85] |
| T. mucrosquamatus | P-3 | 23 | | No | I | Sugihara et al.[85] |
| **Vipera** | | | | | | |
| V. ammodytes ammodytes | HT-1 | 60 | | Yes | III | Fox et al.[86] |
| V. ammodytes ammodytes | HT-2 | 60 | | Yes | III | Fox et al.[86] |
| V. ammodytes ammodytes | HT-3 | 60 | | Yes | III | Fox et al.[86] |
| V. aspis aspis | HT-1 | 67 | | Yes | III | Komori and Sugihara[87] |
| V. berus berus | HMP | 56 | 10, 14, 16 | Yes | III | Samel and Siigur[88] |
| V. lebetina | Lebetase | 23.7 | | Yes | I-B | Siigur and Siigur[89] |
| V. lebetina | Fibrinogenase | 26 | | No | I | Gasmi et al.[90] |
| V. palaestinae | HR-1 | 60 | | Yes | III | Ovadia[91] |
| V. palaestinae | HR-2 | 60 | | Yes | III | Ovadia[91] |
| V. palaestinae | HR-3 | 60 | | Yes | III | Ovadia[91] |
| V. russelli russelli | Russellysin | 93 | | Yes | IV | Gowda et al.[10] |
| **Elapid snake venoms** | | | | | | |
| Naja nigricollis | Proteinase F1 | 58 | | n.d. | III | Evans[92]; Evans and Barrett[93] |
| Ophsophagus hannah | Hannahtoxin | 66 | | Yes | III | Tan and Saifuddin[94] |
| Ophsophagus hannah | Ophiolysin | 70 | 3, 4, 10, 14, 16 | No | III | Yamakowa and Omori-Satoh[95] |
| **Atractaspidid snake venoms** | | | | | | |
| Atractaspis engaddesis | HT-1 | 50 | | Yes | II | Ovadia[96] |

[a] Numbered insulin B-chain sequence: $F^1$-$V^2$-$N^3$-$Q^4$-$H^5$-$L^6$-$C^7$-$G^8$-$S^9$-$H^{10}$-$L^{11}$-$V^{12}$-$E^{13}$-$A^{14}$-$L^{15}$-$Y^{16}$-$L^{17}$-$V^{18}$-$C^{19}$-$G^{20}$-$E^{21}$-$R^{22}$-$G^{23}$-$F^{24}$-$F^{25}$-$Y^{26}$-$T^{27}$-$P^{28}$-$K^{29}$-$A^{30}$.

[b] n.d., Not determined.

species of snakes, as is the case with bovine and porcine trypsins. Thus, atrolysin C (and B) and adamalysin (proteinases I and II) appear to be homologous enzymes from *C. atrox* and *C. adamanteus* venoms as previously suggested.[2] They have similar amino acid compositions,[2] similar insulin B-chain specificities (Table II), and have X-ray crystal structures showing identical topologies thought to be characteristic of the two-disulfide bond reprolysins.[6,7] Obviously, more data on most of the 101 reprolysins are needed, in particular sequence and specificity data and preferably structural information, before reliable assignments of relationships can be made.

In spite of a wealth of diverse information on the numerous reprolysin toxins, it is not feasible for meaningful comparisons of the toxins to be made, primarily owing to the lack of appropriate data on their sequences and their proteolytic functions on specific protein substrates. Also, in some cases, data on similar phenomena from different laboratories are not readily comparable, for various reasons.

Regardless of the inherent difficulties, we feel that an attempt should be made to draw conclusions and form a generalized portraiture of the reprolysins from the information available. We also acknowledge that this should be refined, modified, and corrected as more detailed information becomes available. We thus suggest that the reprolysins should be assigned to four main categories or classes based on their sizes as expressed by the published molecular masses. The relevance of this could naturally be challenged, but as more recent results indicate it does not appear to be a totally futile exercise because sizes should relate to domain composition.[5] In light of the most recent sequence data available, we are thus proposing the assignment of the hemorrhagic toxins into four classes based on size: class I, the small reprolysins, having molecular masses of 20 to 30 kDa and containing only the protease domain; class II, the medium-size enzymes with molecular masses of 30 to 50 kDa, having also a disintegrin-like domain; class III, the reprolysins and most potent hemorrhagic toxins, having molecular masses of 50 to 80 kDa and a third domain; and class IV of molecular mass 80–100 kDa, composed of four domains (see "Structure of Reprolysins" section, p. 362, and Refs. 4 and 5). Ultimately, the best method of classification will be based on sequence comparisons, which are now becoming feasible and indeed substantiate our previously proposed classification of hemorrhagic proteinases.[2,4]

The important developments since the early 1980s have in our view not been the proliferation in the number of reprolysins, but rather the closer scrutiny of a few of them, yielding important insights into the structure and function of the enzymes and opening vistas previously unseen.[2,5–7] The metalloendopeptidases from various snake venoms (the reprolysins) have been assigned a multitude of different names by the researchers that purified them, such as hemorrhagic toxins, hemorrhagic proteases, hemorrhagins, hemorrhagic principles, hemorrhagic factors, fibrinogenases, or simply proteases, along with designating numbers or letters. It would be beneficial if a consensus could be reached on nomenclature by researchers in the field. We would thus suggest that they be termed as suggested by the International Union of Biochemistry and Molecular Biology (IUBMB). The better known and extensively studied reprolysins have been assigned common names and EC numbers by the enzyme nomenclature committee of the IUBMB (Table I). The names, such as atrolysin for hemorrhagic metalloproteinases from *C. atrox* venom and trimerelysin for enzymes from *Trimeresurus flavoviridis*, are not yet in wide use but should with time gain increased acceptance and usage, owing to their simplicity, convenience, and efficacy in classification (Table I).

We are now aware of 102 reprolysins from venoms of 35 species of snakes. Most of these, or 70, have been found to cause hemorrhage (Table II). Another 18 have been determined to be nonhemorrhagic, and 13 have not been assayed for hemorrhagic activity. In Table II the 102 reprolysins have been assigned into four classes according to molecular mass, as previously suggested. It should, however, be borne in mind that it is impossible to classify the toxins with certainty according to domain structure, on either the protein or genetic level, without knowing their amino acid or nucleic acid sequences.

Of the 102 reprolysins counted in Table II about half of them, or 54 enzymes, belong to group I (the small reprolysins containing only a protease domain), 9 fall into group II (the medium size proteins), 30 are placed in group III (the three-domain enzymes), and 6 are assigned to group IV (Table II). It should be recognized and clearly stated that the assignments are based solely on the molecular masses of the reprolysins except for those with known sequences (14 cases).

Many of the large reprolysins are very potent hemorrhagic agents according to their published minimum hemorrhagic doses and have primarily been studied as such. Indeed, all of the most potent hemorrhagic toxins fall into class III. These include HR-1 from *Agkistrodon halys blomhoffii* venom, proteinase H from *C. adamanteus* venom, atrolysin A (HT-a) from *C. atrox* venom, and HR-1A and HR-1B from *T. flavoviridis* venom. From the amino acid compositions, among other data, it seems plausible to suggest that proteinase H and atrolysin A (HT-a) are homologous enzymes from different venoms which might define a subclass termed III-A. Also, HR-1 from *A. halis blomhoffii* and HR-1A and HR-1B from *T. flavoviridis* could be members of the reprolysin subclass (III-A). However, more information, especially on amino acid sequences, is needed to corroborate this suggested subclassification; it is not presented in Table II. One might also speculate that the very large enzymes of group IV with low hemorrhagic potency, such as HR-II from *A. halys blomhoffii* venom and mucrotoxin A from

*Trimeresurus mucrosquamatus,* are four-domain proteins and related to russellysin (RVV-X), coagulation factor X activating enzyme, from *Vipera russelli russelli* venom, suggestive of this fourth group in this classification scheme. Again, more sequence data for the enzymes are needed to substantiate these assignments.

Some of the most hemorrhagically potent small enzymes of class I, with similar amino acid compositions, isoelectric points in the weakly acidic range, and three disulfide bonds, appear to define another subclass of homologous enzymes which we term I-A. The most obvious members of this subclass seem to be atrolysin E (HT-e) from *C. atrox* venom (sequence known), which would be the reference for the subclass, and the apparent homologs from *Agkistrodon acutus* venom, Ac-1, Ac-2, and AaH-I. Other potential members of this subclass are Ac-5 and AaH-II also from *A. acutus* venom, HR-1 from *Trimeresurus gramineus* venom, and HR-a from *T. mucrosquamatus* venom. The basicity of some of the enzymes in class I is not necessarily suggestive of a separate subclass for these proteinases. Thus the basic yet potent hemorrhagic proteinases such as HR-2a (sequence known) and HR-2b from *T. flavoviridis* venom and LHF-II from *Lachesis muta muta* venom (sequence known) may also belong to subclass A of class I.

Some of the enzymes in class I are so weakly hemorrhagic that it is questionable whether they should be considered hemorrhagic toxins at all. We now assign these enzymes to a family designated I-B. Indeed, different biological activities have been associated with some of these toxins. On the basis of their molecular masses, amino acid compositions, low hemorrhagic activities, substrate specificities, amino acid sequences, and X-ray crystal structure, atrolysin C (HT-c and HT-d) from *C. atrox* venom defines this subclass.[1,7,50,52] The basic hemorrhagic toxin atrolysin B, also from *C. atrox* venom, clearly belongs to this subclass of metalloproteinases based on the same criteria as above, in particular the substrate specificity on insulin B chain (Table II) and the amino acid sequence as deduced from the cDNA sequence (see [22] in this volume).[5,49] Proteinase I and proteinase II (adamalysin) from *C. adamanteus* venom are obviously members of this subclass of two-disulfide bond reprolysins.[6] It is possible that the basic hemorrhagic toxin from *A. acutus* venom with low hemorrhagic potency, termed AaH-III, and the so-called fibrinolytic protease (FP), as well as MP-A from *Bothrops moojeni* venom, also belong to the I-B subclass of metalloproteinases.

All the reprolysins in Table II have shown metal dependency when assayed with metal chelators. Of the 102 reprolysins, at least 12 have been analyzed for metal content and all found to contain zinc. The others are inhibited by metal chelators. Both the hemorrhagic and proteolytic activities of the atrolysins have been shown to be dependent on the presence of the zinc.[1,4] Also, calcium ions appear to stabilize the hemorrhagic toxins in aqueous solutions. One calcium binding site has been found in adamalysin and atrolysin C.[6,7] Unfortunately, the metal composition of the majority of the reprolysins has not yet been determined. It can, however, reasonably be assumed that all the reprolysins are metalloproteinases, and they are most probably all zinc enzymes (reviewed in Refs. 2–4). Some of the hemorrhagic reprolysins have been demonstrated to hydrolyze basement membrane preparations.[49,63,97] Results have shown that HT-a, HT-b, HT-c, HT-d, and HT-e from *C. atrox* venom hydrolyze type IV collagen, fibronectin, nidogen, and laminin, all of which are components of basement membranes.[49,98–100] It is also noteworthy that these hemorrhagic toxins do not cleave type I, type III, or type V collagens in the native form but rapidly digest gelatin.

## Purification Procedures

Early investigations of reprolysins (hemorrhagic toxins and other metalloproteinases from snake venoms) were greatly hampered by the lack of sufficiently purified enzymes. This was probably due in large part to the complexity of the venoms coupled with the limited availability of sophisticated and powerful purification techniques. However, since the 1970s there has been a proliferation in the number of purified reprolysins, such that today they count over 100. Furthermore, it is evident that the degree of purity of the enzymes isolated in the 1960s and early 1970s is in many cases suspect, as exemplified by the 20 years of development of the purification of HR-1A and HR-1B, two large hemorrhagic components from the venom of *Trimeresurus flavoviridis.*[76–78]

A case in point is provided by the research history of "old" α-protease (previously EC3.4.24.1), first isolated in 1961 as the basic protease from *C. atrox* venom. Fractionation of the venom on DEAE-cellulose[101] or DEAE-Sephadex[102] revealed three distinct peaks containing caseinolytic activity. The authors referred to the proteolytic fractions as α-, β-, and γ-proteases, respectively, according to the order in which they were eluted from the

[97] A. Ohsaka, M. Just, and E. Habermann, *Biochim. Biophys. Acta.* **323**, 415 (1973).
[98] E. N. Baramova, J. D. Shannon, J. B. Bjarnason, and J. W. Fox, *Arch. Biochem. Biophys.* **275**, 63 (1989).
[99] E. N. Baramova, J. D. Shannon, J. B. Bjarnason, and J. W. Fox, *Matrix* **10**, 91 (1990).
[100] J. B. Bjarnason, L. A. Hite, J. D. Shannon, S. J. Tapaninaho, and J. W. Fox, *Toxicon* **31**, 513 (1993).
[101] G. Pfleiderer and G. Sumyk, *Biochim. Biophys. Acta* **51**, 482 (1961).
[102] G. Pfleiderer and A. Krauss, *Biochem. Z.* **342**, 85 (1965).

DEAE columns. $\alpha$-Protease was unabsorbed at pH 7.2, whereas the $\beta$- and $\gamma$-protease fractions were removed by stepwise increases in buffer concentration[101] or with a NaCl gradient.[102] Subsequently, the $\alpha$-protease fraction was further purified by gel filtration on Sephadex G-75, and two protein peaks were observed.[103] Inactive material was eluted with the void volume, and the second peak contained proteolytic activity and was designated $\alpha$-protease, with a molecular weight of 23,000 estimated from the gel-filtration experiment. No convincing evidence of homogeneity of this or other preparations of the old $\alpha$-protease has been presented. On the contrary, it has more recently been shown, by preparation of the enzyme according to the published procedures, that the old $\alpha$-protease is heterogeneous,[60] even after gel filtration on BioGel P-150 (Bio-Rad, Richmond, CA), comparable to the Sephadex G-75 fractionation reported by Zwilling and Pfleiderer.[103] The proteolytic fraction ($\alpha$-protease) still contains three bands (one major and two minor) on sodium dodecyl sulfate–polyacrylamide gel electrophoresis (SDS–PAGE) and had hemorrhagic activity.[60] Thus it is now clear that the major proteolytic component of the old $\alpha$-protease preparation[103] is atrolysin B or HT-b,[1] whereas the "new" $\alpha$-protease[60] and atroxase[56] are the minor proteolytic components (or component) of the old $\alpha$-protease preparation. Furthermore, from careful scrutiny of the purification methods and other characteristics, it seems reasonable to suggest that new $\alpha$-protease and atroxase are one and the same enzyme.

Most of the reprolysins mentioned in Table II have been purified by standard liquid chromatography techniques, using ion-exchange and gel-filtration methods. For a detailed description of the purification methodology, see [22] in this volume on the atrolysins and the original references mentioned in Table II for other reprolysins.

## Assays for Activity

The proteolytic assay most commonly used in the early period of reprolysin research was the method developed by Kunitz using the milk protein casein as the protein substrate[104] and detection of trichloroacetic acid (TCA)-soluble peptide products after timed incubation. In this assay, the proteinase is incubated with the casein, and after a set reaction time the mixture is subjected to acid precipitation with TCA. The resultant digestion fragments which remained in solution were then quantitated as absorbance at 280 nm, and the value used as a measure of the proteolytic action of the enzyme on the substrate. There are, however, two important limitations to

---

[103] R. Zwilling and G. Pfleiderer, *Hoppe-Seyler's Z. Physiol. Chem.* **38,** 519 (1967).
[104] M. Kunitz, *J. Gen. Physiol.* **30,** 291 (1947).

this technique as it affects snake venom metalloendopeptidases. In our experience, these enzymes are rather specific with respect to their sites of action and substrate preferences, and consequently few peptide bonds of casein may be cleaved. Furthermore, the bonds that are cleaved may not give rise to TCA-soluble peptides, particularly if only a few bonds of casein are cleaved. The second limitation with the Kunitz method is that the monitoring of soluble peptides via absorbance at 280 nm can yield rather low sensitivity which is dependent on the amino acid composition of the peptides released. For enzymes with relatively weak activity on casein, TCA precipitation followed by 280 nm absorbance monitoring may, in fact, be so insensitive that the enzyme may not appear to be active on the substrate and may lead to the false conclusion that it is not proteolytic.

The attempt to establish with reasonable certainty whether a purified enzyme contains proteolytic activity requires the application of well-established, sensitive proteolytic activity assay methods. It should, however, be understood that it is unfeasible to prove that an enzyme is not proteolytic, owing to the varied and specialized specificities toward different proteins and the multitude of peptide bonds available for cleavage. Several methods have been used with success since the late 1970s for the detection of the proteolytic activities of reprolysins. These can generally be divided into two classes: protein substrates and peptide substrates.

The commonly used protein substrates which give a rather qualitative assay of proteolysis include azocoll, azoalbumin, azocasein, and hide powder azure. Various hemorrhagic proteases have been shown to be relatively active toward these substrates.[2,52,105] The assays are simple to perform and sensitive owing to the release of a strongly absorbing dye upon proteolysis of the substrate. Unfortunately, because of the nature of the binding of the dye to the substrate, the insolubility of some of the substrates, and the diversity of peptide bonds cleaved, meaningful kinetic analysis of the digestions cannot be performed. In 1969 Lin et al.[106] introduced a very sensitive and more quantitative assay for proteolytic enzymes which used $N,N$-dimethylhemoglobin or $N,N$-dimethylcasein as substrates. They estimated an approximately 100-fold increase in the sensitivity of the assay over that of the Kunitz method. The sensitivity increase is based on the use of trinitrobenzenesulfonic acid to detect the appearance of new terminal amino groups. The assay was used with success in the isolation and kinetic characterization of the five hemorrhagic reprolysins from *C. atrox* venom, the atrolysins.[1,4]

---

[105] T. Nikai, E. Oguri, M. Kishida, H. Sugihara, N. Mori, and A. T. Tu, *Int. J. Biochem.* **18**, 103 (1986).

[106] Y. Lin, G. E. Means, and R. E. Feeney, *J. Biol. Chem.* **244**, 789 (1969).

The use of peptides and small chromogenic substrates has proved advantageous with the reprolysins in certain instances, particularly where the substrate specificities are known and kinetic data are sought. The oxidized B chain of insulin has been used with many of the enzymes for analysis of the specificity of peptide bond hydrolysis. If properly applied, the method can yield kinetic data.[49,50,52,63] The chromogenic method has been used with success to show proteolytic activity of a hemorrhagic toxin that had previously been determined to be devoid of proteolytic activity.[72,105] Use of the oxidized B chain of insulin has, owing to traditional reasons and convenience, been the initial choice of peptide substrate in these studies. To date, many of the reprolysins have been assayed for proteolysis of the B chain of insulin, and useful comparisons of the cleavage sites among these proteases can be made (Table II).

Other small peptides and chromogenic substrates have been tested for use in proteolytic assays of the reprolysins.[1,63,105] In general, because of the apparent high degree of substrate specificity of the enzymes and the substrate length requirements that they demonstrate, most of them have not shown significant activity on the small substrates. These requirements have been determined by the authors of this review with *C. atrox* atrolysin C (hemorrhagic toxins c and d).[50] These toxins were demonstrated to have both size and sequence specificity in the hydrolysis of small peptides.

One small substrate which has been useful in certain applications is the fluorogenic peptide 2-aminobenzoyl-Ala-Gly-Leu-Ala-4-nitrobenzylamide, on which five of the atrolysins from *C. atrox* are active.[50] All of these enzymes cleave the Gly-Leu bond to yield a change in fluorescence in the reaction mixture. The peptide has proved to be extremely useful for routine quantitative analysis of the proteolytic activity of the atrolysins as well as for assays of inhibitors of the enzymes. It may indeed be a useful general substrate for all reprolysins.

Properties

The first reprolysins to be clearly purified to homogeneity and shown to be zinc-dependent metalloproteinases were *Leucostoma* peptidase A (leucolysin) from *Agkistrodon piscivorus leucostoma*[26] and five hemorrhagic metalloproteinases (atrolysins A–E) from *C. atrox* venom.[1] The atrolysins have been extensively studied since the first report appeared. The zinc ion has been exchanged for cobaltous ion.[107] The specificities on the B chain of insulin have been determined,[49,50,107] and their substrate binding sites

---

[107] E. N. Baramova, J. D. Shannon, J. W. Fox, and J. B. Bjarnason, *Biomed. Biochim. Acta* **50**, 763 (1991).

have been probed with kinetic analysis using a multitude of synthetic substrates and inhibitors.[50] The cleavage of basement membrane matrix components has been described.[98–100,107] The interaction with serum inhibitors has been studied,[108] and the cDNA and amino acid sequences have been determined.[5,51,53] The X-ray crystal structure of atrolysin C has been analyzed,[7] as well as that of adamalysin from *C. adamanteus* venom.[6] Thus, the following chapter (see [22] in this volume) is devoted to a brief review of these enzymes and could serve as a general description of the properties of the reprolysins as a family of enzymes. However, there are subtle differences in properties that need to be addressed in this section.

At least 102 reprolysins have been purified from the venom of 35 species of snakes, ranging in molecular masses from 15 to 100 kDa. It now appears that the enzymes are multimodular proteins, containing up to four modules which have been termed proteinase domain, disintegrin-like domain, high-Cys domain, and a fourth (lectin-like) domain.[4,5] A majority of the reprolysins, or 54 of the 102 in Table II, have molecular masses smaller than 30 kDa, and thus they appear to be composed of the proteinase domain alone. Cleavage specificities on insulin B chain of 18 of those enzymes have been determined, and all but one were found to cleave the Ala[14]-Leu[15] bond, or bond 14 (numbered from the amino terminus of the B chain). The other most commonly cleaved bonds are 10 (15 enzymes), 16 (11 enzymes), 5 (8 enzymes), and 23 (7 enzymes). However, other bonds are cleaved by 4 or fewer enzymes. Thus, numerically, the most common average pattern of cleavage is 5, 10, 14, 16, 23 (His-Leu, His-Leu, Ala-Leu, Tyr-Leu, and Gly-Phe, respectively), which is indeed the actual pattern of cleavage of 4 small reprolysins, namely, atrolysin B (HT-b), atrolysin C (HT-c and HT-d), and HT-3 from *Crotalus ruber ruber* venom. An additional 6 enzymes share four of these cleavage sites, namely, *Leucostoma* peptidase A (leucolysin), proteinases I and II (adamalysin), atroxase, ruberlysin (HT-2) from *C. ruber ruber* venom, and basic proteinase from *T. flavoviridis* venom. It should be borne in mind that conflicting results from cleavage experiments have come from different laboratories, probably because of the many variables in the experiments, such as duration of the assay, concentration of components, temperature of the assay system, purity of the enzyme, methodology, and instrumentation.

Another nine enzymes appear to contain a disintegrin-like domain in addition to the proteinase domain; they have molecular masses in the range between 35 and 50 kDa and belong to class II of this family of

[108] E. N. Baramova, J. D. Shannon, J. B. Bjarnason, S. L. Gonias, and J. W. Fox, *Biochemistry* **29**, 1069 (1990).

metalloproteinases. None of the enzymes have been sequenced, but from the cDNA sequence of atrolysin E (HT-e) it can be concluded that a protein precursor of the enzyme is synthesized as a class II protein and subsequently processed to the class I atrolysin E. Results also suggest that bilitoxin is such a two-module reprolysin.[109] Cleavage specificities of 3 of the 9 enzymes in the class II category have been determined (Table II). They reveal that bond 14 is cleaved by all three enzymes and that bonds 10, 14, 16, and 24 are most commonly cleaved.

A total of 30 reprolysins appear to have molecular masses in the range of approximately 50–80 kDa; hence, they are probably composed of three modules, with a high-Cys domain in addition to the two domains previously mentioned. The sequences of 3 of these enzymes have been determined, that is, the protein sequence of HR-1B (trimerelysin I),[79] and the cDNA sequences of jarahagin[37] and atrolysin A,[4] composed of the three modules already mentioned. Most of the enzymes of this class appear, in general, to be rather potent hemorrhagic toxins. When the cleavage specificities of this group of enzymes are analyzed and compared to those of the class I enzymes, an interesting difference is observed. First, the large reprolysins seem to cleave fewer bonds on average than the small enzymes, or 3.6 bonds as compared to 4.2 bonds for the small ones. Second, they all cleave at bond 14 (Ala-Leu) and many at bond 10 (8 of 10 enzymes) and bond 16 (7 of 10 enzymes). However, in contrast to the small enzymes, few of the large reprolysins appear to cleave at bonds 5 (1 enzyme) and 23 (2 enzymes). Thus, the most common average pattern of cleavage seen for these enzymes is 10, 14, and 16 (His-Leu, Ala-Leu, and Tyr-Leu, respectively), which is the actual pattern for HT-1 from *C. ruber ruber* venom and HMP from *Vipera berus berus* venom (Table II). Another five enzymes of this class have these three cleavage sites in addition to one, two, or three more cleavages. It is not as yet possible to access the significance of these observations. Data on the specificities of more enzymes are needed, preferably obtained by one and the same laboratory for the sake of consistency.

## Structure of Reprolysins and Relationship to Other Metalloproteinases

Since the appearance of sequence data on the snake venom metalloproteinases,[51,73,110] it has become clear that they define a new family of metal-

---

[109] T. Nikai, unpublished results (1994).
[110] T. Miyata, H. Takeya, Y. Ozeki, M. Arakawa, F. Tokunaga, S. Iwanaga, and T. Omori-Satoh, *J. Biochem. (Tokyo)* **105**, 847 (1989).

loendopeptidases[111] which we have suggested be called reprolysins. From protein and cDNA sequence data on the reprolysins generated during the last five years, we can now with confidence define a classification scheme for the enzymes (see Table II). This task, until relatively recently, was a questionable one fraught with uncertainty requiring constant modification as new data became available.[2,3]

Since data are now available on cDNA sequences of some venom metalloproteinases[5,37,53] as well as protein sequences, we have developed a two-part, interrelated classification scheme based on these two sets of data (see [22], this volume, for details). The first part of the scheme (N-I to N-IV) is based on the cDNA sequences of the venom metalloproteinases. The first class, termed N-I (nucleotide class I), is the class of metalloproteinases which at the level of the cDNA (mRNA) codes for a signal sequence, pro (zymogen) sequence, and metalloproteinase sequence. The second nucleotide class, N-II, has signal, pro, proteinase, and disintegrin-like cDNA sequences. N-III is the third nucleotide class, and these sequences represent information for signal, pro, proteinase, disintegrin-like, and high-cysteine domains. The cDNA sequence for atrolysin A belongs to this class.[5] We also propose a putative N-IV class which has sequence information for signal, pro, proteinase, disintegrin-like, high-cysteine, and lectin-like domains, based on the protein sequence of russellysin (RVV-X), a high molecular mass metalloproteinase from *V. russelli russelli*.[112]

When one surveys the protein sequence data of the venom metalloproteinases, one can propose a second, parallel and interrelated classification based on the structure of the mature protein as it occurs after protein purification. The P-I class (protein class I) has only a proteinase domain in its mature form, that is, following proteolytic processing of the signal and zymogen structures (Fig. 1). There is no venom metalloproteinase which has been sufficiently characterized (sequenced) to allow it to be classified definitively as a P-II proteinase. However, there are venom proteinases for which there are sufficient data available to suggest that they belong to the P-II class (Table II[40]). There are three members of the P-III class with known protein or cDNA sequences: atrolysin A from *C. atrox* (Fig. 1), HR-1B from *T. flavoviridis,* and jarahagin from *Bothrops jararaca.*[5,37,79] In the mature form all comprise a proteinase domain, a disintegrin-like domain, and a high-cysteine domain. All of these proteinases are potent hemorrhagic toxins, more so than the lower molecular weight toxins,

---

[111] L. A. Hite, J. W. Fox, and J. B. Bjarnason, *Biol. Chem. Hoppe-Seyler* **373**, 381 (1992).
[112] H. Takeya, S. Nishida, T. Miyata, S.-I. Kawada, Y. Saisaka, T. Morita, and S. Iwanaga, *J. Biol. Chem.* **267**, 14109 (1992).

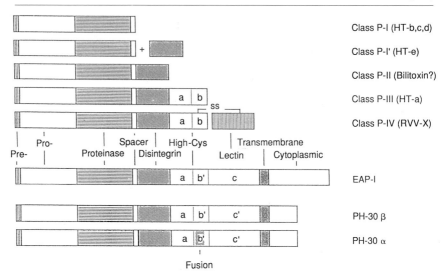

FIG. 1. Schematic representation of the class structures and domain compositions of the reprolysins. Domain structures shared between the snake venom metalloproteinases and mammalian reproductive proteins are indicated.

strongly suggesting the importance of the disintegrin-like and high-cysteine domains to the hemorrhagic activity of the toxins.

The final protein class, P-IV, has one member definably identified to date by amino acid sequence data, namely, russellysin (RVV-X) from *V. russelli russelli*,[112] a 79-kDa protein composed of two disulfide-bonded protein chains (59 and 21 kDa) which functions as a coagulation factor X activating enzyme (Fig. 1). The heavy chain is homologous to the structure of the high molecular weight hemorrhagic toxins such as HT-a and HR-1B, although russellysin is not hemorrhagic. The light chain is disulfide-bonded to the heavy chain through an extra cysteine present in the second subdomain of the high-cysteine domain. Mucrotoxin A is a hemorrhagic metalloproteinase from *T. mucrosquamatus*[83] with a molecular mass of approximately 94 kDa. The amino acid composition of this toxin includes 39 cysteinyl residues, similar to the number found in russellysin (RVV-X) (44 cysteines), which more recently is considered to have a molecular mass of 93 kDa.[10] There will probably be more examples of reprolysins of the P-IV class appearing with time in the literature (Table II).

In the discussion above we have presented a new classification scheme for the venom metalloproteinases based on either the nucleotide sequences or the mature, processed forms of the proteinases. We feel that the two interrelated classifications are necessary since there is clear evidence that the protein forms can be proteolytically processed, producing a product different in structure (and perhaps function) from the originally translated precursor. This scheme should allow for a clearer understanding of the ontogeny of the metalloproteinases as they are found in the crude venom and thus ameliorate some of the confusion surrounding the structural relationships of the various venom metalloproteinases (reprolysins).

The snake venom metalloproteinases were thought to be representatives of a family of metalloproteinases with no homologs of nonvenom origin.[112] However, there are now reports of several mammalian proteins which have clear sequence similarities to the venom metalloproteinases. PH-30 is a protein from guinea pig sperm, composed of two similar chains, $\alpha$ and $\beta$.[113] It is bound to the membrane of guinea pig sperm and is thought to be involved in the sperm–egg fusion process because one chain of the protein contains a virus-like fusion sequence. When the protein sequences of the chains were first reported, the protein was observed to be composed of only a disintegrin-like domain (lacking the signature RGD sequence found in disintegrin peptides) followed by a high-cysteine domain ending with a transmembrane sequence and cytoplasmic domain. From more recent cDNA sequence studies on this protein there appears to be a metalloproteinase or metalloproteinase-like sequence amino terminal to the disintegrin-like domain of the chains; thus, PH-30 presents a structural organization not unlike that of the N-III/N-IV class of venom metalloproteinases (Fig. 1).

Another example of a similar mammalian reproductive protein is the epididymal apical protein I (EAP-I) from rat.[114] This is an androgen-induced, epididymis-specific protein which is homologous to PH-30. Although it does not appear to have a fusion sequence, it does have a transmembrane domain and overall domain organization similar to that of PH-30 (Fig. 1). A mouse cDNA sequence coding for a protein named cyritestin has also been reported which shows the same structural organization as PH-30 and EAP-I.[115]

The sequence similarity shared between the proteinase domains of the mammalian reproductive proteins (MRPs) and the venom proteinase domains is over 30%.[6] The homology in the disintegrin-like domains is more striking. Interestingly, the disintegrin-like domains in the mammalian proteases also lack the signature RGD sequence found in the disintegrin peptides and therefore are more similar to the venom metalloproteinase disintegrin-

[113] C. P. Blobel, T. G. Wolfsberg, C. W. Turck, D. G. Myles, P. Primakoff, and J. M. White, *Nature* (*London*) **356,** 248 (1992).

[114] A. C. F. Perry, R. Jones, P. J. Barker, and L. Hall, *Biochem. J.* **286,** 671 (1992).

[115] A. Senftleben, S. Wallat, L. Lemaire, and U. A. Heinlein, unpublished (1992).

like domains. Finally, the cysteine repeat pattern in the first high-cysteine subdomain of the venom metalloproteinases is nearly identical to the same region at the beginning of the high-cysteine domain of the mammalian proteases.[4,5] The second subdomain of the venom high-cysteine domain is different from those of the mammalian reproductive proteins, suggesting a different function.

On the basis of the structural similarities of the venom metalloproteinases and the mammalian reproductive proteins, it is tempting to speculate on the evolutionary relationship of their functions. Unfortunately, proteolytic activity for the mammalian proteins has yet to be demonstrated, so whether the metalloproteinase-like sequences can actually function as proteinases is unknown. It is not unreasonable to expect the proteinase function and thus the biological activity of these related venom and mammalian enzymes to be modulated by the domains following the proteinase domain. For example, in the case of the hemorrhagic toxins the disintegrin-like domain may function to alter platelet aggregation via interaction with the GPIIb IIa receptor and also target the enzyme to regions where platelets may adhere (i.e., disruptions in the capillary basement membrane). In the PH-30 chains the disintegrin-like domains may also serve to target the protein and therefore the sperm to receptors present on the egg. In the case of the PH-30 $\alpha$ chain, the second high-cysteine subdomain contains within it the virus-like fusion sequence, and one can speculate that this subdomain is important in presenting that fusion sequence in a recognizable form to a cell membrane. With the venom protein RVV-X the second subdomain serves as the connecting site for the light chain (C-lectin).

It is thus clear that the venom metalloproteinases, regardless of size and function, are structurally related via the classification scheme based on domain structure described above. Furthermore, the venom proteinases, particularly of the P-III and P-IV classes, are structurally related to the mammalian reproductive proteins EAP-I, PH-30, and cyritestin. It is expected that additional proteins will be found from snakes, mammals, as well as other organisms, which will strengthen the foundation for the proposal of an encompassing class of unique proteins forming a family of the metalloproteinases which are related by the structural homologies we have outlined above. We have proposed that this family be called the reprolysins, a term derived from the reptilian and reproductive tract origin of the proteins.

An informative development in the field of reprolysin research has been the analysis of the X-ray crystal structures of adamalysin (proteinase II) and atrolysin C (HT-d), homologous enzymes from the venoms of *C. adamanteus* and *C. atrox,* respectively.[6,7] Their structures are identical, revealing an oblate ellipsoidal molecule with an active-site cleft separating a fairly irregularly folded small "lower" subdomain from the "upper" main domain which is composed of an open twisted $\alpha/\beta$ structure having a five-stranded $\beta$ sheet and four $\alpha$ helices. The peptide chain of the enzymes starts at the "lower" subdomain surface and forms the regularly folded main domain with residues 6 to 150 (Fig. 2). The topology of the upper domain begins with a $\beta\alpha\beta$ motif followed by a $\alpha$-helix link of the C helix to a $\beta\beta\beta$ motif and ending in the D helix, which is terminated by Gly-149. The lower subdomain, consisting of the last 50 residues, is organized in multiple turns cross-connected by a disulfide bridge linking Cys-157 and Cys-164. The chain ends in a long C-terminal $\alpha$ helix and an extended segment clamped to the upper domain with a disulfide bridge connecting Cys-117 to Cys-197. The disulfide bridges are typical of the so-called two-disulfide reprolysins. The majority of the reprolysins, however, are three-disulfide proteinases, having disulfide connections from Cys-157 to Cys-181 and from Cys-159 to Cys-164 in the lower subdomain, as well as the Cys-117 to Cys-197 connection.

As previously suggested,[51] the catalytic zinc ion in the active-site cleft is coordinated by His-142, His-146, His-152, and a water molecule anchored to Glu-143 in a nearly tetrahedral manner, as predicted for atrolysin E by

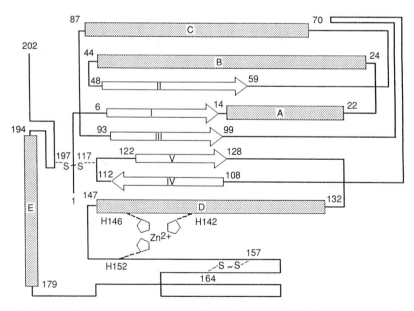

FIG. 2. Secondary structure plot of a reprolysin of the two-disulfide bond type, drawn from X-ray crystal structure data for atrolysin C and adamalysin. The upper and lower domains are arranged above and below the active-site cleft. Helices are labeled with capital letters (A and E) and $\beta$ strands with Roman numerals (I to V).

spectroscopic studies of cobaltous HT-e.[100] Residues His-142, Glu-143, and His-146 are part of the "long active site" D helix which extends to Gly-149, where it turns sharply toward His-152. The substrate binding site is bordered by the Met turn (Met-166) which forms a "basement" of the Zn locus and thus helps anchor the essential 168–172 strand. This strand forms one "wall" of the extended binding site. The segment between residues 109 and 114 of the antiparallel $\beta$ strand of the upper main domain defines the "upper wall" of the extended binding site. A calcium ion is observed liganded by the side chains of Glu-9 and Asp-93 of the main upper domain and Asn-200 and the carbonyl group of Cys-197 of the C-terminal "tail" as well as by internal and external water molecules.

The crystal structures of zinc endopeptidases have hitherto been analyzed for only three thermolysin-like bacterial proteinases[116,117] and astacin.[118] A comparison of the two reprolysin structures with those of astasin and thermolysin reveal a high degree of topological equivalence with the main domain of astacin, as well as the zinc-binding region and the Met turn, but a much lesser extent of similarity with thermolysin. Thus, it has been suggested that the reprolysins should be grouped into a common superfamily along with the families of astacins, matrixins, and serralysins but distinct from the thermolysin family.[118]

---

[116] B. W. Matthews, *Acc. Chem. Res.* **21,** 333 (1988).
[117] D. R. Holland, D. R. Tronwud, H. W. Pley, K. M. Flaherty, W. Stark, J. N. Jansonius, D. B. McKay, and B. W. Matthews, *Biochemistry* **31,** 11310 (1992).
[118] W. Bode, F.-X. Gomis-Rüth, and W. Stöcker, *FEBS Lett.* **331,** 134 (1993).

---

## [22] Atrolysins: Metalloproteinases from *Crotalus atrox* Venom

*By* Jay W. Fox and Jón B. Bjarnason

### Introduction

Atrolysin is the trivial name given to the homologous zinc metalloproteinases from the western diamondback rattlesnake, *Crotalus atrox.* Currently there are six *C. atrox* metalloproteinases which are considered to be atrolysins by nature of their biochemical properties. Atrolysin A (formerly known as Ht-a) and atrolysin F (Ht-f) are class P-III hemorrhagic toxins. Atrolysins B (Ht-b), Cc and Cd (the isoenzymes Ht-c and Ht-d), and E

(Ht-e) are all members of the class P-I hemorrhagic toxins (see [21] in this volume). Other metalloproteinases have been isolated from *C. atrox* venom, but their characterization has been insufficient to classify them definitively as atrolysins. The isolation and biochemical properties of the six atrolysins are described in this chapter along with the structural details that characterize these proteins.

## Purification

Isolation of the atrolysins begins with lyophilized crude venom which is dialyzed against 10 m$M$ borate, pH 9.0, containing 0.1 $M$ NaCl and 2 m$M$ CaCl$_2$. The first chromatography step is ion-exchange on DEAE-cellulose that yields five fractions, A-1 through A-5.[1] Fraction A-1 when subjected to four additional chromatography steps (gel filtration on Sephadex G-75, ion-exchange on DEAE-cellulose, and two gel-filtration steps on AcA 44) yields atrolysin A. Atrolysin B also is derived from fraction A-1 following three additional chromatographies (gel filtration on Sephadex G-75 and two ion-exchange chromatographies on CM-cellulose). Atrolysin C made up of the isoenzymes Ht-c and Ht-d is purified from fraction A-2 by gel filtration on Sephadex G-75 and subsequently two ion-exchange chromatographies on DEAE-cellulose. Atrolysin E is purified from fraction A-4 by gel filtration on Sephadex G-75 followed by ion-exchange chromatography on DEAE-cellulose. Atrolysin F is isolated from fraction A-1 by sequential chromatography on Sephadex G-75, DEAE-cellulose, and two CM-cellulose columns.[2] The purity of the atrolysins following the above chromatography steps is typically assessed by sodium dodecyl sulfate (SDS), native, and isoelectric focusing polyacrylamide gel electrophoresis (PAGE). The approximate yields of atrolysins A, B, Cc, Cd, E, and F from 10 g of lyophilized crude venom are 32, 370, 17, 31, 160, and 150 mg, respectively.

## Biochemical Properties of Atrolysins

*Atrolysin E*

Atrolysin E is the most extensively characterized of all of the *C. atrox* metalloproteinases. It is a class P-Ia hemorrhagic toxin with moderate

---

[1] J. B. Bjarnason and A. T. Tu, *Biochemistry* **17,** 3395 (1978).
[2] T. Nikai, N. Mori, M. Kishida, H. Sugihara, and A. T. Tu, *Arch. Biochem. Biophys.* **231,** 309 (1984).

potency (Table I). The estimated molecular mass of atrolysin E by SDS–PAGE is 27,500 Da, and the proteinase has an isoelectric point of 5.6. The minimum hemorrhagic dose for atrolysin E is 1 $\mu$g.[1] Zinc is found in a 1:1 molar ratio and is critical for the proteolytic and hemorrhagic activities of the protein.[3] Cobalt(II) can be directly exchanged for the zinc in the protein with no significant loss of secondary structure or proteolytic specificity or activity.[3] Examination of the cobalt protein with circular dichroism and absorption spectrometry in the visible region (400–650 nm) indicates a distorted tetrahedral coordination of the metal liganded by oxygen and nitrogen ligands. The assignment of coordination geometry has been corroborated by X-ray crystallography studies on a homologous protein, adamalysin (EC 3.4.24.47).[4]

The sites of cleavage of oxidized insulin B chain by atrolysin E are generally representative for the venom metalloproteinases (see [21] in this volume). The standard method for determining cleavage sites is by timed digestions of insulin B chain by the proteinase, with the products isolated by reversed-phase ($C_{18}$) high-performance liquid chromatography (HPLC) and subjected to amino acid analysis for identification.[3] A novel method being developed in our laboratory is to follow progression of the digestion with matrix-assisted laser desorption mass spectrometry. This allows for simultaneous detection and identification of the digestion products. Limitations of the technique are the background peaks arising from the matrix at the low mass end of the spectrum.

Microscopic studies on the nature of the biological activity of atrolysin E pointed to the relationship of proteolytic degradation of basement membranes surrounding capillaries to the hemorrhagic activity of the protein, which was confirmed using isolated preparations of basement membranes from Englebreth-Holm–Swarm tumors.[5–8] Using isolated components of extracellular matrix and basement membranes, further details of specificity can be elucidated. Atrolysin E readily degrades fibrinogen into nonclotting fragments and fibronectin into a variety of lower molecular mass fragments.[6] Interestingly atrolysin E, like atrolysins A, C, and B, does not significantly digest fibrin. Atrolysin E can also degrade the gelatins of collagen types I, II, III, and V to limited digestion products; however, the native forms of

[3] J. B. Bjarnason and J. W. Fox, *Biochemistry* **33**, 3770 (1983).
[4] F.-X. Gomis-Rüth, L. F. Kress, E. Meyer, and W. Bode, *EMBO J.* **12**, 4151 (1993).
[5] C. L. Ownby, J. B. Bjarnason, and A. T. Tu, *Am. J. Pathol.* **93**, 201 (1978).
[6] J. B. Bjarnason, D. Hamilton, and J. W. Fox, *Biol. Chem. Hoppe-Seyler* **369**, 121 (1988).
[7] E. N. Baramova, J. D. Shannon, J. W. Fox, and J. B. Bjarnason, *Biomed. Biochim. Acta* **50**, 763 (1991).
[8] E. N. Baramova, J. D. Shannon, J. B. Bjarnason, and J. W. Fox, *Arch. Biochem. Biophys.* **275**, 63 (1989).

## TABLE I
### Properties of Atrolysins A, B, Cc, Cd, E, and F

| Property | Atrolysin | | | | | |
| --- | --- | --- | --- | --- | --- | --- |
| | A | B | Cc | Cd | E | F |
| Molecular mass (kDa) by SDS–PAGE | 68 | 24 | 24 | 24 | 25.7 | 64 |
| Isoelectric point (pI) | Weakly acidic | 8.2 | 6.0 | 6.1 | 5.6 | 7.7 |
| Minimum hemorrhagic dose | 0.04 | 3 | 8 | 11 | 1 | 0.5 |
| Metal content | 1 Zn | 1 Zn | 1 Zn | 1 Zn | 1 Zn | 1 Zn |
| Hemorrhagic | + | + | + | + | + | + |
| Proteolytic | + | + | + | + | + | + |
| Carbohydrate (molar ratio) | Fucose (2.3), glucosamine (9.0), galactose (2.0), mannose (4.8), sialic acid (2.2) | None detected | None detected | None detected | None detected | None determined |
| Other activities | Digests basement membrane preparations, collagen IV, laminin, nidogen, fibronectin, fibrinogen, gelatins I, III, IV | Digests basement membrane preparations, laminin, nidogen, fibrinogen | Digests basement membrane preparations, collagen IV, laminin, nidogen, fibronectin, fibrinogen, gelatins I, III, IV | Digests basement membrane preparations, collagen IV, laminin, nidogen, fibronectin, fibrinogen, gelatins I, III, IV | Digests basement membrane preparations, collagen IV, laminin, nidogen, fibronectin, fibrinogen, gelatins I, III, IV | Digests basement membrane preparations, collagen IV, laminin, nidogen, fibronectin, fibrinogen, gelatins, I, III, IV |
| Inhibitors | EDTA, 1,10-phenanthroline, HTI from *C. adamanteus*, Oprin | EDTA, 1,10-phenanthroline, HTI from *C. adamanteus* | EDTA, 1,10-phenanthroline, amino acid hydroxamates, $\alpha_2$-macroglobulin | EDTA, 1,10-phenanthroline, amino acid hydroxamates, $\alpha_2$-macroglobulin | EDTA, 1,10-phenanthroline, $\alpha_2$-macroglobulin | EDTA, 1,10-phenanthroline |

these collagens are not affected.[8] Collagen IV, the collegen typically found in basement membrane structures, is digested to limited products by atrolysin E. Incubation of collagen IV with atrolysin E followed by resolution of the digestion fragments by SDS–PAGE (7% polyacrylamide gel) and blotting on a polyvinyl difluoride (PVDF) membrane and subsequent Edman degradation sequencing of the blotted collegen IV fragments allows for the identification of the sites of cleavage of collagen IV by atrolysin E.[9] The collagen IV $\alpha$I(IV) chains are cleaved in the triplet interruption at Ala$^{219}$-Gln$^{220}$, and the $\alpha$II(IV) chain is cleaved in the same region at the Thr$^{228}$-Leu$^{229}$ bond.

Similar studies using the basement membrane components laminin and nidogen and the laminin/nidogen 1 : 1 complexes as substrates for atrolysin E have been performed.[7] Laminin and nidogen are important components in basement membrane, with laminin capable of forming a homopolymeric network and bound nidogen serving to link laminin to the collagen IV network and heparan sulfate proteoglycans (HSPG).[10] Although laminin is relatively insensitive to digestion by atrolysin E, isolated nidogen is cleaved at positions 75, 296, 336, 402, 478, 625, 702, and 920.[7] Nidogen present in the laminin/nidogen complex is cleaved in a more restricted manner at positions 322, 336, 351, 840, and 953. This difference in cleavage patterns by atrolysin E for nidogen in the complex and isolated nidogen is interpreted to be due to the masking of regions of nidogen as a result of its interaction with laminin. Most of the cleavages of nidogen in both circumstances occur in the proteinase-sensitive flexible link region of the molecule between the G1 and G2 globules. However, some of the cleavages identified are in the rodlike region linking the G2 (collagen IV/HSPG binding) globule to the G3 (laminin binding) globule, which suggests that nidogen cleavage by atrolysin E could be effective in dissociating the laminin network from the collagen IV network in basement membranes.

Several small, synthetic inhibitors have been developed for atrolysin E taking advantage of its substrate specificity and mechanism of catalysis.[11] Two groups of compounds, one with a N-[1-(R,S)-3-dicarboxypropyl] and the other with a N-[1-(R,S)-carboxypentyl] moiety coupled to short peptide structures, have been assayed for atrolysin E inhibition. None of the carboxypropyl-based compounds is inhibitory, whereas the carboxypentyl compounds have $K_i$ values against atrolysin E in the low micromolar range.

[9] E. N. Baramova, J. D. Shannon, J. B. Bjarnason, and J. W. Fox, *Matrix* **10**, 91 (1990).

[10] J. W. Fox, U. Mayer, R. Nischt, M. Aumailley, D. Reinhardt, H. Wiedmann, K. Mann, R. Timpl, T. Kreig, J. Engel, and M.-L. Chu, *EMBO J.* **10**, (1991).

[11] A. Robeva, V. Politi, J. D. Shannon, J. B. Bjarnason, and J. W. Fox, *Biomed. Biochim. Acta* **50**, 769 (1991).

Several short peptides have been isolated from the crude venom of *C. atrox* which are also capable of acting as inhibitors of atrolysin E.[11] Pyro-Glu-Asn-Trp and Pyro-Glu-Glu-Trp inhibit atrolysin E with $K_i$ values of 11 and 19 $\mu M$, respectively. The concentration of the pyroglutamate peptides in the crude venom is in the millimolar range and thus may serve as potent inhibitors of the atrolysins in the venom gland. On envenomation into the tissues, the effective concentration of the inhibitors would likely drop to levels such that the atrolysins would no longer be inhibited.

The lower molecular weight atrolysins are inhibited by human plasma glycoprotein $\alpha_2$-macroglobulin.[12] Atrolysin E binds the inhibitor at 22° in a 2:1 molar ratio. Inhibition of atrolysin E by $\alpha_2$-macroglobulin occurs at a rate of $2.3 \times 10^{-5} M^{-1} \sec^{-1}$ as measured by the change of intrinsic fluorescence of $\alpha_2$-macroglobulin on interaction with the proteinase. Following binding to $\alpha_2$-macroglobulin, atrolysin E cleaves the inhibitor at two sites, a primary one at Val[689]-Met[690] and a secondary site Gly[693]-His[694], both in the bait region of the molecule.

*Atrolysins B and C*

From 10 g of lyophilized crude *C. atrox* venom there is an approximate yield of 370 mg of pure atrolysin B, making it the most abundant hemorrhagic toxin found in the venom.[1] The molecular weight as estimated from SDS–PAGE is 24,000, and it contains zinc in a 1:1 molar ratio. The minimum hemorrhagic dose for the proteinase is 3 $\mu g$, which is moderately potent for a class P-1 hemorrhagic toxin (Table I). Atrolysin B is the only atrolysin having a basic isoelectric point, which is estimated from the protein sequence to be 8.2.[13] This significant difference in net charge when compared to the other atrolysins is a result of sequence differences disseminated throughout the structure rather than localized to a specific region (Fig. 1). Atrolysin B also possesses myonecrotic activity, a functional characteristic not shared by the other atrolysins.[5]

Atrolysin C is the trivial name given to the two minimally hemorrhagic isoenzymes Ht-c and Ht-d (Table I). Both proteinases share similar biochemical properties, as would be expected for proteins which differ from one another by a single amino acid substitution.[13–15] Peptide mapping and protein sequencing of atrolysins Cc and Cd were used to identify the Asp-

[12] E. N. Baramova, J. D. Shannon, J. B. Bjarnason, S. L. Gonias, and J. W. Fox, *Biochemistry* **29**, 1069 (1990).

[13] L. A. Hite, L.-G. Jia, J. B. Bjarnason, and J. W. Fox, *Arch. Biochem. Biophys.* **308**, 182 (1994).

[14] J. D. Shannon, E. N. Baramova, J. B. Bjarnason, and J. W. Fox, *J. Biol. Chem.* **264**, 11575 (1989).

[15] J. B. Bjarnason and J. W. Fox, *Biochim. Biophys. Acta* **911**, 356 (1987).

```
Atrolysin      10        20        30        40        50        60        70        80        90        100
          Pre-      Pro-
E    MIQVLLVTICLAAFPYQGSSIILESGNVNDYEVIYPRKVTALPKGAVQPKYEDTMQYELKVNGEPVVLHLEKNKGLFSKDYISETHYSFDGRKITTNPSVE
B    MIEVLLVTICLAVFPYQGSSIILESGNVNDYEVVYPRKVTALPKGAVQPKYEDAMQYELKVNGEPVVLHLEKNKELFSKDYSETHYSPDGRKITTNPSVE
Cc   MIEVVLVTICLAVFPYQGSSIILESGNVNDYEVVYPRKVTALPKGAVQPKYEDAMQYELKVNGEPVVLHLEKNKELFSKDYSETHYSPDGRKITTNPSVE
Dd   MIEVLLVTICLAVFPYQGSSIILESGNVNDYEVVYPRKVTALPKGAVQPKYEDAMQYELKVNGEPVVLHLEKNKELFSKDYSETHYSPDGRKITTNPSVE

               110       120       130       140       150       160       170       180       190       200
                                                                                                  Proteinase
A                                                                                         ...ERLTKRYVE
E    DHCYYHGRIENDADSTASISACNGLKGHFKLQGEMYLIEPLKLSDSEAHAVFLKNVEKEDEAPKMCGVTQNWESYEPIKKASDLNINPEHQ---RYVE
B    DHCYRGRIENDADSTASISACNGLKGHFKLQGEMYLIEPLELSDSEAHAVFKYENVEKEDEAPKMCGVTQNWESYEPIKKASDLNLNPDQQNLPQRYIE
Cc   DHCYRGRIENDADSTASISACNGLKGHFKLQGEMYLIEPLIEPLELSDSEAHAVFKLENVEKEDEAPKMCGVTQNWESYEPIKKASDLNLNPDQQNLPQRYIE
Cd   DHCYRGRIENDADSTASISACNGLKGHFKLQGEMYLIEPLELELSDSEAHAVFKLENVEKEDEAPKMCGVTQNWESYEPIKKASDLNLNPDQQNLPQRYIE

               210       220       230       240       250       260       270       280       290       300
A    LVIVADHRMFTKYNGNLKKIRKWIYQIVNTINEIYIIPLNIRVALVRLEIWSNGDLIDVTSAANVTLKSFGNWRVTNLLRRKSHDNAQLTAIDLDEETLG
E    LFIVVDHGMVTKYNGDSDKIRQRVHQMVNIMKESYTYMYIDILLAGIEIWSNGDLINVQPASPNTLNSFGEWRETDLLKRKSHDNAQLLTSIAFDEQIIG
B    LVVVADHRVFMKYNSDLNTIRRTRVHEIVNFINGFYRSLNIHVSLTDLEIWSDDFITVQSSAKNTLNSFGEWREADLLRRKSHDAQLLTAINFEGKIIG
Cc   LVVVADHRVFMKYNSDLNTIRRTVHEIVNFINGFYRSLNIHVSLTDLEIWSNEDQINIQSASSDTLNAFAEWRETDLLNRKSHDNAQLTAIELDEETLG
Cd   LVVVADHRVFMKYNSDLNTIRRTVHEIVNEINGFYRSLNIHVSLTDLEIWSNEDQINIQSASSDTLNAFAEWRETDLLNRKSHDNAQLLTAIELDEETLG

               310       320       330       340       350       360       370       380       390       400
                                                                                                    Spacer
A    LAPLGTMCDFKLSIGIVQDHSPINLLVAVTMAELGHNLGMVHDENRCHSCSTPACVMCAVLRQRSYEFSDCLSNHYRTFIINYNPQCILNEPLQTDIIS
E    RAYIGGICDFKRSTGVVQDHSEINLRVAVTMTHELGHNLGIIHDTDSCSCGGYISCIMSPVISDEPSKYFSDCSYIQCWEFIMNQKPQCILKKPLRTDTVS
B    RAVTSSMCNFPRKSVGIVKDHSFINLLVGVTMAHELGHNLGMNHDGKCLRGASLCIMRPGLTPGRSYEFSDDSMGYIQSFLNQYKPQCILNKPLRIDPVS
Cc   LAPLGTMCDFKLSIGIVQDHSFINLLMGVTMAHELGHNLGMEHDGKDCLRGASLCIMRPGLTKGRSYEFSADSMHYYERFLKQYKPQCILNKPLRIDPVS
Cd   LAPLGTMCDFKLSIGIVQDHSFINLLMGVTMAHELGHNLGMEHDGKDCLRGASLCIMRPGLTKGRSYEFSDDSMHYYERFLKQYKPQCILNKPLRIDPVS

               410       420       430       440       450       460       470       480       490       500
               Disintegrin-like                                                            High Cys
A    PPVCGNELLEVGEECDCGSPRTCRDPCCDAATCKLHSWVECESGECCQQCKFTSAGNVCRPARSECDIAESCTGQSADCPTDDFHRNGKPCLHNFGYCYN
E    TPVSGNELLEAGIECDCGSL---ENPCCYATTCKMRPGSQCAEGLCCDQCRFMKKGTVCRVSMVDRN-DDTCTGQSADCP-----RNGL
B    TPVSGNELLEAGEE
Cc   TPVSGNELLEAGEE
Cd   TPVSGNELLEAGEE

               510       520       530       540       550       560       570       580       590       600
A    GNCPIMYHQCYALWGSNVTVAPDACFDINQSGNNSFYCRKENGVNIFCAQEDVKCGRLFCNVNDFLCRHKYSDDGMVDHGTKCADGKVCKNRQCVDVTTA

               610
A    YKSTSGFSQI
```

Fig. 1. Translated cDNA sequences for atrolysins A, B, Cc, Cd, and E. The pre-, pro-, proteinase, disintegrin-like, and high-Cys domains are noted above the sequences. A full-length clone for atrolysin A has not been found, hence the lack of pre- and prosequences.[13]

181 in atrolysin Cd which is substituted by an alanyl residue in atrolysin Cc. This is the only sequence difference detected between the two isoenzymes, a finding which is confirmed by cDNA sequencing of full-length cDNA clones of the proteins.[13] The proteins have pyroglutamyl-blocked N termini and are 202 amino acid residues in length. Atrolysin C secondary structure has been investigated using circular dichroism spectropolarimetry from which estimates of 42% $\alpha$ helix, 37% $\beta$ sheet, and 11% turn structures were made.[15] These estimates have been corroborated by the crystallographic studies on atrolysin Cd.[16]

Atrolysin C is capable of proteolytically degrading fibronectin, laminin, nidogen, collagen type I, III, and V gelatins, as well as basement membrane preparations. The digestion patterns of these substrates, as observed by SDS–PAGE, were identical for atrolysins Cc and Cd but were different for those produced by atrolysins E and A.[8] The overall activity of atrolysin C on basement membrane components and basement membrane preparations is somewhat slower then that observed for atrolysins E and A, a characteristic which is reflected by the lower hemorrhagic activity of these proteinases.[6]

Synthetic peptide substrates have been used to define further the peptide bond specificities of atrolysin C.[17] From a suite of peptides, $NH_2$-Leu-Val-Glu-Ala-Leu-Tyr-Leu-COOH was determined to be an excellent substrate, with the Ala-Leu bond being hydrolyzed with a $k_{cat}/K_m$ value of 134 min$^{-1}$ $M^{-1}$. The kinetics for the cleavages are determined by detecting the production of free amino groups via fluorescamine, with the fluorescence followed by $\lambda_{ex}$ of 390 nm, $\lambda_{em}$ of 475 nm. The results are correlated to standard fluorescence concentration curves for the cleavage products. Using progressively shorter peptides, the active site of atrolysin C was shown to require a minimum of four amino acid residues for peptide bond cleavage. Substitution of amino acid residues in positions P4–P3' in synthetic peptide substrates is a useful method to determine the extended substrate binding sites of the atrolysins. In the case of atrolysin C the most critical features appear to be the need for an extended substrate without charged side chains near the scissile peptide bond, a smaller residue in the P1 position, and a large aliphatic residue in P1.

The fluorogenic substrate 2-aminobenzoyl-Ala-Gly-Leu-Ala-4-nitro-benzylamide is a useful compound for assessing atrolysin kinetics and inhibition.[17] Continuous reaction monitoring can be accomplished by measuring the change of fluorescence ($\lambda_{ex}$ 340 nm, $\lambda_{em}$ 415 nm) using the fluorogenic

[16] D. Zhang, I. Botos, F.-X. Gomis-Rüth, R. Doll, C. Blood, F. G. Njoroge, J. W. Fox, W. Bode, and E. Meyer, *Proc. Natl. Acad. Sci. USA* **91**, 8447 (1994).
[17] J. W. Fox, R. Campbell, L. Beggerly, and J. B. Bjarnason, *Eur. J. Biochem.* **156**, 65 (1986).

substrate and varying concentrations of inhibitors. Dixon plot analysis of the data allows for the calculation of $K_i$ values for the inhibitors. Using this method, the $K_i$ values for a variety of synthetic inhibitors of atrolysin C have been determined, one of the most potent being the hydroxamic dipeptide Z-L-Leu-Gly-NHOH with a $K_i$ of $60 \times 10^{-6}$ $M$.[18]

The inhibition of atrolysin C with human $\alpha_2$-macroglobulin can be analyzed by observing changes in the intrinsic fluorescence of the plasma inhibitor as discussed with atrolysin E. Atrolysin C is rapidly inhibited by $\alpha_2$-macroglobulin with a rate of $6.2 \times 10^{-6}$ $M^{-1}$ $\sec^{-1}$.[12] The molar ratio of binding of atrolysin C with $\alpha_2$-macroglobulin is 1.1 : 1, and the site cleaved in the bait region by the proteinase is Arg$^{696}$-Leu$^{697}$ as determined by Edman degradation sequencing of the electroblotted digestion fragments on PVDF membranes.

*Atrolysin A*

Atrolysin A when first isolated was remarkable for its relatively high hemorrhagic potency (0.04 $\mu$g) as compared to the other atrolysins. From its molecular mass (68 kDa), the proteinase is classified as a P-III atrolysin, meaning that there is in addition to the proteinase domain a disintegrin-like domain and a high-cysteine containing domain C-terminal to the proteinase domain. Zinc is present in a 1 : 1 molar ratio in the protein as in the other atrolysins. The weakly acidic toxin has a blocked N terminus as evidenced by a lack of sequence data resulting from Edman degradation of protein. The digestion pattern of the insulin B chain by atrolysin A is most similar to the other P-III atrolysin, atrolysin F.[18]

Of the five atrolysins which have been examined for the presence of carbohydrate, atrolysin A is the only proteinase which is clearly glycosylated (Table I). Following hydrolysis of the carbohydrate structures from atrolysin A and identification and quantitation of the resultant products by ion-exchange chromatography, we were able to detect fucose, glucosamine, galactose, mannose, and $N$-acetylneuraminic acid (sialic acid) on atrolysin A. Consensus sites for N-glycosylation are found at positions 263, 517, 529, and 533 in atrolysin A (Fig. 1). However, protein sequence data from atrolysin A suggest that residues Asn-517 and Asn-533 are actually glycosylated as they do not yield identifiable Edman degradation products, whereas position Asn-529 is identified as an unmodified residue.[13]

As is seen with atrolysin E, atrolysin A is very efficient at the proteolytic degradation of basement membrane preparations, even more so than atrolysins B and C, thus suggesting a correlation between basement membrane

---

[18] J. B. Bjarnason and J. W. Fox, *J. Pharmacol. Exp. Ther.* **62,** 321 (1994).

degradation and hemorrhagic potency.[6] Atrolysin A cleaves fibrinogen to nonclotting products and does not significantly degrade fibrin, as is similar for atrolysins E, B, and C. With regard to isolated basement membrane component proteins, atrolysin A degrades fibronectin, collagen IV, and nidogen in a manner similar to that which has been reported for the matrix metalloproteinase (MMP) stromelysin.[8]

The structural basis for the high hemorrhagic potency of atrolysin A has been a major point of experimental emphasis. Superficially, one would anticipate that the presence of additional structure information found in the form of the disintegrin-like and high-cysteine domains is the mechanism for potentiating the intrinsic hemorrhagic activity of the proteinase domain. Another possibility is that this particular atrolysin is less effectively inhibited by $\alpha_2$-macroglobulin. This is what is experimentally observed when examining the rate and stoichiometry of atrolysin A interaction with $\alpha_2$-macroglobulin. Atrolysin A interacts only very slowly with human $\alpha_2$-macroglobulin at 22°, and only at a higher temperature (37°) did the proteinase successfully cleave and become covalently bound to the inhibitor.[12] $\alpha_2$-Macroglobulin is primarily cleaved in the bait region by atrolysin A at the $\mathrm{Arg}^{696}\text{-}\mathrm{Leu}^{697}$ bond, the same bond cleaved by atrolysin C and plasmin, thrombin, trypsin, and thermolysin, and at a second bond at $\mathrm{His}^{694}\text{-}\mathrm{Ala}^{695}$ as determined by Edman degradation analysis of the digestion products. These results suggest that under normal conditions the inhibition of atrolysin A by $\alpha_2$-macroglobulin is less effective than is observed for the lower mass atrolysins and hence may have some influence on the hemorrhagic potency of atrolysin A.

*Atrolysin F*

The atrolysin F proteinase is the least well characterized of all of the atrolysins; however, sufficient data have been accumulated to term it an atrolysin confidently. Biochemical characterization of atrolysin F suggests that it is similar to atrolysin A in that both are members of the class P-III sharing similar molecular masses (~64–68 kDa). Zinc is found in a 1 : 1 molar ratio, and the proteinase demonstrates potent hemorrhagic activity, having a minimum hemorrhagic dose of 0.53 $\mu$g.[2] There is a question as to whether atrolysin F shares the same domain structure as atrolysin A as the sequence of atrolysin F has not been determined. From Ouchterlony radial diffusion it appears that the two proteinases are related but not identical.

The proteolytic specificity on oxidized insulin B chain is similar to that of atrolysin A, but additional sites of cleavage of the B chain by atrolysin F are seen. Biological activity of atrolysin F has also been investigated by digestion studies with fibrinogen. Fibrinogen is not cleaved by atrolysin F to produce a clot. When the fibrinogen digestion products are examined by SDS–PAGE, it is seen that the γ chain is hydrolyzed while the Aα and Bβ chains remain intact.[2] Therefore, in addition to producing hemorrhage, the proteinase also functions as a defibrinogenating agent.

## Other Crotalus atrox Metalloproteinases

As observed in [21] in this volume,[19] a total of 13 metalloproteinases have been isolated from C. atrox venom. Of these 6 have been sufficiently characterized such that they can be considered atrolysins and have been discussed above. The remaining 7 proteinases, namely, HT-g, atroxase, protease I, protease IV, collagenase, fibrinogenase, and new α-protease, will require additional characterization before being termed atrolysins. For most of these metalloproteinases we lack data on their insulin B chain cleavage specificities.

## Structure of Atrolysins

### General Structural Characteristics of Atrolysins

As described above, all of the atrolysins are zinc metalloproteinases containing zinc in a 1 : 1 molar ratio and share similar digestion specificities on insulin B chain. Furthermore, the atrolysins all share homologous pre-, pro-, and proteinase domains. Although minor sequence differences can be observed in these domains, the primary difference among the atrolysins (which gives rise to the organization of the atrolysins into the four protein classes, P-I through P-IV) is the presence of additional domains. The function of the additional domains appears to be modulation of the biological activity of the proteins.[13]

### Pre- and Prosequences

Until relatively recently the nature of the nascent structure of venom metalloproteinases as synthesized in the venom gland was unclear. The protein sequence of the proteinases as isolated from the crude venom always began in the proteinase domain, thus suggesting that the proteinases were not synthesized as proenzymes. However, with the first report of a cDNA sequence for a venom proteinase, atrolysin E, it became clear that the venom proteinases are synthesized with both pre- and prodomains which are processed off prior to storage of the proteins in the venom

---

[19] J. B. Bjarnason and J. W. Fox, this volume [21].

gland.[20] The signal sequences for atrolysins B, Cc, Cd, and E are all 18 residues in length, highly conserved, and predicted to be cleaved between Gly-18 and Ser-19 (Fig. 1).

From the cDNA sequences of the atrolysins there is a region of 169 amino acids following the signal sequence which comprises the prodomain of the proteins (Fig. 1). These sequences have 95% shared identity, which suggests an important function for this domain. Approximately 20 residues from the end of the prodomain there is a highly conserved sequence, PKMCGVT, which is absolutely conserved in all of the atrolysin precursors as well as in other venom metalloproteinase sequences such as trigramin, jararhagin, and rhodostomin.[20] We have proposed that this sequence may function in a fashion reminiscent to the "cysteine switch" or "Velcro mechanism" postulated for the matrix metalloproteinases.[20] The function of this region is to bind to the active-site zinc via the cysteinyl sulfur and prevent proteolytic activity. This feature can be tested with the use of synthetic peptides containing the consensus sequence. In the case of the MMPs the analogous consensus sequence is PRCGV/NPD, and peptides which contain that sequence are effective inhibitors of 72-kDa type IV collagenase.[21] Similar peptides containing the venom consensus sequence have been used in our laboratory and show specificity of atrolysin inhibition with regard to peptide sequence, size, and the oxidation state of the cysteinyl sulfur. In the case of the MMPs, the zymogen is activated by several proteolytic cleavages in the prodomain, the final one being autocatalytic to produce the active form of the enzyme.[22] The nature of atrolysin activation is as yet unclear but is likely to be analogous to MMP activation.

## Proteinase Domains and Spacer Region

Initial protein sequencing studies on venom metalloproteinases were often frustrated owing to the prevalence of blocked amino termini primarily the result of cyclization of glutamine residues to form the intractable pyroglutamyl N terminus.[14] This difficulty can be overcome by enzymatic removal of the residue with pyroglutamase, but the yields of resultant free amino-terminal amino acid are generally low. The identification of the nature of the blocked N terminus can be deduced from cDNA sequences of the protein or by mass spectroscopic analysis of a peptide fragment containing the blocked N terminus. From protein sequence analyses of the

[20] L. A. Hite, J. D. Shannon, J. B. Bjarnason, and J. W. Fox, *Biochemistry* **31,** 6203 (1992).
[21] W. G. Stetler-Stevenson, J. A. Talano, M. E. Gallagher, H. C. Krutzsch, and L. A. Liotta, *Am. J. Med. Sci.* **302,** 163 (1991).
[22] T. Crabbe, F. Willenbrock, D. Eaton, P. Hynds, A. F. Carne, G. Murphy, and A. J. Docherty, *Biochemistry* **31,** 8500 (1992).

atrolysins there is seen some difference in the processing site between the pro- and proteinase domains (Fig. 1). Atrolysins B, Cc, and Cd all have pyroglutamate-blocked N termini resulting from a cleavage between conserved glutamines. Atrolysin E has a slightly different structure in this region and is processed to give an asparaginyl residue as a free N terminus. The situation for atrolysin A is somewhat uncertain in that protein sequence data for the mature protein are not available (likely owing to a pyroglutaminyl residue), and therefore inference from the cDNA sequence must be made for the putative N terminus of the protein.

The proteinase domains of the atrolysins range from 202 to 204 amino acids in length, and all contain the signature active-site consensus sequence, $HELGHNLGX_2HDX_{10}CI/VM$. From crystallographic studies on a homologous venom proteinase, adamalysin, it is seen that the zinc atom is complexed by the three histidine residues in the active-site sequence along with a water molecule, giving rise to a tetrahedral coordination sphere of the metal.[4] The conserved CI/VM sequence observed C-terminal to the zinc binding site appears to be important in providing a hydrophobic "basement" for substrate binding. Interestingly, the zinc-binding ligands are mutually conserved in the venom metalloproteinases, the MMPs, and some of the mammalian metalloproteinases.[23] The methionyl residue C-terminal to the active site is also seen in all the venom metalloproteinases, the mammalian metalloproteinases, and the MMPs.

From protein sequence comparative analysis of the atrolysins, some details of the relationships among the homologous domains can be gained. It appears that in general atrolysin proteinase domains can be organized into two groups, those with three disulfide bonds and those with two. The disulfide bond arrangement for the atrolysins has been elucidated using standard protocols of proteolytic digestion of the unreduced protein, isolation of disulfide-bonded fragments, and subsequent sequence analysis of the fragments with the data compared to the known primary structure of the protein.[14] Atrolysins E and A have three disulfide bonds within the proteinase domain with a nested structure of $Cys^{308}$-$Cys^{388}$, $Cys^{348}$-$Cys^{372}$, and $Cys^{350}$-$Cys^{355}$. Although the primary structure of atrolysin F is not known, from the amino acid composition it can be deduced that it also has a three-disulfide bond proteinase domain. Atrolysin A has two additional cysteinyl residues, one at position 358 in the proteinase domain and another at position 404 in the spacer region. Amino acid compositional studies have not indicated the presence of a free sulfhydryl group, and therefore those two residues must be involved in disulfide bonds, either with one another

[23] T. G. Wolfsberg, J. F. Bazan, C. P. Blobel, D. G. Myles, P. Primakoff, and J. M. White, *Proc. Natl. Acad. Sci. U.S.A.* **90**, 10783 (1993).

or perhaps with the following disintegrin-like domain. The two-disulfide bond arrangement for atrolysins B, Cc, and Cd is $Cys^{308}$-$Cys^{388}$ and $Cys^{348}$-$Cys^{355}$.

The proteinase domain for the atrolysins ends at position 393 (Fig. 1) as confirmed by the protein sequence determination of atrolysins E and C, which have in their mature form only the proteinase domain comprising their structure.[14,20] In all the atrolysins, following the proteinase domain is a short "spacer" region of 21 amino acids, which in the case of the class P-I toxins is absent in the mature protein form. The spacer region serves, in the P-II through P-IV class atrolysins, as a splice site for an additional domain C-terminal to the proteinase domain.[13] The spacer sequences are very highly conserved. The spacer of atrolysin A has a cysteine residue at position 404 that we postulate is critical for the maintenance of its P-III class structure by forming a disulfide bond to a free cysteine residue in the succeeding disintegrin-like domain. In the case of atrolysin E this cysteine is absent, and the spacer region and disintegrin-like domain observed from the cDNA sequence are processed away from the nascent protein, leaving only the proteinase domain and consequently a P-Ia class structure for the mature protein. From the cDNA sequences of atrolysins B, Cc, and Cd there is observed neither a cysteine in the spacer region nor a following disintegrin-like domain. Therefore, translation of these structures from the mRNA to the nascent protein structure and subsequently the mature proteins would not require any proteolytic processing at the C-terminal region of the proteins.

*Disintegrin-like Domains*

Disintegrins are peptides isolated from the venoms of Crotalidae and Viperidae snakes. The peptides range from 49 to 83 amino acids in length and can be divided structurally into three classes based on the number of disulfide bonds present; 4, 16, or 7 (Fig. 2). All appear to have the general biological property of inhibiting the interaction of the platelet integrin receptor GP IIb/IIIa with fibrinogen and hence disrupting platelet aggregation.[24] In addition to the high density of disulfide bonds in these peptides, they are also distinguished by the presence of the signature Arg-Gly-Asp sequence often found in sequences of certain integrin ligands. The only disintegrin isolated to date which lacks the Arg-Gly-Asp sequence is barbourin from the venom of the crotalid snake *Sistrurus miliarius barbouri*. Instead of the Arg-Gly-Asp sequence, barbourin has the homologous Lys-

---

[24] C.-M. Teng and T.-F. Huang, *Thromb. Haemostasis* **65,** 624 (1991).

At-E    LRTDTVSTPVSGNELLEAGIECDCGSL----ENPCCYATTCKMRPGSQCAEGLCCDQ---CRFMKKGTVCRVSMVD-RNDDTCTGQSADCP-----RNG
At-A    LQTDIISPPVCGNELLEVGEECDCGSPRTC-RDPCCDAATCKLHSWVECESGECCQQ---CKFTSAGNVCRPARSECDIAESCTGQSADCPTDDFHRNG
HR1B    SKTDIVSPPVCGNELLEAGEECDCGSPENC-QYQCCDAASCKLHSWVKCESGECCDQ---CRFRTAGTECRAAESECDIPESCTGQSADCPTDRFHRNG
Jara    LGTDIISPPVCGNELLEVGEECDCGTPENC-QNECCDAATCKLKSGSQCGHGDCCEQ---CKFSKSGTECRASMSECDPAEHCTGQSSECPADVFHKNG
RVVX    LRKDIVSPPVCGNEIWEEGEECDCGSPANC-QNPCCDAATCKLKPGAECCGNGLCCYQ---CKIKTAGTVCRRARDECDVPEHCTGQSAECPRDQLQQNG

EAPI    FPCNFDDFQFCGNKKLDEGEECDCGPPQEC-TNPCCDAHTCVLKPGFTCAEGECCES---CQIKKAGSICRPAEDECDFPEMCTGHSPACPKDQFRV
PH30β   SNPVCGNNRVEQEDCDCGSQEEC-QDTCCDAATCRLKSTSRCAQGPCCNQ---CEFKTKGEVCRESTDECDLFEYCNGSSGAC-QEDLY

Bitistat3  VSPPVCGNKILEQEDCDCGSPANC-QDQCCNAATCKLTPGSQCNHGECCDQ---CKFKKARTVCRIARGD-WNDDYCTGKSSDCPWNH
Trigramin       EAGEDCDCGSP----ANPCCDAATCKLIPGAQCGEGLCCDQ---CSFIEEGTVCRIARGD-DLDDYCNGRSAGCPRNPFH
Rhodostomin         GKECDCSSP----ENPCCDAATCKLRPGAQCGEGLCCEQ---CKFSRAGKICRIPRGD-MPDDRCTGQSADCPRYH
Eristicophin                LRQEEPCATGPCCRR---CKFKRAGKVCRVARGD-WNNDYCTGKSCDCPRNPWNG
Echistatin                ECESGPCCRN---CKFLKEGTICKRARGD-DMDDYCNGKTCDCPRNPHKGPAT
                                                                    ***

FIG. 2. Comparison of atrolysin A and E disintegrin-like domain sequences with sequences from mammalian reproductive proteins and true disintegrins. The RGD tripeptide region is noted by asterisks. References for sequences are as follows: HR1B, H. Takeya, K. Oda, T. Miyata, T. Omori-Satoh, and S. Iwanaga. *J. Biol. Chem.* **265,** 16068 (1990); Jara (jarahagin), M. J. I. Paine, H. P. Desmond, R. D. G. Theakson, and J. M. Crampton, *J. Biol. Chem.* **267,** 22869 (1992); EAP-I, A. C. F. Perry, R. Jones, P. J. Barker, and L. Hall, *Biochem. J.* **286,** 671 (1992); PH30B, Wolfsberg *et al.*[23]; Bitistat3 (bitistatin 3), R. J. Shebuski, D. R. Ramjit, G. H. Bencen, and M. A. Polokoff, *J. Biol. Chem.* **264,** 21550 (1989); trigramin, L.-C. Au, Y.-B. Huang, T.-F. Huang, G.-W. Teh, H.-H. Lin, and K.-B. Choo, *Biochem. Biophys. Res. Commun.* **181,** 585 (1989); rhodostomin, M. S. Dennis, W. J. Henzel, R. M. Pitti, M. T. Lipari, M. A. Napier, T. A. Deisher, S. Bunting, and R. A. Lazarus, *Proc. Natl. Acad. Sci. U.S.A.* **87,** 9359 (1989); eristocophin, L.-C. Au, Y.-B. Hwang, T.-F. Huang, G.-W. Teh, H.-H. Lin, and K.-B. Choo, *Biochem. Biophys. Res. Commun.* **181,** 585 (1991); and echistatin, Z.-R. Gan, R. J. Gould, J. W. Jacobs, P. A. Friedman, and M. A. Polokoff, *J. Biol. Chem.* **263,** 19827 (1988). (Adapted with permission from Hite *et al.*[13])

Gly-Asp sequence and has been shown to have a certain degree of specificity for the GP IIb/IIIa integrin.[25]

From the translated cDNA sequence for atrolysins E and A there is a domain homologous to the disintegrins present immediately following the spacer region (Figs. 1 and 2). These domains, however, differ from the majority of the disintegrins in that they lack the Arg-Gly-Asp sequence and in fact bear little conservation to that sequence in the region. We have termed these domains "disintegrin-like" as they lack the Arg-Gly-Asp sequence and are found in the context of a metalloproteinase domain.[20] Based on sequence similarity with the disintegrins, it is likely the disulfide bond arrangement for the disintegrin-like domains is identical to that determined for the disintegrins,[26] such that the arrangement in the disintegrin-like domains would be $Cys^{415}$-$Cys^{433}$, $Cys^{417}$-$Cys^{428}$, $Cys^{427}$-$Cys^{450}$, $Cys^{441}$-$Cys^{447}$, $Cys^{446}$-$Cys^{472}$, and $Cys^{460}$-$Cys^{479}$. In atrolysin A there is an "odd" cysteinyl residue at position 358 that is likely disulfide bonded to one of the three "odd" cysteinyl residues in the proteinase domain, spacer region, or high-cysteine domain.

The role of the disintegrin-like domains in modulating the biological activity of atrolysins E, A, and presumably F is uncertain. However, preliminary studies in our laboratory clearly indicate that recombinant disintegrin-like domains, produced by polymerase chain reaction (PCR) amplification of that region of the cDNA, insertion into a bacterial expression vector, and subsequent production of the recombinant protein, are very potent inhibitors of platelet aggregation and cell adhesion. Furthermore, many of these activities can be produced by peptide sequences identical to those in the region in the disintegrin-like domain which are at the same position of the Arg-Gly-Asp sequence in the disintegrins. Nevertheless, it is uncertain how the disintegrin-like domain modulates the activity of the atrolysin. Perhaps the platelet aggregation inhibitory activity of the disintegrin-like domain functions to potentiate the hemorrhagic activity directly by preventing subsequent thrombus formation, or the domain may act by binding to a cellular integrin to localize the proteinase to a critical region such as the endothelial cells of capillaries or fibroblasts in the stroma for concentrated disruption of the extracellular matrix. The method our laboratory is using to address these issues is recombinant DNA technology to make chimeric atrolysins and then examine them for altered biological activities.

Finally, it should be noted that some of the disintegrin peptides have been demonstrated to be the products of proteolytic processing of precursor

[25] R. M. Scarborough, J. W. Rose, M. A. Hsu, D. R. Phillips, V. A. Fried, A. M. Campbell, L. Nannizzi, and I. F. Charo, *J. Biol. Chem.* **266,** 9359 (1991).

[26] M. Adler, P. Carter, R. A. Lazarus, and G. Wagner, *Biochemistry* **32,** 282 (1993).

FIG. 3. Comparison of the high-Cys domain of atrolysin A with homologous regions from other venom metalloproteinases. References for sequences are as follows: HR1B from *Trimeresurus flavoviridis*, H. Takeya, K. Oda, T. Miyata, T. Omori-Satoh, and S. Iwanaga, *J. Biol. Chem.* **265,** 16068 (1990); jarahagin from *Bothrops jararaca*, M. J. I. Paine, H. P. Desmond, R. D. G. Theakson, and J. M. Crampton, *J. Biol. Chem.* **267,** 22869 (1992); and RVVX from *Vipera russelli russelli*, Takeya et al.[28] **C,** Absolutely conserved Cys residues; ^, other Cys residues.

proteins homologous to the atrolysins,[27] and thus it is probable that all of the disintegrins are produced by such a mechanism.

*High-Cysteine Domain*

The high-cysteine domain, which is characterized by the high number of cysteinyl residues, is found in the atrolysins of the P-III (and presumably P-IV) class as well as in the mammalian reproductive proteins.[13] It is of great interest in that very little is known about its function. In atrolysin A the high-cysteine domain is 121 residues in length and has 13 cysteinyl residues, of which all are probably involved in disulfide bonds. As previously discussed, in the case of atrolysin A one of the cysteinyl residues is likely disulfide bonded to one of the other "odd" cysteinyl residues in either the proteinase domain, spacer region, or disintegrin-like domain. As seen in Fig. 3[28] the positions of the cysteinyl residues in the domain are strictly conserved among the venom metalloproteinases that have the domain and relatively conserved in the homologous domain of the mammalian reproductive metalloproteinases. Comparison of the venom high-cysteine domain to the homologous domain in the mammalian reproductive metalloproteinases leads to the suggestion of a two-subdomain organization, with the N-terminal subdomain a of the venom structure more highly conserved to the like region in the mammalian proteinase (Fig. 3).

[27] L.-C. Au, Y.-B. Huang, T.-F. Huang, G.-W. Teh, H.-H. Lin, and K.-B. Choo, *Biochem. Biophys. Res. Commun.* **181,** 585 (1991).

[28] H. Takeya, S. Nishida, T. Miyata, S.-I. Kawada, Y. Saisaka, and S. Iwanaga, *J. Biol. Chem.* **267,** 14109 (1992).

We have postulated that the high-cystein domain is critical for maintenance of a structural supraorganization of the other domains to maximize
the composite biological activity of the protein.[18] From comparison to other
venom metalloproteinases, it can be suggested that the lack of the high-
cysteine domain C-terminal to the disintegrin-like domain often leads to a
proteolytic processing that removes the disintegrin-like domain from the
proteinase domain, such as the case with atrolysin E. However, that is not
always true as there is experimental evidence for stable venom metalloproteinases which lack a high-cysteine domain but have the characteristic
proteinase plus disintegrin-like P-II structure.[28] Further experimentation
with chimeric venom metalloproteinases with various domain constructs
should lead to the solution of these outstanding structure–function questions.

*Biosynthesis of Atrolysins*

Many features of the biosynthesis of the atrolysins have been discussed
in the context of the domain structure of the proteinases. However, several
critical points should be individually summarized (Fig. 4). From the cDNA
structures for atrolysins A, B, Cc, Cd, and E, the proteinase transcripts can
be classified into one of three groups, N-I through N-III, based on the
presence of particular nucleotide sequences which code for various domains.[13] For example, atrolysins B, Cc, and Cd all are transcribed having
only the pre-, pro-, and proteinase domains and represent the N-I class,
whereas atrolysin E has those domains and the additional disintegrin-like
domain and thus is a member of the N-II class. The atrolysin A transcript
has homologous coding sequence for all of the domains seen in the N-II
members plus coding sequence for the high-cysteine domain and hence is
an N-III member. No atrolysin or coding sequence for an atrolysin which
can be classified as a N-IV member has yet been found, although there
have been proteinases isolated from other venoms whose DNA transcripts
are most likely of the N-IV class.[28]

The presence of so many different metalloproteinases in crotalid and
viperid venoms is in part due to the apparent proteolytic processing of the
nascent proteins translated from the four nucleotide classes of structures.[18]
The P-I protein class of venom metalloproteinases as exemplified by atrolysins B, Cc, and Cd do not undergo any further proteolytic processing at the
C-terminal regions other than the removal of the spacer region sequence.
Processing of the proteins however, does occur to remove the pre- and
prodomains to give rise to the active, mature proteins as isolated from the
venom. P-II proteins apparently may or may not be processed to eliminate
the disintegrin-like domain. Atrolysin E lacks its disintegrin-like domain in

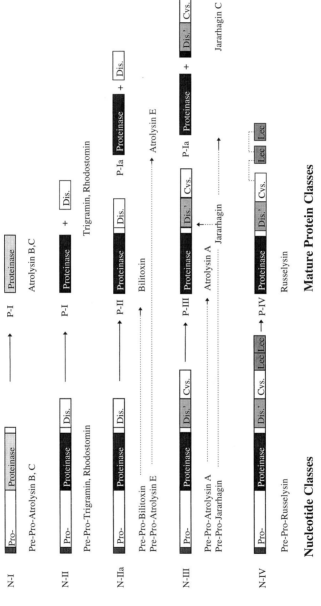

FIG. 4. Schematic representation of translation and posttranslational proteolytic processing of the nucleotide and protein class I–IV structures for the venom metalloproteinases. The schematic indicates the three known nucleotide classes (I–III) and the fourth hypothesized class (IV). The right-hand side of the schematic shows the possible protein products resulting from posttranslational processing, including how most (if not all) disintegrins are likely produced. Hi-cys, High cysteine. (Adapted with permission from Hite *et al.*[13])

the mature protein, whereas a homologous P-II venom metalloproteinase, bilitoxin, maintains the disintegrin-like domain in the mature form.[29] Whether the P-III and P-IV classes *in vivo* undergo proteolytic processing to give rise to one of the smaller protein classes is unknown. However, there are reports that under certain conditions the P-III proteinases, including atrolysin A, can demonstrate autocatalytic processing to produce truncated forms of the proteinases lacking the disintegrin-like and high-cysteine domains.[18,30] For these processing events to be totally understood, more data representing the DNA structures of the venom proteinases and their mature protein sequences must be gathered. Furthermore, investigations as to the nature of the proteinases responsible for the proteolytic processing of the nascent forms of these proteinases should also be undertaken.

## Conclusions

This and the preceding chapter[19] have illustrated the complexity of the venom metalloproteinases in general and the atrolysins in particular. Studies on the reprolysin subfamily of the metalloproteinases, especially the atrolysins, have generated a new understanding of the biochemical and structural characteristics of these metalloproteinases. Insight has also been gained on the processes involved in the biological synthesis of the proteinases from the level of the DNA to the mature protein. There are many venom metalloproteinases which have yet to be characterized in the depth seen for the atrolysins, and certainly there are many more venom metalloproteinases which have yet to be isolated. The values of such studies have become even more obvious with the discovery of the mammalian metalloproteinases that share many of the structural and functional characteristics of the venom metalloproteinases. Together, studies on the atrolysins and other members of the reprolysin subfamily have strengthened the field of proteinase research and should continue to do so in the future.

[29] K. Imai, T. Nikai, H. Sugihara, and C. L. Ownby, *Int. J. Biochem.* **21**, 667 (1989).
[30] H. Takeya, S. Nishida, N. Nishino, Y. Makinose, T. Omori-Satoh, T. Nikai, H. Sugihara, and S. Iwanaga, *J. Biochem.* (*Tokyo*) **113**, 474 (1993).

## [23] Membrane-Associated Metalloproteinase Recognized by Characteristic Cleavage of Myelin Basic Protein: Assay and Isolation

*By* LINDA HOWARD *and* PAUL GLYNN

### Introduction

Artifactual proteolysis following tissue homogenization and cell breakage is a common problem in biochemistry. However, it was the observation of a pattern of protein degradation likely to reflect an *in vitro* artifact that serendipitously allowed the identification of a novel metalloproteinase in brain myelin membrane preparations.[1,2] This proteinase, subsequently demonstrated in a variety of tissues, has been cloned and shown to be a member of a newly recognized family of mammalian proteins related to the disintegrin–metalloproteinase precursors of certain snake venoms (see also [13] in this volume).[3] The physiological substrates and functions of the metalloproteinase are unknown. At present, the only generally applicable assay for its detection is the original one, namely, the generation of a characteristic fragment from myelin basic protein. Here, the observations leading to the identification of the metalloproteinase are briefly summarized, and methods for its isolation and assay are described in detail.

A particulate fraction enriched in myelin membranes can be isolated easily from brain homogenates.[4] Two polypeptides, proteolipid protein (PLP; an integral membrane protein) and myelin basic protein (MBP; an extrinsic membrane protein), comprise over 75% of the total protein in the membrane preparations. When myelin preparations are incubated in neutral pH isotonic salt solutions, MBP dissociates from the cytoplasmic membrane surface to which it is normally attached by electrostatic interactions. The dissociated MBP becomes accessible to, and is degraded by, proteinases in the myelin preparation.[1] Among the standard inhibitors of the four classes of proteinases, 1,10-phenanthroline (PA) and dithiothreitol (DTT) are the most effective in reducing the overall breakdown of MBP, suggesting that the dominant degradative activity is due to a metalloproteinase.[2] Degradation of MBP by the myelin-associated metalloproteinase leads to the formation of a characteristic 10-kDa fragment easily detected

[1] P. Glynn, A. Chantry, N. Groome, M. L. Cuzner, *J. Neurochem.* **48,** 752 (1987).
[2] A. Chantry, C. Earl, N. Groome, and P. Glynn, *J. Neurochem.* **50,** 688 (1988).
[3] L. Howard, X. Lu, S. Mitchell, S. Griffiths, and P. Glynn, in preparation.
[4] W. T. Norton and S. E. Poduslo, *J. Neurochem.* **21,** 749 (1973).

on sodium dodecyl sulfate–polyacrylamide gel electrophoresis (SDS–PAGE).[1,2] This results from a single cleavage of the 170-residue protein between Pro-73 and Gln-74, giving rise to the C-terminal 74–170 (10-kDa) fragment. [MBP(74–170)].[5]

The metalloproteinase has been isolated from bovine brain myelin preparations and partially characterized.[6] It is a single-chain N-glycosylated polypeptide of about 58 kDa, and endoglycosidase F treatment reduces the apparent size to about 52 kDa. The purified enzyme cleaves many peptide bonds in MBP, but MBP(1–73) and MBP(74–170) are the most prominent degradation products. The proteinase is active between pH 7 and pH 9, is inhibited by PA and DTT but not by phosphoramidon, and is not significantly activated by addition of millimolar amounts of $Ca^{2+}$. Histones, but not casein, are also degraded by the purified enzyme; however, its peptide bond and substrate preferences have not been investigated in any detail.[6]

A monoclonal antibody, designated CG4, has been raised to the proteinase isolated from bovine brain myelin. CG4 detects a 58-kDa band on Western blots of glycoproteins extracted from various bovine tissues including peripheral nerve, lung, spleen, heart, kidney, adrenal, and skeletal muscle.[7] By immunohistochemistry, CG4 reacts with the myelin-forming cells of the central (oligodendrocyte) and peripheral (Schwann cell) nervous systems.[8] CG4 does not cross-react with the metalloproteinase in rat and human myelin, but an enzyme activity which degrades MBP to the characteristic product MBP(74–170) is present in particulate fractions from the brains of mammals, birds, reptiles, and amphibians.[8]

Assay Procedure

Detection of the PA-sensitive formation of MBP(74–170) from MBP, monitored by SDS–PAGE, remains the only general assay method for this proteinase. In myelin preparations the breakdown of endogenous MBP can be monitored by SDS–PAGE, but, in studying the enzyme in other tissues, exogenous MBP must be added. In the latter situation it is necessary to extract a tissue glycoprotein fraction as low molecular weight tissue proteins will obscure visualization of MBP(74–170) on SDS–PAGE. Furthermore, in tissues such as kidney in which other proteinases are abundant, a cocktail of inhibitors for serine and cysteine class proteinases and phosphoramidon-

[5] N. Groome, A. Chantry, C. Earl, J. Newcombe, J. Keen, J. Findlay, and P. Glynn, *J. Neuroimmunol.* **19**, 77 (1988).
[6] A. Chantry, N. A. Gregson, and P. Glynn, *J. Biol. Chem.* **264**, 21603 (1989).
[7] A. Chantry and P. Glynn, *Biochem. J.* **268**, 245 (1990).
[8] A. Chantry, N. A. Gregson, and P. Glynn, *Neurochem. Res.* **17**, 861 (1992).

sensitive metalloproteases is added to reduce degradation of both exogenously added MBP and of any MBP(74–170) product that is formed. Even then, it may be necessary to fractionate the glycoproteins on DEAE-Sephacel and analyze the fractions by SDS–PAGE, in order to verify that formation of MBP(74–170) copurifies with a 58-kDa polypeptide.

The assay procedure is rather cumbersome but does provide a means to identify and then isolate the metalloproteinase in extracts from any tissue or cultured cells in which it is present. The assay can be quantified by densitometry of Coomassie blue-stained gels (provided a relatively purified enzyme preparation is used) or by using [125]I-labeled MBP and counting [125]I-MBP(74–170) excized from the gel. [Incubation of [125]I-MBP with the purified metalloproteinase does not result in the formation of trichloroacetic acid (TCA)-soluble [125]I-labeled MBP fragments.[6]] To exemplify the general methods for studying this metalloproteinase, its assay and isolation from bovine kidney are detailed below.

*Isolation of Myelin Basic Protein for Use as Metalloproteinase Substrate*

The MBP preparations used as substrate in the metalloproteinase assay must be free of autolytic degradation products. We have found the MBP preparations available from Sigma (St. Louis, MO) to be unsatisfactory in this respect. Several different methods for MBP isolation have been described[9]; the one we have selected involves initial isolation of myelin membranes from brain, followed by acid extraction and gel filtration. In the past, bovine brain tissue has been the starting material of choice, but its availability in Britain has been severely limited by events subsequent to the outbreak of bovine spongiform encephalomyelitis in the late 1980's. However, porcine brain is also an acceptable starting material for the isolation of MBP.

*Isolation of Myelin Membranes.* The procedure is essentially that of Norton and Poduslo[4] and is outlined only briefly. Brain tissue (fresh or frozen) is homogenized in 10 volumes of 0.32 $M$ sucrose in medium A (10 m$M$ Tris-HCl, pH 7.4/0.02% (w/v) sodium azide) and then carefully layered on 0.8 $M$ sucrose in medium A. After centrifugation (40,000 $g$; 40 min; 4°) the floating layer of crude myelinated axons is recovered from the interface. This material is hypotonically shocked by dilution in medium A to remove soluble axonal components. Crude myelin membranes are recovered by centrifugation, and aliquots are stored at $-20°$.

*Isolation of Myelin Basic Protein from Myelin Membranes.* Membranes equivalent to 100 mg protein are thawed and pelleted by centrifugation (40,000 $g$; 40 min; 4°) in medium A. The pellets are resuspended by homogenizing in 50 ml of 0.1 $M$ HCl/5% methanol, then stirred or shaken for 40

[9] P. R. Dunkley and P. R. Carnegie, *Res. Methods Neurochem.* **2**, 219 (1974).

min at 37°. After centrifugation (40,000 $g$; 15 min; 4°) the clear supernatant is saved and concentrated to less than 5 ml by ultrafiltration through an Amicon (Danvers, MA) YM10 membrane. This amount of concentrate is sufficient for two runs on a column (1.6 × 97 cm) of Sephadex G-50 equilibrated in 10 m$M$ HCl. The column is run at 10 ml/h (4°), and 5 ml fractions are collected. The MBP elutes at about 0.42 of the total column volume, and fractions around this value are analyzed by SDS–PAGE on a 15% polyacrylamide minigel. Fractions containing only intact MBP (18.4 kDa) are pooled and then freeze-dried in aliquots equivalent to 0.2–0.5 mg protein for long-term storage (at −20°). Note that MBP migrates anomalously on SDS–PAGE so that the intact polypeptide (18.4 kDa) runs at about the same level as the standard 21-kDa molecular mass marker protein.

*Assay for Myelin Basic Protein-Degrading Metalloproteinase Activity*

Stock solutions (100× concentrates) of the following proteinase inhibitors (obtainable from Sigma) are prepared: (1) 100 m$M$ benzamidine; (2) 0.1 m$M$ leupeptin; (3) 1 m$M$ E-64 [*trans*-epoxysuccinyl-L-lencylamido-(4-guanidino)butane]; (4) 5 m$M$ phosphoramidon; (5) 100 m$M$ phenylmethylsulfonyl fluoride (PMSF); (6) 1$M$ $N$-ethylmaleimide; and (7) 200 m$M$ 1,10-phenanthroline (PA). (Solutions 1 and 2 are made in water, 3 in 50% (v/v) methanol, and the remainder in methanol.) One volume of each of the inhibitors 1–6 are mixed with 1 volume of either methanol or PA to produce a 14.3× concentrated inhibitor cocktail. The PA-containing cocktail (inhibitor 7) should only be used when attempting to inhibit the required metalloproteinase activity (e.g., Fig. 2).

Aliquots (37.2 μl) of the proteinase-containing solution (glycoprotein or more purified fraction) are incubated (37°; 30 min) in Eppendorf tubes with 2.8 μl of the 14.3× concentrated inhibitor cocktail. Freeze-dried aliquots of MBP are dissolved at 2.3 mg/ml in water, then 4 volumes of this solution are mixed with 1 volume of 750 m$M$ Tris-HCl, pH 7.4. Aliquots (20 μl) of this buffered MBP solution are then added to the reaction mix and incubation (37°) continued for 3 hr. The reaction is stopped by the addition of 15 μl of 10% SDS/10% sucrose/0.01% bromphenol blue (all w/v), and 25-μl aliquots are run on a 15% polyacrylamide-SDS minigel and stained with Coomassie blue. The required metalloproteinase activity is indicated by the disappearance of the intact MBP band and appearance of the characteristic 10-kDa MBP fragment.

Isolation of Metalloproteinase from Bovine Kidney

*Step 1: Solubilization and Isolation of Kidney Glycoproteins.* For convenience, the procedure described uses a 6 × 250 ml rotor (Beckman, Palo Alto, CA, Type 19) capable of generating at least 50,000 $g_{max}$ in order to process relatively large volumes. The method can, of course, be applied to smaller volume rotors.

Bovine kidneys (fresh or frozen) are homogenized in 10 volumes of medium A (10 m$M$ Tris-HCl, pH 7.4/0.02% sodium azide) and a crude particulate fraction prepared by centrifugation (54,000 $g$) for 30 min. The particulate fraction may be used immediately or stored at $-20°$. The particulate fraction (1200–1500 mg protein) is suspended in 600 ml medium A, then 300 ml of 9% (w/v) Triton X-100 in medium A is added, and the mixture is stirred for 10 min at room temperature. Four hundred fifty milliliters of medium B (450 m$M$ Tris-HCl, pH 7.4/6 m$M$ CaCl$_2$/6 m$M$ MgCl$_2$) is then added and stirring continued for a further 1 hr at room temperature followed by centrifugation (54,000 $g$; 1 hr; 4°). The supernatant is saved, and two further 1200- to 1500-mg batches of particulate fraction are processed identicially.

The three Triton extracts (~3.5 liters) are pooled and pumped (overnight; 4°) through a concanavalin A (Con A)-Sepharose column (5 × 5.5 cm) which has been equilibrated with 0.1% Triton X-100 in Con A buffer (50 m$M$ Tris-HCl, pH 7.4/100 m$M$ NaCl/2 m$M$ CaCl$_2$/2 m$M$ MgCl$_2$/0.02% sodium azide). The next day, at room temperature, the column is washed sequentially with the following: (i) 300 ml Con A buffer containing 0.1% Triton X-100; (ii) 300 ml Con A buffer alone; and (iii) 300 ml Con A buffer containing 0.3% CHAPS 3-[(3-cholamidopropyl) dimethylammonio]-1-propane sulfonate. Finally, glycoproteins are eluted from the column with 400 ml of 0.5 $M$ methylmannoside in 0.3% CHAPS/Con A buffer; the first 50 ml of eluate is discarded and the following 350 ml saved. (The column is regenerated by washing in 0.1% Triton X-100/Con A buffer.) The eluate is ultraconcentrated to less than 50 ml in a stirred Amicon cell (YM10 membrane) and then dialyzed overnight (4°) against 1 liter of 10 m$M$ ammonium bicarbonate/0.3% CHAPS.

*Step 2: DEAE-Sephacel Chromatography.* At room temperature, the dialyzed Con A eluate is pumped (at about 130 ml/hr) onto a column of DEAE-Sephacel (1.6 × 12 cm) equilibrated in 10 m$M$ ammonium bicarbonate/0.3% CHAPS. The column is washed with 80 ml of 10 m$M$ ammonium bicarbonate/0.3% CHAPS, and then a linear gradient comprising 100 ml each of 10 m$M$ and 300 m$M$ ammonium bicarbonate/0.3% CHAPS is applied. Ten-milliliter fractions are collected and aliquots of each assayed for MBP-degrading activity as described above. At least three distinct MBP-degrading activities are resolved on DEAE-Sephacel, with peaks occurring around fractions 6, 11, and 16 (Fig. 1). When the three fractions are reassayed in the presence PA, only the early peak (fraction 6) is clearly seen to contain a metalloproteinase activity which generates the characteristic 10-kDa MBP fragment (Fig. 2). Note in both Figs. 1 and

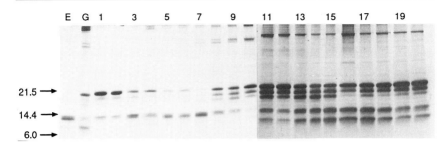

FIG. 1. DEAE-Sepharose fractionation of MBP-degrading activity in glycoproteins solubilized from bovine kidney. Solubilized bovine kidney glycoproteins were chromatographed on DEAE-Sephacel, and aliquots from the 20 DEAE fractions and the crude glycoprotein fraction (G) were assayed for MBP-degrading activity as described in the text. For comparison, MBP degradation by a sample (0.2 μg) of the pure metalloproteinase isolated from bovine brain myelin is shown in lane E.

2 that it is difficult to discern formation of the 10-kDa MBP fragment when the crude glycoprotein fraction is assayed.

*Step 3: CM-Sepharose Chromatography.* The pooled metalloproteinase-containing DEAE fractions (30 ml) are dialyzed overnight (4°) against 1 liter of 10 m$M$ sodium acetate, pH 5.0/0.3% CHAPS. After brief centrifuga-

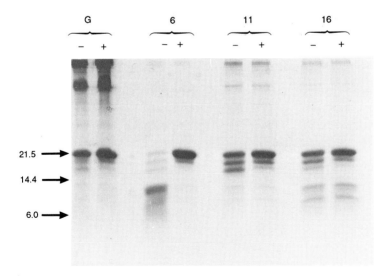

FIG. 2. Degradation of MBP by DEAE-Sephacel-fractionated bovine kidney glycoproteins: effect of 1,10-phenanthroline. Crude solubilized kidney glycoproteins (G) and DEAE-Sephacel fractions 6, 11, and 16 (see Fig. 1) were assayed for MBP degradation in the absence (−) or presence (+) of 2 m$M$ PA.

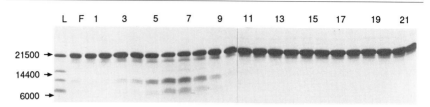

FIG. 3. CM-Sepharose fractionation of DEAE-Sephacel-purified MBP-degrading metallo-proteinase. DEAE-Sephacel fractions 5–7 (see Fig. 1) were pooled and chromatographed on CM-Sepharose. The load (L), flow-through (F), and 22 gradient fractions were assayed for MBP-degrading activity.

tion (2000 $g$; 10 min) to remove precipitated protein, the pool is pumped (at 50 ml/hr; room temperature) onto a small column (1.6 × 1.5 cm) of CM-Sepharose equilibrated in 10 m$M$ sodium acetate, pH 5.0/0.3% CHAPS. The column is washed with 25 ml of the starting buffer, and then a gradient comprising 25 ml each of starting buffer and starting buffer containing 0.3 $M$ NaCl is applied. Fractions (2.5 ml) are collected and 37.5 $\mu$l of each assayed as described above. A single peak of MBP-degrading activity elutes

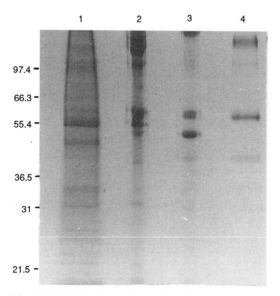

FIG. 4. Analysis by SDS–PAGE of metalloproteinase isolation from bovine kidney. Fractions from successive stages of the isolation protocol were subjected to SDS–PAGE on a 10% polyacrylamide and stained with Coomassie blue. Lane 1, Crude particulate fraction (13 $\mu$g protein of 4500 mg); 2, crude glycoprotein (13 $\mu$g of 96 mg); 3, DEAE fractions 5–7 (6 $\mu$g of 2.6 mg); 4, CM fractions 5–8 (3 $\mu$g of 0.184 mg).

at about 80–120 m$M$ NaCl (fractions 5–8) (Fig. 3). Analysis of the pooled peak fractions by SDS–PAGE on a 10% minigel (Fig. 4) indicates that the main component of this material is a 58-kDa polypeptide, the expected size for the metalloproteinase.[6] The 58-kDa polypeptide can be finally isolated by gel-filtration chromatography on Superose (cf. Ref. 6). However, this is not generally necessary; in fact, the high molecular mass component in lane 4 (Fig. 4) appears to be a disulfide-linked aggregate of the 58-kDa monomer (data not shown). Figure 4 also indicates the amounts of total protein at each stage of the purification; overall, 184 $\mu$g of the CM-purified protein is obtained from 4500 mg protein of the crude particular fraction.

### Acknowledgment

We thank SmithKline Beecham for financial support of L. H.

# [24] Serralysin and Related Bacterial Proteinases

*By* Hiroshi Maeda and Kazuyuki Morihara

### Introduction

*Serratia marcescens* and *Pseudomonas aeruginosa* produce zinc metalloproteinases that exhibit specificity in the cleavage of peptide bonds completely different from that of the thermolysin family, showing similar, broad specificities in the cleavage of oxidized insulin B chain. The enzymes form the serralysin subfamily of family M12, as defined by Rawlings and Barrett (in this volume [13]). Members of the subfamily are the alkaline proteinases from *Pseudomonas aeruginosa*[1,2] and *Serratia marcescens*,[3] as well as enzymes from *Pseudomonas fragi, Serratia* sp. E-15,[1] *Proteus mirabilis*[1] and *Escherichia freundii*.[1] There has been confusion over the identity in physicochemical and enzymatic properties[3–6] of the proteinases of *S. marcescens*,

[1] K. Morihara and K. Oda, *in* "Microbial Degradation of Natural Products" (G. Winkelmann, ed.), p. 293. VCH, Weinheim, 1992.
[2] K. Morihara, *Biochim. Biophys. Acta* **73**, 113 (1963).
[3] C. J. Decedue, E. A. Broussand II, A. D. Lansor, and H. D. Braymer, *Biochim. Biophys. Acta* **569**, 293 (1979).
[4] K. Matsumoto, H. Maeda, K. Takata, R. Kamata, and R. Okamura, *J. Bacteriol.* **157**, 225 (1984).
[5] D. Lyerly and A. Kreger, *Infect. Immun.* **24**, 411 (1979).
[6] K. Miyata, K. Maejima, K. Tomoda, and M. Isono, *Agric. Biol. Chem.* **34**, 310 (1970).

many of which were from clinical isolates, and with few exceptions[4] the studies focus on only one peak in each *Serratia* strain. Early studies reported a wide range of $M_r$ values (from 30,000 to over 60,000).

The primary structures of metalloproteinases of the serralysin subfamily from *P. aeruginosa*,[7,8] *Serratia* sp. E-15,[9] and *Erwinia chrysanthemi*[10,11] are very similar. The tertiary structures of metalloproteinases from *P. aeruginosa*[12] and *Serratia* sp. E-15.[13] have been also reported.

*Serratia marcescens*, as represented by the clinically isolated strain kums 3958, is a nonpigment-forming strain, and it produces three, or perhaps four, discrete proteinases of which the 56K proteinase is the most abundant. The 56K proteinase is immunologically identical with the E-15 proteinase and has been the most extensively studied of the three *Serratia* proteinases because it is obtained from pathogenic clinical isolates and is produced in the largest amounts. Both the 56K and 60K proteinases are inhibited by 10 m$M$ EDTA and 10 m$M$ tetramethylenepentamine,[4,14] and they are similar zinc proteinases. Each contains one zinc atom per mole as revealed by neutron activation analysis,[14] and they have somewhat similar amino acid compositions.[4] However, there are two distinct differences between them. Antibody raised against the 56K proteinase inhibited 96% of its activity but did not affect 60K proteinase activity. Conversely, antibody raised against the 60K proteinase inhibited its activity but did not affect the 56K proteinase at all. The substrate specificity is slightly different, as discussed later. Antibody raised against a crude culture filtrate of *S. marcescens* kums 3958 showed immunoprecipitation reactions against all the proteinases, with the three distinct precipitin lines showing no spurs.[4] Therefore, we can conclude that there are at least three immunologically distinct proteinases, two of which can be considered serralysins.[1,13] Metal-depleted 56K and 60K proteinases, which have no proteolytic activity, can be reactivated by $Zn^{2+}$. However, only the 56K enzyme can be reactivated by $Fe^{2+}$, and $Ca^{2+}$ activates both 56K and 60K proteinases, to 10 and 23%, respectively. Among other metals, $Cu^{2+}$ showed very weak activation of 9 and 4%,

[7] K. Okuda, K. Morihara, Y. Atsumi, H. Takeuchi, S. Kawamoto, H. Kawasaki, K. Suzuki, and J. Fukushima, *Infect. Immun.* **58**, 4083 (1990).

[8] F. Duong, A. Lazdunski, B. Cami, and M. Murgier, *Gene* **121**, 47 (1992).

[9] K. Nakahama, K. Yoshimura, R. Marumoto, M. Kikuchi, I. S. Lee, T. Hase, and H. Matsubara, *Nucleic Acids Res.* **14**, 5843 (1986).

[10] P. Delepelaire and C. Wandersman, *J. Biol. Chem.* **264**, 9083 (1989).

[11] G. S. Dahler, F. Barras, and N. T. Keen, *J. Bacteriol.* **172**, 5803 (1990).

[12] U. Baumann, S. Wu, K. M. Flaherty, and D. B. McKay, *EMBO J.* **12**, 3357 (1993).

[13] K. Hamada, H. Hiramatsu, R. Hujiwara, Y. Katsuya, Y. Hata, and Y. Katsube, *in* "Protein Structure" (*In Japanese*), Proc. 45th Symposium. Sept. 11, 1994, Osaka, p. 81.

[14] K. Matsumoto, H. Maeda, and R. Okamura, *J. Biochem.* (*Tokyo*) **96**, 1165 (1984).

respectively.[14] $Mn^{2+}$ and $Co^{2+}$ gave only 0–5% reactivation, whereas they reactivated the apoenzyme of the E-16 enzyme almost completely.[15]

Serratia marcescens also produces a 73-kDa proteinase that is not a metalloenzyme but is thiol-dependent. Some preparations of the Serratia 73K proteinase run with two different mobilities in polyacrylamide gel electrophoresis without sodium dodecyl sulfate (SDS) and in isoelectric focusing.[16] This heterogeneity may arise by deamidation during purification.[17]

## Assay Methods

### Caseinolysis

The casein digestion method based on that of Kunitz[18] has long been used for the assay of many proteinases including serralysin. At the end of the incubatin, the digestion mixture is treated with trichloroacetic acid (TCA) to precipitate undigested casein, and the quantity of soluble products is determined by the method of Lowry et al.[19] The method is not specific for serralysin, but it can usually be used for routine assays.

### Reagents

Substrate: 2% casein solution is made by dissolving 2 g of casein (Hammersten, Merck, Darmstadt, Germany) in 70 ml of 0.1 $M$ NaOH; the solution is adjusted to pH 7.4 with 1 $M$ $KH_2PO_4$, and the total volume is adjusted to 100 ml with distilled water

10 m$M$ Tris-HCl buffer, pH 7.4, containing 10 m$M$ $CaCl_2$

0.4 $M$ $Na_2CO_3$

10% (w/v) Trichloroacetic acid (TCA)

Folin and Ciocalteu's phenol reagent (Sigma, St. Louis, MO), diluted with an equal volume of distilled water as suggested by the supplier

Procedure. The 2% casein solution and the enzyme solution which is appropriately diluted with 10 m$M$ Tris-HCl buffer are preincubated separately in a water bath at 40° for at least 10 min prior to the assay. The reaction is then carried out with 1 ml of casein solution plus 1 ml of enzyme solution at 40° for 10 min. The reaction is stopped by the addition of 2 ml

[15] K. Miyata, K. Tomoda, and H. Isono, Agric. Biol. Chem. **35**, 460 (1971).
[16] A. Molla, T. Yamamoto, and H. Maeda, J. Biochem. (Tokyo) **104**, 616 (1988).
[17] H. Maeda and K. Kuromizu, J. Biochem (Tokyo) **81**, 25 (1977).
[18] M. Kunitz, J. Gen. Physiol. **30**, 291 (1947).
[19] O. H. Lowry, N. J. Rosebrough, A. L. Farrand, and R. J. Randall, J. Biol. Chem. **193**, 265 (1951).

of 10% TCA, and, after shaking, the mixtures are kept in the bath for about 30 min and then filtered through Whatman (Clifton, NJ) No. 1 paper.

One milliliter of the filtrate is mixed with 5 ml of the $Na_2CO_3$ solution, followed by 1 ml of phenol reagent. After 15 min, the absorbance at 670 nm is read for determination of the amount of liberated tyrosine. The blank is prepared by mixing the casein solution first with TCA and then adding the enzyme solution to the casein–TCA mixture. The specific activity (unit) is expressed as milligrams tyrosine per minute per milligram enzyme. The purified alkaline proteinase of *P. aeruginosa* has a specificity activity of about 1 unit/mg, which is about one-tenth that of pseudolysin.

The tyrosine and tryptophan content of the TCA-soluble products can alternatively be quantitated directly by absorption at 280 nm. Also, a chromophoric substrate such as azocasein (Sigma, St. Louis, MO) may be used, with which the extent of digestion is determined by reading the absorbance of TCA-soluble products at 370 nm.

*Fluorescence Polarization Method*

Proteolytic activity can be measured as a change in fluorescence polarization (FP) when fluorescein isothiocyanate (FITC)-labeled gelatin is used as substrate.[4,20] The final concentration of the substrate is about 10–100 n$M$ of fluorescein equivalents. A small enzyme sample (e.g., from a chromatographic fraction) is placed in a cuvette containing the substrate in 1.0 ml of 50 m$M$ Tris-HCl buffer (pH 8.0), and FP values are measured automatically at 30° by a fluorescence spectropolarimeter. Because the FP value reflects the molecular volume, any decrease in the FP value during incubation indicates the decrease in molecular mass due to proteolysis.[20,21]

Any substrate protein can be used in the method after FITC labeling, although a molecular mass greater than 10 kDa is preferable. The assay requires precise temperature control, within ±0.03°.

Labeling of substrate proteins with FITC can readily be accomplished at a slightly alkaline pH, usually in 0.1–0.5 $M$ sodium carbonate–bicarbonate buffer, pH 8.5. The higher the pH, the more rapid and extensive will be the labeling.[22] Most frequently, about a 20-fold molar excess of FITC (highest grade available) over amino groups is added to the protein solution (1–10 mg/ml in 0.1 $M$ sodium carbonate–bicarbonate buffer, pH 8.5) with stirring at room temperature. Free FITC and FITC-labeled protein can be separated by extensive dialysis and/or Sephadex G-50 column chromatography. Following removal of free FITC and its degradation product, the FP

[20] H. Maeda, *Anal. Biochem.* **92**, 222 (1979).
[21] H. Maeda, *Clin. Chem.* **24**, 2139 (1978).
[22] H. Maeda, H. Kawaguchi, K. Tsujimura, and N. Ishida, *J. Biochem.* (*Tokyo*) **65**, 777 (1969).

value of the FITC-labeled protein becomes high and remains constant during any further dialysis or gel chromatography.

## Fluorometric Assay with Fluorogenic Peptide Substrates

Serralysin can be assayed with peptidyl-methylcoumarylamide (MCA) substrates.[23] The principle of the use of such substrates is described in [2] in this volume. For serralysin the best substrates are Pro-Phe-Arg-MCA and Z-Phe-Arg-MCA (available from Peptide Institute Inc., Ina, Mino City, Osaka 562, Japan).

The reaction is carried out in 0.5 ml of 50 m$M$ Tris-HCl buffer (pH 8.0) containing 0.1 $M$ NaCl and 10 m$M$ CaCl$_2$. The peptidyl-MCA substrate is dissolved in dimethyl sulfoxide at 30° to give a stock concentration of 0.1 m$M$. The reaction is started by the addition of about 5 $\mu$g (in 5 $\mu$l) of enzyme. 7-Amino-4-methylcoumarin (AMC) liberated from the substrate fluoresces at 440 nm when excited at 380 nm, and the increase in fluorescence is monitored continuously with a recorder linked to the spectrofluorimeter. The rate of increase of fluorescence intensity is the measure of enzyme activity.

## Production and Purification Procedures

### Serralysin and Related Proteinases from Serratia marcescens

*Culture and Ammonium Sulfate Precipitation.* A 0.5-ml sample of a stationary-phase overnight culture of *S. marcescens* is introduced into 250 ml of tryptosoy broth (Difco, Detroit, MI) in a 1-liter flask, with reciprocal shaking at 1.75 Hz, and incubated at 30° for 36 hr.[4] Three liters of culture from 12 such flasks is centrifuged at 8000 $g$ for 30 min and the supernatant filtered through a membrane filter (0.22 $\mu$m), both steps being conducted at 4°. Ammonium sulfate is added slowly to the filtrate with gentle stirring to 90% saturation. After standing for 24 hr at 4°, the precipitate is collected by centrifugation at 8000 $g$ for 45 min, dissolved in about 500 ml of water, and dialyzed for 24 hr against 15 liters of water and finally twice against 15 liters of 10 m$M$ Tris-HCl buffer (pH 8.3), both at 4°. It can be concentrated either with a rotary evaporator at room temperature under reduced pressure or in an ultrafiltration cell fitted with an Amicon (Danvers, MA) YM10 membrane under nitrogen gas (50 psi).

*Column Chromatography.* The dialyzed sample (630 ml) is applied to a column (4 × 40 cm) of DEAE-cellulose (DE52) equilibrated with 10 m$M$

---

[23] T. Morita, H. Kato, S. Iwanaga, K. Takada, T. Kimura, and S. Sakakibara, *J. Biochem. (Tokyo)* **82**, 1495 (1977).

Tris-HCl buffer (pH 8.3) and washed with 1 liter of the buffer at a flow rate of 125 ml/hr. A linear gradient formed from 2 liters of buffer and 2 liters of the same buffer with 0.3 $M$ NaCl is then applied. The eluate is collected in 5.0-ml fractions for the measurement of absorbance at 280 nm and enzyme activity. The active fractions are combined and P-I, P-II, and P-III are obtained (see Table I), dialyzed against distilled water at 4°, and lyophilized. The lyophilized material is redissolved and applied to a column (4.0 × 68 cm) of Sephadex G-100 equilibrated and eluted with 10 m$M$ sodium phosphate buffer (pH 6.8). The active fractions are combined and applied to a column (2.2 × 30 cm) of DEAE-cellulose equilibrated with 10 m$M$ sodium phosphate buffer (pH 6.8) for rechromatography. The column is eluted with 600 ml of the phosphate buffer at a flow rate of 40 ml/hr, after which a linear gradient consisting of 400 ml of 10 m$M$ phosphate buffer (pH 6.8) and 400 ml of the same buffer with 0.1 $M$ NaCl is applied. From this column, the proteinase is eluted as a symmetrical peak, and the product exhibits a single band on electrophoresis. Fractions (3.9 ml) of the eluate are collected as before, dialyzed against distilled water, and lyophilized. The lyophilized powder is stored at −70° until required. The results of a representative purification are summarized in Table I.

TABLE I
PURIFICATION OF Serratia marcescens PROTEINASES[a]

| Purification step | Volume (ml) | Protein (mg) | Total activity (FP: %)[b] | Specific activity (FP: %/mg)[b] | Purification (-fold) | Recovery (%) |
|---|---|---|---|---|---|---|
| Culture supernatant | 2700 | 10,152[c] | 4,389,700 | 432 | 1 | 100 |
| 90% saturated (NH₄)₂SO₄ precipitate | 620 | 794 | 4,333,300 | 5458 | 12.6 | 99 |
| DEAE (pH 8.3) | | | | | | |
| 73K (active P-I)[d] | 273 | 66.6 | 133,500 | 2005 | 4.6 | 3.1 |
| 60K (active P-II)[d] | 483 | 76.3 | 630,000 | 8257 | 19.1 | 14.4 |
| 56K (active P-III)[d] | 716 | 188.5 | 1,626,800 | 8630 | 20.0 | 37 |
| Sephadex G-100 | | | | | | |
| 73K | 103 | 0.7 | 79,900 | 114,143 | 264.2 | 1.8 |
| 60K | 147 | 24.1 | 625,500 | 25,954 | 60.1 | 14.3 |
| 56K | 129 | 109.0 | 1,457,000 | 13,367 | 30.9 | 33.2 |
| Rechromatography of 60K from previous step on DEAE (pH 6.8) | | | | | | |
| 60K | 156 | 14.8 | 285,200 | 19,270 | 44.6 | 6.5 |
| 56K | 91 | 4.4 | 87,600 | 19,909 | 46.1 | 2.0 |

[a] Adapted from Matsumoto et al.[4]
[b] Proteolytic activity was measured with gelatin as the substrate by fluorescence polarimetry.
[c] A part of this value might have been derived from the culture medium itself.
[d] Notations correspond to the peaks in Ref. 4.

## Serralysin-Type Proteinase from Pseudomonas aeruginosa

### Strains and Type of Proteinase Produced

*Pseudomonas aeruginosa* strains can be classified into three groups by their ability to produce elastase (E) activity (due to pseudolysin) and alkaline proteinase (P).[24] These are the elastase- and alkaline proteinase-producing (E+P+) IFO 3455 strain, the elastase-negative and alkaline proteinase-producing (E−P+) IFO 3080 strain, and the strain producing neither elastase nor alkaline proteinase (E−P−), strain PA-103. Alkaline proteinase can be produced by the former two strains in either synthetic or semisynthetic medium. Generally, the E−P+ strain, IFO 3080, is used for the production of *Pseudomonas* serralysin, because of the convenience of purification.[2]

*Culture Medium and Culture.* The semisynthetic medium contains 7% (w/v) glucose, 1% $(NH_4)_2HPO_4$, 1% $Na_2HPO_4 \cdot 12H_2O$, 0.2% $KH_2PO_4$, 0.05% $MgSO_4 \cdot 7H_2O$, 0.2% yeast extract, and 2.5% $CaCO_3$, pH 7.0. *Pseudomonas aeruginosa* IFO 3080 is cultured with shaking (1.3 Hz, 10 cm amplitude) in a 500-ml flask containing 100 ml of the medium for 3 to 4 days at 28°. Proteinase production depends on the high (7%) glucose concentration in the medium, and extra $CaCO_3$ must also be included for neutralization. A complex medium such as nutrient broth (Difco, Detroit, MI) is not suitable. The yield of alkaline proteinase in the semisynthetic medium is commonly about 500 mg/liter culture.

*Purification and Column Chromatography.* The purification procedure[2] consists of ammonium sulfate precipitation (0.3 to 0.6 saturation) and acetone (60%, v/v) precipitation at 4°. The precipitates are dissolved in 20 m$M$ sodium phosphate buffer (pH 8.0), and the solution is dialyzed against the same buffer for 2 days (2 changes a day) at 4°. The retentate, after removing the insoluble material by centrifugation, is applied to a column of DEAE-cellulose which has been equilibrated with 20 m$M$ phosphate buffer, pH 8.0, and is then eluted with 20 m$M$ phosphate buffer containing 50 m$M$ and then 0.2 $M$ NaCl.[2]

When the IFO 3455 strain (E+P+) is used for production, elastase (pseudolysin) can be eluted first with the buffer containing 50 m$M$ NaCl, followed by elution of alkaline proteinase with the second buffer containing 0.2 $M$ NaCl. When the IFO 3080 strain (E−P+) is used for production, elution with the buffer containing 50 m$M$ NaCl is not necessary. Further purification is achieved by rechromatography.

*Crystallization.* The fractions containing alkaline proteinase are dialyzed against 10 m$M$ $CaCl_2$ for 2 days (4 changes) in the cold, followed by

[24] K. Morihara, *J. Bacteriol.* **88,** 745 (1964).

concentration over an Amicon YM10 membrane to less than 2% of the initial volume. After centrifugation, cold acetone is added dropwise to the clear supernatant with stirring in the cold until a slight turbidity appears, then the suspension is stored in a refrigerator. Needlelike crystals form overnight at 4°. Further acetone is added to the suspension of crystals, and the suspension is then stored in the refrigerator for 24 hr or longer. More than 85% of the proteinase in the suspension crystallizes. Recrystallization is carried out as follows. The crude crystals are collected by centrifugation and are washed with cold 50% (v/v) acetone in the cold. They are then dissolved in 10 m$M$ CaCl$_2$ adjusted to pH 10 with 0.1 $M$ NaOH (the crystals are only slightly soluble in neutral solution). After adjustment of the pH to neutral with acetic acid, insoluble material is discarded, and acetone is added to the clear solution as described above. Under these conditions, crystallization commences almost immediately. After two or three recrystallizations, the crystals are suspended in water and lyophilized. The yield is about 400–500 mg from 5 liters of broth (about 20% of the theoretical yield based on the total proteolytic activity of the broth).

## Properties of Serralysin and Related Proteinases

### Enzymatic Properties

The optimum pH for hydrolysis of casein is broad, in the range pH 6–10. The enzymes are inhibited by EDTA and $o$-phenanthroline, indicating that they are metalloproteinases. It has been shown that the metalloproteinase from *Serratia piscatram* (or *Serratia* sp. E-15 proteinase) and from *S. marcescens* contain one gram atom of Zn per molecule as an essential component (see Introduction).[14,15]

### Substrate Specificity

The specificity of peptide bond cleavage in the oxidized insulin B chain has been examined with the metalloproteinases from *P. aeruginosa, Pseudomonas fragi, Serratia* sp. E-15, and *Proteus mirabilis*. They all cleave the peptide bonds Asn$^3$-Gln$^4$, CysSO$_3$H$^7$-Gly$^8$, Arg$^{22}$-Gly$^{23}$, and Tyr$^{26}$-Thr$^{27}$. Other peptide bonds such as His$^{10}$-Leu$^{11}$, Glu$^{13}$-Ala$^{14}$, Ala$^{14}$-Leu$^{15}$, Tyr$^{16}$-Leu$^{17}$, Gly$^{20}$-Gly$^{21}$, Phe$^{25}$-Tyr$^{26}$, and Tyr$^{26}$-Thr$^{27}$ are also sensitive to most of the enzymes. Therefore, we can say that the enzymes belonging to the serralysin group show broad substrate side-chain specificity.

Table II gives kinetic parameters of the hydrolysis of various peptide substrates. The specificity at either side of the scissile bond is not stringent, but the molecular size of the peptide substrates is important, and hydrophobic amino acid residues at the P2 and P2' positions are very favorable.

TABLE II
KINETIC PARAMETERS FOR HYDROLYSIS OF SYNTHETIC PEPTIDES BY
ALKALINE PROTEINASE OF *Pseudomonas aeruginosa* IFO 3080[a]

| Substrate P3 P2 P1 P1' P2' P3' | $K_m$ (m$M$) | $k_{cat}$ (sec$^{-1}$) |
|---|---|---|
| Z-Gly-Ala-Ala | 7.7 | 0.62 |
| Z-Ala-Ala-Ala | 2.7 | 0.92 |
| Z-Phe-Ala-Ala | 0.4 (1.7) | 0.62 (0.30) |
| Z-Phe-Ala-Ala | 1.8 | 0.09 |
| Z-Phe-Ala-Ala | 0.4 | 0.04 |
| Z-Ala-Phe-Gly-Ala | 2.5 (13.0) | 11.4 (7.7) |
| Z-Ala-Phe-Leu-Ala | 0.6 | 5.3 |
| Z-Gly-Leu-Gly-Gly | 11.0 | <0.05 |
| Z-Gly-Leu-Gly-Gly-Ala | 4.3 | |
| Z-Ala-Gly-Gly-Leu | 5.4 (25.0) | 6.8 (5.0) |
| Z-Ala-Gly-Gly-Leu-Ala | 9.0 | 50.3 |
| Z-Gly-Pro-Gly-Gly-Pro-Ala | 5.3 | 0.6 |

[a] The reaction was carried out at pH 7.0 and 40° in the presence of 10 m$M$ CoCl$_2$. Parameters in parentheses were determined in the absence of CoCl$_2$. From K. Morihara, H. Tsuzuki, and T. Oka, *Biochim. Biophys. Acta* **309**, 414 (1973); K. Morihara and H. Tsuzki, *Agric. Biol. Chem.* **38**, 621 (1974).

Thus, Z-Ala-Gly+Gly-Leu-Ala is the most susceptible of the peptides tested (+marks the peptide bond cleaved). Z-Gly-Pro-Gly+Gly-Pro-Ala, a substrate of clostridial collagenase, is hydrolyzed moderately at the same site by *Pseudomonas* alkaline proteinase. Among the various metal ions tested, cobalt stimulated activity, and a severalfold increase in the rate of hydrolysis of synthetic peptides and also of gelatin was observed at concentrations higher than 5 m$M$ (Table II).

Contrary to earlier reports, both *Serratia* and *Pseudomonas* enzymes efficiently cleave peptide bonds after Arg in Pro-Phe-Arg+MCA and Z-Phe-Arg+MCA.[25-31] Tables III and IV show the hydrolysis of synthetic

[25] N. Nishino, W. Shimizu, M. Hirotsuka, T. Fujimoto, and H. Maeda, in "Peptide Chemistry" (T. Miyazawa, ed.), p. 233. Protein Research Foundation, Osaka, 1987.

[26] T. Morimoto, N. Nishino, T. Fujimoto, and T. Yamamoto, in "Peptide Chemistry" (N. Yanaihara, ed.), p. 387. Protein Research Foundation, Osaka, 1990.

[27] K. Matsumoto, T. Yamamoto, R. Kamata, and H. Maeda, *J. Biochem. (Tokyo)* **96**, 739 (1984).

[28] A. Molla, T. Yamamoto, T. Akaike, S. Miyoshi, and H. Maeda, *J. Biol. Chem.* **264**, 10589 (1989).

[29] H. Maeda, K. Maruo, T. Akaike, H. Kaminishi, and Y. Hagiwara, in *Agents Actions* **38**(3), 362 (1992).

[30] H. Maeda and A. Molla, *Clin. Chim. Acta* **185**, 357 (1989).

TABLE III
SUBSTRATE SPECIFICITY OF *Pseudomonas aeruginosa*
ALKALINE PROTEINASE AND *Serratia 56K* PROTEINASE
AGAINST SYNTHETIC PEPTIDES[a]

| Substrate | | | | | | | $k_{cat}/K_m$ ($M^{-1}$ $sec^{-1}$) | |
|---|---|---|---|---|---|---|---|---|
| P4 | P3 | P2 | P1 | P1' | P2' | P3' | PA | 56K |
| | | | ↓ | | | | | |
| Z-Phe-Arg-MCA | | | | | | | 20 | 380 |
| Abz-Gly-Phe-Arg-Gly-Nba[b] | | | | | | | — | 150 |
| Abz-Gly-Phe-Arg-Ala-Nba | | | | | | | 510 | — |
| Abz-Gly-Phe-Arg-Val-Nba | | | | | | | 120 | 140 |
| Abz-Gly-Phe-Arg-Leu-Nba | | | | | | | 510 | 5300 |
| Abz-Gly-Phe-Arg-Phe-Nba | | | | | | | 1100 | 1700 |
| Abz-Gly-Phe-Arg-Leu-Leu-Nba | | | | | | | 7500 | 3200 |

[a] Reactions were carried out at pH 8.0 and 25°. PA, Alkaline proteinase of *P. aeruginosa*. From H. Tanaka, T. Yamamoto, Y. Shibuya, N. Nishino, S. Tanase, Y. Miyauchi, and T. Kambara, *Biochim. Biophys. Acta* **1138**, 243 (1992), and U. Senba, T. Yamamoto, T. Kunisada, Y. Shibuya, S. Tanase, T. Kambara, and H. Okabe, *Biochim. Biophys. Acta* **1159**, 113 (1992).
[b] Nba, 4-nitrobenzyl amide.

substrates in which Arg occupies the P1 position by *S. marcescens* 56K and 60K proteinases and *P. aeruginosa* alkaline proteinase.

Molla *et al.*[32] also examined the substrate specificity of *Serratia* 56K proteinase against a more complex protein, porcine plasma fibronectin, which is a dimer of 220 kDa. There were only four major sites of cleavage: 2 of the 9 Arg-Thr bonds, 1 of 11 Leu-Ser bonds, and 1 of 7 Gln-Glu bonds. There are a total of 118, 124, and 125 Arg, Leu, and Gln residues, respectively, in porcine fibronectin. Sequence analysis showed that all the sites of cleavage were located between domains. Specifically, the cleavages were between the heparin-binding domain and the gelatin/collagen-binding domain (Arg[259]-Thr[260]), between the gelatin/collagen-binding domain and the DNA/heparin/cell-binding domain (Leu[687]-Ser[688]), and between the DNA/heparin/cell-binding domain and the C-terminal cluster (Arg[2239]-Thr[2240]). There was a less complete cleavage between the inactive portion

[31] K. Matsumoto, T. Yamamoto, R. Kamata, and H. Maeda, *in* "Kinins IV" (L. M. Greenbaum and H. S. Margolius, eds.), p. 71. Plenum, New York, 1986.
[32] A. Molla, S. Tanase, Y. M. Hong, and H. Maeda, *Biochim. Biophys. Acta* **955**, 77 (1988).

TABLE IV
SUBSTRATE SPECIFICITY OF *Serratia Proteinase*[a]

| Substrate | Cleaving enzyme | Enzyme activity[b] $\Delta f/\text{min}$ | % |
|---|---|---|---|
| 1. Z-Phe-Arg-MCA | Plasma kallikrein and cathepsin B | 23.3 | 100 |
| 2. Pro-Phe-Arg-MCA | Pancreatic and urinary kallikrein | 10.9 | 46.8 |
| 3. Boc-Val-Leu-Lys-MCA | Plasmin | 7.1 | 30.4 |
| 4. Boc-Phe-Ser-Arg-MCA | Trypsin | 6.6 | 28.5 |
| 5. Boc-Val-Pro-Arg-MCA | $\alpha$-Thrombin | 2.4 | 10.3 |
| 6. Bz-Arg-MCA | Trypsin | 2.4 | 10.1 |
| 7. Boc-Ile-Glu-Gly-Arg-MCA | Factor Xa | 1.3 | 5.5 |
| 8. Boc-Leu-Gly-Arg-MCA | Horshoe crab clotting enzyme | 0.4 | 1.6 |
| 9. Suc-Ala-Pro-Ala-MCA | Elastase | 0.3 | 1.3 |
| 10. Gly-Pro-MCA | X-Prolyl-dipeptidyl-aminopeptidase | 0.2 | 0.9 |
| 11. Glt-Gly-Arg-MCA | Urokinase | 0.04 | 0.2 |
| 12. Suc-Ala-Ala-Pro-Phe-MCA | Chymotrypsin | 0.04 | 0.2 |
| 13. Arg-MCA | Cathepsin H | 0.02 | 0.1 |
| 14. Suc-Gly-Pro-MCA | Post-proline-cleaving enzyme | 0.02 | 0.1 |
| 15. Suc-Gly-Pro-Leu-Gly-Pro-MCA | Collagenase-like peptidase | 0.02 | 0.1 |

[a] Z, Benzyloxycarbonyl; MCA, 4-methylcoumaryl-7-amide; Bz, benzoyl; Boc, *tert*-butyloxycarbonyl; Suc, succinyl; Glt, glutaryl. From Matsumoto *et al.*[27]

[b] $\Delta f/\text{min}$, Increase of relative fluorescence intensity of liberated AMC at 441 nm; %, relative velocity with respect to substrate 1.

of the intradomain and the DNA/heparin/cell-binding domain ($\text{Gln}^{865}$-$\text{Glu}^{866}$). Therefore, the efficiency of peptide bond cleavage by the proteinase is primarily dominated by conformational integrity, and the specific amino acid sequence seems to be of secondary importance.

Another example is the cleavage of immunoglobulins $G_1$, $G_2$, $G_3$, and $G_4$ and immunoglobulins $A_1$ and $A_2$ by the *Serratia* 56K proteinase, in which interdomain cleavage at the hinge region is predominant.[33] Tanaka *et al.* also showed that the site of cleavage of prekallikrein by *P. aeruginosa* elastase is similar to that of trypsin, at $\text{Arg}^{371}$-$\text{Ile}^{372}$ among many Arg and Ile residues.[34] The cleavage site of guinea pig Hageman factor by bovine trypsin is believed to be at the extradomain region $\text{Arg}^{340}$-$\text{Ile}^{341}$ among numerous Arg residues in the Hageman factor.[35] Therefore, physiological

[33] A. Molla, T. Kagimoto, and H. Maeda, *Infect. Immun.* **56**, 916 (1988).
[34] H. Tanaka, T. Yamamoto, Y. Shibuya, N. Nishino, S. Tanase, Y. Miyauchi, and T. Kambara, *Biochim. Biophys. Acta* **1138**, 243 (1992).
[35] U. Senba, T. Yamamoto, T. Kunisada, Y. Shibuya, S. Tanase, T. Kambara, and H. Okabe, *Biochim. Biophys. Acta* **1159**, 113 (1992).

TABLE V
PROTEINASE ACTIVITY OF SERRALYSIN AND THERMOLYSIN FAMILIES USING SYNTHETIC
Z-Phe-Arg-MCA AND EFFECT OF PHOSPHORAMIDON AND ZINCOV

| Proteinase | Relative activity[a] (%): $\Delta f/\text{min}^b$ | $IC_{50}^{c,d}$ ($\mu M$) | |
| | | Phosphoramidon | Zincov |
| --- | --- | --- | --- |
| *Serratia marcescens* 56K | 100 | Not inhibited | Not inhibited |
| *Serratia piscatrum* E-15 | 92.5 | Not inhibited | Not inhibited |
| *Pseudomonas aeruginosa* alkaline proteinase | 12.5 | Not inhibited[e] | Not inhibited[e] |
| Subtilisin | 12.5 | Not inhibited[e] | Not inhibited[e] |
| Thermolysin | 4.5 | 64.7 (1.88)[f] | 631.6 (18.4)[f] |
| *Pseudomonas aeruginosa* elastase | 0 | 97.4 (2.95)[f] | 674.9 (20.5)[f] |

[a] Z-Phe-Arg-MCA is the substrate for plasma kallikrein. Reactions were conducted at pH 7.0 with 10 $\mu g/\text{ml}$ proteinase (final) and 5$\mu g/\text{ml}$ MCA substrate.

[b] See Table III.

[c] Substrate was azocasein at 2.5 $\mu g/\text{ml}$, pH 7.0; enzyme solution was 3.0 $\mu g/\text{ml}$.

[d] $IC_{50}$ value is the 50% inhibitory concentration of each inhibitor at an enzyme concentration of 1.0 $\mu M$.

[e] Activity increased rather than inhibition.

[f] Values in the parentheses show those of the enzyme concentration at 1.0 $\mu g/\text{ml}$.

and pathological functions of the proteinases, especially *in vivo*, should be interpreted carefully. Furthermore, the size of the proteinase molecule may restrict its access to folded peptide chains in most native proteins through steric hindrance. For these reasons, the results obtained with oxidized insulin chains should be extrapolated to native proteins with caution.

As shown in Tables III, IV, and V and previous publications, it is interesting that the *Serratia* 56K proteinase cleaves Z-Phe-Arg-MCA and Pro-Phe-Arg-MCA, which are the optimal substrates for plasma kallikrein and pancreatic–urinary kallikrein, respectively. This led us to examine the effect of the proteinases on the Hageman factor–kallikrein–kinin system as discussed later.[27-31,36-44]

[36] A. Molla, K. Matsumoto, I. Oyamada, T. Katsuki, and H. Maeda, *Infect. Immun.* **53**, 522 (1986).

[37] A. Molla, T. Akaike, and H. Maeda, *Infect. Immun.* **57**, 1868 (1989).

[38] H. Kaminishi, M. Tanaka, T. Cho, H. Maeda, and Y. Hagiwara, *Infect. Immun.* **58**, 2139 (1990).

[39] K. Maruo, T. Akaike, Y. Inada, I. Ohkubo, T. Ono, and H. Maeda, *J. Biol. Chem.* **268**, 17711 (1993).

[40] H. Kaminishi, T. Cho, T. Ito, A. Iwata, K. Kawasaki, Y. Hagiwara, and H. Maeda, *FEMS Microbiol. Lett.* **114**, 109 (1993).

[41] R. Kamata, K. Matsumoto, R. Okamura, T. Yamamoto, and H. Maeda, *Ophthalmology* (*Philadelphia*) **92**, 1452 (1985).

TABLE VI
EFFECTS OF ENZYME INHIBITORS ON PROTEINASES OF
*Serratia marcescens*[a]

| Reagents[b] | Residual proteinase activity (%)[c] | | |
|---|---|---|---|
| | 56K | 60K | 73K |
| None (control) | 100 | 100 | 100 |
| EDTA | 0 | 0 | 85 |
| EGTA | 95 | 98 | 65 |
| L-Cysteine | 92 | 94 | 157 |
| Glutathione (SH) | 98 | 97 | 168 |
| Dithiothreitol | 71 | 85 | 161 |
| 2-Mercaptoethanol | 99 | 94 | 121 |
| Iodoacetic acid | 92 | 100 | 31 (85) |
| N-Ethylmaleimide | 99 | 100 | 72 (100) |
| p-Chloromercuribenzoic acid | — | — | 0 (0) |
| Antipain | 60 | 95 | 89 |
| Diisopropyl fluorophosphate | 103 | 97 | 100 |
| Phenylmethylsulfonyl fluoride | 88 | 94 | 58 |
| Leupeptin | 90 | 102 | 77 |

[a] The *S. marcescens* proteinases are designated 56K, 60K, and 73K, from Matsumoto *et al.*[4]
[b] Enzyme inhibitors were used at a concentration of 1.92 m$M$.
[c] The assay procedure was as described in the text and was conducted at pH 5.0. The results at pH 8.0 were essentially similar to those at pH 5.0. The ratio of inhibitor to enzyme (mole/mole) was 5600 for both 56K and 60K and 37,000 for the 73K enzyme. The values in parentheses were obtained after treatment at pH 8.0.

The metalloproteinases of this group are generally inhibited by conventional metal-chelating agents (Tables V and VI), but they are insensitive to phosphoramidon and Zincov (2-(N-hydroxycarboxamido)-4-methylpentanoyl-L-alanyl-glycine amide; Calbiochem, La Jolla, CA) (Table V), which are specific inhibitors of metalloproteinases of the thermolysin family. After

[42] A. Molla, Y. Matsumura, T. Yamamoto, R. Okamura, and H. Maeda, *Infect. Immun.* **55**, 2509 (1987).
[43] S. Miyagawa, R. Kamata, K. Matsumoto, R. Okamura, and H. Maeda, *Albrect von Graefes Arch. Ophthalmol.* **229**, 281 (1991).
[44] S. Miyagawa, K. Matsumoto, R. Kamata, R. Okamura, and H. Maeda, *Jpn. J. Ophthalmol.* **35**, 402 (1991).

TABLE VII
REACTIVATION OF EDTA-INACTIVATED *Serratia* 56K AND
60K PROTEINASES BY VARIOUS DIVALENT METAL IONS[a]

|                                | Restored activity (%) | |
|--------------------------------|:---------:|:---------:|
| Reagent                        | 56K       | 60K       |
| ZnCl$_2$                       | 72        | 104       |
| CaCl$_2$                       | 10        | 23        |
| MgSO$_4$                       | 0         | 0         |
| MnCl$_2$                       | 5         | 0         |
| FeSO$_4$                       | 90        | 7         |
| CoCl$_2$                       | 3         | 0         |
| CuSO$_4$                       | 9         | 4         |
| Controls                       |           |           |
| Native enzyme + EDTA (2 m$M$)  | 0         | 0         |
| Native enzyme alone            | 100       | 100       |

[a] To the EDTA-inactivated enzyme solution, metal ions
at 2 m$M$ were added and incubated in 50 m$M$ sodium
acetate buffer (pH 5.0) at 30° for 20 min. Substrate was
then added. The enzyme concentration was 0.35 $\mu M$ in
the assay. From Matsumoto *et al.*[4,14]

EDTA inactivation, the E-16, 56K, and 60K proteinases can be reactivated
with Zn$^{2+}$, Fe$^{2+}$, and other divalent cations (Table VII).[4,14,15] Nishino *et al.*
discovered competitive inhibitors of both the *S. marcescens* 56K proteinase
and *P. aeruginosa* proteinase, as shown in Table VIII.[25,26,45] The inhibitors
are more potent against 56K proteinase than *P. aeruginosa* alkaline proteinase.

*Structures*

The primary structures of the proteinases from *Serratia* sp. E-15[9] and
*P. aeruginosa*[7,8] have been deduced from DNA sequences of the structural
genes. The molecular mass of the E-15 enzyme is 50,600 (470 amino acid
residues), and that of *P. aeruginosa* proteinase is 49,500 (470 residues).
Alignment of the primary structures is shown in Fig. 1, which demonstrates
that the sequences are very similar to one another (55% identical).

*Erwinia chrysanthemi*, a phytopathogenic gram-negative bacterium, can
produce three or four metalloendopeptidases. Amino acid sequences of

[45] N. Nishino, W. Shimizu, T. Fujimoto, and H. Maeda, *in* "Peptide Chemistry" (M. Ueki,
ed.), p. 21. Protein Research Foundation, Osaka, 1989.

TABLE VIII
INHIBITION OF *Pseudomonas aeruginosa* ALKALINE PROTEINASE AND
*Serratia* 56K PROTEINASE BY SUBSTRATE ANALOGS[a]

| Inhibitor | $K_i$ $(\mu M)^b$ PA | 56K |
|---|---|---|
| Bz-Phe-Arg-N(H)—[benzene ring with SH ortho] | 7.0 (5.5) | 18 (1.9) |
| Bz-Phe-Arg-N(H)—[benzene ring]—SH | 7.5 (7.1) | 1.6 (1.4) |
| Bz-Phe-Arg-N(H)—[benzene ring with SBzl ortho] | 50 (17) | 0.7 (0.4) |
| Bz-Phe-Arg-N(H)—[benzene ring]—SBzl | 55 (7.0) | 2.0 (0.6) |

[a] Reactions with *P. aeruginosa* alkaline proteinase (PA) and *Serratia* 56K proteinase were conducted at pH 7.9 and 25°. From Tanaka *et al.*[34] and Nishino *et al.*[25]
[b] Values in parentheses were obtained after 12 hr of incubation.

proteinases from two strains of *E. chrysanthemi*, strain B374[10] and strain EC16,[11] have been deduced from cDNA sequences of the structural genes. The results indicate that both amino acid sequences are very similar to that of *Serratia* sp. E-15 proteinase, the percentage of identities for proteinase B and EC16 proteinase being 61 and 49%, respectively. The identity of proteinase B with the EC proteinase is 65%.

It has not been shown directly that the metalloproteinases from *P. aeruginosa* and *E. chrysanthemi* contain a zinc ion as an essential component, but the consensus sequence that is responsible for zinc binding in astacin,[46] namely, HEXXHXUGUXH (in which X represents an arbitrary

[46] W. Bode, F.-X. Gomis-Rüth, R. Huber, R. Zwilling, and W. Stöcker, *Nature* (*London*) **358,** 164 (1992).

```
AP    1   GRSDAYTQVDNFLHAYARGGDELVNGHPSYTVDQAAEQILREQASWQ--KAPGDSVLTLS    58
          *  **  **   **      **  *    **  *  **    *    *  *  *  *
SP    1   AATTGYDAVDDLLHYHERGNGIQINGKDSFSNEQAGLFITRENQTWNGYKVFGQPV-KLT    59

AP   59   YSFLTKPNDFFNTPWKYVSDIYSLGKFSAFSAQQQAQAKLSLQSWSDVTNIHFVDAGQGD   118
          **   *   *     *    **  ** ********** **  ** *    *
SP   60   FSF---PDYKFSST--NVAGDTGLSKF---SAEQQQQAKLSLQSWADVANITFTEVAAGQ   111

AP  119   QGDLTFGNFS------SSVGGAAFAFLPD---VPDALKGQSWYLINSSYSANVNPANGNY   169
          ****  *     *  * ****    *  * **  *  * *    *      **  *
SP  112   KANITFGNYSQDRPGHYDYGTQAYAFLPNTIWQGQDLGGQTWYNVNQS--NVKHPATEDY   169

AP  170   GRQTLTHEIGHTLGLSHPGDYNAGEGDPTYADATYAEDTRAYSVMSYWEEQNTGQDFKGA   229
          ****  ****** *************** *** * ******* *  **** * *** *  *
SP  170   GRQTFTHEIGHALGLSHPGDYNAGEGNPTYRDVTYAEDTRQFSLMSYWSETNTGGDNGGH   229

AP  230   YSSAPLLDDIAAIQKLYGANLTTRTGDTVYGFNSNTERDFYSATSSSSKLVFSVWDAGGN   289
          *  *********** ****** ************** *** * ** * * * * ******
SP  230   YAAAPLLDDIAAIQHLYGANLSTRTGDTVYGFNSNTGRDFLSTTSNSQKVIFAAWDAGGN   289

AP  290   DTLDFSGFSQNQKINLNEKALSDVGGLKGNVSIAAGVTVENAIGGSGSDLLIGNDVANVL   349
          ** ****   **  ****** *************** *****      *   **    ***
SP  290   DTFDFSGYTANQRINLNEKSFSDVGGLKGNVSIAAGVTIENAIGFRQR-LIVGNAANNVL   348

AP  350   KGGAGNDILYGGLGADQLWGGAGADTFVYGDIAESS-AAAPDTLRDFVSGQDKIDLSGLD   408
          ******* *  **  *** ****** *  **   *  *  *  * *   *** * ******
SP  349   KGGAGNDVLFGGGGADELWGGAGKDIFVF-SAASDSAPGASDWIRDFQKGIDKIDLSFFN   407

AP  410   AFVNGGLVLQYVDAFAGKAGQAILSYDAASKAGSLAIDFSGDAHADFAINLIGQATQADIVV  470
          **  *  **  * ***  *      *       *       **      **  *
SP  408   KEAQSSDFIHFVDHFSGAAGEALLSYNASNNVTDLSVNIGGHQAPDFLVKIVGQVDVATDFIV  470
```

FIG. 1. Amino acid sequence of *Pseudomonas* alkaline proteinase (AP) and *Serratia* sp. E-15 protease (SP), both deduced from cDNA sequences based on refs. 7–9.

amino acid and U is a bulky hydrophobic residue) can be seen in the proteinases from *Serratia* sp. E-15, *P. aeruginosa*, and *E. chrysanthemi* (Fig. 2). The consensus sequence is different from that of the thermolysin family (see [13] in this volume).

The tertiary structure of the metalloproteinase from *P. aeruginosa* serralysin has been solved by X-ray analysis to a resolution of 1.6 Å (Fig. 3).[12] The enzyme has two distinct structural domains, the N- and C-terminal domains. The N-terminal domain is the proteolytic domain; it has an overall tertiary fold and active-site zinc ligation similar to that of astacin. The C-terminal domain consists of a 21-strand β sandwich. Within this domain is a novel so-called parallel β roll structure in which successive β strands are wound in a right-handed spiral, and in which $Ca^{2+}$ is bound within the turns between strands by a repeated GGXGXD sequence motif that is found in a diverse group of proteins by gram-negative bacteria. A preliminary report on the tertiary structure of the E-15 proteinase[13] indicates that the structure is similar to that of the *Pseudomonas* alkaline proteinase.

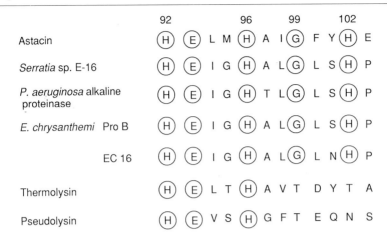

FIG. 2. Amino acid sequence alignment around the zinc-binding consensus sequence for astacin and the related serralysin or pseudolysin family proteinases.

nary report on the tertiary structure of the E-15 proteinase[13] indicates that the structure is similar to that of the *Pseudomonas* alkaline proteinase.

*Pathogenic activity.* As previously discussed elsewhere,[27–31,36–44,47] microbial proteinases can cause a wide range of pathogenic effects in infected hosts. Effects include (1) inactivation of endogenous proteinase inhibitors such as $\alpha_1$-proteinase inhibitor, C1-esterase inhibitor, and antithrombin III, resulting in uncontrolled clotting, fibrinolysis, and kinin release[27–31,36–39]; (2) activation of the kinin-generating cascade[27–31,36–39]: Hageman factor → (pre)kallikrein → kininogen → kinin; (3) activation of the coagulation cascade (factors XII, X, II, etc.)[27–31,36–40]; (4) inactivation of the complement system and chemotaxis (C5a, etc.)[31,36,47]; (5) inactivation of defense-oriented proteins including immunoglobulins G and A, lysozyme, and transferrin[33,36]; (6) tissue degeneration,[41–44] leading to spreading of infection and septicemia[48]; (7) cytotoxicity to cells: $\alpha_2$-macroglobulin complex formation followed by attachment, endocytosis via $\alpha_2$-macroglobulin receptor, and regeneration of proteinase activity in cells[49,50]; (8) activation of factor X and plasminogen, leading to the activation of influenza and paramyxovirus in-

[47] T. Oda, Y. Kojima, T. Akaike, S. Ijiri, A. Molla, and H. Maeda, *Infect. Immun.* **58**, 1269 (1990).

[48] H. Maeda, T. Akaike, Y. Sakata, and K. Maruo, *in* "Proteases, Protease Inhibitors, and Protease-Derived Peptides" (J. Chironis and J. E. Repine, eds.), p. 159. Birkhaeuser, Basel, 1993.

[49] H. Maeda, A. Molla, T. Oda, and T. Katsuki, *J. Biol. Chem.* **262**, 10946 (1987).

[50] H. Maeda, A. Molla, K. Sakamoto, A. Murakami, and Y. Matsumura, *Cancer Res.* **49**, 660 (1989).

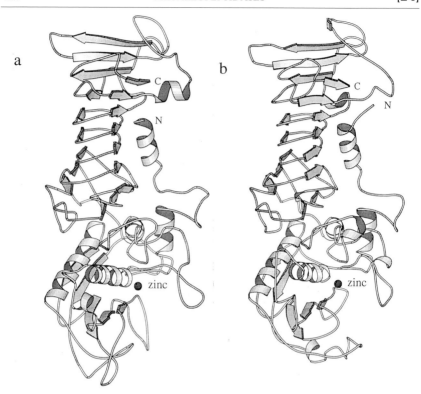

Fig. 3. A Ribbon representations of the molecular structures of (a) *Serratia* proteinase from *Serratia* sp. E-15[13] and (b) alkaline proteinase from *Pseudomonas aeruginosa* IFO 3455.[54] The overall structures of both enzymes are quite similar. (With permission and courtesy of Dr. Y. Hata, Kyoto University.)

fectivity[51-53]; (9) activation of bacterial toxin and/or enhancement of spreading of both exotoxin and endotoxin mediated through kinin activation.[48,53]

Readers interested in pathogenic effects are advised to consult the references cited. It should be emphasized that the activation of the kinin-generating cascade and the direct liberation of bradykinin from H-kininogen are typical activities of endogenous serine proteinases *in vivo*. However, 14 microbial metalloproteinases have also been shown to activate this process in one way or another, even in the presence of blood plasma, so

[51] T. Akaike, A. Molla, M. Ando, S. Araki, and H. Maeda, *J. Virol.* **63,** 2252 (1989).
[52] H. D. Klenk and W. Garten, *Trends Microbiol.* **2,** 39 (1994).
[53] V. A. Gordon and S. M. Leppla, *Infect. Immun.* **62,** 333 (1994).
[54] H. Miyake, Y. Hata, T. Fujii, T. Akutagawa, K. Morihara, and Y. Katsube, *Bull. Inst. Chem. Res., Kyoto Univ.*, **72,** 373 (1994).

it appears that activation of the kinin-generating system is a universal mechanism accompanying cleavage of uniquely scissile bonds in zymogens by microbial proteinases, regardless of their diverse substrate specificities.[27,28,39,40]

# [25] Vertebrate Collagenases

By Marianna Dioszegi, Paul Cannon, and Harold E. Van Wart

## Introduction and Background

The most abundant structural macromolecules present in the extracellular matrix are the collagens, a class of molecules derived from several multigene families.[1,2] The individual collagen types differ with respect to details of their structures in ways that critically influence the manner by which they are proteolytically degraded. Thus, a brief review of the structural diversity present in collagens is a necessary prerequisite to a definition of the different "collagenases" that are responsible for their metabolic turnover.

All members of the collagen family share the property that they are composed of three polypeptide chains referred to as α chains that form a tightly coiled right-handed triple helix, the so-called collagen fold, in part of the molecule. The triple-helical conformation endows such sections of collagen molecules with a stiff rodlike conformation that confers a marked resistance to proteolysis. In addition to the triple-helical domains, all collagens also contain globular domains[3] which are generally much more susceptible to proteolysis. Interactions between the different domains present in collagen molecules lead to their association into either fibrillar or nonfibrillar aggregates in the extracellular matrix. The aggregates function as supporting elements of tissue and also serve other developmental and physiological functions. There are at least 15 different collagen types with unique structures, functions, and spatial and temporal tissue distributions.

Although space does not permit a detailed discussion of the structures of the different collagen types, it is appropriate to elaborate on the relationship between proteolytic susceptibility and structure for the substrates of interest

---

[1] E. J. Miller and S. Gay, this series, Vol. 144, p. 3.
[2] T. F. Linsenmayer, in "Cell Biology of Extracellular Matrix" (E. D. Hay, ed.), p. 7. Plenum, New York, 1991.
[3] R. E. Burgeson, Annu. Rev. Cell Biol. 4, 551 (1988).

in this chapter. The most abundant collagens are the "classic" interstitial collagen types I, II, and III. They consist of homologous $\alpha$ chains with molecular masses of approximately 95 kDa, each of which contains an uninterrupted Gly-X-Y repeating sequence throughout 90% or more of its length. The resultant collagen molecules are stiff triple-helical rods that polymerize *in vivo* to form fibrils. There are short peptides at the N and C termini referred to as telopeptides that do not have this repeating sequence and that are not triple helical. Type V and XI collagens also form fibrils. However, other collagen types adopt a variety of alternate supramolecular structures.[1-3] For example, type IV collagen consists of helical domains separated by globular domains. The flexibility conferred by the non-helical domains enables type IV collagen monomers to associate to form sheetlike networks that are found almost exclusively in basement membranes. The flexibility conferred by this structure also makes basement membranes susceptible to attack by a broad spectrum of proteinases.[4] Thus, the fibril-forming collagens are generally much more resistant to proteolysis than the non-fibril-forming types.

The interstitial collagenases are endopeptidases that catalyze the hydrolytic cleavage of the undenatured, triple-helical portions of the classic interstitial type I, II, and III collagens under physiological conditions. Because those collagens exist as fibrils in tissue, this definition requires that an interstitial collagenase be able to utilize these fibrils as substrates. The definition excludes proteinases capable of hydrolyzing only the non-triple-helical telopeptide portions of the collagens, as well as enzymes capable of degrading the triple-helical domains of collagens only in solution or at extremes of pH. However, it includes collagenases that make single or multiple scissions within triple helices. For example, a number of bacterial collagenases have been identified that cleave the triple-helical regions of the interstitial collagens at multiple loci and that ultimately digest the substrates into a mixture of small peptides.[5,6] In contrast, the vertebrate collagenases are a subset of interstitial collagenases that are distinguished by their ability to dissolve fibrils of types I, II, and III collagens by making a single scission across all three $\alpha$ chains at a specific sensitive locus of the exposed tropocollagen (TC) monomers approximately three-quarters from the N terminus to produce the so-called TC$^A$ and TC$^B$ fragments. This mode of attack was first elucidated by Gross and co-workers with tadpole collagenase.[7,8] The known interstitial collagenases do not hydrolyze type

[4] C. Smith, H. E. Van Wart, and D. E. Schwartz, *Anal. Biochem.* **139,** 448 (1984)

[5] S. Seifter and E. Harper, *in* "The Enzymes" (P. D. Boyer, ed.), 3rd Ed., Vol. 3, p. 649. Academic Press, New York, 1971.

[6] M. F. French, A. Bhown, and H. E. Van Wart, *J. Protein Chem.* **11,** 83 (1992).

[7] J. Gross and C. M. Lapiere, *Proc. Natl. Acad. Sci. U.S.A.* **48,** 1014 (1962).

[8] J. Gross and Y. Nagai, *Proc. Natl. Acad. Sci. U.S.A.* **54,** 1197 (1965).

IV and V collagens in a similar manenr. Those collagens are believed to be hydrolyzed *in vivo* by other proteinases that may be termed type IV and type V collagenases, respectively.[9]

At present, only two distinct vertebrate interstitial collagenases (see note added in proof) have been identified, which will be referred to as the fibroblast-type (EC 3.4.24.7) and neutrophil-type (EC 3.4.24.34) collagenases on the basis of the source from which each was originally identified. Both are members of the matrix metalloproteinase (MMP) family[10] and have also been designated MMP-1 and MMP-8, respectively.[11] In the subsequent discussion, the human fibroblast- and neutrophil-type collagenases will be designated as HFC and HNC, respectively. A brief review of the properties of the two enzymes will facilitate the topics presented below. Both HFC and HNC are monomeric metalloenzymes that are believed to contain two atoms of zinc and two atoms of calcium per polypeptide chain.[12] One zinc atom resides at the active site and participates directly in catalysis, whereas the second zinc probably serves a structural role. The two calcium atoms are also believed to stabilize the tertiary structure of the proteins. The sequences of HFC[13,14] and HNC[15,16] deduced from the cDNAs reveal that the two enzymes are very similar. The fibroblast-type collagenases from rabbit,[17] pig,[18] rat,[19] and mouse[20] have also been cloned.

The HFC and HNC enzymes are synthesized as preproenzymes of 469 and 467 amino acids, of which the first 19 and 20, respectively, are the signal peptides. A schematic comparison of the two preprocollagenases

[9] H. Birkedal-Hansen, this series, Vol. 144, p. 140.

[10] H. Birkedal-Hansen, W. G. I. Moore, M. K. Bodden, L. J. Windsor, B. Birkedal-Hansen, A. DeCarlo, and J. A. Engler, *Crit. Rev. Oral Biol. Med.* **4**, 197 (1993).

[11] H. Nagase, A. J. Barrett, and J. F. Woessner, Jr., in "Matrix Metalloproteinases and Inhibitors" (H. Birkedal-Hansen, Z. Werb, H. Welgus, and H. Van Wart, eds.), Matrix Spec. Suppl. No. 1, p. 421. Gustav Fischer, Stuttgart, 1992.

[12] This is based on the homology of HFC and HNC to two other MMP, matrilysin and stromelysin, whose X-ray structures (K. Appelt and M. Browner, personal communications) reveal the presence of these metal atoms.

[13] G. I. Goldberg, S. M. Wilhelm, A. Kronberger, E. A. Bauer, G. A. Grant, and A. Z. Eisen, *J. Biol. Chem.* **261**, 6600 (1986).

[14] S. E. Whitham, G. Murphy, P. Angel, H.-J. Rhamsdorf, B. J. Smith, A. Lyons, T. J. R. Harris, J. J. Reynolds, P. Herrlich, and A. J. P. Docherty, *Biochem. J.* **240**, 913 (1986).

[15] K. A. Hasty, T. F. Pourmotabbed, G. I. Goldberg, J. P. Thompson, D. G. Spinella, R. M. Stevens, and C. L. Mainardi, *J. Biol. Chem.* **265**, 11421 (1990).

[16] P. Devarajan, K. A. Mookhtiar, H. E. Van Wart, and N. Berliner, *Blood* **77**, 2731 (1991).

[17] M. E. Fini, I. M. Plucinska, A. S. Mayer, R. H. Gross, and C. E. Brinckerhoff, *Biochemistry* **26**, 6156 (1987).

[18] N. J. Clarke, M. C. O'Hare, T. E. Cawston, and G. P. Harper, *Nucleic Acids Res.* **18**, 6703 (1990).

[19] C. O. Quinn, D. K. Scott, C. E. Brinckerhoff, L. M. Matrisian, J. J. Jeffrey, and N. C. Partridge, *J. Biol. Chem.* **265**, 22342 (1990).

[20] P. Henriet, G. G. Rousseau, and Y. Eeckhout, *FEBS Lett.* **310**, 175 (1992).

illustrating several of the features discussed below is shown in Fig. 1. After removal of the signal peptides, the collagenases are converted to the respective proenzymes, pro-HFC and pro-HNC, both with molecular masses (predicted from the protein sequences) of approximately 52 kDa. Although the zymogens are single polypeptide chains, they can be viewed as consisting of three domains. The N-terminal portions of HFC and HNC consist of a propeptide of approximately 80 residues that is autolytically lost subsequent to activation, leading to a reduction in molecular mass of approximately 9 kDa. The central portions of the enzymes consist of a catalytic domain with an approximate molecular mass of 19 kDa that contains all four metal atoms, including the active-site zinc atom.[12] The C-terminal portion, referred to as the hemopexin- or vitronectin-like domain, is critically important for collagenase activity. Both HFC and HNC are known to undergo an autolytic cleavage at a site very close to the boundary between the

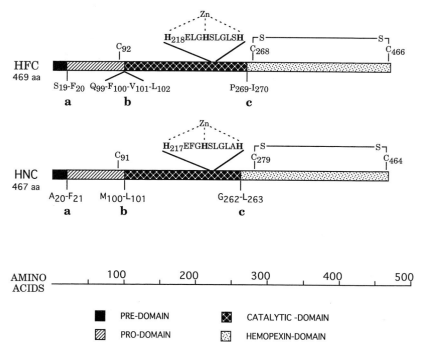

FIG. 1. Schematic comparison of the structures of prepro-HFC and prepro-HNC. The three histidine residues responsible for binding the catalytic zinc as well as the three cysteine residues present in each protein are indicated. The two cysteines in the hemopexin domain are disulfide bonded. The third, unpaired cysteine is part of the conserved "cysteine switch" region in the propeptide domain. The sites of proteolysis between enzyme domains are indicated as follows: **a,** signal sequence cleavage site; **b,** activation-associated autolytic cleavage site(s); **c,** autolytic degradation site.

catalytic and hemopexin-like domains. Loss of the hemopexin-like domain causes the two enzymes to lose the ability to hydrolyze types I, II, and III collagens, but not the ability to hydrolyze peptide substrates.[21–23]

As noted above, both pro-HFC and pro-HNC (and the other pro-MMPs) are secreted by cells as zymogens that require activation before acting on their substrates. According to the "cysteine switch" model, the sulfhydryl group of a cysteine residue located in the propeptide domains of both procollagenases is coordinated to the active-site zinc atom in the catalytic domain in a manner that covers the active site and renders the zymogens latent.[24,25] Release of the cysteine switch residue from the zinc atom with the concomitant formation of a catalytically competent active site is required for zymogen activation. The cysteine residue is dissociated from the zinc atom on autolytic activation, or via activation by trypsin or other proteases, by proteolytic loss of the propeptide domain. Dissociation can also be brought about by chaotropic agents such as NaSCN or NaI, and surfactants such as sodium dodecyl sulfate (SDS). In addition, pro-HFC can be activated by sulfhydryl reagents that react with the cysteine switch residue during moments when it is transiently dissociated from the zinc atom, thereby preventing reassociation and shifting the equilibrium to the active form by mass action. These reactions [alkylation, oxidation, Hg(II) complexation, disulfide exchange] all convert the sulfhydryl group of the cysteine switch residue to a species that cannot serve as a good zinc ligand.

Although HFC and HNC are homologous (58% at the protein sequence level) and share many properties, there are several key differences between the two collagenases. The first pertains to the location and extent of carbohydrate processing. Some of the pro-HFC molecules synthesized by cells contain N-linked carbohydrate arising from glycosylation at one of two possible Asn residues, raising the apparent molecular mass of these chains by approximately 5 kDa.[26] The glycosylation does not alter the function of the enzyme molecules in any known way. Thus, glycosylated and nonglycosylated pro-HFC species are synthesized by fibroblasts and other cell types that have molecular masses of approximately 52 and 57 kDa, respectively.

[21] I. M. Clark and T. E. Cawston, *Biochem. J.* **263**, 21 (1989).
[22] B. Birkedal-Hansen, W. G. I. Moore, R. E. Taylor, A. S. Bhown, and H. Birkedal-Hansen, *Biochemistry* **27**, 6751 (1988).
[23] V. Knauper, A. Osthues, Y. A. DeClerk, K. E. Langley, J. Blaser, and H. Tschesche, *Biochem. J.* **291**, 847 (1993).
[24] E. B. Springman, E. L. Angleton, H. Birkedal-Hansen, and H. E. Van Wart, *Proc. Natl. Acad. Sci. U.S.A.* **87**, 364 (1990).
[25] H. E. Van Wart and H. Birkedal-Hansen, *Proc. Natl. Acad. Sci. U.S.A.* **87**, 5578 (1990).
[26] S. M. Wilhelm, Z. Z. Eisen, M. Teter, S. D. Clark, A. Kronberger, and G. I. Goldberg, *Proc. Natl. Acad. Sci. U.S.A.* **83**, 3756 (1986).

In contrast, pro-HNC has only been detected as a glycoprotein with an intact apparent molecular mass of 75 kDa. The 23-kDa difference between the observed molecular mass and that predicted from the protein sequence is due to glycosylation at multiple Asn residues.[27] Pro-HNC is easily proteolyzed during purification to a 58-kDa form by removal of the N-terminal 64 residues; this species retains the latency of the zymogen.[27,28] HFC is produced by a variety of cell types, but only by *de novo* synthesis in response to specific stimuli. It is not stored intracellularly, but is immediately secreted as the pro-HFC zymogen. In contrast, HNC is only known to be produced by neutrophils and only during their maturation in bone marrow. In addition, pro-HNC is stored intracellularly in secondary granules and only released when neutrophils are stimulated to degranulate.

The HFC and HNC enzymes also have different substrate specificities. Both hydrolyze all three $\alpha$ chains of types I, II, and III collagens following the Gly residue of the partial sequences Gly-(Ile or Leu)-(Ala or Leu) in the collagens at a site that corresponds to Gly-775 in type I collagen.[29,30] However, the two collagenases hydrolyze types I, II, and III collagens at markedly different rates, with HNC exhibiting a marked preference for type I[27,31,32] and HFC a preference for type III.[27,33] HFC has also been reported to hydrolyze types VII[34] and X[35–37] collagens, but at loci different from the cleavage site in the interstitial collagens. As equivalent studies have not been carried out with HNC, it is not clear whether those collagens are also substrates for both collagenases. A systematic investigation of the peptide substrate specificities of the two collagenases has also been carried out.[38,39] The results indicate that the specificities of the two enzymes are

[27] S. K. Mallya, K. A. Mookhtiar, Y. Gao, K. Brew, M. Dioszegi, H. Birkedal-Hansen, and H. E. Van Wart, *Biochemistry* **29**, 10628 (1990).

[28] K. A. Mookhtiar and H. E. Van Wart, *Biochemistry* **29**, 10620 (1990).

[29] J. Gross, E. Harper, E. D. Harris, Jr., P. A. McCroskery, J. H. Highberger, C. Corbett, and A. H. Kang, *Biochem. Biophys. Res. Commun.* **61**, 605 (1974).

[30] E. J. Miller, E. D. Harris, Jr., E. Chung, J. E. Finch, Jr., P. A. McCroskery, and W. T. Butler, *Biochemistry* **15**, 787 (1976).

[31] A. L. Horwitz, A. J. Hance, and R. G. Crystal, *Proc. Natl. Acad. Sci. U.S.A.* **74**, 897 (1977).

[32] K. A. Hasty, J. J. Jeffrey, M. S. Hibbs, and H. G. Welgus, *J. Biol. Chem.* **262**, 10048 (1987).

[33] H. G. Welgus, J. J. Jeffrey, and A. Z. Eisen, *J. Biol. Chem.* **256**, 9511 (1981).

[34] J. L. Seltzer, A. Z. Eisen, E. A. Bauer, N. P. Morris, R. W. Glanville, and R. E. Burgeson, *J. Biol. Chem.* **264**, 3822 (1989).

[35] T. M. Schmid, R. Mayne, J. J. Jeffrey, and T. F. Linsenmayer, *J. Biol. Chem.* **261**, 4184 (1986).

[36] S. J. Gadher, D. R. Eyre, V. C. Duance, S. F. Wotton, L. W. Heck, T. M. Schmid, and D. E. Woolley, *Eur. J. Biochem.* **175**, 1 (1988).

[37] S. J. Gadher, T. M. Schmid, L. W. Heck, and D. E. Woolley, *Matrix* **9**, 109 (1989).

[38] G. B. Fields, H. E. Van Wart, and H. Birkedal-Hansen, *J. Biol. Chem.* **262**, 6221 (1987).

[39] S. J. Netzel-Arnett, G. B. Fields, H. Birkedal-Hansen, and H. E. Van Wart, *J. Biol. Chem.* **266**, 6747 (1991).

similar, yet there are certain differences in the preference for amino acids at individual subsites. These differences have been used to design optimal fluorogenic substrates for the two enzymes.[40]

Many of the important procedures for producing, assaying, and purifying vertebrate collagenases have been delineated in earlier chapters in this series.[9,41,42] Accordingly, this contribution will focus on recent progress in these areas. In particular, there has been considerable progress in the cloning and expression of collagenase enzymes, as well as in the development of more convenient assays. The information presented below pertains specifically to HFC and HNC; however, most of what is presented is expected to be relevant to the analogous collagenases from other species.

## Assays for Vertebrate Collagenases

The choice of an assay for the study of a vertebrate collagenase is critically dependent on the nature of the study. In general, there are two types of assays that are based on the use of either native collagens or peptides as substrates. If the assay is needed to follow the purification of a collagenase, then it is essential that a collagen-based assay be used. This is because none of the available peptide substrates are totally specific for collagenases; most of the substrates can be hydrolyzed by other proteinases. Thus, the measurement of true collagen hydrolysis cannot be avoided. In contrast, for studies of pure or highly purified collagenases, it is much more convenient to utilize one of several chromogenic peptides as substrates, because the assays are sensitive, rapid, and convenient.

### Collagen-Based Assays

A wide variety of methods has been reported for the measurement of vertebrate collagenase activity that range from qualitative screening assays to those that attempt to quantitate the rate of hydrolysis of the collagen substrates. The interstitial collagens that are used as substrates can exist in a variety of forms, including insoluble irreversibly cross-linked fibrils, fibrils or gels reconstituted from acid- or salt-soluble collagens, and soluble collagens. Assays have been developed that use all of these forms in either the native or radiolabeled states as substrates. A number of the assays have been well described in earlier chapters in this series, including a viscometric assay,[42] procedures based on the dissolution of collagen films or plaques,[42]

[40] S. Netzel-Arnett, S. K. Mallya, H. Nagase, H. Birkedal-Hansen, and H. E. Van Wart, *Anal. Biochem.* **195,** 86 (1991).
[41] T. E. Cawston and G. Murphy, this series, Vol. 80, p. 711.
[42] E. D. Harris, Jr., and C. A. Vater, this series, Vol. 82, p. 423.

and several assays based on the hydrolysis of radiolabeled soluble collagen or collagen fibrils.[9,41,42]

The most convenient and sensitive collagenase assays are those that use soluble radiolabeled collagens as substrates. After separation of the $TC^A$ and $TC^B$ hydrolysis fragments from the intact collagens, the concentration of fragments is determined by scintillation counting. The use of radiolabeled collagens in such assays is not straightforward and is complicated by numerous subtleties. The state that a collagen will adopt is markedly dependent on the temperature, pH, ionic strength, the collagen concentration, source, and type, and also whether the molecule has intact telopeptides or has been chemically modified. The preparation and properties of radiolabeled rat tendon type I, bovine cartilage type II, and human amnion type III collagens have been systematically investigated in order to provide an information base with which to choose conditions for collagenase assays that utilize collagens in different states as substrates.[43] On the basis of these data, rapid, sensitive, and convenient assays for the hydrolysis of soluble $^3$H-acetylated types I, II, and III collagens by vertebrate collagenases have been developed.[44]

*Isolation of Types I, II, and III Collagens.* The isolation and characterization of different collagen types have been described in detail earlier.[45] The reader is also referred to procedures specific for rat tendon type I,[46] bovine cartilage type II,[47] and human amnion type III[47] collagens.

*Radiolabeling of Collagens.* Taking into account economy and ease of reaction, $^3$H-acetylation is the recommended method for preparing radiolabeled collagens with high specific activities [counts per minute (cpm) per milligram] and labeling indices (number of Lys + hydroxylysine (Hyl) residues acetylated per tropocollagen molecule) for general use in soluble collagenase assays. Both acid-soluble (e.g., rat tendon type I) and pepsinized (e.g., bovine cartilage type II or human amnion type III) or pronase-treated radiolabeled collagens can be used as soluble substrates under the appropriate conditions. The acid-soluble collagens remain fully soluble in buffers at pH 7.5 containing 0.2–0.4 $M$ NaCl in the 4°–35° range as long as they have labeling indices of 40 Lys + Hyl, or greater.[43] Thus, acetylation serves not only to introduce the radiolabel, but also to prevent fibrillogenesis. Pepsinized or pronase-treated collagens have shortened telopeptides. As intact telopeptides are necessary for efficient fibrillogenesis, the treated collagens are much more soluble under the conditions specified above, and only modest acetylation levels ensure full solubility. The procedure

---

[43] K. A. Mookhtiar, S. K. Mallya, and H. E. Van Wart, *Anal. Biochem.* **158,** 322 (1986).
[44] S. K. Mallya, K. A. Mookhtiar, and H. E. Van Wart, *Anal. Biochem.* **158,** 334 (1986).
[45] E. J. Miller and R. K. Rhodes, this series. Vol. 82, p. 33.
[46] G. Chandrakasan, D. A. Torchia, and K. A. Piez, *J. Biol. Chem.* **251,** 6062 (1976).
[47] R. K. Rhodes and E. J. Miller, *Biochemistry* **17,** 3442 (1978).

described below produces high labeling indices (~40 Lys + Hyl) suitable for soluble assays. A labeling procedure to produce collagens with low labeling indices that are suitable as substrates for fibril assays has been described earlier.[9]

Prior to the radiolabeling, stock solutions of the collagen (2–5 mg/ml in 10 m$M$ acetic acid), 2 $M$ CaCl$_2$, and 0.1 $M$ sodium borate, pH 11, are made. Dry benzene is also prepared by treatment with a small quantity of sodium metal. To start the labeling procedure, the ampoule containing the [$^3$H]acetic anhydride (New England Nuclear, Boston, MA, or Amersham, Arlington Heights, IL; 5 mCi of reagent with a specific activity of 50 mCi/mmol) is immersed in liquid nitrogen for 10 min in a fume hood to condense the reagent. The ampoule is then withdrawn halfway, 1 ml of dry benzene is added, and the seal broken with a steel rod. As the benzene enters the ampoule, it freezes to a snowy white appearance. The solution is then thawed and used for the labeling reaction as soon as possible. The concentration of $^3$H should be determined at this stage by counting a small aliquot in order to ensure that the solubilization of reagent has proceded correctly.

For each 50 mg of collagen, 1 mCi of $^3$H should be used in the labeling reaction. A 250-ml silanized glass beaker is set up in an ice bath on a magnetic stirrer with access to a pH meter. With constant stirring, 50 ml of collagen stock solution is slowly mixed first with 50 ml of cold water and then 12.5 ml of 2 $M$ CaCl$_2$. Addition of CaCl$_2$ first is avoided, because it will cause the collagen to precipitate. Last, 12.5 ml of 0.1 $M$ sodium borate, pH 11, is added to give a solution of pH 9.0. If necessary, the pH should be adjusted to pH 9.0 with a few drops of 1 $N$ NaOH. The [$^3$H]acetic anhydride (1 ml) is added in five aliquots over a period of 1 hr with gentle stirring at 5°–10°. It is important to maintain the temperature in that range. The solution is covered and stirred for an additional hour. The solution is then acidified with 250 $\mu$l of glacial acetic acid and dialyzed extensively against 10 m$M$ acetic acid at 4° until the dialyzate contains essentially no $^3$H. The specific activity of the collagen should be approximately 2 × 10$^6$ cpm/mg. The solution is stored at 4°.

*Collagenase Assay Procedure.* Routine assays with all three collagen types are carried out at 30° at a collagen concentration of 100 $\mu$g/ml. $^3$H-acetylated collagen (100 $\mu$g; 40 $\mu$l of a 2.5 mg/ml stock solution in 10 m$M$ acetic acid) is added to a sample of activated collagenase (see below) dissolved in assay buffer (50 m$M$ Tricine, 0.2 $M$ NaCl, 10 m$M$ CaCl$_2$, pH 7.5) at 30° to give a total volume of 1 ml. As a function of time, 100-$\mu$l aliquots of the reaction mixture are transferred to microcentrifuge tubes containing 50 $\mu$l of 20 m$M$ 1,10-phenanthroline dissolved in assay buffer at 23° to quench the reaction. Aliquots are removed at 0, 0.5, 1, 2, 3, 4, and 5 hr, immediately vortexed, and incubated for 5 min at 30° for rat type I and at 34° for bovine type II and human type III collagens to denature

TABLE I
KINETIC PARAMETERS FOR HYDROLYSIS OF SOLUBLE COLLAGENS BY HUMAN FIBROBLAST
AND NEUTROPHIL COLLAGENASES[a]

| Collagen | $k_{cat}$ ($hr^{-1}$) | | $K_m$ ($\mu M$) | | $k_{cat}/K_m$ ($\mu M^{-1} hr^{-1}$) | |
|---|---|---|---|---|---|---|
| | HFC | HNC | HFC | HNC | HFC | HNC |
| Rat type I | 16 | 690 | 0.83 | 1.0 | 19 | 690 |
| Bovine type II | 3.2 | 130 | 2.4 | 2.3 | 1.3 | 57 |
| Human type III | 350 | 200 | 1.7 | 2.5 | 210 | 80 |

[a] Kinetic parameters were determined from standard collagen-based assays using trypsin-activated latent HFC or PCMB-activated latent 58-kDa HNC.

the TC$^A$ and TC$^B$ fragments. Next, 150 $\mu$l of 23° dioxane is added while the tubes are still in the water bath to precipitate the intact collagen. The tubes are vortexed, chilled on ice for 10 min, and centrifuged at 4° for 10 min at 12,000 $g$ in an Eppendorf microcentrifuge. Aliquots of the supernatant (150 $\mu$l) are transferred to a vial containing 5 ml of scintillation cocktail and the $^3$H concentration determined. The counts per minute from each sample are plotted versus time and an initial slope determined (in units of cpm/hr). The counts per minute value corresponding to full hydrolysis is obtained by counting 50 $\mu$l of the reaction mixture directly. The initial rate of the reaction, $v_i$ (in units of $\mu$g/hr), is calculated from the initial slope by dividing by the counts per minute corresponding to total hydrolysis and multiplying by the mass of collagen in the assay.

If desired, the assays can be carried out at any temperature in the range 1°–30°. However, it is essential that the 5-min incubation step at 30° be included to ensure full denaturation of the collagen fragments before addition of the dioxane. Failure to incubate the quench tubes lowers the sensitivity of the assays markedly. It is also essential that the quench tubes not be removed from the water bath until after the addition of the dioxane. Otherwise, the fragments will renature, and a similar loss in sensitivity will be encountered. By varying the collagen concentration in the standard assay, $v_i$ values can be measured as a function of substrate concentration. From such data, double-reciprocal plots of $[E_0]/v_i$ or $1/v_i$ versus $1/[S]$ can be constructed and the kinetic parameters $K_m$ and $k_{cat}$ or $V_{max}$, respectively, determined. The kinetic parameters for the hydrolysis of types I, II, and III collagens by trypsin-activated HFC and the $p$-chloromercuribenzoate (PCMB)-activated 58-kDa form of HNC are shown in Table I. In carrying out the assays at variable collagen concentrations, the only requisite change from the standard assay procedure is that the collagen in the quench tubes must be diluted to less than 70 $\mu g/ml$ prior to incubation. These assays can also be used to quantitate the $K_I$ values of inhibitors and to assess their mode of inhibition, although the peptide-based assays are better suited for that purpose.

*Peptide-Based Assays*

Two types of chromogenic peptide-based assays are available for vertebrate collagenases. The first type is based on the hydrolysis of thioester substrates[48] such as acetyl-Pro-Leu-Gly-(S-Leu)-Leu-Gly-O-ethyl, where S-Leu is a Leu residue in which the NH group has been replaced by a S atom.[49–51] On hydrolysis of the scissile Gly-(S-Leu) bond, the free thiol HS-Leu is liberated. By carrying out the reaction in the presence of a reagent that reacts with the thiol product to give a spectral response, the progress of the reaction can be monitored continuously. Two reagents that can be used for spectrophotometric assays are dithionitrobenzoate (DTNB) and 4,4′-dithiopyridine (PDS), both of which react with mercaptans to give chromophoric products with $\Delta\varepsilon_{412}$ and $\Delta\varepsilon_{323}$ values of 0.01415 and 0.0198 $\mu M^{-1} cm^{-1}$, respectively. Alternatively, $N$-(7-dimethylamino-4-methylcoumarinyl)maleimide (DACM; Sigma, St. Louis, MO, or Fluka, Ronkonkoma, NY) can be used in a fluorometric version of the assay. The DACM reagent reacts with mercaptans to give a fluorescent product ($\lambda_{ex}$ 395 nm, $\lambda_{em}$ 478 nm).[52]

Routine assays are carried out in 50 m$M$ HEPES or Tricine, 0.2 $M$ NaCl, 10 m$M$ CaCl$_2$, pH 7.5, in the presence of 0.5 m$M$ PDS or DTNB at a substrate concentration of 20 $\mu M$ in a total volume of 0.4 ml. The assays are initiated by addition of preactivated collagenase (see below) to a final enzyme concentration of approximately 4 n$M$. The slope of the absorbance change ($\Delta A/\Delta t$) is measured using less than 5% of full hydrolysis, and the initial rates, $v_i$ (in units of $\mu M$/min), are calculated using the extinction coefficient changes given above. The fluorescent version of the assay is carried out in a similar fashion in the presence of 0.5 m$M$ DACM. Because thioester substrates are inherently easy to hydrolyze, there is usually a background rate of hydrolysis (observed in the absence of enzyme) that must be subtracted from the assays to obtain the true collagenase-catalyzed rate. The kinetic parameters for hydrolysis of DACM by HFC at pH 7.5, 37°, are $k_{cat}$ = 370,000 hr$^{-1}$ and $K_m$ = 3.9 m$M$.[50]

The thiopeptide assay is the fastest and most convenient assay available.

[48] J. C. Powers and C.-M. Kam, this volume [1].
[49] H. Weingarten and J. Feder, *Anal. Biochem.* **147,** 437 (1985).
[50] H. Weingarten, R. Martin, and J. Feder, *Biochemistry* **24,** 6730 (1985).
[51] H. Weingarten and J. Feder, *Biochem. Biophys. Res. Commun.* **139,** 1184 (1986).
[52] A. Zantema, J. A. Maasen, J. Kriek, and W. Möller, *Biochemistry* **21,** 3069 (1982).

However, the substrate is inherently unstable, with a rate of decomposition that is higher in the presence of nucleophilic buffers and above pH 7.0. There is also interference in the assay from substances that absorb in the range 300–400 nm, and the assay cannot be used with samples containing mercaptans. Because the $K_m$ value for HFC is high and exceeds the solubility of the substrate, the assay is not well suited for measuring kinetic parameters or evaluating $K_I$ values for inhibitors.

A second type of assay is a continuously recording fluorescence assay that is based on the hydrolysis of peptides with intramolecularly quenched fluorescence. The first such assay was developed by Stack and Gray for the substrate DNP-Pro-Leu-Gly-Leu-Trp-Ala-D-Arg, where DNP stands for the 2,4-dinitrophenyl group.[53] The key features of the peptide are the presence of the fluorescent Trp residue in subsite $P'_2$ and the DNP quenching group at the N terminus. Hydrolysis of the Gly-Leu bond by collagenases relieves the quenching in the intact substrate, and the resulting fluorescence increase serves as the basis for the assay (see [2] in this volume). HFC hydrolyzes the peptide at pH 7.6, 37°, with $k_{cat}$ = 38 hr$^{-1}$ and $K_m$ = 7.1 $\mu M$.[53] In a related study, the substrates DNP-Pro-Leu-Ala-Leu-Trp-Ala-Arg and DNP-Pro-Leu-Ala-Tyr-Trp-Ala-Arg have been developed as substrates individually optimized for HFC ($k_{cat}$ = 4300 hr$^{-1}$ and $K_m$ = 130 $\mu M$ at pH 7.5, 23°) and HNC ($k_{cat}$ = 11,000 hr$^{-1}$ and $K_m$ = 7.7 $\mu M$ at pH 7.5, 23°), respectively.[40] Knight and associates have reported a more sensitive collagenase assay ($k_{cat}/K_m$ = 14,800 $M^{-1}$ sec$^{-1}$ at pH 7.5, 37°, for HFC) that is based on the hydrolysis of MCA-Pro-Leu-Gly-Leu-Dpa-Ala-Arg-NH$_2$, where MCA stands for 7-methoxycoumarin-4-yl and Dpa for $N$-3-(2,4-dinitrophenyl)-L-2,3-diaminopropionyl.[54] As the assays are described in detail elsewhere in this volume,[55] the procedures are not repeated here.

*Activation of Collagenases*

The activation of vertebrate procollagenases can be achieved by a variety of means. As discussed above, pro-MMP zymogens can be activated either proteolytically or nonproteolytically. In general, the optimal conditions for activation by any treatment must be explored empirically. In choosing a method of activation, one should make sure that it is compatible with the assay to be used. Thus, organomercurial activation should be avoided if the enzyme is to be used with the thioester assay. If proteolytic activation is used, the activating protease should be inactivated if it could also compete with the collagenase for the substrate.

[53] M. S. Stack and R. D. Gray, *J. Biol. Chem.* **264,** 4277 (1989).
[54] C. G. Knight, F. Willenbrock, and G. Murphy, *FEBS Lett.* **296,** 263 (1992).
[55] C. G. Knight, this volume [2].

Pro-HFC can be activated by treatment with 50 $\mu$g/ml trypsin for 15 min at 23°, after which the trypsin can be inactivated by the addition of a 4-fold excess of soybean trypsin inhibitor (SBTI) or removed with a SBTI-agarose resin. Alternatively, full activation can be achieved by incubation with 100 $\mu M$ PCMB for 30 min at 23°. Activation can also be achieved by similar treatments with other organomercurials and other classes of activators.[24] The 58- and 75-kDa forms of pro-HNC must be activated differently. The 58-kDa form is easily activated by treatment with 100 $\mu M$ PCMB for 15 min at 23°. In contrast, full activation of the 75-kDa form requires treatment with both trypsin (10 $\mu$g/ml for 15 min followed by a 4-fold excess of SBTI) and PCMB (100 $\mu M$ for 10 min) at 23°.[56]

## Sources of Vertebrate Collagenases

Vertebrate collagenases have been purified in their naturally occurring forms from a wide variety of sources including different tissues, tissue culture media, and cells. Those materials continue to represent one type of source of the enzymes. In addition, cDNA clones have been obtained for a number of the collagenases,[13–20] and the availability of the clones has made possible the expression of recombinant collagenases. A number of systems are available for the expression of proteins from cDNA clones.[57] This can be an attractive alternative to isolation from natural sources for a variety of reasons. First, a recombinant enzyme can often be obtained in larger quantities. Second, there may be a safety issue associated with the natural source (e.g., the isolation of HNC from human blood). Third, expression in bacteria can produce a collagenase that is not posttranslationally modified. This can be advantageous if the enzyme is to be used for crystallography, because many glycoproteins are difficult to crystallize. A fourth potential advantage of using a recombinant enzyme as a source for purification is that there are often fewer contaminating activities than in native enzyme preparations. Last, recombinant systems make possible the expression of truncated, chimeric, and specifically mutated enzymes.

This section discusses sources of both naturally occurring and recombinant full-length (including the hemopexin-like domain) forms of the fibroblast- and neutrophil-type collagenases, since only these retain collagenase activity. The numbering of amino acid residues is based on the initiating methionine being designated as the first residue. The relation of the various truncations to the domain structure of the enzymes is indicated in Fig. 1.

---

[56] H. E. Van Wart, *in* "Matrix Metalloproteinases and Inhibitors" (H. Birkedal-Hansen, Z. Werb, H. Welgus, and H. Van Wart, eds.), *Matrix* Spec. Suppl. No. 1, p. 31. Gustav Fischer, Stuttgart, 1992.

[57] D. V. Goedel (ed.), this series, Vol. 185.

*Fibroblast-Type Collagenase*

*Naturally Occurring Enzyme.* Vertebrate collagenases have been isolated from a number of different organisms and sources. These have been extensively reviewed in previous volumes of this series.[9,41,42] The conditioned medium from cultured connective tissue fibroblasts generally contains low levels (~100–200 ng/ml) of pro-HFC. Stimulation with cytokines or phorbol esters results, following a lag period, in a significant increase in the amount of enzyme secreted by the cells. For example, medium conditioned for 7 days by culturing human gingival fibroblasts in the presence of interleukin-1$\beta$ contains in excess of 5 $\mu$g/ml of pro-HFC.[58] This medium can be used as the starting material for purification of the enzyme.

*Recombinant Enzyme.* Full-length recombinant prepro-HFC has been expressed in stably transfected C127 cells (ATCC CRL 1616)[59] and NOS mouse myeloma cells (ECACC cont. no. 85110503),[60] both of which secrete pro-HFC into the culture medium. Transiently transfected COS-7 cells also produce small quantities of the enzyme.[59] Purified recombinant pro-HFC from the myeloma cell expression system resembles the naturally occurring enzyme in that it migrates as a doublet (glycosylated and unglycosylated forms) with the expected apparent molecular weights during SDS–polyacrylamide gel electrophoresis. The enzyme also shows full latency until activation with organomercurials or trypsin. After activation, the resultant HFC degrades fibrillar collagen and is inhibited by tissue inhibitor of metalloproteinase-1 (TIMP-1). Similar results have been obtained for rabbit fibroblast-type collagenase that was stably expressed in baby hamster kidney cells at a level of 3 $\mu$g/ml.[61]

The availability of a cDNA clone allows for the addition of exogenous protein sequences to form fusion proteins which may facilitate detection and purification of the desired collagenase. For example, a chimeric protein consisting of a signal sequence from rat stromelysin followed by a sequence derived from protein A fused to amino acids 24–469 of HFC has been expressed transiently in COS-7 cells.[62] The protein A portion of the molecule allows the facile affinity purification of the chimeric protein using immunoglobulin G (IgG)-Sepharose. The fusion protein is expressed in

[58] M. W. Lark, L. A. Walakovits, T. K. Shah, J. Vanmiddlesworth, P. M. Cameron, and T. Y. Lin, *Connect. Tissue Res.* **25**, 49 (1990).
[59] G. Murphy, M. I. Cockett, P. E. Stephens, B. J. Smith, and A. J. P. Docherty, *Biochem. J.* **248**, 265 (1987).
[60] G. Murphy, J. A. Allan, F. Willenbrock, M. I. Cockett, J. P. O'Connell, and A. J. Docherty, *J. Biol. Chem.* **267**, 9612 (1992).
[61] C. E. Brinckerhoff, K. Suzuki, T. I. Mitchell, F. Oram, C. I. Coon, R. D. Palmiter, and H. Nagase, *J. Biol. Chem.* **265**, 22262 (1990).
[62] R. Sanchez-Lopez, C. M. Alexander, O. Behrendtsen, R. Breathnach, and Z. Werb. *J. Biol. Chem.* **268**, 7238 (1993).

latent form and can be eluted from the affinity column by activation with organomercurials, which leads to autolytic release of HFC (catalytic plus hemopexin domains). The resultant enzyme is capable of cleaving fibrillar collagen.

Pro-HFC (amino acids 20 to 469) has also been expressed in *Escherichia coli*.[63] The enzyme is insoluble but can be extracted with guanidine hydrochloride and refolded to produce a species which, with the exception of glycosylation, is functionally equivalent to the native enzyme in terms of a number of properties. The lower relative specific activity of the recombinant enzyme toward fibrillar collagen versus casein compared to the naturally occurring protein, however, suggests that refolding of the C-terminal domain may not be correct. In another study, the cDNA for porcine fibroblast-type collagenase has been used to create a fusion protein with $\beta$-galactosidase which was expressed in *E. coli*.[64] The majority of that fusion protein is also insoluble. However, the protein can be extracted, refolded, and cleaved with factor Xa to generate an active collagenase with the authentic N terminus of the activated naturally occurring collagenase.

*Neutrophil-Type Collagenase*

*Naturally Occurring Enzyme.* Pro-HNC is only known to be located in the secondary granules of mature human neutrophils. Thus, that material is the sole current source of naturally occurring HNC.

*Recombinant Enzyme.* Pro-HNC has been expressed transiently in COS-7 cells from a cDNA clone in the pcDNA I expression vector[15] and in a cell-free *in vitro* translation system.[16] In both cases, the expressed proteins are able to cleave specifically collagen fibrils to produce the characteristic $TC^A$ and $TC^B$ fragments following activation. Interestingly, in the case of the COS-7 cell transient transfections, pro-HNC is secreted from the cells. This contrasts with the normal trafficking of the neutrophil, where the newly synthesized enzyme is stored in granules.

Pro-HNC has also been expressed in our laboratory in insect (Sf9) cells using a baculovirus vector.[65] An *Eco*RI cDNA fragment containing the complete coding region of prepro-HNC was subcloned into a transfer vector and cotransfected with linearized viral DNA into Sf9 cells. Recombinant viruses were cloned, and those directing a high level of expression of secreted protein were selected by analysis of culture medium with a polyclonal antibody that recognizes HNC.[27] On organomercurial activation, the me-

[63] L. J. Windsor, H. Birkedal-Hansen, B. Birkedal-Hansen, and J. A. Engler, *Biochemistry* **30**, 641 (1991).

[64] M. C. O'Hare, N. J. Clarke, and T. E. Cawston, *Gene* **111**, 245 (1992).

[65] M. Dioszegi, P. Cannon, and H. E. Van Wart, unpublished data (1994).

dium from cells infected with recombinant virus, but not that from cells infected with a control virus, exhibited collagenase activity.

## Purification of Vertebrate Collagenases

### Fibroblast-Type Collagenase

The purification of pro-HFC and of homologous fibroblast-type collagenases from other species from cell culture media and tissues has been described previously.[9,41,42] The same protocols should be useful for purification of the recombinant enzyme from bacterial and mammalian expression systems. Thus, no additional attention is directed to the purification of the fibroblast-type collagenase here.

### Neutrophil-Type Collagenase

A procedure for purifying pro-HNC was presented earlier.[9] However, there have been substantial improvements in the purification of pro-HNC that are described below. Progress in the study of HNC has been slow relative to that of HFC, largely because of the difficulty in purifying the former. The HNC enzyme is more unstable than HFC, and, unless certain precautions are taken, there are large losses during purification. Several purification protocols have been developed to isolate various active and latent forms of HNC in high yield.

Three different starting materials, each with a different level of purity of HNC, can be used for the purification. The first potential starting material is the supernatant from freshly harvested human neutrophils that have been treated with phorbol myristate acetate (PMA) to stimulate enzyme release in the presence of phenylmethylsulfonyl fluoride (PMSF), $\alpha_1$-protease inhibitor, and the antioxidants methionine and catalase.[66] The supernatants contain predominantly the latent 75-kDa form of HNC. A second source is an extract from fresh buffy coats prepared in the presence of PMSF (1 mM) and benzamidine (50 mM).[28] The extract, which is considerably more crude, also contains predominantly intact, latent pro-HNC. However, variable amounts of a latent 58-kDa form are present if smaller quantities of benzamidine are used, owing to cleavage by a contaminating proteinase. If the PMSF and benzamidine are both omitted, an active 58-kDa form of HNC is the predominant species. A third starting material is the cell extract from buffy coats obtained from outdated blood.[28] This is the most readily available source that contains as much HNC as fresh buffy coats. However, the resultant extract is more heavily contaminated by hemoglobin.

---

[66] S. T. Test and S. F. Weiss, J. Biol. Chem. 259, 399 (1984).

Buffy coats from outdated or fresh blood are diluted with an equal volume of 0.9% (w/v) NaCl. For buffy coats from fresh blood, dextran T500 is added to a final concentration of 1.1% and allowed to sit for 1 hr at 25°. The upper yellow layer is then removed and centrifuged at 2000 $g$ for 20 min. When outdated blood is the source, the dextran step is impractical and is omitted. Instead, the cell suspension is centrifuged for 10 min at 1000 $g$ at 4°. The cell pellets from either procedure are then subjected to hypotonic hemolysis for 1 min. The resulting suspension is made isotonic with 3.6% NaCl, and the cells are pelleted by centrifugation for 10 min at 3000 $g$. After washing the cells twice with 0.9% NaCl, they are suspended in an equal volume of 10 m$M$ Tris, 1 $M$ NaCl, 5 m$M$ CaCl$_2$, pH 7.5, containing 0.05% (v/v) Brij 35 and 50 $\mu M$ ZnSO$_4$, homogenized in a blender, and diluted to 300 ml with the same buffer. If pro-HNC is to be isolated, PMSF (100 m$M$ in 2-propanol) is added to a final concentration of 1 m$M$. The suspension is then freeze–thawed seven times in dry ice/acetone and in a 35° water bath, respectively. After centrifugation at 16,000 $g$ for 1 hr to remove cell debris, the supernatant is diluted with 10 m$M$ Tris, 5 m$M$ CaCl$_2$, pH 7.5, containing 0.05% Brij 35 and 50 $\mu M$ ZnSO$_4$ so that the final NaCl concentration is 0.5 $M$. A small amount of solid deoxyribonuclease I and enough MgSO$_4$ to make its final concentration 0.5 m$M$ are added to degrade the DNA and reduce the viscosity of the solution. After standing overnight at 4°, the solution is clarified by centrifugation at 16,000 $g$ for 3 hr. The supernatant serves as the starting material for the chromatographic purification.

In general, attempts to purify HNC chromatographically will be unsuccessful unless there is strict adherence to several procedures. First, solutions of HNC containing less than 0.5 $M$ NaCl lose activity rapidly, probably because of adsorption of the enzyme to glass and plastic surfaces. A similar loss in activity is encountered on dialysis and concentration. Even in the presence of 0.5 $M$ NaCl, the addition of the surfactant Brij 35 is also required to stabilize the enzyme. Third, HNC is stabilized by 50 $\mu M$ zinc ions. As HNC contains two intrinsic zinc atoms, the exogenous zinc could prevent the loss of intrinsic zinc during the purification. Fourth, HNC is more stable to almost all manipulations when kept in the latent form. Last, freeze–thawing cycles generally have a deleterious effect on the activity of HNC. The purified enzyme retains activity best when quick-frozen in dry ice/acetone and stored at −70°.

The first chromatographic step is designed to separate the collagenase activity from the bulk of the hemoglobin. The crude buffy coat extract dissolved in 10 m$M$ Tris, 0.5 $M$ NaCl, 5 m$M$ CaCl$_2$, pH 7.5, containing 0.05% Brij 35 and 50 $\mu M$ ZnSO$_4$ is applied to a Reactive Red 120-agarose column (2.5 × 60 cm) that had previously been equilibrated with the same buffer, except devoid of Brij 35, at a flow rate of 25 ml/hr. After washing with this buffer until the absorbance reaches the baseline, the collagenase activity is eluted with this buffer containing 1 $M$ NaCl and 0.05% Brij 35. This step removes the majority of the hemoglobin, aminopeptidase, cathepsin G, myeloperoxidase, and elastase and 80% of the gelatinase with a typical collagenase yield of 60–100%.

The HNC-rich fraction from the Reactive Red 120-agarose column is concentrated 35-fold. In the case of a latent preparation, the collagenase is activated by incubation with 0.5 m$M$ PCMB for 30 min at 4°. The sample is then applied to a Sepharose-CH-Pro-Leu-Gly-NHOH affinity column (1.5 × 12 cm) at a flow rate of 15 ml/hr. Prior to application of the sample, the column is equilibrated with 10 m$M$ Tris, 10 m$M$ CaCl$_2$, 1 $M$ NaCl, pH 7.5, containing 0.05% Brij 35 and 0.1 m$M$ PCMB. For the isolation of active HNC, the activation step is unnecessary. The column is then washed with the same buffer until the absorbance returns to the baseline. In the case of a latent sample, the column is next washed with this same buffer devoid of PCMB until all of the PCMB is removed. The active or latent HNC is then eluted with 100 m$M$ Tris, 100 m$M$ CaCl$_2$, 1 $M$ NaCl, pH 9, containing 0.05% Brij 35. Zinc ions are omitted in this buffer so that they will not compete with HNC for the hydroxamate ligand. Immediately after elution, the fraction is adjusted to pH 7.5 with 0.5 $M$ MES, pH 6, containing 1 $M$ NaCl and 0.05% Brij 35, and the ZnSO$_4$ concentration is restored to 50 $\mu M$. This step typically results in a 60- and 160-fold purification of the latent and active enzymes with yields of 65 and 120%, respectively. This step removes all remaining impurities except for a trace of gelatinase.

The last step entails chromatography over gelatin-Sepharose to remove the contaminating neutrophil gelatinase. The HNC fraction from the previous step is diluted 2-fold with a 1 $M$ NaCl solution containing 0.05% Brij 35 to lower the concentratons of Tris and CaCl$_2$. The latent or active samples are then applied to a gelatin-Sepharose column (1 × 10 cm) at a flow rate of 15 ml/hr, and the column is washed with 50 m$M$ Tris, 1 $M$ NaCl, 5 m$M$ CaCl$_2$, pH 7.5, containing 0.05% Brij 35 and 50 $\mu M$ ZnSO$_4$ until the absorbance reaches baseline to give a fraction that contains pure HNC. Overall, the three-step purification described above gives chromatographically pure 58-kDa latent or active HNC from outdated blood with a typical cumulative yield of 40–80%.

Other workers have also developed protocols for purifying HNC, and two of these are mentioned briefly. Tschesche and associates (see [26], this volume) have published a procedure that begins with the homogenization of buffy coats in 20 m$M$ Tris, 5 m$M$ CaCl$_2$, 20 m$M$ benzamidine, 0.5 m$M$ ZnCl$_2$, pH 8.5, in the presence of PMSF and diisopropyl fluorophosphate.[67] The sample is centrifuged at 48,000 $g$ for 30 min and the supernatant dialyzed against 20 m$M$ Tris, 5 m$M$ CaCl$_2$, 0.5 m$M$ ZnCl$_2$, pH 8.5, and applied to a Q-Sepharose Fast Flow column (Pharmacia, Uppsala, Sweden) (4.5 × 19 cm) in the same buffer. Pro-HNC is eluted with a 0–0.15 $M$ NaCl gradient. The collagenase-rich fractions are pooled and applied to a zinc-chelate Sepharose column (1.4 × 10 cm). The pro-HNC passes through the column and is dialyzed against 20 m$M$ Tris, 5 m$M$ CaCl$_2$, 0.5 m$M$ ZnCl$_2$, 0.1 $M$ NaCl, pH 7.5, and applied to an Orange-Sepharose column (1.4 × 10 cm). The pro-HNC is eluted with a 0.1–2 $M$ NaCl gradient. Last, the enzyme is concentrated and chromatographed on a Sephacryl S-300 column (1.8 × 80 cm) equilibrated with 20 m$M$ Tris, 10 m$M$ CaCl$_2$, 0.5 m$M$ ZnCl$_2$, 0.3 $M$ NaCl, pH 7.5.

Sorsa has reported a purification scheme in which the crude starting extract is first applied to a Sepharose 6B gel-filtration column (Pharmacia, Uppsala, Sweden) (1.5 × 70 cm) in 50 m$M$ Tris, 10 m$M$ CaCl$_2$, 0.2 $M$ NaCl, pH 7.5.[68] The fractions containing HNC are then applied to a QAE-Sephadex A-50 (Pharmacia, Uppsala, Sweden) column. The last purification step consists of chromatography over Cibacron Blue Sepharose (Pharmacia, Uppsala, Sweden) (1 × 10 cm). The sample is applied in 5 m$M$ Tris, 0.1 $M$ NaCl, 1 m$M$ CaCl$_2$, pH 7.5. HNC is eluted with a 0.1–1.0 $M$ NaCl gradient at an NaCl concentration of 0.5 $M$.

*Note Added in Proof.* A third human interstitial collagenase has recently been cloned from a breast tumor cDNA library (J. M. Freije, I. Diez-Itza, M. M. Balbin, L. M. Sanchez, R. Blasco, J. Tolivia, and C. Lopex-Otin, *J. Biol. Chem.* **269,** 16766 (1994)). This enzyme appears to be the human homolog of the previously described rat[19] and mouse[20] collagenases with which it shows 86% amino acid identity. Human fibroblast-type collagenase shows only 50% amino acid identity with the rodent enzymes. Expression of this cDNA in a vaccinia virus system results in the synthesis of a protein which, upon activation with APMA, shows collagenase activity against type I collagen fibrils. This new enzyme has been designated collagenase-3 or MMP-13.

[67] V. Knäuper, S. Krämer, H. Reinke, and H. Tschesche, *Biol. Chem. Hoppe-Seyler* **371** (Suppl.), 295 (1990).
[68] T. A. Sorsa, *Scand. J. Rheum.* **16,** 167 (1987).

# [26] Human Neutrophil Collagenase

*By* HARALD TSCHESCHE

## Introduction

Neutrophil collagenase (EC 3.4.24.34) is a zinc-containing matrix-degrading metalloproteinase (MMP) contained as a proenzyme in the specific granules of polymorphonuclear leukocytes (PMNL).[1] It is assumed to be synthesized during differentiation of bone marrow stem cells in the myelocyte stage[2] and is encoded by a gene different from that of the fibroblast-type interstitial collagenase.[3,4] The enzyme is highly glycosylated, with about one-third of the molecular weight arising from carbohydrate.[4,5] The proenzyme, also referred to as latent enzyme, is released on stimulation of the cells by various agents, such as chemotactic formylpeptides,[6,7] human C3a and C5a anaphylatoxins,[8,9] tumor necrosis factor (TNF$a$),[10] interleukin-1,[11] interleukin-8,[12] fibrinogen and fibrin-derived products,[13,14] granulocyte/macrophage colony-stimulating factor (GM-CSF),[15] calcium ionophore A23187,[16] and leukotrienes and prostaglandins.[7] The interstitial procollagenase is a three-domain protein composed of propeptide and catalytic domains followed by a C-terminal hemopexin-like domain[17] (see [25] in this volume).

Activation of the proenzyme occurs in the extracellular space after secretion and can be achieved *in vitro* by various proteolytic enzymes such as the serine proteinases trypsin, chymotrypsin, cathepsin G, and tissue

[1] G. Murphy, J. J. Reynolds, U. Bretz, and M. Baggiolini, *Biochem. J.* **162,** 195 (1977).

[2] D. F. Bainton and M. G. Farquhar, *J. Cell Biol.* **28,** 277 (1966).

[3] K. A. Hasty, T. F. Pourmotabbed, G. I. Goldberg, J. P. Thompson, D. G. Spinella, R. M. Stevens, and C. L. Mainardi, *J. Biol. Chem.* **265,** 11421 (1990).

[4] V. Knäuper, S. Krämer, H. Reinke, and H. Tschesche, *Eur. J. Biochem.* **189,** 295 (1990).

[5] H. Tschesche, V. Knäuper, S. Krämer, J. Michaelis, R. Oberhoff, and H. Reinke, *in* "Matrix Metalloproteinases and Inhibitors" (H. Birkedal-Hansen, Z. Werb, H. Welgus, and H. Van Wart, eds.), *Matrix* Suppl. No. 1, p. 245. Gustav Fischer, Stuttgart, 1992.

[6] H. Tschesche, A. Schettler, H. Thorn, B. Bakowski, V. Knäuper, H. Reinke, and B. M. Jockusch, *in* "Proteinases and Their Inhibitors—Recent Developments" (E. Auerswald, H. Fritz, and V. Turk, eds.), Proc. 8th Winter School, KFA Jülich GmbH, p. 31. 1989.

[7] A. Schettler, H. Thorn, B. M. Jockusch, and H. Tschesche, *Eur. J. Biochem.* **197,** 197 (1991).

[8] H. N. Fernandez, P. M. Henson, A. Otani, and T. E. Hugli, *J. Immunol.* **120,** 109 (1978).

[9] P. A. Ward and J. H. Hill, *J. Immunol.* **104,** 535 (1970).

[10] S. J. Klebanoff, M. A. Vadas, J. M. Harlan, L. H. Sparkes, J. R. Gamble, J. M. Agosti, and A. M. Waltersdorph, *J. Immunol.* **136,** 4220 (1986).

[11] T. A. Luger, J. A. Charon, M. Colot, M. Micksche, and J. J. Oppenheim, *J. Immunol.* **131,** 816 (1983).

[12] T. Oshimura, K. Matsushima, S. Tanaka, E. A. Robinson, E. Apella, J. J. Oppenheim, and E. J. Leonard, *Proc. Natl. Acad. Sci. U.S.A.* **84,** 9233 (1987).

[13] A. B. Kay, D. S. Pepper, and M. R. Ewart, *Nature (London) New Biol.* **243,** 56 (1973).

[14] R. McKenzie, D. S. Pepper, and A. B. Kay, *Thromb. Res.* **6,** 1 (1975).

[15] A. Yuo, K. Kitagawa, A. Ohsaka, M. Ohta, K. Miyazono, T. Okabe, A. Urabe, M. Saito, and F. Takaku, *Blood* **74,** 2144 (1989).

[16] V. Knäuper, S. M. Wilhelm, P. K. Separack, Y. A. DeClerk, V. E. Langley, A. Osthues, and H. Tschesche, *Biochem. J.* **581** (1993).

[17] H. Birkedal-Hansen, W. G. I. Moore, M. K. Bodden, L. J. Windsor, B. Birkedal-Hansen, A. DeCarlo, and J. A. Engler, *Crit. Rev. Oral Biol. Med.* **4**(2), 197 (1993).

kallikrein,[4,5,18] but not by leukocyte elastase or plasmin,[4,5] but also by metalloproteinases such as stromelysin 1 (MMP-3)[16] or 2 (MMP-10).[23] Activation is also initiated by autocatalysis,[5,18,19] by mercury compounds,[5,20,21,47] by oxidative agents such as oxygen radicals, hydrogen peroxide, or hypochlorite,[22,47] or sodium gold thiomalate.[23,47] The molecular mass is reduced by about 20 kDa on activation owing to loss of the N-terminal propeptide.[4,5,18] The active enzyme cleaves type I, II, and III triple helical collagen into one-quarter and three-quarter fragments well below the denaturation temperature with preference for type I (i.e., it hydrolyzes the $Gly^{775}$-$Leu^{776}$ and $Gly^{775}$-$Ile^{776}$ peptide bonds).[5,18,24–26] It thus differs in specificity from the fibroblast-type enzyme, which prefers type III collagen over type I.[18,24] Studies with synthetic oligopeptides covering the P4 to P4' subsites of the substrate demonstrate similar but distinct substrate specificities[27] of the two enzymes.

The leukocyte collagenase has more general proteolytic properties. Besides promoting cleavage of triple-helical collagen and self-degradation during autocatalytic activation[4,5,18,19,28,29] and autolysis,[30] it cleaves gelatin peptides, fibronectin,[5] proteoglycans,[5] cartilage aggrecan,[31] serpins,[30,32–34] and peptides like angiotensin and substance P.[34a]

[18] H. Van Wart, in "Matrix Metalloproteinases and Inhibitors" (H. Birkedal-Hansen, Z. Werb, H. Welgus, and H. Van Wart, eds.), *Matrix* Suppl. No. 1, p. 31. Gustav Fischer, Stuttgart, 1992.

[19] K. A. Hasty, M. S. Hibbs, A. H. Kang, and C. L. Mainardi, *J. Biol. Chem.* **261,** 5645 (1986).

[20] J. Bläser, V. Knäuper, A. Osthues, H. Reinke, and H. Tschesche, *Eur. J. Biochem.* **202,** 1223 (1991).

[21] J. Bläser, S. Triebel, H. Reinke, and H. Tschesche, *FEBS Lett.* **313,** 59 (1992).

[22] S. J. Weiss, G. Peppin, X. Ortiz, C. Ragsdale, and S. T. Test, *Science* **227,** 747 (1985).

[23] S. Lindy, T. Sorsa, K. Suomalainen, and H. Turto, *FEBS Lett.* **208,** 23 (1986).

[24] S. K. Mallya, K. A. Mookhtiar, Y. Gao, K. Brew, M. Dioszegi, H. Birkedal-Hansen, and H. E. Van Wart, *Biochemistry* **29,** 10628 (1990).

[25] A. I. Horwitz, A. J. Hance, and R. G. Crystal, *Proc. Natl. Acad. Sci. U.S.A.* **74,** 894 (1977).

[26] K. A. Hasty, J. J. Jeffrey, M. S. Hibbs, and H. G. Welgus, *J. Biol. Chem.* **262,** 10048 (1987).

[27] S. Netzel-Arnett, G. Fields, H. Birkedal-Hansen, and H. E. Van Wart, *J. Biol. Chem.* **266,** 6747 (1991).

[28] K. A. Mookhtiar and H. E. Van Wart, *Biochemistry* **29,** 10620 (1990).

[29] T. Sorsa, K. Suomalainem, H. Turto, and S. Lindy, *Med. Biol.* **63,** 66 (1985).

[30] V. Knäuper, A. Osthues, Y. A. DeClerck, K. E. Langley, J. Bläser, and H. Tschesche, *Biochem. J.* **291,** 847 (1993).

[31] A. J. Fosang, K. Last, V. Knäuper, P. J. Neame, G. Murphy, T. E. Hardingham, H. Tschesche, and J. A. Hamilton, *Biochem. J.* **295,** 273 (1993).

[32] V. Knäuper, H. Reinke, and H. Tschesche, *FEBS Lett.* **263,** 355 (1990).

[33] V. Knäuper, S. Triebel, H. Reinke, and H. Tschesche, *FEBS Lett.* **290,** 99 (1991).

[34] J. Michaelis, M. C. M. Vissers, and C. C. Winterbourn, in "Matrix Metalloproteinases and Inhibitors" (H. Birkedal-Hansen, Z. Werb, H. Welgus, and H. Van Wart, eds.), *Matrix* Suppl. No. 1, p. 80. Gustav Fischer, Stuttgart, 1992.

Purification of Procollagenase from Polymorphonuclear
  Leukocyte Granules

Neutrophil collagenase was detected by Lazarus *et al.*[35] in 1968 and
assigned to the specific granules by Murphy *et al.*[36] Purification has been
described by several groups with differing results.[4,19,23,28,29,37,38] To obtain
predominantly homogeneous, nontruncated latent enzyme, it is essential
to conduct the granule extraction in the presence of the serine proteinase
inhibitor diisopropyl fluorophosphate (DFP) or phenylmethylsulfonyl fluoride (PMSF).

*Reagents and Materials*

Hypotonic NaCl solution: 0.6 m$M$ KCl, 22 m$M$ NaCl
Ringer's solution: 4 m$M$ KCl, 152 m$M$ NaCl
DNP-peptide solution: 5 × 10$^{-4}$ $M$ 2,4-dinitrophenyl (DNP)-peptide
    (see below), 50 m$M$ Tris-HCl, pH 7.5, 150 m$M$ NaCl, 5 m$M$ CaCl$_2$,
    0.02% (w/v) bovine serum albumin (BSA), 0.02% NaN$_3$
Coupling buffer: 0.1 $M$ NaHCO$_3$ pH 8.0, 0.5 $M$ NaCl
Buffer 1: 20 m$M$ Tris-HCl, pH 9.0, 5 m$M$ CaCl$_2$, 0.5 m$M$ ZnCl$_2$, 0.05%
    PMSF, 0.05% DFP, 0.05% benzamidine hydrochloride
Buffer 2: 20 m$M$ Tris-HCl, pH 8.5, 5 m$M$ CaCl$_2$, 5 m$M$ ZnCl$_2$,
    0.02% NaN$_3$
Buffer 3: 20 m$M$ Tris-HCl, pH 8.5, 5 m$M$ CaCl$_2$, 0.5 m$M$ ZnCl$_2$,
    0.02% NaN$_3$
Buffer 4: 20 m$M$ Tris-HCl, pH 7.5, 5 m$M$ CaCl$_2$, 0.5 m$M$ ZnCl$_2$, 100
    m$M$ NaCl
Buffer 5: 20 m$M$ Tris-HCl, pH 7.5, 5 m$M$ CaCl$_2$, 0.5 $M$ NaCl
Buffer 6: 100 m$M$ Tris-HCl, pH 9.0, 20 m$M$ CaCl$_2$, 0.5 $M$ NaCl
Buffer 7: 600 m$M$ Tris-HCl, pH 7.5, 20 m$M$ CaCl$_2$, 0.5 $M$ NaCl
Buffer 8: 50 m$M$ Tris-HCl, pH 7.5, 5 m$M$ CaCl$_2$, 0.5 $M$ NaCl, 0.5 m$M$
    ZnCl$_2$, 0.02% NaN$_3$

1,4-Butanediol diglycidyl ether, disodium iminodiacetate, and Reactive
Orange 14 are purchased from Sigma (Deisenhofen, Germany), Pro-Leu-
Gly-NHOH and the DNP-peptide from Bachem (Bubendorf, Switzerland),
and Sepharose CL-6B, CH-Sepharose 4B, DEAE-Sepharose fast flow, and
Q-Sepharose fast flow from Pharmacia (Freiburg, Germany). All other
reagents are of the purest grade available.

[35] G. S. Lazarus, R. S. Brown, J. R. Daniels, and H. M. Fullmer, *Science* **159,** 1483 (1968).
[36] G. Murphy, J. J. Reynolds, U. Bretz, and M. Baggiolini, *Biochem. J.* **162,** 195 (1977).
[37] P. Christner, D. Damato, M. Reinhart, and W. Abrams, *Biochemisry* **21,** 6005 (1982).
[38] J. E. Callaway, J. A. Garcia, C. L. Hersch, R. K. Yeh, and M. Gilmore-Hebert, *Biochemistry*
    **25,** 4757 (1986).

*Preparation of Chromatography Resins*

*Preparation of Zinc-Chelate-Sepharose.* A mixture of 250 ml Sepharose CL-6B, 250 ml 1,4-butanediol diglycidyl ether, 250 ml 0.6 $M$ NaOH, and 500 mg NaBH$_4$ is stirrd overnight and washed extensively with water. The epoxy-activated Sepharose is stirred with a solution of 34 g disodium iminodiacetate in 170 ml of 2 $M$ Na$_2$CO$_3$ for 24 hr at 65°. The material is washed again with water extensively and then loaded with Zn$^{2+}$ by washing with an aqueous solution of 1% ZnCl$_2$ (w/v).

*Preparation of Orange-Sepharose CL-6B.* A suspension of 250 ml Sepharose CL-6B in 250 ml water is heated to 60°.[39] After the addition of 10 g Reactive Orange 14, the dark red suspension is stirred at 30° for 30 min. Then 100 g NaCl is added, and the suspension is stirred for 1 hr at 80°. After the addition of 8 g Na$_2$CO$_3$ stirring is continued for 2 hr at 80°. The material is washed with water until no dye is detected in the eluate.

*Preparation of Hydroxamate-Sepharose.* Hydroxamate-Sepharose is prepared with slight modification to the manufacturer's suggestions (Pharmacia). Fifteen grams of activated CH-Sepharose 4B is allowed to swell in 1 m$M$ HCl. To remove additives the swollen gel is washed with 3 liters of 1 m$M$ HCl. After 250 mg Pro-Leu-Gly-NHOH is dissolved in 75 ml coupling buffer, the gel is added to this solution and shaken at room temperature for 2 days. Excess ligand is washed away with coupling buffer. To block remaining reactive groups the gel is shaken with 0.1 $M$ Tris-HCl, pH 8.0, for 1 hr. The material is then washed three times with solutions of alternating pH: with 0.1 $M$ acetate, pH 4.0, 0.5 $M$ NaCl, and with 0.1 $M$ Tris-HCl, pH 8.0, 0.5 $M$ NaCl.

*Assay Procedure*

The collagenase-containing fractions can be identified in every chromatographic step using the synthetic peptide DNP-Pro-Gln-Gly-Ile-Ala-Gly-Gln-D-Arg-OH as the substrate, which is hydrolyzed between Gly and Ile.[40] The reaction is stopped by the addition of 500 $\mu$l of 1 $N$ HCl, which leads to denaturation of the enzyme. The DNP-labeled tripeptide is extracted with 1.6 ml ethyl acetate/$n$-butanol (1:0.15, v/v), and the organic phase can be measured photometrically at 366 nm ($\varepsilon = 1.7$ mol$^{-1}$ cm$^{-1}$). Degradation of 1 $\mu$mol DNP-labeled peptide per minute is defined as 1 unit.

For activation of the proenzyme 100 $\mu$l of the collagenase-containing solution is incubated with 2 $\mu$l of 20 m$M$ HgCl$_2$ at 37° for 1 hr. After the

---

[39] H. J. Böhme, G. Kopperschläger, J. Schulz, and E. Hofmann, *J. Chromatogr.* **69**, 209 (1972).
[40] Y. Masui, T. Takemoto, H. Hori, and Y. Nagai, *Biochem. Med.* **17**, 215 (1977).

addition of 100 $\mu$l solution of the DNP-labeled octapeptide, incubation is continued for 1 hr.

### Purification Procedures

All purification steps should be carried out at 4°.[4] The columns must be equilibrated with starting buffer before use.

*Step 1: Preparation of Granule Pellet.* Buffy coat (2 liters from approximately 20 liters of human blood) is mixed with plasmatonin (Plasmatonin : buffy coat ratio, 1 : 4, v/v).[41] Within 75 min the erythrocytes settle to some extent, which allows raw separation of erythrocytes in the sediment and leukocytes (PMNL) in the supernatant. The supernatant is centrifuged at 150 $g$ for 20 min. To disrupt remaining erythrocytes, the leukocyte pellet is suspended in a hypotonic NaCl solution. After 30 sec two parts of 3.5% NaCl solution is added to make the solution isotonic again. The procedure provides a hypotonic lysis of the erythrocytes, whereas leukocytes remain essentially intact. After centrifugation at 150 $g$ for 10 min the supernatant is discarded. The pellet is washed twice with Ringer's phosphate buffer (buffer : pellet ratio, 5 : 1) (0.5 $M$ potassium phosphate, pH 7.0 : Ringer's solution ratio, 55 : 500, v/v). Each washing is followed by centrifugation at 150 $g$ for 15 min. The pellet is then suspended in buffer 1, and the resulting suspension is subjected to 5 MPa for 15 min in a Parr bomb. After decompression the leukocytes are disrupted. The pellet (approximately 10 g net weight) obtained by centrifugation at 48,000 $g$ for 4 hr contains granules, debris, nuclei, mitochondria, etc. The pellet can be stored at −75°.

*Step 2: Extraction of Granule Pellet.* Five hundred grams of the frozen granule pellet is suspended in 300 ml buffer 2 and 3 ml of 2 $M$ DFP solution. After homogenization with an Ultraturrax homogenizer (Braun, Melsungen, Germany), the suspension is centrifuged at 48,000 $g$ for 30 min. The red-colored supernatant contains PMNL procollagenase which can be purified in latent form in five chromatographic steps.

*Step 3: Affinity Chromatography on Zinc-Chelate-Sepharose CL-6B.* In the first purification step the leukocyte extract is applied to a zinc-chelate-Sepharose CL-6B column (column size 3.5 × 25 cm).[4] The column is washed with buffer 3 until no further protein can be detected in the eluate. PMNL procollagenase passes unretarded.

*Step 4: Anion-Exchange Chromatography on DEAE-Sepharose Fast Flow.* The yellow eluate is directly applied to a DEAE-Sepharose fast flow column (column size 3.5 × 20 cm) and washed with buffer 3 until no further

[41] S. Engelbrecht, E. Pieper, H. W. Macartney, W. Rautenberg, H. R. Wenzel, and H. Tschesche, *Hoppe-Seyler's Z. Phys. Chem.* **363,** 305 (1982).

protein can be detected in the eluate.[4] The unretarded fractions containing PMNL procollagenase are dialyzed against buffer 3 overnight and then centrifuged at 48,000 $g$ for 30 min to eliminate precipitated protein.

*Step 5: Anion-Exchange Chromatography on Q-Sepharose Fast Flow.* The supernatant is then applied to a Q-Sepharose fast flow column (3.5 × 25 cm).[4] The column is washed with buffer 3 until all unbound protein is washed out. Then the bound proteins, including PMNL procollagenase, are eluted with a linear NaCl gradient (0–0.15 $M$ NaCl, 2 × 600 ml). Fractions with peptidolytic activity after activation are pooled to a conductivity of 4.5 mS. The elution profile is shown in Fig. 1. Fractions with higher conductivities contain contaminating proteins which cannot be removed in the following purification steps. They are pooled separately and used for the next purification or are purified in activated form on hydroxamate-Sepharose (see below, step 7a) after separate purification on Orange-Sepharose CL-6B.

*Step 6: Affinity Chromatography on Orange-Sepharose CL-6B.* The low conductivity pool from the Q-Sepharose step is dialyzed overnight against buffer 4 and then applied to Orange-Sepharose CL-6B (column size 4 × 25 cm).[4] The column is washed with buffer 4 until all unbound proteins are removed. Then the bound proteins are eluted with a nonlinear NaCl gradient (0.1–2 $M$ NaCl; 600 ml, 250 ml). The elution profile is shown in

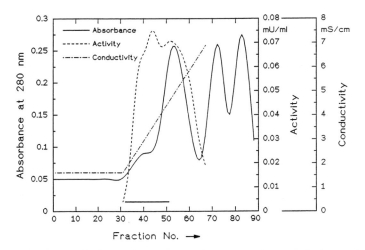

FIG. 1. Elution profile from anion-exchange chromatography on Q-Sepharose fast flow (elution of unbound protein not shown). Protein was eluted from the column (column size 3.5 × 25 cm, fraction size 10 ml) with a linear NaCl gradient (0–0.15 $M$ NaCl in buffer 3, 2 × 600 ml). The pooled fractions are marked with a bar.

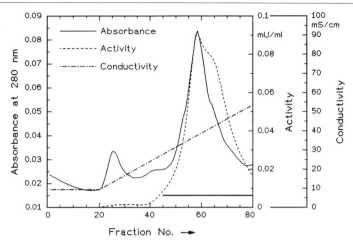

FIG. 2. Elution profile from affinity chromatography on Orange-Sepharose CL-6B (elution of unbound protein not shown). Protein was eluted from the column (column size 4 × 25 cm, fraction size 10 ml) with a nonlinear NaCl gradient (0.1–2 *M* NaCl in buffer 4; 600 ml, 250 ml). The pooled fractions are marked with a bar.

Fig. 2. Fractions with peptidolytic activity after activation (35–50 mS) are pooled. The pool is concentrated to 5 ml by ultrafiltration using an Amicon (Danvers, MA) XM50 membrane.

*Step 7: Gel Filtration on Sephacryl S-300.* Purification to homogeneity is finally achieved by gel filtration on a Sephacryl S-300 column (column size 120 cm, flow rate 20 ml/hr) with buffer 8. The elution profile is shown in Fig. 3.

*Step 7a: Affinity Chromatography on Hydroxamate-Sepharose.* The high conductivity pool of the Q-Sepharose fractions (step 5) is applied to Orange-Sepharose CL-6B in an analogous way as described for the low conductivity pool in step 6.[42] The collagenase-containing fractions are pooled, concentrated to 100 ml by ultrafiltration using an Amicon XM50 membrane, and activated by incubation with 1 m*M* HgCl₂ at 37°. The activated collagenase is dialyzed overnight against buffer 5, applied to a hydroxamate-Sepharose column (1.2 × 15 cm), and washed with buffer 5 to remove all unbound protein. Under these conditions active collagenase binds selectively to hydroxamate-Sepharose. The enzyme is eluted with buffer 6. To achieve an appropriate pH, 6-ml fractions are collected and 2 ml buffer 7 is added to each fraction. The elution profile is shown in Fig. 4. The collagenase-containing fractions are pooled and dialyzed overnight against buffer 8.

[42] W. M. Moore and C. A. Spilburg, *Biochemistry* **25,** 5189 (1986).

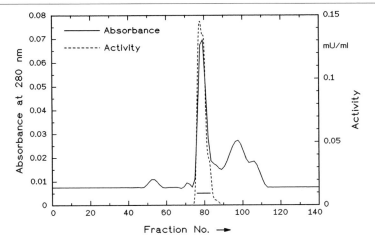

FIG. 3. Elution profile from gel filtration on Sephacryl S-300 (elution of unbound protein not shown; column size 120 cm, flow rate 20 ml/hr, fraction size 5 ml). The pooled fractions are marked with a bar.

The purity of PMN collagenase can be proved by sodium dodecyl sulfate–polyacrylamide gel electrophoresis (SDS–PAGE). The latent form has an apparent $M_r$ of 85,000, the $M_r$ of the active form is 64,000. The extent of purification by the individual steps is given in Table I.

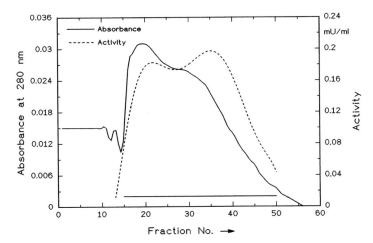

FIG. 4. Elution profile from affinity chromatography on hydroxamate-Sepharose (elution of unbound protein not shown). Protein was eluted from the column (column size 1.2 × 15 cm, fraction size 6 ml) with buffer 6. The pooled fractions are marked with a bar.

TABLE I
PURIFICATION OF POLYMORPHONUCLEAR LEUKOCYTE COLLAGENASE

| Step | Total units (mU) | Specific activity (mU/mg) | Purification (-fold) | Recovery (%) |
|---|---|---|---|---|
| Extract | 5500 | 3.9 | — | — |
| Q-Sepharose FF | 2400 | 6.6 | 1 | 100 |
| Zinc Chelate Sepharose | 1509 | 7.3 | 1.2 | 63 |
| Orange Sepharose | 1604 | 86.7 | 13.1 | 66 |
| Sephacryl S-300 | 332 | 102.6 | 15.5 | 14 |

*Note.* Taken from ref. 49, where performed in a variant order of steps.

## Preparation of Recombinant Enzyme

The full-length PMNL procollagenase is cloned and expressed in *E. coli* and found stored intracellularly in inclusion bodies. On refolding, however, it becomes autocatalytically degraded, which leads to extremely low yields on purification. The C-terminally truncated procollagenase, lacking the hemopexin-like domain, is also mainly degraded during the refolding process involving extensive dialysis. However, alternative renaturation on Q-Sepharose fast flow prevents excessive degradation. Only the catalytic domain of PMNL collagenase is stable enough to be effectively refolded by dialysis against a zinc-containing buffer.

*Procedure for Cloning cDNAs Coding for Polymorphonuclear Leukocyte Collagenase Variants*

Total RNA isolated from bone marrow suspensions of leukemia patients[42] is reverse transcribed with reverse transcriptase (BRL, Gaithersburg, MD).[43] After reverse transcription, polymerase chain reaction (PCR) amplification is performed, and the cDNA coding for full-length PMNL collagenase is selectively amplified using two oligonucleotide primers: "colstart," 5'-TTCCCATATGTTTCCTGTATCTTCTAAAGA-3', and "colend," 5'-TGATGGATCCTCAGCCATATCTACAGTTAGG-3. The colstart primer incorporates sequences for a unique*Nde*I site and an initiating methionine. Colend introduces sequences for a stop codon and a unique *Bam*HI site. For amplification of the full-length PMNL collagenase cDNA, 30 cycles of PCR (94°, 75 sec; 53°, 75 sec; 72°, 110 sec) are carried out using *Taq* DNA polymerase. The amplified fragment is treated with Klenow

[43] S. Schnierer, T. Kleine, T. Gote, A. Hillemann, V. Knäuper, and H. Tschesche, *Biochem. Biophys. Res. Commun.* **191,** 319 (1993).

DNA polymerase, digested with *Nde*I and *Bam*HI, and ligated into the *Nde*I and *Bam*HI site of the T7 expression vector pET11a.[44]

The truncated form of the PMNL collagenase cDNA, lacking the coding region for the hemopexin-like domain, is amplified by PCR using the above program and the primers colstart and "colcatend" (5'-GTTGGATCCT-CATCCATAGATGGCCTGAAT-3'). Like colend, colcatend incorporates sequences for a stop codon and a unique *Bam*HI site. The resulting PCR fragment is also cloned into the *Nde*I/*Bam*HI site of the pET11a vector. The strategy used for amplification and cloning of the truncated PMNL procollagenase cDNA is also employed for the coding region of the catalytic domain. To amplify the cDNA corresponding to the catalytic domain the primers "colcatstart" (5'-TGGTCATATGTTAACCCCAG-GAAACCCC-3') and colcatend are used. Colcatstart, as all the other forward primers, carries a *Nde*I recognition site.

*Methods for Expression, Extraction, and Purification*

*Reagents*

Lysis buffer: 100 m$M$ Tris-HCl, pH 8.5, 5 m$M$ benzamidine, 5 m$M$ 2-mercaptoethanol

Buffer 1: 20 m$M$ Tris-HCl, pH 8.5, 100 m$M$ 2-mercaptoethanol, 8 $M$ urea

Buffer R1: 10 m$M$ Tris-HCl, pH 8.5, 5 m$M$ CaCl$_2$, 100 m$M$ 2-mercapto-ethanol

Buffer R2: 20 m$M$ Tris-HCl, pH 7.5, 5 m$M$ CaCl$_2$, 0.5 m$M$ ZnCl$_2$

Buffer R3: 20 m$M$ Tris-HCl, pH 7.5, 5 m$M$ CaCl$_2$, 100 m$M$ NaCl, 0.5 m$M$ ZnCl$_2$

Buffer H1: 20 m$M$ Tris-HCl, pH 7.5, 10 m$M$ CaCl$_2$, 0.5 $M$ NaCl

Buffer H2: 100 m$M$ Tris-HCl, pH 9.0, 0.5 m$M$ CaCl$_2$, 0.5 $M$ NaCl

*Expression of Polymorphonuclear Leukocyte Collagenase Variants.* For expression of the full-length PMNL procollagenase, the C-terminally truncated enzyme, or the catalytic domain, *Escherichia coli* strain Bl21 (DE3) is transformed with the corresponding plasmids to allow T7 RNA polymerase-mediated transcription of the cloned cDNAs. Cultures of the transformed *E. coli* strains are grown overnight at 37° in medium containing 10 g Bacto-tryptone, 5 g yeast extract, and 10 g NaCl per liter and 200 $\mu$g/ml ampicillin. Two liters of the same medium is inoculated with 20 ml of the overnight culture at 37°. When a cell density corresponding to an OD$_{578}$ of 0.6 is reached, isopropylthiogalactoside (IPTG) is added to a final concentration

[44] F. W. Studier, A. H. Rosenberg, J. J. Dunn, and J. W. Dubendorff, this series, Vol. 185, p. 60.

of 0.4 m$M$, and the incubation is continued for another 4 hr at 37° to maximize the production of the different PMNL collagenase variants.

*Extraction from Inclusion Bodies.* The cells are pelleted at 5000 $g$ for 20 min at 4° and resuspended in 70 ml lysis buffer per gram cell pellet. After the addition of 0.3 mg lysozyme per milliliter lysis buffer, the mixture is incubated for 30 min at 37° under shaking and afterward stored for 15 min on ice. The lysate is centrifuged for 10 min at 10,000 $g$, and the resulting pellet is suspended in lysis buffer containing 1 $M$ urea for washing. After centrifugation of the suspension at 10,000 $g$ for 10 min, solubilization of the resulting pellet is achieved in buffer 1. The mixture is incubated for 1 hr at 50° under shaking, and insoluble material is subsequently removed by centrifugation at 12,000 $g$ for 25 min. The supernatant (crude extract) containing the solubilized inclusion bodies is immediately frozen and stored at $-20°$ until further purification. The buffers used for isolation of the inclusion bodies containing the catalytic domain only do not require 2-mercaptoethanol as there is no cysteine in that portion of the enzyme.

*Refolding and Purification of C-Terminally Truncated Polymorphonuclear Leukocyte Collagenase.* The crude extract containing the C-terminally truncated PMNL procollagenase is mixed with 100 g Q-Sepharose and incubated for 8 hr at 4° under shaking. A column is loaded with the suspension and equilibrated with buffer R1. Refolding of the truncated proenzyme is performed by a further equilibration step in buffer R2, and afterward partial purification is achieved by applying a 1000-ml linear gradient of 0–2 $M$ NaCl. Fractions are analyzed by SDS–PAGE and immunoblotting. Those containing the truncated PMNL-procollagenase are pooled and dialyzed against buffer R2. Final purification is performed by FPLC (fast protein liquid chromatography) using Mono Q-Sepharose equilibrated with buffer R1 and loaded with the dialyzed sample. Elution of the truncated enzyme is carried out by applying a linear gradient of 0–200 m$M$ NaCl in buffer R2 for 60 min at a flow rate of 1 ml/min. The C-terminally truncated PMNL procollagenase is eluted at a salt concentration of 50 m$M$ NaCl and is thus purified to apparent homogeneity. The procedure allows the isolation of 1 mg purified protein from 1 liter of *E. coli* suspension.

*Refolding and Purification of Polymorphonuclear Leukocyte Collagenase Catalytic Domain.* Refolding of the catalytic domain from the crude extract is performed by dialysis at 4° against renaturation buffer R3. Proteins that precipitate during the procedure are removed by centrifugation at 10,000 $g$ for 10 min. Before final purification the zinc ions have to be removed. The supernatant is therefore dialyzed against buffer H1. Purification to homogeneity is then achieved by affinity chromatography on hydroxamate-Sepharose.[42] The protein mixture is loaded onto the hydroxamate column equilibrated in buffer H1, the column is subsequently washed

M_r x 10^-3   1   2   3   4

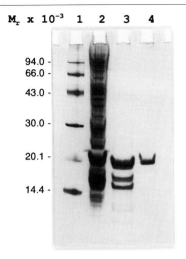

FIG. 5. Silver-stained SDS–polyacrylamide gel demonstrating the purification of the PMNL collagenase catalytic domain: lane 1, molecular weight markers; lane 2, *E. coli* crude extract 4 hr after induction of catalytic domain expression with IPTG; lane 3, refolded catalytic domain after dialysis against renaturation buffer; lane 4, catalytic domain purified to homogeneity by hydroxamate affinity chromatography.

with 3 volumes of buffer H1, and the PMNL collagenase catalytic domain is eluted with buffer H2. The catalytic domain is fractionated into a volume of 5 ml, 2 ml of which is 600 m$M$ Tris-HCl, pH 6.5, 5 m$M$ CaCl$_2$, and 0.5 $M$ NaCl and 3 ml of which is eluate. By this method, 25 mg highly purified enzyme can be isolated from 1 liter of *E. coli* suspension (Fig. 5).

Proenzyme Activation

The proenzyme becomes activated by proteolytic release of the propeptide domain, which involves cleavages in the region of residues 71 to 81 either by various proteolytic enzymes or by autocatalysis (see Introduction). Activation of the virgin proenzyme requires a conformational change with disruption of the putative Cys-Zn$^{2+}$ bond, which ensures latency of the enzyme by keeping the active-site Zn$^{2+}$ covered by the propeptide domain. Displacing the cryptic Cys-71 of the conserved PRCVGPD sequence motif by a water molecule, which is necessary for catalysis,[5,45] exposes the catalytic

[45] B. L. Vallee and S. S. Auld, *in* "Matrix Metalloproteinases and Inhibitors" (H. Birkedal-Hansen, Z. Werb, H. G. Welgus, and H. E. Van Wart, eds.), *Matrix* Suppl. No. 1, p. 5. Gustav Fischer, Stuttgart, 1992.

site. This process, referred to as the "cysteine switch" mechanism,[46,47] can be induced by various proteinases which cleave the propeptide via one or more successive clips, by reagents which directly interact with the propeptide Cys[71]-residue such as organomercurials, thiol reagents, or oxidants, or by chaotropic agents and detergents which shift the equilibrium of the conformations to the open, autocatalytically activated form.

*Procedures*

*Activation by Mercurials.* Procollagenase (0.23 $\mu M$) dissolved in 20 m$M$ Tris-HCl, pH 7.5, 5 m$M$ CaCl$_2$, and 200 m$M$ NaCl is treated with HgCl$_2$ or 4-aminophenylmercuric acetate (APMA) in a concentration up to 1 m$M$ for 1 hr at 37°.[4,5,20]

*Truncation by Trypsin to Latent Enzyme.* Procollagenase (0.23 $\mu M$) dissolved in 20 m$M$ Tris-HCl, pH 7.5, 5 m$M$ CaCl$_2$, 200 m$M$ NaCl is incubated with 4.6 n$M$ porcine trypsin (N-tosyl-L-phenylalanine chloromethane treated) for 5 min at 37°.[48] The reaction is terminated by the addition of a 10-fold molar excess of bovine pancreatic trypsin inhibitor (BPTI) or diisopropyl fluorophosphate (DFP). The procedure leads to latent enzyme with Phe-49 at the N terminus (see Fig. 6).

*Activation by Trypsin.* For full activation the above procedure is extended to 15–120 min at 37° before termination.[4,5] With trypsin the active collagenase with Cys-71 at the N terminus is obtained; chymotrypsin or cathepsin G produces the form with Met-80.[3–5,7,30]

*Activation by Stromelysin.* For activation by stromelysin,[48] procollagenase (2 $\mu M$) is incubated in 20 m$M$ Tris-HCl, pH 7.5, 5 m$M$ CaCl$_2$, and 200 m$M$ NaCl with 0.2 $\mu M$ active stromelysin I or II (or activated recombinant stromelysin) for 6 to 8 hr at 37°. The "superactivated form" with Phe-79 as the N terminus is obtained (Figs. 6 and 7). This Phe form is about two to three times as active as the trypsin- or HgCl$_2$-activated forms against peptides or triple-helical type I collagen[48] (see below). Specific activities of 11,000 units/mg are obtained for stromelysin-activated collagenase compared to 3000 and 3300 units/mg for trypsin- and HgCl$_2$-activated enzymes against soluble type I collagen as the substrate.[48] In addition, the Phe form is more stable against autoproteolytic fragmentation than the trypsin- or HgCl$_2$-activated forms.

[46] H. E. Van Wart and H. Birkedal-Hansen, *Proc. Natl. Acad. Sci. U.S.A.* **87,** 5578 (1990).
[47] E. B. Springman, E. L. Angleton, H. Birkedal-Hansen, and H. E. Van Wart, *Proc. Natl. Acad. Sci. U.S.A.* **87,** 364 (1990).
[48] V. Knäuper, S. M. Wilhelm, P. K. Seperack, Y. A. DeClerck, K. E. Langley, A. Osthues, and H. Tschesche, *Biochem. J.* **295,** 581 (1993).

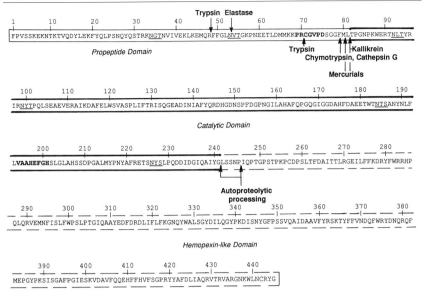

Fig. 6. Amino acid sequence of human PMNL collagenase [3]. Indicated are the propeptide, catalytic, and hemopexin-like domains and the processing site for truncation [4, 30] and activation [20, 21] by serine proteinases and mercurials. The autoproteolytic processing sites for cleavage of the hemopexin-like domain are also indicated. Potential glycosylation sites are underlined.

## Properties and Structure

### Catalytic Activity

The procollagenase consists of three domains: the propeptide, catalytic, and hemopexin-like domains.[17] Proteolytic or autolytic cleavage of the N-terminal propeptide domain generates the active enzyme. The residue at the newly generated N terminus depends on the individual substrate specificity of the activating proteinase, and several variants have been reported[4,5,18,20,28,29,30,49] (see Fig. 6). The variant with N-terminal Phe-79 seems to be the most stable one and represents the "superactivated" form with the highest specific activity.[48] This property is obviously explained by the compact structure in which the N-terminal segment from Pro-86 is fixed against the globular core (see below).[50]

[49] V. Knäuper, S. Krämer, H. Reinke, and H. Tschesche, Biol. Chem. Hoppe-Seyler 371(Suppl.), 295 (1990).
[50] P. Reinemer, F. Grams, R. Huber, T. Kleine, S. Schnierer, M. Pieper, H. Tschesche, and W. Bode, FEBS Lett. 338, 227 (1994).

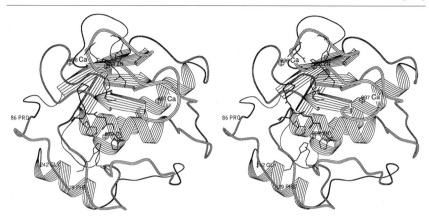

FIG. 7. Structure of the catalytic domain of human neutrophil collagenase shown as a ribbon model.[50,51] An upper (N-terminal) and a lower (C-terminal) part are separated by a relatively flat active-site cleft with the "catalytic" zinc ion near a very deep S1' pocket. The upper part harbors a second "structural" zinc ion and two calcium ions. The Pro-Leu-Gly-NHOH inhibitor lies antiparallel to the edge strand on the unprimed S pocket with the Pro residing in a hydrophobic groove formed by the side chains of His-162, Phe-164, and Ser-151, whereas the Leu forms two inter-main-chain hydrogen bonds to Ala-163 with its side chain situated in a small opening lined by His-201, Ala-206, and His-207. The hydroxamate group is oriented toward Glu-198 with the carbonyl oxygen and the hydroxyl oxygen complexing the "catalytic" zinc.[50]

The catalytic domain without the hemopexin-like domain is generated as an autolytic fragment from the active collagenase[30] or can be obtained from the recombinant C-terminally truncated proenzyme after activation by trypsin or, preferably, by stromelysin.[17] It is also conveniently prepared as a recombinant inactive protein in *E. coli* inclusion bodies and then activated by dialysis as described above.

The catalytic domain is a fully active enzyme with, however, diminished turnover rates against various peptides, for example, the assay DNP-peptide, substance P, or angiotensin I.[50a] However, it exhibits no activity against triple-helical type I collagen, so this activity presumably depends on the presence of the hemopexin-like domain.[43] As a hypothesis, we assume some type of clamping mechanism in which the hemopexin-like domain arrests the triple-helical collagen strands at the active site. By rotation around the strand axis all three individual peptide chains could be cleaved at the same locus to yield strand cleavage into two triple-helical fragments.

*Overall Structure of Catalytic Domain*

The recombinant catalytic domains, residues $Met^{80}$–$Gly^{242}$ (proenzyme numbering, Fig. 6) and residues $Phe^{79}$–$Gly^{242}$, cloned and expressed in *E.*

[50a] O. Diekmann and H. Tschesche, *Braz. J. Med. Biol. Res.* **27,** 1877 (1994).

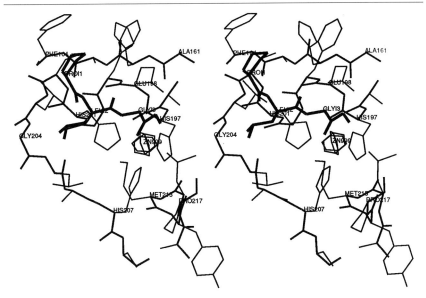

FIG. 8. Detailed model of the environment of the catalytic zinc ion (Zn[999]) coordinated by the three His residues of the His[197]-Glu[198]-X-X-His[201]-X-X-Gly[204]-X-X-His[207] zinc-binding consensus sequence and by the hydroxamic acid moiety of the inhibitor.[50,51]

*coli*,[43] have been purified to homogeneity and cocrystallized with the substrate analog inhibitor Pro-Leu-Gly-NHOH.[51] The X-ray crystal structures of the noncovalent complexes have been solved at 2.0 and 2.5 Å resolution, respectively[51,52] (Fig. 7). The structure shows a spherical molecule with a relatively flat active-site cleft, which separates a small "lower" part from the "upper" main body. The lower part comprising the last 38 residues is formed by multiple turns and ends in a long C-terminal α helix at the surface of the molecule. The "upper" part consists of a central, highly twisted five-stranded β-pleated sheet, which is flanked by an S-shaped double loop and two other bridging loops on its convex side and two long α helices on its concave side. The sheet covers the large hydrophobic core.

The "catalytic" zinc ion (Zn[999]) is pentacoordinated by three imidazole $N_{\varepsilon 2}$ atoms of His-197 (2.0 Å), His-201 (2.2 Å), and His-207 (1.9 Å) of the HEXXHXXGXXH conserved sequence motif at the bottom of the active-site cleft (Fig. 8). The hydroxamic acid moiety of the inhibitor coordinates with its carbonyl (2.0 Å) and its hydroxyl oxygen (2.2 Å) to the zinc ion.

[51] W. Bode, P. Reinemer, R. Huber, T. Kleine, S. Schnierer, and H. Tschesche, *EMBO J.* **13**, 1263 (1994).
[52] W. Bode, F.-X. Gomis-Rüth, R. Huber, R. Zwilling, and W. Stöcker, *Nature (London)* **358**, 164 (1992).

In addition to those two oxygen–zinc contacts, the hydroxamate group with its oxygen and nitrogen is close to the O of Glu-198 and forms a hydrogen bond via its nitrogen to the carbonyl group of Ala-161 within the extended edge strand Ala[161]–Phe[164] of the $\beta$ sheet.

The structure bears some similarity to crayfish astacin[52,53] and invertebrate venom adamalysin II and to the *Pseudomonas aeruginosa* alkaline proteinase,[54,55] all of which also contain the "Met turn"[56] which forms the base of the active-site residues. In those enzymes the C-terminal chain exhibits a largely irregular folding with a wide right-hand loop followed by the $\alpha$ helix. The loop is stabilized by a tight 1.4 turn Ala[213]-Leu-Met-Tyr[216], the Met turn, which provides a hydrophobic base for the three liganding His residues of the catalytic zinc ion. This conserved topological element is characteristic of the "metzincins".[57]

The structure of the PMNL collagenase catalytic domain harbors a second structural zinc ion (Zn[998]) and two calcium ions (Ca[996] and Ca[997]). The structural zinc ion and one calcium are sandwiched between the surface S-shaped loop and the surface of the $\beta$ sheet. This second zinc is tetrahedrally coordinated by His-147, Asp-149, His-162, and His-175, whereas the calcium ion (Ca[997]) is nearly octahedrally coordinated by Asp-154, Gly-155, Asn-157, Ile-159, Asp-177, and Glu-180. The second calcium ion is located on the convex side of the $\beta$ sheet at the end of the glycine-rich loop from Phe-164 to Gly-172. It is octahedrally coordinated with the carbonyl groups of Asp-137, Gly-169, Gly-171, and Asp-173 and two solvent molecules.

The strictly conserved residues Asp-232 and Asp-233 are contained within the C-terminal helix. The side chain of Asp-233 is buried and forms a hydrogen bond to the Met turn residues Leu-214 and Met-215, thus stabilizing the active-site base and proving essential for catalytic activity.[57] The Asp-232 residue is positioned at the side of a hydrophobic groove formed by the C-terminal helix residues Gly-236, Ala-238, and Ile-240 closed by Leu-205 and Met-215. This residue serves as a contact site for the side chain of residue Phe-79 in the superactivated enzyme form,[48] whereas in the structural model of the Met[80]–Gly[242] form the N terminus from Pro-86 is disordered[51,52] (Fig. 7) and could not be traced in the X-ray crystallographic structure.

[53] F.-X. Gomis-Rüth, W. Stöcker, R. Huber, R. Zwilling, and W. Bode, *J. Mol. Biol.* **229**, 945 (1993).
[54] F.-X. Gomis-Rüth, L. F. Kress, and W. Bode, *EMBO J.* **12**, 4151 (1993).
[55] U. Baumann, S. Wu, K. M. Flaherty, and D. B. MacKay, *EMBO J.* **12**, 3357 (1993).
[56] W. Bode, F.-X. Gomis-Rüth, and W. Stöcker, *FEBS Lett.* **331**, 134 (1993).
[57] T. Hirose, C. Patterson, T. Pourmotabbed, C. L. Mainardi, and K. A. Hasty, *Proc. Natl. Acad. Sci. U.S.A.* **90**, 2569 (1993).

The structure of the superactive Phe[79]–Gly[242] catalytic domain reveals that the N-terminal heptapeptide segment Phe[79]-Met-Leu-Thr-Pro-Gly-Asn[85] binds to the concave surface at the bottom of the molecule located between Pro-86 and Ser-209 and formed by the C-terminal $\alpha$ helix and the extended chain element Gly[204]–Ser[209] following the active-site helix. N-Terminal to Thr-82 the chain turns downward to the C-terminal helix crossing over with the strand following the active-site helix and residue Leu-205 to form the only regular inter-main-chain hydrogen bond. The N-terminal ammonium group of Phe-79 forms a salt bridge with the carboxylate moiety of the strictly conserved Asp-232. Another strictly conserved residue is Gly-236, one turn away from Asp-232 in the C-terminal helix, which serves as a contact site for the side chain of residue 79. A more bulky amino acid replacing it would cause stereochemical strain for correct orientation of the N terminus.

The structural models of the Phe[79]–Gly[242] form and the Met[80]–Gly[242] form of human PMNL collagenase superimpose well. In particular, the geometry of the active-site residue remains unchanged. The activity enhancement of superactivation seems to be linked to the disorder–order transition of the N-terminal segment and the formation of the salt bridge between the Phe-79 N-terminal ammonium group and the side chain of Asp-232. This seems to lead to stabilization of the active site and, hence, to the transitions states in the superactive form. In contrast, the mobile N-terminal segment of the truncated forms may also interfere with binding of a peptide substrate, as the shortened N-terminal segment cannot form the salt bridge and interlock with the main body of the enzyme.[50]

Acknowledgments

This work was supported by the Deutsche Forschungsgemeinschaft, SFB 223 of the Universität Bielefeld and SFB 207 of the Universität München, as well as by the Fonds der Chemischen Industrie.

# [27] Human Stromelysins 1 and 2

By HIDEAKI NAGASE

## Introduction

Stromelysin 1 (EC 3.4.24.17) and stromelysin 2 (EC 3.4.24.22) are zinc metalloendopeptidases belonging to the matrix metalloproteinase (MMP) family. Stromelysin 1 is referred to as matrix metalloproteinase 3 (MMP-

3) according to the numerical designation of MMPs.[1] A metalloproteinase called "proteoglycanase" from rabbit bone culture medium,[2] "procollagenase activator" from rabbit synovial fibroblasts,[3] "collagen-activating protein" from bovine cartilage,[4] and "transin" from transformed rat cells[5] is stromelysin 1 in the respective species. An acid metalloproteinase isolated from human cartilage[6] which digests aggrecan core protein optimally at pH 5.3 has also proved to be stromelysin 1.[7] Stromelysin 1 is not readily detected in normal cells in culture or tissue homogenates, but its synthesis is greatly enhanced in mesenchymal cells when they are treated with cytokines such as interleukin 1 (IL-1) and tumor necrosis factor $\alpha$ (TNF$\alpha$), growth factors, phorbol myristate acetate (PMA), and several other stimuli.[8,9] Stromelysin 1 is secreted from the cells as inactive zymogen (proMMP-3).[8,9] On activation the enzyme is capable of degrading a number of extracellular matrix components,[8,9] and it participates in activation of procollagenases (proMMP-1 and proMMP-8)[10,11] and progelatinase B (proMMP-9).[12]

Stromelysin 2 (MMP-10) was first identified by molecular cloning.[13,14] It is also secreted from the cell as an inactive zymogen. Human prostromelysins 1 and 2 are 78% identical in amino acid sequence. Stromelysin 2 is expressed in human keratinocytes when they are treated with PMA, transforming growth factor $\alpha$, epidermal growth factor, or TNF$\alpha$ but not

[1] H. Nagase, A. J. Barrett, and J. F. Woessner, Jr., *Matrix* Suppl. 1, 421 (1992).

[2] W. A. Galloway, G. Murphy, J. D. Sandy, J. Gavrilovic, T. E. Cawston, and J. J. Reynolds, *Biochem. J.* **209,** 741 (1983).

[3] C. A. Vater, H. Nagase, and E. D. Harris, Jr., *J. Biol. Chem.* **258,** 9374 (1983).

[4] B. V. Treadwell, J. Neidel, M. Pavia, C. A. Towle, M. E. Trice, and H. J. Mankin, *Arch. Biochem. Biophys.* **251,** 715 (1986).

[5] L. M. Matrisian, G. T. Bowden, P. Kreig, G. Fürstenberger, J.-P. Briand, P. Leroy, and R. Breathnach, *Proc. Natl. Acad. Sci. U.S.A.* **83,** 9413 (1986).

[6] W. Azzo and J. F. Woessner, Jr., *J. Biol. Chem.* **261,** 5434 (1986).

[7] S. M. Wilhelm, Z. H. Shao, T. J. Housley, P. K. Seperack, A. P. Baumann, Z. Gunja-Smith, and J. F. Woessner, Jr., *J. Biol. Chem.* **268,** 21906 (1993).

[8] J. F. Woessner, Jr., *FASEB J.* **5,** 2145 (1991).

[9] H. Birkedal-Hansen, W. G. I. Moore, M. K. Bodden, L. J. Windsor, B. Birkedal-Hansen, A. DeCarlo, and J. A. Engler, *Crit. Rev. Oral Biol. Med.* **4,** 197 (1993).

[10] H. Nagase, Y. Ogata, K. Suzuki, J. J. Enghild, and G. Salvesen, *Biochem. Soc. Trans.* **19,** 715 (1991).

[11] V. Knäuper, S. M. Wilhelm, P. K. Seperack, Y. A. DeClerck, K. E. Langley, A. Osthues, and H. Tschesche, *Biochem. J.* **295,** 581 (1993).

[12] Y. Ogata, J. J. Enghild, and H. Nagase, *J. Biol. Chem.* **267,** 3581 (1992).

[13] D. Muller, B. Quantin, M. C. Gesnel, R. Millon-Collard, J. Abecassis, and R. Breathnach, *Biochem. J.* **253,** 187 (1988).

[14] R. Breathnach, L. M. Matrisian, M. C. Gesnel, A. Staub, and P. Leroy, *Nucleic Acids Res.* **15,** 1139 (1987).

in gingival fibroblasts under similar conditions.[15] Some enzymatic properties of stromelysin 2 have been characterized using the recombinant enzyme expressed in mammalian cells.[16–18]

## Assay Methods for Human Stromelysin 1

The enzymatic activity of stromelysin 1 can be measured using aggrecan entrapped in polyacrylamide beads,[19] Azocoll,[20] and radiolabeled proteins such as [14]C-acetylated casein[21] and reduced, [3]H-carboxymethylated transferrin ([[3]H]Cm-Tf)[22] as protein substrates. A number of fluorogenic synthetic substrates are also described.[23] In this section the methods using [[3]H]Cm-Tf and Mca-Arg-Pro-Lys-Pro-Val-Glu-Nva-Trp-Arg-Lys(Dnp)-NH$_2$[24,25] are described as they provide simple, sensitive assays. The latter synthetic substrate is readily hydrolyzed by stromelysin 1 (MMP-3), but not by interstitial collagenase (MMP-1) or gelatinase A (MMP-2).[25]

### Transferrin Radioassay

*Principle.* Transferrin is reduced and SH groups are carboxymethylated with [[3]H]iodoacetic acid.[22] This exposes hydrophobic residues which renders transferrin readily susceptible to stromelysin 1 as the enzyme prefers hydrophobic side chains at the P1' site. Digestion of [[3]H]Cm-Tf yields peptides soluble in a dilute trichloroacetic acid (TCA) solution, and the products are measured by released radioactivity. Transferrin is chosen as it contains a large number of cysteine residues (37 Cys in 678 residues).

*Preparation of Substrate.* Human transferrin (Sigma, St. Louis, MO) (200 mg) is dissolved in 20 ml of 0.2 $M$ Tris-HCl (pH 8.6)–8 $M$ urea containing 20 m$M$ dithiothreitol (DTT) and incubated at 37° for 4 hr. For effective [3]H-carboxymethylation of the reduced protein, a rapid and

[15] L. J. Windsor, H. Grenett, B. Birkedal-Hansen, M. K. Bodden, J. A. Engler, and H. Birkedal-Hansen, *J. Biol. Chem.* **268,** 17341 (1993).

[16] R. Nicholson, G. Murphy, and R. Breathnach, *Biochemistry* **28,** 5195 (1989).

[17] G. Murphy, M. I. Cockett, R. V. Ward, and A. J. P. Docherty, *Biochem. J.* **277,** 277 (1991).

[18] Q. Nguyen, G. Murphy, C. E. Hughes, J. S. Mort, and P. J. Roughley, *Biochem. J.* **295,** 595 (1993).

[19] H. Nagase and J. F. Woessner, Jr., *Anal. Biochem.* **107,** 385 (1980).

[20] Z. Gunja-Smith, H. Nagase, and J. F. Woessner, Jr., *Biochem. J.* **258,** 115 (1989).

[21] T. E. Cawston, W. A. Galloway, E. Mercer, G. Murphy, and J. J. Reynolds, *Biochem. J.* **195,** 159 (1981).

[22] T. Yee and H. Nagase, unpublished work (1983).

[23] C. G. Knight, this volume [2].

[24] Mca, (7-Methylcoumarin-4-yl)acetyl; Nva, norvaline; Dnp, 2,4-dinitrophenyl.

[25] H. Nagase, C. G. Fields, and G. B. Fields, *J. Biol. Chem.* **269,** 20952 (1994).

complete removal of DTT is essential. This is accomplished by applying the sample to a column of Sephadex G-50 (bed volume 200 ml) equilibrated with 0.2 $M$ Tris-HCl (pH 8.6)–8 $M$ urea at room temperature. The separation of reduced transferrin and DTT is monitored by the absorbance at 280 nm. The reduced transferrin fractions recovered in the void volume (about 48 ml) are pooled and immediately reacted with 1 mCi of [$^3$H]iodoacetic acid (Amersham, Arlington Heights, IL; 90 mCi/mmol) for 30 min at 23°, then chased with 30 m$M$ iodoacetic acid for 30 min at 23° to complete carboxymethylation. Incorporation of [$^3$H]iodoacetic acid is greater than 90%. The reduced $^3$H-carboxymethylated transferrin ([$^3$H]Cm-Tf) is then mixed with an equal volume of cold 10% (w/v) TCA, and the precipitate formed is collected by brief centrifugation using a benchtop clinical centrifuge. The [$^3$H]Cm-Tf pellet is washed with cold 10% (w/v) TCA 4 times to remove the free label and TCA-soluble peptides and dissolved into 50 m$M$ Tris-HCl (pH 7.5)–0.15 $M$ NaCl or an appropriate buffer at a concentration of 3 mg/ml. The [$^3$H]Cm-Tf precipitates below pH 5.5. The $A_{280\,nm,cm}^{1\%}$ value of 8.13 is used to measure the concentration of [$^3$H]Cm-Tf. The substrate is stored at $-20°$ in 1-ml portions.

*Reagents*

TNC buffer: 50 m$M$ Tris-HCl (pH 7.5)–0.15 $M$ NaCl–10 m$M$ CaCl$_2$–0.02% (w/v) NaN$_3$–0.05% (w/v) Brij 35
[$^3$H]Cm-Tf solution: [$^3$H]Cm-Tf dissolved in TNC buffer at 3 mg/ml
Stopping reagent: 3.3% (w/v) TCA
*Procedure.* A 10-$\mu$l portion of [$^3$H]Cm-Tf solution is placed in a 1.5-ml microcentrifuge tube and mixed with 10 $\mu$l of the enzyme (0.2–2.0 $\mu$g/ml of stromelysin 1) in TNC buffer. After incubation at 37° for 1–4 hr, the reactions are stopped by adding 200 $\mu$l of 3.3% (w/v) TCA, and the tubes are allowed to stand for 15 min at room temperature. The precipitates formed are then removed by centrifugation for 5 min using an Eppendorf microcentrifuge, and 100 $\mu$l of the supernatant is mixed with 3 ml of a suitable phase-combined scintillant and the radioactivity measured. The assay also includes (a) TNC buffer alone for determination of background radioactivity and (b) 10 $\mu$l of trypsin solution (100 $\mu$g/ml) for the total digestion of [$^3$H]Cm-Tf. The total radioactivity of 30 $\mu$g [$^3$H]Cm-Tf may vary from preparation to preparation [20,000–35,000 counts/min (cpm)/30 $\mu$g [$^3$H]Cm-Tf].

The assay is linear up to about 10% digestion of [$^3$H]Cm-Tf (Fig. 1, inset). A second phase of apparent linearity of the assay can be also observed above 10% digestion (Fig. 1). The estimation of enzymatic activity based on the second phase of linearity gives a false (2- to 3-fold lower) specific activity

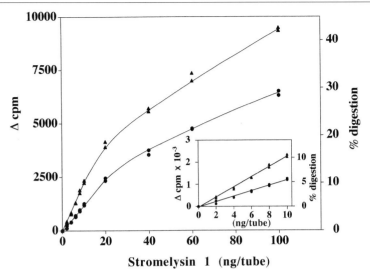

FIG. 1. Transferrin radioassay. Various concentrations of stromelysin 1 (10 µl) were mixed with [³H]Cm-Tf (10 µl) in TNC buffer and incubated for 2 hr (●) and 4 hr (▲) at 37°. The total radioactivity of 10 µl of [³H]Cm-Tf digested by 10 µl of trypsin (100 µg/ml) was 23,000 cpm.

of the enzyme. The linearity of the assay can be tested by incubating with trypsin (0.2–2.0 ng/assay) in TNC buffer at 37° for 1 hr.

*Calculation.* One unit (U) of [³H]Cm-Tf digestion is defined as the production of 1 µg [³H]Cm-Tf fragments soluble in 3.3% TCA in 1 min at 37°. The units of an enzyme solution are determined as follows:

$$\text{U/ml enzyme solution} = 30 \times \frac{\text{cpm}_{\text{sample}} - \text{cpm}_{\text{buffer}}}{\text{cpm}_{\text{trypsin}} - \text{cpm}_{\text{buffer}}} \times 100 \times \frac{1}{t\,(\text{min})}$$

The specific activity of the fully activated stromelysin 1 against [³H]Cm-Tf is 840 U/mg prostromelysin 1.

## Assay with Fluorogenic Peptide

*Principle.* The peptide substrate Mca-Arg-Pro-Lys-Pro-Val-Glu-Nva-Trp-Arg-Lys(Dnp)-NH₂ containing Mca [(7-methoxycoumarin-4-yl)acetyl] as a fluorophore group and Dnp (2,4-dinitrophenyl) as a quencher group is barely fluorescent. An increase in fluorescence is observed on hydrolysis of the peptide. Synthesis of the substrate is described elsewhere.[25] Stromelysin 1 exclusively cleaves the Glu-Nva bonds.[25]

*Reagents*

Substrate stock solution: A 10 m$M$ solution of substrate in dimethyl sulfoxide (DMSO) is stored at $-20°$, and the working substrate solution (10 $\mu M$) is prepared by dilution with TNC buffer; the substrate is soluble at 300 $\mu M$ in TNC buffer containing 3% (v/v) DMSO.

Stopping reagent: 3% (v/v) glacial acetic acid

Mca-Arg-Pro-Lys-Pro-Val-Glu standard is generated by complete hydrolysis of the substrate by stromelysin 1 and dilution to 1.0–0.1 $\mu M$ in the stopping reagent; the fluorescence values of the standard solutions are used to determine the amount of the substrate hydrolysis

*Procedure.* A typical assay is carried out by incubating a 100-$\mu$l portion of 10 $\mu M$ substrate with 10 $\mu$l of the enzyme solution (0.2–2.0 n$M$) in TNC buffer at 37°. After incubation at 37° for an exact period of time (20–30 min), the reaction is stopped by the addition of 900 $\mu$l of the stopping reagent. The fluorescence of the product, Mca-Arg-Pro-Lys-Pro-Val-Glu, is measured by excitation at 325 nm and emission at 393 nm. The amount of substrate hydrolysis is calculated based on the fluorescence values of the Mca-Arg-Pro-Lys-Pro-Val-Glu standard solutions after subtraction of the reaction blank value (stopping solution added before the enzyme). Continuous rate assays can also be made with a fluorimeter equipped with a temperature-controlled cell and a recorder. The $k_{cat}/K_m$ value for stromelysin 1 is 218,000 $M^{-1}$ sec$^{-1}$ (see Table III).

## Purification of Human Stromelysin 1

Stromelysin 1 is found as prostromelysin 1 in the conditioned culture medium or tissue extracts in most cases. Purification of stromelysin 1 as a zymogen is easier and gives a better yield than that of the activated enzyme because the latter consists of several molecular species.

### Sources of Prostromelysin 1

A number of fibroblasts from various species in culture can be stimulated with PMA and/or IL-1 to produce prostromelysin 1. Under these conditions, the production of interstitial procollagenase (proMMP-1), and in some cases progelatinase B (proMMP-9), is also enhanced. Progelatinase A (proMMP-2) is constitutively produced by fibroblasts in culture. Human synovial,[26,27] gingival,[28] and dermal[29] fibroblasts are good sources, as relatively high amounts of prostromelysin 1 are produced when the cells are stimulated. A human rheumatoid synovial cell culture system is described as an example.[26]

Synovial tissues removed at arthroplasty from patients with rheumatoid arthritis are minced and cells are dissociated by digestion with bacterial collagenase (Sigma) (4 mg/ml) at 37° for 1 hr and then with 0.25% trypsin (GIBCO, Grand Island, NY) for 1 hr at room temperature with gentle agitation. After removal of the undigested tissue debris by filtering through several layers of sterile gauze, cells are plated in 35-mm tissue culture plates and grown in Dulbecco's modified Eagle's medium (DMEM) containing 20% (v/v) fetal calf serum and penicillin and streptomycin (50 $\mu$g/ml each) at 37° with 5% $CO_2$ and air. Cells after the third and fourth passages grown to confluency in DMEM containing 10% (v/v) fetal calf serum are washed with Hanks' balanced salt solution (HBSS) and treated with 1 n$M$ IL-1$\alpha$ or IL-1$\beta$ in DMEM containing 0.2% (w/v) lactalbumin hydrolyzate for 3 to 4 days. The conditioned medium is harvested, centrifuged to remove cellular debris, and stored at $-20°$ until use. The production of prostromelysin 1 is synergistically enhanced by the combination of 1 n$M$ IL-1 and 1 n$M$ TNF$\alpha$.[27] The addition of PMA (10–50 ng/ml) further augments the production of prostromelysin 1.[27] Human gingival and dermal fibroblasts grown from the explants of minced tissues are stimulated in the same manner as for rheumatoid synovial cells.

## Purification Procedures

All purification procedures are carried out at room temperature with TC buffer [50 m$M$ Tris-HCl (pH 7.5)–10 m$M$ $CaCl_2$] containing 0.05% (w/v) Brij 35 and 0.02% (w/v) $NaN_3$ unless otherwise stated.

## Method 1

Method 1 is based on previously published procedures.[26,30] Seven chromatographic steps are required.

[26] Y. Okada, H. Nagase, and E. D. Harris, Jr., *J. Biol. Chem.* **261**, 14245 (1986).
[27] K. L. MacNaul, N. Chartrain, M. Lark, M. J. Tocci, and N. I. Hutchinson, *J. Biol. Chem.* **265**, 17238 (1990).
[28] M. W. Lark, L. A. Walakovits, T. K. Shah, J. Vanmiddlesworth, P. M. Cameron, and T. Y. Lin, *Connect. Tissue Res.* **25**, 49 (1990).
[29] S. M. Wilhelm, I. E. Collier, A. Kronberger, A. Z. Eisen, B. L. Marmer, G. A. Grant, E. A. Bauer, and G. I. Goldberg, *Proc. Natl. Acad. Sci. U.S.A.* **84**, 6725 (1987).
[30] Y. Okada, E. D. Harris, Jr., and H. Nagase, *Biochem. J.* **254**, 731 (1988).

*Step 1: DEAE-Cellulose Chromatography.* The crude conditioned culture medium (500 ml) is concentrated about 10- to 20-fold using an Amicon (Danvers, MA) Diaflo apparatus with a YM10 membrane and applied to a column of DEAE-cellulose (25 ml packed resin) equilibrated with 50 m$M$ Tris-HCl (pH 8.0)–0.15 $M$ NaCl–10 m$M$ CaCl$_2$ without Brij 35. All proMMPs are recovered in the unbound fraction. This step, which removes glycosaminoglycans and phenol red, is essential for reproducible binding of proMMP-1, proMMP-2, and proMMP-3 in the next Green A Matrex gel (Amicon) step.

*Step 2: Green A Matrex Gel Chromatography.* The DEAE nonbinding fraction is applied to a column of Green A Matrex gel (2.5 × 5 cm) equilibrated with TNC buffer [50 m$M$ Tris-HCl (pH 7.5)–0.15 $M$ NaCl– 10 m$M$ CaCl$_2$–0.05% Brij 35–0.02% NaN$_3$]. Most proteins do not bind to the column, whereas all three proMMPs bind. After washing the column with TNC buffer, prostromelysin 1 (proMMP-3) is eluted stepwise with TC buffer containing 0.3 $M$ NaCl. The recovery of prostromelysin 1 is about 50%. This fraction also contains progelatinase A (proMMP-2), whereas about 75% of interstitial procollagenase (proMMP-1) is recovered with TC buffer containing 2.0 $M$ NaCl.

*Step 3: Gelatin-Sepharose Affinity Chromatography.* Progelatinase A present in the 0.3 $M$ NaCl fraction is removed by a column of gelatin-Sepharose (Pharmacia LKB, Piscataway, NJ) (2.5 × 4 cm) equilibrated with TC buffer containing 0.3 $M$ NaCl. Prostromelysin 1 is recovered in the unbound fraction. The bound progelatinase A can be eluted from the column with 5% (v/v) DMSO in TC buffer containing 1 $M$ NaCl.

*Step 4: Gel-Permeation Chromatography.* The unbound prostromelysin 1 fraction is concentrated and subjected to gel-permeation chromatography on Sephacryl S-200 equilibrated with TC buffer containing 0.4 $M$ NaCl. The purified prostromelysin 1 exhibits a doublet of 57 kDa (major) and 59 kDa (minor) bands on sodium dodecyl sulfate–polyacrylamide gel electrophoresis (SDS–PAGE).

*Step 5: Further Purification of Prostromelysin 1.* The material obtained in step 4 may contain a small amount of proMMP-1 of 52 and 56 kDa. ProMMP-1 can be removed by an anti-human MMP-1 immunoadsorbent column equilibrated with TC buffer containing 0.4 $M$ NaCl. Alternatively, application of the sample in 20 m$M$ Tris-HCl (pH 7.5)–0.15 $M$ NaCl–10 m$M$ CaCl$_2$ to a CM-cellulose column equilibrated with the same buffer removes proMMP-1, and prostromelysin 1 is recovered in the unbound fraction.[28] The CM-cellulose step may be used after step 1 as described by Lark *et al.*[28]

*Method 2*

Prostromelysin 1 can be purified more efficiently by affinity chromatography using polyclonal antibodies raised against activated stromelysin 1.[31] This method is useful for isolating prostromelysin 1 from tissue extracts, for example, human articular cartilage.[20] The crude conditioned culture medium (500 ml) concentrated 10- to 20-fold over an Amicon YM10 membrane is applied to a column (2.5 × 10 cm) of sheep polyclonal anti-human stromelysin 1 immunoglobulin G (IgG)-Affi-Gel 10 (6 mg IgG fraction coupled to 1 ml of Affi-Gel 10) equilibrated with TNC buffer without Brij 35 at a flow rate of 5 ml/min. After washing the column, the bound prostromelysin 1 is eluted with 6 $M$ urea in TNC buffer without Brij 35 at the same flow rate, and the pooled fractions are dialyzed against 200 volumes of TNC buffer without Brij 35 three times. The material is concentrated by a YM10 membrane and applied to Sephacryl S-200 as in step 4 in Method 1. Prostromelysin 1 is stable to 6 $M$ urea treatment, and the majority is recovered as a proform. About 1.5 mg of prostromelysin 1 can be obtained from the medium (500 ml) of stimulated human rheumatoid synovial cells with 90% recovery.

*Separation of Glycosylated Form of Prostromelysin 1*

Prostromelysin 1 secreted from fibroblasts contains about 20% glycosylated 59-kDa form. This form can be separated from the unglycosylated prostromelysin 1 (57 kDa) by using a concanavalin A-Sepharose column (Pharmacia LKB) equilibrated with TC buffer containing 0.4 $M$ NaCl.[30] Only the 59-kDa prostromelysin 1 binds to the column, and it is eluted with 1 $M$ $\alpha$-methyl-D-mannoside. Both unbound (unglycosylated) and eluted (glycosylated) fractions require further purification by gel-permeation chromatography on Sephacryl S-200 as in step 4 to remove a small amount of concanavalin A and other proteins from the concanavalin A-Sepharose.

*Separation of Active Stromelysin 1 and Prostromelysin 1*

During storage or purification procedures prostromelysin 1 may be partially converted to the active forms of 45 and 28 kDa. Active forms can be separated from the proenzyme by chromatography on Green A Matrex gel.[26,32] A typical example is shown for stromelysin 1 in Fig. 2.

The mixture of stromelysin 1 and prostromelysin 1 (total of ~0.5 mg protein) in TNC buffer is applied to a column of Green A Matrex gel (1

[31] A. Ito and H. Nagase, *Arch. Biochem. Biophys.* **267,** 211 (1988).
[32] L.-C. Chen, M. E. Noelken, and H. Nagase, *Biochemistry* **32,** 10289 (1993).

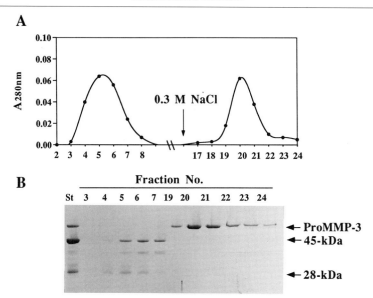

FIG. 2. Separation of prostromelysin 1 and active forms of stromelysin 1. (A) Partially activated prostromelysin 1 (proMMP-3) (0.5 mg) was applied to a column of Green A Matrex gel (0.75 × 1.2 cm) equilibrated with TNC buffer. The bound prostromelysin 1 was eluted with 0.3 $M$ NaCl TC buffer containing 0.05% Brij 35 and 0.02% NaN$_3$ at a flow rate of 0.5 ml/min, and fractions of 1 ml were collected. (B) Analysis by SDS–PAGE of fractions eluted from the Green A Matrex gel column. St, Starting material.

ml) equilibrated with TNC buffer. The active forms of stromelysin 1 (45 and 28 kDa) do not bind to the resin, whereas prostromelysin 1 binds and is eluted with TC buffer containing 0.3 $M$ NaCl, 0.05% Brij 35, and 0.02% NaN$_3$.

## Activation of Prostromelysin 1

Human prostromelysin 1 is activated by mercurial compounds, endopeptidases, and heat treatment. On activation the zymogen is processed to the 45- and 28-kDa forms, both of which have indistinguishable specific activity and substrate specificity.

### Activation by 4-Aminophenylmercuric Acetate

4-Aminophenylmercuric acetate (APMA) stock solution is made by dissolving APMA (final concentration 20 m$M$) in 0.2 $M$ NaOH and titrating

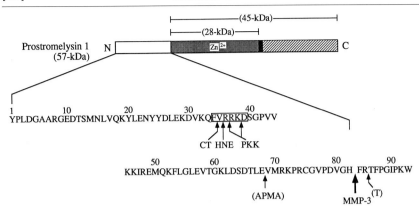

FIG. 3. Steps involved in prostromelysin 1 activation by proteinases and APMA. Initial cleavage sites by chymotrypsin (CT), human neutrophil elastase (HNE), and plasma kallikrein (PKK) as well as the autolytic site of cleavage promoted by APMA in the propeptide are indicated. The final activation is mediated by the action of stromelysin 1 intermediates generated by proteinases or APMA, and a 45-kDa stromelysin 1 is produced. The 45-kDa form undergoes autolysis to the 28-kDa enzyme. Trypsin (T) cleaves the Arg[84]-Thr[85] bond and generates a 45-kDa species which exhibits 20% of the activity against [³H]Cm-Tf. The open bar denotes the propeptide domain, the stippled bar the catalytic domain, the solid bar the proline-rich region, and the hatched bar the C-terminal domain.

with concentrated HCl to pH 11.0. The solution is stable at room temperature for at least 4 weeks. Below pH 11.0 APMA precipitates at that concentration. The stock solution can be diluted to 3 m$M$ with TNC buffer without precipitation.

The rate of prostromelysin 1 activation depends on the concentration of APMA. Full activation with 1.5 m$M$ APMA requires a 22-hr incubation at 37° in TNC buffer.[30] The proenzyme of 57 kDa is initially processed to 46 kDa and then to 45 and 28 kDa. Generation of the 46-kDa form is considered to be an intramolecular reaction as the rate of proMMP-3 conversion to the 46-kDa intermediate follows first-order kinetics ($t_{1/2}$ of 2.0 hr with 1.5 m$M$ APMA at 37°), and it is independent of the initial concentration of the zymogen or the presence of an excess amount of substrate.[30] Subsequent conversion of the 46-kDa intermediate to the 45- and 28-kDa species is a bimolecular reaction.[30] Both the 45- and the 28-kDa forms have the same N terminus, Phe-83, indicating that the 28-kDa form lacks the C-terminal domain[33] (Fig. 3).

[33] H. Nagase, J. J. Enghild, K. Suzuki, and G. Salvesen, *Biochemistry* **29**, 5783 (1990).

*Activation by Endopeptidases*

A number of endopeptidases activate prostromelysin 1. These include trypsin,[30] chymotrypsin,[30] tryptase,[34] rat chymases I and II,[35] plasmin,[30] plasma kallikrein,[30] neutrophil elastase,[33] *Pseudomonas* elastase,[36] and thermolysin.[30] However, stromelysin 1 cannot activate its own proenzyme.[30] Studies with representative endopeptidases have indicated that activator proteinases attack the Phe[34]-Val-Arg-Arg-Lys-Asp[39] region located at the middle of the propeptide and generate a 53-kDa intermediate[33] (Fig. 3). The sequence around that "bait" region dictates which proteinases become activators of prostromelysin 1. The complete removal of propeptide is conducted by cleaving the His[82]-Phe[83] bond via a bimolecular reacton of partially activated intermediates.[33] Prolonged incubation with activator proteinases may result in further processing or degradation of the enzyme (e.g., by thermolysin or plasma kallikrein).[30]

A stable, fully active 45-kDa stromelysin 1 can be generated by treatment with chymotrypsin or neutrophil elastase at a concentration of 5 $\mu$g/ml for 1–4 hr at 37° following inactivation of the serine proteinase by 2 m$M$ diisopropyl fluorophosphate or 2 m$M$ phenylmethylsulfonyl fluoride. When prostromelysin 1 is treated with 5–10 $\mu$g/ml of trypsin for more than 5 min at 37°, the Arg[84]-Thr[85] bond is cleaved by trypsin (Fig. 3) and only about 20% of the full activity is detected against [$^3$H]Cm-Tf.[30] Thus, trypsin is not suitable for generation of the fully active stromelysin 1.

*Heat Activation*

Prostromelysin 1 is self-activated to the 45- and 28-kDa active forms by incubation at 55°.[37] The rate of activation is influenced by the salt concentrations of the buffer: prostromelysin 1 in 50 m$M$ Tris-HCl (pH 7.5)–10 m$M$ Ca$^{2+}$ is almost completely converted to 45- and 28-kDa species by heating at 55° after 60 min, whereas the conversion rate is lower with 0.4 $M$ NaCl.[36] Phe-83 is found as the N terminus in the heat-activated forms.[37]

*Titration of Active Site*

Tissue inhibitors of metalloproteinases (TIMPs) and $\alpha_2$-macroglobulin ($\alpha_2$M) form 1:1 complexes with active stromelysin 1 and inhibit the enzy-

[34] B. L. Gruber, M. J. Marchese, K. Suzuki, L. B. Schwartz, Y. Okada, H. Nagase, and N. S. Ramamurthy, *J. Clin. Invest.* **84**, 1657 (1989).
[35] K. Suzuki, M. Lees, G. Newlands, H. Nagase, and D. E. Woolley, *Biochem. J.* **305**, 301 (1995).
[36] K. Suzuki and H. Nagase, unpublished work (1993).
[37] P. A. Koklitis, G. Murphy, C. Sutton, and S. Angal, *Biochem. J.* **276**, 217 (1991).

matic activity. The two groups of inhibitors are useful to titrate the active site of stromelysin 1. Titration with $\alpha_2M$ requires large protein substrates in the assay.

Ten microliters of approximately 100 n$M$ activated stromelysin is mixed with 10 $\mu$l of 20, 40, 60, . . . , 200 n$M$ TIMP-1 or $\alpha_2M$ solution and incubated at 37° for 1 hr. A 10-$\mu$l portion of each mixture is then assayed for residual activity against [$^3$H]Cm-Tf. A linear plot of activity against the inhibitor molarity reaches, or is extrapolated to be, zero activity at the molarity of the enzyme solution. On activation of prostromelysin 1, two active forms (45 and 28 kDa) are generated, but both bind to the inhibitor with a 1:1 stoichiometry.[38]

## Properties of Human Stromelysin 1

### Structural Features

Human stromelysin 1 is synthesized as a preproenzyme of 477 amino acids and is secreted from cells as a proenzyme of 460 amino acids which consists of the propeptide (82 amino acids), the catalytic domain (165 amino acids), the proline-rich region (25 amino acids), and the C-terminal domain (188 amino acids).[29,39,40] There are three cysteines in prostromelysin 1. The propeptide contains the R$^{73}$RCGVPD$^{79}$ sequence, a motif found in all proMMPs, and the catalytic domain has the zinc-binding signature HEXXHXXGXXH found in a number of metalloendopeptidases. The C-terminal domain shares some sequence similarity with hemopexin and vitronectin. Two cysteines in the C-terminal domain are disulfide bonded. X-Ray crystallographic studies of C-terminal domain-truncated pro-stromelysin 1 have indicated that the proenzyme contains two Zn$^{2+}$ and three Ca$^{2+}$ and that the Zn$^{2+}$ ion at the active site is bound to His-201, His-205, and His-211 and Cys-75 in the propeptide.[41] Interaction of Cys-75 and the Zn$^{2+}$ of the active site has also been demonstrated via spectroscopic evidence of sulfur ligation to Co$^{2+}$ in the cobalt-substituted pro-stromelysin 1[42] and by extended X-ray absorption fine structure spectros-

---

[38] Y. Okada, S. Watanabe, I. Nakahishi, J. Kishi, T. Hayakawa, W. Watorek, J. Travis, and H. Nagase, *FEBS Lett.* **229**, 157 (1988).

[39] S. E. Whitham, G. Murphy, P. Angel, H.-J. Rahmsdorf, B. J. Smith, A. Lyons, T. J. R. Harris, J. J. Reynolds, P. Herrlich, and A. J. P. Docherty, *Biochem. J.* **240**, 913 (1986).

[40] J. Saus, S. Quinones, Y. Otani, H. Nagase, E. D. Harris, Jr., and M. Kurkinen, *J. Biol. Chem.* **263**, 6742 (1988).

[41] K. Appelt, personal communication (1993).

[42] S. P. Salowe, A. I. Marcy, G. C. Cuca, C. K. Smith, I. E. Kopka, W. K. Hagmann, and J. D. Hermes, *Biochemistry* **31**, 4535 (1992).

copy.[43] The $Cys^{75}$–$Zn^{2+}$ interaction maintains the latency of proMMPs as it prevents the formation of a water–zinc complex that is required for catalytic reaction.[44] Two potential glycosylation sites are found at Asn-101 in the catalytic domain and at Asn-381 in the C-terminal domain. Both prostromelysin 1 and the active 45-kDa stromelysin 1 bind to collagen I through the C-terminal domain.[45]

*Stability*

Prostromelysin 1 in TNC buffer is stable to storage at −20° at least for 1 year, at 4° for 4 months, and at 37° for 48 hr. Activated stromelysin 1 in TNC buffer is also stable at −20° and 4°. Little loss of activity is observed with the sample kept at room temperature (~23°) for at least 4 weeks. Heating the enzyme at 80° for 15 min results in 35% loss of activity.[26] The enzyme is stable in the range pH 5.5–9.5, but about 50% of the activity is lost at pH 5.0 and pH 10.0, and more than 80% at pH 4.5 and pH 10.5, after 15 hr.[46]

*pH Optimum*

Stromelysin 1 exhibits an acid pH optimum of pH 5.3–5.5 for digestion of aggrecan, but it retains about 30% of the activity at pH 7.5 and about 50% at pH 8.0.[7] The acid pH optimum of stromelysin 1 and a shoulder of activity extending from pH 7.5 to 8.0 with three $pK_a$ values estimated to be 5.4, 6.1, and 9.5 were clearly demonstrated by hydrolysis of a substance P analog, Arg-Pro-Lys-Pro-Gln-Gln-Phe-Phe-Gly-Leu-Nle-NH$_2$ (Fig. 4).[46] Digestion of gelatin or Azocoll by stromelysin 1 exhibits a similar pH dependence,[20] but the optimal activity against [$^3$H]Cm-Tf is shifted to pH 7.5 with a shoulder of activity around pH 6.0–6.5.[36]

*Requirement for Calcium Ion*

Stromelysin 1 placed in $Ca^{2+}$-free buffer exhibits no enzymatic activity. The $Ca^{2+}$ is required to maintain the active conformation of the enzyme. Full activity is detected by the addition of 1 m$M$ $Ca^{2+}$ at pH 7.5 and is retained at even higher concentrations of $Ca^{2+}$.[26] At lower concentrations

[43] R. C. Holz, S. P. Salowe, C. K. Smith, G. C. Cuca, and L. Que, Jr., *J. Am. Chem. Soc.* **114,** 9611 (1992).
[44] H. E. Van Wart and H. Birkedal-Hansen, *Proc. Natl. Acad. Sci. U.S.A.* **87,** 5578 (1990).
[45] G. Murphy, J. A. Allan, F. Willenbrock, M. I. Cockett, J. P. O'Connell, and A. J. P. Docherty, *J. Biol. Chem.* **267,** 9612 (1992).
[46] R. K. Harrison, B. Chang, L. Niedzwiecki, and R. L. Stein, *Biochemistry* **31,** 10757 (1992).

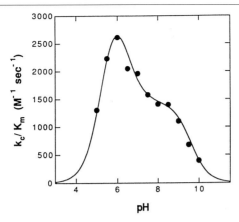

FIG. 4. Dependence of $k_{cat}/K_m$ on pH for the stromelysin 1-catalyzed hydrolysis of Arg-Pro-Lys-Pro-Gln-Gln-Phe-Gly-Leu-Nle-NH$_2$. (Reprinted with permission from Harrison *et al.*[46] Copyright 1992 American Chemical Society.)

of Ca$^{2+}$ the enzyme is autolyzed.[47] At pH 5.3 in 0.1 $M$ MES–Tris buffer, the requirement of Ca$^{2+}$ increases to 5–6 m$M$, and only 20–30% of activity is detected with 1 m$M$ Ca$^{2+}$ at pH 5.3.[7] The decreased affinity for Ca$^{2+}$ at acidic pH, therefore, increases the rate of stromelysin 1 autolysis at a low Ca$^{2+}$ concentration.[7] Little or no activity is detected with Zn$^{2+}$, Cd$^{2+}$, Mg$^{2+}$, Mn$^{2+}$, or Hg$^{2+}$.[26] The ion Co$^{2+}$ gives about 30% of the activity at concentrations of 5–10 m$M$.[26]

*Reactivation of Apostromelysin 1*

Treatment of stromelysin 1 in TNC buffer with 20 m$M$ EDTA for 15 min at 20° and subsequent removal of the EDTA–metal complexes by using a spin column of Sephadex G-10 equilibrated with metal-free buffer result in almost complete loss of enzymatic activity.[26] Addition of Ca$^{2+}$, Hg$^{2+}$, Zn$^{2+}$, or Mn$^{2+}$ alone up to a concentration of 10 m$M$ at pH 7.5 does not restore activity, but the addition of 10 m$M$ Co$^{2+}$ restores 40% of the original activity.[26] In the presence of 5 m$M$ Ca$^{2+}$, 0.5 m$M$ Zn$^{2+}$ restores 60% of the activity[26]; Co$^{2+}$ also restores the original activity about 60% but requires a higher concentration ($>5$ m$M$).[26] Extensive dialysis of stromelysin 1 against 2 m$M$ 1,10-phenanthroline in 50 m$M$ HEPES–5 m$M$ Ca$^{2+}$ (pH 7.0) at 4° followed by dialysis against the buffer without 1,10-phenanthroline reduces the activity by 85%.[42] The enzymatic activity restored is

[47] T. J. Housley, A. P. Baumann, I. D. Braun, G. Davis, P. K. Seperack, and S. M. Wilhelm, *J. Biol. Chem.* **268,** 4481 (1993).

TABLE I
SPECIFIC ACTIVITIES OF HUMAN STROMELYSIN 1 ON PROTEIN SUBSTRATES AT pH 7.5

| Substrate | Source | Temperature (°C) | Specific activity ($\mu$g digested min$^{-1}$ mg$^{-1}$)$^a$ | $k_{cat}/K_m$ ($M^{-1}$ sec$^{-1}$) | Ref |
|---|---|---|---|---|---|
| Aggrecan core protein | Bovine | 37 | 315 | | b |
| Collagen IV | Mouse EHS tumor | 32 | 140 | | c |
| Fibronectin | Human | 37 | 150 | | c |
| Gelatin I | Guinea pig | 37 | 48 | | c |
| Elastin | | 37 | 1.8 | | d |
| [$^3$H]Cm-Tf | Human | 37 | 840 | | c |
| Casein | | 37 | 270 | | d |
| $\alpha_1$-Proteinase inhibitor | | 37 | | 5400 | e |
| $\alpha_1$-Antichymotrypsin | | 37 | | 35,000 | e |
| Antithrombin | | 37 | | <50 | e |
| $\alpha_2$-Macroglobulin | | 25 | | 56,000 | f |
| Aggrecan core protein | Human | 22 | | 4000 | g |

$^a$ Milligram protein is determined as prostromelysin 1.
$^b$ Y. Okada, H. Nagase, and E. D. Harris, Jr., *J. Biol. Chem.* **261,** 14245 (1986).
$^c$ H. Nagase, Y. Ogata, K. Suzuki, J. J. Enghild, and G. Salvesen, *Biochem. Soc. Trans.* **19,** 715 (1991).
$^d$ P. A. Koklitis, G. Murphy, C. Satton, and S. Angal, *Biochem. J.* **276,** 217 (1991).
$^e$ A. E. Mast, J. J. Enghild, H. Nagase, K. Suzuki, S. V. Pizzo, and G. Salvesen, *J. Biol. Chem.* **266,** 15810 (1991)
$^f$ J. J. Enghild, G. Salvesen, K. Brew, and H. Nagase, *J. Biol. Chem.* **264,** 8779 (1989).
$^g$ M. Poe, R. L. Stein, and J. K. Wu, *Arch. Biochem. Biophys.* **298,** 757 (1992).

proportional to the amount of Zn$^{2+}$ added.[42] Again, Co$^{2+}$ restores the activity about 80% but binds to the enzyme more weakly, and it is displaced by the subsequent addition of Zn$^{2+}$, which increases the activity.[42]

## Substrate Specificity

*Protein Substrates.* Specific activities of stromelysin 1 on some protein substrates and the cleavage sites identified are summarized in Tables I and II, respectively. The enzyme digests a number of extracellular matrix components including aggrecan core protein,[26,48] cartilage link protein,[49] fibronectin,[7,26] collagens III,[20] IV,[17,26] IX,[50] and X,[50] telopeptides of collagens I,[26] II,[50] and XI,[50] laminin A chain,[26] gelatins,[10,26] and elastin,[17] but activity on collagen III and elastin is weak. Aggrecan core protein, fibronectin, collagen IV, laminin A chain, and gelatins are cleaved into several fragments. The cleavage site of collagen IX is located in the noncollagenous domain 2 (NC 2 domain) of $\alpha1$(IX), $\alpha2$(IX), and $\alpha3$(IX) chains.[50] Plasma serine proteinase inhibitors (serpins), $\alpha_1$-chymotrypsin, $\alpha_1$-proteinase inhib-

[48] C. R. Flannery, M. W. Lark, and J. D. Sandy, *J. Biol. Chem.* **267,** 1008 (1992).
[49] Q. Nguyen, G. Murphy, P. J. Roughley, and J. S. Mort, *Biochem. J.* **259,** 61 (1989).
[50] J.-J. Wu, M. W. Lark, L. E. Chun, and D. R. Eyre, *J. Biol. Chem.* **266,** 5625 (1991).

TABLE II
HUMAN STROMELYSIN 1 CLEAVAGE SITES IN NATURAL PROTEINS

| Substrate | Source | P4 | P3 | P2 | P1 | + | $P_1'$ | $P_2'$ | $P_3'$ | $P_4'$ | Ref. |
|---|---|---|---|---|---|---|---|---|---|---|---|
| Aggrecan core protein | Human | I | P | E | $N^{341}$ | — | $F^{342}$ | F | G | V | a |
| Cartilage link | Human | R | A | I | $H^{16}$ | — | $I^{17}$ | Q | A | E | b |
| Collagen $\alpha$1 (II) | Bovine | A | G | G | $A^{115}$ | — | $Q^{116}$ | M | G | V | c |
| Collagen $\alpha$1 (IX) | Bovine | (L | A | A | $S^l)^{597}$ | — | $L^{598}$ | K | R | P | c |
| Collagen $\alpha$2 (IX) | Bovine | (E | V | A | $S^l)^{597}$ | — | $A^{598}$ | K | R | E | c |
| Fibronectin | Human | P | F | S | $P^{689}$ | — | $L^{690}$ | V | A | T | d |
| $\alpha_2$-Macroglobulin | Human | G | P | E | $G^{679}$ | — | $L^{680}$ | R | V | G | e |
| | | R | V | G | $F^{684}$ | — | $Y^{685}$ | E | S | D | e |
| Ovostatin | Chicken | L | N | A | $G^{677}$ | — | $F^{678}$ | T | A | S | e |
| $\alpha_1$-Proteinases inhibitor | Human | E | A | I | $P^{357}$ | — | $M^{358}$ | S | I | P | f |
| $\alpha_1$-Antichymotrypsin | Human | L | L | S | $A^{360}$ | — | $L^{361}$ | V | E | T | f |
| Antithrobin III | Human | I | A | G | $R^{393}$ | — | $S^{394}$ | L | N | P | f |
| ProMMP-1 | Human | D | V | A | $Q^{80}$ | — | $F^{81}$ | V | L | T | g |
| ProMMP-3 | Human | D | T | L | $E^{68}$ | — | $V^{69}$ | M | R | K | h |
| | | D | V | G | $H^{82}$ | — | $F^{83}$ | R | T | F | h |
| ProMMP-8 | Human | D | S | G | $G^{78}$ | — | $F^{79}$ | M | L | T | i |
| ProMMP-9 | Human | R | V | A | $E^{40}$ | — | $M^{41}$ | R | G | E | j |
| | | D | L | G | $R^{87}$ | — | $F^{88}$ | Q | T | F | j |
| Substance P | | K | P | Q | $Q^6$ | — | $F^7$ | F | G | L | k |
| Insulin B chain | | L | V | E | $A^{14}$ | — | $L^{15}$ | Y | L | V | d |
| | | E | A | L | $Y^{16}$ | — | $L^{17}$ | V | C | G | d |
| | | | | | | | | | SO_3 | | |

[a] C. R. Flannery, M. W. Lark, and J. D. Sandy, *J. Biol. Chem.* **267,** 1008 (1992).

[b] Q. Nguyen, G. Murphy, P. J. Roughley, and J. S. Mort, *Biochem. J.* **259,** 61 (1989).

[c] J.-J. Wu, M. W. Lark, L. E. Chun, and D. R. Eyre, *J. Biol. Chem.* **266,** 5625 (1991).

[d] S. M. Wilhelm, Z.-H. Shao, T. J. Housley, P. K. Seperack, A. P. Baumann, Z. Gunja-Smith, and J. F. Woessner, Jr., *J. Biol. Chem.* **268,** 21906 (1993).

[e] J. J. Enghild, G. Salvesen, K. Brew, and H. Nagase, *J. Biol. Chem.* **264,** 8779 (1989).

[f] A. E. Mast, J. J. Enghild, H. Nagase, K. Suzuki, S. V. Pizzo, and G. Salvesen, *J. Biol. Chem.* **266,** 15810 (1991).

[g] K. Suzuki, J. J. Enghild, T. Morodomi, G. Salvesen, and H. Nagase, *Biochemistry* **29,** 10261 (1990).

[h] H. Nagase, J. J. Enghild, K. Suzuki, and G. Salvesen, *Biochemistry* **29,** 5783 (1990).

[i] V. Knäuper, S. M. Wilhelm, P. K. Seperack, Y. A. DeClerck, K. E. Langley, A. Ostheus, and H. Tschesche, *Biochem. J.* **295,** 581 (1993).

[j] Y. Ogata, J. J. Enghild, and H. Nagase, *J. Biol. Chem.* **267,** 3581 (1992).

[k] R. Harrison, J. Teahan, and R. Stein, *Anal. Biochem.* **180,** 110 (1989).

[l] Based on chicken $\alpha$1(IX) and $\alpha$2(IX) sequences; Y. Ninomiya, M. van der Rest, R. Mayne, G. Lozano, and B. R. Olsen, *Biochemistry* **24,** 4223 (1985).

TABLE III
Kinetic Constants of Human Stromelysin 1 for Synthetic Substrates at pH 7.5

| Sequence[a] | Temperature (°C) | $k_{cat}$ (sec$^{-1}$) | $K_m$ ($\mu M$) | $k_{cat}/K_m$ ($M^{-1}$ sec$^{-1}$) | Ref |
|---|---|---|---|---|---|
| Dnp-Pro-Leu-Gly + Leu-Trp-Ala-D-Arg-NH$_2$ | 37 | | | 2200 | b |
| Dnp-Pro-Tyr-Ala + Tyr-Trp-Met-Arg-NH$_2$ | 23 | 0.24 | 100 | 2400 | c |
| Mca-Pro-Leu-Gly + Leu-Dpa-Ala-Arg-NH$_2$ | 37 | | | 23,000 | b |
| Dnp-Arg-Pro-Lys-Pro-Leu-Ala + Nva-Trp-NH$_2$ | 25 | | | 45,000 | d |
| Mca-Arg-Pro-Lys-Pro-Val-Glu + Nva-Trp-Arg-Lys(Dnp)-NH$_2$ | 37 | 5.4 | 25 | 218,000 | e |

[a] Dnp, 2,4-Dinitrophenyl; Mca, (7-methoxycoumarin-4-yl)acetyl; Dpa, N-3-(2,4-dinitrophenyl)-L-2,3-diaminopropio
nyl; Nva, norvaline.
[b] C. G. Knight, F. Willenbrock, and J. Murphy, FEBS Lett. **296**, 263 (1992).
[c] S. Netzel-Arnett, S. K. Mallya, H. Nagase, H. Birkedal-Hansen, and H. E. Van Wart, Anal. Biochem. **195**, 86 (1991).
[d] L. Niedzwiecki, J. Teahan, R. K. Harrison, and R. L. Stein, Biochemistry **31**, 12618 (1992).
[e] H. Nagase, C. G. Fields, and G. B. Fields, J. Biol. Chem. **269**, 20952 (1994).

itor, and antithrombin III are cleaved by stromelysin 1 at their reactive loops and are inactivated.[51] The bait region of $\alpha_2$-macroglobulin is rapidly hydrolyzed by stromelysin 1, but the reaction with ovastatin is slow.[52] Stromelysin 1 also cleaves propeptides of a number of proMMPs and participates in their activation.[10–12]

Alignment of the cleavage sites in proteins indicates that stromelysin 1 readily hydrolyzes a peptide bond with hydrophobic side chains at the P1' position, but there is no well-defined P1 specificity (Table II). The lack of strict requirements at the P1 position is also evident from studies with synthetic substrates.[53]

*Synthetic Substrates.* A number of synthetic substrates for stromelysin 1 have been reported, and representatives are shown in Table III. The peptide Mca-Arg-Pro-Lys-Pro-Val-Glu + Nva-Trp-Arg-Lys(Dnp)-NH$_2$ is shown to be the most susceptible synthetic substrate for stromelysin 1, with a $k_{cat}/K_m$ ratio of 218,000 at 37°, pH 7.5.[25] Studies by Niedzwiecki et al.[53] based on substance P and its analogs demonstrated that catalytic efficiency is dependent on the length of peptide, suggesting an extended substrate binding site in stromelysin 1. This is particularly prominent with N-terminal truncation. The peptide containing only 3 residues on the N-terminal side of the scissile bond is no longer hydrolyzed unless the P3 site is blocked. Two residues are sufficient at the P' side when the carboxyl group of P2' is blocked. Stromelysin 1 hydrolyzes substrates with Phe, Tyr, Leu, Ile, and

[51] A. E. Mast, J. J. Enghild, H. Nagase, K. Suzuki, S. V. Pizzo, and G. Salvesen, J. Biol. Chem. **266**, 15810 (1991).
[52] J. J. Enghild, G. Salvesen, K. Brew, and H. Nagase, J. Biol. Chem. **264**, 8779 (1989).
[53] L. Niedzwiecki, J. Teahan, R. K. Harrison, and R. L. Stein, Biochemistry **31**, 12618 (1992).

Val at the P1' position, but the P1 position is not specified. Preferred residues at P2 and P3 sites are shown to be Leu and Pro, respectively.[53] Preferred residues at P2' are aromatic side chains.[53]

*Inhibitors*

Almost complete inhibition of the enzymic activity is observed with 10 m$M$ EDTA or 10 m$M$ EGTA, or 5 m$M$ 1,10-phenanthroline in TNC buffer which contains 10 m$M$ CaCl$_2$. About 50% inhibition is observed with 0.5 m$M$ cysteine or dithiotheitol. Phosphoramidon is not an effective inhibitor: only about 10% of inhibition is seen at 0.1 m$M$. Natural inhibitors include tissue inhibitors of metalloproteinases (TIMP-1 and TIMP-2), $\alpha_2$-macro-globulin and ovostatin. They inhibit stromelysin 1 by forming 1:1 enzyme-inhibitor complexes. The $K_1$ value of TIMP-1 and stromelysin 1 is estimated to be <0.2 n$M$.[45] The rate of binding to $\alpha_2 M$ is rapid ($k_{cat}/K_m$ = 56,000 $M^{-1}\text{sec}^{-1}$).[52] Synthetic inhibitors characterized with $K_i$ values include a hydroxamate compound, HONH-CO-CH$_2$-CH(n-pentyl)-CO-Leu-Phe-NH$_2$ ($K_i$ = 120 n$M$ at pH 7.5, 25°),[54] and a phosphonamidate compound, phthaloyl-$N$-(CH$_2$)$_4$-P(O$_2^-$)-Ile-($\beta$-naphthyl)-Ala-NH-CH$_3$ ($K_i$ = 7 n$M$; $k_{on}$ = 2.7 × 10$^4$ $M^{-1}\text{sec}^{-1}$; $k_{off}$ = 1.9 × 10$^{-4}$ sec$^{-1}$ at pH 5, 25°).[55]

## Activation of Other Promatrix Metalloproteinases by Stromelysin 1

*Activation of Procollagenases*

Stromelysin 1 plays a critical role in activation of procollagenases (proMMP-1 and proMMP-8) by specifically cleaving the Gln$^{80}$-Phe$^{81}$ bond of proMMP-1[56] and the Gly$^{78}$-Phe$^{79}$ bond of proMMP-8.[11] The N-terminal Phe in both collagenases is essential for expression of the full collagenolytic activity.[11,56] Although a catalytic amount of stromelysin 1 cleaves the Gly$^{78}$-Phe$^{79}$ bond of proMMP-8,[11] cleavage of the Gln$^{80}$-Phe$^{81}$ bond of proMMP-1 requires prior removal of a portion of propeptide by treatment with proteinases or APMA.[56] Proteinases that participate in this reaction in proMMP-1 are trypsin, plasmin, and plasma kallikrein.[56] Without stro-melysin 1, both MMP-1 and MMP-8 undergo autolytic cleavage of the Phe$^{81}$-Val$^{82}$ and Val$^{82}$-Leu$^{83}$ bonds and of the Phe$^{79}$-Met$^{80}$ and Met$^{80}$-Leu$^{81}$ bonds, respectively. The products exhibit only 40–50% of collagenolytic ac-tivity.[11,56]

[54] A. I. Marcy, L. L. Eiberger, R. Harrison, H. K. Chan, N. I. Hutchinson, W. K. Hagmann, P. M. Cameron, D. A. Boulton, and J. D. Hermes, *Biochemistry* **30**, 6476 (1991).

[55] M. Izquierdo-Martin and R. L. Stein, *Bioorg. Med. Chem.* **1**, 19 (1993).

[56] K. Suzuki, J. J. Enghild, T. Morodomi, G. Salvesen, and H. Nagase, *Biochemistry* **29**, 10261 (1990).

Full activation of proMMP-1 can be attained by incubation of proMMP-1 with 10 $\mu$g/ml of trypsin for 1 hr, or with 1 m$M$ APMA for 4 hr, at 37° in the presence of an equimolar amount of stromelysin 1. Trypsin is inactivated by 2 m$M$ diisopropyl fluorophosphate. Activation of proMMP-8 in TNC buffer is attained by incubation with stromelysin 1 at a molar ratio of 1:0.1 for 6 hr at 37°.[11] To examine the action of MMP-1 and MMP-8 on collagens I and II, the removal of stromelysin 1 is not necessary because stromelysin 1 does not digest collagens I and II, but it is required for testing the action on other substrates. Stromelysin 1 is removed by an antistromelysin 1 IgG-affinity column.

## Activation of Progelatinase B

A catalytic amount of stromelysin 1 activates progelatinase B (proMMP-9) by sequential cleavage of the Glu$^{40}$-Met$^{41}$ and Ala$^{87}$-Phe$^{88}$ bonds.[12] Full activation can be achieved after incubation with a 1/10 molar amount of stromelysin 1 at 37° for 4 hr. The activity of gelatinase B can be checked on gelatin without removing stromelysin 1 because the activity of stromelysin 1 on gelatin is negligible under these conditions. Examination of gelatinase B activity on other substrates requires the removal of stromelysin 1.

## Assay for Human Stromelysin 2

The enzymatic activity of stromelysin 2 is assayed by the same methods described for stromelysin 1. The enzyme readily hydrolyzes Mca-Arg-Pro-Lys-Pro-Val-Glu-Nva-Trp-Arg-Lys(Dnp)-NH$_2$.[25] The general proteolytic activity of stromelysin 2 is, however, weaker than that of stromelysin 1: the specific activity of stromelysin 2 on [$^3$H]Cm-Tf is 340 U/mg prostromelysin 2 at pH 7.5 when fully activated.

## Purification of Human Stromelysin 2

*Sources of Prostromelysin 2.* Windsor *et al.*[15] have reported that human keratinocytes in culture stimulated with PMA, transforming growth factor $\alpha$, epidermal growth factor, or TNF$\alpha$ produce prostromelysin 2, but gingival fibroblasts do not. Recombinant prostromelysin 2 has been expressed in mouse myeloma NSO cells (ECACC Cat. No. 85110503) stably transfected with the stromelysin 2 cDNA ligated into the expression vector pEE-12[17] that contains a glutamine synthetase gene as an amplifiable marker.[57]

[57] M. I. Cockett, C. R. Bebbington, and G. T. Yarranton, *Bio/Technology* **8**, 662 (1990).

*Purification Procedure.* Prostromelysin 2 behaves similarly in purification to prostromelysin 1.[58] Prostromelysin 2 binds to Green A Matrex gel equilibrated with TNC buffer and eluted with 0.3 $M$ NaCl. Thus, a similar purification procedure to that described for prostromelysin 1 is applicable when the starting material is serum-free conditioned medium from stimulated cells.

The conditioned medium of transfected NSO cells, however, contains a large amount of bovine serum albumin that is required for culturing the cells. Thus, Green A Matrex gel is not effective as serum albumin binds to the resin. Instead, $Zn^{2+}$-chelate chromatography is substituted for this step. The $Zn^{2+}$-chelate Sepharose resin is prepared by successive washing of a column of chelating Sepharose fast flow (Pharmacia LKB) (2.5 × 5 cm) with 2 volumes of 20 m$M$ EDTA (pH 8.0), 3–5 volumes water, and 20 ml of 35 m$M$ ZnCl$_2$. After equilibration of the $Zn^{2+}$-chelate Sepharose column with TC buffer containing 0.4 $M$ NaCl, the crude medium (250 ml) of transfected NSO cells (containing ~1 mg of prostromelysin 2) concentrated about 10-fold with a YM10 membrane is applied to the column. After extensive washing with 50 m$M$ cacodylate hydrochloride (pH 6.0)–0.8 $M$ NaCl–10 m$M$ CaCl$_2$, the bound prostromelysin 2 is eluted with 50 m$M$ imidazole hydrochloride (pH 7.5)–1 $M$ NaCl–10 m$M$ CaCl$_2$. The fractions containing prostromelysin 2 (assayed with [$^3$H]Cm-Tf after activation with 1 m$M$ APMA for 2 hr at 37°) are combined and dialyzed against TNC buffer. The purified prostromelysin 2 exhibits a single band at 57 kDa on SDS–PAGE under reducing conditions. During purification or storage of the culture medium, prostromelysin 2 is often partially converted to the active 45-kDa stromelysin 2. Separation of the two forms can be carried out effectively by Green A Matrex gel chromatography as described for stromelysin 1.

## Activation of Prostromelysin 2

Prostromelysin 2 can be activated by APMA, endopeptidases, and heat treatment. It is processed to the 45- and 28-kDa enzymes in a manner similar to that for prostromelysin 1.[58] However, the activation rate of prostromelysin 2 by APMA is significantly faster than that of prostromelysin 1. A 2-hr incubation with 1.5 m$M$ APMA at 37° is sufficient for full activation of prostromelysin 2. The activation rate by heat treatment is affected by the salt concentrations as in the case of prostromelysin 1.

Only a limited number of endopeptidases have been tested for activation of prostromelysin 2. Trypsin, chymotrypsin, and plasmin activate the zymo-

[58] K. Suzuki, H. Nagase, G. Murphy, and A. J. P. Docherty, unpublished results (1994).

gen fully. Neutrophil elastase does not activate prostromelysin 2.[58] This is probably due to the lack of Val[35] in the proteinase-susceptible "bait" region sequence, Phe[34]-Arg-Arg-Lys-Asp[38]. In contrast to the case for prostromelysin 1, prostromelysin 2 is fully activated by trypsin. This is explained by the fact that Arg[84] in prostromelysin 1 is replaced by Ser in prostromelysin 2.

## Properties of Human Stromelysin 2

Human prostromelysin 2 is synthesized as preproenzyme of 476 amino acids and is secreted from cells as the proenzyme. It has one potential glycosylation site at Asn[100], but it is unlikely to be a glycoprotein as it does not bind to concanavalin A.[58] Prostromelysin 2 in TNC buffer is stable at $-20°$ for 4 weeks, or at $4°$ for 2 weeks, but it tends to undergo autoactivation on longer storage at $-20°$ or on repeated freezing and thawing.

Stromelysin 2 digests aggrecan core protein,[17] cartilage link protein,[18] collagen IV,[17] gelatin,[16] and elastin.[17] Activities on those substrates are about one-half that of stromelysin 1.[17] Cartilage link protein is cleaved at the His[16]-Ile[17] bond, the same site cleaved by stromelysin 1.[18] In contrast to stromelysin 1, the activity on fibronectin of stromelysin 2 is negligible.[58] The enzyme has optimal activity against [³H]Cm-Tf and Azocoll at pH 7.5–8.0. Only about 25% of the activity is found at pH 6.0. However, the activity against aggrecan at pH 5.5 is double that at pH 7.5.[17] Stromelysin 2 participates in proMMP-1 activation in a similar manner to stromelysin 1. The rate of proMMP-1 activation by stromelysin 2 is similar to that by stromelysin 1.[16,58] The enzyme hydrolyzes Mca-Arg-Pro-Lys-Pro-Val-Glu-Nva-Trp-Arg-Lys(Dnp)-NH$_2$ more slowly than stromelysin 1 at pH 7.5 ($k_{cat}$ 3.8 sec$^{-1}$, $K_m$ 23 $\mu M$, $k_{cat}/K_m$ 17 $\times$ 10$^4$ $M^{-1}$ sec$^{-1}$). The enzymatic activity is inhibited by EDTA, EGTA, and 1,10-phenanthroline,[17] and by TIMP-1 at a 1:1 stoichiometry.[17,58] About 50% inhibition is observed with 0.5 m$M$ cysteine or dithiothreitol.[58]

# [28] Gelatinases A and B

By GILLIAN MURPHY and THOMAS CRABBE

## Introduction

Two members of the matrix metalloendopeptidase (MMP) family have been described that are active in the cleavage of denatured collagens, type V collagen, elastin, and other matrix proteins and execute a specific cleavage of native type IV collagen chains. They are $M_r$ 72,000 gelatinase A (MMP2; EC 3.4.24.35) and $M_r$ 92,000 gelatinase B (MMP9; EC 3.24.4.35). Both enzymes are also referred to as type IV collagenases[1,2] and are encoded by genes located on human chromosome 16.[1]

Gelatinase A is the most widely distributed MMP, being produced constitutively by many cell types in culture. Gelatinase B can also be secreted by mesenchymal cells in culture after induction by cytokines and other agents and is a major product of monocytes and tumor cells. It is also found packaged in a granule fraction in polymorphonuclear leukocytes. Both gelatinases may be detected in plasma, and gelatinase B has been shown to be present in saliva. Their potential role, particularly in basement membrane turnover, has been discussed in a number of reviews.[3–5] Although the two gelatinases have overlapping substrate specificities and localization patterns, it is noteworthy that they are differently regulated at both the transcriptional and extracellular level.[6] A unique feature of the gelatinases is the binding of their proforms to TIMP-1 (gelatinase B) and TIMP-2 (gelatinase A).

Assay Methods

Gelatinases A and B have most commonly been assayed with gelatin as the substrate. Labeled denatured type I collagen is most frequently used and the release of trichloroacetic acid-soluble fragments monitored. This is an essentially inaccurate assay because the generation of large molecular mass fragments is not monitored. Furthermore, the substrate is not specific and will be turned over at a relatively lower rate by most proteinases. The presence of contaminating TIMPs in crude gelatinase preparations such as conditioned media from cell cultures adds to the inaccuracies. Because of their specific gelatinase-binding properties, the TIMPs cannot be completely removed from the system (e.g., by reduction techniques suitable for the assay of MMPs such as stromelysin in the presence of TIMP) without denaturation of the enzymes. The gelatinases are relatively efficient pepti-

[1] I. E. Collier, S. M. Wilhelm, A. Z. Eisen, B. L. Marmer, G. A. Grant, J. L. Seltzer, A. Kronberger, C. He, E. A. Bauer, and G. I. Goldberg, *J. Biol. Chem.* **263,** 6579 (1988).
[2] S. M. Wilhelm, I. E. Collier, B. L. Marmer, A. Z. Eisen, G. A. Grant, and G. I. Goldberg, *J. Biol. Chem.* **264,** 17213 (1989).
[3] K. Tryggvason, M. Hoyhtya, and T. Salo, *Biochim. Biophys. Acta* **907,** 191 (1987).
[4] H. Birkedal-Hansen, W. G. I. Moore, M. K. Bodden, L. J. Windsor, B. Birkedal-Hansen, A. DeCarlo, and J. A. Engler, *Crit. Rev. Oral Biol. Med.* **4,** 197 (1993).
[5] G. Murphy and J. J. Reynolds, *in* "Connective Tissue and Its Heritable Disorders" (P. M. Royce and B. Steinmann, eds.), p. 287. Wiley–Liss, New York, 1993.
[6] P. Huhtala, A. Tuuttila, L. T. Chow, J. Lohi, J. Keski-Oja, and K. Tryggvason, *J. Biol. Chem.* **266,** 16485 (1991).

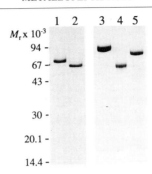

Fig. 1. SDS–polyacrylamide gel electrophoresis of latent and active gelatinases A and B. Recombinant human progelatinases A and B secreted by transfected mouse myeloma cells were purified and activated as described in the text [J. P. O'Connell, F. Willenbrock, A. J. P. Docherty, D. Eaton, and G. Murphy, *J. Biol. Chem.* **269**, 14967 (1994); T. Crabbe, C. Ioannou, and A. J. P. Docherty, *Eur. J. Biochem.* **218**, 431 (1993)]. Samples were reduced with 2-mercaptoethanol and run on a 10–20% polyacrylamide gel. Lane 1, Progelatinase A; lane 2, APMA-activated gelatinase A; lane 3, progelatinase B; lane 4, APMA-activated gelatinase B; lane 5, stromelysin 1-activated gelatinase B.

dases, and a number of good peptide substrates have been developed.[7,8] These are not completely specific and are still subject to TIMP interference, but they are particularly convenient for kinetic studies. A high throughput assay suitable for use with ELISA (enzyme-linked immunosorbent assay) microplates has been described by Bickett *et al.*[9]

*Activation of Progelatinases A and B*

Activation of progelatinase A is best accomplished by treatment with 4-aminophenylmercuric acetate (APMA). It is believed that the organomercurial reagent causes a conformational alteration that allows a stepwise autolytic cleavage of the propeptide. Crude preparations are maximally activated by incubation with 2 m$M$ APMA for 1 hr at 37°; purified progelatinase A is fully activated by incubation with 2 m$M$ APMA for 1 hr at 25°. Longer incubations may be required for the activation of diluted (<1 $\mu M$) purified progelatinase A. Activation can be followed by enzyme assay or sodium dodecyl sulfate (SDS)–polyacrylamide gel electrophoresis (Fig. 1), which will show cleavage of the $M_r$ 72,000 proenzyme to the $M_r$ 66000 active form. Owing to a marked tendency to autolyze, active gelatinase A may be stored at 4° for only 4–5 days.

[7] M. S. Stack and R. D. Gray, *J. Biol. Chem.* **264**, 4277 (1989).
[8] C. G. Knight, F. Willenbrock, and G. Murphy, *FEBS Lett.* **296**, 263 (1992).
[9] D. M. Bickett, M. D. Green, J. Berman, M. Dezube, A. S. Howe, P. J. Brown, J. T. Roth, and G. M. McGeehan, *Anal. Biochem.* **212**, 58 (1993).

Although the $M_r$ 92,000 progelatinase B is fully activated by incubation with 2 m$M$ APMA for 2 hr at 37°, the procedure results in an additional autolytic cleavage that removes the C-terminal domain and generates an $M_r$ 65,000 enzyme (Fig. 1). As this will affect the interactions of the enzyme with TIMP-1, it is recommended that activated stromelysin 1 is used instead.[10,11] Incubation at 37° for 2 hr at a 40:1 molar ratio of progelatinase B to stromelysin 1 will remove only the propeptide to give the $M_r$ 82,000 active gelatinase A (Fig. 1). Active gelatinase B is less prone to activity loss and may be stored at 4° for several weeks.

*Gelatinolytic Assay*

*Reagents*

Substrate: Radiolabeled type I collagen[12,13] is prepared by acetylation with [14]C- or [3]H-labeled acetic anhydride and dialyzed into 50 m$M$ Tris-HCl, pH 7.5, 0.15 $M$ NaCl, 0.02% w/v sodium azide. The material is stable at 4° for several months. The collagen should be denatured just before use because the resulting gelatin is highly susceptible to proteolysis by bacteria. Denaturation is achieved by heating at 60° for 20 min

Buffer: 50 m$M$ Tris-HCl, pH 7.5, 0.15 $M$ NaCl, 10 m$M$ CaCl$_2$, 0.02% NaN$_3$

*Procedure.* The assay is carried out at 37° in a total volume of 250 $\mu$l with 100 $\mu$g of radiolabeled gelatin. Assays of 1–20 hr show a reasonably linear response between 10 and 60% lysis of the gelatin. The incubation is terminated by the addition of 50 $\mu$l of cold 90% (w/v) trichloroacetic acid. Samples should be kept in ice for at least 10 min to allow complete precipitation of undegraded gelatin and then spun at 10,000 $g$ for 15 min. The radioactivity of portions of the supernatant is determined. Blank values (gelatin in the presence of buffer alone) of the counts soluble in trichloroacetic acid should be deducted. The precise specific activity of the 100-$\mu$g gelatin samples under the conditions of the assay are obtained from the counts released by total lysis with 10 $\mu$g of trypsin. One unit of gelatinase activity hydrolyzes 1 $\mu$g gelatin/min.

*Fluorometric Assay*

The most accurate assays for gelatinases A and B involve the use of quenched fluorescent peptide substrates. Dinitrophenyl-Pro-Leu-Gly-Leu-

[10] Y. Ogata, J. J. Enghild, and H. J. Nagase, *J. Biol. Chem.* **267**, 3581 (1992).
[11] J. P. O'Connell, F. Willenbrock, A. J. P. Docherty, D. Eaton, and G. Murphy, *J. Biol. Chem.* **269**, 14967 (1994).
[12] T. E. Cawston and A. J. Barrett, *Anal. Biochem.* **99**, 340 (1979).
[13] Sigma (St. Louis, MO) markets collagens from several species.

Trp-Ala-D-Arg-NH$_2$[14] was first described as a suitable substrate,[7] and this was modified to Mca-Pro-Leu-Gly-Leu-Dpa-Ala-Arg[15] to give greater sensitivity.[8] This low fluorescing substrate liberates the intensely fluorescent Mca-Pro-Leu-Gly on cleavage which can be quantified fluorometrically. The assays are not suitable for gelatinase assays in crude biological samples as the background fluorescence interferes with the product fluorescence.

*Reagents*

Substrate: Mca-Pro-Leu-Gly-Leu-Dpa-Ala-Arg (obtainable from Calbiochem/Novabiochem, La Jolla, CA) is dissolved in dimethyl sulfoxide (DMSO). The concentration of stock solutions may be determined from the absorbance at 410 nm, assuming $\varepsilon = 7500$ $M^{-1}$ cm$^{-1}$

Buffer: 0.1 $M$ Tris-HCl, pH 7.5, 0.1 $M$ NaCl, 10 m$M$ CaCl$_2$, and 0.05% v/v Brij 35

*Procedure.* The assay is carried out at 25° with the enzyme (10–100 p$M$) in 2.5 ml of buffer containing 1.5 $\mu M$ substrate by continuous monitoring using a fluorometer ($\lambda_{ex}$ 328 nm, $\lambda_{em}$ 393 nm). The instrument is set to zero with substrate buffer and calibrated with Mca-Pro-Leu. Assays should be made with substrate concentrations well below the $K_m$ to avoid absorptive quenching effects. The initial rate of cleavage of substrate measured over 10 min should be proportional to substrate concentration in the range 1–8 $\mu M$, but it is important that less than 10% of the substrate is degraded during the assay (see [2] in this volume).

*Active-Site Titration*

Where the active-site molarity of the gelatinases is required, a procedure using TIMPs as titrants has been developed and is described in [30] in this volume. It is recommended that TIMP-1 be used to titrate gelatinase A and TIMP-2 for gelatinase B.

*Zymography*

As little as 1 pg of gelatinase can be detected using the polyacrylamide gel electrophoresis technique with gelatin incorporated into the gel as described in detail in [30] in this volume. Samples for assessment are electrophoresed and the gel washed in Triton X-100 before incubation in 50 m$M$ Tris-HCl, pH 7.5, 20 m$M$ CaCl$_2$, 0.02% NaN$_3$ at 37° for 1–20 hr, as necessary.

---

[14] Standard abbreviations are used for the amino acids (L-configuration).
[15] Available from Novabiochem: Mca, (7-methoxycoumarin-4-yl)acetyl; Dpa, *N*-3-(2,4-dinitrophenyl)-L-2,3-diaminopropionyl.

The gel is finally stained with Coomassie blue. The technique is not strictly quantitative, but some comparison between samples may be made by gel scanning at low levels of activity when the gel clearing is not diffuse. It is advisable to run a range of dilutions of each sample to be sure that the gelatin lysis is in the linear range. A particularly sensitive version of this technique has been developed by Overall and Limeback.[16]

Purification

Cell-conditioned culture media are the usual source of starting material for progelatinase A and B purification. The former has been purified from skin explant, primary skin, gingival or synovial fibroblast, as well as tumor cell sources. Progelatinase B has largely been purified from U937 cell cultures or the supernatant from selectively degranulated neutrophils. The gelatinases are particularly prone to autolytic degradation once activated, so they are best maintained as the latent proenzyme during purification and storage. Pretreatment of culture medium by chromatography on DEAE-Sepharose[17] or Green A agarose[18] has been employed to separate gelatinases partially from other MMPs. Gelatinase purification is then largely accomplished by affinity chromatography using gelatin-Sepharose 4B or fast flow (Pharmacia, Piscataway, NJ). The gelatinases have a high affinity for this ligand, which, in combination with the high binding capacity of the matrix (about 10 mg protein/ml), allows the use of a small column with a large cross-sectional area and makes prior desalting of the culture supernatant unnecessary. When large quantities of culture supernatant are to be processed, it may be preferable to employ a batch purification method. Other gelatin-binding proteins can be removed by a 1 $M$ salt wash before elution of the gelatinases by 10% (v/v) dimethyl sulfoxide (DMSO).

It is common for progelatinase A and progelatinase B to be present in the same culture supernatant, and, although a gradient of 0 to 10% DMSO has been reported to help their separation,[19] a second chromatography step is usually required. Concanavalin A (Con A)-Sepharose is often employed at this stage because it will bind the glycosylated progelatinase B but not progelatinase A. Progelatinase B can then be eluted using 0.5 $M$ methyl-$\alpha$-D-mannoside. Another concern is that often a proportion of the progelatinase A is secreted as a 1 : 1 stoichiometric complex with TIMP-2, whereas

[16] C. M. Overall and H. Limeback, *Biochem. J.* **256,** 965 (1988).
[17] R. V. Ward, R. M. Hembry, J. J. Reynolds, and G. Murphy, *Biochem. J.* **278,** 179 (1991).
[18] T. Morodomi, Y. Ogata, Y. Sasaguri, M. Morimatsu, and H. Nagase, *Biochem. J.* **285,** 603 (1992).
[19] D. Moutsiakis, P. Mancuso, H. Krutzsch, W. Stetler-Stevenson, and S. Zucker, *Connect. Tissue Res.* **28,** 213 (1992).

progelatinase B is similarly complexed with TIMP-1. The TIMP cannot be removed without prior denaturation of the complex, and activation of the gelatinase in its presence leads to low specific activities.[20] It is preferable, therefore, simply to separate the complex from the free proenzyme. In the case of progelatinase A this can be accomplished using heparin Sepharose.[21] The complex and the free proenzyme both bind but are separated on elution with a 0 to 0.2 $M$ NaCl gradient. The progelatinase B–TIMP-1 complex can be separated from free proenzyme by immunoaffinity chromatography using an antibody raised against TIMP-1[17] or by chromatography on Green A agarose.[18] In some cases progelatinase B may also be contaminated with collagenase bound to the C-terminal domain, but this can be separated from the free (or TIMP bound) progelatinase B by the use of a DMSO gradient during gelatin Sepharose chromatography.[22]

All purification steps should be performed at 4° and can be monitored by assaying the degradation of denatured type I collagen that results from the treatment of samples with APMA. Zymography on SDS–polyacrylamide gels and N-terminal amino acid sequencing are more specific tests that can be carried out to identify progelatinase A and progelatinase B. The concentration of the enzymes can be determined by absorbance at 280 nm [$\varepsilon$(progelatinase A) = 122,800 $M^{-1}$ cm$^{-1}$; $\varepsilon$ (progelatinase B) = 114,360 $M^{-1}$ cm$^{-1}$].

*Storage of Progelatinase A and B*

Once purified, progelatinases A and B can be stored in aliquots at −70°. Even under these conditions, however, the samples will gradually become activated and then degraded by autolysis catalyzed by the low levels of active gelatinase that copurify with the proenzyme. Degradation of progelatinase B can largely be prevented by storage at low concentrations (<1 $\mu M$) in the presence of the detergent Brij 35, but progelatinase A is more susceptible to autolysis and must be frozen in buffers that limit any activity. This can be accomplished by storage in 25 m$M$ 2-($N$-morpholino)ethanesulfonic acid–NaOH, pH 6.0. The presence of Brij 35 also confers some stability at pH 7.5.

[20] D. E. Kleiner, A. Tuuttila, K. Tryggvason, and W. G. Stetler-Stevenson, *Biochemistry* **32,** 1583 (1993).
[21] H. Kolkenbrock, D. Orgel, A. Hecker-Kia, W. Noack, and N. Ulbrich, *Eur. J. Biochem.* **198,** 775 (1991).
[22] G. I. Goldberg, A. Strongin, I. E. Collier, L. T. Genrich, and B. L. Marmer, *J. Biol. Chem.* **267,** 4583 (1992).

Properties of Progelatinases A and B

Progelatinase A cloned from mouse[23] and human sources[1] has been found to be 97% identical. Analysis by reduced SDS–polyacrylamide gel electrophoresis gives human progelatinase A an $M_r$ of 72,000, in agreement with the amino acid sequence. Progelatinase B has also been cloned from mouse[24] and human[2] cells, and the two forms show 72% similarity. Human progelatinase B has an $M_r$ of 92,000 on SDS–polyacrylamide gel electrophoresis, although the amino acid sequence predicts a protein of $M_r$ 76,240. The extra mass has been shown to be due to N- and O-linked glycosylation.[2] The heterogeneity of glycosylation results in slightly different $M_r$ proteins from different sources. Mouse progelatinase B has an $M_r$ of 105,000 owing to differences in glycosylation as well as the insertion of a 16-amino acid segment in the collagen-like region between the N- and C-terminal domains (Fig. 2). Both enzymes show the modular structure typical of the MMPs, comprising a signal sequence, an N-terminal propeptide which confers latency on the proenzymes, a catalytic domain containing $Zn^{2+}$ and $Ca^{2+}$ binding sites, and a hemopexin- and vitronectin-like C-terminal domain (Fig. 2). They differ from other MMPs in the possession of another domain inserted within the catalytic domain that has homology with the three type II repeats of the gelatin-binding region of fibronectin.[1,2,25] Progelatinase B has an additional 54-amino acid extension in the proline-rich "hinge" sequence that lies between the catalytic and C-terminal domains and shows sequence similarity to a number of collagens (Figs. 2 and 3).

As previously mentioned, the proenzymes are often isolated complexed with TIMP-1 (progelatinase B)[22] or TIMP-2 (progelatinase A).[26,27] The complexes result from interactions between the C-terminal domains of both proenzyme and inhibitor and can also be formed when purified TIMP is added to purified progelatinase. As the reconstituted complexes are more easily dissociated than those isolated from cell cultures, it has been proposed that the latter are formed during intracellular protein folding. In the absence of TIMP-1, progelatinase B is able to form a stable homodimer or a complex with interstitial procollagenase.[22] Instances of ternary complexes between

[23] P. Reponen, C. Sahlberg, P. Huhtala, T. Hurskainen, I. Thesleff, and K. Tryggvason, *J. Biol. Chem.* **267,** 7856 (1992).

[24] H. Tanaka, K. Hojo, H. Yoshida, T. Yoshioka, and K. Sugita, *Biochem. Biophys. Res. Commun.* **190,** 732 (1993).

[25] I. E. Collier, P. A. Krasnov, A. Y. Strongin, H. Birkedal-Hansen, and G. I. Goldberg, *J. Biol. Chem.* **267,** 6776 (1992).

[26] E. W. Howard and M. J. Banda, *J. Biol. Chem.* **266,** 17972 (1991).

[27] D. E. Kleiner, E. J. Unsworth, H. C. Krutzsch, and W. G. Stetler-Stevenson, *Biochemistry* **31,** 1665 (1992).

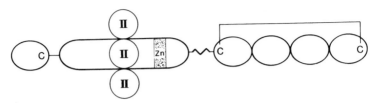

FIG. 2. Domain structure of gelatinases A and B. Analysis of the primary structure of gelatinases A and B in relation to other members of the matrix metalloproteinase (MMP) family shows that the mature proteins consist of four protein domains. The propeptide domain (1) is responsible for the maintenance of latency. Exogenous and endogenous proteolytic cleavages (see Fig. 3) remove this region to reveal the active enzyme forms. The catalytic domain (2) contains the active-site zinc (and a potential structural zinc) as well as stabilizing calcium ligands. Domains 1 and 2 (and the signal peptide) are common to all the MMPs and may be considered as components of a primordial gene. Boundaries between the domains do not always correspond to exon–intron junctions [I. E. Collier, G. A. P. Bruns, G. I. Goldberg, and D. S. Gerhard, *Genomics* 9, 429 (1991); L. Matrisian, *BioEssays* 14, 455 (1992)]. The C-terminal domain (3) is similar to the tandem repeat structure of the hemopexin family and is coded for by a discrete set of exons. The presence of this domain in most of the MMPs confers specific properties on each individual enzyme. In the case of gelatinase A it determines cell binding properties and TIMP interactions. For gelatinase B it has TIMP and collagenase binding properties. The gelatin-binding domain (4) is unique to the gelatinases and is coded for by three separate exons determining three repeats of the type II module defined in fibronectin [I. E. Collier, G. A. P. Bruns, G. I. Goldberg, and D. S. Gerhard, *Genomics* 9, 429 (1991)]. This domain confers the properties of matrix and specific macromolecular substrate binding on the gelatinases. Finally gelatinase B has acquired an additional 54-amino acid, proline-rich extension (5) to the short proline-rich "hinge" region between the catalytic and C-terminal domains of the other MMPs, including gelatinase A. The sequence is related to that of a number of helical collagens.[2]

```
        ▶ start of proenzyme
1) APSPIIKFPG.DVAPKT.DKELAVQYLNTFYGCPKESCNLFVLK........DTLKKMQKFFG  53
2) APRQRQSTLVLFPGDLRTNLTDRQ..LAEEY.LYRYGYTRVAEMRGESKSLGP.ALLLLQKQLS  60

             ▶ start of active enzyme
1) LPQTGDLDQNTIETMRKPRCGNPDVAN YNFFPRKPKWDKNQITYRIIGYTPDLDPETVDDAFARAFQV 121
2) LPETGELDSATLKAMRTPRCGVPDLGR FQTFEGDLKWHHHNITYWIQNYSEDLPRAVIDDAFARAFAL 128


1) WSDVTPLRFSRIHDGEADIMINFGRWEHGDGYPFDGKDGLLAHAFAPGTGVGGDSHFDDDELWTLGEGQ 190
2) WSAVTPLTFTRVYSRDADIVIQFGVAEHGDGYPFDGKDGLLAHAFPPGPGIQGDAHFDDDELWSLGKGV 197


1) VVRVKY.GNADGEYCKFPFLFNGKEYNSCTDTGRSDGFLWCSTTYNFEKDGKYGFCPHE 248
2) VVPTRF.GNADGAACHFPFIFEGRSYSACTTDGRSDGLPWCSTTANYDTDDRFGFCPSE 255
```

gelatin
binding
repeats
```
1) ALFT.MGGNAEGQPCKFPFRFQGTSYDSCTTEGRTDGYRWCGTTEDYDRDKKYGFCPET 306
2) RLYTRD.GNADGKPCQFPFIFQGQSYSACTTDGRSDGYRWCATTANYDRDKLFGFCPTR 313

1) AMSTV.GGNSEGAPCVFPFTFLGNKYESCTSAGRSDGKMWCATTANYDDDRKWGFCPDQ 364
2) ADSTVMGGNSAGELCVFPFTFLGKEYSTCTSEGRGDGRLWCATTSNFDSDKKWGFCPDQ 372
```

```
         zinc binding site                      collagen-like hinge
1) GYSLFLVAAHEFGHAMGLEHSQDPGALMAPIYTYTKNFRLSQDDIKGIQELYG............... 417
2) GYSLFLVAAHEFGHALGLDHSSVPEALMYPMYRFTEGPPLHKDDVNGIRHLYGPRPEPEPRPPTTTTPQ 441


1) .....................................ASPDID..LGT.GPTPTLGPVT.PEI...CK 441
2) PTAPPTVCPTGPPTVHPSERPTAGPTGPPSAGPTGPPTAGP...ST.ATTVPL.......SPV.DDACN 498


1) QDIVFDGIAQIRGE.IFFFKDRFIWRTVTPR.DKPMGPL.LVATFWPELPEKIDAVYEAPQEEKAVFFA 507
2) VNI.FDAIAEI.GNQLYLFKDGKYWRFSEGRGSRPQGP.FLIADKWPALPRKLDSVFEEPLSKKLFFFS 564


1) GNEYWIYS.ASTLERGYPKPLTSLGL..PPDV.QRVDAA.FNWSKNKKTYIFAGDKFWRYNEVKKKMDP 571
2) GRQVWVYTGASVL..G.PRRLDKLGLGA..DVAQ.VTGAL..RSGRGKMLLFSGRRLWRFDVKAQMVDP 624


1) GFPKLIADAWNNAIPDNL.DAVVDLEGGGHSYFFKGAYY.....LKLENQSLKSVKFGSIKSDWLGC   631
2) RSASEVDRMFPGVPLDTHD.VFQYREK..AYFCQDRFYWRVSSRSELNQVDQVGYVTYD...ILQCPED 688
```

Fig. 3. Sequence alignment of progelatinases A and B. The predicted amio acid sequences of progelatinase A (1)[1] and progelatinase B (2)[2] have been aligned to demonstrate the modular arrangement of the enzymes shown in Fig. 2.

progelatinases, TIMPs, and active MMPs have also been cited.[21] The physiological significance of the various complexes is as yet poorly defined.

Deletion mutagenesis and biochemical studies have shown that the C-terminal domain of progelatinase A allows it to bind to the cell membranes of a number of stimulated mesenchymal and tumor cells.[28,29] Such binding may promote a progelatinase A activation mechanism that has not yet been fully characterized.

[28] G. Murphy, F. Willenbrock, R. V. Ward, M. I. Cockett, D. Eaton, and A. J. P. Docherty, *Biochem. J.* **283,** 637 (1992).
[29] A. Y. Strongin, B. L. Marmer, G. A. Grant, and G. I. Goldberg, *J. Biol. Chem.* **268,** 14033 (1993).

## Properties of Activated Gelatinases A and B

Gelatinase A and B activation is accompanied by the loss of amino acids at the N terminus by proteolytic processing. Although this can be accomplished by incubation with APMA, which initiates autocatalytic cleavages, the physiological mechanisms of activation are as yet unclear. Whereas it is possible that *in vivo* progelatinase B is activated by other endopeptidases (e.g., cathepsin G, tissue kallikrein, or plasmin),[11,30,31] progelatinase A is not activated by any of the endopeptidases so far tested.[32] Data demonstrating that progelatinase A, but not progelatinase B, can be activated by fibroblast cell membranes have led to the proposal that *in vivo* the former might be activated by a specific cell surface event.[28,29,33,34] When exposed to APMA, progelatinase A cleavage occurs at the Asn[80]-Tyr[81] peptide bond (Fig. 4)[32,35] which, by sequence homology alignment, is equivalent to the final autolytic cleavage site in the activation of the other MMPs. It has been proposed that the aromatic residue that forms the N terminus of each of the activated MMPs is essential to the expression of full activity.[36] The Asn[80]-Tyr[81] bond can be cleaved by active gelatinase A in the absence of APMA, but the rate of reaction is very slow.[37] The presence of the organomercurial greatly accelerates the reaction, presumably by causing conformational alterations to the proenzyme that lead to autolytic cleavages at sites upstream of the Asn[80]-Tyr[81] bond.[38] In the absence of a complete propeptide, the susceptibility of the Asn[80]-Tyr[81] bond to autolysis is increased. A similar mechanism seems to occur during cell membrane-induced processing. A progelatinase A intermediate cleaved at Asn[37]-Leu[38] has been identified, which is followed by an Asn[80]-Tyr[81] cleavage (Fig. 4).[29]

Activation of purified gelatinase A is followed by numerous slow secondary autolytic cleavages that give rise to an $M_r$ 42,000 active species that

[30] Y. Okada, Y. Gonoji, K. Naka, K. Tomita, I. Nakanishi, K. Iwata, K. Yamashita, and T. Hayakawa, *J. Biol. Chem.* **267,** 21712 (1992).

[31] S. Desrivières, H. Lu, N. Peyri, C. Soria, Y. Legrand, and S. Menashi, *J. Cell. Physiol.* **157,** 587 (1993).

[32] Y. Okada, T. Morodomi, J. J. Enghild, K. Suzuki, A. Yasui, I. Nakanishi, G. Salvesen, and H. Nagase, *Eur. J. Biochem.* **194,** 721 (1990).

[33] P. D. Brown, D. E. Kleiner, E. J. Unsworth, and W. G. Stetler-Stevenson, *Kidney Int.* **43,** 163 (1993).

[34] R. V. Ward, S. J. Atkinson, P. M. Slocombe, A. J. P. Docherty, J. J. Reynolds, and G. Murphy, *Biochim. Biophys. Acta* **1079,** 242 (1991).

[35] W. G. Stetler-Stevenson, H. C. Krutzsch, M. P. Wacher, I. M. K. Margulies, and L. A. Liotta, *J. Biol. Chem.* **264,** 1353 (1989).

[36] K. Suzuki, J. J. Enghild, T. Morodomi, G. Salvesen, and H. Nagase, *Biochemistry* **29,** 10261 (1990).

[37] T. Crabbe, C. Ioannou, and A. J. P. Docherty, *Eur. J. Biochem.* **218,** 431 (1993).

[38] T. Crabbe and C. Sutton, unpublished (1992).

**Progelatinase A**

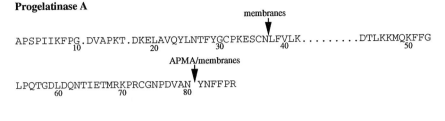

**Progelatinase B**

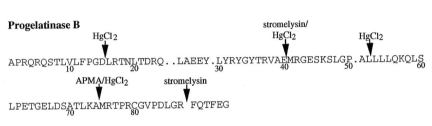

Fig. 4. Propeptide cleavages occurring during the activation processing of progelatinases A and B. Progelatinase A has been shown to be processed autolytically in the presence of 4-aminophenylmercuric acetate (APMA). Although transient intermediate cleavages occur, only the final cleavage has been defined.[32,35] In the presence of some cells or membrane preparations derived from them, progelatinase A activation occurs with two definable propeptide cleavages.[29] Progelatinase B is also autolytically processed in the presence of HgCl₂ or APMA with definable intermediates [Ref. 30; S. Triebel, J. Bläser, H. Reinke, V. Knäuper, and H. Tschesche, *FEBS Lett.* **298**, 280 (1992)]. Stromelysin can be shown to cleave at two sites in the propeptide to yield active enzyme.[10,11]

has lost the noncatalytic C-terminal domain and several inactive breakdown products. Most conspicuous among the latter is an $M_r$ 32,000 fragment that has been identified as the gelatinase A C-terminal domain.[26] With time, gelatinase A activity is lost, and the $M_r$ 31,000 fragment builds up in the absence of the $M_r$ 42,000 active species. It would appear, therefore, that a series of cleavages within the N-terminal catalytic domain eventually causes the loss of gelatinase A activity.

In contrast to the case of gelatinase A, the APMA-induced (or HgCl₂) activation of gelatinase B gives rise to an active species that retains part of the propeptide (Figs. 1 and 4). This includes the PRCGVP sequence thought to confer latency on the proforms of the MMPs.[39] A number of autolytic cleavages occur transiently (Fig. 4), and the final N-terminal autolytic cleavage occurs at Ala⁷⁴-Met⁷⁵, leaving 13 amino acids of the propeptide still attached.[30,40] These can be cleaved away by stromelysin 1[10]

[39] E. B. Springman, E. L. Angleton, H. Birkedal-Hansen, and H. E. Van Wart, *Proc. Natl. Acad. Sci. U.S.A.* **87**, 364 (1990).
[40] S. Triebel, J. Bläser, H. Reinke, V. Knäuper, and H. Tschesche, *FEBS Lett.* **298**, 280 (1992).

TABLE I
RELATIVE ACTIVITIES OF GELATINASES A AND B AGAINST MACROMOLECULAR
AND PEPTIDE SUBSTRATES[a]

| Substance | Gelatinase A | Gelatinase B |
|---|---|---|
| Type I gelatin ($\mu$g/min per nmol) | 544[b] | 695[b] |
| Elastin ($\mu$g/hr per nmol) | 270[b] | 88[b] |
| Type IV collagen ($\mu$g/hr per nmol) | 270[b] | 499[b] |
| Aggrecan ($\mu$g chondroitin sulfate/hr/nmol) | 1.02[b] | 0.45[b] |
| Mca-Pro-Leu-Gly-Leu-Dpa-Ala-Arg ($k_{cat}/K_m$, $M^{-1}$ sec$^{-1}$ $\times$ 10$^{-6}$) | 3.2[c] | 0.89[d] |

[a] The assay conditions are described in the references.
[b] G. Murphy, M. I. Cockett, R. V. Ward, and A. J. P. Docherty, *Biochem. J.* **277**, 277 (1991).
[c] Murphy *et al.*[28]
[d] O'Connell *et al.*[11]

(Fig. 1), suggesting that gelatinase B is unable to cleave the Arg$^{87}$-Phe$^{88}$ peptide bond that defines the end of the propeptide. In the presence of mercurials, considerable C-terminal self-processing of gelatinase B accompanies the N-terminal cleavages. This results in a fall in $M_r$ to about 65,000 (cleavage of Ala$^{398}$-Leu$^{399}$).[40] The active forms of gelatinase B generated by APMA or stromelysin 1 have the same catalytic efficiency against a peptide or a gelatin substrate in agreement with the deletion mutagenesis studies of the gelatinase B C-terminal domain.[11]

The two gelatinases have similar substrate specificities (Table I). In addition to their gelatinolytic activities, which appear to be identical for type I gelatin,[18] they are also both able to degrade native collagens of types IV and V, and the extent and approximate sites of cleavage catalyzed by the two gelatinases appear to be identical. Cleavage of type IV collagen has been of particular interest since the native protein is specifically cleaved at a single site into approximately one-quarter and three-quarter fragments (Fig. 5).[41,42] The precise site of cleavage has not been identified. The rate of cleavage of solubilized native type IV collagen (25°–30°) by the gelatinases is very low, but it is markedly accelerated by incubation at 37°, when the collagen structure is less stable. Gelatinases A and B are both potent elastin-degrading enzymes.[43,44] Gelatinase A is additionally able to degrade

[41] L. I. Fessler, K. G. Duncan, J. H. Fessler, T. Salo, and K. Tryggvason, *J. Biol. Chem.* **259**, 9783 (1984).
[42] G. Murphy, R. Ward, R. M. Hembry, J. J. Reynolds, K. Kuhn, and K. Tryggvason, *Biochem. J.* **258**, 463 (1989).
[43] G. Murphy, M. I. Cockett, R. V. Ward, and A. J. P. Docherty, *Biochem. J.* **277**, 277 (1991).
[44] R. M. Senior, G. L. Griffin, C. J. Fliszar, S. D. Shapiro, G. I. Goldberg, and H. G. Welgus, *J. Biol. Chem.* **266**, 7870 (1991).

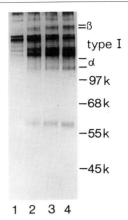

Fig. 5. Cleavage of soluble EHS type IV collagen by gelatinase A. Type IV collagen prepared from Engelbreth-Holm–Swarm (EHS) tumor [H. K. Kleinman, M. L. McGarvey, L. A. Liotta, P. G. Robey, K. Tryggrason, and G. R. Martin, *Biochemistry* **21,** 6188 (1982)] was incubated with active gelatinase A at 30° for 12 hr, and the products were analyzed by electrophoresis on a reduced SDS–polyacrylamide gel (10%) and visualized by silver staining. Substrates were EHS type IV collagen alone (lane 1), collagen plus 0.001 μg of gelatinase A (lane 2), collagen plus 0.0025 μg of gelatinase A (lane 3), and collagen plus 0.005 μg of gelatinase A (lane 4). The pattern of gelatinase B cleavage appears to be very similar (not shown).[30,42] Molecular weights of standard proteins and type I collagen are shown at the right-hand side of the gel.

collagen types VII and XI, aggrecan, fibronectin, laminin A and B chains, and myelin basic protein, whereas gelatinase B has been shown to degrade type I, III, and XI collagens, aggrecan, laminin A chain, and myelin basic protein. The pH optimum of gelatinase A has been determined against Azocoll[32] and a synthetic substrate[45] and found to be pH 8.5 and pH 7.0, respectively.

A study on synthetic peptide substrates has shown that the two gelatinases have a similar site of action specificity.[46] Both prefer a small amino acid (e.g., Gly or Ala) at the P1 site and an aliphatic or hydrophobic residue at P1' site. Another study investigating denatured type I collagen cleavage indicated that gelatinase A might require a hydroxyproline residue 5 amino acids either upstream or downstream of the cleavage site.[47] Minor variations in the substrate specificity of gelatinases A and B suggest that they have different subsequent subsite preferences.

[45] P. Hynds, unpublished (1992).
[46] S. Netzel-Arnett, Q.-X. Sang, W. G. I. Moore, M. Navre, H. Birkedal-Hansen, and H. E. Van Wart, *Biochemistry* **32,** 6427 (1993).
[47] J. L. Seltzer, K. T. Akers, H. Weingarten, G. A. Grant, D. W. McCourt, and A. Z. Eisen, *J. Biol. Chem.* **265,** 20409 (1990).

The particular ability of these two matrix metalloproteinases to degrade denatured collagen rapidly has been assigned to their possession of the three type II repeats of the gelatin-binding region of fibronectin. Expression of the individual repeats from gelatinase B have shown that one of the modules has significantly more gelatin-binding ability than the other two.[25] It was noted that 10% DMSO, which prevents progelatinase A or B binding to gelatin-Sepharose, had little effect on the gelatinolytic activity of active gelatinase B. In contrast, active gelatinase A is 70% inhibited by 10% DMSO using gelatin or peptide substrate assays.[48] In accordance with this, a deletion mutant of gelatinase A that lacks this domain was only weakly active on gelatin but retained full activity against a synthetic substrate.[49]

The activity of both gelatinases A and B is inhibited by the classic inhibitors of zinc-dependent metalloenzymes, for example, $o$-phenanthroline. EDTA also inhibits activity, but it is not certain whether it acts by removing the $Zn^{2+}$ that participates in the hydrolytic reaction mechanism or the more loosely bound $Ca^{2+}$ that is required to maintain a catalytically active conformation. Neither enzyme is inhibited by inhibitors specific to serine or cysteine proteinases. Both gelatinases are inhibited by the general protease inhibitor $\alpha_2$-macroglobulin and the specific inhibitors of metalloproteinases, TIMP-1 and TIMP-2. The TIMPs form 1:1 stoichiometric complexes with the enzymes and appear to be tight-binding competitive inhibitors, inhibiting with $K_i$ values in the picomolar range.[11,50,51] In the case of gelatinase A, TIMP interactions occur with both the active site in the N-terminal domain and sites on the enzyme C-terminal domain. TIMP-2 specifically interacts tightly with the C-terminal domain of both pro- and active gelatinase A.[26,27,50] Active gelatinase B binds more rapidly to TIMP-1 than to TIMP-2, largely owing to C-terminal domain interactions which may be similar to those determining the specific ability of TIMP-1 to bind to progelatinase B. The final active gelatinase B-TIMP complexes involve binding of the N-terminal domains as well as the C-terminal domains of each protein[11] (see [30], this volume).

[48] G. Murphy, unpublished (1993).
[49] G. Murphy, Q. Nguyen, M. I. Cockett, S. J. Atkinson, J. A. Allan, G. C. Knight, F. Willenbrock and A. J. P. Docherty, *J. Biol. Chem.* **269**, 6632 (1994).
[50] F. Willenbrock, T. Crabbe, P. M. Slocombe, C. W. Sutton, A. J. P. Docherty, M. I. Cockett, M. O'Shea, K. Brocklehurst, I. R. Phillips, and G. Murphy, *Biochemistry* **32**, 4330 (1993).
[51] G. Murphy and F. Willenbrock, this volume, [30].

# [29] Matrilysin

### By J. Frederick Woessner, Jr.

## Introduction

Matrilysin is distinguished from the other matrix metalloproteinases by its small size: $M_r$ 28,000 for the latent form and $M_r$ 19,000 for the active form. The small size is due to the absence of the hemopexin-like domain found in all other members of the matrixin family.[1] The enzyme was first described in 1980 as a metalloproteinase found in the involuting uterus of the rat[2]; however, only modest purification was achieved, and characterization was not sufficient to permit unambiguous distinction from related enzymes such as stromelysin. It was not until 1988 that the enzyme was fully purified and characterized in more detail.[3] It was referred to as uterine metalloproteinase or ump. In that same year, a human cDNA clone of corresponding size was expressed in COS cells and shown to have similar properties[4]; the human enzyme was named putative (later punctuated) metalloproteinase 1 or pump-1. Because the enzyme is neither putative nor limited to the uterus, the Enzyme Commission name matrilysin (EC 3.4.24.23) is to be preferred. The shorthand designation MMP-7 (matrix metalloproteinase 7) is in common use.

## Assay Procedures

Impure preparations of matrilysin are readily assayed by digestion of Azocoll[3] or transferrin; however, those substrates are digested by a wide variety of proteinases. Pure enzyme may be assayed by these methods as well, but the fluorescent peptide assay is more sensitive and rapid. Immunoassays have not yet been published for matrilysin; in any event, such assays will not quantitate the active form.

### Assay with Azocoll

*Principle.* Azocoll is a general proteinase substrate prepared by diazotization of finely ground dried cowhide. Although the major constituent of hide is collagen, that protein becomes denatured so that nonspecific en-

---

[1] J. F. Woessner, Jr., *FASEB J.* **5,** 2145 (1991).
[2] A. Sellers and J. F. Woessner, Jr., *Biochem. J.* **189,** 521 (1980).
[3] J. F. Woessner, Jr., and C. J. Taplin, *J. Biol. Chem.* **263,** 16918 (1988).
[4] B. Quantin, G. Murphy, and R. Breathnach, *Biochemistry* **28,** 5327 (1989).

zymes attack it. The released fragments are soluble (although of a large size that barely enter polyacrylamide gels). After incubation the colored material is separated from the insoluble particles of hide that settle to the bottom of the incubation tube. Unfortunately, it is difficult to find good commercial preparations of Azocoll. Calbiochem/Novabiochem (La Jolla, CA) supplies a preparation of 100 mesh, but all batches are not equal. The color production by 0.8 mg/ml of Azocoll, completely digested by trypsin, should be 0.5–0.7 at 520 nm. In the assay to be described 20 ng rat matrilysin should give an absorbance of about 0.24 after 18 hr of digestion. It may be necessary to increase the substrate concentration for batches that do not meet these specifications.

*Reagents*

Assay buffer: 50 m$M$ Tris-HCl, 0.2 $M$ NaCl, 10 m$M$ CaCl$_2$, 0.02% NaN$_3$ (w/v), 0.05% Brij 35 (w/v), adjusted to pH 7.5

Substrate: Azocoll is suspended at 4 mg/ml in buffer and kept briskly stirring in a small beaker; a pipette tip with the end cut off sufficient to pass the Azocoll particles is used for dispensing

Activator: Aminophenylmercuric acetate (APMA) is dissolved in 0.1 $M$ NaOH to a final concentration of 50 m$M$

Inhibitor: Dissolve 1,10-phenanthroline in 95% (v/v) ethanol to make 0.1 $M$

*Procedure.* Add enzyme preparation (2–20 ng, 0.3–3 units, in assay buffer) to a 12 × 75 mm polypropylene Chemtube (Bio-Rad, Richmond, CA). Add buffer to a final volume of 1 ml. These tubes will be used to measure active enzyme. For total enzyme activity, add 12.5 $\mu$l APMA to each tube for crude enzymes or 1 $\mu$l for pure enzyme. Blanks are prepared by the addition of 12.5 $\mu$l of phenanthroline. Each sample should be prepared in duplicate. In the final step, 0.25 ml of suspended Azocoll is added to give a final volume of 1.25 ml. The tubes are capped with plug-type closures (USA/Scientific Plastics, Ocala, FL) and laid flat on a glass or metal plate. The rows are anchored by passing labeling tape over the tubes and around the plate. The plate is then placed in a Dubnoff metabolic shaker bath with the tube axes parallel to the direction of shaking. Incubation is continued for 18 hr at 37°. The tubes are removed from the bath, placed upright and centrifuged briefly at low speed. The absorbance of the supernatant is then read in a spectrophotometer at 520 nm.

Tubes with enzyme alone measure active enzyme; tubes with added APMA measure total activity. Subtraction of the first from the second measures latent enzyme. In each case, the phenanthroline blank must first be subtracted. This corrects not only for spontaneous release of dye from substrate but also for nonspecific digestion by any proteinase not in the metalloproteinase class. A unit is defined as the digestion of 1 μg Azocoll/min; this is based on the color yield of trypsin-digested Azocoll. Pure rat matrilysin has a specific activity of 17,000 μg Azocoll digested/min/mg zymogen weight.

*Comment.* The long incubation time is selected because it permits assays that begin at the end of the day (useful during purification steps) and because of the sensitivity achieved. If a rapid assay is needed, the transferrin method should be used (see below). The assay is a simple colorimetric method requiring only common equipment. It is important that the Azocoll remain in suspension during pipetting from the stock and during incubation; avoid glassware in which there is clumping at the liquid surface or on the walls. The assay is not perfectly linear with enzyme concentration but shows some falling off at higher levels. It is best not to exceed absorbancies of 0.25; below this point a calibration curve can be constructed for greatest accuracy, but in practice this is rarely necessary. If the time of assay is reduced (by increasing the enzyme levels) it may become necessary to activate the enzyme with APMA before adding the substrate; in the 18-hr assay this is not necessary. For rapid activation, APMA should be used at 1 m$M$ concentration.

With crude enzyme preparations there may be many other enzymes that cleave Azocoll. Members of the cysteine class will be blocked by APMA, and only the metalloproteinases will be blocked by phenanthroline. Among the metalloproteinases, interference is to be expected mainly from other members of the matrix metalloproteinase family such as gelatinase. Although it is likely that specific gelatinase inhibitors will come on the market shortly, at the present time one can distinguish matrilysin from gelatinases only by passing them through a molecular sieve column such as Ultrogel AcA 54 to separate the very small matrilysin from the others. The presence of matrilysin can be further verified by zymography on transferrin gels (see below).

## Assay with Transferrin

*Principle.* The transferrin assay method is presented in detail by Nagase.[5] Transferrin has a high content of disulfide bonds. These are cleaved by reduction followed by carboxymethylation with [$^3$H]iodoacetic acid to give a completely denatured protein of high specific activity. Following digestion by proteinase, the undigested transferrin is precipitated with trichloroacetic acid and the supernatant radioactivity is determined.

[5] H. Nagase, this volume [27].

*Reagents*

Buffer: Tris buffer as described for Azocoll assay above

Substrate: Tritiated transferrin is prepared as described,[5] and its specific activity is typically 0.1 $\mu$Ci/mg; the product is suspended in buffer at a final concentration of 2.5 mg/ml

Activator: APMA is dissolved at a level of 15 m$M$ in 0.1 $N$ NaOH

Inhibitor: Prepare a 0.2 $M$ solution of sodium EDTA in assay buffer and adjust to pH 7.8 (to allow for a decrease when mixed with CaCl$_2$)

Precipitant: 5% (w/v) aqueous solution of trichloroacetic acid (TCA)

*Procedure.* To a microcentrifuge tube add 20 $\mu$l (50 $\mu$g) transferrin solution, 20 $\mu$l enzyme preparation (in buffer, 0–60 ng or 0–0.06 unit), and buffer to a total of 60 $\mu$l. To activate latent enzyme add 2 $\mu$l APMA, and to prepare blanks add 9 $\mu$l EDTA. Incubate for 2 hr at 37°. Stop the reaction by the addition of 400 $\mu$l TCA and hold the tubes in an ice–water bath for 10 min to allow precipitates to form. Centrifuge for 4 min in a benchtop microcentrifuge. Remove 150 $\mu$l supernatant to a scintillation vial (7 ml) and add 5 ml scintillant (Cytoscint, ICN, Irvine, CA). Mix and count for 1 min. Determine the extent of digestion by comparison to standards of transferrin added directly to the scintillation fluid. Do not exceed 30% digestion of the substrate. One unit of enzyme digests 1 $\mu$g of transferrin/ min. The specific activity of pure rat matrilysin is 1150 units/mg zymogen weight.

*Comment.* The method is quite sensitive, but the TCA precipitation step is readily perturbed by various substances that might be in the enzyme preparation, such as salts, buffers, and proteins. If the method is applied to crude preparations it is advisable to incubate blanks containing only transferrin and buffer, then add the enzyme after the incubation and immediately before the TCA. In our experience, several peaks of "enzyme activity" found by molecular sieving disappeared after this treatment. The comments on crude enzymes under the Azocoll assay are also pertinent here.

*Assay with Fluorescent Mca-Peptide*

Knight *et al.*[6] introduced the use of (7-methoxycoumarin-4-yl)acetyl (Mca)-Pro-Leu-Gly-Leu-[3-(2,4-dinitrophenyl)-L-2,3-diaminopropionyl]- Ala-Arg-NH$_2$ as a general substrate for the assay of matrix metalloproteinases. The principles are discussed by Knight.[7] The original method employed continuous recording of fluorescence, but the present method calls for sampling at short intervals.

[6] C. G. Knight, F. Willenbrock, and G. Murphy, *FEBS Lett.* **296,** 263 (1992).
[7] C. G. Knight, this volume [2].

*Reagents*

Assay buffer: As for the Azocoll assay

Substrate: The Mca peptide ($M_r$ 1093) is purchased from Bachem Bioscience, Inc. (Philadelphia, PA). It is dissolved in 15% (v/v) dimethyl sulfoxide (DMSO) at a concentration of 53 $\mu M$ and stored in aliquots at $-80°$. The peptide is light sensitive, and appropriate precautions must be taken in the preparation of stock solutions, storage, and assay

Stop solution: 0.1 $M$ sodium acetate buffer, pH 4.0

Standard: 7-Methoxycoumarin-4-acetic acid is dissolved in 2 $\mu M$ Mca peptide solution in buffer at a final concentration of 200 n$M$. Aliquots of 70 $\mu$l are then mixed with 0.75 ml stop solution for fluorescence determination

*Procedure.* Enzyme preparation (2–20 ng) plus 24 $\mu$l substrate stock (2 $\mu M$ final concentration) is made to a total volume of 0.63 ml with buffer (sufficient for 9 aliquots). The mixture is incubated at 37°, and duplicate aliquots of 70 $\mu$l are removed at 0, 5, 10, and 20 min and placed into tubes containing 0.75 ml of stop solution. Fluorescence is then determined in a 1-ml cuvette using an excitation wavelength of 328 nm and emission at 393 nm. The substrate and buffer alone are used to zero the instrument, and the standard is used to set the instrument to 0.8 of the maximum range. This amount of fluorescence is equivalent to 10% digestion of the substrate (14 pmol). Use of the standard permits calculation of the picomoles of peptide cleaved per minute (equals 1 unit). Rat matrilysin cleaves 390 nmol/min/mg zymogen.

*Comment.* The assay is very sensitive and rapid. It requires 1/10 as much enzyme as the Azocoll assay and 1/50 the time. However, the peptide is an excellent substrate for other matrixins, particularly for gelatinase A. Therefore, the assay is suitable for purified matrilysin but not for crude extracts or culture media containing other matrixins. The substrate concentration is about 1/20 of the $K_m$ value; therefore, the initial rate data may also be used to calculate $k_{cat}/K_m$.[7] A comparison of rat matrilysin and human recombinant matrilysin as assayed by the three methods is shown in Table I.

Zymography

Matrilysin preparations can be examined for enzyme activity by zymography. The details for gelatin zymography are given elsewhere in this volume.[8] Gelatin is not a suitable substrate for matrilysin. However, if one prepares gels in the same manner as described, but substituting 0.4 mg/

---

[8] J. F. Woessner, Jr., this volume [31].

TABLE I
COMPARISON OF HUMAN AND RAT MATRILYSIN ACTIVITIES

| | | Matrilysin | |
| --- | --- | --- | --- |
| Substrate | Unit | Rat | Human |
| Azocoll | μg/min/mg enzyme[a] | 17,000 | 8000 |
| CM-transferrin | μg/min/mg enzyme | 1150 | 1900 |
| Mca-peptide | nmol/min/mg enzyme | 390 | 1160 |

[a] Enzyme protein content is based on zymogen of $M_r$ 28,000. The protein content for rat matrilysin was estimated by titration with TIMP-1.

ml carboxymethylated transferrin for the gelatin, excellent results can be obtained. The transferrin is prepared as described,[5] but nonradioactive iodoacetic acid is used for carboxymethylation of the reduced protein. The method can detect 1–30 ng of matrilysin. Prestained markers (GIBCO–BRL, Gaithersburg, MD) permit identification of the $M_r$ 28,000 zymogen and the $M_r$ 19,000 active form. Intermediate activation forms of 21–24 kDa may also be detected. The method is good for detecting matrilysin in crude extracts. One must keep in mind that stromelysin and macrophage elastase may also appear. Both should show bands at 54 kDa, with elastase showing a band of active enzyme at 18 kDa and stromelysin showing bands of active enzyme at 48, 28, and 24 kDa. Crude extracts will often contain serine proteinases in this region so that gels must be incubated with 30 m$M$ EDTA to prove that a given band is due to a metal-dependent proteinase.

## Purification of Matrilysin from Rat Uterus

The only example of matrilysin purification from a tissue is that of the rat uterine enzyme. The drawbacks to this purification are that the starting material is expensive and the yields are only a few micrograms. On the other hand, no rodent cell lines have yet been reported to produce matrilysin. The method given follows Ref. 3 closely with some modifications. (All steps are carried out in the cold unless otherwise noted.)

*Extraction.* Pregnant rats of the Sprague-Dawley strain are commercially available from several suppliers. At 1 day postpartum the uterus is dissected free of mesentery and cervix, rinsed, minced, and homogenized in 10 volumes of 10 m$M$ CaCl$_2$ and 0.25% (w/v) Triton X-100 in glass homogenizers. A total of 12 g of tissue (about 9 uteri) is used. The homogenate is centrifuged at 6000 $g$ for 30 min. The precipitate is rehomogenized in 10 volumes of 50 m$M$ Tris-HCl, pH 7.5, containing 0.1 $M$ CaCl$_2$ and 0.15 $M$ NaCl. The suspension is placed in metal centrifuge tubes and heated to 60° for 4 min.

After centrifugation in the cold at 17,500 $g$ for 30 min, the supernatant is retained for enzyme purification.

*Chromatographies.* The supernatant is first concentrated in an ice bath by slowly adding solid $(NH_4)_2SO_4$ to 25% saturation and then to 75% saturation, with centrifugation at 20,000 $g$ for 40 min after each addition. The second pellet is resuspended in 40 ml of assay buffer (see Azocoll assay above). This is applied to a 5 × 50 cm column of Ultrogel AcA 54 molecular sieve resin (Sepracor, Marlborough, MA) equilibrated in the same buffer. The latent enzyme emerges at about 680 ml, and the volume between 630 and 730 ml is collected. The exact position is determined by assay with Azocoll and APMA activation. This position is close to that of a chymotrypsinogen marker. The active enzyme emerges later but is discarded because it is superimposed on a peak of mast cell chymase which complicates its purification. Gelatinases emerge near the void volume and are discarded.

The enzyme is next applied to a 1.5 ml column of Blue Sepharose (Pharmacia, Piscataway, NJ) and washed with an additional 30 ml of assay buffer. The enzyme is then eluted with a linear gradient formed by 30 ml of 0.2 $M$ and 30 ml of 2.0 $M$ NaCl in assay buffer. The enzyme emerges at about 0.8 $M$ salt. At this stage the enzyme is about 4–5% pure. However, the preparation is usually free of any detectable contaminating proteinases and is suitable for many routine purposes.

*Final Steps.* Two further steps are required to produce homogeneous enzyme. First, a 1.5-ml Chelating-Sepharose column (Pharmacia) is prepared and charged with zinc according to the manufacturer's instructions. It is then equilibrated with 25 m$M$ boric acid, 1 m$M$ $CaCl_2$, 0.05% Brij 35, and 0.03% (v/v) toluene brought to pH 8.0 with NaOH. The eluate from Blue Sepharose is dialyzed against this borate buffer and applied to the column. The column is then washed with 25 m$M$ sodium cacodylate buffer, pH 6.5, and the enzyme is eluted with a linear gradient of 30 ml cacodylate buffer and 30 ml of this buffer containing 0.5 $M$ NaCl. The enzyme emerges at about 50 m$M$ NaCl. In the final step, a smaller 1.6 × 80 cm column of Ultrogel AcA 54 is used; it is equilibrated as described above. However, the enzyme is dialyzed into a buffer containing 25 m$M$ Tris-HCl, pH 7.5, 5 m$M$ $CaCl_2$, 0.02% $NaN_3$, and 0.05% Brij 35. Under these conditions, the enzyme interacts with the column and emerges later than normal ($K_{av} = 0.65$). The purification is summarized in Table II.

*Product.* The final product is homogeneous and shows a single band at $M_r$ 28,000 on sodium dodecyl sulfate–polyacrylamide gel electrophoresis (SDS–PAGE) and silver staining. A minor band at $M_r$ 19,000 may also appear owing to partial activation of the latent enzyme. It should be noted that there is no adequate method for the measurement of protein in these

TABLE II
PURIFICATION OF MATRILYSIN FROM RAT UTERUS

| Procedure | Protein | Units | Units/mg protein | Yield (%) | Purification (-fold) |
|---|---|---|---|---|---|
| Crude heat extract | 260 mg | 1170 | 4.5 | 100 | 1 |
| Ammonium sulfate precipitate | 110 mg | 1030 | 9.4 | 88 | 2 |
| Ultrogel AcA 54 | 16 mg | 360 | 22.5 | 31 | 5 |
| Blue Sepharose | 420 μg | 320 | 760 | 27 | 170 |
| Zinc chelate column | (100 μg)[a] | 185 | (1850) | 16 | (410) |
| Ultrogel AcA 5 | 3.3 μg | 51 | 17,000 | 4 | 3780 |

[a] Protein content is only approximate.

last two steps. In the original report,[3] protein was estimated by the intensity of silver staining. More recently an estimate of final specific activity has been based on a titration with known amounts of TIMP-1 (tissue inhibitor of metalloproteinase), which is assumed to bind in 1:1 ratio. This indicates that the overall yield is about 3 μg of enzyme with a specific activity of 17,000 Azocoll units/mg.

## Preparation of Matrilysin by Other Methods

### Purification from Cell Culture Media

Human matrilysin has been purified from cell culture in two instances. Miyazaki et al.[9] purified enzyme secreted by a human rectal carcinoma cell line CaR-1. Eight liters of conditioned media were concentrated by ammonium sulfate and passed through a Cellulofine GCL2000 molecular sieve column (Chisso Ltd., Tokyo). The active fractions were subjected to anion-exchange high-performance liquid chromatography (HPLC) on a Shodex QA-824 column (Showa Denko Ltd., Tokyo), followed by molecular sieve HPLC on TSK gel G3000SW (Tosoh Ltd., Tokyo). Overall purification was 110-fold with a yield of 0.7 mg. A purification table was not presented.

Human matrilysin has also been purified from cultured human kidney cells by Marcotte et al.[10] Cultures were maintained for 21 days without medium change. The medium was purified by S-Sepharose (Pharmacia) chromatography, salt-gradient elution from Cibacron Blue-agarose (Sigma), and reversed-phase HPLC on C$_4$ (Vydac, Hesperia, CA) with

[9] K. Miyazaki, K. Funahashi, Y. Numata, N. Koshikawa, K. Akaogi, Y. Kikkawa, H. Yasumitsu, and M. Umeda, *J. Biol. Chem.* **268**, 14387 (1993).

[10] P. S. Marcotte, I. M. Kozan, S. A. Dorwin, and J. M. Ryan, *J. Biol. Chem.* **267,** 13803 (1992).

acetonitrile gradient. The extent of purification is not discussed, but the final product is homogeneous.

*Purification Following Expression of Recombinant cDNA*

The classic preparation of matrilysin was detailed by Quantin *et al.*[4] COS cells were transfected with a partial cDNA for matrilysin in the pPROTA vector. The secreted product was a fusion protein between staphylococcal protein A and matrilysin. Purification was achieved by interaction between the expressed immunoglobulin G (IgG)-binding domain of staphylococcal protein A and an IgG-Sepharose column. On addition of APMA, the enzyme underwent autodigestion to release the active form of matrilysin.

More recently Murphy *et al.*[11] have expressed the cDNA in NSO mouse myeloma cells using the pEE12 vector (by a proprietary method). The proenzyme was purified from the medium by binding to S-Sepharose and elution in 0.2 $M$ NaCl, followed by Sephacryl S200 (Pharmacia) chromatography.

Expression of human matrilysin in *Escherichia coli* has been achieved by Ye *et al.*[12]; however, the enzyme appears in the inclusion bodies of the cell and has not been successfully refolded.

Physical Properties of Matrilysin

The rat and human matrilysin preparations are quite stable. Enzyme in the tissue can be stored for years at $-20°$. Purified enzyme may be kept for months in the refrigerator in latent or active form. The latent form tends to gradually become active on prolonged storage.

The $M_r$ for the latent proenzyme is 27,772 for the rat and 27,938 for the human; $M_r$ for the active form is 18,934 for the rat and 19,133 for the human. These values are based on the amino acid sequences deduced from the respective cDNA sequences.[4,13] Both enzymes have Leu-Pro-Leu at the amino terminus, and the active forms have Tyr-Ser-Leu (human)[14] and Phe-Ser-Leu (rat).[13]

Activation of matrilysin can be accomplished by the addition of APMA or by cleavage with trypsin or other proteinases. The enzyme is sensitive to APMA, so, in an overnight assay with Azocoll, the APMA concentration should be reduced to 50 $\mu M$ or less. With crude enzyme, 0.5 m$M$ APMA may be required.

[11] G. Murphy, M. I. Cockett, R. V. Ward, and A. J. P. Docherty, *Biochem. J.* **277**, 277 (1991).
[12] Q-Z. Ye, L. L. Johnson, and V. Baragi, *Biochem. Biophys. Res. Commun.* **186**, 143 (1992).
[13] S. R. Abramson, G. E. Conner, H. Nagase, I. Neuhaus, and J. F. Woessner, Jr., *J. Biol. Chem.* submitted
[14] T. Crabbe, F. Willenbrock, D. Eaton, P. Hynds, A. F. Carne, G. Murphy, and A. J. P. Docherty, *Biochemistry* **31**, 8500 (1992).

TABLE III
SPECIFICITY OF MATRILYSINS

| Substrate | P5 | P4 | P3 | P2 | P1 | | P1' | P2' | P3' | P4' | P5' | Ref. |
|---|---|---|---|---|---|---|---|---|---|---|---|---|
| Rat matrilysin | | | | | | | | | | | | |
| Insulin B chain | His | Leu | Val | Glu | Ala | + | **Leu** | Tyr | Leu | Val | Cys-19 | a |
| Insulin B chain | Val | Glu | Ala | Leu | Tyr | + | **Leu** | Val | Cys | Gly | Glu-21 | a |
| Collagenase peptide | | Dnp | Pro | Leu | Gly | + | **Ile** | Ala | Gly | D-Arg | NH₂ | a |
| Self-activation | Pro | Asp | Val | Ala | Glu | + | **Phe** | Ser | Leu | Met | Pro-82 | b |
| α2(I) chain | Pro | Gly | Pro | Ser | Gly | + | **Ile** | Thr | Gly | Pro | Hyp-717 | b |
| α2(I) chain | Ala | Gly | Phe | Gln | Gly | + | **Leu** | Leu | Gly | Ala | Hyp-780 | b |
| α2(I) chain | Pro | Gly | Pro | Leu | Gly | + | **Ile** | Ala | Gly | Pro | Hyp-813 | b |
| Human matrilysin | | | | | | | | | | | | |
| Self-activation | Pro | Asp | Val | Ala | Glu | + | **Tyr** | Ser | Leu | Phe | Pro-82 | c |
| Mca-peptide | | Mca | Pro | Leu | Gly | + | **Leu** | Dpa | Ala | Arg | NH₂ | d |
| Aggrecan | Phe | Thr | Ser | Glu | Asp | + | **Leu** | Val | Val | Gln | Val-446 | e |
| Aggrecan | Asp | Ile | Asp | Glu | Asn | + | **Phe** | Phe | Gly | Val | Gly-346 | e |
| Prourokinase | Ser | Pro | Pro | Glu | Glu | + | **Leu** | Lys | Phe | Gln | Cys-148 | f |
| Entactin | Glu | Leu | Ile | Gly | Glu | + | **Leu** | Ser | Phe | Tyr | Asp-39 | g |
| Entactin | Thr | Ser | His | Leu | Gly | + | **Leu** | Glu | Asp | Val | Ala-280 | g |
| Entactin | Gly | Ser | Pro | Asp | Ala | + | **Leu** | Gln | Asn | Pro | Cys-642 | g |
| Entactin | Gln | Arg | Ala | Gln | Cys | + | **Ile** | Tyr | Met | Gly | Gly-752 | g |
| Octapeptide | | Gly | Pro | Gln | Ala | + | **Ile** | Ala | Gly | Gln | | h |
| Best amino acid | | Pro | Leu | Ala | + | **Leu** | Trp | Met | Gln | | h |

[a] J. F. Woessner, Jr., and C. J. Taplin, *J. Biol. Chem.* **263,** 16918 (1988).

[b] S. R. Abramson, G. E. Conner, H. Nagase, I. Neuhaus, and J. F. Woessner, Jr., *J. Biol. Chem.* submitted.

[c] T. Crabbe, F. Willenbrock, D. Eaton, P. Hynds, A. F. Carne, G. Murphy, and A. J. P. Docherty, *Biochemistry* **31,** 8500 (1992).

[d] C. G. Knight, F. Willenbrock, and G. Murphy, *FEBS Lett.* **296,** 263 (1992).

[e] A. J. Fosang, P. J. Neame, K. Last, T. E. Hardingham, G. Murphy, and J. A. Hamilton, *J. Biol. Chem.* **267,** 19470 (1992).

[f] P. S. Marcotte, I. M. Kozan, S. A. Dorwin, and J. M. Ryan, *J. Biol. Chem.* **267,** 13803 (1992).

[g] U. I. Sires, G. L. Griffin, T. J. Broekelmann, R. P. Mecham, G. Murphy, A. E. Chung, H. E. Welgus, and R. M. Senior, *J. Biol. Chem.* **268,** 2069 (1993).

[h] S. Netzel-Arnett, Q.-X. Sang, W. G. I. Moore, M. Navre, H. Birkedal-Hansen, and H. E. Van Wart, *Biochemistry* **32,** 6427 (1993).

The p$I$ for the rat enzyme is 5.9. The pH optimum for Azocoll and aggrecan digestion is pH 7.0. Matrilysin is a zinc-dependent enzyme that also requires 1 m$M$ CaCl₂ for optimal activity in overnight assays.

## Enzymatic Properties of Matrilysin

The human and rat enzymes have very similar properties. The only major difference found to date is that the human enzyme digests elastin at a reasonably good rate (190 $\mu$g/hr/nmol enzyme),[11] whereas the rat enzyme has no detectable action on this substrate.[3]

Matrilysins digest gelatins of types I, II, III, and IV (but not the corresponding collagens), fibronectin, laminin, casein, Azocoll, 2,4-dinitrophenyl-Pro-Leu-Gly-Ile-Ala-Gly-Pro-D-Arg (a collagenase substrate), and Mca-peptide. Rat matrilysin has an unusual specificity on gelatin of type I collagen prepared from rat. Action on the $\alpha 1$ chain is low, and on the $\alpha 2$ chain there are three main cleavage sites at and near the site of collagenase action.[13] These are illustrated in Table III. Bond cleavages in other substrates are also shown. The pattern of specificities is not very revealing; the main consistent feature is the presence of a hydrophobic group (of large size) in position P1′. A detailed study of the action of human matrilysin on a series of 58 octapeptides has been performed by Netzel-Arnett et al.[15] The best peptide of the series was Gly-Pro-Gln-Ala+Ile-Ala-Gly-Gln; the best amino acid found at each given position is shown in the last entry of Table III.

Matrilysin is inhibited by a wide variety of chelating agents including EDTA, 1,10-phenanthroline, and dithiothreitol, but it is only poorly inhibited by the thermolysin inhibitor phosphoramidon.[3] More specific matrix metalloproteinase inhibitors such as SC 40827 {N-[(3-N-benzyloxycarbonyl)amino-1-(R)-carboxypropyl]-L-Leu-O-methyl-L-Tyr-N-methylamide} and SC 44463 ($N^4$-hydroxyl-$N^1$-{1S-[(4-methoxphenyl)methyl]-2-(methylamino)-2-oxoethyl}-2R-(2-methylpropyl)butanediamide) inhibit Azocoll digestion with $IC_{50}$ values of $3 \times 10^{-5}$ and $1 \times 10^{-8}$ $M$, respectively.[13]

Natural inhibitors of matrilysin include $\alpha_2$-macroglobulin and $\alpha_1$-inhibitor$_3$[16] and the corresponding chicken egg inhibitor ovostatin.[3] The tissue inhibitors TIMP-1[14] and TIMP-2[17] also inhibit in stoichiometric fashion.

### Acknowledgments

Work in the author's laboratory was supported by National Institutes of Health Grant HD-06773. The assistance of Ms. Carolyn Taplin and Susan Abramson in the preparation of this chapter is gratefully acknowledged. Human recombinant matrilysin was generously provided by A. J. P. Docherty of Celltech, Inc., Slough, UK.

[15] S. Netzel-Arnett, Q-X. Sang, W. G. I. Moore, M. Navre, H. Birkedal-Hansen, and H. E. Van Wart, *Biochemistry* **32**, 6427 (1993).

[16] C. Zhu and J. F. Woessner, Jr., *Biol. Reprod.* **45**, 334 (1991).

[17] K. Miyazaki, K. Funahashi, Y. Numata, N. Koshikawa, K. Akaogi, Y. Kikkawa, H. Yasumitsu, and M. Umeda, *J. Biol. Chem.* **268**, 14387 (1993).

## [30] Tissue Inhibitors of Matrix Metalloendopeptidases

By GILLIAN MURPHY and FRANCES WILLENBROCK

### Introduction

Three members of the tissue inhibitor of metalloendopeptidase family, TIMP-1, TIMP-2, and TIMP-3, have been cloned from a number of species. The individual TIMPs show about 40% sequence identity, but they share considerable higher structural similarity, notably because of the conservation of 12 cysteine residues that have been shown to form disulfide bonds in TIMP-1,[1] giving a six-loop structure (Fig. 1). Human TIMP-1 is a 184-amino acid complex glycoprotein with a molecular mass of 28.5 kDa.[2] Glycosylation occurs at two sites and is heterogeneous, with variable sialic acid substitution leading to a heterogeneous p$I$,[3,4] but it is not necessary for function.[5,6] TIMP-2 is a nonglycosylated 194-amino acid protein of 21 kDa molecular mass which is characterized by an extended negatively charged C terminus.[7] Human TIMP-3, a 188-amino acid protein, has a theoretical molecular mass of 21 kDa.[8] The sequence of TIMP-3 contains a C-terminal potential glycosylation site, but a glycosylated form has not been described. All the TIMPs inhibit active matrix metalloendopeptidases (MMPs), binding in a 1 : 1 molar ratio to form tight, noncovalent complexes. Exceptionally strong C-terminal domain interactions between TIMP-1 and gelatinase B and TIMP-2 and gelatinase A mean that complexes between the respective proenzymes and inhibitors can also occur.

Tissue inhibitors of metalloproteinases have been shown to be produced by many cell types in culture,[9] and they are found in body fluids and tissue

[1] R. A. Williamson, F. A. O. Marston, S. Angal, P. Koklitis, M. Panico, H. R. Morris, A. F. Carne, B. J. Smith, T. J. R. Harris, and R. B. Freedman, *Biochem. J.* **268**, 267 (1990).

[2] A. J. P. Docherty, A. Lyons, B. J. Smith, E. M. Wright, P. E. Stephens, T. J. R. Harris, G. Murphy, and J. J. Reynolds, *Nature (London)* **318**, 66 (1985).

[3] G. Murphy and Z. Werb, *Biochim. Biophys. Acta* **839**, 214 (1985).

[4] J.-I. Kishi and T. J. Hayakawa, *Biochemistry* **96**, 395 (1984).

[5] S. P. Tolley, G. J. Davies, M. O'Shea, M. I. Cockett, A. J. P. Docherty, and G. Murphy, *Proteins; Struct. Funct. Genet.* **17**, 435 (1993).

[6] G. P. Stricklin, *Collagen Relat. Res.* **6**, 219 (1986).

[7] T. C. Boone, M. J. Johnson, Y. A. De Clerck, and K. E. Langley, *Proc. Natl. Acad. Sci. U.S.A.* **87**, 2800 (1990).

[8] S. S. Apte, M.-G. Mattei, and B. R. Olsen, *Genomics* **19**, 86 (1994).

[9] H. Birkedal-Hansen, W. G. I. Moore, M. K. Bodden, L. J. Windsor, B. Birkedal-Hansen, A. DeCarlo, and J. A. Engler, *Crit. Rev. Oral Biol. Med.* **4**, 197 (1993).

extracts.[10,11] They are thought to play a role in the regulation of the diverse degradative processes involving MMPs that occur during morphogenesis, growth, and destructive pathologies. Growth factor activity has been demonstrated for TIMP-1[12,13] and TIMP-2.[14]

## Assay Methods

The activity of TIMPs has been routinely assayed in many laboratories by the assessment of inhibition of crude or purified activated MMPs in macromolecular substrate assays. Measurements of the inhibition of collagenase (MMP1) in the [14]C-labeled collagen fibril assay constitute the most common variation.[15] One unit of TIMP activity is defined as the amount required to give 50% inhibition of 2 units (2 $\mu$g collagen degraded/min) of collagenase. The assay does not account for TIMPs already complexed to active MMPs, but it has been found that TIMPs complexed to progelatinases are freely available for assay. The former may be assayed by dissociation of MMP–TIMP complexes at pH 3 in the presence of EDTA. The dissociation process destroys the MMP activity, and free TIMP is available for assay.[16,17] Proenzyme forms of MMPs present in crude preparations of TIMPs do not interfere with the assay. Distinction between glycosylated TIMP-1 and unglycosylated TIMPs (generally TIMP-2) in crude preparations may be made by passage through minicolumns of concanavalin A (Con A)-Sepharose, which bind TIMP-1 only.[18]

## *Fluorometric Assay*

The most accurate method for determination of TIMP concentration involves the use of the internally quenched fluorescent peptide assays of MMPs.[19–21] The assays are not suitable for crude TIMPs from biological

[10] G. Murphy, T. E. Cawston, and J. J. Reynolds, *Biochem. J.* **195,** 167 (1981).
[11] H. G. Welgus, E. A. Bauer, and G. P. Stricklin, *J. Invest. Dermatol.* **87,** 592 (1986).
[12] J. C. Gasson, D. W. Golde, S. E. Kaufman, C. A. Westbrook, R. M. Hewick, R. J. Kaufman, G. C. Wong, P. A. Temple, A. C. Leary, E. L. Brown, E. C. Orr, and S. C. Clark, *Nature (London)* **315,** 768 (1985).
[13] T. Hayakawa, K. Yamashita, J.-I. Kishi, and K. Harigaya, *FEBS Lett.* **268,** 125 (1990).
[14] T. Hayakawa, K. Yamashita, K. Tanzawa, E. Uchijima, and K. Iwata, *FEBS Lett.* **298,** 29 (1992).
[15] T. E. Cawston and G. Murphy, this series, Vol. 80, p. 711.
[16] G. Murphy, P. Koklitis, and A. F. Carne, *Biochem. J.* **261,** 1031 (1989).
[17] V. Lefebvre and G. Vaes, *Biochim. Biophys. Acta* **992,** 355 (1989).
[18] C. M. Overall, J. L. Wrana, and J. Sodek, *J. Biol. Chem.* **264,** 1860 (1989).
[19] M. S. Stack and R. D. Gray, *J. Biol. Chem.* **264,** 4277 (1989).
[20] C. G. Knight, F. Willenbrock, and G. Murphy, *FEBS Lett.* **296,** 263 (1992).
[21] L. Niedzwiecki, J. Teahan, R. K. Harrison, and R. L. Stein, *Biochemistry* **31,** 12618 (1992).

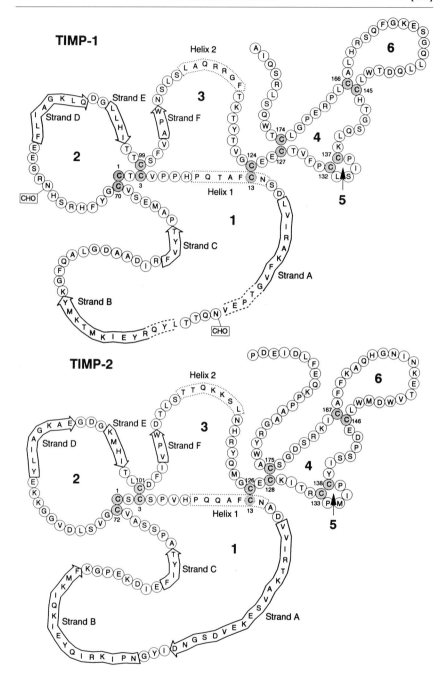

samples, however, as the background fluorescence interferes with the product fluorescence.

*Principle.* The method exploits the high affinity of TIMPs for the MMPs so that, under the conditions used, the interaction is essentially irreversible and the TIMP is titrated by the enzyme. The preferred enzyme for the measurement is stromelysin owing to the stability of both pro and active forms. Stromelysin is assayed using the substrates Mca-Pro-Leu-Ala-Nva-Dpa-Ala-Arg[22] or Mca-Pro-Leu-Gly-Leu-Dpa-Ala-Arg.[23] Hydrolysis of the internally quenched substrates results in the release of the intensely fluorescent methoxycoumarin peptide, Mca-Pro-Leu-Ala or Mca-Pro-Leu-Gly ($\lambda_{ex}$ 328 nm, $\lambda_{em}$ 393 nm).

*Reagents*

Substrate: Mca-Pro-Leu-Ala-Nva-Dpa-Ala-Arg or Mca-Pro-Leu-Gly-Leu-Dpa-Ala-Arg is dissolved in dimethyl sulfoxide (DMSO) as a 100 $\mu M$ solution (stored at 4°)

Incubation buffer: 50 m$M$ Tris-HCl (pH 7.5), 0.1 $M$ NaCl, 10 m$M$ CaCl$_2$, and 0.05% v/v Brij 35

Assay buffer: 50 m$M$ 2-($N$-morpholino)ethanesulfonic acid or sodium cacodylate, pH 6.0, 0.1 $M$ NaCl, 10 m$M$ CaCl$_2$, and 0.05% Brij 35

Stromelysin: Prostromelysin is activated with 10 $\mu$g/ml trypsin at 37° for 30 min and the reaction terminated with 100 $\mu$g/ml soybean trypsin inhibitor. The concentration of stromelysin can be determined from absorbance at 280 nm using the extinction coefficient of 59,100 $M^{-1}$ cm$^{-1}$ for prostromelysin and by titration of the active enzyme against a known concentration of standard TIMP-1. The activated stromelysin can be divided into aliquots and frozen; it retains full activity for several months

*Procedure.* The TIMP samples are preincubated with active stromelysin (at a concentration of 25–100 n$M$) for 30 min at room temperature prior

---

[22] Standard abbreviations are used for the amino acid residues (L configuration). Mca, 7-Methoxycoumarin; Dpa, 2,4-dinitrophenyl-L-2,3,-diaminopropionyl.

[23] Available from Novabiochem (La Jolla, CA).

---

Fig. 1. Loop structure of TIMP-1 and TIMP-2. The cDNA-predicted amino acid sequences of human TIMP-1 and TIMP-2 are depicted as six-loop structures on the basis of the assignment of the disulfide bonds in TIMP-1.[1] In TIMP-1, loops 1 and 2 contain carbohydrate binding sites (CHO); TIMP-1 is a complex glycoprotein and is variably substituted with sialic acid,[4] giving it a heterogeneous charge.[3] TIMP-2 differs from TIMP-1 in all species studied in that it is unglycosylated, loop 1 has a different amino acid sequence alignment, and the C terminus is extended and highly negatively charged. The residues contributing to the 6 $\beta$ sheets, strands A–F ($\Rightarrow$) and 2 $\alpha$-helices (▯▯▯), were determined by N.M.R. analysis of loops 1–3 of TIMP-2 and subsequently deduced for N-TIMP-1,[35] and are shown courtesy of R. Williamson, University of Kent.

to assay. The assay is started by the addition of 50 $\mu$l of the stromelysin–TIMP reaction mixture to assay buffer containing 0.5–1 $\mu M$ substrate and a final concentration of 1% dimethyl sulfoxide. Initial velocities (increased fluorescence units per unit time) are plotted against the volume of stock TIMP solution. The volume of TIMP required to give complete inhibition (i.e., equal to the enzyme) is determined from the intercept on the abscissa.

*Reverse Zymography.* The MMP inhibitory activities of TIMPs may conveniently be demonstrated and semiquantified using a substrate gel electrophoresis technique (reverse zymography). Samples for assessment are prepared as for conventional sodium dodecyl sulfate (SDS)–polyacrylamide gel electrophoresis,[24] except that reducing agents are omitted and samples are not boiled. Polyacrylamide gels (up to 11% polyacrylamide or a 4–15% gradient) are prepared with 0.5 mg/ml gelatin (Sigma, St. Louis, MO) incorporated. After electrophoresis the gel is washed twice for 15 min each time in 2.5% Triton X-100, rinsed with water, and incubated in a crude preparation of activated MMPs. Crude culture media from rabbit or human skin fibroblasts stimulated with phorbol myristate acetate which contain gelatinolytic, collagenolytic, and caseinolytic activities in a latent form may be used. Treatment of such culture media with 2 m$M$ 4-aminophenylmercuric acetate (APMA) at 37° for 2–4 hr will activate the proMMP content. The medium may be dialyzed against 25 m$M$ Tris-HCl, pH 7.5, containing 0.1 $M$ NaCl, 10 m$M$ CaCl$_2$, and 0.02% w/v sodium azide and stored at $-20$° for long periods. The amount of MMP activity in the medium will determine the length of incubation with the Triton-treated polyacrylamide gel that is required to clear the gelatin partially. A period of about 1 hr at 37° is convenient. The gel may then be stained with Coomassie blue and destained. The MMP inhibitor band will be revealed as darker areas of the gel where the gelatin has not been cleared. Partial quantitation of the activity present is possible by scanning densitometry and comparison with standard preparations of TIMPs. It is important to run a number of concentrations of each sample to be sure that the data are in the linear range. The TIMPs may be tentatively identified by the apparent molecular masses using this system: TIMP-1 runs as a diffuse band of about 29 kDa, TIMP-2 at 22 kDa, and TIMP-3 at about 23–24 kDa.

Purification

Both TIMP-1 and TIMP-2 are generally purified from the conditioned media of cultured cells. Fibroblasts (both primary and cell lines) produce relatively high levels of TIMP-1 when stimulated with phorbol myristate acetate or cytokines such as interleukin-1. TIMP-1 may be purified from

---

[24] U. K. Laemmli and M. Favre, *J. Mol. Biol.* **80**, 575 (1973).

the culture medium of fetal lung or other fibroblasts using a modification of the method of Cawston *et al.*[25] The medium (buffered with 10 m$M$ Tris-HCl, pH 7.5) is depleted of MMPs by initial passage through an iminodiacetic acid Sepharose column saturated with zinc acetate and equilibrated with 25 m$M$ Tris-HCl, pH 7.5, containing 0.15 $M$ NaCl, 10 m$M$ CaCl$_2$, 0.05% Brij 35, and 0.02% NaN$_3$. The inhibitory activity does not bind to the matrix, but the MMPs bind and may be eluted if required with 50 m$M$ sodium acetate, pH 4.7, containing 10 m$M$ CaCl$_2$. The inhibitor pool should be dialyzed against 25 m$M$ Tris-HCl, pH 7.5, containing 0.05% Brij 35, 0.02% NaN$_3$ and then applied to heparin Sepharose equilibrated with the same buffer. After washing with this buffer to the background $A_{280}$, TIMP-1 is eluted with a gradient of NaCl, eluting at 0.20–0.4 $M$ NaCl. Further purification of TIMP-1 into forms of relative homogeneity of glycosylation can be effected using concanavalin A-Sepharose equilibrated with the same buffer. Heterogeneously glycosylated material does not bind to the column, but the majority of TIMP-1 will bind through exposed mannose residues and is eluted with the buffer containing 0.5 $M$ $\alpha$-methylmannoside.

The inhibitor TIMP-2 has been purified from the culture media from both tumor cells and fibroblasts. In some cultures all the TIMP-2 is found complexed with progelatinase A. Culture media may then be applied to a gelatin-agarose affinity column equilibrated with 25 m$M$ Tris-HCl, pH 75, 0.5 $M$ NaCl, 10 m$M$ CaCl$_2$, 0.02% Brij 35, and 0.025% sodium azide (buffer A) to which the progelatinase–TIMP-2 complex will bind. After washing with buffer A the complex can be eluted with buffer A containing 10% dimethyl sulfoxide. Contamination of the preparation with progelatinase B–TIMP-1 complexes will vary, but these may be removed by concanavalin A-Sepharose chromatography using buffer A. The progelatinase A–TIMP-2 complex can be dissociated and the TIMP-2 recovered using reversed-phase high-performance liquid chromatography (HPLC) in the presence of 0.1% trifluoroacetic acid.[26] Alternatively, the complex may be treated with 10 m$M$ EDTA, pH 3, at 37° for 1 hr prior to separation of the partially denatured gelatinase A and free TIMP-2 by gel filtration on Sephacryl S-200 using buffer A.[27] If free progelatinase A is present in the culture medium and is required in a native form, alternative chromatographic procedures are required, as described in [28] in this volume.

Both TIMP-1 and TIMP-2 are very stable and may be stored in neutral pH buffer or freeze-dried at −70° for long periods without loss of activity. Molar extinction coefficients at 280 nm of 26,500 $M^{-1}$ cm$^{-1}$ for TIMP-

[25] T. E. Cawston, W. A. Galloway, E. Mercer, G. Murphy, and J. J. Reynolds, *Biochem. J.* **195,** 159 (1981).

[26] E. W. Howard, E. C. Bullen, and M. J. Banda, *J. Biol. Chem.* **266,** 13070 (1991).

[27] R. V. Ward, R. M. Hembry, J. J. Reynolds, and G. Murphy, *Biochem. J.* **278,** 179 (1991).

$1^{28}$ and 39,600 $M^{-1}$ cm$^{-1}$ for TIMP-2[29] have been calculated from amino acid analyses.

## Titration of Active Matrix Metalloproteinase Concentration with Tissue Inhibitors of Metalloproteinases

The technique of TIMP titration has been used routinely to determine the concentration of active gelatinases A and B, collagenase, stromelysin 1, and matrilysin. Enzyme activity is monitored using Mca-Pro-Leu-Gly-Leu-Dpa-Ala-Arg or Mca-Pro-Leu-Ala-Nva-Dpa-Ala-Arg as described above.

*Enzyme Preparation.* As the method involves a preincubation of enzyme and TIMP, precautions must be taken to ensure that from the time that the titration is started the enzyme does not undergo further activation or degradation. Enzymes that are activated by organomercurial compounds such as 4-aminophenylmercuric acetate are separated from the activator by gel-filtration spin dialysis or diluted 1000-fold. Enzymes that are activated by other matrix metalloproteinases are incubated in the presence of the lowest concentration of activating enzyme required to give optimal activation. Activation by other proteinases such as trypsin is stopped at the appropriate time by specific inhibitors such as soybean trypsin inhibitor. All active enzyme preparations are stored on ice throughout.

*Inhibitor Concentration.* The concentration of stock TIMP solutions is determined both by amino acid analysis and by absorbance at 280 nm.

*Method.* The method is essentially the same as the titration method described above for the determination of TIMP concentration. Concentrations and conditions that have been used routinely to titrate matrix metalloproteinases are summarized in Table I. Enzyme and inhibitor are preincubated at concentrations that are much greater than the $K_i$ for the interaction to ensure complete complex formation. The use of a preincubation at high concentrations also increases the rate of complex formation. The latter consideration is particularly important for the titration of gelatinase B with TIMP-1 and when titrating matrilysin or forms of the enzymes which lack a C-terminal domain, many of which interact slowly with either TIMP.[30] Assay conditions are chosen such that the enzyme concentration is much greater than $K_i$ and so that a rapid rate of reaction is observed in the absence of inhibitor. The reaction is started by the addition of the enzyme–inhibitor reaction mixture to a solution of substrate (0.5 $\mu M$ Mca-Pro-Leu-Gly-Leu-Dpa-Ala-Arg or Mca-Pro-Leu-Ala-Nva-Dpa-Ala-Arg) in buffer

[28] G. Murphy, A. Houbrechts, M. I. Cockett, R. A. Williamson, M. O'Shea, and A. J. P. Docherty, *Biochemistry* **30.** 8097 (1991).

[29] Y. A. DeClerck, T.-D. Yean, H. S. Lu, J. Ting, and K. F. Langley, *J. Biol. Chem.* **266,** 3893 (1991).

[30] F. Willenbrock, T. Crabbe, P. M. Slocombe, C. W. Sutton, A. J. P. Docherty, M. I. Cockett, M. O'Shea, K. Brocklehurst, I. R. Phillips, and G. Murphy, *Biochemistry* **32,** 4330 (1993).

TABLE I
CONDITIONS RECOMMENDED FOR ACTIVE-SITE TITRATION OF MATRIX METALLOPROTEINASES BY
TISSUE INHIBITORS

| | | Enzyme concentration (n$M$) | | |
| | | Minimum, | | |
| Enzyme | TIMP | in preincubation | In assay | Conditions |
|---|---|---|---|---|
| atrilysin | TIMP-1 | 500 | 10 | After 4-hr preincubation, assay at pH 7.5, 25° |
| romelysin | TIMP-1 or TIMP-2 | 50 | 5 | Assay at pH 6, 37° |
| llagenase | TIMP-1 | 250 | 25 | After 1-hr preincubation, assay at pH 7.5, 37° |
| elatinase A | TIMP-1 or TIMP-2 | 10 | 1 | Assay at pH 7.5, 25° |
| elatinase B | TIMP-1 | 100 | 1 | After 6-hr preincubation, assay at pH 7.5, 25° |
| elatinase B | TIMP-2 | 10 | 1 | Assay at pH 7.5, 25° |

containing 10 m$M$ CaCl$_2$, 0.1 $M$ NaCl, 0.05% Brij 35, and a final dimethyl sulfoxide concentration of 1%. The initial rate of substrate hydrolysis is recorded over 2–5 min. Under those conditions no dissociation of the enzyme–TIMP complexes is observed. Initial velocities are plotted against the TIMP concentration, and the enzyme concentration is determined from the intercept on the abscissa.

Properties

*Stability.* Both TIMP-1 and TIMP-2 require an intact disulfide-bonded structure to inhibit the MMPs and are therefore inactivated by reduction with thiol compounds such as dithiothreitol. TIMP-1 is stable to acid pH and heat, retaining full activity after incubation at pH 2 or 100° for 30 min.[31] TIMP-1 is inactivated by proteolytic cleavage by human neutrophil elastase, trypsin, and $\alpha$-chymotrypsin.[32]

*Structure.* The TIMP molecule consists of six loops defined by disulfide bonds. The correct tertiary structure is essential for activity, as reduced TIMP and limit digestion proteolytic fragments of TIMP-1 are inactive. Tryptic peptide mapping has revealed that the C-terminal four amino acids and a region in loop 1 consisting of Lys[23]–Arg[59] are exposed, whereas loop 2 may be buried.[33] Tryptic digests under nonreducing conditions result in the formation of an insoluble core peptide with a molecular mass of approximately 8 kDa.

[31] A. Osthues, V. Knäuper, R. Oberhoff, H. Reinke, and H. Tschesche, *FEBS Lett.* **296,** 16 (1992).

[32] Y. Okada, S. Watanabe, I. Nakanishi, J.-I. Kishi, T. Hayakawa, W. Watorek, J. Travis, and H. Nagase, *FEBS Lett.* **229,** 157 (1988).

[33] R. A. Williamson, B. J. Smith, S. Angal, G. Murphy, and R. B. Freedman, *Biochim. Biophys. Acta* **1164,** 8 (1993).

Mutagenesis of both TIMP-1 and TIMP-2 to remove sections of the molecule has demonstrated that two structurally distinct domains exist; the N-terminal domain (N-TIMP) consists of loops 1–3 and the C-terminal domain (C-TIMP) of loops 4–6 (Fig. 1). The N-terminal domain can fold independently of the C-terminal domain to give a functional MMP inhibitor. A mutant of TIMP-1 consisting of the N-terminal domain only (N-TIMP-1) has been compared with wild-type TIMP-1 by chemical modification studies and susceptibility to trypsin. The inhibitory activities of both forms of TIMP-1 are affected to a similar extent by modification with diethyl pyrocarbonate.[34] Trypsin digestion releases the same peptides with sequences common to both forms of TIMPs, and the order in which they are released is also the same.[33] Therefore, it has been concluded that the conformation adopted by the mutant is the same as that of the N-terminal domain of the wild-type inhibitor. The C-terminal domain has not been expressed in sufficient quantities to allow its characterization.

Homonuclear two-dimensional $^1$H nuclear magnetic resonance spectroscopy has been used to obtain essentially complete sequence-specific assignments for 123 of the 127 amino acid residues present in the active N-terminal domain of TIMP-2 (N-TIMP-2).[35] Analysis of the through-space nuclear Overhauser effect data obtained for N-TIMP-2 allowed determination of both secondary structure of the domain (Fig. 1) and also a low-resolution tertiary structure defining the protein backbone topology. The protein contains a five-stranded antiparallel $\beta$-sheet that is rolled over on itself to form a closed $\beta$-barrel, and two short helices which pack close to one another on the same barrel face. A comparison of the N-TIMP-2 structure with other known protein folds reveals that the $\beta$-barrel topology is homologous to that seen in proteins of the oligosaccharide/oligonucleotide binding (OB) fold family. The common structural features include the number of $\beta$-strands and their arrangement, the $\beta$-barrel shear number, an interstrand hydrogen bond network, the packing of the hydrophobic core, and a conserved $\beta$-bulge. Superpositions of the $\beta$-barrels from N-TIMP-2 and two previously known members of the OB protein fold family (Staphylococcal nuclease and *Escherichia coli* heat-labile enterotoxin) confirmed the similarity in $\beta$-barrel topography. Comparison of sequence conservation in the TIMPs with the active site of other family members indicates that they do not share a common ligand binding site.[35]

The highly conserved sequence motifs $V_{18}IRA/TK_{22}$ in strand A and $YIF_{84}LI/L/VA/TG_{88}$ in strand D of N-TIMP-2 (Fig. 1) both form an integral

[34] R. A. Williamson, B. J. Smith, S. Angal, and R. B. Freedman, *Biochim. Biophys. Acta* **1203**, 147 (1993).
[35] R. A. Williamson, G. Martorell, M. D. Carr, G. Murphy, A. J. P. Docherty, R. B. Freedman, and J. Feeney, *Biochemistry* **33**, 11745 (1994).

TABLE II
APPARENT ASSOCIATION RATE CONSTANTS FOR INTERACTION OF MATRIX
METALLOPROTEINASES WITH TISSUE INHIBITORS[a]

| | $10^{-6} \times k_{on}$ $(M^{-1}\ sec^{-1})$ | | | |
| Enzyme | TIMP-1 | N-TIMP-1 | TIMP-2 | N-TIMP-2 |
|---|---|---|---|---|
| Stromelysin 1 | 1.9 | 1.2 | 0.3 | 0.05 |
| Gelatinase A | 3.4 | 0.03 | 21 | 0.13 |
| Gelatinase B | 11.1 | 0.08 | 0.26 | 0.29 |
| Matrilysin | 0.3 | 0.4 | n.d. | n.d. |

[a] Determined at pH 7.5, ionic strength $I$ of 0.2, and 25°, except for stromelysin 1 which was determined at 37°. n.d., Not determined.

part of the closed $\beta$-barrel structure and are involved in the packing of the two $\alpha$ helices onto the $\beta$-barrel core and therefore have essential structure roles in the protein backbone. However, conserved surface regions of potential functional importance could also be identified.[35]

*Binding of Inhibitor to Proenzyme.* In addition to TIMP binding to active MMPs, strong interactions have been detected between TIMP-1 and progelatinase B and between TIMP-2 and progelatinase A. The TIMP-2 binding site for progelatinase A consists primarily of the charged C-terminal extension, which is unique to TIMP-2, although other low affinity interactions are probably present in the final complex.[30] Binding is to the C-terminal domain of the enzyme, and the reconstituted complex has an approximate $K_d$ of 50 p$M$. There appear to be differences between the stability of native and reconstituted complexes, the latter being considerably more stable and dissociable only by severe treatments that denature the enzyme.[27,35a–37] The proenzyme in the native complex can be activated and the active form can be inhibited by additional TIMP-1 and TIMP-2. This inhibition appears to be slow, however, possibly owing to the masking of important binding sites on the C-terminal domain of the enzyme by the first TIMP-1 molecule. The progelatinase B–TIMP-1 complex is also formed by interactions between the C-terminal domains of enzyme and inhibitor.

*Chemical Modification and Mutagenesis.* The role of the two domains of the TIMP-1 and TIMP-2 molecules has been studied by comparing wild-type TIMPs and genetically engineered forms that consist of the N-terminal domain only. The truncated TIMPs inhibit all the MMPs including genetically engineered forms of the enzymes in which the C-terminal domains have been removed. Thus the TIMP N-terminal domain interacts with the enzyme cata-

[35a] G. I. Goldberg, B. L. Marmer, G. A. Grant, A. Z. Eisen, S. Wilhelm, and C. He, *Proc. Natl. Acad. Sci. U.S.A.* **86,** 8207 (1989).
[36] W. G. Stetler-Stevenson, H. C. Krutzsch, and L. A. Liotta, *J. Biol. Chem.* **264,** 17374 (1989).
[37] E. W. Howard and M. J. Banda, *J. Biol. Chem.* **266,** 17972 (1991).

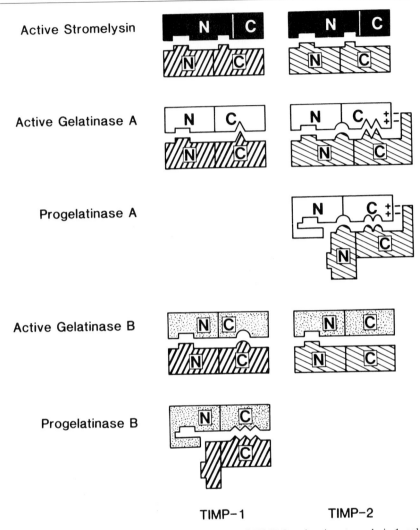

Active Stromelysin

Active Gelatinase A

Progelatinase A

Active Gelatinase B

Progelatinase B

TIMP-1                TIMP-2

FIG. 2. Domain interactions between TIMP-1 and TIMP-2 and active stromelysin 1, gelatinase A, progelatinase A, gelatinase B, and progelatinase B during complex formation. Both inhibitors (lower hatched boxes) are depicted as consisting of an N-terminal domain (N) of three disulfide-bonded loops 1–3 (Fig. 1) and a C-terminal domain (C) consisting of loops 4–6. The enzymes are also shown as N-terminal and C-terminal domains. The N-terminal domains of both TIMP-1 and TIMP-2 contain the inhibitory sites and interact with the catalytic N-terminal domain of each of the active MMPs. The C-terminal domain interactions probably involve different sites for different MMPs. In the case of stromelysin 1, which binds TIMP-1 more rapidly than TIMP-2, the TIMP C-terminal domain interacts with the catalytic domain of the enzyme.[39] For gelatinase A, binding of TIMP-2 is more rapid than the binding of TIMP-1. This appears to be due in part to the ability of the 9-amino acid negatively charged C-terminal motif to bind specifically within the gelatinase A C-terminal domain. Other charged

lytic domain. It is not possible to assess the effect of the removal of the TIMP C-terminal domain on the $K_i$ for the interaction, as the $K_i$ values are too low to determine accurately using the assay systems presently available. However, in many cases, the rate of inhibition is dramatically decreased by the removal of the C-terminal domain (Table II), suggesting that this domain also binds to the enzyme. Chemical cross-linking of TIMP-2 and gelatinase A has also demonstrated that there are two binding sites on TIMP-2,[38] and it has been suggested that binding to the second site stabilizes the proenzyme.[37] A kinetic analysis of the inhibition of gelatinase A, stromelysin 1, and hybrid enzymes in which the C-terminal domains of the enzymes have been exchanged has demonstrated that the TIMP C-terminal domain contains at least two different binding sites. One of the binding sites interacts with the C-terminal domain of gelatinase A, whereas the other binds to the N-terminal domain of stromelysin 1[39] (Fig. 2). In both cases the interactions increase the rate of inhibition, although it is not clear whether they play a role in stabilizing the final form of the complex.

Both chemical modification and site-directed mutagenesis studies have been performed on TIMP-1 in an attempt to define the roles of individual amino acid residues.[1,34] Modification of tyrosine, arginine, and lysine residues had no effect on TIMP activity, whereas modification with an excess of diethyl pyrocarbonate (DEPC) abolishes TIMP activity. The residues modified are His-95, His-144, and His-164, and it was proposed that His-95 is essential as this is within the N-terminal domain and is highly conserved. Systematic site-directed mutagenesis of each of the conserved histidine residues in the TIMP N-terminal domain, however, has suggested a role for His-7 and residues in the region between Cys-3 and Cys-13 but not for His-95.[40] Wild-type TIMP-1, the H7A and H95A mutants, and the N-terminal domain of TIMP-1 are all equally sensitive to DEPC modifica-

[38] D. E. Kleiner, E. J. Unsworth, H. C. Krutzsch, and W. G. Stetler-Stevenson, *Biochemistry* **31**, 1665 (1992).
[39] Q. Nguyen, F. Willenbrock, M. I. Cockett, M. O'Shea, A. J. P. Docherty, and G. Murphy, *Biochemistry* **33**, 2089 (1994).
[40] M. O'Shea, F. Willenbrock, R. A. Williamson, M. I. Cockett, R. B. Freedman, J. J. Reynolds, A. J. P. Docherty, and G. Murphy, *Biochemistry* **31**, 10146 (1992).

interactions occur between the gelatinase A and TIMP-2 C-terminal domains,[30] and these may also be responsible for the fact that TIMP-2 can bind to progelatinase A,[36,37] where N-terminal domain interactions cannot occur. However, to emphasize that neither set of binding sites has been characterized they are represented differently. Gelatinase B binds TIMP-1 more rapidly than TIMP-2, and the C-terminal domain interactions of both inhibitor and enzyme appear to determine this property. As for the progelatinase A–TIMP-2 interaction, the ability of TIMP-1 to bind to progelatinase B[35a] in the absence of N-terminal interactions may be due to the C-terminal domain interactions defined between the active gelatinase B and TIMP-1, which are depicted using different symbols.

TABLE III
APPARENT $K_i$ VALUES FOR TIMP-1–MATRIX
METALLOPROTEINASE INTERACTIONS

| Enzyme[a] | $K_i$ (pM) | Ref. |
|---|---|---|
| Matrilysin | 370 | b |
| Stromelysin 1 | <200 | c |
| Stromelysin 1 | 200 | d |
| Rabbit bone collagenase | 140 | e |
| Human skin fibroblast collagenase | <1000 | f |
| Collagenase | <200 | c |
| Pig synovial collagenase (cow TIMP-1) | 300 | g |
| Gelatinase A | <2 | h |

[a] All enzymes are recombinant human enzymes unless otherwise
stated.
[b] O'Shea et al.[40]
[c] G. Murphy, J. A. Allan, F. Willenbrock, M. I. Cockett, J. P. O'Connell, and A. J. P. Docherty, J. Biol. Chem. **267**, 9612 (1992).
[d] J. A. Teahan and R. L. Stein, FASEB J. **4**, (Abstract 1652),
A 1977 (1991).
[e] T. E. Cawston, G. Murphy, E. Mercer, W. A. Galloway, B. L. Hazleman, and J. J. Reynolds, Biochem. J. **211**, 313 (1983).
[f] H. G. Welgus, J. J. Jeffrey, A. Z. Eisen, W. T. Roswit, and
G. P. Stricklin, Collagen Relat. Res. **5**, 167 (1985).
[g] Lelievre et al.[41]
[h] G. Murphy, F. Willenbrock, R. V. Ward, M. I. Cockett, D. Eaton,
and A. J. P. Docherty, Biochem. J. **283**, 637 (1992).

tion with a similar loss of inhibitory activity. There is at present no explanation for the effect of DEPC modification on TIMP-1. It is possible that
there is a local modification-induced unfolding of TIMP-1 or that each of the
modified histidines contributes to the binding of TIMP-1 and that disruption
of any one of these interactions is sufficient to alter TIMP-1 activity.

*Mechanism of Inhibition.* The inhibitors TIMP-1 and TIMP-2 specifically
inhibit the activated mammalian MMPs and, with the exception of a soybean
leaf metalloendopeptidase (also a member of family M10), have no known
activity on other metalloendopeptidases. The specificity of inhibition has
led to the proposal that inhibition by TIMPs be taken as a criterion for
inclusion of a metalloendopeptidase in the MMP family. Inhibition is reversible with a 1 : 1 stoichiometry, and TIMP is not modified by the enzyme.[28]
Very tight complexes are formed between TIMP and each MMP, with $K_i$
values of less than 1 nM (Table III).[41] Unfortunately, accurate determination of $K_i$ values is prevented by the lack of sensitivity of the enzyme assays
(i.e., the necessary condition [E] < $K_i$ cannot be fulfilled) and the long

[41] Y. Lelievre, R. Bouboutou, J. Boiziau, D. Faucher, D. Achard, and T. Cartwright, Matrix
**10**, 292 (1990).
[42] J. O'Connell, T. Crabbe, F. Willenbrock, and G. Murphy, unpublished studies (1993).

incubation times required to ensure complete complex formation at these enzyme concentrations. Apparent association rate constants have been determined for the inhibition of several enzymes by TIMP-1 and TIMP-2 (Table II). Inhibition of full-length enzymes is rapid, with rate constants in the range of $10^5-10^7$ $M^{-1}$ sec$^{-1}$, whereas inhibition of matrilysin is somewhat slower. The highest values of the rate constants are for the interaction of gelatinase A with TIMP-2 and gelatinase B with TIMP-1. The additional binding site on the latent proform of gelatinase A for TIMP-2 increases the rate of formation of the active gelatinase A–TIMP-2 complex, and it is possible that a similar mechanism exists in the interaction of active gelatinase B with TIMP-1.[42]

Inhibition probably occurs as a result of binding at the enzyme active site, as competition has been observed between TIMP-1 and low molecular weight synthetic inhibitors that are directed at the catalytic zinc.[39,40] However, the catalytically essential glutamic acid of the MMP active site does not appear to be involved in the interaction as mutation of this residue in gelatinase A does not affect TIMP binding.[43] The amino acid residues in TIMP that are responsible for binding at the enzyme active site are not known. Two sequences in TIMP-1 (Asp-16 to Lys-22 and Glu-82 to Gly-87) which could fulfill the specificity requirements of the enzymes have been identified.[44] It has been proposed that the enzyme S1'–S3' subsites interact with the hydrophobic residues of these sequences, aligning the carboxyl group of Asp-16 or Glu-82 to interact with the catalytically essential zinc and thus inactivate the enzyme. However, mutations within both the regions did not alter the affinity of TIMP-1 for matrilysin,[40] or the rate of association of TIMP-1 and gelatinase A,[45] indicating that the sequences are not essential for TIMP activity. Both sequences have recently been shown to form an integral part of the $\beta$ barrel core of N-TIMP-2 and therefore have essential structural roles.[35] The results from site-directed mutagenesis studies suggested that residues His-7 and Gln-9 might interact at the enzyme active site, as mutations at these positions decreased the affinity of TIMP-1 for matrilysin.[40] However, further studies have shown that these mutations result in a form of TIMP-1 that has the kinetic characteristics of the N-terminal domain mutant.[45] The residues are situated in a hydrophobic patch on the surface of N-TIMP-2 and are adjacent to its C-terminus.[35] It is possible, therefore, that the region between Cys-3 and Cys-13 is responsible for maintaining the overall conformation of the full-length TIMP so that the two domains are correctly orientated for docking to the enzyme.

[43] T. Crabbe, S. Zucker, M. I. Cockett, F. Willenbrock, S. Tickle, J. P. O'Connell, J. M. Scothern, G. Murphy, and A. J. P. Docherty, *Biochemistry* **33**, 6684 (1994).
[44] J. F. Woessner, *FASEB J.* **5**, 2145 (1991).
[45] F. Willenbrock, unpublished (1993).

The picture that is emerging of the mechanism of inhibition by TIMP is complex, with numerous points of interaction occurring between the enzyme and inhibitor. The TIMP C-terminal domain has more than one enzyme binding site and acts to increase the rate of inhibition of many enzymes. The mechanism for the rate enhancement is probably simply by increasing the probability of interaction between the enzyme and inhibitor N-terminal domains. Recent advances in determining the structures of both the MMPs and the TIMPs will allow the use of a rational approach to site-directed mutagenesis studies and, therefore, the identification of residues involved in the binding of TIMP to the enzyme active site.

# [31] Quantification of Matrix Metalloproteinases in Tissue Samples

By J. FREDERICK WOESSNER, JR.

## Introduction

A number of papers have reported on the assay of matrix metalloproteinases (MMPs) in tissue extracts or homogenates. Many reports contain one or more serious errors: the extraction of enzyme is incomplete, latent enzymes are not activated, and inhibitors of MMPs are ignored. If a comparison is made between enzyme levels in two different situations, for example, normal and disease state, then the errors may become serious. If the extraction is incomplete, changes in the extracellular matrix may render the enzyme more readily extractable. If an activating factor (enzyme) is increased, more enzyme may become active. If an inhibitor is diminished, more enzyme will be detected. In all of those cases there may be no change at all in the total enzyme content of the tissue. It is the purpose of this chapter to deal with such common problems by showing how to extract the proteinases, how to check for completeness of extraction, how to activate latent forms, and how to remove or destroy inhibitors.

*Tissue Extract versus Culture Medium Assays*

Most studies of MMPs do not concern the assay of extracts. Of some 3000 reports on metalloproteinases and their inhibitors,[1] about 250 deal with the direct determination of enzyme in tissue homogenates and extracts. The vast majority of work in this field has been concerned with cell, tissue,

---

[1] J. F. Woessner, Jr., *Matrix* (Suppl.) **1**, 425 (1992).

and organ culture and the determination of enzyme activity in the media produced by such cultures. This is because the proteinases are difficult to extract from tissue and because the amounts found in culture are typically about 50 times higher than those in tissues. However, even in the study of culture fluid the problems of inhibition and activation remain. Furthermore, it is often desirable to check what is actually going on in the living animal. Although the presence of a particular metalloproteinase in a tissue may be detected by immunohistochemistry and quantified by enzyme-linked immunosorbent assay (ELISA) methods, those methods do not generally provide information about the fraction of enzyme that is in the active state as opposed to that in latent proenzyme form. Moreover, the ELISA methods require a prior extraction of the enzyme.

## Matrixin Family

To date nine distinct mammalian matrixins[2] are known: interstitial collagenase (MMP-1), neutrophil collagenase (MMP-8), stromelysin 1 (MMP-3), gelatinase A (MMP-2), gelatinase B (MMP-9), matrilysin (MMP-7), macrophage elastase (MMP-12), and stromelysins 2 and 3 (MMP-10 and MMP-11). Of these the first six are frequently found to be tightly bound to the extracellular matrix, and the other three have not been examined. The neutrophil collagenase (MMP-8) and macrophage elastase (MMP-12) are stored in the cells as granules awaiting discharge. Extraction from such cells is straightforward. However, on discharge the enzymes may become attached to unknown components of the matrix.

## Binding of Matrixins to Extracellular Matrix

In early studies, it was postulated that collagenase might be bound to collagen fibers in the matrix and that heating to denature the collagen might release the enzyme.[3] The methods based on this idea were successful, but subsequent work with many other matrixins has given similar results. It seems unlikely that collagen is the anchor in all cases. In fact, Welgus et al.[4] present data showing a failure of latent MMP-1 to bind to collagen at all! Thus, in fact, we do not really know which components of the extracellular matrix bind the matrixins. A striking illustration of binding is found in the case of perfusion of the rat ovary.[5] An ovary is perfused with

[2] H. Nagase, A. J. Barrett, and J. F. Woessner, Jr., *Matrix* (Suppl.) **1**, 421 (1992).
[3] J. G. Weeks, J. Halme, and J. F. Woessner, Jr., *Biochim. Biophys. Acta* **445**, 205 (1976).
[4] H. G. Welgus, J. J. Jeffrey, A. Z. Eisen, W. T. Roswit, and G. P. Stricklin, *Collagen Relat. Res.* **5**, 167 (1985).
[5] T. A. Butler, C. Zhu, R. A. Mueller, G. C. Fuller, W. J. LeMaire, and J. F. Woessner, Jr., *Biol. Reprod.* **44**, 1183 (1991).

20 times its own volume of fluid every minute, yet the collagenases and gelatinases known to be present in the ovary and required for ovulation do not wash out of the ovary over a period of 12 hr. Because collagenase was the first matrixin to be discovered and because most of the principles of tissue assay can be described for this example, the first part of this chapter deals with that enzyme. Additional metalloproteinases are then taken up as special cases.

### Strategy for Collagenase Assay in Tissue Samples

Collagenase is perhaps the simplest of the matrixins to assay in tissue and extracts because it requires a very specific substrate: collagen. More importantly, the triple helix of collagen is resistant to most of the known proteinases expected in mammalian tissues. The strategy for assaying collagenase is outlined in Fig. 1. Tissue collagenase (MMP-1, MMP-8) is bound to the extracellular matrix (part of MMP-8 may still be in cells). The enzyme is in two distinguishable forms, namely, active and latent. There may also be enzyme bound to inhibitors such as tissue inhibitor of metalloproteinases (TIMP); this presents a separate problem not covered in the outline (Fig. 1). Both active and latent forms are largely insoluble but not completely so (typically 80–85%). Most tissues contain several known inhibitors of MMP-1 including TIMP-1, TIMP-2, and $\alpha_2$-macroglobulin, and, in rodents, other members of the macroglobulin family including $\alpha_1$-macroglobulin and $\alpha_1$-inhibitor$_3$ ($\alpha_1 I_3$).[6] If one attempts to measure MMP-1 directly in tissue homogenates, the various inhibitors will interfere and the collagen of the tissue will compete with any added collagen substrate. For those reasons it is desirable to extract the collagenase.

### Triton Extraction

If one extracts tissues with Triton X-100 and 10 m$M$ CaCl$_2$, cells are disrupted and intracellular and membrane-bound enzyme should be released. This step also extracts most of the TIMPs and macroglobulins. The macroglobulins are large ($M_r$ 720,000) compared to collagenase ($M_r$ 44,000–54,000) and the TIMPs are small ($M_r$ 20,000–28,000). Therefore, a molecular sieve column will usually suffice to separate enzyme from inhibitors and permit valid assay. Assay with and without aminophenylmercuric acetate (APMA) will distinguish latent and active forms of MMP-1; use of phenanthroline will block collagenase and provide an appropriate correction for other proteinases that might affect collagen. In general, APMA will block

[6] C. Zhu and J. F. Woessner, Jr., *Biol. Reprod.* **45**, 334 (1991).

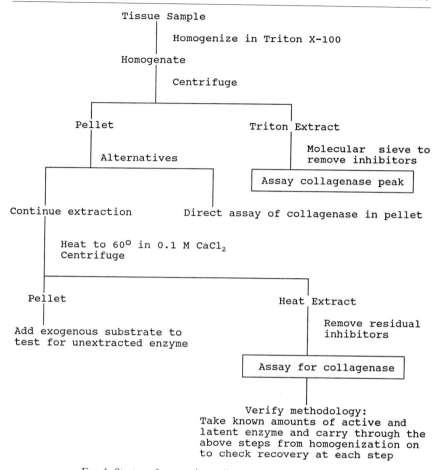

Fig. 1. Strategy for assaying collagenase in tissue samples.

any cysteine enzyme such as cathepsin B, and 4-(2-aminoethyl)benzenesulfonyl fluoride (AEBSF) may be added to block serine enzymes.

## Heat Extraction

The remaining enzyme is extracted with a heating step in the presence of a relatively high calcium concentration. This brings out most of the remaining collagenase. By this time, inhibitors are largely absent, but it may be prudent to check the ability of reduction and alkylation to destroy any residual TIMPs. Active and latent enzyme recovered here may be used to test the efficacy and recovery at each of the preceding steps by adding these back to tissue minces prior to the homogenization step.

## Residual Enzyme

The final residue can now be checked for remaining enzyme. The endogenous collagen has been at least partly denatured by heating, so exogenous substrate must be added. One can measure the amount of denatured collagen by means of its hydroxyproline content and test such amounts in the collagenase assay to determine the extent of interference that might be expected. Typically, one finds about 5–10% of enzyme in the residue. So the total enzyme is distributed about 15% in Triton extract, 80% in heat extract, and 5% in residue. Of course, each of the above steps may require modification for different types of tissue. A final problem is that the methods do not distinguish interstitial collagenase (MMP-1) from neutrophil collagenase (MMP-8). Molecular sieving and specific antibodies will help to resolve this aspect.

The overall approach may therefore be broken down into three segments: methods of collagenase assay, extraction of enzyme from the tissue, and destruction or removal of inhibitors.

## Assay of Collagenase Activity

There are almost as many methods for collagenase assay as there are groups working in the field. This points up the generally unsatisfactory nature of these assays caused by difficulties in preparing substrate, in distinguishing digestion products from substrate, and in reducing high blanks. One common assay in wide use is the fibril assay of Cawston and Barrett, which is described elsewhere in this series by Cawston and Murphy[7] and is not repeated here. The direct pellet assay[8] is described here as a method employing endogenous collagen substrate already present in the tissue sample. This is shown in Fig. 1 as an alternative to the heat extraction step. The telopeptide-free collagenase assay[9] is recommended for assaying extracts because it avoids a difficulty that may be encountered in using the fibril assay for crude systems, namely, enzymes may digest the telopeptide domains of collagen, freeing molecules from the fibrils without cleaving the triple-helical domain. It is always wise to check the digestion products by sodium dodecyl sulfate–polyacrylamide gel electrophoresis (SDS–

---

[7] T. E. Cawston and G. Murphy, this series, Vol. 80, p. 711.
[8] J. F. Woessner, Jr., and J. N. Ryan, *Biochim. Biophys. Acta* **309,** 397 (1973).
[9] D. D. Dean and J. F. Woessner, Jr., *Anal. Biochem.* **148,** 174 (1985).

PAGE) to be certain the typical three-quarter and one-quarter fragments of tropocollagen are produced by the enzyme.[10]

## Pellet Assay

In spite of the foregoing cautions, it is possible in many cases to obtain a reasonable estimate of collagenase directly in tissue homogenates.[8] To accomplish this, one takes advantage of the relative solubility of macroglobulins and TIMP and the insolubility of collagenase in the Triton X-100 extraction. This first extraction removes most of the inhibitory materials, and the enzyme can then be assayed by its action on endogenous collagen or on added radioactive substrate.

### Reagents

Homogenizing fluid: 0.25% (v/v) Triton X-100 dissolved in 10 m$M$ CaCl$_2$

Assay buffer: 50 m$M$ Tris-HCl, pH 7.5, containing 0.15 $M$ NaCl, 10 m$M$ CaCl$_2$, 0.03% (w/v) Brij 35, and 0.02% (w/v) NaN$_3$

*Procedure.* The tissue to be studied is first homogenized in cold Triton fluid. A tissue such as uterus is thoroughly rinsed, minced, and homogenized with a Ten-Broeck glass homogenizer. A tougher tissue such as cartilage requires a high-speed device such as the Polytron (Brinkmann, Westbury, NY). Skin and wound tissue can be disrupted in liquid nitrogen with a Bessman Tissue Pulverizer (Spectrum, Los Angeles, CA). In each case, 20 ml of fluid is added for each gram of wet tissue. Collagenase activity is not affected by freezing; tissues frozen for years at $-20°$ will retain full activity. When the tissue is thoroughly dispersed, the homogenate is centrifuged at 6000 $g$ for 30 min at 4°.

The resultant pellet is resuspended in 10 volumes of assay buffer and distributed to Beckman 12-ml polyallomer centrifuge tubes (2 ml/tube). Latent enzyme can be activated by adding 1 m$M$ APMA; active enzyme is measured without addition. Both types of measurements require that blanks be prepared with 30 m$M$ EDTA or 2 m$M$ 1,10-phenanthroline. The tubes are tightly capped with rubber stoppers or with centrifuge tube closures. They are incubated for 18 hr at 37° in a Dubnoff metabolic shaker at a shaking rate sufficient to keep the insoluble pellet material in suspension. At the end of incubation the tubes are chilled and centrifuged for 30 min at 30,000 $g$. The supernatant will contain collagen digestion products, and the pellets contain undigested collagen.

[10] M. Ågren, C. J. Taplin, J. F. Woessner, Jr., W. H. Eaglstein, and P. M. Mertz, *J. Invest. Dermatol.* **99,** 709 (1992).

The collagen content of both supernatant and pellet is determined by hydrolysis and hydroxyproline assay. The pellet and the supernatant are each transferred to Pyrex tubes with Teflon-lined caps. Two milliliters of 12 $N$ HCl is added to the supernatant and 2 ml of 6 $N$ HCl to the pellet. (The addition can be made by way of the centrifuge tubes to rinse out remnants of pellet.) The tubes are tightly capped and heated at 110° for 18 hr to hydrolyze the protein. Aliquots of hydrolysate are neutralized with 10 $N$ and 1 $N$ NaOH and neutral red indicator. Hydroxyproline is determined by any convenient method such as the Ehrlich's reagent colorimetric method.[11] The total hydroxyproline in pellet and supernatant are added to give total collagen content; the portion in the supernatant measures percent digestion. Hydroxyproline content can be converted to approximate collagen equivalents by multiplication by the factor 7.46.

Comments. The method has been widely applied to a variety of tissues in many laboratories. It has also been adapted to very small samples of tissue (30 mg).[12] However, it is cumbersome in that incubation, hydrolysis, and hydroxyproline assay are time-consuming. A further difficulty is that one cannot be certain that all inhibitors have been removed by the first extraction in Triton. Also, in the uterus it is found that a serine proteinase is present that activates part of the latent MMP-1 during the assay.[13] Weeks et al.[3] have shown that one can add radioactive soluble collagen to the pellet and achieve digestion of the exogenous collagen. That approach may be simpler than determining hydroxyproline content. However, if the endogenous collagen content varies in different samples to be compared, that collagen will compete to different extents with the added substrate. A way to correct for this is presented by Ågren et al.,[10] but one must again determine hydroxyproline content. In general, it would probably be best to add labeled collagen and use the results as a general indication of the presence of significant amounts of enzyme. Detailed study of changing enzyme levels would be better accomplished by extraction and use of the following assay.[9]

*Telopeptide-Free Collagen Assay*

*Reagents*

Carrier collagen: Acid soluble collagen is prepared from rat skins by a lengthy procedure given in detail elsewhere in this series.[7] The

---

[11] J. F. Woessner, Jr., *Arch. Biochem. Biophys.* **93**, 440 (1961).

[12] T. I. Morales, J. F. Woessner, D. S. Howell, J. M. Marsh, and W. J. LeMaire, *Biochim. Biophys. Acta* **524**, 428 (1978).

[13] J. F. Woessner, Jr., *Biochem. J.* **161**, 535 (1977).

final product in 0.5 $M$ acetic acid may be dialyzed into 50 m$M$ Tris-HCl, pH 7.5, containing 0.3 $M$ NaCl and 0.02% (w/v) NaN$_3$. This may be frozen in small aliquots at $-70°$; once frozen, the collagen should not be thawed and refrozen. The collagen content is measured by hydroxyproline determination[11]

Radioactive collagen substrate: Collagen as prepared above is labeled at the last stage before freezing. Collagen in 0.5 $M$ acetic acid is dialyzed into 10 m$M$ disodium tetraborate, pH 9.0, containing 0.2 $M$ CaCl$_2$. The concentration should be 2.5–4 mg/ml. Acetylation is performed using 1 mCi of [H$^3$]acetic anhydride for each 20 mg collagen. The label is transferred in dry dioxane and stirred for 30 min at 4°. The collagen is dialyzed against 50 m$M$ Tris, 0.3 $M$ NaCl, 5 m$M$ calcium acetate until free label is removed. The product is precipitated by 30% saturated ammonium sulfate, followed by dialysis against 0.5 $M$ acetic acid. The solution is pepsinized by adding bovine pepsin A (3000 U/mg) in a ratio of 300 U pepsin:1 mg collagen. The mixture is stirred for 6 hr at 24° and overnight in the cold. Collagen is precipitated with 1.7 $M$ NaCl, collected, and dissolved in 50 m$M$ Tris, pH 7.5, 0.3 $M$ NaCl, and 2 $M$ urea. It is then dialyzed as in the final step for carrier collagen and frozen. Analysis by SDS–PAGE is used to determine that telopeptides have been completely cleaved (no $\beta$ bands or dimers seen); if not, repeat the digestion cycle

Assay buffer: 50 m$M$ Tris, 0.2 $M$ NaCl, 10 m$M$ CaCl$_2$, 0.05% (w/v) Brij 35, 0.02% (w/v) NaN$_3$, pH 7.5

1,10-Phenanthroline: 0.1 $M$ in ethanol

Aminophenylmercuric acetate: 25 m$M$ APMA in 0.1 $N$ NaOH

Secondary proteinases: 2.5 mg trypsin and 2.5 mg chymotrypsin in 2 ml assay buffer

Buffered EDTA: 50 m$M$ Tris, 0.2 $M$ NaCl, 0.2 $M$ Na$_4$EDTA, pH 7.8

Trichloroacetic acid (TCA): 20% (w/v) solution

*Procedure.* Pipette 17 $\mu$g substrate collagen into a 1.5-ml microcentrifuge tube using a positive displacement pipette (Microman, Rainin, Woburn, MA). Add enzyme preparation and buffer to a total volume of 110 $\mu$l. To measure latent enzyme, add 2.2 $\mu$l APMA. For blanks add 2.2 $\mu$l phenanthroline stock. Incubate for 18 hr at 30.0° ± 0.1°. Cool tubes on ice for 10 min and add the following to each tube: 200 $\mu$g carrier collagen (about 65 $\mu$l), 35 $\mu$l EDTA, and 17 $\mu$l secondary proteinase mixture. Conduct a secondary digestion for 90 min at 31.5°. Chill tubes on ice and add an equal volume of ice-cold 20% TCA, vortex, let stand for 20 min, and then centrifuge in microcentrifuge. Transfer 100 $\mu$l of supernatant into a 7-ml scintillation vial containing 5 ml Cytoscint (ICN, Irvine, CA), mix, and count in a liquid scintillation counter. A standard is prepared by pipetting 17 $\mu$g substrate directly into a scintillation vial. One unit of collagenase digests 1 $\mu$g collagen/min. Purified human collagenase has a specific activity of about 5000 U/mg in the assay.

*Comments.* This method avoids complications arising from the effects of proteinases on the telopeptide region of the collagen molecule. However, there are still difficulties with the assay. A trypsin control can be run by adding collagen, buffer, and 1 $\mu$g trypsin per tube. This typically gives about 7% digestion owing to the fact that even at 30° there is a slow denaturation of telopeptide-free collagen. There is some digestion of collagen in every tube during the secondary digestion, giving an appreciable blank which is corrected by the phenanthroline control. Collagen is viscous and sticks to glass, so it is important to use a positive displacement pipette and plastic tips. The temperature of the bath is also critical and should not be allowed to vary more than 0.1°. In the study of crude tissue extracts one may want to add inhibitors of serine proteinases such as soybean trypsin inhibitor (SBTI) to block unwanted enzymes and to prevent activation of latent collagenase. However, this will also block the enzymes of secondary digestion which must then be added in excess. By the same token, use of trypsin to activate latent enzyme, followed by SBTI to block trypsin, is less desirable than activation by APMA. To be certain that collagenase is actually present, the digests (before secondary treatment) can be analyzed by SDS–PAGE to show the typical three-quarter and one-quarter fragments of the collagen $\alpha$ chains ($M_r$ 75,000 and 25,000).[10]

### Extraction of Collagenase from Tissue Samples

In many tissues examined to date about 80% of interstitial collagenase is not readily soluble in aqueous buffers or detergents. A number of papers have been published in which simple buffers and salt solutions are used to extract measurable amounts of collagenase. However, in no case was it shown how much of the total enzyme was recovered. Therefore, changes in amount of enzyme recovered in various diseased states or following various treatments may be suspect because of possible changes in extractability rather than changes in total enzyme.

Two methods of extraction have found wide applicability in the literature: heat extraction[3] and the use of 4–5 $M$ urea in buffer as described originally by Wirl.[14] Although Wirl tried a number of extraction methods, he did not check for unextracted enzyme as Weeks *et al.* did.[3] We have compared the use of 5 $M$ urea versus heat in a study of pig skin wounds

---

[14] G. Wirl, *Connect. Tissue Res.* **5,** 171 (1977).

and found that the heating method recovered twice as much enzyme as the urea method.[10] Therefore, the former is described here.

*Reagents*

Triton solution: 0.25% (v/v) Triton X-100 dissolved in 10 m$M$ CaCl$_2$
Heat extract solution: 50 m$M$ Tris, pH 7.5, containing 0.1 $M$ CaCl$_2$ and 0.15 $M$ NaCl
Assay buffer: 50 m$M$ Tris-HCl, pH 7.5, containing 0.2 $M$ NaCl, 10 m$M$ CaCl$_2$, 0.03% (w/v) Brij 35, and 0.02% (w/v) NaN$_3$

*Procedure.* For collagenase extraction,[3] the tissue is weighed and homogenized as described under "Pellet Assay" above. Twenty milliliters of Triton solution is used for each gram of wet tissue. The homogenates are centrifuged at 2° for 30 min at 6000 g. The supernatants are reserved (Triton extract), and the pellets are then resuspended by homogenizing in 20 volumes of heat-extraction fluid. The suspension is distributed to 50-ml metal tubes (to promote rapid heat exchange). The tubes are placed in a 60° bath for 4 min with constant agitation, then chilled in an ice–water bath. After centrifugation at 20,000 g for 30 min at 4°, the supernatants are collected (heat extract). Both fractions could be assayed at this stage, but it would be best to remove inhibitors (see next section) before proceeding with the assay.

*Comments.* The method outlined has been applied to a wide variety of tissues. However, it will not be suitable in every case. For example, Eeckhout *et al.* found that heating destroyed collagenase activity in the case of bone tissue.[15] In the case of human cartilage, reducing the CaCl$_2$ concentration from 0.1 $M$ to 10 m$M$ in the heating step resulted in greater recovery of enzyme (J. F. Woessner, Jr., unpublished, 1982). In the case of pig wounds, the heating step was extended to 6 min. This improved recovery of total enzyme, but at the expense of loss of much of the active enzyme.[10] The optimum conditions must be worked out for each tissue anew. Active and latent enzyme can be added to check recovery during homogenization and heating. The final pellet can be checked for completeness of extraction by adding radioactive substrate and correcting for endogenous collagen interference as we have described.[10] The pellet can also be suspended in sample buffer and subjected to collagen zymography as described by Birkedal-Hansen.[16]

Removal of Interfering Inhibitors from Extracts

As mentioned above, the major tissue inhibitors are the macroglobulins and the TIMPs. All members of the macroglobulin family are susceptible

[15] Y. Eeckhout, J. M. Delaissé, and G. Vaes, *Biochem. J.* **239**, 793 (1986).
[16] H. Birkedal-Hansen, this series, Vol. 144, p. 140.

to inactivation by treatment with methylamine to disrupt the thiol ester bond. All except $\alpha_1I_3$ are also susceptible to reduction, which dissociates the subunits. The TIMPs are susceptible to reduction and alkylation, which disrupts the six critical disulfide bridges in the molecules.

## Molecular Sieve Method

*Procedure.* Ultrogel AcA 54 (Sepracor, Marlborough, MA) has been found to be suitable for removing inhibitors from extracts. A typical separation is illustrated in Fig. 2: the void volume material includes $\alpha_2$-macroglobulin and $\alpha_1I_3$, gelatinases barely enter the column, collagenase is somewhat retarded, and TIMPs come shortly after. Because the fractions are already in an assay buffer, collagenase can be determined on aliquots as large as 90 $\mu$l in the assay described above. The APMA reagent should be added to detect latent forms of MMP-1. Active forms should be detectable without further addition or treatment. Inhibitor fractions can be identified by adding them back to fractions containing collagenase that has been previously activated with APMA. After preincubation for 1 hr, the assay for collagenase is carried out.

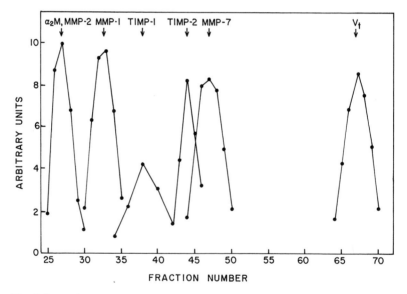

FIG. 2. Separation of collagenase from inhibitors on an Ultrogel AcA 54 column. A sample of 3 ml was applied to a $1 \times 115$ cm column of AcA 54 and eluted with sample buffer; fractions of 1.5 ml were collected. The position of peaks is indicated along the top. $\alpha_2$M, $\alpha_2$-Macroglobulin; MMP-2, position of gelatinases A and B; MMP-1, collagenase; MMP-7, matrilysin.

If large numbers of samples are to be examined, small columns of AcA 54 may be employed, for example, 0.7 × 17 cm disposable columns (Bio-Rad, Richmond, CA), with a sample volume of 0.3 ml. After the relative positions are established for all components from the large column, it is usually possible to adjust the collected volume of effluent from such small columns into three pools (void, collagenase, TIMPs) so that only one fraction needs to be assayed to determine the total enzyme content.

## Chemical Destruction of Inhibitors

*Procedure.* The use of methylamine is combined with reduction/alkylation to destroy all presently known inhibitors. For human tissue extracts, reduction/alkylation should suffice; for rodents, both steps are needed because $\alpha_1 I_3$ is present. However, studies by Rubenstein *et al.* suggest that other mammals may also have this inhibitor.[17] Extracts (Triton or heat) are treated by adding sufficient 100 m$M$ dithiothreitol (DTT) to give a final concentration of 2 m$M$. Extracts are incubated at 37° for 30 min followed by addition of sufficient 100 m$M$ iodoacetamide to give a final concentration of 5 m$M$. After a further incubation of 30 min the reagents may be removed by dialysis. However, if $\alpha_1 I_3$ is present or if one wishes to destroy macroglobulins independently, one should add 2 $M$ methylamine (base) (Sigma, St. Louis, MO) to a final concentration of 0.2 $M$ and allow the mixture to stand at 24° for 30 min. At this point all reagents are removed by dialysis overnight against cold assay buffer.

For small samples (0.3 ml), it is convenient to remove the reagents by use of a spin column as follows. Small columns of Sephadex G-10 (Pharmacia, Piscataway, NJ) are prepared by equilibrating the beads with assay buffer and packing 1.6 ml into a 2-ml disposable glass column (Bio-Rad). The columns are washed with buffer, which is then drained by gravity. The columns are placed inside tubes and centrifuged at 800 $g$ for 4 min at room temperature. An enzyme sample of 0.3 ml is added, and centrifugation is repeated under the same conditions of speed and time. Enzyme should appear in the effluent, and the various reagents should remain in the gel. The volume of enzyme or gel may have to be adjusted slightly to improve the separation (check with Blue Dextran and methyl red markers). Known amounts of enzyme are spun under the same conditions to check recovery. The treated samples are then assayed by the telopeptide-free assay and corrected for volume changes caused by adding reagents, conducting dialysis, etc.

*Comments.* The method is quicker than the use of a sieve column. It is advisable to add known amounts of active enzyme to extracts to be certain

---

[17] D. S. Rubenstein, I. B. Thogersen, S. V. Pizzo, and J. J. Enghild, *Biochem. J.* **290,** 85 (1993).

no destruction occurs in that step. In general, there is so much protein in crude extracts that the low levels of DTT do not appear to disrupt enzyme disulfide bridges; however, some losses have been noted with tissue culture media.[18] There is a further problem that has not yet been studied in detail. Gelatinase A is frequently found with TIMP-2 bound to its C-domain; gelatinase B similarly may have associated TIMP-1. It is not clear that these forms of TIMPs will be completely destroyed by reduction and alkylation, owing to protection by the attached protein. Later, during assay, some TIMP might be transferred to, or interact with, the active center of collagenase, causing a reduction in activity.[19] With the use of an AcA 54 column, the gelatinases will be close to the void volume and will not interfere with the assay of collagenase in later fractions. An alternative approach is to titrate the inhibitors by addition of a noncollagenolytic metalloproteinase such as stromelysin.[18] This has the disadvantage of activating latent collagenase if added in excess.

It is beyond the scope of this chapter to discuss the measurement of inhibitors. However, details of measurements of $\alpha_2$-macroglobulin, $\alpha_1 I_3$, and TIMP in the rat ovary are described by Zhu and Woessner.[6] Similarly, TIMP measurements have been made in extracts of human cartilage, together with measurements of metalloproteinases.[20]

Yamashita et al.[21] show that collagenase–TIMP complexes may be dissociated by use of an anti-TIMP antibody column. Clark et al.[22] have developed an immunoassay which determines the amount of complex. Methods to dissociate complexes of collagenase with $\alpha_2$-macroglobulin are presented by Abe and Nagai.[23]

## Measurement of Stromelysin in Tissue Samples

When we go beyond the measurement of collagenase to consider other matrixins, we encounter a further difficulty, namely, the lack of a specific substrate. Work has been directed to the development of substrates that will distinguish among the various metalloproteinases, just as we now can differentiate various blood-clotting enzymes. However, it is too early to

---

[18] V. Lefebvre and G. Vaes, *Biochim. Biophys. Acta* **992,** 355 (1989).

[19] Y. Itoh, Y. Ogata, and H. Nagase, *FASEB J.* **7,** A1210 (1993).

[20] D. D. Dean, J. Martel-Pelletier, J.-P. Pelletier, D. S. Howell, and J. F. Woessner, Jr., *J. Invest. Dermatol.* **84,** 678 (1989).

[21] K. Yamashita, J. Zhang, L. Zou, H. Hayakawa, T. Noguchi, I. Kondo, O. Narita, N. Fujimoto, K. Iwata, and T. Hayakawa, *Matrix* **12,** 481 (1992).

[22] I. M. Clark, J. K. Wright, T. E. Cawston, and B. L. Hazleman, *Matrix* **12,** 108 (1982).

[23] S. Abe and Y. Nagai, *J. Biochem.* (*Tokyo*) **73,** 897 (1973).

describe the accurate measurement of crude tissue extracts with such substrates. Similarly, specific inhibitors are currently under development which would also greatly facilitate such assays.

In our laboratory, we have used cartilage proteoglycan incorporated in polyacrylamide beads as a sensitive substrate for stromelysin.[24] However, this is not a specific substrate, and it is relatively complicated to prepare and to distribute the substrate to the assay tubes. Therefore, the transferrin assay is recommended for the study of stromelysin. The details of the method are given in full by Nagase,[25] so they need not be repeated here. Stromelysin may be extracted by the same methods as outlined above for collagenase (e.g., by Triton and heat extraction). However, we have found in the case of cartilage that a reasonably good recovery can be made using 2 $M$ guanidine hydrochloride (GuHCl).[20]

*Procedure.* Pieces of cartilage are diced into small fragments of 1–2 mm$^3$ and homogenized in a Polytron with 10 volumes of 50 m$M$ Tris, pH 7.5, containing 10 m$M$ CaCl$_2$ and 2 $M$ guanidine hydrochloride. The supernatant is collected by centrifugation at 21,000 $g$ for 30 min at 4°. The resultant extract contains large amounts of proteoglycan which interfere with the transferrin assay. Therefore, the extract is diluted 4-fold with DEAE buffer (20 m$M$ Tris-HCl, pH 8.0, 5 m$M$ CaCl$_2$). DEAE-Sephacel (Pharmacia) is equilibrated with the same buffer. For each 40 ml of extract add 20 ml settled DEAE gel and stir gently at room temperature for 2 hr. The slurry is then packed into a wide column. The fall through and 1 volume of buffer rinse are combined; this should contain the enzyme, and the proteoglycan should remain bound to the column. The supernatant is then treated with dithiothreitol and iodoacetamide as described above, followed by overnight dialysis against a large volume of assay buffer. Samples are then concentrated 8-fold on an Amicon (Danvers, MA) YM5 membrane and assayed with transferrin.[25]

*Comments.* The use of guanidine was found to give better recoveries of stromelysin than the classic heating method used for collagenase. At the same time, there is good recovery of TIMP. Cartilage does not have appreciable amounts of $\alpha_2$-macroglobulin, but this could be important in other tissues and may warrant further treatment with methylamine. Stromelysin can be further identified in such extracts by zymography on casein or transferrin substrate gels. There are at least two active forms at $M_r$ 48,000 and 28,000 as well as the latent form visualized at 54,000. No other enzyme gives such a pattern.

[24] H. Nagase and J. F. Woessner, Jr., *Anal. Biochem.* **107,** 385 (1980).
[25] H. Nagase, this volume [27].

## Measurement of Matrilysin in Tissue Extracts

Methods for the extraction of matrilysin from rat uterus are given elsewhere in this volume.[26] The methods are the same as those used for collagenase. We have not found the enzyme in extracts of rat ovary. Matrilysin may be assayed on transferrin or on Azocoll. The latent form of the enzyme is somewhat retarded on Ultrogel AcA 54 columns run under the conditions described above, so it should emerge well after latent collagenase and stromelysin (Fig. 2, $K_{av}$ of 0.5). This is probably the best way to distinguish the enzyme in crude preparations. It can also be identified (but not unambiguously) on transferrin zymograms.[26]

## Measurement of Gelatinases A and B in Tissue Extracts

Measurement of gelatinases A and B in tissue extracts is an area that is still under development. Assays for gelatinase using denatured collagen substrates have been available for a long time. Details of the assay methods are given elsewhere.[27] However, as mentioned above, gelatinase A (72-kDa gelatinase, MMP-2) often has a molecule of TIMP-2 attached to its C-terminal domain, and gelatinase B (92-kDa gelatinase, MMP-9) similarly is found in association with TIMP-1. This produces complications such that there is no clear agreement on how to assay pure complexed gelatinases, let alone crude extracts. It is known that the TIMP molecule is bound through its C-terminal domain to the C-terminal domain of gelatinases and that on activation of latent enzyme the TIMP molecule may rearrange itself in relation to the enzyme so that it now blocks the active center.[28] Other matrixins that become active may also interact with the TIMP bound to the C-terminal domain of neighboring molecules.[19] Furthermore, active enzyme on incubation at 37° may become inactivated but when examined by zymography still show full activity (J. F. Woessner, Jr., unpublished, 1993).

In the face of the various complications, crude extracts should be examined by zymography. This will reveal latent and active forms at the proper molecular weight positions, even if they have bound TIMP prior to the SDS–PAGE step. Zymography indicates whether the enzyme has undergone an activation step *in vivo*, even if it is uncertain how much activity such a form will display in the presence of TIMP. Zymography can be made at least partially quantitative by the use of gel scanners and the

---

[26] J. F. Woessner, Jr., this volume [29].
[27] G. Murphy and T. Crabbe, this volume [28].
[28] D. E. Kleiner, Jr., A. Tuuttila, K. Tryggvason, and W. G. Stetler-Stevenson, *Biochemistry* **32,** 1583 (1993).

inclusion of reference enzyme with each gel. The following method is based on the work of Hibbs and co-workers.[29]

## Zymographic Assay of Gelatinases

### Reagents

Gelatin: Mix gelatin (Sigma, porcine skin, 300 bloom) at a concentration of 5 mg/ml in water and heat to 60° until dissolved

Polyacrylamide gel: The standard method of Laemmli[30] is followed. Gels of 55 × 80 × 0.75 mm are cast using 5 ml of gel. The gel comprises 1.25 ml aqueous acrylamide (375 mg)/bis-acrylamide (10 mg); 1.25 ml 1.4 $M$ Tris–HCl, pH 8.8, containing 0.4% w/v SDS; and water; 0.5 ml gelatin solution; 25 $\mu$l of 10% w/v ammonium persulfate; and 2 ml water, 0.5 ml gelatin solution, 25 $\mu$l of 10% ammonium persulfate, and 25 $\mu$l $N,N,N',N'$-tetramethylethylenediamine (TEMED). The final gel strength is 7.5%, and the gelatin content is 0.5 mg/ml. A 5-mm stacking gel is added at the top

Sample buffer: 4× sample buffer is prepared according to Laemmli but omitting any reducing agent

Incubation buffer: 50 m$M$ Tris-HCl, pH 7.5, containing 10 m$M$ CaCl$_2$, 1 $\mu M$ ZnCl$_2$, 1% (v/v) Triton X-100, and 0.02% (w/v) NaN$_3$

*Procedure.* Enzyme samples of 30 $\mu$l are mixed with 10 $\mu$l sample buffer and held at room temperature for at least 30 min to allow coating of the protein with SDS. Boiling is not recommended, and reducing agents are omitted because of possible interference with subsequent refolding of gelatinase. Aliquots of 25 $\mu$l are applied to each lane of the gel. Gelatinase standards must be included on each gel. Molecular weight markers may also be included if desired (but the gelatinase standards provide adequate reference markers). The gels are run in a PROTEAN II apparatus (Bio-Rad) at 200 V constant voltage until the dye front nears the end. It is important to keep the gels cold during the run.

When electrophoresis is complete, the gels are separated from the glass plates and washed in Pyrex staining dishes (10 × 12 cm) with 100 ml of 0.25% Triton X-100 (two times, 15 min each). The washing, incubation, and staining are all done with gentle shaking of the gels in a Dubnoff shaker or on a rotatory shaker. This first step removes SDS from the gel and proteins. The gels are then incubated overnight (16 hr) with incubation buffer. The buffer is decanted, and the gels are stained with 0.1% Coomassie blue in 40% 2-propanol for a minimum of 1 hr. They are then destained

[29] J. A. Lorenzo, C. C. Pilbeam, J. F. Kalinowski, and M. S. Hibbs, *Matrix* **12,** 282 (1992).
[30] U. K. Laemmli, *Nature* (*London*) **277,** 680 (1970).

in 7% acetic acid. The gels may be measured when wet, but they are easier to handle and can readily be remeasured if dried between two sheets of BioGelWrap (BioDesign, Carmel, NY). When SDS is removed, the gelatinases refold (at least partially) and digest away the gelatin, leaving a clear band against a purple background. The band intensity is determined by scanning densitometry using any of a variety of instruments such as the Zeineh soft laser scanning densitometer (Biomed Instruments, Chicago, IL) or Image Quant Personal Densitometer (Molecular Dynamics, Sunnyvale, CA). The instrument may be zeroed on the supporting glass plate, and the maximum absorption can be set on a uniformly stained portion of the gel.

A standard curve gel and tracing are illustrated in Figs. 3 and 4. The curve will vary from gel to gel and is not linear; thus, it is desirable to apply at least two concentrations of reference enzyme to each gel. Reference gelatinases are not commercially available at this time.

*Comments.* Zymography is based on the ability of SDS-denatured metalloproteinases to refold themselves after SDS is removed. They then digest the gelatin substrate. Even the latent forms develop activity. About 35% of the enzyme activity is recovered during the refolding (Woessner, unpublished, 1994); the latent form may still retain its propeptide domain. The propeptide reduces the digestion of gelatin somewhat so that latent enzyme produces only 60% of the digestion produced by active enzyme on a mole basis. Therefore, the standard curve for latent MMP-9 (Fig. 4) would correspond to a curve of active enzyme in which only 60% as much enzyme was applied. The appropriate factor for latent MMP-2 is 33%.

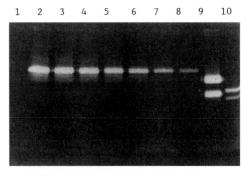

FIG. 3. Gelatin zymography of standards of latent gelatinase B. Lane 1 contained 300 pg of enzyme; lane 2, 250 pg; lane 3, 200 pg; lane 4, 150 pg; lane 5, 150 pg; lane 6, 80 pg; lane 7, 50 pg; lane 8, 25 pg; lane 9, sample of active gelatinase B after storage for 2 months in a refrigerator; lane 10, gelatinase A, latent and active forms (72 and 64 kDa). It should be noted that gelatinase B was complexed with TIMP-1 prior to electrophoresis.

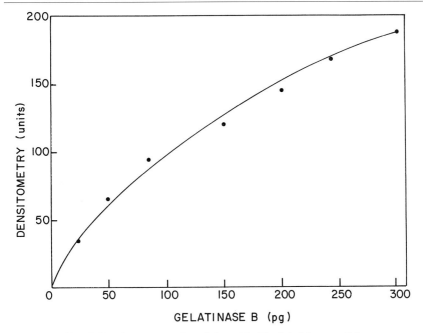

FIG. 4. Densitometry tracing of the gel in Fig. 3 (arbitrary units).

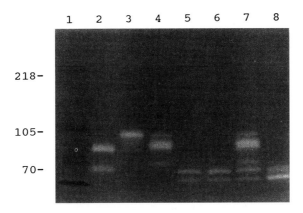

FIG. 5. Gelatin zymography of various gelatinase preparations using a 5% polyacrylamide gel. Lane 1 contained reduced stained markers; lane 2, preparation of active gelatinase B (150 pg) stored for several months in refrigerator; lane 3, gelatinase B (300 pg) largely in latent form and complexed with TIMP-1; lane 4, gelatinase B after treatment with 1 m$M$ APMA for 24 hr; lanes 5 and 6, gelatinase A (1.5 ng), latent and active; lane 7, mixture of lanes 4 and 6; lane 8, extract of involuting rat uterus.

In spite of the widespread use of zymography to quantify gelatinases, it must be recognized that the method is still very crude. The associated TIMP molecules are dissociated, so the activity that is seen on the gels may be much higher than that expressed *in vivo*. A sample gel is illustrated in Fig. 5. Here we see standards of authentic gelatinases. Latent gelatinase A is seen at 72 kDa and the active form at 64 kDa. It is readily apparent that the proteins move anomalously in relation to the markers; this is because the enzymes are not reduced (whereas the markers are). Latent gelatinase B (lane 3, Fig. 5) is at 92 kDa and on activation with APMA gives bands at about 85 and 80 kDa. This preparation contained bound TIMP-1, so according to Morodomi et al.[31] it would not digest gelatin prior to electrophoresis. Lane 1 (Fig. 5) shows a TIMP-free preparation of gelatinase B that was stored for months in the refrigerator. It has become completely active and shows a further band at 68 kDa. It will be seen that this band overlaps those for gelatinase A and would complicate measurement of both enzymes. Additional bands will be seen in crude extracts. Gelatinase B has an associated 25-kDa protein (microglobulin) giving a band at 125 kDa and a dimer at 215 kDa[32]; these complexes depend on disulfide bridges which are not dissociated in zymography. Gelatinase A frequently gives a band at 46 kDa.

An important observation that can be made, particularly if a tissue contains only a single species of gelatinase, is whether the enzyme is completely in the latent form or if it has undergone at least partial activation to lower $M_r$ forms. Even if the enzyme contains bound TIMP it is likely to express some activity *in vivo,* typically about one-tenth that expected in the absence of bound TIMP.

### Acknowledgments

Work in the author's laboratory was supported by National Institutes of Health Grants HD-06773 and AR-16940. The assistance of Ms. Carolyn Taplin is gratefully acknowledged. Human gelatinase A and B preparations were generously supplied by Dr. Hideaki Nagase, University of Kansas Medical Center, Kansas City, Kansas.

[31] T. Morodomi, Y. Ogata, Y. Sasaguri, M. Morimatsu, and H. Nagase, *Biochem. J.* **285,** 603 (1992).
[32] S. Triebel, J. Bläser, H. Reinke, and H. Tschesche, *FEBS Lett.* **314,** 386 (1992).

# [32] Thimet Oligopeptidase and Oligopeptidase M or Neurolysin*

*By* Alan J. Barrett, Molly A. Brown, Pamela M. Dando,
C. Graham Knight, Norman McKie, Neil D. Rawlings,
and Atsushi Serizawa

## Introduction

Thimet oligopeptidase (TOP) has been "discovered" independently on a number of different occasions, and only recently have the several separate identities of the enzyme been unified. *Pz-peptidase* was a eukaryotic enzyme that hydrolyzed the Pz-peptide, 4-phenylazobenzyloxycarbonyl-Pro-Leu+Gly-Pro-D-Arg,[1] which had been designed as a synthetic substrate for bacterial collagenase.[2] Much of the work on Pz-peptidase was stimulated by the idea that the enzyme might participate in the physiological turnover of collagen.[3] Other peptide substrates, lacking the Pz blocking group but retaining the Pro-Xaa-Gly sequence reminiscent of the repeating unit of collagen, were also used, and for these the activity was termed *collagenase-like peptidase.*[4] The catalytic type of the peptidase was unclear, as it had some characteristics of both cysteine peptidases and metallopeptidases, and accordingly it was for a time catalogued as of unknown catalytic type, EC 3.4.99.31.

*Endo-oligopeptidase A* is the latest in a series of names (including kininase A) given to an enzyme from brain (especially rabbit) that was

---

* *Note added in proof:* At the time of writing the present chapter, it seemed probable that oligopeptidase M was identical with neurolysin, but the term oligopeptidase M was nevertheless retained pending further confirmation. Additional data now leave almost no doubt that the two are the same, and in these circumstances the name neurolysin, recommended by IUBMB, is to be preferred.

[1] Abbreviations and conventions: Cpp, *N*-[1-(*RS*)-carboxy-3-phenylpropyl]-; DFP, diisopropyl fluorophosphate; DTT, dithiothreitol; Mca, (7-methoxycoumarin-4-yl)acetyl; Mcc, 7-methoxycoumarin-3-carboxylyl-; NHMec, 7-(4-methyl)coumarylamide; Nle, L-norleucine; pAb, *p*-aminobenzoate; Pz, 4-phenylazobenzyloxycarbonyl; QF01, Dnp-Pro-Leu-Gly-Pro-Trp-D-Lys; QF02, Mcc-Pro-Leu-Gly-Pro-D-Lys(Dnp). The symbol + is used to mark the bond hydrolyzed in a peptide sequence (Ref. 17, p. 372).

[2] E. Wünsch and H. G. Heidrich, *Hoppe-Seyler's Z. Physiol. Chem.* **333,** 149 (1963).

[3] T. I. Morales and J. F. Woessner, Jr., *J. Biol. Chem.* **252,** 4855 (1977).

[4] K. Kojima, H. Kinoshita, T. Kato, T. Nagatsu, K. Takada, and S. Sakakibara, *Anal. Biochem.* **100,** 43 (1979).

discovered by Camargo and co-workers.[5,6] The enzyme cleaved bradykinin and a variety of other biologically active peptides, and it was suggested to be significant in the physiological degradation of such peptides, and perhaps also in the generation of enkephalins. Endo-oligopeptidase A showed clear thiol dependence, and it was considered to be a cysteine-type peptidase, EC 3.4.22.19.[7,8]

*Soluble metalloendopeptidase*, or *endopeptidase 24.15*, was found in rat brain and other tissues by Orlowski and co-workers.[9] These workers noted the similarities in specificity to endooligopeptidase A, but their clear evidence of metallopeptidase character of the rat soluble metalloendopeptidase contrasted with the marked thiol dependence of the rabbit enzyme as described by the Camargo group, and justified the number EC 3.4.24.15.

Research on Alzheimer's amyloidosis has led to the detection of a peptidase in brain that cleaves oligopeptide substrates modeling the site of amyloidogenic cleavage of the Alzheimer's precursor protein at the appropriate bond.[10,11] The enzyme from human brain was named *amyloidin* when it was cloned and sequenced.[12] Following the isolation of human TOP from erythrocytes, Dando *et al.* found that the N-terminal sequence of human TOP is identical to residues 19–24 of amyloidin. Many other properties described for amyloidin agree wtih those of thimet oligopeptidase, and there is now no doubt that the enzymes are one and the same.[13]

When it was shown that activities which had previously been attributed to different enzymes are in fact due to a single enzyme that unusually combines thiol and metal dependence,[14,15] none of the existing names seemed entirely appropriate, so thimet peptidase was proposed, as an acronym for the thiol-dependent metallopeptidase character of the enzyme.[16] The name subsequently recommended in the IUBMB list (1992) was thimet

[5] A. C. M. Camargo, R. Shapanka, and L. J. Greene, *Biochemistry* **12**, 1838 (1973).

[6] A. C. M. Camargo, H. Caldo, and M. L. Reis, *J. Biol. Chem.* **254**, 5304 (1979).

[7] M. D. Gomes, L. Juliano, E. S. Ferro, R. Matsueda, and A. C. M. Camargo, *Biochem. Biophys. Res. Commun.* **197**, 501 (1993).

[8] H. Taguchi, Y. Nishiyama, A. C. M. Camargo, and Y. Okada, *Chem. Pharm. Bull.* **41**, 2038 (1993).

[9] M. Orlowski, C. Michaud, and T. G. Chu, *Eur. J. Biochem.* **135**, 81 (1983).

[10] J. R. McDermott, J. A. Biggins, and A. M. Gibson, *Biochem. Biophys. Res. Commun.* **185**, 746 (1992).

[11] G. Papastoitsis, R. Siman, R. Scott, and C. R. Abraham, *Biochemistry* **33**, 192 (1994).

[12] H. F. Dovey, P. A. Seubert, S. Sinha, L. Conlogue, S. P. Little, and E. M. Johnstone, Patent WO 92/07068 (1992).

[13] P. M. Dando, M. A. Brown, and A. J. Barrett, *Biochem. J.* **294**, 451 (1993).

[14] A. J. Barrett and U. Tisljar, *Biochem. J.* **261**, 1047 (1989).

[15] U. Tisljar, A. C. M. de Camargo, C. A. da Costa, and A. J. Barrett, *Biochem. Biophys. Res. Commun.* **162**, 1460 (1989).

[16] U. Tisljar and A. J. Barrett, *Biochem. J.* **267**, 531 (1990).

oligopeptidase, EC 3.4.24.15,[17] which also takes note of the unusual substrate-size dependence of the enzyme (see below). Here, we abbreviate thimet oligopeptidase as TOP, for convenience.

*Oligopeptidase M or Neurolysin.* Like that of TOP, the identity of oligopeptidase M (MOP) has emerged by the interweaving of several apparently unrelated lines of research. In early work with the Pz-peptide, activity was detected in mitochondria, and the question was raised whether there might be a mitochondrial collagenase.[18] Work by Tisljar and Barrett subsequently established that the enzyme is a mitochondrial metallopeptidase that is distinct from both collagenase and TOP, despite similarities in substrate specificity.[19]

Quite separately, research directed at the identification of receptors for angiotensin II led to the detection and cloning of a soluble angiotensin II-binding protein, and the raising of a monoclonal antibody, M3C.[20,21] The sequence of the soluble angiotensin-binding protein showed it to be closely related to TOP, but even more closely related to a microsomal processing endopeptidase.[22] The microsomal metalloendopeptidase was believed to be responsible for the posttranslational processing of γ-carboxyglutamic acid-containing blood coagulation factors.[23,24]

More work on the mitochondrial peptidase in our laboratory has unified these lines of research. The isolation of the rat enzyme has led to the discovery that it has an N-terminal, 20-amino acid sequence in which 19 of the residues are identical to the sequence from Ser-14 in the rabbit "microsomal" enzyme (numbered as in Fig. 1; Ser-38 in the mitochondrial oligopeptidase sequence), and thus it is clearly the same protein. Moreover, the rat mitochondrial enzyme is recognized by the antibody against the rabbit soluble angiotensin-binding protein.[25] We can conclude that the mitochondrial enzyme is a metallooligopeptidase that is identical with both the soluble angiotensin-binding protein and the microsomal processing endopeptidase. The enzyme was initially termed "thimet peptidase II,"[19] but we here use the name "oligopeptidase M" (abbreviated MOP, for mitochondrial oligopeptidase). MOP is quite distinct from the homologous

[17] Nomenclature Committee of the International Union of Biochemistry and Molecular Biology, "Enzyme Nomenclature 1992." Academic Press, Orlando, Florida, 1992.
[18] H. G. Heidrich, D. Prokopova, and K. Hannig, *Hoppe-Seyler's Z. Physiol. Chem.* **350,** 1430 (1969).
[19] U. Tisljar and A. J. Barrett, *FEBS Lett.* **264,** 84 (1990).
[20] N. Sugiura, H. Hagiwara, and S. Hirose, *J. Biol. Chem.* **267,** 18067 (1992).
[21] R. L. Soffer, M. A. R. Kiron, A. Mitra, and S. J. Fluharty, *Methods Neurosci.* **5,** 192 (1991).
[22] N. McKie, P. M. Dando, N. D. Rawlings, and A. J. Barrett, *Biochem. J.* **295,** 57 (1993).
[23] S. Kawabata and E. W. Davie, *J. Biol. Chem.* **267,** 10331 (1992).
[24] S. Kawabata, K. Nakagawa, T. Muta, S. Iwanaga, and E. W. Davie, *J. Biol. Chem.* **268,** 12498 (1993).
[25] A. Serizawa, P. M. Dando, and A. J. Barrett, *J. Biol. Chem.* **270,** 2092 (1995).

mitochondrial enzyme, mitochondrial intermediate peptidase, which is a highly specific endopeptidase and not an oligopeptidase (see [33] in this volume).

Neurolysin (EC 3.4.24.16) ([36], this volume) is extremely similar in biochemical properties to MOP, and the probability is that again the two enzymes are identical.[25]

Assay of Thimet Oligopeptidase

The assay with Pz-Pro-Leu+Gly-Pro-D-Arg is probably only of historical interest, now. After the incubation, the reaction was stopped by acidification, and the Pz-Pro-Leu product was extracted from the acidic solution into ethyl acetate for determination by spectrophotometry.[3] Alternatively, the product could be determined by high-performance liquid chromatography (HPLC).[26] Hydrolysis of natural peptides, notably bradykinin and dynorphin-related peptides, was long the standard assay in the laboratory of Camargo.[27] The best of the discontinuous assay procedures are those with Bz-Gly+Ala-Ala-Phe-pAb and the quenched fluorescence substrates, described in detail below.

For kinetic analyses, it is highly desirable to have an assay in which the formation of product can be monitored continuously during the reaction. The first such assay for TOP was that of Kojima et al.[4] in which the substrate was Suc-Gly-Pro-Leu+Gly-Pro-NHMec, but release of the fluorescent $NH_2Mec$ required the action of the serine peptidase dipeptidyl-peptidase IV included in the mixture as a reagent. This complication made the assay unsatisfactory for many types of kinetic experiments. The direct fluorimetric assays with Dnp-Pro-Leu+Gly-Pro-Trp-D-Lys (QF01)[28] and Mcc-Pro-Leu+Gly-Pro-D-Lys(Dnp) (QF02)[29] are sensitive, convenient, and suited to most types of experiment. These assays also are described in detail below.

Since the introduction of QF01 and QF02, an enkephalin-based quenched fluorescence substrate has been introduced by Juliano and colleagues.[30] The substrate, o-aminobenzoyl-Gly-Gly-Phe-Leu+Arg-Arg-Val-[N-(2,4-dinitrophenyl)ethylenediamine], is a very sensitive one for TOP, but it is also readily hydrolyzed by peptidases with trypsin-like, chymotrypsin-like, and other specificities.[31] Some analytical procedures that have been

[26] T. Chikuma, Y. Ishii, and T. Kato, J. Chromatogr. 348, 205 (1985).
[27] M. A. Cicilini, M. J. F. Ribeiro, E. B. Oliveira, R. A. Mortara, and A. C. M. Camargo, Peptides 9, 945 (1988).
[28] A. J. Barrett, C. G. Knight, M. A. Brown, and U. Tisljar, Biochem. J. 260, 259 (1989).
[29] U. Tisljar, C. G. Knight, and A. J. Barrett, Anal. Biochem. 186, 112 (1990).
[30] L. Juliano, J. R. Chagas, I. Y. Hirata, E. Carmona, M. Sucupira, E. S. Oliveira, E. B. Oliveira, and A. C. M. Camargo, Biochem. Biophys. Res. Commun. 173, 647 (1990).
[31] P. M. Dando and A. J. Barrett, unpublished work (1994).

developed for the bacterial oligopeptidase A ([34] in this volume) might well be applicable to TOP, also.

*Assay with Bz-Gly-Ala-Ala-Phe-p-Aminobenzoate*

The substrate Bz-Gly-Ala-Ala-Phe-*p*-aminobenzoate was introduced for TOP (then termed soluble metalloendopeptidase) in a two-stage assay, in which cleavage of the -Gly+Ala- bond by TOP was followed by the action of an aminopeptidase that hydrolyzed the Ala-Ala-Phe-*p*-aminobenzoate product to yield free *p*-aminobenzoic acid, which was quantified in a diazo-coupling procedure.[9] A simpler procedure is one in which the aminopepti-dase is present from the start of the assay.[32] This is satisfactory for many purposes, including assays during purification, but may not be appropriate for determination of pH dependence or the activities of inhibitors, because effects on the aminopeptidase may be difficult to disentangle from those on TOP.

The aminopeptidase used corresponds to membrane alanyl aminopepti-dase (EC 3.4.11.2) and is not the cytosolic leucine aminopeptidase (EC 3.4.11.1). Commercial alanyl aminopeptidase tends to be contaminated with neprilysin, which cleaves the substrate, giving high blank values. Neprilysin may be removed by chromatography on phenyl-Sepharose, as in its purifi-cation,[33] but we find it more convenient to include phosphoramidon in the assay mixture as an inhibitor of neprilysin. The procedure for the simplified assay is as follows.

*Reagents*

Dithiothreitol solution: 2 m$M$ in water, prepared on the day

Aminopeptidase: Leucine aminopeptidase, microsomal (Sigma, St. Louis, MO, L5006)

Phosphoramidon: 0.5 m$M$ in water; store at $-20°$

Buffer: Tris-HCl, 0.4 $M$, pH 7.0

Substrate, Bz-Gly-Ala-Ala-Phe-*p*-aminobenzoate: A 10 m$M$ stock so-lution is prepared by dissolving the acid in water containing 1 equiva-lent of NaOH. Preparation of the substrate has been described[9]

Trichloroacetic acid solution: 25% (w/v)

Sodium nitrite solution: 0.25%, dissolved freshly in water

Ammonium sulfamate solution: 1.25% (w/v) in water

*N*-(1-Naphthyl)ethylenediamine dihydrochloride: 0.1% in ethanol

Reference standard: *p*-Aminobenzoate, potassium salt (Sigma, A0254), 1 m$M$ in water

---

[32] M. Orlowski, S. Reznik, J. Ayala, and A. R. Pierotti, *Biochem. J.* **261,** 951 (1989).

[33] J. Almenoff and M. Orlowski, *Biochemistry* **22,** 590 (1983).

*Procedure.* Each reaction mixture initially has the following composition:

| | |
|---|---|
| Enzyme solution | 5–20 $\mu$l |
| Dithiothreitol (2 m$M$) | 20 $\mu$l |
| Aminopeptidase (2 mg/ml) | 10 $\mu$l |
| Phosphoramidon (0.5 m$M$) | 10 $\mu$l |
| Buffer (0.4 $M$ Tris-HCl, pH 7.0) | 100 $\mu$l |
| Water to 190 $\mu$l total volume | |

Warm the mixture to 37°, start the reaction by adding substrate solution (20 $\mu$l), and incubate for 30 min. Then add 0.1 ml of the trichloroacetic acid solution, followed by 0.2 ml of sodium nitrite solution. Let stand for 3 min and add 0.2 ml ammonium sulfamate solution. After a further 2 min, add 0.5 ml $N$-(1-naphthyl)ethylenediamine. Measure $A_{555}$ and calculate the concentration of aminobenzoate from a standard curve prepared with pure $p$-aminobenzoate in place of enzyme. One unit is defined as the amount of activity that catalyzes the release of 1 $\mu$mol product/hr.

### Quenched Fluorescence Assays with QF01 and QF02

*Principle.* Thimet oligopeptidase cleaves a peptide bond in the substrate so that the fluorophore is separated from the quenching dinitrophenyl group and fluorescence appears (see [2] in this volume for more information on quenched fluorescence assays). Normally, the increase in fluorescence is monitored continuously, but stopped assays also are possible. With Perkin-Elmer (Norwalk, CT) fluorimeters, the experiment can conveniently be controlled right through to the calculation of results by use of the freely distributed FLUSYS software running on an IBM-compatible computer.[34] Either QF01 or QF02 may be used as the substrate. QF01 has the advantage of being water-soluble, so that no dimethyl sulfoxide (DMSO) is introduced into the reaction mixture; on the other hand, QF02 is rather more sensitive to hydrolysis by TOP, and the higher wavelengths used greatly decrease the likelihood of interference by proteins or colored materials in crude samples.[29] The substrates and standards are available from Calbiochem/ Novabiochem (La Jolla, CA).

### Reagents

Buffer: 50 m$M$ Tris-HCl, pH 7.8, containing 0.1% Brij 35
QF01 substrate stock solution: Dnp-Pro-Leu+Gly-Pro-Trp-D-Lys dissolved in water at 1.0 m$M$ ($A_{365}$ = 17.0); store at −20°

---

[34] N. D. Rawlings and A. J. Barrett, *Comput. Appl. Biosci.* **6**, 118 (1990).

QF02 substrate stock solution: Mcc-Pro-Leu+Gly-Pro-D-Lys(Dnp) dissolved in DMSO at 1.0 m$M$ ($A_{347}$ = 32)

Fluorescence standards: For QF01, the standard is Ac-Trp-NH$_2$, which is prepared as a 0.5 m$M$ stock solution in water and stored at 4°; for QF02, the standard is Mcc-Pro-Leu, a 5 m$M$ solution of which in DMSO is stored at 4°

ZnCl$_2$ stock solution: 100 m$M$ ZnCl$_2$ in water

*Preactivation of enzyme.* Some enzyme samples require preactivation in 5 m$M$ dithiothreitol (DTT) for full activity, so this is done as a matter of routine. The enzyme solution is mixed with an equal volume of 10 m$M$ DTT and allowed to stand at 0° for 10 min before dilution into the assay at a final concentration of 0.1 m$M$ DTT.

*Setting up Fluorimeter.* For QF01, $\lambda_{ex}$ is 283 nm and $\lambda_{em}$ is 350 nm, whereas for QF02, $\lambda_{ex}$ is 345 nm and $\lambda_{em}$ is 405 nm. Standards corresponding to the fluorescent products of hydrolysis of the QF substrates must be used. The appropriate standard (see above) is diluted to 1 $\mu M$ in assay buffer to set the fluorimeter to 1000 arbitrary units. Fluorescence is very much affected by temperature, so that $\Delta F$ values fall by about 10% as stopped assay samples cool from 30° to room temperature. It is therefore essential to set up the fluorimeter with standard at exactly the temperature of the samples. For continuous assays, this is simple, because everything is done in the thermostatted cell of the fluorimeter, usually at 30°.

*Procedure for Continuous Assays.* The quartz cuvette in the fluorimeter is thermostatted to 30° and set up for efficient stirring. The fluorimeter is controlled by the FLUSYS software. Calibration is as described above, but with the standard solution at 30°. The buffer is prewarmed to 30°. Buffer (2.45 ml) is pipetted into the cuvette, followed by 25 $\mu$l of substrate stock solution. When the baseline fluorescence is stable, the enzyme solution (25 $\mu$l) is introduced, and the increase in fluorescence is followed. A linear rate of increase in fluorescence is expected.

*Procedure for Stopped Assays.* Mix in a plastic tube 2.45 ml of buffer and 25 $\mu$l of sample. Add 25 $\mu$l of substrate stock solution and incubate for 60 min at 30°. Stop the reaction by adding 25 $\mu$l of ZnCl$_2$ solution. Prepare a blank for each sample by mixing sample, buffer, ZnCl$_2$, and substrate. Incubate for 60 min at 30°.

## Assay of Oligopeptidase M

Like TOP, MOP hydrolyzes Pz-Pro-Leu+Gly-Pro-D-Arg, Bz-Gly+Ala-Ala-Phe-$p$-aminobenzoate, Dnp-Pro-Leu+Gly-Pro-Trp-D-Lys, and Mcc-Pro-Leu+Gly-Pro-D-Lys(Dnp) (although with different kinetic parameters), so that the assays described above are applicable. However,

MOP is not activated by thiols, so the pretreatment with DTT may be omitted, and no thiol compound need be added to the assay buffer.

### Specificity of Assays

The substrates recommended above show a good measure of selectivity for the two oligopeptidases in the presence of other enzymes, but they are not totally resistant to the action of other peptidases that may be present in some samples, as has been discussed.[35] Any contribution from prolyl oligopeptidase or other serine peptidases can be excluded by the use of 1 m$M$ diisopropyl fluorophosphate (DFP), and the action of neprilysin can be eliminated by 1 $\mu M$ phosphoramidon. Products should be analyzed by reversed-phase HPLC whenever interference by any other enzyme is suspected, because different products will generally be formed. QF02 is cleaved by neurolysin just as it is by TOP and MOP,[36] and, at the time of writing, we are unable to distinguish neurolysin from MOP.

The assays can be made to distinguish between TOP and MOP by use of inhibitors. Thus, the dipeptide Pro-Ile (10 m$M$) is a selective inhibitor of MOP, whereas Cpp-Ala-Ala-Phe-pAb (0.5 $\mu M$) is selective for TOP. There are also differences in the specificity of action of the two enzymes on oligopeptides that might be used in HPLC assays (see below). Suitable antibodies can also provide complete selectivity in the assays of the two enzymes.[25]

### Purification of Oligopeptidases

Among the sources from which TOP has been purified to homogeneity are rat testis,[13,32] human erythrocytes,[13] and *Escherichia coli* expressing the rat enzyme under the control of the plasmid pEXTOP.[37] The TOP enzyme is very difficult to separate quantitatively from plasma albumin, so that preparations from testis and other tissues tend to retain some contamination, but with washed erythrocytes there is no such problem. The recombinant enzyme is readily isolated in much larger quantities than are feasible with the natural enzyme. Accordingly, methods for purification of the natural enzyme from human erythrocytes and the recombinant rat enzyme have been selected for detailed description. For MOP, rat liver mitochondria form a suitable source.

[35] U. Tisljar, *Biol. Chem. Hoppe-Seyler* **374,** 91 (1993).

[36] P. Dauch, H. Barelli, J.-P. Vincent, and F. Checler, *Biochem. J.* **280,** 421 (1991).

[37] N. McKie, M. A. Brown, P. M. Dando, and A. J. Barrett, *Biochem J.* (in press).

## Human Erythrocyte Thimet Oligopeptidase

*Extraction and Batch Purification.* Red blood cells (1320 ml) obtained from a blood transfusion service are washed three times by suspension and centrifugation (3500 $g$ for 15 min, 4°C) in 4 liters of 1% NaCl containing 0.05% trisodium citrate. The cells are finally resuspended in 2 liters of 10 m$M$ sodium phosphate buffer, pH 7.4, containing 5 m$M$ 2-mercaptoethanol, 0.05% (w/v) Brij 35, and 25 $\mu M$ phenylmethylsulfonyl fluoride (buffer A). Mercaptoethanol and Brij 35 at the stated concentrations are included in all solutions used subsequently, unless otherwise stated. The results of work with the recombinant enzyme suggest that 0.1 m$M$ ZnCl$_2$ may also be beneficial. The suspension is shaken with 500 ml of CCl$_4$, and the emulsion centrifuged. The clear red supernatant is retained and diluted with 2 liters of buffer A. Whatman (Clifton, NJ) DE-52 DEAE-cellulose is equilibrated with buffer A, and 150 g (damp weight) of this is stirred with the lysate for 30 min, by which time essentially all of the TOP activity is adsorbed. The DEAE-cellulose is collected by centrifugation and washed twice with 2 liters of buffer A in the centrifuge, and further on a sintered glass filter. The adsorbed enzyme is eluted with two treatments (250 ml each) of 500 m$M$ KCl in 100 m$M$ sodium phosphate, pH 7.4, containing Brij 35 and 2-mercaptoethanol in the centrifuge. The eluates are combined and dialyzed against 30 m$M$ Tris-HCl, pH 7.8, containing the usual mercaptoethanol and Brij 35 (buffer B).

*DEAE-Cellulose Chromatography.* The sample is run on a column (200 ml bed volume) of DEAE-cellulose in buffer B, elution being achieved with a linear gradient to 0.30 $M$ NaCl in 75 m$M$ Tris-HCl buffer over 5 bed volumes. Fractions containing activity are combined and transferred into 10 m$M$ sodium phosphate buffer, pH 7.0, and concentrated in an Amicon (Danvers, MA) ultrafiltration cell.

*Hydroxyapatite Chromatography.* The preparation is next run on a column (40 ml bed volume) of hydroxyapatite (Bio-Rad, Richmond, CA, BioGel HT) in 10 m$M$ sodium phosphate buffer, pH 7.0, with a gradient rising to 150 m$M$ sodium phosphate, pH 7.0, over 6 bed volumes. Active fractions are combined and transferred into 20 m$M$ triethanolamine, 10 m$M$ HCl, 200 m$M$ NaCl, pH 7.8, by ultrafiltration, and at the same time concentrated to 2 ml.

*Sephacryl S-200 Chromatography.* The sample is run on a column (92 cm × 16 mm, 185 ml) of Sephacryl S-200 HR (Pharmacia, Piscataway, NJ) in 20 m$M$ triethanolamine–NaCl buffer, pH 7.8. Active fractions are combined and run on a Mono Q column [HR 5/5 Pharmacia FPLC (fast protein liquid chromatography)] with a gradient to 300 m$M$ NaCl.

*Mono P Fast Protein Liquid Chromatography.* The solution is dialyzed into 25 m$M$ Bis–Tris-HCl, pH 6.3, for running on Mono P (HR 5/5, Pharmacia FPLC). Elution is with Polybuffer 74 diluted 10-fold and adjusted to pH 4.0 with HCl. Active fractions are combined to form the final product.

## Recombinant Rat Thimet Oligopeptidase

*Induced Expression of Thimet Oligopeptidase in Escherichia coli Strain BL21 DE3 (pLysS).* The construction of the pEXTOP plasmid is as described.[37] For induced expression of TOP in *E. coli*,[38] the *E. coli* strains are grown at 37° in 2TY broth (1% yeast extract, 1% tryptone, and 0.5% NaCl) unless otherwise noted. To enrich for ampicillin-resistant and hence plasmid-bearing clones, the *E. coli* BL21 DE3 (pLysS)/pEXTOP cells are streaked onto 2TY plates containing chloramphenicol (33 $\mu$g/ml) and ampicillin (100 $\mu$g/ml) and grown overnight at 37°. A single colony from the plate is inoculated into 3 ml of 2TY medium with chloramphenicol (33 $\mu$g/ml) and ampicillin (100 $\mu$g/ml) and grown until the cells are just visible in the tube. They are then restreaked on a fresh 2TY plate with antibiotics as above and incubated for a further 12 hr at 37°. A single colony from the plate is used to inoculate 100 ml of 2TY, and the culture is grown overnight at 37°. The overnight culture is spun down and washed in 200 ml of fresh 2TY medium and finally resuspended in 20 ml of 2TY medium. The cells serve as an inoculum for the large-scale cultures used for isolation of the enzyme. In each experiment 200 $\mu$l of the cell suspension is used to inoculate a 1-liter culture in TB medium,[39] which is grown until $D_{600}$ exceeds 1.5. At this point the production of TOP is induced by the addition of isopropyl-$\beta$-D-thiogalactopyranoside to a final concentration of 0.4 m$M$, and cells are grown for 16 hr at 37° before harvest.

*Extraction of Cells.* Cells from 20 liters of culture are collected by centrifugation (4000 $g$), and washed twice in 50 m$M$ Tris-HCl buffer, pH 7.8, containing 100 m$M$ NaCl, and twice more in buffer alone. The cell pellet (200 g) is frozen at $-70°$ and then thawed at 35°, with the addition of 3.2 liters of 50 m$M$ buffer A. The cell suspension is homogenized in a Waring blendor (30 sec), then frozen and thawed once more. The mixture is made 0.1 mg/ml in protamine sulfate and centrifuged at 10,000 $g$. All steps in the purification of recombinant TOP from the extract are at 4°. Buffer A contains Tris-HCl at the given molarity (of HCl), pH 7.8, with

---

[38] F. W. Studier, A. H. Rosenberg, J. J. Dunn, and J. W. Dubendorff, this series, Vol. 185, p. 60.

[39] J. Sambrook, E. F. Fritsch, and T. Maniatis (eds.) "Molecular Cloning: A Laboratory Manual." Cold Spring Harbor Laboratory Press, New York, 1989.

the addition of 0.05% (w/v) Brij 35, 5 m$M$ 2-mercaptoethanol, and 100 $\mu M$ ZnCl$_2$, which have been found to stabilize the enzyme.

*Batch DEAE-Cellulose Extraction.* The supernatant is diluted with 1.5 volumes of water and stirred overnight at 4° with 400 g of damp DE-52 DEAE-cellulose (Whatman) that has been equilibrated with 20 m$M$ buffer A. The ion exchanger is collected by centrifugation (15 min at 3000 g) and washed three times with 20 m$M$ buffer A. The enzyme is then eluted by two treatments of the exchanger with 0.4 $M$ NaCl in the same buffer (total volume of washings about 1 liter) in the centrifuge bottle.

*Ammonium Sulfate Fractionation.* The solution (4°C) is made 1.4 $M$ in ammonium sulfate by addition of the solid, and after 30 min the precipitate is removed by centrifugation (8000 g, 15 min, 4°C) and discarded. The supernatant is made 2.4 $M$ in ammonium sulfate, and the precipitate, collected as above, is dissolved in the minimum volume of 50 m$M$ buffer A and dialyzed against the same overnight.

*DEAE-Cellulose Chromatography.* The crude enzyme sample is run onto a bed (2.5 × 60 cm) of DE-52 equilibrated with 20 m$M$ buffer A. The column is washed with the 20 m$M$ buffer and then eluted with a gradient (0.9 plus 0.9 liter) to 0.3 $M$ NaCl in the buffer. Active fractions are combined. The enzyme is concentrated by precipitation from 2.4 $M$ ammonium sulfate, and the pellet is redissolved in 20 m$M$ buffer A modified by the omission of Brij 35 and the addition of 1.25 $M$ ammonium sulfate.

*Pentyl Agarose Chromatography.* The solution in 1.23 $M$ ammonium sulfate is run onto a 100-ml bed of pentyl agarose (Sigma, P5393). The column is washed with 2 bed volumes of 1.23 $M$ ammonium sulfate and then eluted with a gradient (400 plus 400 ml) descending from 1.23 $M$ to buffer alone. Active fractions are combined and dialyzed into 20 m$M$ buffer A.

*Resource Q Chromatography.* The preparation is divided into two equal parts, to be run separately on a 6-ml Resource Q column (Pharmacia, 17-1179-01). Elution is with a gradient (100 ml) to 0.2 $M$ NaCl in buffer A. Fractions are assayed and also subjected to sodium dodecyl sulfate (SDS)–polyacrylamide gel electrophoresis, and those having a specific activity of 3–6 units/mg (in the QF02 assay) that show a single protein band are combined as the final product. The pure enzyme is stored as a solution of at least 5 mg/ml in 20 m$M$ buffer A plus 100 m$M$ NaCl and 0.1% (w/v) NaN$_3$ at 4°.

*Purification of Oligopeptidase M from Rat Liver Mitochondria*

*Preparation of Mitoplasts.* The method is based on that described by Isaya and Kalousek ([33] in this volume). All operations are carried out at 0°–4°. Rat liver (450 g) is rinsed with saline, chopped with a pair of scissors, and homogenized in 4 volumes of 10 m$M$ Tris-HCl, pH 8.0, containing 250 m$M$ sucrose, 5 m$M$ 2-mercaptoethanol, and 0.1 m$M$ phenylmethylsulfonyl fluoride (TS buffer). The tissue is homogenized with two up-and-down strokes in a motor-driven Potter–Elvehjem homogenizer running at 1000 rpm. The homogenate is centrifuged at 1000 $g$ for 5 min, and the debris pellet discarded.

The first supernatant is again centrifuged, this time for 15 min at 10,000 $g$. The second supernatant is discarded, and the pellet (crude mitochondria) is resuspended in 1 volume of TS buffer. Digitonin stock solution (5%, w/v) is added to 0.045% to lyse lysosomes and peroxisomes. After 5 min, with occasional stirring, a further 3 volumes of TS buffer is added, and the mitochondria are washed twice by centrifugation (10,000 $g$, 15 min) and resuspension in TS buffer.

The washes are discarded, and the supernatant (purified mitochondria) is diluted to a protein concentration of 100 mg/ml with TS buffer. Digitonin stock solution is added to make 0.55% digitonin, and after 15 min with occasional stirring 3 volumes of TS buffer is added, and the mitoplasts are washed twice by centrifugation (25,000 $g$, 15 min) and resuspension in TS buffer.

The mitoplast pellet is diluted to a protein concentration of 50 mg/ml in TS buffer, and Lubrol WX stock solution (10%, w/v) is added to make 1.1% (i.e., 0.18 mg Lubrol/mg protein). After 30 min with occasional stirring the suspension is diluted with an equal volume of TS buffer and centrifuged at 100,000 $g$ for 1 hr. The supernatant, representing the mitochondrial matrix extract, forms the source for the chromatographic isolation of MOP. Activity is assayed with QF02; TOP seems to be absent from mitoplasts, so all activity is attributed to MOP.

*DEAE-Cellulose Chromatography.* A column (26 × 380 mm) of DEAE-cellulose (Whatman DE-52) is equilibrated with 10 m$M$ Tris-HCl, pH 8.0, containing 0.05% (w/v) Brij 35, 5 m$M$ 2-mercaptoethanol, 0.1 m$M$ ZnCl$_2$, and 0.1 m$M$ phenylmethylsulfonyl fluoride. The sample is run on the column with a 2-liter linear gradient from 0 to 0.3 $M$ NaCl in the buffer. Fractions of 15 ml are collected at a flow rate of 1 ml/min. Fractions containing activity are combined, concentrated to about 40 ml by ultrafiltration over an Amicon PM30 membrane, and dialyzed against 10 m$M$ sodium phosphate buffer, pH 6.8, containing 5 m$M$ 2-mercaptoethanol, 0.5% Brij 35, and 0.1 m$M$ ZnCl$_2$ (PB buffer).

*Hydroxyapatite Chromatography.* A column (16 × 80 mm) of hydroxyapatite (Bio-Rad, BioGel HT) is equilibrated with PB buffer, and the sample is run with a 400-ml linear gradient from 10 to 300 m$M$ sodium phosphate. Fractions (3.0 ml) are collected at 7.5 ml/hr. The active fractions are combined and concentrated by ultrafiltration to 2–3 ml.

Oligopeptidase M is eluted from hydroxyapatite at a phosphate concentration distinctly greater than that for TOP, and this is the only step discovered so far in which the two enzymes are readily separated. With mitoplasts as the source of enzyme, however, no peak attributable to TOP is detected.

*Sephacryl S-300 and HiTrap Blue Chromatography.* The preparation is next run on a column (26 × 650 ml) of Sephacryl S-300 in 10 m$M$ Tris-HCl, pH 8.0, containing 0.15 $M$ NaCl, 0.05% Brij 35, 5 m$M$ 2-mercaptoethanol, and 0.1 m$M$ ZnCl$_2$. Fractions (5 ml) are collected at 24 ml/hr. The active fractions are combined and diluted with an equal volume of 10 m$M$ Tris-HCl, pH 8.0, containing 0.05% Brij 35, 5 m$M$ 2-mercaptoethanol, and 0.1 m$M$ ZnCl$_2$. The solution is passed through a 1-ml column of Pharmacia HiTrap Blue in 10 m$M$ Tris-HCl, pH 8.0, containing 0.1 $M$ NaCl, 0.05% Brij 35, 5 m$M$ 2-mercaptoethanol, and 0.1 m$M$ ZnCl$_2$.

*Mono Q Fast Protein Liquid Chromatography.* The final step is Pharmacia FPLC on a Mono Q (HR 5/5) column equilibrated in 10 m$M$ Tris-HCl, pH 8.0, containing 0.1 $M$ NaCl, 0.05% Brij 35, 5 m$M$ 2-mercaptoethanol, and 0.1 m$M$ ZnCl$_2$. Elution is with a linear gradient from 0.1 to 0.2 $M$ NaCl run at 0.5 ml/min over 30 min. Fractions of 0.5 ml are collected.

## Typical Results from Purification Procedures

Peak tubes from the final columns in the purification of TOP commonly show specific activities (in the QF02 assay) up to 6 units/mg protein, but values rapidly fall to 3 units/mg or below, and then become stable during storage. The enzyme is stable for weeks at 4° in 50 m$M$ Tris-HCl, pH 7.8, 100 m$M$ NaCl, 0.1% NaN$_3$, preferably at a concentration of at least 1 mg/ml.

Laboratory-scale preparations of TOP from chicken liver,[40] rat testis, or human red cells[13] typically yield several hundred micrograms of the homogeneous enzyme. In subsequent work with recombinant rat TOP, it has been found that inclusion of 0.1 m$M$ ZnCl$_2$ in the buffers used during the procedure markedly stabilizes the enzyme, increasing yields. Approximately 40 mg (20% yield) of the recombinant enzyme could be obtained from a single preparation.[37] For MOP, a preparation on the scale described typically yields 0.5 mg of the pure enzyme, with a specific activity of 3.4 units/mg in the assay with QF02.[25]

[40] A. J. Barrett and M. A. Brown, *Biochem. J.* **271,** 701 (1990).

TABLE I
AMINO ACID COMPOSITIONS OF THIMET OLIGOPEPTIDASE AND
OLIGOPEPTIDASE M FROM VARIOUS MAMMALIAN SPECIES[a]

| Amino acid | TOP | | | MOP | |
|---|---|---|---|---|---|
| | Rat | Human | Pig | Pig | Rabbit |
| Ala | 52 | 54 | 55 | 41 | 43 |
| Cys | 15 | 16 | 15 | 13 | 13 |
| Asp | 41 | 38 | 37 | 44 | 47 |
| Glu | 60 | 64 | 60 | 57 | 54 |
| Phe | 28 | 28 | 26 | 29 | 32 |
| Gly | 39 | 40 | 40 | 40 | 37 |
| His | 18 | 20 | 17 | 18 | 16 |
| Ile | 20 | 19 | 22 | 29 | 29 |
| Lys | 42 | 38 | 44 | 51 | 44 |
| Leu | 80 | 77 | 81 | 81 | 78 |
| Met | 19 | 19 | 18 | 28 | 22 |
| Asn | 20 | 18 | 19 | 26 | 30 |
| Pro | 33 | 30 | 30 | 25 | 26 |
| Gln | 37 | 38 | 38 | 25 | 31 |
| Arg | 40 | 47 | 40 | 40 | 43 |
| Ser | 32 | 33 | 34 | 40 | 43 |
| Thr | 35 | 34 | 34 | 39 | 37 |
| Val | 43 | 43 | 43 | 43 | 44 |
| Trp | 7 | 7 | 7 | 6 | 6 |
| Tyr | 26 | 26 | 27 | 29 | 29 |
| Res | 687 | 689 | 687 | 704 | 704 |
| $M_r$ | 78,343 | 78,814 | 78,145 | 80,758 | 80,779 |

[a] The $M_r$ values are calculated simply from the amino acid compositions on the assumption of no disulfide bonds, and thus make no provision for any possible nonpeptide prosthetic groups or the metal atoms. References to the original publications of the amino acid sequences are given in the legend to Fig. 3. Res stands for total amino acid residues, and $M_r$ is relative molecular weight.

## Composition and Structure

cDNA encoding rat TOP was first cloned and sequenced by Pierotti *et al.*,[41] but some significant errors were detected by McKie *et al.*, which were corrected by use of a clone provided by the original workers.[22] The revised sequence contains 687 amino acid residues and gives a calculated molecular mass of 78.3 kDa (Table I). Subsequently, closely related sequences have

[41] A. Pierotti, K.-W. Dong, M. J. Glucksman, M. Orlowski, and J. L. Roberts, *Biochemistry* **29**, 10323 (1990).

been described for human TOP (as amyloidin)[12] and pig TOP.[42] The compositions are summarized in Table I.

Deduced amino acid sequences have also been reported for MOP from pig (as soluble angiotensin-binding protein)[20] and rabbit (as microsomal endopeptidase),[24] and again the compositions are summarized in Table I. Both SDS–polyacrylamide gel electrophoresis and gel permeation chromatography of TOP and MOP give apparent molecular mass values of 75–80 kDa, in reasonable agreement with the amino acid compositions (but see below). We conclude that both enzymes are monomeric, single-chain proteins. The value of $A_{280}^{1\%}$ has been determined as 6.25 for recombinant rat TOP, giving a value of 48,960 for the molar extinction coefficient.[37] Values for the isoelectric point of TOP from a variety of sources have generally been 5.0 ± 0.2 (reviewed in Ref. 35).

An alignment of the available sequences of TOP and MOP (Fig. 1) shows that they are closely related, with about 60% of residues being totally conserved among the five sequences. Assuming that the initiator Met residues are as shown on Fig. 1,[20] MOP contains 24 N-terminal residues not present in TOP. These may contain a mitochondrial targeting sequence: cleavage at -Arg-Glu⁵+ would conform to the common occurrence of Arg in position P2 for mitochondrial processing peptidase,[44] and removal of a further octapeptide by mitochondrial intermediate peptidase would give rise to the N-terminal Ser-38 residue (MOP numbering) identified in the enzyme from mitochondria.[25] The shortened N terminus of MOP, as isolated from rat liver mitochondria, is probably responsible for our observation that the enzyme migrates in SDS–polyacrylamide gel electrophoresis as if of a molecular mass about 2 kDa less than TOP, contrary to the deduced $M_r$ values given for MOP in Table I.

Preparations of active TOP from rat testis[41] and the recombinant enzyme from *E. coli*[37] have N-terminal Lys-2 (Fig. 1 numbering), so it can be concluded that the enzyme does not require cleavage of an N-terminal propeptide for activity. However, N-terminal truncation of the molecule does occur, as a preparation of human TOP had N-terminal Ser-19.[13]

The His⁴⁷³-Glu-Phe-Gly-His sequence is assumed to represent a partial zinc-binding site, and to contain the catalytic Glu, by analogy with other metallopeptidases (see [13], this volume). No third ligand has been identified, but Glu-502 and Glu-509 are both candidates.

The sequences of TOP and MOP contain 13–16 cysteine residues (Table I), 8 of which are conserved (Fig. 1). Recombinant rat TOP gives strong

[42] A. Kato, N. Sugiura, H. Hagiwara, and S. Hirose, *Eur. J. Biochem.* **221,** 159 (1994).
[43] Deleted in proof.
[44] B. S. Glick, E. M. Beasley, and G. Schatz, *Trends Biochem. Sci.* **17,** 453 (1992).

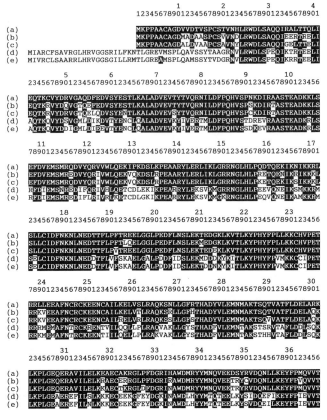

FIG. 1. Alignment of the amino acid sequences of species variants of TOP and MOP. Residues identical to those in rat TOP are printed in white on black, and putative zinc-binding His residues are marked with an asterisk (*). The sequences are those of (a) rat TOP, (b) human TOP, (c) pig TOP, (d) rabbit MOP, and (e) pig MOP. For original references to the sequences, see the legend to Fig. 3.

reactions with 5,5′-dithiobis(2-nitrobenzoic acid) (DTNB) and 2,2′-dithiodipyridine (PDS), corresponding to at least 10 of the cysteine thiol groups being free to react.[37] One of the conserved Cys residues, Cys-483, has been suggested to be responsible for the thiol dependence of TOP,[41] but the fact that it is also present in MOP, which shows no activation by thiols, raises a doubt about that. It might seem more likely that one or more of the 5 cysteine residues that are conserved in TOP but absent from MOP would be responsible. Of these, 3 seem too close to the termini to be strong candidates, and the closely located Cys-246 and Cys-253 may seem the most attractive.

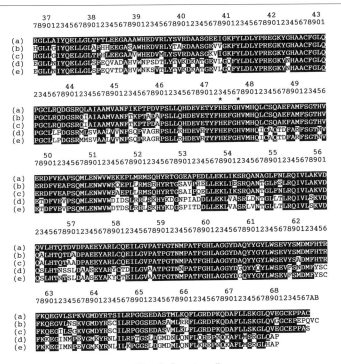

FIG. 1. (*continued*)

It has been suggested that TOP occurs partially associated with the plasma membrane or the membranes of synaptic vesicles (see below). However, the hydrophobicity profiles of the TOP and MOP sequences show no markedly hydrophobic regions suggestive of transmembrane segments. Periodate–Schiff staining of bands of TOP and MOP on polyacrylamide gels has given negative results, so there is no indication of any glycosylation of either molecule.[25]

Analysis of several preparations of recombinant rat TOP showed the presence of about 1.8 g · atom/mol of zinc.[37] In view of the presence of a typical consensus sequence for binding of a catalytic zinc atom, and stabilization of the enzyme by zinc ions, we suggest that there is one catalytic and one structural zinc atom per molecule.

## Enzymatic Activity

### Requirements for Activity

*pH Optimum.* In assays with the quenched fluorescence substrates, the activity of rat TOP is greater at pH 8.2 than at lower pH values (down to pH 6.5). The stability of the enzyme falls off above pH 8, however, so that pH 7.8 is optimal for routine assays of 10 min or so. Activity is much the same in 50 m$M$ Tris-HCl and sodium phosphate buffers. Thus we have not been able to confirm the earlier report of an optimum at pH 7.0 obtained in assays that also depended on the activity of alanyl aminopeptidase as a reagent.[9]

*Thiol Activation.* Preparations of TOP that have been exposed to air in solutions lacking thiol compounds typically show very marked stimulation of activity by low concentrations of dithiothreitol or 2-mercaptoethanol. The effect was characterized in detail for chicken TOP (as Pz-peptidase), for which it was found that maximal activity was obtained in 50 $\mu M$ dithiothreitol or 1 m$M$ mercaptoethanol. The activity fell off at higher concentrations of either reagent, such that activity was about halved at 10 times those concentrations.[40] The explanation for the biphasic effect of thiol compounds on the activity of TOP is probably that one or more of the thiol groups of the molecule tend to hinder activity in some way when substituted by large groups in mixed disulfides that form under oxidizing conditions. High concentrations of thiol compounds bind significantly to the catalytic zinc atom, however, causing inhibition. We were able to show that the inhibition at high thiol concentrations is related to the thiophilicity of the metal, because it was almost absent in the Mn$^{2+}$ form of the enzyme.[40] The inhibition of TOP by 10 m$M$ DTT is instantaneously reversible on dilution of the thiol, so that pretreatment of the enzyme with 10 m$M$ DTT is a good form of preactivation of the enzyme.

Work with MOP that had been partially purified without the use of hydroxyapatite gave the indication that MOP is activated by thiols.[19] Subsequent work has not confirmed this, however,[25] and we now attribute the activation seen in the early work to contamination of the MOP preparation by TOP.

*Metal Ions.* The effects of divalent metal ions in the reactivation of TOP after dialysis against EDTA were studied with the chicken enzyme.[40] The metal ions were used at 50 $\mu M$ concentration, and there was essentially complete reactivation by Zn$^{2+}$. The percentages of reactivation by other metals were as follows: Mn$^{2+}$, 71%; Ca$^{2+}$, 35%; Co$^{2+}$, 31%; Cd$^{2+}$, 22%. The ions Mg$^{2+}$, Cu$^{2+}$, Ni$^{2+}$, and Fe$^{2+}$ were without effect. We have not been able to confirm the reports[9,32] of superactivation of rat TOP by Co$^{2+}$. Inhibition by higher concentrations of Zn$^{2+}$ begins to be apparent at 100 $\mu M$. An excess of Mn$^{2+}$ is much less inhibitory, however.[40]

*Substrate Specificity*

All substrates that have clearly been shown to be cleaved by TOP are of 18 or fewer amino acid residues, and this is the basis of the designation "oligopeptidase." Bonds that are cleaved in, for example, fragments 1–12 or 16–28 of vasoactive intestinal peptide are totally unaffected in the complete peptide of 28 residues.[13] All attempts to show cleavage of proteins by TOP have been unsuccessful, although a report of hydrolysis of the Alzheimer's amyloid precursor protein[11] will be important if confirmed.

Data for the specificity of cleavage of peptides by TOP agree closely, whether the enzyme is from rat testis,[13,32] human erythrocytes,[13] or rabbit brain.[45] $Ac\text{-}Ala_4$ is perhaps the simplest substrate yet described for TOP.[40] A few examples of positions of hydrolysis of larger peptides are shown in Fig. 2. Despite the considerable amount of work that has been done, it has not been possible to formulate a description of the specificity of the enzyme such that one can predict the point of cleavage of a peptide before it is determined experimentally. As has been discussed,[13] hydrophobic side chains in the P1 and P3' positions tend to be favorable, as is proline in P2. Subsequent work has revealed selectivity among hydrophobic P1 residues, in that bonds formed by those with branched side chains (isoleucine, valine) are not readily hydrolyzed. The free C-terminal carboxyl group has a favorable influence on activity, and it is probably responsible for the fact that many of the recorded scissile bonds are three or four residues in from the C terminus.[13,45a] The characteristic of the enzyme of cleaving most enkephalin-containing peptides in such a way as to generate the free enkephalin, and not to destroy the enkephalin,[45] distinguishes it from many other metallopeptidases. In addition to the peptides shown in Fig. 2, TOP is known to hydrolyze angiotensin I, angiotensin II, luliberin, and substance P.[13]

Information on the specificity of peptide bond hydrolysis by MOP has only relatively recently been obtained.[25] As can be seen in Fig. 2a, a number of cleavage positions are as by TOP. Cleavage of the Pz-peptide that led to the discovery of the enzyme[18] is also the same, as is action on QF01, QF02, and bradykinin. Other natural and synthetic substrates are cleaved differently, however (Fig. 2b). The hydrolysis of neurotensin solely at the -Pro+Tyr- bond is particularly striking, having been reported previously only for neurolysin (EC 3.4.24.16, also known as neurotensin-degrading enzyme; see [36], this volume). The hydrolysis of dynorphin A(1–17) by MOP is also distinctive, because the peptide is a potent inhibitor, but not a substrate, for TOP.[13] Moreover, the $-Leu^5+Arg-$ bond cleaved in the 1–8 fragment is resistant in dynorphin A(1–17).

Some of the $K_m$ values that have been determined for TOP and MOP are shown in Table II. For reasons that are as yet unexplained, TOP that

[45] A. C. M. Camargo, M. D. Gomes, O. Toffoletto, M. J. F. Ribeiro, E. S. Ferro, B. L. Fernandes, K. Suzuki, Y. Sasaki, and L. Juliano, *Neuropeptides* **26,** 281 (1994).
[45a] C. G. Knight, P. M. Dando, and A. J. Barrett, *Biochem. J.* (in press).

| Peptide | Cleavage |
|---------|----------|

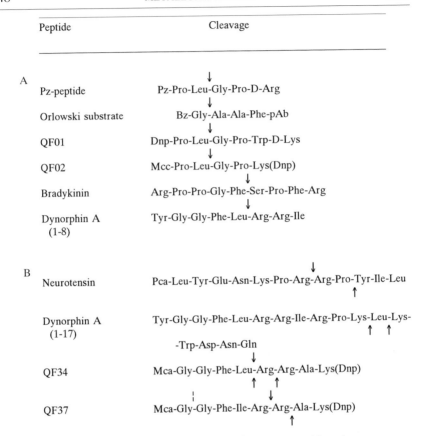

FIG. 2. Substrate specificities of TOP and MOP. (A) Cleavage positions that are common to both enzymes (marked ↓ ). (B) Differences between the enzymes, the points of hydrolysis by TOP being marked ↓ (or ⁞ for a weak cleavage) and those by MOP ↑.

is freshly diluted from solution in 100 $\mu M$ ZnCl$_2$ shows $K_i$ values for two Cpp-peptide inhibitors that are severalfold higher than those for the enzyme stored without zinc.

*Inhibitors*

As metallopeptidases, TOP and MOP are inactivated by exposure to metal-chelating agents. In early work, confusion over the question of whether TOP (as endo-oligopeptidase A) is a metallopeptidase seems to have arisen from the fact that EDTA is very slow to remove the essential zinc at above pH 7, and there are indications that this may be particularly

TABLE II
KINETIC PARAMETERS FOR THIMET OLIGOPEPTIDASE AND OLIGOPEPTIDASE M

| Component | $K_m$ ($\mu M$) | | | |
|---|---|---|---|---|
| | Human erythrocyte TOP | Rat testis TOP | Rat recombinant TOP from *E. coli* | Rat liver MOP |
| Substrate | | | | |
| QF01 | 13.3 | 12.0 | 16.1 | >50 |
| QF02 | 9.0 | 23.0 | 34.3 | 49.8 |
| QF34[a] | nd[b] | nd | 1.4 | 4.3 |
| QF37[a] | nd | nd | 2.5 | 2.2 |

| | $K_i$ (n$M$) | | | |
|---|---|---|---|---|
| Inhibitor | | | | |
| Cpp-AAF-pAb | 9.8 | 30 (124[c]) | 186[c] | 4480 |
| Cpp-AAY-pAb | nd | 16 | 40[c] | 700 |
| Cpp-APF-pAb | 1.3 | 7 (55[c]) | 31[c] | 3820 |
| Pro-Ile | nd | nd | ≫10 m$M$ | 0.54 m$M$ |
| Dynorphin A(1–13) | 11.2 | 48 (36[c]) | 83 | 1400 |

[a] The structures of QF34 and QF37 are shown in Fig. 2.
[b] nd, Not determined.
[c] Assayed in 25 n$M$ ZnCl$_2$, immediately after dilution from 100 $\mu M$ ZnCl$_2$.

true of the enzyme from species such as rabbit. (The pH dependence of the formation of apoenzymes from metallopeptidases generally is discussed in [14] of this volume.) MOP also is inhibited by EDTA and 1,10-phenanthroline.[25]

Compounds that react covalently with thiol groups produce forms of TOP that range widely in activity. In a typical experiment, rat recombinant TOP was fully activated with DTT in a small volume of solution and was then diluted into a large volume of 1 m$M$ solution of alkylating agent in large excess over the total thiols. Assays were done 5 and 10 min later and gave similar results, indicating that the reactions with the alkylating agents had gone to completion. Activities relative to the control without alkylating agent were as follows: iodoacetate, 116%; iodoacetamide, 96%; N-ethylmaleimide, 46%; and N-phenylmaleimide, 12%. It was concluded that alkylation of all readily reactive thiol groups of TOP (perhaps as many as the 10 that react rapidly with DTNB and PDS) affects activity in a way that depends on the nature of the alkylating groups. The small, negatively charged carboxymethyl groups cause an increase in activity under the assay conditions, whereas the progressively larger and more hydrophobic carboxamidomethyl, ethylsuccinimido, and phenylsuccinimido substituents cause increasing degrees of inhibition. Both TOP and MOP (from rat) are completely inhibited in the presence of 50 $\mu M$ Hg$^{2+}$ or p-hydroxymercuribenzoate.

Excellent reversible inhibitors of TOP were discovered by Orlowski and co-workers.[46,47] These were analogs of the substrate Bz-Gly-Ala-Ala-Phe-pAb in which the Bz-Gly- group was replaced by the N-[1-(RS)-carboxy-3-phenylpropyl]- (or Cpp-) group. The most potent of the compounds was Cpp-Ala-Ala-Tyr-pAb ($K_i$ 16 nM for the rat enzyme). Further analogs were synthesized by Knight in a study of the structure–function relationships of such inhibitors.[48] The most potent compound was Cpp-Ala-Pro-Phe-pAb ($K_i$ 7 nM). It has been shown that Cpp-Ala-Ala-Phe-pAb is a very selective inhibitor of TOP, with little effect on neprilysin and peptidyl-dipeptidase A. However, it is cleaved by neprilysin to form Cpp-Ala-Ala, which is a potent inhibitor of peptidyl-dipeptidase A. This occurs in enzyme preparations containing even traces of neprilysin,[49,50] which complicates the interpretation of in vivo experiments with the inhibitors (see below). Our work has shown that MOP also is inhibited by the Cpp-tripeptidyl-pAb compounds, although much less potently (Table II).[25]

A further potent series of inhibitors of TOP arose from work directed at the inhibition of bacterial collagenases.[51,52] The compound N-(phenylethylphosphonyl)-Gly-Pro-Nle inhibited TOP with $K_i$ values of the order of 7 nM, and it was a 10-fold more potent inhibitor of neurolysin (probably identical with MOP). Like the Cpp-peptides, the phosphonyl peptides were highly selective for TOP and neurolysin, having little effect on a range of other metallopeptidases.

A peculiarity of the specificity of TOP is the fact that although dynorphin A(1–8) is an excellent substrate of TOP, being cleaved as shown in Fig. 2, the longer peptide dynorphin A(1–13) is not hydrolyzed at all but is a very powerful inhibitor, with a $K_i$ about 48 nM.[13,30,40,53] MOP also is significantly inhibited by dynorphin A(1–13), although more weakly (Table II). Inhibitors of various other metallopeptidases, including phosphoramidon, thiorphan, and captopril, have little action on TOP[40] and MOP.

[46] T. G. Chu and M. Orlowski, *Biochemistry* **23,** 3598 (1984).

[47] M. Orlowski, C. Michaud, and C. J. Molineaux, *Biochemistry* **27,** 597 (1988).

[48] C. G. Knight and A. J. Barrett, *FEBS Lett.* **294,** 183 (1991).

[49] C. H. Williams, T. Yamamoto, D. M. Walsh, and D. Allsop, *Biochem. J.* **294,** 681 (1993).

[50] C. Cardozo and M. Orlowski, *Peptides* **14,** 1259 (1993).

[51] V. Dive, A. Yiotakis, C. Roumestand, B. Gilquin, J. Labadie, and F. Toma, *Int. J. Pept. Protein Res.* **39,** 506 (1992).

[52] H. Barelli, V. Dive, A. Yiotakis, J.-P. Vincent, and F. Checler, *Biochem. J.* **287,** 621 (1992).

[53] J. R. McDermott, *Biochem. Soc. Trans.* **12,** 307 (1984).

*Homologs and Evolution*

The enzymes TOP and MOP are members of the single family (M3) of metallopeptidases that is known to include both exopeptidases and endopeptidases, and which is thus of exceptional diversity ([13], this volume). Although no sequence data are available for neurolysin ([36], this volume), its properties would be consistent with its being a member of family M3, or even being identical to MOP. The main discrepancy between the data for MOP and neurolysin seems to be the synaptosomal localization reported for neurolysin, as contrasted with the mitochondrial localization of MOP. The characteristic cleavage of neurotensin at the -Pro$^{10}$+Tyr- bond, together with behavior on hydroxyapatite, the high activity on QF02, and inhibition by Pro-Ile, clearly links the enzymes, however.[25]

The other known homologs of TOP and MOP in eukaryotes are mitochondrial intermediate peptidase ([33], this volume), which is known from rat, and saccharolysin (EC 3.4.24.37; formerly peptidase *yscD*), known from *Saccharomyces cerevisiae*.[54] In prokaryotes, two homologs, oligopeptidase A and peptidyl-dipeptidase A, have been identified in both *E. coli* and *Salmonella typhimurium* ([34], this volume). Also, the oligopeptidase termed peptidase F of *Lactococcus lactis* has been shown to be a homolog ([35], this volume).

A phylogenetic tree showing the relationships of the proteins is presented in Fig. 3. The dating of the dendrogram suggests that TOP and MOP diverged from a common ancestor in relatively recent times, during the evolution of animals. The ancestor may have resembled saccharolysin, which is present in both the cytoplasm and mitochondria of yeast.[54]

Biological Aspects

*Distribution*

*Thimet Oligopeptidase.* The indications are that TOP is ubiquitously distributed across animal species,[35] and the chicken enzyme is remarkably similar in properties to that from rat.[40] An enzyme hydrolyzing QF02, and strongly inhibited by Cpp-Ala-Ala-Phe-pAb (and therefore probably TOP), has been detected in the muscle of bony fish and in tissues of invertebrates including octopus, earthworm, and garden snail.[35] In sexually immature mammals, the distribution among organs is fairly uniform; however, in

---

[54] M. Büchler, U. Tisljar, and D. H. Wolf, *Eur. J. Biochem.* **219,** 627 (1994).

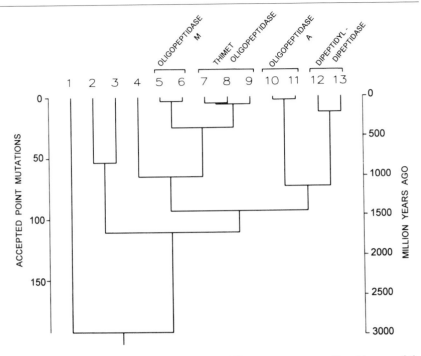

FIG. 3. Phylogenetic tree for the TOP family. The sequences were aligned by use of the
PILEUP program (Genetics Computer Group, in "Program Manual for the GCG Package."
University of Wisconsin, Madison, 1991), and the tree was constructed by use of the KITSCH
program of the PHYLIP package [J. Felsenstein, *Evolution* **39,** 783 (1985)], which assumes
a constant evolutionary rate. The tree was dated by taking the divergence between fungi and
mammals as 1000 million years ago (m.y.a.). It can be seen that on this basis the divergence
of the eukaryote and prokaryote proteins occurred about 1450 m.y.a. Key to sequences: (1)
*Lactococcus* peptidase F [V. Monnet, M. Nardi, A. Chopin, M.-C. Chopin, and J.-C. Gripon,
*J. Biol. Chem.* **269,** 32070 (1994)]; (2) *Schizophylum* sequence [L. Giasson, C. A. Specht, C.
Milgrim, C. P. Novotny, and R. C. Ullrich, *Mol. Gen. Genet.* **218,** 72 (1989)]; (3) rat mitochon-
drial intermediate peptidase [G. Isaya, F. Kalousek, and L. E. Rosenberg, *Proc. Natl. Acad.
Sci. U.S.A.* **89,** 8317 (1992)]; (4) saccharolysin [M. Büchler, U. Tisljar, and D. H. Wolf, *Eur.
J. Biochem.* **219,** 627 (1994)]; (5) rabbit MOP (as microsomal endopeptidase) [S. Kawabata,
K. Nakagawa, T. Muta, S. Iwanaga, and E. W. Davie, *J. Biol. Chem.* **268,** 12498 (1993)]; (6)
pig MOP (as soluble angiotensin binding protein) [N. Sugiura, H. Hagiwara, and S. Hirose,
*J. Biol. Chem.* **267,** 18067 (1992)]; (7) human TOP [H. F. Dovey, P. A. Seubert, S. Sinha, L.
Conlogue, S. P. Little, and E. M. Johnstone, Patent WO 92/07068 (1992)]; (8) pig TOP [A.
Kato, N. Sugiura, H. Hagiwara, and S. Hirose, *Eur. J. Biochem.* **221,** 159 (1994)]; (9) rat TOP
[N. McKie, P. M. Dando, N. D. Rawlings, and A. J. Barrett, *Biochem. J.* **295,** 57 (1993)]; (10)
*Salmonella typhimurium* oligopeptidase A [C. A. Conlin and C. G. Miller, *J. Bacteriol.* **174,** 1631
(1992)]; (11) *Escherichia coli* oligopeptidase A [C. A. Conlin, N. J. Trun, T. J. Silhavy, and
C. G. Miller *J. Bacteriol.* **174,** 5881 (1992)]; (12) *S. typhimurium* peptidyl-dipeptidase [S. Hamil-
ton and C. G. Miller, *J. Bacteriol.* **174,** 1626 (1992)]; (13) *E. coli* peptidyl-dipeptidase [B.
Henrich, S. Becker, U. Schroeder, R. Plapp, *J. Bacteriol.* **175,** 7290 (1993)].

mature male rats there is particularly high activity in testis,[55] and the same way well apply to other species.

The few immunolocalization studies on TOP in tissues and cells tend to be marred by the use of incompletely specific antibodies, reflected in multiple bands on immunoblots. In rat lung, strong reactions were obtained from bronchial epithelial cells and alveolar macrophages.[56] For rat brain, results are conflicting, as one report locates TOP in the cerebellum, especially in Purkinje cells,[57] whereas another finds it absent.[58] A study of the distribution of TOP in the vicinity of the hypothalamus by biochemical assay of punch samples revealed that the enzyme was broadly distributed and was particularly abundant where potential substrates such as dynorphins and luliberin are located.[59] The colocalization of TOP and [Met⁵]enkephalin in rat brain seemed consistent with a proposed role in the formation of the enkephalins.[58]

Strong nuclear staining for TOP in brain cells was reported by Healy and Orlowski,[57] whereas intense reaction from mast cell granules has also been described.[60] In chicken fibroblasts in monolayer culture, a particulate distribution consistent with endosomal localization has been seen.[61]

In subcellular fractionation studies, TOP is found predominantly in the cytosolic fraction of homogenates, implying that it is mainly distributed in the soluble phase of the cytoplasm, or possibly in fragile organelles such as endosomes. A certain proportion of the activity, sometimes up to 25%, has been reported to be associated with membrane fractions including synaptosomes, however.[58,62,63] These studies were made before the recognition that MOP acts on all the conventional substrates of TOP, and that neurolysin may do so, too (if it is distinct). Because membrane fractions might easily have been contaminated with fragments of mitochondria, further work is needed.

[55] A. Lasdun, S. Reznik, C. J. Molineaux, and M. Orlowski, *J. Pharmacol. Exp. Ther.* **251,** 439 (1989).

[56] H.-S. H. Choi, M. Lesser, C. Cardozo, and M. Orlowski, *Am. J. Respir. Cell Mol. Biol.* **3,** 619 (1990).

[57] D. P. Healy and M. Orlowski, *Brain Res.* **571,** 121 (1992).

[58] E. S. Oliveira, P. E. P. Leite, M. G. Spillantini, A. C. M. Camargo, and S. P. Hunt, *J. Neurochem.* **55,** 1114 (1990).

[59] C. J. Molineaux, B. Yu, and J. M. Ayala, *Neuropeptides* **18,** 49 (1991).

[60] S. H. Paik, F. Betti, A. C. M. Camargo, and E. S. De Oliveira, *J. Neuroimmunol.* **38,** 35 (1992).

[61] J.-M. Chen, A. Changco, M. A. Brown, and A. J. Barrett, *Exp. Cell Res.* **216,** 80 (1995).

[62] G. R. Acker, C. Molineaux, and M. Orlowski, *J. Neurochem.* **48,** 284 (1987).

[63] C. J. Molineaux and J. M. Ayala, *J. Neurochem.* **55,** 611 (1990).

*Oligopeptidase M.* In subcellular fractionations of rat liver following Potter–Elvehjem homogenization,[64,65] we have found that MOP is primarily localized in the mitochondrial fraction, and to a lesser extent in the cytosol. None was detected in microsomes, lysosomes, or peroxisomes.[25] Elsewhere, a microsomal fraction prepared following a 5-min treatment of the tissue with a Waring blender[23] may well have been contaminated with fragments of mitochondria, and, as was suggested above, the same may be true of some synaptosomal fractions containing the MOP-like neurolysin. MOP can be eluted from mitoplasts at acidic or alkaline pH values, and it is apparently associated noncovalently with the outer surface of the inner mitochondrial membrane, perhaps through an interaction with a transmembrane protein.[25]

## Possible Physiological Functions

As Pz-peptidase or collagenase-like peptidase, TOP was for some time thought to be involved in the physiological catabolism of collagen. The evidence for this was not strong, depending mainly on the ability of the enzyme to cleave synthetic substrates of bacterial collagenase, occasional reports of increased activity correlated with collagen degradation,[66] and a report of synergism with gelatinase in action on gelatin.[67] The enzyme is able to hydrolyze synthetic analogs of collagen-derived peptides of the composition $(Gly-Pro-Leu)_n$ and $(Gly-Hypro-Leu)_n$, in which $n$ is in the range 2–5.[68] However, the enzyme does not have the characteristics of a member of the collagenase family of matrix metalloendopeptidases,[69] and an attempt to confirm the synergistic action with gelatinase was unsuccessful.[70] There is little evidence that the enzyme is ever secreted by cells,[71] and the small amounts that are detectable extracellularly (where they might potentially act on connective tissue matrix components) may well arise simply from cell death. Our view, therefore, is that TOP is unlikely to play a significant part in connective tissue matrix resorption.

In the test tube, TOP efficiently degrades many peptides that have important biological activities (see above). The inhibitor Cpp-Ala-Ala-Phe-

[64] S. Watabe and T. Kimura, *J. Biol. Chem.* **260,** 5511 (1985).
[65] H. Beaufay, P. Jacques, P. Baudhuin, O. Z. Sellinger, J. Berthet, and C. de Duve, *Biochem. J.* **92,** 184 (1964).
[66] M. Rajabi and J. F. Woessner, Jr., *Am. J. Obstet. Gynecol.* **150,** 821 (1984).
[67] K. Sakyo, J.-I. Kobayashi, A. Ito, and Y. Mori, *J. Biochem. (Tokyo)* **94,** 1913 (1983).
[68] C. G. Knight, P. M. Dando, and A. J. Barrett, *Biochem. J.,* in press.
[69] A. J. Barrett and U. Tisljar, *Matrix* **1**(Suppl.), 95 (1992).
[70] P. M. Dando, G. Murphy, and A. J. Barrett, unpublished work (1994).
[71] E. S. Ferro, D. V. Tambourgi, F. Gobersztejn, M. D. Gomes, M. Sucupira, M. C. S. Armelin, T. L. Kipnis, and A. C. M. Camargo, *Biochem. Biophys. Res. Commun.* **191,** 275 (1993).

pAb has been used in *in vivo* experiments with rats that seemed to indicate that TOP is indeed responsible for biological degradation of enkephalin-containing peptides, luliberin, and bradykinin.[55,63,72] In some studies the inhibitor of TOP was administered together with an inhibitor of neprilysin, because it was known that neprilysin could cleave the inhibitor at the -Ala+Phe- bond. The cleavage by neprilysin inactivates the compound as an inhibitor of TOP; however, one of the products, Cpp-Ala-Ala, is a potent inhibitor of peptidyl-dipeptidase A (angiotensin-converting enzyme),[49] and it may well be that the effects obtained with Cpp-Ala-Ala-Phe-pAb *in vivo* were, in fact, due to indirect inhibition of peptidyl-dipeptidase A.[50,73] It has been suggested that the abundant TOP in testis may function to destroy luliberin and maintain an appropriate hormonal balance.[74]

Because TOP is mostly located inside cells and is probably not secreted or presented on the outer surface of most cells to any appreciable extent, one must consider the possibility that its main functions are intracellular. If the indications of endosomal localization mentioned above are confirmed, the intracellular enzyme may still act on exogenous substrates. There are also many potential substrates formed within cells, however, and the thiol dependence of TOP would certainly suit the enzyme well to activity in the cytoplasmic environment. The possibility that TOP participates in the intracellular catabolism of oligopeptides would be consistent with evidence that homologous enzymes function in this way in bacteria and yeast. Thus, mutant strains of bacteria deficient in oligopeptidase A ([34], this volume) and yeast deficient in saccharolysin[54] show accumulation of peptides. MOP could equally well function in the mitochondrion to hydrolyze cleaved targeting peptides, or peptide intermediates in protein catabolism.

Work with the mutant bacteria deficient in oligopeptidase A has given evidence that oligopeptidase A has restricted endopeptidase activity ([34], this volume). This gives additional weight to suggestions that TOP also may have true endopeptidase activity.[11]

Somewhat surprisingly, the biological activity of TOP seems to be essentially unregulated. As has been mentioned above, it seems to be fully active as synthesized, with no propeptide, and also there is little evidence of natural inhibitors, apart from, possibly, dynorphin A(1–13). The activity of the enzyme still might be regulated by control of its intracellular location, of course. Otherwise, we seem to be forced to the conclusion that its activities are entirely benign, and no control is necessary.

[72] J. A. Schriefer and C. J. Molineaux, *J. Pharmacol. Exp. Ther.* **266,** 700 (1993).

[73] X.-P. Yang, S. Saitoh, A. G. Scicli, E. Mascha, M. Orlowski, and O. A. Carretero, *Hypertension (Dallas)* **23**(Suppl.), I235 (1994).

[74] A. R. Pierotti, A. Lasdun, J. M. Ayala, J. L. Roberts, and C. J. Molineaux, *Mol. Cell. Endocrinol.* **76,** 95 (1991).

Acknowledgments

We are most grateful to Dr. R. L. Soffer for the gift of monoclonal antibody M3C and Dr. Luiz Juliano for the gift of o-aminobenzoyl-Gly-Gly-Phe-Leu-Arg-Arg-Val-[N-(2,4,-dinitrophenyl)ethylenediamine]. We also thank Drs. Véronique Monnet, Shigehisa Hirose, and Dieter H. Wolf for sharing data prior to publication. Financial support was received from the Medical Research Council (UK) and Snow Brand Milk Products Company, Ltd., Sapporo, Japan.

# [33] Mitochondrial Intermediate Peptidase

By GRAZIA ISAYA and FRANTISEK KALOUSEK

## Introduction

Precursors cleaved by two matrix peptidases represent a distinct group among the proteins that are imported into mitochondria. They are targeted to the matrix and the inner membrane and correlated by the presence of the octapeptide motif (F, I, or L)-X-X-(S, T, or G)-X-X-X-X at the carboxyl terminus of the targeting signal.[1,2] In vitro studies showed that the precursors are first cleaved by a general mitochondrial processing peptidase (MPP, EC 3.4.99.41)[3] that removes most of the leader peptide, leaving the octapeptide sequence at the protein amino terminus.[4] The resulting intermediate-size polypeptides are then processed to mature subunits by mitochondrial intermediate peptidase (MIP, EC 3.4.24.59), which specifically cleaves off the octapeptide.[5]

To define the structure–function relationships of MIP, we have purified the enzyme to homogeneity from rat liver mitochondrial matrix[6] and have analyzed the requirements for MIP activity[6] and substrate specificity.[4,5] A full-length rat liver MIP cDNA has been isolated,[7] and sequence analysis has revealed that MIP is structurally related to a new family of intracellular metallopeptidases.[7,8]

[1] J. P. Hendrick, P. E. Hodges, and L. E. Rosenberg, Proc. Natl. Acad. Sci. U.S.A. 86, 4056 (1988).
[2] Y. Gavel and G. von Heijne, Protein Eng. 4, 33 (1990).
[3] M. Brunner and W. Neupert, this volume [46].
[4] G. Isaya, F. Kalousek, W. A. Fenton, and L. E. Rosenberg, J. Cell Biol. 113, 65 (1991).
[5] G. Isaya, F. Kalousek, and L. E. Rosenberg, J. Biol. Chem. 267, 7904 (1992).
[6] F. Kalousek, G. Isaya, and L. E. Rosenberg, EMBO J. 11, 2803 (1992).
[7] G. Isaya, F. Kalousek, and L. E. Rosenberg, Proc. Natl. Acad. Sci. U.S.A. 89, 8317 (1992).
[8] N. D. Rawlings and A. J. Barrett, this volume [7].

## Purification of Rat Liver Mitochondrial Intermediate Peptidase

*Preparation of Mitochondrial Matrix.* For isolation and subfractionation of mitochondria, we use a modification of the procedure originally described by Schnaitman and Greenawalt.[9] Liver tissue from 20 male Sprague-Dawley rats, weighing 90 to 130 g, is used at one time, and all operations are carried out at 0°–5°. Rats are euthanized by cervical dislocation; liver tissue is rapidly removed and carefully minced with a pair of scissors. Four volumes of HMS buffer (2 m$M$ HEPES, pH 7.4, 220 m$M$ mannitol, 70 m$M$ sucrose) is added, and homogenization is carried out with a motor-driven Teflon pestle in a glass tissue grinder at a speed of 1000 rpm. The homogenate is centrifuged for exactly 1 min at 5000 $g$; the supernatant is collected and centrifuged at 10,000 $g$ for 6 min.

Contamination of mitochondria by lysosomes is minimized by treating the pellet with a diluted digitonin solution, according to the procedure of Loewenstein *et al.*[10] The pellet is first resuspended in a small volume of HMS, followed by the addition of an equal volume of a 1.75 mg/ml digitonin (Fisher Scientific, Fair Lawn, NJ) solution for 7 min with occasional stirring. Three volumes of HMS is added to dilute the digitonin, and the mitochondrial suspension is washed twice by a few passes in a Teflon–glass tissue grinder followed by centrifugation at 10,000 $g$ for 6 min.

For mitoplast preparation, the isolated mitochondria are diluted to a protein concentration of 100 mg/ml and incubated with an equal volume of a solution containing 12 mg/ml digitonin in HMS for 15 min with occasional stirring. Three volumes of HMS is added, and the mitoplast suspension is washed twice by gentle homogenization and centrifugation at 25,000 $g$ for 15 min. The mitoplast pellet is diluted to 50 mg protein/ml, incubated with 0.18 mg Lubrol WX (Sigma, St. Louis, MO) per milligram protein for 30 min, diluted with 1 volume of HMS, and centrifuged at 100,000 $g$ for 30 min. The soluble matrix fraction is divided into aliquots and stored at −70°.

Mitochondrial intermediate peptidase is purified from mitochondrial matrix in five steps (Table I), as described.[6] The first three steps are carried out using mitochondrial matrix prepared from two batches of 20 rats. Attempts to use larger volumes of matrix as the starting material result in a lower yield because greater time is required and the enzyme is rapidly inactivated in matrix (50% after 24 hr at 4°) and after the first step of purification. The first three steps should be carried out within 72 hr and repeated until the matrix fractions from 320 rats have been processed to this point and kept at −70°. The final two steps can be carried out using

[9] C. Schnaitman and J. W. Greenawalt, *J. Cell Biol.* **38,** 158 (1968).
[10] J. Loewenstein, H. R. Scholte, and E. M. Wit-Peeters, *Biochim. Biophys. Acta* **223,** 432 (1970).

TABLE I

PURIFICATION OF MITOCHONDRIAL INTERMEDIATE PEPTIDASE FROM RAT
LIVER MITOCHONDRIA[a]

| Fraction | Total protein (mg) | Total activity (U × 10⁻³) | Specific activity (U/mg) | Purification factor (-fold) | Yield (%) |
|---|---|---|---|---|---|
| Matrix | 15,750 | 210 | 13.3 | 1.0 | 100 |
| DEAE BioGel A | 1395 | 159 | 114 | 8.5 | 76 |
| Heparin-agarose | 217 | 72 | 331 | 24.8 | 34 |
| Hydroxyapatite | 28.7 | 54 | 1881 | 141 | 25 |
| ω-Aminooctyl-agarose | 1.3 | 9.4 | 7230 | 543 | 4.5 |
| Mono Q | 0.14 | 4.4 | 29,950 | 2251 | 2.1 |

[a] One unit of activity is defined as the amount of enzyme that catalyzes the conversion of 50% of the M-iOTC in 1 $\mu$l of *in vitro* translation mixture to its mature form in 10 min at 27°. Reprinted by permission of Oxford University Press from Kalousek *et al.*[6]

eight-fraction pools from the third step combined together. All steps of purification are monitored by sodium dodecyl sulfate–polyacrylamide gel electrophoresis (SDS–PAGE) and Coomassie blue staining of protein gels, and by determination of MIP activity.

*DEAE BioGel A Chromatography.* The mitochondrial matrix fraction (~2000 mg) is diluted with 10 m$M$ HEPES, pH 7.4, 1 m$M$ dithiothreitol (DTT), brought to 20 m$M$ NaCl in a final volume of 400 ml, and loaded on a DEAE BioGel A column (2.5 × 9 cm Bio-Rad, Richmond, CA), equilibrated with the same buffer, at a flow rate of 120 ml/hr. After absorption, the column is washed with 2 column volumes of the same buffer, followed by washing with 16 column volumes of 30 m$M$ NaCl, 10 m$M$ HEPES, pH 7.4, 1 m$M$ DTT. The peptidase is then eluted with a 500-ml linear gradient from 30 m$M$ NaCl to 150 m$M$ NaCl in 10 m$M$ HEPES, pH 7.4, 1 m$M$ DTT, at a flow rate of 60 ml/hr. Fractions with a NaCl concentration between 70 and 100 m$M$ are pooled, concentrated on a Centriprep 30 centrifugal microconcentrator (Amicon, Danvers, MA) to 20 ml, diluted to 100 ml with 10 m$M$ HEPES, pH 7.4, 1 m$M$ DTT, and reconcentrated to 25 ml. The concentration of NaCl is then adjusted to 20 m$M$.

*Heparin-Agarose Chromatography.* The concentrated sample from the previous step is loaded on a heparin-agarose column (2.5 × 6 cm, Sigma, St. Louis, MO), equilibrated with 20 m$M$ NaCl, 10 m$M$ HEPES, pH 7.4, 1 m$M$ DTT, at a flow rate of 24 ml/hr. The MIP activity is eluted with the same buffer at the same flow rate. The first 50 ml of the eluate is used in the subsequent step.

*Hydroxyapatite Chromatography.* The heparin eluate is diluted with 10 m$M$ HEPES, pH 7.4, 1 m$M$ DTT, to 100 ml and applied to a hydroxyapatite column (1.5 × 4.6 cm prepared as described[6]), equilibrated with 10 m$M$ NaCl, 10 m$M$ HEPES, pH 7.4, 1 m$M$ DTT, at a flow rate of 24 ml/hr. After absorption, the column is washed extensively with 10 m$M$ potassium phosphate, pH 7.0, 1 m$M$ DTT. About 40% of the total protein and less than 10% of MIP elute and are discarded. The major portion of MIP activity is eluted by a 150-ml linear gradient from 10 m$M$ potassium phosphate, pH 7.0, 1 m$M$ DTT, to 60 m$M$ potassium phosphate, pH 7.6, 1 m$M$ DTT (3.5-ml fractions). Proteins from every other fraction are analyzed by SDS–PAGE using both 8.3% T, 4% C and 6.1% T, 2.5% C gels (T denotes the total concentration of acrylamide and bisacrylamide; C denotes the percentage concentration of the cross-linker relative to T). Using a 8.3% T, 4% C gel, it is possible to distinguish MIP from a protein which moves slightly faster and starts to be eluted earlier than MIP. Using a 6.1% T, 2.5% C gel, a slower moving impurity is detected, which is present in at least 10-fold excess over MIP. Only fractions that are virtually free of both contaminants are pooled (usually between 26 and 42 m$M$ potassium phosphate), concentrated to 5 ml on a Centriprep 10, frozen on dry ice, and kept at $-70°$.

*ω-Aminooctyl-Agarose.* Pools from eight hydroxyapatite columns are combined and diluted with 50 m$M$ NaCl, 10 m$M$ HEPES, pH 7.4, 1 m$M$ DTT, to 100 ml and loaded on a 1.5 × 6 cm ω-aminooctyl-agarose column (Sigma, St. Louis, MO), equilibrated with the same buffer, at a flow rate of 24 ml/hr. The column is washed with 12 column volumes of 150 m$M$ NaCl, 10 m$M$ HEPES, pH 7.4, 1 m$M$ DTT. The MIP activity is eluted with a 500-ml linear gradient from 150 to 500 m$M$ NaCl in 10 m$M$ HEPES, pH 7.4, 1 m$M$ DTT, at a flow rate of 24 ml/hr. Fractions with MIP activity and free of most MPP activity (300–340 m$M$ NaCl) are pooled, concentrated, diluted with 10 m$M$ HEPES, pH 7.4, 1 m$M$ DTT, and reconcentrated on a Centriprep 10 to a final NaCl concentration of 20 m$M$ and a final volume of 3 ml.

*Mono Q Fast Protein Liquid Chromatography.* One-half of the pool from the previous step is loaded on a Mono Q anion-exchange column (HR 5/5) (Pharmacia Biotech, Piscataway, NJ) which has been equilibrated with the same buffer. After washing the column with 2 ml of the same buffer and 6 ml of 120 m$M$ NaCl, 10 m$M$ HEPES, pH 7.4, 1 m$M$ DTT, MIP activity is eluted with a 30-ml linear gradient between 120 and 280 m$M$ NaCl in 10 m$M$ HEPES, pH 7.4, 1 m$M$ DTT. The MIP activity elutes between 165 and 205 m$M$ NaCl. Active fractions are pooled, concentrated, diluted with 10 m$M$ HEPES, pH 7.4, 1 m$M$ DTT, reconcentrated with a Centricon 10 to a final volume of 300 $\mu$l (20 m$M$ NaCl), and stored at $-70°$ without loss of activity for at least 6 months.

With this procedure, a 2250-fold MIP purification is achieved with a final yield of about 2%. All steps are reproducible in purity, recovery, and specific activity.

*Properties of Purified Enzyme.* Purified, active MIP is a monomer of 75 kDa. It should be noted that about 20% of the final MIP preparation consists of equimolar amounts of two peptides of 47 and 28 kDa which are not separable under nondenaturing conditions and represent the products of a single cleavage in the 75-kDa protein. The enzyme has a broad pH optimum between pH 6.6 and pH 8.9, and it is sensitive to thiol-blocking reagents. At a concentration of 0.2 m$M$, inactivation is essentially complete with $N$-ethylmaleimide or $p$-hydroxymercuribenzoate, whereas approximately 50% inactivation is obtained with iodoacetic acid and iodoacetamide. Optimum activity requires divalent cations: addition of 1 m$M$ Mn$^{2+}$, Mg$^{2+}$, or Ca$^{2+}$ stimulates the activity 4.5-, 4.8-, and 2.2-fold, respectively, whereas 1 m$M$ Zn$^{2+}$, Co$^{2+}$, or Fe$^{2+}$ inhibits MIP activity completely. Also, the enzyme is reversibly inhibited by 50 $\mu M$ EDTA. To test for these requirements, Mg$^{2+}$ and other cations normally present in the reticulocyte lysate used for the assay (see below) should be removed from the translation mixture containing the radiolabeled substrate. This can be easily achieved by diluting the translation mixture with 10 m$M$ HEPES, pH 7.4, and ultrafiltering the sample through a Centricon 30.

Determination of Mitochondrial Intermediate Peptidase Activity

Similar to other known leader peptidases, MIP has a complex specificity. A characteristic octapeptide sequence is always present at the MIP substrate amino termini, including, in particular, a large hydrophobic amino acid (normally phenylalanine, isoleucine, or leucine) at P8 and a small hydroxylated residue (serine or threonine), or glycine at P5.[1,2] Although there is no obvious consensus sequence on the carboxyl-terminal side of the MIP cleavage sites,[1,2] only certain mitochondrial proteins can be cleaved by MIP,[5] and mutations as distant as the P70' position may severely affect cleavage.[11]

These requirements constitute the main problem in the use of short synthetic substrates and a colorimetric assay for determination of MIP activity. Alternatively, we have developed an assay that utilizes *in vitro* translated mitochondrial precursor or intermediate proteins. The precursor for ornithine transcarbamylase (pOTC) is routinely used in our laboratory, and several other cloned precursors with a typical MIP processing site in the targeting signal are available. When the precursors are imported into intact mitochondria, the cleavage catalyzed by MIP is part of a general

[11] L. A. Graham, U. Brandt, J. S. Sargent, and B. L. Trumpower, *J. Bioenerg. Biomembr.* **25**, 245 (1993).

protein import reaction involving binding of the precursor to surface receptors, energy-dependent translocation across the outer and inner mitochondrial membranes, interaction with molecular chaperones, and initial cleavage by the general processing peptidase, MPP.[12] Many of the steps are circumvented when the precursor is incubated directly with purified MIP. Under those conditions, however, initial cleavage by MPP is still required to expose the octapeptide at the amino terminus of the substrate and provide an optimized MIP processing site, representing a rate-limiting step.

Thus, to be able to measure MIP activity independently of the presence of MPP, we have modified the pOTC cDNA via polymerase chain reaction (PCR) amplification and constructed an artificial intermediate-size polypeptide (iOTC) that is translated *in vitro* from a methionine in the P8 position (M-iOTC).[4] *In vitro* transcriptions are performed from the pOTC and M-iOTC cDNAs cloned in transcription vectors under the control of the phage SP6 polymerase promoter (many systems for standard *in vitro* transcription and translation are commercially available, e.g., from Promega, Madison, WI, or GIBCO–BRL, Gaithersburg, MD). Messenger RNAs are translated in rabbit reticulocyte lysate in the presence of [35S]methionine. Translation mixtures containing radiolabeled substrates are divided into aliquots and stored at $-70°$ for up to 2 months. One-half to one unit of purified enzyme is added to 1 $\mu$l of translation mixture in a total volume of 10 $\mu$l in the presence of 1 m$M$ Mn$^{2+}$. Reactions are incubated at 27° for 10 to 20 min and analyzed directly by SDS–PAGE and fluorography; quantitations of protein bands are carried out by densitometric analysis of fluorograms, and cleavage sites are determined by amino-terminal radiosequence analysis conducted directly on the reaction products electroblotted onto polyvinylidene difluoride membranes (Immobilon P from Millipore, Bedford, MA).

A typical processing reaction using purified rat liver MIP, with or without addition of purified MPP, is shown in Fig. 1. When MPP alone is added to the reaction, pOTC is processed to a characteristic intermediate-size polypeptide (iOTC) (lane 2, Fig. 1); there is no detectable proteolytic activity on incubation of pOTC with MIP alone (lane 3, Fig. 1), and simultaneous incubation with MPP plus MIP is required to yield mature OTC (lane 4, Fig. 1). On the other hand, M-iOTC is processed to mature OTC by MIP alone (lane 7, Fig. 1), and addition of MPP has no further effects (lane 8, Fig. 1).

---

[12] F. U. Hartl, N. Pfanner, D. W. Nicholson, and W. Neupert, *Biochim. Biophys. Acta* **988,** 1 (1989).

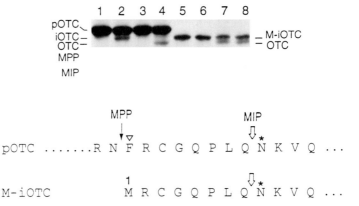

FIG. 1. *In vitro* assay for MIP activity determination. [35S]Methionine-labeled pOTC and M-iOTC were incubated with MPP and/or MIP, as indicated, for 20 min at 27°. Reaction products were analyzed directly by SDS–PAGE and fluorography. The amino acid sequences of pOTC and M-iOTC in the region surrounding the MPP (solid arrow) and MIP (open arrow) cleavage sites are shown. The triangle marks the iOTC amino terminus, an asterisk marks the mature OTC amino terminus, and 1 marks the initiator methionine in M-iOTC. Reprinted with permission of ASBMB from Isaya *et al.*[5]

In general, processing by the purified matrix peptidases is consistent with processing by intact mitochondria.[4] If purified MIP and/or MPP are not available, crude mitochondrial matrix may represent a good alternative; a matrix fraction with MPP activity and essentially free of detectable MIP activity can be obtained by treatment of crude matrix with 0.2 m$M$ $N$-ethylmaleimide.

## Inhibition of Mitochondrial Intermediate Peptidase by Synthetic Octapeptides

Synthetic octapeptides inhibit MIP activity *in vitro.*[5] This apparently depends on the ability of MIP to bind to a specific octapeptide whether or not the peptide is joined to the remaining portion of the corresponding intermediate protein.

In the following experiment, MIP inhibition by synthetic iOTC octapeptides is used as a means to define the main structural features that direct recognition by MIP. Peptides are synthesized on an Applied Biosystems (Foster City, CA) solid-phase peptide synthesizer and purified by reversed-phase high-performance liquid chromatography. Peptide characterization is by amino acid analysis and fast atom bombardment mass spectrometry. Each peptide is resuspended in 10 m$M$ HEPES buffer, pH 7.4, and is added to the processing reactions at the required concentrations in the presence of 1 $\mu$l of *in vitro* translated M-iOTC and 1 U of purified MIP. The following peptides are used: OTC(25–32), corresponding to the natural iOTC octapeptide, OTC(25Y–32), containing a tyrosine that substitutes for the amino-terminal phenylalanine, OCT(26–32), not including the amino-terminal phenylalanine, and OTC(25–31), not including the carboxyl-terminal glutamine. Three random peptides are used in control reactions: OTC(31–40), corresponding to the 2 carboxyl-terminal residues of the iOTC octapeptide joined to the 8 amino-terminal residues of mature OTC and including the MIP cleavage site; random octapeptide A (anaphylatoxin C3a fragment 70–77) (Sigma); and random decapeptide B (adrenocorticotropic hormone fragment 1–10) (Sigma). Table II summarizes the results of three independent determinations of MIP inhibition by increasing concentrations of each peptide.

That and similar experiments with other peptides[5] have led to the conclusion that synthetic octapeptides are specifically recognized by MIP. They appear to bind to a site that is crucial for MIP activity, competing with the octapeptide at the amino terminus of the substrate. In this sense, it is noteworthy that the decapeptide OTC(25–34), containing the octapeptide sequence joined to the MIP cleavage site, did not inhibit MIP more effi-

TABLE II

INHIBITION OF MITOCHONDRIAL INTERMEDIATE PEPTIDASE BY SYNTHETIC PEPTIDES[a]

| Peptide | Peptide concentration ($\mu M$) | | | | | | | |
| --- | --- | --- | --- | --- | --- | --- | --- | --- |
| | 0 | 0.05 | 0.5 | 2 | 5 | 20 | 50 | 100 |
| OTC(25–32), FRCGQPLQ | 100 | 93 ± 15 | 61 ± 12 | 47 ± 1 | 36 ± 1 | 3 ± 2 | | |
| OTC(25Y–32), YRCGQPLQ | 100 | | | | 73 ± 11 | 71 ± 15 | 47 ± 22 | 12 ± 18 |
| OTC(26–32), RCGQPLQ | 100 | | | | 89 ± 4 | 95 ± 7 | 94 ± 12 | 86 ± 9 |
| OTC(25–31), FRCGQPL | 100 | | | | 92 ± 4 | 74 ± 18 | 43 ± 26 | 19 ± 8 |
| OTC(31–40), LQNKVQLKGR | 100 | | | | | | | 77 ± 10 |
| Random A, ASHLGLAR | 100 | | | | | | | 95 ± 10 |
| Random B, SYSMDHFRWG | 100 | | | | | | | 86 ± 14 |

[a] M-iOTC was incubated with purified MIP *in vitro* in the presence of the indicated peptides. Three independent determinations were performed for each peptide at the indicated concentrations. The amounts of OTC formed by MIP from M-iOTC in the presence of the peptides are expressed as the percent of OTC formed in the absence of peptides. Values were determined by densitometry of fluorograms and are expressed as means ± SD. Reprinted with permission of ASBMB from Isaya *et al.*[5]

ciently than OTC(25–32).[5] Positioning of a large hydrophobic amino acid at the peptide amino terminus is essential but not sufficient for MIP inhibition. Although none of the remaining residues in the octapeptide is individually as crucial as the amino-terminal residue,[4] a characteristic overall amino acid composition (consisting of predominantly hydrophobic and hydroxylated amino acids with no acidic residues) and a minimum length of 8 amino acids are also important features. In general, we have found that MIP inhibition by a certain octapeptide is reproducible with different substrates.[5] On the other hand, MIP exhibits a wide range of affinity for different octapeptides. For example, with synthetic peptide Fe/S(25–32) (LTTSTALQ), corresponding to the *Neurospora crassa* Rieske iron-sulfur protein intermediate octapeptide,[13] the concentration of peptide required for MIP inhibition is 5-fold higher than with OTC(25–32).

## Molecular Characterization of Mitochondrial Intermediate Peptidase

*Isolation of Full-Length Rat Liver cDNA.* To isolate a rat liver MIP cDNA, purified MIP is used to obtain amino acid sequence information from the amino terminus and tryptic peptides. Degenerate oligonucleotide primers are synthesized based on the underlined amino acids from the sequences of two peptides: F-D-F-E-I-S-G-I-H-L-D-E-E-K, and H-L-V-I-D-G-L-H-A-E-A-S-D-D-L-V. To generate a partial cDNA, poly(A)$^+$ RNA is prepared from rat cardiac muscle, first-strand cDNA is generated with reverse transcriptase and random hexamers, and PCR amplifications are carried out on the single-stranded cDNA pool. Because at the time of the experiment the positions of the two tryptic peptides in the intact protein were not known, the PCR amplifications are carried out with various combinations of sense and antisense primers from both peptides, and several amplification products of different sizes are obtained in each reaction. Only one among such products should be derived from specific amplification of the MIP cDNA. The expected size of the authentic product is not known, however, and no internal oligonucleotide probe is available to screen the products by Southern blot hybridization. Therefore, PCR products are analyzed by sequencing.

The deduced amino acid sequence of one fragment of 468 bp is found to contain the sequence of the two tryptic peptides used for primer design. This partial cDNA clone is then used to screen a rat liver cDNA library, and 11 positive clones are isolated after three rounds of screening. The largest of the clones is found to contain an open reading frame of 2130 bp

[13] F. U. Hartl, B. Schmidt, E. Wachter, H. Weiss, and W. Neupert, *Cell* (*Cambridge, Mass.*) **47,** 939 (1986).

specifying a protein of 710 amino acids that includes the sequence of four tryptic peptides and the amino-terminal sequence derived from purified MIP (GenBank Accession No. M96633). The protein contains a predominance of acidic over basic amino acids (82 versus 60), and it is generally hydrophilic. Eighteen cysteine residues are distributed over the length of the protein, and a putative zinc-binding motif, H-E-M-G-H, is found toward the carboxyl terminus.

*Mitochondrial Import and Processing of Enzyme Precursor.* A 33-residue putative mitochondrial leader peptide is encoded by the cDNA sequence upstream from the mature MIP amino terminus, as determined by amino acid sequencing of purified MIP (Fig. 2). The sequence surrounding the mature amino terminus shows the motif X-R-X+X-S, a cleavage site typically recognized by MPP, but no octapeptide motif is found in this region, suggesting that MIP may not be involved in its own proteolytic maturation.

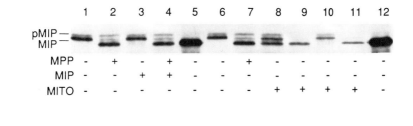

FIG. 2. Processing of *in vitro* translated MIP precursor by purified MPP and MIP and import by isolated rat liver mitochondria. [35S]Methionine-labeled MIP precursor (lanes 1 and 6) was incubated with purified MPP (lanes 2 and 7), MIP (lane 3), MPP plus MIP (lane 4), or freshly isolated, intact rat liver mitochondria (lane 8) for 20 min at 27°. After import, an aliquot was treated with trypsin (lane 9), whereas a second aliquot was separated into supernatant (lane 10) and pellet (lane 11) by centrifugation. Products of the processing and import reactions were analyzed directly by SDS–PAGE and fluorography. In lanes 5 and 12, a protein translated from an initiating methionine positioned at the mature MIP amino terminus indicates the migration position of the mature-size protein. The band immediately above the MIP precursor band is a translation artifact; that product is not processed by purified MPP nor imported by intact mitochondria. The amino acid sequence of the MIP precursor leader peptide is shown at bottom, and the site of MPP cleavage is indicated by the arrow. An asterisk shows the mature MIP amino terminus as determined by amino acid sequence analysis of purified rat liver MIP. Reprinted from Isaya *et al.*[7]

To analyze the biogenesis of the MIP precursor (pMIP), two cDNAs encoding pMIP and mature-size MIP are cloned in *in vitro* transcription vectors. Translation of the cDNAs yields two major products of 78 and 75 kDa, respectively (Fig. 2, lanes 1 and 5). *In vitro* processing of pMIP by the two purified matrix peptidases is performed as described above (Fig. 2). Consistent with the absence of a site of possible MIP cleavage, pMIP is cleaved to mature-size MIP by MPP (Fig. 2, lane 2), whereas MIP does not have any effect on pMIP processing (Fig. 2, lanes 3 and 4). Radiolabeled pMIP is also incubated with freshly isolated rat liver mitochondria. Six microliters of pMIP translation mixture is incubated with 4 $\mu$l of a mitochondrial suspension at a final protein concentration of 8 mg/ml for 20 min at 27°. After import, an aliquot is treated with trypsin (40 $\mu$g/ml final concentration) for 5 min on ice and then supplemented with soybean trypsin inhibitor (100 $\mu$g/ml). Mitochondria from a second aliquot are reisolated by centrifugation, and the resulting supernatant and pellet are analyzed separately. The MIP precursor is processed to mature MIP by intact mitochondria (Fig. 2, lane 8) identically to processing by purified MPP (Fig. 2, lane 7). Most of the newly formed mature MIP is protected from externally added trypsin (Fig. 2, lane 9) and is associated with the mitochondrial pellet (Fig. 2, lane 11), whereas most of the unprocessed pMIP remains in the supernatant (Fig. 2, lane 10) and is degraded by trypsin treatment (Fig. 2, lane 9).

*Sequence Analysis of Rat Mitochondrial Intermediate Peptidase.* The predicted MIP protein is used as a query peptide sequence to search for similarities in sequence databases using the program TFASTA[14] (October 1993). The top scoring sequences are *Schizophyllum commune* hypothetical metallopeptidase (GenBank Accession No. M97179) (40% identity over 204 amino acids), *Salmonella typhimurium* (M84574) and *Escherichia coli* (M93984) oligopeptidase A (30% identity over 230 amino acids), *Saccharomyces cerevisiae* YCL57w hypothetical metallopeptidase (X59720) (21% identity over 440 amino acids), pig soluble angiotensin-binding protein (D11336) (26% identity over 395 amino acids), rat testes EC 3.4.24.15 (M61142) (28% identity over 166 amino acids), and *E. coli* dipeptidyl carboxypeptidase (X57947) (29% identity over 219 amino acids). The predicted molecular masses of members of this group of proteins are very close, ranging between 70 and 75 kDa. Except for the presence of a putative zinc-binding motif, they show no homology to any other proteins in the sequence databases.

In addition to the structural similarity, MIP shares with thimet oligopeptidase the characteristic of being a thiol-dependent metallopepti-

---

[14] University of Wisconsin, Genetics Computer Group, Madison.

dase.[15] Also, the affinity of EC 3.4.24.15[15] and oligopeptidase A[16] for small peptide substrates is consistent with the inhibition of MIP activity by synthetic octapeptides.

## Acknowledgments

This work was supported by Grants GM48076, to G. I., and DK09527, to F. K., from the National Institutes of Health.

[15] A. J. Barrett, M. A. Brown, P. M. Dando, C. G. Knight, N. McKie, N. D. Rawlings, and A. Serigawa, this volume [32].
[16] C. A. Conlin and C. G. Miller, this volume [34].

# [34] Dipeptidyl Carboxypeptidase and Oligopeptidase A from *Escherichia coli* and *Salmonella typhimurium*

By Christopher A. Conlin and Charles G. Miller

## Introduction

Dipeptidyl carboxypeptidase (Dcp), an exopeptidase which removes dipeptides from the C termini of its substrates, was first purified and characterized from extracts of *Escherichia coli* by Yaron and co-workers.[1] The enzyme was the subject of an earlier chapter in this series.[2] Oligopeptidase A (OpdA) was originally identified in extracts of *Salmonella typhimurium* as an activity able to hydrolyze some of the same small peptide substrates as Dcp but still present in mutants lacking Dcp.[3,4] It is now clear that although the two activities differ in specificity they are structurally related: both are members of a proteinase family that includes not only bacterial enzymes but also enzymes from fungi and from higher eukaryotes.

## Assay and Detection of Dipeptidyl Carboxypeptidase

*N*-Acetyl-Ala$_3$ (AcAla$_3$) is a convenient and specific substrate for Dcp. It is not hydrolyzed by any other peptidases present in crude extracts of *E. coli* or *S. typhimurium*. Extracts of *dcp* mutants do not hydrolyze this

[1] A. Yaron, D. Mlynar, and A. Berger, *Biochem. Biophys. Res. Commun.* **47**, 897 (1972).
[2] A. Yaron, this series, Vol. 45, p. 599.
[3] E. R. Vimr, L. Green, and C. G. Miller, *J. Bacteriol.* **153**, 1259 (1983).
[4] E. R. Vimr and C. G. Miller, *J. Bacteriol.* **153**, 1252 (1983).

substrate. The reaction products can be detected by the ninhydrin method[4] or by high-performance liquid chromatography (HPLC).[3] In a crude extract of a wild-type strain of *S. typhimurium* or *E. coli*, the product of the reaction, $Ala_2$, is rapidly converted by dipeptidases to free Ala. In extracts of a strain lacking peptidases N, A, B, and D, $Ala_2$ is stable and is the observed product.

*Ninhydrin Method*

*Reagents*

Assay buffer: 50 m$M$ sodium barbital, pH 8.1

Substrate: 3.66 m$M$ AcAla$_3$ in assay buffer containing 0.1 m$M$ CoCl$_2$

Enzyme: crude soluble protein extract (diluted to ~2 mg/ml) of *S. typhimurium* LT2 from cells grown to late exponential phase in rich medium, washed, and resuspended in assay buffer

Ninhydrin reagent: Ninsol (Pierce Chemical Co., Rockford, IL)

Trichloroacetic acid (TCA), 50% (v/v) in water

Methanol, 50% (v/v) in water

*Procedure.* Two parts of substrate solution is added to one part extract and the reaction mixture incubated at 37°. At various times, 0.3-ml samples are withdrawn into plastic centrifuge tubes (Eppendorf) containing 30 $\mu$l of 50% TCA. After centrifugation to remove precipitated protein, a 0.1-ml sample of the supernatant fluid is added to 0.4 ml of water, followed by 0.5 ml Ninsol reagent. After 15 min of incubation in a boiling water bath, the tubes are cooled with running tap water. Next, 50% methanol is added to dilute the blue color into a readable range, and the absorbance at 570 nm is read on a spectrophotometer. The values are converted to alanine equivalents from a standard curve constructed with L-alanine. One unit of activity is defined as the amount of enzyme able to produce 1 $\mu$mol of alanine per minute. The specific activity of a crude extract of *S. typhimurium* LT2 is approximately 0.09 $\mu$mol alanine per minute per milligram of protein.

## Assay and Detection of Oligopeptidase A

*Ninhydrin Method*

We have used AcAla$_4$ for assaying OpdA in a manner essentially identical to the ninhydrin method described above (with an AcAla$_4$ concentration in the stock solution of 1.45 m$M$). The substrate is hydrolyzed by both Dcp and OpdA but is a specific substrate for OpdA in mutant strains lacking Dcp. The product of OpdA hydrolysis, $Ala_3$, is rapidly hydrolyzed to free alanine by peptidases present in the crude extract but accumulates in extracts of peptidase-deficient mutants.

*Spectrophotometric Aminopeptidase-Coupled Assay*

Benzyloxycarbonyl-Ala-Ala-Leu-*p*-nitroanilide (Z-AAL-pNA) can be used as the substrate in a relatively specific spectrophotometric assay for OpdA.[5] Although OpdA does not liberate the chromogen (*p*-nitroaniline), it hydrolyzes the substrate to produce Leu-*p*-nitroanilide (LNA) as the product. In the presence of peptidase N, a broad specificity aminopeptidase with naphthylamidase activity, the LNA is rapidly converted to *p*-nitroaniline, and the conversion can be followed by monitoring the increase in absorbance at 410 nm. The method is conveniently adaptable to microassays in plastic depression plates and has been used to screen for strains carrying plasmids that overproduce OpdA.[6] The gene encoding peptidase N has been cloned from both *E. coli*[7] and *S. typhimurium* (C. G. Miller, unpublished). It should be noted that this sort of assay may be adapted to other enzymes which do not release the chromogen from chromogenic peptide substrates.

*Reagents*

Substrate: Z-AAL-pNA (Sigma, St. Louis, MO), 0.5 mg/ml (0.07 m$M$) in dimethylformamide
Buffer: 50 m$M$ N-methyldiethanolamine, pH 8.0, 0.5 m$M$ CoCl$_2$
Partially purified peptidase N, 5 mg protein/ml, specific activity approximately 40 units/mg protein[8] [the enzyme can be conveniently purified from a *S. typhimurium* strain (TN1727/pJG2) lacking other peptidases including OpdA and containing a plasmid overproducing peptidase N by chromatography on DEAE-cellulose (Whatman, Clifton, NJ, DE-52)]
*Procedure.* To 1 ml of a mixture containing buffer and peptidase N (1 unit), add sample containing oligopeptidase A. The reaction is initiated by addition of substrate to a final concentration of 0.12 m$M$, and the absorbance change is followed at 410 nm in a recording spectrophotometer. The rate of hydrolysis is determined using a $\Delta\varepsilon_{410}$ of 8900.

*Detection of Oligopeptidase A Activity in Microcultures*

It is frequently useful to be able to screen a large number of cultures for peptidase activity. The availability of a convenient procedure for such screening allows the isolation of mutants lacking or overproducing the activity. Strains carrying peptidase plasmids can also be identified using

[5] C. A. Conlin, Ph.D. Thesis, Case Western Reserve University, Cleveland, Ohio (1991).
[6] C. A. Conlin and C. G. Miller, *J. Bacteriol.* **174**, 1631 (1992).
[7] M. Foglino, S. Gharbi, and A. Lazdunski, *Gene* **49**, 303 (1986).
[8] H. J. Lee, J. N. LaRue, and I. B. Wilson, *Anal. Biochem.* **41**, 397 (1971).

such screens. Microcultures grown in the wells of plastic depression plates can be used to carry out such screens. The procedure used to screen a plasmid library for strains carrying *opdA* plasmids is described.

The wells of plastic depression plates (Linbro Scientific) containing 0.2 ml of a suitable medium [LB plus chloramphenicol (20 $\mu$g/ml) when screening a pBR328 library] are inoculated with cells from the individual colonies to be tested (transformants containing plasmids from a library prepared on an *opdA*$^+$ strain) using any convenient method (e.g., sterile toothpicks). The plates are incubated overnight, and a replica is made using a multipronged transfer device to transfer a few cells from each microculture to the surface of an 150 × 50 mm agar plate. Cells in the depression plates are harvested by centrifugation of the plastic depression plates using the plastic depression plate adapters supplied by most centrifuge manufacturers and washed once in 200 $\mu$l of 50 m$M$ Tris-HCl, pH 7.6 (Tris buffer). The pellet is resuspended in 30 $\mu$l lysis solution (lysozyme, 1 mg/ml in Tris buffer) and the plates subjected to three freeze–thaw cycles ($-70°/+37°$). After addition of 150 $\mu$l of Tris buffer containing 0.5 m$M$ CoCl$_2$ to each well, 17 $\mu$l of substrate solution (Z-AAL-*p*NA, 0.5 mg/ml in dimethylformamide) is added and the plates incubated (37°, 1 hr). To release the chromogen from the product of OpdA hydrolysis, 5 $\mu$l of peptidase N solution (as above) is added, and the plates are incubated for an additional 10 min. The liberated *p*-nitroaniline can be observed directly or after diazotization, which produces a more intense purple color.[9] If diazotization is used, 10 $\mu$l of NaNO$_2$ solution (0.1% in 1 $N$ HCl) and 10 $\mu$l ammonium sulfamate (0.5% v/v in 1 $N$ HCl) are added successively to each well. After 5 min of incubation at room temperature, 10 $\mu$l *N*-(1-naphthyl)ethylenediamine solution (0.05% v/v in 95% ethanol) is added. Positive wells are identified by the presence of significantly darker color, and the strains corresponding to the positive wells can be recovered from the agar plate replica.

The use of coupled assay methods is not limited to Dcp and OpdA but can be applied quite generally to assay peptidases which produce peptide products. It is necessary to have available a source of a peptidase which hydrolyzes the product peptide(s) but not the substrate peptide. Mutants lacking potentially interfering peptidases but overproducing the desired peptidase make it possible to use crude extracts for this purpose, but any source of a sufficiently pure peptidase of appropriate specificity should suffice.

The use of microculture screening is also quite general. Variants of the procedure have been used to isolate mutants lacking *opdA*[3] or *pepE*,[10] or

[9] B. G. Ohlsson, B. R. Westrom, and B. W. Karlsson, *Anal. Biochem.* **152,** 239 (1986).
[10] T. H. Carter and C. G. Miller, *J. Bacteriol.* **159,** 453 (1984).

lacking or overproducing *pepT*.[11,12] The growth and lysis of the microcultures are similar for all the procedures. Plastic depression plates can also be used in convenient qualitative assays for screening for the presence of a particular activity in a variety of organisms or for assaying fractions during purification.[13]

### Detection of Peptidase Activity after Gel Electrophoresis

One of the major problems in characterizing the peptidases of an organism or tissue is the frequent presence of multiple activities able to hydrolyze the same substrate under similar conditions. Detection of peptidase activity after electrophoresis of crude extracts frequently provides a convenient method for determining the number and specificities of the peptide hydrolases present.[14] Because both Dcp and Opd produce peptides rather than free amino acids, it is not possible to make direct use of coupled assays based on the detection of free amino acids.[14,15]

In the procedure, the nondenaturing gel through which the peptidase-containing sample has been electrophoresed is overlayed with agar containing a peptide substrate, L-amino-acid oxidase, peroxidase, and *o*-dianisidine. Peptidase activity in the gel produces free amino acids which are oxidized by L-amino-acid oxidase, producing $H_2O_2$ which serves to oxidize *o*-dianisidine to a brown insoluble pigment in a reaction catalyzed by peroxidase. By adding a second peptidase which is able to hydrolyze the products formed by Dcp or Opd to free amino acids but which is unable to attack the substrate directly, the method can be conveniently used to detect peptidases that produce peptide products. For Dcp, peptidase D (PepD) of *E. coli* and *S. typhimurium*, a broad specificity dipeptidase, can be added to the gel overlay reaction mixture to hydrolyze the dipeptide products to free amino acids.[13] The *pepD* genes from both *E. coli*[16,17] and *S. typhimurium*[13] have been cloned on high copy number plasmids, and strains carrying such plasmids significantly overexpress PepD. Crude extracts of mutant strains of *S. typhimurium* lacking other peptidases but carrying *pepD* on a plasmid can be used as a source of PepD for *in situ* detection of Dcp activity.[13] A similar approach allows identification of OpdA activity after

[11] K. L. Strauch, T. H. Carter, and C. G. Miller, *J. Bacteriol.* **156**, 743 (1983).
[12] K. L. Strauch and C. G. Miller, *J. Bacteriol.* **154**, 763 (1983).
[13] S. Hamilton and C. G. Miller, *J. Bacteriol.* **174**, 1626 (1992).
[14] C. G. Miller and K. Mackinnon, *J. Bacteriol.* **120**, 355 (1974).
[15] W. H. Lewis and H. Harris, *Nature (London)* **215**, 351 (1967).
[16] B. Henrich, U. Monnerjahn, and R. Plapp, *J. Bacteriol.* **172**, 4641 (1990).
[17] J. Klein, B. Henrich, and R. Plapp, *J. Gen. Microbiol.* **132**, 2337 (1986).

gel electrophoresis. The simplest OpdA substrates are blocked peptides. Such substrates lack a free N terminus and are therefore resistant to the action of aminopeptidases. Hydrolysis of such a peptide produces a free N terminus[3] and a potential substrate for aminopeptidases which can liberate free amino acids from the unblocked peptide product but cannot attack the blocked substrate.

Purification

Purification of Dcp[2] and OpdA[18] from *E. coli* strains carrying single copies of the genes encoding the proteins has been described. With the isolation of overproducing strains, purification of both Dcp and OpdA becomes relatively simple. The methods used in our laboratory to obtain pure material for N-terminal sequencing are presented below, but it seems likely that given the levels of these activities in the crude extracts many other equally effective purification schemes could be devised.

*Purification of Dipeptidyl Carboxypeptidase.* Cells (34 g wet weight, 4°C) from strain TN2676[13] grown to stationary phase in LB plus chloramphenicol (20 μg/ml) medium are collected by centrifugation, resuspended in 30 ml of 50 mM N-methyldiethanolamine buffer, pH 8.1 (DEA), and disrupted by sonication. After removal of debris by centrifugation (1 hr at 26,000 g, 4°C), the supernatant (24 ml) is loaded on a Q-Sepharose (Pharmacia, Piscataway, NJ) column (100 ml volume), which is washed with 300 ml of DEA and 350 ml of DEA containing 0.2 M NaCl, and eluted with a linear gradient (2 liters total volume, 0.22 to 0.42 M NaCl in DEA). Active fractions are identified by a microtiter well assay using Met-Gly-Met-Met as the substrate,[13] pooled (140 ml), and concentrated to 2.3 ml with a Centriprep concentrator (Amicon, Danvers, MA). The concentrated material is applied to a gel-filtration column (Ultrogel AcA 34, 500 ml total volume) which is eluted with DEA. The activity elutes at a position corresponding to a molecular mass of approximately 80 kDa, and the active fractions (25 ml) are pooled and concentrated to 2.7 ml as described above. The material (1 ml) is further purified on a Mono Q column (Pharmacia) and Superose 12 (Pharmacia). Approximately 360 μg of protein is obtained. The material shows a single band after sodium dodecyl sulfate–polyacrylamide gel electrophoresis (SDS–PAGE) and staining with Coomassie blue.

*Purification of Oligopeptidase A.* Cells from a 10-liter overnight LB plus chloramphenicol (20 μg/ml) culture of TN3080 [*leuBCD485 pepN90 pepA16 pepB11 pepP1 pepQ1 supQ302*(*proAB pepD*)/*pCM128*] are har-

---

[18] P. Novak, P. H. Ray, and I. K. Dev, *J. Biol. Chem.* **261,** 420 (1986).

vested by centrifugation, resuspended in Tris buffer (10 m$M$ Tris-HCl, pH 7.9, 5 m$M$ MgCl$_2$, 0.1 $M$ NaCl), lysed by sonication, and the extract cleared by centrifugation (200,000 $g$, 2 hr at 4°C). The cleared extract is applied to a DEAE-cellulose column (Whatman DE-52, 50 × 2.5 cm) equilibrated in Tris buffer, and the column is eluted with a linear NaCl gradient (0.1–0.35 $M$). Active fractions, detected using the microassay method with Z-AAL-$p$NA as the substrate, are pooled and concentrated by ultrafiltration using an Amicon YM30 membrane. The concentrated material is applied to an Ultrogel AcA 34 gel-filtration column which is eluted with Tris buffer, and the active fractions are pooled and concentrated with an Amicon Centriprep-10. The material is chromatographed on a chloramphenicol caproate-agarose affinity column (Sigma) to remove chloramphenicol ace-tyltransferase encoded by the vector (pBR328). (This step would not be necessary if a construct carrying the *opdA* gene in a vector lacking the *cat* gene were used.) The material obtained from the column is again concentrated and applied to a Mono Q (Pharmacia) column which is eluted with a linear NaCl gradient. Peak fractions are concentrated and subjected to a second Mono Q column chromatography. The most active fractions from the column are pooled, concentrated, and chromatographed on a Superose 6 column (Pharmacia) to yield material with a specific activity of 22 U/mg protein with an estimated purity of approximately 95%.

### Specificity

Little new information on the specificity of Dcp has been obtained since 1976 when the then available knowledge was summarized.[2] The enzyme requires a free C terminus, and it cannot hydrolyze bonds in which the peptide nitrogen is donated by proline or in which both amino acids are Gly.

Oligopeptidase A was originally detected as an activity able to hydrolyze AcAla$_4$ in extracts of a $dcp^-$ *S. typhimurium* strain. The original studies of the specificity of the enzyme indicated that OpdA could hydrolyze N-blocked peptides with at least four amino acids (e.g., AcAla$_4$), but at least five amino acids were required in order for an unblocked peptide to be hydrolyzed (e.g., Ala$_5$).[3] The ability of a substrate to be hydrolyzed is influenced by amino acids other than those contributing to the scissile bond, as Z-Gly$_4$ or Z-Gly$_5$ were not substrates whereas Z-Gly-Pro-Gly-Gly-Pro-Ala was hydrolyzed at the Gly-Gly bond. The presence of Ala or Gly on either side of the scissile bond was permissive for hydrolysis, but the number of peptides tested was not sufficient to determine the range of allowable amino acids at the scissile bond.

Novak and co-workers[18,19] found that *in vitro* OpdA was the major soluble activity able to hydrolyze the prolipoprotein signal peptide. Analysis of the products of OpdA-catalyzed hydrolysis of the 20-amino acid signal peptide revealed six cleavage sites, each one involving either Gly or Ala on one side or the other of the scissile bond. The smallest product produced was a tripeptide. The enzyme did not attack the prolipoprotein itself, suggesting that it recognizes an altered conformation of the peptide achieved only after release from the precursor, not the primary sequence itself. An alternative hypothesis, that the enzyme is able to attack only peptides below a certain minimum size, appears to be excluded by studies of the OpdA-catalyzed hydrolysis of the bacteriophage P22 gp7 protein (see below).

Mutants of *S. typhimurium* lacking OpdA are unable to support normal growth of bacteriophage P22.[20] In the wild-type cell, the phage protein gp7 is produced by the removal of 20 amino acids from the N terminus of its precursor. Such processing does not take place in an *opdA* mutant. No other phage or bacterial protein is required for the processing reaction. Purified OpdA can efficiently process the precursor protein synthesized in an *in vitro* transcription/translation system derived from rabbit reticulocytes (A. Toguchi and C. Conlin, unpublished, 1993). In that case, the OpdA substrate is a protein of 229 amino acids, not a small peptide. The bond apparently cleaved by OpdA (Glu-Lys) does not fit the rules deduced from either the small peptide studies or the prolipoprotein signal peptide cleavage studies. Although none of the substrates previously tested contained such a bond, OpdA may recognize something other than the nature of the amino acids forming the scissile bond.

Dipeptidyl carboxypeptidase is inhibited by the antihypertensive drug captopril [(2S)-1-(3-mercapto-2-methylpropionyl)-l-proline],[4,21] but OpdA is not.[4] The Dcp enzyme is also inhibited by dipeptides, especially those containing positively charged amino acids.[22]

Metal Ion Activation

Dipeptidyl carboxypeptidase is not inhibited by dialysis against 0.1 m$M$ EDTA although it is activated by $Co^{2+}$.[2] Given the presence of a $Zn^{2+}$ binding site in the amino acid sequence and the fact that other members of the same structural family require much higher concentrations of EDTA for rapid inactivation, it seems reasonable to guess that Dcp contains a metal ion, probably $Zn^{2+}$. No metal ion analysis has been reported, however.

[19] P. Novak and I. K. Dev, *J. Bacteriol.* **170**, 5067 (1988).
[20] C. A. Conlin, E. R. Vimr, and C. G. Miller, *J. Bacteriol.* **174**, 5869 (1992).
[21] C. E. Deutch and R. L. Soffer, *Proc. Natl. Acad. Sci. U.S.A.* **75**, 5998 (1978).
[22] C. A. Conlin, N. J. Trun, T. J. Silhavy, and C. G. Miller, *J. Bacteriol.* **174**, 5881 (1992).

Like Dcp, OpdA retains activity through purification in buffers to which no metal ion has been added. The purified enzyme is activated by 10 $\mu M$ $Co^{2+}$ and to a lesser extent by 0.1 m$M$ $Mn^{2+}$.[3] In addition, OpdA is completely inhibited by 0.1 m$M$ $Zn^{2+}$, 1 m$M$ $Ni^{2+}$, 0.1 m$M$ $Cu^{2+}$, and 1 m$M$ EDTA.[3,5]

## Cloning and Nucleotide Sequence Analysis

The *dcp* gene has been cloned from *S. typhimurium*[13] by selection for complementation of the *N*-acetyl-Ala$_3$ utilization defect conferred by *dcp* mutations,[21] and the nucleotide sequence has been determined (GenBank Accession No. M84575). The *E. coli* gene has also been cloned and sequenced (GenBank Accession No. X57947). The *opdA* genes of both *S. typhimurium* (GenBank Accession No. M84574) and *E. coli* (GenBank Accession No. M93984) have been cloned and sequenced.[6,22] Analysis of the sequences has produced several interesting results.

*Dipeptidyl Carboxypeptidase and Oligopeptidase A as Related Enzymes.* Comparison of the *E. coli* and *Salmonella* amino acid sequences predicted for the two corresponding proteins shows them to be very similar (94% identical for OpdA, 78% identical for Dcp). The similarities are typical for the two closely related organisms. Surprisingly, the amino acid sequences of Dcp and OpdA are also quite similar to one another. Both proteins are 680 amino acids in length and share (comparing the two *Salmonella* proteins) 33% identical amino acids along their entire lengths. One region, centered at about amino acid 470, is particularly strongly conserved and appears to be a $Zn^{2+}$ binding site.[23] The primary sequence similarity between Dcp and OpdA is greater than that between any of the other *E. coli* or *S. typhimurium* peptidases for which sequence information is available.

*Dipeptidyl Carboxypeptidase and Oligopeptidase A as Members of a Group of Related Proteases.* The amino acid sequences of Dcp and OpdA show striking similarities to a group of proteases found in a variety of organisms. As shown in Fig. 1,[24–26] the similarities define a family of enzymes with examples from eubacteria, fungi, and animals. Although the sequence similarity is particularly striking in the vicinity of the putative $Zn^{2+}$ binding site (Fig. 2),[27] there is clear similarity over large regions of the sequence.

[23] C. V. Jongeneel, J. Bouvier, and A. Bairoch, *FEBS Lett.* **242**, 211 (1989).
[24] N. Sugiura, H. Hagiwara, and S. Hirose, *J. Biol. Chem.* **267**, 18067 (1992).
[25] A. Pierotti, K.-W. Dong, M. J. Glucksman, M. Orlowski, and J. L. Roberts, *Biochemistry* **29**, 10323 (1990).
[26] F. Kalousek, G. Isaya, and L. E. Rosenberg, *EMBO J.* **11**, 2803 (1992).
[27] B. Henrich, S. Becker, U. Schroeder, and R. Plapp, *J. Bacteriol.* **175**, 7290 (1993).

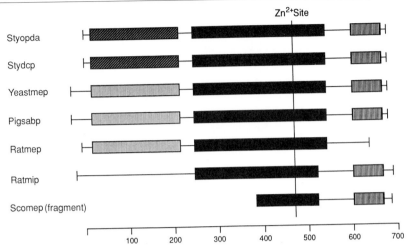

FIG. 1. Schematic comparison of amino acid sequence similarities among Dcp, OpdA, and structurally related enzymes. A block of sequence similarity that includes the putative $Zn^{2+}$ binding site is shared by all of the enzymes. All enzymes except rat metalloendopeptidase 24.15 also share a block of sequence similarity near the C termini. Blocks of sequence similarity in the N-terminal region divide the group into three families: (1) OpdA and Dcp; (2) yeast metalloendopeptidase, pig angiotensin binding protein, and rat metallopeptidase EP 24.15; and (3) rat mitochondrial intermediate peptidase. The *Schizophyllum commune* metalloendopeptidase sequence (Scomep) is incomplete. Styopda, *Salmonella typhimurium* oligopeptidase A[6]; Stydcp, *Salmonella typhimurium* dipeptidyl carboxypeptidase[13]; Yeastmep, *Saccharomyces cerevisiae* metallopeptidase (deduced from chromosome III complete sequence, EMBL Accession No. X59720); Pigsabp, pig angiotensin-binding protein[24]; Ratmep, rat metalloendopeptidase[25]; Ratmip, rat mitochondrial intermediate peptidase.[26]

Comparison of the sequences shows a central sequence domain including the putative $Zn^{2+}$ binding site which is similar in all members of the family. Three different types of N-terminal domains are apparent. A shorter region of C-terminal similarity is present in all but one of the sequences. It is interesting that three of the enzymes [OpdA, rat metalloendopeptidase (Ratmep), and rat mitochondrial intermediate peptidase (Ratmip)] are known or thought to play a role in protein processing. Although angiotensin-converting enzyme (dipeptidyl carboxypeptidase I)[28] is a dipeptidyl carboxypeptidase which, like Dcp, is inhibited by the drug captopril,[21] there appears to be no significant amino acid sequence similarity to Dcp,[13,29] and angiotensin-converting enzyme is not a member of this family.

[28] J. W. Ryan, this series, Vol. 163, p. 194.
[29] F. Soubrier, G. F. Alhenc, C. Hubert, J. Allegrini, M. John, G. Tregear, and P. Corvol, *Proc. Natl. Acad. Sci. U.S.A.* **85,** 9386 (1988).

```
          430
Ratpmip   RLKEDGSYQL PVVVLMLNLP HASRDFPTLL TPGMMENLFH EMGHAMHSML
Scomep    IPGLPGTYQQ PLVVLLCEFA RPSLGAAVLE WHEVM.TLFH EMGHAMHSMI
Pigsabp   CLLPDGSRMM SVAALVVNFS QPRAGRPSLL RHDEVRTYFH EFGHVMHQIC
Ratmep    CLRQDGSRQL AIAAMVANFT KPTPDVPSLL QHDEVETYFH EFGHVMHQLC
Yeastmep  FMIDDTTRSY PVTALVCNFS KSTKDKPSLL KHNEIVTFFH ELGHGIHDLV
Styopda   MRKADGTLQK PVAYLTCNFN RPVNGKPALF THDEVITLFH EFGHGLHHML
Ecoprlc   MRKADGSLQK PVAYLTCNFN RPVNGKPALF THDEVITLFH EFGHGLHHML
Stydcp    STLNET...R PVIYNVCNYQ KPVDGQPALL LWDDVITLFH EFGHTLHGLF
Ecodcp    STLNKT...H PVIYNVCNYQ KPAAGEPALL LWDDVITLFH EFGHTLHGLF
                     L                     FH E GH  H
```

```
          480
Ratpmip   GRTRYQHVTG .TRCPTDFAE VPSILMEYFS NDYRVVSQFA KHYQTGQPLP
Scomep    GRTGYQNVSG .TRCPTDFVE LPSILMEHFL NSRQVLSLFH ADSTSSSSQP
Pigsabp   AQTDFARFSG .TNVEIDFVE VPSQMLENVV WDIDSLRRLS KHYKDGSPIT
Ratmep    SQAEFAMFSG .THVERDFVE APSQMLENVV WEKEPLMRMS QHYRTGGEAP
Yeastmep  GQNKESRFNG PGSVPWDFVE APSQMLEFWT WNKNELINLS SHYKTGEKIP
Styopda   TRIETAGVSG ISGVPWDAVE LPSQFMENWC WEPEALAFIS GHYETGEPLP
Ecoprlc   TRIETAGVSG ISGVPWDAVE LPSQFMENWC WEPEALGFIS GHYETGEPLP
Stydcp    AVQRYATLSG .TNTPRDFVE FPSQINEHWA SHPRVFERYA RHVDSGEKMP
Ecodcp    ARQRYATLSG .TNTPRDFVE FPSQINEHWA THPQVFARYA RHYQSGAAMP
                G       D E  PS    E
```

FIG. 2. Amino acid sequence comparison of the region surrounding the putative $Zn^{2+}$ binding site. The amino acid sequence from residues 430 to 529 of OpdA is compared to the corresponding regions of structurally related enzymes. The positions identical for all sequences are indicated below the alignment. Many other sites show a strong preference for a particular amino acid or a requirement for similar amino acids. Abbreviations are as in Fig. 1 and as follows: Ecoprlc, *Escherichia coli* OpdA (PrlC)[22]; Ecodcp, *Escherichia coli* Dcp.[27]

*Oligopeptidase A as Heat-Shock Protein.* Inspection of the region 5′ to the *opdA* coding region revealed a near consensus $\sigma^{32}$ promoter.[6] Primer extension analysis has shown that this promoter is actually used, and results with mutants defective in $\sigma^{32}$ (*htpR*) have shown that $\sigma^{32}$ is required for *opdA* transcription.[30] The *opdA* gene is therefore a member of the heat-shock regulon. The *dcp* gene shows no heat-shock promoter, and there is no evidence that it is regulated.

## Physiological Roles of Dipeptidyl Carboxypeptidase and Oligopeptidase A

Clearly Dcp can play a catabolic role: it is required for the utilization of AcAla$_3$ as a sole nitrogen source. It seems unlikely that catabolism

[30] C. A. Conlin, manuscript in preparation.

is its major function, however, because N-blocked peptides are poorly transported into the cell and most unblocked substrates of Dcp (four amino acids or larger) would also be expected to be transported poorly. It has been proposed that a fraction (<10%) of the Dcp in the cell is located in the periplasmic space and that such localization may play a role in the utilization of N-blocked peptides.[21] The Dcp protein does not contain a signal peptide, however, so it is not a typical periplasmic protein. Studies of the uptake of radiolabeled AcAla$_3$ by S. typhimurium dcp mutants show that the compound can be transported into the cell intact and that transport depends on a functional Opp (oligopeptide permease transport) system.[31] These results make it unnecessary to postulate periplasmic hydrolysis of the substrate to explain its utilization.

It is likely that Dcp has a more important function of participating in the degradation of intracellular proteins. It is clear that small peptides are intermediates in the degradation of a variety of types of intracellular proteins.[32,33] Several aminopeptidases have been implicated in the degradation of these peptides.[33] Data on protein degradation in strains lacking Dcp in addition to the aminopeptidases suggest that Dcp, the only known C-terminal exopeptidase in S. typhimurium and E. coli, also participates in this process.[4]

The results of in vitro studies suggest that OpdA may play a special role in the degradation of signal peptides. Most of the proteins that are exported from the cytoplasm undergo an N-terminal processing as part of the export pathway. One product of the processing, typically a 20- to 30-amino acid peptide (the "signal peptide") is so rapidly degraded that it can almost never be observed in vivo. The major cytoplasmic activity able to degrade one such signal peptide (lipoprotein) is OpdA.[18] A further indication that OpdA interacts in some special way with signal peptides comes from the discovery that prlC mutations are alleles of opdA.[22] The prlC mutations suppress the localization defect conferred by certain mutations in the signal sequence of the LamB protein.[34,35] Although the basis of the suppressor phenotype is not understood, the isolation of such mutations in opdA suggests the possibility that the protein can be mutationally altered to allow it to recognize the signal peptide before it is released from the precursor. As noted above, OpdA is required for the processing of the precursor of phage protein gp7 in the infection of S. typhimurium by phage P22. This observation seems especially interesting in view of the fact that

[31] E. R. Vimr, Ph.D. Thesis, Case Western Reserve University, Cleveland, Ohio (1981).
[32] C. G. Miller and L. Green, J. Bacteriol. 147, 925 (1981).
[33] C. Yen, L. Green, and C. G. Miller, J. Mol. Biol. 143, 21 (1980).
[34] N. J. Trun and T. J. Silhavy, Genetics 116, 513 (1987).
[35] N. J. Trun and T. J. Silhavy, J. Mol. Biol. 205, 665 (1989).

some of the other enzymes structurally related to OpdA are thought to play roles in the generation of peptide hormones from their precursors. It will be interesting to see if OpdA plays a role in the processing of any cellular proteins.

# [35] Oligopeptidases from *Lactococcus lactis*

By VÉRONIQUE MONNET

## Introduction

Lactic acid bacteria are auxotrophic bacteria which must hydrolyze milk proteins to grow to high cell densities and consequently to acidify milk quickly during the cheese-making process. The proteolytic systems of lactic acid bacteria are complex and have been extensively studied for several years.[1–3] A cell envelope-associated proteinase essential for the optimal development of bacteria in milk is the first enzyme to act in the degradation of caseins. The proteinase hydrolyzes proteins into peptides which can, provided they have suitable chain length, be transported through the plasma membrane using an oligopeptide transport system. The peptides are further degraded intracellularly by different peptidases. For *Lactococcus lactis,* a dozen different peptidases have been described and characterized. All seem to have an intracellular location and potentially play a role in the nutrition of the cells. Furthermore, when used in cheese manufacture, lactic acid bacteria tend to lyse, and the lactococcal peptidases liberated in the curd are probably involved in the cheese-ripening process. Most results have been reported for the cell envelope-located proteinase and aminopeptidases, but there has also been interest in at least two oligopeptidases present in *Lactococcus lactis.* These enzymes do not hydrolyze proteins (or hydrolyze them at a very slow rate), di-, tri-, or tetrapeptides. This chapter focuses on the properties of these enzymes.

The oligopeptidases purified to date are listed in Table I.[4–10] They

[1] V. Monnet, M. P. Chapot-Chartier, and J. C. Gripon, *Lait* **73,** 97 (1993).
[2] P. S. T. Tan, B. Poolman, and W. N. Konings, *J. Dairy Res.* **60,** 269 (1993).
[3] J. Kok, *Food Biotechnol.* **19,** 670 (1991).
[4] M. J. Desmazeaud and C. Zevaco, *Ann. Biol. Anim. Biochim. Biophys.* **16,** 851 (1976).
[5] T. R. Yan, N. Azuma, S. Kaminogawa, and K. Yamauchi, *Appl. Environ. Microbiol.* **53,** 2296 (1987).
[6] T. R. Yan, N. Azuma, S. Kaminogawa, and K. Yamauchi, *Eur. J. Biochem.* **163,** 259 (1987).
[7] R. Baankreis, Ph.D. Thesis (1992).
[8] P. S. T. Tan, K. M. Pos, and W. N. Konings, *Appl. Environ. Microbiol.* **57,** 3593 (1991).

TABLE I

OLIGOPEPTIDASES PURIFIED FROM *Lactococcus lactis*

| Organism | Name of enzyme | Optimum temperature | pH optimum | $M_r$ | Inhibitors[a] | Reactivation after EDTA treatment | Substrate routinely used | Gene cloned and sequenced | Homology | Refs. |
|---|---|---|---|---|---|---|---|---|---|---|
| *L. lactis* biovar. *diacetylactis* | Neutral endopeptidase | 45° | 7 | 49,500 | EDTA, o-phenanthroline | $Mn^{2+}$, $Zn^{2+}$ | Insulin B chain | No | | Desmazeaud and Zevaco[4] |
| *L. lactis* subsp. *cremoris* H61 | LEP II | 40° | 6 | 80,000 (dimer) | EDTA | $Zn^{2+}$, $Mn^{2+}$ | Fragment 1–23 $\alpha_{s1}$-casein | No | | Yan et al.[6] |
| *L. lactis* subsp. *cremoris* C13 | NEP (neutral endopeptidase) | 35° | 6 | 67,000 | EDTA, phosphoramidon, thiorphan | $Zn^{2+}$, $Co^{2+}$ | Fragment 1–23 $\alpha_{s1}$-casein | No | | Baankreis[7] |
| *L. lactis* subsp. *cremoris* Wg2 | PepO | 30°–38° | 6–6.5 | 70,000 | EDTA, PMSF, mercaptoethanol, $Zn^{2+}$ | $Co^{2+}$ | Met-enkephalin | Yes | Family M13 | Tan et al.,[8] Mierau et al.[9] |
| *L. lactis* subsp. *cremoris* H61 | LEP I | 40° | 7–7.5 | 98,000 | EDTA, o-phenanthroline | $Mn^{2+}$, $Co^{2+}$ | Fragment 1–23 $\alpha_{s1}$-casein | No | | Yan et al.[5] |
| *L. lactis* subsp. NCDO 763 | PepF | 40° | 8 | 70,000 | EDTA, o-phenanthroline, $Cu^{2+}$ | $Mn^{2+}$, $Co^{2+}$ | Bradykinin | Yes | Family M3 | Monnet et al.[10] |

[a] PMSF, Phenylmethylsulfonyl fluoride.

share common features such as metal content and the ability to cleave only peptides. However, some differences which will be detailed later (i.e., pH optimum or substrate specificity) allow a separation into two groups of similar enzymes. Further work on the peptidases may demonstrate the existence of only two different oligopeptidases in *L. lactis*. Focusing on recently published investigations, we now describe the methodology used to purify the enzymes and characterize their properties.

Enzyme Assays

The main difficulty encountered in enzyme assays is the choice of a substrate which allows for the detection of an oligopeptidase and the measurement of its activity in a cell-free extract. The substrate must be specific for only one oligopeptidase, and its hydrolysis products must be measurable (i.e., they must not be degraded further). Nonmodified peptides can routinely be used as substrates. However, they usually are degraded by various proteolytic enzymes which sometimes prevent the specific detection of one oligopeptidase in cell-free extracts.

The 1–23 fragment of $\alpha_{s1}$-casein is used as a substrate.[5,6] The peptide is related to the cheese-ripening process as it is the main salt-soluble peptide resulting from the action of chymosin (a constituent of rennet) on $\alpha_{s1}$-casein. Therefore, it can be considered as a "natural" substrate for the lactococcal oligopeptidases but must be prepared and purified from milk. Its degradation is followed by reversed-phase high-performance liquid chromatography (RP-HPLC) using standard conditions. The reaction mixture contains 50 $\mu$l of substrate solution (0.1%, w/v, in 10 m$M$ sodium phosphate buffer, pH 8), 50 $\mu$l of the enzyme solution, and 400 $\mu$l of 50 m$M$ sodium phosphate buffer, pH 6. After incubation at 37° for 60 min, 0.1 ml of acetic acid is added to the mixture to stop the reaction. The hydrolyzate is analyzed by RP-HPLC with a Zorbax ODS column (4.6 × 150 mm). The mobile phase is 0.1% trifluoroacetic acid (TFA), and acetonitrile is increased linearly from 0 to 40% in 40 min at a flow rate of 1 ml/min. The effluent is monitored at 230 nm.

Commercially available peptides can also be used.[8] Met-enkephalin (Tyr-Gly-Gly-Phe-Met) has been used for the detection of PepO, the oligopeptidase purified from *Lactococcus lactis* subsp. *cremoris* Wg2 (Table I). The reaction mixture containing 1.25 m$M$ Met-enkephalin in 20 m$M$

[9] I. Mierau, P. S. T. Tan, A. Haandrikman, J. Kok, K. J. Leenhouts, W. N. Konings, and G. Venema, *J. Bacteriol.* **175,** 2087 (1993).
[10] V. Monnet, M. Nardi, A. Chopin, M. C. Chopin, and J. C. Gripon, *J. Biol. Chem.* **269,** 32070 (1994).

Tris-HCl, pH 7, is incubated with various amounts of enzyme at 30° for 15 min. Samples of 50 $\mu$l each are taken every minute, and the reaction is stopped by addition of 10 $\mu$l of 30% acetic acid and by cooling the mixture to 4°. Ten microliters of the mixture is spotted onto a precoated 0.25-cm-thick silica gel 60 plate, and thin-layer chromatography is performed. Peptides and amino acids are visible in UV light after staining of the gels with 0.05% fluorescamine in 99% acetone. The time point at which complete hydrolysis is observed is determined. On the assumption that hydrolysis proceeds linearly with time, the activity is calculated and expressed in micromoles of Met-enkephalin hydrolyzed per minute.

Bradykinin (Arg-Pro-Pro-Gly-Phe-Ser-Pro-Phe-Arg) is routinely used for PepF detection.[10] It was chosen because it is similar in size, proline content, and hydrophobicity to the peptides liberated from caseins by the cell envelope proteinase. The activity is measured by incubating the enzyme solution with bradykinin (0.23 m$M$ final concentration) in 50 m$M$ Tris-HCl buffer, pH 8, at 40°. The digestion is stopped by the addition of TFA to 1% final concentration. The reaction is followed by RP-HPLC using a $C_{18}$ Nucleosil column (250 $\times$ 4.6 mm) in a TFA/acetonitrile solvent system (solvent A: 0.115% TFA; solvent B: 0.1% TFA, 60% $CH_3CN$) with a linear gradient from 20 to 80% solvent B in 15 min. The absorbance at 214 nm allows the measurement of peak areas, which are converted to amounts of peptide. One unit of endopeptidase activity is defined as the amount of enzyme releasing 1 nmol of the peptide Arg-Pro-Pro-Gly-Phe per hour.

### Preparation of Cell-Free Extracts

To avoid too long a description of purification procedures, only the purification procedures of the two peptidases on which genetic studies have been made are described.

For the purification of PepO, cells are grown in 5 liters MRS medium at a controlled pH of 6.3, and growth of the culture is stopped when the $OD_{660\,nm}$ is 1.9.[8] Cells are harvested by centrifugation at 7000 $g$ for 15 min at 4° and washed with 400 ml of 50 m$M$ potassium phosphate, pH 6. The cells are suspended in 50 ml of 20 m$M$ potassium phosphate buffer and disrupted using a French press at 4°. The disrupted cells are centrifuged for 20 min at 17,500 $g$.

For the preparation of PepF, cells of *Lactococcus lactis* subsp. *lactis* NCDO 763 are grown in 5 liters of M17 medium.[10] The culture is stopped at the end of the logarithmic phase when the $OD_{480\,nm}$ is about 6. Cells are harvested by centrifugation and washed twice with 50 m$M$ sodium $\beta$-glycerophosphate, pH 7, and resuspended in 250 ml of 50 m$M$ Bis–Tris buffer, pH 7, and incubated for 90 min at 30° with 0.5 mg/ml lysozyme,

1 U/ml mutanolysin, and 1 m$M$ MgCl$_2$. The lysate is centrifuged for 30 min at 20,000 $g$, and cellular debris is discarded. RNase (6.87 mg) and DNase (0.125 mg) are added to the supernatant, which is then incubated for 2 hr at 30° in the presence of 1 m$M$ MgCl$_2$. Nucleic acids are then precipitated with 62 m$M$ MnSO$_4$ and removed by centrifugation for 30 min at 10,000 $g$ and 4°.

Sonication is also used to break cells for the purification of LEP I and LEP II.[5,6] The LEP I and LEP II oligopeptidases are purified from *Lactococcus lactis* subsp. *cremoris* H61.

## Purification of Oligopeptidases

Ion-exchange, hydrophobic interaction, hydroxyapatite, and gel-filtration chromatographies have been the main methods used to purify the oligopeptidases. The first step is generally an anion-exchange chromatography which resolves the activity toward fragment 1–23 of $\alpha_{s1}$-casein into two distinct peaks as observed for the purification of LEP I and LEP II corresponding to the two oligopeptidases (Fig. 1). Similar observations have been made during the purification of PepF with bradykinin as a substrate.

### *Purification of PepO*

A five-step procedure is used to purify PepO oligopeptidase.[8] As shown in Table II, the yield of activity is 17.4%, and the enzyme is purified 678-fold. The first step is an anion-exchange chromatography on a DEAE-Sephacel column (1.6 × 20 cm) equilibrated with 20 m$M$ potassium phosphate buffer (pH 6) containing 0.1 $M$ NaCl. Four hundred milliliters of cell extract is diluted with potassium phosphate buffer to the same ionic strength and is applied to the column. Proteins are eluted at 54 ml/hr with a linear gradient from 0.1 to 0.4 $M$ NaCl in the same buffer. Fractions containing Met-enkephalin-hydrolyzing activities are pooled and applied, after addition of sufficient NaCl, to a phenyl-Sepharose CL-4B column (1.8 × 8.5 cm) equilibrated with 20 m$M$ Tris-HCl (pH 6) containing 4 $M$ NaCl. The enzyme is eluted at 30 ml/hr with a linear gradient of 4 to 0 $M$ NaCl in Tris-HCl buffer. The active fraction is found between 0.9 and 0.6 $M$ NaCl and then applied to a hydroxyapatite column (1.5 × 6.3 cm) equilibrated with 10 m$M$ sodium phosphate, pH 6. The endopeptidase is eluted at 40 ml/hr with a linear gradient of sodium phosphate from 0.01 to 0.4 $M$ at pH 6. The activity is eluted between 0.21 and 0.25 $M$ sodium phosphate and applied to a second anion-exchange column (Mono Q HR 5/5) equilibrated with 20 m$M$ Tris-HCl (pH 6). The enzyme is eluted at 1 ml/min with

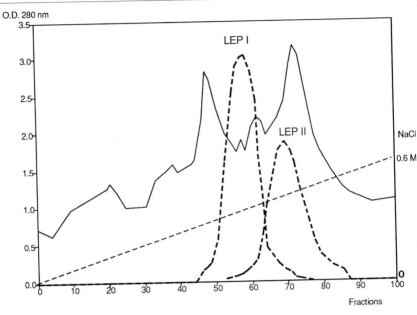

Fig. 1. DEAE-Sephacel chromatography of a cell-free extract from *L. lactis* subsp. *cremoris* H61 after ammonium sulfate precipitation and resuspension in 50 m*M* sodium phosphate buffer pH 6.[5,6] Elution was achieved with a linear gradient from 0 to 0.6 *M* NaCl, and two peaks of activity hydrolyzing fragment 1–23 of $\alpha_{s1}$-casein were observed. The flow rate was 25 ml/hr, and 7-ml fractions were collected. Protein was monitored at 280 nm.

TABLE II

PURIFICATION OF LACTOCOCCAL ENDOPEPTIDASE PepO FROM *Lactococcus lactis* subsp. *cremoris* Wg2[a]

| Purification step | Total protein (mg) | Total activity (units) | Specific activity (units/mg) | Yield (%) | Purification (-fold) |
|---|---|---|---|---|---|
| Cell extract | 6640 | 278.9 | 0.042 | 100 | 1 |
| DEAE-Sephacel | 902.1 | 387.9 | 0.43 | 139 | 10 |
| Phenyl-Sepharose | 102.6 | 112.4 | 1.096 | 40.3 | 26 |
| Hydroxyapatite | 38.5 | 96.3 | 2.5 | 34.5 | 60 |
| Mono Q | 2.4 | 48.5 | 20.208 | 17.4 | 481 |
| Phenyl-Superose | 1.7 | 48.4 | 28.47 | 17.4 | 678 |

[a] Total activity is expressed as micromoles of Met-enkephalin hydrolyzed per minute. Specific activity is expressed as micromoles of Met-enkephalin hydrolyzed per milligram of protein per minute. From Tan *et al.*[8]

a linear gradient from 0.1 to 0.3 $M$ NaCl and found between 0.23 and 0.27 $M$ NaCl. The last step is a second hydrophobic interaction chromatography on a phenyl-Superose column equilibrated with 20 m$M$ Tris-HCl (pH 6) containing 4 $M$ NaCl. After the addition of NaCl, the active fraction is applied to the column and eluted at a flow rate of 0.5 ml/min with a linear gradient from 4 to 0 $M$ NaCl. The enzyme elutes as a single peak at 1.6 $M$ NaCl.

*Purification of PepF*

The PepF peptidase is purified in a four-step procedure.[10] The yield of purification and purification factors are given in Table III. Determination of the endopeptidase activity in the crude extract using bradykinin as the substrate is not possible because several activities degrade that peptide. The first purification step is an anion-exchange chromatography. Two hundred fifty milliliters of crude extract is applied to a Q-Sepharose fast flow column (4 × 15 cm) equilibrated with 25 m$M$ Bis–Tris-HCl buffer, pH 6.5. Proteins are eluted at 5 ml/min with a linear gradient from 0 to 0.2 $M$ NaCl in 200 min. The fractions containing the enzyme are dialyzed against 25 m$M$ Bis–Tris-HCl buffer, pH 7, and then applied to another ion-exchange column (Mono Q HR 10/10, Pharmacia, Piscataway, NJ). The elution is achieved using a linear gradient from 0 to 0.12 $M$ in 10 min and then from 0.12 to 0.22 $M$ NaCl in 60 min in the same Bis–Tris-HCl buffer. The active fractions are eluted around 0.17 $M$, pooled, and dialyzed against 10 m$M$ sodium phosphate buffer, pH 6.8, and applied to a hydroxyapatite column (1 × 5 cm). Elution is achieved using a linear gradient from 0.01 to 0.5 $M$ sodium phosphate in 10 min and then from 0.1 to 0.4 $M$ sodium phosphate

TABLE III
PURIFICATION OF LACTOCOCCAL ENDOPEPTIDASE PepF FROM *Lactococcus lactis* subsp. *lactis* NCDO 763[a]

| Purification step | Total protein (mg) | Total activity (units) | Specific activity (units/mg) | Yield (%) | Purification (-fold) |
|---|---|---|---|---|---|
| Lysate | 464 | nd | nd | nd | nd |
| Anion exchange, pH 6.5 (fast flow Sepharose) | 53 | 157,000 | 2962 | 100 | 1 |
| Anion exchange, pH 7 (Mono Q) | 9.5 | 94,080 | 9903 | 60 | 3.3 |
| Hydroxyapatite | 0.16 | 12,384 | 77,400 | 7.8 | 26 |
| Gel filtration | 0.048 | 10,761 | 224,187 | 6.8 | 75 |

[a] Total activity is expressed as nanomoles of bradykinin hydrolyzed per hour. Specific activity is expressed as nanomoles of bradykinin hydrolyzed per milligram of protein per hour. nd, Not determined. Data from Monnet *et al.*[10]

at a 0.8 ml/min flow rate. The active fractions are dialyzed against water for 1 hr and concentrated 10-fold in a vacuum concentrator. Purification is achieved by injection of 250-$\mu$l fractions on a Superose 12 column, and elution is performed with 15 m$M$ Tris-HCl buffer, pH 8, containing 0.15 $M$ NaCl at a flow rate of 0.25 ml/min.

The purified fraction exhibits a major band (A) and a minor band (B) which cannot be totally eliminated during the purification procedure (Fig. 2). We have shown that the two bands exhibit the same activity toward bradykinin. After a longer migration time allowing a better separation of the bands from one another, the lane of the gel is cut into 2 mm slices. The slices are washed twice with 50 m$M$ Tris-HCl buffer containing 2.5% Triton X-100 to eliminate sodium dodecyl sulfate (SDS). Gel slices are then incubated in 50 m$M$ Tris-HCl, pH 8, in the presence of 0.23 m$M$ bradykinin, and under these conditions the activity can be measured by analysis of the supernatant by RP-HPLC as described for the enzyme assays.

## Physicochemical Properties of Oligopeptidases

*Molecular Mass.* Molecular masses are estimated by gel filtration and by SDS–polyacrylamide gel electrophoresis (SDS–PAGE). Except for the

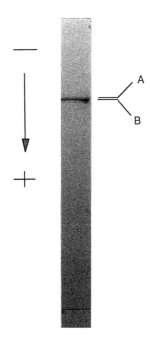

FIG. 2. Analysis by SDS–PAGE of the PepF purified fraction. Two bands are visible: a major one (A) and a minor one (B).

oligopeptidase from *L. lactis* biovar. *diacetylactis,* the molecular masses of all the oligopeptidases are between 67 and 98 kDa. Only one dimeric structure was reported for LEP II.

*pH and Temperature Activity and Stability Optima.* The enzymes described are all optimally active between 35° and 45°. The best pH conditions are more variable, and we can distinguish a first group of peptidases active at pH 6–7 and a second one active at pH 7–8 (Table I). Stability was tested for LEP I and LEP II, and it appears that the enzymes are fairly stable between pH 6 and 9 and below 50°. One-half of the activity remains after 10 min at 55° for LEP I and after 10 min at 65° for LEP II.

*Isoelectric Point.* Isoelectric points are determined by chromatofocusing. Values of 5.1 and 4.3 are reported for LEP I and PepO, respectively.

*Inhibitors.* All the lactococcal oligopeptidases are inhibited by EDTA and *o*-phenanthroline. After EDTA treatment, reactivation is achieved by divalent cations: $Mn^{2+}$, $Co^{2+}$, and $Zn^{2+}$ for the first group of enzymes, and $Mn^{2+}$ and $Co^{2+}$ for the second group (Table I).

*Amino Acid Composition.* Amino acid compositions are given for LEP I and LEP II and are compared to those deduced from the gene sequences of PepO and PepF (Fig. 3). Contradictory results are found for LEP II and PepO, whereas the amino acid compositions of LEP I and PepF correlate rather well (Fig. 3). This additional finding confirms that LEP I and PepF are most probably identical.

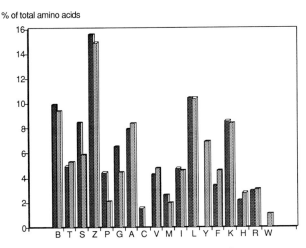

FIG. 3. Comparison of the amino acid composition of LEP I[5] (cross-hatched bars) to that deduced from the gene sequence of PepF[10] (stippled bars). Columns B represent D + N, and Z is E + Q. (Data from Yan *et al.*[5] and Monnet *et al.*[10])

*Substrate Specificity.* The oligopeptidases are generally unable to hydrolyze proteins (caseins were mainly used as the substrate) or di-, tri-, and tetrapeptides. Consequently these peptidases can be described as size-specific endopeptidases or oligopeptidases. For the first group of enzymes [neprilysin (NEP), LEP II, PepO], the smallest peptide hydrolyzed is Met-enkephalin (5 amino acids) and the largest is insulin B chain (30 amino acids) (Table IV). The cleavage site in fragment 1–23 of $\alpha_{s1}$-casein is not recognized in the whole protein, demonstrating the importance of substrate size. Hydrolysis of casein has been reported for the neutral endopeptidase from *L. lactis* biovar. *diacetylactis*.[4] However, the extent of casein hydrolysis is only 1% of that observed with insulin B chain. Another endopeptidase with similar properties has been isolated from *L. lactis* subsp. *lactis* NCDO 763 and partially purified.[11] It is also able to hydrolyze caseins but at a very low rate compared to insulin B chain. The observations suggest that the endopeptidases from the first group are perhaps able to cleave proteins but show that the reaction is very unfavorable.

Except for two of the cleavage sites of fragment 1–23 from $\alpha_{s1}$-casein, the oligopeptidases prefer hydrolyzing peptide bonds involving the amino group of hydrophobic amino acids, and in this respect they are similar to thermolysin and to mammalian neprilysin (EC 3.4.24.11). Bonds involving the imino group of proline can be cleaved by LEP II as shown with bradykinin. A neutral endopeptidase with similar properties and specificity was also purified from a lactic acid bacterium from another genus: *Streptococcus thermophilus*.[12]

The second group of enzymes (LEP I and PepF) has a restricted substrate size specificity; the smallest substrate cleaved is $\beta$-casomorphin (7 residues) and the largest fragment 1–23 from $\alpha_{s1}$-casein. We have demonstrated the importance of the size of the substrate for PepF using fragment 2–7 of bradykinin, which is not hydrolyzed even though it contains the Phe-Ser bond cleaved in bradykinin. Hydrophobic and basic amino acids are involved in the observed cleaved bonds. Moreover, proline is often found in position P2′ of the cleaved bonds. We have observed with PepF that Pro-X bonds can be cleaved, as observed with $\beta$-casomorphin, but at a very low rate.

*Kinetic Data.* Kinetic data have been calculated from the hydrolysis of fragment 1–23 of $\alpha_{s1}$-casein. The $K_m$ values were reported to be 14.2 and 270 p$M$ for LEP I and LEP II, respectively, showing that LEP I has a better affinity for the substrate.

[11] G. Muset, V. Monnet, and J. C. Gripon, *J. Dairy Res.* **56**, 765 (1989).
[12] M. J. Desmazeaud, *Biochimie* **56**, 1173 (1974).

## TABLE IV
### SUBSTRATE SPECIFICITY OF OLIGOPEPTIDASES FROM *Lactococcus lactis*

| Substrate | Hydrolyzed by | Not hydrolyzed by | Cleavage sites observed with LEP II, PepO, or NEP | Cleavage sites observed with LEP I or PepF |
|---|---|---|---|---|
| Met-enkephalin | LEP II, PepO | LEP I, PepF | Y-G-G-F-M | |
| β-Casomorphin | PepF | LEP I, LEP II, PepO | | Y-P-F-P-G-P-I |
| Bradykinin | LEP I, LEP II, PepO, PepF | | R-P-P-G-F-S-P-F-R | R-P-P-G-F-S-P-F-R |
| Substance P | LEP I, LEP II, PepO, PepF | | R-P-K-P-Q-Q-F-F-G-L-M-NH₂ | R-P-K-P-Q-Q-F-F-G-L-M-NH₂ |
| Neurotensin | LEP I, LEP II, PepO, PepF | | Pyr-L-Y-E-N-K-P-R-R-P-Y-I-L | Pyr-L-Y-E-N-K-P-R-R-P-Y-I-L |
| Fragment 1–23 $\alpha_{s1}$-casein | LEP I, LEP II, PepO, PepF | | R-P-K-H-P-I-K-H-Q-G-L-P-Q-E-V-L-N-E-N-L-L-R-F | R-P-K-H-P-I-K-H-Q-G-L-P-Q-E-V-L-N-E-N-L-L-R-F |
| Glucagon | LEP II, PepO, NEP | LEP I, PepF | H-S-Q-G-T-F-T-S-D-Y-S-K-Y-L-D-S-R-R-A-Q-D-F-V-Q-W-L-M-N-T | |
| Insulin B chain | LEP II, PepO, NEP | LEP I, PepF | F-V-N-Q-H-L-C-G-S-H-L-V-E-A-L-Y-L-V-C-G-E-R-G-F-F-Y-T-P-K-A | |

```
PepO       - M-------------------------------------------------------- -1

NEP/HUMAN  - GKSESQMDITDINTPKPKKKQRWTPLEISLSVLVLLLTIIAVTMIALYATYDDGI -55

                                                          A
PepO       - ------------------TRIQDDLFATVNAEWLENAEIPADKHRISAFDELV -36
                               *   * *     **    **    *    ** *
NEP/HUMAN  - CKSSDCIKSAARLIQNMDATTEPCTDFFKYACGGWLKRNVIPETSRYGNFDILR -110

PepO       - LKNEKNLAKDLADLSQNLPTDNPELLEAIKFYNKAGDWQAREKADFSAVKNELAK -91
              * * *.  .        *. *.      * * *       * *
NEP/HUMAN  - DELEVVLKDVLQEPK-------TEDIVAVQKAKALYRSCINESAIDSRGGEPLLK -158

PepO       - ---------VETLNTFEDFKNNLT------QLVFHSQAPLPFSFSVEPDMKDAIH -131
              * * *       .    *      **    * * *..
NEP/HUMAN  - LLPDIYGWPVATENWEQKYGASWTAEKAIAQLNSKYGKKVLINLFVGTDDKNSVN -213

PepO       - YSLGFSGPGLILPDTTYYNDEHPRKKELLDFWAKNTSEILKTFDVENAEEIAKSA -186
              .    * * **   **    *           * *.    .*.
NEP/HUMAN  - HVIHIDQPRLGLPSRDYYECTGIYKEAC-------------TAYVDFMISVARLI -255

PepO       - LKFDALLVPSANTSEEWAKYAELYHPISTDSFVSKVKNLDLKSLIKDLVKTEPDK -241
              .*.  ..* * **   *.       .*      .*    *   .
NEP/HUMAN  - RQEERLPIDENQLALEMNKVMELEKEIANATAKPEDRNDPMLLYNKMTLAQIQNN -310

PepO       - VIVYEDRFYESFD-----------SLINEENWSLIKAWMLTKIARGATSFFNED -284
              .          *.    *. ***   ***.    * . *
NEP/HUMAN  - FSLEINGKPFSWLNFTNEIMSTVNISITNEEDVVVYAPEYLTKLKPILTKYSARD -365

PepO       - LRILGGAYGRFLSNVQ------EARSQEKHQLDLTESYFSQVI----------- -321
              * *           *.*   * * *
NEP/HUMAN  - LQNLMSWRFIMDLVSSLSRTYKESRNAFRKALYGTTSETATWRRCANYVNGNMEN -420

PepO       - --GLFYGKKYFGEAGKADVKRMVTAMIKVYQARLSKNEWLSQETAEKAIEKLDAI -374
              * *   *   * *  ... .*.   *    *. **  .* ** **
NEP/HUMAN  - AVGRLYVEAAFAGESKHVVEDLIAQIREVFIQTLDDLTWMDAETKKRAEEKALAI -475

PepO       - TPFIGFPD-------KLPEIYSRLKTTSGSLYEDALKFDEILTARTFEKFSEDVD -422
              **.**   **  **  * *      .*   .   ..    *  * **
NEP/HUMAN  - KERIGYPDDIVSNDNKLNNEYLELNYKEDEYFENIIQNLKFSQSKQLKKLREKVD -530

                                                              A   ZA
PepO       - KTSWHMPAHMVNAYYSPDSNTIVFPAAILQAPFYSLEQSSSQNYGGIGTVIAHEI -477
              * *    * .*. .**   * ***** *** **.* **  ****** ** ***
NEP/HUMAN  - KDEWISGAAVVNAFYSSGRNQIVFPAGILQPPFFSAQQSNSLNYGGIGMVIGHEI -585

              Z
PepO       - SHAFDNNGAQFDKEGNLNKWWLDEDYEAFEEKQKEMIAL---FDGVETEAGPANG -529
             .* ** ** .* *.*.** **    * *   *.    *    .       .*
NEP/HUMAN  - THGFDDNGRNFNKDGDLVDWWTQQSASNFKEQSQCMVYQYGNFSWDLAGGQHLNG -640

              Z
PepO       - KLIVSENIADQGGITAALTAAKDEKDVDLKA------------FFSQWAKIWRM -571
              . ****.**. *  *      **  ..* .*
NEP/HUMAN  - INTLGENIADNGGLGQAYRAYQNYIKKNGEEKLLPGLDLNHKQLFFLNFAQVWCG -695

                                A                              A
PepO       - KASKEFQQMLLSMDFHAPAKLRANIPPTNLEEFYDTFDVKETDKMYRAPENRLKI -626
              *.    .   * *.* *      *   **..* .     *   **  .  ..
NEP/HUMAN  - TYRPEYAVNSIKTDVHSPGNFRIIGTLQNSAEFSEAFHCRKNSYMN--PEKKQRV -748

PepO       - W -627
              *
NEP/HUMAN  - W -749
```

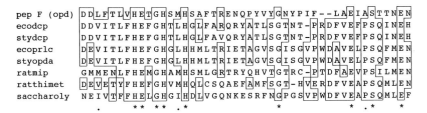

```
pep F (opd)  D D L F T L V H E T G H S M H S A F T R E N Q P Y V Y G N Y P I F - - L A E I A S T T N E N
ecodcp       D D V I T L F H E F G H T L H G L F A R Q R Y A T L S G T N T - P R D F V E F P S Q I N E H
stydcp       D D V I T L F H E F G H T L H G L F A V Q R Y A T L S G T N T - P R D F V E F P S Q I N E H
ecoprlc      D E V I T L F H E F G H G L H H M L T R I E T A G V S G I S G V P W D A V E L P S Q F M E N
styopda      D E V I T L F H E F G H G L H H M L T R I E T A G V S G I S G V P W D A V E L P S Q F M E N
ratmip       G M M E N L F H E M G H A M H S M L G R T R Y Q H V T G T R C - P T D F A E V P S I L M E N
ratthimet    D E V E T Y F H E F G H V M H Q L C S Q A E F A M F S G T - H V E R D F V E A P S Q M L E N
saccharoly   N E I V T F F H E L G H G I H D L V G Q N K E S R F N G P G S V P W D F V E A P S Q M L E F
             .       * *   * *   . *                           *             *   . *       *
```

FIG. 5. Alignment of the amino acid sequences around the catalytic site of the lactococcal endopeptidase PepF, a dipeptidyl carboxypeptidase from *E. coli* (ecodcp), a dipeptidyl carboxypeptidase from *S. typhimurium* (stydcp), an oligoendopeptidase from *E. coli* (ecoprlc), an oligoendopeptidase from *S. typhimurium* (styopda), mitochondrial intermediate peptidase from rat (ratmip), thimet oligopeptidase from rat (ratthimet), and saccharolysin from *S. cerevisiae* (saccharoly).[10] Asterisks (*) indicate identical amino acids; dots (·), similar amino acids. Identical amino acids are boxed. From Monnet *et al*.[10]

## Cloning and Sequencing of Oligopeptidase Genes

The genes of two oligopeptidases have been cloned. A lactococcal genome library in phage λ was screened using antibodies for *pepO*.[9] For *pepF* a lactococcal genome library in *Escherichia coli* was screened using a probe deduced from protein sequences.[10] In both cases, open reading frames were identified that encode proteins of 627 and 602 amino acids for *pepO* and *pepF*, respectively. Neither typical bacterial leader peptides nor possible membrane-spanning domains were present for the two genes. No consensus promoter sequences were found, suggesting that the genes are part of an operon. Motifs characteristic of Zn-dependent metallopeptidases (H-E-X-X-H) were found for both peptidases.

A plasmid location of the *pepF* gene was demonstrated. The gene was part of the large 55-kb plasmid carrying the cell envelope proteinase gene and the lactose operon. Another gene which shows some similarity was detected on the chromosome by hybridization with the plasmid gene under nonstringent conditions. The product of the chromosomal gene is probably the minor protein observed in the purified fraction. To check this, the gene is currently being cloned and sequenced.

---

FIG. 4. Alignment of the amino acid sequences of the lactococcal endopeptidase PepO and human neprilysin (NEP). Asterisks (*) indicate identical amino acids; dots (·), similar amino acids; double underline, membrane-spanning domain of NEP. Amino acids of the active site (A) or involved in $Zn^{2+}$ binding (Z) of NEP are boxed. (From Mierau *et al*.[9])

Homology

No obvious homology between PepO and any bacterial enzyme was found, but a striking similarity was observed with the neutral endopeptidase of mammals (neprilysin; EC 3.4.24.11) (Fig. 4).[9] The overall identity of PepO with human neprilysin is 27.1% but increases to 71% for a stretch of 42 amino acids surrounding the zinc-binding site of the enzymes. Consequently, the oligopeptidase PepO belongs to family M13 according to the classification of Rawlings and Barrett[13] (see also [13] in this volume).

For PepF, homologies were found in the region of the active site with two bacterial oligopeptidases (OpdA from *Salmonella typhimurium* and prlC from *E. coli*), two bacterial dipeptidyl carboxypeptidases (dcp from *E. coli* and from *S. typhimurium*), a peptidase from *Saccharomyces* (saccharolysin), and also two rat peptidases (rat mitochondrial intermediate peptidase and rat thimet oligopeptidase, EC 3.4.24.15) (Fig. 5).[10] Consequently, the oligopeptidase PepF can be assigned to the family M3 according to the classification of Rawlings and Barrett ([13], this volume).

Role of Lactococcal Oligopeptidases

The role of the enzymes has not yet been clearly demonstrated, and several hypotheses can be proposed. It has been shown that the gene *pepO* is located in the same operon as the genes encoding an oligopeptide transport system.[9] The oligopeptidase most probably plays a role in the nutrition of the cell by hydrolyzing peptides which have been transported through the plasma membrane. However, that activity is not essential for the growth of lactococci in milk, as has been demonstrated by the normal development of a mutant deficient in PepO activity in milk.

A second possible role for the endopeptidases is to participate in the turnover of cytoplasmic proteins and to degrade signal peptides. A signal peptide peptidase activity has been attributed to an endopeptidase (OpdA from *E. coli*[14]) which displays homology at the active site with PepF. Finally, we know that lactic acid bacteria lyse in curd and liberate the peptidases into the cheese. Oligopeptidases potentially act together with other peptidases to hydrolyze peptides during the proteolysis of the curd.

---

[13] N. D. Rawlings and A. J. Barrett, *Biochem. J.* **290**, 205 (1993).
[14] P. Novak and I. K. Dev, *J. Bacteriol.* **170**, 5067 (1988).

# [36] Neurolysin: Purification and Assays

*By* F. Checler, H. Barelli, P. Dauch, V. Dive, B. Vincent,
and J. P. Vincent

## Introduction

Neurolysin (EC 3.4.24.16, also known as neurotensin-degrading neutral metalloendopeptidase) inactivates neurotensin by cleaving the peptide at the $Pro^{10}$-$Tyr^{11}$ bond, leading to the biologically inert catabolites neurotensin 1–10 and neurotensin 11–13.[1] Previous studies indicated that substitution of the Tyr-11 residue by an aromatic D-amino acid led to highly potent neurotensin analogs *in vivo*,[2-4] although they behaved as poor agonists of neurotensin receptors in *in vitro* systems virtually devoid of efficient proteolysis.[5] Altogether, this suggested that the D-amino acid substitution at position 11 had conferred resistance to degradation. Such a hypothesis was confirmed by our studies showing that D-Phe[11] and D-Tyr[11] neurotensin indeed totally resisted proteolysis, *in vitro* and *in vivo* after intracerebroventricular administration in the rat.[6] This reinforced the hypothesis of a key role of the $Pro^{10}$-$Tyr^{11}$ bond in the processes of neurotensin inactivation. Moreover, it was striking to observe that cleavage at the $Pro^{10}$-$Tyr^{11}$ bond was ubiquitously detected during the course of studies on neurotensin catabolism by cells and tissues of central and peripheral origin.[7]

The development of a specific polyclonal antiserum directed toward rat brain neurolysin allowed us to establish that the distribution of the endopeptidase in the brain paralleled that of neurotensin receptors.[8] Furthermore, it was possible to show that the receptors of neurotensin appeared on a very discrete population of primary cultured neurons from mouse embryos that coexpressed neurolysin.[9] The data are in good agreement

[1] F. Checler, J. P. Vincent, and P. Kitabgi, *J. Biol. Chem.* **261,** 11274 (1986).
[2] F. B. Jolicoeur, A. Barbeau, F. Rioux, R. Quirion, and S. Saint-Pierre, *Peptides* **2,** 171 (1981).
[3] P. T. Loosen, C. B. Nemeroff, G. Bissette, G. Burnett, A. J. Prange, Jr., and M. A. Lipton, *Neuropharmacology* **17,** 109 (1978).
[4] J. E. Rivier, L. H. Lazarus, M. H. Perrin, and M. R. Brown, *J. Med. Chem.* **20,** 1409 (1977).
[5] P. Kitabgi, F. Checler, J. Mazella, and J. P. Vincent, *in* "Reviews in Basic and Clinical Pharmacology," (R. P. Ebstein and R. H. Belmaker, eds.), p. 397. Freund Publishing House Ltd, London 1985.
[6] F. Checler, J. P. Vincent, and P. Kitabgi, *J. Pharmacol. Exp. Ther.* **227,** 743 (1983).
[7] F. Checler, H. Barelli, P. Kitabgi, and J. P. Vincent, *Biochimie* **70,** 75 (1988).
[8] J. Woulfe, F. Checler, and A. Beaudet, *Eur. J. Neurosci,* **4,** 1309 (1992).
[9] J. Chabry, F. Checler, J. P. Vincent, and J. Mazella, *J. Neurosci.* **10,** 3916 (1990).

with the idea that a peptidase involved in the physiological degradation of a given peptide should be located in the vicinity of the specific receptors mediating the message in order to interrupt it efficiently.

More direct evidence of the involvement of neurolysin in the physiological degradation of neurotensin came from the development of the specific inhibitors Pro-Ile and phosphodiepryl 03. Those agents allowed us to show that, in a model of vascularly perfused dog ileum *in vivo*, both inhibitors prevented neurotensin degradation as well as neurotensin 1–10 formation.[10] The data are not antagonistic with the possibility that neurolysin participates in other proteolytic processes as suggested by its ability to cleave other neuropeptides besides neurotensin, at least *in vitro*.[1] The two specific neurolysin inhibitors should prove useful in assessing the pharmacological significance of these observations.

In this chapter, we focus on procedures for purification of neurolysin not only from isolated rat brain synaptic membranes but also from more crude tissue preparations. Furthermore, we describe a fluorimetric substrate and specific inhibitors that can be used in several neurolysin assays. Finally, we document some characteristics of the enzyme that were established by means of the new tools.

## Materials

Tritiated neurotensin, [3,11-*tyrosyl*-3,5-³H(N)]neurotensin, is from New England Nuclear (Boston, MA). Neurotensin, neuromedin N, and peptidyl-AMC* fluorimetric substrates are purchased from Peninsula Laboratories (Belmont, CA). Bz-Gly-Ala-Ala-Phe-pAB, Cpe-Ala-Ala-Phe-pAB, Cpp-Ala-Ala-Tyr-pAB, Z-Pro-prolinal, and pyroglutamyl ketone ester are kindly given by Drs. M. Orlowski and S. Wilk (Department of Pharmacology, Mount Sinai School of Medicine, New York, NY). Arphamenine B is from Interchim (Asnières, France). Phosphoramidon [*N*-(α-L-rhamnopyranosyloxydihydroxyphosphinyl)-L-leucyl-L-tryptophan] is from Boehringer (Mannheim, Germany). Bestatin, Pro-Ile, and leucine aminopeptidase (microsomal, type IV) are from Sigma (St. Louis, MO). Amastatin, diprotin A, and Mcc-Pro-Leu-Gly-Pro-D-Lys-Dnp are from Novabiochem (La Jolla, CA). Phosphodiepryl 03 [*N*-(phenylethylphosphonyl)glycyl-L-prolyl-L-

---

[10] F. Checler, H. Barelli, P. Dauch, B. Vincent, V. Dive, A. Beaudet, E. E. Daniel, J. E. T. Fox-Threlkeld, Y. Masuo, and J. P. Vincent, *Biochem. Soc. Trans.* **21**, 692 (1993).

* CPE, *N*-[1-(*RS*)-carboxy-2-phenylethyl]-; CPP, *N*-[1-(*RS*)-carboxy-3-phenylpropyl]-; Mcc, 7-methoxycoumarin-3-carboxylyl-; AMC, 7-(4-methyl)coumarylamide; pAB, *p*-aminobenzoate; Dnp, 2,4-dinitrophenyl-; Z, benzyl-; Bz, benzoyl-.

aminohexanoic acid] is synthesized by the method of Dive et al.[11] All resins are from IBF-Pharmindustrie (Villeneuve la Garenne, France).

### Detection of Neurotensin-Degrading Neutral Metallopeptidase

The first evidence that a novel peptidase likely contributed to the catabolism of neurotensin in the central nervous system came from initial studies which examined the proteolytic events accounting for neurotensin breakdown by purified rat brain synaptic membranes[12-14] We have shown that three major N-terminal neurotensin fragments devoid of biological activity can be derived from primary cleavages of the neurotensin molecule, namely, neurotensin 1-8, neurotensin 1-10, and neurotensin 1-11 (Fig. 1A). The formation of neurotensin 1-8 and neurotensin 1-11 is ascribed to thimet oligopeptidase (EC 3.4.24.15) and neprilysin (EC 3.4.24.11), respectively[12, 13] (see model in Fig. 1B). However, besides neprilysin, a major contribution of neurotensin 1-10 formation appears totally resistant to a series of specific inhibitors developed toward various exo- and endopeptidases.[14] This was the first clue to the presence of a novel proteolytic activity capable of inactivating neurotensin. The peptidase has been purified from rat brain synaptic membranes according to the following procedure.

### Purification of Neurolysin from Rat Brain Synaptic Membranes

The choice of purified rat brain synaptic membranes as starting material is of crucial importance to achieve the preparation of pure neurolysin. Indeed, isolation of the membranes includes extensive washings with a hypotonic buffer (5 m$M$ Tris-HCl, pH 7.5) and high-speed centrifugations that allow the removal of abundant neurotensin 1-10-forming enzymes such as carboxypeptidase A and prolyl oligopeptidase (EC 3.4.21.26) that will hinder the detection of neurolysin.

### *Preparation of Rat Brain Synaptic Membranes*

Male Sprague-Dawley rats (200–250 g) are decapitated. Whole brains are removed and synaptic membranes are prepared following the procedure of Jones and Matus.[15] The resulting pellets should be stored at −70°.

[11] V. Dive, A. Yiotakis, A. Nicolaou, and F. Toma, *Eur. J. Biochem.* **191,** 685 (1990).
[12] F. Checler, J. P. Vincent, and P. Kitabgi, *J. Neurochem.* **41,** 375 (1983).
[13] F. Checler, J. P. Vincent, and P. Kitabgi, *J. Neurochem.* **45,** 1509 (1985).
[14] F. Checler, P. C. Emson, J. P. Vincent, and P. Kitabgi, *J. Neurochem.* **43,** 1295 (1984).
[15] D. H. Jones and A. I. Matus, *Biochim. Biophys. Acta* **356,** 276 (1974).

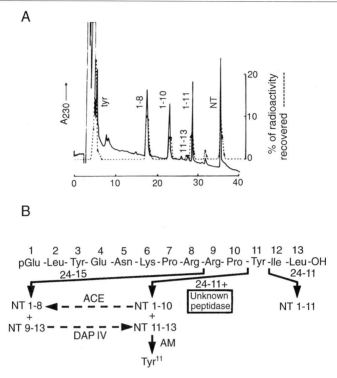

Fig. 1. Analysis by HPLC of neurotensin degradation by rat brain synaptic membranes and proposed model of inactivation. Neurotensin (5 nmol) and [$^3$H]neurotensin (100,000 cpm, 80 Ci/mmol) were incubated with rat brain synaptic membranes for 30 min at 37°. Catabolites were separated by HPLC (A) as described in the text, using buffer/solvent system 1. Degradation products were identified by amino acid analysis and by the retention times compared with synthetic fragments. The diagram of neurotensin (B) represents the primary inactivating cleavages (solid arrows) of neurotensin and the secondary cleavages taking place on degradation products. Peptidases were identified by means of the following specific inhibitors: 24.11 (neprilysin) was inhibited by thiorphan, $10^{-6}$ $M$; 24.15 (thimet oligopeptidase) by CPE-Ala-Ala-Phe-pAB, $10^{-5}$ $M$; ACE (angiotensin-converting enzyme) by captopril, $10^{-6}$ $M$; AM (alanyl aminopeptidase) by bestatin, $10^{-5}$ $M$; and DAP IV (dipeptidyl-peptidase IV) by diprotin A, $10^{-3}$ $M$.

## Purification of Enzyme

*Preparation and Solubilization.* All steps are carried out at 4° unless otherwise indicated. Synaptic membranes (540 mg protein) corresponding to 80 rat brains are resuspended in 70 ml of 5 m$M$ Tris-HCl, pH 7.5, and centrifuged at 170,000 $g$ for 45 min. The supernatant is discarded, and the resulting pellet is treated twice following the same procedure. The final pellet is resuspended in 20 m$M$ Tris-HCl, pH 7.5, containing 1% w/w Triton X-100 and gently stirred for 45 min. The suspension is then centrifuged at 170,000 $g$ for 45 min and the supernatant kept for further purification.

*Ion-Exchange Chromatography on DEAE-Trisacryl.* The supernatant is applied to a DEAE-Trisacryl (IBF, Villeneuve la Garenne, France) column (25 × 175 mm) previously equilibrated with 20 m$M$ Tris-HCl, pH 7.5, buffer. The resin is washed with the same buffer and then eluted with a linear gradient established between 200 ml of 20 m$M$ Tris-HCl, pH 7.5, and 200 ml of the same buffer containing 0.15 $M$ NaCl. The column is run at a flow rate of 2.5 ml/min, and fractions of 4 ml are collected. All fractions are tested for neurotensin-degrading activity and analyzed by high-performance liquid chromatography (HPLC) (see below, Analysis of Neurotensin Degradation by HPLC).

Neurolysin can be quantitatively retained on DEAE-Trisacryl and eluted as a single peak of neurotensin 1–10-generating activity by increasing the salt concentration of the elution buffer (Fig. 2). The pooled fractions exhibiting neurolysin activity (pool D1, Fig. 2) are contaminated at this stage by an activity that generates neurotensin 1–8 and neurotensin 9–13 (later identified as thimet oligopeptidase). The only chromatographic step that eliminates this activity is use of a hydroxyapatite column.

*Hydroxyapatite Chromatography.* Active fractions (pool D1) are dialyzed overnight against two changes of 3 liters each of 1 m$M$ KH$_2$PO$_4$, pH

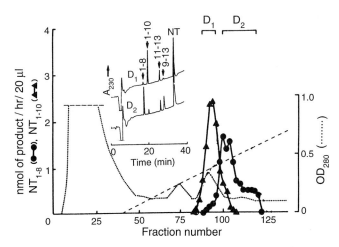

FIG. 2. DEAE chromatography of neurolysin purification. DEAE chromatography was performed as described in the text, and all fractions were assayed for the ability to cleave neurotensin. Neurotensin 1–8 and 1–10 formation was monitored by HPLC according to the procedure described in the text (system 2). Insets correspond to HPLC analysis of neurotensin hydrolysis by D1 and D2 pooled proteins.

7.4, and loaded at a flow rate of 10 ml/hr onto a hydroxyapatite column (10 × 60 mm) previously equilibrated with 1 m$M$ KH$_2$PO$_4$, pH 7.4. The column is then washed with the same buffer, and activity is eluted with a linear gradient established between 75 ml of 1 m$M$ KH$_2$PO$_4$, pH 7.4, and 75 ml of 0.25 $M$ KH$_2$PO$_4$, pH 7.4. Fractions of 1.5 ml are collected and assayed for neurotensin degradation. As illustrated in the Fig. 3, at this step, the enzyme emerges as a single symetrical absorbing peak (pool H1) that is fully resolved from the thimet oligopeptidase which is recovered at lower phosphate concentrations (pool H2) (Fig. 3). Furthermore, the purified activity appears totally devoid of contaminating peptidases. These two

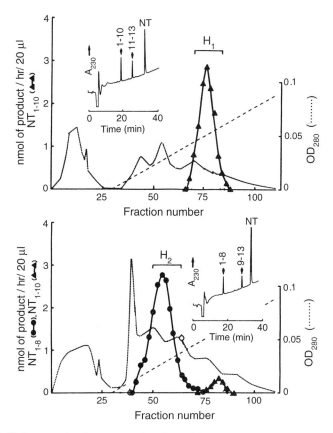

FIG. 3. Hydroxyapatite chromatography of post-DEAE D1 and D2 pooled proteins. Post-DEAE pooled proteins (D1, top profile; D2, bottom profile) were concentrated, dialyzed, and loaded on the hydroxyapatite column. All fractions were tested for neurotensin degradation and analyzed by HPLC as described in the text (system 2). Insets correspond to HPLC analysis of neurotensin degradation by H1 or H2 pooled proteins.

steps give a purification factor of about 550-fold with an overall recovery of 15%.[1]

*Gel-Filtration Chromatography.* An aliquot (5 ml) of the active pool H1 is concentrated to 940 $\mu$l with an ultrafiltration cell (Amicon, Danvers, MA) equipped with a YM10 membrane (Diaflo). Three hundred $\mu$l is injected onto two connected columns (Lichrospher 100 diol, Merck, Darmstadt, Germany) previously equilibrated with 100 m$M$ Tris-HCl, pH 7.5, containing 0.15 $M$ NaCl. Elution is carried out isocratically with the equilibrating buffer at a flow rate of 0.3 ml/min. All fractions (300 $\mu$l) are tested for neurotensin degradation. Molecular weight is deduced from a standard curve established from separate injections of cytochrome $c$, chymotrypsinogen, ovalbumin, and bovine serum albumin.

*Purity of Enzyme*

*One-Dimensional SDS–PAGE.* Aliquots of homogenate and posthydroxyapatite pooled proteins are evaporated (SpeedVac), resuspended in 100 $\mu$l of sodium phosphate buffer (pH 7.5) containing 2% sodium dodecyl sulfate, and 5% 2-mercaptoethanol, and then boiled for 5 min. Samples are subjected to sodium dodecyl sulfate–polyacrylamide gel electrophoresis (SDS–PAGE). Electrophoresis is carried out according to Laemmli[16] in an 8% acrylamide gel for 5 hr at room temperature (40 mA per gel). Standards (phosphorylase $b$, 92,500; bovine serum albumin, 66,500; ovalbumin, 45,000; carbonate dehydratase, 31,000; soybean trypsin inhibitor, 21,500) and proteins are revealed by staining with Coomassie Brillant Blue R-250 (Bio-Rad, Richmond, CA) and destaining in acetic acid. The analysis of the purity of a batch of neurolysin (pool H1 of Fig. 3) indicates a single band with an apparent molecular mass of 70–75 kDa (Fig. 4). Further assessment of the extent of purity is performed by two-dimensional SDS–PAGE analysis.

*Two-Dimensional SDS–PAGE.* Isoelectric focusing is carried out in 3 × 100 mm tubes according to the method described by O'Farrell[17] and modified by Cabral and Schatz.[18] A current of 0.2 mA per gel is applied for 40 min, then the voltage is set up to 280 V and maintained constant overnight. Standards (bovine serum albumin, ovalbumin, and carbonate dehydratase) are run under the same conditions. The second dimension in sodium dodecyl sulfate is performed in slab gels with 10% acrylamide as described above for the one-dimensional SDS–PAGE. The analysis (Fig. 4)

---

[16] U. K. Laemmli, *Nature (London)* **227**, 680 (1970).
[17] P. Z. O'Farrell, H. M. Goodman, and P. O'Farrell, *Cell (Cambridge, Mass.)* **12**, 1133 (1977).
[18] F. Cabral and G. Schatz, this series, Vol. 56, p. 602.

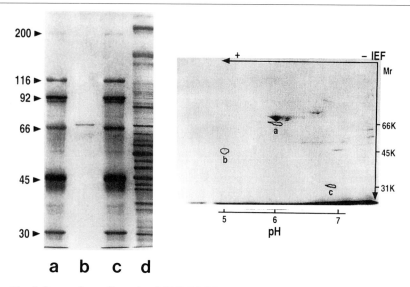

FIG. 4. One- and two-dimensional SDS–PAGE analysis of purified neurolysin. One-dimensional SDS–PAGE is shown at left. Aliquots of homogenate (100 μg) or posthydroxyapatite pooled proteins (pool H1, 7 μg protein) were subjected to electrophoresis in a 8% sodium dodecyl sulfate-polyacrylamide gel under the conditions described in the text. Lanes a and c contain standards; lane b, posthydroxyapatite proteins; lane d, homogenate. Two-dimensional SDS–PAGE is shown at right. Thirty micrograms of pool H1 proteins was loaded in 4% acrylamide gels containing 2% ampholines (pH 5–7) and 8 M urea. Isoelectric focusing was carried out as described in the text, and then the second-dimension separation was performed in slab gels with 10% acrylamide (see text). Drawn circles indicate the migrating pattern of standards (a, bovine serum albumin; b, ovalbumin; c, carbonate dehydratase) run in parallel gels under the same conditions.

revealed faint contaminations by proteins of lower molecular mass that appeared totally devoid of enzymatic activities.[1]

## Purification of Neurolysin from Whole Tissue Homogenates

The preparation of rat brain synaptic membranes is a tedious procedure, and therefore it has been important to devise a protocol that could be applied to whole tissue homogenates as the starting material. Such a procedure did not lead to absolute purity, but that was compensated for by higher yields of the peptidase. The purification procedure starts with the two efficient DEAE and hydroxyapatite chromatography steps described above and involves the following two additional steps.

*Additional Purification Steps*

*Concanavalin A Chromatography.* Posthydroxyapatite activity is concentrated, dialyzed (final volume, ~15 ml; final phosphate concentration, 60 $\mu M$) and is loaded onto a 7-ml column of concanavalin A previously equilibrated at a flow rate of 10 ml/hr in 20 m$M$ Tris-HCl, pH 7.5, containing 150 m$M$ NaCl, 1 m$M$ MnCl$_2$, and 1 m$M$ CaCl$_2$. The resin is then washed with the same buffer, and a gradient is established between 75 ml of the equilibrating buffer and 75 ml of the same eluent containing 0.3 $M$ methyl-$\alpha$-D-glucopyranoside. Active fractions (1.5 ml) are pooled, concentrated, and dialyzed in an Amicon cell (YM10 membrane).

*Gel-Filtration Chromatography.* Pooled fractions (3 ml) from the concanavalin A step are loaded onto an 80 × 2 cm AcA 34 column equilibrated with 20 m$M$ Tris-HCl, pH 7.5, containing 0.15 $M$ NaCl, at a flow rate of 15 ml/hr. Fractions (13 min) are tested for neurotensin degradation. The column is calibrated in parallel with the following standards: chymotrypsinogen, 25 kDa; ovalbumin, 45 kDa; bovine serum albumin, 67 kDa; bovine $\gamma$-globulin, 169 kDa; catalase, 240 kDa. The $V_0$ and $V_T$ values are determined with ferritin and Dnp-alanine, respectively.

*Fluorimetric Assays*

Because the starting material represents a crude fraction, it is absolutely crucial to check for possible contamination by other exo- or endopeptidases. This can easily be performed by means of the following fluorimetric or chromogenic assays.

Unless otherwise stated, all fluorimetric substrates are incubated at 37° in a final volume of 100 $\mu$l of 50 m$M$ Tris-HCl, pH 7.5, generally with about 10–20 $\mu$l of posthydroxyapatite or post-AcA 34 pooled proteins. Hydrolysis of all the following 7-AMC-containing substrates is monitored by fluorimetry (excitation and emission wavelengths of 380 and 460 nm, respectively), and quantitation is deduced from a standard curve established with known amounts of 7-amino-4-methylcoumarin.[19]

*Glutamyl Aminopeptidase, Arginyl Aminopeptidase, Prolyl Oligopeptidase, and Pyroglutamyl Peptidase.* Glutamyl aminopeptidase (EC 3.4.11.7), arginyl aminopeptidase (EC 3.4.11.6), prolyl oligopeptidase (EC 3.4.21.26), and pyroglutamyl peptidase (EC 3.4.19.3) are assayed with $\alpha$-Glu-AMC (0.1 m$M$), Arg-AMC (0.5 m$M$), Z-Gly-Pro-AMC (0.5 m$M$) and pGlu-AMC (0.5 m$M$), respectively. Inhibition studies are carried out in the presence of 10 $\mu M$ amastatin (glutamyl aminopeptidase), 0.5 $\mu M$ arphamenine B

---

[19] F. Checler, *in* "Methods in Neurotransmitter and Neuropeptide Research, Part 2" (T. Nagatsu, H. Parvez, eds.), Vol. 11, pp. 375–418, Elsevier, Amsterdam, 1993.

(arginyl aminopeptidase), 1 $\mu M$ Z-Pro-prolinal (prolyl oligopeptidase), or 1 $\mu M$ pyroglutamyl ketone ester (pyroglutamyl peptidase), respectively.

*Alanyl Aminopeptidase.* Alanyl aminopeptidase (EC 3.4.11.2) is measured with Leu-AMC (0.5 m$M$), and inhibition is achieved by prior incubation of the enzyme with 50 $\mu M$ bestatin.

*Dipeptidyl-peptidase II and Dipeptidyl-peptidase IV.* Dipeptidyl-peptidase II and dipeptidyl-peptidase IV are assayed with Lys-Ala-AMC (1 m$M$) and Gly-Pro-AMC (0.5 m$M$), respectively. Inhibition studies are carried out in the presence of 1 $\mu M$ kelatorphan (dipeptidyl-peptidase II) or 1 m$M$ diprotin A (dipeptidyl-peptidase IV).

*Neprilysin.* Activity of neprilysin is assayed with 1 m$M$ Suc-Ala-Ala-Phe-AMC, and incubations are ended by the addition of 1 $\mu M$ phosphoramidon. The same concentration of the inhibitor is added at the beginning for incubations under inhibitory conditions. Because neprilysin attacks the substrate at the Ala-Phe bond, 5 $\mu$g of exogenous alanyl aminopeptidase (EC 3.4.11.2) is added after the 2-hr incubation to release AMC from Phe-AMC.

*Angiotensin Converting Enzyme.* Angiotensin-converting enzyme (EC 3.4.15.1) is monitored by incubating Bz-Gly-His-Leu (2 m$M$) in the above Tris buffer containing 0.3 $M$ NaCl and 10 $\mu M$ 2-mercaptoethanol. Inhibition studies are performed with 1 $\mu M$ captopril that is used in the presence of 10 $\mu M$ 2-mercaptoethanol to protect the reactive sulfhydryl group. Reactions are stopped by the addition of 0.4 ml of 2 $M$ NaOH followed by 3 ml of water. Liberated His-Leu is measured by adding 0.1 ml of 1% (w/v) *o*-phthalaldehyde in methanol, followed 4 min later by 0.2 ml of 6 $M$ HCl. Quantitation of His-Leu released is deduced from a standard curve established with known amounts of His-Leu (fluorimeter setting of 365 and 495 nm as excitation and emission wavelengths, respectively).

### Chromogenic Assays

*Thimet Oligopeptidase.* Thimet oligopeptidase (EC 3.4.24.15) is measured with 1 m$M$ Bz-Gly-Ala-Ala-Phe-pAB.[20] The inhibitors Cpe-Ala-Ala-Phe-pAB (50 $\mu M$) or Cpp-Ala-Ala-Tyr-pAB (1 $\mu M$) are added either at the beginning or at the end of the incubations. The degradation product Ala-Ala-Phe-pAB is subjected to further processing by 5 $\mu$g of alanyl aminopeptidase to generate the free pAB (*p*-aminobenzoate) chromophore. Owing to contaminating neprilysin present in the commercial microsomal kidney alanyl aminopeptidase (Sigma), the former enzyme is always used in the presence of 1 $\mu M$ of the neprilysin inhibitor phosphoramidon.

[20] M. Orlowski, C. Michaud, and T. G. Chu, *Eur. J. Biochem.* **135**, 81 (1983).

After termination of the incubations by the addition of 0.1 ml of 25% w/v trichloroacetic or glacial acetic acid, *p*-aminobenzoate is determined by the addition of 0.2 ml of an aqueous solution of 0.25% sodium nitrite. After 3 min, 0.2 ml of 1.25% ammonium sulfamate is added followed, 2 minutes later, by 0.5 ml of 0.1% *N*-(1-naphthyl)ethylenediamine dihydrochloride (in ethanol).[20] Absorbance is monitored at 555 nm, after allowing 10 min for color development, and compared with the absorbance obtained with known amounts of standard *p*-aminobenzoate. The azo dye is water-soluble and stable at room temperature and at 4°.

By means of the above enzymatic assays, we have shown that after the hydroxyapatite step of the purification of neurolysin from homogenates of rat brain,[21] ileum,[22] and kidney,[23] the pooled fractions are totally devoid of the above other peptidases.

## Assays of Neurolysin

### Neurotensin as Substrate

During the course of the first purification of neurolysin, all the fractions were monitored for the ability to degrade neurotensin by HPLC analysis[1] according to the following procedures.

*Analysis of Neurotensin Degradation by HPLC.* Chromatography is carried out on a Waters Associates (Milford, MA) apparatus equipped with a Model 481 detector, two Model 510 pumps, a WISP injector, and an automated gradient controller. Elutions are carried out at room temperature onto an RP18 Lichrosorb column (Merck), at a flow rate of 1 ml/min, and the absorbance is monitored at 230 nm with a detector setting of 0.1 full scale. Samples are chromatographed with the two following buffer/solvent systems:

System 1: 35-min nonlinear gradient (gradient 7) of 20 m*M* ammonium acetate, pH 6.4/acetonitrile from 90:10 v/v to 65:35 (v/v) (a typical chromatogram and the retention times of neurotensin and a series of neurotensin synthetic fragments are given in Fig. 5A)

System 2: 42-min linear gradient of 0.1% TFA (trifluoroacetic acid), 0.05% TEA (triethylamine)/0.1% TFA, 0.05% TEA in acetonitrile from 90:10 to 60:40 (the retention times of neurotensin and synthetic peptides are given in Fig. 5B)

[21] H. Barelli, F. Girard, S. St. Pierre, P. Kitabgi, J. P. Vincent, and F. Checler, *Neurochem. Int.* **12,** 351 (1988).
[22] H. Barelli, J. P. Vincent, and F. Checler, *Eur. J. Biochem.* **175,** 481 (1988).
[23] H. Barelli, J. P. Vincent, and F. Checler, *Eur. J. Biochem.* **211,** 79 (1993).

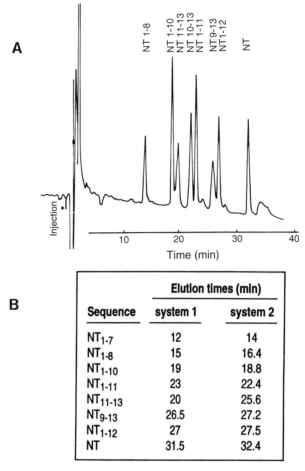

Fig. 5. Separation by HPLC and retention times of synthetic neurotensin fragments. (A) Resolution by HPLC of the indicated neurotensin (NT) fragments run in solvent/buffer system 1 (see text). (B) Retention times of a series of neurotensin fragments run in systems 1 and 2 (see text).

Analysis of neurotensin degradation by HPLC has the advantage of allowing the direct identification of the catabolite but is too laborious for repetitive assays such as those necessary to examine the physicochemical, biochemical, or specificity properties of the enzyme. For such work, we have developed, a rapid, reproducible, and routine assay based on the chromatographic properties of tritiated neurotensin and its labeled catabolites [$^3$H]neurotensin 1–10 and [$^3$H]neurotensin 11–13.

*SPC25 Chromatographic Assay of Tritiated Neurotensin Degradation.* Tritiated neurotensin [3,11-*tyrosyl*-3,5-$^3$H(N)]neurotensin (obtained from New England Nuclear at a specific radioactivity of about 50–100 Ci/mmol) is probably the most useful radiolabeled substrate for neurotensin degradation studies. We have previously showed that the peptide is equally labeled on both Tyr-3 and Tyr-11 of the neurotensin sequence. Accordingly, both of the two catabolites neurotensin 1–10 and neurotensin 11–13 are radioactive and display equal specific radioactivities.

The routine assay based on the chromatographic properties of [$^3$H]neurotensin 1–10, [$^3$H]neurotensin 11–13, and [$^3$H]neurotensin is summarized in Fig. 6. In the procedure, [$^3$H]neurotensin [100,000 counts/min (cpm)] is incubated with 5–100 ng of purified enzyme for 30 min at 37°, in a final volume of 100 $\mu$l of 50 m$M$ Tris-HCl, pH 7.5. Incubations are terminated by the addition of 5 $\mu$l of 0.5 $M$ HCl and diluted with 900 $\mu$l distilled water. The mixture is then loaded on mini-SPC25 Sephadex columns (500 $\mu$l) preequilibrated with 10 m$M$ acetic acid, pH 5.5. Step 1 of the elution consists of 4 ml of the equilibration buffer in order to elute [$^3$H]neurotensin 11–13, neutral under the present conditions of pH. In step 2, the same buffer containing 0.2 $M$ NaCl is applied to elute the basic peptides [$^3$H]neurotensin 1–10 and [$^3$H]neurotensin. Separate reversed-phase HPLC of both eluates (Fig. 6) confirms the expected identity of the tritiated peptides recovered in each step.[1] Thus, more than 95% of the radioactivity recovered after the first elution step consists of [$^3$H]neurotensin 11–13, whereas more than 92% of the second eluate consists of a mixture of [$^3$H]neurotensin 1–10 and [$^3$H]neurotensin. The fact that the amount of radioactivity identified as [$^3$H]neurotensin 11–13 matches that recovered as [$^3$H]neurotensin 1–10, together with the equal distribution of Tyr$^3$ and Tyr$^{11}$ of the [$^3$H] neurotensin molecule, allows us to calculate the amount of degraded [$^3$H]neurotensin according to the following equation:

$$\%[^3\text{H}]\text{neurotensin } 1\text{--}10 = \%[^3\text{H}]\text{neurotensin } 11\text{--}13$$
$$= \frac{\text{radioactivity recovered in step 1} \times 100}{\text{radioactivity recovered in steps 1 and 2}}$$

The above technique establishes that neurolysin (i) belongs to the class of metallopeptidases and is maximally active at neutral pH, (ii) exhibits endopeptidase activity because a free C terminus is not an absolute requirement of potential substrates, (iii) displays a high affinity for neurotensin ($K_m$ of ~2 $\mu M$) and recognizes the C-terminal pentapeptide neurotensin 9–13 as the shortest full substrate, and (iv) hydrolyzes biologically active peptides belonging to other families such as angiotensin I, bradykinin, substance P, dynorphin 1–8 but appears poorly active toward other neurotensin congeners such as neuromedin N[1] or Kinetensin.[23]

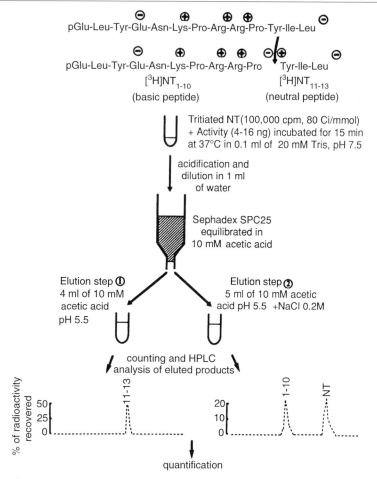

FIG. 6. Quantification of tritiated neurotensin degradation on Sephadex SPC25. The scheme summarizes the Sephadex SPC25 procedure described in the text.

## Mcc-Pro-Leu-Gly-Pro-D-Lys-Dnp as Substrate

The peptide Mcc-Pro-Leu-Gly-Pro-D-Lys-Dnp (QFS) was first reported as a substrate of Pz-peptidase,[24] a peptidase widely distributed in mammals, that is now termed thimet oligopeptidase (EC 3.4.24.15).[25] The first indication that, in the central nervous system, an additional peptidase likely contributes to QFS hydrolysis came from the very partial inhibition of

[24] U. Tisljar, C. G. Knight, and A. J. Barrett, *Anal. Biochem.* **186**, 112 (1990).
[25] U. Tisljar and A. J. Barrett, *Biochem. J.* **267**, 531 (1990).

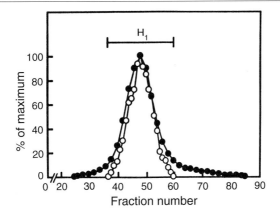

FIG. 7. Elution of neurolysin and Mcc-Pro-Leu-Gly-Pro-D-Lys-Dnp activities from hydroxy-apatite. All fractions from the hydroxyapatite step of neurolysin purification were assayed for neurotensin 1–10 formation by HPLC (●) or fluorimetrically recorded for Mcc-Pro-Leu-Gly-Pro-D-Lys-Dnp hydrolysis (○) as described in the text. Values are expressed as the percentage of maximal activity recovered.

QFS degradation by saturating concentrations of the thimet oligopeptidase inhibitor[26] Cpp-Ala-Ala-Tyr-pAB.[27] This led us to examine the possibility that neurolysin could cleave QFS. This can be performed by the two following procedures.

*Fluorimetric Assay.* The QFS substrate (50 $\mu M$) is incubated at 37° with the enzyme in a final volume of 100 $\mu$l of 20 m$M$ Tris-HCl, pH 7.5. Incubations are stopped by the addition of 2.5 ml of 80 m$M$ sodium formate, pH 3.7, and then activity is monitored with a fluorimeter set at 345 and 405 nm as excitation and emission wavelengths, respectively. Quantification of the Mcc-Pro-Leu released is deduced from a standard curve established with known amounts of synthetic Mcc-Pro-Leu.

*Assay by HPLC.* The substrate QFS (4 nmol, 20 $\mu M$) is incubated at 37° with the enzyme in a final volume of 200 $\mu$l of 20 m$M$ Tris-HCl, pH 7.5. Acidified samples are analyzed in buffer/solvent system 2 (see above) according to two distinct gradients:

Gradient 1: 42-min linear gradient of 0.1% TFA, 0.05% TEA/0.1% TFA, 0.05% TEA in acetonitrile from 75:25 to 20:80 (v/v)

Gradient 2: biphasic gradient of 0.1% TFA, 0.05% TEA/0.1% TFA, 0.05% TEA in acetonitrile from 90:10 to 75:25 (v/v) in 15 min then from 75:25 to 20:80 (v/v) in 40 min

[26] P. Dauch, H. Barelli, J. P. Vincent, and F. Checler, *Biochem. J.* **280,** 421 (1991).
[27] T. G. Chu and M. Orlowski, *Biochemistry* **23,** 3598 (1984).

Figure 7 clearly shows that, during the hydroxyapatite step of neurolysin purification (see above), a QFS-hydrolyzing activity appears superimposable on the neurotensin 1–10-forming enzyme.[26] The possibility that QFS indeed behaves as a substrate of neurolysin was further reinforced by the observation that QFS hydrolysis can be dose-dependently inhibited by neurotensin, with a $K_i$ value (2.6 $\mu M$) that closely corresponds to the $K_m$ of the enzyme for neurotensin.[26]

Initial velocity measurements carried out with purified neurolysin (Fig. 8A) allow us to derive a $K_m$ value of about 25 $\mu M$.[26] Kinetic analysis carried out at virtual saturating concentrations of QFS indicates that the rate of

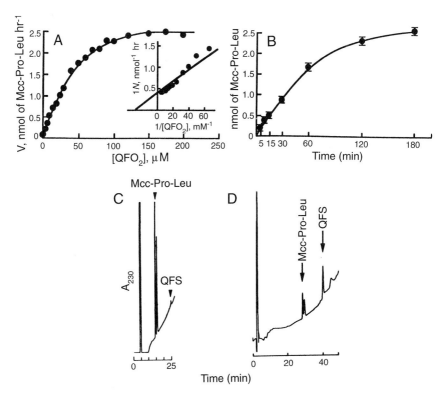

FIG. 8. Kinetic parameters and HPLC analysis of QFS hydrolysis by purified rat brain neurolysin. For fluorimetry, QFS was incubated at 37° with purified neurolysin. Hydrolysis was fluorimetrically recorded and quantified as described in the text. (A) Initial velocity as a function of substrate concentration. *Inset*: Lineweaver–Burk plot of the data. (B) Formation of Mcc-Pro-Leu as a function of time at a 100 $\mu M$ concentration of QFS. For HPLC; QFS (4–5 nmol) was incubated with purified neurolysin, and analysis was performed by HPLC according to the systems 1 (C) and 2 (D).

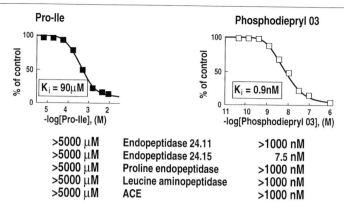

| Pro-Ile | | Phosphodiepryl 03 |
|---|---|---|
| >5000 μM | Endopeptidase 24.11 | >1000 nM |
| >5000 μM | Endopeptidase 24.15 | 7.5 nM |
| >5000 μM | Proline endopeptidase | >1000 nM |
| >5000 μM | Leucine aminopeptidase | >1000 nM |
| >5000 μM | ACE | >1000 nM |

FIG. 9. Inhibition of neurolysin by Pro-Ile and phosphodiepryl 03. Neurolysin was incubated with 20 $\mu M$ neurotensin in the absence (control) or presence of increasing concentrations of Pro-Ile or phosphodiepryl 03. The $K_i$ values were calculated from $IC_{50}$ data with the formula $IC_{50} = K_i(1 + [S]/K_m)$, with the $K_m$ value for neurotensin taken as 2 $\mu M$. At bottom is indicated the $K_i$ values of Pro-Ile (left-hand column) and phosphodiepryl 03 (right-hand column) for the indicated peptidases.

appearance of the product remains linear within the first 60 min of incubation (Fig. 8B). The specific activity of neurolysin toward QFS is about 215 nmol/hr/mg protein.[26] As shown in Fig. 8C, D, HPLC analysis carried out according to the two different gradient systems allows us to monitor two degradation products derived from a single cleavage taking place at the Leu-Gly peptide bond.[26]

The QFS peptide can be used as the substrate during the course of neurolysin purification. The fluorimetric assay permits the quickest purification of the enzyme and is now routinely used in our laboratory.

## Inhibitors of Neurolysin

### Dipeptides

We previously screened a series of dipeptides as potential inhibitors of neurolysin. Among several Pro-X dipeptides mimicking the Pro-X peptide bond of neurotensin that is cleaved by neurolysin, we showed that Pro-Ile was the most potent blocking agent.[28] Figure 9 indicates that the dipeptide half-inhibited the brain enzyme with a $K_i$ value of 90 $\mu M$. Interestingly, we have established that Pro-Ile displays an exclusive selectivity toward neurolysin: a 5 m$M$ concentration of the agent did not affect various exo-

[28] P. Dauch, J. P. Vincent, and F. Checler, *Eur. J. Biochem.* **202,** 269 (1991).

TABLE I
DISTRIBUTION OF NEUROLYSIN IN RAT ORGANS[a]

| Tissue | Specific activity (nmol Mcc-Pro-Leu/hr/mg protein) |
|---|---|
| Liver | 57.5 + 2.7 (100) |
| Kidney | 55.1 ± 3.8 (95.8) |
| Bladder | 46.2 ± 2.4 (80.3) |
| Ileum | 38.9 ± 2 (67.6) |
| Testis | 38.4 ± 2 (66.8) |
| Duodenum | 34 ± 2.1 (59.1) |
| Spleen | 28.4 ± 0.7 (49.4) |
| Jejunum | 26.4 ± 1.2 (45.9) |
| Cecum | 26 ± 1 (45.2) |
| Stomach | 24 ± 0.9 (41.7) |
| Esophagus | 21.1 ± 1.9 (36.7) |
| Colon | 17.9 ± 1.1 (31.1) |
| Brain | 15.5 ± 0.4 (26.9) |
| Lung | 13.7 ± 0.7 (23.8) |
| Heart | 8.5 ± 0.5 (14.8) |

[a] Neurolysin activity was fluorimetrically measured as described in the text in the absence or presence of 10 m$M$ Pro-Ile. Specific activities represent the means ± SEM of 3 to 7 independent determinations.

and endopeptidases (see Fig. 9), most of them also belonging to the class of zinc metallopeptidases.[28]

*Phosphodiepryl 03*

The development of specific inhibitors of zinc metallopeptidases has often been achieved by introducing metal-coordinating groups, which could be carboalkyls, thiols, and hydroxamate, or by mimicking the transition state of the catalytic reaction by phospho groups. Phosphonamide peptides have been previously proved to behave as potent inhibitors of zinc bacterial collagenases from *Clostridium histolyticum*.[11] It is noteworthy that the structure of one of them, namely, phosphodiepryl 03, displayed the Gly-Pro sequence that corresponds to the N terminus of one of the degradation products of QFS (Gly-Pro-D-Lys-Dnp, see above) generated by neurolysin. We have therefore examined phosphodiepryl 03 as a putative inhibitor of neurolysin.[29] The compound dose-dependently and potently ($K_i$ 0.9 n$M$) inhibits the peptidase (Fig. 9). Furthermore, phosphodiepryl 03 displays a rather satisfactory selectivity toward neurolysin (Fig. 9), although it should

[29] H. Barelli, V. Dive, A. Yiotakis, J. P. Vincent, and F. Checler, *Biochem. J.* **287,** 621 (1992).

TABLE II
CEREBRAL DISTRIBUTION OF NEUROLYSIN[a]

| Brain zone | Specific activity |
| --- | --- |
| Olfactory bulb | |
| External plexiform layer | 32.3 ± 2 |
| Internal granular layer | 29 ± 0.9 |
| Cerebral cortex | |
| Frontal | 20.1 ± 2.5 |
| Cingulate | 30 ± 2.6 |
| Parietal | 12.2 ± 3.1 |
| Piriform | 23.3 ± 3 |
| Enthorinal | 10.6 ± 2.8 |
| Basal forebrain | |
| Lateral septal nucleus | |
| Dorsal | 21.6 ± 1.3 |
| Intermediate | 16.8 ± 3.3 |
| Medial septal nucleus | 10.3 ± 2.7 |
| Diagonal band | 16.7 ± 4.3 |
| Olfactory tubercule | 23.3 ± 5.3 |
| Basal ganglia | |
| Accumbens nucleus | 14.8 ± 3.4 |
| Striatum | |
| Anterior | 19 ± 1.5 |
| Medial | 28.2 ± 0.8 |
| Posterior | 14.3 ± 1 |
| Globus pallidus | 24.4 ± 3.4 |
| Habenular nucleus | 12 ± 2.1 |
| Amygdaloid complex | |
| Central nucleus | 9.4 ± 0.5 |
| Medial nucleus | 10.5 ± 1.8 |
| Brain stem | |
| Superior colliculus | 8.8 ± 1.5 |
| Central gray | 8.4 ± 1.9 |
| Substantia nigra | |
| Compact part | 6.8 ± 1.7 |
| Reticular part | 6.9 ± 2.5 |
| Ventral tegmental area | 6.4 ± 2.8 |
| Interfascicular nucleus | 16.2 ± 2 |
| Interpeduncular nucleus | 19.3 ± 1.1 |
| Dorsal raphe nucleus | 9.4 ± 2.6 |
| Locus coeruleus | 13.8 ± 0.9 |
| Pontine reticular nucleus | 3.4 ± 0.8 |
| White matter | |
| Corpus callosum | 11.7 ± 0.7 |
| Internal capsule | 4 ± 1.3 |
| Ventral hippocampal commissure | 5.7 ± 0.3 |
| Cerebellum | 3.9 ± 0.8 |

[a] Neurolysin activity was fluorimetrically measured as described in the text in the absence or presence of 10 m$M$ Pro-Ile. Specific activities represent the means ± SEM of 4 to 10 determinations carried out with two to five independent homogenates prepared from pooled or individual brain micropunches.

TABLE III
DISTRIBUTION OF NEUROLYSIN IN THALAMUS,
HYPOTHALAMUS, HIPPOCAMPUS, PINEAL, AND
PITUITARY GLANDS[a]

| Area | Specific activity |
|------|-------------------|
| Thalamus | |
| Paratenial nucleus | 6.3 ± 2.7 |
| Paraventricular nucleus | 11.2 ± 2.9 |
| Mediodorsal nucleus | 11.6 ± 3 |
| Laterodorsal nucleus | 14.5 ± 0.5 |
| Ventrolateral nucleus | 10.9 ± 2.7 |
| Ventromedial nucleus | 8.2 ± 1.4 |
| Ventral posterior nucleus | 4.9 ± 2 |
| Gelatinosus nucleus | 16.9 ± 2.1 |
| Rhomboid nucleus | 12.8 ± 3.6 |
| Reuniens nucleus | 9.6 ± 2.5 |
| Posterior thalamus nuclear group | 9.7 ± 2.5 |
| Dorsolateral geniculate nucleus | 12.2 ± 2.3 |
| Medial geniculate nucleus | 7.6 ± 1.2 |
| Hippocampal formation | |
| $CA_1$ | 2.2 ± 0.4 |
| $CA_2$ | 5 ± 0.5 |
| $CA_3$ | 2 ± 0.5 |
| Dentate gyrus | 16 ± 1.1 |
| Hypothalamus | |
| Suprachiasmatic nucleus | 18 ± 0.4 |
| Supraoptic nucleus | 8 ± 1.9 |
| Periventricular nucleus | 6.1 ± 1.3 |
| Arcuate nucleus | 18.1 ± 4.2 |
| Lateral hypothalamic area | 7.1 ± 2.4 |
| Mammillary nucleus | 8.6 ± 1.2 |
| Pituitary gland | |
| Anterior lobe | 11.4 ± 4.1 |
| Posterior lobe | 10.6 ± 5.7 |
| Pineal gland | 12 ± 4 |

[a] Neurolysin activity was fluorimetrically measured as described in the text in the absence or presence of 10 m$M$ Pro-Ile. Specific activities represent the means ± SEM of 4 to 10 determinations carried out with two to five independent homogenates prepared from pooled or individual brain micropunches.

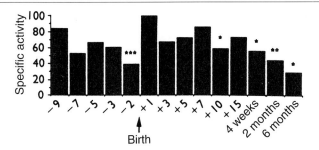

FIG. 10. Neurolysin activity was fluorimetrically measured as described in the text in the absence or presence of 10 m$M$ Pro-Ile. Specific activities represent the means $\pm$ SEM of 4 to 18 determinations. Statistical analyses were performed by the Student's $t$ test (*$p$ < 0.05, **$p$ < 0.005, ***$p$ < 0.001, compared with day 1 after birth).

be noted that it also blocked thimet oligopeptidase (with an about 10-fold higher $K_i$ value, Fig. 9). The inhibitor phosphodiepryl 03, which is hydrophilic and potent, has allowed us to establish for the first time the involvement of neurolysin in the physiological inactivation of neurotensin, *in vivo*, in a vascularly perfused dog ileum model (Barelli *et al.*[30]).

### Fluorimetric Assay of Neurolysin in Whole Tissue Homogenates

As stated above QFS also behaves as a substrate of thimet oligopeptidase. Therefore, hydrolysis of QFS in a complex mixture such as whole tissue homogenates could be ascribed to two components, a Pro-Ile-sensitive cleavage elicited by neurolysin and a Cpp-Ala-Ala-Tyr-pAB-sensitive hydrolysis arising from thimet oligopeptidase. Therefore, a selective inhibition of thimet oligopeptidase would make it possible to measure neurolysin activity in crude extracts. This could be achieved by preincubation of the enzymatic source with 0.5 $\mu M$ Cpp-Ala-Ala-Tyr-pAB.[26] This procedure has allowed us to establish the tissue distribution, cerebral regionalization, and ontogeny of neurolysin in murines.[31]

### *Tissue Distribution*

Table I indicates that neurolysin could be detected ubiquitously in rat organs. The enzyme is particularly abundant in liver and kidney, whereas heart is the tissue that displays the lowest specific activity.

---

[30] H. Barelli, J. E. T. Fox-Threlkeld, V. Dive, E. E. Daniel, J. P. Vincent, and F. Checler, *Br. J. Pharmacol* **112**, 127–132 (1994).

[31] P. Dauch, Y. Masuo, J. P. Vincent, and F. Checler, *J. Neurochem.* **59**, 1862 (1992).

*Cerebral Regionalization*

There is a 16-fold difference between the highest and lowest activities in various parts of the brain (Tables II and III). Neurolysin activity appears particularly enriched in the olfactory bulb and tubercule, cingulate cortex, medial striatum, and globus pallidus (Table II). In contrast, the enzyme activity is particularly low in the cerebellum (Table II) as well as in the $CA_1$, $CA_2$, and $CA_3$ parts of the hippocampal formation (Table III). It should be noted that the distribution of neurotensin receptors established by means of various radiolabeled neurotensins indicates that there exists a good parallel between neurotensin receptor densities and neurolysin activity in the rat brain.[32,33]

*Ontogeny*

Neurolysin is developmentally regulated in mice brain (Fig. 10). The enzyme is detected *in utero* as early as 9 days before parturition, increases transiently just after birth, then regularly decreases until adulthood (Fig. 10). Here again, it is striking that high-affinity neurotensin binding sites increase in number during the postnatal period.[34,35]

Altogether, the parallel localization and similar postnatal regulation of neurolysin and neurotensin receptors further suggest the functional importance of neurolysin in the termination of the neurotensinergic message.

[32] L. H. Lazarus, M. R. Brown, and M. H. Perrin, *Neuropharmacology* **16**, 625 (1977).
[33] E. Moyse, W. Rostène, M. Vial, K. Leonard, J. Mazella, P. Kitabgi, J. P. Vincent, and A. Beaudet, *Neuroscience* (*Oxford*) **22**, (1987).
[34] A. Schotte and P. M. Laduron, *Brain Res.* **408**, 326 (1987).
[35] N. Zsürger, J. Chabry, A. Coquerel, and J. P. Vincent, *Brain Res.* **586**, 303 (1992).

# [37] Leishmanolysin: Surface Metalloproteinase of *Leishmania*

By JACQUES BOUVIER, PASCAL SCHNEIDER, and ROBERT ETGES

## Introduction

*General Introduction*

The genus *Leishmania* includes kinetoplastid protozoan parasites responsible for a spectrum of zoonotic and anthroponotic diseases collectively called the leishmaniases. Leishmaniasis is classified by the World Health Organization (WHO) as one of the six major parasitic diseases, affecting over 12 million people worldwide. The wide spectrum of clinical manifestations in humans ranges from self-healing cutaneous ulcers to nonresolving mucocutaneous lesions, and visceral disease with hepatosplenomegaly and bone marrow infection leading, without treatment, to death.[1,2] As for many other parasitic diseases, there is no effective vaccine available. Current chemotherapeutic agents are not only obsolete, but are also suboptimal in that they display significant toxicity and have selected drug-resistant parasites in some parts of the world.[3] Further, significant resistance of the vector to currently acceptable insecticides has been reported.

Despite the diversity of clinical features, the life cycles and nutritional requirements of all species of *Leishmania* are remarkably similar. The parasite proliferates alternately as attached or motile flagellated promastigotes in the midgut of the hematophagous dipteran insect vectors[4,5] and as sessile amastigotes in the acidified phagolysosomes of the macrophages of mammalian hosts.[6] The immediate environments encountered by the different species of *Leishmania* during the life cycles are therefore likely to be quite similar.

*Biochemical Aspects*

Promastigotes of Old and New World species of *Leishmania* express numerous surface molecules, including transporters and ectoenzymes,[7,8] among them an abundant membrane glycoprotein, known as gp63, at high density ($>5 \times 10^5$ molecules per cell) at the surface of the parasite.[9,10] The gp63 molecule is bound to the membrane by a glycosylphosphatidylinositol

[1] Leishmaniasis, *in* "Tropical Diseases, Progress in Research, 1989–1990," p. 79. World Health Organization, Geneva, 1991.
[2] Kala-azar surges on two fronts. *TDR News* **37**, 1 (1991).
[3] M. Ouellette and B. Papadopoulou, *Parasitol. Today* **9**, 150 (1993).
[4] D. H. Molyneux and R. Killick-Kendrick, *in* "The Leishmaniases in Biology and Medicine" (W. Peters and R. Killick-Kendrick, eds.), Vol. 1, p. 121. Academic Press, London, 1987.
[5] L. L. Walters, *J. Eukaryot. Microbiol.* **40**, 196 (1993).
[6] J.-C. Antoine, E. Prina, C. Jouanne, and P. Bongrand, *Infect. Immun.* **58**, 779 (1990).
[7] K.-P. Chang, G. Chaudhuri, and D. Fong, *Annu. Rev. Microbiol.* **44**, 499 (1990).
[8] P. Schneider, C. Bordier, and R. J. Etges, *in* "Intracellular Parasites" (J. L. Avila and J. R. Harris, eds.), Subcellular Biochemistry, Vol. 18, p. 39. Plenum, New York and London, 1992.
[9] C. Bordier, *Parasitol. Today* **3**, 151 (1987).
[10] R. J. Etges and J. Bouvier, *in* "Biochemical Protozoology" (G. Coombs and M. North, eds.), p. 221. Taylor & Francis, London, 1991.

(GPI) anchor[11,12] similar to that which attaches a wide variety of proteins to membranes of eukaryotes.[13,14] The surface protein has been demonstrated on promastigotes residing in the midgut of the phlebotomine sandfly vector,[15,16] and it occurs at the surface of the promastigotes of all *Leishmania* species examined so far except for *L. tarentolae.*[17,18] The "surface antigen" or major surface glycoprotein of the promastigote was shown to be a proteolytic enzyme,[19] designated the promastigote surface protease, or PSP, and later characterized as a zinc endopeptidase (EC 3.4.24.36).[20-22] Its predominant $\beta$ structure, determined by Raman spectroscopy[23] and circular dichroism,[20] distinguishes it from the well-characterized low molecular weight soluble metalloendopeptidases thermolysin[24] and astacin.[25,26]

Surface metalloproteinase activity is not only a highly conserved feature of the genus *Leishmania,* but also occurs at the surface of the monogenetic trypanosomatids *Crithidia* and *Herpetomonas.*[18,27,28] Peptide-mapping analyses showed that the related proteinases are structurally conserved among the species of *Leishmania.*[29,30] Genetic studies have shown that several closely related families of genes express the metalloproteinase, some in a stage-specific manner. Intriguingly, a soluble metalloproteinase has been

[11] R. J. Etges, J. Bouvier, and C. Bordier, *EMBO J.* **5,** 597 (1986).

[12] P. Schneider, M. A. J. Ferguson, M. J. McConville, A. Mehlert, S. W. Homans, and C. Bordier, *J. Biol. Chem.* **265,** 16955 (1990).

[13] G. A. M. Cross, *Annu. Rev. Cell Biol.* **6,** 1 (1990).

[14] M. J. McConville and M. A. J. Ferguson, *Biochem. J.* **294,** 305 (1993).

[15] F. Grimm, L. Jenni, J. Bouvier, R. J. Etges, and C. Bordier, *Acta Trop.* **44,** 375 (1987).

[16] C. R. Davies, A. M. Cooper, C. Peacock, R. P. Lane, and J. M. Blackwell, *Parasitology* **101,** 337 (1990).

[17] J. Bouvier, R. J. Etges, and C. Bordier, *Mol. Biochem. Parasitol,* **24,** 73 (1987).

[18] R. J. Etges, *Acta Trop.* **50,** 205 (1992).

[19] R. J. Etges, J. Bouvier, and C. Bordier, *J. Biol. Chem.* **261,** 9099 (1986).

[20] J. Bouvier, C. Bordier, H. Vogel, R. Reichelt, and R. J. Etges, *Mol. Biochem. Parasitol.* **37,** 235 (1989).

[21] C. V. Jongeneel, J. Bouvier, and A. Bairoch, *FEBS Lett.* **242,** 211 (1989).

[22] G. Chaudhuri, M. Chaudhuri, A. Pan, and K.-P. Chang, *J. Biol. Chem.* **264,** 7483 (1989).

[23] F. Jähnig, and R. J. Etges, *FEBS Lett.* **241,** 79 (1988).

[24] M. Levitt and J. Greer, *J. Mol. Biol.* **114,** 181 (1977).

[25] W. Stöcker, F.-X. Gomis-Rüth, W. Bode, and R. Zwilling, *Eur. J. Biochem.* **214,** 215 (1993).

[26] F.-X. Gomis-Rüth, W. Stöcker, R. Huber, R. Zwilling, and W. Bode, *J. Mol. Biol.* **229,** 945 (1993).

[27] J. A. Inverso, E. Medina-Acosta, J. O'Connor, D. G. Russell, and G. A. M. Cross, *Mol. Biochem. Parasitol.* **57,** 47 (1993).

[28] P. Schneider and T. A. Glaser, *Mol. Biochem. Parasitol.* **58,** 277 (1993).

[29] V. Colomer-Gould, L. Galvao-Quintao, J. Keithly, and N. Noguiera, *J. Exp. Med.* **162,** 902 (1985).

[30] R. J. Etges, J. Bouvier, R. Hoffman, and C. Bordier, *Mol. Biochem. Parasitol.* **14,** 141 (1985).

isolated from the amastigotes of *L. mexicana* which appears to be a component of the lysosomal compartment of the intracellular parasite.[31] Compared to the membrane-bound enzyme of the promastigote, the amastigote homolog has a lower pH optimum (pH 5.5–6 versus pH 8–9). This member of the *Leishmania* metalloproteinase family may be encoded by a gene designated C1,[32] which diverges in sequence at the carboxyl terminus compared to the sequences of the GPI-anchored forms. Because the amastigote enzyme is neither associated with the promastigote nor membrane bound, the designation promastigote surface protease is no longer appropriate: we therefore propose to call the growing family of homologous kinetoplastid metalloproteinases, whether soluble or membrane bound, the leishmanolysins.[33]

In spite of the numerous investigations devoted to the characterization of leishmanolysin, the role of the enzyme and the identity of its substrates *in vivo* remain the subject of intense speculation. Whether the proteolytic activity performs a nutritional role by providing heme, amino acids, and small peptides to membrane amino acid transporters of the promastigote, or whether it plays a defensive role against host or vector proteinases, remains to be demonstrated. It has also been speculated that leishmanolysin could protect the promastigote by cleaving the potentially membranolytic components of serum complement[34] or microbicidal proteins of the macrophage.[22]

Clearly, more precise knowledge of the peptide bond specificity and active site accessibility is essential to our understanding of the role played by leishmanolysin. To date, the peptide bond preference of *L. major* leishmanolysin has been examined in detail using a variety of natural and synthetic substrates. Like many other metalloendoproteinases, the specificity of leishmanolysin is defined essentially by the P′ subsites of the substrate.[35,36] These findings have allowed the design of a synthetic peptide hydroxamate inhibitor,[35] and the kinetic parameters obtained with the model peptide substrate $H_2N$-L-I-A-Y-+L-K-K-A-T-COOH permitted the rational design of a fluorogenic substrate for leishmanolysin based on radiationless energy transfer (RET) methodology.[37]

---

[31] T. Ilg, D. Harbecke, and P. Overath, *FEBS Lett.* **327,** 103 (1993).

[32] E. Medina-Acosta, S. M. Beverley, and D. G. Russell, *Infect. Agents Dis.* **2,** 25 (1993).

[33] N. D. Rawlings and A. J. Barrett, *Biochem. J.* **290,** 205 (1993).

[34] S. M. Puentes, D. M. Dwyer, P. A. Bates, and K. A. Joiner, *J. Immunol.* **143,** 3743 (1989).

[35] J. Bouvier, P. Schneider, R. J. Etges, and C. Bordier, *Biochemistry* **29,** 10113 (1990).

[36] H. S. Ip, D. G. Russell, and G. A. M. Cross, *Mol. Biochem. Parasitol.* **40,** 163 (1990).

[37] J. Bouvier, P. Schneider, and B. Malcolm, *Exp. Parasitol.* **76,** 146 (1993).

*Molecular Aspects*

The complete nucleotide sequences of representative leishmanolysins have been deduced in several species of *Leishmania* and *Crithidia*.[27,38-41] The enzymes are encoded by three families of multiple, heterogeneous genes which map to a single chromosome.[32,42,43] Although messenger RNA was reported to be constitutively transcribed in both stages of the parasite[42] and a polypeptide cross-reacting with promastigote leishmanolysin was detected in amastigotes of several *Leishmania* species,[22,29,44,45] the enzymatic activity, amount, and solubility behavior of amastigote leishmanolysin were only relatively recently investigated in *L. mexicana*.[31] Regulated expression of leishmanolysin appears to occur at specific stages of the life cycle. Correspondingly distinct mRNA populations for the enzyme have been shown to be differentially expressed during development of promastigotes to an infectious form.[41,46,47] In amastigotes of *L. major* LV39, however, the expression of leishmanolysin mRNA was virtually undetectable, and protein expression (both as proteolytic activity and immunoreactive protein) was shown to be down-regulated by a factor of 300 times or more compared to the promastigote.[48]

*Immunological Aspects*

The involvement of leishmanolysin in the early phases of infection as a ligand for the mannosyl-fucosyl receptor, as an acceptor for complement component C3b deposition, or simply as the "major surface antigen"[34,49,50]

[38] L. L. Button and W. R. McMaster, *J. Exp. Med.* **167,** 724 (1988); correction: *J. Exp. Med.* **171,** 589 (1990).

[39] R. A. Miller, S. G. Reed, and M. Parsons, *Mol. Biochem. Parasitol.* **39,** 267 (1990).

[40] H. B. Steinkraus and P. J. Langer, *Mol. Biochem. Parasitol.* **52,** 141 (1992).

[41] S. C. Roberts, K. G. Swihart, M. W. Agey, R. Ramamoorthy, M. E. Wilson, and J. E. Donelson, *Mol. Biochem. Parasitol.* **62,** 157 (1993).

[42] L. L. Button, D. G. Russell, H. L. Klein, E. Medina-Acosta, R. Karess, and W. R. McMaster, *Mol. Biochem. Parasitol.* **32,** 271 (1989).

[43] J. R. Webb, L. L. Button, and W. R. McMaster, *Mol. Biochem. Parasitol.* **48,** 173 (1991).

[44] E. Medina-Acosta, R. E. Karess, H. Schwarz, and D. G. Russell, *Mol. Biochem. Parasitol.* **37,** 263 (1989).

[45] T. O. Frommel, L. L. Button, Y. Fujikura, and W. R. McMaster, *Mol. Biochem. Parasitol.* **38,** 25 (1990).

[46] R. Ramamoorthy, J. E. Donelson, K. E. Paetz, M. Maybodi, S. C. Roberts, and M. E. Wilson, *J. Biol. Chem.* **267,** 1888 (1992).

[47] M. E. Wilson, K. E. Paetz, R. Ramamoorthy, and J. E. Donelson, *J. Biol. Chem.* **268,** 15731 (1993).

[48] P. Schneider, J.-P. Rosat, J. Bouvier, J. Louis, and C. Bordier, *Exp. Parasitol.* **75,** 196 (1992).

[49] D. M. Mosser and P. J. Edelson, *Nature (London)* **327,** 329 (1987).

[50] D. G. Russell and P. Talamas-Rohana, *Immunol. Today* **10,** 328 (1989).

has led to considerable effort to exploit the surface metalloproteinase as a molecularly defined vaccine. Indeed, intraperitoneal vaccination of inbred mice with purified liposome-reconstituted *L. mexicana* leishmanolysin in Freund's complete adjuvant confers significant protection to CBA mice on challenge infection[51]; however, attempts to protect BALB/c mice with recombinant *L. major* leishmanolysin were unsuccessful.[52] More recent vaccination results with leishmanolysin expressed cytoplasmically by genetically engineered *Mycobacterium bovis* are more encouraging.[53] Synthetic peptides of predicted T-cell epitopes from *L. major* leishmanolysin were shown to protect mice from subsequent low-dose challenge infection with promastigotes from different species of *Leishmania*.[54,55] However, the poor capacity of T cells from human cutaneous leishmaniasis patients to recognize purified leishmanolysin *in vitro* suggests that leishmanolysin or leishmanolysin-derived peptides alone are inadequate to vaccinate human populations.[56,57] Indeed, the ability of T cells from many different strains of mice to recognize purified *L. mexicana* promastigote leishmanolysin *in vitro* was not correlated with susceptibility or resistance to challenge infection with that species.[58]

## Methodology

### Organisms and Culture Conditions

Promastigotes of *Leishmania* can be grown axenically to high density. In contrast, studies on the amastigote have been hampered by the difficulty of isolating the intracellular organism free of host macrophage components. The time-consuming gradient centrifugation techniques used to isolate amastigotes from infected mice can allow the partial transformation to the promastigote stage. Now, it is possible to isolate amastigotes rapidly essentially free of contaminating macrophage and lesion debris using gentle

[51] D. G. Russell and J. Alexander, *J. Immunol.* **140**, 1274 (1988).

[52] E. Handman, L. L. Button, and R. W. McMaster, *Exp. Parasitol.* **70**, 427 (1990).

[53] N. D. Connell, E. Medina-Acosta, W. R. McMaster, B. R. Bloom, and D. G. Russell, *Proc. Natl. Acad. Sci. U.S.A.* **90**, 11473 (1993).

[54] A. Jardim, J. Alexander, H. S. Teh, D. Ou, and R. Olafson, *J. Exp. Parasitol.* **172**, 645 (1990).

[55] D. M. Russo, A. Jardim, E. M. Carvalho, P. R. Sleath, R. J. Armitage, R. W. Olafson, and S. G. Reed, *J. Immunol.* **150**, 932 (1993).

[56] C. L. Jaffe, R. Shor, H. Trau, and J. H. Passwell, *Clin. Exp. Immunol.* **80**, 77 (1990).

[57] S. C. F. Mendonça, D. G. Russell, and S. G. Coutinho, *Clin. Exp. Immunol.* **83**, 472 (1991).

[58] J. A. Lopez, H.-A. Reins, R. J. Etges, L. L. Button, W. R. McMaster, P. Overath, and J. Klein, *J. Immunol.* **146**, 1328 (1991).

filtration through polycarbonate filters.[59] Filter-purified amastigotes of *L. major* show no signs of expression of promastigote-specific messenger RNA or proteins.[48] These techniques do, however, require the infection of mice with the parasite several weeks or months prior to isolation. Several reports have described the establishment of axenic amastigotes.[60–63]

We routinely maintain promastigotes at 26° in HOSMEM II medium[64] containing 5% heat-inactivated fetal calf serum. (*Note:* It is essential to screen many batches of serum for optimum growth; those suitable for mammalian cell culture may not support high-density promastigote growth.) For large-scale (50 ml–1 liter) cultivation of promastigotes, a modification of the medium described by Schaefer *et al.*[65] is used (*vide infra*). For some strains of *Leishmania*, fetal bovine serum can be replaced by 300 mg liter$^{-1}$ bovine serum albumin and 4 mg liter$^{-1}$ polyoxyethylenesorbitan monooleate (Tween 80). It is worth noting that all the species of human-isolated *Leishmania* from the reference collection of the London School of Tropical Medicine and Hygiene have been grown in HOSMEM II medium as described.[18]

For the production of leishmanolysin used in our studies, three different strains of *L. major* have been used: the infective strain *L. major* MRHO/ SU/59/Neal P (LV39) and the noninfective *L. major* LEM 513 and *L. major* LRC-L119. The last strain is particularly valuable as a source of homogeneous leishmanolysin in that it does not produce the anionic glycolipid lipophosphoglycan (LPG), which complicates anion-exchange chromatographic procedures, and expresses approximately four times more leishmanolysin at the cell surface than the infective strains.

For the large-scale preparation of leishmanolysin for crystallization trials, the following detailed protocol was developed to produce 30 to 50 mg of leishmanolysin per purification. For smaller amounts of protein, washed promastigotes are lysed directly in Triton X-114.

*Purification of Leishmanolysin*

*Cells and Culture Conditions.* Lipophosphoglycan (LPG)-deficient *Leishmania major* LRC-L119 promastigotes are grown in 1 liter of modified

[59] T. A. Glaser, S. J. Wells, T. W. Spithill, J. M. Pettit, D. C. Humphris, and A. J. Mukkada, *Exp. Parasitol.* **71**, 343 (1990).

[60] P. S. Doyle, J. C. Engel, P. F. Pimenta, P. P. da Silva, and D. M. Dwyer, *Exp. Parasitol.* **73**, 326 (1991).

[61] P. M. Rainey, T. W. Spithill, D. McMahon-Pratt, and A. A. Pan, *Mol. Biochem. Parasitol.* **49**, 111 (1991).

[62] P. A. Bates, C. D. Robertson, L. Tetley, and G. H. Coombs, *Parasitology* **105**, 193 (1992).

[63] S. Eperon and D. McMahon-Pratt, *J. Protozool.* **36**, 502 (1989).

[64] R. L. Berens and J. J. Marr, *J. Parasitol.* **64**, 160 (1978).

[65] F. W. Schaefer III, E. J. Bell, and F. J. Etges, *Exp. Parasitol.* **28**, 465 (1970).

Schaefer's medium[65] containing 5% heat-inactivated fetal bovine serum, 12.5 mg ml$^{-1}$ folic acid, 6.25 mg ml$^{-1}$ hemin (replacing the rabbit erythrocyte lysate), and 20 mg ml$^{-1}$ gentamycin sulfate, in 2-liter baffled Erlenmeyer flasks. The cultures, inoculated to 1 to 2 × 10$^6$ cells ml$^{-1}$, are shaken at approximately 100 rpm on an orbital platform at 26° until maximum cell density is attained (7 × 10$^7$ to 1 × 10$^8$ cells ml$^{-1}$ in 4 to 6 days depending on ambient temperature and inoculum).

Stationary-phase promastigotes are harvested by centrifugation (1000 *g*, 10 min, 4°, Sorvall GS3 rotor) and washed twice in ice-cold 20 m*M* Tris, pH 7.4, 140 m*M* NaCl (Tris-buffered saline or TBS).

*Protein Concentration Determination.* Protein concentrations are determined with the bicinchoninic acid reagent with bovine serum albumin as standard (Pierce, Rockford, IL) using the manufacturer's instructions.

*Isolation of Leishmania Promastigote Membranes.* Washed, pelleted promastigotes are gently resuspended in hypotonic 10 m*M* Tris, pH 8, containing 5 $\mu$g ml$^{-1}$ of the cysteine proteinase inhibitor leupeptin, at 30°, allowed to swell for approximately 15 min, then cooled on ice. The cells are disrupted in an ice-cold 40 ml Dounce homogenizer (10 to 20 strokes, tight glass pestle; Kontes, Vineland, NY). The cells are examined by phase-contrast microscopy during the homogenization procedure to ensure that more than 95% of the cells are disrupted. The homogenate is centrifuged at 10,000 *g* (15 min, 4°, Sorvall SS-34 or GSA rotor). Pelleted material is resuspended in cold TBS by Dounce homogenization, pelleted as described, then stored at −20°.

*Detergent Extraction of Leishmanolysin.* Promastigote membranes obtained from 2–4 × 10$^{12}$ promastigotes, representing 8 to 16 g of total cellular protein, or 4 to 8 g of membrane-associated protein, are resuspended in ice-cold TBS containing 5 $\mu$g ml$^{-1}$ leupeptin to approximately 2 × 10$^9$ cells ml$^{-1}$. Precondensed, antioxidant-treated Triton X-114[66] (see miniprint section of Ref. 66) is then added to a final concentration of 2%, and the suspension stirred at 4° for 1 hr.

*Isolation of Leishmanolysin by Phase Separation in Triton X-114 Solution.* The Triton X-114 extract is centrifuged at 8000 *g* (15 min, 4°, Sorvall GS3 rotor) to pellet insoluble material, composed in *Leishmania* primarily of $\alpha$ and $\beta$ tubulins from the membrane-anchored cold-stable microtubular cytoskeleton and flagellar axoneme. The supernatant is recovered and warmed to 30° to promote clouding of the polyoxyethylene nonionic detergent,[67] then centrifuged at 7000 *g* (20 min, 25°, Sorvall GS3 rotor) both to promote phase separation and to pellet further insoluble material that

[66] C. Bordier, *J. Biol. Chem.* **256,** 1604 (1981).

[67] C. Bordier, *in* "Post-translational Modifications by Lipids" (U. Brodbeck and C. Bordier, eds.), p. 29. Springer-Verlag, Berlin, 1988.

appears after warming. Following centrifugation, the upper detergent-depleted aqueous phase is removed and cooled, and the detergent-enriched lower phase, containing the amphiphilic molecules extracted from the pro-mastigote membranes, is carefully separated from the pelleted material. The aqueous phase is then made to 1% Triton X-114, the detergent phase diluted to 2% detergent with ice-cold TBS, and the phase separation repeated to ensure optimum enrichment of amphiphilic components. Leish-manolysin is recovered nearly quantitatively in the pooled detergent phases (enzymatic activity is not detectable in the aqueous phase or pellet fractions).

*Detergent Exchange and Anion-Exchange Chromatographic Enrichment of Leishmanolysin.* The final detergent phase containing leishmanolysin is diluted to approximately 3% Triton X-114 with 20 m$M$ Tris, pH 8, and subjected to a final round of phase separation as described above. Although less efficient in the absence of salt, this separation step reduces the amount of NaCl present in the sample sufficiently for the leishmanolysin to interact with the anion exchanger. The detergent phase is again diluted to 2% Triton X-114 and made to 1% lauryldodecylamine *N*-oxide (LDAO; Serva, Heidelberg, Germany) to prevent clouding of the Triton X-114.[68]

The sample is immediately applied to a 50-ml radial flow column (Superflow 50, Sepragen, Inc., San Leandro, CA) of Q-Sepharose fast flow (Pharmacia, Freiburg, Germany) equilibrated in 20 m$M$ Tris, pH 8, 3 m$M$ LDAO at 4 to 6 ml min$^{-1}$ at room temperature. The extremely low back pressure of the apparatus allows rapid application of the viscous Triton X-114-containing sample. The column is washed with 7 to 10 bed volumes of 20 m$M$ Tris, pH 8, 10 m$M$ LDAO to remove residual Triton X-114, until the absorbance of the flowthrough at 280 nm reaches a stable baseline (Triton X-114 absorbs strongly at 280 nm, whereas LDAO does not).

Bound macromolecules are then eluted from the equilibrated, washed column with a 3-liter gradient of 50 to 300 m$M$ NaCl in 20 m$M$ Tris, pH 8, 3 m$M$ LDAO at a flow rate of 8 ml min$^{-1}$, collecting fractions of 30 ml. Leishmanolysin is recovered in a relatively narrow peak, well separated from the bulk of the eluted protein. Fractions containing leishmanolysin are identified by enzymatic activity as described below.

Pooled fractions containing leishmanolysin are concentrated by ultrafil-tration (Amicon, Danvers, MA, PM30 membrane) and applied at 1 ml min$^{-1}$ to a 10-ml (1 × 10 cm) column of the strong anion exchanger Mono Q (Pharmacia) equilibrated with 20 m$M$ Tris, pH 8, 3 m$M$ LDAO, at room temperature. Column-bound proteins are eluted at 4 ml min$^{-1}$ with a 290-ml gradient of 50 to 300 m$M$ NaCl in 20 m$M$ Tris, pH 8, 3 m$M$ LDAO,

[68] J. Bouvier, R. J. Etges, and C. Bordier, *J. Biol. Chem.* **260,** 15504 (1985).

collecting fractions of 10 ml. Leishmanolysin elutes in a single well-defined peak at approximately 200 m$M$ NaCl.

Peak fractions are pooled and concentrated by ultrafiltration, then applied to a 1-ml (0.5 × 5 cm) column of Mono Q equilibrated in the same buffer. Bound leishmanolysin is eluted at 1 ml min$^{-1}$ with a 20-ml gradient of 50 to 300 m$M$ NaCl as described for the 10-ml column, collecting fractions of 0.5 ml. Active fractions are recovered, pooled, and concentrated to approximately 1 ml in 20 m$M$ Tris-HCl, pH 7.2.

*Lipase-Mediated Solubilization of Leishmanolysin.* The diacylglycerol of the glycosylphosphatidylinositol anchor of leishmanolysin is enzymatically removed by incubation with the phosphatidylinositol-specific phospholipase C (PIPLC) of *Bacillus thuringiensis* (American Radiolabeled Chemicals, St. Louis, MO). In a microcentrifuge tube, 50 mU of lipase is added to the concentrated, chromatographically purified amphiphilic leishmanolysin. The mixture is incubated at 37° for 60 min, then placed on ice and made to 2% Triton X-114. The tube is warmed to 30° until clouding occurs (2–3 min), then centrifuged (13,000 $g$, 30 sec, room temperature). The upper aqueous phase is made to 1% Triton X-114, the detergent phase diluted to the original volume with ice-cold TBS, and the process repeated. Approximately 70% of the initially amphiphilic leishmanolysin is solubilized by PIPLC treatment (the remaining 30% remains resistant to further lipase digestion).

The aqueous phases are then pooled, diluted 4-fold with 20 m$M$ Tris, pH 8 (without detergent), and applied as previously described to a detergent-free 1-ml (0.5 × 5 cm) column of Mono Q equilibrated in the same buffer. The column is washed with a sufficient volume of buffer to remove the Triton X-114 (monitored at 280 nm), and the bound hydrophilic leishmanolysin is eluted as described for the amphiphilic form.

*Yield.* In a representative purification, 2.3 × 10$^{12}$ L119 promastigotes, representing 9.2 g of cellular protein (of which 2% is represented by the surface metalloproteinase, or 184 mg), yielded 61.4 mg of purified leishmanolysin, of which 70%, or 43 mg, is susceptible to PIPLC solubilization. The overall yield of leishmanolysin is 33.4%, and that of the soluble, PIPLC-digested form, 24%.

*Isotopic Labeling*

Promastigote leishmanolysin is a glycosylphosphatidylinositol (GPI)-anchored proteinase.[11,12,68] Biosynthetic labeling with specific radiolabeled anchor precursors such as ethanolamine, inositol, phosphate, and myristic (or palmitic) acid, in addition to phosphatidylinositol-specific phospholipase C sensitivity, demonstrated that the membrane-bound protein was not anchored to the membrane of the promastigote by a transmembrane sequence of hydrophobic amino acid residues, but rather by a GPI anchor similar to that described for the variant surface glycoproteins of the African trypanosomes.

Log-phase promastigotes are washed twice in cold TBS and resuspended to a concentration of $10^7$ cells ml$^{-1}$ in methionine-free RPMI 1640 medium (GIBCO, Grand Island, NY) containing 10 $\mu$Ci of [$^{35}$S]methionine (Amersham Radiochemical Center, Amersham, UK) or to a concentration of $3 \times 10^7$ cells ml$^{-1}$ in serum-free HOSMEM II containing 100 $\mu$Ci of [9,10($n$)-$^3$H]myristic (tetradecanoic) or [$^3$H]palmitic (dodecanoic) acid (Amersham), which is prepared just before use by adding an appropriate volume of the isotopically labeled fatty acid in toluene to a sterile glass culture vessel, drying it with a stream of filtered N$_2$, and adding sterile medium. Twenty minutes after the inoculation of the radioactive fatty acid-containing medium with Leishmania promastigotes ($\sim$1 $\times$ 10$^6$ ml$^{-1}$), sterile heat-inactivated fetal calf serum is added to a final concentration of 5%, and the incubation is allowed to proceed overnight at 26°.[11,68]

Biosynthetic incorporation of [$^3$H]glucosamine (10 $\mu$Ci ml$^{-1}$) or [$^{32}$P]phosphoric acid (20 $\mu$Ci ml$^{-1}$) is accomplished in serum-free RPMI 1640 medium minus glucose or phosphate, respectively. Biosynthetic incorporation of $^{65}$ZnCl$_2$ (Amersham) is performed in 50 ml of HOSMEM II containing 100 $\mu$Ci (60 Ci mol$^{-1}$) of the isotope.[20] For each substrate, incorporation is allowed to occur for 18 hr or more, during which time the cell density doubles. Conditions for surface radioiodination of Leishmania promastigotes using IodoGen (Pierce) have been described.[18]

Labeled cells are washed 5 times in cold TBS, then lysed directly in 2% Triton X-114 in TBS and subjected to phase separation. Aliquots of the different detergent phases obtained for each specific labeling are analyzed by sodium dodecyl sulfate–polyacrylamide gel electrophoresis (SDS–PAGE) and fluorography.

*Enzymatic Assays*

Several assays to detect the proteolytic activity of leishmanolysin have been developed. Zymographic techniques that exploit immobilized or insoluble substrates (generally high molecular weight proteins such as gelatin or fibrinogen copolymerized in a polyacrylamide or agarose gel matrix) and the detergent stability of leishmanolysin allow the detection of the enzyme following separation by SDS–PAGE. Such qualitative analysis allows the detection of proteolytic activity in complex samples[17,18] and the assay has been standardized with known quantities of purified leishmanolysin to generate quantitative results.[48]

*Zymographic Techniques.* Based on the method of Lacks and Springhorn,[69] ovine fibrinogen (240 $\mu$g ml$^{-1}$ in the gel) or gelatin (0.08%) is copolymerized in the polyacrylamide gel matrix. Electrophoresis in the presence of SDS is performed without sample denaturation by heating or reduction. After electrophoresis, the gels are washed to remove excess detergent and then allowed to incubate in a buffer of appropriate pH for the proteinase of interest, for several hours or overnight at room temperature, with constant agitation. The gels are stained in Coomassie blue and destained to reveal clear bands generated by proteinase activity on a blue-stained gelatin or fibrinogen background. The inclusion of a sample containing a known amount of enzyme allows quantitative densitometry to be performed. Substrate-free gels can, after electrophoresis, be incubated in diffusible fluorogenic substrate solutions (i.e., 7-amino-4-methyl-coumarin-derivatized peptides) and photographed under ultraviolet illumination. This powerful and, in some cases, highly specific technique has been applied to numerous enzymes.[70]

*Assays in Solution.* Numerous soluble protein or peptide substrates have been used to detect proteolytic activity in solution. Azocasein (Sigma, St. Louis, MO) at 20 mg ml$^{-1}$ in an appropriate buffer is an economical and readily available substrate for many proteolytic enzymes. After incubation with the proteinase, low molecular weight acid-soluble azopeptides are separated from intact azocasein by precipitation of the latter with 2.5% (final concentration) trichloroacetic acid (TCA). After centrifugation, the absorbance of the acid-soluble peptides is determined at 366 nm.[17,19] However, the technique is relatively insensitive, is not enzyme or peptide-bond specific, and cannot be used at acidic pH owing to the precipitation of azocasein below pH 5.

The convenient milk assay, using intact casein micelles as a substrate, is compared with substrate-gel analysis in Fig. 1. Nonfat powdered milk is dissolved at 1.5% (w/v) in TBS, pH 7.5, and the solution is centrifuged for 5 min at 10,000 $g$ prior to use. The assay is then performed by pipetting 10-$\mu$l aliquots of proteinase solution into the wells of a 96-well microtiter plate containing 0.1 ml per well of the milk solution and incubating at 37°. The decrease in light scattering (turbidity) of the casein micelles is measured at 405 nm in an ELISA (enzyme-linked immunosorbent assay) reader after different periods of time. The milk assay, however, suffers from the same limitations as assays with azocasein.

Although the milk assay has been applied to the investigation of different proteinases (P. Schneider and C. Bordier, unpublished), it proved to

---

[69] S. A. Lacks and S. S. Springhorn, *J. Biol. Chem.* **255,** 7467 (1980).
[70] O. Gabriel and D. M. Gersten, *Anal. Biochem.* **203,** 1 (1992).

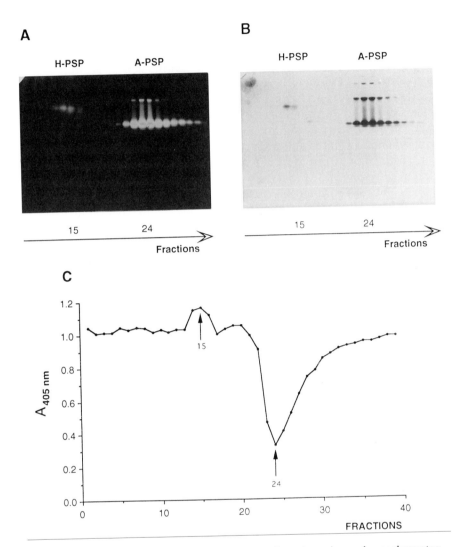

FIG. 1. Enzymatic detection of leishmanolysin in fractions after anion-exchange chromatography, showing a comparison of fibrinogen–SDS–PAGE analysis and the milk assay. Fractions collected during elution of the strong anion exchanger Mono Q are assayed by substrate-containing and normal SDS–polyacrylamide gels. Ten microliters from each fraction is mixed with 3 $\mu$l of 5× concentrated sample buffer and loaded onto 12% polyacrylamide gels containing or not 240 $\mu$g ml$^{-1}$ copolymerized fibrinogen. The samples are neither heated nor reduced. After electrophoresis, the fibrinogen-containing gel is incubated overnight (16 hr) in 50 ml of TBS, and later stained with Coomassie brilliant blue and destained. The enzymatic activity is revealed as clear bands on a blue-stained fibrinogen background (A). The second gel, which does not contain fibrinogen, is simply stained with Coomassie blue and destained (B). For the milk assay, 10-$\mu$l aliquots of the same fractions are pipetted into the wells of a 96-well microtiter plate containing 0.1 ml of 1.5% (15 mg ml$^{-1}$) nonfat milk in TBS. The remaining turbidity of the casein micelles is measured after 60 min of incubation at 405 nm in an ELISA reader (C).

be very specific for leishmanolysin, presumably owing to the ability of the GPI anchor of the enzyme to insert into the casein micelle. *Leishmania* cysteine proteinases are not detected by this method, although they can be readily demonstrated in the same fractions with suitable chromogenic or fluorogenic substrates, or gelatin–SDS–PAGE. In practice, this simple assay rapidly identifies leishmanolysin-containing fractions during chromatographic purification. It is quite sensitive: as little as 2 $\mu$g ml$^{-1}$ of leishmanolysin, which represents 0.32 pmol of enzyme in the assay, can be detected. Amphiphilic (A) and hydrophilic (H) forms of leishmanolysin (the latter generated by the action of phosphatidylinositol-specific phospholipase C) do not act identically on casein micelles (Fig. 1C): fractions 14–16, which contain the H-leishmanolysin (H-PSP) induce an initial increase in the absorbance at 405 nm. In some cases, it was even possible to observe precipitation of the casein. After prolonged incubation, however, the precipitated casein is eventually cleared, indicating that it is hydrolyzed by the soluble proteinase. Precipitation of casein was never observed with the amphiphilic form of the enzyme (A-PSP).

To design and synthesize a better substrate for leishmanolysin, the peptide bond preference was investigated using natural and synthetic peptides of known amino acid sequence. Peptides 11–25, 39–53, and 81–104 from horse cytochrome *c* (Cyt *c*) together with peptides 170–182 from CW3 HLA antigen and 83–94 from the $\beta$ chain of HLA-DO were described elsewhere.[35] Peptides 94–102 and 94–102 with a serine in position 98 from Cyt *c*, as well as the nonapeptide 97–105 and peptide [−4]–[+5] derived from the leishmanolysin sequence,[38] were synthesized using the solid-phase method.[71] The other peptides (glucagon, oxidized insulin B chain) were obtained commercially.

All peptide substrates presented in Fig. 2, except for the VSG peptide 36,[36] are incubated at a concentration of 250 $\mu$M in TBS with the soluble form of leishmanolysin (HPSP) at a concentration of 8 n$M$ (molar ratio of enzyme to substrate, 1:3 $\times$ 10$^4$). At 0, 1, 2, 4, 8, and 16 min after the addition of the enzyme, 20-$\mu$l aliquots are removed from the reaction vial, mixed with 10 $\mu$l of 0.1 $M$ HCl, and heated for 2 min at 95° to inactivate the enzyme. Hydrolysis products are chromatographically separated by high-performance liquid chromatography (HPLC) on a Waters $\mu$Bondapack C$_{18}$ reversed-phase column (Waters Associates, Milford, MA) equilibrated with 0.1% (v/v) trifluoroacetic acid in distilled water. From each sample, 15 $\mu$l is applied to the column, and the retained peptides are eluted with a 25-ml, 0 to 40% linear gradient of acetonitrile containing 0.1% (v/v)

---

[71] E. Atherton and R. C. Sheppard, *in* "Solid Phase Peptide Synthesis, a Practical Approach" (D. Rickwood and B. D. Hames, eds.), IRL Press, Oxford, 1989.

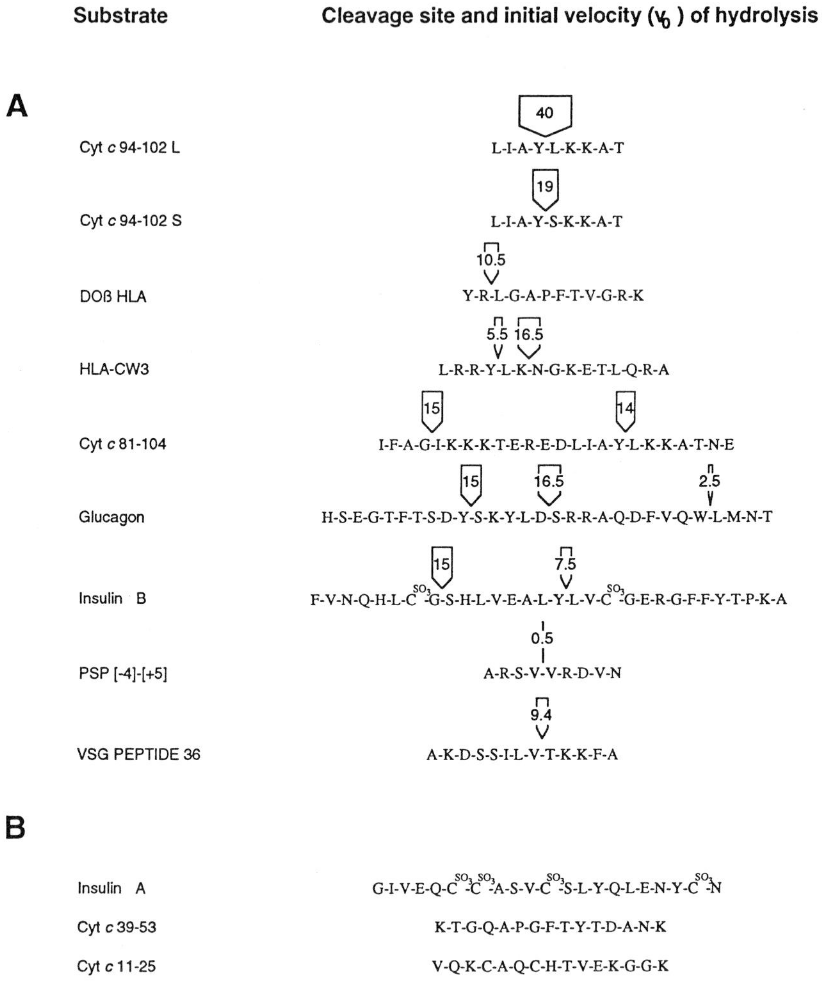

| Substrate | Cleavage site and initial velocity ($v_0$ ) of hydrolysis |
|-----------|------------------------------------------------------------|

**A**

Cyt *c* 94-102 L

$\boxed{40}$

L-I-A-Y-L-K-K-A-T

Cyt *c* 94-102 S

$\boxed{19}$

L-I-A-Y-S-K-K-A-T

DOβ HLA

10.5

Y-R-L-G-A-P-F-T-V-G-R-K

HLA-CW3

5.5  16.5

L-R-R-Y-L-K-N-G-K-E-T-L-Q-R-A

Cyt *c* 81-104

$\boxed{15}$                         $\boxed{14}$

I-F-A-G-I-K-K-K-T-E-R-E-D-L-I-A-Y-L-K-K-A-T-N-E

Glucagon

$\boxed{15}$        16.5                    2.5

H-S-E-G-T-F-T-S-D-Y-S-K-Y-L-D-S-R-R-A-Q-D-F-V-Q-W-L-M-N-T

Insulin  B

$\boxed{15}$            7.5

F-V-N-Q-H-L-C$^{SO_3}$-G-S-H-L-V-E-A-L-Y-L-V-C$^{SO_3}$-G-E-R-G-F-F-Y-T-P-K-A

PSP [-4]-[+5]

0.5

A-R-S-V-V-R-D-V-N

VSG PEPTIDE 36

9.4

A-K-D-S-S-I-L-V-T-K-K-F-A

**B**

Insulin  A

G-I-V-E-Q-C$^{SO_3}$-C$^{SO_3}$-A-S-V-C$^{SO_3}$-S-L-Y-Q-L-E-N-Y-C$^{SO_3}$-N

Cyt *c* 39-53

K-T-G-Q-A-P-G-F-T-Y-T-D-A-N-K

Cyt *c* 11-25

V-Q-K-C-A-Q-C-H-T-V-E-K-G-G-K

Fig. 2. Proteolytic activity of leishmanolysin on peptides. Time course experiments were performed using 250 $\mu M$ of each substrate and 8 n$M$ of leishmanolysin at 37° in TBS at pH 7.5. Sites of hydrolysis are indicated by arrows, the widths of which are proportional to the initial rate of hydrolysis $v_0$. The numbers within the arrows represent $v_0$ and are given in moles of peptide bonds cleaved per second per mole of enzyme (A). Peptides used as substrates that were not hydrolyzed, even after extended incubation with leishmanolysin, are also shown (B).

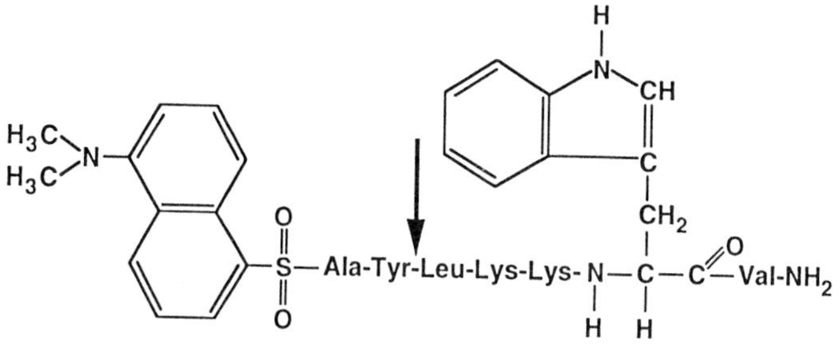

Fig. 3. Structure of fluorogenic peptide substrate designed for leishmanolysin. The indole fluorescence of the tryptophan residue is efficiently quenched through resonance energy transfer by an N-terminal dansyl group located five amino acid residues away. The heptapeptide is cleaved by leishmanolysin between the tyrosine and residues (arrow) with a $k_{cat}/K_m$ ratio of $8.8 \times 10^6$ mol$^{-1}$ sec$^{-1}$.

trifluoroacetic acid, at a flow rate of 1 ml min$^{-1}$. The eluate is monitored at 214 nm, and single elution peaks are collected for subsequent amino acid composition analyses. Peak areas are recorded with a D-2000 Hitachi integrator (Merck, Darmstadt, Germany).

Because the rate of appearance of the products is constant over the first 4 to 8 min of the reaction when less than 20% of the substrate is hydrolyzed, the initial velocity ($v_0$) of appearance for each product can be measured in time course experiments. For amino acid analyses, isolated products are lyophilized, then hydrolyzed in 100 $\mu$l of constant-boiling 6 $M$ HCl (Pierce), containing 1% (v/v) phenol, for 16 hr at 110° under vacuum. Analyses of phenylisothiocyanate-derivatized amino acids are performed as described.[35,72] Although this strategy allowed us to investigate the substrate specificity of leishmanolysin, it is time-consuming and requires the synthesis and analysis of numerous peptides of known primary structure. The results of the study, illustrated in Fig. 2, identified an efficiently cleaved peptide substrate, H$_2$N-L-I-A-Y-L-K-K-A-T-COOH, derived from the cytochrome *c* peptide 81–104. The peptide is cleaved by leishmanolysin between the tyrosine and leucine residues with a $k_{cat}/K_m$ ratio of $1.8 \times 10^6$ mol$^{-1}$ sec$^{-1}$. The promising kinetic values obtained with the peptide[35] led us to design and synthesize a fluorescent peptide substrate to exploit the radiationless energy transfer (RET) methodology (Fig. 3).[37] Such improved methods have been applied to a variety of proteolytic enzymes.[73,74]

[72] R. L. Heinrickson and S. C. Meredith, *Anal. Biochem.* **136**, 65 (1984).
[73] J. R. Petithory, F. R. Masiarz, J. F. Kirsch, D. V. Santi, and B. A. Malcolm, *Proc. Natl. Acad. Sci. U.S.A.* **88**, 11510 (1991).
[74] D. J. Matthews and J. A. Wells, *Science* **260**, 1113 (1993).

## Fluorogenic peptide substrate assay on Thin Layer Chromatography

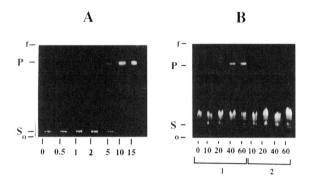

## [¹²⁵I]-peptide substrate assay on Thin Layer Chromatography

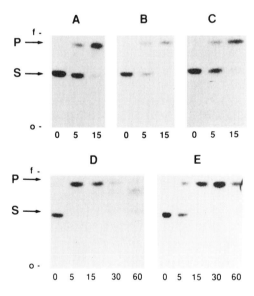

Fig. 4. Peptide substrate assays analyzed by thin-layer chromatography. (Top) Glutaralde-hyde-fixed promastigotes of *L. major* LV39 (B) were incubated at a concentration of $10^7$ cells $ml^{-1}$ with the fluorogenic substrate. Aliquots were collected at the indicated times (minutes) and mixed with 0.1 *M* HCl to stop the enzymatic reaction (group 1). After 10 min, cells were removed form one-half of the reaction mixture (group 2), and the supernatant was incubated as the other half with cells. Hydrolysis products were then analyzed by TLC on silica gel plates as described in the text. A parallel control experiment was performed with purified leishmanolysin (A). o, Origin; S, substrate (Dns-AYLKKWV-NH₂); P, product of hydrolysis

Both the internally quenched fluorogenic substrate and the underivatized peptide can be used to compare the proteolytic activities that are specifically due to leishmanolysin at the surface of intact, fixed, or living promastigotes of several species of *Leishmania,* or purified leishmanolysin in solution. The products of proteolysis are separated by thin-layer chromatography (TLC) using either the iodinated form of the synthetic peptide substrate or the fluorescent oligopeptide. Interestingly, the presence of the bulky iodine atom on the tyrosine residue contributing to the scissile peptide bond in the peptides does not affect the susceptibility of the bond to cleavage by leishmanolysin.[28]

*Thin-Layer Chromatographic Analyses.* Substrates [100 $\mu M$; $7 \times 10^5$ counts min$^{-1}$ (cpm) ml$^{-1}$ of iodinated peptide mixed with cold peptide] and leishmanolysin (40 n$M$) are mixed in a final volume of 100 $\mu$l. At the times indicated in Fig. 4, 10-$\mu$l aliquots are removed from the reaction vial and mixed with 3 $\mu$l of 0.1 $M$ HCl to inactivate the enzyme.

Washed living promastigotes or promastigotes which have been fixed for 30 min in 0.1% glutaraldehyde in Hanks' balanced salt solution[19] are resuspended at a concentration of $10^7$ cells ml$^{-1}$ in Hanks' balanced salt solution containing 100 $\mu M$ peptide substrate and then incubated at 26° with constant agitation. After 10 min, one-half of the reaction was removed and centrifuged for 3 min at 13,000 $g$ to eliminate the promastigotes, and the supernatant was further incubated at 26°. At appropriate times, 20-$\mu$l aliquots are removed and mixed with 10 $\mu$l of 0.1 $M$ HCl to stop the reaction.

Thin-layer chromatographic separations are performed on silica gel 60 plates (Merck) using 1-butanol/acetic acid/water in a 8 : 1 : 1 ratio (by volume) as solvent. The fluorescent dansyl group is visualized by ultraviolet radiation at 362 nm and photographed on Polaroid 667 film with a yellow filter (Kodak, Rochester, NY, No. 12) (Fig. 4, top). For analysis of the leishmanolysins of different species of *Leishmania,* iodinated peptides generated by the surface metalloproteinase are analyzed by TLC on cellulose plates and resolved in butanol/pyridine/acetic acid/water (97 : 75 : 15 : 60, by volume).[75] The results of the analyses are shown in Fig. 4. Interestingly, all species of

[75] A. Takeda and A. L. Maizel, *Science* **250,** 676 (1990).

---

(Dns-AY); f, migration front. (Bottom) The model peptide Cyt *c* 94–102 [98]L was iodinated and incubated with glutaraldehyde-fixed promastigotes ($10^8$ cells ml$^{-1}$) from several species of *Leishmania* and *Herpetomonas samuelpessoai.* Aliquots were taken at different time points (0, 5, 15, 30, and 60 min), and the products of hydrolysis were analyzed by TLC as described in the text. o, Origin; f, migration front; S, substrate (LIA[$^{125}$I]YLKKAT); P, product (LIA[$^{125}$I]Y). (A) *Leishmania amazonensis;* (B) *L. infantum;* (C) *L. mexicana;* (D) *L. major;* (E) *H. samuelpessoai.*

*Leishmania* tested so far, as well as *Herpetomonas samuelpessoai*,[28] hydrolyzed the peptide substrates at the same peptide bond, suggesting that all leishmanolysins have the same substrate specificity (Fig. 4, bottom).

## Concluding Remarks

Leishmanolysin has been and continues to be the subject of many investigations. Despite the large amount of information collected to date,[10] the precise role of the enzyme in the life cycle of *Leishmania* remains a subject of speculation. Like many other metalloendopeptidases, leishmanolysin has a rather defined, but not strict, substrate specificity. Although leishmanolysin is encoded by three families of heterogeneous genes which are differentially expressed during the life cycle of the parasite, the relevance of the regulated expression remains largely unknown. Antigenic variation is unknown in *Leishmania*. Do different proteinases encoded by the different gene families have distinct substrate specificities? This would seem unlikely, given the results shown in Fig. 4. Perhaps subtle differences in structure around the active site broaden the pH range over which the enzyme functions, which would be a clear advantage to an organism exposed to the alkaline midgut of the insect vector and the acidified phagolysosome of its mammalian host macrophage.

Little is known concerning the activity of leishmanolysin in the amastigote stage, and there is still debate regarding the pH optimum of the enzyme. Leishmanolysin expression in amastigotes is clearly reduced compared to that in the promastigote stage.[31,48,68] The identification of an apparently soluble leishmanolysin in the amastigotes of *L. mexicana*[31] may provide confirmation for the predicted absence of a GPI anchor from the gene product encoded by the C1 leishmanolysin gene of that species.[32] The amastigote enzyme has a lower optimum of pH 5.5–6 and is located intracellularly within the lysosomal compartment of the parasite (megasome).[31] The lower pH optimum reported for the *L. mexicana* amastigote leishmanolysin is similar to that of leishmanial cysteine proteinases, located in the same subcellular compartment, and is still higher than the predicted lower pH limit for metalloproteinase activity determined by the $pK_a$ value of 5.3 of the catalytically critical glutamic acid residue in the active site, which must remain unprotonated in order to polarize the molecule of water bound to the active-site zinc atom. The pH optimum for the soluble amastigote leishmanolysin is still much higher than the value of pH 4 reported for the *L. amazonensis* promastigote enzyme.[7,22] The presence of leishmanolysin homologs at the surface of the monogenetic kinetoplastid parasites of insects *Herpetomonas samuelpessoai* and *Crithidia fasciculata*[18,27,28] suggests that leishmanolysin is likely to play a role in the insect vector, and not in the mammalian host. Further studies on the biosynthesis and substrate specificity of the soluble amastigote leishmanolysin are necessary to clarify its role in the parasite life cycle.

The observation that all leishmanolysins so far examined have the same peptide bond preference in a limited range of substrates does not reflect the heterogeneity observed in the sequences of the different genes analyzed to date. Moreover, the highest degree of similarity observed among the different genes is located in the part encoding the putative active site of the enzyme, although, even there, some nonconservative amino acid substitutions were observed.[41] Unrelated metalloproteinases share few similarities in sequence or secondary structure outside of the highly conserved and extended zinc-binding domain of the active site; indeed, several members have even inverted the conserved catalytic site![25] The addition of the three-dimensional structures of the crayfish digestive enzyme astacin,[33] the snake venom metalloproteinase adamalysin II,[76] *Pseudomonas* alkaline protease,[77] and human collagenase[26] to the existing solved structures of small, soluble bacterial metalloproteinases has allowed the description of distinct families of metalloproteinases[25,26,33] which differ in the mode of zinc coordination from that of thermolysin. Currently, refinement of the three-dimensional structure of *L. major* L119 leishmanolysin is in progress.[78]

The detailed three-dimensional structure of the *Leishmania* metalloproteinase should allow us to improve existing synthetic fluorescent oligopeptide substrates and inhibitors to help explore the role of the surface metalloproteinase *in vivo*. Molecular replacement studies using the sequences of different leishmanolysin genes of *Leishmania* and *Crithidia*[27,38–41] may help to explain the differences in substrate specificities and pH optima reported for the enzymes. Additionally, it will represent the first of the family of large membrane-bound metalloproteinases to be crystallized and characterized, permitting comparison with the pharmacologically significant mammalian enkephalinases and matrix metalloproteinases.

## Acknowledgments

We thank Dr. Theresa Glaser for careful reading of the manuscript, and Thierry Bornand and Marcel Allegrini for helping in the preparation of figures. This work received financial support from UNDP/World Bank/World Health Organization Special Programme for Research and Training in Tropical Diseases (ID 900447) and from the Swiss National Fund (31-30857.91). Special thanks are due to the Structural Biology Summer Visitors Programme of the European Molecular Biology Laboratory, Heidelberg, for providing the opportunity for R.E. to participate with Edith Schlagenhauf and Peter Metcalf in the crystallographic analysis of leishmanolysin.

[76] F.-X. Gomis-Rüth, L. F. Kress, and W. Bode, *EMBO J.* **12,** 4151 (1993).
[77] U. Baumann, S. Wu, K. M. Flaherty, and D. B. McKay, *EMBO J.* **12,** 3357 (1993).
[78] E. Schlagenhauf, R. Etges, and P. Metcalf, *Proteins: Structure, Function, and Genetics* **21,** (in press).

# [38] Immunoglobulin A-Metallo-Type Specific Prolyl Endopeptidases

By ANDREW G. PLAUT and ANDREW WRIGHT

## Introduction

Immunoglobulin A (IgA) proteinases are bacterial endopeptidases produced by infectious agents in the genera *Streptococcus, Neisseria, Haemophilus, Ureaplasma, Clostridium, Capnocytophaga,* and *Bacteroides.* The only known substrate is human IgA, the antibody in secretions that bathe human mucus membranes. More specifically, nearly all IgA proteinases cleave only the IgA1 subclass of secretory and serum IgA. The proteinases cleave one peptide bond in the IgA1 heavy polypeptide chain, yielding intact antigen-binding Fab domains and Fc domains of the antibody protein. Because bacterial colonization possibly leading to infection usually begins on mucosal surfaces, proteolytic disassembly of mucosal antibodies is likely to be a factor in virulence. Elsewhere in this series (Volume 244, [10]) we describe the serine-type IgA-specific prolyl endopeptidases, the mechanism by which the enzymes are secreted from gram-negative bacterial cells, the characteristics of IgA1 that are the basis for its enzyme susceptibility, and the diseases caused by IgA proteinase-positive bacteria. Here we describe only the metallo-type IgA proteinases, IUBMB classification EC 3.4.24.13 and peptidase evolutionary family M27, as classified by Rawlings and Barrett[1] (see [13] in this volume).

A definitive assignment of the IgA proteinase of *Streptococcus sanguis* [American Type Culture Collection (ATCC), Rockville, MD, Cat. No. 10556] as a metalloproteinase is based on analysis of the deduced primary amino acid sequence, as discussed below. Metallo-type IgA proteinases of other bacteria are so classified on the basis of their sensitivity to inhibition by metal chelators (Table I).

Several microorganisms that cocolonize the oral cavity with IgA proteinase-positive species are themselves enzyme-negative; these include *Streptococcus gordonii, S. mitis* (biovar 2), *S. salivarius,* and *S. mutans.* The taxonomy of these and other viridans-type streptococci in the oral cavity and pharynx has been materially advanced by a newer classification scheme of Kilian *et al.,*[2] which should be consulted. *Proteus mirabilis* produces an

---

[1] N. Rawlings and A. J. Barrett, *Biochem. J.* **290,** 205 (1993).
[2] M. Kilian, L. Mikkelsen, and J. Henrichsen, *Int. J. Syst. Bacteriol.* **39,** 471 (1989).

TABLE I

BACTERIA PRODUCING METALLO-TYPE IMMUNOGLOBULIN A PROTEINASES

| Bacteria | Comments | Refs. |
|---|---|---|
| *Streptococcus sanguis* | All strains enzyme-positive; metal binding site defined | *a* |
| *Streptococcus pneumoniae* | All strains positive; inhibition by 5 m$M$ EDTA$^d$ | *b, c* |
| *Streptococcus oralis* | All strains positive | *e* |
| *Streptococcus mitis* | Biovar 1 is 20% enzyme-positive; biovar 2 is negative | *e* |
| *Capnocytophaga ochracea, C. gingivalis, C. sputigena* | All enzymes inhibited by 6 m$M$ EDTA and 10 m$M$ bathocuproine disulfonate (BCDS) | *f* |

[a] J. V. Gilbert, A. G. Plaut, and A. Wright, *Infect. Immun.* **59**, 7 (1991).
[b] C. Male, *Infect. Immun.* **26**, 254 (1979).
[c] M. Kilian, J. Mestecky, R. Kulhavy, M. Tomana, and W. T. Butler, *J. Immunol.* **124**, 2596 (1980).
[d] M. Proctor and P. J. Manning, *Infect. Immun.* **58**, 2733 (1990).
[e] M. Kilian, L. Mikkelsen, and J. Henrichsen, *Int. J. Syst. Bacteriol.* **39**, 471 (1989).
[f] V. G. Frandsen, J. Reinholdt, and M. Kilian, *Infect. Immun.* **55**, 631 (1987).

EDTA-sensitive proteinase that cleaves immunoglobulins, but this is not regarded as an IgA proteinase because it has substrates other than IgA.[3]

## Cleavage Sites of Immunoglobulin A Proteinases in Human IgA1

Each IgA1 proteinase attacks a single peptide bond in the hinge region of the human IgA1 heavy polypeptide chain (Fig. 1). All streptococcal species including *S. sanguis*,[4] *S. pneumoniae*,[5] and other oral streptococci cleave the same peptide bond, Pro$^{227}$-Thr$^{228}$ (numbering according to Tsukida *et al.*[6]), as shown in Fig. 1. The IgA proteinase of *Capnocytophaga* species cleaves Pro$^{223}$-Ser$^{224}$, the same bond known to be cleaved by *Bacteroides melaninogenicus* IgA proteinase.[7] Cleavage of IgA1 can be accompanied by deglycosylation of the substrate by associated bacterial glycosidases,[8,9] most notably seen with *S. pneumoniae* and *S. oralis* proteinases.

[3] M. Kerr, *Biochem. J.* **271**, 285 (1990).
[4] A. G. Plaut, F. V. Gilbert, M. S. Artenstein, and J. D. Capra, *Science* **190**, 1103 (1975).
[5] M. Kilian, J. Mestecky, R. Kulhavy, M. Tomana, and W. T. Butler, *J. Immunol.* **124**, 2596 (1980).
[6] Y. Tsuzukida, C. C. Wang, and F. W. Putnam, *Proc. Natl. Acad. Sci. U.S.A.* **76**, 1104 (1979).
[7] S. B. Mortensen and M. Kilian, *Infect. Immun.* **45**, 550 (1984).
[8] E. V. G. Frandsen, J. Reinholdt, and M. Kilian, *Infect. Immun.* **55**, 631 (1987).
[9] J. Reinholdt, M. Tomana, S. B. Mortensen, and M. Kilian, *Infect. Immun.* **58**, 1186 (1990).

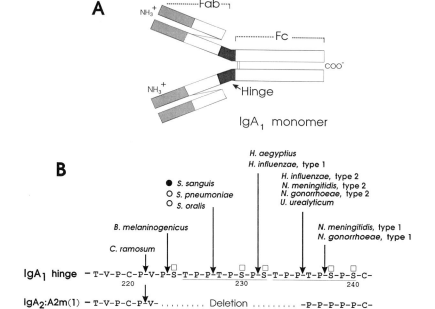

Fig. 1. (A) Diagram of monomeric human IgA1 protein showing variable regions (light shading) and constant regions (nonshaded and hinge) of the heavy and light polypeptide chains. The Fab and Fc fragments are produced by IgA1 proteinase cleavage at the heavy chain hinge. (B) Primary sequence of the IgA1 hinge [deep shaded in (A)] showing peptide bonds cleaved by each IgA1 proteinase. A solid circle denotes the *Streptococcus sanguis* enzyme whose assignment to the metalloproteinase class is most clearly documented; open circles identify proteinases provisionally assigned to that class. The duplicated octapeptide in IgA1 is underlined, and open squares at serine residues represent O-linked glycosylation sites. Both IgA2 allotypes have the hinge deletion that confers resistance to all but the *Clostridium ramosum* proteinase. Amino acid numbers are those previously published [Y. Tsuzukida, C. C. Wang, and F. W. Putnam, *Proc. Natl. Acad. Sci. U.S.A.* **76,** 1104 (1979)].

## Synthesis and Secretion of Metallo-Type Immunoglobulin A Proteinases

The secretion mechanism of the serine-type IgA proteinases in gram-negative species is outlined elsewhere in this series (Volume 244, [10]). The means by which metallo-type streptococcal enzymes are secreted from the wild-type cell is not yet known.

## Specificity

The IgA proteinases cleave both serum and secretory human IgA, and their specificity is, with one exception, confined to IgA1 proteins. The allotypes of $IgA_2$ [A2m(1) and A2m(2)] are not substrates because they lack the critical 13-residue stretch containing most cleavage sites in IgA1 (Fig. 1). IgA1 and $IgA_2$ proteins are, however, found in serum and secretions of all normal persons. With the exception of IgA1 proteins of great apes (gorilla, chimpanzee), there have been no natural substrates identified for the metallo-type IgA proteinases. The $K_m$ of the *S. sanguis* proteinase for human IgA1 is $5.5 \times 10^{-6}\,M$.[10] The enzyme is active over a relatively broad pH range, pH 5.5–7.5; the cleavage rate of human IgA1 is maximal at pH 6–7.[10]

As seen in Fig. 1, proline invariably contributes the carboxyl group to the bond cleaved by each proteinase. The occurrence of alternative cleavage sites among serine-type IgA proteinases of *Haemophilus* and *Neisseria* species and the structural basis for enzyme type specificity are discussed elsewhere in this series (Volume 244, [10]). The five O-linked oligosaccharide side chains of the hinge region in IgA (Fig. 1) do not influence IgA proteinase specificity, but there is some suggestion that initial substrate recognition or the rate of IgA1 cleavage may be dependent on glycosylation; removal of hinge sialic acids increases the rate of IgA cleavage, whereas more extensive substrate deglycosylation appears to decrease the cleavage rate.[9] All IgA1 proteins are substrates despite known carbohydrate heterogeneity, suggesting that a major influence on IgA proteinase cleavage is unlikely.

### Assay of Immunoglobulin A Proteinases

Assays of the highly specific IgA proteinases are dependent on human IgA proteins as substrates. The purification of human serum and/or secretory IgA for this purpose and methods of qualitative and quantitative assay are provided elsewhere (this series, Volume 244, [10]). Our laboratory uses the same assay techniques for all IgA proteinases, regardless of the catalytic mechanism. A novel assay based on the ability of IgA-binding proteins on group A streptococci to capture IgA1 cleavage fragments has also been reported.[11]

### Purification of *Streptococcus sanguis* Immunoglobulin A Proteinase

Purification strategies for the IgA proteinases typically begin with culture of the bacterial cells in liquid media and isolation of the proteinase from the cell-free supernatant. Although extracellular secretion and the large size of IgA proteinases are advantages in purification, protein levels

---

[10] A. G. Plaut, J. V. Gilbert, and I. Heller, *Exp. Med. Biol.* **107**, 489 (1978).
[11] L. Lindahl, C. Schalen, and P. Christensen, *J. Clin. Microbiol.* **13**, 991 (1981).

in media are low. Enzymes can also be purified from recombinant *iga*[+] bacteria, with the important caveat that even active enzyme from a foreign environment may be structurally incomplete relative to the natural form.[12]

A purification scheme for the *S. sanguis* ATCC 10556 metallo-type IgA proteinase has been published[9]; our laboratory uses that method with minor modifications. Bacterial cells are cultured in liquid Todd–Hewitt medium for 24 hr at 37°, and cell-free medium is obtained by centrifugation. Crude precipitates of IgA proteinase are made by 60% ammonium sulfate precipitation (390 g/liter). After 2 hr to allow complete salting out, the precipitate is redissolved in 0.1 $M$ phosphate buffer, pH 7.0, containing 0.1% (w/v) sodium azide and applied to a 1.6 × 50 cm column for size-exclusion chromatography on Superose 12 (Pharmacia, Uppsala, Sweden) or equivalent chromatographic material. Proteins are eluted with the same buffer at 1 ml/min, and eluted enzyme is monitored by qualitative assay.

We have purified analytical amounts of *S. sanguis* IgA proteinases using an antibody raised to a recombinant tandem repeat segment of the enzyme protein (see below). The IgG is purified from the rabbit serum by affinity binding to protein A-Sepharose (Bio-Rad, Richmond, CA). The IgG is eluted with 0.1 $M$ sodium citrate buffer, pH 3.0, concentrated by positive pressure dialysis, and linked to Affi-Prep 10 Gel (Bio-Rad) according to the manufacturer's protocol. Crude *S. sanguis* proteinase obtained from culture supernatants as described above is resuspended in 50 m$M$ Tris-HCl buffer, pH 7.5, containing 0.1% (w/v) sodium azide. After dialysis against that buffer, enzyme is applied to the antibody column and washed with 10 bed volumes 10 m$M$ sodium phosphate buffer, pH 6.8. Enzyme can be eluted from the column with 100 m$M$ glycine buffer, pH 2.5. The enzyme activity is restored by bringing the solution to pH 7–8 by adding 1 $M$ sodium phosphate buffer, pH 8.0 (usually about 1/20, v/v).

Storage

Reduction of activity of purified and crude IgA proteinase stored in assay and purification buffers occurs slowly (weeks to months) at 4°; frozen enzymes are stable for at least 1 year.

Structure and Catalytic Mechanism

A complete nucleotide and deduced amino acid sequence for the IgA proteinase of *Streptococcus sanguis* (ATCC strain 10556) has been reported.[12] The *iga* gene of that strain had earlier been cloned into *Escherichia*

---

[12] J. V. Gilbert, A. G. Plaut, and A. Wright, *Infect. Immun.* **59,** 7 (1991).

*coli,* where it encoded a protein whose substrate specificity was identical to that of the wild-type enzyme.[13] The amino acid sequence deduced from the 5634-nucleotide open reading frame of the cloned gene has a calculated molecular mass of 208 kDa, with no cysteine residues. The first methionine of the open reading frame precedes a putative signal sequence, followed by a hydrophobic stretch which could be a membrane anchor. The amino-terminal sequence of the recombinant enzyme recovered from the *E. coli* periplasm indicates a translational start site in that organism downstream of both the putative signal sequence and the hydrophobic stretch. The protein, 186 kDa, lacks a signal sequence at the amino terminus, and the mechanism by which it is secreted into the *E. coli* periplasm is unknown. The translational start point of the protein has not yet been determined for the enzyme produced in *S. sanguis;* however, preliminary indications are that it is upstream of the translational start site used in *E. coli* (A. G. Plaut and A. Wright, unpublished). It is also not known if a precursor enzyme processed to a mature form is produced by *S. sanguis,* similar to the serine IgA proteinases of *Haemophilus* and *Neisseria*[14] (this series, Volume 244, [10]).

The deduced amino acid sequence of the *S. sanguis* protease has several important features. Most important, there is no homology with the IgA proteinases of *Neisseria gonorrhoeae* and *Haemophilus influenzae.* This confirmed earlier findings that probes from within the *S. sanguis* gene failed to hybridize to *iga* DNA of *Haemophilus* and *Neisseria* species, even under conditions of low stringency. Second, the protein has a pentapeptide sequence HEMTH, analogous to the HxxTH (where x is any amino acid) zinc-binding signature in other metalloproteinases.[15] Site-specific mutagenesis which alters that segment to encode for FKMTH leads to production of a full-length mutant polypeptide in *E. coli* that completely lacks IgA proteinase activity.[12] There is as yet no crystallographic information localizing the pentapeptide to the enzyme active site. Third, near the amino terminus there is a 198-residue stretch which consists of 10 very similar (but not identical) tandem repeats to 20 amino acids. Of the 198 residues in the region, 48 are either glycine or proline, and 28 are glutamic acid. No repeats of any kind have been found in the serine-type IgA proteinases of gram-negative bacteria, but repeated regions are relatively common in other extracellular and surface proteins of pathogenic streptococci (referenced in Gilbert *et al.*[13]) Repeats often confer structural and biological properties that may be important in pathogenesis, and tandemly repeated regions of

[13] J. V. Gilbert, A. G. Plaut, Y. Fishman, and A. Wright, *Infect. Immun.* **56,** 1961 (1988).
[14] J. Pohlner, R. Halter, K. Beyreuther, and T. F. Meyer, *Nature (London)* **325,** 458 (1987).
[15] C.-V. Jongeneel, J. Bouvier, and A. Bairoch, *FEBS Lett.* **242,** 211 (1989).

DNA can lead to genetic variability through high frequency homologous recombination. It is not known if such variability takes place in the tandem repeat of the *S. sanguis* proteinase.

## Chemical Inhibition

The IgA proteinases of *S. sanguis*, *S. pneumoniae*, and *Capnocytophaga* species are inhibited by millimolar concentrations of EDTA.[8,9,16,17] In addition, the *S. sanguis* proteinase is inhibited by 67 m$M$ $o$-phenanthroline and the *Capnocytophaga* proteinases by 10 m$M$ bathocuproine disulfonate (BCDS).[8]

Although prolylboronic acids are potent inhibitors of the IgA proteinases of *Neisseriae gonorrhoeae* and *Haemophilus influenzae*,[18] they are not effective against the proteinase from *Streptococcus sanguis*, even though the sequence in IgA1 cleaved by the enzyme is the same as that cleaved by the type 2 *Haemophilus* and *Neisseria* proteinases. This is not altogether surprising because peptide boronic acids, although generally very potent aganst serine proteinases, are much less effective against metalloproteinases. Efforts to synthesize inhibitors of the streptococcal metallo-type IgA proteinase have thus focused on strategies for zinc enzymes. We have synthesized appropriate derivatives of the ketomethylene, phosphonate, and phosphoramidon classes of inhibitors, none of which show any ability to inhibit the *S. sanguis* IgA proteinase under our usual assay conditions. Several inhibitors, all analogs of the IgA1 hinge region substrate, weakly inhibit serine-type *N. gonorrhoeae* IgA proteinases ($K_i$ ~$10^{-4}$ $M$). The inability of the compounds to inhibit even weakly the *S. sanguis* and *S. pneumoniae* proteinases is surprising and suggests either that these are not members of the zinc-dependent metalloproteinase class of enzymes or that they have unusual specificity requirements that are not yet understood.

Physiological proteinase inhibitors such as $\alpha_2$-macroglobulin and $\alpha_1$-proteinase inhibitor do not inhibit the *S. sanguis* or *Capnocytophaga* IgA proteinases.[8] All inhibition by human serum and secretions is mediated by antiproteinase antibodies, as discussed below.

## Antibody Inhibition of Immunoglobulin A Proteinases

Antibodies to IgA proteinases capable of inhibiting activity have been found in human secretions and serum, as discussed with reference to the

[16] M. Proctor and P. J. Manning, *Infect. Immun.* **58**, 2733 (1990).

[17] J. Biewenga and F. Daus, *Immunol. Commun.* **12**, 491 (1983).

[18] W. W. Bachovchin, A. G. Plaut, G. R. Flentke, M. Lynch, and C. A. Kettner, *J. Biol. Chem.* **265**, 3738 (1990).

serine-type IgA proteinases (this series, Volume 244, [10]). Interestingly, naturally occurring inhibiting antibodies to the *S. sanguis* proteinase in serum and colostrum are at substantially lower titers than those against the *Haemophilus* and *Neisseria* serine-type enzymes.[9] Also, normal mixed human saliva and parotid saliva have weak inhibiting activity. Inhibiting anti-IgA proteinases in human serum are typically of the IgG class, whereas in secretions such antibodies are of the IgA isotype. Antibodies raised in rabbits to the *Capnocytophaga* proteinases do inhibit activity and have been used to classify the enzymes.[8]

Measurement of serum antibody can be accomplished using ELISAs (enzyme-linked immunosorbent assays),[19,20] but quantitation of inhibiting antibodies involves assay. Detailed methods for this are provided elsewhere in this series (Volume 244, [10]).

Biological Consequences of Streptococcal Immunoglobulin A Proteinase Production

As discussed elsewhere (this series, Volume 244, [10]), locally produced IgA is the principal antibody in secretions that bathe mucosal surfaces,[3,21] and IgA hydrolysis by microbial IgA proteinases may allow microbes to circumvent immunity or to recruit antibodies or their fragments as a step in the infectious process. Because metallo-type IgA proteinases are primarily produced by oral microorganisms, considerations of salivary secretory IgA are most relevant for enzymes of this catalytic type.

The tissue distribution of IgA proteinase-positive bacteria in the human oral cavity has been clearly defined.[20,22,23] Production of IgA proteinase is characteristic of those species that initiate bacterial colonization of the recently cleaned tooth surface to begin formation of dental plaque, the precursor to caries development.[20,22] For example, in one study proteinase production was found among 88% of streptococci colonizing plaque at 4 hr after cleaning, in contrast to 10% of streptococci on the dorsum of the tongue or on the oropharynx[23] in the same subject. There is, however, uncertainty as to the importance of the ability of a microorganism to cleave IgA as a prerequisite to colonization of the tooth. A study[24] of nine persons

[19] G. F. Brooks, C. J. Lammel, M. S. Blake, B. Kusecek, and M. Achtman, *J. Infect. Dis.* **166,** 1316 (1992).

[20] M. Kilian and K. Holmgren, *Infect. Immun.* **31,** 868 (1981).

[21] M. H. Mulks, *in* "Bacterial Enzymes and Virulence" (I. A. Holder, ed.), p. 82. CRC Press, Boca Raton, Florida, 1985.

[22] M. Kilian, J. Mestecky, and M. W. Russell, *Microbiol.. Rev.* **52,** 296 (1988).

[23] M. Kilian, J. Reinholdt, B. Nyvad, E. V. G. Frandsen, and L. Mikkelsen, *Immunol. Invest.* **18,** 161 (1989).

[24] J. Reinholdt, V. Friman, and M. Kilian, *Infect. Immun.* **61,** 3998 (1993).

with profound serum and salivary IgA deficiency (but with compensatory elevations of IgM in saliva) showed that the ratio of IgA proteinase-producing bacteria to total colony-forming units (cfu) on the tooth surface was the same as in normal subjects. This indicates that the ability to cleave salivary IgA is not essential for successful tissue colonization in an environment rich in IgA antibody, and that microbial IgA proteinases may have substrates other than IgA. The role of salivary antibody in defense of the oral cavity against bacterial pathogens is also not clear; salivary secretory IgA responses to specific streptococcal surface antigens have been shown not to differ in titer, $IgA1/IgA_2$ distribution, or antigen-binding affinity among caries-susceptible and caries-resistant individuals.[25]

Despite uncertainties as to the biological activities of IgA proteinases, Fab fragments can be found on bacteria in newly forming human dental plaque, and incubation of proteinase-positive streptococci with salivary secretory IgA (sIgA) produces such fragments.[26-28] Enzyme-negative bacteria can also passively acquire Fab fragments seemingly generated by extracellular proteinases of species with which they cocolonize.[26] The Fab fragments on surface antigens of bacteria not only mask antigenic determinants to prevent binding by other antibodies (and their associated inflammatory mediators[28]), but may also cause a relative increase in microbial hydrophobicity, an important element in binding of cells to dental-acquired pellicle and hydroxyapatite.[27] Fab fragments are relatively more hydrophobic than intact sIgA, possibly because they lack the carbohydrate-rich Fc and SC components of the intact antibody protein.

The role of the IgA proteinase of *Streptococcus pneumoniae* (although not yet clearly confirmed as a metalloproteinase) has not been investigated in similar detail, although this species causes both mucosal and systemic infection and is one of the most important human pathogens.

Acknowledgments

Work was supported by National Institutes of Health Grant DE 09677, NIH Grant AI 20337, and the GRASP Digestive Disease Research Center (NIH Grant DK-34928).

[25] H. Hocini, S. Iscaki, J.-P. Bouvet, and J. Pillot, *Infect. Immun.* **61,** 3597 (1993).
[26] T. Ahl and J. Reinholdt, *Infect. Immun.* **59,** 563 (1991).
[27] G. Hajishengallis, E. Nikolova, and M. W. Russell, *Infect. Immun.* **60,** 5057 (1992).
[28] M. W. Russell, J. Reinholdt, and M. Kilian, *Eur. J. Immunol.* **19,** 2243 (1989).

# [39] Tetanus and Botulism Neurotoxins: Isolation and Assay

By Giampietro Schiavo and Cesare Montecucco

## Introduction

The paralysis associated with tetanus and botulism is caused by neurotoxins produced by bacteria of the genus *Clostridium*. These are the most powerful toxins known, the $LD_{50}$ in mice being in the range 0.1–1 ng/kg. Tetanus neurotoxin (TeNT) is produced by toxigenic strains of *Clostridium tetani* in a single type, whereas seven different serotypes (A, B, C, D, E, F, and G) of botulism neurotoxin (BoNT) are made by *Clostridium botulinum* and other clostridia.[1–3]

The neurotoxins have a similar structural organization. They are synthesized as a single polypeptide chain of 150 kDa, which accumulates in the cytosol until bacterial lysis. When released the toxin is rapidly cleaved by various proteinases, at a single exposed loop.[4] This generates the two-chain neurotoxin, composed of a heavy chain (H, 100 kDa) and a light chain (L, 50 kDa), held together by a single disulfide bond, as depicted in Fig. 1.

These toxins, as well as all bacterial protein toxins with intracellular targets, penetrate cells with a four-step mechanism: (a) binding, (b) internalization, (c) membrane translocation, and (d) cytosolic target modification. The 50-kDa carboxyl-terminal part of the H chain appears to be responsible for the neurospecific binding at the presynaptic membrane of the neuromuscular junction.[1–3] Neuronal receptors involved in such a specific and high-affinity binding have not been identified, but different receptors are involved in the binding of the different neurotoxins.[5,6] After binding, the neurotoxins are internalized into endosome-like vesicular organelles, and at this stage they cannot be neutralized by antitoxin antibodies.[6,7] For the toxins to reach their cytosolic targets, they must move across the intracellular vesicle membrane, and there is evidence that this step requires acidifica-

[1] L. L. Simpson (ed.), "Botulinum Neurotoxins and Tetanus Toxin." Academic Press, New York, 1989.

[2] C. Montecucco, E. Papini, and G. Schiavo, *FEBS Lett.* **346,** 92 (1994).

[3] C. Montecucco and G. Schiavo, *Mol. Microbiol.* **13,** 1 (1994).

[4] K. G. Krieglstein, A. H. Henschen, U. Weller, and E. Habermann, *Eur. J. Biochem.* **202,** 41 (1991).

[5] J. D. Black and J. O. Dolly, *J. Cell Biol.* **103,** 521 (1986).

[6] C. Montecucco, *Trends Biochem. Sci.* **11,** 314 (1986).

[7] J. D. Black and J. O. Dolly, *J. Cell Biol.* **103,** 535 (1986).

METHODS IN ENZYMOLOGY, VOL. 248

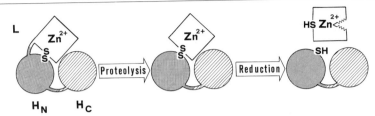

FIG. 1. Scheme of the structure and activation of tetanus and botulism neurotoxins. The single polypeptide chain of 150 kDa is folded in three distinct 50-kDa domains: L is the zinc endopeptidase, $H_N$ is responsible for cell penetration, and $H_C$ is responsible for the neurospecific binding. The proteinase activity of each toxin is revealed only after the generation of free L chain by selective proteolysis and reduction.

tion of the vesicle lumen.[8] The 50 kDa amino-terminal part of the H chain is mainly responsible for the translocation of the L chain.[9,10] Finally, once inside the neuronal cytosol, each clostridial neurotoxin L chain recognizes and cleaves at a unique site a single protein component of the neuroexocytosis apparatus.[3,11–18]

The L chains of TeNT and BoNT represent a new group of zinc-dependent endopeptidases.[3] A single molecule of L chain is capable of catalyzing the proteolysis of many substrate molecules before being neutralized by the cell.[3] This property, together with the neurospecificity, accounts for the high potency of the neurotoxins. This chapter deals with the isolation of

[8] L. C. Williamson and E. A. Neale, *J. Neurochem.* **63**, 2342 (1994).
[9] D. H. Hoch, M. Romero-Mira, B. E. Ehrlich, A. Finkelstein, B. R. DasGupta, and L. L. Simpson, *Proc. Natl. Acad. Sci. U.S.A.* **82**, 1692 (1985).
[10] C. C. Shone, P. Hambleton, and J. Melling, *Eur. J. Biochem.* **167**, 175 (1987).
[11] G. Schiavo, B. Poulain, O. Rossetto, F. Benfenati, L. Tauc, and C. Montecucco, *EMBO J.* **11**, 3577 (1992).
[12] G. Schiavo, F. Benfenati, B. Poulain, O. Rossetto, P. Polverino de Laureto, B. R. DasGupta, and C. Montecucco, *Nature (London)* **359**, 832 (1992).
[13] Deleted in press.
[14] G. Schiavo, C. C. Shone, O. Rossetto, F. C. G. Alexander, and C. Montecucco, *J. Biol. Chem.* **268**, 11516 (1993).
[15] J. Blasi, E. R. Chapman, E. Link, T. Binz, S. Yamasaki, P. De Camilli, T. Sudhof, H. Nieman, and R. Jahn, *Nature (London)* **365**, 160 (1993).
[16] G. Schiavo, O. Rossetto, S. Catsicas, P. Polverino De Laureto, B. R. DasGupta, F. Benfenati, and C. Montecucco, *J. Biol. Chem.* **268**, 23784 (1993).
[17] G. Schiavo, A. Santucci, B. R. DasGupta, P. P. Metha, J. Jontes, F. Benfenati, M. C. Wilson, and C. Montecucco, *FEBS Lett.* **335**, 99 (1993).
[18] J. Blasi, E. R. Chapman, S. Yamasaki, T. Binz, H. Nieman, and R. Jahn, *EMBO J.* **12**, 4821 (1993).
[18a] G. Schiavo, C. Malizio, W. S. Trimble, P. Polverino de Laureto, H. Sugiyama, E. Johnson, and C. Montecucco, *J. Biol. Chem.* **269**, 20213 (1994).

neurotoxins free from contaminant proteinase activities, their characterization, and assay of the zinc endopeptidase activities.

### Safety Precautions

Clostridial neurotoxins are very toxic. However, they do not affect individuals immunized with the corresponding toxoids (toxins detoxified by treatment with paraformaldehyde). In most countries children are vaccinated with tetanus toxoid, and this is sufficient to provide full protection against tetanus for decades. A booster injection of tetanus toxoid (available from pharmacies and health authorities) before starting research with tetanus toxin is advisable. One dose is sufficient to bring the serum antitetanus toxin titer to a full protection level.

In contrast, the vaccine for BoNT serotypes A, B, C, D, and E is not commercially available and can be obtained from the Centers for Disease Control (CDC, Atlanta, GA). Only after the third injection (usually performed 2 months after the first) is a protective serum anti-BoNT titer generally, but not always, achieved. This can be checked by incubating various dilutions of the serum with the toxin and then injecting into mice. Human anti-TeNT antibodies and horse anti-BoNT antibodies are also available from health authorities, and their injection immediately after accidental penetration of the toxin into the circulatory system is sufficient to prevent the disease.

Work with the toxins should be performed in a contained space. Every tool should be washed at the end of the experiment with dilute sodium hypochlorite (the toxins are extremely sensitive to oxidants).

### Purification of Tetanus Neurotoxin

*Principle.* Tetanus neurotoxin is an abundant protein of the supernatants obtained from cultures of toxigenic strains of *Clostridium tetani.* The TeNT gene is contained in a plasmid and codes for a single polypeptide chain protein (s-TeNT), composed of 1315 amino acids ($M_r$ 150,700).[3] s-TeNT is released by bacterial lysis and is rapidly converted by bacterial proteinases into the two-chain TeNT form.[4] The most suitable growth medium for the preparation of TeNT does not contain proteins, thus simplifying the purification procedures.[19] Different growth conditions and extraction procedures are followed for the preparation of the single-chain TeNT (s-TeNT), two-chain TeNT (TeNT), and the L chain.

---

[19] W. C. Latham, D. F. Bent, and L. Levine, *Appl. Microbiol.* **10,** 146 (1962).

*Reagents*

Buffer A: 2 m$M$ Sodium citrate, 100 m$M$ sodium potassium phosphate, 1 m$M$ EDTA, 1 m$M$ sodium azide, 1 m$M$ benzamidine, pH 7.5

Buffer B: 10 m$M$ Sodium phosphate, pH 7.4

Buffer C: 10 m$M$ Sodium HEPES, 150 m$M$ NaCl, pH 7.4

Buffer D: 320 m$M$ Sucrose, 4 m$M$ sodium HEPES, pH 7.3

Buffer E: 4 m$M$ Sodium HEPES, 300 m$M$ glycine, 0.02% w/w sodium azide, pH 7.3

*Single-Chain Tetanus Neurotoxin.* *Clostridium tetani* (Harvard strain) is harvested before the end of the exponential growth phase[20] by centrifugation for 10 min at 10,000 *g*, and the resulting bacterial pellet is washed twice with 10 m$M$ sodium phosphate, 145 m$M$ NaCl, pH 7.5. The cells are lysed by overnight extraction under stirring at 4° with 100 m$M$ sodium citrate, 1 $M$ NaCl, 2 m$M$ benzamidine, 1 m$M$ diisopropyl fluorophosphate (DFP), pH 7.5. The supernatant, clarified by centrifugation at 10,000 *g* for 30 min, is fractionated by the addition of a saturated ammonium sulfate solution to 43% saturation at 4°; the solution is kept at pH 7.0 by the addition of 1 $M$ NaH$_2$PO$_4$. The toxin precipitate is centrifuged at 22,000 *g* for 30 min and washed with 40% saturated ammonium sulfate in buffer A. After centrifugation as above, the pellet is dissolved in 50 ml of buffer A containing 1 m$M$ diisopropyl fluorophosphate, dialyzed against 10 m$M$ sodium phosphate, pH 6.8, and then clarified by centrifugation at 40,000 *g* for 30 min. The resulting supernatant, containing 80% pure single-chain TeNT, is processed with DEAE-cellulose and Aca 34 chromatography as described for two-chain TeNT.[20,21]

*Two-Chain Tetanus Neurotoxin and L Chain.* A 6- to 8-day-old sterile culture supernatant of *Clostridium tetani* (Harvard strain), obtained from a vaccine company (Sclavo, Siena, Italy), is precipitated with 250 g/liter of finely ground solid ammonium sulfate at 4°. The pellet is recovered by centrifugation at 16,000 *g* for 10 min at 4° and dissolved in 10 m$M$ sodium phosphate, pH 7.4 (buffer B). The precipitation step is repeated with the final percentage of ammonium sulfate increased to 46% of saturation. The resulting pellet is dissolved in buffer B and dialyzed extensively against the same buffer. After clarification of the dialyzed toxin solution by centrifugation (16,000 *g* for 10 min at 4°), the dark-brown supernatant is applied to a diethylaminoethyl-derivatized cellulose (DEAE-cellulose) column (DE-52; Whatman, Clifton, NJ; 15 mg protein/ml resin), previously equilibrated with buffer B. The toxin is eluted with a linear gradient of 10–100 m$M$

---

[20] K. Ozutsumi, N. Sugimoto, and M. Matsuda, *Appl. Environ. Microbiol.* **49,** 939 (1985).

[21] U. Weller, F. Mauler, and E. Habermann, *Naunyn-Schmiedeberg's Arch. Pharmacol.* **338,** 99 (1988).

sodium phosphate, pH 7.4, and fractions of 4 ml are collected. After checking the protein composition by sodium dodecyl sulfate–polyacrylamide gel electrophoresis (SDS–PAGE) on 10% polyacrylamide gels, the toxin-containing fractions are pooled and precipitated with 60% saturated ammonium sulfate.

The ammonium sulfate suspension is spun at 27,000 $g$ for 15 min at 4°, and the resulting pellet is dissolved in a minimum volume of buffer B and applied to an Aca 34 gel-filtration column (3 × 130 cm), equilibrated with the same buffer and eluted at a 25 ml/hr constant flow. Fractions of 8 ml are collected. Two peaks, corresponding to TeNT (95% pure) and L chain (70% pure), are obtained. The proportion of L chain increases with the aging of the bacterial culture. The fractions corresponding to TeNT and L chain are pooled from the eluate, precipitated with 60% saturated ammonium sulfate at pH 7.0, and stored at 4° until further purification is performed.

*Nicking of Single-Chain Tetanus Neurotoxin.* The percentage of single-chain and two-chain TeNT in the final toxin preparation may vary depending on the growth conditions and the age of the bacterial culture. To obtain the two-chain toxin, the solution containing s-TeNT is treated with a proteinase to promote cleavage at a single site located within an exposed loop.[4] It is advisable to perform an analytical test to optimize nicking conditions first. s-TeNT is cleaved with tosylsulfonyl phenylalanyl chloromethyl ketone-treated trypsin (Serva, Heidelberg, Germany) at 25° for 60 min using a toxin/proteinase ratio of 1000 : 1 (w/w). Proteolysis is terminated by the addition of soybean trypsin inhibitor at a proteinase-to-inhibitor ratio of 1 : 4 (w/w). The procedure is also applicable with minor modifications to the nicking of single-chain BoNTs.[22]

*Immobilized Metal Ion Affinity Chromatography.* The TeNT, isolated as above, and BoNT serotypes A, B, C, E, and F isolated as described elsewhere[22] contain traces of contaminant clostridial proteinases. The neurotoxins commercially available from Calbiochem (La Jolla, CA), Sigma (St. Louis, MO), or Wako (Neuss, Germany) also contain additional proteins. The neurotoxins can be rapidly freed from contaminant proteases using an immobilized metal ion affinity chromatography (IMAC) step, which exploits the coordination between electron donor groups of the protein surface and the chelated metal ions bound to the iminodiacetic acid groups attached to the agar-based matrix.[23,24] Among the protein

[22] L. L. Simpson, J. J. Schmidt, and J. L. Middlebrook, this series, Vol. 165, p. 76.
[23] F. H. Arnold, *Bio Technology* **9**, 151 (1991).
[24] E. S. Hemdan, Y. Zhao, E. Sulkowski, and J. Porath, *Proc. Natl. Acad. Sci. U.S.A.* **86**, 1881 (1989).

surface residues able to interact with the transition metal ions in IMAC separations, histidines and cysteines are the predominant ligands.[23,24] The experimental procedure may be divided into two sections: preparation of the metal-bound column and purification of the toxin.[25]

A chelating Superose HR 10/2 column (Pharmacia, Piscataway, NJ), connected with a Beckman (Palo Alto, CA) System Gold high-performance liquid chromatography (HPLC) apparatus, is washed at a constant flow rate of 0.5 ml/min with 7.5 ml of 50 m$M$ EDTA, 500 m$M$ NaCl, pH 6.0, and equilibrated with 10 ml of 50 m$M$ sodium acetate, 200 m$M$ NaCl, pH 6.0. The column is loaded with zinc ions by flowing 7.5 ml of 100 m$M$ zinc chloride through it. Unbound metal ions are removed with 10 ml of 50 m$M$ sodium acetate, 200 m$M$ NaCl, pH 6.0. Finally, the column is equilibrated with 10 m$M$ sodium HEPES, 150 m$M$ NaCl, pH 7.4 (buffer C). The TeNT and BoNTs (0.1–10 mg), previously dialyzed into buffer C, are loaded onto the column at room temperature, and unbound material is washed away with the same buffer. Toxins are eluted with a linear gradient of 0–25 m$M$ imidazole in buffer C at a constant flow rate of 0.5 ml/min (Fig. 2A).

This procedure is also useful for the purification of $H_C$, the 50-kDa carboxy-terminal part of the heavy chain of TeNT, which shows an identical retention time. This is due to the fact that histidine residues located in $H_C$ are responsible for TeNT binding to the immobilized zinc. The same procedure is followed to purify BoNTs, as depicted for BoNT/A in Fig. 2A. Purified TeNT, its $H_C$ fragment, and BoNTs are dialyzed against 10 m$M$ sodium HEPES, 50 m$M$ NaCl, pH 7.2, and, after freezing in liquid nitrogen, are stored at $-80°$.

This procedure cannot be used with BoNT/D because that serotype is not retained by the zinc-IMAC column. To obtain a proteinase-free BoNT/D preparation, a commercially available source (0.5–10 mg; Wako) is used. The material is dialyzed against 20 m$M$ Tris-Cl, pH 8.0, and loaded onto a Mono Q HR 5/5 column, previously equilibrated with the same buffer. Proteins are eluted with a step gradient of NaCl, as shown in Fig. 2B: BoNT/D is present in peak 1, as determined by SDS–PAGE (Fig. 2C); peak 2 contains a nontoxic unidentified protein.

*Purification of L Chain of Tetanus Toxin.* The L chain, obtained as above, is contaminated by TeNT and small molecular weight peptides. A three-step HPLC purification procedure is followed to remove the contaminants. The L chain ammonium sulfate precipitate is dissolved in 100 m$M$ sodium phosphate, pH 6.8, and loaded in 20-mg aliquots onto a preparative G 3000 SW column (21.5 × 600 mm, Pharmacia). The peak enriched in L

---

[25] O. Rossetto, G. Schiavo, P. Polverino de Laureto, S. Fabbiani, and C. Montecucco, *Biochem. J.* **285,** 9 (1992).

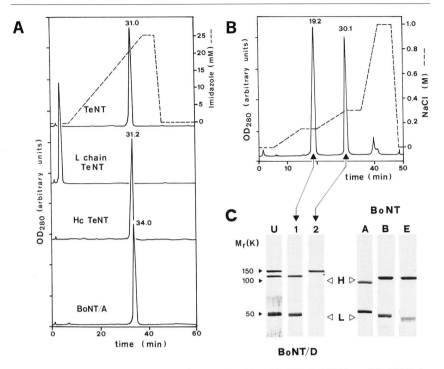

FIG. 2. Purification of TeNT, L chain of TeNT, $H_C$ of TeNT, BoNT/A, and BoNT/D by HPLC. (A) Chelating Superose-Zn(II) chromatography: samples were eluted with the imidazole gradient shown in the top graph, and the retention time for each peak is indicated. (B) BoNT/D was eluted from a Mono Q HR HPLC column with the NaCl step gradient depicted as a dashed line. (C) The SDS–PAGE profiles of BoNT/D before chromatography (U), peak 1 (neurotoxin), and peak 2 (nontoxic peptide). BoNT serotypes A, B, and E, obtained as in (A), are also shown.

chain is loaded onto a zinc-IMAC column to eliminate residual TeNT. The L chain is collected in the void volume, dialyzed against 20 m$M$ Tris-Cl, pH 8.0, and applied to a Mono Q HR 5/5 column previously equilibrated with the same buffer. The L chain is eluted with a linear gradient of NaCl (0–200 m$M$) at a 1.0 ml/min flow rate. Fractions containing the purified protein are collected and precipitated with ammonium sulfate (60% saturation). The resulting L chain is homogeneous on both SDS–PAGE and reversed-phase chromatography on a BU-300 column (100 × 46 mm; Pierce, Rockford, IL). The L chain is kept at 4°; alternatively, after freezing in liquid nitrogen, it is stored at −80° as indicated above.

*Zinc Content Determination.* Routinely the zinc content and the animal toxicity of the purified toxins are measured as a quality control of the

TABLE I
TARGET AND PEPTIDE BOND SPECIFICITIES OF TETANUS AND BOTULISM NEUROTOXINS

| Toxin type | Neuronal target | Peptide bond cleaved | Ref. |
|---|---|---|---|
| TeNT | VAMP/synaptobrevin | Gln-Phe | 11–12 |
| BoNT/A | SNAP-25 | Gln-Arg | 15–17 |
| BoNT/B | VAMP/synaptobrevin | Gln-Phe | 12 |
| BoNT/C | Syntaxin | Lys-Ala | 18, 26a |
| BoNT/D | VAMP/synaptobrevin | Lys-Leu | 16 |
| BoNT/E | SNAP-25 | Arg-Ile | 16, 17 |
| BoNT/F | VAMP/synaptobrevin | Gln-Lye | 14 |
| BoNT/G | VAMP/synaptobrevin | Ala-Ala | 18a |

preparation. The metal content is determined by atomic absorption spectroscopy, after extensive dialysis of the samples at 4° in 10 m$M$ sodium HEPES, 100 m$M$ NaCl, pH 7.0, pretreated with Amberlite MB-3 (Sigma).[11,14,26] The TeNT, its L chain, and BoNTs have an average contents of 0.8–1.0 gram atoms zinc per mole toxin.[11,14,26] For toxicity determination, the neurotoxins are diluted in phosphate-buffered saline containing 0.1% w/w bovine serum albumin and injected intraperitoneally in 40-day-old mice. Time before death is recorded, and a mouse lethal dose is calculated.[21]

Functional Assay of Clostridial Neurotoxins

Both TeNT and BoNT are zinc endopeptidases highly specific for their respective substrates: vesicle associated membrane protein (VAMP)/synaptobrevin or synaptosomal associated membrane protein of 25 kDa (SNAP-25) or syntaxin, as reported in Table I. The targets are cleaved at a unique peptide bond (Table I). The TeNT and BoNTs do not cleave short peptides spanning the respective cleavage sites. They appear to require the recognition of additional portions of the substrate protein, still to be identified. As a consequence, a simple spectrophotometric or fluorometric assay of their activity is not yet available. The zinc-dependent activity must be assayed with the target protein or a substantial portion of it. The target protein may be obtained by subcellular fractionation of the nervous tissue. Alternatively, suitable substrates may be obtained by recombinant techniques or by peptide synthesis.

*Preparation of Small Synaptic Vesicles.* The procedure for the prepara-

[26] G. Schiavo, O. Rossetto, A. Santucci, B. R. DasGupta, and C. Montecucco, *J. Biol. Chem.* **267**, 23479 (1992).

[26a] G. Schiavo, C. C. Shone, M. K. Bennett, R. H. Scheller, and C. Montecucco, *J. Biol. Chem.* **270**, in press (1995).

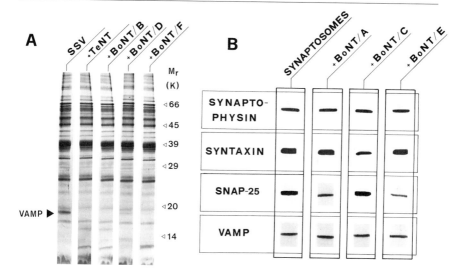

FIG. 3. Electrophoretic assay of the proteolytic activity of tetanus and botulism neurotoxins. (A) Electrophoretic profile of untreated rat brain small synaptic vesicles (SSV), SSV plus TeNT, and SSV plus BoNT/B, D, and F with the cleavage of VAMP/synaptobrevin (whose position in the gel is marked by arrow) into two fragments. (B) Immunoblotting analysis of the proteolysis of syntaxin by BoNT/C and of SNAP-25 by BoNT/A and E, as detailed in the text.

tion of small synaptic vesicles (SSV) from brain follows established methods.[27] Minor modifications extend its applicability to nonmammalian sources, such as chicken, easily obtainable from abattoirs. In a typical preparation, brains from 10 chickens or 7 rats (Wistar) are obtained by decapitation. The brains are placed into 320 m$M$ sucrose, 4 m$M$ sodium HEPES, pH 7.3 (buffer D), at 4°. The cortexes are removed, pooled, and homogenized in 20 ml of buffer D in a Teflon–glass homogenizer using 12 up-and-down strokes at 900 rpm. The homogenate is diluted to 75 ml with the same buffer containing 100 $\mu M$ phenylmethylsulfonyl fluoride (PMSF) and 2 $\mu$/ml pepstatin (final concentrations), then centrifuged at 900 $g$ for 10 min. The supernatant is collected and centrifuged at 9200 $g$ for 15 min. The resulting pellet is washed in 60 ml of buffer D, centrifuged at 10,500 for 15 min, and finally resuspended in 4.5 ml of buffer D.

This suspension, referred to as crude synaptosomes, is diluted with 50 ml of ice-cold distilled water and homogenized in a 50-ml Teflon–glass homogenizer using 8 strokes at 900 rpm. The suspension is immediately buffered with 0.5 ml of 1 $M$ sodium HEPES, pH 7.4, and 100 $\mu M$ PMSF and 2 $\mu$g/ml pepstatin are added (final concentrations). After 30 min at 4°,

[27] W. B. Huttner, W. Schiebler, P. Greengard, and P. De Camilli, *J. Cell Biol.* **96**, 1374 (1983).

the lysed synaptosomes are centrifuged at 25,000 $g$ for 20 min at 4° to yield a pellet (fraction LP1) and a clear supernatant, enriched in small synaptic vesicles. This fraction is centrifuged at 165,000 $g$ for 120 min, and the transparent pellet (LP2) is resuspended in 1 ml of buffer D containing proteinase inhibitors. The suspension is homogenized with 8 strokes in a 5-ml Teflon–glass homogenizer, followed by forcing it through a 25-gauge needle eight times. The suspension is then layered onto a continuous linear sucrose gradient generated in a 40-ml Ultraclear tube (Beckman) from 18 ml of 50 m$M$ sucrose and 16.5 ml of 800 m$M$ sucrose at 4°. After centrifugation on a Beckman SW28 rotor at 26,000 rpm for 4 hr at 4°, the gradient reveals a broad and light turbid band in the 250–450 m$M$ sucrose gradient, containing small synaptic vesicles. The band is collected with a peristaltic pump and then diluted with 4 m$M$ sodium HEPES, 300 m$M$ glycine, 0.02% sodium azide, pH 7.3 (buffer E), to 50 ml. The small synaptic vesicles are recovered by centrifugation at 165,000 $g$ for 2 hr at 4°, and the transparent pellet (SG-V) is resuspended, as previously indicated for the LP2 fraction, in 1.0 ml of glycine buffer plus 100 $\mu M$ PMSF and 2 $\mu$g/ml pepstatin. The protein concentration is determined with the Bradford assay. The preparation remains viable for at least 3 days.

*Proteolytic Assay.* Forty-five micrograms of LP2 or 30 $\mu$g of SV-G in 30 $\mu$l of buffer E plus 150 m$M$ NaCl is incubated with 3 $\mu$l of 100–200 n$M$ TeNT or BoNT serotypes B, D, F, or G (previously reduced for 30 min at 37° in the presence of 10 m$M$ dithiothreitol) for 2 hr at 37°. Samples are adjusted to 2% SDS, 5% 2-mercaptoethanol, 5 m$M$ EDTA, and 5% v/v glycerol and then boiled for 2 min. Proteins are analyzed in 13–18% polyacrylamide linear gradient gels via SDS–PAGE and stained with Coomassie blue or, after transfer onto nitrocellulose membrane, with specific anti-VAMP/synaptobrevin antibodies (Fig. 3A).

Alternatively 50 $\mu$g of LP1 fraction in 30 $\mu$l of glycine buffer, 150 m$M$ NaCl, 0.4% octylglucoside is incubated with 3 $\mu$l of 200 n$M$ BoNT/A, C, or E (previously reduced as above) for 2 hr at 37°. After SDS–PAGE on 13–18% linear polyacrylamide gradient gels, proteins are analyzed by Western blotting with antibodies against SNAP-25 (for BoNT/A or E) or syntaxin (for BoNT/C) (Fig. 3B). Quantitative analysis of the proteolytic cleavage is performed by densitometric analysis of the stained gel and/or Western blot.

Acknowledgments

Supported by grants from the National Research Center (CNR) Target Project on Biotechnology and Bioinstrumentation and from Telethon-Italia (Grant 473).

# [40] Human Carboxypeptidase N: Lysine Carboxypeptidase

*By* Randal A. Skidgel

## Introduction

Human plasma carboxypeptidase N (lysine carboxypeptidase; EC 3.4.17.3) was discovered by Erdös and Sloane in the early 1960s[1] as an enzyme that inactivates bradykinin by cleaving its C-terminal arginine. It was rediscovered and renamed many times as it was found to cleave C-terminal Arg or Lys from a variety of peptide and protein substrates.[2,3] Carboxypeptidase N is an important plasma regulator of peptide hormone activity and protects against the deleterious actions of peptides released into the circulation (e.g., anaphylatoxins C3a, C4a, and C5a).[2,3] As one indication of its importance, no person has ever been found to lack the enzyme completely, and patients with genetically determined low levels are very rare.[3-5] In one patient, low blood levels (about 20% of normal) were associated with recurrent attacks of angioneurotic edema.[5]

## Methods of Assay

A variety of assays have been described for carboxypeptidase N.[1,6-12] Most methods are based on measuring an increase or decrease in UV absorbance after removal of the C-terminal basic amino acid from N-terminally blocked substrates such as benzoyl(Bz)-Gly-Arg, Bz-Gly-Lys, Bz-Ala-Lys, furylacryloyl (FA)-Ala-Lys, FA-Ala-Arg, and the ester sub-

---

[1] E. G. Erdös and E. M. Sloane, *Biochem. Pharmacol.* **11**, 585 (1962).
[2] E. G. Erdös, *Handb. Exp. Pharmacol.* **25**(Suppl.), 427 (1979).
[3] R. A. Skidgel, *Trends Pharmacol. Sci.* **9**, 299 (1988).
[4] E. G. Erdös, I. M. Wohler, M. I. Levine, and M. P. Westerman, *Clin. Chim. Acta* **11**, 39 (1965).
[5] K. P. Mathews, P. M. Pan, N. J. Gardner, and T. E. Hugli, *Ann. Int. Med.* **93**, 443 (1980).
[6] E. G. Erdös, E. M. Sloane, and I. M. Wohler, *Biochem. Pharmacol.* **13**, 893 (1964).
[7] E. G. Erdös, H. Y. T. Yang, L. L. Tague, and N. Manning, *Biochem. Pharmacol.* **16**, 1287 (1967).
[8] T. J. McKay, A. W. Phelan, and T. H. Plummer, Jr., *Arch. Biochem. Biophys.* **197**, 487 (1979).
[9] T. H. Plummer and M. T. Kimmel, *Anal. Biochem.* **108**, 348 (1980).
[10] T. H. Plummer, Jr., and E. G. Erdös, this series, Vol. 80, p. 442.
[11] R. A. Skidgel and E. G. Erdös, *in* "Methods of Enzymatic Analysis" (H. U. Bergmeyer, ed.), Vol. 5, p. 60. Verlag Chemie, Weinheim, 1984.
[12] R. A. Skidgel, *in* "Methods in Neurosciences: Peptide Technology" (P. M. Conn, ed.), Vol. 6, p. 373. Academic Press, Orlando, Florida, 1991.

strate Bz-Gly-argininic acid.[6-11] A more sensitive technique uses a fluorescent substrate, dansyl-Ala-Arg, in which the fluorescent product is preferentially extracted into chloroform after acidification of the reaction mixture.[12]

A convenient method for monitoring the purification of carboxypeptidase N is that originally described by Plummer and Kimmel[9] using FA-Ala-Lys.[10,11] The substrate FA-Ala-Lys incorporates two features that previous studies have shown to be optimal for carboxypeptidase N activity, namely, a C-terminal lysine (which is turned over faster than C-terminal arginine) and a penultimate alanine. In addition, because the decrease in absorbance is monitored at 336 nm, there is little interference from protein absorbance in impure samples.

*Reagents*

Buffer: 0.1 $M$ HEPES (pH 7.75) containing 0.5 $M$ NaCl

Substrate: 5 m$M$ FA-Ala-Lys (18.23 mg/10 ml water) is stable for about 2 weeks at 4° or can be stored frozen in aliquots for several months; FA-Ala-Lys is commercially available from several manufacturers (e.g., Sigma, St. Louis, MO)

*Procedure.* To a test tube is added 0.5 ml of buffer, 0.1 ml of substrate, and enough water to give a final volume (including sample) of 1.0 ml. The mixture is warmed to 37° in a water bath, enzyme sample is added with brief mixing, and then the solution is rapidly transferred to a prewarmed cuvette (10 mm path) in a thermostatted (37°) chamber of a recording spectrophotometer. (The sample can be read against a reference cuvette containing only substrate and buffer or a neutral density filter to compensate for the initial absorbance of the substrate.) The instrument is adjusted to give a $\Delta A$ of 0.1 full scale, and the change in absorbance at 336 nm is recorded continuously for about 2–3 min. As with any spectrophotometric assay, the possibility that particulate matter and other substances may cause nonspecific changes in absorbance must be considered. Enzyme and substrate blanks should be run under the same conditions to rule out these possibilities.

*Calculation of Activity Units.* One unit is defined as the amount of enzyme required to hydrolyze 1 $\mu$mol of substrate per minute at 37°. The $\Delta\varepsilon$ ($M^{-1}$ cm$^{-1}$) was reported to be $-1300$ for FA-Ala-Lys at 336 nm;[9] however, the total $\Delta A$ for commercial FA-Ala-Lys should be determined for each batch of substrate by comparing the absorbance of the substrate to that of the split products (i.e., by allowing the reaction to go to completion). For example, in our hands, one batch of commercial substrate gave a total $\Delta A$ of 0.545 for 0.5 m$M$ substrate (0.5 $\mu$mol in 1 ml).[11]

Purification

*General Comments.* The procedure outlined below has been used successfully in our laboratory for many years.[13] It is a relatively simple two-step procedure involving DE-52 ion-exchange chromatography and affinity chromatography on arginine-Sepharose. To expedite the purification process, we use batchwise adsorption for the chromatographic steps, and dialysis is not required except after the last step. Thus, the whole purification can easily be completed within 2–3 days. The use of arginine-Sepharose to purify carboxypeptidase N was originally described by Oshima *et al.*,[14] and a procedure using the same principles as outlined here and giving similar results was published by Plummer and Hurwitz[15] and detailed in an earlier volume in this series.[10] The major differences in the latter method are (1) use of *p*-aminobenzoyl-L-arginine-Sepharose as the affinity gel, (2) dialysis before and after each step, and (3) application of the samples to resins already packed in a column.

*Materials.* Phenylmethylsulfonyl fluoride (PMSF), diisopropyl fluorophosphate (DFP), aprotinin, 3-[(3-cholamidopropyl)dimethylammonio]-1-propane sulfonate (CHAPS), and guanidinoethylmercaptosuccinic acid (GEMSA) are available from commercial sources (e.g., Sigma). GEMSA can also be synthesized as described.[10]

DE-52 (Whatman, Clifton, NJ) can be reused several times and is recycled using the following procedure (eliminate the first step if using new resin). One liter of settled gel is mixed with 2 liters of 2 *M* NaCl and allowed to stand for a few minutes. The solution is removed with a fritted glass funnel, and the gel is washed successively with 2 liters of water, 2 liters of 2 *N* NaOH, several liters of water until the pH drops to approximately pH 8, 2 liters of 1 *N* HCl, and finally water again until the pH is approximately pH 5–6. The gel can be stored at 4° in the presence of 0.005% (w/v) thimerosal to inhibit bacterial growth.

Arginine-Sepharose is prepared from epichlorohydrin-activated Sepharose 6B[16] or can be purchased from Pharmacia (Piscataway, NJ). To activate the Sepharose 6B, 100 ml of settled gel is mixed with 85 ml water and 130 ml of 1 *N* NaOH. After warming to 40° in a water bath, 15 ml of epichlorohydrin (Sigma) is added, and the mixture is stirred constantly (with an overhead stirrer) for 2 hr at 40°. The activated gel is washed with 5 liters of water. For the coupling reaction, arginine (20 g) is dissolved in 100 ml water, and

[13] Y. Levin, R. A. Skidgel, and E. G. Erdös, *Proc. Natl. Acad. Sci. U.S.A.* **79,** 4618 (1982).
[14] G. Oshima, J. Kato, and E. G. Erdös, *Biochim. Biophys. Acta* **365,** 344 (1974).
[15] T. H. Plummer and M. Y. Hurwitz, *J. Biol. Chem.* **253,** 3907 (1978).
[16] J. Porath and N. Fornstedt, *J. Chromatogr.* **51,** 479 (1970).

the mixture is adjusted to pH 12.4 with 10 $N$ NaOH. The arginine solution and the activated gel are mixed, readjusted to pH 12.3–12.4, and then allowed to stir overnight at room temperature. The gel is then washed with 1–2 liters of water, 1 liter of dilute acetic acid (pH 3–4), 1 liter of 2 $M$ NaCl, and then extensively with water until no salt remains. (This can be tested easily by dropping the eluent into a test tube containing a solution of AgNO$_3$; the absence of a white AgCl precipitate indicates that the washing is complete.) Once prepared, the arginine-Sepharose can be reused several times using the following washing procedure. Used arginine-Sepharose (100 ml settled gel) is suspended in a solution of 1 $M$ arginine and 2 $M$ NaCl and allowed to stand at room temperature for about 30 min. The liquid is removed by filtration and the gel washed twice more with 2 $M$ NaCl (100 ml each time). The gel is then washed extensively with water (at least 2 liters) until no salt remains and is stored in the presence of 0.02% (w/v) sodium azide at 4°.

### Purification Procedure

*Plasma Preparation.* Pooled outdated human plasma (2000–2500 ml) is stirred for 1 hr at 4° in the presence of 1 m$M$ DFP, 1 m$M$ PMSF, and 200 units/ml aprotinin. The plasma is then either filtered through Whatman filter paper or centrifuged to remove any precipitates. The starting carboxypeptidase N activity is measured with the FA-Ala-Lys substrate as described above.

*Ion-Exchange Chromatography.* The pooled plasma, diluted to 4 liters with 50 m$M$ Tris-HCl, pH 7.2, is mixed with 1 liter of settled DE-52 ion-exchange resin (prepared as described above) and the mixture adjusted to pH 7.2 with 1 $N$ HCl, if necessary. After stirring for 1–2 hr, the gel is allowed to settle, and the supernatant is tested for carboxypeptidase activity to determine how much is bound (should be about 70–85%). The diluted plasma is removed by vacuum filtration on a large fritted glass funnel, and the gel is washed with 1 liter of 50 m$M$ Tris-HCl, pH 7.2, containing 50 m$M$ NaCl, by resuspension and filtration. The gel is again resuspended in 1 liter of the same buffer and then poured into a 5 × 60 cm column. Proteins are eluted with a 4 liter plus 4 liter gradient of 50 m$M$ to 0.25 $M$ NaCl in 50 m$M$ Tris-HCl, pH 7.2, at a flow rate of 175 ml/hr, and 19-ml fractions are collected and tested for enzyme activity (Fig. 1). The carboxypeptidase activity elutes near the middle of the gradient (at about 0.13–0.15 $M$ NaCl).

*Arginine-Sepharose Affinity Chromatography.* To the pooled active fractions (about 1600–2300 ml) from the DE-52 column are added 0.5 m$M$ PMSF, 0.5 m$M$ DFP, 100 units/ml aprotinin, 1% (v/v) $n$-butanol, and finally arginine-Sepharose (100 ml settled gel). The mixture is stirred for 1 hr, the

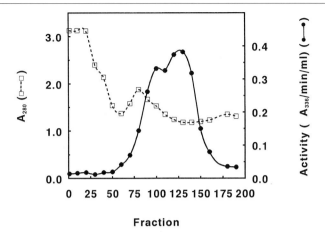

Fig. 1. Ion-exchange chromatography of carboxypeptidase N. Carboxypeptidase N in diluted human plasma was adsorbed batchwise onto DE-52 ion-exchange resin. The resin was washed with 50 mM Tris-HCl, pH 7.2, containing 50 mM NaCl and then poured into a 5 × 60 cm column. The enzyme was eluted with a 4 liter plus 4 liter gradient of 50 mM to 0.25 M NaCl in 50 mM Tris-HCl, pH 7.2 (starting with fraction 1), and 19-ml fractions were collected. The apparent presence of two components in the activity peak is probably due to heterogeneity in the glycosylation of the 83-kDa subunit of carboxypeptidase N. Sometimes only a single broad peak of activity is eluted.

supernatant is tested for carboxypeptidase activity to determine the percent binding (usually about 80%), and then the solution is removed by vacuum filtration. The gel is resuspended in 200 ml of 10 mM sodium phosphate, pH 7.0, containing 0.1 M NaCl and 1% n-butanol and poured into a 5 × 5 cm column. The column is washed with the same buffer containing 0.2 M NaCl until the protein concentration reaches baseline and then with the same buffer containing 0.25 M NaCl. The fractions are monitored closely, and when a small protein peak (with a little carboxypeptidase activity) is eluted with the 0.25 M NaCl, the eluant is immediately switched to the same buffer containing 1 mM GEMSA to elute carboxypeptidase N (Fig. 2). Active fractions are pooled (about 60–100 ml) and then dialyzed overnight against 2–3 changes (4 liters each) of 0.1 M NH$_4$HCO$_3$ to remove the GEMSA, which is an enzyme inhibitor.

Using this procedure, the enzyme is purified over 4000-fold with a specific activity of 25.7 units/mg with FA-Ala-Lys substrate (Table I). This corresponds to a specific activity of 103 units/mg with hippurylargininic acid (which is cleaved 4-fold faster), essentially identical with the value of 104 units/mg reported by Plummer and Hurwitz.[15] Although the latter reported a slightly lower 2608-fold purification, this can be attributed to

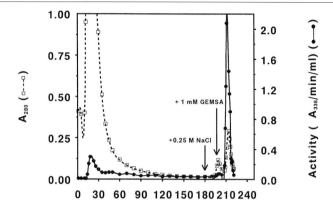

**Fraction**

Fig. 2. Affinity chromatography of carboxypeptidase N on arginine-Sepharose. The pooled active fractions from the DE-52 column (Fig. 1) were adsorbed batchwise onto 100 ml of settled arginine-Sepharose gel. The supernatant was filtered off, and the gel was resuspended in 200 ml of 10 m$M$ sodium phosphate, pH 7.0, containing 0.1 $M$ NaCl and 1% $n$-butanol and then poured into a 5 × 5 cm column. After packing, the column was washed with the same buffer containing 0.2 $M$ NaCl (fractions 1–179) and then with buffer containing 0.25 $M$ NaCl (first arrow). Carboxypeptidase N was eluted with 0.25 $M$ NaCl in the same buffer containing 1 m$M$ GEMSA (second arrow).

TABLE I

PURIFICATION OF HUMAN PLASMA CARBOXYPEPTIDASE N

| Purification step | Volume (ml) | Activity[a] (units) | Protein[b] (mg) | Specific activity (units/mg) | Yield (%) | Purification (-fold) |
|---|---|---|---|---|---|---|
| Plasma | 2250 | 982 | 155,700 | 0.0063 | 100 | 1 |
| DE-52 column | 2350 | 526 | 2192 | 0.240 | 54 | 38 |
| Arginine-Sepharose column[c] | 7.2 | 149 | 5.8 | 25.7 | 15 | 4079 |

[a] One unit is 1 $\mu$mol of FA-Ala-Lys cleaved per minute at 37°.

[b] Protein was determined by the method of M. M. Bradford, *Anal. Biochem.* **72**, 248 (1976), using bovine serum albumin as the standard. For the purified protein, the final value was calculated using a conversion factor based on quantitative amino acid analysis of purified enzyme on which the Bradford protein assay was also done.

[c] Values given are for pooled fractions after dialysis and concentration.

our present use of the Bradford protein assay (Table I). We previously reported an average purification of 2665-fold for our method using the Lowry protein assay.[13] The overall yield of 15% shown in Table I is a minimum value and varies from preparation to preparation to as high as 30–35%.

*Concentration and Storage.* As with most enzymes, carboxypeptidase N is more stable when stored at concentrations of at least 1 mg/ml. The enzyme can be concentrated with high recovery in an Amicon (Danvers, MA) stirred cell concentrator equipped with a YM10 membrane. The enzyme is most stable when stored at −20° in 50% glycerol, which prevents the solution from freezing. Care must be taken to assure that the volume of glycerol added is accurate (the viscous glycerol is best dispensed using a positive displacement pipette) and that the freezer temperature is not below −20° (a 50% glycerol solution will freeze at −23°). Under these conditions, the enzyme will maintain its activity for several years, although some degradation of the subunits will be noticed when analyzed by sodium dodecyl sulfate–polyacrylamide gel electrophoresis (SDS–PAGE) after long-term storage. The addition of 0.5 m$M$ DFP and 0.5 m$M$ PMSF can help to stabilize the enzyme by inhibiting any trace proteinase contamination in the preparation. For situations in which glycerol might interfere with subsequent analyses, carboxypeptidase N can be quick-frozen and stored at −70°, although repeated freeze–thawing is to be avoided.

*Separation and Isolation of Subunits of Carboxypeptidase N.* Purified carboxypeptidase N runs as a single band on native polyacrylamide gel electrophoresis, but it dissociates into three major bands on denaturing gels (with or without reduction): a diffuse band of 83 kDa (the noncatalytic subunit) and two major bands of 55 and 48 kDa representing two forms of the active subunit.[13,15] (The calculated molecular mass of the active subunit based on the cDNA sequence is 50 kDa, with reported values from SDS–PAGE analysis varying from 44 to 55 kDa; historically, however, the most frequently cited numbers are 55 and 48 kDa, so those values will be used here.)

We originally reported a technique to dissociate the subunits of carboxypeptidase N using 3 $M$ guanidine and separation by chromatography on Sephadex G-75 superfine.[13] That procedure results in the isolation of the 48-kDa form of the active subunit, as the 55-kDa form either aggregates or does not elute from the column, possibly because of higher hydrophobicity.[17] A modification of the procedure,[17] using CHAPS detergent in the

[17] R. A. Skidgel, C. D. Bennett, J. W. Schilling, F. Tan, D. K. Weerasinghe, and E. G. Erdös, *Biochem. Biophys. Res. Commun.* **154**, 1323 (1988).

buffers, allows the separation and isolation of both major forms of the active subunit.

Purified carboxypeptidase N (1 to 5 mg at a concentration of 1 to 2 mg/ml) is mixed on ice with an equal volume of 0.1 $M$ Tris-HCl, pH 7.5, containing 6 $M$ guanidine hydrochloride and 1% CHAPS and stirred for 30 min. The mixture is applied to a column (2.6 × 96 cm) of Sephacryl S-300 (Sephacryl S-200 has also been used successfully), preequilibrated with 0.05 $M$ Tris-HCl, pH 7.5, containing 0.1 $M$ NaCl and 0.5% CHAPS, and is eluted at 12 ml/hr. Three major protein peaks are eluted: the first peak contains the noncatalytic 83-kDa subunit, the second peak primarily the 48-kDa form of the active subunit, and the last peak the 55-kDa form of the active subunit. The CHAPS detergent can be removed by dialysis, and the 83-kDa subunit fractions can be concentrated on an Amicon concentrator equipped with a YM10 membrane.

The active subunit fractions are more difficult to concentrate because, owing to its hydrophobic nature, the active subunit binds to YM10 ultrafiltration membranes and the activity must be recovered by washing the membrane with a 50% glycerol solution.[13] Generally, better results have been obtained by concentrating in the dialysis bag itself. The pooled fractions are placed in a dialysis bag, and the bag is then buried in a container containing dry Sephadex G-200 (other high molecular weight hygroscopic materials may also be used, e.g., carboxymethylcellulose or polyethylene glycol) and allowed to stand at 4° for several hours, until the volume is reduced to the desired level. The process can be shortened somewhat by rocking the sample and periodically removing the wet Sephadex and exposing the bag to fresh dry material. As with the intact enzyme, activity of the isolated catalytic subunit is best preserved if the protein is stored at −20° in 50% glycerol.

Properties

*Molecular Weight.* The size estimates of purified human carboxypeptidase N range from about 270 to 330 kDa, depending on the method used, although 280 kDa is the value most frequently cited.[10,13–15] It is considered to be a tetrameric enzyme composed of a dimer of heterodimers, with each heterodimer containing one catalytic subunit ($M_r$ 44,000–55,000 in SDS–PAGE) and one noncatalytic 83-kDa subunit.[3,13,15] The 83-kDa subunit is heavily glycosylated[13,15] (about 28% by weight), with a carbohydrate composition typical of proteins containing Asn-linked complex carbohydrate chains.[15] In contrast, the active subunit lacks carbohydrate.[13,15]

*Enzymatic Properties.* The lack of carbohydrate, along with its small size and relative instability at 37°,[13] would not allow the active subunit to survive in the circulation for very long. Thus, the primary role of the 83-kDa subunit is to carry and protect the active subunit in the blood.[3,13] Although the 83-kDa subunit might modulate the ability of the enzyme to interact with certain substrates and inhibitors,[18,19] the majority of the enzymatic properties of the intact 280-kDa tetramer are preserved in the isolated catalytic subunit.[13,20]

Carboxypeptidase N contains zinc in the active center as a required cofactor and therefore is inhibited by chelating agents such as EDTA and $o$-phenanthroline.[2,6,7,15] Specific and high-affinity bi-product analog inhibitors of carboxypeptidase N and other B-type carboxypeptidases were developed by the Plummer group.[21,22] These include GEMSA ($K_i$ of 1 $\mu M$ for carboxypeptidase N) and DL-2-mercaptomethyl-3-guanidinoethylthiopropanoic acid ($K_i$ of 2 n$M$).

Replacement of $Zn^{2+}$ in the active center by $Co^{2+}$ activates carboxypeptidase N by about 2- to 6-fold at neutral pH (depending on the substrate),[6,9,20,23] owing to an increase in the value of $k_{cat}$.[20] In contrast, $Cd^{2+}$ inhibits the enzyme.[6,7] The pH optimum of carboxypeptidase N is in the neutral range, and it retains little activity at pH 5.5[6,23] (about 7% of that measured at pH 7.5 with dansyl-Ala-Arg). However, $Co^{2+}$ is still able to activate the enzyme at pH 5.5 to 156% of the value measured at pH 7.5 without $Co^{2+}$.[23]

Carboxypeptidase N cleaves a variety of synthetic and naturally occurring substrates containing C-terminal Arg or Lys. In general, the enzyme cleaves Lys faster than Arg (owing to a higher $k_{cat}$), and the penultimate residue also plays an important role, with alanine being preferred in many cases. These principles are reflected in the catalytic constants reported for various substrates[8–10,20,24] (Table II).

*Sequence.* The cDNA sequences for both subunits of carboxypeptidase N were determined from clones isolated from human liver cDNA libraries.[25,26] The cDNA sequence of the 50-kDa catalytic subunit encodes a putative 20-amino acid signal peptide and a mature protein of 438 amino acids that has only low sequence identity with pancreatic carboxypeptidase

[18] R. A. Skidgel, M. S. Kawahara, and T. E. Hugli, *Adv. Exp. Med. Biol.* **198A**, 375 (1986).
[19] F. Tan, H. Jackman, R. A. Skidgel, E. K. Zsigmond, and E. G. Erdös, *Anesthesiology* **70**, 267 (1989).
[20] R. A. Skidgel, A. R. Johnson, and E. G. Erdös, *Biochem. Pharmacol.* **33**, 3471 (1984).
[21] T. J. McKay and T. H. Plummer, Jr., *Biochemistry* **17**, 401 (1978).
[22] T. H. Plummer, Jr., and T. J. Ryan, *Biochem. Biophys. Res. Commun.* **98**, 448 (1981).
[23] P. A. Deddish, R. A. Skidgel, and E. G. Erdös, *Biochem. J.* **261**, 289 (1989).
[24] G. Oshima, J. Kato, and E. G. Erdös, *Arch. Biochem. Biophys.* **170**, 132 (1975).
[25] F. Tan, D. K. Weerasinghe, R. A. Skidgel, H. Tamei, R. K. Kaul, I. B. Roninson, J. W. Schilling, and E. G. Erdös, *J. Biol. Chem.* **265**, 13 (1990).
[26] W. Gebhard, M. Schube, and M. Eulitz, *Eur. J. Biochem.* **178**, 603 (1989).

TABLE II

KINETIC CONSTANTS FOR SUBSTRATES OF CARBOXYPEPTIDASE N

| Substrate | $K_m$ ($\mu M$) | $k_{cat}$ (min$^{-1}$) | $k_{cat}/K_m$ ($\mu M^{-1}$ min$^{-1}$) | Ref. |
|---|---|---|---|---|
| Bz-Gly-Lys | 1400 | 960 | 0.7 | a, b |
| Bz-Gly-Arg | 650 | 240 | 0.4 | a |
| Bz-Ala-Lys | 350 | 21,120 | 60.3 | a |
| Bz-Ala-Arg | 280 | 8340 | 29.8 | a |
| FA-Ala-Lys | 340 | 5820 | 17.1 | c |
| FA-Ala-Arg | 260 | 1860 | 7.1 | c |
| [Leu$^5$]Enkephalin-Arg$^6$ | 57 | 375 | 6.6 | d |
| [Met$^5$]Enkephalin-Arg$^6$ | 49 | 1024 | 20.9 | d |
| [Met$^5$]Enkephalin-Lys$^6$ | 216 | 6204 | 28.7 | d |
| Bradykinin | 19 | 58 | 3.1 | d |

[a] T. J. McKay, A. W. Phelan, and T. H. Plummer, Jr., *Arch. Biochem. Biophys.* **197**, 487 (1979).
[b] G. Oshima, J. Kato, and E. G. Erdös, *Arch. Biochem. Biophys.* **170**, 132 (1975).
[c] T. H. Plummer and M. T. Kimmel, *Anal. Biochem.* **108**, 348 (1980).
[d] R. A. Skidgel, A. R. Johnson, and E. G. Erdös, *Biochem. Pharmacol.* **33**, 3471 (1984).

A (19%) or B (17%).[26] However, higher identity was found with membrane-bound carboxypeptidase M (41%) and the secretory granule prohormone processing carboxypeptidase H or E (49%), especially in regions containing putative active site residues, where identities range from 69 to 93%.[26,27] Carboxypeptidases N, M, and H are longer than carboxypeptidases A and B, and their extensions at the C terminus have no counterpart in the pancreatic enzymes and no similarities to one another.[26,27] The 50-kDa subunit of carboxypeptidase N contains several dibasic sites in the C-terminal region that could potentially be cleaved to give rise to the heterogeneity found in the size of the purified subunit.

The cDNA sequence of the 83-kDa subunit encodes a 59-kDa protein with no sequence similarity to the 50-kDa active subunit or any other carboxypeptidase.[25] As expected from the high carbohydrate content, the sequence contains seven potential Asn-linked glycosylation sites and, in addition, a region rich in threonine and serine which may be a site for attachment of O-linked carbohydrate.[25]

The most striking feature of the deduced sequence is a domain comprising about 54% of the protein which contains 12 leucine-rich tandem repeats of 24 amino acids each.[25] The consensus repeated sequence contains conserved residues in 10 of 24 positions in each repeat, primarily consisting of

[27] F. Tan, S. Chan, D. F. Steiner, J. W. Schilling, and R. A. Skidgel, *J. Biol. Chem.* **264**, 13165 (1989).

leucine in 6 positions, leucine and other hydrophobic amino acids in positions 4 and 5, proline in position 1, and asparagine in position 19. This repeating pattern was first discovered in the leucine-rich $\alpha_2$-glycoprotein and has been found in a variety of other mammalian proteins (e.g., platelet glycoprotein (GP)Ib$_\alpha$, GP V, and GP IX, proteoglycans, RNase inhibitor, luteinizing hormone receptor, oligodendrocyte/myelin glycoprotein, and U2 snRNP-A') and even in *Drosophila*, yeast, bacterial, and viral proteins.[25,28] Although the X-ray crystal structure is known for only the RNase inhibitor, the leucine-rich repeat region likely forms a common structural element that is critical for the functioning of the proteins, all of which participate in some sort of binding interaction (e.g., protein–protein or protein–membrane). We have proposed that the leucine-rich repeat region in the 83-kDa subunit mediates its interaction with the 50-kDa active subunit to form a heterodimer and that the N- and C-terminal domains of the 83-kDa subunit are responsible for association of the two heterodimers to form the tetramer.[25,28] Elucidation of the exact structure and binding interactions of the two subunits of carboxypeptidase N will require further study.

### Acknowledgment

Some of the studies reported here were supported by National Institutes of Health Grants HL 36473 and HL 36082.

[28] R. A. Skidgel and F. Tan, *Agents Actions* **38**(Suppl.), 359 (1992).

# [41] Human Carboxypeptidase M

*By* FULONG TAN, PETER A. DEDDISH, and RANDAL A. SKIDGEL

## Introduction

Investigations in the late 1950s and early 1960s led to the discovery of two mammalian carboxypeptidases that specifically cleave C-terminal arginine or lysine: pancreatic carboxypeptidase B[1] (EC 3.4.17.2) and human plasma carboxypeptidase N[2] (lysine carboxypeptidase; EC 3.4.17.3). Although a few reports[3] indicated the possible existence of a "tissue" carboxy-

[1] J. E. Folk and J. A. Gladner, *J. Biol. Chem.* **231**, 379 (1958).
[2] E. G. Erdös and E. M. Sloane, *Biochem. Pharmacol.* **11**, 585 (1962).
[3] E. G. Erdös, *Handb. Exp. Pharmacol.* **25**(Suppl.), 427 (1979).

peptidase B-type enzyme, it was not clear whether the activity differed from carboxypeptidase N or carboxypeptidase B. After purifying a unique carboxypeptidase from human urine,[4] we found a membrane-bound carboxypeptidase B-type enzyme to be present in a variety of tissues, including kidney, lung, placenta, and blood vessels, and in cultured endothelial cells and fibroblasts.[5,6] The enzyme was purified to homogeneity from human placenta,[7] and analysis of the enzymatic and structural characteristics and determination of the cDNA sequence proved it to be a unique enzyme.[7,8] We named it carboxypeptidase M to denote the fact that it is membrane-bound.[7]

## Methods of Assay

Many of the assays described for other B-type carboxypeptidases, such as carboxypeptidase N,[2,9–12] have been used for carboxypeptidase M.[5,7,13] As with carboxypeptidase N,[9,10] substrates containing a penultimate alanine are among the best for carboxypeptidase M.[7] However, carboxypeptidase M prefers substrates with C-terminal arginine[7] whereas carboxypeptidase N prefers C-terminal lysine.[3,5,9,10]

The majority of the assays developed for these enzymes rely on measuring changes in absorbance on removal of the C-terminal amino acid from a dipeptide substrate containing a UV-absorbing N-terminal blocking group.[3,9–12] Some examples of this type of substrate are benzoyl(Bz)-Gly-Lys, Bz-Gly-Arg, Bz-Gly-argininic acid, Bz-Ala-Lys, furylacryloyl-Ala-Lys, and furylacryloyl-Ala-Arg. The most convenient way of using these substrates is to measure the change in absorbance over time with a continuously recording spectrophotometer. However, the method is not sensitive enough for some samples and also requires an optically clear solution, which can be difficult to obtain with tissue homogenates or membrane fractions. Modifications of these techniques can eliminate some of the deficiencies but

---

[4] R. A. Skidgel, R. M. Davis, and E. G. Erdös, *Anal. Biochem.* **140**, 520 (1984).

[5] R. A. Skidgel, A. R. Johnson, and E. G. Erdös, *Biochem. Pharmacol.* **33**, 3471 (1984).

[6] R. A. Skidgel, *Trends Pharmacol. Sci.* **9**, 299 (1988).

[7] R. A. Skidgel, R. M. Davis, and F. Tan, *J. Biol. Chem.* **264**, 2236 (1989).

[8] F. Tan, S. Chan, D. F. Steiner, J. W. Schilling, and R. A. Skidgel, *J. Biol. Chem.* **264**, 13165 (1989).

[9] T. J. McKay, A. W. Phelan, and T. H. Plummer, Jr., *Arch. Biochem. Biophys.* **197**, 487 (1979).

[10] R. A. Skidgel, this volume [40].

[11] T. H. Plummer, Jr., and E. G. Erdös, this series, Vol. 80, p. 442.

[12] R. A. Skidgel and E. G. Erdös, *in* "Methods of Enzymatic Analysis" (H. U. Bergmeyer, ed.), Vol. 5, p. 60. Verlag Chemie, Weinheim, 1984.

[13] R. A. Skidgel, *in* "Methods in Neurosciences: Peptide Technology" (P. M. Conn, ed.), Vol. 6, p. 373. Academic Press, Orlando, Florida, 1991.

involve longer incubation times followed by extraction of the product[5,12] or separation by high-performance liquid chromatography (HPLC).[14] Indeed, we initially used Bz-Gly-Lys as a substrate (measuring the product, Bz-Gly, after extraction) to measure carboxypeptidase M in tissue fractions and to follow its purification.[5,7] However, this type of assay is somewhat tedious to carry out. Detailed procedures for using UV-absorbing dipeptide substrates have been reported elsewhere.[10–12]

Assays based on fluorescent substrates are generally more sensitive than those using the UV-absorbing substrates. We now routinely assay carboxypeptidase M activity using the fluorescent substrate dansyl-Ala-Arg.[13] Because the substrate and product are both equally fluorescent, the assay utilizes the ability to extract the product, dansyl-Ala, preferentially into chloroform while the remaining substrate stays in the aqueous phase at acid pH. It is similar in principle to the assay technique developed for carboxypeptidase H.[15] The assay is very convenient because the reaction, extraction, and fluorescence measurements are all done in the same tube.

*Synthesis of Dansyl-Ala-Arg.* The substrate is readily synthesized by dansylating the dipeptide Ala-Arg.[13] The synthesis outlined below can be easily scaled up or down to match the anticipated need.

Although the absolute exclusion of light is not necessary, it is preferable to limit light exposure (e.g., by use of aluminum foil) whenever possible. To 1.0 g of Ala-Arg (Bachem Bioscience Inc., Philadelphia, PA) dissolved in 150 ml of 0.2 $M$ NaHCO$_3$, pH 8.7, is added slowly (with constant stirring) 10.0 g of dansyl chloride (Sigma Chemical Company, St. Louis, MO) dissolved in 150 ml of acetone. The solution is stirred for 2 hr at room temperature, during which time it is normal for some of the excess dansyl chloride to precipitate. The precipitate is removed by filtration, and the acetone is evaporated on a rotary evaporator. During evaporation, more precipitate forms and is likewise removed. The remaining aqueous solution (about 150 ml) is acidified to pH 3.5 with concentrated HCl, clarified by filtration, and extracted 2 times with 700 ml of chloroform (which is discarded). The aqueous solution is neutralized to about pH 7.2 with 10 $N$ NaOH, and residual chloroform is removed by rotary evaporation.

The dansyl-Ala-Arg can be purified by preparative thin-layer chromatography or HPLC as previously described.[13] Alternatively, we have developed a simpler procedure using a Sep-Pak C$_{18}$ cartridge (Millipore, Bedford, MA). To the aqueous solution containing crude dansyl-Ala-Arg is added acetonitrile to a final concentration of 10% (v/v). A Sep-Pak Vac 35-ml cartridge is washed first with 150 ml of acetonitrile and then equilibrated

[14] B. G. Grimwood, A. L. Tarentino, and T. H. Plummer, Jr., *Anal. Biochem.* **170**, 264 (1988).
[15] L. D. Fricker and S. H. Snyder, *Proc. Natl. Acad. Sci. U.S.A.* **79**, 3886 (1982).

with 150 ml of 10% (v/v) acetonitrile in water. The sample is applied and allowed to load by gravity. The cartridge is then washed with 100 ml of 15% acetonitrile in water to elute impurities remaining after the extraction (e.g., dansyl hydroxide). The column can be visually monitored with a hand-held UV light (long wavelength). The impurities eluted in the first wash fluoresce with a bright blue color. Dansyl-Ala-Arg, which fluoresces bright yellow, is then eluted with 70% acetonitrile in water until no more fluorescence is seen in the eluate (about 60 ml). The solvent is removed by rotary evaporation, and the dansyl-Ala-Arg is redissolved in water to a final concentration of 10 m$M$ and stored in 1-ml aliquots, frozen at $-70°$ and protected from light. Using this procedure, we obtained a 61% yield of dansyl-Ala-Arg, which gives enough substrate for about 45,000 determinations. Analytical HPLC analysis showed the substrate to have a purity of 98.8%.

*Enzyme Assay*

*Reagents*

Buffer: 0.2 $M$ HEPES [4-(2-hydroxyethyl)-1-piperazineethanesulfonic acid], pH 7.0, containing 0.2% (v/v) Triton X-100

Substrate: 1.0 m$M$ Dansyl-Ala-Arg (4.64 mg/10 ml water or dilute 10 m$M$ stock solution 1:10). Divide into 1-ml aliquots and store protected from light at $-20°$ to $-70°$. It is stable for at least 1 year under these conditions, but repeated freeze–thawing (more than 2 or 3 times) should be avoided

Inhibitor: 100 $\mu M$ MGTA (DL-2-mercaptomethyl-3-guanidinoethyl-thiopropanoic acid; Calbiochem, La Jolla, CA). Dissolve 0.237 mg in 10 ml of degassed water, divide into aliquots, and store frozen. Thaw only as much as needed and discard after use. Freeze–thawing should be avoided. Because MGTA has a free sulfhydryl group that is essential for inhibitory activity, oxidizing agents or mercurial compounds should not be used in the same reaction mixture

Stop solution: 1.0 $M$ citric acid adjusted to pH 3.1 with NaOH

*Procedure.* To a 10 × 75 mm glass test tube on ice is added 125 $\mu$l of buffer, 5–50 $\mu$l of enzyme sample, 0 or 25 $\mu$l of 100 $\mu M$ MGTA, and 0–70 $\mu$l water to give a final volume of 200 $\mu$l. For each set of reactions, one enzyme blank (no substrate) and one substrate blank (no enzyme) are prepared. To assure the specificity of the reaction, samples can be preincubated with and without the MGTA inhibitor. (Although this should be always done with unknown samples, it is not necessary to do it with each sample when screening activity in fractions from a purification.) Samples are preincubated for 5–10 min on ice, and then 50 $\mu$l of 1.0 m$M$ dansyl-

Ala-Arg substrate (200 $\mu M$ final) is added to start the reaction. Samples are incubated at 37° for 15 min to 3 hr, depending on activity, and the reaction is stopped by adding 150 $\mu$l of the stop solution (1.0 $M$ citrate, pH 3.1). Chloroform (1.0 ml) is added to each tube, mixed vigorously for 15 sec (to extract the dansyl-Ala product), and then centrifuged at about 800 $g$ for 10 min to separate the phases. The fluorescence in the chloroform layer (bottom layer) is measured relative to a chloroform blank at 340 nm excitation wavelength and 495 nm emission. A simple test tube adapter can be used which allows the fluorescence to be read directly in the 10 × 75 mm tubes. The position of the tube should be checked to assure that the chloroform layer is high enough so that the excitation beam does not hit the aqueous phase containing the unreacted fluorescent substrate. Alternatively, the chloroform layer can be transferred to a cuvette for measurement of the fluorescence.

*Calculation of Activity Units.* Carboxypeptidase activity is defined as the difference in fluorescence between the uninhibited sample and the sample inhibited with 10 $\mu M$ MGTA. Usually over 95% of the activity is inhibited with MGTA, even in crude tissue fractions, indicating the specificity of the substrate. Fluorescence units (FU) are converted to nanomoles of substrate by constructing a standard curve of FU versus concentration of dansyl-Ala (commercially available from Sigma and other vendors) in the range of 1 to 10 $\mu M$. Tubes are set up as described above except dansyl-Ala is substituted for dansyl-Ala-Arg and extra water substituted for the enzyme sample. The dansyl-Ala soutions are extracted as above and the fluorescence measured. The extraction step automatically corrects for extraction efficiency so that the concentration of dansyl-Ala in a reaction mixture can be determined directly from the standard curve. The reaction is linear until approximately 6 nmol of dansyl-Ala has been produced (75 FU on our Perkin-Elmer, Norwalk, CT, instrument). Thus, enzyme concentration or reaction time should be adjusted to keep FU readings in the linear range of the assay. Because assay conditions and fluorometers will vary from laboratory to laboratory, a time course should be done and the amount of dansyl-Ala produced versus time plotted to determine the linearity of the assay under conditions that will be used routinely.

## Purification

*General Comments.* The purification of human carboxypeptidase M outlined below contains some modifications of the original procedure that was published.[7] These procedures have been used successfully in our laboratory for several years. Because carboxypeptidase M is anchored to plasma membranes by covalent attachment to a phosphatidylinositol–glycan

(PI-G) anchor,[16,17] it first must be solubilized. The original procedure used the detergent 3-[(3-cholamidopropyl)dimethylammonio]-1-propane sulfonate (CHAPS) to solubilize the enzyme. More recently, we have used bacterial phosphatidylinositol-specific phospholipase C (PI-PLC) to release it in soluble form. The latter method eliminates the need to use detergents in the buffers during purification. Whichever method of solubilization is used, the subsequent steps are the same: ion-exchange chromatography on a column of Q-Sepharose (Pharmacia, Piscataway, NJ), followed by affinity chromatography on arginine-Sepharose. At this stage of purification, there is no detectable contaminating enzyme activity, although one or two additional protein bands may be seen when the preparation is analyzed by sodium dodecyl sulfate–polyacrylamide gel electrophoresis (SDS–PAGE). An additional one or two steps of purification can be performed, depending on the purity required. Separation by HPLC on a Mono Q column (Pharmacia, Piscataway, NJ) and/or size-exclusion chromatography on TSK-G3000SW (Toso Haas, Montgomeryville, PA) or Superdex 75 (Pharmacia, Piscataway, NJ) columns have been used successfully. For unknown reasons, varying amounts of transferrin often copurify with carboxypeptidase M, as determined by sequencing the approximately 80-kDa contaminant band seen in SDS–PAGE analysis of a carboxypeptidase M preparation.[18] This impurity can be efficiently and specifically removed by affinity adsorption with immobilized anti-human transferrin antibody (which is commercially available).

*Materials.* L-Arginine-Sepharose is prepared using epichlorohydrin-activated Sepharose 6B as described.[10] The affinity gel can also be obtained from Pharmacia. The following reagents are commercially available: CHAPS, Brij 35, aprotinin, *p*-chloromercuriphenyl sulfonate (PCMS), phenylmethylsulfonyl fluoride (PMSF), diisopropyl fluorophosphate (DFP) from Sigma; guanidinoethylmercaptosuccinic acid (GEMSA) from Calbiochem; and PI-PLC (from *Bacillus thuringiensis*) from ICN Biochemicals (Costa Mesa, CA). GEMSA can also be synthesized as described.[11]

*Procedure*

Unless otherwise stated, all centrifugations and chromatographic separations are carried out at 4°.

*Isolation of Placental Microvilli.* A microvillous membrane fraction from human placenta is prepared as described[7,19] with some modifications. Pieces

[16] P. A. Deddish, R. A. Skidgel, V. B. Kriho, X.-Y. Li, R. P. Becker, and E. G. Erdös, *J. Biol. Chem.* **265**, 15083 (1990).
[17] R. A. Skidgel, F. Tan, P. A. Deddish, and X.-Y. Li, *Biomed. Biochim. Acta* **50**, 815 (1991).
[18] H. S. Lu, F. Tan, and R. A. Skidgel, unpublished (1988).
[19] A. R. Johnson, R. A. Skidgel, J. T. Gafford, and E. G. Erdös, *Peptides* **5**, 789 (1984).

of villous tissue (about 1000 g total) are dissected from the fetal surface of 6–8 placentas and rinsed three times with cold normal (0.9% w/v) saline (500 ml each wash). The tissue is finely minced with a pair of scissors or processed through a meat grinder and then resuspended in 5 liters of cold normal saline containing proteinase inhibitors (1 m$M$ PMSF, 0.2 m$M$ PCMS, 0.5 m$M$ DFP, and 2500 units of aprotinin). The mixture is stirred for 45 min at 4° during which time microvillar vesicles bud off into the solution. The suspension is filtered through four layers of gauze and then centrifuged for 11 min at 10,000 g. The microvilli are obtained by centrifuging the supernatant fraction at 10,000 g for 120–180 min. If carboxypeptidase M is to be solubilized by detergent extraction, the pellet is resuspended in 50 ml of 0.1 $M$ Tris-HCl, pH 8.5, containing the proteinase inhibitors listed above. If the enzyme is to be released with PI-PLC, the pellet is suspended in 100 ml of 25 m$M$ Tris-HCl, pH 7.5, containing 0.25 $M$ sucrose, 1 m$M$ PMSF, 0.1 m$M$ PCMS, 0.1 m$M$ leupeptin, and 10 $\mu$g/ml aprotinin. If necessary, the resuspended pellet can be quickly frozen and stored at −70° before the solubilization step.

*Solubilization of Carboxypeptidase M with Detergent.* To 50 ml of the suspended microvillar fraction is added 50 ml of 1.6% (w/v) CHAPS in water. The suspension is briefly sonicated, stirred for 1 hr at 4°, and then centrifuged at 100,000 g for 1 hr. The supernatant, containing solubilized carboxypeptidase M, is collected, and the activity is measured (along with the starting material) to determine the amount of enzyme solubilized (usually at least 80%).

*Solubilization of Carboxypeptidase M with Phosphatidylinositol-Specific Phospholipase C.* To the microvillar pellet suspended in 100 ml of buffer (as described above) is added 3–4 units of PI-PLC (1 unit is the amount of enzyme that cleaves 1 $\mu$mol of phosphatidylinositol per minute at pH 7.5 and 37°). The mixture is stirred at 37° for 45 min and then centrifuged at 100,000 g for 1 hr. The supernatant, containing the released carboxypeptidase M, is collected and tested for activity. Usually about 80% of the total carboxypeptidase M is released by this procedure. Because the quality of the PI-PLC can vary from batch to batch, it can first be tested on a small aliquot of the microvillar pellet before proceeding. If the large-scale release is unsatisfactory, the pellet can be resuspended and treated a second time.

The purification procedures outlined below are those used for the CHAPS detergent-solubilized enzyme. The conditions used for purification of the PI-PLC-released enzyme are identical except no detergent is used in any of the buffers. In addition, before the first ion-exchange column, the supernatant from the PI-PLC treatment should be adjusted to pH 8.5 with NaOH.

*Chromatography on Q-Sepharose.* To the CHAPS-solubilized membrane preparation is added Brij 35 detergent to a final concentration of 1%

(w/v). The solution is then applied to a Q-Sepharose column (2.6 × 10 cm or 1.6 × 15 cm) preequilibrated with 50 m$M$ Tris-HCl, pH 8.5, containing 1% Brij 35. The column is washed with 2 bed volumes of the equilibration buffer, and the bound proteins are then eluted with a linear gradient (500 ml plus 500 ml) of 0 to 0.5 $M$ NaCl in the same buffer at a flow rate of 60 ml/hr. Fractions (8 ml each) are collected and assayed for carboxypeptidase activity. The active fractions are pooled, concentrated to 50 ml on an Amicon (Danvers, MA) stirred cell concentrator equipped with a YM10 membrane, and then dialyzed overnight against 4 liters of 50 m$M$ Tris-HCl, pH 7.2, with one change.

*Affinity Chromatography on Arginine-Sepharose.* A column of arginine-Sepharose (1.6 × 20 cm) is packed and equilibrated with 50 m$M$ Tris-HCl, pH 7.2, containing 0.5% (w/v) CHAPS. The dialyzed concentrate from the Q-Sepharose column is applied, and the column is washed with the equilibration buffer until the absorbance (at 280 nm) of the eluate reaches baseline (about 150 ml). The column is then eluted with 100 ml of the same buffer containing 1 m$M$ GEMSA and finally buffer containing 0.5 $M$ NaCl (Fig. 1). Fractions of 2.5–4.5 ml are collected and assayed for carboxypepti-

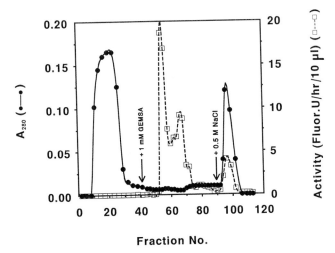

**Fraction No.**

Fig. 1. Arginine-Sepharose affinity chromatography of human carboxypeptidase M. In this preparation, carboxypeptidase M was released from the microvillar fraction with PI-PLC, and thus detergent was not used in the buffer. The pooled active fractions from a Q-Sepharose column were dialyzed, concentrated to 60 ml, and then applied to a column (1.6 × 20 cm) of arginine-Sepharose which was preequilibrated with 50 m$M$ Tris-HCl, pH 7.2. (Fraction 1 marks the start of column loading; 4-ml-fractions were collected.) The column was then eluted with 100 ml of the equilibration buffer, 200 ml of buffer containing 1 m$M$ GEMSA (first arrow), and then 150 ml of buffer containing 0.5 $M$ NaCl (second arrow). For further details, see text.

dase M activity. The activity profile is not an accurate representation of the elution of carboxypeptidase M because the GEMSA used to elute the enzyme also inhibits it in the assay. Thus, after an initial sharp rise in the activity peak, the apparent activity decreases rapidly owing to the increased amount of GEMSA eluting in the later tubes (Fig. 1). For this reason, fractions after the apparent activity peak (about an additional 50 ml) are also pooled with the active fractions. For example, in the purification shown in Fig. 1, fractions 53 to 86 were pooled together. (Alternatively, 0.1 m$M$ MGTA can be used to elute the enzyme. Because MGTA contains a free sulfhydryl group, its inhibitory activity can be abolished by including 0.1 m$M$ PCMS in the assay mixture. This allows more precise definition of the activity peak.) The pooled fractions are concentrated on an Amicon stirred cell concentrator equipped with a YM10 membrane and then dialyzed against 3 changes (1 liter each) of 50 m$M$ Tris-HCl, pH 7.2, containing 0.1% CHAPS. After dialysis, CHAPS is added to a final concentration of 0.4%. After every chromatographic run, the arginine-Sepharose resin is regenerated as described[10] by washing with a solution containing 1 $M$ arginine and 2 $M$ NaCl followed by extensive washing with 2 $M$ NaCl and then water.

Using this purification procedure, carboxypeptidase M is purified almost 300-fold with a 38% yield (Table I). As mentioned above, at this stage of purification some contaminating bands remain when the preparation is analyzed by SDS–PAGE. One or both of the following steps can be carried out to remove the remaining contaminants, depending on the purity required.

*Chromatography on a Mono-Q Column.* A Mono-Q HR 5/5 column (5 × 50 mm) attached to a Waters gradient HPLC system is equilibrated

TABLE I
PURIFICATION OF HUMAN PLACENTAL CARBOXYPEPTIDASE M

| Purification step | Volume (ml) | Activity[a] (units) | Protein[b] (mg) | Specific activity (units/mg) | Yield (%) | Purification (-fold) |
|---|---|---|---|---|---|---|
| Placental microvilli | 230 | 10,549 | 2254 | 4.7 | 100 | 1.0 |
| CHAPS supernatant | 230 | 10,378 | 552 | 18.8 | 98 | 4.0 |
| Q-Sepharose | 50 | 7062 | 145 | 48.7 | 67 | 10.4 |
| Arginine-Sepharose | 24 | 4032 | 3 | 1344 | 38 | 286 |

[a] One unit equals 1 nmol of dansyl-Ala-Arg hydrolyzed per minute at 37°, assayed as described.

[b] Protein was assayed by the method of M. M. Bradford, *Anal. Biochem.* **72,** 248 (1976), using bovine serum albumin as standard.

with 50 m$M$ Tris-HCl, pH 8.0, containing 0.2% CHAPS. About 500 $\mu$g of the concentrated, dialyzed carboxypeptidase M preparation after arginine-Sepharose chromatography is injected into the column and eluted at room temperature at a flow rate of 1 ml/min. The column is washed for 5 min with equilibration buffer and then eluted with either a 50-min linear gradient of 0.0 to 1.0 $M$ NaCl in the same buffer or a linear gradient of 0.0 to 0.2 $M$ NaCl for 5 min followed by a linear gradient of 0.2 to 0.5 $M$ NaCl in the same buffer and at the same flow rate for 60 min. Proteins are detected by UV absorption at 280 nm, and 0.5-ml fractions are collected and assayed for carboxypeptidase M activity. The active fractions are pooled.

*HPLC Size-Exclusion Chromatography.* The carboxypeptidase M preparation from the Mono Q column is concentrated 5- to 10-fold on an Amicon concentrator (YM10 membrane) and then injected into a Superdex 75 HR 10/30 column (Pharmacia, Piscataway, NJ) (preequilibrated with 50 m$M$ sodium phosphate, pH 7.0, containing 0.15 $M$ NaCl). The column is eluted with the equilibration buffer at a flow rate of 0.5 ml/min, and 0.25-ml fractions are collected and assayed for carboxypeptidase M activity. Active fractions are pooled and concentrated as above. Satisfactory results have also been obtained using a TSK G3000SW column in the same HPLC system.[7]

*Storage.* Carboxypeptidase M is relatively stable if it is concentrated, quick-frozen, and then stored at $-70°$. Alternatively, it can be stored at $-20°$ in 50% (v/v) glycerol, which prevents freezing of the solution. Under these conditions, the activity is stable for at least 1 year.

Properties

*Molecular Weight.* When analyzed by SDS–PAGE, the enzyme runs as a single band of 62 kDa with or without reduction, showing that it is a single-chain protein.[7] In gel filtration on a TSK G3000SW column in the presence of CHAPS detergent, carboxypeptidase M elutes as a single peak corresponding to a molecular mass of 73 kDa.[7] The slightly higher value is probably due to the binding of the enzyme to detergent micelles and/or its glycoprotein nature. That carboxypeptidase M is a glycoprotein was proved by two methods. When applied to a column of concanavalin A-Sepharose, purified carboxypeptidase M bound tightly and could only be eluted with buffer containing 0.5 $M$ $\alpha$-methyl-D-glucoside or $\alpha$-methyl-D-mannoside (Fig. 2). Chemical deglycosylation with trifluoromethanesulfonic acid reduced the mass of carboxypeptidase M to 47.6 kDa in SDS–PAGE, indicating that the enzyme is a glycoprotein containing about 23% carbohydrate by weight.[7]

*Enzymatic Properties.* Carboxypeptidase M is activated by CoCl$_2$ and

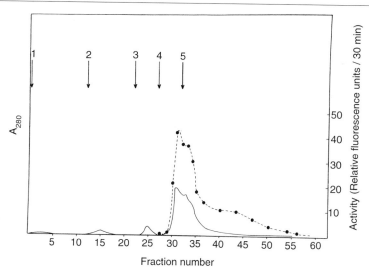

FIG. 2. Chromatography of human carboxypeptidase M on concanavalin A-Sepharose. Purified human carboxypeptidase M (about 600 μg in 200 μl) was applied to a column (1.8 × 2.5 cm) of concanavalin A-Sepharose preequilibrated with buffer containing 50 mM Tris-HCl, pH 7.0, 1.0 mM CaCl₂, 1.0 mM MnCl₂, and 1% CHAPS. The column was washed with 11 ml of equilibration buffer (arrow 1) and then eluted stepwise (at 8 ml/hr) with the same buffer containing 0.25 $M$ NaCl (10 ml; arrow 2), 0.25 $M$ NaCl plus 0.25 $M$ α-methyl-D-glucoside (5 ml; arrow 3), 0.25 $M$ NaCl plus 0.5 $M$ α-methyl-D-glucoside (5 ml; arrow 4), and 0.25 $M$ NaCl plus 0.5 $M$ α-methyl-D-mannoside (25 ml; arrow 5). Fractions of 1 ml were collected and assayed: $A_{280}$ (solid line), carboxypeptidase M activity (dashed line).

inhibited by $o$-phenanthroline, MGTA, GEMSA, and cadmium acetate, demonstrating that it is a metallopeptidase.[5,7] It has a neutral pH optimum and cleaves only C-terminal Arg or Lys from a variety of synthetic peptide substrates as well as naturally occurring peptides such as bradykinin, Arg[6]- and Lys[6]-enkephalins, and dynorphin A(1–13).[7] Carboxypeptidase M prefers C-terminal arginine over lysine as demonstrated by faster cleavage of 5 mM Bz-Gly-Arg (2.2 μmol/min/mg) than 5 mM Bz-Gly-Lys (1.3 μmol/min/mg).[7] However, the penultimate residue can also dramatically affect the rate of hydrolysis, as 1 mM Bz-Ala-Lys is cleaved about 17 times faster than 5 mM Bz-Gly-Lys. Of all the synethetic substrates tested, the ester substrate (1 mM Bz-Gly-argininic acid) was cleaved fastest (102 μmol/min/mg).[7]

Naturally occurring peptides were also tested as substrates for carboxypeptidase M, and the following kinetic constants were determined[7]: [Leu⁵]enkephalin-Arg⁶, $K_m$ = 63 μM, $k_{cat}$ = 106 min⁻¹; [Met⁵]enkephalin-Arg⁶, $K_m$ = 46 μM, $k_{cat}$ = 934 min⁻¹; [Met⁵]enkephalin-Lys⁶, $K_m$ = 375 μM, $k_{cat}$ = 663 min⁻¹; and bradykinin, $K_m$ = 16 μM, $k_{cat}$ = 147 min⁻¹.

Thus, [Met[5]]enkephalin-Arg[6] has the highest specificity constant ($k_{cat}/K_m$), and changing the C-terminal amino acid to lysine ([Met[5]]enkephalin-Lys[6]) decreases it over 10-fold owing to a large increase in the $K_m$. As with the synthetic substrates, changing the penultimate residue also markedly affected the kinetics. Although the $K_m$ of [Leu[5]]enkephalin-Arg[6] is similar to that of [Met[5]]enkephalin-Arg[6], the $k_{cat}$ is almost 9-fold lower. Of the substrates tested, bradykinin (with C-terminal Phe[8]-Arg[9]) has the lowest $K_m$ and the second highest $k_{cat}/K_m$.

*Cloning and Sequencing of Carboxypeptidase M.* A carboxypeptidase M clone was isolated from a human placental cDNA library.[8] The 2009-base pair insert contained an open reading frame of 1317 base pairs coding for a mature protein of 426 residues (after removal of the signal peptide).[8] The deduced sequence contains six potential Asn-linked glycosylation sites, consistent with its 23% carbohydrate content. The amino acid sequence of carboxypeptidase M is 41% identical with that of the active subunit of human plasma carboxypeptidase N, 43% identical with human and 41% with bovine carboxypeptidase H, but only 15% identical with bovine pancreatic carboxypeptidase A or B.[8] Although the overall sequence identity with the pancreatic carboxypeptidases is not high, many of the active site residues identified in carboxypeptidases A and B[20,21] have been conserved in carboxypeptidase M.[8] These include the three zinc-binding residues (His-66, Glu-69, and His-173), two of the substrate binding residues (Arg-137 and Tyr-242), and the catalytic Glu-264. The arginine at position 127 in carboxypeptidase A, thought to polarize the scissile carbonyl group in the substrate, corresponds to Lys-118 in carboxypeptidase M and the putative residue in the side-chain binding pocket (Ile-255 in carboxypeptidase A and Asp-253 in carboxypeptidase B) is Gln-249.[8] Definitive proof of the actual active-site residues in carboxypeptidase M will require the use of site-directed mutagenesis and/or X-ray crystallography.

*Membrane Anchoring of Carboxypeptidase M.* Carboxypeptidase M is firmly anchored to cell membranes, yet the primary sequence does not contain a true transmembrane spanning region.[8] The extreme C terminus contains a mildly hydrophobic stretch of 15 amino acids that is similar in nature to those of other proteins that are attached to the membrane via a phosphatidylinositol–glycan (PI-G) anchor.[22] In this type of membrane attachment, the hydrophobic C terminus acts as a temporary anchor in the Golgi that is then cleaved off, leading to attachment of the new C terminus

---

[20] D. W. Christianson and W. N. Lipscomb, *Acc. Chem. Res.* **22**, 62 (1989).
[21] K. Titani, L. H. Ericsson, K. A. Walsh, and H. Neurath, *Proc. Natl. Acad. Sci. U.S.A.* **72**, 1666 (1975).
[22] M. G. Low, *FASEB J.* **3**, 1600 (1989).

to the ethanolamine moiety of a preformed PI-G anchor.[22] Consistent with this mechanism of membrane anchoring, carboxypeptidase M can be released from membrane preparations by bacterial PI-PLC[16] (see above). Direct evidence for the presence of a PI-G anchor was obtained by labeling Madin-Darby canine kidney (MDCK) cells (which have high carboxypeptidase M activity) with [³H]ethanolamine. Antiserum specific for carboxypeptidase M immunoprecipitated a single radiolabeled band from the solubilized membrane fraction, corresponding in size to that of carboxypeptidase M, as revealed by SDS–PAGE and auroradiography.[16]

*Distribution and Function.* Membrane-bound carboxypeptidase M is a widely distributed ectoenzyme, present in many human and animal organs and tissues such as kidney, lung, intestine, brain, peripheral nerves, blood vessels, placenta, and in cultured fibroblasts, endothelial cells, and MDCK cells.[5,6,16,23,24] As such, it can participate in a variety of functions including controlling peptide hormone activity at the cell surface, extracellular protein and peptide degradation, and extracellular prohormone processing.[6] After its release from cell membranes into body fluids it may function in soluble form. For example, significant carboxypeptidase M-like activity has been found in amniotic fluid, urine, and seminal plasma.[4,5,25] Future studies should help to define the important role(s) of carboxypeptidase M in the regulation of peptide hormone activity.

### Acknowledgment

Some of the reported studies were supported in part by National Institutes of Health Grants DK41431, HL36473, and HL36082.

[23] A. Nagae, P. A. Deddish, R. P. Becker, C. H. Anderson, M. Abe, R. A. Skidgel, and E. G. Erdös, *J. Neurochem.* **59**, 2201 (1992).
[24] A. Nagae, M. Abe, R. P. Becker, P. A. Deddish, R. A. Skidgel, and E. G. Erdös, *Am. J. Respir. Cell Mol. Biol.* **9**, 221 (1993).
[25] R. A. Skidgel, P. A. Deddish, and R. M. Davis, *Arch. Biochem. Biophys.* **267**, 660 (1988).

# [42] Carboxypeptidase T

*By* Valentin M. Stepanov

### Introduction

Whereas exocellular endopeptidases are rather common among the enzymes produced by prokaryotic microorganisms, secretion of exopeptidases seems to be a rare event. Carboxypeptidase T was found to be secreted by *Thermoactinomyces vulgaris,* strain INMI,[1] along with thermitase, an endopeptidase that belongs to the subtilisin family, small amounts of a metalloproteinase, and a Glu, Asp-specific serine proteinase. The functional characteristics of the extracellular enzyme, named carboxypeptidase T (EC 3.4.17.18), and, more importantly, its amino acid sequence[2] and tertiary structure[3,4] clearly indicate that it belongs to the metallocarboxypeptidase family, being homologous to the well-characterized carboxypeptidases A and B. Remarkably, carboxypeptidase T specificity toward peptide substrates combines the characteristics of carboxypeptidases A and B, that is, the enzyme cleaves off C-terminal neutral, preferably hydrophobic, amino acids, like carboxypeptidase A, and also arginine and lysine residues that bear cationic groups in their side chains. Some other prokaryotic metallocarboxypeptidases such as carboxypeptidase SG produced by *Streptomyces griseus,* strain K-1,[5-7] and carboxypeptidase L,[8] isolated from *Streptomyces spheroides,* strain 35, culture filtrate, share this remarkable dual specificity pattern.

## Molecular Characteristics

A molecular mass of 38,000 Da was assessed by sodium dodecyl sulfate (SDS) electrophoresis in 10% polyacrylamide gel, whereas a mass of 36,928 Da was calculated from the amino acid sequence. The isoelectric point (p$I$) of 5.3, a value which agrees with the enzyme amino acid composition, was determined by isoelectrofocusing on polyacrylamide plates in pH 3.5–9.0 gradient.

[1] A. L. Osterman, V. M. Stepanov, G. N. Rudenskaya, O. M. Khodova, I. A. Tsaplina, M. B. Yakovleva, and L. G. Loginova, *Biokhimiya* **49,** 292 (1984).
[2] S. V. Smulevich, A. L. Osterman, O. V. Galperina, M. V. Matz, O. P. Zagnitko, R. M. Kadyrov, I. A. Tsaplina, N. V. Grishin, G. G. Chestukhina, and V. M. Stepanov, *FEBS Lett.* **291,** 75 (1991).
[3] K. M. Polyakov, G. V. Obmolova, I. P. Kuranova, B. V. Strokopytov, B. K. Vainshtein, O. V. Mosolova, V. M. Stepanov, and G. N. Rudenskaya, *Mol. Biol. (Moscow)* **23,** 266 (1989).
[4] A. Teplyakov, K. Polyakov, G. Obmolova, B. Strokopytov, I. Kuranova, A. Osterman, N. Grishin, S. Smulevich, O. Zagnitko, O. Galperina, M. Matz, and V. Stepanov, *Eur. J. Biochem.* **208,** 281 (1992).
[5] Y. F. Seber, T. P. Toomey, Y. T. Powell, K. Brew, and W. M. Awad, *J. Biol. Chem.* **251,** 204 (1976).
[6] K. Breddam, T. Y. Bazzone, B. Holmquist, and B. L. Vallee, *Biochemistry* **18,** 1563 (1979).
[7] Y. Narahachi, *J. Biochem. (Tokyo)* **107,** 879 (1990).
[8] G. N. Rudenskaya, V. G. Kreier, N. S. Landau, N. I. Tarasova, E. A. Timokhina, N. S. Egorov, and V. M. Stepanov, *Biokhimiya* **52,** 2002 (1986).

*Primary Structure.* The primary structure of carboxypeptidase T was determined by DNA sequencing of the cloned structural gene.[2] The N-terminal sequence of 31 amino acid residues, established by automated Edman degradation of the mature enzyme, as well as the sequences of five tetra- to heptapeptides isolated from the proteolytic digest, coincided with those predicted by DNA sequencing.

The carboxypeptidase T gene coding for the open reading frame contains 1269 bp. The mature enzyme sequence consists of 326 amino acid residues, preceded by 98 residues that comprise a prepro region. The N-terminal stretch of the latter sequence starts with a motif characteristic for a signal peptide, namely, three positively charged residues followed by a prolonged hydrophobic stretch. The boundary between pre- and prosequences cannot be specified yet. The propart of the carboxypeptidase T precursor contains about 70 amino acid residues and does not reveal any homology with the propeptides of mammalian metallocarboxypeptidases. The mature enzyme sequence starts from an aspartic acid residue. It follows that a Gln-Asp bond is split to convert the precursor to the mature carboxypeptidase T. Hence, a proteinase with a rather unusual specificity pattern, capable of hydrolyzing a Gln-Asp peptide bond, seems to be involved in procarboxypeptidase T activation. The homologous carboxypeptidase SG from *Streptomyces griseus* also possesses Asp as the N-terminal amino acid. Procarboxypeptidase T has not been observed in *Thermoactinomyces vulgaris* cultures.

Sequence alignment of carboxypeptidase T and 15 other metallocarboxypeptidases[9] has shown that the enzyme belongs to a subfamily of "short" metallocarboxypeptidases referred to as "digestive" enzymes owing to their apparently simple degradative function. These carboxypeptidases lack several loops or even small domains that are inserted in the substantially longer sequences of their "regulatory" homologs, like carboxypeptidases E, N, and M.

Carboxypeptidase SG from *Streptomyces griseus,* the nearest homolog of carboxypeptidase T, revealed 63.9% identical residues. A level of 27% identity was found when the enzyme sequence was compared with that of carboxypeptidase A, and 26% with carboxypeptidase B. These figures indicate a rather remote relation of the prokaryotic and eukaryotic enzymes, although their tertiary folds and the structural elements responsible for the catalytic mechanisms are remarkably similar.

*Tertiary Structure.* The crystal structure of carboxypeptidase T was determined by X-ray diffraction at 2.35 Å resolution.[4] The core of the molecule is composed of a twisted eight-stranded $\beta$ sheet and six $\alpha$ helices antiparallel

---

[9] A. L. Osterman, N. V. Grishin, S. V. Smulevitch, M. V. Matz, O. P. Zagnitko, L. P. Revina, and V. M. Stepanov, *J. Protein Chem.* **11**, 561 (1992).

TABLE I
AMINO ACID RESIDUES FORMING S1′ SPECIFICITY POCKETS IN CARBOXYPEPTIDASES
A, B, AND T

| Enzyme | Residue | | | | | | |
|---|---|---|---|---|---|---|---|
|  | 203 | 243 | 248 | 250 | 253 | 255 | 268 |
| Carboxypeptidase A[a] | Leu | Ile | Tyr | Ala | Gly | Ile | Thr |
| Carboxypeptidase B[a] | Leu | **Gly** | Tyr | Ala | Gly | **Asp** | Thr |
|  | 211 | 251 | 255 | 257 | 260 | 262 | 275 |
| Carboxypeptidase T | Leu | **Ala** | Tyr | **Thr** | **Asp** | **Thr** | Thr |

[a] Numbering according to carboxypeptidase A sequence.

to the β strands. Superposition of carboxypeptidase A and T tertiary structures gave an average deviation 0.8 Å for 235 compared $C_\alpha$ atoms, whereas certain differences were detected in the flexible regions of the molecule distant from the active site.

*Substrate Binding Site and Specificity.* Crystallographic data allow us to discuss the most prominent feature of the enzyme specificity, namely, its capacity to digest peptides with hydrophobic amino acid residues at P1′ position (like carboxypeptidase A) as well as to cleave off the C-terminal arginine or lysine residues (similar to carboxypeptidase B). Such a dual specificity pattern has to be explained by the structure of the carboxypeptidase T primary specificity (S1′) pocket. Topologically equivalent amino acid residues whose side chains line the S1′ pockets in carboxypeptidases A, B, and T are compared in the Table I.

The presence of Asp-255 in carboxypeptidase B and Asp-253 (Asp-260[10]) in carboxypeptidase T explain the specificity of the two enzymes toward arginine or lysine residues at P1′ of the peptide substrates. On the other hand, it seems plausible that the Asp-253 (Asp-260) carboxylate might form a hydrogen bond with threonine-255 (Thr-262) also located in the carboxypeptidase T binding pocket, which would promote the escape of water molecules from the P1′ pocket and thus facilitate the binding of the hydrophobic side chains. This hydrogen bond is absent in the carboxypeptidase B structure.

As shown on model peptides, carboxypeptidase T cleaves off C-terminal Arg, Lys, His, Phe, Tyr, Trp, Leu, Met, and Ala residues. A free α-amino group hampers the hydrolysis; hence, only acylated dipeptides are hy-

[10] Carboxypeptidase A numbering is used for amino acid residues; carboxypeptidase T numbering is given in parentheses.

TABLE II
KINETIC PARAMETERS FOR HYDROLYSIS OF PEPTIDE SUBSTRATES Z-Ala-Ala-Xaa-OH BY
CARBOXYPEPTIDASES A, B, AND T[a]

| | Carboxypeptidase | | | | | | | | |
| | A | | | B | | | T | | |
| Xaa | $K_m$ ($\mu M$) | $k_{cat}$ (Sec$^{-1}$) | $k_{cat}/K_m$ ($\mu M^{-1}$ Sec$^{-1}$) | $K_m$ ($\mu M$) | $k_{cat}$ (Sec$^{-1}$) | $k_{cat}/K_m$ ($\mu M^{-1}$ Sec$^{-1}$) | $K_m$ ($\mu M$) | $k_{cat}$ (Sec$^{-1}$) | $k_{cat}/K_m$ ($\mu M^{-1}$ Sec$^{-1}$) |
|---|---|---|---|---|---|---|---|---|---|
| Phe | 3 | 42 | 14 | 430 | 11 | 0.026 | 13 | 2.6 | 0.2 |
| Leu | 12 | 35 | 3 | 1000 | 14 | 0.014 | 26 | 12.5 | 0.48 |
| Trp | 1.8 | 2 | 1.1 | 240 | 0.8 | 0.003 | 10 | 0.06 | 0.006 |
| Arg | — | — | — | 10 | 4.5 | 0.45 | 920 | 57 | 0.061 |
| Lys | — | — | — | 15 | 8.7 | 0.58 | 780 | 5 | 0.006 |

[a] Data provided by N. V. Grishin and V. V. Morkovtsev in this laboratory.

drolyzed. The P1 amino acid residue, preceding the C terminus, influences the rate of hydrolysis; for example, glycine and glutamic acid at P1 retard the cleavage, whereas proline blocks it completely. Kinetic parameters for hydrolysis of a series of tripeptide substrates by carboxypeptidases A, B, and T (Table II) indicate that the hydrophobic amino acids phenylalanine and leucine are cleaved by carboxypeptidases A and T with comparable efficiency. On the other hand, carboxypeptidase B splits off the C-terminal arginine and lysine far better than carboxypeptidase T; $K_m$ values for the latter enzyme correspond to a rather loose binding of the respective substrates, only partially compensated for by relatively high $k_{cat}$ values.

The carboxylate group of Asp-253 (Asp-260) is positioned within the S1 subsite of carboxypeptidase T differently from its functional analog Asp-255 in carboxypeptidase B. Although it allows the enzyme to attack peptide bonds formed by arginine or lysine, it fails to confer a level of efficiency characteristic of a highly specialized enzyme. It appears that the loss of efficiency represents a functionally acceptable cost of the broad specificity of carboxypeptidase T, insofar as the subtilisin-like serine proteinase produced concomitantly with carboxypeptidase T, would degrade protein substrates predominantly to peptides with hydrophobic C-terminal residues. Rather limited activity of the enzyme toward arginine and lysine residues would nevertheless be sufficient for protein degradation.

*Functional Groups Involved in Catalytic Mechanism.* Although there are still uncertainties concerning the catalytic mechanism of metallocarboxypeptidases, it appears that the amino acid residues involved have been identified fairly well.[11] As shown by the sequence alignment and the tertiary

[11] D. W. Christianson and W. N. Lipscomb, *Acc. Chem. Res.* **22**, 62 (1989).

TABLE III

TOPOLOGICALLY AND FUNCTIONALLY EQUIVALENT RESIDUES IN ACTIVE SITES OF
CARBOXYPEPTIDASES A AND T

| | Residue in carboxypeptidase | |
| Presumed functional role[a] | A | T |
| --- | --- | --- |
| $Zn^{2+}$ coordination | His-69 | His-69 |
| | His-196 | His-204 |
| | Glu-72 | Glu-72 |
| Fixation of P1' residue carboxylate | Asn-144 | Asn-146 |
| | Arg-145 | Arg-147 |
| | Tyr-248 | Tyr-255 |
| Polarization of P1' residue carbonyl group | Arg-127 | Arg-129 |
| $H_2O$ polarization, H-donation to the leaving P1' residue NH group | Glu-270 | Glu-277 |

[a] D. W. Christianson and W. N. Lipscomb, *Acc. Chem. Res.* **22,** 62 (1989).

structures, the residues that directly participate in catalysis are strictly conserved in all metallocarboxypeptidases, including carboxypeptidase T. Eight residues of carboxypeptidase A, for which specific roles in catalysis have been defined, are retained in carboxypeptidase T, and their spatial positions are in line with their presumed functions (Table III). The enzyme contains one $Zn^{2+}$ ion per molecule, bound in the same manner as in carboxypeptidase A. The enzyme is completely inhibited by chelating agents (25 m$M$ Na$_2$EDTA or 10 m$M$ $o$-phenanthroline) but proved insensitive to the inhibitors of serine (1 m$M$ phenylmethylsulfonyl fluoride or Z-Ala-Ala-Phe-CH$_2$Cl) or cysteine (1 m$M$ Hg$^{2+}$ or $p$-hydroxymercuribenzoate) proteinases. The enzyme hydrolyzes substrates optimally at pH 7–8.

One functionally important feature distinguishes the carboxypeptidase T structure from those of carboxypeptidases A and B: it contains four binding sites for Ca$^{2+}$ ions. The Ca$^{2+}$ seems to be important for the stability of the enzyme, especially for temperature stability. In the presence of 1 m$M$ Ca$^{2+}$ the enzyme retains at least 80% of activity after 8 hr of incubation at 37°, whereas in the absence of added calcium ions activity drops to 50–60% after 3 hr at 20°.

## Activity Assay

The activity of carboxypeptidase T is measured against the chromogenic substrates 2,4-dinitrophenyl(Dnp)-Ala-Ala-Arg-OH or Dnp-Ala-Ala-Leu-

Arg-OH or their analogs.[12] The method serves also for the assay of carboxy-peptidase B or other carboxypeptidases (e.g., carboxypeptidase N) capable of cleaving C-terminal arginine residues.

$$\text{Dnp-Ala-Ala-Arg-OH} \xrightarrow{\text{carboxypeptidase T}} \text{Dnp-Ala-Ala-OH} + \text{H-Arg-OH}$$

Splitting of C-terminal arginine dramatically changes the physicochemical properties of the molecule bearing the 2,4-dinitrophenyl chromophore group. Thus, Dnp-Ala-Ala-OH can be extracted into an organic solvent (e.g., ethyl acetate from acidified water solution), whereas the uncleaved substrate Dnp-Ala-Ala-Arg-OH remains in the water phase. This simple procedure allows one to measure the extent of substrate hydrolysis and to assess the enzyme activity.[12] An even simpler protocol, especially suitable for carboxypeptidase T assay in column fractions, depends on an ion-exchange separation of the colored hydrolysis product from the excess substrate. Dnp-Ala-Ala-Arg-OH in weakly acidic solution is bound by a cation exchanger (e.g., sulfoethyl or sulfopropyl-Sephadex) owing to the positively charged guanidino group of the arginine residue, whereas the product, Dnp-Ala-Ala-OH, is unbound and appears in the breakthrough, where its concentration can be easily assessed spectrophotometrically. The substrate synthesis and detailed assay protocol are given below.

*Procedures*

*Synthesis of Dnp-Ala-Ala-Arg-OH.* Synthesis of the Dnp-Ala-Ala-Arg-OH substrate[13] is mediated by the enzymes subtilisin and trypsin. To a

$$\text{Dnp-Ala-Ala-OCH}_3 + \text{H-Arg-NH}_2 \xrightarrow{\text{subtilisin}} \text{Dnp-Ala-Ala-Arg-NH}_2$$

$$\text{Dnp-Ala-Ala-Arg-NH}_2 + \text{H}_2\text{O} \xrightarrow{\text{trypsin}} \text{Dnp-Ala-Ala-Arg-OH}$$

solution of 294 mg (1.2 mmol) of H-Arg-NH$_2$·2HCl in 3.5 ml dimethyl sulfoxide, 0.2 ml water, and 0.5 ml triethylamine is added 272 mg (0.8 mmol) of Dnp-Ala-Ala-OCH$_3$ in 6.5 ml, followed by 1 g of catalyst. To prepare the catalyst 1 g of macroporous glass (CPG-10, Serva, Heidelberg, Germany) is soaked in a solution of 193 mg subtilisin in 2 ml of 50 m$M$ Tris-HCl, pH 7.5, buffer, then dried in a desiccator. The mixture is shaken for 48 hr at 20°, and then the catalyst is filtered off, washed with a mixture of dimethyl sulfoxide, water, and acetonitrile (35:2:63; v/v). The catalyst can be used repeatedly. To the filtrate and washings is added 1 ml of 5.7

---

[12] L. A. Lyublinskaya, T. I. Vaganova, T. S. Paskhina, and V. M. Stepanov, *Biokhimiya* **38,** 790 (1973).
[13] Procedure developed by M. P. Yusupova in this laboratory (1993).

$M$ HCl, and the acetonitrile is evaporated under reduced pressure. The residue is diluted with 150 ml water and extracted with ethyl acetate (50 ml, three times). To the water layer is added 0.5 $M$ NaOH to pH 7.4; then 2 mg of trypsin is introduced, and the mixture is stirred for 2 hr at 32°. After acidification with 5.7 $M$ HCl to pH 3.0 the product is extracted with water-saturated $n$-butanol (150 ml, four times). The butanol layer is washed with water (10 ml, 2 times) and evaporated under reduced pressure. The residue is dissolved in 5 ml $CH_3OH$ and precipitated with 45 ml of ethyl ether. The crystalline yellow product is filtered off and dried over KOH. The yield of Dnp-Ala-Ala-Arg-OH is routinely 63%. The peptide gives one peak, with a retention time of 7.6 min, when subjected to high-performance liquid chromatography (HPLC) on Beckman (Palo Alto, CA) Ultrasphere ODS column (4.6 × 45 mm) in a methanol concentration gradient [solution A: 5% $CH_3OH$, solution B: 90% $CH_3OH$, both containing 0.05% $CF_3COOH$ and $(C_2H_5)_3N$; gradient from 15 to 50% B in 15 min, detection at 360 nm].

*Assay of Carboxypeptidase T activity.* To 1 ml of 0.5 m$M$ substrate solution in 0.1 $M$ TRIS-HCl buffer, pH 7.5, 10–100 μl of the enzyme solution is added.[1] The mixture is incubated for 10–60 min at 37°, and then 0.2 ml of 50% $CH_3COOH$ is added to stop the reaction. The mixture is quantitatively transferred to a microcolumn (plastic cone from an Eppendorf automatic pipette plugged with cotton) that contains 2 ml of SP-Sephadex C-25, preequilibrated with 1 $M$ $CH_3COOH$. The column is washed with 1 $M$ $CH_3COOH$ (two times, 1 ml). The washings are combined, and the $A_{360}$ of the solution is measured. To calculate Dnp-Ala-Ala-OH concentration, a molar extinction value ($\varepsilon_{360}$) of 15,000 is used. One activity unit is equal to the amount of enzyme that hydrolyzes 1 μmol of the substrate in 1 min under the specified conditions.

TABLE IV

PURIFICATION OF CARBOXYPEPTIDASE T

| Purification step | Protein content ($A_{280}$) | Specific activity (units/$A_{280}$) | Yield (%) | Purification (-fold) |
|---|---|---|---|---|
| Dried culture filtrate of *Thermoactinomyces vulgaris* | 25,000 | 0.025 | 100 | 1 |
| Chromatography on bacitracin-silochrome | 1150 | 0.38 | 74 | 14.5 |
| Chromatography on bacitracin-Sepharose | 40 | 14.3 | 85 | 550 |

## Isolation of Carboxypeptidase T

The isolation procedure[1] is based on application of affinity chromatography on bacitracin-silochrome to isolate the proteolytic enzymes contained in the culture filtrate. In further chromatography on a bacitracin-Sepharose 4B column, specific elution of carboxypeptidase T was made possible by its inhibition with the chelating agent $Na_2EDTA$. The activity was then restored by extensive dialysis of the preparation against calcium chloride solution. Bacitracin-containing ligands have been repeatedly used for affinity separation of many proteolytic enzymes,[13–15] and their preparation has been previously described.[14,15] The adsorption of carboxypeptidase T is apparently due to the enzyme-binding site interacting with the bacitracin peptide chain, perhaps with an additional interaction between the $Zn^{2+}$ and the cyclopeptide. Good results on the final purification step were also obtained using as the adsorbent Sepharose 4B with covalently bound $p$-aminobenzyl succinate that served as a substrate analog (CABS-Sepharose).

*Procedure.* Table IV summarizes a typical purification procedure. To 50 g of dried *Thermoactinomyces vulgaris* culture filtrate is added 150 ml of 50 m$M$ Tris-HCl buffer, pH 7.2, that contained 1 m$M$ $CaCl_2$. The insoluble material is removed by centrifugation, and the solution is applied to a 350-ml bacitracin-silochrome column. The column is washed with the same buffer, then eluted with 20% (v/v) 2-propanol in 1 $M$ NaCl and 50 m$M$ Tris-HCl buffer, pH 7.2. The eluate is dialyzed against water, then applied to a bacitracin-Sepharose 4B column (0.8 × 20 cm), preequilibrated with pH 7.2 buffer. The column is washed with 50 m$M$ Tris-HCl, pH 7.2, buffer, containing 1 m$M$ $CaCl_2$, and then carboxypeptidase T is eluted with 18 m$M$ $Na_2EDTA$ in pH 7.2 buffer. The column is washed with 0.5 $M$ NaCl in the same buffer, and the serine proteinase is eluted with 20% isopropyl alcohol in 1 $M$ NaCl and 50 m$M$ Tris-HCl buffer, pH 7.2. To the fraction that contains carboxypeptidase T is added $CaCl_2$ to 50 m$M$, and the solution is dialyzed against 1 m$M$ $CaCl_2$ then against water and lyophilized, which gives about 20 mg of purified enzyme.

[14] V. M. Stepanov and G. N. Rudenskaya, *J. Appl. Biochem.* **5,** 420 (1983).
[15] V. M. Stepanov, G. N. Rudenskaya, A. V. Gaida, and A. L. Osterman, *J. Biochem. Biophys. Methods* **5,** 177 (1981).

# [43] Pitrilysin

By Angela Anastasi and Alan J. Barrett

## Introduction

Pitrilysin is a peptidase so far known only from *Escherichia coli*, which was originally discovered as an activity hydrolyzing polypeptide fragments of $\beta$-galactosidase, and termed protease III.[1] The enzyme was also encountered as an insulin-degrading activity from the periplasmic space of *E. coli*, and referred to as protease Pi.[2] The name pitrilysin (EC 3.4.24.55, formerly EC 3.4.99.44) is now recommended[3]; the name is entirely trivial but may call to mind the fact that the enzyme is the product of the *ptr* gene in *E. coli*. The lysin ending is commonly used in the names of metalloendopeptidases. Much of the interest in pitrilysin has stemmed from the fact that it is a member of a group of metallopeptidases described as family M16 in [13] in this volume. Other members of this group of proteins are eukaryotic endopeptidases, insulysin (formerly known as insulinase; see [44]), the two subunits of the mitochondrial processing peptidase (see [46]), and a paired-basic processing enzyme (see [45] in this volume).

## Assay Methods

Cheng and Zipser assayed pitrilysin in a complementation assay, in which the catalytic activity of incomplete molecules of $\beta$-galactosidase was restored by fragments derived from the N terminus of the intact form of the enzyme.[1] The activating fragments (termed auto $\alpha$) served as substrates for pitrilysin, so that the auto $\alpha$ activity was destroyed, and correspondingly little $\beta$-galactosidase activity was regenerated, when pitrilysin activity was high.

In the majority of subsequent work with the enzyme, radiolabeled insulin has been used as substrate.[2,4] However, the insulin assay has the disadvantage that it does not allow continuous monitoring of enzymatic activity, which is not ideal for kinetic work. This disadvantage has been overcome

---

[1] Y.-S. E. Cheng and D. Zipser, *J. Biol. Chem.* **254**, 4698 (1979).
[2] A. L. Goldberg, K. H. S. Swamy, C. H. Chung, and F. S. Larimore, this series, Vol. 80, p. 680.
[3] Nomenclature Committee of the International Union of Biochemistry and Molecular Biology, "Enzyme Nomenclature 1992." Academic Press, Orlando, Florida, 1992.
[4] C. C. Dykstra and S. R. Kushner, *J. Bacteriol.* **163**, 1055 (1985).

by the development of a fluorimetric assay.[5] Both insulin-degrading and fluorimetric assays are described below.

*Insulin Degradation Assay*

To quantify pitrilysin activity by use of insulin as the substrate, reaction mixtures are set up containing 5–15 μg of insulin, a trace amount of [125]I-labeled insulin [about 12,000–15,000 counts/min (cpm)], and the enzyme, in a final volume of 0.5 ml of assay buffer (50 m$M$ Tris-HCl, pH 7.5, containing 0.05% Brij 35). The radiolabeled insulin is obtainable from Lise Screen Ltd. (Watford, UK). After incubation for 1 hr at 37°, the reaction is stopped by the addition of 60 μl of 100% (w/v) trichloroacetic acid solution and 40 μl of bovine serum albumin solution (30 mg/ml) as carrier, with good mixing. The assay tubes are kept in ice for 30 min, and, after centrifugation, 0.4-ml samples of supernatant containing acid-soluble products are removed for gamma counting. The response of the assay is linear to about 35% degradation of the substrate. Another form of assay for insulin degradation, the plate assay, is described in [44].

*Quenched Fluorescence Assay*

As is described in [2] in this volume, a quenched fluorescence substrate is one in which a short sequence of amino acids containing the scissile peptide bond links a potentially fluorescent group to a group that acts as an internal quencher of the fluorescence. Cleavage of the scissile bond results in the appearance of fluorescence, which is monitored.

The quenched fluorescence substrate for pitrilysin and insulysin, QF27, was designed on the basis of the structure of fragment 16–28 of the vasoactive intestinal peptide (VIP).[5,6] The potential fluorophore is a derivative of 7-methoxycoumarin, and the quencher is a 2,4-dinitrophenyl group. Thus, the structure of QF27 is Mca-Nle-Ala-Val-Lys-Lys-Tyr-Leu-Asn-Ser-Lys(Dnp)-Leu-Asp-D-Lys. The enhancement of fluorescence resulting from hydrolysis is approximately 20-fold, and the sensitivity of the assay is simlar to that with insulin as substrate.

*Reagents.* The syntheses of QF27 and the reference standard, $N^2$-acetyl-$N^6$-Mca-L-Lys, have been described,[5] and the compounds are commercially available from Calbiochem NovaBiochem (Nottingham, UK).

[5] A. Anastasi, C. G. Knight, and A. J. Barrett, *Biochem. J.* **290**, 601 (1993).
[6] Abbreviations: Abz, *o*-aminobenzoyl; Dpa, DL-2-amino-3-(7-methoxy-4-coumaryl)propionic acid; Lys(Dnp), $N^6$-(2′,4′-dinitrophenyl)lysine; Mca, (7-methoxycoumarin-4-yl)acetyl; Nle, norleucine; VIP, vasoactive intestinal peptide. For amino acids, the standard three-letter or one-letter codes are used, except that k is D-lysine and in the oxidized B chain of insulin C is cysteic acid.

Substrate stock solution: QF27 is dissolved in dimethyl sulfoxide at 2.5 m$M$, and stored at 4°

Standard stock solution: $N^2$-Ac-$N^6$-Mca-L-Lys is dissolved in dimethyl sulfoxide at 5 m$M$, and stored at 4°

Assay buffer: 50 m$M$ Tris-HCl, pH 7.5, containing 0.05% Brij 35

*Procedure.* Fluorimetric assays with QF27 are conveniently made in a Perkin-Elmer (Norwalk, CT) spectrofluorimeter controlled by an IBM-compatible computer running the FLUSYS software.[7] Other fluorimeters can be used, but software for the collection of kinetic data is highly desirable. The sample is contained in a quartz cuvette accurately thermostated at 37°, and is stirred continuously.

Excitation and emission wavelengths are set to 328 and 393 nm, respectively. The fluorimeter is zeroed against substrate in assay buffer and calibrated to read 1000 arbitrary units of fluorescence for a solution containing both reference standard at a concentration 10% that of substrate and working strength substrate. In routine assays, the substrate concentration is 10 $\mu M$ (10 $\mu$l of the stock solution added), and the reference standard concentration is 1 $\mu M$ (50 $\mu$l of 1/100 diluted stock solution). Because the substrate causes some quenching of the fluorescence of the product and standard, it is essential that the instrument is restandardized whenever the substrate concentration is changed, as in the determination of $K_m$ values (see [2], this volume).

Continuous assays are performed in a total volume of 2.5 ml of assay buffer with 10 $\mu$l of substrate stock solution (final 10 $\mu M$), and the reaction is started by the addition of about 0.05 milliunit (mU) of enzyme, 1 mU of activity being defined as that hydrolyzing 1 nmol of substrate per minute at 37°. The fluorescence is recorded, and the rate of increase in fluorescence over a suitable time interval (5–30 min) is calculated by linear regression analysis of the values against time.

Assays with QF27 have been found to show accurately measurable rates of hydrolysis, linear with time, for enzyme concentrations in the range 0.01–0.1 mU/ml (17–170 ng/ml). QF27 is not susceptible to hydrolysis only by pitrilysin, and it is also the first described synthetic substrate for insulysin.[5]

## Purification

Pitrilysin has been purified from wild-type *E. coli*.[2,4] However, since the preparation of plasmids containing the gene for pitrilysin,[8,9] the preferred

[7] N. D. Rawlings and A. J. Barrett, *Comput. Appl. Biosci.* **6,** 118 (1990).
[8] C. C. Dykstra, D. Prasher, and S. R. Kushner, *J. Bacteriol.* **157,** 21 (1984).
[9] P. W. Finch, R. E. Wilson, K. Brown, I. D. Hickson, and P. T. Emmerson, *Nucleic Acids Res.* **14,** 7695 (1986).

starting material has become one of the *E. coli* strains that overexpress the enzyme. Anastasi *et al.*[5] compared two strains of *E. coli*: the strain PE004 that harbors the plasmid pPF307 (given by Professor P. T. Emmerson, Department of Biochemistry, University of Newcastle upon Tyne, Newcastle NE1 7RU, UK) and strain SK7814 containing plasmid pCDK35 (given by Dr. S. R. Kushner, Department of Genetics, Life Sciences Building, University of Georgia, Athens, GA 30602), both of which were grown in Luria–Bertani medium.[10] For the PE004 strain, the medium was supplemented with 50 $\mu$g/ml kanamycin, and the culture was grown initially at 30°; at $D_{660}$ = 0.05–0.1, the temperature was shifted to 37° for 4–5 hr, for temperature induction.[11] The SK7814 strain was grown in medium supplemented with 170 $\mu$g of chloramphenicol/ml for 5–6 hr to a $D_{660}$ value of approximately 2.0.

Pitrilysin was identifiable in both cytoplasmic and periplasmic fractions of the overexpressing cells by assay and by immunoblot, two-thirds of the activity being present in the cytoplasmic fraction, but the specific activity of the periplasmic fraction was 6-fold greater; that fraction was therefore preferred as the source for purification. The periplasmic extract of strain PE004 had an initial specific activity one-third that of strain SK7814, and accordingly the latter strain was used for further work.

*Procedure*

Approximately 12 g (wet weight) of SK7814 cells grown as above is harvested from culture (3 liters) by centrifugation at 5000 $g$ for 30 min, then washed with 10 m$M$ Tris-HCl, pH 7.5, containing 0.1 $M$ NaCl, followed by 10 m$M$ Tris-HCl, pH 7.5. All buffers used subsequently in the purification of pitrilysin contain 0.05% Brij 35. For the extraction of the periplasmic proteins, the pellet is mixed with 20 ml of chloroform, allowed to stand for 15 min at room temperature, and suspended in 180 ml of 10 m$M$ Tris-HCl, pH 7.5.[12] The suspension is centrifuged for 20 min at 6500 $g$. The supernatant is removed, recentrifuged at 8000 $g$ for 20 min, dialyzed against 20 m$M$ Tris-HCl, pH 8.0, and run on a column (1.5 $\times$ 11.5 cm, 20 ml) of Q-Sepharose (Sigma, St. Louis, MO) equilibrated in the same buffer. The column is eluted with a gradient to 250 m$M$ NaCl in the Tris buffer at a flow rate of 90 ml/hr. The fractions (2 ml) are assayed, activity being found at about 100 m$M$ NaCl. Active fractions are combined and dialyzed against 20 m$M$ potassium phosphate buffer, pH 7.0. The sample is run on a column (1 $\times$ 3 cm, 2.3 ml) of hydroxyapatite (Bio-Rad, Richmond, CA) and eluted

[10] J. Sambrook, E. F. Fritsch, and T. Maniatis, "Molecular Cloning: A Laboratory Manual." Cold Spring Harbor Laboratory, Cold Spring Harbor, New York, 1989.
[11] S. Yasuda and T. Takagi, *J. Bacteriol.* **154,** 1153 (1983).
[12] G. F.-L. Ames, C. Prody, and S. Kustu, *J. Bacteriol.* **160,** 1181 (1984).

with a gradient (60 ml) of 20–250 mM potassium phosphate buffer, pH 7.0. Active fractions are combined, dialyzed against 25 mM Bis–Tris-HCl, pH 6.3, and run on the Pharmacia (Piscataway, NJ) FPLC (fast protein liquid chromatography) Mono P (HR 5/20) column at a flow rate of 0.5 ml/min. The column is preequilibrated with the Bis–Tris buffer and is developed with Polybuffer 74 diluted 10-fold (pH 4.0). Active fractions are combined, dialyzed into 50 mM Tris-HCl, pH 7.5, and supplemented with glycerol to 40% (v/v) before the preparation is stored at −20°.

Typically, the final preparation of pitrilysin is 145-fold purified, in 4% yield.[5] The protein appears as a single band in sodium dodecyl sulfate (SDS)–polyacrylamide gel electrophoresis, with mobility slightly higher than that of β-galactosidase (116 kDa), consist with the molecular mass of 107 kDa calculated from the amino acid sequence.

## Structure

The ptr gene encoding pitrilysin was cloned as a part of a 19-kb BamHI fragment and was found to be physically located between the recB and recC genes, whose products are subunits of exonuclease V. When the genes were amplified for the overexpression of exonuclease V, an increase in pitrilysin activity was obtained incidentally.[8] The complete nucleotide sequence of the ptr gene[9] shows that pitrilysin is a member of family M16 of metallopeptidases ([13], this volume).

Pitrilysin contains zinc,[13] and site-directed mutagenesis has implicated the two histidine residues in the sequence -His-Tyr-Leu-Glu-His- in the binding of zinc. Moreover, both of the His residues and the Glu in the same sequence are required for the catalytic activity of the enzyme.[14]

## Requirements for Activity

The pH optimum for hydrolysis of QF27 by pitrilysin is about pH 7.5. The enzyme shows no activation by thiol compounds (unlike the homologous insulysin; see [44]), merely being inhibited at high concentrations, as is normal for metallopeptidases. The metal dependence of pitrilysin is described below, but no addition of divalent cations to the solution is normally required for full activity.

## Substrate Specificity

Little has been known about the specificity of pitrilysin for cleavage of peptide bonds, although Cheng and Zipser showed cleavage of insulin B

---

[13] L. Ding, A. B. Becker, A. Suzuki, and R. A. Roth, J. Biol. Chem. 267, 2414 (1992).
[14] A. B. Becker and R. A. Roth, Proc. Natl. Acad. Sci. U.S.A. 89, 3835 (1992).

chain at the -Tyr[16]+Leu[17]-bond and more slowly at -Phe[24]+Tyr[25]-.[1] The points of cleavage of a number of oligopeptides by pitrilysin were determined by use of high-performance liquid chromatography analysis, by Anastasi *et al.*,[5] and the results are summarized in Fig. 1. With each of the substrates shown, all susceptible bonds were cleaved at similar rates so that the relative heights of HPLC peaks for the products did not change appreciably with time of incubation. There was no further degradation of the products. In that small set of scissile bonds, no clear preference for amino acids around the bond hydrolyzed by pitrilysin is apparent, although the residue in the P1 position is commonly Leu or Tyr.

*Kinetics.* Kinetic parameters for the hydrolysis of the synthetic substrate QF27 by pitrilysin are $K_m = 7.7 \pm 0.75$ $\mu M$ and $k_{cat} = 2.9$ sec$^{-1}$ (taking an $M_r$ of 107,000, and assuming that a preparation of 580 mU/mg was fully active). The value of $k_{cat}/K_m$ was thus $3.8 \times 10^5$ $M^{-1}$ sec$^{-1}$. It should be noted that two bonds are cleaved in QF27 at approximately equal rates, so the rate of cleavage of each would be about one-half the total rate of hydrolysis.

QF27 has also been found to be a substrate for *Drosophila* insulysin, which forms products identical to those resulting from the action of pitrilysin. The $K_m$ of *Drosophila* insulysin for QF27 is $18 \pm 3.3$ $\mu M$.

Rates of hydrolysis of peptides were determined by HPLC analysis after timed incubations and expressed relative to the rate for insulin B chain (Table I). The rate of disappearance of intact insulin B chain was the greatest in the series, despite the fact that most of the peptides were cleaved at two or more bonds. Substrates of pitrilysin so far discovered range in

```
Insulin B chain (oxidized)
                                    ↓
    F-V-N-Q-H-L-C-G-S-H-L-V-E-A-L-Y-L-V-C-G-E-R-G-F-F-Y-T-P-K-A

VIP (pig)
                                  ↓
    H-S-D-A-V-F-T-D-N-Y-T-R-L-R-K-Q-M-A-V-K-K-Y-L-N-S-I-L-N

VIP(10-28) (pig)
                          ↓ ↓                    ↓ ↓
              Y-T-R-L-R-K-Q-M-A-V-K-K-Y-L-N-S-I-L-N

VIP(16-28) (chicken)
                                      ↓ ↓ ↓
                          E-M-A-V-K-K-Y-L-N-S-V-L-T

                                              Dnp
QF27                                  ↓   ↓    |
                    Mca-Nle-A-V-K-K-Y-L-N-S-K-L-D-k
```

FIG. 1. Points of cleavage of various peptides by pitrilysin. The sequences of the polypeptides are shown in the single-letter code, and points of hydrolysis are marked by arrows.

TABLE I
RELATIVE RATES OF DEGRADATION OF PEPTIDES BY PITRILYSIN[a]

| Peptide | Relative rate of degradation | Bonds cleaved |
|---------|------------------------------|---------------|
| Insulin B chain (oxidized) (FVNQHLCGSHLVEALYLVCGERGFFYTPKA) | 1 | 1 |
| Secretin (HSDGTFTSELSRLREGARLQRLLQGLV) | 0.3 | Multiple |
| VIP(1–28) (HSDAVFTDNYTRLRKQMAVKKYLNSILN) | 0.23 | 1 |
| VIP(10–28) (YTRLRKQMAVKKYLNSILN) | 0.2 | 4 |
| QF27 [Mca-Nle-AVKKYLNSK(Dnp)LDk] | 0.2 | 2 |
| Thyrocalcitonin (CSNLSTCVLSAYWRNLNNFHRFSGMGFGPETP) | 0.13 | Multiple |
| VIP(16–28) (EMAVKKYLNSVLT) | 0.07 | 3 |
| Substance P (RPKPQQFFGLM) | 0.07 | 2 |
| Angiotensinogen(1–14) (DRVYIHPFHLLVYS) | 0.03 | 2 |
| LH-RH (EHWSYGLRPG) | 0.02 | 2 |
| VIP(1–12) (HSDAVFTDNYTR) | 0 | — |
| Angiotensin I (DRVYIHPFHL) | 0 | — |
| Angiotensin II (DRVYIHPF) | 0 | — |
| Dynorphin A(1–13) (YGGFLRRIRPKLK) | 0 | — |
| Dynorphin A(1–8) (YGGFLRRI) | 0 | — |
| Glucagon(22–29) (FVQWLMNT) | 0 | — |
| Insulin A chain (GIVEQCCASVCSLYQLENYCN) | 0 | — |
| EGF[b] receptor(1005–1016) (DVVDADEYLIPQ) | 0 | — |
| α-Casein(90–96) (RYLGYLD) | 0 | — |
| Bradykinin (RPPGFSPFR) | 0 | — |

[a] The values of the apparent first-order rate constants for hydrolysis of the peptides were determined as described in the text and have been normalized by taking the value for insulin B chain as 1.
[b] EGF, Epidermal growth factor.

size from 10 to 51 amino acid residues. The smallest of the peptides for which cleavage was detectable were luliberin (luteinizing hormone-releasing hormone, LH-RH) (10 residues) and substance P (11 residues). A number of potential substrates of less than 10 residues we all found not to be cleaved (see Table I). In addition to those shown were Abz-Arg-Ala-Leu-Tyr-Leu-Val-Lys(Dnp), containing the bond rapidly cleaved by pitrilysin in the insulin B chain, and Mca-Pro-Leu-Gly-Leu-Dpa-Ala-Arg-NH$_2$ (known as QF24), a substrate for mammalian matrix metalloendopeptidases.[15] These results indicate that there is a lower limit of the size of a substrate for pitrilysin close to 10 residues.

[15] C. G. Knight, F. Willenbrock, and G. Murphy, FEBS Lett. 296, 263 (1992).

Insulin, glucagon, and insulin B chain added to assays of pitrilysin with QF27 as substrate were found to behave as inhibitors with $K_i$ values of 7.1, 1.1, and 17.7 $\mu M$, respectively. These results were consistent with the three polypeptides being substrates of pitrilysin, with these very low $K_m$ values. Only at very low concentrations of insulin or glucagon can appreciable rates of degradation by pitrilysin be detected by HPLC analysis, however. For example, at a concentration of 50 $\mu M$ no cleavage of either insulin or glucagon was detected after 2 hr, whereas when the substrate concentrations were reduced to 1–5 $\mu M$ cleavage was readily apparent. These observations probably show that insulin and glucagon have very low $k_{cat}$ values, as well as low $K_m$ values, with pitrilysin. However, an alternative possiblity that has yet to be excluded is that both polypeptides aggregate in solution at the higher concentrations, decreasing their susceptibility to hydrolysis.[16,17]

In summary, the data for the specificity of action of pitrilysin on polypeptides indicate a preference for a bulky hydrophobic residue in the P1 position, but otherwise there is no readily definable specificity for amino acid residues around a scissile bond in the small set of cleavage sites so far identified. No substrate of less than 10 residues has yet been discovered, and no hydrolysis of any peptide bond less than four residues from an N or C terminus has been detected.

*Lack of Activity on Proteins*

Goldberg *et al.*[2] were unable to detect cleavage of casein or globin by pitrilysin, but Baneyx and Georgiou[18] suggested that pitrilysin hydrolyzed a fusion protein in living *E. coli.* To determine whether proteins are cleaved by pitrilysin, Anastasi *et al.*[5] made tests with oxyhemoglobin (subunit 16 kDa), casein (23 kDa), phosphorylase *a* (97 kDa), and gelatin (110 kDa) by SDS–polyacrylamide gel electrophoresis. None of the proteins was seen to be cleaved, however. Further tests were made with three of the proteins by including them in assays with QF27, in case they might be behaving like insulin and glucagon, but no significant inhibition of the hydrolysis of the substrate was detected.

The great majority of endopeptidases are bound by $\alpha_2$-macroglobulin in a characteristic "trapping" reaction. Experiments with pitrilysin failed to show any such reaction, however.[5] This was consistent with the evidence that the enzyme does not act on large proteins, since the interaction with

---

[16] Y. Pocker and S. B. Biswas, *Biochemistry* **20**, 4354 (1981).

[17] K. Rose, L.-A. Savoy, A. V. Muir, J. G. Davies, R. E. Offord, and G. Turcatti, *Biochem. J.* **256**, 847 (1988).

[18] F. Baneyx and G. Georgiou, *J. Bacteriol.* **173**, 2696 (1991).

$\alpha_2$-macroglobulin requires the cleavage of the macroglobulin in the "bait region."[19]

## Metal Dependence of Pitrilysin

Like other metallopeptidases, pitrilysin is readily inhibited by chelating agents.[5] The relative effectiveness of various divalent metal ions in reactivating pitrilysin after its exposure to 0.3 m$M$ 2,6-pyridinedicarboxylic acid has been reported.[5] The most effective ion in reactivating the enzyme was $Zn^{2+}$, giving 96% activity at a concentration of 0.1 m$M$, consistent with the finding that pitrilysin contains zinc.[13] Less effective were $Co^{2+}$, $Ca^{2+}$, and $Mn^{2+}$ which gave 75, 70, and 52% activity, respectively, at a concentration of 0.5 m$M$. Activation appeared instantaneous with $Zn^{2+}$, $Co^{2+}$, and $Mn^{2+}$, but not with $Ca^{2+}$, for which the activity increased progressively over 2–3 min. Further activation by $Zn^{2+}$ was observed following treatment with $Mn^{2+}$ and $Ca^{2+}$, reaching 96 and 90% activity, respectively. Although $Cu^{2+}$, $Cd^{2+}$, and $Mg^{2+}$ did not reactivate the enzyme, they did not prevent subsequent activation by 0.1 m$M$ $Zn^{2+}$, which resulted in 88, 49, and 105% activity, respectively.

In excess, metal ions tend to inhibit pitrilysin that contains its endogenous metal. Thus 1 m$M$ $Zn^{2+}$, $Co^{2+}$, $Mn^{2+}$, and $Ca^{2+}$ inhibited by 90, 32, 15, and 9%, respectively. Inhibition by excess zinc and other metal ions has been reported for many other metallopeptidases, and a mechanism for the effect of zinc on carboxypeptidase A has been proposed.[20]

## Inhibitors

No potent and selective inhibitor of pitrilysin has been described, and the enzyme is little affected by many inhibitors of other metallopeptidases that have been tested.[5] The antibiotic bacitracin was found to inhibit pitrilysin, especially in the presence of $Zn^{2+}$.[5] The most potent inhibition was achieved with a commercial preparation of zinc bacitracin, for which $K_i$ was 12.9 $\mu M$. Bacitracin is known to form a complex with zinc, and an interpretation of the effect of zinc on inhibition by pitrilysin would be that the conformation of the zinc–bacitracin complex is such that it is a more potent inhibitor than free bacitracin.

Pitrilysin is not inhibited by normal serum but was 69% inhibited by a rabbit antiserum to the enzyme.[5] Dynorphin A(1–13) behaved as a competitive inhibitor of pitrilysin in the quenched fluorescence assay, with a $K_i$ of 13.3 $\mu M$, and no hydrolysis of the peptide was detected by HPLC.[5]

[19] A. J. Barrett and P. M. Starkey, *Biochem. J.* **133,** 709 (1973).
[20] K. S. Larsen and D. S. Auld, *Biochemistry* **28,** 9620 (1989).

# [44] Insulysin and Pitrilysin: Insulin-Degrading Enzymes of Mammals and Bacteria

By ANDREW B. BECKER and RICHARD A. ROTH

## Introduction

Insulin regulates a wide variety of metabolic and cellular processes in target tissues. Most, if not all, of the varied effects of insulin appear to be mediated through a specific cell-surface receptor.[1,2] After insulin binds to its receptor, the receptor–hormone complex is internalized in endocytic vesicles.[3–5] When the vesicles are acidified, the complex dissociates, with most of the insulin receptors being recycled to the cell surface and the insulin being degraded inside of the cell.

Although the process of insulin degradation and the role the process plays in cellular regulation has been studied for approximately 40 years, significant progress on understanding the enzymes responsible for hormone degradation has only occurred since the 1980s. One enzyme which has been postulated to play a role in the process of insulin degradation has been given the name insulin-degrading enzyme (to be referred to as IDE), insulinase, and, most recently, insulysin (EC 3.4.24.56).[6,7] There are numerous lines of evidence implicating IDE in the degradation of insulin *in vivo*, including the following: (1) mice with elevated levels of IDE are resistant to elevated levels of insulin[8]; (2) the degradation products of insulin from intact cells are nearly identical to those seen with purified IDE[6,7,9–12]; (3) inhibitors of IDE also inhibit insulin degradation in intact cells[13]; (4) micro-

---

[1] J. M. Tavare and K. Siddle, *Biochim. Biophys. Acta* **1178,** 21 (1993).
[2] R. A. Roth, *Handb. Exp. Pharmacol.* **92,** 169 (1990).
[3] J. R. Levy and J. M. Olefsky, *Handb. Exp. Pharmacol.* **92,** 237 (1990).
[4] J.-J. I. Doherty, D. G. Kay, W. H. Lai, B. I. Posner, and J. M. Bergeron, *J. Cell Biol.* **110,** 35 (1990).
[5] R. J. Pease, G. D. Smith, and T. J. Peters, *Biochem. J.* **228,** 137 (1985).
[6] W. C. Duckworth, *Handb. Exp. Pharmacol.* **92,** 143 (1990).
[7] W. C. Duckworth, F. G. Hamel, D. E. Peavy, J. J. Liepnieieks, M. P. Ryan, M. A. Hermodson, and B. H. Frank, *J. Biol. Chem.* **263,** 1826 (1988).
[8] R. E. Beyer, *Acta Endocrinol.* **19,** 309 (1955).
[9] F. G. Williams, D. E. Johnson, and G. E. Bauer, *Metab. Clin. Exp.* **39,** 231 (1990).
[10] W. C. Duckworth, *Endocr. Rev.* **9,** 319 (1988).
[11] F. G. Hamel, B. I. Posner, J. J. Bergeron, B. H. Frank, and W. C. Duckworth, *J. Biol. Chem.* **263,** 6703 (1988).
[12] R. K. Assoian and H. S. Tager, *J. Biol. Chem.* **257,** 9078 (1981).
[13] C. Kayalar and W. T. Wong, *J. Biol. Chem.* **264,** 8928 (1989).

injection of antibodies directed against IDE inhibit the degradation of insulin in intact cells[14]; (5) insulin can be cross-linked to IDE in intact cells[15]; (6) IDE has been reported to be present in endosomes[16,17]; and (7) overexpression of IDE has been found to increase the rate of insulin degradation in intact cells.[18] The above data are all consistent with a role for IDE in insulin metabolism. However, it is possible that other, as yet unidentified, enzymes may also contribute to the process of insulin degradation *in vivo*.

In addition to its role in insulin degradation, IDE has also been shown to degrade a number of other cellular proteins, including glucagon,[19] atrial natriuretic factor,[20] transforming growth factor $\alpha$,[21] and oxidatively damaged hemoglobin.[22] IDE has also been implicated in the processes of membrane fusion and cell development. For example, it has been shown that inhibitors of differentiation in myoblasts, which function by preventing membrane fusion, also inhibit degradation of insulin by the endogenous insulin-degrading enzyme.[13] There also appear to be cellular enzymes with similar biochemical properties to IDE which are important in the processes of exocytosis in mast cells[23] and intracellular membrane fusion in the fertilization of sea urchin eggs.[24] These findings imply that IDE may play a more general role in hormone metabolism and cellular regulation.

A more general role for the enzyme is also supported by its broad tissue distribution. Although the levels of IDE are high in liver and kidney, which are major sites of insulin degradation, the enzyme is also present at high levels in other tissues, including brain and testis. Radioimmunoassays revealed high levels of IDE in all tissues tested, including brain, muscle, liver, and kidney, with the highest levels in liver.[25] Northern blots of rat tissues also revealed a large range of tissues containing IDE mRNA, with highest levels in adult testis, tongue, and brain; moderate levels in kidney, prostate, heart, muscle, liver, intestine, and skin; and low levels in the spleen, lung,

---

[14] K. Shii and R. A. Roth, *Proc. Natl. Acad. Sci. U.S.A.* **83,** 4147 (1986).

[15] J. Hari, K. Shii, and R. A. Roth, *Endocrinology (Baltimore)* **120,** 829 (1987).

[16] J. Fawcett and R. Rabkin, *Endocrinology (Baltimore)* **133,** 1539 (1993).

[17] F. G. Hamel, M. J. Mahone, and W. C. Duckworth, *Diabetes* **40,** 436 (1991).

[18] W.-L. Kuo, B. D. Gehm, and M. R. Rosner, *Mol. Endocrinol.* **5,** 1467 (1991).

[19] R. J. Kischner and A. L. Goldberg, *J. Biol. Chem.* **258,** 967 (1983).

[20] D. Muller, H. Baumeister, F. Buck, and D. Richter, *Eur. J. Biochem.* **202,** 285 (1991).

[21] B. D. Gehm and M. R. Rosner, *Endocrinology (Baltimore)* **128,** 1603 (1991).

[22] J. M. Fagan and L. Waxman, *Biochem J.* **277,** 779 (1991).

[23] D. I. Mundy, *Cell (Cambridge, Mass.)* **40,** 645 (1985).

[24] H. A. Farach, D. I. Mundy, W. J. Srittmatter, and W. J. Lennarz, *J. Biol. Chem.* **262,** 5483 (1987).

[25] H. Akiyama, K. Yokono, K. Shii, W. Ogawa, H. Taniguchi, S. Baba, and M. Kasuga, *Biochem. Biophys. Res. Commun.* **170,** 1325 (1990).

thymus, and uterus.[26,27] Studies using *in situ* hybridization revealed high levels of IDE mRNA in the liver and kidney of rats.[28] There was also a high level of message in the adrenal cortex from birth to adulthood in the rat and in the brown fat of rat pups. On a per cell basis, the highest levels of IDE mRNA were found in germ cells, in the egg cell in females and spermatocytes in males.

Although the existence of a specific insulin-degrading enzyme was reported over 30 years ago,[29] progress in the study of IDE was limited by the low levels of the molecule in cells and its lack of stability *in vitro*. Small amounts of IDE were purified from human red blood cells using conventional column chromatography.[19,30] These studies revealed the presence of a single polypeptide of 110 kDa on denaturing polyacrylamide gels. More recently, monoclonal antibodies against the enzyme have been generated,[14] and affinity columns composed of the antibodies have yielded enough homogeneous human IDE for partial sequence analysis.[31] The protein sequence was used to isolate a cDNA clone that encodes human IDE.

Although IDE had previously been shown to be a thiol-dependent metalloproteinase based on its inhibitor profile,[19] the amino acid sequence of the human enzyme, deduced from the cDNA clone, showed no significant homology to any known mammalian proteinases.[31] Interestingly, the protein sequence of human IDE did show 46% sequence similarity to an *Escherichia coli* proteinase, called protease III, Pi, or, most recently, pitrilysin (EC 3.4.24.55).[32] In addition, three regions in the sequences of the two enzymes are between 54 and 80% identical, implying that these regions may be important for the enzymatic functions of IDE and protease III. Subsequently, the homologous IDE enzymes were cloned from *Drosophila*,[33] rat,[27] and *Eimeria bovis*.[34] All of the enzymes contain a high level of sequence identity with human IDE, most markedly in the three highly conserved domains. In addition, the cloned sequences for all the enzymes encode proteins of approximately 100 kDa. These findings indicate that

[26] W.-L. Kuo, A. G. Montag, and M. R. Rosner, *Endocrinology (Baltimore)* **132**, 604 (1993).
[27] H. Baumeister, D. Muller, M. Rehbein, and D. Richter, *FEBS Lett.* **317**, 250 (1993).
[28] C. A. Bondy, J. Zhou, E. Chin, R. R. Reinhardt, L. Ding, and R. A. Roth, *J. Clin. Invest.* **93**, 966 (1994).
[29] I. A. Mirsky, *Recent Prog. Horm. Res.* **13**, 429 (1957).
[30] K. Shii, K. Yokono, S. Baba, and R. A. Roth, *Diabetes* **35**, 675 (1986).
[31] J. A. Affholter, V. A. Fried, and R. A. Roth, *Science* **242**, 1415 (1988).
[32] P. W. Finch, R. E. Wilson, K. Brown, I. D. Hickson, and P. T. Emmerson, *Nucleic Acids Res.* **14**, 7695 (1986).
[33] W.-L. Kuo, B. D. Gehm, and M. R. Rosner, *Mol. Endocrinol.* **4**, 1580 (1990).
[34] M. A. Abrahamsen, T. G. Clark, P. Mascolo, C. A. Speer, and M. W. White, *Mol. Biochem. Parasitol.* **57**, 1 (1993).

the enzyme is highly conserved throughout evolution and lends further support to the physiological importance of the protein.

It has also been noted that the two subunits of the mitochondrial processing proteinase (MPP) also contain significant sequence identity to members of the IDE family of enzymes, particularly in the first highly conserved domain of the protein family.[35] The two subunits ($\alpha$ and $\beta$) form a 110-kDa complex and function to process proteins being transported into mitochondria.[36] It has been shown that both subunits need to be present in order to form an active complex.[37]

All the members of the IDE family of proteinases have been shown to have an absolute requirement for divalent cations for activity, characterizing them as metalloproteinases.[19,38,39] In addition, purified protease III has been shown to contain stoichiometric amounts of zinc,[38] whereas partially purified human IDE contains high levels of zinc and significantly lower levels of manganese.[40] Interestingly, none of the members of this proteinase family contain the consensus active site (HEXXH) described in the major family of zinc metalloendopeptidases represented by thermolysin in which the two histidines have been shown to be required to coordinate the active-site zinc and the glutamate required for the nucleophilic attack of the peptide bond.[41,42] However, within the first highly conserved domain of the insulin-degrading enzymes is the sequence HXXEH. (This sequence is found in the $\beta$ but not the $\alpha$ subunit of the mitochondrial processing proteinase, consistent with the need for both subunits to obtain an active complex.) The sequence HXXEH was proposed to be an inversion of the active-site sequence found in the other family of metalloendopeptidases.[43]

Site-directed mutagenesis of the two histidines to arginine and the glutamate to glutamine in protease III indicated that the two histidines but not the glutamate were required to maintain the presence of zinc in the purified protein, whereas all three residues were found to be critical for enzymatic activity.[43] These findings are consistent with the hypothesis that the sequence HXXEH serves as an inverted active site of more traditional metalloproteinases. Subsequent work on human IDE showed that site-directed

[35] N. D. Rawlings and A. J. Barrett, *Biochem J.* **275**, 389 (1991).
[36] B. S. Glick, E. M. Beasley, and G. Schatz, *Trends Biochem. Sci.* **17**, 453 (1992).
[37] V. Geli, *Proc. Natl. Acad. Sci. U.S.A.* **90**, 6247 (1993).
[38] L. Ding, A. B. Becker, A. Suzuki, and R. A. Roth, *J. Biol. Chem.* **267**, 2414 (1992).
[39] Y.-S. Cheng and D. Zipser, *J. Biol. Chem.* **254**, 4698 (1979).
[40] A. Ebrahim, F. G. Hamel, R. G. Bennett, and W. C. Duckworth, *Biochem. Biophys. Res. Commun.* **181**, 1398 (1991).
[41] B. L. Vallee and D. S. Auld, *Biochemistry* **29**, 5647 (1990).
[42] B. W. Matthews, *Acc. Chem. Res.* **21**, 333 (1988).
[43] A. B. Becker and R. A. Roth, *Proc. Natl. Acad. Sci. U.S.A.* **89**, 3835 (1992).

mutagenesis of the first histidine in the sequence and the glutamate resulted in loss of proteolytic activity with retention of insulin-binding properties.[44,45] This work confirms the importance of the two residues in the sequence for proteolytic activity; However, it fails to address the critical question of whether the histidine and/or glutamate are zinc-binding residues in human IDE as the affinities for zinc of the wild-type and mutant proteins were not measured. Further studies on protease III using site-directed mutagenesis showed that a downstream glutamate when mutated to glutamine resulted in loss of catalytic activity as well as a significant decrease in zinc binding.[46] The results from these studies indicate that the active-site and third zinc-binding residue in protease III comprises the sequence $HXXEH(X)_{76}E$. This sequence is found in all IDE family members as well as the $\beta$ subunit of the mitochondrial processing proteinase. Further studies need to be performed on other members of this protein family to determine if the sequence represents the active site in these proteinases.

The active sites of the *Drosophila*, human, and rat enzymes all contain a cysteine (HXCEH), whereas the bacterial enzyme does not. In addition, the first three enzymes are inhibited by sulfhydryl-modifying reagents, whereas the bacterial enzyme is not. This led to the hypothesis that the cysteine is responsible for the sensitivity to these agents.[38] However, site-directed mutagenesis of the bacterial protease III to add the cysteine did not confer on the enzyme a sensitivity to sulfhydryl-modifying reagents.[47] Also, site-directed mutagenesis of the human enzyme to remove the cysteine did not change the sensitivity of the enzyme to sulfhydryl-modifying agents.[45] Thus, it presently is not known which cysteine is responsible for the sensitivity of the eukaryotic enzymes to sulfhydryl-modifying reagents.

In addition to a putative active site, the amino acid sequence of IDE revealed a potential peroxisomal targeting sequence in the carboxyl terminus of the enzyme.[27] The sequence, A/S-K-L, is found in the rat, *Drosophila*, and human homologs of the enzyme.[27,33,48] In contrast, the bacterial enzyme protease IIII contains a signal sequence which directs it to the periplasmic space.[32,49] The cellular localization of IDE is therefore open to question. Prior biochemical studies support a primary cytoplasmic localization (with a small fraction of the enzyme being found in membrane fractions and endosomes), whereas the presence of the targeting sequence would suggest a peroxisomal localization for the enzyme. In addition, there may be

[44] B. D. Gehm, W. L. Kuo, R. K. Perlman, and M. R. Rosner, *J. Biol. Chem.* **268,** 7943 (1993).
[45] R. K. Perlman, B. D. Gehm, W. L. Kuo, and M. R. Rosner, *J. Biol. Chem.* **268,** 21538 (1993).
[46] A. B. Becker and R. A. Roth, *Biochem. J.* **292,** 137 (1993).
[47] L. Ding and R. A. Roth, unpublished studies (1992).
[48] J. A. Affholter, C. L. Hsieh, U. Francke, and R. A. Roth, *Mol. Endocrinol.* **4,** 1125 (1990).
[49] K. H. Swamy and A. L. Goldberg, *J. Bacteriol.* **149,** 1027 (1982).

multiple forms of IDE as its mRNA exhibits two major forms (a 3.4- and a 6.3-kb form) and as well as minor variants of these on Northern blots.[26,27] However, it has not yet been shown that the different IDE mRNAs differ in the coding region and yield different protein products.

Another issue which remains to be resolved based on the IDE sequence is the initiation site for translation. Originally, translation was proposed to start at the first ATG in the open reading frame of the human IDE clone at nucleotides 16–18.[31] However, this ATG does not match closely to the consensus start site described by Kozak.[50] The flanking region surrounding the second ATG in the open reading frame, located 123 nucleotides downstream from the first ATG, more closely resembles the Kozak consensus start site. In support of the second ATG being the initiation codon, a truncation mutant of human IDE, which eliminated the first ATG, expressed a protein of the same molecular weight as the untruncated clone.[47]

Although there has been a great deal of progress in characterizing the enzymatic properties of the insulin-degrading enzymes over the past decade, no definitive role for the enzymes in cellular processes has been identified. To define clearly the role of the enzymes in cellular regulation, two types of experiments need to be performed, namely, overexpression and knockouts of insulin-degrading enzyme in intact cells. To date, there have been two reports of the successful expression of a full-length IDE protein in eukaryotic cells.[18,48] In the first, the expressed human enzyme was stably expressed in Chinese hamster ovary cells, allowing the expressed human enzyme to be distinguished from the endogenous hamster enzyme with a human-specific monoclonal antibody. The expressed human enzyme was shown to have the same specific activity in degrading insulin as the native human enzyme. No net increase in insulin-degrading activity was observed, however, in either cell lysates or intact cells as a result of the expression of the human cDNA. The second group found a 2- to 3-fold increase in the amount of insulin-degrading activity in the lysates of transiently transfected COS cells and a 3-fold increase in the degradation of insulin by the intact cells, providing further evidence that IDE is involved in insulin metabolism *in situ*.[18] Unfortunately, no evidence was presented in the studies that the increased degradation of insulin by the transfected cells occurred through an insulin receptor-mediated process as has previously been shown for normal cells.

One critical issue which remains to be addressed is the assays used for the study of insulin degradation. The assay which is most commonly used relies on the precipitation of intact insulin by trichloroacetic acid (TCA). As the human, *Drosophila,* and bacterial proteinases are known to make

---

[50] M. Kozak, *J. Cell Biol.* **115**, 887 (1991).

only a limited number of cleavages in the insulin molecule, the TCA assay is not an optimal assay for studying the enzymes. In the next section, we describe highly sensitive and specific assays for the study of insulin degradation by the enzymes.

## Plate Assays for Measuring Insulin Degradation

Any assay used for studying the catalytic activity of an enzyme should ideally meet three requirements: (1) high sensitivity, (2) high specificity, and (3) reproducibility. Because a number of proteinases can presumably degrade at least part of the insulin molecule, it is difficult to develop a specific assay for insulin degradation which relies on insulin as a substrate. Therefore, at least one group is working on the development of synthetic substrates which will be highly specific for bacterial protease III in an effort to confront this difficulty[51] (see [43], this volume). Alternatively, assays which isolate the proteinase before measuring its enzymatic activity can be utilized, and a method for such an assay is described below.

It has been shown that the human and *Drosophila* insulin-degrading enzymes make a limited number of cleavages in the insulin molecule.[52,53] Similarly, bacterial protease III cleaves insulin B chain in a limited number of sites.[39] The most commonly used assay in the study of insulin degradation involves the precipitation of insulin or insulin B chain by trichloroacetic acid (TCA). The assay is typically performed by incubating $^{125}$I-labeled insulin or insulin B chain with either whole cells, cell extracts, or purified enzyme for a specified amount of time. The reaction is terminated by adding TCA to a final concentration of 7.5 to 10% w/v, incubating on ice for 10 min in the presence of bovine serum albumin (BSA), and then microcentrifuging the sample to separate the supernatant from the precipitate. The amount of degradation is then calculated from the increase in TCA-soluble $^{125}$I in comparison with control incubations without enzyme. It has been shown using this assay that purified red blood cell IDE will degrade $^{125}$I-labeled insulin in a linear fashion for up to 60 min and maximizes at 20% of the insulin being degraded to a non-TCA-precipitable form.[30] This low level of degradation has consistently been shown using this assay with other preparations of IDE as well as bacterial protease III.[38]

However, other assays of insulin degradation, including gel filtration and the receptor binding assay described below, indicate that nearly 100%

[51] A. Anastasi, C. G. Knight, and A. J. Barrett, *Biochem. J.* **290,** 601 (1993).

[52] W. C. Duckworth, F. G. Hamel, R. Bennett, M. P. Ryan, and R. A. Roth, *J. Biol. Chem.* **265,** 2984 (1990).

[53] W. C. Duckwork, J. V. Garcia, J. J. Liepnieks, F. G. Hamel, M. A. Hermodson, B. H. Frank, and M. R. Rosner, *Biochemistry* **28,** 2471 (1989).

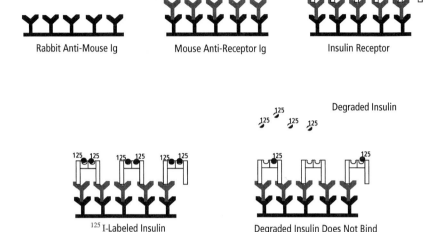

FIG. 1. Plate binding assay for insulin degradation. Microtiter plates are incubated sequentially with anti-mouse immunoglobulin (Ig), a monoclonal antireceptor antibody, a source of insulin receptor, and the labeled insulin which is to be tested for the extent of degradation. Parallel plates are used to assess the amount of labeled insulin bound without prior treatment with the proteinase (normally around 10 to 20% of the input counts). The decrease in binding as a result of incubation with the proteinase is used to calculate the percent degradation.

of the $^{125}$I-labeled insulin is actually degraded within 60 min, with a plateau being reached in under 30 min.[30] This discrepancy presumably is due to the low sensitivity of the TCA assay, probably as a result of the limited number of cleavages made by the insulin-degrading enzymes in the insulin molecule. This would result in only minor changes in the insulin structure and thereby cause only limited changes in precipitability. Furthermore, $^{125}$I-labeled insulin stored at $-20°$ or $-70°$ has an increasing fraction of TCA-soluble $^{125}$I in as little as several days and may reach as high as 10–20% solubility within 2–4 weeks. This high level of nonprecipitable label makes these preparations unsuitable for the TCA assay and requires frequent iodinations of the hormone.

To monitor insulin degradation, we have therefore developed a plate binding assay which relies on the ability of insulin to bind to the insulin receptor with high affinity (Fig. 1).[54] Even small changes in the insulin molecule will affect the affinity of the hormone for its receptor, especially

[54] D. O. Morgan and R. A. Roth, *Endocrinology* (*Baltimore*) **116,** 1224 (1985).

in light of the fact that the part of the insulin molecule recognized by human IDE is also responsible for binding of the hormone to its receptor.[55]

The assay is performed on a 96-well polyvinyl chloride microtiter plate (plates can be bought from Dynatech or Falcon). The wells in the plate are coated with 50 $\mu$l of an affinity-purified rabbit anti-mouse antibody solution (20 $\mu$g/ml in 20 m$M$ NaHCO$_3$, pH 9.6) (antibody can be purchased from PelFreeze or Cappel). The plates are then incubated for 2–4 hr at room temperature, and the wells are washed twice with ice-cold WGBT buffer (50 m$M$ HEPES, pH 7.6, 150 m$M$ NaCl, 0.2% w/v BSA, and 0.1% w/v Triton X-100). Forty-five microliters of a monoclonal anti-insulin receptor antibody (29B4) (the antibody can be purchased from Oncogene Sciences or Santa Cruz Biotechnology) is then applied to the wells (1 $\mu$g/ml in WGBT buffer). The wells are again incubated at room temperature for 2–4 hr and washed twice with WGBT buffer. Forty microliters of a lysate of CHO-T cells (Chinese hamster ovary cells overexpressing human insulin receptor at approximately $1 \times 10^6$ per cell[56]) is then applied to the wells. [The cells are lysed and harvested by taking a confluent 100-mm plate of CHO-T cells, washing them twice with a phosphate-buffered saline (PBS) solution, and then lysing them for 3 min on ice in 3 ml of a solution consisting of HEPES-buffered saline, pH 7.6, and 2% Triton X-100. The solution is removed from the tissue culture plate and applied to the wells.] The plate is then incubated overnight at 4° and subsequently washed twice with WGBT buffer. [125]I-Labeled insulin can then be applied to the wells. The plates are stored at 4° for 4 hr, and then the wells are washed three times with WGBT. The wells are subsequently cut from the plate and counted directly in a gamma counter. Typically 3000 to 10,000 counts/min (cpm) are bound specifically to the receptor (i.g., displaceable by unlabeled insulin).

The plate assay is simple to use, highly sensitive, and very reproducible. The source of the insulin-degrading activity used in the assay can be whole cells, cell extracts, purified IDE, or, as described below, IDE immunoimmobilized on a plate. The assay also relies on the absolute number of counts bound to the immunoimmobilized receptor and does not appear to be as affected by the age of the iodinated hormone as is the TCA assay. In addition, the assay can also be modified for the study of the degradation of other hormones by using the appropriate receptor.

As described above, studies of the insulin-degrading activity in cell extracts can be complicated by the presence of multiple enzymes which

[55] J. A. Affholter, M. A. Cascieri, M. L. Bayne, J. Brange, M. Casaretto, and R. A. Roth, *Biochemistry* **29**, 7727 (1990).

[56] L. Ellis, E. Clauser, D. O. Morgan, M. Edery, R. A. Roth, and W. J. Rutter, *Cell (Cambridge, Mass.)* **45**, 721 (1986).

degrade the substrate. To overcome this, one can purify the specific enzyme from the extract. This can be a time-consuming process. Moreover, in the case of IDE the enzyme is extremely labile (thereby resulting in a preparation with low specific activity), and there is the possibility that the isolated enzyme differs from the native enzyme in some properties. One solution to such problems is to use microtiter plates coated with a specific antibody to immunoisolate the enzyme immediately before the assay. Instead of using an anti-insulin receptor antibody as described above, we have used the anti-IDE monoclonal antibody (9B12) and an antiprotease III monoclonal antibody (2B4D9) to immunocapture the respective enzymes from cell lysates.[38] Once the enzyme is captured on the plate and the other cellular proteins are washed away, $^{125}$I-labeled insulin can be applied to the enzyme for various amounts of time at 37° or 24°. The reaction is terminated by removing the substrate, adding EDTA, and placing the material on ice. The extent of degradation of the insulin can then be assessed by measuring its ability to bind to the insulin receptor, as described above.

In addition to measuring the degradation of insulin, the binding of insulin to the enzyme can also be measured by incubating the immunoimmobilized enzyme at 4° with $^{125}$I-labeled insulin. By including different concentrations of other proteins in the assay, one can assess the relative affinity of the enzyme for different substrates.[55] Similarly, one can determine the effects of mutations of the enzyme or proteinase inhibitors on the ability of the immobilized enzymes to bind insulin. As in the receptor binding assay, the assay is simple to use, highly sensitive, and very reproducible. Finally, the assay can be modified for the study of a large number of enzymes as long as a noninhibitory mono- or polyclonal antibody exists. For example, we have utilized the assay to measure the protein kinase activity of immunoimmobilized insulin receptor[57] and protein kinase C.

Conclusion

Insulin-degrading enzyme (IDE) has been implicated in the physiological process of insulin degradation after internalization. Since the 1980s, a great deal has been learned at both the DNA and protein level about the enzyme as well as other members of this unique family. However, many questions remain unanswered, including the localization of the enzyme in the body and in the cell, the physiological substrates of the enzyme, the effects of overexpression or gene knock-outs, and the regulation of enzyme expression. In this chapter we review much of what is known about the

---

[57] G. Steele-Perkins and R. A. Roth, *J. Biol. Chem.* **265**, 9458 (1990).

enzyme as well as describe the development of a series of highly sensitive and specific assays for studying insulin-degrading enzymes.

## Acknowledgment

Work in the authors' laboratory is supported by National Institutes of Health Grant DK34926.

# [45] N-Arginine Dibasic Convertase

*By* PAUL COHEN, ADRIAN R. PIEROTTI, VALÉRIE CHESNEAU,
THIERRY FOULON, and ANNIK PRAT

## Introduction

The search for endopeptidase(s) responsible for processing of somato-statin precursor(s) by selective proteolysis at the Arg-Lys dibasic site[1-3] has led to the discovery of a new metalloendopeptidase that is able to cleave the somatostatin-28 Arg-Lys doublet on the N-terminal side of the arginine residue.[4-7] Because the endopeptidase generates $[\text{Arg}^{-2}\text{-Lys}^{-1}]$ somatostatin-14, a substrate for the copurifying aminopeptidase(s) B, the complex of both activities was originally called somatostatin-28 convertase. The endopeptidase, first characterized in rat brain cortex extracts and then in the testis, is moreover able to perform selective cleavage on the N-terminal side of Arg from dibasic sites in a number of prohormone fragments.[8] Therefore, it has been renamed NRD convertase for N-arginine

[1] P. Cohen, A. Morel, P. Gluschankof, S. Gomez, and P. Nicolas, *Adv. Exp. Med. Biol.* **188,** 109 (1985).

[2] P. Cohen, P. Kuks, S. Gomez, and A. Morel, *in* "Peptide Hormones as Prohormones" (J. Martinez, ed.), p. 83. Ellis Horwood Chichester, Sussex G-B., 1989.

[3] A. Lepage-Lezin, P. Joseph-Bravo, G. Devilliers, L. Benedetti, J.-M. Launay, S. Gomez, and P. Cohen, *J. Biol. Chem.* **266,** 1679 (1991).

[4] P. Gluschankof, A. Morel, S. Gomez, P. Nicolas, C. Fahy, and P. Cohen, *Proc. Natl. Acad. Sci. U.S.A.* **81,** 6662 (1984).

[5] S. Gomez, P. Gluschankof, A. Morel, and P. Cohen, *J. Biol. Chem.* **260,** 10541 (1985).

[6] P. Gluschankof, S. Gomez, A. Morel, and P. Cohen, *J. Biol. Chem.* **262,** 9615 (1987).

[7] S. Gomez, P. Gluschankof, A. Lepage, and P. Cohen, *Proc. Natl. Acad. Sci. U.S.A.* **85,** 5468 (1988).

[8] V. Chesneau, A. R. Pierotti, N. Barré, C. Créminon, C. Tougard, and P. Cohen, *J. Biol. Chem.* **269,** 2056 (1994).

dibasic convertase. Its complete amino acid sequence has now been deduced by cDNA cloning and sequencing.[9]

## Assay Procedure

### Principle

A simple and rapid assay was originally designed to allow for the reliable detection in tissue extracts of an endopeptidase activity capable of cleaving selectively at the Arg-Lys moiety of a synthetic peptide substrate mimicking the prosomatostatin sequence.[4] Starting from the prosomatostatin sequence[1,2] around the Arg-Lys doublet (Pro$^{-5}$-Arg-Glu-*Arg*$^{-2}$-*Lys*$^{-1}$-Ala-Gly-Cys-Lys-Asn-Phe$^6$), the cleavage site for somatostatin-14 production from either the full-length precursor or from somatostatin-28 (the entire 28-mer C-terminal domain of prosomatostatin), an undecapeptide was designed as follows. In this precursor segment, Cys$^3$ and Phe$^6$ were replaced by Ala and Tyr-NH$_2$, respectively, in order to avoid SH oxidation reactions and to provide the peptide sequence with a site for $^{125}$I labeling. Therefore the following undecapeptide was synthesized and used as a substrate in further studies, as an alternative to somatostatin-28 (S-28)[4]:

Pro-Arg-Glu-*Arg-Lys*-Ala-Gly-Ala-Lys-Asn-Tyr-NH$_2$

that is,

[Ala$^{17}$,Tyr$^{20}$]S-28(10–20)-NH$_2$     Peptide I

To identify the generated products, a number of peptide fragments were also synthesized.[4,6] These included the following:

[Ala$^{17}$,Tyr$^{20}$]S-28(15–20)-NH$_2$:

Ala-Gly-Ala-Lys-Asn-Tyr-NH$_2$     Peptide II

[Ala$^{17}$,Tyr$^{20}$]S-28(14–20)-NH$_2$:

*Lys*-Ala-Gly-Ala-Lys-Asn-Tyr-NH$_2$     Peptide VI

[Ala$^{17}$,Tyr$^{20}$]S-28(13–20)-NH$_2$:

*Arg-Lys*-Ala-Gly-Ala-Lys-Asn-Tyr-NH$_2$     Peptide V

The assay can be performed by using either unlabeled or labeled substrate(s).

### Labeled Substrate

The assay,[4] performed using $^{125}$I-labeled peptide I as substrate, is recommended for a rapid screening of fractions obtained by molecular sieve

[9] A. R. Pierotti, A. Prat, V. Chesneau, F. Gaudoux, A.-M. Leseney, T. Foulon, and P. Cohen, *Proc. Natl. Acad. Sci. U.S.A.* **91**, 6078 (1994).

chromatography of tissue extracts. The Tyr residue side chain in peptide I provides a site for [125]I labeling which allows rapid identification of the generated reaction product(s) without altering the possible requirement of the proteinase for an aromatic residue (i.e., Phe[20] in somatostatin-28).

The [125]I-labeling of peptides I and II is done by the lactoperoxidase method,[10] and the [125]I-labeled peptides I and II are purified on a carboxymethyl cellulose (CM52) resin. Labeled peptide I is retained on the resin after washing with 100 m$M$ ammonium acetate buffer, pH 6.0, whereas peptide II is eluted with the same buffer. Control experiments indicate that the degradation product (*vide infra*) does not bind to the resin in 30 m$M$ ammonium acetate.

The convertase activity is assayed by using an aliquot of [125]I-labeled peptide I [20,000 counts/min (cpm)] dissolved in 10 $\mu$l of 250 m$M$ Tris-HCl, pH 7.0, and incubated at 37° with an aliquot either of crude organ extract or of fractions from the molecular sieve column (final volume is 300 $\mu$l in 0.25 $M$ Tris-HCl, pH 7.0). At various reaction times, 50 $\mu$l of the mixture is removed, and the reaction is stopped by the addition of 600 $\mu$l of CM52 resin equilibrated in 30 m$M$ ammonium acetate, pH 6.0 (1 : 10, v/v). The mixture is then centrifuged at 10,000 $g$ for 30 sec. The resulting pellet is washed once with the equilibration buffer and then twice with 100 m$M$ ammonium acetate buffer, pH 6.0. The iodinated peptides retained on the resin are counted after each salt concentration wash. Results can be calculated as follows:

$$\text{Specific activity} = \frac{(x_1 - x_2) - (y_1 - y_2)}{y_2} \times 100 \tag{1}$$

$$\text{Degrading activity} = \frac{(t - x_1) - (t - y_1)}{y_1} \times 100 \tag{2}$$

Specific activity [Eq. (1)] represents percent peptide II generated, whereas degrading activity [Eq. (2)] represents percent Asn-[[125]I]Tyr-NH$_2$ generated. In the expressions, $t$ is total cpm in the 50-$\mu$l reaction mixture, $x_1$ is cpm recovered bound to the resin after the 30 m$M$ ammonium acetate wash, $x_2$ is the remaining cpm bound to the resin after the last wash, and $y_1$, and $y_2$ are the cpm recovered after the 30 and 100 m$M$ washes of the resin with peptide I alone.

In tissue extracts, this method allows the Arg-Lys specific endopeptidase activity to be discriminated from other proteolytic actions cleaving the Lys-Asn bond. This, combined with other selective properties of the enzyme (*vide infra*), permits the detection and discrimination of the endopeptidase

---

[10] J. J. Marchalonis, *Biochem. J.* **113,** 299 (1969).

from the large range of proteolytic activities present in crude extracts at the early stages of purification. Given the large number of such contaminating enzymes, the addition of inhibitors in the NRD convertase assay may help in identifying more precisely the endopeptidase responsible for the observed cleavage. Results in Table I (*vide infra*) indicate that reagents like amastatin could be useful.

### Unlabeled Substrates

Peptide substrate (either peptide I or S-28)(5 $\mu$g) is incubated at 37° with an enzyme sample (quantity of protein depending on the purification step) in a 300-$\mu$l final volume of 0.25 $M$ Tris-HCl buffer, pH 7.0. The reaction is stopped by the addition of 30 $\mu$l of concentrated acetic acid, and the resulting mixture is subjected to high-performance liquid chromatography (HPLC) analysis on a $\mu$Bondapak $C_{18}$ column (300 $\times$ 3.0 mm; Waters, Milford, MA). For incubations with peptide I, a gradient of acetonitrile in 1-hexanesulfonic acid (1 g/liter, 20 m$M$ acetic acid) from 20 to 28% in 12 min at a flow rate of 1 ml min$^{-1}$ is used. Substrates and generated products can be unequivocally identified by the HPLC retention times relative to standards. When S-28 is used as a substrate the generated C-terminal Arg$^{13}$–Cys$^{28}$ fragment, called [Arg$^{-2}$-Lys$^{-1}$]S-14, is identified by N-terminal sequencing. In that case substrate is separated from product by a linear gradient of 25% acetonitrile in 0.1 $M$ phosphate triethylamine, pH 3.0, at 1 ml min$^{-1}$. The S-28(1–12) N-terminal fragment is identified both by its retention time and by radioimmunoassay using a specific antiserum.

### Purification Procedure

Purification can be achieved by using either rat brain cortex or testis as a source of peptidase.[7,8] However, because the latter tissue is found to be a predominant source of activity (*vide infra*), it is recommended for enzymes studies requiring larger quantities of material.

Early work had shown that the endopeptidase activity was largely contaminated by aminopeptidases of the B type which copurified during the enzyme preparation. Purification of the two activities is achieved by a succession of differential precipitation, molecular sieve chromatography (Sephadex G-150), and ion-exchange chromatography (DEAE-Trisacryl). The ion-exchange step is efficient in separating the endopeptidase from the aminopeptidase activities.[6] Aminopeptidases elute around 50 and 100 m$M$ KCl, respectively, whereas the endopeptidase is recovered between

290 and 350 m$M$ KCl (Fig. 1). The endopeptidase activity is then completely purified in a fourth step involving an adsorption column (hydroxyapatite). Purification steps are carried out at 4°. Frozen testes from Wistar rats (10 testes, 13 g; Pel-freeze, Rogers, AR) are defrosted, homogenized (20%, w/v) in 50 m$M$ phosphate buffer, pH 7.4, 100 m$M$ KCl, and centrifuged at 2000 $g$ for 15 min. The supernatant is precipitated at pH 4.7 with 1 $M$ ammonium acetate buffer, pH 4.5, 5 m$M$ 2-mercaptoethanol, and centrifuged at 7800 $g$ for 30 min. The supernatant is readjusted to pH 7.5, the enzyme preparation is concentrated to one-tenth of its initial volume in an ultrafiltration cell (Amicon, Danvers, MA, Diaflo membrane YM30), and run on a Sephadex G-150 superfine column (Pharmacia, Piscataway, NJ) in 250 m$M$ Tris-HCl buffer, pH 7.5, 5 m$M$ 2-mercaptoethanol. The active fractions, diluted 5 times with deionized water, are applied to a DEAE-Trisacryl M ion-exchange column (IBF, Villeneuve la Garenne, France) (100 ml bed volume) preequilibrated in 50 m$M$ Tris-HCl buffer, pH 7.5, and eluted by two successive linear salt gradients from 0 to 150 m$M$ KCl in 400 ml and from 150 to 400 m$M$ KCl in 400 ml (Fig. 1). Fractions (3 ml) are collected and tested for enzyme activity as described above. Fractions within the central part of the activity peak are pooled and submitted to an hydroxyapatite-Ultrogel column (30 ml bed volume; IBF) preequilibrated with 50 m$M$ Tris-HCl, buffer, pH 7.5, 200 m$M$ KCl, 1 m$M$ potassium phosphate (buffer A). Elution is carried out with a 200-ml linear

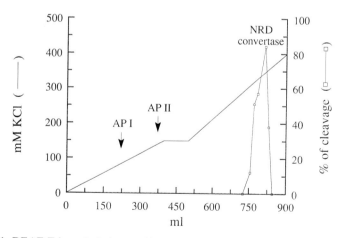

Fig. 1. DEAE-Trisacryl elution profile. Elution fractions (3 ml) were tested for NRD convertase specific activity using peptide I as the substrate as described in the text. Results are expressed as the percentage of cleaved substrate. Elution positions of the two contaminating B-type aminopeptidases (AP I, aminopeptidase I; AP II, aminopeptidase II) are indicated by arrows. (Adapted from Gluschankof *et al.*[6])

gradient of buffer A to the same buffer with 250 m$M$ potassium phosphate. The pool of active fractions is concentrated and desalted on ultrafiltration cartridges (Amicon, Centriflo membrane cones, type CF25) and can be further analyzed by polyacrylamide gel electrophoresis (PAGE).

The purification procedure originally used to extract the endopeptidase activity from rat brain[7] differs from the protocol described above[8] in the following points. First, fresh tissues initially used did not appear to give better yields than frozen organs. Second, two concentration steps on a Speed-Vac system were avantageously replaced by ultrafiltration systems because they allow elimination of molecules of $M_r$ less than 30,000 and avoid concentration of salts. The last main modification was replacement of the original HPLC gel-filtration column (TSK G 3000 SW, LKB, Uppsala, Sweden) by an adsorption hydroxyapatite column which is more efficient in eliminating the last protein contaminants.

Properties

*pH and Inhibitor Profiles*

Both the brain cortex and testis enzymes exhibit a similar pH dependence with a broad curve showing a maximum activity at basic pH values (around pH 8.5).[7,8]

The sensitivity of NRD convertase toward classic peptidase inhibitors and divalent cations is tested as follows: enzyme (250 ng) is preincubated with either inhibitor or divalent cation in a total volume of 250 $\mu$l for 30 min at 37° in 250 m$M$ Tris-HCl buffer, pH 7.5. The reaction is initiated by the addition of dynorphin A at 6.5 $\mu M$ final concentration and is carried out for 15 min at 37°. Results are expressed as the percent activity remaining, taking the incubation control as 100% cleavage (Table I). When tested with peptide I and/or dynorphin A as substrates, both enzymes from cortex and testis show the typical behavior of metallopeptidases with a marked sensitivity to divalent cation chelators (Table I) and, in addition, a sensitivity to some sulfhydryl reagents like $N$-ethylmaleimide (NEM). Interestingly, the convertase activity is inhibited by classic aminopeptidase inhibitors like amastatin and bestatin. In contrast, epibestatin and the usual serine and aspartic proteinase inhibitors have no effect.

Reactivation of NRD convertase after inhibition by cation chelators can be achieved as follows: enzyme (420 ng) is first preincubated for 30 min with 10 $\mu M$ EDTA at 37° in 250 m$M$ Tris-HCl buffer, pH 7.5. Dynorphin A (6.5 $\mu M$ final concentration) and the divalent cation (15 or 25 $\mu M$ final concentration) are then added, and the reaction mixture is incubated for

TABLE I
INHIBITOR PROFILE FOR N-ARGININE DIBASIC CONVERTASE[a]

| Inhibitor | Concentration ($\mu M$) | Activity (% control) |
|---|---|---|
| EDTA | 10 | 0 |
| EGTA | 1 | 26 |
| | 500 | 7 |
| 1,10-Phenanthroline | 100 | 73 |
| | 500 | 0 |
| Phosphoramidon | 100 | 21 |
| Captopril | 10 | 87 |
| Tos-Phe-CH$_2$Cl | 10 | 120 |
| | 250 | 50 |
| Phenylmethylsulfonyl fluoride (PMSF) | 100 | 74 |
| Leupeptin | 100 | 95 |
| Dithiothreitol (DTT) | 100 | 116 |
| Iodoacetamide | 100 | 78 |
| N-Ethylmaleimide (NEM) | 1000 | 92 |
| | 100 | 0 |
| (2-Guanidinoethylmercapto)succinic acid (GEMSA) | 10 | 63 |
| Amastatin | 100 | 100 |
| | 20 | 0 |
| Bestatin | 10 | 78 |
| | 50 | 13 |
| | 20 | 31 |
| | 10 | 100 |
| Cations | | |
| Ca$^{2+}$ | 100 | 112 |
| Mg$^{2+}$ | 100 | 105 |
| Cu$^{2+}$ | 100 | 44.5 |
| Co$^{2+}$ | 100 | 36 |
| Mn$^{2+}$ | 100 | 57 |
| Zn$^{2+}$ | 100 | 11 |

[a]Adapted from Chesneau et al.[8]

30 min at 37°. Results are expressed as the percent activity recovered, taking the incubation control as 100% cleavage (Table II).

Both cation inhibition and reactivation profiles confirm the metalloenzyme character of the convertase (Tables I and II), and Zn$^{2+}$ ions exhibit both the most inhibitory and the most reactivatory effects. In contrast to other known processing endopeptidases specific for dibasic sites (KEX2 gene product family),[11] NRD convertase does not show any Ca$^{2+}$ dependence under routine assay conditions.

[11] R. S. Fuller, A. Brake, and J. Thorner, Proc. Natl. Acad. Sci. U.S.A. 86, 1434 (1989).

TABLE II
CATION REACTIVATION PROFILE FOR N-ARGININE
DIBASIC CONVERTASE[a]

| Reagent | Concentration ($\mu M$) | Activity (% control) |
|---|---|---|
| Inhibition control | — | 22 |
| CaCl$_2$ | 15 | 74 |
| | 25 | 90 |
| MgCl$_2$ | 15 | 21 |
| | 25 | 43 |
| CoCl$_2$ | 15 | 74 |
| | 25 | 74 |
| MnCl$_2$ | 15 | 94 |
| | 25 | 114 |
| CuSO$_4$ | 15 | 14 |
| | 25 | 18 |
| ZnSO$_4$ | 15 | 122 |
| | 25 | 90 |

[a] Adapted from Chesneau et al.[8]

*Substrate Specificity*

The substrate selectivity of the convertase was investigated by using *in vitro* tests with either synthetic peptides mimicking dibasic maturation sites of prohormonal sequences or/and physiologically important peptides which can constitute hormonal precursors for further maturation steps.[7,8]

The assay for potential substrates is conducted as follows: a 40-$\mu M$ final concentration of peptide substrate is incubated with an aliquot of the enzyme pool recovered after the DEAE-Trisacryl M fractionation step (*vide supra*) in 250 m$M$ Tris-HCl buffer, pH 7.5, for 1, 3, and 15 hr at 37°. At the end of the reaction time, the mixture is analyzed by reversed-phase HPLC fractionation on a $\mu$Bondapack C$_{18}$ column (3.9 × 300 mm; Waters). Various elution systems can be used to separate the substrates from the cleavage reaction products. The absorbance at 215 nm ($A_{215\,nm}$) is monitored, and fragments can be identified by their retention times by reference to standard peptides and by amino acid composition.

The following gradients of acetonitrile in buffer B (1 g/liter 1-hexanesulfonic acid, 20 m$M$ acetic acid) at a flow rate of 1 ml min$^{-1}$ can be used: 20 to 28% in 12 min for peptide I and derivatives, peptide V, [Arg$^0$,Ala$^{17}$]S-28(1–19)-NH$_2$, and prosomatostatin(56–68)-NH$_2$; 20 to 35% in 30 min for $\alpha$-neoendorphin and dynorphin B; 20 to 40% in 20 min for other opioid peptides, preproneurotensin 154–170, and proocytocin/neurophysin 1–20. Atrial natriuretic factor (ANF) 1–28 is analyzed using an isocratic gradient of 29% acetonitrile in buffer B. Somatostatin-28 is analyzed as described earlier (see Assay Procedure). The kinetic parameters are determined from initial velocity measurements at various substrate concentrations (six or seven points in a range from 5 times below to 5 times above the $K_m$ are taken to draw plots) and calculated from Hanes–Woolf plots by a linear regression program. Because incubation conditions may vary, pseudo-$k_{cat}$ values can be evaluated by taking arbitrarily 1 UE = 4.5 ng of enzyme preparation/$\mu$l. Values are given $\pm 10\%$ (Table III).

It is found that all cleavages occur strictly on the amino-terminal side of the arginine residue in dibasic sites independently of the nature of the doublet (Table III). In the case of opioid peptides, the Arg-Arg site is predominantly cleaved in the middle. A second minor cleavage upstream of the maturation site is observed in some peptides (not shown). The only exception to the rule is $\alpha$-neoendorphin, whose Arg-Lys processing site is cleaved in the middle like the other opioid peptides; however, comparison of the $K_m$ values reveals a very low affinity of the enzyme for that substrate.

Selectivity for basic doublets is confirmed by the absence of cleavage at the level of single Arg residues [like prosomatostatin(56–68)-NH$_2$; a peptide mimicking the sequence surrounding the monobasic maturation site of prosomatostatin] and in the Arg singlets present in several peptides. In addition, the enzyme does not cleave peptide I when either Arg or Lys is replaced by D-*Arg* or D-*Lys*.

Another interesting feature of the catalytic behavior of NRD convertase is its higher affinity for larger peptides. Comparison of the kinetic data obtained with the somatostatin peptide series and the dynorphin A-related peptides show that lower $K_m$ values are obtained with the larger substrates, namely, 43 $\mu M$ for somatostatin-28 and 6.45 $\mu M$ for dynorphin A. ANF(1–28) and preproneurotensin (154–170) also exhibit low $K_m$ values. Finally, it is noteworthy that the presence of a dibasic site in a peptide is not a sufficient condition to achieve enzyme cleavage. Indeed, the Lys-Arg doublet in preproenkephalin (128–140) and in proocytocin/neurophysin (1–20) is not cleaved, whereas the Lys-Arg moiety in preproneurotensin (154–170) is cleaved with high efficiency. This peptide, which also contains an Arg-Arg moiety in the middle of the neurotensin sequence, is only cleaved at the prohormonal maturation site. This behavior suggests that other factors including secondary structure and specific subsite requirements are important for specific recognition and/or catalysis by the endopeptidase.

Convertase Structure

Molecular sieve chromatography of the enzyme originally indicated a molecular mass of approximately 90 kDa for the cortex enzyme.[4,6,7] How-

## TABLE III
### Substrate Dependence Profile for N-Arginine Dibasic Convertase[a]

| Peptide | Sequence | $K_m$ ($\mu M$) | Pseudo-$k_{cat}$ (pmol/min/UE) | Pseudo-$k_{cat}/K_m$ ($\mu l$/min/UE) |
|---|---|---|---|---|
| **Somatostatin and derivatives** | | | | |
| [Tyr20]S-28(13–20)-NH2 (peptide V) | R **K** AGAKNY-NH2 | | Not cleaved | |
| [Ala17,Tyr20]S-28(10–20)-NH2 (peptide 1) | PRE⊥R **K** AGAKNY-NH2 | 428 | 400 | 0.93 |
| [D-Arg13,Ala17,Tyr20]S-28(10–20)-NH2 | PRE **R** **K** AGAKNY-NH2 | | Not cleaved | |
| [D-Lys14,Ala17,Tyr20]S-28(10–20)-NH2 | PRE R **K** AGAKNY-NH2 | | Not cleaved | |
| [Gln12,Ala17,Tyr20]S-28(10–20)-NH2 | PRQ⊥R **K** AGAKNY-NH2 | | Not determined | |
| [Gly12,Ala17,Tyr20]S-28(10–20)-NH2 | PRG⊥R **K** AGAKNY-NH2 | | Not determined | |
| [Arg0,Ala17]S-28(1–19)-NH2 | RSANSNPAMAPRE⊥R **K** AGAKN-NH2 | 227 | 74.8 | 0.33 |
| Somatostatin-28 | SANSNPAMAPRE⊥R **K** AGCKNFFWKTFTSC | 43 | 15.8 | 0.37 |
| Prosomatostatin(56–68)-NH2 | DEMRLELQRSANS-NH2 | | Not cleaved | |
| **Opioid** | | | | |
| Dynorphin A | YGGFL **R**⊥R IRPKLKWDNQ | 6.45 | 59.9 | 9.3 |
| Dynorphin A(1–13) | YGGFL **R**⊥R IRPKLK | 24 | 625 | 26 |
| Dynorphin A(1–10) | YGGFL **R**⊥R IRP | 127 | 240 | 1.89 |
| Dynorphin A(1–9) | YGGFL **R**⊥R IR | 162 | 105 | 0.65 |
| Dynorphin A(1–7) | YGGFL **R** R | | Not cleaved | |
| Dynorphin B | YGGFL **R**⊥R QFKVVT | 209 | 101 | 0.48 |
| α-Neoendorphin | YGGFL **R**⊥K YPL | 974 | 92.2 | 0.095 |
| Bovine adrenal medulla dodecapeptide (BAM-12P) | YGGFM **R**⊥R VGRPE | 103 | 365 | 3.54 |
| Preproenkephalin (128–140) | GGEVLA **K** R YGGFM | | Not cleaved | |
| **Miscellaneous neuropeptides** | | | | |
| Preproneurotensin (154–170) | ENKRRPYIL **K**⊥R ASYYY | 17.3 | 85.6 | 4.90 |
| ANF(1–28) | SL **R**⊥R SSCFGGRIDRIGAQSGLGCNSFRY | 6.25 | 114 | 18.3 |
| Proocytocin/neurophysin (1–20) | CYIQNCPLGG **K** R AVLDLDVR | | Not cleaved | |

*Note.* **R** and **K** correspond to DR and DK, respectively. ⊥ means the scissile bond.
[a] Values are given ± 10%. Adapted from Ref. 8.

ever, electrophoretic analysis of testis enzyme under denaturing conditions reveals two separated forms of $M_r$ 140,000 and 110,000[8] (Fig. 2, lane a). Microsequencing of endolysin C fragments generated from the two proteins indicates a close relationship, although the impossibility of N-terminal sequencing of the full-length forms prevented us from establishing the exact nature of the relationship between the two enzymes.[8] Limited posttranslational (or postextraction) proteolysis could account for this feature. Indeed, Northern blotting performed on RNA preparations from testis did not shown the presence of more than one form of mRNA.

Immunoblotting using antibodies raised against the purified testis enzyme revealed two bands in both brain cortex and testis extracts[8] (Fig. 2, lanes b–e). Because the endopeptidase characterized from either tissue exhibits a similar selectivity for peptide substrates and a comparable sensitivity to peptidase inhibitors (*vide supra*), it can be inferred that both are functionally and immunochemically similar.

## cDNA Cloning and Sequencing

The purification of adequate amounts of the rat testis enzyme allowed us to obtain the sequence of four tryptic fragments and to raise polyclonal antibodies.[8] The antibodies were used to screen a rat testis expression cDNA library in the λ Zap II vector (1.5 × 10[6] recombinants; Stratagene,

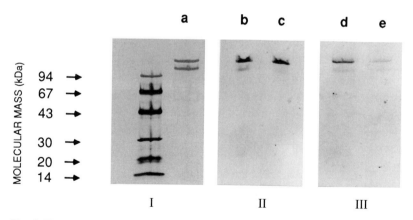

FIG. 2. Electrophoretic analysis under denaturing conditions of the rat testes enzyme and its relationship to that from brain cortex. (I) Molecular mass markers; lane a, purified testis enzyme. The gel was stained using the silver staining technique. (II) Immunoblot of purified testis enzyme (lane b) and of brain cortex enzyme (lane c). (III) Immunoblot of the corresponding crude extracts from testis (2 μg protein) (lane d) and brain cortex (7 μg) (lane e). (Adapted from Chesneau *et al.*[8])

La Jolla, CA). In addition, one of the convertase fragments, a 15-mer peptide, was used to design a 26-mer degenerate inosine-containing oligonucleotide probe. Standard methods[12] were used to screen by both approaches a total of 3 × 10$^6$ clones. From the 40 selected clones, 32 appeared to correspond to the NRD convertase cDNA. The entire sequence of the largest cDNA obtained exhibits the following features: a 5' untranslated region of 64 nucleotides, an open reading frame of 1161 codons, and a 3' region of 34 nucleotides. This is in good agreement with the Northern blot analysis of testis RNA, which reveals a convertase messenger with an approximate size of 3.7 kb.[9]

The deduced amino acid sequence encodes a 1161-residue protein with a calculated molecular mass of 133 kDa, consistent with the largest form of the purified enzyme.[8,9] The sequence shows the following main features (Fig. 3)[13–19]: (1) a putative signal peptide, of about 20 residues; (2) a 71-residue acidic stretch (residues 139 to 209) exceptionally rich in aspartate and glutamate (79%); (3) the putative pentapeptide zinc-binding motif (HXXEH; residues 244 to 248) in which XX is Phe-Leu (this motif belongs to the insulinase signature[20] and has been identified as being the functional binding site for a zinc atom in the human insulin-degrading enzyme[21] and protease III from *Escherichia coli*[22]; these observations clearly confirm the metallopeptidase character of the enzyme); and (4) no clearly identifiable transmembrane domain.

The presence of an acidic stretch in the NH$_2$-terminal region of the protein is an interesting feature and may be responsible for the acidic p*I* of 4.9 (V. Chesneau *et al.*, 1993, unpublished data). This kind of acidic region is observed in a number of proteins like nucleolins, the avian progesterone receptor, transcription factors, calcium-binding proteins, and, notably, the carboxypeptidase Kex1 from *Saccharomyces cerevisiae*. The function of

[12] J. Sambrook, E. F. Fritsch, and T. Maniatis, *in* "Molecular Cloning: A Laboratory Manual." Cold Spring Harbor Laboratory, Cold Spring Harbor, New York, 1989.

[13] J. A. Affholter, V. A. Fried, and R. A. Roth, *Science* **242**, 1415 (1988).

[14] W.-L. Kuo, B. D. Gehm, and M. R. Rosner, *Mol. Endocrinol.* **4**, 1580 (1990).

[15] H. Baumeister, D. Müller, M. Rehbein, and D. Richter, *FEBS Lett.* **317**, 250 (1993).

[16] P. W. Finch, R. E. Wilson, K. Brown, I. D. Hickson, and P. T. Emmerson, *Nucleic Acids Res.* **14**, 7695 (1986).

[17] J. J. M. Meulenberg, E. Sellink, N. H. Riegman, and P. W. Postma, *Mol. Gen. Genet.* **232**, 284 (1992).

[18] N.-Y. Chen, S.-Q. Jiang, D. A. Klein, and H. Paulus, *J. Biol. Chem.* **268**, 9448 (1992).

[19] J. Kleiber, F. Kalousek, M. Swaroop, and L. E. Rosenberg, *Proc. Natl. Acad. Sci. U.S.A.* **87**, 7978 (1990).

[20] N. D. Rawlings and A. J. Barrett, *Biochem. J.* **275**, 389 (1991).

[21] R. K. Perlman, B. D. Gehm, W.-L. Kuo, and M. R. Rosner, *J. Biol. Chem.* **268**, 21538 (1993).

[22] A. B. Becker and R. A. Roth, *Proc. Natl. Acad. Sci. U.S.A.* **89**, 3835 (1993).

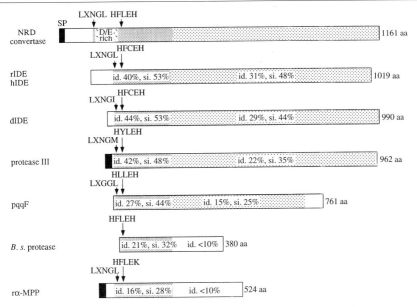

FIG. 3. Schematic comparison of NRD convertase with related proteins. rIDE, Rat insulin-degrading enzyme[13]; hIDE, human insulin-degrading enzyme[14]; dIDE, *Drosophila* insulin-degrading enzyme[15]; protease III, *E. coli* protease III[16]; pqqF, open reading frame F of the pyrroloquinoline quinone operon of *Klebsiella pneumoniae*[17]; B.s.protease, *Bacillus subtilis* processing proteinase[18]; rα-MPP, rat mitochondrial matrix processing peptidase α subunit.[19] Putative signal peptides are shown as solid boxes; nonhomologous regions are represented by open boxes; homologous regions are shown as shaded boxes, with percentages of identity (id.) and similarity (si.) indicated for each conserved region. The consensus motif LXNG(I/L/M) zinc-binding site, and protein sizes are indicated. The NRD convertase D/E-rich domain is cross-hatched.

these highly charged protein domains remains unidentified. However, it is hypothesized that the domain may participate in the interaction of the enzyme with subcellular membranes and/or in substrate binding and recognition via electrostatic interactions.

## Enzyme Distribution

Determination of the convertase activity in brain indicates that the enzyme is detected in at least three regions, namely, hypothalamus, pituitary gland, and neocortex.[4] Further measurements reveal that the testis is a major source of enzyme as that organ contains 5 to 10 times as much activity as the brain cortex[8]; indeed, a clear signal in the brain sample could not be detected by Northern blot analysis. In testis, immunocytochemical studies show that the protein is present in the last stage spermatids.[8] On the other hand, *in situ* hybridization with a ribonucleic acid probe reveals that NRD convertase transcripts appear earlier during spermatogenesis.[9] In addition, NRD convertase activity can also be detected in adrenal gland and heart tissues.[8]

### N-Arginine Dibasic Convertase as Member of M16 Pitrilysin Family

A FASTA search in the GenPro databank (version 79) has been performed on the NRD convertase amino acid sequence deleted of its acidic stretch (−71 residues). From 48 sequences showing scores higher than 100, 7 significantly homologous proteins were selected: the human, rat, and *Drosophila* insulin-degrading enzymes (IDEs or insulysins,[13–15] protease III from *E. coli* known as pitrilysin,[16] the open reading frame F of the pyrroloquinoline quinone operon of *Klebsiella pneumoniae* (pqqF[17]), the rat mitochondrial matrix processing peptidase (MPP) $\alpha$ subunit,[19] and the *Bacillus subtilis* processing proteinase, a putative gene flanking the diaminopimelate operon.[18] A schematic representation of NRD convertase and the homologous proteins, which are all members of the M16 pitrilysin family,[23] is shown in Fig. 3. The highest similarly is observed in the region of the zinc-binding motif and reaches 50% with insulin-degrading enzyme from three species and protease III from *E. coli*. With those proteins, the similarity can be extended to the entire sequence, although it decreases (35 to 48%). Surprisingly, the NRD convertase acidic stretch is inserted in the highly conserved region, about 25 residues upstream from the HXXEH zinc-binding site.

The physiological substrate(s) of the convertase has not been identified as yet. The *in vitro* catalytic properties of the enzyme are consistent with a role in processing of dibasic sites in propeptides and proteins, and the presence of a putative signal peptide argues in favor of its association with the machinery of the cell secretory pathways.

### Acknowledgments

Work from authors' laboratory was supported in part by funds from the Centre National de le Recherche Scientifique to Unité de Recherche Associée 1682 and the Université Pierre et Marie Curie (Paris VI).

[23] N. D. Rawlings and A. J. Barrett, *Biochem. J.* **290,** 205 (1993).

## [46] Purification and Characterization of Mitochondrial Processing Peptidase of *Neurospora crassa*

### By MICHAEL BRUNNER and WALTER NEUPERT

### Introduction

Many nuclear encoded mitochondrial proteins are synthesized with a cleavable N-terminal presequence which targets the protein to the mitochondrial matrix.[1,2] This presequence is proteolytically removed in the mitochondrial matrix during or shortly after import of the precursor.[3] The processing reaction is carried out by the mitochondrial processing peptidase (MPP).[4–8] The enzyme is a heterodimeric protein which is composed of the two related polypeptides $\alpha$-MPP and $\beta$-MPP.[7–9] This chapter focuses on the purification and characterization of the two components of MPP from mitochondria of the filamentous fungus *Neurospora crassa*. $\alpha$-MPP is found as a soluble monomeric protein.[7,9] $\beta$-MPP of *N. crassa* is identical to the core I protein of the $bc_1$ complex of the respiratory chain.[10] Some $\beta$-MPP is found soluble in the matrix, but the majority of the protein is associated with the mitochondrial inner membrane.[9–11] Both components, $\alpha$-MPP and $\beta$-MPP, are required for the proteolytic action on precursor proteins.[11] The $\alpha$- and $\beta$-MPP themselves are made as

[1] G. Attardi and G. Schatz, *Annu. Rev. Cell Biol.* **4**, 289 (1988).

[2] F.-U. Hartl and W. Neupert, *Science* **247**, 930 (1990).

[3] P. Böhni, S. Gasser, C. Leaver, and G. Schatz, *in* "The organisation of the mitochondrial genome" (Kroon, A. M., Saccone, C., eds.) p. 423. Elsevier/North-Holland, Amsterdam, 1980.

[4] B. Schmidt, E. Wachter, W. Sebald, and W. Neupert, *Eur. J. Biochem.* **144**, 581 (1984).

[5] S. D. Emr, A. Vassarotti, J. Garrett, B. L. Geller, M. Takeda, and Douglas, M. G., *J. Cell Biol.* **102**, 523–533.

[6] P. Böhni, G. Daum, and G. Schatz, *J. Biol. Chem.* **258**, 4937 (1983).

[7] G. Hawlitschek, H. Schneider, B. Schmidt, M. Tropschug, F.-U. Hartl, and W. Neupert, *Cell (Cambridge, Mass.)* **53**, 795 (1988).

[8] M. Yang, R. E. Jensen, M. P. Yaffe, W. Oppliger, and G. Schatz, *EMBO J.* **7**, 3857 (1988).

[9] R. A. Pollock, F.-U. Hartl, M. Y. Cheng, J. Ostermann, A. Horwich, and W. Neupert, *EMBO J.* **7**, 3493 (1988).

[10] U. Schulte, M. Arretz, H. Schneider, M. Tropschug, E. Wachter, W. Neupert, and H. Weiss, *Nature (London)* **339**, 147 (1989).

[11] M. Arretz, H. Schneider, B. Guiard, M. Brunner, and W. Neupert, *J. Biol. Chem.* **269**, 4959 (1994).

preproteins with amino-terminal presequences which are cleaved by mature MPP.[7]

In rat and yeast, $\alpha$- and $\beta$-MPP form a stable complex which is soluble in the matrix,[8,12,13] and both subunits of yeast MPP are required for proteolytic activity.[14] By contrast, all processing activity in mitochondria from potato is associated with the $bc_1$ complex of the mitochondrial inner membrane.[15] It remains to be established which of the three corelike proteins constitute the active peptidase in mitochondria of potato tubers.

## Purification Procedure for *Neurospora crassa* Mitochondrial Processing Peptidase

### Expression of Mature Form of α-MPP in Escherichia coli and Purification of Native Protein

*Construction of Recombinant Expression System for Mature α-MPP.* In *Neurospora crassa,* $\alpha$-MPP comprises about 0.03% of the mitochondrial protein, and purification yields only minor amounts of the protein.[7] Therefore, $\alpha$-MPP has been expressed in *E. coli.* To obtain the mature form of $\alpha$-MPP, the leucine codon corresponding to the mature N terminus of $\alpha$-MPP is replaced by an ATG codon using polymerase chain reaction (PCR) mutagenesis. As the DNA template for the PCR, pGEM3 DNA is used that carries a full-length preMPP cDNA inserted in an SP6 promoter orientation. The upstream oligonucleotide for PCR is GGGCAT**ATG**GC-TACGAGAGCCGCTGCTGTC, which contains an *Nde*I (CATATG) site overlapping the ATG codon. As downstream primer an oligonucleotide specific to the T7 promoter is used. The resulting DNA fragment is digested with *Nde*I and *Eco*RI, which cleaves in the polylinker of pGEM 3, and subcloned into the expression vector pJLA503,[16] yielding pRM.[11] Positive clones are verified by DNA sequencing.

*Preparation of Bacterial Extracts Containing α-MPP.* The plasmid pRM is transformed into *Escherichia coli* Ec538. For large-scale production of $\alpha$-MPP, four 1.5-liter cultures of the recombinant *E. coli.* cells are grown in 5-liter Erlenmeyer flasks at 30° in Luria broth (LB) (100 $\mu$g/ml ampicillin) to a cell density corresponding to an $OD_{578}$ of 1.0. For heat induction of $\alpha$-MPP expression the temperature is rapidly shifted to 42° by adding to

[12] W. Ou, A. Ito, H. Okazaki, and T. Omura, *EMBO J.* **8,** 2605 (1989).

[13] J. Kleiber, F. Kalousek, M. Swaroop, and L. E. Rosenberg, *Proc. Natl. Acad. Sci. U.S.A.* **87,** 7978 (1990).

[14] V. Geli, *Proc. Natl. Acad. Sci. U.S.A.* **90,** 6247 (1993).

[15] H. P. Braun, M. Emmermann, V. Kruft, and U. K. Schmitz, *EMBO J.* **11,** 3219 (1992).

[16] B. Schauder, H. Blöcker, R. Frank, and J. E. G. McCarthy, *Gene* **52,** 279 (1987).

the cultures an equal volume of LB medium at 56°. The cultures are incubated for 2 hr at 42°, and subsequently the cells are harvested by centrifugation for 10 min at 7000 rpm in a Sorvall JA-10 rotor. The bacterial pellet is resuspended in 200 ml of 30 m$M$ Tris-HCl, pH 7.5, 5 m$M$ EDTA, 1 m$M$ phenylmethylsulfonyl fluoride (PMSF). After mixing for 15 min at 4°, cells are pelleted (10 min, 7000 rpm, Sorvall JA-10 rotor) and resuspended in 200 ml of 30 m$M$ Tris-HCl, pH 7.5, 1 m$M$ MnCl$_2$, 1 m$M$ PMSF, and 1.25% (w/v) Triton X-100. The sample is extensively sonicated to disrupt the cells and break the chromosomal DNA (Branson sonifier, 10 pulses of 10 sec, setting 8 and 50% duty). Subsequently, the debris is removed by centrifugation (30 min, 9000 rpm, JA-10 rotor).

*DEAE-Cellulose Chromatography.* The supernatant is loaded at a flow rate of 4 ml/min on a 5 × 20 cm DEAE-cellulose column equilibrated with buffer A [30 m$M$ Tris-HCl, pH 7.5, 1 m$M$ dithiothreitol (DTT), 0.25% Triton X-100, 2.5% glycerol]. The column is washed with 1000 ml buffer A containing 60 m$M$ NaCl. $\alpha$-MPP is eluted in buffer A with a 1500-ml linear gradient of 60 to 90 m$M$ NaCl. Fractions of 30 ml are collected, and $\alpha$-MPP is found in fractions 12 through 24.

*Hydroxyapatite Chromatography.* The fractions containing $\alpha$-MPP are pooled, and 1/20 volume of 0.5 $M$ sodium phosphate, pH 6.9, is added. The sample is loaded at a flow rate of 1.25 ml/min on a 5 × 8 cm hydroxyapatite column equilibrated with buffer B [20 m$M$ 3-($N$-morpholino)propanesulfonic acid (MOPS)–KOH, pH 7.1, 50 m$M$ NaCl, pH 7.1, 1 m$M$ DTT, 5% glycerol] containing 5 m$M$ sodium-phosphate. The column is first washed with 125 ml equilibration buffer and then with a 250-ml linear gradient (in buffer B) of 5 to 132 m$M$ sodium phosphate. Subsequently, $\alpha$-MPP is eluted with buffer B containing 132 m$M$ sodium phosphate. Fractions of 10 ml are collected, and $\alpha$-MPP is found in fractions 40 to 68.

*Concentration of $\alpha$-MPP.* The pooled fractions are dialyzed 2 times against 5 liters each time of 30 m$M$ Tris-HCl, pH 7.5, 1 m$M$ DTT and loaded on a 1.6 × 5 cm DEAE-cellulose column equilibrated with 30 m$M$ Tris-HCl, pH 7.5, 1 m$M$ DTT, 2.5% glycerol. The column is washed with 60 ml buffer A containing 30 m$M$ NaCl, and $\alpha$-MPP is eluted with 20 ml buffer A containing 100 m$M$ NaCl.

The purification of $\alpha$-MPP from 12 liters of *E. coli* culture yields about 15 mg protein at a concentration of 0.75 mg/ml. The protein is divided into aliquots, frozen in liquid nitrogen, and stored at −70°. All chromatography steps are carried out at 4°. The purity of $\alpha$-MPP can be monitored by sodium dodecyl sulfate–polyacrylamide gel electrophoresis (SDS–PAGE) and Western blotting.[17] Immune decoration can be carried out with a rabbit

---

[17] W. N. Burnette, *Anal. Biochem.* **121**, 195 (1981).

anti-$\alpha$-MPP serum and visualized using goat anti-rabbit immunoglobulin G (IgG) coupled to alkaline phosphatase.[18]

### Purification of $\beta$-MPP from $bc_1$ Complex of Mitochondrial Inner Membrane

In *Neurospora crassa* $\beta$-MPP is identical to the core I protein of the $bc_1$ complex.[10] Mitochondrial membranes are used for the purification of $\beta$-MPP because the membrane-derived $\beta$ subunit is much more abundant than the soluble form.[7,11,19,20]

*Isolation of Mitochondria.* Mitochondria from *N. crassa* hyphae are isolated according to Sebald *et al.*[21] Briefly, hyphae (1000 g) are resuspended in 4 liters SET buffer (250 m$M$ sucrose, 1 m$M$ EDTA, 10 m$M$ Tris-HCl, pH 7.2, 1 m$M$ PMSF) by homogenization in a kitchen blender. The cells are disrupted in a grind mill, and the debris is removed by centrifugation (15 min, 3000 rpm, JA-10 rotor). The supernatant is recovered, and mitochondria are pelleted by centrifugation for 15 min at 10,000 rpm (JA-10 rotor). The mitochondrial pellet is resuspended in 200 ml NATES buffer [200 m$M$ NaCl, 1 m$M$ EDTA, 5% (w/v) sucrose, 1 m$M$ PMSF, 50 m$M$ Tris, pH 7.1], and the protein concentration is adjusted to 5 mg/ml.

*Preparation and Solubilization of Mitochondrial Membranes.* Mitochondria are broken by sonication (Branson sonifier, 10 pulses of 10 sec, setting 8 and 50% (w/v) duty), and debris is removed by centrifugation (30 min, 9000 rpm, JA-10 rotor). The supernatant is recovered, and mitochondrial membranes are sedimented at 100,000 $g$ for 1 hr (37,500 rpm, 70-Ti rotor). The membranes are resuspended in 20 m$M$ Tris, pH 7.5, 20 m$M$ NaCl, 1 m$M$ EDTA, 5% sucrose, 1 m$M$ PMSF, and the protein concentration is adjusted to 5 mg/ml. To solubilize the membranes Triton X-100 [20% (w/v) stock solution] is added to a final concentration of 1.5%, and the sample is kept for 15 min on ice with gentle stirring. Nonsolubilized material is removed by a centrifugation (30 min, 15,000 rpm, JA-20 rotor).

*DEAE-Cellulose Chromatography.* The membrane extract is loaded at a flow rate of 2 ml/min on a 2.6 × 30 cm DEAE-cellulose column equilibrated with buffer D (30 m$M$ Tris-HCl, pH 7.5, 0.5% Triton X-100). The column is washed with 200 ml buffer D containing 30 m$M$ NaCl. $\beta$-MPP is eluted with a 1000-ml linear gradient of 30 to 300 m$M$ NaCl. Fractions

[18] M. S. Blake, K. H. Johnston, G. J. Russel-Jones, and E. C. Gotschlick, *Anal. Biochem.* **136,** 175 (1984).

[19] P. Linke and H. Weiss, this series, Vol. 126, p. 201.

[20] H. Weiss, *Curr. Top. Bioenerg.* **15,** 67 (1987).

[21] W. Sebald, W. Neupert, and H. Weiss, this series, Vol. 55, p. 144.

of 20 ml are collected. The majority of $\beta$-MPP is found in fractions 20 through 28.

*Hydroxyapatite Chromatography.* The pooled fractions are diluted with an equal volume of buffer E (30 m$M$ MOPS–KOH, pH 7.1, 50 m$M$ NaCl, 5 m$M$ sodium phosphate, pH 7.1, 0.25% Triton X-100), and the sample is loaded on a 2.6 × 8 cm hydroxyapatite column (equilibrated with buffer E) at a flow rate of 1 ml/min. The column is washed with 80 ml buffer E containing 60 m$M$ sodium phosphate and with a 120-ml gradient (in buffer E) of 60 to 80 m$M$ sodium phosphate. $\beta$-MPP is eluted with a 230-ml linear gradient of 80 to 200 m$M$ sodium phosphate and 270 ml buffer E containing 200 m$M$ sodium phosphate. Fractions of 10 ml are collected. Some $\beta$-MPP (<20%) elutes around fraction 10. This material corresponds to the monomeric form of $\beta$-MPP.[11] Most of the $\beta$-MPP, however, elutes in a complex with the core II protein[11] in fractions 28 through 40. The pooled fractions are dialyzed twice for 2 hr against 5 liters of 30 m$M$ Tris-HCl, pH 8.4. Under these conditions $\beta$-MPP dissociates from the core II protein,[11,19,20] and the monomeric protein is purified to homogeneity by rechromatography on an anion-exchange column and a hydroxyapatite column as follows.

*Chromatography on Q-Sepharose.* The dialyzed sample is loaded at a flow rate of 1 ml/min on a 1.6 × 13 cm Q-Sepharose column, equilibrated with buffer F (30 m$M$ Tris-HCl, pH 8.4, 0.25% Triton X-100). The column is washed with 70 ml buffer F containing 150 m$M$ NaCl. $\beta$-MPP is eluted with a 90-ml linear gradient of 150 to 700 m$M$ NaCl. Fractions of 2 ml are collected, and $\beta$-MPP is recovered in fractions 15–19.

*Rechromatography on Hydroxyapatite.* The pooled fractions are diluted with 3 volumes of buffer E (see above) and loaded at a flow rate of 1 ml/ min on a 1.6 × 4 cm hydroxyapatite column (equilibrated with buffer E). The column is washed with 30 ml buffer E containing 20 m$M$ sodium phosphate, pH 7.1. $\beta$-MPP is eluted with a 90-ml linear gradient of 40 to 200 m$M$ sodium phosphate. Fractions of 2 ml are collected, and $\beta$-MPP is recovered in fractions 11–19.

At this stage $\beta$-MPP is purified to homogeneity. The fractions are pooled and dialyzed twice for 2 hr against 30 m$M$ Tris-HCl, pH 8.4. The sample still contains detergent, as Triton X-100 is not removed under these conditions. The protein solution is divided into aliquots frozen in liquid nitrogen, and stored at −70°. The purification yields about 5 mg $\beta$-MPP, and the protein concentration is 0.5–0.8 mg/ml.

All steps are carried out at 4°. The purification is monitored at each step by SDS–PAGE, Western blotting,[17] and immune decoration[18] using a polyclonal rabbit anti-$\beta$-MPP serum.

## Expression in Escherichia coli and Purification of Chimeric Preprotein $pb_2\Delta19(167)$-DHFR

*Expression of $pb_2\Delta19(167)$-DHFR in Escherichia coli.* $pb_2(167)\Delta19$-DHFR is a chimeric preprotein where the N-terminal portion of precytochrome $b_2$ (up to amino acid residue 167) is fused to the mouse dihydrofolate reductase (DHFR).[22] The protein carries a deletion of 19 amino acids in the sorting signal for the intermembrane space. The DNA encoding the chimeric preprotein has been subcloned into the *E. coli* expression vector pUHE 24[23] under the control of an isopropylthiogolactoside (IPTG)-inducible promoter. The resulting plasmid, pUHE 73-1, was transformed into *E. coli* XL Blue cells (any *E. coli* strain carrying the *lac* $I^Q$ mutation can be used to maintain the expression plasmid in a stable manner).

For expression of $pb_2\Delta19(167)$-DHFR, a 400-ml culture of *E. coli* cells harboring pUHE 73-1 is grown at 37° in LB medium (100 mg/ml ampicillin) to an $OD_{598}$ of 0.6. IPTG is added to a final concentration of 1 m$M$, and the culture is incubated for 2 hr at 37°. Under these conditions $pb_2(167)\Delta19$-DHFR is efficiently expressed and is found in inclusion bodies.

*Purification of Fusion Protein $pb_2\Delta19(167)$-DHFR from Inclusion Bodies.* To purify the inclusion bodies the cells are harvested by centrifugation (10 min, 7000 rpm, 4° Sorvall JA-10 rotor). The pellet is resuspended in 5 ml of 50 m$M$ Tris-HCl, pH 8, 1 mg/ml lysozyme, and the suspension is incubated for 30 min at room temperature. Then 30 ml IB buffer [50 m$M$ Tris-HCl, pH 8, 1 m$M$ EDTA, 10 m$M$ DTT, 1 m$M$ PMSF, 25% (w/v) sucrose] containing 1% Triton X-100 is added. The sample is sonicated (Branson sonifier, 10 pulses of 10 sec, setting 7 and 40% duty), and inclusion bodies are collected by centrifugation (30 min, 20,000 rpm, JA-20 rotor). To wash the inclusion bodies, the pellet is resuspended 35 ml IB buffer containing 1% Triton X-100. The suspension is sonicated and the inclusion bodies collected by centrifugation as described above. This wash step is repeated twice with IP buffer containing 0.1% Triton X-100 in the second wash and no detergent in the third wash step. Finally, the inclusion bodies are solubilized in 2 ml of 30 m$M$ Tris-HCl, pH 8.7, 7 $M$ urea. The solution is added to DEAE-cellulose resin (2.0 ml bed volume) equilibrated with 30 m$M$ Tris-HCl, pH 8.7, 7 $M$ urea and incubated for 30 min at 4° on a rolling rocker. The precursor protein does not bind to the anion-exchange resin under these conditions; however, an inhibitor of the processing reaction is removed at this step. To recover $pb_2\Delta19(167)$-DHFR in a concen-

[22] H. Koll, B. Guiard, J. Rassow, J. Ostermann, A. L. Horwich, W. Neupert, and F.-U. Hartl, *Cell (Cambridge, Mass.)* **68,** 1163 (1992).
[23] H. Bujard, R. Gentz, M. Lanzer, D. Stueber, M. Mueller, I. Ibrahimi, M.-T. Haeuptle, and B. Dobberstein, this series, Vol. 155, p. 416.

trated form, the suspension is packed into 1-ml spin columns, and the flow through is collected by a brief spin (5 to 10 sec) in an Eppendorf centrifuge. The purification of $pb_2\Delta19(167)$-DHFR from a 400-ml culture yields about 120 mg preprotein at a concentration of 50 mg/ml. The preprotein is divided into aliquots and stored at $-70°$.

## Enzyme Activity Determinations

*Principle*

Purified *N. crassa* MPP is capable of specifically processing radiochemical and chemical quantities of mitochondrial preproteins.[11,24] Protocols for both assays are given below. Thus, in combination with preprotein import into isolated mitochondria, it can be established whether a precursor is processed by MPP, how often it is processed by the enzyme, and, in addition, whether maturation of a preprotein in mitochondria involves processing by other peptidases such as the mitochondrial intermediate peptidase (MIP)[25] or the intermembrane space peptidase (IMP1).[26] A major aim of analyzing processing of chemical quantities of preprotein is to characterize the processing peptidase and its interaction with preproteins.

*Method A: Processing of Radiochemical Amounts of [$^{35}$S]Methionine-Labeled Precursor Proteins*

Radiolabeled preproteins are synthesized in a rabbit reticulocyte lysate in the presence of [$^{35}$S]methionine by coupled transcription/translation.[27,28] *In vitro* processing reactions of preproteins are performed at 25° in 30 $\mu$l assay buffer (30 m$M$ Tris-HCl, pH 7.7, 1 m$M$ MnCl$_2$, 0.1 m$M$ DTT) containing 1% Triton X-100. The $\alpha$- and $\beta$-MPP forms, 4 pmol each, are incubated with reticulocyte lysate containing the radiolabeled preprotein. The lysate may comprise up to 10% of the assay volume without affecting the processing reaction. Usually preproteins are efficiently processed within less than 3 min. However, with some preproteins the kinetics of processing are very slow, and an incubation for up to 60 min may be required to detect a significant fraction of processed protein. Processing assays are terminated by the addition of 10 $\mu$l Laemmli buffer (4-fold concentrated). Aliquots are analyzed by SDS–PAGE and fluorography.

[24] M. Arretz, H. Schneider, U. Wienhues, and W. Neupert, *Biomed. Biochem. Acta* **50**, 403 (1991).
[25] F. Kalousek, J. P. Hendrik, and L. E. Rosenberg, *Proc. Natl. Acad. Sci. U.S.A.* **85**, 7536 (1988).
[26] M. Behrns, G. Michaelis, and E. Pratje, *Mol. Gen. Genet.* **228**, 167 (1991).
[27] H. R. B. Pelham and R. J. Jackson, *Eur. J. Biochem.* **67**, 247 (1976).
[28] D. Stüber, I. Ibrahimi, D. Culter, B. Dobberstein, and H. Bujard, *EMBO J.* **3**, 3143 (1984).

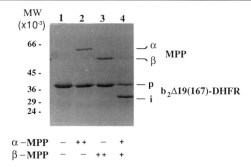

*Method B: Processing of Chemical Amounts of pb₂Δ19(167)-DHFR*

Standard *in vitro* processing reactions of chemical quantities of prepro-tein are performed at 25°. The α- and β-MPP (4 pmol each) are prewarmed to 25° in 95 μl assay buffer (see above). The processing reaction is started by the addition of 200 pmol (7.5 μg) pb₂Δ19(167)-DHFR in 5 μl of 7 *M* urea, 30 m*M* Tris-HCl, pH 8.7. The reaction is terminated by the addition of 30 μl Laemmli buffer (4-fold concentrated) containing 1% Triton X-100. When diluted out of the denaturant, pb₂Δ19(167)-DHFR has the tendency to bind to the plastic walls of Eppendorf tubes. The addition of Triton X-100 (in addition to the SDS of the sample buffer) facilitates the quantita-tive recovery of the protein.[29] The processing products are analyzed by SDS–PAGE as shown in Fig. 1 and quantified by densitometry (Pharmacia-LKB, Piscataway, NJ) of Coomassie blue-stained gels.

*Variation of Assay Conditions*

To process chemical quantities of preprotein, the precursor has to be diluted out of 7 *M* urea. Urea does not affect the processing reaction over a broad range of concentrations. The dependence of processing of pb₂Δ19(167)-DHFR on urea concentration, pH, and temperature is shown

---

[29] M. Arretz, Dissertation, Universität München (1993) (Ph.D thesis).

in Fig. 2A–C. The kinetics of processing of $pb_2\Delta19(167)$-DHFR under standard conditions (25°, 350 m$M$ urea, assay buffer, pH 7.7) are shown in Fig. 2D. Processing of 2 $\mu M$ $pb_2\Delta19(167)$-DHFR by 40 n$M$ $\alpha$- and $\beta$-MPP was half-maximal after 2 min.

## Enzyme Kinetics

Purified MPP and purified precursor can be used to analyze the kinetics and specificity of interaction between MPP and preprotein substrates. Processing of $pb_2\Delta19(167)$-DHFR by MPP follows Michaelis–Menten kinetics.[11] The $K_m$ and $V_{max}$ values have been determined according to the following protocol. A stock solution of preprotein (1.3 m$M$) is prediluted in 7 $M$ urea, 30 m$M$ Tris-HCl, pH 8.7, to concentrations between 5 and 100 $\mu M$, and 5 $\mu$l of each dilution is prewarmed to 25°. The processing reactions are started by the addition of 4 pmol MPP in 95 $\mu$l assay buffer (prewarmed to 25°). After 2 min the reactions are terminated by the addition of 30 $\mu$l Laemmli buffer (4-fold concentrated), 1% Triton X-100. Samples are analyzed by SDS–PAGE and quantified by densitometry. Analysis of kinetic data (Fig. 3A) reveals an apparent $K_m$ of 1.27 $\mu M$ and a $V_{max}$ of 0.27 $\mu M$/min. Under the assumption that the purified enzyme is fully active, the molar activity of MPP is given by $k_{cat} = (V_{max}/[MPP]) = 6.75$ min$^{-1}$.

## Competitive Inhibition of Processing by
## Presequence-Derived Peptides

The purpose of competition assays is to use presequence peptides or portions thereof to characterize and to delineate the MPP binding site in a preprotein. Such peptides act as competitive inhibitors of preprotein processing.[11] The affinity of MPP for presequence-derived peptides can be low, with dissociation constants in the 1 to 100 $\mu M$ range. Accordingly, it is difficult to detect and quantify this interaction. The affinity of MPP for the peptides can be determined in competition assays in which $pb_2\Delta19(167)$-DHFR is used as a reporter substrate. This is exemplified in Fig. 3B for peptide $pb_2(1–31)$ which is derived from the matrix targeting signal of cytochrome $b_2$. Processing of $pb_2\Delta19(167)$-DHFR was monitored in the absence and presence of 40 $\mu M$ peptide. Michaelis–Menten kinetics were analyzed as described above.

In the presence of presequence peptide the apparent $K_m$ for processing is increased. The maximal rate of substrate turnover, $V_{max}$, is not affected by the peptide. This indicates that the presequence peptide acts as competitive inhibitor. In the presence of 40 $\mu M$ peptide the apparent $K_{m(+peptide)}$ for processing was 2.47 $\mu M$. This is 2-fold higher than in the absence of prepep-

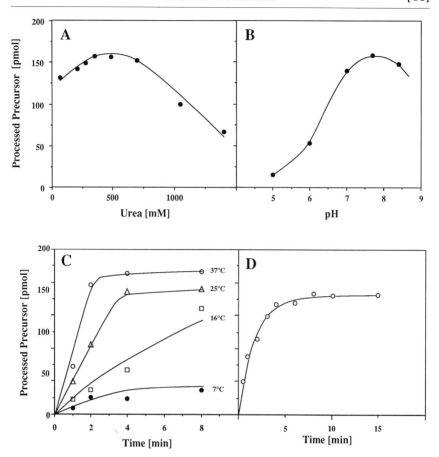

Fig. 2. Dependence of processing of pb$_2$Δ19(167)-DHFR on urea concentration, pH, temperature, and time. Processing reactions were carried out with 2 μM pb$_2$Δ19(167)-DHFR and 40 nM MPP. Processing was monitored by SDS–PAGE and quantified by densitometry. (A) Dependence on concentration of urea. Aliquots of 1 μl containing 200 pmol pb$_2$Δ19(167)-DHFR in 7 M urea, 30 mM Tris-HCl, pH 8.4, were diluted into 99 μl assay buffer containing sufficient urea to give the final concentrations indicated. (B) pH dependence. The pH of the assay buffer was adjusted with 30 mM Bis–Tris–NaOH to between pH 5.5 and 7.0 and with Tris-HCl between pH 7.0 and 8.5. (C) Temperature dependence. Processing was monitored as a function of time at the temperatures indicated. (D) Kinetics of processing under standard conditions. pb$_2$Δ19(167)-DHFR (200 pmol) was incubated under standard conditions with 5 pmol MPP for the times indicated. Standard assays were performed at 25° in 100 μl of 30 mM Tris-HCl, pH 7.7, 0.1 mM DTT, 0.1 mM MnCl$_2$, and 350 mM urea.

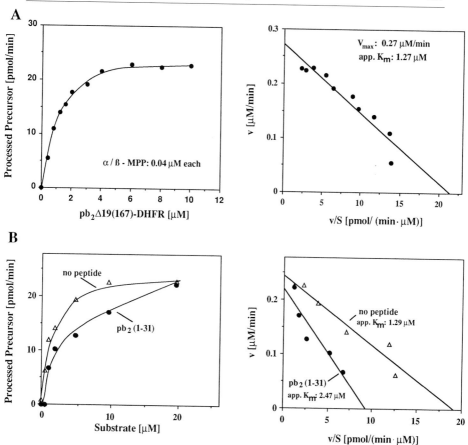

FIG. 3. Michaelis–Menten kinetics and inhibition of processing by prepeptides. Processing of pb$_2$Δ19(167)-DHFR by purified MPP was carried out under standard conditions (see text) for 2 min with 40 n$M$ MPP and at the substrate concentrations indicated. Processing was monitored by SDS–PAGE and quantified by densitometry of Coomassie blue-stained gels. (A) The amount of pb$_2$Δ19(167)-DHFR processed per minute is plotted versus substrate concentration, and data are analyzed according to Eadie–Hofstee plots (right-hand side). The slope of the least-squares fit through the data points corresponds to an apparent $K_m$ of 1.27 $\mu M$ and the $y$-intercept to a $V_{max}$ of 0.27 $\mu M$/min. (B) Inhibition of processing by presequence peptides. Michaelis–Menten kinetics of processing of pb$_2$Δ19(167)-DHFR were determined in the absence and presence of 40 $\mu M$ of pb$_2$(1–31) derived from the presequence of cytochrome $b_2$. Samples were analyzed by SDS–PAGE and densitometry. Analysis of data according to Eadie–Hofstee plots (right-hand side) revealed apparent $K_m$ values of 1.29 $\mu M$ in the absence and 2.47 $\mu M$ in the presence of peptide pb$_2$(1–31).

tide ($K_{m(-peptide)}$ = 1.29 $\mu M$). The expected increase of the Michaelis–Menten constant in the presence of the inhibitory peptide is given by $K_{m(+peptide)}$ = $K_{m(-peptide)}$ × (1 + [peptide]/$K_i$). Accordingly, the affinity of MPP for the presequence peptide is characterized by an apparent $K_i$ of 37 $\mu M$ [i.e., 1.29 $\mu M$ × 40 $\mu M$/(2.47 − 1.29) $\mu M$]. Thus, the interaction of MPP with the presequence peptide is about 30-fold weaker than that with the intact preprotein. The competition assay is a sensitive method to detect and quantify such weak interactions with the peptidase.

# [47] O-Sialoglycoprotease from *Pasteurella haemolytica*

*By* ALAN MELLORS and REGGIE Y. C. LO

## Introduction

A novel endopeptidase, which hydrolyzes peptide bonds within glycoprotein substrates, has been described in culture supernatants of the bacterium *Pasteurella haemolytica*.[1] The *P. haemolytica* enzyme is designated as an *O*-sialoglycoprotein endopeptidase (E.C. 3.4.24.57) or glycoprotease because of its specificity for the proteolytic cleavage of proteins which have a significant number of *O*-sialoglycans attached to serine and threonine residues.[2] The amino acid sequence predicted from the nucleotide sequence of the *P. haemolytica* serotype A1 glycoprotease gene, and partially confirmed by N-terminal sequencing of the gene product, shows no significant homology with any other proteolytic enzymes but suggests that the enzyme is a metal ion-dependent protease.[3]

### Occurrence

*Pasteurella haemolytica* is a gram-negative coccobacillus found as a commensal in the nasopharynx of cattle and sheep. There are at least 16 serotypes: serotype A1 is commonly isolated from the lungs of cattle during outbreaks of bovine pneumonic pasteurellosis (bovine shipping fever).[4] The gene *gcp* which encodes the glycoprotease enzyme is found in *P. haemolytica* serotypes which ferment arabinose (A biotypes) but not in

[1] G. L. Otulakowski, P. E. Shewen, E. A. Udoh, A. Mellors, and B. N. Wilkie, *Infect. Immun.* **42**, 64 (1983).
[2] K. M. Abdullah, E. A. Udoh, P. E. Shewen, and A. Mellors, *Infect. Immun.* **60**, 56 (1992).
[3] K. M. Abdullah, R. Y. C. Lo, and A. Mellors, *J. Bacteriol.* **173**, 5597 (1991).
[4] P. E. Shewen, A. Sharp, and B. N. Wilkie, *Vet. Med.* **83**, 1078 (1988).

those which ferment trehalose (T biotypes).[5,5a] Similarly, the glycoprotease enzyme activity has been found to be secreted only by A biovars of *P. haemolytica* and has been characterized from culture supernatants of the serotype A1. The gene for the glycoprotease of *P. haemolytica* A1 has been cloned, isolated, and sequenced: it codes for a protein of 325 amino acids, with a predicted molecular mass of 35.2 kDa, isoelectric point of pH 5.1, and a metal ion-binding region.[3] The amino acid sequence deduced from the *gcp* gene, and partially confirmed by N-terminal sequencing, shows no homology with any known class of proteolytic enzymes.

*Substrate Specificity*

The novel feature of the *P. haemolytica* glycoprotease is its marked specificity for sialylated glycoproteins. Many substrate proteins and synthetic peptides have been tested, but proteolytic cleavage has been found only for those bearing sialoglycan chains attached to serine or threonine side chains. The best characterized substrate is human glycophorin A, a transmembrane cell surface protein of human erythrocytes, which contains 15 *O*-linked glycans and 1 *N*-linked glycan.[6] The removal of sialate residues from glycophorin A by treatment with *Clostridium perfringens* neuraminidase destroys the susceptibility of the glycoprotein to hydrolysis by the enzyme.[2] Cleavage of glycophorin A will occur when the substrate is present in solution or if it is present *in situ* in the erythrocyte membrane.

Similar cell surface glycoproteins that are cleaved by the glycoprotease[7] include human CD34 (an antigen expressed on panhematopoietic stem cells in the bone marrow), CD43/leukosialin (a sialomucin that has been implicated in immune cell function and cell-signaling phenomena), CD44 (a receptor for hyaluronate and probably other components of the extracellular matrix), and CD45/leukocyte common antigen (a family of molecules involved in leukocyte activation phenomena whose intracellular domain encodes a tyrosine phosphatase). Other *O*-glycosylated substrates have been identified including the counterreceptor or ligand for the platelet selectin (P-selectin),[8,9] the mucin-like counterreceptor (sgp50) for the leu-

[5] K. M. Abdullah, R. Y. C. Lo, and A. Mellors, *Biochem. Soc. Trans.* **18**, 901 (1990).
[5a] C. W. Lee, R. Y. C. Lo, P. E. Shewen, and A. Mellors, *FEMS Microbiol. Lett.* **121**, 199 (1994).
[6] M. Tomita, H. Furthmayr, and V. T. Marchesi, *Biochemistry* **17**, 4756 (1978).
[7] D. R. Sutherland, K. M. Abdullah, P. Cyopick, and A. Mellors, *J. Immunol.* **148**, 1458 (1992).
[8] C. N. Steininger, C. A. Eddy, R. M. Leimgruber, A. Mellors, and J. K. Welply, *Biochem. Biophys. Res. Commun.* **188**, 760 (1992).
[9] K. E. Norgard, K. L. Moore, S. Diaz, N. L. Stults, S. Ushiyama, R. P. McEver, R. D. Cummings, and A. Varki, *J. Biol. Chem.* **268**, 12764 (1993).

kocyte selectin (L-selectin),[9] the receptor for interleukin 7,[10] and epitectin (a human tumor cell surface antigen).[11] In general the substrates contain extracellular domains in which the serine/threonine residues account for 30% of the amino acids. Although several of these structures also bear $N$-linked glycans, most of the glycans in the mucin-like domains are attached via $O$-linkage to serine or threonine residues. On the other hand, several $N$-glycoprotein human leukocyte cell surface antigens are not cleaved by the glycoprotease, including CD18/11[a,b,c] (leukocyte integrins), CD71 (transferrin receptor), HLA class I, and 8A3 antigens.[7] Soluble glycoproteins of the $N$-linked type, including human immunoglobulins $A_1$ and $A_2$, bovine $\alpha_1$-acid glycoprotein, bovine $\alpha$-lactalbumin, and hen ovalbumin, are not cleaved.[2] A soluble $O$-glycosylated glycoprotein, bovine milk $\kappa$-casein, has also been characterized as a substrate for the glycoprotease. $\kappa$-Casein is not a cell surface molecule but has amphipathic properties, and its use in a rapid assay for the glycoprotease is described below.

## Isolation

The glycoprotease enzyme can be partially purified from culture supernatants of $P.$ $haemolytica$ A1 (strain 43270, ATCC, Rockville, MD), and the following is a modification of the previously published procedure.[2] A brain–heart infusion broth culture grown for 2–3 hr is used to inoculate 6 liters of RPMI 1640 medium containing 5% (v/v) heat-inactivated fetal calf serum. For serum-free preparations, 0.1% (w/v) polyoxyethylene glycol (average $M_r$ of 6000) can be substituted for serum proteins as a glycoprotease stabilizer. After 3–4 hr of aerobic culture at 37° with shaking, the culture supernatant is collected by centrifugation at 17,000 $g$ for 15 min, sterilized by filtration through a 0.45-$\mu$m filter, and then concentrated 20-fold by ultrafiltration with an Amicon (Danvers, MA) PM10 filter (10,000 $M_r$ average cutoff). An equal volume of 10 m$M$ acetate buffer, pH 4.0, is added to the concentrate, and the solution is adjusted to pH 4.5 by the addition of 1 $M$ HCl. After 30 min at 0°, the precipitate is removed by centrifugation at 15,000 $g$ and 4° for 15 min and resuspended in 50 m$M$ HEPES [4-(2-hydroxyethyl)-1-piperazineethane sulfonate] buffer, pH 7.4, frozen, and lyophilized for storage. A 70-fold increase in specific activity can be obtained by gel-exclusion chromatography of the redissolved lyophilate on Bio-Gel P-300 (Bio-Rad, Richmond, CA).[2] When purified to apparent homogeneity by high-performance liquid chromatography (HPLC), using gel-exclusion and anion-exchange columns, the enzyme

[10] L. Healy and I. Titley, personal communication (1993).
[11] R.-H. Hu, A. Mellors, and V. P. Bhavanandan, $Arch.$ $Biochem.$ $Biophys.$ **310,** 300 (1994).

TABLE I
PARTIAL PURIFICATION OF *Pasturella haemolytica* GLYCOPROTEASE[a]

| Fraction | Volume (ml) | Protein (g) | Total activity ($\mu$mol/min) | Recovery (%) | Specific activity (pmol/min/mg protein) |
|---|---|---|---|---|---|
| Culture supernatant | 6000 | 14.40 | 6.78 | 100 | 0.47 |
| Concentrated supernatant | 350 | 7.52 | 1.56 | 22.4 | 0.21 |
| pH 4.5 supernatant | 890 | 6.56 | 0.18 | 2.5 | 0.03 |
| pH 4.5 precipitate | 233 | 1.03 | 3.06 | 45.4 | 2.99 |

[a] A 6-liter culture of *P. haemolytica* A1 grown in RPMI 1640 with 5% fetal calf serum was concentrated by ultrafiltration over a 10-kDa filter, and the pH was lowered to pH 4.5 as described in the text. The precipitate formed in 30 min at 0° was resuspended in 50 m$M$ HEPES, pH 7.4. *O*-Sialoglycoprotein endopeptidase activity in the fractions was determined by the hydrolysis of [125]I-labeled glycophorin A in a 30-min incubation, followed by SDS–PAGE analysis of the products and autoradiography, as described in the text.

shows a molecular mass of 35 kDa by sodium dodecyl sulfate–polyacrylamide gel electrophoresis (SDS–PAGE) analysis, but it is very low in specific activity. The use of fetal calf serum or bovine serum albumin in the culture medium is required for the protein precipitation at pH 4.6 and yields a stabilized glycoprotease-enriched preparation, low in other *P. haemolytica* proteinase or peptidase activities. Although the preparation contains some *P. haemolytica* A1 sialidase activity,[12] as measured against 2'-(4-methylumbelliferyl)-$\alpha$-D-*N*-acetylneuraminic acid,[13] the sialidase activity against sialoglycoproteins is below detection limits when measured by SDS–PAGE analysis, under the conditions used for the proteolysis of glycophorin A. The preparation is also free from *P. haemolytica* leukotoxin activity.[14]

The glycoprotease activity of the lyophilate has been shown to be stable for over 1 year when kept refrigerated at 4°, or frozen at $-12°$. Dilute solutions will lose activity at 0°–4°, or on freezing and thawing repeatedly. Table I shows the specific activities against glycophorin A of various preparations of the glycoprotease-enriched fractions from *P. haemolytica* A1.

Assay

Two methods are described here for the assay of the *P. haemolytica* glycoprotease, both based on the hydrolysis of [125]I-labeled substrates. The first method is based on the proteolysis of [125]I-labeled glycophorin A but

[12] G. H. Frank and L. B. Tabatabai, *Infect. Immun.* **32,** 1119 (1981).

[13] M. Potier, L. Mameli, M. Belisle, L. Dallaire, and S. B. Melancon, *Anal. Biochem.* **94,** 287 (1979).

[14] P. E. Shewen and B. N. Wilkie, *Infect. Immun.* **35,** 91 (1982).

is applicable to other purified *O*-sialoglycoproteins. It is the more specific of the two assay methods, because the substrate is relatively resistant to proteolysis by trace amounts of other proteinases present in biological materials, such as the bovine serum proteins used to precipitate and stabilize the glycoprotease activity. The second method, which is more rapid and convenient, is based on the generation of trichloroacetic acid-soluble (TCA-soluble) peptides when [125]I-labeled κ-casein, a bovine milk glycoprotein, is degraded by the glycoprotease. However, κ-casein is susceptible to hydrolysis by proteinases other than the *P. haemolytica* glycoprotease, such as contaminating proteinases present in the biological sample. Therefore, the rapid κ-casein method should always be checked by SDS–PAGE analysis of the products[15] to determine that the products result from glycoprotease action. The more specific substrate glycophorin A cannot be readily used in a rapid assay because neither it nor its digestion products can be precipitated by protein precipitants such as TCA.

*Glycophorin A Hydrolysis Assay*

Glycophorin A is prepared from human erythrocyte ghosts[16] and radioiodinated by standard methods.[17] Each enzyme assay tube contains 3.5 μg [125]I-labeled glycophorin A and enzyme protein in 50 m$M$ HEPES buffer, pH 7.4, in a total volume of 25 μl. Incubation of the mixture is at 37° for 5–60 min. For the assay of extracts containing low levels of the *P. haemolytica* glycoprotease, which require prolonged incubations (18 hr), a mixture of proteinase inhibitors can be added. For example, pepstatin A, antipain, leupeptin, and aprotinin can be used at final concentrations of 1 μg/ml each, without effect on the glycoprotease enzyme. Hydrolysis of the substrate is terminated by the addition of Laemmli sample buffer[15] and boiling for 3 min. An aliquot of the SDS-treated sample, equal in volume to that loaded onto the gel, is counted to determine the total radioactivity per lane (**A** dpm). Analysis by SDS–PAGE of the substrate and products is then carried out on 12% polyacrylamide gels under reducing conditions, the gels are stained with Coomassie dye, and the radioactive substrate and product bands are located by autoradiography of the dried gel. Unlike κ-casein, glycophorin A stains poorly with Coomassie dye and the autoradiographs are essential for the location of substrate and product bands. Low molecular weight products of glycophorin hydrolysis will migrate with the dye front and be lost from the gel. Therefore quantitation of hydrolysis should be based on the disappearance of substrate and, under conditions of extensive

---

[15] U. K. Laemmli, *Nature* (*London*) **227**, 680 (1970).
[16] J. P. Segrest, T. M. Wilkinson, and L. Sheng, *Biochim. Biophys. Acta* **554**, 533 (1979).
[17] M. A. K. Markwell, *Anal. Biochem.* **125**, 427 (1982).

glycoprotease action, will always exceed the appearance of product bands on the gel. Uncleaved glycophorin A migrates on SDS–PAGE as a dimer and a monomer, and both bands are cut from the gel, combined and counted in a gamma counter (**B** dpm). The percentage hydrolysis, corrected for variable recovery in each lane, is calculated by comparing enzyme assays with unhydrolyzed controls containing no enzyme. The calculation is as follows:

$$\mathbf{N} = \mathbf{B} \div \mathbf{A}$$
$$\% \text{ Hydrolysis} = 100 \times [1 - (\mathbf{N} \text{ for assay} \div \mathbf{N} \text{ for control})]$$

### κ-*Casein Hydrolysis Assay*

The bovine milk protein κ-casein differs from glycophorin A and other transmembrane protein substrates in that it is not a cell surface antigen. It is an amphipathic protein with six O-linked glycans per molecule.[18] Native unreduced κ-casein is known to exist in high molecular weight aggregates, of 10 or more monomer subunits, so that each aggregate may behave in solution like a large glycoprotein extensively covered with O-sialoglycans.[19] Unlike the cell surface antigens, it is a substrate which is available in quantity and which can be readily separated from much of its cleavage products by precipitation with cold 10% (w/v) TCA. κ-Casein, however, is more sensitive than glycophorin A to hydrolysis by other proteinases, and control assays to measure any nonspecific activity should always be performed. Furthermore, the degree of solubility of κ-casein and its cleavage products in cold 10% TCA is affected by the presence of other proteins, detergents, and salts. Therefore, the effects of those variables should be determined in control experiments in which the hydrolysis of κ-casein is also estimated by SDS–PAGE analysis of products as described for glycophorin A. κ-Casein (commercially available from Sigma, St. Louis, MO; samples used in our laboratory were the gift of D. Dalgleish, Food Science, Univ. of Guelph, ON, Canada) can be prepared from bovine milk[18] and radioiodinated by standard methods.[17]

In the rapid assay of *P. haemolytica* glycoprotease, each enzyme assay tube contains 3.5 μg [125]I-labeled κ-casein and enzyme protein in 50 m*M* HEPES buffer, pH 7.4 (total volume of 25 μl). Incubation of the mixture is carried out at 37° for 5–60 min. As for the glycophorin A hydrolysis, low levels of the *P. haemolytica* glycoprotease may require prolonged incubations (18 hr). A mixture of proteinase inhibitors, as described above, should be added to prevent proteolysis by other proteolytic enzymes. At the end

---

[18] J. Mercier, G. Brignon, and B. Ribadeau-Dumas, *Eur. J. Biochem.* **35**, 222 (1973).
[19] L. K. Rasmussen, P. Hojrup, and T. E. Petersen, *Eur. J. Biochem.* **207**, 215 (1992).

of the reaction period, 75 $\mu$l of water is added, a 10-$\mu$l aliquot is removed for counting of total [125]I radioactivity in a gamma counter ($x$ dpm), and the enzyme action in the remainder is stopped by the addition of 10 $\mu$l of 100% (w/v) TCA. The mixture is kept on ice for 30 min and then centrifuged at 9000 $g$ for 10 min, followed by removal of 11 $\mu$l of the supernatant for gamma counting ($y$ dpm). Controls should include enzyme samples that have been inactivated by heating at 100° for 3 min. The apparent percentage hydrolysis of the 3.5 $\mu$g $\kappa$-casein substrate is given by the formula % hydrolysis = 101($y/x$).

Figure 1 compares the apparent hydrolysis rates for $\kappa$-casein measured as above for TCA-soluble products in the rapid assay with the hydrolysis rate as measured by SDS–PAGE analysis. Because not all the products of $\kappa$-casein digestion are soluble in 10% TCA, the amount of hydrolysis observed in the TCA assay is less than that measured by SDS–PAGE analysis of the disappearance of $\kappa$-casein substrate. Furthermore, the proportion of $\kappa$-casein degradation products that appear in the 10% TCA supernatant is affected by the presence of detergents or other solutes in the assay system, so that whenever reaction conditions are changed the rapid assay should be validated by SDS–PAGE analysis of the rate of disappearance of substrate, as described for glycophorin A digestion above. For first-order kinet-

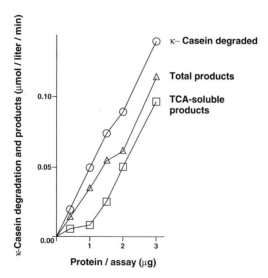

FIG. 1. Hydrolysis of [125]I-labeled $\kappa$-casein as measured by the disappearance of the substrate from SDS–polyacrylamide gels and the appearance of product bands on the gel (total products), compared with that proportion of total products that remain soluble in 10% TCA. The assay consisted of 3.5 $\mu$g [125]I-labeled $\kappa$-casein and varying amounts of enzyme protein in 50 m$M$ HEPES buffer, pH 7.4 (total volume 25 $\mu$l). Incubation of the enzyme and substrate was at 37° for 15 min.

ics with respect to enzyme concentration, the percentage hydrolysis of [125]I-labeled κ-casein in the assay should not exceed 15%. Blank values, determined in the absence of enzyme, should be subtracted from the percentage hydrolysis values determined in the presence of enzyme. Blank values will depend on the degree of purity of the κ-casein and should not exceed 6%.

Figure 2a shows the hydrolysis of glycophorin A, measured as described in the previous section, compared with that of κ-casein measured by the TCA assay. Figure 2b shows the effect of substrate concentration on the rates of hydrolysis. At substrate concentrations higher than those used in the assays above, there is a relatively reduced activity of the glycoprotease, for both glycophorin A and κ-casein, which may be due to substrate inhibition but is more likely to be caused by increased aggregation of substrate molecules at higher concentrations.

*Positional Specificity, Glycophorin A, and κ-Casein*

For the best characterized substrates of the glycoprotease, the main cleavage sites are $Arg^{31}$-$Asp^{32}$ for glycophorin $A^2$ and $Lys^{24}$-$Tyr^{25}$ for

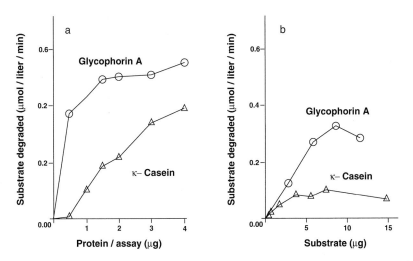

FIG. 2. Hydrolysis of human glycophorin A compared with that of bovine κ-casein: (a) Hydrolysis of the two substrates in a 5 min incubation with equivalent amounts of *P. haemolytica* glycoprotease pH 4.5 precipitate, as measured by the disappearance of TCA-soluble products for [125]I-labeled glycophorin A bands from SDS–PAGE gels and by the appearance of TCA-soluble products for [125]I-labeled κ-casein. (b) Hydrolysis of the two [125]I-labeled substrates as a function of substrate concentration, in a 5 min incubation with 3.5 μg glycophorin A and a 15-min incubation with 3.5 μg κ-casein. Conditions were as described in the assay methods.

κ-casein.[20] For glycophorin A, the susceptible peptide bonds appear to be in the extensively O-sialoglycosylated extracellular domain, so that the enzyme may require neighboring sialate residues for binding and hydrolysis of the peptide backbone.[2] However, on prolonged incubation of glycophorin A with the glycoprotease, there are many apparent sites of cleavage within the extracellular domain,[21] and further characterization of the peptide bond specificity is needed. In the hydrolysis of glycophorin by the glycoprotease, there are a number of apparent cleavage sites which do not correlate with basic, acidic, or aromatic amino acid residues, so that it is not possible to assign the enzyme to any known class of proteinases, and this is borne out by the lack of sequence similarity with any other prokaryote or eukaryote proteinase. At substrate concentrations less than 10 $\mu M$, Michaelis–Menten kinetics are observed for these two substrates, and the $K_m$ values are similar, being about 5 $\mu M$ for the cleavage of glycophorin A and 3 $\mu M$ for bovine κ-casein, based on $M_r$ values of 31,000 (glycosylated glycophorin A) and 19,000 (unglycosylated κ-casein), respectively. However, in assessing these $K_m$ values, it is assumed that all the glycophorin A or κ-casein substrate is chemically homogeneous and in solution in the monomeric state, whereas both substrates are present in solution as aggregates and are chemically heterogeneous.

The complete hydrolysis of glycophorin A by the glycoprotease can be readily accomplished by increasing the enzyme concentration. However, for κ-casein, hydrolysis never exceeds 80%, and this is probably due to heterogeneity of the substrate, which exists not only in differentially glycosylated and phosphorylated forms but also as a range of aggregates, crosslinked by intermolecular disulfide bridges.[19] The primary structure of bovine κ-casein is closely homologous with that of another mammalian soluble glycoprotein, γ-fibrinogen, from a variety of species,[22] but the γ-fibrinogens from 12 mammalian species were tested and found not to be substrates for the glycoprotease.

## Cleavage of O-Sialoglycoproteins from Surface of Live Cells

The P. haemolytica glycoprotease is active in cell culture media and can be used to cleave cell surface antigens, with no loss of cell viability. Treated cells show the loss of specific epitopes, revealed by immunofluorescence using monoclonal antibodies. Fluorescence-activated cell sorting (FACS)

---

[20] S. Simpson and A. Mellors, unpublished work (1993).

[21] H. Nakada, M. Inoue, Y. Numata, N. Tanaka, I. Funakoshi, S. Fukui, A. Mellors, and I. Yamashina, Proc. Natl. Acad. Sci. U.S.A. 90, 2495 (1993).

[22] P. Jolles, M. H. Loucheux-Lefebvre, and A. Henschen, J. Mol. Evol. 11, 271 (1978).

analysis enables the identification of subsets of mixed cell populations, differentiated from one another by the presence on the cell surface of glycoprotease-susceptible molecules. Alternatively, the glycoproteins on the live cell surface can be metabolically labeled by the incorporation of radiolabeled glycoprotein precursor molecules, such as [$^{35}$S]cysteine, or surface-labeled by $^{125}$I-iodination by the use of lactoperoxidase.[23] By such methods, live cells have been labeled *in vitro*, and the *P. haemolytica* glycoprotease has been shown to cleave specifically a number of cell surface O-sialoglycoproteins of functional significance, without loss of cell viability.[7,8] For the treatment of mammalian cells in culture, about $5 \times 10^6$ cells are incubated at 37° with sufficient *P. haemolytica* glycoprotease activity to cleave 0.1–1.0 $\mu$g glycophorin A per second (about 12–20 $\mu$g protein in the lyophilate prepared as above). The cells are incubated with enzyme for 30 min, in cell culture medium which can include 10% fetal calf serum, and then the enzyme is removed by washing the cells with cell culture medium.

The enzyme was shown to cleave the human leukocyte antigens CD34, CD43, CD44, and CD45 on the surface of live KG-la (acute myelogenous leukemia) cells.[7] The cleavage of CD34, without affecting cell viability, results in the removal of certain CD34 epitopes while other CD34 epitopes are retained on the cell surface.[24] This differentiation of epitopes is the basis for an improved method of isolation of human bone marrow stem cells, namely, a subpopulation of human bone marrow cells (1–3% of the total cells) that can give rise to all known blood cell lineages and is therefore useful for autologous or allogeneic bone marrow transplantation. Specific antibodies coupled to paramagnetic microspheres are used to tag CD34$^+$ cells, which are then magnetically isolated from the heterogeneous populations of bone marrow cells. The cells are subsequently released from the paramagnetic spheres by short-term treatment (30 min) with the glycoprotease to cleave the CD34 antigen, with no effect on cell viability.[25] Previous protocols for the immunomagnetic selection of CD34$^+$ cells have used a nonspecific proteinase (chymopapain) or prolonged incubations (18 hr) of the cells to remove the magnetic beads. The specificity and noncytotoxic nature of the *P. haemolytica* glycoprotease thus provide advantages for the isolation of highly purified CD34$^+$ hematopoietic stem cells,[25] with therapeutic potential in cancer treatment and gene therapy.

[23] D. R. Phillips, this series. Vol. 215, p. 412.
[24] D. R. Sutherland, J. C. W. Marsh, J. Davidson, M. A. Baker, A. Keating, and A. Mellors, *Exp. Hematol.* **20,** 590 (1992).
[25] J. C. W. Marsh, D. R. Sutherland, J. Davidson, A. Mellors, and A. Keating, *Leukemia* **6,** 926 (1992).

The glycoprotease will cleave the P-selectin ligand, on the surface of human lymphoid cells.[8,9] This cell surface $O$-sialoglycoprotein bears the sialyl-Lewis[X] antigen and is the specific site of attachment of the platelet P-selectin adhesion molecule. The glycoprotease removes only a few radio-labeled glycoproteins from the surface of [[35]S]cysteine metabolically labeled HL-60 (promyelocytic leukemia) cells, as revealed by two-dimensional protein electrophoresis and autoradiography.[8] Cleavage by the enzyme of the P-selectin ligand on cells that express it abrogates the ability of these cells to bind to soluble P-selectin, though the enzyme does not remove the sialyl-Lewis[X] epitope of the P-selectin ligand.[9] This suggests that the sialyl-Lewis[X] antigen is not a specific requirement for cell adhesion. A similar adhesion molecule, the L-selectin ligand or Sgp40, which is found on lymphoid cells and acts as the site of adhesion of L-selectin, is also cleaved by the enzyme.[9] It is likely that the physiological targets of the enzyme, in the bovine respiratory tract, are sialoglycoproteins on the surfaces of endothelial cells, alveolar macrophages, alveolar neutrophils, or other leukocytes. The properties of the *P. haemolytica* glycoprotease and its occurrence in virulent serotypes of *P. haemolytica* suggest that the enzyme may have a specific pathogenic role in bovine shipping fever. The glycoprotease may interfere with the host immune response perhaps by the specific cleavage of glycoproteins on immune cell surfaces or by aiding the bacterium in the colonization of the lung.

## Enzyme Properties

The *P. haemolytica* glycoprotease is optimally active at about pH 7.4 and is stable over the range pH 4.5–8.0. The enzyme can be inhibited by treatment with relatively high concentrations of EDTA ($K_i$ is about 50 m$M$), so that it is classified as a metalloproteinase. This is supported by the presence of a putative metal ion-binding site in the protein structure predicted from the nucleotide sequence for *gcp*. The metal ion cofactor is thought to be tightly bound to the enzyme, as the EDTA-inactivated glycoprotease can be reactivated by dialysis.[2] Zinc-dependent bacterial metalloproteinases of the thermolysin family are inhibited by much lower concentrations of EDTA. The glycoprotease does not cleave thermolysin substrates such as furylacryloylglycylleucinamide, nor is it inhibited by thermolysin inhibitors such as phosphoramidon. The glycoprotease is irreversibly inactivated by high salt concentrations, which precludes its purification by ammonium sulfate fractionation. Sodium chloride concentrations of 0.3 $M$ will inactivate the enzyme within a few hours. DEAE-cellulose ion-exchange chromatography can be carried out for isolation of the enzyme, by elution with 0.015-0.500 $M$ HEPES buffer pH 7.4 containing 0.025% (v/v) polyoxyethylene glycol ($M_r$ 6,000).

It has been shown that a number of *P. haemolytica* glycoprotease-susceptible substrates become resistant to cleavage by the glycoprotease if they are first treated with sialidase from *Clostridium perfringens* or *Vibrio cholerae*. Thus, the proteolysis seems to be dependent on the presence of terminal sialate residues. Abolition or reduction of glycoprotease suscepti-bility by desialylation has been shown for glycophorin A[2], CD34, CD44[7], the L-selectin ligand, the P-selectin ligand,[8,9] and κ-casein.

The gene *gcp* which codes for the enzyme has been isolated, sequenced, and expressed in *Escherichia coli*.[3] As for the native protein, the recombi-nant enzyme rGcp has a molecular mass of 35 kDa. The rGcp protein shows no sequence similarity with other proteinases but shares 76% homology with the putative product of an *E. coli* open reading frame of unknown function, *orfX*, which is part of the macromolecular synthesis operon and which includes the *rpsU, dnaG,* and *rpsD* genes.[26] Another homologous open reading frame of unknown function has been found recently on chromo-some IV of *Saccharomyces cerevisiae* where it is associated with genes essential for cellular function.[27] The native enzyme is secreted into the medium of *P. haemolytica* A1 cultures; however, when it is expressed heterologously in *E. coli*, rGcp is not fully secreted but is trapped in the periplasmic space. Therefore, the recombinant gene product can be readily isolated free from most host proteins by osmotic shock treatment.[3] A recom-binant fusion protein has been constructed between Gcp and HlyA, the secretion signal of *E. coli* α-hemolysin.[28] The fusion protein, rGcp-F, is secreted into the culture supernatant of *E. coli* when the α-hemolysin secretion functions are supplied *in trans* in a separate plasmid. Murine monoclonal immunoglobulin G antibodies have been raised against the recombinant protein rGcp and can be used to identify the *P. haemolytica* glycoprotease and the recombinant *gcp* products by SDS–PAGE analysis and immunoblotting.[29] Bovine antisera from cattle exposed to *P. haemolyt-ica* in the field are able to neutralize the glycoprotease activity against glycophorin A. The polyclonal antibodies will also immunoprecipitate the enzyme from solution, when used in conjunction with *Staphylococcus* pro-tein A-Sepharose beads.[30]

[26] M. Nesin, J. R. Lupski, P. Svec, and G. N. Godson, *Gene* **51,** 149 (1987).
[27] M. Simon, P. Benit, A. Vassal, C. Dubois, and G. Faye, *Yeast* **10,** in press (1994).
[28] N. Mackman, K. Baker, L. Gray, R. Haigh, J.-M. Nicaud, and I. B. Holland, *EMBO J.* **6,** 2835 (1987).
[29] R. Y. C. Lo, M. A. Watt, S. Gyroffy, and A. Mellors, *FEMS Microbiol. Lett.* **116,** 225 (1994).
[30] C. W. Lee, P. E. Shewen, W. M. Cladman, J. A. R. Conlon, A. Mellors, and R. Y. C. Lo, *Can. J. Vet. Res.* **58,** 93 (1994).

## Acknowledgments

The authors are grateful to W. Cladman for assistance in the preparation of this chapter. The contributions of J. R. Scott, G. L. Otulakowski, P. E. Shewen, B. N. Wilkie, A. E. Udoh, K. M. Abdullah, S. S. Lo, D. R. Sutherland, J. C. Marsh, S. Gyroffy, S. Simpson, Y. Shad, and M. A. Watt are gratefully acknowledged. Research support came from the Natural Sciences and Engineering Council, and the National Cancer Institution, Canada.

# [48] β-Lytic Endopeptidases

By Efrat Kessler

## Introduction

β-Lytic endopeptidases are a family of evolutionarily related bacterial extracellular metalloendopeptidases which cause cell lysis of other bacteria by cleaving specific peptide bonds within the cell wall peptidoglycan network. Based on amino acid sequence homology and similarity of physicochemical and enzymatic properties, four endopeptidases may now be classified as such. These are *Lysobacter enzymogenes* (previously, *Sorangium* or *Myxobacter* 495)[1] β-lytic endopeptidase, the first such enzyme to be described,[2] *Achromobacter lyticus* β-lytic endopeptidase,[3] *Pseudomonas aeruginosa* LasA (staphylolytic proteinase),[4,5] and *Aeromonas hydrophila* proteinase (AhP).[6] Although the enzymes were discovered in the 1960s, relatively little is known about the substrate specificity, active site, tertiary structure, and regulation of the β-lytic endopeptidases. Their potential as antimicrobial agents and tools in biochemical research, as well as the clinical importance of some (e.g., *P. aeruginosa* LasA) have, however, initiated renewed interest in the endopeptidases, three of which have been the subject of recent investigations.[3,5,6–8]

[1] D. R. Whitaker, this series, Vol. 19, p. 599.
[2] D. R. Whitaker, *Can. J. Biochem.* **43**, 1935 (1965).
[3] S. L. Li, S. Norioka, and F. Sakiyama, *J. Bacteriol.* **172**, 6506 (1990).
[4] M. E. Burke and P. A. Pattee, *J. Bacteriol.* **93**, 860 (1967).
[5] E. Kessler, M. Safrin, J. C. Olson, and D. E. Ohman, *J. Biol. Chem.* **268**, 7503 (1993).
[6] A. G. Loewy, U. V. Santer, M. Wieczorek, J. K. Blodgett, S. W. Jones, and J. C. Cheronis, *J. Biol. Chem.* **268**, 9071 (1993).
[7] E. Kessler, M. Safrin, and D. E. Ohman, manuscript in preparation.
[8] J. E. Peters, S. J. Park, A. Darzins, L. C. Freck, J. M. Saulnier, J. M. Wallach, and D. R. Galloway, *Mol. Microbiol.* **6**, 1155 (1992).

Assay Methods

Owing to a lack of suitable synthetic substrates and low activity toward standard protein substrates, most assays rely on bacteriolytic activity. Two such assays are described below. Also described is a spectrophotometric assay with a synthetic substrate and a method for identifying staphylolytic activity in polyacrylamide gels.

### Staphylolytic Activity of Pseudomonas aeruginosa LasA

*Substrate.* The staphylolytic activity of *P. aeruginosa* LasA is assayed[5] with lyophilized *Staphylococcus aureus* cells as the substrate. We use strain $D_2C$ from Sigma (St. Louis, MO), but any laboratory strain is equally suitable. The lyophilized cells are suspended in 20 m$M$ Tris-HCl, pH 8.5 (10 mg/ml in Tris buffer), and the suspension is heated at 100° for 10 min. If stored at 4°, this (stock) suspension is stable for at least 2 weeks.

*Procedure.* A concentrated suspension of *S. aureus* cells is diluted 30-fold in Tris buffer to yield an approximate $OD_{595}$ of 0.8. Then 0.9 ml of the diluted cell suspension is placed in a 1-ml optical cell, and, after equilibration to 25°, the reaction is initiated by mixing with enzyme (100 $\mu$l or less, in which case the volume is preadjusted to 1 ml with Tris buffer). The rate of decrease in absorbance at 595 nm caused by cell lysis is monitored spectrophotometrically at 25°.

*Kinetics.* After a short lag, the apparent rate of decrease in absorbance is linear up to a $\Delta A_{595}$ of about $-0.1$. With pure LasA, cell lysis is partial, and the decrease in absorbance levels off when $A_{595}$ is 0.3–0.4 (~50% drop in absorbance).

*Activity.* One unit of activity is the amount of enzyme which causes an $A_{595}$ decrease of 1 unit/min. Enzyme aliquots in the range of 0.01 to 0.2 units (0.07 to 1.4 $\mu$g of purified enzyme) are recommended.

*Notes.* Proteinases in the crude culture filtrate of *P. aeruginosa*, especially elastase and alkaline proteinase, interfere with the assay.[7] Alone, these proteinases are not staphylolytic. However, cell walls partially degraded by LasA are rendered susceptible to the action of other proteinases, resulting in enhanced reaction rates and complete cell lysis. Slightly different versions of this procedure have also been described.[4,9,10]

[9] M. Lache, W. R. Hearn, J. W. Zyskind, D. J. Tipper, and J. L. Strominger, *J. Bacteriol.* **100**, 254 (1969).
[10] N. Brito, M. A. Falcon, A. Carnicero, A. M. Gutierrez-Navarro, and T. B. Mansito, *Res. Microbiol.* **140**, 125 (1989).

*Bacteriolytic Activity of Lysobacter enzymogenes β-Lytic Endopeptidase*

*Substrate.* Acetone-dried *Arthrobacter globiformis* cells are used to assay the *L. enzymogenes* β-lytic endopeptidase.[11,12] A thick suspension of cells from a 24-hr culture is mixed with 11 volumes of acetone chilled to −15°. The cells are allowed to settle at room temperature, washed in ether, and dried *in vacuo.* The dried cells are resuspended in distilled water, and the drying process is repeated.

*Procedure.* A fresh *A. globiformis* cell suspension (0.3 mg/ml) in 25m*M* Tris-HCl, pH 9.0, is prepared 1 hr before the assay.[12] If kept at 4°, this suspension is stable for at least 5 hr. Five milliliters of the fresh cell suspension is mixed with 200 μl of enzyme, and the rate of decrease of absorbance at 660 nm owing to cell lysis is monitored spectrophotometrically at 25°.

*Kinetics.* The apparent kinetics of the reaction is zero order. The reaction rate is linear up to a $\Delta A_{660}$ of approximately −0.2.

*Activity.* One unit of activity is the amount of enzyme which causes a decrease in absorbance of 0.03 unit/min.

*Spectrophotometric Assay of Lysobacter enzymogenes*
*β-Lytic Endopeptidase*

*Substrate.* 3-(2-furylacryloyl)-Gly-Leu-amide (FAGLA; Sigma) is the substrate in the spectrophotometric assay of *L. enzymogenes* β-lytic endopeptidase.[13,14]

*Procedure.* Three milliliters of substrate solution (2.5 m*M* in 0.2 *M* Tris-HCl, pH 7.2) is mixed with purified β-lytic endopeptidase (final concentration 20–120 μg/ml), and the rate of decrease in absorbance at 345 nm arising from hydrolysis of the Gly-Leu peptide bond is recorded spectrophotometrically at 25°.

*Notes.* The substrate has a considerable absorption at 345 nm ($\varepsilon$ = 766). The use of a double-beam spectrophotometer with a substrate blank is recommended. The $\Delta\varepsilon_{345}$ for complete hydrolysis is 317. The substrate is stable in solutions of neutral to slightly acid pH if stored at −20°. The assay requires enzyme concentrations one to two orders of magnitude higher than those used in assays of bacteriolytic activity.

*Detection of Pseudomonas aeruginosa LasA Activity in Acrylamide Gels*

Samples containing staphylolytic proteinase (1–10 μg) are subjected to electrophoresis under nondenaturing conditions in a 12% polyacrylamide

[11] D. C. Gillespie and F. D. Cook, *Can. J. Microbiol.* **11**, 109 (1965).
[12] D. R. Whitaker, F. D. Cook, and D. C. Gillespie, *Can. J. Biochem.* **43**, 1927 (1965).
[13] N. B. Oza, *Int. J. Pept. Protein Res.* **5**, 365 (1973).
[14] J. Feder, *Biochem. Biophys. Res. Commun.* **32**, 326 (1968).

gel.[15] After electrophoresis, the gel is placed over an indicator agar gel (1% w/v in 20 m$M$ Tris-HCl, 0.05% w/v sodium azide, pH 8.5) containing 3 mg/ ml of heat-killed *Staphylococci* and incubated in a humid atmosphere at 37°. Clearing zones caused by cell lysis are seen within 6 to 16 hr.[10]

## Purification of Enzymes

β-Lytic endopeptidases are highly basic proteins. Thus, cation-exchange chromatography is a useful step in the purification. Most published protocols take advantage of this and are often based on no more than three steps: concentration of medium proteins by ammonium sulfate or acetone precipitation, removal of a major portion of the contaminating proteins by adsorption to an anion-exchanger at pH 7.5 to 8.0, and chromatography of the effluent on a cation-exchanger at pH 6 to 7.5.

### *Lysobacter enzymogenes β-Lytic Endopeptidase*

A detailed, large-scale purification procedure for *L. enzymogenes* β-lytic endopeptidase has been described before in this series.[1] The β-lytic enzyme was obtained as a by-product in the purification of *Lysobacter α*-lytic endopeptidase. Briefly, enzyme production was achieved in a fermentor containing 100–130 liters of medium. The β- and α-lytic endopeptidases were adsorbed on Amberlite CG50 and fractionated by stepwise elution with citrate buffers.[1] Final purification of the β-lytic endopeptidase (which eluted from the resin earlier than the α-lytic endopeptidase) was achieved by rechromatography. Approximately 2 g of enzyme was obtained from 130 liters of medium, and the purified fraction was estimated to contain no more than 0.2% of α-lytic endopeptidase.[16]

### *Achromobacter lyticus β-Lytic Endopeptidase*

One-half gram of a commercial achromopeptidase preparation (available from Sigma) is dissolved in 120 ml of 10 m$M$ Tris-HCl, pH 8, and chromatographed on a 1.5 × 30 cm column of CM-Sepharose CL-6B, with a linear gradient of 0 to 0.4 $M$ NaCl in the same buffer.[3] The active peak (lytic activity is determined toward *Micrococcus luteus* cells, essentially as described above for staphylolytic activity) is chromatographed on a 1.5 × 80 cm Sephadex G-75 column, equilibrated with 10 m$M$ Tris-HCl, pH 8. This is followed by cation-exchange high-performance liquid chromatography (HPLC) on an MCI Gel CQK31S (CM) column (7.5 × 75 mm; Mitsubishi

[15] B. J. Davis, *Ann. N.Y. Acad. Sci.* **121,** 404 (1964).
[16] D. R. Whitaker, *Can J. Biochem.* **45,** 991 (1967).

Kasei) with a linear gradient of 0 to 0.5 $M$ NaCl in 50 m$M$ Tris-HCl, pH 7.0. The active peak contains a mixture of $\alpha$- and $\beta$-lytic endopeptidases (800 $\mu$g protein; purification factor 46), which, after lyophilization, are separated by reversed-phase HPLC on a $C_4$ column ($\mu$Bondasphere, 15 $\mu$m, 30 nm, 0.39 × 30 cm; Waters, Milford, MA). The column is equilibrated with 0.1% v/v trifluoroacetic acid (TFA) and eluted with a linear gradient of acetonitrile (0 to 80% v/v in 0.08% v/v TFA for 40 min) at a flow rate of 1.0 ml/min. The enzyme thus obtained is homogeneous on sodium dodecyl sulfate–polyacrylamide gel electrophoresis (SDS–PAGE), but inactive.

## Pseudomonas aeruginosa LasA (Staphylolytic Proteinase)

Two procedures are described below. Method I, used in our laboratory, is simple and associated with high recovery, but the purified enzyme fraction is only 95 to 98% pure.[5] The initial steps in both method I and method II are basically the same. Method II, however, includes an extra chromatofocusing step, yielding an enzyme preparation that is homogeneous on SDS–PAGE.[17] In principle, both methods may be combined.

*Method I. Pseudomonas aeruginosa* strain FRD2 is grown (37°, 24 hr) with shaking (~110 strokes/min) in four 2-liter conical flasks containing 400 ml of tryptic soy broth without dextrose (Difco, Detroit, MI).[5] Cells are removed by centrifugation (6000 $g$, 20 min), the clear supernatant is chilled, and the proteins in it are precipitated with ammonium sulfate (ultrapure, Schwarz-Mann, Orangeburg, NY, 80% saturation). After standing at 4° for 16 hr, the precipitate is collected by centrifugation (4°, 60 min, 20,000 $g$), dissolved in approximately 15 ml of 20 m$M$ Tris-HCl, 0.5 m$M$ CaCl$_2$, pH 7.5 (buffer A), and dialyzed against the same buffer. The dialyzed solution (35–40 ml) is clarified by centrifugation (12,000 $g$, 10 min) and (may be) stored at −20° until further processing.

For purification, the concentrated crude culture filtrate (34 ml; 238 mg protein) is chromatographed at 4° on a 4.4 × 28 cm column of DEAE-cellulose (DE-52, Whatman, Clifton, NJ) equilibrated and washed with 3 to 4 volumes of buffer A (isocratic conditions). The flow rate is 80 ml/hr, and 8.7-ml fractions are collected. The column is monitored by measuring the absorbance at 280 nm and assaying staphylolytic, proteolytic, and elastolytic activities. A major portion of the contaminating proteins adsorb to the column. Proteins which do not bind are resolved into four distinct peaks that emerge from the column at elution volumes of 139 (I), 209 (II), 452 (III), and 1087 (IV) ml. The staphylolytic endopeptidase is eluted in peak I, whereas elastase is the last protein to emerge from the column (peak

[17] J. E. Peters and D. R. Galloway, *J. Bacteriol.* **172**, 2236 (1990).

IV). The trailing edge of peak I (which contains staphylolytic activity) is only partially resolved from peak II.

An improved resolution of the two proteins is achieved by rechromatography. The active fractions in peak I are concentrated with Aquacide II (Calbiochem, La Jolla, CA) to 2.5 ml, dialyzed against buffer A, and run on a second, 2.2 × 28 cm column of DEAE-cellulose, equilibrated and eluted with buffer A at a flow rate of 20 ml/hr; 2.2-ml fractions are collected. By absorbance at 280 nm, the staphylolytic proteinase emerges from the column in the pass-through fractions as a major symmetrical peak. It is followed by a barely detectable peak of protein II. On SDS–PAGE, protein II migrates slightly more slowly than LasA. As judged by SDS–PAGE analysis and silver staining, the purified enzyme fraction is free of protein II. It contains, however, up to 5% of a smaller (molecular weight 15,000) contaminating protein, which apparently does not interfere with the assays. The activity yield of each step is 90%; thus, the overall yield is about 80%.

Because of interference by elastase and alkaline proteinase, the specific activity of LasA in the crude culture filtrate cannot be determined with accuracy. We assume that the experimental value (3 units/mg) measured for the sample is about 3-fold higher than the theoretical one. Based on this, the theoretical increase in specific activity (from 1 to 137 units/mg of the purified enzyme) suggests a purification factor of the order of 140. The yield is approximately 4 mg per liter of culture medium.

*Method II.* LasA is purified from 3 liters of culture filtrate of *P. aeruginosa* strain PAO1 or PA220.[17] Proteins in the filtrate are precipitated twice with ammonium sulfate (60% saturation). The precipitate formed (4°, 16 hr) after each of the precipitation steps is dissolved in water. The protein solution from the second precipitation step is made 65% in acetone, and the precipitate formed is collected by centrifugation and then dissolved in and dialyzed against 20 m*M* sodium phosphate, pH 8. The dialyzed solution is run on a 2.5 × 80 cm column of DEAE-Sephacel (Pharmacia, Piscataway, NJ) equilibrated and washed with the same buffer. LasA runs through the column, whereas elastase as well as a large portion of other contaminating proteins adsorb to the column. The run-through LasA containing fractions are pooled and concentrated by ammonium sulfate precipitation. The precipitate is dissolved in and dialyzed against 25 m*M* diethanolamine, pH 9.5, and subjected to chromatofocusing by FPLC (fast protein liquid chromatography) on a Mono P column (Pharmacia) at a flow rate of 1 ml/min. LasA runs through the column. The active fractions are pooled, dialyzed against 20 m*M* sodium phosphate, pH 8.0, and stored at −80°. The purified LasA fraction is homogeneous in Coomassie Blue-stained SDS–polyacrylamide gels.

*Notes.* No information was given as to the purification factor and yield.[17] The ammonium sulfate and acetone precipitation steps as well as DEAE-Sephacel chromatography followed procedures originally introduced by Morihara *et al.*[18] for the purification of *Pseudomonas* elastase.

### *Aeromonas hydrophila β-Lytic Endopeptidase*

A strain of *A. hydrophila* isolated from the digestive system of the leech *Hirudo medicinalis* is grown in buffered tryptone yeast extract broth (tryptone, 8 g/liter; yeast extract, 5 g/liter; $NaH_2PO_4$, 10 g/liter, adjusted to pH 7.0) for 16–18 hr at 30°.[6] The cell suspension is chilled and centrifuged to remove the bacteria. EDTA titrated to pH 7.4 is added to a final concentration of 20 m$M$ to reduce the total proteolytic activity in the culture filtrate. The filtrate is concentrated 20-fold with an Amicon (Danvers, MA) CH2 system and S1Y10 membrane, and proteins in the medium are precipitated with ammonium sulfate (80% saturation). The precipitate is dissolved in and dialyzed against 20 m$M$ HEPES, 0.1 $M$ NaCl, 20 m$M$ EDTA, pH 7.4 (buffer A), and 38 ml of the dialyzed solution is applied to a Baker-bond WP-PEI column (2.5 × 20 cm) equilibrated and washed with buffer A. The pass-through fractions containing AhP are dialyzed (4°; 1–3 hr; longer dialysis at this point results in some loss of AhP activity) against 20 m$M$ HEPES, 30 m$M$ NaCl, pH 6.8 (buffer B), and run on a CM-cellulose column (2.5 × 20 cm) equilibrated with the same buffer. After initial washing with buffer B, the bound enzyme is eluted from the column with a linear gradient from the equilibration buffer to 20 m$M$ HEPES, 0.4 $M$ NaCl, pH 7.4. The enzyme peak emerges at 0.2 $M$ NaCl. The active fractions are concentrated by precipitation with ammonium sulfate (80% saturation), dialyzed against buffer A, and stored frozen. The yield is 2 to 4 mg/liter of culture medium.

The purified enzyme migrates as a single band on SDS–PAGE using Coomassie blue or silver staining, and as a single tailing peak on $C_{18}$ (Waters) reversed-phase HPLC. Isoelectric focusing (in 4 $M$ urea) followed by silver staining shows a major band at pH 8.5 and two minor, closely migrating bands at pH 8.0.

*Notes.* Two assays are used to follow activity during purification. (1) In the Fibrin assay, SDS–PAGE is used to assay conversion of the γ-chain dimer of fibrin to a modified γ monomer caused by cleavage of a peptide bond in the vicinity of the cross-link. (2) The Azocoll assay[6] is the alternative. Both assay methods suffer from interference by other proteinases. In

---

[18] K. Morihara, H. Tsuzuki, T. Oka, H. Inoue, and M. Ebata, *J. Biol. Chem.* **240** 3295 (1965).

addition, the fibrin assay is cumbersome, whereas the Azocoll assay seems to be insensitive with purified AhP.

## Properties of Enzymes

In addition to the detailed discussion of the individual β-lytic endopeptidases below, an overview of the enzymatic properties is given in Tables I and II. A striking similarity in properties of the enzymes may be noted. Furthermore, although still somewhat sporadic, current information on the specificity of cleavage of the enzymes suggests the same primary biological

TABLE I

PHYSICOCHEMICAL PROPERTIES OF β-LYTIC ENDOPEPTIDASES

| Parameter | Lyso[a] | Achr[b,c] | LasA | AhP[d] |
|---|---|---|---|---|
| Molecular mass (kDa) | | | | |
| By sequence | 19,100[e] | 19,286 | 19,950[f] | |
| By SDS–PAGE | | 22,000 | 21,000[g] | 19,000 |
| By sedimentation | 19,000[h] | | | |
| By gel filtration | 9500[h] | | 15,000[i] | |
| Number of residues | 178[e] | 179 | 182[f] | |
| Percentage sequence identity relative to Lyso[a] | 100 | >96 | 38[f,g] | 46[j] |
| Ratio of basic/acidic residues | 16/3 | 16/4 | 17/7 | |
| p$l$ | | | 9.24[f] | 8.5 |
| Zinc atom | +[e] | Likely | +[k,l] | + |
| Preproenzyme (~370 residues) | | + | +[f,m] | |

[a] *Lysobacter enzymogenes* β-lytic endopeptidase.
[b] *Achromobacter lyticus* β-lytic endopeptidase.
[c] S. L. Li, S. Norioka, and F. Sakiyama, *J. Bacteriol.* **172,** 6506 (1990).
[d] A. G. Loewy, U. V. Santer, M. Wieczorek, J. K. Blodgett, S. W. Jones, and J. C. Cheronis, *J. Biol. Chem.* **268,** 9071 (1993).
[e] A. P. Damaglou, L. C. Allen, and D. R. Whitaker, *in* "Atlas of Protein Sequence and Structure" (M. O. Dayhoff, ed.), Vol. 5(Suppl. 2), p. 198. National Biomedical Research Foundation, Washington, D.C., 1976.
[f] A. Darzins, J. E. Peters, and D. R. Galloway, *Nucleic Acids Res.* **18,** 6444 (1990).
[g] E. Kessler, M. Safrin, J. C. Olson, and D. E. Ohman, *J. Biol. Chem.* **268,** 7503 (1993).
[h] L. Jurasek and D. R. Whitaker, *Can. J. Biochem.* **43,** 1955 (1965).
[i] N. Brito, M. A. Falcon, A. Carnicero, A. M. Gutierrez-Navarro, and T. B. Mansito, *Res. Microbiol.* **140,** 125 (1989).
[j] Based on sequence of the first 40 residues.
[k] Deduced from inhibition by the Zn-specific chelator tetraethylenepentamine (TEP).
[l] E. Kessler, M. Safrin, and D. E. Ohman, manuscript in preparation.
[m] P. A. Schad and B. H. Iglewski, *J. Bacteriol.* **170,** 2784 (1988).

TABLE II
Enzymatic Properties of β-Lytic Endopeptidases[a]

| Parameter | Lyso[b] | LasA | AhP[c] |
|---|---|---|---|
| pH optimum | | | |
|   Bacteriolysis | 8.5–9.0[d] | 8.5[e] | |
|   Synthetic substrates cleavage | 6.5[f] | | 7.5 |
| Proteolytic activity | Low[g] | Negative[h] | Low |
| Elastolytic activity | Yes[f] | Yes[h] | Likely |
| Staphylolytic activity | Likely | Yes[h] | |
| Activity on bacterial cell walls | | | |
|   Release of C-terminal Gly | Yes[i] | | |
|   Release of di- and triglycine | | Yes[j] | |
| Specificity of cleavage | | | |
|   Gly$_5$ | | Yes[h,k] | |
|   Gly$_6$ | | Yes[h,k] | |
|   Internal Gly-X (X = neutral residue) | Yes[f,l] | Yes[h,j,k] | Yes |
|   Internal Gly-Pro | No[f] | No[m] | No |
|   Dipeptides, tripeptides, or amide, ester, and p-nitroanilide substrates | No[f] | | No |
| Preferred cleavage: internal Gly-Gly+X (X = Ala, Gly, Phe) | | Yes[m] | Yes |
| Interference by charge | Yes[f,l] | Yes[h,k] | Yes |
| Inhibition by | | | |
|   Reducing agents | | Yes[m] | Yes |
|   Zinc chelators | Yes[f] | Yes[m] | Yes |
|   EDTA | | Weak[m] | Weak |
|   Phosphoramidon | | No[m] | Weak |
|   Serine proteinase inhibitors | No[n] | No[m] | No |
|   Thiol proteinase inhibitors | | No[m] | No |

[a] *Achromobacter lyticus* β-lytic endopeptidase is excluded because its enzymatic properties are not known.

[b] *Lysobacter enzymogenes* β-lytic endopeptidase.

[c] A. G. Loewy, U. V. Santer, M. Wieczorek, J. K. Blodgett, S. W. Jones, and J. C. Cheronis, *J. Biol. Chem.* **268**, 9071 (1993).

[d] D. C. Gillespie and F. D. Cook, *Can. J. Microbiol.* **11**, 109 (1965).

[e] M. E. Burke and P. A. Pattee, *J. Bacteriol.* **93**, 860 (1967).

[f] N. B. Oza, *Int. J. Pept. Protein Res.* **5**, 365 (1973).

[g] D. R. Whitaker, *Can. J. Biochem.* **43**, 1935 (1965).

[h] E. Kessler, M. Safrin, J. C. Olson, and D. E. Ohman, *J. Biol. Chem.* **268**, 7503 (1993).

[i] C. S. Tsai, D. R. Whitaker, L. Jurasek, and D. C. Gillespie, *Can. J. Biochem.* **43**, 1971 (1965).

[j] M. Lache, W. R. Hearn, J. W. Zyskind, D. J. Tipper, and J. L. Strominger, *J. Bacteriol.* **100**, 254 (1969).

[k] N. Brito, M. A. Falcon, A. Carnicero, A. M. Gutierrez-Navarro, and T. B. Mansito, *Res. Microbiol.* **140**, 125 (1989).

[l] D. R. Whitaker, C. Roy, C. S. Tsai, and L. Jurasek, *Can. J. Biochem.* **43**, 1961 (1965).

[m] E. Kessler, M. Safrin, and D. E. Ohman, manuscript in preparation.

[n] D. R. Whitaker, *Can. J. Biochem.* **45**, 991 (1967).

role for all, namely, defense against bacteria in the environment, in particular, against species of *Staphylococcus* or closely related organisms.

## *Lysobacter enzymogenes β-Lytic Endopeptidase*

*Physicochemical Properties.* Based on the amino acid sequence[19] (Fig. 1A), the molecular weight of the enzyme is 19,100. Molecular weight by gel filtration is around 10,000, an anomaly resulting from weak adsorption to the resin.[20] The enzyme contains 178 amino acid residues, two disulfide bonds linking residues 65–111 and 155–168,[19] and one atom of zinc.[21] It does not, however, contain the consensus zinc binding sequence HEXXH typical of most zinc metalloendopeptidases. Of 19 ionizable residues, 16 are basic. At pH 8, electrophoretic mobility toward the cathode is slightly slower than that of lysozyme.[2] Other data are as follows: $A_{280}^{1\%} = 20.5^2$; $s_{20,\omega}^{\circ} = 2.2$ Svedberg units.[20]

The enzyme has been crystallized.[22] The crystals were rhombic prisms with space group $P2_1$, $2_1$, $2_1$ and unit cell parameters $a = 54.1$, $b = 99.6$, and $c = 53.9$ Å. The asymmetric unit contained two molecules of β-lytic endopeptidase.

*Bacteriolytic Activity.* The purified enzyme lyses both *Arthrobacter globiformis* and *Micrococcus lysodeikticus* cells.[2] Lysis of the former occurs more readily: 2–6 μg of purified enzyme causes complete lysis of *A. globiformis* cells within 1 hr, whereas 16–30 μg of enzyme and several hours are required to effect complete lysis of *M. lysodeikticus* cells.[2] Lysis of *S. aureus* cells at a rate faster than that of either *M. lysodeikticus* or *A. globiformis* was observed with *L. enzymogenes* culture filtrate.[11] Based on the specificity of cleavage within defined substrates and by analogy to *P. aeruginosa* and *A. hydrophila* β-lytic endopeptidases (see below), this may largely reflect β-lytic endopeptidase action. The optimum pH for bacteriolysis is pH 9.0.[11]

*Action on Isolated Cell Walls.* Hydrolysis of *M. lysodeikticus* cell walls is accompanied by an increase in C-terminal glycine and (to a lesser extent) alanine, as well as an increase in free ε-amino groups of lysine and N-terminal alanine residues.[23]

[19] A. P. Damaglou, L. C. Allen, and D. R. Whitaker, *in* "Atlas of Protein Sequence and Structure" (M. O. Dayhoff, ed.), Vol. 5 (Suppl. 2), p. 198. National Biomedical Research Foundation, Washington, D.C., 1976.

[20] L. Jurasek and D. R. Whitaker, *Can. J. Biochem.* **43,** 1955 (1965).

[21] L. Jurasek and D. R. Whitaker, *Can. J. Biochem.* **45,** 917 (1967).

[22] W. B. T. Cruse and D. R. Whitaker, *J. Mol. Biol.* **102,** 173 (1976).

[23] C. S. Tsai, D. R. Whitaker, L. Jurasek, and D. C. Gillespie, *Can. J. Biochem.* **43,** 1971 (1965).

**A**

```
LasA  APPSNLMQLPWRQGYSWQPNGAHSNTGSG-YPYSSFDASYDWPRWGS--ATY   49
      | | | |    | ||   |||:|||||  || || | |      |||
Achr  SPNGLLQFPFPRGASWHVGGAHTNTGSGNYPMSSLDMSRG-GGWGSNQNGN     50
      |||||||||||||||||||||||||||||||||||||||||||  ||||||
Lyso  SPNGLLQFPFPRGASWHVGGAHTNTGSGNYPMSSLDMSRG-GG--SNQNGN     48


LasA  SVVAAHAG-TVRVLSRCQVRVTHPSGWATNYYHMDQIQVSNGQQVSADTKLG   100
      | |  || :  :  | |   : |  || | ||| :||   | :||  | ::
Achr  WVSASAAG-SFKRHSSCFAEIVHTGGWSTTYYHLMNIQYNTGANVSMNTAIA   101
      |||||||| |||||||||||||||||||||||||||||||||||||||||||
Lyso  WVSASAAGGSFKRHSSCFAEIVHTGGWSTTYYHLMNIQYNTGANVSMNTAIA   100


LasA  VYAGNINTALCEGGSSTGPHLHFSLLYNGAFVSLQGASFGPYRINVGTSNYD   152
      |     ||| || |||||| | ||  ||| | |  |:|   ||| | ||
Achr  NPANTQAQALCNGGQSTGPHEHWSLKQNGSFYHLNGTYLSGYRITATGSSYD   153
      |  |||||||||||||||||| |||||||||||||||||||||||||||||
Lyso  NAPNTQAQALCNGGQSTGPHQHWSLKQNGSFYHLNGTYLSGYRITATGSSYD   152


LasA  NDCRRYYFYNQSAGTTHCAFRPLYNPGLAL   182
      | |:|   :  |   | :     |||
Achr  TNCSRFYLTKN--GQNYC-YGYYVNPGPN   179
      |||||||||||  |||||  |||||||||
Lyso  TNCSRFYLTKN--GQNYC-YGYYVNPGPN   178
```

**B**

```
LasA  APPSNLMQLPWRQGYSWQPNGAHSNTGSG-YPYSSFDASYDWP   42
      :::. :|||||||||||.:|||||:||||  |||||:|.||| |
AhP   AGGQFQLPWRQGYSWKANGAHSHTGSG-YPYSSIDVSYDXP    40
      :.| :|:|:..| ||...|||.:|||| || ||:|:| : .
Lyso  SPNGLLQFPFPRGASWHVGGAHTNTGSGNYPMSSLDMSRGGG   42
```

FIG. 1. Comparison of amino acid sequences of β-lytic endopeptidases. Numbers on the right-hand side correspond to the positions in the individual proteins. (A) Alignment of the amino acid sequences of *P. aeruginosa* LasA, *A. lyticus* β-lytic endopeptidase (Achr), and *L. enzymogenes* β-lytic endopeptidase (Lyso). Vertical lines indicate identical residues, and colons indicate amino acid residues with similar properties. [From E. Kessler, M. Safrin, J. C. Olson, and D. E. Ohman, *J. Biol. Chem.* **268,** 7503 (1993).] (B) Alignment of the amino-terminal sequences of LasA, *A. hydrophila* β-lytic endopeptidase (AhP), and *L. enzymogenes* β-lytic endopeptidase (Lyso). Vertical lines indicate identical residues, colons indicate residues with a similarity index [M. Gribskov and R. R. Burgess, *Nucleic Acids Res.* **14,** 6745 (1986)] of 0.5 or greater, and one dot indicates residues with similarity index of 0.1 or greater. [From A. G. Loewy, U. V. Santer, M. Wieczorek, J. K. Blodgett, S. W. Jones, and J. C. Cheronis, *J. Biol. Chem.* **268,** 9071 (1993).]

*Proteolytic and Elastolytic Activities.* Activity toward casein (at pH 7.4) is limited. To be detected, 50–100 μg of the purified enzyme was required.[2] The enzyme is reported to also act on elastin-orcein, although no quantitative information was given.[13] The enzyme does not act on the A chain of (performic acid) oxidized insulin. It cleaves the B chain readily at the $Gly^{23}$-$Phe^{24}$ peptide bond and more slowly at the $Val^{18}$-$Cys^{19}$-$SO_3H$ bond.[24] Of a long list of synthetic substrates, including dipeptides, tripeptides, N-blocked amino acid esters, and amides that have been tested,[13,24] only benzyloxycarbonyl(Z)-Gly-Phe-amide (but not Z-Gly-Phe) and FAGLA were hydrolyzed.[13] The cleavage occurred on the carboxyl-terminal side of glycine and required enzyme concentrations in the range of 75–300 μg/ml. The pH optimum for FAGLA hydrolysis is pH 6.5.

*Inhibitors.* The enzyme activity toward FAGLA is inhibited by 1,10-phenanthroline. Activity is fully restored on addition of $Zn^{2+}$ at a zinc-to-protein molar ratio of 1.0.[13] The enzyme is insensitive to diisopropyl fluorophosphate.[16]

### *Achromobacter lyticus* β-Lytic Endopeptidase

Based on amino acid sequence (Fig. 1A), the *A. lyticus* enzyme consists of 179 amino acid residues and its molecular weight is 19,286.[3] Although the p*I* has not been reported, the basic nature of the enzyme is evident from the high ratio (16:4) of basic to acidic residues. The amino acid sequence is over 96% identical to that of *L. enzymogenes* β-lytic endopeptidase (Fig. 1A). Thus, although, enzymatic properties of *Achromobacter* β-lytic endopeptidase have not yet been studied, it is reasonable to assume that, once elucidated, they will prove essentially the same as those of *Lysobacter* β-lytic endopeptidase.

The nucleotide sequence of the structural gene indicates that *A. lyticus* β-lytic endopeptidase is synthesized as a 374-residue preproenzyme with an extra amino-terminal sequence of 195 amino acid residues. In addition to the signal peptide, this includes a long propeptide (presumably 171 residues long) characterized by its high content of basic amino acids. Several clusters of arginines occupy positions −170 to −169, −164 to −163, −102 to −101, −86 to −85, −46 to −45, and −36 to −33. No significant sequence similarity has been found to propeptides of other proteinases, but similar clusters of basic amino acids exist in the propeptides of *L. enzymogenes*

---

[24] D. R. Whitaker, C. Roy, C. S. Tsai, and L. Jurasek, *Can. J. Biochem.* **43,** 1961 (1965).

α-lytic endopeptidase,[25] *Achromobacter* lysyl endopeptidase (*Achromobacter* protease I),[26] and *P. aeruginosa* LasA.[27,28]

## Pseudomonas aeruginosa LasA

*Physicochemical Properties.* By analysis of the amino acid sequence[17,27,28] (Figure 1A), LasA consists of 182 amino acid residues, and its molecular weight is 19,950. The molecular weight by gel filtration on Sephadex G-50 is only 15,000 owing to weak adsorption.[10] The predicted p*I* is 9.24.[28] The ratio of basic to acidic residues is 17:7. The amino acid sequence of LasA shows approximately 40% identity with those of *L. enzymogenes* and *A. lyticus* β-lytic endopeptidases (Fig. 1A).[5,27,28]

*Biosynthesis and Structural Gene.* The nucleotide sequence of the structural gene[27,28] indicates that LasA is synthesized as a 370-residue preproenzyme with an amino-terminal, 188-residue prepro sequence. The cell-associated proenzyme is inactive.[29]

*Bacteriolytic Activity.* LasA lyses rapidly various strains of *S. aureus*[4,5,9,10] and a number of other bacteria of the same genus such as *S. saprophyticus, S. epidermidis,* and *S. warnerri.*[10] Rapid lysis of *Streptomyces griseus* was also reported.[10] *Micrococcus lysodeikticus* and *Gaffkia tetragena* cells, in which the interpeptide bridges contain some glycine residues, are lysed more slowly, whereas bacteria containing no glycine interpeptides are not lysed.[9,10] The sensitive bacteria have oligoglycine (pentaglycine in *Staphylococcus*) interpeptide bridges within their peptidoglycan[9,10,30] that are cleaved by LasA.[9,10] Cell lysis depends on ionic strength. The rate of lysis of *S. aureus* cells is maximal at buffer concentrations of 10 to 20 m*M*. It drops to almost none as the buffer concentration approaches 0.1 *M*.[4] Lysis is also prevented in buffers containing 1 *M* sucrose and in undiluted human serum. Under these conditions, *S. aureus* cells are converted to spheroplasts, but these lyse rapidly on dilution. The optimum pH for bacteriolysis is pH 8.5.[4]

*Action on Isolated Cell Walls.* Lysis of isolated cell walls of *S. aureus* is associated with the release of di-, tri-, and tetraglycine and a slight increase of N-terminal alanine[9] (the C-terminal glycine residue of the pentaglycine cross-bridge is attached via a peptide bond to the amino group of D-Ala[30]).

[25] D. M. Epstein and P. C. Wensink, *J. Biol. Chem.* **263,** 16586 (1988).
[26] T. Ohara, K. Makino, H. Shinagawa, A. Nakata, S. Norioka, and F. Sakiyama, *J. Biol. Chem.* **264,** 20625 (1989).
[27] P. A. Schad and B. H. Iglewski, *J. Bacteriol.* **170,** 2784 (1988).
[28] A. Darzins, J. E. Peters, and D. R. Galloway, *Nucleic Acids Res.* **18,** 6444 (1990).
[29] A. Carnicero, T. B. Mansito, J. M. Roldan, and M. A. Falcon, *Arch. Microbiol.* **154,** 37 (1990).
[30] K. H. Schleifer and O. Kandler, *Bacteriol. Rev.* **36,** 407 (1972).

Isolated cell walls of *M. radiodurans* and *G. tetragena*, which contain (intra-tetrapeptide) diglycine sequences and three glycines and a serine residue, respectively, in place of the pentaglycine bridge of *S. aureus*, are lysed at rates comparable to that of *S. aureus* cell walls.[9]

*Proteolytic and Elastolytic Activities.* LasA is practically inactive against azocasein, but it exhibits basal elastolytic activity. In assays with insoluble elastin–Congo red, 5 μg of enzyme produces an increase in absorbance at 495 nm (owing to elastin solubilization) of 0.06 in 5 hr.[5,7,8] This renders elastin more susceptible to the action of other proteinases. The elastolytic activity of *Pseudomonas* elastase may be enhanced by one order of magnitude, with the degree of enhancement depending on LasA/elastase ratio.[7,17] The elastolytic activity of thermolysin, proteinase K, or human neutrophil elastase (10 μg each) is increased 2-fold in the presence of 3 μg LasA.[17] The alkaline proteinase of *P. aeruginosa*, which by itself does not degrade elastin, acquires elastolytic activity.[31]

*Specificity of Cleavage.* The preferred cleavage sites in elastin are Gly-Ala peptide bonds within internal Gly-Gly-Ala sequences. There is also evidence for cleavages of Gly-Gly and Gly-Phe bonds, presumably within Gly-Gly-Gly and Gly-Gly-Phe sequences, respectively.[7] Synthetic penta-glycine is cleaved readily into the respective di- and tripeptides,[5,10] whereas hexaglycine is cleaved symmetrically to generate triglycine.[5] Di-, tri-, or tetraglycine are not cleaved.[5,10] Other synthetic substrates resistant to LasA are Gly-Ser, Gly-Ala, and Lys-Ala,[10] Z-Gly-Pro-Gly-Gly-Pro-Ala and Ac-Ala$_4$,[7] as well as tetraalanine and tosyl-Gly-Pro-Lys-*p*-nitroanilide.[8] Thus, LasA appears to be selective for Gly-Gly-X sequences located internally within oligo- or polypeptides, with the preferred X residue in position $P_1'$[32] being Gly, Ala, or Phe. Charged side chains and terminal amino or carboxyl groups in the vicinity of the scissile bond interfere with LasA activity. A proline residue in the X position is not acceptable, and polypeptides containing no Gly-Gly sequences are virtually resistant to LasA.[7]

*Biological Activities.* LasA inhibits growth of *S. aureus*. Inhibition is observed with purified enzyme and in mixed cultures with *P. aeruginosa*, where maximal LasA production and maximal growth inhibition of *S. aureus* coincide.[33]

Full expression of the elastolytic phenotype of *P. aeruginosa* depends on LasA expression. By comparison to parental strains, elastolytic activity

[31] C. Wolz, E. Hellstern, M. Haug, D. R. Galloway, M. L. Vasil, and G. Doring, *Mol. Microbiol.* **5,** 2125 (1991).

[32] I. Schecter and A. Berger, *Biochem. Biophys. Res. Commun.* **27,** 157 (1967).

[33] F. R. Perestelo, M. T. Blanco, A. M. Gutierrez-Navarro, and M. A. Falcon, *Microbios Lett.* **30,** 85 (1985).

in culture filtrates of *lasA P. aeruginosa* mutants is reduced.[5,27,34,35] LasA contributes to virulence of *P. aeruginosa* in experimental models of acute[36] and chronic[37] lung infections.

*Inhibitors.* The staphylolytic activity of LasA is inhibited by several metal chelators including (the zinc-specific chelator) tetraethylenepentamine (TEP), 1,10-phenanthroline, EDTA (at concentrations higher than 20 m$M$), and EGTA ($\geq$20 m$M$). Dithiothreitol (DTT) and $ZnCl_2$ ($\geq$10$^{-4}$ $M$) are also inhibitory to the enzyme. The staphylolytic activity of LasA is insensitive to TLCK (tosyl-lysine chloromethyl ketone), $N$-ethylmaleimide, DFP (diisopropyl fluorophosphate), PMSF (phenylmethylsulfonyl fluoride), leupeptin, and phosphoramidon. Inhibition by TEP indicates that LasA is a zinc metalloproteinase. Like *L. enzymogenes* and *A. lyticus* $\beta$-lytic endopeptidases, it does not contain the consensus $Zn^{2+}$-binding sequence HEXXH.

*Application of Inhibitors to LasA Studies: Relevance to other $\beta$-Lytic Endopeptidases.* Purified LasA preparations that appear homogeneous on SDS–PAGE may contain undetectable traces of other extracellular proteinases, sufficient to interfere with specificity studies. For example, LasA was reported to cleave $\beta$-casein at the Lys$^{29}$-Ile$^{30}$ peptide bond,[30] located within the highly charged sequence -Asn-Lys-Lys-Ile-Glu-Lys-Phe.[8] The cleavage is fully prevented by TLCK (which is not inhibitory to LasA) but not by the LasA inhibitor 1,10-phenanthroline.[7] Also, cleavage of Ac-Ala$_4$, observed in our laboratory in the absence of inhibitors,[7] was completely prevented in the presence of TLCK and phosphoramidon, neither of which inhibit LasA. As all $\beta$-lytic endopeptidases are isolated from culture filtrates rich in proteolytic enzymes, such interference by traces of extracellular proteinases is not unlikely with any $\beta$-lytic endopeptidase preparation. To avoid such complications it is advisable to conduct activity and specificity studies of $\beta$-lytic endopeptidases in the presence of proteinase inhibitors.

*Stability.* LasA staphylolytic activity is stable at 4° in 20 m$M$ Tris-HCl, 0.5 m$M$ CaCl$_2$, pH 7.5, for at least 2 days. At $-20$°, the staphylolytic activity of LasA is stable for several months. After 1 year, however, a 20–30% loss in activity is observed. Calcium is not essential, but we added it during purification to stabilize elastase. At 37°, the enzyme is stable for 2 hr, but 40% of the activity is lost after 1 day. Full inactivation occurs in 15 min at 65°.[4]

[34] D. E. Ohman, S. J. Cryz, and B. H. Iglewski, *J. Bacteriol.* **142**, 836 (1980).
[35] J. B. Goldberg and D. E. Ohman, *J. Bacteriol.* **169**, 4532 (1987).
[36] L. L. Blackwood, R. M. Stone, B. H. Iglewski, and J. E. Pennington, *Infect. Immun.* **39**, 198 (1983).
[37] D. E. Woods, S. J. Cryz, R. L. Friedman, and B. H. Iglewski, *Infect. Immun.* **36**, 1223 (1982).

## Aeromonas hydrophila β-Lytic Endopeptidase

*Physiochemical Properties.* By SDS–PAGE, the molecular weight of AhP is 19,000.[6] Migration without reduction is slightly greater, suggesting the presence of intramolecular disulfide bonds. From isoelectric focusing, the p*l* is 8.5. The enzyme contains one atom of zinc, and the amino-terminal sequence of the first 40 residues (Fig. 1B) shows 46% identity with *L. enzymogenes* β-lytic endopeptidase and 69% identity with *P. aeruginosa* LasA.

*Bacteriolytic Activity.* The bacteriolytic activity of AhP has not been examined. Based on its specificity of cleavage (see below) and similarity to LasA, AhP is likely to possess staphylolytic activity. A staphylolytic proteinase which shares several properties with other β-lytic endopeptidases has been isolated from culture filtrates of *A. hydrophila*.[38] That enzyme was described as a small basic protein (p*l* 9.0–9.5) with a pH optimum for lytic activity of pH 8.5. Lysis of *S. aureus* cell walls was accompanied by the liberation of di- and triglycine. Synthetic glycine oligopeptides larger than triglycine but not polyglycine were hydrolyzed. The staphylolytic activity was inhibited by $Zn^{2+}$ but not by iodoacetate or *p*-chloromercuribenzoate. Inhibition of 30% was observed with 1 m*M* EDTA. The enzyme was retarded on Sephadex of various sieving capacities. Sedimentation analysis gave a $s^0_{20,w}$ of 1.1 S. It is not clear whether AhP and the staphylolytic proteinase are the same. A staphylolytic proteinase from *Myxobacter* AL1, which, based on current information, may not be classified as β-lytic endopeptidase ($M_r$ by sedimentation, 14,000; no sequence data are available), has been documented,[39] suggesting that the same organism may produce two distinct extracellular proteinases with staphylolytic activity.

*Specificity of Cleavage.* AhP hydrolyzes the Gly-Ala peptide bond within the Gly-Gly-Ala sequence located near the cross-link site in the γ-chain dimer of fibrin, converting the dimer to a modified γ-monomer. A synthetic fragment of the same region in the γ chain of fibrin, namely, Ac-EGQQH-HLGGAK(εTfa, trifluoroacetyl)QA-CONH₂ is cleaved at a single site, between the P₁ Gly and P₁′ Ala residues (in the terminology of Ref. 32). The fragment Ac-LGGAK(εTfa)QA-CONH₂ is even a better substrate, but Ac-LGAK(εTfa)QA-CONH₂, with one of the glycine residues missing, is hydrolyzed 200 times more slowly. Short synthetic substrates with the general structure Ac- and succinyl(Suc)-Gly-Gly-X-amide containing Ala, Phe, or *p*-nitrophenylalanine (Nph) in the X position are readily cleaved at the Gly-X bond. However, cleavages of the Gly-Gly bonds also occur,

---

[38] N. W. Coles, C. M. Gilbo, and A. J. Broad, *Biochem. J.* **111,** 7 (1969).
[39] R. L. Jackson and G. R. Matsueda, this series, Vol. 19, p. 591.

which are faster with the Ac- than with the Suc-N-substituted substrates. Polar residues, especially those with free $\alpha$-amino or $\alpha$-carboxyl groups located in close proximity to the scissile bonds, are not tolerated well, and Suc-Gly-Gly-$p$-nitroanilide is not cleaved. The optimal pH for activity, assessed with Suc-Gly-Gly-Nph-CONH$_2$, is pH 7.5.

*Kinetic Constants.* At pH 7.0, the $k_{cat}/K_m$ value for Ac-Gly-Nph-CONH$_2$ cleavage is 240 $M^{-1}$ sec$^{-1}$. At pH 7.5, the individual $k_{cat}$ and $K_m$ constants for hydrolysis of the substrate are 3.4 sec$^{-1}$ and 17 m$M$, respectively (i.e., $k_{cat}/K_m$ = 200 $M^{-1}$ sec$^{-1}$). The $k_{cat}/K_m$ values for the cleavages (at pH 7.0) of the Gly-Gly and Gly-Nph peptide bonds in Suc-Gly-Gly-Nph-CONH$_2$ are 170 and 1030 $M^{-1}$ sec$^{-1}$, respectively.

*Inhibitors.* The enzyme is inhibited by 1,10-phenanthroline and 8-hydroxyquinoline 5-sulfonic acid (both at 1 m$M$) but is insensitive to 5 m$M$ EDTA. DTT and 2-mercaptoethanol are also inhibitory to the enzyme. Iodoacetate, diisopropyl fluorophosphate, PMSF, pepstatin, and $\alpha_2$-macroglobulin have no effect on the enzyme activity.

*Stability.* Enzyme activity is stable in 5% methanol, 2% acetone, and 2% all v/v dimethyl sulfoxide, when incubated with these agents 30 min before and during the fibrinolytic assay. The enzyme activity is also retained after 1 hr exposure to 4 $M$ urea, 4.5 $M$ ammonium sulfate, or 2 $M$ NaCl, followed by dialysis against 20 m$M$ HEPES, 0.1 $M$ NaCl, pH 7.4. Loss of activity (5–8% and 30%) is observed after 1 hr of incubation in 50 m$M$ HEPES, 50 m$M$ NaCl, pH 7.0, at room temperature and 40°, respectively. Boiling for 15 sec in the same buffer causes complete inactivation.

# [49] Procollagen N-Peptidases: Procollagen N-Proteinases

*By* Karl E. Kadler, Samantha J. Lightfoot, and Rod B. Watson

## Introduction

The procollagen N-peptidases cleave the N-propeptides of types I, II, and III procollagen at specific sites within the substrates. These endopeptidases are fundamental to the conversion of procollagen to collagen that self-assembles into the cross-striated fibrils characteristic of the extracellular matrix of connective tissues. The peptidases are present in active forms in only minute quantities even in developing tissues during active collagen synthesis. Procedures have been developed for their purification from chick embryo leg tendons, bovine tendons and skin (type I/II procollagen

N-peptidase), and human placenta (type III procollagen N-peptidase). This chapter describes the purification of the two known procollagen N-peptidases and discusses observations on the substrate requirements of the type I/II procollagen N-peptidase.

The following reactions are catalyzed by the procollagen N-peptidases:

Type I/II procollagen N-peptidase (EC 3.4.24.14)
      Type I/II procollagen → type I/II pCcollagen[1] + N-propeptides
      Type I/II pNcollagen → type I/II collagen + N-propeptides
Type III procollagen N-peptidase
      Type III procollagen → type III pCcollagen + N-propeptides
      Type III pNcollagen → type III collagen + N-propeptides

Collagen is, arguably, the most abundant structural protein in the animal kingdom. It occurs in the extracellular matrix of connective tissue as insoluble assemblies including sheetlike laminas (in basement membrane), hexagonal lattices (in Descemet's membrane), fine filaments, and fibrils (for review, see Kielty et al.[2]). More than 17 genetically distinct types of collagen occur, the most abundant of which are the type I, III, and V collagens that occur as the cross-striated fibrils in tendon, ligament, skin, and bone, and type II and XI collagen that occurs as fibrils in cartilage and vitreous humor. A distinguishing feature of fibril-forming collagens is that they are synthesized as soluble precursor procollagens. The conversion of procollagen to collagen is fundamental to the synthesis of the mature collagen that spontaneously self-assembles into fibrils. The procollagen N-peptidases catalyze the removal of the N-propeptides of procollagen during the conversion of procollagen to collagen; C-peptidase catalyzes the removal of the C-propeptides of procollagen. Two procollagen N-peptidases have been identified. The same enzyme cleaves the N-propeptides of both types I and II procollagen. A separate enzyme cleaves the N-propeptides of type III procollagen. The peptidases that cleave the N- and C-propeptides of type V and XI procollagen have not been identified. The N-peptidases have a

---

[1] N-, Amino; C-, carboxyl; NEM, N-ethylmaleimide; PMSF, phenylmethylsulfonyl fluoride; PAB, p-aminobenzamidine; proα chain, α chain of procollagen; α chain, α chain of collagen; pCcollagen, an intermediate in the processing of procollagen to collagen containing the carboxyl propeptides but not the amino propeptides; pNcollagen, an intermediate in the processing of procollagen to collagen containing the amino propeptides but not the carboxyl propeptides; pCα chain, α chain of pCcollagen; pNα chain, α chain of pNcollagen; EDTA, ethylenediaminetetraacetic acid; Brij 35, polyoxyethylene lauryl ether; Tris, tris(hydroxymethyl)aminomethane; SDS, sodium dodecyl sulfate; PAGE, polyacrylamide gel electrophoresis; U, units.

[2] C. M. Kielty, I. Hopkinson, and M. E. Grant, in "Connective Tissue and Its Heritable Disorders; Molecular, Genetic and Medical Aspects" (P. M. Royce and B. Steinmann, eds.), p. 103. Wiley–Liss, New York, 1993.

number of special features, many of which depend largely on the structure of the procollagen substrate.

The procollagens comprise three polypeptide chains, each having a C-propeptide, a domain of repeating Gly-X-Y triplets (in which X and Y can be any residue but are frequently proline and hydroxyproline, respectively), and an N-propeptide that itself contains repeats of Gly-X-Y triplets (for review, see Kadler[3]). Type I procollagen comprises two proα1(I) and one proα2(I) chains, whereas types II and III are both homotrimers consisting of [proα1(II)]$_3$ and [proα1(III)]$_3$, respectively. Assembly of procollagen proceeds to a C to N direction starting with the association of three C-propeptides, followed by zipperlike assembly of the Gly-X-Y domains into a triple helix, and is concluded by the assembly of the N-propeptides which themselves assemble into a minor triple helix (for review of triple helix folding, see Kadler[3] and Engel and Prockop[4]). Electron microscopy of isolated procollagen molecules[5] and pNcollagen in fibrils[6] and sheets[7] suggests that the N-propeptides are folded back onto the procollagen molecule (see Fig. 1). Sequence analysis suggests that hydrophobic and electrostatic interactions stabilize the hairpin conformation.[8] The biological requirements of a folded-back conformation of the N-propeptide are unclear but may include presentation to the N-peptidase of the bonds cleaved by the N-peptidase and regulation of collagen fibril morphology.

Some of the earliest evidence to suggest the existence of an enzyme that specifically excises the N-propeptides of type I procollagen came from studies of procollagen processing in chick embryo calvaria[9] and from studies of dermatosparaxis, a heritable disorder of animals characterized by fragility of skin.[10-12] Here, lack of N-peptidase activity results in the accumulation of pNcollagen, an intermediate in the processing of procollagen to collagen that contains the N-propeptides but not the C-propeptides. An enzyme that specifically removed the N-propeptides of type I procollagen was later identified in calf tendon,[13] in cell culture,[14-17] and in chick embryo leg

[3] K. E. Kadler, *Protein Profile* **1**, 519 (1994).
[4] J. Engel and D. J. Prockop, *Annu. Rev. Biophys. Biophys. Chem.* **20**, 137 (1991).
[5] A. P. Mould and D. J. S. Hulmes, *J. Mol. Biol.* **195**, 543 (1987).
[6] D. F. Holmes, R. B. Watson, B. Steinmann, and K. E. Kadler, *J. Biol. Chem.* **268**, 15758 (1993).
[7] D. F. Holmes, A. P. Mould, and J. A. Chapman, *J. Mol. Biol.* **220**, 111 (1991).
[8] R. B. Watson, S. J. Lightfoot, and K. E. Kadler, *Int. J. Exp. Pathol.* **74**, A18 (1993).
[9] L. I. Fessler, N. P. Morris, and J. H. Fessler, *Proc. Natl. Acad. Sci. U.S.A.* **72**, 4905 (1975).
[10] A. Lenaers, M. Ansay, B. V. Nusgens, and M. Lapiere, *Eur. J. Biochem.* **23**, 533 (1971).
[11] C. M. Lapiere, A. Lenaers, and L. D. Kohn, *Proc. Natl. Acad. Sci. U.S.A.* **68**, 3054 (1971).
[12] C. M. Lapiere and G. Pierard, *J. Invest. Dermatol.* **62**, 582 (1974).
[13] L. D. Kohn, C. Isersky, J. Zupnik, A. Lenaers, G. Lee, and C. M. Lapiere, *Proc. Natl. Acad. Sci. U.S.A.* **71**, 40 (1974).
[14] D. L. Layman and R. Ross, *Arch. Biochem. Biophys.* **157**, 451 (1973).

N-propeptide

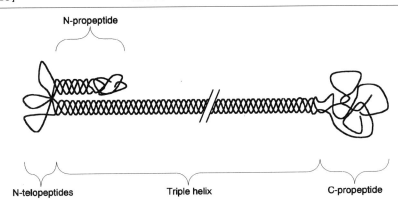

N-telopeptides                    Triple helix                    C-propeptide

FIG. 1. Diagrammatic representation of the extracellular steps in collagen biosynthesis. Procollagen consists of a central 300-nm-long triple-helical region (comprising three chains each with ~1000 residues wound into a triple helix) flanked by a trimeric globular C-propeptide domain and a timeric N-propeptide domain, consisting of a short triple helix and a globular region, bent back onto the main triple helix. The molecule has a molecular mass of 450 kDa, the triple-helical region accounting for 300 kDa, the C-propeptide domain 100 kDa, and the N-propeptide domain about 45 kDa. Although types II and III procollagen are very similar in structure to type I, it is not known whether the N-propeptides of these procollagens are in a bent-back conformation. The secreted procollagen molecules are processed to collagen in the extracellular space by procollagen N- and C-peptidases. The collagen molecules generated then spontaneously self-assemble into cross-striated fibrils, and the growing fibrils are stabilized by the formation of intermolecular cross-links.

tendons[18–23] with an apparent molecular mass of 260 kDa,[18] and from whole chick embryos[24–27] with an apparent molecular mass of 320 kDa. It is a neutral metalloproteinase (pH optimum pH 7.0–8.0), requiring $Zn^{2+}$ and $Ca^{2+}$ for full activity,[19] and has been purified to homogeneity by Hojima

[15] B. Goldberg, M. B. Taubman, and A. Radin, Cell (Cambridge, Mass.) 4, (1975).
[16] D. L. Layman, Proc. Soc. Exp. Biol. Med. 166, 325 (1981).
[17] L. Gerstenfeld, J. C. Beldekas, G. E. Sonenshein, and C. Franzblau, J. Biol. Chem. 259, 9158 (1984).
[18] J. Uitto and J. R. Lichtenstein, Biochem. Biophys. Res. Commun. 71, 60 (1976).
[19] L. Tuderman, K. I. Kivirikko, and D. J. Prockop, Biochemistry 17, 2948 (1978).
[20] M. K. K. Leung, L. I. Fessler, D. B. Greenberg, and J. H. Fessler, J. Biol. Chem. 254, 224 (1979).
[21] T. Morikawa, L. Tuderman, and D. J. Prockop, Biochemistry 19, 2646 (1980).
[22] L. Tuderman and D. J. Prockop, Eur. J. Biochem. 125, 545 (1982).
[23] Y. Hojima, J. McKenzie, M. van der Rest, and D. J. Prockop, J. Biol. Chem. 264, 11336 (1989).
[24] K. Tanzawa, J. Berger, and D. J. Prockop, J. Biol. Chem. 260, 1120 (1985).
[25] J. Berger, K. Tanzawa, and D. J. Prockop, Biochemistry 24, 600 (1985).
[26] K. E. Dombrowski, J. E. Sheats, and D. J. Prockop, Biochemistry 25, 4302 (1986).
[27] K. E. Dombrowski and D. J. Prockop, J. Biol. Chem. 263, 16545 (1988).

*et al.* from chick embryo leg tendons[23] and bovine tendons.[28] In gel-filtration chromatography the enzyme has an apparent molecular mass of 500 kDa, and silver nitrate staining of the most pure preparations of the chick peptidase suggests that it is composed of four subunits of molecular mass 161, 135, 120, and 61 kDa; catalytic activity was attributed to the two larger subunits. Storage of the enzyme can result in the proteolytic degradation of some of the subunits to produce a 300-kDa form of the enzyme that retains full proteolytic activity. This smaller enzyme was most probably the form previously extracted from whole chick embryos.[24] The *N*-peptidase that cleaves type I procollagen also cleaves the *N*-propeptides of type II procollagen[29] and has been named, accordingly, the type I/II procollagen *N*-peptidase.

Type I and II procollagens are cleaved by the type I/II *N*-peptidase at -Pro-Gln- bonds in the proα1(I) (at position 161–162 in the human chain), -Ala-Gln- bonds in the proα2(I) (at position 79–80 in the human chain), and -Ala-Gln- bonds in the proα1(II) chains of type II procollagen (at position 181–182 in the human chain) that are located at the junction of the *N*-telopeptide and the *N*-propeptide domains.[30] Properties of the enzyme include the following: the $K_m$ for chick type I procollagen is 54 n$M$ at pH 7.5 and 35° (for human type I procollagen the $K_m$ is in the range 0.8–1.2 $\mu M$,[31]), and the $k_{cat}$ for chick type I procollagen is 350 hr$^{-1}$; the activation energy for the reaction with chick type I procollagen is 30 kJ mol$^{-1}$; and the isoelectric point of the enzyme is 3.6 (values from Hojima *et al.*[23]).

Synthetic peptides with amino acid sequences identical to the sequence surrounding the proα1(I) cleavage site are not cleaved by the type I/II enzyme.[21] However, such peptides were effective inhibitors of the enzyme, especially those peptides that contained L-phenylalanine 3 residues amino-terminal to the cleavage site. The type I procollagen is cleaved sequentially by the enzyme in that one proα1(I) chain and the proα2(I) chain are cleaved rapidly and the remaining proα1(1) chain is cleaved slowly. In contrast, no partially cleaved intermediates were found when type I homotrimer or type II procollagen was used as the substrate.[25]

Type III procollagen *N*-peptidase specifically removes the *N*-propeptides from type III procollagen (at -Pro-Gln- bonds) and will not cleave type I and II procollagen. The enzyme was partially purified from calf

---

[28] Y. Hojima, M. M. Mörgelin, J. Engel, M.-M. Boutillon, M. van der Rest, J. McKenzie, G. C. Chon, N. Rafi, A. M. Romanic, and D. J. Prockop, *J. Biol. Chem.* **269**, 11381 (1994).

[29] J. Uitto, *Biochemistry* **16**, 3421 (1977).

[30] D. Horlein, P. P. Fietzek, and K. Kühn, *FEBS Lett.* **89**, 279 (1978).

[31] S. J. Lightfoot, D. F. Holmes, A. Brass, M. E. Grant, P. H. Byers, and K. E. Kadler, *J. Biol. Chem.* **267**, 25521 (1992).

tendon fibroblast culture medium,[32] from bovine aorta smooth muscle cells,[33] and human placenta.[34] Type III $N$-peptidase is a neutral metalloproteinase; it requires $Zn^{2+}$ and $Ca^{2+}$ and a native procollagen structure[33] for full activity. It cleaves at a slower rate than does the type I/II procollagen $N$-peptidase[35]; the $K_m$ value for the human enzyme against bovine type III procollagen is 2 $\mu M$.

Of special interest, mutations in the genes for type I procollagen (COL1A1 and COL1A2) that result in either the Ehlers–Danlos syndrome (EDS) type VII or in osteogenesis imperfecta (OI, "brittle bone disease") can alter the structure of the procollagen and slow the rate of cleavage by the $N$-peptidase. The biochemical basis of EDS type VII is a failure to remove the $N$-propeptides of type I procollagen[36,37] (for review, see Steinmann et al.[38] and Kadler[39]). Individuals with EDS type VII have mutations in COL1A1 or COL1A2 that cause skipping of exon 6 during processing of the pre-mRNAs for either the proα1(I) or proα2(I) chains. Exon 6 in both genes encodes the 18 residues that include the site for cleavage by the type I/II procollagen $N$-peptidase. Studies of an individual with EDS showed a G to A transition at the 5' donor site of exon 6 of the COL1A2 that caused the synthesis of proα2$^{-ex6}$ chains that lacked the $N$-peptidase cleavage site.[40] Procollagen molecules synthesized by the individual's dermal fibroblasts in culture were a 1:1 mixture of normal and abnormal molecules. Incubation of the procollagen with $N$-peptidase resulted in a 1:1 mixture of pCcollagen and $N$-peptidase-resistant procollagen. Notably, the normal proα1(I) chains in abnormal molecules were resistant to cleavage by the $N$-peptidase. Incubation of the mixture with $C$-peptidase generated collagen and abnormal pNcollagen (pNcollagen$^{-ex6}$) that readily copolymerized into fibrils. Fibrils generated in vitro by cleavage of the mixture with $C$-peptidase[41] initially formed hieroglyphic fibrils that could be re-

[32] B. V. Nusgens, Y. Goebels, H. Shinkai, and C. M. Lapiere, Biochem. J. **191,** 699 (1980).

[33] R. Halila and L. Peltonen, Biochemistry **23,** 1251 (1984).

[34] R. Halila and L. Peltonen, Biochem. J. **239,** 47 (1986).

[35] L. I. Fessler, R. Timpl, and J. H. Fessler, J. Biol. Chem. **256,** 2531 (1981).

[36] J. R. Lichtenstein, L. D. Kohn, G. R. Martin, P. Byers, and V. A. McKusick, Trans. Assoc. Am. Physicians **86,** 333 (1973).

[37] J. R. Lichtenstein, G. R. Martin, L. D. Kohn, P. H. Byers, and V. A. McKusick, Science **182,** 298 (1973).

[38] B. Steinmann, P. M. Royce, and A. Superti-Furga, in "Connective Tissue and Its Heritable Disorders: Molecular, Genetic and Medical Aspects" (P. M. Royce and B. Steinmann, eds.), p. 351. Wiley–Liss. New York, 1993.

[39] K. E. Kadler, Int. J. Exp. Pathol. **71,** 319 (1993).

[40] R. B. Watson, G. A. Wallis, D. F. Holmes, D. Viljoen, P. H. Byers, and K. E. Kadler, J. Biol. Chem. **267,** 9093 (1992).

[41] K. E. Kadler, Y. Hojima, and D. J. Prockop, J. Biol. Chem. **262,** 15696 (1987).

solved to fibrils with near-circular cross sections with additional $N$-pepti-dase. Further incubation of the hieroglyphic fibrils with the $N$-peptidase resulted in partial cleavage of the pNcollagen$^{-ex6}$ in which the abnormal pN$\alpha$2(I) chains remained intact. The $N$-propeptides were in a bent-back conformation and restricted to the surface of the hieroglyphs. Cleavage of the pro$\alpha$1(I) chains but not the pro$\alpha$2(I)$^{-ex6}$ chains in abnormal molecules resulted in a change in conformation of the $N$-propeptides such that some were in an extended conformation. Examination of fibrils in the skin of an individual with EDS type VII showed that the $N$-propeptides were in both a hair-pin conformation and extended in the fibrils.[6] The results showed that the multiple actions of the $N$-peptidase have a vital role in determining the morphology of collagen fibrils and the phenotype of some heritable connective tissue diseases.

Other work has shown that mutations in COL1A1 and COL1A2, which cause OI and result in substitutions of other residues for glycine, slow the rate of cleavage of type I procollagen by the type I/II procollagen $N$-peptidase.[42] Vogel *et al.* identified an individual with OI whose cells in culture secreted a type I procollagen that contained a substitution of cyste-ine for glycine $\alpha$1(I)748 in one or both pro$\alpha$1(I) chains.[43] The procollagen molecules were cleaved slowly by $N$-peptidase, and rotary shadowing elec-tron microscopy showed a flexible kink at the site of the substitution. It was proposed that the substitution was accommodated within a loop in the defective chain and that the stagger of the three chains in the triple helix was altered amino-terminal to the substitution. The phase shift thus generated in the triple helix would be propagated to the N-terminal end of the molecule during normal folding and, presumably, would disrupt the conformation of the $N$-peptidase cleavage site. Lightfoot *et al.* showed that OI type I procollagens containing substitutions of aspartate and arginine for glycine were cleaved slowly by the $N$-peptidase but did not have kinks in the triple helix.[31] The slow cleavage of a fragment of an abnormal procollagen that did not contain the substitution demonstrated directly that an altered con-formation is generated at the site of the substitution and propagated to the N end of the triple helix of the procollagen molecule and disrupts the conformation of the $N$-peptidase cleavage site. Wallis *et al.* showed that an OI type I procollagen containing a 3-amino acid deletion in the triple helix was cleaved normally by the type I/II $N$-peptidase.[44] Also, a type I procollagen containing a substitution of serine for glycine $\alpha$1(I)883 in one

[42] K. E. Dombrowski, B. E. Vogel, and D. J. Prockop, *Biochemistry* **28,** 7107 (1989).

[43] B. E. Vogel, R. Doelz, K. E. Kadler, Y. Hojima, J. Engel, and D. J. Prockop, *J. Biol. Chem.* **263,** 19249 (1988).

[44] G. A. Wallis, K. E. Kadler, B. J. Starman, and P. H. Byers, *J. Biol. Chem.* **267,** 25529 (1992).

or both proα1(I) chains was cleaved normally by the N-peptidase.[45] The results suggest that the type I/II N-peptidase is sensitive to changes in the conformation of its cleavage site in the procollagen molecule caused by phase shifts of 3 amino acids and greater.


Nomenclature

Enzymes that cleave the amino propeptides of procollagen have been referred to in the literature as amino procollagen peptidase, procollagen amino-terminal endopeptidase, amino-terminal procollagen peptidase, procollagen amino-terminal proteinase, and procollagen amino (N-) proteinase. Historically, the enzyme was called a peptidase, presumably because it was thought to act on the telopeptides of type I collagen. When it became clear that the enzyme cleaved a larger procollagen molecule to a smaller intermediate by specific enzymatic cleavage of internal bonds, the term proteinase was adopted. In current literature the enzyme is frequently referred to as procollagen N-proteinase. However, in general terms, a peptide hydrolase is called a peptidase. In agreement with the guidelines in *Enzyme Nomenclature* (1992) and *Biochemical Nomenclature and Related Documents: A Compendium* (1992), the enzymes shall be referred to here as procollagen N-peptidases.


Assay Methods

*Type I/II Procollagen N-Peptidase*

Two assay methods have been described for type I/II procollagen N-peptidase using radiolabeled type I procollagen as substrate. In the gel assay, the substrate is incubated with enzyme, the cleavage products are separated in sodium dodecyl sulfate (SDS)–polyacrylamide gels under reducing conditions, and the cleavage of proα chains to pCα chains is quantitated by fluorography or phosphoimaging of dried gels. In the rapid assay,[19] radiolabeled procollagen is incubated with the enzyme, the triple-helical molecules (procollagen substrate and pCcollagen product) are precipitated by 27% (v/v) ethanol, and the soluble N-propeptides are determined by liquid scintillation counting. The rapid assay gives no measure of specificity of propeptide cleavage and can be used definitively only in the absence of other proteinase activity. In contrast, the gel assay readily distinguishes the

---

[45] S. J. Lightfoot, M. S. Atkinson, G. Murphy, P. H. Byers, and K. E. Kadler, *J. Biol. Chem.* **269**, 30352 (1994).

cleavage of N-propeptides from that of C-propeptides, and it is the assay of choice at early stages of purification or for crude extracts.

*Definition of Enzyme Unit.* One unit of activity is the amount that converts 1 μg of procollagen to pCcollagen in 1 hr at 34° at an initial substrate concentration of 10 μg/ml.

*Assay Procedure.* To a 1.5-ml microcentrifuge tube add 1 μg $^{14}$C-labeled type I procollagen (the substrate may be prepared as described previously[19,22]), 5 m$M$ CaCl$_2$, and 0.01% (w/v) Brij 35 contained within a volume of 10 μl. Up to 90 μl of the solution to be assayed for N-peptidase activity is added. The sodium chloride and Tris buffer concentrations should be adjusted to 0.12–0.15 and 0.05–0.2 $M$, respectively. For gel assays, samples are incubated at 34° for times of 0, 0.5, 1, 2, and 4 hr. At the end of the incubation period, the reaction is stopped by the addition of 10 μl of TE buffer [1 $M$ Tris-HCl buffer, pH 7.4 (20°), 250 m$M$ EDTA, and 0.1% (w/v) NaN$_3$]. Samples are analyzed by SDS–PAGE in the presence of a reducing agent, and the procollagen bands are displayed either by fluorography and exposure of the dried gels to film or by phosphoimaging. The initial rate of cleavage of procollagen to pCcollagen is estimated by plotting [pCα1(I) chains]/[proα1(I) + pCα1(I) chains] against time of incubation.

For rapid assays, the samples are incubated at 34° for 1 and 3 hr. The reaction is stopped by adding 100 μl of 0.2× TE buffer and 100 μl of 78% (v/v) ethanol that was at −20°. The solution is vortexed repeatedly, held on ice (0°) for 45 min, and centrifuged at 15,000 $g$ for 15 min, and then 150 μl of the supernatant is taken for liquid scintillation counting. The units of activity of the N-peptidase preparation (U/ml) are readily calculated from the disintegrations per minute (dpm) of free N-propeptides cleaved by consideration of the molecular masses of the N-propeptides (45 kDa) and the procollagen (450 kDa), the specific activity of the $^{14}$C-labeled procollagen (dpm/g), the time of incubation (hr), and the volumes of sample used.

*Type III Procollagen N-Peptidase*

Two assay methods have been described using radiolabeled type III procollagen as the substrate.[46] In the gel assay, the substrate is incubated with enzyme, the cleavage products are separated in SDS–polyacrylamide gels under reducing conditions, and the cleavage of proα1(III) chains of pCα1(III) chains is quantitated by fluorography of the dried gels. In the rapid assay, radiolabeled procollagen is incubated with the enzyme, the

---

[46] R. Halila, *Anal. Biochem.* **145**, 205 (1985).

triple-helical molecules (procollagen substrate and pCcollagen product) are precipitated by 30% (w/v) ammonium sulfate, and the soluble N-propeptides are determined by liquid scintillation counting.

*Definition of Enzyme Unit.* One unit of activity has been defined as the amount of enzyme present in 1 mg of original tissue.[32,33] However, because of the potential variation in yields of the proteinase obtained using different tissues, and different source animals, a better definition of 1 unit of activity, and one which would be in line with the assay of type I/II N-peptidase, would be the amount that converts 1 $\mu$g of procollagen to pCcollagen in 1 hr at 34° at an initial substrate concentration of 10 $\mu$g/ml.

*Assay Procedure.* To a 1.5-ml microcentrifuge tube add 10 $\mu$g [14]C-labeled type III procollagen (the substrate may be prepared as previously described[46,47]) and 2 m$M$ CaCl$_2$ contained within a volume of 40 $\mu$l. Up to 100 $\mu$l of the solution to be assayed for N-peptidase activity is added. The concentrations of sodium chloride and sodium cacodylate buffer, pH 7.4, should be adjusted to 0.15 $M$ and 0.1$M$, respectively. For gel assays, samples are incubated at 30° for times of 0, 0.5, 1, 2, and 4 hr. At the end of the incubation period, the reaction is stopped by cooling the reaction tubes in an ice bath and adding 20 $\mu$l of 250 m$M$ EDTA, 10 m$M$ p-aminobenzamidine (PAB), 10 m$M$ phenylmethylsulfonyl fluoride (PMSF), and 100 m$M$ N-ethylmaleimide (NEM). Samples are analyzed by SDS–PAGE in the presence of a reducing agent, and the procollagen bands are displayed either by fluorography and exposure of the dried gels to film or by phosphoimaging. The initial rate of cleavage of procollagen to pCcollagen is estimated by plotting [pC$\alpha$1(III) chains]/[pro$\alpha$1(III) + pC$\alpha$1(III) chains] against time of incubation.

For rapid assays, the samples are incubated at 30° for 3 hr. The reaction is stopped by cooling the reaction tubes on ice and adding 20 $\mu$l of 250 m$M$ EDTA, 10 m$M$ PAB, 10 m$M$ PMSF, and 100 m$M$ NEM. Forty microliters of newborn calf serum and 100 $\mu$l of 0.1 $M$ sodium cacodylate buffer, pH 7.4, containing 90% (w/v) ammonium sulfate are added for a minimum of 2 hr at 4°, and the samples are centrifuged at 13,500 $g$ for 15 min. One hundred microliters of the supernatant is taken for liquid scintillation counting. The units of activity of the N-peptidase preparation (U/ml) are readily calculated as above.

## Purification of Type I/II Procollagen N-Peptidase

Although the importance of the procollagen N-peptidases has been recognized since the 1980s, studies have been restricted by low levels of

---

[47] B. Nusgens and C. M. Lapiere, *Anal. Biochem.* **95**, 406 (1979).

the proteins in tissues. The richest source of the type I/II enzyme is 17-day chick embryo leg tendons, but estimates suggest that the enzyme may be present in amounts as minute as 1 $\mu$g per gram of tendon. The methods described are those developed[24] and refined[23] previously, with minor modifications by us for the enzyme in chick leg tendon.

Because of the small amounts of enzyme available in tissue, the purification is best performed with tendons from 3600–4800 embryos. The purification procedure below need not be adjusted if the number of embryos used lies within this range. The recovery and purity figures obtained in our laboratory agree very well with those obtained earlier.[23] The values stated below are taken from Hojima et al.[23] Tendons are harvested at the rate of 600 per week and stored at $-70°$ until required. All subsequent steps are performed at 4° or on ice. Sodium azide at 0.02% (w/v) is used as a bacteriostat and is added to all buffers described below. The pH of buffers is adjusted at 20°. Ten-milliliter fractions are collected at all steps.

*Step 1: Extraction.* Tendons are thawed, pooled, and washed 3 times in 600 ml, each time, of 25 m$M$ Tris-HCl buffer, pH 7.5, contaning 50 m$M$ NaCl. The tendons are divided into three equal parts and separately suspended in 750 ml of 0.2 $M$ Tris-HCl buffer, pH 7.5, containing 1 $M$ NaCl and 0.05% (w/v) Brij 35. The sealed containers are rotated for 16 hr, after which the tendons are collected by centrifugation at 17,000 $g$ for 30 min. The enzyme is recovered in the supernatant. Protein (8040 mg) is recovered containing 55,770 U of enzyme at a specific activity of 6.9 U/mg.

*Step 2: Ammonium Sulfate Precipitation.* Proteins in the supernatant are precipitated by ammonium sulfate (176 mg/ml) for a minimum of 1.5 hr and collected by centrifugation at 17,000 $g$ for 60 min. The precipitate is resuspended in 360 ml of the extraction buffer, stirred for a minimum of 1.5 hr, and centrifuged at 17,000 $g$ for 60 min, and the supernatant is recovered. Protein (3540 mg) is recovered containing 37,500 U of enzyme at a specific activity of 10.6 U/mg, giving a 2-fold purification.

*Step 3: Adsorption on Concanavalin A-Sepharose.* The sample is mixed with 30 ml of concanavalin A-Sepharose that has previously been equilibrated with 0.2 $M$ Tris-HCl buffer, pH 7.5, containing 2 $M$ NaCl and 0.05% (w/v) Brij 35. The resin is packed into a column of size 1.6 × 20 cm and equilibrated with 250 ml of buffer at a flow rate of 30 ml/hr. Adsorbed protein is eluted with 500 ml of the equilibration buffer to which has been added 0.5 $M$ $\alpha$-methyl-D-mannoside (prefiltered through a Whatman, Clifton, NJ, No. 1 filter) and the pH adjusted to pH 7.0. The eluted sample is pressure concentrated to 50 ml using an Amicon (Danvers, MA) YM30 membrane in a 400-ml chamber. The sample is allowed to stir for 1 hr after the concentration is complete, and the membrane is washed with 10 ml of the elution buffer for a further 2 hr. The samples are combined and dialyzed for a total of 20 hr against two changes of 4 liters each of 0.1 $M$ Tris-HCl buffer, pH 7.5, containing 0.2 $M$ NaCl and 0.02% (w/v) Brij 35. The dialyzed sample is centrifuged at 17,000 $g$ for 30 min and the supernatant recovered. Protein (169 mg) is recovered containing 30,200 U of enzyme at a specific activity of 179 U/mg giving a 26-fold purification.

*Step 4: Heparin-Sepharose Chromatography.* The sample is applied to a 1.5 × 10.5 cm column of heparin-Sepharose, equilibrated with the dialysis buffer, at 20 ml/hr. The column is washed with 100 ml of the buffer and the *N*-peptidase is eluted with a linear NaCl gradient between 0.2 and 1.4 $M$ NaCl at a flow rate of 20 ml/hr. The gradient is prepared with 250 ml of equilibration buffer and 250 ml of the buffer containing 1.4 $M$ NaCl. Rapid assay of alternate fractions shows that the procollagen *N*-peptidase activity is eluted between 0.6 and 0.8 $M$ NaCl. Fractions containing activity are pooled and concentrated to 4 ml using an Amicon YM30 membrane and a 200-ml chamber. The sample is allowed to stir for 1 hr after concentration is complete, and the membrane is washed for a further 2 hr with 2 ml of the buffer used in the next step (see below). The samples are combined. Protein (7.68 mg) is recovered containing 27,700 U of enzyme at a specific activity of 3610 U/mg, giving a 520-fold purification.

*Step 5: Sephacryl S-300 Gel Filtration.* The combined sample is applied to a 5 × 60 cm Sephacryl S-300 column, equilibrated with 50 m$M$ Tris-HCl buffer, pH 7.5, containing 0.75 $M$ NaCl and 0.02% (w/v) Brij 35, at a flow rate of 15–20 ml/hr. Two peaks of enzymatic activity are seen corresponding to apparent molecular masses of 500 and 300 kDa. The fractions containing the 500-kDa peak (the first activity peak) are pooled and concentrated by ultrafiltration to a volume of 4 ml, as before. Protein (1.74 mg) is recovered containing 19,000 U of enzyme at a specific activity of 10,900 U/mg, giving a 1580-fold purification. The *N*-peptidase is sufficiently pure at this point to generate pCcollagen for subsequent use in a cell-free system of generating collagen fibrils by cleavage of the pCcollagen with procollagen *C*-peptidase.[41]

*Step 6: Sephacryl S-300 Gel Filtration.* The sample is applied to a 2.5 × 100 cm column of Sephacryl S-300, equilibrated in the same buffer used in the previous step. The flow rate is 15–20 ml/hr. The rapid assay is used to detect the *N*-peptidase activity, and tubes containing the enzyme are pooled and concentrated to a final volume of 10 ml, as before. The concentrated sample is dialyzed against two changes of 5 liters each of 50 m$M$ Tris-HCl buffer, pH 7.5, containing 0.15 $M$ NaCl and 0.02% (w/v) Brij 35 for a total of 15 hr. Protein (0.824 mg) is recovered containing 15,500 U of enzyme at a specific activity of 18,800 U/mg, representing a 2720-fold purification.

*Step 7: CM-Sephadex C-50 Chromatography.* The sample is applied to a 1.5 × 9 cm column of CM-Sephadex C-50, equilibrated with the dialysis buffer, at a flow rate of 30 ml/hr. The column is washed with 50 ml of dialysis buffer, and the sample is eluted with a linear NaCl gradient between 0.15 and 0.55 *M* NaCl, prepared with 200 ml of dialysis buffer and 200 ml of dialysis buffer containing 0.55 *M* NaCl. Enzyme activity elutes between approximately 0.3 and 0.4 *M* NaCl. Fractions containing enzyme actvity are pooled and solid NaCl added to 0.75 *M* before concentration to 4 ml. The added salt prevents adsorption to the YM30 membrane. Protein (0.335 mg) is recovered containing 10,600 U of enzyme at a specific activity of 31,700 U/mg, giving a 4590-fold purification.

*Step 8: Sephacryl S-300 Gel Filtration.* The sample is applid to a 2.5 × 114 cm column of Sephacryl S-300, equilibrated with a 50 m*M* Tris-HCl buffer, pH 7.5, containing 0.75 *M* NaCl and 0.02% (w/v) Brij 35, at a flow rate of 15–20 ml/hr. Two activity peaks are seen: a main peak with an apparent molecular mass of 500 kDa and another smaller peak with an apparent molecular mass of 300 kDa. The peak fractions from the main activity peak are pooled. The fractions from the small actvity peak are pooled separately. Protein (0.042 mg) was recovered containing 6430 U of enzyme at a specific activity of 153,000 U/mg giving a 22,200-fold purification.

*Storage.* Impure preparations of the enzyme from steps early in the purification scheme can be left overnight at 4° without significant loss of activity. The highest purity preparations should be kept at −70°. Working stocks of the enzyme can be kept at −20°.

## Purification of Type I/II Procollagen *N*-Peptidase from Bovine Sources

We have found that a mixture of freshly isolated tendons and ligaments from fetal calves can be a very rich source of the type I/II *N*-peptidase. After finely chopping 44 g of tissue from a 210-day calf fetus, and using the same extraction method as for the chick tendons, we obtained 42,000 U of enzyme (K. E. Kadler, S. J. Lightfoot, and R. B. Watson, unpublished observations, 1992) compared with 55,770 U of the enzyme from approximately 150 g of chick tendons.

The values stated below are taken from Hojima *et al.*[28] Tendons taken from fetal calves are stored at −20° for 1–2 months until required. All subsequent steps are performed at 4° or on ice. Sodium azide at 0.02% (w/v) is used as a bacteriostat and is added to all buffers described below.

The pH of buffers is adjusted at 20°. Ten-milliliter fractions are collected at all steps unless otherwise stated.

*Step 1: Extraction.* Two hundred grams of tendon material is thawed, pooled, and washed 3 times in 600 ml each time of 25 m$M$ Tris-HCl buffer, pH 7.5, containing 75 m$M$ NaCl. Fatty tissue is removed, and the tendons are cut into approximately 0.5-cm pieces. The tendon pieces are divided into three equal parts and extracted as for chick tendons. The tendons are extracted again under the same conditions and the supernatants combined. Protein (4250 mg) is recovered containing 42,000 U of enzyme at a specific activity of 9.8 U/mg.

*Step 2: Ammonium Sulfate Precipitation.* Proteins in the supernatant are precipitated by ammonium sulfate as for the chick enzyme. Precipitates floating on the supernatant are collected by filtration through 24-cm-diameter Whatman No. 1 filter papers. All precipitates obtained are resuspended in 350 ml of the extraction buffer as for the chick enzyme. Protein (701 mg) is recovered containing 27,000 U of enzyme at a specific activity of 38.5 U/mg, giving a 4-fold purification.

*Step 3: Adsorption on Concanavalin A-Sepharose.* This step is the same as step 3 for the chick enzyme with the exception that the α-methyl-D-mannoside-containing elution buffer is adjusted to pH 6.8. Protein (96 mg) is recovered containing 13,500 U of enzyme at a specific activity of 140 U/mg, representing a 14-fold purification.

*Step 4: Heparin-Sepharose Chromatography.* This step is the same as step 4 for the chick enzyme with the exception that a 1.5 × 16 cm column of Heparin-Sepharose is used. Protein (1.82 mg) is recovered containing 10,300 U of enzyme at a specific activity of 5660 U/mg, giving a 580-fold purification.

*Step 5: Sephacryl S-300 Gel Filtration.* This step is the same as step 5 for the chick enzyme with the exceptions that the concentration of NaCl in the elution buffer is increased to 1 $M$ (from 0.75 $M$) and fraction volumes of 7.5 ml are collected (instead of 10 ml). The concentrated sample is dialyzed against two changes of 2 liters each time of 50 m$M$ Tris-HCl buffer, pH 7.5, containing 0.15 $M$ NaCl and 0.02% Brij 35 for a total of 15 hr. Protein (0.355 mg) is recovered containing 6,250 U of enzyme at a specific activity of 17,600 U/mg, giving a 1800-fold purification.

*Step 6: CM-Sephadex C-50 Chromatography.* This step is the same as step 7 for the chick enzyme with the exceptions that the flow rate is reduced to 16 ml/hr (from 30 ml/hr) and 5-ml fractions (instead of 10 ml) are collected. Protein (0.173 mg) is recovered containing 6100 U of enzyme at a specific activity of 35,300 U/mg, giving a 3600-fold purification.

*Step 7: Sephacryl S-300 Gel Filtration.* This step is the same as step 8 for the chick enzyme. Only one activity peak is seen and fractions containing the peak of the activity and of high specific activity are pooled. Protein (0.021 mg) is recovered containing 3340 U of enzyme at a specific activity of 157,000 U/mg, giving a 16,000-fold purification.

## Purification of Type III Procollagen N-Peptidase

The methods described are those developed earlier[33,34] with minor modifications by us. All steps are performed at 4° or on ice.

*Step 1: Homogenization.* Two human placentas are obtained from normal deliveries and the fetal membranes and umbilical cord removed. Fetal blood is rinsed from the placental vessels by thorough washing in cold distilled water, and the placental tissue is cut into small pieces and further washed in deionized water (5 ml/g tissue). The tissue pieces are suspended in 0.1 $M$ sodium cacodylate buffer, pH 7.4, containing 2 $M$ KCl and 0.5% (v/v) Triton X-100, then homogenized with a glass–Teflon homogenizer at 100 rpm for 5 min. The homogenate is stirred for 2 hr and centrifuged at 8000 $g$ for 1 hr and the supernatant and pellet recovered separately. The pellet is homogenized again in the same volume of buffer, stirred overnight, and centrifuged at 8000 $g$ for 1 hr. The pellet and supernatant are collected separately and the homogenization repeated. The supernatants are pooled. Protein (4200 mg) is recovered containing 4200 U of enzyme.

*Step 2: Ammonium Sulfate Precipitation.* Proteins in the supernatant are precipitated by the addition of 390 mg/ml ammonium sulfate for a minimum of 1.5 hr and then collected by centrifugation at 17,000 $g$ for 60 min. The precipitate is resuspended in 2 times the minimum volume of 0.1 $M$ sodium cacodylate buffer, pH 7.4, containing 1 $M$ KCl and dialyzed against two changes of 35 volumes each of the same buffer for a total of 20 hr. Protein (2256 mg) is recovered containing 2797 U of enzyme at a specific activity of 1.24 U/mg, giving a 1.24-fold purification.

*Step 3: Concanavalin A-Ultrogel Chromatography.* The sample is applied to a 1.6 × 10 cm column of concanavalin A-Ultrogel, equilibrated with the dialysis buffer, at a flow rate of 5 ml/hr. The column is washed with 200 ml of equilibration buffer, and proteins are eluted with 500 ml of the equilibration buffer to which has been added 0.3 $M$ α-methyl-D-mannoside (prefiltered through a Whatman No. 1 filter). Three-milliliter fractions are collected. Alternate tubes are assayed for enzyme activity, and peak activity fractions are pooled and dialyzed against two changes of 35 volumes each of 0.1 $M$ sodium cacodylate buffer, pH 7.4, containing 0.1 $M$ KCl for a total of 20 hr. Protein (73 mg) is recovered containing 1234 U of enzyme at a specific activity of 16.9 U/mg, giving a 16.9-fold purification.

*Step 4: Heparin-Sepharose Chromatography.* The sample is applied to 1 × 10 cm column of heparin-Sepharose 6B, equilibrated in the same buffer, at a flow rate of 10 ml/hr. The column is washed with 50 ml of equilibration buffer, and the N-peptidase is eluted with a linear KCl gradient between 0.1 and 1 M KCl at the same flow rate. The gradient is prepared with 100 ml of equilibration buffer and 100 ml of the buffer containing 1 M KCl. Two-milliliter fractions are collected. Highest activity fractions are pooled and dialyzed against two changes of 35 volumes each of sodium cacodylate buffer, pH 7.4, containing 0.1 M KCl for a total of 20 hr. Protein (1.52 mg) is recovered containing 584 U of enzyme at a specific activity of 384 U/mg, giving a 384-fold purification.

*Step 5: pNcollagen-Sepharose 4B Chromatography.* The sample is applied to a 1 × 4 cm column of type III pNcollagen covalently coupled to CNBr-activated Sepharose 4B, equilibrated with dialysis buffer, at a flow rate of 1–2 ml/hr. The column is washed with 10 column volumes of equilibration buffer, and N-peptidase is eluted with 0.1 M cacodylate buffer, pH 7.4, containing 1 M KCl. Peak activity fractions are pooled, the N-peptidase is precipitated by the addition of 390 mg/ml ammonium sulfate for a minimum of 1.5 hr and the material is collected by centrifugation at 17,000 g for 60 min. The pellet is resuspended in a minimum volume of 0.1 M sodium cacodylate buffer, pH 7.4, and dialyzed against two changes of 35 volumes each of the same buffer for a total of 12 hr. Protein (0.055 mg) was recovered containing 319 U of enzyme at a specific activity of 5800 U/mg, giving a 5800-fold purification.

*Storage.* As a general rule the enzyme is kept frozen at −70°C in 0.5-ml volumes. Working stocks of the enzyme can be kept at −20°C.

### Acknowledgments

The authors thank the Wellcome Trust, Nuffield Foundation (Oliver Bird Fund), and the Osteogenesis Imperfecta Foundation (USA) for supporting research carried out in their own laboratory. K.E.K. is a Senior Research Fellow in Basic Biomedical Science with the Wellcome Trust.

# [50] Procollagen C-Peptidase: Procollagen C-Proteinase

*By* KARL E. KADLER and ROD B. WATSON

## Introduction

Procollagen C-peptidase cleaves the carboxyl propeptides of procollagen, the soluble precursor of collagen that self-assembles into the fibrils that are the major source of mechanical strength of connective tissues.

Removal of the carboxyl propeptides lowers the solubility of procollagen by at least 10,000-fold and initiates the self-assembly of the fibrils. This chapter describes the work that led up to the identification of the C-peptidase and its purification from chick and mouse sources. It also describes in detail the purification of the peptidase from chick embryo leg tendons,[1] which by any standards was a tour de force in enzyme purification and characterization.

*Reactions catalyzed by procollagen C-peptidase*

Type I/II/III procollagen → type I/II/III pNcollagen[2] + C-propeptides
Type I/II/III pCcollagen → type I/II/III collagen + C-propeptides

Collagen is a major structural protein in animals. It occurs as the cross-striated fibrils and larger fibers and fiber bundles that form a molecular scaffold in the extracellular matrix of connective tissues (for review, see Kielty et al.[3]). More than 17 genetically distinct types of collagen occur, the most abundant of which are the type I and III collagens that occur as fibrils in tendon, ligament, skin, and bone and the type II collagen that occurs as fibrils in cartilage and vitreous humor. Collagens of types I, II, and III are produced as precursor procollagens. The procollagens comprise three polypeptide chains each having a C-propeptide, a domain of repeating Gly-X-Y triplets (in which X and Y can be any residue but are frequently proline and hydroxyproline, respectively), and an N-propeptide that itself contains repeats of Gly-X-Y triplets (for review, see Prockop et al.[4]). Type I procollagen comprises two proα1(I) chains and one proα2(I) chain, whereas types II and III are both homotrimers consisting of [proα1(II)]₃ and [pro-α1(III)]₃, respectively. Conversion of procollagen to collagen is fundamental to the synthesis of the collagen that self-assembles into fibrils. Removal of the N-propeptide is catalyzed by procollagen N-peptidase, and removal of the C-propeptide is catalyzed by procollagen C-peptidase (Fig. 1).

[1] Y. Hojima, M. van Der Rest, and D. J. Prockop, *J. Biol. Chem.* **260**, 15996 (1985).

[2] N-, Amino; C-, carboxyl; proα chain, α chain of procollagen; α chain, α chain of collagen; pCcollagen, an intermediate in the processing of procollagen to collagen containing the carboxyl propeptides but not the amino propeptides; pNcollagen, an intermediate in the processing of procollagen to collagen containing the amino propeptides but not the carboxyl propeptides; pCα chain, α chain of pCcollagen, pNα chain, α chain of pNcollagen; EDTA, ethylenediaminetetraacetic acid; Brij 35, polyoxyethylene lauryl ether; Tris, tris(hydroxymethyl)aminomethane; SDS, sodium dodecyl sulfate; PAGE, polyacrylamide gel electrophoresis.

[3] C. M. Kielty, I. Hopkinson, and M. E. Grant, *in* "Connective Tissue and Its Heritable Disorders: Molecular, Genetic and Medical Aspects" (P. M. Royce and B. Steinmann, eds.), p. 103. Wiley–Liss, New York, 1993.

[4] D. J. Prockop, K. I. Kivirikko, L. Tuderman, and N. A. Guzman, *N. Engl. J. Med.* **301**, 13 (1979).

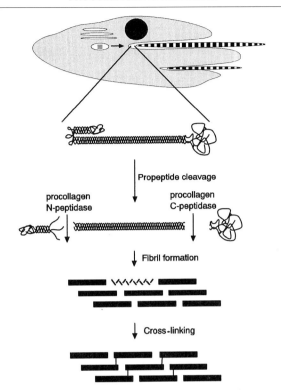

FIG. 1. Cleavage of types I, II, and III procollagen by the procollagen *N*- and *C*-peptidases to collagen. Procollagen consists of a central 300-nm-long triple-helical region (comprising three chains each with ~1000 residues) flanked by a trimeric globular *C*-propeptide domain (right-hand side) and a trimeric *N*-propeptide domain (left-hand side). The procollagen molecule has a molecular mass of 450 kDa. The triple-helical region accounts for 300 kDa, the *C*-propeptide domain around 100 kDa, and the *N*-propeptide domain approximately 45 kDa. Procollagen is secreted from cells and is converted to collagen by the removal of the *N*- and *C*-propeptides by procollagen *N*-peptidase and the procollagen *C*-peptidase, respectively. The collagen generated in the reaction spontaneously self-assembles into cross-striated fibrils that occur in the extracellular matrix of connective tissues.

Evidence for the existence of a specific procollagen *C*-peptidase first came from observations of human and mouse fibroblast cultures in which *C*-propeptides of type I procollagen are rapidly cleaved at neutral pH.[5,6] Similar activity was seen in cultures of chick calvaria,[7,8] chick tendon fibro-

[5] B. D. Goldberg, M. B. Taubman, and A. Radin, *Cell* (*Cambridge, Mass.*) **4,** 45 (1975).

[6] E. Kessler and B. Goldberg, *Anal. Biochem.* **86,** 463 (1978).

[7] L. I. Fessler, N. P. Morris, and J. H. Fessler, *Proc. Natl. Acad. Sci. U.S.A.* **72,** 4905 (1975).

[8] J. M. Davidson, L. S. McEneany, and P. Bornstein, *Eur. J. Biochem.* **81,** 349 (1977).

blasts,[9–11] and chick embryo cartilage cells[12] and in extracts of chick calvaria.[13] There was initial skepticism over the existence of a true $C$-peptidase because acidic, cathepsin-like, proteinases were shown to cleave the $C$-propeptides from procollagen.[14,15] However, cathepsins do not cleave at the physiological site recognized at the C end of the collagen molecule, and they are only weakly active at neutral pH. The best candidate molecule for the $C$-peptidase was identified and purified to apparent homogeneity from chick leg tendon culture medium.[1] It acts at neutral pH, it is a neutral metalloproteinase (and therefore similar to other extracellular matrix proteinases such as the procollagen $N$-peptidases), and it specifically cleaves the –X–Asp– where X bonds that mark the junction between the $C$-telopeptides and the $C$-propeptides, in the pro$\alpha$1(I) and pro$\alpha$2(I) chains of type I procollagen, the pro$\alpha$1(II) chains of type II procollagen, and in the pro$\alpha$1(III) chains of type III procollagen. Site-directed substitution of the conserved Asp at the cleavage site in the pro$\alpha$2(I) chain blocked cleavage by $C$-proteinase *in vitro*.[16] Ala-Asp bonds are found at the junction between the $C$-telopeptides and the $C$-propeptides in the chains for human pro$\alpha$1(V) and human pro$\alpha$1(XI) (the sequences are reviewed by Ayad *et al.*[17]), and presumably the same procollagen $C$-peptidase may cleave the $C$-propeptides of types V and XI procollagen.

Two forms of the procollagen $C$-peptidase are obtained from chick tissues and have been termed EA and EB. Both forms specifically cleave the $C$-propeptides from type I procollagen with equal efficiency.[1] The two forms of the enzyme have apparent molecular masses of 110 kDa for the E-A form and 95 kDa for the E-B form as determined by gel-filtration chromatography. It is not known how E-A and E-B are related to one another. In general, the two forms occur in a 2 : 1 ratio of E-A to E-B; the ratio is flipped during long incubations of the chick leg tendons. E-B may, therefore, be a partially degraded form of E-A or may be a processed form of E-A. All subsequent reference to the $C$-peptidase is equally applicable to both E-A and E-B forms unless stated otherwise.

[9] J. Uitto and J. R. Lichtenstein, *Biochem. Biophys. Res. Commun.* **71,** 60 (1976).

[10] D. Duskin, M. J. Davidson, and P. Bornstein, *Arch. Biochem. Biophys.* **185,** 326 (1978).

[11] M. K. K. Leung, L. I. Fessler, D. B. Greenberg, and J. H. Fessler, *J. Biol. Chem.* **254,** 224 (1979).

[12] Uitto, J. *Biochemistry* **16,** 3421 (1977).

[13] F. K. Njieha, T. Morikawa, L. Tuderman, and D. J. Prockop, *Biochemistry* **21,** 757 (1982).

[14] J. M. Davidson, L. S. G. McEneany, and P. Bornstein, *Eur. J. Biochem.* **100,** 551 (1979).

[15] D. L. Helseth and A. Veis, *Proc. Natl. Acad. Sci. U.S.A.* **81,** 3302 (1984).

[16] S. T. Lee, E. Kessler, and D. S. Greenspan, *J. Biol. Chem.* **265,** 21992 (1990).

[17] S. Ayad, R. P. Boot-Handford, M. J. Humphries, K. E. Kadler, and A. Shuttleworth, "The Extracellular Matrix Facts Book." Academic Press, London, 1993.

The procollagen C-peptidase is highly pH dependent, with an optimum of pH 8.0–8.5. No activity was seen at pH 6 or below.[1] Approximately 5 to 10 m$M$ CaCl$_2$ was required for optimum activity, and calcium was found to have a stabilizing affect on the enzyme at physiological temperatures. The $K_m$ for the enzyme was 75–96 n$M$, and the $k_{cat}$ was 32–41 h$^{-1}$, depending on the method of analysis. The activation energy of the enzyme is 21,000 cal mol$^{-1}$. The enzyme does not appear to require a native conformation of the substrate for activity and will cleave heat-denatured procollagens as effectively as native procollagens. A C-peptidase secreted by human cells in culture cleaves a recombinant protein of proα1(III) at the –Gly–Asp– bond between the C-telopeptide and the C-propeptide (S. Harris, R. B. Watson, N. Bulleid, and K. Kadler, unpublished observations). A variety of divalent metal ion chelators, such as EDTA, inhibit the enzyme completely. 1,10-Phenanthroline was the most effective inhibitor tested.[1]

A procollagen C-peptidase has been isolated from mouse fibroblast cultures.[18] The molecular mass of the active enzyme was 125 ± 5 kDa in gel filtration (compared with 110 kDa for the chick enzyme). The pH dependence of the enzyme and the substrate specificity are the same as those for the chick enzyme, with the same specific -Ala-Asp- bond being cleaved in proα1(I) and proα2(I) chains of type I procollagen. Complete inhibition of the enzyme was achieved by metal chelators but not by inhibitors of serine and cysteine proteinases.

An enhancer protein from mouse fibroblasts that increases the activity of C-peptidase has also been described.[19–21] This 55-kDa glycoprotein enhances the activity of C-peptidase on type I procollagen by about 4-fold, increasing the apparent $K_m$ and $V_{max}$ values 16- and 20-fold, respectively.

Nomenclature

The enzyme that cleaves the carboxyl propeptides of type I procollagen has been referred to in the literature as procollagen C-proteinase, procollagen peptidase, procollagen carboxy-terminal proteinase, carboxy-terminal endopeptidase, and carboxyl procollagen peptidase. Historically, the enzyme was called a peptidase, presumably because it was thought to act on

[18] E. Kessler, R. Adar, B. Goldberg, and R. Niece, *Collagen Relat. Res.* **6,** 249 (1986).
[19] R. Adar, E. Kessler, and B. Goldberg, *Collagen Relat. Res.* **6,** 267 (1986).
[20] E. Kessler and R. Adar, *Eur. J. Biochem.* **186,** 115 (1989).
[21] E. Kessler, A. P. Mould, and D. J. S. Hulmes, *Biochem. Biophys. Res. Commun.* **173,** 81 (1990).

the telopeptides of type I collagen and was similar in this regard to the procollagen aminopeptidase. When it became clear that the enzyme cleaved a larger procollagen molecule to a smaller intermediate by specific enzymatic cleavage of internal bonds, the term proteinase was adopted. In current literature the enzyme is frequently referred to as procollagen *C*-proteinase. However, modern nomenclature states that a peptide hydrolase is a peptidase. We tentatively refer to the enzyme that cleaves the *C*-propeptides of types I, II, and III procollagen (and possibly types V and XI) as the procollagen *C*-peptidase in agreement with the guidelines of *Enzyme Nomenclature* (1992) and *Biochemical Nomenclature and Related Documents: A Compendium* (1992). The procollagen *C*-peptidase has not been assigned an EC number at this time.

Assay Method

To a 1.5-ml microcentrifuge tube add 1 $\mu$g [14]C-labeled type I procollagen (the substrate may be prepared as described previously[22,23]), 5 m$M$ CaCl$_2$, and 0.01% Brij 35 contained within a volume of 10 $\mu$l. Up to 90 $\mu$l of the solution to be assayed for *C*-peptidase activity is added. The sodium chloride and Tris buffer concentrations should be adjusted to 0.12–0.15 $M$ and 0.05–0.2 $M$, respectively and the final reaction volume to 100 $\mu$l. For gel assays, samples are incubated at 34° for times of 0, 0.5, 1, 2, and 4 hr. At the end of the incubation period, the reaction is stopped by the addition of 10 $\mu$l of TE buffer (1 $M$ Tris-HCl, 250 m$M$ EDTA, 0.2% NaN$_3$, pH 7.4). Samples are analyzed by sodium dodecyl sulfate–polyacrylamide gel electrophoresis (SDS–PAGE) in the presence of a reducing agent, and the procollagen bands are displayed either by fluorography and exposure of the dried gels to film or by phosphoimaging. The initial rate of cleavage of procollagen to pNcollagen is estimated by plotting the ratio [pN$\alpha$1(I) chains]/[pro$\alpha$1(I) + pN$\alpha$1(I) chains] against the time of incubation.

For rapid assays, the samples are incubated at 34° for 1 and 3 hr. The reaction is stopped by adding 100 $\mu$l of 0.2× TE buffer and 100 $\mu$l of 78% ethanol that is at $-20°$. The solution is vortexed repeatedly, held on ice (0°) for 45 min, and centrifuged at 15,000 $g$ for 15 min, and 150 $\mu$l of the supernatant is taken for liquid scintillation counting. The units of activity of the *C*-peptidase preparation (U/ml) are readily calculated from the disintegrations per minute (dpm) of free *C*-propeptides cleaved by consideration of the relative molecular weights of the *C*-propeptides (100 kDa) and the procollagen (450 kDa), the specific activity of the [14]C-labeled

[22] L. Tuderman and D. J. Prockop, *Eur. J. Biochem.* **125,** 545 (1982).
[23] L. Tuderman, K. I. Kivirikko, and D. J. Prockop, *Biochemistry* **17,** 2948 (1978).

procollagen (dpm/$\mu$g), the time of incubation (hr), and the volumes of sample used.

*Definition of Enzyme Unit.* One unit of activity is the amount that converts 1 $\mu$g of procollagen to pNcollagen in 1 hr at 34° at an initial substrate concentration of 10 $\mu$g/ml.

## Purification of Procollagen C-Peptidase from Chick Embryo Tendons

The methods described are those developed by Hojima *et al.*,[1] with minor modifications, for the enzyme in chick leg tendon. With the exception of step 1, all procedures are performed at 4° or on ice. Plastic containers are used throughout, and glass columns are siliconized with Sigmacote (Sigma, St. Louis, MO). All buffers contain 0.02% (w/v) $NaN_3$ as a bacteriostat, and the pH is adjusted at room temperature. The recovery and purity figures obtained in our laboratory agree very well with those obtained by Hojima *et al.*[1] The values stated below are taken from Hojima *et al.*[1] Ten-milliliter fractions are collected at all steps.

*Step 1: Organ Culture of Leg Tendons.* Leg tendons are removed from approximately 50 dozen 17-day-old chick embryos. The tendons (~25–30 g) are thoroughly washed in Kreb's glucose solution to remove all traces of blood and then cultured for 3 × 12 hr at 30° in a total of 1000 ml Dulbecco's modified Eagle's medium (DMEM) containing 100 units/ml penicillin and 100 $\mu$g/ml streptomycin, with gentle shaking. The culture is performed in two 1-liter Erlenmeyer flasks, each containing half of the tendons, which are gassed for several minutes with a stream of 5% $CO_2$/95% air prior to culture and tightly sealed. The culture medium is harvested by gentle centrifugation, to avoid damaging the tendons, and stored at −20°.

*Step 2: Preparation of Culture Medium.* Frozen culture medium from a total of 300–400 dozen embryos is thawed at 4° and pooled (6–8 liters). The pH of the culture medium is adjusted to 7.5 by the addition of concentrated hydrochloric acid. Solid Tris base is added to a final concentration of 50 m$M$ and solid NaCl added to a final concentration of 0.3 $M$. $NaN_3$ is added to a concentration of 0.02%, and the solution is adjusted to pH 8.0. The sample is centrifuged at 17,000 $g$ for 30 min at 4° and the supernatant collected. Protein (4060 mg) is recovered containing 5570 units of enzyme at a specific activity of 1.37 units/mg.

*Step 3: Green A Matrex Gel Chromatography.* The sample is applied to a 5 × 10 cm column of Green A Matrex gel (Amicon, Danvers, MA), equilibrated with 50 m$M$ Tris-HCl buffer, pH 7.5, containing 0.3 $M$ NaCl, at a flow rate of 100 ml/hr. The column is washed with 500 ml of equilibration buffer followed by 500 ml of 50 m$M$ Tris-HCl buffer, pH 7.5, containing

1 $M$ NaCl, to remove adsorbed impurities. The $C$-peptidase is eluted from the column with 50 m$M$ Tris-HCl buffer, pH 7.5, containing 3 $M$ NaCl and 2 $M$ urea, at the same flow rate. Fractions containing enzyme activity, as determined by the rapid assay, are pooled and pressure concentrated to approximately 30 ml using an Amicon YM30 membrane and a 200-ml chamber. The sample is allowed to stir for 1 hr after concentration is complete, and the membrane is washed with 10 ml of the elution buffer for a further 2 hr. The samples are combined. The sample is diluted with an equal volume of 0.15 $M$ Tris-HCl buffer, pH 7.5, containing 2 m$M$ CaCl$_2$ to give final concentrations of 0.1 $M$ Tris-HCl, 1.5 $M$ NaCl, 1 $M$ urea, and 1 m$M$ CaCl$_2$ and a final volume of approximately 80 ml. The sample is centrifuged at 48,000 $g$ for 30 min and the supernatant recovered. Protein (233 mg) is recovered containing 5590 units of enzyme at a specific activity of 24 units/mg, giving a purification of 18-fold.

*Step 4: Concanavalin A-Sepharose Chromatography.* The sample is applied to a 1.5 × 10 cm column of concanavalin A-Sepharose, equilibrated with 0.1 $M$ Tris-HCl buffer, pH 7.5, containing 1.5 $M$ NaCl, 1 $M$ urea, and 1 m$M$ CaCl$_2$, at a flow rate of 20 ml/hr. The column is washed with 100 ml of the equilibration buffer and the adsorbed protein eluted with 500 ml of equilibration buffer to which has been added 1 $M$ $\alpha$-methyl-D-mannoside and the pH adjusted to 7.0. The eluted sample is pressure concentrated to about 15 ml using an Amicon YM30 membrane and a 200-ml chamber. The sample is allowed to stir for 1 hr after concentration is complete, and the membrane is washed for a further 2 hr with 2 × 2 ml of the dialysis buffer used in the next step. The samples are combined, Brij 35 is added to a final concentration of 0.005%, and the sample is dialyzed against two changes of 2 liters each of 0.3 $M$ Tris-HCl buffer, pH 7.5, containing 50 m$M$ NaCl and 0.005% Brij 35 for a total of 20 hr. The dialyzed sample is centrifuged at 48,000 $g$ for 30 min and the supernatant collected. Protein (37.2 mg) is recovered containing 5630 units of enzyme at a specific activity of 151 units/mg, giving a 110-fold purification.

*Step 5: Heparin-Sepharose Chromatography.* The sample is applied to a 2.5 × 8 cm column of heparin-Sepharose, equilibrated with the dialysis buffer, at 20 ml/hr, and the column is washed with 250 ml of the equilibration buffer. The column is further washed with 500 ml of 0.3 $M$ Tris-HCl buffer, pH 7.5, containing 0.2 $M$ NaCl and 0.005% Brij 35 and then with 500 ml of 0.3 $M$ Tris-HCl buffer, pH 7.5, containing 1.5 $M$ NaCl, 1 $M$ urea, and 0.005% Brij 35. Rapid assay of alternate fractions shows two peaks of enzyme activity. The major peak (containing E-A) appears just after the void volume, and fractions containing activity are pooled. The second peak (containing E-B), which appears during the second wash, is also pooled and stored separately. It is not known how the enzymes in the first and second activity peaks differ, but only fractions from the first activity peak are purified further. The final wash is to clean the column for reuse.

Fractions from the first activity peak are pooled, and NaCl, urea, α-methyl-D-mannoside, and $CaCl_2$ are added to final concentrations of 1 $M$, 1 $M$, 0.2 $M$, and 5 m$M$, respectively. The sample is concentrated to approximately 2 ml using an Amicon YM30 membrane and a 50-ml chamber. The sample is allowed to stir for 1 hr after concentration is complete, and the membrane is washed for a further 2 hr with 2 × 1 ml of the final wash buffer used on the column. The samples are combined. Protein (6.11 mg) is recovered containing 4030 units of enzyme at a specific activity of 659 units/mg, representing a 480-fold purification.

*Step 6: Sephacryl S-300 Gel Filtration.* The sample is applied to a 2.5 × 112 cm column of Sephacryl S-300, equilibrated with 0.3 $M$ Tris-HCl buffer, pH 7.5, containing 1.5 $M$ NaCl, 5 m$M$ $CaCl_2$, and 0.005% Brij 35 at a flow rate of 15 ml/hr. The rapid assay is used to detect the *C*-peptidase activity, and tubes containing most of the enzyme are pooled. Urea is added to a final concentration of 1 $M$ and α-methyl-D-mannoside added to a final concentration of 0.2 $M$. The sample is concentrated to approximately 2 ml using an Amicon YM30 membrane and a 50-ml chamber. The sample is allowed to stir for 1 hr after concentration is complete, and the membrane is washed for a further 2 hr with 2 × 0.5 ml of the final wash buffer used on the column in step 5. The samples are combined. Protein (0.438 mg) is recovered containing 2680 units of enzyme at a specific activity of 6130 units/mg, giving a 4470-fold purification. At this point the *C*-peptidase is sufficiently pure to cleave pCcollagen in a cell-free system of generating collagen fibrils.

*Step 7: Sephacryl S-200 Gel Filtration.* The sample is applied to a 2.5 × 100 cm column of Sephacryl S-200, equilibrated with 0.3 $M$ Tris-HCl buffer, pH 7.5, containing 1 $M$ NaCl, 5 m$M$ $CaCl_2$, and 0.005% Brij 35 at a flow rate of 15 ml/hr. The rapid assay is used to detect the *C*-peptidase activity, and tubes containing most of the enzyme are pooled. Protein (0.048 mg) is recovered containing 1380 units of enzyme at a specific activity of 29,100 units/mg, giving a 21,200-fold purification.

*Storage.* The preparations should be kept at −70°. Working stocks of the enzyme can be kept at −20°.

## Purification of Procollagen *C*-Peptidase from Mouse Fibroblast Culture Medium

The methods described are those developed by Kessler *et al.*,[18] with minor modifications, for the enzyme from mouse fibroblast culture medium.

With the exception of step 1, all procedures are performed at 4° or on ice. Plastic containers are used throughout, and glass columns siliconized with Sigmacote (Sigma). All buffers contain 0.02% $NaN_3$ as a bacteriostat, and the pH was adjusted at room temperature. The values stated below are taken from Kessler et al.[18]

*Step 1: Culture of 3T6 Mouse Fibroblasts.* Mouse 3T6 fibroblasts are grown to confluency in 30 850-$cm^2$ roller bottles in DMEM supplemented with 10% fetal calf serum and 75 $\mu$g/ml sodium ascorbate. The confluent cells are washed twice with phosphate-buffered saline and incubated with serum-free medium for 4 hr. The medium is removed and the cells incubated in fresh serum-free medium for 24 hr. The medium is collected and the cells incubated for 24 hr in serum-supplemented medium followed by a further serum-free incubation cycle. In total the cells go through four rounds of serum-free incubations. Protein (419 mg) is recovered containing 419 units of enzyme at a specific activity of 1 unit/mg.

*Step 2: Ammonium Sulfate Precipitation.* Immediately after collection each portion of the serum-free medium is cooled on ice for 1 hr, and proteins in the medium are precipitated with ammonium sulfate (229 mg/ml) for 16 hr and collected by centrifugation at 17,000 $g$ for 60 min. The precipitate is resuspended in a minimum volume of 50 m$M$ Tris-HCl buffer, pH 7.5, containing 0.15 $M$ NaCl and 5 m$M$ $CaCl_2$ and dialyzed against two changes of 35 volumes each of the same buffer for a total of 20 hr. The dialyzed samples are stored at $-20°$. Protein recovery is not assessed at this stage.

*Step 3: Ammonium Sulfate Fractionation.* The dialyzed samples from all four rounds of cell incubations are thawed and pooled. Ammonium sulfate is added to 30% (176 mg/ml) for a minimum of 1.5 hr and the sample centrifuged at 17,000 $g$ for 60 min. The supernatant is recovered, ammonium sulfate added to a final concentration of 45% (an additional 86 mg/ml) for a minimum of 1.5 hr, and precipitated proteins collected by centrifugation at 17,000 $g$ for 60 min. The precipitate is resuspended in a minimum volume of 50 m$M$ Tris-HCl buffer, pH 7.5, containing 0.15 $M$ NaCl and 5 m$M$ $CaCl_2$ and dialyzed against two changes of 35 volumes each of the same buffer for a total of 20 hr. Protein (102 mg) is recovered containing 206.5 units of enzyme at a specific activity of 2 units/mg, representing a 2-fold purification.

*Step 4: Sephadex G-150 Chromatography.* Solid guanidine hydrochloride is added to the dialyzed sample to a concentration of 0.5 $M$. After standing for 3 hr the sample is applied to a 2.2 × 130 cm column of Sephadex G-150, equilibrated in 50 m$M$ Tris-HCl buffer, pH 7.5, containing 0.15 $M$ NaCl and 5 m$M$ $CaCl_2$ at a flow rate of 16 ml/hr. The column is eluted with the equilibration buffer. Ten-milliliter fractions are collected, and all fractions are dialyzed against two changes of 500 ml each of 50 m$M$ Tris-HCl buffer, pH 7.5, containing 0.15 $M$ NaCl and 5 m$M$ $CaCl_2$ for a total of 20 hr and assayed for enzyme activity. Peak activity fractions are pooled and concentrated by ammonium sulfate precipitation; proteins are precipitated by the addition of ammonium sulfate (229 mg/ml) for 16 hr and collected by centrifugation at 17,000 $g$ for 60 min. The precipitate is resuspended in a minimum volume of 50 m$M$ HEPES buffer, pH 7.5, containing 0.15 $M$ NaCl, 5 m$M$ CaCl$_2$, and 0.05% Brij 35 and dialyzed against two changes of 35 volumes each of the same buffer for a total of 20 hr. Protein (8.2 mg) is recovered containing 93.3 units of enzyme at a specific activity of 11.4 units/mg, giving a purification of 11.4-fold.

*Step 5: Carboxyl Propeptide-Sepharose 4B Chromatography.* The sample is applied to a 1 × 5 cm column of the carboxyl propeptide of type I procollagen covalently coupled to CNBr-activated Sepharose 4B (the substrate may be prepared as described earlier[24]), equilibrated in 50 m$M$ HEPES buffer, pH 7.5, containing 0.15 $M$ NaCl, 5 m$M$ CaCl$_2$, and 0.05% Brij 35, and allowed to adsorb for 18 hr. The column is washed with 4 column volumes of the equilibration buffer, and the enzyme is eluted with the equilibration buffer to which has been added 1 $M$ NaCl and 1 $M$ guanidine hydrochloride, at a flow rate of 12 ml/hr. One-milliliter fractions are collected and are dialyzed against two changes of 35 volumes each of 50 m$M$ Tris-HCl buffer, pH 7.5, containing 0.15 $M$ NaCl and 5 m$M$ CaCl$_2$ for a total of 20 hr and assayed for enzyme activity. Peak activity fractions are pooled. Protein (0.2 mg) is recovered containing 19.6 units of enzyme at a specific activity of 98 units/mg, giving a 98-fold purification.

*Storage.* The preparations should be kept at $-70°$. Working stocks of the enzyme can be kept at $-20°$.

## Acknowledgments

The authors thank the Wellcome Trust for supporting research carried out in their own laboratory. K.E.K. is a Senior Research Fellow in Basic Biomedical Science with the Wellcome Trust.

---

[24] B. D. Goldberg, R. G. Phelps, E. Kessler, M. J. Klein, and M. B. Taubman, *Collagen Relat. Res.* **5,** 393 (1985).

## [51] Peptidyl-Asp Metalloendopeptidase

*By* Marie-Luise Hagmann, Ursula Geuss, Stephan Fischer, and Georg-Burkhard Kresse

### Introduction

Peptidyl-Asp Metalloendopeptidase (EC 3.4.24.33, also called endopeptidase Asp-N) is an extracellular metalloproteinase secreted by a *Pseudomonas fragi* mutant. This chapter summarizes the purification and properties of the enzyme.

### Assay Methods

The activity of endopeptidase Asp-N is assayed by the procedure of Kunitz,[1] using casein or hemoglobin as the substrate. A unit of activity is defined as the amount of enzyme which yields a 0.001 optical density unit of change per minute at 280 nm.[1] A detailed procedure is described by Laskowski[2] in an earlier volume of this series. With endopeptidase Asp-N the assay is linear up to an absorbance of 0.8.[1]

Alternatively, activity can be measured with Azocoll[3] or resorufin-labeled casein as the substrate.[4] Fifty microliters of 0.4% (w/v) resorufin-labeled casein in distilled water and 50 $\mu$l of 0.2 $M$ Tris buffer, pH 7.8, 20 m$M$ CaCl$_2$, are incubated with 100 $\mu$l endopeptidase solution for 15 min at 37°. A blank containing 100 $\mu$l of distilled water instead of enzyme solution is incubated in parallel. The reaction is stopped by the addition of 480 $\mu$l of 5% (w/v) trichloroacetic acid. Precipitation of undigested labeled casein is completed by a further incubation at 37° for 10 min. After a 5-min centrifugation, 400 $\mu$l of the clear supernatant is withdrawn and mixed with 600 $\mu$l of 0.5 $M$ Tris buffer, pH 8.8, and the absorbance at 574 nm is immediately read against the blank. One-tenth microgram of purified endopeptidase Asp-N gives an absorbance difference of 0.09.

Both methods are general proteinase assays. No assay procedure using a specific chromogenic substrate for endopeptidase Asp-N has yet been described. However, the cleavage specificity can be checked with peptide

---

[1] J. Noreau and G. R. Drapeau, *J. Bacteriol.* **140**, 911 (1979).

[2] M. Laskowski, this series, Vol. 2, p. 26.

[3] G. L. Moore, *Anal. Biochem.* **32**, 122 (1969).

[4] E. Schickaneder, U. Geuss, W. Hösel, and H. v. d. Eltz, *Fresenius Z. Anal. Chem.* **330**, 360 (1988).

substrates, for example, glucagon. The assay is carried out as follows: 100 $\mu$g of glucagon, dissolved in distilled water, is incubated with 0.2 $\mu$g of endopeptidase Asp-N in 50 m$M$ sodium phosphate buffer, pH 8.0, for 60 min at 37° (final volume, 100 $\mu$l). A control incubation is performed without the enzyme. Digestion is stopped by heating the incubation mixture for 10 min in a boiling water bath. After addition of 100 $\mu$l of 0.1% trifluoroacetic acid, an aliquot is subjected to high-performance liquid chromatography (HPLC) analysis for separation of the cleavage products. The Hypersil 5-$\mu$m (53 × 4.6 mm) reversed-phase column is run with solvent A, 0.1% (v/v) trifluoroacetic acid in distilled water, and solvent B, 70% (v/v) acetonitrile, 0.1% (v/v) trifluoroacetic acid in distilled water. The gradient from 0 to 100% solvent B is run for 30 min at a flow rate of 0.5 ml/min, with detection at 215 nm. Fractions containing the cleavage products are further characterized by N-terminal sequencing using an automated sequencer. To check for the presence of proteinase impurities prolonged incubation times up to 18 hr may have to be used.

Purification Procedure

Endopeptidase Asp-N was first isolated from the culture supernatant of *Pseudomonas fragi* which, in contrast to the majority of microorganisms, was thought to produce only a single extracellular proteinase.[1,5] However, owing to the low yield of the enzyme from this organism, the properties of the proteinase could only be studied when a derepressed mutant derived from *P. fragi* ATCC 4973 (American Type Culture Collection, Rockville, MD) became available which produces 40 times higher enzyme levels than the parental organism.[1] In these studies, it also became clear that *P. fragi* in fact also produces more than one proteinase species. One of the mutant proteinases was found[6] to possess an unusual cleavage specificity (see below). The following purification procedure was described by the group of Drapeau.[1,6]

*Step 1: Growth of Pseudomonas fragi.* The culture medium used for growing the *P. fragi* mutant consists of 10.0 g/liter hydrolyzed casein (salt-free), 0.5% (w/v) yeast extract, 0.30 g/liter L-tryptophan, 0.10 g/liter cysteine hydrochloride, 0.20 g/liter MgSO$_4$, 1.4 g/liter CaCl$_2$·2H$_2$O, 1.0 g/liter KH$_2$PO$_4$, 0.50 g/liter KNO$_3$, 0.50 g/liter NaCl, and 4.0 mg/liter Zn-EDTA; it is adjusted to pH 7.5 with 1 $M$ NaOH and autoclaved at 120° for 20 min. Flasks of 2000 ml containing 500 ml of medium are inoculatd with 25 ml of an overnight culture grown in brain–heart infusion broth. The flasks are

[5] M. A. Porzio and A. M. Pearson, *Biochim. Biophys. Acta.* **384,** 235 (1975).
[6] G. R. Drapeau, *J. Biol. Chem.* **255,** 839 (1980).

placed on a rotatory shaker for incubation at 17° for 24 to 36 hr until the culture reaches a turbidity of 450 Klett units.

*Step 2: Ammonium Sulfate Fractionation.* The cells are removed by centrifugation, and to the spent medium is added, with stirring, 660 g of $(NH_4)_2SO_4$ per liter. The suspension is stirred for an additional 2 hr at 4°, and the precipitate is collected by centrifugation. The material is suspended in a small volume of 5 m$M$ sodium acetate buffer, pH 3.8, containing 5 m$M$ CaCl$_2$, and dialyzed overnight against the same buffer.

*Step 3: SP-Sephadex Chromatography.* The dialyzate is centrifuged to remove insoluble material and then run on a column of SP-Sephadex C-25 (5 × 20 cm) previously equilibrated with the acetate–calcium buffer. Elution is performed using the same buffer containing a linear gradient of 0 to 0.3 $M$ NaCl followed by a step elution with buffer containing 0.6 $M$ NaCl. Fractions showing proteolytic activity are pooled. Endopeptidase Asp-N activity is recovered in the last peak of activity which elutes from the column. It is rechromatographed on a second column of SP-Sephadex to ensure removal of possible contamination from other proteinases secreted by the mutant. Finally, the sample is neutralized by adding 1 $M$ Tris-HCl buffer, pH 7.5, and lyophilized.

The enzyme is available commercially in highly purified form (for use in protein sequencing) as endoproteinase Asp-N from several suppliers.

## Properties

*Stability of Enzyme.* Endopeptidase Asp-N lyophilized in the presence of Tris-HCl buffer, pH 7.5, is stable for at least 2 years when stored at 4°. After reconstitution of the lyophilizate with distilled water (2 $\mu$g enzyme with 50 $\mu$l water, giving a buffer concentration of 10 m$M$ Tris-HCl), the resulting enzyme solution is stable for 1 week at 4° without detectable loss of activity.

Endopeptidase Asp-N is frequently used for digestion of proteins, which often has to be performed under denaturing conditions. Therefore, stability of the proteinase in the presence of denaturants has been studied.[7] Its stability in the presence of several protein-denaturing additives is shown in Table I; endopeptidase Asp-N is also active in up to 2 $M$ urea.[6]

*Purity. Pseudomonas fragi* endopeptidase purified as described above gave a single band in sodium dodecyl sulfate–polyacrylamide gel electrophoresis (SDS–PAGE).[1] However, there are indications that the preparation may still contain nonprotein impurities.[8]

---

[7] U. Geuss, unpublished (1990).
[8] G. R. Drapeau, unpublished (1986).

TABLE I

INFLUENCE OF DENATURANTS ON ENDOPEPTIDASE Asp-N[a]

| Denaturing agent | Concentration | Relative enzyme activity (%) |
|---|---|---|
| Control | — | 100 |
| Sodium dodecyl sulfate | 0.001% (w/v) | 113 |
| | 0.01% (w/v) | 122 |
| | 0.1% (w/v) | 10 |
| Urea | 0.1 $M$ | 100 |
| | 0.5 $M$ | 108 |
| | 1.0 $M$ | 105 |
| Guanidine hydrochloride | 0.1 $M$ | 100 |
| | 0.5 $M$ | 85 |
| | 1.0 $M$ | 80 |
| Acetonitrile | 1% (v/v) | 90 |
| | 5% (v/v) | 115 |
| | 10% (v/v) | 125 |

[a] Endopeptidase Asp-N was incubated, in 25 m$M$ sodium phosphate buffer, pH 7.8, for 6 hr at 25°, in the presence of the denaturants listed. Activity was assayed with azocoll as the substrate.

In protein sequencing, even traces of impurities absorbing in the 206–230 nm range used for peptide detection in HPLC fingerprinting may interfere. Therefore, endopeptidase Asp-N may have to be further purified by ion-exchange chromatography for this application.

*Molecular Weight.* The molecular weight of the wild-type *P. fragi* proteinase has been reported to be 52,000[1] as determined by SDS–PAGE. However, in more recent studies[7,9] with highly purified endopeptidase Asp-N from the mutant strain, the molecular weight was found to be 27,000 in SDS–PAGE as well as HPLC gel filtration (TSK G 2000 SW column). The reason for the discrepancy is not clear but may be indicative of a difference between the wild-type and mutant proteinases. The finding that the mutant proteinase shows a changed cleavage pattern toward insulin[1] supports this assumption.

*Amino Acid Sequence.* Only the partial amino acid sequences shown in Table II are available at present.[10] These peptides may occur as autolysis products during prolonged incubations with the endopeptidase.

*pH Dependence.* With casein, hemoglobin, or Azocoll as the substrate, maximum activity is found in the range pH 7.0–8.5.[6,7] More than 75% of

[9] J. Tschakert, Diploma Thesis, University of Munich (1989).
[10] K. A. Walsh, unpublished (1992).

TABLE II
Partial Amino Acid Sequences of Endopeptidase Asp-N

| Partial sequence[a] | Number of amino acids |
|---|---|
| ... DIATDSSTSPYAYGHGYRYEPATGwRTIMAY NCTRSCPRLNYWSNPNISYdigp ... | 54 |
| ... DCATGYYSFAHEIGHLQsar ... | 20 |
| ... DNQRVLVNTKATIAAFr ... | 17 |
| ... ELARYETTNYTESGSF ... | 16 |
| ... DSIHTSRNTYTAA ... | 13 |
| ... ESNQGYVNSNVGI ... | 13 |
| ... DTDLARFRGTS ... | 11 |

[a] Amino acids in lowercase letters are tentative.

the maximum activity is measured between pH 6.5 and 9.0. There is no significant difference when various buffer systems are used (Tris, phosphate, acetate, or bicarbonate buffers).

*Inhibitors and Activators.* The data obtained from inhibition experiments support the metalloproteinase character of endopeptidase Asp-N. The proteinase is completely inhibited by chelating agents like 1,10-phenanthroline,[1] EDTA, and EGTA (at 0.76–1.25 m$M$ concentration).[9] No inhibition was observed with the serine proteinase inhibitor phenylmethylsulfonyl fluoride (PMSF).[5,9] Inhibition by $\alpha_2$-macroglobulin occurred at a molar ratio (inhibitor : proteinase) of about 18 : 1 when incubated for 1 hr at 37°.[9] 1,10-Phenanthroline-inhibited wild-type proteinase could be reactivated by the addition of $Zn^{2+}$ or $Co^{2+}$, whereas the addition of $Ca^{2+}$ or $Mg^{2+}$ had no effect.[1] However, no reactivation of the mutant enzyme by $Zn^{2+}$ could be obtained after prolonged incubation (1 hr at 37°) with 1,10-phenanthroline,[9] suggesting irreversible modification or conformational alteration of the apoenzyme under these conditions.

*Specificity.* In 1980, it was shown by Drapeau[6] using oxidized pancreatic ribonuclease and myoglobin as substrates, that endopeptidase Asp-N specifically cleaves peptide bonds at the N-terminal side of either aspartic or cysteic acid residues. This specificity was confirmed by Ponstingl *et al.*[11] using $\alpha$-tubulin as well as by Fischer *et al.*[12] with peptides (e.g., glucagon) as the substrate.

[11] H. Ponstingl, G. Maier, M. Little, and E. Krauhs, *in* "Advanced Methods in Protein Microsequence Analysis" (B. Wittmann-Liebold, ed.), p. 316. Springer-Verlag Berlin and Heidelberg, 1986.

[12] S. Fischer, U. Geuss, M. Schäffer, G.-B. Kresse, and G. R. Drapeau, *J. Protein Chem.* **7**, 225 (1988).

Cysteic acid does not normally occur in proteins but is formed from cysteine by oxidation with performic acid, a procedure often used in protein sequencing to eliminate disulfide bridges. Bonds with cysteic acid in P1' can be cleaved by endopeptidase Asp-N, so to restrict cleavage to aspartic residues, blockage of the cysteine residues by pyridylethylation or carboxymethylation rather than oxidation is recommended. No cleavage by endopeptidase Asp-N at such modified residues could be detected.[9]

Studies by Ingrosso *et al.*[13] and Tetaz *et al.*[14] have shown that, under certain conditions, endopeptidase Asp-N possesses the ability to cleave peptide bonds also at the N-terminal side of some (but not all) glutamic acid residues. Although these studies even suggested cleavage at glutamyl residues at a similar rate to cleavage at aspartate residues, later kinetic studies[9,15] with glucagon and insulin A and B chain as the substrates demonstrated a high kinetic preference for aspartic residues: the rate of aspartic specific cleavage, in these cases, was at least 2000-fold higher that the glutamic activity of the endopeptidase. No crucial structure or sequence requirements for hydrolysis could be found.

*Application.* The main application of endopeptidase Asp-N is in the digestion of peptides and proteins prior to amino acid sequence analysis,[11] of which many examples have been published since 1986. Owing to the mostly low amount of aspartic residues within a polypeptide chain, endoproteinase Asp-N is appropriate to generate larger peptide fragments as compared to tryptic or chymotryptic cleavage. In combination with glycosidases, endopeptidase Asp-N may also be used to detect glycosylation sites within a protein molecule[16]: enzymatic removal of Asn-bound carbohydrate results in the unmasking of aspartic acid residues, thus creating new potential cleavage sites for the proteinase. By comparison of the endopeptidase Asp-N cleavage patterns of unmodified and deglycosylated target protein, glycosylation sites can be identified.

*Distribution.* Endopeptidase Asp-N is unique among the known microbial proteinases in its cleavage specificity. There are indications of the presence of similar proteinases in higher organisms, for example, the proteinase cleaving thymosin-$\beta$4 in murine bone marrow cells.[17] Until now, these are the only known proteinases specific for preaspartate cleavage.

[13] D. Ingrosso, A. V. Fowler, J. Bleibaum, and S. Clarke, *Biochem. Biophys. Res. Commun.* **162,** 1528 (1989).

[14] T. Tetaz, J. R. Morrison, J. Andreou, and N. H. Fidge, *Biochem. Int.* **22,** 561 (1990).

[15] U. Geuss, M. Schäffer, J. Tschakert, and G.-B. Kresse, *J. Protein Chem.* **9,** 299 (1990).

[16] C. K. Leonard, M. W. Spellman, L. Riddle, R. J. Harris, J. N. Thomas, and T. J. Gregory, *J. Biol. Chem.* **265,** 10373 (1990).

[17] C. Grillon, K. Rieger, J. Bakala, D. Schott, J.-L. Morgat, E. Hannappel, W. Voelter, and M. Lenfant, *FEBS Lett.* **274,** 30 (1990).

# Author Index

Numbers in parentheses are footnote reference numbers and indicate that an author's work is referred to although the name is not cited in the text.

# Subject Index

globomycin treatment, 171–172
*iap* gene product, 117–118
methionyl aminopeptidase, 220–221
murein endopeptidase, 118–119
protease III, *See* Pitrilysin
*sohB* gene product, 117–118
tail-specific protease, 117–118
*x-ileS-lsp-orf149-lytB* operon, 174
X-Pro aminopeptidase, 213, 220–221
*Escherichia freundii*, proteinase, 395
Esterase, thioester substrate, 16
Ethylenediamine tetraacetic acid
astacin inhibition, 318
binding to transition metals, stability constants, 230
calcium binding, equilibrium constant for, 232–233
magnesium binding, equilibrium constant for, 232
meprin inhibition, 341
metallopeptidase inhibition
mechanism of action, 233–235
time dependence, 234–236
zinc binding, equilibrium constant for, 232
Ethylene glycol bis($\beta$-aminoethyl ether) tetraacetic acid
binding to transition metals, stability constants, 230
calcium binding, equilibrium constant for, 232–233
magnesium binding, equilibrium constant for, 232
zinc binding, equilibrium constant for, 232

## F

Factor I, thioester substrates, 13
Factor VIIa, thioester substrates, 13
subsite mapping studies, 17
Factor IXa, thioester substrates, 13
subsite mapping studies, 17
Factor Xa, active-site titration, benzyl *p*-guanidinothiobenzoate in, 14
Factor XIa, substrate, fluorogenic, 24
Factor XIIa, active-site titration, benzyl *p*-guanidinothiobenzoate in, 14
Factor Bb, thioester substrates, 13
Factor C2, thioester substrates, 13
Factor D, thioester substrates, 13

Fibrinogenase, *Crotalus atrox*, 351, 378
$\alpha$-Fibrinogenase, properties, 194
Fibrolase, southern copperhead snake, properties, 192–194
*Flavobacterium*, creatinase, 221
Fluorescein mono-*p*-guanidinobenzoate, in active-site titration of serine proteinases, 87, 89
Fluorescence quenching, 19
assay methods, principles, 25–27
by resonance energy transfer, 19, 24–25
Fluorogenic substrate
in active-site titration of serine proteinases, 87, 89, 91
amine peptidyl derivatives, 19–23
applications, in proteinase specificity mapping, 35–44
macromolecular, 33–34
peptide
contact-quenched, 19, 23–31
quenched, 19, 23
assay methods, 25–27
calibration, 26
Dabcyl-Edans, 27–28
dansyl–tryptophan, 28
design, 31–32
2,4-dinitrophenyl–*o*-aminobenzoyl, 30
2,4-dinitrophenyl–7-methoxycoumarin, 29–30
2,4-dinitrophenyl–tryptophan, 28–29
3-nitrotyrosine–*o*-aminobenzoyl, 31
pitfalls with, 32–33
precautions with, 32–33
synthetic peptide library approach, 33

## G

Gelatin, degradation, by gelatinases A and B, 482–483
Gelatinase
active-site titration, 100–101, 474, 502–503
gene, 204–205
peptide thioester substrate, 15–16
polypeptide chain structure, 203–204
Gelatinase A
activation, 480–484
active-site titration, 474, 502–503
activity, 470–471

# S

Saccharolysin
  amino acid sequence, 208
  evolution, 552
  homologs, 551, 591–592
  properties, 206–207
*Salmonella*
  aspartyl dipeptidase, 118, 120
  oligopeptidase A, *See* Oligopeptidase A
SCH 39,370
  development, 267
  properties, 268
SCH 32,615, development, 267
Scytalidopepsin B
  amino acid sequence, 151
  and aspergillopepsin II, comparison, 151–
    152
  properties, 109, 116
Scytalidopepsin family
  acidic pH optimum, 107
  properties, 109, 115–116
Serine carboxypeptidase, acidic pH opti-
    mum, 107
Serine peptidase
  assay, 7–9
  peptide thioester substrates, 3–18
    kinetic constants, 10–14
    specificity, 14–15
    thioester substrates, 10–15
Serine proteinase
  active-site titration
    burst, 86–91
      sensitivity, 88–89
    with tritiated diisopropyl fluorophos-
      phate, 92
  substrate turnover, kinetics, 86
Serralysin
  assay
    caseinolysis method, 397–398
    fluorescence polarization method, 398–
      399
    fluorimetric, with fluorogenic peptide
      substrates, 399
  glycine-rich repeats, 203
  pathogenic effects, 411–413
  production, 399–402
  properties, 192–194, 201–205, 402–413
    enzymatic, 402

*Pseudomonas*, 202–203, 395
  purification, 399–402
*Serratia*, 395
  polypeptide chain structure, 204
  structure, 191, 396, 408–413
  substrate specificity, 402–408
Serralysin G, *Erwinia*, 203
*Serratia marcescens*
  culture, 399
  hemolysin activator, 222
  proteinases, 395–397
    inhibition, 407–409
    pathogenic effects, 411–413
    production, 399–400
    properties, 402–413
    proteinase activity, 406
    purification, 399–400
    reactivation, by divalent metal ions,
      408
    73-kDa thiol-dependent, 397
*Serratia* sp. E.-15, proteinase, 395–396
  proteinase activity, 406
  structure, 408–412
  substrate specificity, 402–408
Sheep pulmonary adenomatosis virus, ret-
    ropepsin, 113
*O*-Sialoglycoprotease, *Pasteurella*, *See* *O*-Si-
    aloglycoprotein endopeptidase
*O*-Sialoglycoprotein endopeptidase, *Pasteur-
    ella*, 223–224, 728–740
  activity, 728
  gene, 739
  sialate residue requirement, 739
  amino acid sequence, 728
  assay, 731–736
    κ-casein hydrolysis method, 731–736
    glycophorin A hydrolysis method, 731–
      733
  cleavage of *O*-sialoglycoproteins from sur-
    face of live cells, 736–738
  cleavage specificity, 728
  gene, 728–729
  glycophorin A hydrolysis, 735–736
  homologs, 224
  inhibition, 738–739
  isolation, 730–731
  metal ion cofactor, 738
  occurrence, 728–729

# Z